■ TABLE 1.5

Approximate Physical Properties of Some Common Liquids (BG Units)

Liquid	Temperature (°F)	Density, ρ (slugs/ft^3)	Specific Weight, γ (lb/ft^3)	Dynamic Viscosity, μ (lb·s/ft^2)	Kinematic Viscosity, ν (ft^2/s)	Surface Tension,[a] σ (lb/ft)	Vapor Pressure, p_v [lb/in.^{2} (abs)]	Bulk Modulus,[b] E_v (lb/in.^{2})
Carbon tetrachloride	68	3.09	99.5	2.00 E − 5	6.47 E − 6	1.84 E − 3	1.9 E + 0	1.91 E + 5
Ethyl alcohol	68	1.53	49.3	2.49 E − 5	1.63 E − 5	1.56 E − 3	8.5 E − 1	1.54 E + 5
Gasoline[c]	60	1.32	42.5	6.5 E − 6	4.9 E − 6	1.5 E − 3	8.0 E + 0	1.9 E + 5
Glycerin	68	2.44	78.6	3.13 E − 2	1.28 E − 2	4.34 E − 3	2.0 E − 6	6.56 E + 5
Mercury	68	26.3	847	3.28 E − 5	1.25 E − 6	3.19 E − 2	2.3 E − 5	4.14 E + 6
SAE 30 oil[c]	60	1.77	57.0	8.0 E − 3	4.5 E − 3	2.5 E − 3	—	2.2 E + 5
Seawater	60	1.99	64.0	2.51 E − 5	1.26 E − 5	5.03 E − 3	2.56 E − 1	3.39 E + 5
Water	60	1.94	62.4	2.34 E − 5	1.21 E − 5	5.03 E − 3	2.56 E − 1	3.12 E + 5

[a]In contact with air.

[b]Isentropic bulk modulus calculated from speed of sound.

[c]Typical values. Properties of petroleum products vary.

■ TABLE 1.6

Approximate Physical Properties of Some Common Liquids (SI Units)

Liquid	Temperature (°C)	Density, ρ (kg/m^3)	Specific Weight, γ (kN/m^3)	Dynamic Viscosity, μ (N·s/m^2)	Kinematic Viscosity, ν (m^2/s)	Surface Tension,[a] σ (N/m)	Vapor Pressure, p_v [N/m^2 (abs)]	Bulk [illegible]
Carbon tetrachloride	20	1,590	15.6	9.58 E − 4	6.03 E − 7	2.69 E − 2	1.3 E + 4	[illegible]
Ethyl alcohol	20	789	7.74	1.19 E − 3	1.51 E − 6	2.28 E − 2	5.9 E + 3	[illegible]
Gasoline[c]	15.6	680	6.67	3.1 E − 4	4.6 E − 7	2.2 E − 2	5.5 E + 4	[illegible]
Glycerin	20	1,260	12.4	1.50 E + 0	1.19 E − 3	6.33 E − 2	1.4 E − 2	[illegible]
Mercury	20	13,600	133	1.57 E − 3	1.15 E − 7	4.66 E − 1	1.6 E − 1	[illegible]
SAE 30 oil[c]	15.6	912	8.95	3.8 E − 1	4.2 E − 4	3.6 E − 2	—	[illegible]
Seawater	15.6	1,030	10.1	1.20 E − 3	1.17 E − 6	7.34 E − 2	1.77 E + 3	[illegible]
Water	15.6	999	9.80	1.12 E − 3	1.12 E − 6	7.34 E − 2	1.77 E + 3	[illegible]

[a]In contact with air.

[b]Isentropic bulk modulus calculated from speed of sound.

[c]Typical values. Properties of petroleum products vary.

TABLE 1.7
Approximate Physical Properties of Some Common Gases at Standard Atmospheric Pressure (BG Units)

Gas	Temperature (°F)	Density, ρ (slugs/ft³)	Specific Weight, γ (lb/ft³)	Dynamic Viscosity, μ (lb·s/ft²)	Kinematic Viscosity, ν (ft²/s)	Gas Constant,[a] R (ft·lb/slug·°R)	Specific Heat Ratio,[b] k
Air (standard)	59	2.38 E − 3	7.65 E − 2	3.74 E − 7	1.57 E − 4	1.716 E + 3	1.40
Carbon dioxide	68	3.55 E − 3	1.14 E − 1	3.07 E − 7	8.65 E − 5	1.130 E + 3	1.30
Helium	68	3.23 E − 4	1.04 E − 2	4.09 E − 7	1.27 E − 3	1.242 E + 4	1.66
Hydrogen	68	1.63 E − 4	5.25 E − 3	1.85 E − 7	1.13 E − 3	2.466 E + 4	1.41
Methane (natural gas)	68	1.29 E − 3	4.15 E − 2	2.29 E − 7	1.78 E − 4	3.099 E + 3	1.31
Nitrogen	68	2.26 E − 3	7.28 E − 2	3.68 E − 7	1.63 E − 4	1.775 E + 3	1.40
Oxygen	68	2.58 E − 3	8.31 E − 2	4.25 E − 7	1.65 E − 4	1.554 E + 3	1.40

[a]Values of the gas constant are independent of temperature.
[b]Values of the specific heat ratio depend only slightly on temperature.

TABLE 1.8
Approximate Physical Properties of Some Common Gases at Standard Atmospheric Pressure (SI Units)

Gas	Temperature (°C)	Density, ρ (kg/m³)	Specific Weight, γ (N/m³)	Dynamic Viscosity, μ (N·s/m²)	Kinematic Viscosity, ν (m²/s)	Gas Constant,[a] R (J/kg·K)	Specific Heat Ratio,[b] k
Air (standard)	15	1.23 E + 0	1.20 E + 1	1.79 E − 5	1.46 E − 5	2.869 E + 2	1.40
Carbon dioxide	20	1.83 E + 0	1.80 E + 1	1.47 E − 5	8.03 E − 6	1.889 E + 2	1.30
Helium	20	1.66 E − 1	1.63 E + 0	1.94 E − 5	1.15 E − 4	2.077 E + 3	1.66
Hydrogen	20	8.38 E − 2	8.22 E − 1	8.84 E − 6	1.05 E − 4	4.124 E + 3	1.41
Methane (natural gas)	20	6.67 E − 1	6.54 E + 0	1.10 E − 5	1.65 E − 5	5.183 E + 2	1.31
Nitrogen	20	1.16 E + 0	1.14 E + 1	1.76 E − 5	1.52 E − 5	2.968 E + 2	1.40
Oxygen	20	1.33 E + 0	1.30 E + 1	2.04 E − 5	1.53 E − 5	2.598 E + 2	1.40

[a]Values of the gas constant are independent of temperature.
[b]Values of the specific heat ratio depend only slightly on temperature.

Fundamentals of Fluid Mechanics

Update Edition

Third Edition Update

Fundamentals of Fluid Mechanics

BRUCE R. MUNSON
DONALD F. YOUNG
Department of Aerospace Engineering and Engineering Mechanics
THEODORE H. OKIISHI
Department of Mechanical Engineering
Iowa State University
Ames, Iowa, USA

John Wiley & Sons, Inc.
New York Chichester Weinheim Brisbane Singapore Toronto

Cover Photos: Whirlpool, Francoise Sauze/Science Photo Library/Photo Researchers, Inc.
Computer Simulation of a Vortex, Dr. Fred Espenak/Science Photo Library/Photo Researchers, Inc.

ACQUISITIONS EDITOR Charity Robey

MARKETING MANAGER Harper Mooy

PRODUCTION EDITOR Tony VenGraitis

OUTSIDE PRODUCTION MANAGEMENT Ingrao Associates

INTERIOR DESIGNER Madelyn Lesure

ILLUSTRATION EDITOR Sigmund Malinowski

JUNIOR ILLUSTRATION COORDINATOR Gene Aiello

ELECTRONIC ILLUSTRATIONS Radiant Illustration and Design

COVER DESIGNER Madelyn Lesure

This book was set in Times Roman by Ruttle, Shaw & Wetherill, Inc. and printed and bound by Von Hoffmann Press, Inc. The cover was printed by Lehigh Press.

Library of Congress Cataloging-in-Publication Data

Munson, Bruce Roy, 1940-
Fundamentals of fluid mechanics / Bruce R. Munson, Donald F. Young, Theodore H. Okiishi. — 3rd ed. update
p. cm.
Includes bibliographical references and index.
ISBN 0-471-35502-X (cloth : alk. paper and CD-ROM)
1. Fluid mechanics. I. Young, Donald F. II. Okiishi, T. H. (Theodore Hisao), 1939- . III. Title.

10 9 8 7 6 5 4 3

To Erik and all others who possess the curiosity,
patience, and desire to learn

About the Authors

Bruce R. Munson, Professor of Engineering Mechanics at Iowa State University since 1974, received his B.S. and M.S. degrees from Purdue University and his Ph.D. degree from the Aerospace Engineering and Mechanics Department of the University of Minnesota in 1970.

From 1970 to 1974, Dr. Munson was on the mechanical engineering faculty of Duke University. From 1964 to 1966, he worked as an engineer in the jet engine fuel control department of Bendix Aerospace Corporation, South Bend, Indiana.

Dr. Munson's main professional activity has been in the area of fluid mechanics education and research. He has been responsible for the development of many fluid mechanics courses for studies in civil engineering, mechanical engineering, engineering science, and agricultural engineering and is the recipient of an Iowa State University Superior Engineering Teacher Award and the Iowa State University Alumni Association Faculty Citation.

He has authored and coauthored many theoretical and experimental technical papers on hydrodynamic stability, low Reynolds number flow, secondary flow, and the applications of viscous incompressible flow. He is a member of The American Society of Mechanical Engineers, The American Physical Society, and The American Society for Engineering Education.

Donald F. Young, Anson Marston Distinguished Professor in Engineering, is a faculty member in the Department of Aerospace Engineering and Engineering Mechanics at Iowa State University. Dr. Young received his B.S. degree in mechanical engineering, his M.S. and Ph.D. degrees in theoretical and applied mechanics from Iowa State, and has taught both undergraduate and graduate courses in fluid mechanics for many years. In addition to being named a Distinguished Professor in the College of Engineering, Dr. Young has also received the Standard Oil Foundation Outstanding Teacher Award and the Iowa State University Alumni Association Faculty Citation. He has been engaged in fluid mechanics research for more than 35 years, with special interests in similitude and modeling and the interdisciplinary field of biomedical fluid mechanics. Dr. Young has contributed to many technical publications and is the author or coauthor of two textbooks on applied mechanics. He is a Fellow of The American Society of Mechanical Engineers.

Theodore H. Okiishi, Associate Dean of Engineering and past Chair of Mechanical Engineering at Iowa State University, has taught fluid mechanics courses there since 1967. He received his undergraduate and graduate degrees at Iowa State.

From 1965 to 1967, Dr. Okiishi served as a U.S. Army officer with duty assignments at the National Aeronautics and Space Administration Lewis Research Center, Cleveland, Ohio, where he participated in rocket nozzle heat transfer research, and at the Combined Intelligence Center, Saigon, Republic of South Vietnam, where he studied seasonal river flooding problems.

Professor Okiishi is active in research on turbomachinery fluid dynamics. He and his graduate students and other colleagues have written a number of journal articles based on their studies. Some of these projects have involved significant collaboration with government and industrial laboratory researchers with one technical paper winning the ASME Melville Medal.

Dr. Okiishi has received several awards for teaching. He has developed undergraduate and graduate courses in classical fluid dynamics as well as the fluid dynamics of turbomachines.

He is a licensed professional engineer. His technical society activities include having been chair of the board of directors of The American Society of Mechanical Engineers (ASME) International Gas Turbine Institute. He is a Fellow of The American Society of Mechanical Engineers and the technical editor of the *Journal of Turbomachinery*.

Preface

This book is intended for junior and senior engineering students who are interested in learning some fundamental aspects of fluid mechanics. This area of mechanics is mature, and a complete coverage of all aspects of it obviously cannot be accomplished in a single volume. We developed this text to be used as a first course. The principles considered are classical and have been well-established for many years. However, fluid mechanics education has improved with experience in the classroom, and we have brought to bear in this book our own ideas about the teaching of this interesting and important subject. This third edition has been prepared after several years of experience by the authors using the first and second editions for an introductory course in fluid mechanics. Based on this experience, along with suggestions from reviewers, colleagues, and students, we have made a number of changes in this new edition. Many of these changes are minor and have been made to simply clarify, update and expand certain ideas and concepts. Major changes include the addition of new problems, the addition of review problems, a slight modification to compressible flow problem-solving techniques, and the addition of a summary sentence in the margin of each page of text.

One of our aims is to represent fluid mechanics as it really is—an exciting and useful discipline. To this end, we include analyses of numerous everyday examples of fluid-flow phenomena to which students and faculty can easily relate. In the third edition 165 examples are presented that provide detailed solutions to a variety of problems. Also, a generous set of homework problems in each chapter stresses the practical application of principles. Those problems that can be worked best with a programmable calculator or a computer, about 10% of the problems, are so identified. Also included in most chapters are several open-ended problems. These problems require critical thinking in that in order to work them one must make various assumptions and provide the necessary data. Students are thus required to make reasonable estimates or to obtain additional information outside the classroom. These open-ended problems are clearly identified. Another feature is the inclusion of extended, laboratory-type problems in most chapters. Actual experimental data are included in these problems, and the student is asked to perform a detailed analysis of the problem similar to that required for a typical laboratory. It is believed that this type of problem will be particularly useful for fluid mechanics courses that do not have a laboratory as a part of the course. These laboratory-type problems are located at the end of the problems section in most chapters and can be easily recognized. The examples and homework problems illustrate the considerable versatility of fluid mechanical analyses.

A summary or (highlight) sentence is inserted on each page of text.

Since this is an introductory text, we have designed the presentation of material to allow for the gradual development of student confidence in fluid mechanics problem solving. Each important concept or notion is considered in terms of simple and easy-to-understand circumstances before more complicated features are introduced. Two new features have been introduced in the third edition to aid in this understanding. First, a brief summary (or highlight) sentence has been added to each page of text. These sentences serve to prepare or remind the reader about an important concept discussed on that page. The entire page must still be read to understand the material—the summary sentences merely reinforce the comprehension. Second, a set of review problems covering most of the main topics has been added at the end of each chapter. Complete, detailed solutions to these review problems are available in the supplement titled *Student Solution Manual for Fundamentals of Fluid Mechanics*, by Munson, Young, and Okiishi (John Wiley and Sons, Inc., New York, 1997).

Two systems of units continue to be used throughout the text: the British Gravitational System (pounds, slugs, feet, and seconds), and the International System of Units (newtons, kilograms, meters, and seconds). Both systems are widely used, and we believe that students need to be knowledgeable and comfortable with both systems. Approximately one-half of the examples and homework problems use the British System; the other half is based on the International System.

In the first four chapters, the student is made aware of some fundamental aspects of fluid motion, including important fluid properties, regimes of flow, pressure variations in fluids at rest and in motion, fluid kinematics, and methods of flow description and analysis. The Bernoulli equation is introduced in Chapter 3 to draw attention, early on, to some of the interesting effects of fluid motion on the distribution of pressure in a flow field. We believe that this timely consideration of elementary fluid dynamics will increase student enthusiasm for the more complicated material that follows. In Chapter 4, we convey the essential elements of kinematics, including Eulerian and Lagrangian mathematical descriptions of flow phenomena, and indicate the vital relationship between the two views. For teachers who wish to consider kinematics in detail before the material on elementary fluid dynamics, Chapters 3 and 4 can be interchanged without loss of continuity.

Chapters 5, 6, and 7 expand on the basic analysis methods generally used to solve or to begin solving fluid mechanics problems. Emphasis is placed on understanding how flow phenomena are described mathematically and on when and how to use infinitesimal and finite control volumes. Owing to the importance of numerical techniques in fluid mechanics, we have included introductory material on this subject in Chapter 6. The effects of fluid friction on pressure and velocity distributions are also considered in some detail. A formal course in thermodynamics is not required to understand the various portions of the text that consider some elementary aspects of the thermodynamics of fluid flow. Chapter 7 features the advantages of using dimensional analysis and similitude for organizing test data and for planning experiments and the basic techniques involved.

Chapters 8 to 12 offer students opportunities for the further application of the principles learned early in the text. Also, where appropriate, additional important notions such as boundary layers, transition from laminar to turbulent flow, turbulence modeling, chaos, and flow separation are introduced. Practical concerns such as pipe flow, open-channel flow, flow measurement, drag and lift, the effects of compressibility, and the fluid mechanics fundamentals associated with turbomachines are included.

The compressible flow tables found in the previous editions (and in other texts) have been replaced by corresponding graphs. It is felt that in the current era of visual learning, these graphs allow a fuller understanding of the characteristics of the compressible flow functions.

A new supplement is the *Fluid Mechanics Phenomena* CD containing short video segments that illustrate various aspects of "real world" fluid mechanics. Many of the segments show how fluid mechanics is related to familiar devices and everyday experiences. A short text included with each segment indicates the key fluid mechanics topic being demonstrated and provides a brief description of the content. Each video segment is identified in the textbook by an icon of the type shown in the left margin. These icons are located so that the various video segments are associated with the fluid mechanics concepts and theory discussed in the textbook at that location. The number in the icon identifies the segment, e.g., V1.3 refers to video segment number 3 in Chapter 1.

An Instructor's Manual is available to professors who adopt this book for classroom use. This manual contains complete, detailed solutions to all the problems in the text. This Instructor's Manual is available in paper or CD format and may be obtained directly from the publishers.

Students who study this text and who solve a representative set of the exercises provided should acquire a useful knowledge of the fundamentals of fluid mechanics. Faculty who use this text are provided with numerous topics to select from in order to meet the objectives of their own courses. More material is included than can be reasonably covered in one term. All are reminded of the fine collection of supplementary material. Where appropriate, we have cited throughout the text the articles and books that are available for enrichment.

We express our thanks to the many colleagues who have helped in the development of this text, including Professor Bruce Reichert of Kansas State University for help with Chapter 11 and Professor Patrick Kavanagh of Iowa State University for help with Chapter 12. We are indebted to the following reviewers of the third edition for their comments and suggestions:

William Dempster
University of Strathclyde

William Schultz
The University of Michigan

Ana Paula Barros
Penn State University

George D. Catalano
United States Military Academy, West Point

Michael H. Woo
The Citadel

Thomas J. Olenik
New Jersey Institute of Technology

Edward E. O'Brien
SUNY at Stony Brook

V. Dakshina Murty
University of Portland

Patrick V. Farrell
University of Wisconsin-Madison

We wish to express our gratitude to the many persons who supplied the photographs used throughout the text and to Milton Van Dyke for his help in this effort. Finally, we thank our families for their continued encouragement during the writing of this third edition and Barbara Munson for her editing of the manuscript.

Working with students over the years has taught us much about fluid mechanics education. We have tried in earnest to draw from this experience for the benefit of users of this book. Obviously we are still learning, and we welcome any suggestions and comments from you.

BRUCE R. MUNSON
DONALD F. YOUNG
THEODORE H. OKIISHI

Contents

6 DIFFERENTIAL ANALYSIS OF FLUID FLOW 309

7 SIMILITUDE, DIMENSIONAL ANALYSIS, AND MODELING 397

11 COMPRESSIBLE FLOW 699

12 TURBOMACHINES 779

A UNIT CONVERSION TABLES 846

B PHYSICAL PROPERTIES OF FLUIDS 850

C PROPERTIES OF THE U.S. STANDARD ATMOSPHERE 856

D COMPRESSIBLE FLOW DATA FOR AN IDEAL GAS 858

ANSWERS 859

INDEX 869

Fundamentals of Fluid Mechanics

Update Edition

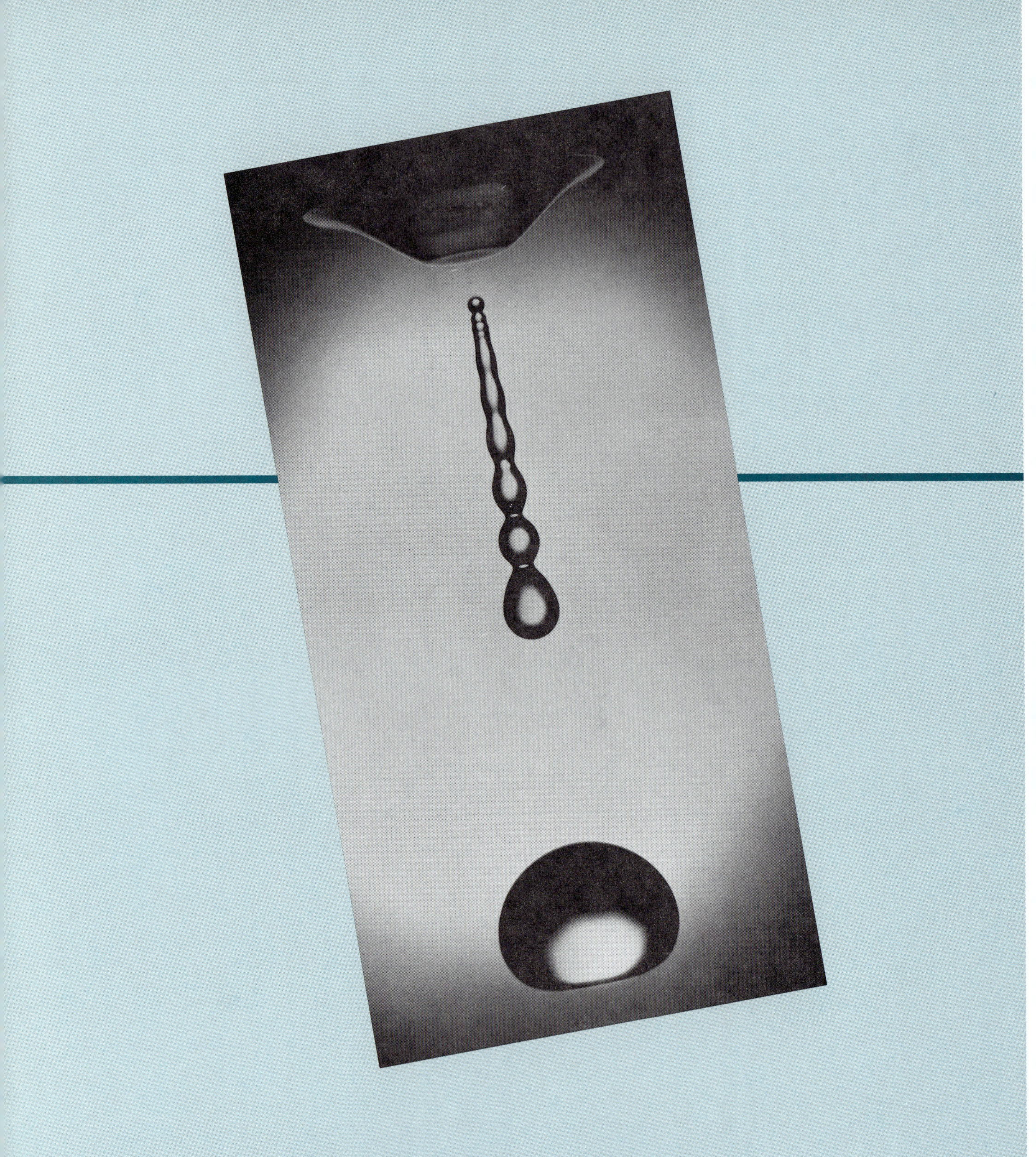

The breakup of a liquid jet into drops is a function of fluid properties such as density, viscosity, and surface tension. [Reprinted with permission from American Institute of Physics (Ref. 6) and the American Association for the Advancement of Science (Ref. 7).]

1 Introduction

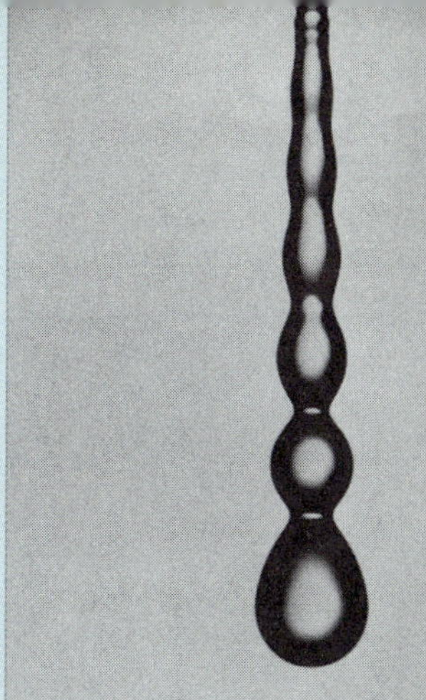

Fluid mechanics is concerned with the behavior of liquids and gases at rest and in motion.

Fluid mechanics is that discipline within the broad field of applied mechanics concerned with the behavior of liquids and gases at rest or in motion. This field of mechanics obviously encompasses a vast array of problems that may vary from the study of blood flow in the capillaries (which are only a few microns in diameter) to the flow of crude oil across Alaska through an 800-mile-long, 4-ft-diameter pipe. Fluid mechanics principles are needed to explain why airplanes are made streamlined with smooth surfaces for the most efficient flight, whereas golf balls are made with rough surfaces (dimpled) to increase their efficiency. Numerous interesting questions can be answered by using relatively simple fluid mechanics ideas. For example:

- How can a rocket generate thrust without having any air to push against in outer space?
- Why can't you hear a supersonic airplane until it has gone past you?
- How can a river flow downstream with a significant velocity even though the slope of the surface is so small that it could not be detected with an ordinary level?
- How can information obtained from model airplanes be used to design the real thing?
- Why does a stream of water from a faucet sometimes appear to have a smooth surface, but sometimes a rough surface?
- How much greater gas mileage can be obtained by improved aerodynamic design of cars and trucks?

The list of applications and questions goes on and on—but you get the point; fluid mechanics is a very important, practical subject. It is very likely that during your career as an engineer you will be involved in the analysis and design of systems that require a good understanding of fluid mechanics. It is hoped that this introductory text will provide a sound foundation of the fundamental aspects of fluid mechanics.

1.1 Some Characteristics of Fluids

One of the first questions we need to explore is—what is a fluid? Or we might ask—what is the difference between a solid and a fluid? We have a general, vague idea of the difference. A solid is "hard" and not easily deformed, whereas a fluid is "soft" and is easily deformed (we can readily move through air). Although quite descriptive, these casual observations of the differences between solids and fluids are not very satisfactory from a scientific or engineering point of view. A closer look at the molecular structure of materials reveals that matter that we commonly think of as a solid (steel, concrete, etc.) has densely spaced molecules with large intermolecular cohesive forces that allow the solid to maintain its shape, and to not be easily deformed. However, for matter that we normally think of as a liquid (water, oil, etc.), the molecules are spaced farther apart, the intermolecular forces are smaller than for solids, and the molecules have more freedom of movement. Thus, liquids can be easily deformed (but not easily compressed) and can be poured into containers or forced through a tube. Gases (air, oxygen, etc.) have even greater molecular spacing and freedom of motion with negligible cohesive intermolecular forces and as a consequence are easily deformed (and compressed) and will completely fill the volume of any container in which they are placed.

A fluid, such as water or air, deforms continuously when acted on by shearing stresses of any magnitude.

Although the differences between solids and fluids can be explained qualitatively on the basis of molecular structure, a more specific distinction is based on how they deform under the action of an external load. Specifically, *a fluid is defined as a substance that deforms continuously when acted on by a shearing stress of any magnitude.* A shearing stress (force per unit area) is created whenever a tangential force acts on a surface. When common solids such as steel or other metals are acted on by a shearing stress, they will initially deform (usually a very small deformation), but they will not continuously deform (flow). However, common fluids such as water, oil, and air satisfy the definition of a fluid—that is, they will flow when acted on by a shearing stress. Some materials, such as slurries, tar, putty, toothpaste, and so on, are not easily classified since they will behave as a solid if the applied shearing stress is small, but if the stress exceeds some critical value, the substance will flow. The study of such materials is called *rheology* and does not fall within the province of classical fluid mechanics. Thus, all the fluids we will be concerned with in this text will conform to the definition of a fluid given previously.

Although the molecular structure of fluids is important in distinguishing one fluid from another, it is not possible to study the behavior of individual molecules when trying to describe the behavior of fluids at rest or in motion. Rather, we characterize the behavior by considering the average, or macroscopic, value of the quantity of interest, where the average is evaluated over a small volume containing a large number of molecules. Thus, when we say that the velocity at a certain point in a fluid is so much, we are really indicating the average velocity of the molecules in a small volume surrounding the point. The volume is small compared with the physical dimensions of the system of interest, but large compared with the average distance between molecules. Is this a reasonable way to describe the behavior of a fluid? The answer is generally yes, since the spacing between molecules is typically very small. For gases at normal pressures and temperatures, the spacing is on the order of 10^{-6} mm, and for liquids it is on the order of 10^{-7} mm. The number of molecules per cubic millimeter is on the order of 10^{18} for gases and 10^{21} for liquids. It is thus clear that the number of molecules in a very tiny volume is huge and the idea of using average values taken over this volume is certainly reasonable. We thus assume that all the fluid characteristics we are interested in (pressure, velocity, etc.) vary continuously throughout the fluid—that is, we treat the fluid as a *continuum*. This concept will certainly be valid for all the circumstances considered in this text. One area of fluid mechanics for which the continuum concept breaks down is in the study of rarefied gases such as would be encountered at very high altitudes. In this case the spacing between air molecules can become large and the continuum concept is no longer acceptable.

1.2 Dimensions, Dimensional Homogeneity, and Units

Fluid characteristics can be described qualitatively in terms of certain basic quantities such as length, time, and mass.

Since in our study of fluid mechanics we will be dealing with a variety of fluid characteristics, it is necessary to develop a system for describing these characteristics both *qualitatively* and *quantitatively*. The qualitative aspect serves to identify the nature, or type, of the characteristics (such as length, time, stress, and velocity), whereas the quantitative aspect provides a numerical measure of the characteristics. The quantitative description requires both a number and a standard by which various quantities can be compared. A standard for length might be a meter or foot, for time an hour or second, and for mass a slug or kilogram. Such standards are called *units*, and several systems of units are in common use as described in the following section. The qualitative description is conveniently given in terms of certain *primary quantities*, such as length, L, time, T, mass, M, and temperature, Θ. These primary quantities can then be used to provide a qualitative description of any other *secondary quantity*: for example, area $\doteq L^2$, velocity $\doteq LT^{-1}$, density $\doteq ML^{-3}$, and so on, where the symbol $\doteq$ is used to indicate the *dimensions* of the secondary quantity in terms of the primary quantities. Thus, to describe qualitatively a velocity, V, we would write

$$V \doteq LT^{-1}$$

and say that "the dimensions of a velocity equal length divided by time." The primary quantities are also referred to as *basic dimensions*.

For a wide variety of problems involving fluid mechanics, only the three basic dimensions, L, T, and M are required. Alternatively, L, T, and F could be used, where F is the basic dimension of force. Since Newton's law states that force is equal to mass times acceleration, it follows that $F \doteq MLT^{-2}$ or $M \doteq FL^{-1}T^2$. Thus, secondary quantities expressed in terms of M can be expressed in terms of F through the relationship above. For example, stress, σ, is a force per unit area, so that $\sigma \doteq FL^{-2}$, but an equivalent dimensional equation is $\sigma \doteq ML^{-1}T^{-2}$. Table 1.1 provides a list of dimensions for a number of common physical quantities.

All theoretically derived equations are *dimensionally homogeneous*—that is, the dimensions of the left side of the equation must be the same as those on the right side, and all additive separate terms must have the same dimensions. We accept as a fundamental premise that all equations describing physical phenomena must be dimensionally homogeneous. If this were not true, we would be attempting to equate or add unlike physical quantities, which would not make sense. For example, the equation for the velocity, V, of a uniformly accelerated body is

$$V = V_0 + at \tag{1.1}$$

where V_0 is the initial velocity, a the acceleration, and t the time interval. In terms of dimensions the equation is

$$LT^{-1} \doteq LT^{-1} + LT^{-1}$$

and thus Eq. 1.1 is dimensionally homogeneous.

Some equations that are known to be valid contain constants having dimensions. The equation for the distance, d, traveled by a freely falling body can be written as

$$d = 16.1t^2 \tag{1.2}$$

and a check of the dimensions reveals that the constant must have the dimensions of LT^{-2} if the equation is to be dimensionally homogeneous. Actually, Eq. 1.2 is a special form of the well-known equation from physics for freely falling bodies,

$$d = \frac{gt^2}{2} \tag{1.3}$$

General homogeneous equations are valid in any system of units.

in which g is the acceleration of gravity. Equation 1.3 is dimensionally homogeneous and valid in any system of units. For $g = 32.2 \text{ ft/s}^2$ the equation reduces to Eq. 1.2 and thus Eq. 1.2 is valid only for the system of units using feet and seconds. Equations that are restricted to a particular system of units can be denoted as *restricted homogeneous equations*, as opposed to equations valid in any system of units, which are *general homogeneous equations*. The preceding discussion indicates one rather elementary, but important, use of the concept of dimensions: the determination of one aspect of the generality of a given equation simply based on a consideration of the dimensions of the various terms in the equation. The concept of dimensions also forms the basis for the powerful tool of *dimensional analysis*, which is considered in detail in Chapter 7.

■ **TABLE 1.1**
Dimensions Associated with Common Physical Quantities

	FLT System	*MLT* System
Acceleration	LT^{-2}	LT^{-2}
Angle	$F^0L^0T^0$	$M^0L^0T^0$
Angular acceleration	T^{-2}	T^{-2}
Angular velocity	T^{-1}	T^{-1}
Area	L^2	L^2
Density	$FL^{-4}T^2$	ML^{-3}
Energy	FL	ML^2T^{-2}
Force	F	MLT^{-2}
Frequency	T^{-1}	T^{-1}
Heat	FL	ML^2T^{-2}
Length	L	L
Mass	$FL^{-1}T^2$	M
Modulus of elasticity	FL^{-2}	$ML^{-1}T^{-2}$
Moment of a force	FL	ML^2T^{-2}
Moment of inertia (area)	L^4	L^4
Moment of inertia (mass)	FLT^2	ML^2
Momentum	FT	MLT^{-1}
Power	FLT^{-1}	ML^2T^{-3}
Pressure	FL^{-2}	$ML^{-1}T^{-2}$
Specific heat	$L^2T^{-2}\Theta^{-1}$	$L^2T^{-2}\Theta^{-1}$
Specific weight	FL^{-3}	$ML^{-2}T^{-2}$
Strain	$F^0L^0T^0$	$M^0L^0T^0$
Stress	FL^{-2}	$ML^{-1}T^{-2}$
Surface tension	FL^{-1}	MT^{-2}
Temperature	Θ	Θ
Time	T	T
Torque	FL	ML^2T^{-2}
Velocity	LT^{-1}	LT^{-1}
Viscosity (dynamic)	$FL^{-2}T$	$ML^{-1}T^{-1}$
Viscosity (kinematic)	L^2T^{-1}	L^2T^{-1}
Volume	L^3	L^3
Work	FL	ML^2T^{-2}

A commonly used equation for determining the volume rate of flow, Q, of a liquid through an orifice located in the side of a tank is

$$Q = 0.61\ A\sqrt{2gh}$$

where A is the area of the orifice, g is the acceleration of gravity, and h is the height of the liquid above the orifice. Investigate the dimensional homogeneity of this formula.

SOLUTION

The dimensions of the various terms in the equation are Q = volume/time $\doteq L^3T^{-1}$, A = area $\doteq L^2$, g = acceleration of gravity $\doteq LT^{-2}$, and h = height $\doteq L$

These terms, when substituted into the equation, yield the dimensional form:

$$(L^3T^{-1}) \doteq (0.61)(L^2)(\sqrt{2})(LT^{-2})^{1/2}(L)^{1/2}$$

or

$$(L^3T^{-1}) \doteq [(0.61)\sqrt{2}](L^3T^{-1})$$

It is clear from this result that the equation is dimensionally homogeneous (both sides of the formula have the same dimensions of L^3T^{-1}), and the numbers (0.61 and $\sqrt{2}$) are dimensionless.

If we were going to use this relationship repeatedly we might be tempted to simplify it by replacing g with its standard value of 32.2 ft/s^2 and rewriting the formula as

$$Q = 4.90\ A\sqrt{h} \tag{1}$$

A quick check of the dimensions reveals that

$$L^3T^{-1} \doteq (4.90)(L^{5/2})$$

and, therefore, the equation expressed as Eq. 1 can only be dimensionally correct if the number 4.90 has the dimensions of $L^{1/2}T^{-1}$. Whenever a number appearing in an equation or formula has dimensions, it means that the specific value of the number will depend on the system of units used. Thus, for the case being considered with feet and seconds used as units, the number 4.90 has units of ft$^{1/2}$/s. Equation 1 will only give the correct value for Q (in ft^3/s) when A is expressed in square feet and h in feet. Thus, Eq. 1 is a *restricted* homogeneous equation, whereas the original equation is a *general* homogeneous equation that would be valid for any consistent system of units. A quick check of the dimensions of the various terms in an equation is a useful practice and will often be helpful in eliminating errors—that is, as noted previously, all physically meaningful equations must be dimensionally homogeneous. We have briefly alluded to units in this example, and this important topic will be considered in more detail in the next section.

1.2.1 Systems of Units

In addition to the qualitative description of the various quantities of interest, it is generally necessary to have a quantitative measure of any given quantity. For example, if we measure the width of this page in the book and say that it is 10 units wide, the statement has no meaning until the unit of length is defined. If we indicate that the unit of length is a meter, and define the meter as some standard length, a unit system for length has been established (and a numerical value can be given to the page width). In addition to length, a unit must be established for each of the remaining basic quantities (force, mass, time, and temperature). There are several systems of units in use and we shall consider three systems that are commonly used in engineering.

British Gravitational (BG) System. In the BG system the unit of length is the foot (ft), the time unit is the second (s), the force unit is the pound (lb), and the temperature unit is the degree Fahrenheit (°F), or the absolute temperature unit is the degree Rankine (°R), where

$$°\text{R} = °\text{F} + 459.67$$

The mass unit, called the *slug*, is defined from Newton's second law (force = mass × acceleration) as

$$1 \text{ lb} = (1 \text{ slug})(1 \text{ ft/s}^2)$$

This relationship indicates that a 1-lb force acting on a mass of 1 slug will give the mass an acceleration of 1 ft/s^2.

The weight, $\mathcal{W}$ (which is the force due to gravity, g) of a mass, m, is given by the equation

Two systems of units that are widely used in engineering are the British Gravitational (BG) System and the International System (SI).

$$\mathcal{W} = mg$$

and in BG units

$$\mathcal{W}(\text{lb}) = m \text{ (slugs) } g \text{ (ft/s}^2)$$

Since the earth's standard gravity is taken as $g = 32.174 \text{ ft/s}^2$ (commonly approximated as 32.2 ft/s^2), it follows that a mass of 1 slug weighs 32.2 lb under standard gravity.

International System (SI). In 1960 the Eleventh General Conference on Weights and Measures, the international organization responsible for maintaining precise uniform standards of measurements, formally adopted the *International System of Units* as the international standard. This system, commonly termed SI, has been widely adopted worldwide and is widely used (although certainly not exclusively) in the United States. It is expected that the long-term trend will be for all countries to accept SI as the accepted standard and it is imperative that engineering students become familiar with this system. In SI the unit of length is the meter (m), the time unit is the second (s), the mass unit is the kilogram (kg), and the temperature unit is the kelvin (K). Note that there is no degree symbol used when expressing a temperature in kelvin units. The Kelvin temperature scale is an absolute scale and is related to the Celsius (centigrade) scale (°C) through the relationship

$$\text{K} = °\text{C} + 273.15$$

Although the Celsius scale is not in itself part of SI, it is common practice to specify temperatures in degrees Celsius when using SI units.

The force unit, called the newton (N), is defined from Newton's second law as

$$1 \text{ N} = (1 \text{ kg})(1 \text{ m/s}^2)$$

Thus, a 1-N force acting on a 1-kg mass will give the mass an acceleration of 1 m/s^2. Standard gravity in SI is 9.807 m/s^2 (commonly approximated as 9.81 m/s^2) so that a 1-kg mass weighs 9.81 N under standard gravity. Note that weight and mass are different, both qualitatively and quantitatively! The unit of *work* in SI is the joule (J), which is the work done

TABLE 1.2
Prefixes for SI Units

Factor by Which Unit Is Multiplied	Prefix	Symbol
10^{12}	tera	T
10^{9}	giga	G
10^{6}	mega	M
10^{3}	kilo	k
10^{2}	hecto	h
10	deka	da
10^{-1}	deci	d
10^{-2}	centi	c
10^{-3}	milli	m
10^{-6}	micro	μ
10^{-9}	nano	n
10^{-12}	pico	p
10^{-15}	femto	f
10^{-18}	atto	a

In mechanics it is very important to distinguish between weight and mass.

when the point of application of a 1-N force is displaced through a 1-m distance in the direction of a force. Thus,

$$1 \text{ J} = 1 \text{ N} \cdot \text{m}$$

The unit of *power* is the watt (W) defined as a joule per second. Thus,

$$1 \text{ W} = 1 \text{ J/s} = 1 \text{ N} \cdot \text{m/s}$$

Prefixes for forming multiples and fractions of SI units are given in Table 1.2. For example, the notation kN would be read as ''kilonewtons'' and stands for 10^3 N. Similarly, mm would be read as ''millimeters'' and stands for 10^{-3} m. The centimeter is not an accepted unit of length in the SI system, so for most problems in fluid mechanics in which SI units are used, lengths will be expressed in millimeters or meters.

English Engineering (EE) System. In the EE system units for force *and* mass are defined independently; thus special care must be exercised when using this system in conjunction with Newton's second law. The basic unit of mass is the pound mass (lbm), the unit of force is the pound (lb).[1] The unit of length is the foot (ft), the unit of time is the second (s), and the absolute temperature scale is the degree Rankine (°R). To make the equation expressing Newton's second law dimensionally homogeneous we write it as

$$\mathbf{F} = \frac{m\mathbf{a}}{g_c} \tag{1.4}$$

where g_c is a constant of proportionality which allows us to define units for both force and mass. For the BG system only the force unit was prescribed and the mass unit defined in a

[1]It is also common practice to use the notation, lbf, to indicate pound force.

consistent manner such that $g_c = 1$. Similarly, for SI the mass unit was prescribed and the force unit defined in a consistent manner such that $g_c = 1$. For the EE system, a 1-lb force is defined as that force which gives a 1 lbm a standard acceleration of gravity which is taken as 32.174 ft/s^2. Thus, for Eq. 1.4 to be both numerically and dimensionally correct

$$1 \text{ lb} = \frac{(1 \text{ lbm})(32.174 \text{ ft/s}^2)}{g_c}$$

so that

$$g_c = \frac{(1 \text{ lbm})(32.174 \text{ ft/s}^2)}{(1 \text{ lb})}$$

With the EE system weight and mass are related through the equation

$$\mathcal{W} = \frac{mg}{g_c}$$

where g is the local acceleration of gravity. Under conditions of standard gravity ($g = g_c$) the weight in pounds and the mass in pound mass are numerically equal. Also, since a 1-lb force gives a mass of 1 lbm an acceleration of 32.174 ft/s^2, and a mass of 1 slug an acceleration of 1 ft/s^2, it follows that

$$1 \text{ slug} = 32.174 \text{ lbm}$$

In this text we will primarily use the BG system and SI for units. The EE system is used very sparingly, and only in those instances where convention dictates its use. Approximately one-half the problems and examples are given in BG units and one-half in SI units. We cannot overemphasize the importance of paying close attention to units when solving problems. It is very easy to introduce huge errors into problem solutions through the use of incorrect units. Get in the habit of using a *consistent* system of units throughout a given solution. It really makes no difference which system you use as long as you are consistent; for example, don't mix slugs and newtons. If problem data are specified in SI units, then use SI units throughout the solution. If the data are specified in BG units, then use BG units throughout the solution. Tables 1.3 and 1.4 provide conversion factors for some quantities that are commonly encountered in fluid mechanics. For convenient reference these tables are also reproduced on the inside of the back cover. Note that in these tables (and others) the numbers are expressed by using computer exponential notation. For example, the number 5.154 E + 2 is equivalent to 5.154×10^2 in scientific notation, and the number 2.832 E − 2 is equivalent to 2.832×10^{-2}. More extensive tables of conversion factors for a large variety of unit systems can be found in Appendix A.

When solving problems it is important to use a consistent system of units, e.g., don't mix BG and SI units.

■ **TABLE 1.3**
Conversion Factors from BG Units to SI Units

(See inside of back cover.)

■ **TABLE 1.4**
Conversion Factors from SI Units to BG Units

(See inside of back cover.)

EXAMPLE 1.2

A tank of water having a total mass of 36 kg rests on the floor of an elevator. Determine the force (in newtons) that the tank exerts on the floor when the elevator is accelerating upward at 7 $\mathrm{ft/s^2}$.

SOLUTION

A free-body diagram of the tank is shown in Fig. E1.2 where $\mathcal{W}$ is the weight of the tank and water, and F_f is the reaction of the floor on the tank. Application of Newton's second law of motion to this body gives

$$\sum \mathbf{F} = m\mathbf{a}$$

or

$$F_f - \mathcal{W} = ma \qquad \textbf{(1)}$$

where we have taken upward as the positive direction. Since $\mathcal{W} = mg$, Eq. 1 can be written as

$$F_f = m(g + a) \qquad \textbf{(2)}$$

Before substituting any number into Eq. 2 we must decide on a system of units, and then be sure all of the data are expressed in these units. Since we want F_f in newtons we will use SI units so that

$$F_f = 36 \text{ kg } [9.81 \text{ m/s}^2 + (7 \text{ ft/s}^2)(0.3048 \text{ m/ft})] = 430 \text{ kg} \cdot \text{m/s}^2$$

Since $1 \text{ N} = 1 \text{ kg} \cdot \text{m/s}^2$ it follows that

$$F_f = 430 \text{ N} \qquad \text{(downward on floor)} \qquad \textbf{(Ans)}$$

The direction is downward since the force shown on the free-body diagram is the force of the floor *on the tank* so that the force the tank exerts *on the floor* is equal in magnitude but opposite in direction.

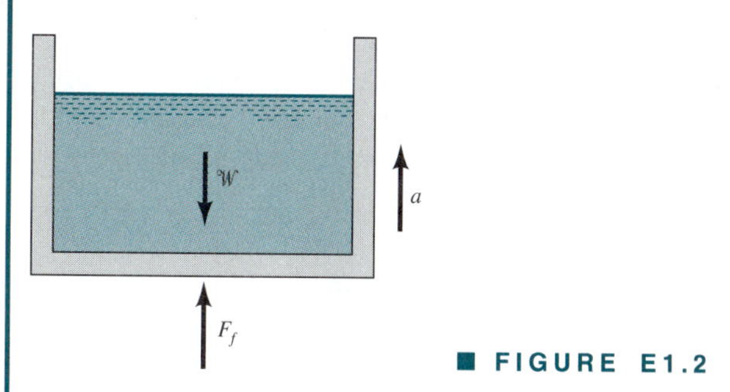

■ FIGURE E1.2

As you work through a large variety of problems in this text, you will find that units play an essential role in arriving at a numerical answer. Be careful! It is easy to mix units and cause large errors. If in the above example the elevator acceleration had been left as 7 $\mathrm{ft/s^2}$ with m and g expressed in SI units, we would have calculated the force as 605 N and the answer would have been 41% too large!

1.3 Analysis of Fluid Behavior

The study of fluid mechanics involves the same fundamental laws you have encountered in physics and other mechanics courses. These laws include Newton's laws of motion, conservation of mass, and the first and second laws of thermodynamics. Thus, there are strong similarities between the general approach to fluid mechanics and to rigid-body and deformable-body solid mechanics. This is indeed helpful since many of the concepts and techniques of analysis used in fluid mechanics will be ones you have encountered before in other courses.

The broad subject of fluid mechanics can be generally subdivided into *fluid statics*, in which the fluid is at rest, and *fluid dynamics*, in which the fluid is moving. In the following chapters we will consider both of these areas in detail. Before we can proceed, however, it will be necessary to define and discuss certain fluid *properties* that are intimately related to fluid behavior. It is obvious that different fluids can have grossly different characteristics. For example, gases are light and compressible, whereas liquids are heavy (by comparison) and relatively incompressible. A syrup flows slowly from a container, but water flows rapidly when poured from the same container. To quantify these differences certain fluid properties are used. In the following several sections the properties that play an important role in the analysis of fluid behavior are considered.

1.4 Measures of Fluid Mass and Weight

1.4.1 Density

The density of a fluid is defined as its mass per unit volume.

The *density* of a fluid, designated by the Greek symbol ρ (rho), is defined as its mass per unit volume. Density is typically used to characterize the mass of a fluid system. In the BG system ρ has units of slugs/ft^3 and in SI the units are kg/m^3.

The value of density can vary widely between different fluids, but for liquids, variations in pressure and temperature generally have only a small effect on the value of ρ. The small change in the density of water with large variations in temperature is illustrated in Fig. 1.1. Tables 1.5 and 1.6 list values of density for several common liquids. The density of water at 60 °F is $1.94\ \text{slugs/ft}^3$ or $999\ \text{kg/m}^3$. The large difference between those two values illustrates the importance of paying attention to units! Unlike liquids, the density of a gas is strongly influenced by both pressure and temperature, and this difference will be discussed in the next section.

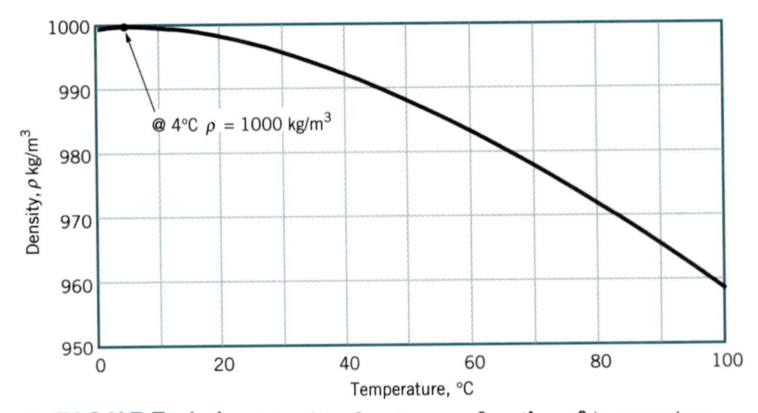

■ FIGURE 1.1 **Density of water as a function of temperature.**

■ **TABLE 1.5**
Approximate Physical Properties of Some Common Liquids (BG Units)

(See inside of front cover.)

■ **TABLE 1.6**
Approximate Physical Properties of Some Common Liquids (SI Units)

(See inside of front cover.)

The *specific volume*, v, is the *volume* per unit mass and is therefore the reciprocal of the density—that is,

$$v = \frac{1}{\rho} \tag{1.5}$$

This property is not commonly used in fluid mechanics but is used in thermodynamics.

1.4.2 Specific Weight

The *specific weight* of a fluid, designated by the Greek symbol γ (gamma), is defined as its *weight* per unit volume. Thus, specific weight is related to density through the equation

$$\gamma = \rho g \tag{1.6}$$

where g is the local acceleration of gravity. Just as density is used to characterize the mass of a fluid system, the specific weight is used to characterize the weight of the system. In the BG system, γ has units of lb/ft^3 and in SI the units are N/m^3. Under conditions of standard gravity ($g = 32.174 \text{ ft/s}^2 = 9.807 \text{ m/s}^2$), water at 60 °F has a specific weight of 62.4 lb/ft^3 and 9.80 kN/m^3. Tables 1.5 and 1.6 list values of specific weight for several common liquids (based on standard gravity). More complete tables for water can be found in Appendix B (Tables B.1 and B.2).

Specific weight is weight per unit volume; specific gravity is the ratio of fluid density to the density of water at a certain temperature.

1.4.3 Specific Gravity

The *specific gravity* of a fluid, designated as *SG*, is defined as the ratio of the density of the fluid to the density of water at some specified temperature. Usually the specified temperature is taken as 4 °C (39.2 °F), and at this temperature the density of water is 1.94 slugs/ft^3 or 1000 kg/m^3. In equation form, specific gravity is expressed as

$$SG = \frac{\rho}{\rho_{H_2O@4°C}} \tag{1.7}$$

and since it is the *ratio* of densities, the value of *SG* does not depend on the system of units used. For example, the specific gravity of mercury at 20 °C is 13.55 and the density of mercury can thus be readily calculated in either BG or SI units through the use of Eq. 1.7 as

$$\rho_{Hg} = (13.55)(1.94 \text{ slugs/ft}^3) = 26.3 \text{ slugs/ft}^3$$

or

$$\rho_{Hg} = (13.55)(1000 \text{ kg/m}^3) = 13.6 \times 10^3 \text{ kg/m}^3$$

It is clear that density, specific weight, and specific gravity are all interrelated, and from a knowledge of any one of the three the others can be calculated.

1.5 Ideal Gas Law

Gases are highly compressible in comparison to liquids, with changes in gas density directly related to changes in pressure and temperature through the equation

$$p = \rho RT \tag{1.8}$$

where p is the absolute pressure, ρ the density, T the absolute temperature,[2] and R is a gas constant. Equation 1.8 is commonly termed the *ideal* or *perfect gas law*, or the *equation of state* for an ideal gas. It is known to closely approximate the behavior of real gases under normal conditions when the gases are not approaching liquefaction.

In the ideal gas law, absolute pressures and temperatures must be used.

Pressure in a fluid at rest is defined as the normal force per unit area exerted on a plane surface (real or imaginary) immersed in a fluid and is created by the bombardment of the surface with the fluid molecules. From the definition, pressure has the dimension of FL^{-2}, and in BG units is expressed as lb/ft^2 (psf) or lb/in.^2 (psi) and in SI units as N/m^2. In SI, 1 N/m^2 is defined as a *pascal*, abbreviated as Pa, and pressures are commonly specified in pascals. The pressure in the ideal gas law must be expressed as an *absolute* pressure, which means that it is measured relative to absolute zero pressure (a pressure that would only occur in a perfect vacuum). Standard sea-level atmospheric pressure (by international agreement) is 14.696 psi (abs) or 101.33 kPa (abs). For most calculations these pressures can be rounded to 14.7 psi and 101 kPa, respectively. In engineering it is common practice to measure pressure relative to the local atmospheric pressure, and when measured in this fashion it is called *gage* pressure. Thus, the absolute pressure can be obtained from the gage pressure by adding the value of the atmospheric pressure. For example, a pressure of 30 psi (gage) in a tire is equal to 44.7 psi (abs) at standard atmospheric pressure. Pressure is a particularly important fluid characteristic and it will be discussed more fully in the next chapter.

The gas constant, R, which appears in Eq. 1.8, depends on the particular gas and is related to the molecular weight of the gas. Values of the gas constant for several common gases are listed in Tables 1.7 and 1.8. Also in these tables the gas density and specific weight are given for standard atmospheric pressure and gravity and for the temperature listed. More complete tables for air at standard atmospheric pressure can be found in Appendix B (Tables B.3 and B.4).

■ TABLE 1.7
Approximate Physical Properties of Some Common Gases at Standard Atmospheric Pressure (BG Units)

(See inside of front cover.)

■ TABLE 1.8
Approximate Physical Properties of Some Common Gases at Standard Atmospheric Pressure (SI Units)

(See inside of front cover.)

[2] We will use T to represent temperature in thermodynamic relationships although T is also used to denote the basic dimension of time.

EXAMPLE 1.3

A compressed air tank has a volume of 0.84 ft^3. When the tank is filled with air at a gage pressure of 50 psi, determine the density of the air and the weight of air in the tank. Assume the temperature is 70 °F and the atmospheric pressure is 14.7 psi (abs).

SOLUTION

The air density can be obtained from the ideal gas law (Eq. 1.8) expressed as

$$\rho = \frac{p}{RT}$$

so that

$$\rho = \frac{(50 \text{ lb/in.}^2 + 14.7 \text{ lb/in.}^2)(144 \text{ in.}^2/\text{ft}^2)}{(1716 \text{ ft}\cdot\text{lb/slug}\cdot{}^\circ\text{R})[(70 + 460)^\circ\text{R}]} = 0.0102 \text{ slugs/ft}^3 \qquad \textbf{(Ans)}$$

Note that both the pressure and temperature were changed to absolute values.

The weight, $\mathcal{W}$, of the air is equal to

$$\mathcal{W} = \rho g \times (\text{volume})$$

$$= (0.0102 \text{ slugs/ft}^3)(32.2 \text{ ft/s}^2)(0.84 \text{ ft}^3)$$

so that since 1 lb = 1 slug·ft/s^2

$$\mathcal{W} = 0.276 \text{ lb} \qquad \textbf{(Ans)}$$

1.6 Viscosity

The properties of density and specific weight are measures of the "heaviness" of a fluid. It is clear, however, that these properties are not sufficient to uniquely characterize how fluids behave since two fluids (such as water and oil) can have approximately the same value of density but behave quite differently when flowing. There is apparently some additional property that is needed to describe the "fluidity" of the fluid.

V1.1

Fluid motion can cause shearing stresses.

To determine this additional property, consider a hypothetical experiment in which a material is placed between two very wide parallel plates as shown in Fig. 1.2*a*. The bottom plate is rigidly fixed, but the upper plate is free to move. If a solid, such as steel, were placed between the two plates and loaded with the force P as shown, the top plate would be displaced through some small distance, δa (assuming the solid was mechanically attached to the plates). The vertical line AB would be rotated through the small angle, $\delta\beta$, to the new position AB'. We note that to resist the applied force, P, a shearing stress, τ, would be developed at the plate-material interface, and for equilibrium to occur $P = \tau A$ where A is the effective upper

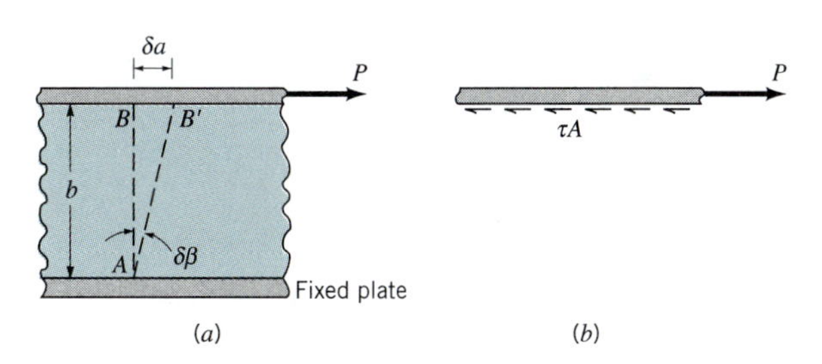

■ **FIGURE 1.2** ***(a)*** **Deformation of material placed between two parallel plates. *(b)* Forces acting on upper plate.**

plate area (Fig. 1.2*b*). It is well known that for elastic solids, such as steel, the small angular displacement, $\delta\beta$ (called the shearing strain), is proportional to the shearing stress, τ, that is developed in the material.

What happens if the solid is replaced with a fluid such as water? We would immediately notice a major difference. When the force P is applied to the upper plate, it will move continuously with a velocity, U (after the initial transient motion has died out) as illustrated in Fig. 1.3. This behavior is consistent with the definition of a fluid—that is, if a shearing stress is applied to a fluid it will deform continuously. A closer inspection of the fluid motion between the two plates would reveal that the fluid in contact with the upper plate moves with the plate velocity, U, and the fluid in contact with the bottom fixed plate has a zero velocity. The fluid between the two plates moves with velocity $u = u(y)$ that would be found to vary linearly, $u = Uy/b$, as illustrated in Fig. 1.3. Thus, a *velocity gradient*, du/dy, is developed in the fluid between the plates. In this particular case the velocity gradient is a constant since $du/dy = U/b$, but in more complex flow situations this would not be true. The experimental observation that the fluid "sticks" to the solid boundaries is a very important one in fluid mechanics and is usually referred to as the *no-slip condition*. All fluids, both liquids and gases, satisfy this condition.

V1.2

Real fluids, even though they may be moving, always "stick" to the solid boundaries that contain them.

In a small time increment, δt, an imaginary vertical line AB in the fluid would rotate through an angle, $\delta\beta$, so that

$$\tan\,\delta\beta \approx \delta\beta = \frac{\delta a}{b}$$

Since $\delta a = U\,\delta t$ it follows that

$$\delta\beta = \frac{U\,\delta t}{b}$$

We note that in this case, $\delta\beta$ is a function not only of the force P (which governs U) but also of time. Thus, it is not reasonable to attempt to relate the shearing stress, τ, to $\delta\beta$ as is done for solids. Rather, we consider the *rate* at which $\delta\beta$ is changing and define the *rate of shearing strain*, $\dot{\gamma}$, as

$$\dot{\gamma} = \lim_{\delta t \to 0} \frac{\delta\beta}{\delta t}$$

which in this instance is equal to

$$\dot{\gamma} = \frac{U}{b} = \frac{du}{dy}$$

A continuation of this experiment would reveal that as the shearing stress, τ, is increased by increasing P (recall that $\tau = P/A$), the rate of shearing strain is increased in direct proportion—that is

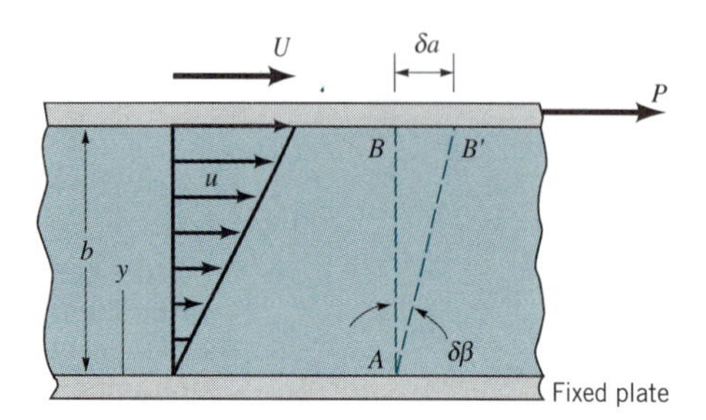

■ **FIGURE 1.3 Behavior of a fluid placed between two parallel plates.**

$$\tau \propto \dot{\gamma}$$

or

$$\tau \propto \frac{du}{dy}$$

This result indicates that for common fluids such as water, oil, gasoline, and air the shearing stress and rate of shearing strain (velocity gradient) can be related with a relationship of the form

V1.3

$$\tau = \mu \frac{du}{dy} \tag{1.9}$$

Dynamic viscosity is the fluid property that relates shearing stress and fluid motion.

where the constant of proportionality is designated by the Greek symbol μ (mu) and is called the *absolute viscosity*, *dynamic viscosity*, or simply the *viscosity* of the fluid. In accordance with Eq. 1.9, plots of τ versus du/dy should be linear with the slope equal to the viscosity as illustrated in Fig. 1.4. The actual value of the viscosity depends on the particular fluid, and for a particular fluid the viscosity is also highly dependent on temperature as illustrated in Fig. 1.4 with the two curves for water. Fluids for which the shearing stress is *linearly* related to the rate of shearing strain (also referred to as rate of angular deformation) are designated as *Newtonian fluids*. Fortunately most common fluids, both liquids and gases, are Newtonian. A more general formulation of Eq. 1.9 which applies to more complex flows of Newtonian fluids is given in Section 6.8.1.

Fluids for which the shearing stress is not linearly related to the rate of shearing strain are designated as *non-Newtonian fluids*. Although there is a variety of types of non-Newtonian fluids, the simplest and most common are shown in Fig. 1.5. The slope of the shearing stress vs rate of shearing strain graph is denoted as the *apparent viscosity*, μ_{ap}. For Newtonian fluids the apparent viscosity is the same as the viscosity and is independent of shear rate.

For *shear thinning fluids* the apparent viscosity decreases with increasing shear rate—the harder the fluid is sheared, the less viscous it becomes. Many colloidal suspensions and polymer solutions are shear thinning. For example, latex paint does not drip from the brush because the shear rate is small and the apparent viscosity is large. However, it flows smoothly

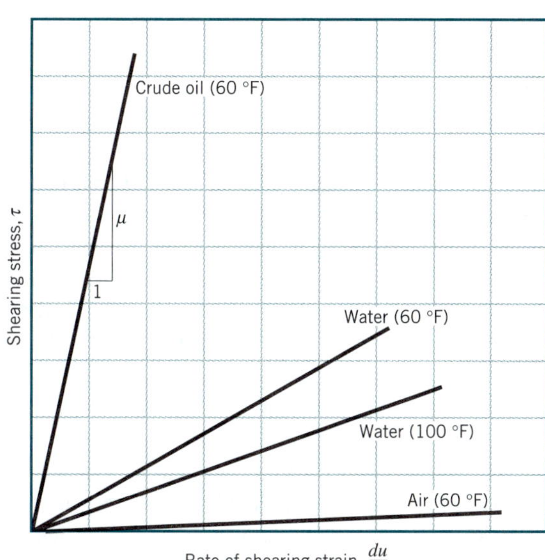

■ **FIGURE 1.4** **Linear variation of shearing stress with rate of shearing strain for common fluids.**

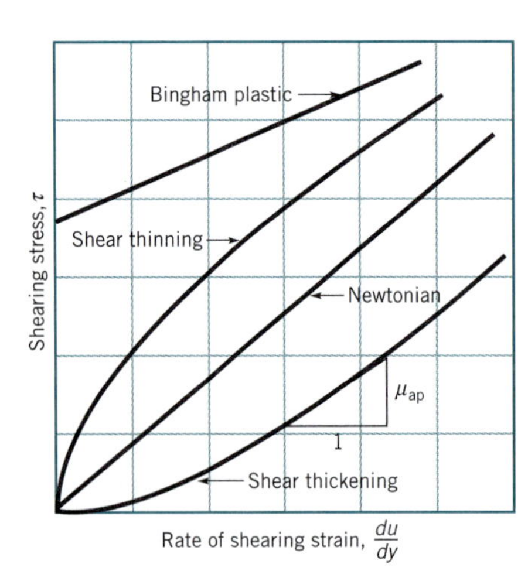

■ FIGURE 1.5 **Variation of shearing stress with rate of shearing strain for several types of fluids, including common non-Newtonian fluids.**

onto the wall because the thin layer of paint between the wall and the brush causes a large shear rate (large du/dy) and a small apparent viscosity.

The various types of non-Newtonian fluids are distinguished by how their apparent viscosity changes with shear rate.

For *shear thickening fluids* the apparent viscosity increases with increasing shear rate—the harder the fluid is sheared, the more viscous it becomes. Common examples of this type of fluid include water-corn starch mixture and water-sand mixture (''quicksand''). Thus, the difficulty in removing an object from quicksand increases dramatically as the speed of removal increases.

The other type of behavior indicated in Fig. 1.5 is that of a *Bingham plastic*, which is neither a fluid nor a solid. Such material can withstand a finite shear stress without motion (therefore, it is not a fluid), but once the yield stress is exceeded it flows like a fluid (hence, it is not a solid). Toothpaste and mayonnaise are common examples of Bingham plastic materials.

V1.4

From Eq. 1.9 it can be readily deduced that the dimensions of viscosity are FTL^{-2}. Thus, in BG units viscosity is given as $\text{lb}\cdot\text{s/ft}^2$ and in SI units as $\text{N}\cdot\text{s/m}^2$. Values of viscosity for several common liquids and gases are listed in Tables 1.5 through 1.8. A quick glance at these tables reveals the wide variation in viscosity among fluids. Viscosity is only mildly dependent on pressure and the effect of pressure is usually neglected. However, as previously mentioned, and as illustrated in Fig. 1.6, viscosity is very sensitive to temperature. For example, as the temperature of water changes from 60 to 100 °F the density decreases by less than 1% but the viscosity decreases by about 40%. It is thus clear that particular attention must be given to temperature when determining viscosity.

Figure 1.6 shows in more detail how the viscosity varies from fluid to fluid and how for a given fluid it varies with temperature. It is to be noted from this figure that the viscosity of liquids decreases with an increase in temperature, whereas for gases an increase in temperature causes an increase in viscosity. This difference in the effect of temperature on the viscosity of liquids and gases can again be traced back to the difference in molecular structure. The liquid molecules are closely spaced, with strong cohesive forces between molecules, and the resistance to relative motion between adjacent layers of fluid is related to these intermolecular forces. As the temperature increases, these cohesive forces are reduced with a corresponding reduction in resistance to motion. Since viscosity is an index of this resistance, it follows that the viscosity is reduced by an increase in temperature. In gases, however, the molecules are widely spaced and intermolecular forces negligible. In this case resistance to relative motion arises due to the exchange of momentum of gas molecules between adjacent layers. As molecules are transported by random motion from a region of low bulk velocity

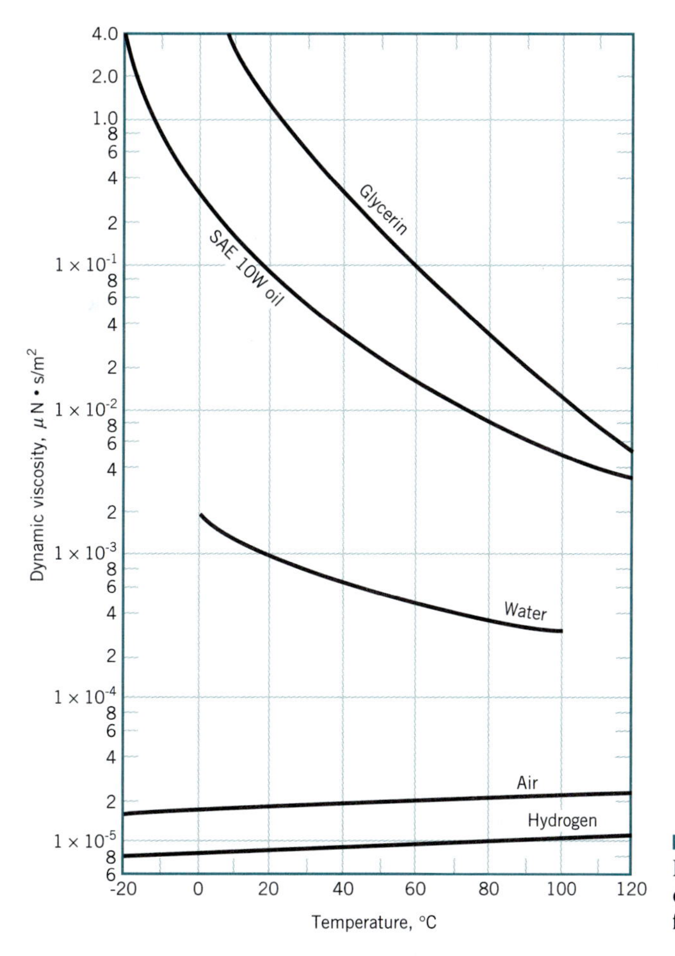

■ FIGURE 1.6 Dynamic (absolute) viscosity of some common fluids as a function of temperature.

to mix with molecules in a region of higher bulk velocity (and vice versa), there is an effective momentum exchange which resists the relative motion between the layers. As the temperature of the gas increases, the random molecular activity increases with a corresponding increase in viscosity.

Viscosity is very sensitive to temperature.

The effect of temperature on viscosity can be closely approximated using two empirical formulas. For gases the *Sutherland equation* can be expressed as

$$\mu = \frac{CT^{3/2}}{T + S} \tag{1.10}$$

where C and S are empirical constants, and T is absolute temperature. Thus, if the viscosity is known at two temperatures, C and S can be determined. Or, if more than two viscosities are known, the data can be correlated with Eq. 1.10 by using some type of curve-fitting scheme.

For liquids an empirical equation that has been used is

$$\mu = De^{B/T} \tag{1.11}$$

where D and B are constants and T is absolute temperature. This equation is often referred to as *Andrade's equation.* As was the case for gases, the viscosity must be known at least for two temperatures so the two constants can be determined. A more detailed discussion of the effect of temperature on fluids can be found in Ref. 1.

Quite often viscosity appears in fluid flow problems combined with the density in the form

$$\nu = \frac{\mu}{\rho}$$

Kinematic viscosity is defined as the ratio of the absolute viscosity to the fluid density.

This ratio is called the *kinematic viscosity* and is denoted with the Greek symbol ν (nu). The dimensions of kinematic viscosity are L^2/T, and the BG units are ft^2/s and SI units are m^2/s. Values of kinematic viscosity for some common liquids and gases are given in Tables 1.5 through 1.8. More extensive tables giving both the dynamic and kinematic viscosities for water and air can be found in Appendix B (Tables B.1 through B.4), and graphs showing the variation in both dynamic and kinematic viscosity with temperature for a variety of fluids are also provided in Appendix B (Figs. B.1 and B.2).

Although in this text we are primarily using BG and SI units, dynamic viscosity is often expressed in the metric CGS (centimeter-gram-second) system with units of dyne·s/cm^2. This combination is called a *poise*, abbreviated P. In the CGS system, kinematic viscosity has units of cm^2/s, and this combination is called a *stoke*, abbreviated St.

A dimensionless combination of variables that is important in the study of viscous flow through pipes is called the *Reynolds number*, Re, defined as $\rho VD/\mu$ where ρ is the fluid density, V the mean fluid velocity, D the pipe diameter, and μ the fluid viscosity. A Newtonian fluid having a viscosity of $0.38\ \text{N·s/m}^2$ and a specific gravity of 0.91 flows through a 25-mm-diameter pipe with a velocity of 2.6 m/s. Determine the value of the Reynolds number using (a) SI units, and (b) BG units.

SOLUTION

(a) The fluid density is calculated from the specific gravity as

$$\rho = SG\ \rho_{\text{H}_2\text{O@4°C}} = 0.91\ (1000\ \text{kg/m}^3) = 910\ \text{kg/m}^3$$

and from the definition of the Reynolds number

$$\text{Re} = \frac{\rho VD}{\mu} = \frac{(910\ \text{kg/m}^3)(2.6\ \text{m/s})(25\ \text{mm})(10^{-3}\ \text{m/mm})}{0.38\ \text{N·s/m}^2}$$

$$= 156\ (\text{kg·m/s}^2)/\text{N}$$

However, since $1\ \text{N} = 1\ \text{kg·m/s}^2$ it follows that the Reynolds number is unitless—that is,

$$\text{Re} = 156 \qquad \textbf{(Ans)}$$

The value of any dimensionless quantity does not depend on the system of units used if all variables that make up the quantity are expressed in a consistent set of units. To check this we will calculate the Reynolds number using BG units.

(b) We first convert all the SI values of the variables appearing in the Reynolds number to BG values by using the conversion factors from Table 1.4. Thus,

$$\rho = (910\ \text{kg/m}^3)(1.940 \times 10^{-3}) = 1.77\ \text{slugs/ft}^3$$

$$V = (2.6\ \text{m/s})(3.281) = 8.53\ \text{ft/s}$$

$$D = (0.025\ \text{m})(3.281) = 8.20 \times 10^{-2}\ \text{ft}$$

$$\mu = (0.38\ \text{N·s/m}^2)(2.089 \times 10^{-2}) = 7.94 \times 10^{-3}\ \text{lb·s/ft}^2$$

and the value of the Reynolds number is

$$\text{Re} = \frac{(1.77 \text{ slugs/ft}^3)(8.53 \text{ ft/s})(8.20 \times 10^{-2} \text{ ft})}{7.94 \times 10^{-3} \text{ lb·s/ft}^2}$$

$$= 156 \text{ (slug·ft/s}^2\text{)/lb} = 156 \qquad \textbf{(Ans)}$$

since 1 lb = 1 slug·ft/s^2. The values from part (a) and part (b) are the same, as expected. Dimensionless quantities play an important role in fluid mechanics and the significance of the Reynolds number as well as other important dimensionless combinations will be discussed in detail in Chapter 7. It should be noted that in the Reynolds number it is actually the ratio μ/ρ that is important, and this is the property that we have defined as the kinematic viscosity.

EXAMPLE 1.5

The velocity distribution for the flow of a Newtonian fluid between two wide, parallel plates (see Fig. E1.5) is given by the equation

$$u = \frac{3V}{2}\left[1 - \left(\frac{y}{h}\right)^2\right]$$

where V is the mean velocity. The fluid has a viscosity of 0.04 lb·s/ft^2. When V = 2 ft/s and h = 0.2 in. determine: (a) the shearing stress acting on the bottom wall, and (b) the shearing stress acting on a plane parallel to the walls and passing through the centerline (midplane).

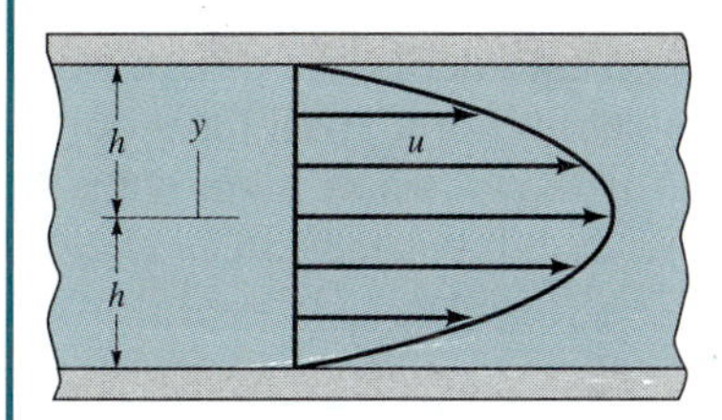

■ FIGURE E1.5

SOLUTION

For this type of parallel flow the shearing stress is obtained from Eq. 1.9,

$$\tau = \mu \frac{du}{dy} \qquad \textbf{(1)}$$

Thus, if the velocity distribution, $u = u(y)$ is known, the shearing stress can be determined at all points by evaluating the velocity gradient, du/dy. For the distribution given

$$\frac{du}{dy} = -\frac{3Vy}{h^2} \qquad \textbf{(2)}$$

(a) Along the bottom wall $y = -h$ so that (from Eq. 2)

$$\frac{du}{dy} = \frac{3V}{h}$$

and therefore the shearing stress is

$$\tau_{\substack{\text{bottom}\\\text{wall}}} = \mu\left(\frac{3V}{h}\right) = \frac{(0.04 \text{ lb}\cdot\text{s/ft}^2)(3)(2 \text{ ft/s})}{(0.2 \text{ in.})(1 \text{ ft/12 in.})}$$

$$= 14.4 \text{ lb/ft}^2 \text{ (in direction of flow)} \qquad \textbf{(Ans)}$$

This stress creates a drag on the wall. Since the velocity distribution is symmetrical, the shearing stress along the upper wall would have the same magnitude and direction.

(b) Along the midplane where $y = 0$ it follows from Eq. 2 that

$$\frac{du}{dy} = 0$$

and thus the shearing stress is

$$\tau_{\text{midplane}} = 0 \qquad \textbf{(Ans)}$$

From Eq. 2 we see that the velocity gradient (and therefore the shearing stress) varies linearly with y and in this particular example varies from 0 at the center of the channel to 14.4 lb/ft^2 at the walls. For the more general case the actual variation will, of course, depend on the nature of the velocity distribution.

1.7 Compressibility of Fluids

1.7.1 Bulk Modulus

An important question to answer when considering the behavior of a particular fluid is how easily can the volume (and thus the density) of a given mass of the fluid be changed when there is a change in pressure? That is, how compressible is the fluid? A property that is commonly used to characterize compressibility is the *bulk modulus*, E_v, defined as

$$E_v = -\frac{dp}{d\mathcal{V}/\mathcal{V}} \qquad \textbf{(1.12)}$$

where dp is the differential change in pressure needed to create a differential change in volume, $d\mathcal{V}$, of a volume $\mathcal{V}$. The negative sign is included since an increase in pressure will cause a decrease in volume. Since a decrease in volume of a given mass, $m = \rho\mathcal{V}$, will result in an increase in density, Eq. 1.12 can also be expressed as

$$E_v = \frac{dp}{d\rho/\rho} \qquad \textbf{(1.13)}$$

Liquids are usually considered to be incompressible, whereas gases are generally considered compressible.

The bulk modulus (also referred to as the *bulk modulus of elasticity*) has dimensions of pressure, FL^{-2}. In BG units values for E_v are usually given as lb/in.^{2} (psi) and in SI units as N/m^2 (Pa). Large values for the bulk modulus indicate that the fluid is relatively incompressible—that is, it takes a large pressure change to create a small change in volume. As expected, values of E_v for common liquids are large (see Tables 1.5 and 1.6). For example, at atmospheric pressure and a temperature of 60 °F it would require a pressure of 3120 psi to compress a unit volume of water 1%. This result is representative of the compressibility of liquids. Since such large pressures are required to effect a change in volume, we conclude that liquids can be considered as *incompressible* for most practical engineering applications.

As liquids are compressed the bulk modulus increases, but the bulk modulus near atmospheric pressure is usually the one of interest. The use of bulk modulus as a property describing compressibility is most prevalent when dealing with liquids, although the bulk modulus can also be determined for gases.

1.7.2 Compression and Expansion of Gases

When gases are compressed (or expanded) the relationship between pressure and density depends on the nature of the process. If the compression or expansion takes place under constant temperature conditions (*isothermal process*), then from Eq. 1.8

$$\frac{p}{\rho} = \text{constant} \tag{1.14}$$

If the compression or expansion is frictionless and no heat is exchanged with the surroundings (*isentropic process*), then

$$\frac{p}{\rho^k} = \text{constant} \tag{1.15}$$

The value of the bulk modulus depends on the type of process involved.

where k is the ratio of the specific heat at constant pressure, c_p, to the specific heat at constant volume, c_v (i.e., $k = c_p/c_v$). The two specific heats are related to the gas constant, R, through the equation $R = c_p - c_v$. As was the case for the ideal gas law, the pressure in both Eqs. 1.14 and 1.15 must be expressed as an absolute pressure. Values of k for some common gases are given in Tables 1.7 and 1.8, and for air over a range of temperatures, in Appendix B (Tables B.3 and B.4).

With explicit equations relating pressure and density the bulk modulus for gases can be determined by obtaining the derivative $dp/d\rho$ from Eq. 1.14 or 1.15 and substituting the results into Eq. 1.13. It follows that for an isothermal process

$$E_v = p \tag{1.16}$$

and for an isentropic process

$$E_v = kp \tag{1.17}$$

Note that in both cases the bulk modulus varies directly with pressure. For air under standard atmospheric conditions with $p = 14.7$ psi (abs) and $k = 1.40$, the isentropic bulk modulus is 20.6 psi. A comparison of this figure with that for water under the same conditions ($E_v =$ 312,000 psi) shows that air is approximately 15,000 times as compressible as water. It is thus clear that in dealing with gases greater attention will need to be given to the effect of compressibility on fluid behavior. However, as will be discussed further in later sections, gases can often be treated as incompressible fluids if the changes in pressure are small.

A cubic foot of helium at an absolute pressure of 14.7 psi is compressed isentropically to $\frac{1}{2}$ ft^3. What is the final pressure?

SOLUTION

For an isentropic compression

$$\frac{p_i}{\rho_i^k} = \frac{p_f}{\rho_f^k}$$

where the subscripts i and f refer to initial and final states, respectively. Since we are interested in the final pressure, p_f, it follows that

$$p_f = \left(\frac{\rho_f}{\rho_i}\right)^k p_i$$

As the volume is reduced by one half, the density must double, since the mass of the gas remains constant. Thus,

$$p_f = (2)^{1.66}(14.7 \text{ psi}) = 46.5 \text{ psi (abs)}$$ **(Ans)**

1.7.3 Speed of Sound

The velocity at which small disturbances propagate in a fluid is called the speed of sound.

Another important consequence of the compressibility of fluids is that disturbances introduced at some point in the fluid propagate at a finite velocity. For example, if a fluid is flowing in a pipe and a valve at the outlet is suddenly closed (thereby creating a localized disturbance), the effect of the valve closure is not felt instantaneously upstream. It takes a finite time for the increased pressure created by the valve closure to propagate to an upstream location. Similarly, a loud speaker diaphragm causes a localized disturbance as it vibrates, and the small change in pressure created by the motion of the diaphragm is propagated through the air with a finite velocity. The velocity at which these small disturbances propagate is called the *acoustic velocity* or the *speed of sound*, c. It will be shown in Chapter 11 that the speed of sound is related to changes in pressure and density of the fluid medium through the equation

$$c = \sqrt{\frac{dp}{d\rho}} \qquad \textbf{(1.18)}$$

or in terms of the bulk modulus defined by Eq. 1.13

$$c = \sqrt{\frac{E_v}{\rho}} \qquad \textbf{(1.19)}$$

Since the disturbance is small, there is negligible heat transfer and the process is assumed to be isentropic. Thus, the pressure-density relationship used in Eq. 1.18 is that for an isentropic process.

For gases undergoing an isentropic process, $E_v = kp$ (Eq. 1.17) so that

$$c = \sqrt{\frac{kp}{\rho}}$$

and making use of the ideal gas law, it follows that

$$c = \sqrt{kRT} \qquad \textbf{(1.20)}$$

Thus, for ideal gases the speed of sound is proportional to the square root of the absolute temperature. For example, for air at 60 °F with $k = 1.40$ and $R = 1716$ ft·lb/slug·°R it follows that $c = 1117$ ft/s. The speed of sound in air at various temperatures can be found in Appendix B (Tables B.3 and B.4). Equation 1.19 is also valid for liquids, and values of E_v can be used to determine the speed of sound in liquids. For water at 20 °C, $E_v = 2.19$ GN/m^2 and $\rho = 998.2$ kg/m^3 so that $c = 1481$ m/s or 4860 ft/s. Note that the speed of sound in water is much higher than in air. If a fluid were truly incompressible ($E_v = \infty$) the speed of sound would be infinite. The speed of sound in water for various temperatures can be found in Appendix B (Tables B.1 and B.2).

A jet aircraft flies at a speed of 550 mph at an altitude of 35,000 ft, where the temperature is −66 °F. Determine the ratio of the speed of the aircraft, V, to that of the speed of sound, c, at the specified altitude. Assume $k = 1.40$.

SOLUTION

From Eq. 1.20 the speed of sound can be calculated as

$$c = \sqrt{kRT} = \sqrt{(1.40)(1716 \text{ ft}\cdot\text{lb/slug}\cdot{}^\circ\text{R})(-66 + 460)\ {}^\circ\text{R}}$$
$$= 973 \text{ ft/s}$$

Since the air speed is

$$V = \frac{(550 \text{ mi/hr})(5280 \text{ ft/mi})}{(3600 \text{ s/hr})} = 807 \text{ ft/s}$$

the ratio is

$$\frac{V}{c} = \frac{807 \text{ ft/s}}{973 \text{ ft/s}} = 0.829 \qquad \textbf{(Ans)}$$

This ratio is called the *Mach number*, Ma. If $\text{Ma} < 1.0$ the aircraft is flying at *subsonic* speeds, whereas for $\text{Ma} > 1.0$ it is flying at *supersonic* speeds. The Mach number is an important dimensionless parameter used in the study of the flow of gases at high speeds and will be further discussed in Chapters 7, 9, and 11.

1.8 Vapor Pressure

It is a common observation that liquids such as water and gasoline will evaporate if they are simply placed in a container open to the atmosphere. Evaporation takes place because some liquid molecules at the surface have sufficient momentum to overcome the intermolecular cohesive forces and escape into the atmosphere. If the container is closed with a small air space left above the surface, and this space evacuated to form a vacuum, a pressure will develop in the space as a result of the vapor that is formed by the escaping molecules. When an equilibrium condition is reached so that the number of molecules leaving the surface is equal to the number entering, the vapor is said to be saturated and the pressure that the vapor exerts on the liquid surface is termed the *vapor pressure*.

Since the development of a vapor pressure is closely associated with molecular activity, the value of vapor pressure for a particular liquid depends on temperature. Values of vapor pressure for water at various temperatures can be found in Appendix B (Tables B.1 and B.2), and the values of vapor pressure for several common liquids at room temperatures are given in Tables 1.5 and 1.6.

A liquid boils when the pressure is reduced to the vapor pressure.

Boiling, which is the formation of vapor bubbles within a fluid mass, is initiated when the absolute pressure in the fluid reaches the vapor pressure. As commonly observed in the kitchen, water at standard atmospheric pressure will boil when the temperature reaches 212 °F (100 °C)—that is, the vapor pressure of water at 212 °F is 14.7 psi (abs). However, if we attempt to boil water at a higher elevation, say 10,000 ft above sea level, where the atmospheric pressure is 10.1 psi (abs), we find that boiling will start when the temperature is about

193 °F. At this temperature the vapor pressure of water is 10.1 psi (abs). Thus, boiling can be induced at a given pressure acting on the fluid by raising the temperature, or at a given fluid temperature by lowering the pressure.

In flowing liquids it is possible for the pressure in localized regions to reach vapor pressure thereby causing cavitation.

An important reason for our interest in vapor pressure and boiling lies in the common observation that in flowing fluids it is possible to develop very low pressure due to the fluid motion, and if the pressure is lowered to the vapor pressure, boiling will occur. For example, this phenomenon may occur in flow through the irregular, narrowed passages of a valve or pump. When vapor bubbles are formed in a flowing fluid they are swept along into regions of higher pressure where they suddenly collapse with sufficient intensity to actually cause structural damage. The formation and subsequent collapse of vapor bubbles in a flowing fluid, called *cavitation*, is an important fluid flow phenomenon to be given further attention in Chapters 3 and 7.

1.9 Surface Tension

V1.5

At the interface between a liquid and a gas, or between two immiscible liquids, forces develop in the liquid surface which cause the surface to behave as if it were a ''skin'' or ''membrane'' stretched over the fluid mass. Although such a skin is not actually present, this conceptual analogy allows us to explain several commonly observed phenomena. For example, a steel needle will float on water if placed gently on the surface because the tension developed in the hypothetical skin supports the needle. Small droplets of mercury will form into spheres when placed on a smooth surface because the cohesive forces in the surface tend to hold all the molecules together in a compact shape. Similarly, discrete water droplets will form when placed on a newly waxed surface.

These various types of surface phenomena are due to the unbalanced cohesive forces acting on the liquid molecules at the fluid surface. Molecules in the interior of the fluid mass are surrounded by molecules that are attracted to each other equally. However, molecules along the surface are subjected to a net force toward the interior. The apparent physical consequence of this unbalanced force along the surface is to create the hypothetical skin or membrane. A tensile force may be considered to be acting in the plane of the surface along any line in the surface. The intensity of the molecular attraction per unit length along any line in the surface is called the *surface tension* and is designated by the Greek symbol σ (sigma). For a given liquid the surface tension depends on temperature as well as the other fluid it is in contact with at the interface. The dimensions of surface tension are FL^{-1} with BG units of lb/ft and SI units of N/m. Values of surface tension for some common liquids (in contact with air) are given in Tables 1.5 and 1.6 and in Appendix B (Tables B.1 and B.2) for water at various temperatures. The value of the surface tension decreases as the temperature increases.

The pressure inside a drop of fluid can be calculated using the free-body diagram in Fig. 1.7. If the spherical drop is cut in half (as shown) the force developed around the edge

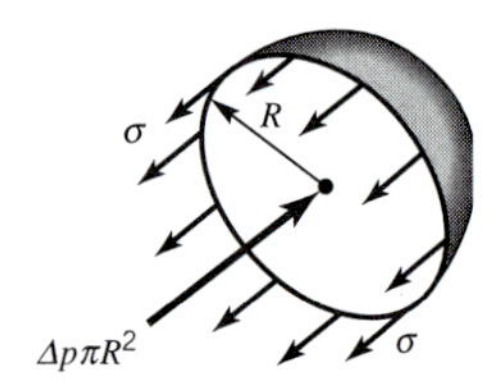

■ FIGURE 1.7 **Forces acting on one-half of a liquid drop.**

due to surface tension is $2\pi R\sigma$. This force must be balanced by the pressure difference, Δp, between the internal pressure, p_i, and the external pressure, p_e, acting over the circular area, πR^2. Thus,

$$2\pi R\sigma = \Delta p \; \pi R^2$$

or

$$\Delta p = p_i - p_e = \frac{2\sigma}{R} \tag{1.21}$$

It is apparent from this result that the pressure inside the drop is greater than the pressure surrounding the drop. (Would the pressure on the inside of a bubble of water be the same as that on the inside of a drop of water of the same diameter and at the same temperature?)

Capillary action in small tubes, which involves a liquid–gas–solid interface, is caused by surface tension.

Among common phenomena associated with surface tension is the rise (or fall) of a liquid in a capillary tube. If a small open tube is inserted into water, the water level in the tube will rise above the water level outside the tube as is illustrated in Fig. 1.8*a*. In this situation we have a liquid–gas–solid interface. For the case illustrated there is an attraction (adhesion) between the wall of the tube and liquid molecules which is strong enough to overcome the mutual attraction (cohesion) of the molecules and pull them up the wall. Hence, the liquid is said to *wet* the solid surface.

The height, h, is governed by the value of the surface tension, σ, the tube radius, R, the specific weight of the liquid, γ, and the *angle of contact*, θ, between the fluid and tube. From the free-body diagram of Fig. 1.8b we see that the vertical force due to the surface tension is equal to $2\pi R\sigma \cos \theta$ and the weight is $\gamma\pi R^2 h$ and these two forces must balance for equilibrium. Thus,

$$\gamma\pi R^2 h = 2\pi R\sigma \cos \theta$$

so that the height is given by the relationship

$$h = \frac{2\sigma \cos \theta}{\gamma R} \tag{1.22}$$

The angle of contact is a function of both the liquid and the surface. For water in contact with clean glass $\theta \approx 0°$. It is clear from Eq. 1.22 that the height is inversely proportional to the tube radius, and therefore the rise of a liquid in a tube as a result of capillary action becomes increasingly pronounced as the tube radius is decreased.

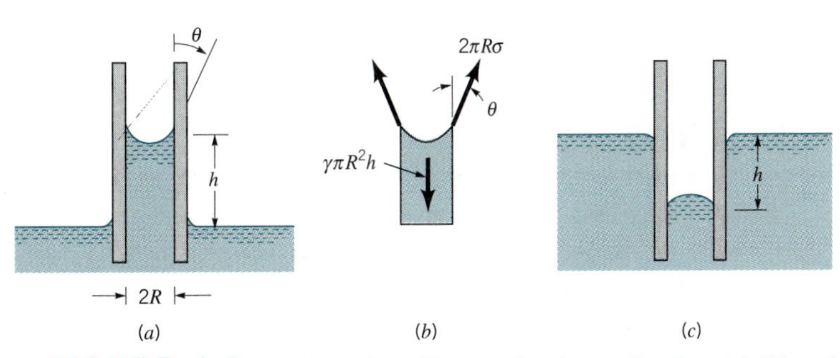

■ **FIGURE 1.8 Effect of capillary action in small tubes. (*a*) Rise of column for a liquid that wets the tube. (*b*) Free-body diagram for calculating column height. (*c*) Depression of column for a nonwetting liquid.**

Pressures are sometimes determined by measuring the height of a column of liquid in a vertical tube. What diameter of clean glass tubing is required so that the rise of water at 20 °C in a tube due to capillary action (as opposed to pressure in the tube) is less than 1.0 mm?

SOLUTION

From Eq. 1.22

$$h = \frac{2\sigma \cos \theta}{\gamma R}$$

so that

$$R = \frac{2\sigma \cos \theta}{\gamma h}$$

For water at 20 °C (from Table B.2), $\sigma = 0.0728$ N/m and $\gamma = 9.789$ kN/m^3. Since $\theta \approx 0°$ it follows that for $h = 1.0$ mm,

$$R = \frac{2(0.0728 \text{ N/m})(1)}{(9.789 \times 10^3 \text{ N/m}^3)(1.0 \text{ mm})(10^{-3} \text{ m/mm})} = 0.0149 \text{ m}$$

and the minimum required tube diameter, D, is

$$D = 2R = 0.0298 \text{ m} = 29.8 \text{ mm}$$ **(Ans)**

If adhesion of molecules to the solid surface is weak compared to the cohesion between molecules, the liquid will not wet the surface and the level in a tube placed in a nonwetting liquid will actually be depressed as shown in Fig. 1.8*c*. Mercury is a good example of a nonwetting liquid when it is in contact with a glass tube. For nonwetting liquids the angle of contact is greater than 90°, and for mercury in contact with clean glass $\theta \approx 130°$.

Surface tension effects play a role in many fluid mechanics problems associated with liquid–gas, liquid–liquid, or liquid–gas–solid interfaces.

Surface tension effects play a role in many fluid mechanics problems including the movement of liquids through soil and other porous media, flow of thin films, formation of drops and bubbles, and the breakup of liquid jets. Surface phenomena associated with liquid-gas, liquid-liquid, liquid-gas-solid interfaces are exceedingly complex, and a more detailed and rigorous discussion of them is beyond the scope of this text. Fortunately, in many fluid mechanics problems, surface phenomena, as characterized by surface tension, are not important, since inertial, gravitational, and viscous forces are much more dominant.

1.10 A Brief Look Back in History

Before proceeding with our study of fluid mechanics, we should pause for a moment to consider the history of this important engineering science. As is true of all basic scientific and engineering disciplines, their actual beginnings are only faintly visible through the haze of early antiquity. But, we know that interest in fluid behavior dates back to the ancient civilizations. Through necessity there was a practical concern about the manner in which spears and arrows could be propelled through the air, in the development of water supply and irrigation systems, and in the design of boats and ships. These developments were of course based on trial and error procedures without any knowledge of mathematics or mechanics.

Some of the earliest writings that pertain to modern fluid mechanics can be traced back to the ancient Greek civilization and subsequent Roman Empire.

However, it was the accumulation of such empirical knowledge that formed the basis for further development during the emergence of the ancient Greek civilization and the subsequent rise of the Roman Empire. Some of the earliest writings that pertain to modern fluid mechanics are those of Archimedes (287–212 B.C.), a Greek mathematician and inventor who first expressed the principles of hydrostatics and flotation. Elaborate water supply systems were built by the Romans during the period from the fourth century B.C. through the early Christian period, and Sextus Julius Frontinus (A.D. 40–103), a Roman engineer, described these systems in detail. However, for the next 1000 years during the Middle Ages (also referred to as the Dark Ages), there appears to have been little added to further understanding of fluid behavior.

Beginning with the Renaissance period (about the fifteenth century) a rather continuous series of contributions began that forms the basis of what we consider to be the science of fluid mechanics. Leonardo da Vinci (1452–1519) described through sketches and writings many different types of flow phenomena. The work of Galileo Galilei (1564–1642) marked the beginning of experimental mechanics. Following the early Renaissance period and during the seventeenth and eighteenth centuries, numerous significant contributions were made. These include theoretical and mathematical advances associated with the famous names of Newton, Bernoulli, Euler, and d'Alembert. Experimental aspects of fluid mechanics were also advanced during this period, but unfortunately the two different approaches, theoretical and experimental, developed along separate paths. *Hydrodynamics* was the term associated with the theoretical or mathematical study of idealized, frictionless fluid behavior, with the term *hydraulics* being used to describe the applied or experimental aspects of real fluid behavior, particularly the behavior of water. Further contributions and refinements were made to both theoretical hydrodynamics and experimental hydraulics during the nineteenth century, with the general differential equations describing fluid motions that are used in modern fluid mechanics being developed in this period. Experimental hydraulics became more of a science, and many of the results of experiments performed during the nineteenth century are still used today.

At the beginning of the twentieth century both the fields of theoretical hydrodynamics and experimental hydraulics were highly developed, and attempts were being made to unify the two. In 1904 a classic paper was presented by a German professor, Ludwig Prandtl (1857–1953), who introduced the concept of a "fluid boundary layer," which laid the foundation for the unification of the theoretical and experimental aspects of fluid mechanics. Prandtl's idea was that for flow next to a solid boundary a thin fluid layer (boundary layer) develops in which friction is very important, but outside this layer the fluid behaves very Gmuch like a frictionless fluid. This relatively simple concept provided the necessary impetus for the resolution of the conflict between the hydrodynamicists and the hydraulicists. Prandtl is generally accepted as the founder of modern fluid mechanics.

Also, during the first decade of the twentieth century, powered flight was first successfully demonstrated with the subsequent vastly increased interest in *aerodynamics*. Because the design of aircraft required a degree of understanding of fluid flow and an ability to make accurate predictions of the effect of air flow on bodies, the field of aerodynamics provided a great stimulus for the many rapid developments in fluid mechanics that have taken place during the twentieth century.

As we proceed with our study of the fundamentals of fluid mechanics, we will continue to note the contributions of many of the pioneers in the field. Table 1.9 provides a chronological listing of some of these contributors and reveals the long journey that makes up the history of fluid mechanics. This list is certainly not comprehensive with regard to all of the past contributors, but includes those who are mentioned in this text. As mention is made in succeeding chapters of the various individuals listed in Table 1.9, a quick glance at this table will reveal where they fit into the historical chain.

■ **TABLE 1.9**

Chronological Listing of Some Contributors to the Science of Fluid Mechanics Noted in the Text[a]

ARCHIMEDES (287–212 B.C.)
Established elementary principles of buoyancy and flotation.

SEXTUS JULIUS FRONTINUS (A.D. 40–103)
Wrote treatise on Roman methods of water distribution.

LEONARDO da VINCI (1452–1519)
Expressed elementary principle of continuity; observed and sketched many basic flow phenomena; suggested designs for hydraulic machinery.

GALILEO GALILEI (1564–1642)
Indirectly stimulated experimental hydraulics; revised Aristotelian concept of vacuum.

EVANGELISTA TORRICELLI (1608–1647)
Related barometric height to weight of atmosphere, and form of liquid jet to trajectory of free fall.

BLAISE PASCAL (1623–1662)
Finally clarified principles of barometer, hydraulic press, and pressure transmissibility.

ISAAC NEWTON (1642–1727)
Explored various aspects of fluid resistance–inertial, viscous, and wave; discovered jet contraction.

HENRI de PITOT (1695–1771)
Constructed double-tube device to indicate water velocity through differential head.

DANIEL BERNOULLI (1700–1782)
Experimented and wrote on many phases of fluid motion, coining name "hydrodynamics"; devised manometry technique and adapted primitive energy principle to explain velocity-head indication; proposed jet propulsion.

LEONHARD EULER (1707–1783)
First explained role of pressure in fluid flow; formulated basic equations of motion and so-called Bernoulli theorem; introduced concept of cavitation and principle of centrifugal machinery.

JEAN le ROND d'ALEMBERT (1717–1783)
Originated notion of velocity and acceleration components, differential expression of continuity, and paradox of zero resistance to steady nonuniform motion.

ANTOINE CHEZY (1718–1798)
Formulated similarity parameter for predicting flow characteristics of one channel from measurements on another.

GIOVANNI BATTISTA VENTURI (1746–1822)
Performed tests on various forms of mouthpieces–in particular, conical contractions and expansions.

LOUIS MARIE HENRI NAVIER (1785–1836)
Extended equations of motion to include "molecular" forces.

AUGUSTIN LOUIS de CAUCHY (1789–1857)
Contributed to the general field of theoretical hydrodynamics and to the study of wave motion.

GOTTHILF HEINRICH LUDWIG HAGEN (1797–1884)
Conducted original studies of resistance in and transition between laminar and turbulent flow.

JEAN LOUIS POISEUILLE (1799–1869)
Performed meticulous tests on resistance of flow through capillary tubes.

HENRI PHILIBERT GASPARD DARCY (1803–1858)
Performed extensive tests on filtration and pipe resistance; initiated open-channel studies carried out by Bazin.

JULIUS WEISBACH (1806–1871)
Incorporated hydraulics in treatise on engineering mechanics, based on original experiments; noteworthy for flow patterns, nondimensional coefficients, weir, and resistance equations.

WILLIAM FROUDE (1810–1879)
Developed many towing-tank techniques, in particular the conversion of wave and boundary layer resistance from model to prototype scale.

ROBERT MANNING (1816–1897)
Proposed several formulas for open-channel resistance.

GEORGE GABRIEL STOKES (1819–1903)
Derived analytically various flow relationships ranging from wave mechanics to viscous resistance—particularly that for the settling of spheres.

ERNST MACH (1838–1916)
One of the pioneers in the field of supersonic aerodynamics.

OSBORNE REYNOLDS (1842–1912)
Described original experiments in many fields—cavitation, river model similarity, pipe resistance—and devised two parameters for viscous flow; adapted equations of motion of a viscous fluid to mean conditions of turbulent flow.

JOHN WILLIAM STRUTT, LORD RAYLEIGH (1842–1919)
Investigated hydrodynamics of bubble collapse, wave motion, jet instability, laminar flow analogies, and dynamic similarity.

VINCENZ STROUHAL (1850–1922)
Investigated the phenomenon of "singing wires."

The rich history of fluid mechanics is fascinating, and many of the contributions of the pioneers in the field are noted in the succeeding chapters.

■ TABLE 1.9 (continued)

EDGAR BUCKINGHAM (1867–1940)
Stimulated interest in the United States in the use of dimensional analysis.

MORITZ WEBER (1871–1951)
Emphasized the use of the principles of similitude in fluid flow studies and formulated a capillarity similarity parameter.

LUDWIG PRANDTL (1875–1953)
Introduced concept of the boundary layer and is generally considered to be the father of present-day fluid mechanics.

LEWIS FERRY MOODY (1880–1953)
Provided many innovations in the field of hydraulic machinery. Proposed a method of correlating pipe resistance data which is widely used.

THEODOR VON KÁRMÁN (1881–1963)
One of the recognized leaders of twentieth century fluid mechanics. Provided major contributions to our understanding of surface resistance, turbulence, and wake phenomena.

PAUL RICHARD HEINRICH BLASIUS (1883–1970)
One of Prandtl's students who provided an analytical solution to the boundary layer equations. Also, demonstrated that pipe resistance was related to the Reynolds number.

[a]Adapted from Ref. 2; used by permission of the Iowa Institute of Hydraulic Research, The University of Iowa.

It is, of course, impossible to summarize the rich history of fluid mechanics in a few paragraphs. Only a brief glimpse is provided, and we hope it will stir your interest. References 2 to 5 are good starting points for further study, and in particular Ref. 2 provides an excellent, broad, easily read history. Try it—you might even enjoy it!

References

1. Reid, R. C., Prausnitz, J. M., and Sherwood, T. K., *The Properties of Gases and Liquids*, 3rd Ed., McGraw-Hill, New York, 1977.
2. Rouse, H. and Ince, S., *History of Hydraulics*, Iowa Institute of Hydraulic Research, Iowa City, 1957, Dover, New York, 1963.
3. Tokaty, G. A., *A History and Philosophy of Fluidmechanics*, G. T. Foulis and Co., Ltd., Oxfordshire, Great Britain, 1971.
4. Rouse, H., *Hydraulics in the United States 1776–1976*, Iowa Institute of Hydraulic Research, Iowa City, Iowa, 1976.
5. Garbrecht, G., ed., *Hydraulics and Hydraulic Research—A Historical Review*, A. A. Balkema, Rotterdam, Netherlands, 1987.
6. Brenner, M. P., Shi, X. D., Eggens, J., and Nagel, S. R., *Physics of Fluids*, Vol. 7, No. 9, 1995.
7. Shi, X. D., Brenner, M. P., and Nagel, S. R., *Science*, Vol. 265, 1994.

Review Problems

Note: Problems designated with (R) are review problems. The phrases within parentheses refer to the main topics to be used in solving the problems. Complete, detailed solutions to these review problems can be found in the supplement titled *Student Solution Manual for Fundamentals of Fluid Mechanics*, by Munson, Young, and Okiishi (John Wiley and Sons, New York, 1997).

1.1R (Dimensions) During a study of a certain flow system the following equation relating the pressures p_1 and p_2 at two points was developed:

$$p_2 = p_1 + \frac{f\ell V}{Dg}$$

In this equation V is a velocity, ℓ the distance between the two points, D a diameter, g the acceleration of gravity, and f a dimensionless coefficient. Is the equation dimensionally consistent?

(ANS: No)

1.2R (Dimensions) If V is a velocity, ℓ a length, $\mathscr{W}$ a weight, and μ a fluid property having dimensions of $FL^{-2}T$, determine the dimensions of: **(a)** $V\ell\mathscr{W}/\mu$, **(b)** $\mathscr{W}\mu\ell$, **(c)** $V\mu/\ell$, and **(d)** $V\ell^2\mu/\mathscr{W}$.

(ANS: L^4T^{-2}; $F^2L^{-1}T$; FL^{-2}; L)

1.3R (Units) Make use of Table 1.4 to express the following quantities in BG units: **(a)** 465 W, **(b)** 92.1 J, **(c)** 536 N/m^2, **(d)** 85.9 mm^3, **(e)** 386 kg/m^2.

(ANS: 3.43×10^2 ft·lb/s; 67.9 ft·lb; 11.2 lb/ft^2; 3.03×10^{-6} ft^3; 2.46 $slugs/ft^2$)

1.4R (Units) A person weighs 165 lb at the earth's surface. Determine the person's mass in slugs, kilograms, and pounds mass.

(ANS: 5.12 slugs; 74.8 kg; 165 lbm)

1.5R (Specific gravity) Make use of Fig. 1.1 to determine the specific gravity of water at 22 and 89 °C. What is the specific volume of water at these two temperatures?

(ANS: 0.998; 0.966; $1.002 \times 10^{-3} m^3/kg$; 1.035×10^{-3} m^3/kg)

1.6R (Specific weight) A 1-ft-diameter cylindrical tank that is 5 ft long weighs 125 lb and is filled with a liquid having a specific weight of 69.6 lb/ft^3. Determine the vertical force required to give the tank an upward acceleration of 9 ft/s^2.

(ANS: 509 lb up)

1.7R (Ideal gas law) Calculate the density and specific weight of air at a gage pressure of 100 psi and a temperature of 100 °F. Assume standard atmospheric pressure.

(ANS: 1.72×10^{-2} $slugs/ft^3$; 0.554 lb/ft^3)

1.8R (Ideal gas law) A large dirigible having a volume of 90,000 m^3 contains helium under standard atmospheric conditions [pressure = 101 kPa (abs) and temperature = 15 °C]. Determine the density and total weight of the helium.

(ANS: 0.169 kg/m^3; $1.49 \times 10^5 N$)

1.9R (Viscosity) A Newtonian fluid having a specific gravity of 0.92 and a kinematic viscosity of 4×10^{-4} m^2/s flows past a fixed surface. The velocity profile near the surface is shown in Fig. P1.9R. Determine the magnitude and direction of the shearing stress developed on the plate. Express your answer in terms of U and δ, with U and δ expressed in units of meters per second and meters, respectively.

(ANS: $0.578\ U/\delta$ N/m^2 acting to right on plate)

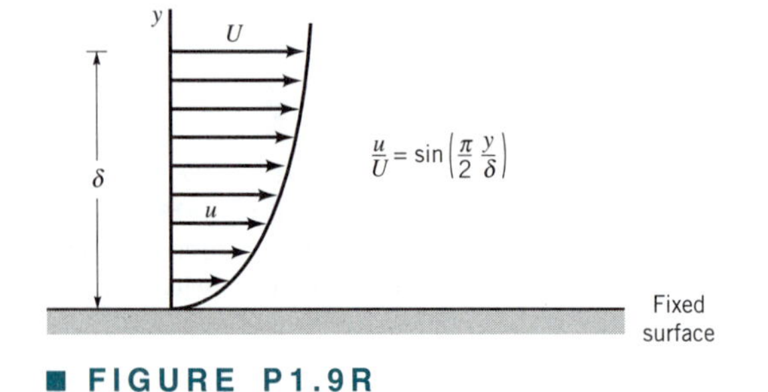

■ **FIGURE P1.9R**

1.10R (Viscosity) A large movable plate is located between two large fixed plates as shown in Fig. P1.10R. Two Newtonian fluids having the viscosities indicated are contained between the plates. Determine the magnitude and direction of the shearing stresses that act on the fixed walls when the moving plate has a velocity of 4 m/s as shown. Assume that the velocity distribution between the plates is linear.

(ANS: 13.3 N/m^2 in direction of moving plate)

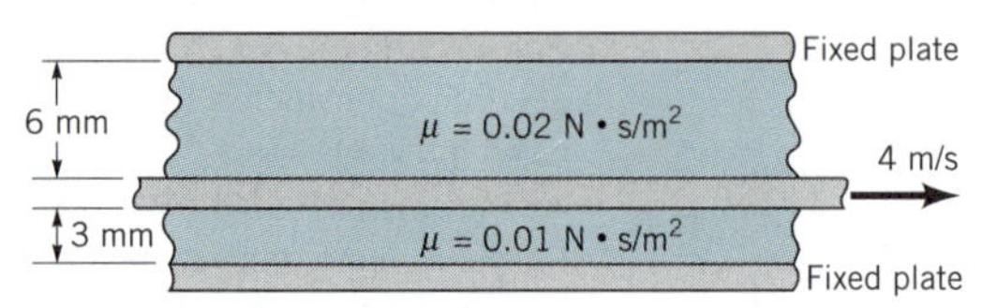

■ **FIGURE P1.10R**

1.11R (Viscosity) Determine the torque required to rotate a 50-mm-diameter vertical cylinder at a constant angular velocity of 30 rad/s inside a fixed outer cylinder that has a diameter of 50.2 mm. The gap between the cylinders is filled with SAE 10 oil at 20 °C. The length of the inner cylinder is 200 mm. Neglect bottom effects and assume the velocity distribution in the gap is linear. If the temperature of the oil increases to 80 °C, what will be the percentage change in the torque?

(ANS: 0.589 N m; 92.0 percent)

1.12R (Bulk modulus) Estimate the increase in pressure (in psi) required to decrease a unit volume of mercury by 0.1%.

(ANS: 4.14×10^3 psi)

1.13R (Bulk modulus) What is the isothermal bulk modulus of nitrogen at a temperature of 90 °F and an absolute pressure of 5600 lb/ft^2?

(ANS: 5600 lb/ft^2)

1.14R (Speed of sound) Compare the speed of sound in mercury and oxygen at 20 °C.

(ANS: $c_{Hg}/c_{O2} = 4.45$)

1.15R (Vapor pressure) At a certain altitude it was found that water boils at 90 °C. What is the atmospheric pressure at this altitude?

(ANS: 70.1 kPa (abs))

Problems

Note: Unless specific values of required fluid properties are given in the statement of the problem, use the values found in the tables on the inside of the front cover. Problems designated with an (*) are intended to be solved with the aid of a programmable calculator or a computer. Problems designated with a (†) are "open-ended" problems and require critical thinking in that to work them one must make various assumptions and provide the necessary data. There is not a unique answer to these problems.

1.1 Determine the dimensions, in both the *FLT* system and the *MLT* system, for **(a)** the product of mass times velocity, **(b)** the product of force times volume, and **(c)** kinetic energy divided by area.

1.2 Verify the dimensions, in both the *FLT* and *MLT* systems, of the following quantities which appear in Table 1.1: **(a)** angular velocity, **(b)** energy, **(c)** moment of inertia (area), **(d)** power, and **(e)** pressure.

1.3 Verify the dimensions, in both the *FLT* system and the *MLT* system, of the following quantities which appear in Table 1.1: **(a)** acceleration, **(b)** stress, **(c)** moment of a force, **(d)** volume, and **(e)** work.

1.4 If P is a force and x a length, what are the dimensions (in the *FLT* system) of **(a)** dP/dx, **(b)** d^3P/dx^3, and **(c)** $\int P\,dx$?

1.5 If u is a velocity, x a length, and t a time, what are the dimensions (in the *MLT* system) of **(a)** $\partial u/\partial t$, **(b)** $\partial^2 u/\partial x \partial t$, and **(c)** $\int (\partial u/\partial t)\,dx$?

1.6 If V is a velocity, ℓ a length, and ν a fluid property having dimensions of L^2T^{-1}, which of the following combinations are dimensionless: **(a)** $V\ell\nu$, **(b)** $V\ell/\nu$, **(c)** $V^2\nu$, **(d)** $V/\ell\nu$?

1.7 Dimensionless combinations of quantities (commonly called dimensionless parameters) play an important role in fluid mechanics. Make up five possible dimensionless parameters by using combinations of some of the quantities listed in Table 1.1.

1.8 The force, P, that is exerted on a spherical particle moving slowly through a liquid is given by the equation

$$P = 3\pi\mu DV$$

where μ is a fluid property (viscosity) having dimensions of $FL^{-2}T$, D is the particle diameter, and V is the particle velocity. What are the dimensions of the constant, 3π? Would you classify this equation as a general homogeneous equation?

1.9 According to information found in an old hydraulics book, the energy loss per unit weight of fluid flowing through a nozzle connected to a hose can be estimated by the formula

$$h = (0.04 \text{ to } 0.09)(D/d)^4V^2/2g$$

where h is the energy loss per unit weight, D the hose diameter, d the nozzle tip diameter, V the fluid velocity in the hose, and g the acceleration of gravity. Do you think this equation is valid in any system of units? Explain.

1.10 The pressure difference, Δp, across a partial blockage in an artery (called a *stenosis*) is approximated by the equation

$$\Delta p = K_v \frac{\mu V}{D} + K_u \left(\frac{A_0}{A_1} - 1\right)^2 \rho V^2$$

where V is the blood velocity, μ the blood viscosity ($FL^{-2}T$), ρ the blood density (ML^{-3}), D the artery diameter, A_0 the area of the unobstructed artery, and A_1 the area of the stenosis. Determine the dimensions of the constants K_v and K_u. Would this equation be valid in any system of units?

1.11 Assume that the speed of sound, c, in a fluid depends on an elastic modulus, E_v, with dimensions FL^{-2}, and the fluid density, ρ, in the form $c = (E_v)^a(\rho)^b$. If this is to be a dimensionally homogeneous equation, what are the values for a and b? Is your result consistent with the standard formula for the speed of sound? (See Eq. 1.19.)

1.12 A formula to estimate the volume rate of flow, Q, flowing over a dam of length, B, is given by the equation

$$Q = 3.09BH^{3/2}$$

where H is the depth of the water above the top of the dam (called the head). This formula gives Q in ft³/s when B and H are in feet. Is the constant 3.09 dimensionless? Would this equation be valid if units other than feet and seconds were used?

† **1.13** Cite an example of a restricted homogeneous equation contained in a technical article found in an engineering journal in your field of interest. Define all terms in the equation, explain why it is a restricted equation, and provide a complete journal citation (title, date, etc.).

1.14 Make use of Table 1.3 to express the following quantities in SI units: **(a)** 10.2 in./min, **(b)** 4.81 slugs, **(c)** 3.02 lb, **(d)** 73.1 ft/s², **(e)** 0.0234 lb·s/ft².

1.15 Make use of Table 1.4 to express the following quantities in BG units: **(a)** 14.2 km, **(b)** 8.14 N/m³, **(c)** 1.61 kg/m³, **(d)** 0.0320 N·m/s, **(e)** 5.67 mm/hr.

1.16 Make use of Appendix A to express the following quantities in SI units: **(a)** 160 acre, **(b)** 742 Btu, **(c)** 240 miles, **(d)** 79.1 hp, **(e)** 60.3 °F.

1.17 Verify the conversion relationships for **(a)** acre, **(b)** bar, and **(c)** U.S. liquid gallon found in Appendix A.

1.18 For Table 1.3 verify the conversion relationships for: **(a)** area, **(b)** density, **(c)** velocity, and **(d)** specific weight. Use the basic conversion relationships: 1 ft = 0.3048 m; 1 lb = 4.4482 N; and 1 slug = 14.594 kg.

1.19 For Table 1.4 verify the conversion relationships for: **(a)** acceleration, **(b)** density, **(c)** pressure, and **(d)** volume flowrate. Use the basic conversion relationships: 1 m = 3.2808 ft; 1 N = 0.22481 lb; and 1 kg = 0.068521 slug.

1.20 Water flows from a large drainage pipe at a rate of 1500 gal/min. What is this volume rate of flow in m³/s and in liters/min?

1.21 A tank of oil has a mass of 30 slugs. **(a)** Determine its weight in pounds and in newtons at the earth's surface. **(b)** What would be its mass (in slugs) and its weight (in pounds) if located on the moon's surface where the gravitational attraction is approximately one-sixth that at the earth's surface?

1.22 A certain object weighs 300 N at the earth's surface. Determine the mass of the object (in kilograms) and its weight (in newtons) when located on a planet with an acceleration of gravity equal to 4.0 ft/s^2.

1.23 An important dimensionless parameter in certain types of fluid flow problems is the *Froude number* defined as $V/\sqrt{g\ell}$, where V is a velocity, g the acceleration of gravity, and ℓ a length. Determine the value of the Froude number for $V = 10$ ft/s, $g = 32.2$ ft/s^2, and $\ell = 2$ ft. Recalculate the Froude number using SI units for V, g, and ℓ. Explain the significance of the results of these calculations.

1.24 The specific weight of a certain liquid is 85.3 lb/ft^3. Determine its density and specific gravity.

1.25 The density of a certain type of jet fuel is 805 kg/m^3. Determine its specific gravity and specific weight.

1.26 An open, rigid-walled, cylindrical tank contains 4 ft^3 of water at 40 °F. Over a 24-hour period of time the water temperature varies from 40 °F to 90 °F. Make use of the data in Appendix B to determine how much the volume of water will change. For a tank diameter of 2 ft, would the corresponding change in water depth be very noticeable? Explain.

† **1.27** Estimate the number of pounds of mercury it would take to fill your bath tub. List all assumptions and show all calculations.

1.28 A liquid when poured into a graduated cylinder is found to weigh 8 N when occupying a volume of 500 ml (milliliters). Determine its specific weight, density, and specific gravity.

1.29 The information on a can of pop indicates that the can contains 355 mL. The mass of a full can of pop is 0.369 kg while an empty can weighs 0.153 N. Determine the specific weight, density, and specific gravity of the pop and compare your results with the corresponding values for water at 20 °C. Express your results in SI units.

***1.30** The variation in the density of water, ρ, with temperature, T, in the range 20 °C $\leq T \leq$ 60 °C, is given in the following table.

Density (kg/m^3)	998.2	997.1	995.7	994.1	992.2	990.2	988.1
Temperature (°C)	20	25	30	35	40	45	50

Use these data to determine an empirical equation of the form $\rho = c_1 + c_2T + c_3T^2$ which can be used to predict the density over the range indicated. Compare the predicted values with the data given. What is the density of water at 42.1 °C?

† **1.31** Estimate the number of kilograms of water consumed per day for household purposes in your city. List all assumptions and show all calculations.

1.32 The density of oxygen contained in a tank is 2.0 kg/m^3 when the temperature is 25 °C. Determine the gage pressure of the gas if the atmospheric pressure is 97 kPa.

1.33 Nitrogen is compressed to a density of 5 kg/m^3 under an absolute pressure of 425 kPa. Determine the temperature in degrees Celsius.

1.34 A closed tank having a volume of 2 ft^3 is filled with 0.30 lb of a gas. A pressure gage attached to the tank reads 12 psi when the gas temperature is 80 °F. There is some question as to whether the gas in the tank is oxygen or helium. Which do you think it is? Explain how you arrived at your answer.

† **1.35** The presence of raindrops in the air during a heavy rainstorm increases the average density of the air/water mixture. Estimate by what percent the average air/water density is greater than that of just still air. State all assumptions and show calculations.

1.36 A tire having a volume of 3 ft^3 contains air at a gage pressure of 26 psi and a temperature of 70 °F. Determine the density of the air and the weight of the air contained in the tire.

1.37 A compressed air tank contains 8 kg of air at a temperature of 80 °C. A gage on the tank reads 300 kPa. Determine the volume of the tank.

***1.38** Develop a computer program for calculating the density of an ideal gas when the gas pressure in pascals (abs), the temperature in degrees Celsius, and the gas constant in J/kg·K are specified.

***1.39** Repeat Problem 1.38 for the case in which the pressure is given in psi (gage), the temperature in degrees Fahrenheit, and the gas constant in ft·lb/slug·°R.

1.40 Make use of the data in Appendix B to determine the dynamic viscosity of mercury at 75 °F. Express your answer in BG units.

1.41 Determine the ratio of the dynamic viscosity of water to air at a temperature of 70 °C. Compare this value with the corresponding ratio of kinematic viscosities. Assume the air is at standard atmospheric pressure.

1.42 The kinematic viscosity and specific gravity of a liquid are 3.5×10^{-4} m^2/s and 0.79, respectively. What is the dynamic viscosity of the liquid in SI units?

1.43 A liquid has a specific weight of 59 lb/ft^3 and a dynamic viscosity of 2.75 lb·s/ft^2. Determine its kinematic viscosity.

1.44 The viscosity of a certain fluid is 5×10^{-4} poise. Determine its viscosity in both SI and BG units.

1.45 The kinematic viscosity of oxygen at 20 °C and a pressure of 150 kPa (abs) is 0.104 stokes. Determine the dynamic viscosity of oxygen at this temperature and pressure.

1.46 Determine from Figs. B.1 and B.2 in Appendix B the dynamic and kinematic viscosity of kerosene at 40 °C. Express your answer in both SI and BG units.

1.47 SAE 30 oil at 60 °F flows through a 2-in.-diameter pipe with a mean velocity of 5 ft/s. Determine the value of the Reynolds number (see Example 1.4).

1.48 Calculate the Reynolds numbers for the flow of water and for air through a 3-mm-diameter tube, if the mean velocity is 3 m/s and the temperature is 30 °C in both cases (see Example 1.4). Assume the air is at standard atmospheric pressure.

1.49 For air at standard atmospheric pressure the values of the constants that appear in the Sutherland equation (Eq. 1.10) are $C = 1.458 \times 10^{-6}$ kg/(m·s·K$^{1/2}$) and $S = 110.4$ K. Use these values to predict the viscosity of air at 10 °C and 90 °C and compare with values given in Table B.4 in Appendix B.

*1.50 Use the values of viscosity of air given in Table B.4 at temperatures of 0, 20, 40, 60, 80, and 100 °C to determine the constants C and S which appear in the Sutherland equation (Eq. 1.10). Compare your results with the values given in Problem 1.49. (*Hint:* Rewrite the equation in the form

$$\frac{T^{3/2}}{\mu} = \left(\frac{1}{C}\right) T + \frac{S}{C}$$

and plot $T^{3/2}/\mu$ versus T. From the slope and intercept of this curve, C and S can be obtained.)

1.51 For a certain liquid $\mu = 7.1 \times 10^{-5}$ lb·s/ft^2 at 40 °F and $\mu = 1.9 \times 10^{-5}$ lb·s/ft^2 at 150 °F. Make use of these data to determine the constants D and B which appear in Andrade's equation (Eq. 1.11). What would be the viscosity at 80 °F?

*1.52 Use the value of the viscosity of water given in Table B.2 at temperatures of 0, 20, 40, 60, 80, and 100 °C to determine the constants D and B which appear in Andrade's equation (Eq. 1.11). Calculate the value of the viscosity at 50 °C and compare with the value given in Table B.2. (*Hint:* Rewrite the equation in the form

$$\ln \mu = (B)\frac{1}{T} + \ln D$$

and plot $\ln \mu$ versus $1/T$. From the slope and intercept of this curve, B and D can be obtained. If a nonlinear curve-fitting program is available the constants can be obtained directly from Eq. 1.11 without rewriting the equation.)

1.53 Crude oil having a viscosity of 9.52×10^{-4} lb·s/ft^2 is contained between parallel plates. The bottom plate is fixed and the upper plate moves when a force P is applied (see Fig. 1.3). If the distance between the two plates is 0.1 in., what value of P is required to translate the plate with a velocity of 3 ft/s? The effective area of the upper plate is 200 in.^{2}

1.54 A thin layer of glycerin flows down an inclined, wide plate with the velocity distribution shown in Fig. P1.54. For h = 0.3 in. and α = 20°, determine the surface velocity, U. Note that for equilibrium, the component of weight acting parallel to the plate surface must be balanced by the shearing force developed along the plate surface. In your analysis assume a unit plate width.

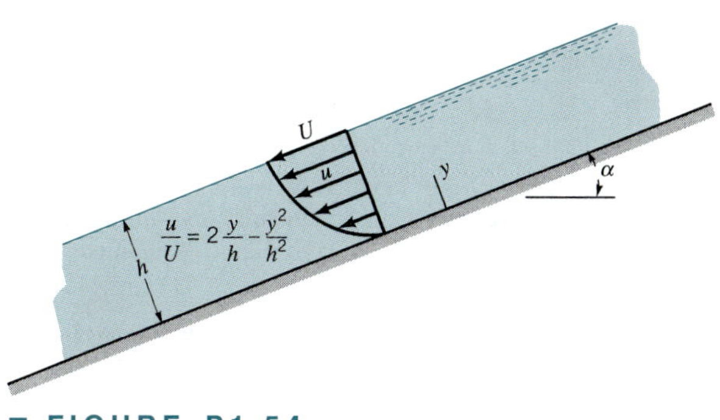

■ FIGURE P1.54

1.55 The viscosity of blood is to be determined from measurements of shear stress, τ, and rate of shearing strain, du/dy, obtained from a small blood sample tested in a suitable viscometer. Based on the data given below determine if the blood is a Newtonian or a non-Newtonian fluid. Explain how you arrived at your answer.

τ (N/m^2)	0.04	0.06	0.12	0.18	0.30	0.52	1.12	2.10
du/dy (s^{-1})	2.25	4.50	11.25	22.5	45.0	90.0	225	450

1.56 A 10-kg block slides down a smooth inclined surface as shown in Fig. P1.56. Determine the terminal velocity of the block if the 0.1-mm gap between the block and the surface contains SAE 30 oil at 60 °F. Assume the velocity distribution in the gap is linear, and the area of the block in contact with the oil is 0.2 m^2.

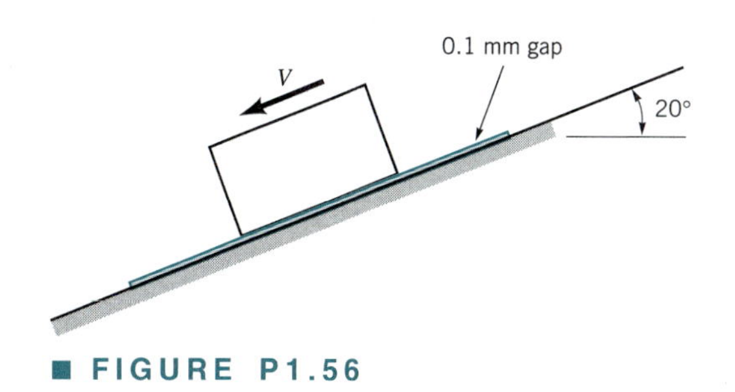

■ FIGURE P1.56

1.57 A piston having a diameter of 5.48 in. and a length of 9.50 in. slides downward with a velocity V through a vertical pipe. The downward motion is resisted by an oil film between the piston and the pipe wall. The film thickness is 0.002 in., and the cylinder weighs 0.5 lb. Estimate V if the oil viscosity is 0.016 lb·s/ft^2. Assume the velocity distribution in the gap is linear.

1.58 A layer of water flows down an inclined fixed surface with the velocity profile shown in Fig. P1.58. Determine the magnitude and direction of the shearing stress that the water exerts on the fixed surface for $U = 3$ m/s and $h = 0.1$ m.

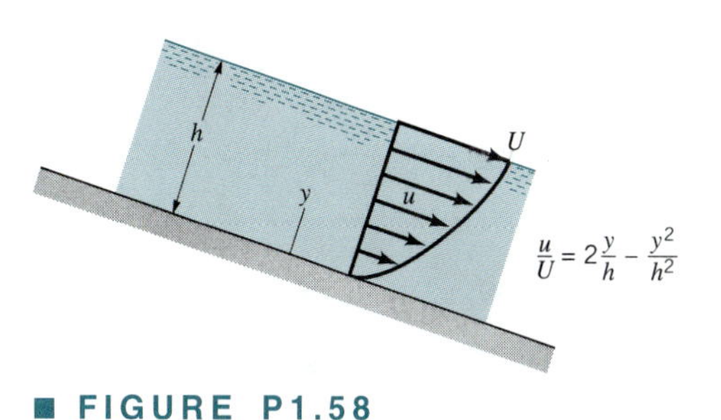

■ FIGURE P1.58

1.59 When a viscous fluid flows past a thin sharp-edged plate, a thin layer adjacent to the plate surface develops in which the velocity, u, changes rapidly from zero to the approach velocity, U, in a small distance, δ. This layer is called a *boundary layer.* The thickness of this layer increases with the distance x along the plate as shown in Fig. P1.59. Assume that $u = U\,y/\delta$

and $\delta = 3.5 \sqrt{\nu x/U}$ where ν is the kinematic viscosity of the fluid. Determine an expression for the force (drag) that would be developed on one side of the plate of length l and width b. Express your answer in terms of l, b, ν, and ρ, where ρ is the fluid density.

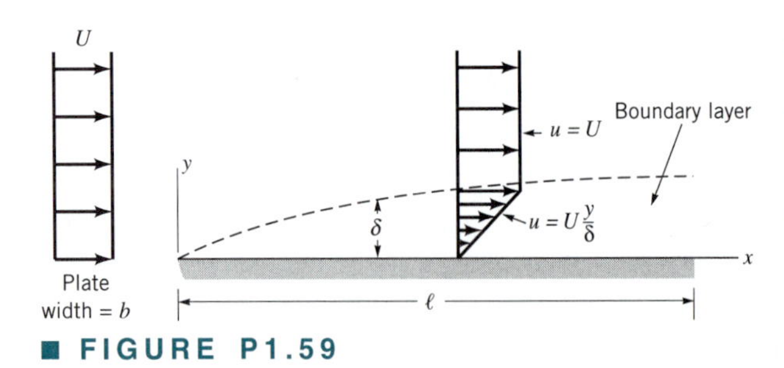

■ FIGURE P1.59

*1.60 Standard air flows past a flat surface and velocity measurements near the surface indicate the following distribution:

y (ft)	0.005	0.01	0.02	0.04	0.06	0.08
u (ft/s)	0.74	1.51	3.03	6.37	10.21	14.43

The coordinate y is measured normal to the surface and u is the velocity parallel to the surface. **(a)** Assume the velocity distribution is of the form

$$u = C_1 y + C_2 y^3$$

and use a standard curve-fitting technique to determine the constants C_1 and C_2. **(b)** Make use of the results of part (a) to determine the magnitude of the shearing stress at the wall ($y = 0$) and at $y = 0.05$ ft.

1.61 The vicosity of liquids can be measured through the use of a *rotating cylinder viscometer* of the type illustrated in Fig. P1.61. In this device the outer cylinder is fixed and the inner cylinder is rotated with an angular velocity, ω. The torque $\mathcal{T}$ required to develop ω is measured and the viscosity is calculated from these two measurements. Develop an equation relating μ, ω, $\mathcal{T}$, ℓ, R_o and R_i. Neglect end effects and assume the velocity distribution in the gap is linear.

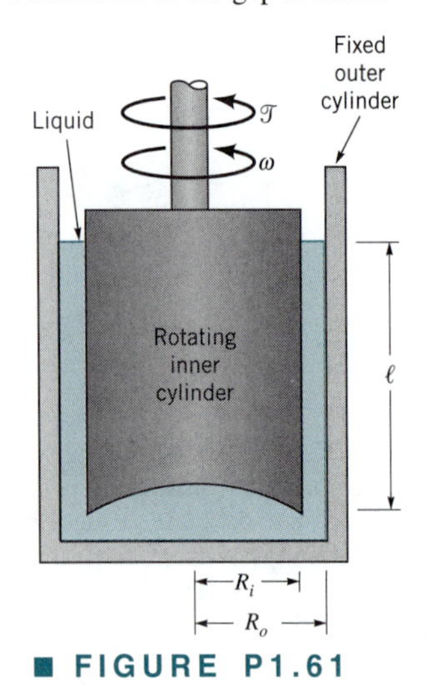

■ FIGURE P1.61

1.62 The space between two 6-in.-long concentric cylinders is filled with glycerin (viscosity $= 8.5 \times 10^{-3}$ lb·s/ft^2). The inner cylinder has a radius of 3 in. and the gap width between cylinders is 0.1 in. Determine the torque and the power required to rotate the inner cylinder at 180 rev/min. The outer cylinder is fixed. Assume the velocity distribution in the gap to be linear.

1.63 One type of rotating cylinder viscometer, called a *Stormer* viscometer, uses a falling weight, $\mathcal{W}$, to cause the cylinder to rotate with an angular velocity, ω, as illustrated in Fig. P1.63. For this device the viscosity, μ, of the liquid is related to $\mathcal{W}$ and ω through the equation $\mathcal{W} = K\mu\omega$, where K is a constant that depends only on the geometry (including the liquid depth) of the viscometer. The value of K is usually determined by using a calibration liquid (a liquid of known viscosity).

(a) Some data for a particular Stormer viscometer, obtained using glycerin at 20 °C as a calibration liquid, are given below. Plot values of the weight as ordinates and values of the angular velocity as abscissae. Draw the best curve through the plotted points and determine K for the viscometer.

$\mathcal{W}$ (lb)	0.22	0.66	1.10	1.54	2.20
ω (rev/s)	0.53	1.59	2.79	3.83	5.49

(b) A liquid of unknown viscosity is placed in the same viscometer used in part (a), and the data given below are obtained. Determine the viscosity of this liquid.

$\mathcal{W}$ (lb)	0.04	0.11	0.22	0.33	0.44
ω (rev/s)	0.72	1.89	3.73	5.44	7.42

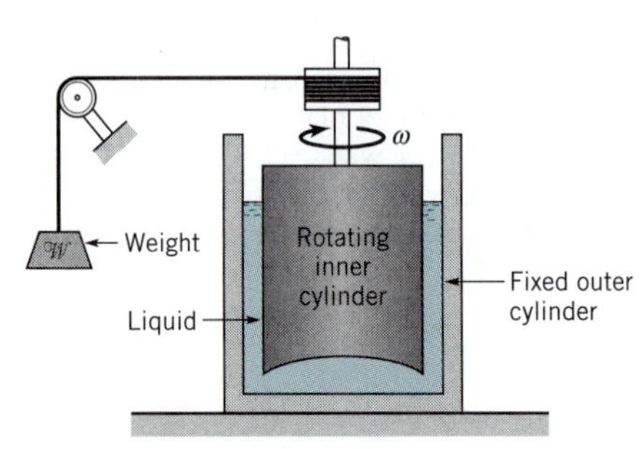

■ FIGURE P1.63

*1.64 The following torque-angular velocity data were obtained with a rotating cylinder viscometer of the type described in Problem 1.61.

Torque (ft·lb)	13.1	26.0	39.5	52.7	64.9	78.6
Angular velocity (rad/s)	1.0	2.0	3.0	4.0	5.0	6.0

For this viscometer $R_o = 2.50$ in., $R_i = 2.45$ in., and $\ell = 5.00$ in. Make use of these data and a standard curve-fitting program to determine the viscosity of the liquid contained in the viscometer.

1.65 A 12-in.-diameter circular plate is placed over a fixed bottom plate with a 0.1-in. gap between the two plates filled with glycerin as shown in Fig. P1.65. Determine the torque required to rotate the circular plate slowly at 2 rpm. Assume that the velocity distribution in the gap is linear and that the shear stress on the edge of the rotating plate is negligible.

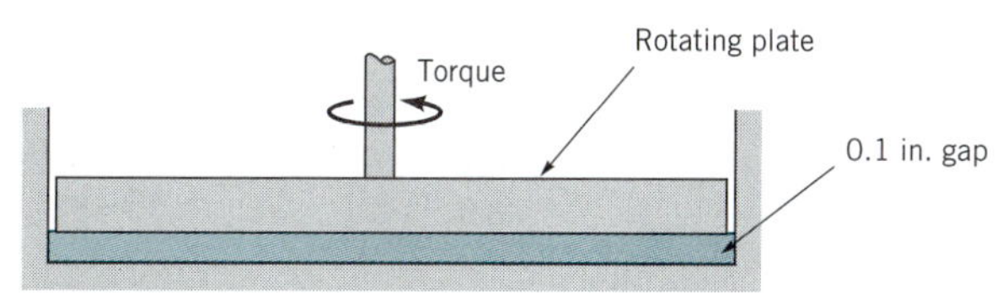

■ **FIGURE P1.65**

† **1.66** Vehicle shock absorbers damp out oscillations caused by road roughness. Describe how a temperature change may affect the operation of a shock absorber.

1.67 A rigid-walled cubical container is completely filled with water at 40 °F and sealed. The water is then heated to 100 °F. Determine the pressure that develops in the container when the water reaches this higher temperature. Assume that the volume of the container remains constant and the value of the bulk modulus of the water remains constant and equal to 300,000 psi.

1.68 In a test to determine the bulk modulus of a liquid it was found that as the absolute pressure was changed from 15 to 3000 psi the volume decreased from 10.240 to 10.138 $in.^3$ Determine the bulk modulus for this liquid.

1.69 Calculate the speed of sound in m/s for **(a)** gasoline, **(b)** mercury, and **(c)** seawater.

1.70 Air is enclosed by a rigid cylinder containing a piston. A pressure gage attached to the cylinder indicates an initial reading of 25 psi. Determine the reading on the gage when the piston has compressed the air to one-third its original volume. Assume the compression process to be isothermal and the local atmospheric pressure to be 14.7 psi.

1.71 Often the assumption is made that the flow of a certain fluid can be considered as incompressible flow if the density of the fluid changes by less than 2%. If air is flowing through a tube such that the air pressure at one section is 9.0 psi and at a downstream section it is 8.6 psi at the same temperature, do you think that this flow could be considered an imcompressible flow? Support your answer with the necessary calculations. Assume standard atmospheric pressure.

1.72 Oxygen at 30 °C and 300 kPa absolute pressure expands isothermally to an absolute pressure of 120 kPa. Determine the final density of the gas.

1.73 Natural gas at 70 °F and standard atmospheric pressure of 14.7 psi is compressed isentropically to a new absolute pressure of 60 psi. Determine the final density and temperature of the gas.

1.74 Compare the isentropic bulk modulus of air at 101 kPa (abs) with that of water at the same pressure.

***1.75** Develop a computer program for calculating the final gage pressure of gas when the initial gage pressure, initial and final volumes, atmospheric pressure, and the type of process (isothermal or isentropic) are specified. Use BG units. Check your program against the results obtained for Problem 1.70.

1.76 Determine the speed of sound at 20 °C in **(a)** air, **(b)** helium, and **(c)** natural gas. Express your answer in m/s.

1.77 Jet airliners typically fly at altitudes between approximately 0 to 40,000 ft. Make use of the data in Appendix C to show on a graph how the speed of sound varies over this range.

1.78 When a fluid flows through a sharp bend, low pressures may develop in localized regions of the bend. Estimate the minimum absolute pressure (in psi) that can develop without causing cavitation if the fluid is water at 160 °F.

1.79 Estimate the minimum absolute pressure (in pascals) that can be developed at the inlet of a pump to avoid cavitation if the fluid is ethyl alcohol at 20 °C.

1.80 When water at 90 °C flows through a converging section of pipe, the pressure is reduced in the direction of flow. Estimate the minimum absolute pressure that can develop without causing cavitation. Express your answer in both BG and SI units.

1.81 A partially filled closed tank contains ethyl alcohol at 68 °F. If the air above the alcohol is evacuated, what is the minimum absolute pressure that develops in the evacuated space?

1.82 Estimate the excess pressure inside a raindrop having a diameter of 3 mm.

1.83 A 12-mm diameter jet of water discharges vertically into the atmosphere. Due to surface tension the pressure inside the jet will be slightly higher than the surrounding atmospheric pressure. Determine this difference in pressure.

1.84 What is the difference between the pressure inside a soap bubble and atmospheric pressure for a 3-in.-diameter bubble? Assume the surface tension of the soap film to be 70% of that of water at 70 °F.

1.85 Determine the height water at 70 °F will rise due to capillary action in a clean, $\frac{1}{4}$-in.-diameter tube. What will be the height if the diameter is reduced to 0.01 in.?

1.86 Two vertical, parallel, clean glass plates are spaced a distance of 2 mm apart. If the plates are placed in water, how high will the water rise between the plates due to capillary action?

1.87 An open, clean glass tube, having a diameter of 3 mm, is inserted vertically into a dish of mercury at 20 °C. How far will the column of mercury in the tube be depressed?

1.88 An open 2-mm-diameter tube is inserted into a pan of ethyl alcohol and a similar 4-mm-diameter tube is inserted into a pan of water. In which tube will the height of the rise of the fluid column due to capillary action be the greatest? Assume the angle of contact is the same for both tubes.

***1.89** The capillary rise in a tube depends on the cleanliness of both the fluid and the tube. Typically, values of h are less than those predicted by Eq. 1.22 using values of σ and θ for clean fluids and tubes. Some measurements of the height, h, to which a water column rises in a vertical open tube of diameter d are given below. The water was tap water at a temperature of 60 °F and no particular effort was made to clean the glass tube.

Fit a curve to these data and estimate the value of the product $\sigma \cos \theta$. If it is assumed that σ has the value given in Table 1.5, what is the value of θ? If it is assumed that θ is equal to 0°, what is the value of σ?

d (in.)	0.3	0.25	0.20	0.15	0.10	0.05
h (in.)	0.133	0.165	0.198	0.273	0.421	0.796

1.90 The capillary tube viscometer device shown in Fig. P1.90 can be used to determine the kinematic viscosity, $\nu = \mu/\rho$, of a liquid. The volume flowrate, Q cubic feet per second or mℓ per second, at which a viscous liquid flows through a small diameter tube (i.e., a capillary tube) depends upon many parameters including the diameter and length of the tube, the acceleration of gravity, the density and viscosity of the liquid, and the head (height) of the liquid above the top of the tube. An advanced analysis of this situation would show that with other parameters held constant, the kinematic viscosity is related to the flowrate as $\nu = K/Q$, where K is a constant. The value of the constant K can be determined by measuring Q for a fluid of known viscosity, in this case water. The flowrate is given by $Q = \forall/t$, where $\forall$ is the volume of water collected in a graduated cylinder in the time period t.

Values of $\forall$ and t determined experimentally when using water at different temperatures are shown in the table below. For each temperature, use the book value of the viscosity of water and the given data to determine the constant K.

It is assumed that the value of K is constant, independent of the viscosity of the fluid used. Do your results support this? Discuss some possible reasons for this not being true.

$\forall$ (mℓ)	t (s)	T (°C)
9.50	15.4	26.3
9.30	17.0	21.3
9.05	20.4	12.3
9.25	13.3	34.3
9.40	9.9	50.4
9.10	8.9	58.0

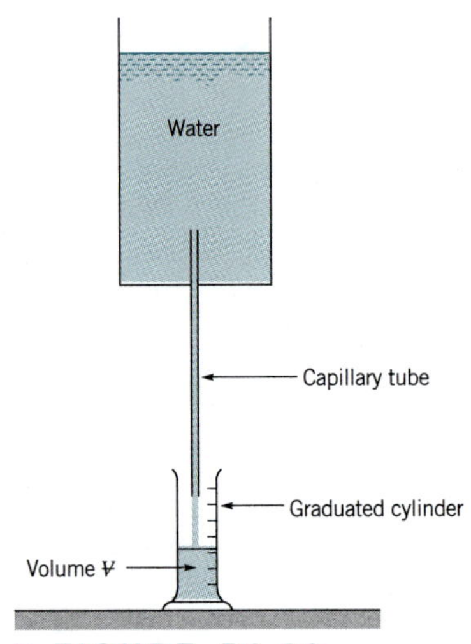

■ FIGURE P1.90

An image of hurricane Allen viewed via satellite: Although there is considerable motion and structure to a hurricane, the pressure variation in the vertical direction is approximated by the pressure-depth relationship for a static fluid. (Visible and infrared image pair from a NOAA satellite using a technique developed at NASA/GSPC.) (Photograph courtesy of A. F. Hasler [Ref. 7].)

2 Fluid Statics

In this chapter we will consider an important class of problems in which the fluid is either at rest or moving in such a manner that there is no relative motion between adjacent particles. In both instances there will be no shearing stresses in the fluid, and the only forces that develop on the surfaces of the particles will be due to the pressure. Thus, our principal concern is to investigate pressure and its variation throughout a fluid and the effect of pressure on submerged surfaces. The absence of shearing stresses greatly simplifies the analysis and, as we will see, allows us to obtain relatively simple solutions to many important practical problems.

2.1 Pressure at a Point

There are no shearing stresses present in a fluid at rest.

As we briefly discussed in Chapter 1, the term pressure is used to indicate the normal force per unit area at a given point acting on a given plane within the fluid mass of interest. A question that immediately arises is how the pressure at a point varies with the orientation of the plane passing through the point. To answer this question, consider the free-body diagram, illustrated in Fig. 2.1, that was obtained by removing a small triangular wedge of fluid from some arbitrary location within a fluid mass. Since we are considering the situation in which there are no shearing stresses, the only external forces acting on the wedge are due to the pressure and the weight. For simplicity the forces in the x direction are not shown, and the z axis is taken as the vertical axis so the weight acts in the negative z direction. Although we are primarily interested in fluids at rest, to make the analysis as general as possible, we will allow the fluid element to have accelerated motion. The assumption of zero shearing stresses will still be valid so long as the fluid element moves as a rigid body; that is, there is no relative motion between adjacent elements.

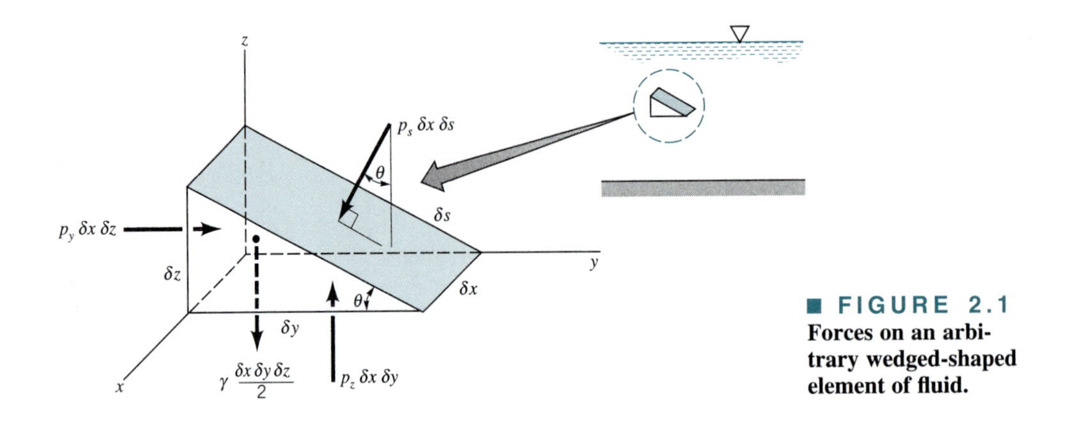

■ **FIGURE 2.1** **Forces on an arbitrary wedged-shaped element of fluid.**

The equations of motion (Newton's second law, $\mathbf{F} = m\mathbf{a}$) in the y and z directions are, respectively,

$$\sum F_y = p_y\,\delta x\,\delta z - p_s\,\delta x\,\delta s\,\sin\theta = \rho\,\frac{\delta x\,\delta y\,\delta z}{2}\,a_y$$

$$\sum F_z = p_z\,\delta x\,\delta y - p_s\,\delta x\,\delta s\,\cos\theta - \gamma\,\frac{\delta x\,\delta y\,\delta z}{2} = \rho\,\frac{\delta x\,\delta y\,\delta z}{2}\,a_z$$

where p_s, p_y, and p_z are the average pressures on the faces, γ and ρ are the fluid specific weight and density, respectively, and a_y, a_z the accelerations. Note that a pressure must be multiplied by an appropriate area to obtain the force generated by the pressure. It follows from the geometry that

$$\delta y = \delta s\,\cos\theta \qquad \delta z = \delta s\,\sin\theta$$

so that the equations of motion can be rewritten as

$$p_y - p_s = \rho a_y\,\frac{\delta y}{2}$$

$$p_z - p_s = (\rho a_z + \gamma)\,\frac{\delta z}{2}$$

Since we are really interested in what is happening at a point, we take the limit as δx, δy, and δz approach zero (while maintaining the angle θ), and it follows that

$$p_y = p_s \qquad p_z = p_s$$

The pressure at a point in a fluid at rest is independent of direction.

or $p_s = p_y = p_z$. The angle θ was arbitrarily chosen so we can conclude that *the pressure at a point in a fluid at rest, or in motion, is independent of direction as long as there are no shearing stresses present.* This important result is known as *Pascal's law* named in honor of Blaise Pascal (1623–1662), a French mathematician who made important contributions in the field of hydrostatics. In Chapter 6 it will be shown that for moving fluids in which there is relative motion between particles (so that shearing stresses develop) the normal stress at a point, which corresponds to pressure in fluids at rest, is not necessarily the same in all directions. In such cases the pressure is defined as the *average* of any three mutually perpendicular normal stresses at the point.

2.2 Basic Equation for Pressure Field

Although we have answered the question of how the pressure at a point varies with direction, we are now faced with an equally important question—how does the pressure in a fluid in which there are no shearing stresses vary from point to point? To answer this question consider a small rectangular element of fluid removed from some arbitrary position within the mass of fluid of interest as illustrated in Fig. 2.2. There are two types of forces acting on this element: *surface forces* due to the pressure, and a *body force* equal to the weight of the element. Other possible types of body forces, such as those due to magnetic fields, will not be considered in this text.

The pressure may vary across a fluid particle.

If we let the pressure at the center of the element be designated as p, then the average pressure on the various faces can be expressed in terms of p and its derivatives as shown in Fig. 2.2. We are actually using a Taylor series expansion of the pressure at the element center to approximate the pressures a short distance away and neglecting higher order terms that will vanish as we let δx, δy, and δz approach zero. For simplicity the surface forces in the x direction are not shown. The resultant surface force in the y direction is

$$\delta F_y = \left(p - \frac{\partial p}{\partial y}\frac{\delta y}{2}\right)\delta x\,\delta z - \left(p + \frac{\partial p}{\partial y}\frac{\delta y}{2}\right)\delta x\,\delta z$$

or

$$\delta F_y = -\frac{\partial p}{\partial y}\,\delta x\,\delta y\,\delta z$$

Similarly, for the x and z directions the resultant surface forces are

$$\delta F_x = -\frac{\partial p}{\partial x}\,\delta x\,\delta y\,\delta z \qquad \delta F_z = -\frac{\partial p}{\partial z}\,\delta x\,\delta y\,\delta z$$

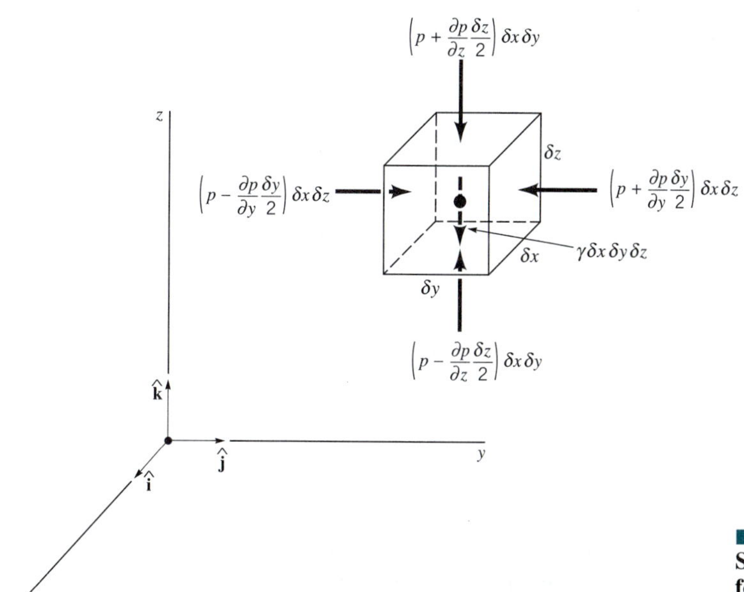

■ **FIGURE 2.2** **Surface and body forces acting on small fluid element.**

The resultant surface force acting on the element can be expressed in vector form as

$$\delta \mathbf{F}_s = \delta F_x \hat{\mathbf{i}} + \delta F_y \hat{\mathbf{j}} + \delta F_z \hat{\mathbf{k}}$$

or

The resultant surface force acting on a small fluid element depends only on the pressure gradient if there are no shearing stresses present.

$$\delta \mathbf{F}_s = -\left(\frac{\partial p}{\partial x} \hat{\mathbf{i}} + \frac{\partial p}{\partial y} \hat{\mathbf{j}} + \frac{\partial p}{\partial z} \hat{\mathbf{k}} \right) \delta x \, \delta y \, \delta z \quad \textbf{(2.1)}$$

where $\hat{\mathbf{i}}$, $\hat{\mathbf{j}}$, and $\hat{\mathbf{k}}$ are the unit vectors along the coordinate axes shown in Fig. 2.2. The group of terms in parentheses in Eq. 2.1 represents in vector form the *pressure gradient* and can be written as

$$\frac{\partial p}{\partial x} \hat{\mathbf{i}} + \frac{\partial p}{\partial y} \hat{\mathbf{j}} + \frac{\partial p}{\partial z} \hat{\mathbf{k}} = \nabla p$$

where

$$\nabla(\ ) = \frac{\partial(\ )}{\partial x} \hat{\mathbf{i}} + \frac{\partial(\ )}{\partial y} \hat{\mathbf{j}} + \frac{\partial(\ )}{\partial z} \hat{\mathbf{k}}$$

and the symbol ∇ is the *gradient* or "del" vector operator. Thus, the resultant surface force per unit volume can be expressed as

$$\frac{\delta \mathbf{F}_s}{\delta x \, \delta y \, \delta z} = -\nabla p$$

Since the z axis is vertical, the weight of the element is

$$-\delta \mathcal{W} \hat{\mathbf{k}} = -\gamma \, \delta x \, \delta y \, \delta z \, \hat{\mathbf{k}}$$

where the negative sign indicates that the force due to the weight is downward (in the negative z direction). Newton's second law, applied to the fluid element, can be expressed as

$$\sum \delta \mathbf{F} = \delta m \, \mathbf{a}$$

where $\Sigma \, \delta \mathbf{F}$ represents the resultant force acting on the element, $\mathbf{a}$ is the acceleration of the element, and δm is the element mass, which can be written as $\rho \, \delta x \, \delta y \, \delta z$. It follows that

$$\sum \delta \mathbf{F} = \delta \mathbf{F}_s - \delta \mathcal{W} \hat{\mathbf{k}} = \delta m \, \mathbf{a}$$

or

$$-\nabla p \, \delta x \, \delta y \, \delta z - \gamma \, \delta x \, \delta y \, \delta z \, \hat{\mathbf{k}} = \rho \, \delta x \, \delta y \, \delta z \, \mathbf{a}$$

and, therefore,

$$-\nabla p - \gamma \hat{\mathbf{k}} = \rho \mathbf{a} \quad \textbf{(2.2)}$$

Equation 2.2 is the general equation of motion for a fluid in which there are no shearing stresses. We will use this equation in Section 2.12 when we consider the pressure distribution in a moving fluid. For the present, however, we will restrict our attention to the special case of a fluid at rest.

2.3 Pressure Variation in a Fluid at Rest

For a fluid at rest **a** = 0 and Eq. 2.2 reduces to

$$\nabla p + \gamma \hat{\mathbf{k}} = 0$$

or in component form

$$\frac{\partial p}{\partial x} = 0 \qquad \frac{\partial p}{\partial y} = 0 \qquad \frac{\partial p}{\partial z} = -\gamma \tag{2.3}$$

These equations show that the pressure does not depend on x or y. Thus, as we move from point to point in a horizontal plane (any plane parallel to the x–y plane), the pressure does not change. Since p depends only on z, the last of Eqs. 2.3 can be written as the ordinary differential equation

For liquids or gases at rest the pressure gradient in the vertical direction at any point in a fluid depends only on the specific weight of the fluid at that point.

$$\boxed{\frac{dp}{dz} = -\gamma} \tag{2.4}$$

Equation 2.4 is the fundamental equation for fluids at rest and can be used to determine how pressure changes with elevation. This equation indicates that the pressure gradient in the vertical direction is negative; that is, the pressure decreases as we move upward in a fluid at rest. There is no requirement that γ be a constant. Thus, it is valid for fluids with constant specific weight, such as liquids, as well as fluids whose specific weight may vary with elevation, such as air or other gases. However, to proceed with the integration of Eq. 2.4 it is necessary to stipulate how the specific weight varies with z.

2.3.1 Incompressible Fluid

Since the specific weight is equal to the product of fluid density and acceleration of gravity ($\gamma = \rho g$), changes in γ are caused either by a change in ρ or g. For most engineering applications the variation in g is negligible, so our main concern is with the possible variation in the fluid density. For liquids the variation in density is usually negligible, even over large vertical distances, so that the assumption of constant specific weight when dealing with liquids is a good one. For this instance, Eq. 2.4 can be directly integrated

$$\int_{p_1}^{p_2} dp = -\gamma \int_{z_1}^{z_2} dz$$

to yield

$$p_2 - p_1 = -\gamma(z_2 - z_1)$$

or

$$p_1 - p_2 = \gamma(z_2 - z_1) \tag{2.5}$$

where p_1 and p_2 are pressures at the vertical elevations z_1 and z_2, as is illustrated in Fig. 2.3.

Equation 2.5 can be written in the compact form

$$p_1 - p_2 = \gamma h \tag{2.6}$$

or

$$p_1 = \gamma h + p_2 \tag{2.7}$$

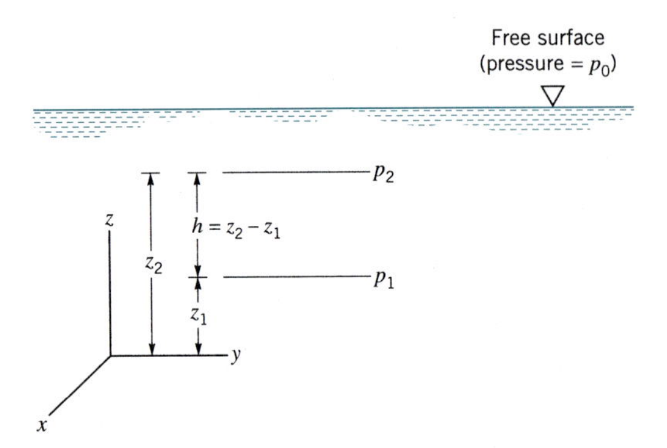

■ **FIGURE 2.3** **Notation for pressure variation in a fluid at rest with a free surface.**

where h is the distance, $z_2 - z_1$, which is the depth of fluid measured downward from the location of p_2. This type of pressure distribution is commonly called a *hydrostatic distribution*, and Eq. 2.7 shows that in an incompressible fluid at rest the pressure varies linearly with depth. The pressure must increase with depth to ''hold up'' the fluid above it.

It can also be observed from Eq. 2.6 that the pressure difference between two points can be specified by the distance h since

$$h = \frac{p_1 - p_2}{\gamma}$$

The pressure head is the height of a column of fluid that would give the specified pressure difference.

In this case h is called the *pressure head* and is interpreted as the height of a column of fluid of specific weight γ required to give a pressure difference $p_1 - p_2$. For example, a pressure difference of 10 psi can be specified in terms of pressure head as 23.1 ft of water ($\gamma = 62.4$ lb/ft^3), or 518 mm of Hg ($\gamma = 133$ kN/m^3).

When one works with liquids there is often a free surface, as is illustrated in Fig. 2.3, and it is convenient to use this surface as a reference plane. The reference pressure p_0 would correspond to the pressure acting on the free surface (which would frequently be atmospheric pressure), and thus if we let $p_2 = p_0$ in Eq. 2.7 it follows that the pressure p at any depth h below the free surface is given by the equation:

$$p = \gamma h + p_0 \qquad \textbf{(2.8)}$$

As is demonstrated by Eq. 2.7 or 2.8, the pressure in a homogeneous, incompressible fluid at rest depends on the depth of the fluid relative to some reference plane, and it is *not* influenced by the *size* or *shape* of the tank or container in which the fluid is held. Thus, in Fig. 2.4 the pressure is the same at all points along the line AB even though the container may have the very irregular shape shown in the figure. The actual value of the pressure along AB depends only on the depth, h, the surface pressure, p_0, and the specific weight, γ, of the liquid in the container.

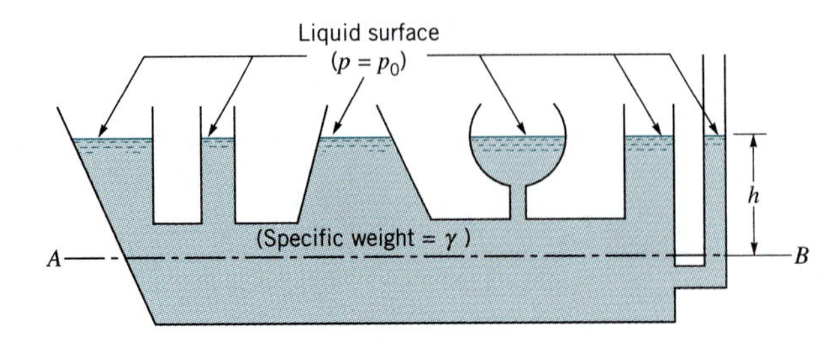

■ **FIGURE 2.4** **Fluid equilibrium in a container of arbitrary shape.**

EXAMPLE 2.1

Because of a leak in a buried gasoline storage tank, water has seeped in to the depth shown in Fig. E2.1. If the specific gravity of the gasoline is $SG = 0.68$, determine the pressure at the gasoline-water interface and at the bottom of the tank. Express the pressure in units of lb/ft^2, lb/in.^2, and as a pressure head in feet of water.

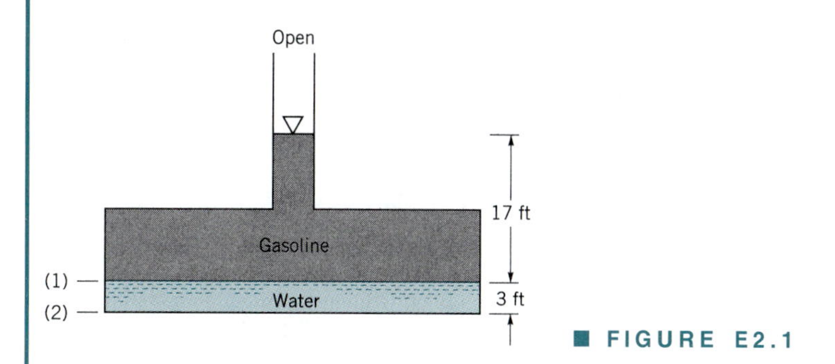

■ FIGURE E2.1

SOLUTION

Since we are dealing with liquids at rest, the pressure distribution will be hydrostatic, and therefore the pressure variation can be found from the equation:

$$p = \gamma h + p_0$$

With p_0 corresponding to the pressure at the free surface of the gasoline, then the pressure at the interface is

$$\begin{aligned} p_1 &= SG\gamma_{H_2O}h + p_0 \\ &= (0.68)(62.4 \text{ lb/ft}^3)(17 \text{ ft}) + p_0 \\ &= 721 + p_0 \ (\text{lb/ft}^2) \end{aligned}$$

If we measure the pressure relative to atmospheric pressure (gage pressure), it follows that $p_0 = 0$, and therefore

$$p_1 = 721 \text{ lb/ft}^2 \qquad \textbf{(Ans)}$$

$$p_1 = \frac{721 \text{ lb/ft}^2}{144 \text{ in.}^2/\text{ft}^2} = 5.01 \text{ lb/in.}^2 \qquad \textbf{(Ans)}$$

$$\frac{p_1}{\gamma_{H_2O}} = \frac{721 \text{ lb/ft}^2}{62.4 \text{ lb/ft}^3} = 11.6 \text{ ft} \qquad \textbf{(Ans)}$$

It is noted that a rectangular column of water 11.6 ft tall and 1 ft^2 in cross section weighs 721 lb. A similar column with a 1-in.^2 cross section weighs 5.01 lb.

We can now apply the same relationship to determine the pressure at the tank bottom; that is,

$$\begin{aligned} p_2 &= \gamma_{H_2O}h_{H_2O} + p_1 \\ &= (62.4 \text{ lb/ft}^3)(3 \text{ ft}) + 721 \text{ lb/ft}^2 \qquad \textbf{(Ans)} \\ &= 908 \text{ lb/ft}^2 \end{aligned}$$

$$p_2 = \frac{908 \text{ lb/ft}^2}{144 \text{ in.}^2/\text{ft}^2} = 6.31 \text{ lb/in.}^2 \qquad \textbf{(Ans)}$$

$$\frac{p_2}{\gamma_{H_2O}} = \frac{908 \text{ lb/ft}^2}{62.4 \text{ lb/ft}^3} = 14.6 \text{ ft} \qquad \textbf{(Ans)}$$

Observe that if we wish to express these pressures in terms of *absolute* pressure, we would have to add the local atmospheric pressure (in appropriate units) to the previous results. A further discussion of gage and absolute pressure is given in Section 2.5.

The transmission of pressure throughout a stationary fluid is the principle upon which many hydraulic devices are based.

The required equality of pressures at equal elevations throughout a system is important for the operation of hydraulic jacks, lifts, and presses, as well as hydraulic controls on aircraft and other types of heavy machinery. The fundamental idea behind such devices and systems is demonstrated in Fig. 2.5. A piston located at one end of a closed system filled with a liquid, such as oil, can be used to change the pressure throughout the system, and thus transmit an applied force F_1 to a second piston where the resulting force is F_2. Since the pressure p acting on the faces of both pistons is the same (the effect of elevation changes is usually negligible for this type of hydraulic device), it follows that $F_2 = (A_2/A_1)F_1$. The piston area A_2 can be made much larger than A_1 and therefore a large mechanical advantage can be developed; that is, a small force applied at the smaller piston can be used to develop a large force at the larger piston. The applied force could be created manually through some type of mechanical device, such as a hydraulic jack, or through compressed air acting directly on the surface of the liquid, as is done in hydraulic lifts commonly found in service stations.

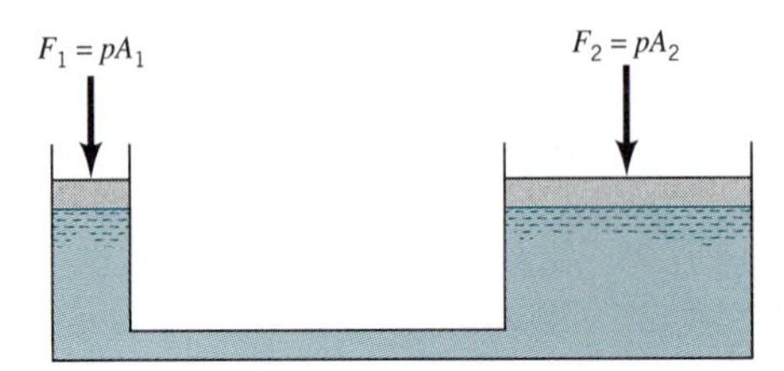

■ **FIGURE 2.5 Transmission of fluid pressure.**

2.3.2 Compressible Fluid

We normally think of gases such as air, oxygen, and nitrogen as being compressible fluids since the density of the gas can change significantly with changes in pressure and temperature. Thus, although Eq. 2.4 applies at a point in a gas, it is necessary to consider the possible variation in γ before the equation can be integrated. However, as was discussed in Chapter 1, the specific weights of common gases are small when compared with those of liquids. For example, the specific weight of air at sea level and 60 °F is 0.0763 lb/ft^3, whereas the specific weight of water under the same conditions is 62.4 lb/ft^3. Since the specific weights of gases are comparatively small, it follows from Eq. 2.4 that the pressure gradient in the vertical direction is correspondingly small, and even over distances of several hundred feet the pressure will remain essentially constant for a gas. This means we can neglect the effect of elevation changes on the pressure in gases in tanks, pipes, and so forth in which the distances involved are small.

For those situations in which the variations in heights are large, on the order of thousands of feet, attention must be given to the variation in the specific weight. As is described in Chapter 1, the equation of state for an ideal (or perfect) gas is

$$p = \rho RT$$

where p is the absolute pressure, R is the gas constant, and T is the absolute temperature. This relationship can be combined with Eq. 2.4 to give

If the specific weight of a fluid varies significantly as we move from point to point, the pressure will no longer vary directly with depth.

$$\frac{dp}{dz} = -\frac{gp}{RT}$$

and by separating variables

$$\int_{p_1}^{p_2} \frac{dp}{p} = \ln \frac{p_2}{p_1} = -\frac{g}{R} \int_{z_1}^{z_2} \frac{dz}{T} \tag{2.9}$$

where g and R are assumed to be constant over the elevation change from z_1 to z_2. Although the acceleration of gravity, g, does vary with elevation, the variation is very small (see Tables C.1 and C.2 in Appendix C), and g is usually assumed constant at some average value for the range of elevation involved.

Before completing the integration, one must specify the nature of the variation of temperature with elevation. For example, if we assume that the temperature has a constant value T_0 over the range z_1 to z_2 (*isothermal* conditions), it then follows from Eq. 2.9 that

$$p_2 = p_1 \exp\left[-\frac{g(z_2 - z_1)}{RT_0}\right] \tag{2.10}$$

This equation provides the desired pressure-elevation relationship for an isothermal layer. For nonisothermal conditions a similar procedure can be followed if the temperature-elevation relationship is known, as is discussed in the following section.

EXAMPLE 2.2

The Empire State Building in New York City, one of the tallest buildings in the world, rises to a height of approximately 1250 ft. Estimate the ratio of the pressure at the top of the building to the pressure at its base, assuming the air to be at a common temperature of 59 °F. Compare this result with that obtained by assuming the air to be incompressible with $\gamma = 0.0765 \text{ lb/ft}^3$ at 14.7 psi(abs) (values for air at standard conditions).

SOLUTION

For the assumed isothermal conditions, and treating air as a compressible fluid, Eq. 2.10 can be applied to yield

$$\frac{p_2}{p_1} = \exp\left[-\frac{g(z_2 - z_1)}{RT_0}\right]$$

$$= \exp\left\{-\frac{(32.2 \text{ ft/s}^2)(1250 \text{ ft})}{(1716 \text{ ft}\cdot\text{lb/slug}\cdot{}^\circ\text{R})[(59 + 460)^\circ\text{R}]}\right\} = 0.956 \qquad \textbf{(Ans)}$$

If the air is treated as an incompressible fluid we can apply Eq. 2.5. In this case

$$p_2 = p_1 - \gamma(z_2 - z_1)$$

or

$$\frac{p_2}{p_1} = 1 - \frac{\gamma(z_2 - z_1)}{p_1}$$

$$= 1 - \frac{(0.0765 \text{ lb/ft}^3)(1250 \text{ ft})}{(14.7 \text{ lb/in.}^2)(144 \text{ in.}^2/\text{ft}^2)} = 0.955 \qquad \textbf{(Ans)}$$

Note that there is little difference between the two results. Since the pressure difference between the bottom and top of the building is small, it follows that the variation in fluid density is small and, therefore, the compressible fluid and incompressible fluid analyses yield essentially the same result.

We see that for both calculations the pressure decreases by less than 5% as we go from ground level to the top of this tall building. It does not require a very large pressure difference to support a 1250-ft-tall column of fluid as light as air. This result supports the earlier statement that the changes in pressures in air and other gases due to elevation changes are very small, even for distances of hundreds of feet. Thus, the pressure differences between the top and bottom of a horizontal pipe carrying a gas, or in a gas storage tank, are negligible since the distances involved are very small.

2.4 Standard Atmosphere

The standard atmosphere is an idealized representation of mean conditions in the earth's atmosphere.

An important application of Eq. 2.9 relates to the variation in pressure in the earth's atmosphere. Ideally, we would like to have measurements of pressure versus altitude over the specific range for the specific conditions (temperature, reference pressure) for which the pressure is to be determined. However, this type of information is usually not available. Thus, a "standard atmosphere" has been determined that can be used in the design of aircraft, missiles, and spacecraft, and in comparing their performance under standard conditions. The concept of a standard atmosphere was first developed in the 1920s, and since that time many national and international committees and organizations have pursued the development of such a standard. The currently accepted standard atmosphere is based on a report published in 1962 and updated in 1976 (see Refs. 1 and 2), defining the so-called *U.S. standard atmosphere*, which is an idealized representation of middle-latitude, year-round mean conditions of the earth's atmosphere. Several important properties for standard atmospheric conditions at *sea level* are listed in Table 2.1, and Fig. 2.6 shows the temperature profile for the U.S. standard atmosphere. As is shown in this figure the temperature decreases with altitude in the

■ **TABLE 2.1**
Properties of U.S. Standard Atmosphere at Sea Level[a]

Property	SI Units	BG Units
Temperature, T	288.15 K (15 °C)	518.67 °R (59.00 °F)
Pressure, p	101.33 kPa (abs)	2116.2 lb/ft^2 (abs) [14.696 lb/in.^{2} (abs)]
Density, ρ	1.225 kg/m^3	0.002377 slugs/ft^3
Specific weight, γ	12.014 N/m^3	0.07647 lb/ft^3
Viscosity, μ	1.789×10^{-5} N·s/m^2	3.737×10^{-7} lb·s/ft^2

[a]Acceleration of gravity at sea level = 9.807 m/s^2 = 32.174 ft/s^2.

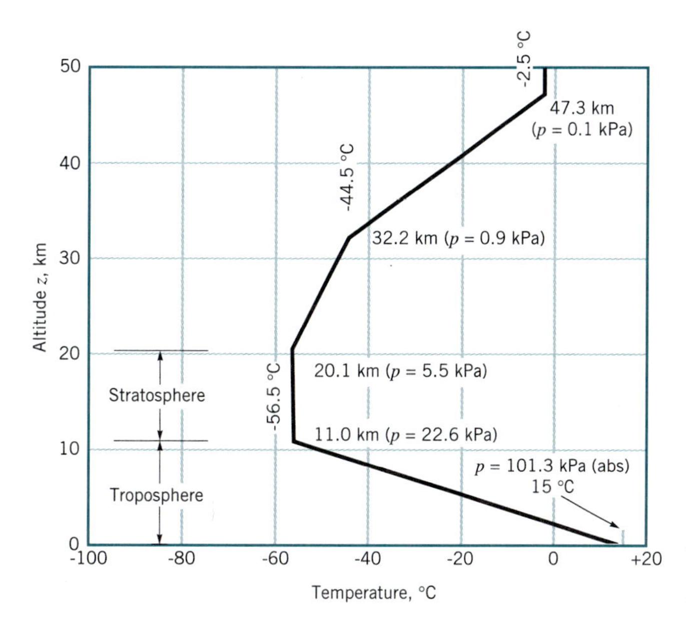

■ **FIGURE 2.6 Variation of temperature with altitude in the U.S. standard atmosphere.**

region nearest the earth's surface (*troposphere*), then becomes essentially constant in the next layer (*stratosphere*), and subsequently starts to increase in the next layer.

Since the temperature variation is represented by a series of linear segments, it is possible to integrate Eq. 2.9 to obtain the corresponding pressure variation. For example, in the troposphere, which extends to an altitude of about 11 km (~36,000 ft), the temperature variation is of the form

$$T = T_a - \beta z \tag{2.11}$$

where T_a is the temperature at sea level ($z = 0$) and β is the *lapse rate* (the rate of change of temperature with elevation). For the standard atmosphere in the troposphere, $\beta = 0.00650$ K/m or 0.00357 °R/ft.

Equation 2.11 used with Eq. 2.9 yields

$$p = p_a\left(1 - \frac{\beta z}{T_a}\right)^{g/R\beta} \tag{2.12}$$

where p_a is the absolute pressure at $z = 0$. With p_a, T_a, and g obtained from Table 2.1, and with the gas constant $R = 286.9$ J/kg·K or 1716 ft·lb/slug·°R, the pressure variation throughout the troposphere can be determined from Eq. 2.12. This calculation shows that at the outer edge of the troposphere, where the temperature is -56.5 °C, the absolute pressure is about 23 kPa (3.3 psia). It is to be noted that modern jetliners cruise at approximately this altitude. Pressures at other altitudes are shown in Fig. 2.6, and tabulated values for temperature, acceleration of gravity, pressure, density, and viscosity for the U.S. standard atmosphere are given in Tables C.1 and C.2 in Appendix C.

2.5 Measurement of Pressure

Pressure is designated as either absolute pressure or gage pressure.

Since pressure is a very important characteristic of a fluid field, it is not surprising that numerous devices and techniques are used in its measurement. As is noted briefly in Chapter 1, the pressure at a point within a fluid mass will be designated as either an *absolute* pressure or a *gage* pressure. Absolute pressure is measured relative to a perfect vacuum (absolute zero

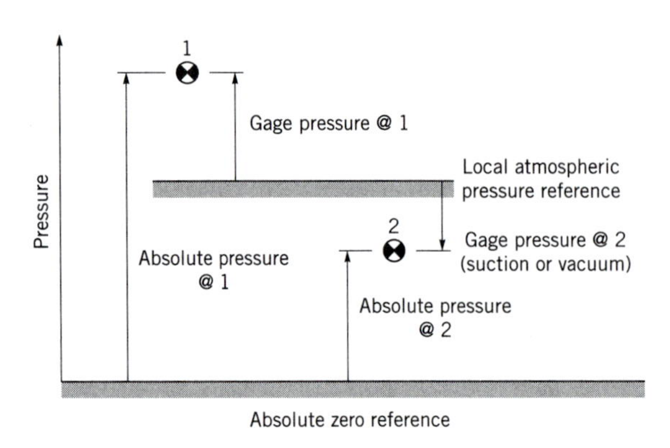

■ FIGURE 2.7 **Graphical representation of gage and absolute pressure.**

pressure), whereas gage pressure is measured relative to the local atmospheric pressure. Thus, a gage pressure of zero corresponds to a pressure that is equal to the local atmospheric pressure. Absolute pressures are always positive, but gage pressures can be either positive or negative depending on whether the pressure is above atmospheric pressure (a positive value) or below atmospheric pressure (a negative value). A negative gage pressure is also referred to as a *suction* or *vacuum* pressure. For example, 10 psi (abs) could be expressed as -4.7 psi (gage), if the local atmospheric pressure is 14.7 psi, or alternatively 4.7 psi suction or 4.7 psi vacuum. The concept of gage and absolute pressure is illustrated graphically in Fig. 2.7 for two typical pressures located at points 1 and 2.

In addition to the reference used for the pressure measurement, the *units* used to express the value are obviously of importance. As is described in Section 1.5, pressure is a force per unit area, and the units in the BG system are lb/ft^2 or lb/in.^2, commonly abbreviated psf or psi, respectively. In the SI system the units are N/m^2; this combination is called the pascal and written as Pa ($1\ \text{N/m}^2 = 1\ \text{Pa}$). As noted earlier, pressure can also be expressed as the height of a column of liquid. Then, the units will refer to the height of the column (in., ft, mm, m, etc.), and in addition, the liquid in the column must be specified (H_2O, Hg, etc.). For example, standard atmospheric pressure can be expressed as 760 mm Hg (abs). *In this text, pressures will be assumed to be gage pressures unless specifically designated absolute.* For example, 10 psi or 100 kPa would be gage pressures, whereas 10 psia or 100 kPa (abs) would refer to absolute pressures. It is to be noted that *pressure differences* are independent of the reference, so that no special notation is required in this case.

A barometer is used to measure atmospheric pressure.

The measurement of atmospheric pressure is usually accomplished with a mercury *barometer*, which in its simplest form consists of a glass tube closed at one end with the open end immersed in a container of mercury as shown in Fig. 2.8. The tube is initially filled with mercury (inverted with its open end up) and then turned upside down (open end down) with the open end in the container of mercury. The column of mercury will come to an equilibrium position where its weight plus the force due to the vapor pressure (which develops in the space above the column) balances the force due to the atmospheric pressure. Thus,

$$p_{\text{atm}} = \gamma h + p_{\text{vapor}} \tag{2.13}$$

where γ is the specific weight of mercury. For most practical purposes the contribution of the vapor pressure can be neglected since it is very small [for mercury, $p_{\text{vapor}} = 0.000023$ lb/in.^2 (abs) at a tempreature of 68 °F] so that $p_{\text{atm}} \approx \gamma h$. It is conventional to specify atmospheric pressure in terms of the height, h, in millimeters or inches of mercury. Note that if water were used instead of mercury, the height of the column would have to be approximately 34 ft rather than 29.9 in. of mercury for an atmospheric pressure of 14.7 psia! The concept of the mercury barometer is an old one, with the invention of this device attributed to Evangelista Torricelli in about 1644.

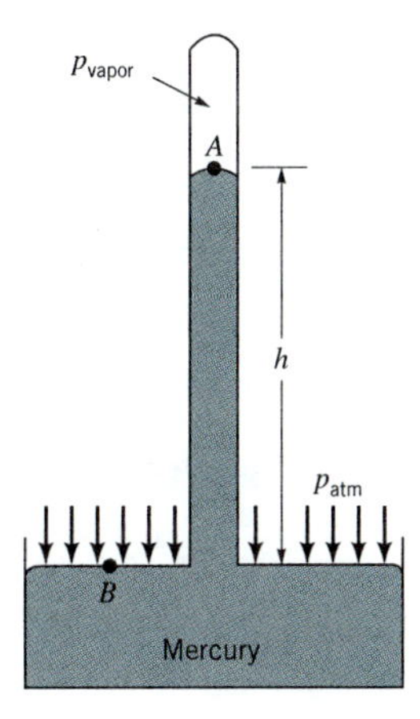

■ FIGURE 2.8 **Mercury barometer.**

EXAMPLE 2.3

A mountain lake has an average temperature of 10 °C and a maximum depth of 40 m. For a barometric pressure of 598 mm Hg, determine the absolute pressure (in pascals) at the deepest part of the lake.

SOLUTION

The pressure in the lake at any depth, h, is given by the equation

$$p = \gamma h + p_0$$

where p_0 is the pressure at the surface. Since we want the absolute pressure, p_0 will be the local barometric pressure expressed in a consistent system of units; that is

$$\frac{p_{\text{barometric}}}{\gamma_{\text{Hg}}} = 598 \text{ mm} = 0.598 \text{ m}$$

and for $\gamma_{\text{Hg}} = 133 \text{ kN/m}^3$

$$p_0 = (0.598 \text{ m})(133 \text{ kN/m}^3) = 79.5 \text{ kN/m}^2$$

From Table B.2, $\gamma_{\text{H}_2\text{O}} = 9.804 \text{ kN/m}^3$ at 10 °C and therefore

$$p = (9.804 \text{ kN/m}^3)(40 \text{ m}) + 79.5 \text{ kN/m}^2$$

$$= 392 \text{ kN/m}^2 + 79.5 \text{ kN/m}^2 = 472 \text{ kPa (abs)} \qquad \textbf{(Ans)}$$

This simple example illustrates the need for close attention to the units used in the calculation of pressure; that is, be sure to use a *consistent* unit system, and be careful not to add a pressure head (m) to a pressure (Pa).

2.6 Manometry

Manometers use vertical or inclined liquid columns to measure pressure.

A standard technique for measuring pressure involves the use of liquid columns in vertical or inclined tubes. Pressure measuring devices based on this technique are called *manometers*. The mercury barometer is an example of one type of manometer, but there are many other configurations possible, depending on the particular application. Three common types of manometers include the piezometer tube, the U-tube manometer, and the inclined-tube manometer.

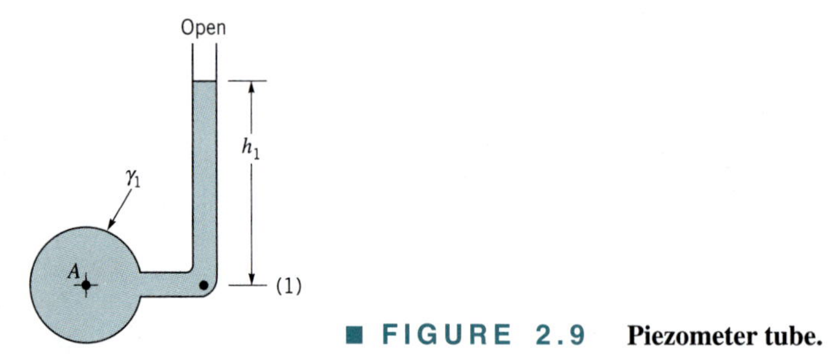

■ FIGURE 2.9 **Piezometer tube.**

2.6.1 Piezometer Tube

The simplest type of manometer consists of a vertical tube, open at the top, and attached to the container in which the pressure is desired, as illustrated in Fig. 2.9. Since manometers involve columns of fluids at rest, the fundamental equation describing their use is Eq. 2.8

$$p = \gamma h + p_0$$

which gives the pressure at any elevation within a homogeneous fluid in terms of a reference pressure p_0 and the vertical distance h between p and p_0. Remember that in a fluid at rest pressure will *increase* as we move *downward* and will decrease as we move *upward.* Application of this equation to the piezometer tube of Fig. 2.9 indicates that the pressure p_A can be determined by a measurement of h_1 through the relationship

$$p_A = \gamma_1 h_1$$

where γ_1 is the specific weight of the liquid in the container. Note that since the tube is open at the top, the pressure p_0 can be set equal to zero (we are now using gage pressure), with the height h_1 measured from the meniscus at the upper surface to point (1). Since point (1) and point A within the container are at the same elevation, $p_A = p_1$.

Although the piezometer tube is a very simple and accurate pressure measuring device, it has several disadvantages. It is only suitable if the pressure in the container is greater than atmospheric pressure (otherwise air would be sucked into the system), and the pressure to be measured must be relatively small so the required height of the column is reasonable. Also, the fluid in the container in which the pressure is to be measured must be a liquid rather than a gas.

2.6.2 U-Tube Manometer

To determine pressure from a manometer, simply use the fact that the pressure in the liquid columns will vary hydrostatically.

To overcome the difficulties noted previously, another type of manometer which is widely used consists of a tube formed into the shape of a U as is shown in Fig. 2.10. The fluid in the manometer is called the *gage fluid.* To find the pressure p_A in terms of the various column heights, we start at one end of the system and work our way around to the other end, simply utilizing Eq. 2.8. Thus, for the U-tube manometer shown in Fig. 2.10, we will start at point A and work around to the open end. The pressure at points A and (1) are the same, and as we move from point (1) to (2) the pressure will increase by $\gamma_1 h_1$. The pressure at point (2) is equal to the pressure at point (3), since the pressures at equal elevations in a continuous mass of fluid at rest must be the same. Note that we could not simply ''jump across'' from point (1) to a point at the same elevation in the right-hand tube since these would not be points within the same continuous mass of fluid. With the pressure at point (3) specified we now move to the open end where the pressure is zero. As we move vertically upward the pressure decreases by an amount $\gamma_2 h_2$. In equation form these various steps can be expressed as

$$p_A + \gamma_1 h_1 - \gamma_2 h_2 = 0$$

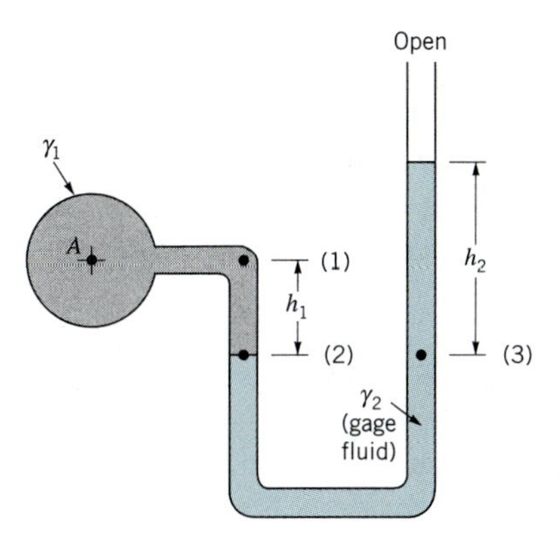

■ FIGURE 2.10 **Simple U-tube manometer.**

and, therefore, the pressure p_A can be written in terms of the column heights as

$$p_A = \gamma_2 h_2 - \gamma_1 h_1 \tag{2.14}$$

The contribution of gas columns in manometers is usually negligible since the weight of the gas is so small.

A major advantage of the U-tube manometer lies in the fact that the gage fluid can be different from the fluid in the container in which the pressure is to be determined. For example, the fluid in A in Fig. 2.10 can be either a liquid or a gas. If A does contain a gas, the contribution of the gas column, $\gamma_1 h_1$, is almost always negligible so that $p_A \approx p_2$ and in this instance Eq. 2.14 becomes

$$p_A = \gamma_2 h_2$$

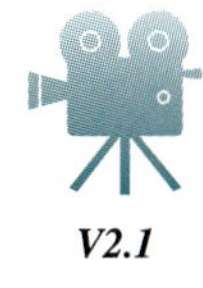

V2.1

Thus, for a given pressure the height, h_2, is governed by the specific weight, γ_2, of the gage fluid used in the manometer. If the pressure p_A is large, then a heavy gage fluid, such as mercury, can be used and a reasonable column height (not too long) can still be maintained. Alternatively, if the pressure p_A is small, a lighter gage fluid, such as water, can be used so that a relatively large column height (which is easily read) can be achieved.

EXAMPLE 2.4

A closed tank contains compressed air and oil ($SG_{\text{oil}} = 0.90$) as is shown in Fig. E2.4. A U-tube manometer using mercury ($SG_{\text{Hg}} = 13.6$) is connected to the tank as shown. For column heights $h_1 = 36$ in., $h_2 = 6$ in., and $h_3 = 9$ in., determine the pressure reading (in psi) of the gage.

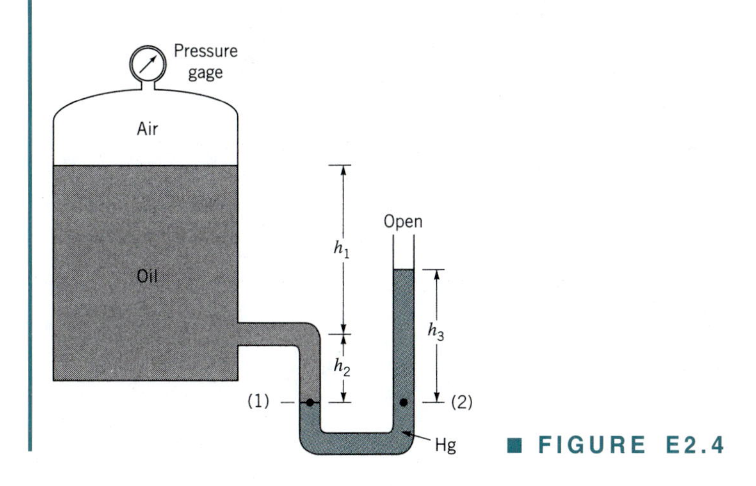

■ FIGURE E2.4

SOLUTION

Following the general procedure of starting at one end of the manometer system and working around to the other, we will start at the air–oil interface in the tank and proceed to the open end where the pressure is zero. The pressure at level (1) is

$$p_1 = p_{\text{air}} + \gamma_{\text{oil}}(h_1 + h_2)$$

This pressure is equal to the pressure at level (2), since these two points are at the same elevation in a homogeneous fluid at rest. As we move from level (2) to the open end, the pressure must decrease by $\gamma_{\text{Hg}}h_3$, and at the open end the pressure is zero. Thus, the manometer equation can be expressed as

$$p_{\text{air}} + \gamma_{\text{oil}}(h_1 + h_2) - \gamma_{\text{Hg}}h_3 = 0$$

or

$$p_{\text{air}} + (SG_{\text{oil}})(\gamma_{\text{H}_2\text{O}})(h_1 + h_2) - (SG_{\text{Hg}})(\gamma_{\text{H}_2\text{O}})h_3 = 0$$

For the values given

$$p_{\text{air}} = -(0.9)(62.4 \text{ lb/ft}^3)\left(\frac{36 + 6}{12} \text{ ft}\right) + (13.6)(62.4 \text{ lb/ft}^3)\left(\frac{9}{12} \text{ ft}\right)$$

so that

$$p_{\text{air}} = 440 \text{ lb/ft}^2$$

Since the specific weight of the air above the oil is much smaller than the specific weight of the oil, the gage should read the pressure we have calculated; that is,

$$p_{\text{gage}} = \frac{440 \text{ lb/ft}^2}{144 \text{ in.}^2/\text{ft}^2} = 3.06 \text{ psi} \qquad \textbf{(Ans)}$$

Manometers are often used to measure the difference in pressure between two points.

The U-tube manometer is also widely used to measure the *difference* in pressure between two containers or two points in a given system. Consider a manometer connected between containers *A* and *B* as is shown in Fig. 2.11. The difference in pressure between *A* and *B* can be found by again starting at one end of the system and working around to the other end. For example, at *A* the pressure is p_A, which is equal to p_1, and as we move to point (2) the pressure increases by $\gamma_1 h_1$. The pressure at p_2 is equal to p_3, and as we move upward

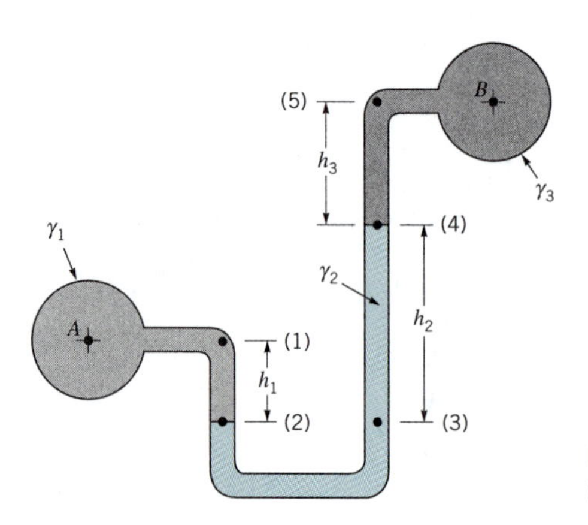

■ **FIGURE 2.11 Differential U-tube manometer.**

to point (4) the pressure decreases by $\gamma_2 h_2$. Similarly, as we continue to move upward from point (4) to (5) the pressure decreases by $\gamma_3 h_3$. Finally, $p_5 = p_B$, since they are at equal elevations. Thus,

$$p_A + \gamma_1 h_1 - \gamma_2 h_2 - \gamma_3 h_3 = p_B$$

and the pressure difference is

$$p_A - p_B = \gamma_2 h_2 + \gamma_3 h_3 - \gamma_1 h_1$$

When the time comes to substitute in numbers, be sure to use a consistent system of units!

Capillary action could affect the manometer reading.

Capillarity due to surface tension at the various fluid interfaces in the manometer is usually not considered, since for a simple U-tube with a meniscus in each leg, the capillary effects cancel (assuming the surface tensions and tube diameters are the same at each meniscus), or we can make the capillary rise negligible by using relatively large bore tubes (with diameters of about 0.5 in. or larger). Two common gage fluids are water and mercury. Both give a well-defined meniscus (a very important characteristic for a gage fluid) and have well-known properties. Of course, the gage fluid must be immiscible with respect to the other fluids in contact with it. For highly accurate measurements, special attention should be given to temperature since the various specific weights of the fluids in the manometer will vary with temperature.

EXAMPLE 2.5

As will be discussed in Chapter 3, the volume rate of flow, Q, through a pipe can be determined by means of a flow nozzle located in the pipe as illustrated in Fig. E2.5. The nozzle creates a pressure drop, $p_A - p_B$, along the pipe which is related to the flow through the equation $Q = K\sqrt{p_A - p_B}$, where K is a constant depending on the pipe and nozzle size. The pressure drop is frequently measured with a differential U-tube manometer of the type illustrated. (a) Determine an equation for $p_A - p_B$ in terms of the specific weight of the flowing fluid, γ_1, the specific weight of the gage fluid, γ_2, and the various heights indicated. (b) For $\gamma_1 = 9.80 \text{ kN/m}^3$, $\gamma_2 = 15.6 \text{ kN/m}^3$, $h_1 = 1.0$ m, and $h_2 = 0.5$ m, what is the value of the pressure drop, $p_A - p_B$?

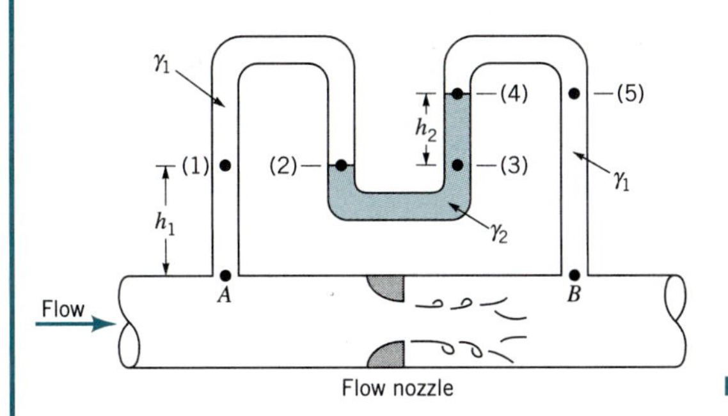

■ FIGURE E2.5

SOLUTION

(a) Although the fluid in the pipe is moving, the fluids in the columns of the manometer are at rest so that the pressure variation in the manometer tubes is hydrostatic. If we start at point A and move vertically upward to level (1), the pressure will decrease by $\gamma_1 h_1$ and will be equal to the pressure at (2) and at (3). We can now move from (3) to (4) where the pressure has been further reduced by $\gamma_2 h_2$. The pressures at levels (4) and (5) are equal, and as we move from (5) to B the pressure will increase by $\gamma_1(h_1 + h_2)$.

Thus, in equation form

$$p_A - \gamma_1 h_1 - \gamma_2 h_2 + \gamma_1(h_1 + h_2) = p_B$$

or

$$p_A - p_B = h_2(\gamma_2 - \gamma_1) \qquad \textbf{(Ans)}$$

It is to be noted that the only column height of importance is the differential reading, h_2. The differential manometer could be placed 0.5 or 5.0 m above the pipe ($h_1 = 0.5$ m or $h_1 = 5.0$ m) and the value of h_2 would remain the same. Relatively large values for the differential reading h_2 can be obtained for small pressure differences, $p_A - p_B$, if the difference between γ_1 and γ_2 is small.

(b) The specific value of the pressure drop for the data given is

$$p_A - p_B = (0.5 \text{ m})(15.6 \text{ kN/m}^3 - 9.80 \text{ kN/m}^3)$$

$$= 2.90 \text{ kPa} \qquad \textbf{(Ans)}$$

2.6.3 Inclined-Tube Manometer

To measure small pressure changes, a manometer of the type shown in Fig. 2.12 is frequently used. One leg of the manometer is inclined at an angle θ, and the differential reading ℓ_2 is measured along the inclined tube. The difference in pressure $p_A - p_B$ can be expressed as

$$p_A + \gamma_1 h_1 - \gamma_2 \ell_2 \sin\theta - \gamma_3 h_3 = p_B$$

or

$$p_A - p_B = \gamma_2 \ell_2 \sin\theta + \gamma_3 h_3 - \gamma_1 h_1 \qquad \textbf{(2.15)}$$

Inclined-tube manometers can be used to measure small pressure differences accurately.

where it is to be noted the pressure difference between points (1) and (2) is due to the *vertical* distance between the points, which can be expressed as $\ell_2 \sin\theta$. Thus, for relatively small angles the differential reading along the inclined tube can be made large even for small pressure differences. The inclined-tube manometer is often used to measure small differences in gas pressures so that if pipes A and B contain a gas then

$$p_A - p_B = \gamma_2 \ell_2 \sin\theta$$

or

$$\ell_2 = \frac{p_A - p_B}{\gamma_2 \sin\theta} \qquad \textbf{(2.16)}$$

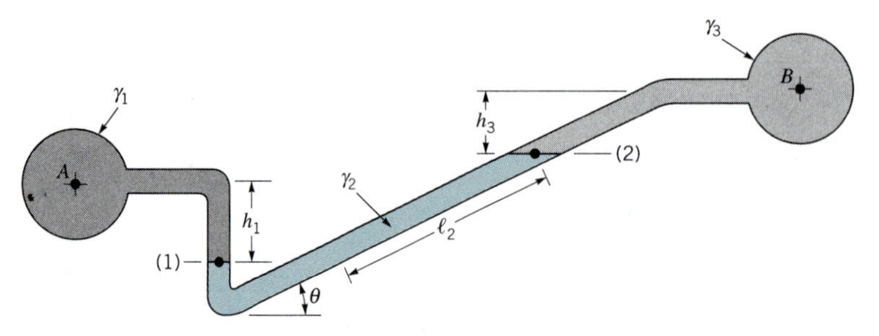

■ **FIGURE 2.12** **Inclined-tube manometer.**

(a) (b)

■ FIGURE 2.13 (a) Liquid-filled Bourdon pressure gages for various pressure ranges. (b) Internal elements of Bourdon gages. The "C-shaped" Bourdon tube is shown on the left, and the "coiled spring" Bourdon tube for high pressures of 1000 psi and above is shown on the right. (Photographs courtesy of Weiss Instruments, Inc.)

where the contributions of the gas columns h_1 and h_3 have been neglected. Equation 2.16 shows that the differential reading ℓ_2 (for a given pressure difference) of the inclined-tube manometer can be increased over that obtained with a conventional U-tube manometer by the factor $1/\sin\theta$. Recall that $\sin\theta \rightarrow 0$ as $\theta \rightarrow 0$.

2.7 Mechanical and Electronic Pressure Measuring Devices

A Bourdon tube pressure gage uses a hollow, elastic, and curved tube to measure pressure.

Although manometers are widely used, they are not well suited for measuring very high pressures, or pressures that are changing rapidly with time. In addition, they require the measurement of one or more column heights, which, although not particularly difficult, can be time consuming. To overcome some of these problems numerous other types of pressure-measuring instruments have been developed. Most of these make use of the idea that when a pressure acts on an elastic structure the structure will deform, and this deformation can be related to the magnitude of the pressure. Probably the most familiar device of this kind is the *Bourdon* pressure gage, which is shown in Fig. 2.13*a*. The essential mechanical element in this gage is the hollow, elastic curved tube (Bourdon tube) which is connected to the pressure source as shown in Fig. 2.13*b*. As the pressure within the tube increases the tube tends to straighten, and although the deformation is small, it can be translated into the motion of a pointer on a dial as illustrated. Since it is the difference in pressure between the outside of the tube (atmospheric pressure) and the inside of the tube that causes the movement of the tube, the indicated pressure is gage pressure. The Bourdon gage must be calibrated so that the dial reading can directly indicate the pressure in suitable units such as psi, psf, or pascals. A zero reading on the gage indicates that the measured pressure is equal to the local atmospheric pressure. This type of gage can be used to measure a negative gage pressure (vacuum) as well as positive pressures.

V2.2

The *aneroid* barometer is another type of mechanical gage that is used for measuring atmospheric pressure. Since atmospheric pressure is specified as an absolute pressure, the conventional Bourdon gage is not suitable for this measurement. The common aneroid barometer contains a hollow, closed, elastic element which is evacuated so that the pressure inside the element is near absolute zero. As the external atmospheric pressure changes, the element deflects, and this motion can be translated into the movement of an attached dial. As with the Bourdon gage, the dial can be calibrated to give atmospheric pressure directly, with the usual units being millimeters or inches of mercury.

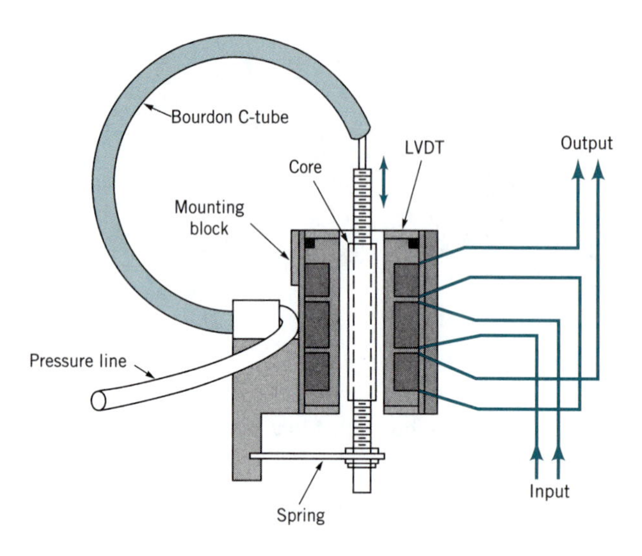

■ **FIGURE 2.14**
Pressure transducer which combines a linear variable differential transformer (LVDT) with a Bourdon gage. (From Ref. 4, used by permission.)

A pressure transducer converts pressure into an electrical output.

For many applications in which pressure measurements are required, the pressure must be measured with a device that converts the pressure into an electrical output. For example, it may be desirable to continuously monitor a pressure that is changing with time. This type of pressure measuring device is called a *pressure transducer*, and many different designs are used. One possible type of transducer is one in which a Bourdon tube is connected to a linear variable differential transformer (LVDT), as is illustrated in Fig. 2.14. The core of the LVDT is connected to the free end of the Bourdon so that as a pressure is applied the resulting motion of the end of the tube moves the core through the coil and an output voltage develops. This voltage is a linear function of the pressure and could be recorded on an oscillograph or digitized for storage or processing on a computer.

One disadvantage of a pressure transducer using a Bourdon tube as the elastic sensing element is that it is limited to the measurement of pressures that are static or only changing slowly (quasistatic). Because of the relatively large mass of the Bourdon tube, it cannot respond to rapid changes in pressure. To overcome this difficulty a different type of transducer is used in which the sensing element is a thin, elastic diaphragm which is in contact with the fluid. As the pressure changes, the diaphragm deflects, and this deflection can be sensed and converted into an electrical voltage. One way to accomplish this is to locate strain gages either on the surface of the diaphragm not in contact with the fluid, or on an element attached to the diaphragm. These gages can accurately sense the small strains induced in the diaphragm and provide an output voltage proportional to pressure. This type of transducer is capable of measuring accurately both small and large pressures, as well as both static and dynamic pressures. For example, strain-gage pressure transducers of the type shown in Fig. 2.15 are used to measure arterial blood pressure, which is a relatively small pressure that varies periodically with a fundamental frequency of about 1 Hz. The transducer is usually connected to the blood vessel by means of a liquid-filled, small diameter tube called a pressure catheter. Although the strain-gage type of transducers can be designed to have very good frequency response (up to approximately 10 kHz), they become less sensitive at the higher frequencies since the diaphragm must be made stiffer to achieve the higher frequency response. As an alternative the diaphragm can be constructed of a piezoelectric crystal to be used as both the elastic element and the sensor. When a pressure is applied to the crystal a voltage develops because of the deformation of the crystal. This voltage is directly related to the applied pressure. Depending on the design, this type of transducer can be used to measure both very low and high pressures (up to approximately 100,000 psi) at high frequencies. Additional information on pressure transducers can be found in Refs. 3, 4, and 5.

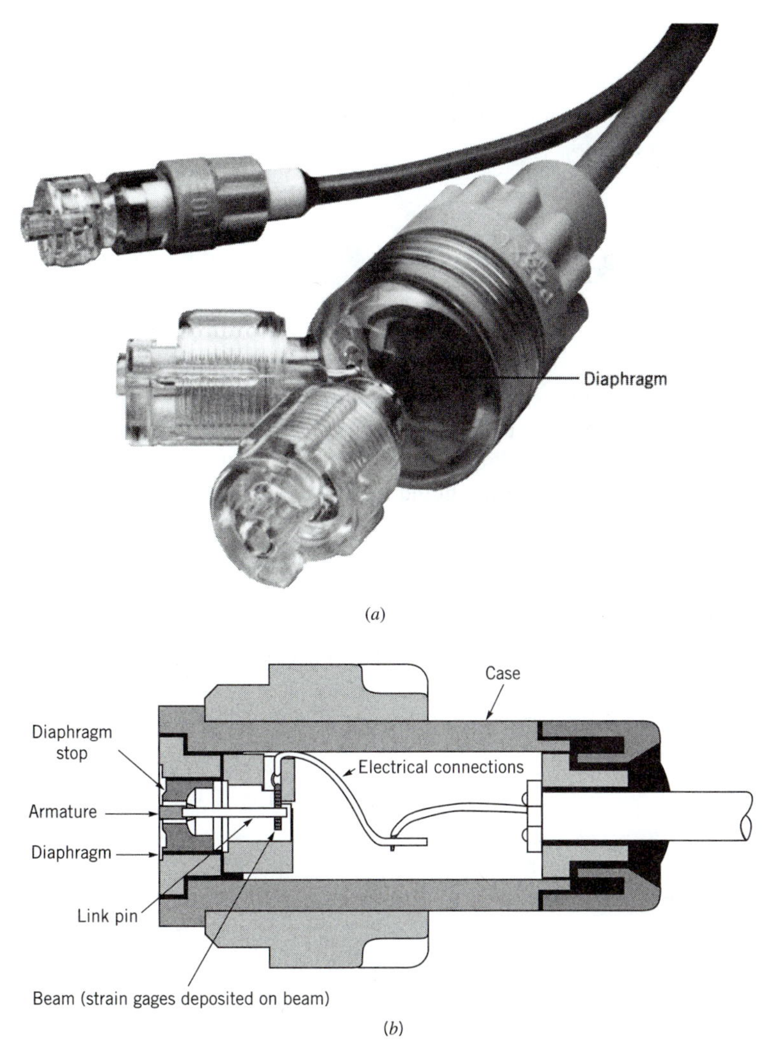

■ **FIGURE 2.15** **(*a*) Two different sized strain-gage pressure transducers (Spectramed Models P10EZ and P23XL) commonly used to measure physiological pressures. Plastic domes are filled with fluid and connected to blood vessels through a needle or catheter. (Photograph courtesy of Spectramed, Inc.) (*b*) Schematic diagram of the P23XL transducer with the dome removed. Deflection of the diaphragm due to pressure is measured with a silicon beam on which strain gages and an associated bridge circuit have been deposited.**

2.8 Hydrostatic Force on a Plane Surface

V2.3

When determining the resultant force on an area, the effect of atmospheric pressure often cancels.

When a surface is submerged in a fluid, forces develop on the surface due to the fluid. The determination of these forces is important in the design of storage tanks, ships, dams, and other hydraulic structures. For fluids at rest we know that the force must be *perpendicular* to the surface since there are no shearing stresses present. We also know that the pressure will vary linearly with depth if the fluid is incompressible. For a horizontal surface, such as the bottom of a liquid-filled tank (Fig. 2.16), the magnitude of the resultant force is simply $F_R = pA$, where p is the uniform pressure on the bottom and A is the area of the bottom. For the open tank shown, $p = \gamma h$. Note that if atmospheric pressure acts on both sides of the bottom, as is illustrated, the *resultant* force on the bottom is simply due to the liquid in the tank. Since the pressure is constant and uniformly distributed over the bottom, the resultant force acts through the centroid of the area as shown in Fig. 2.16.

For the more general case in which a submerged plane surface is inclined, as is illustrated in Fig. 2.17, the determination of the resultant force acting on the surface is more involved. For the present we will assume that the fluid surface is open to the atmosphere. Let the plane in which the surface lies intersect the free surface at 0 and make an angle θ with

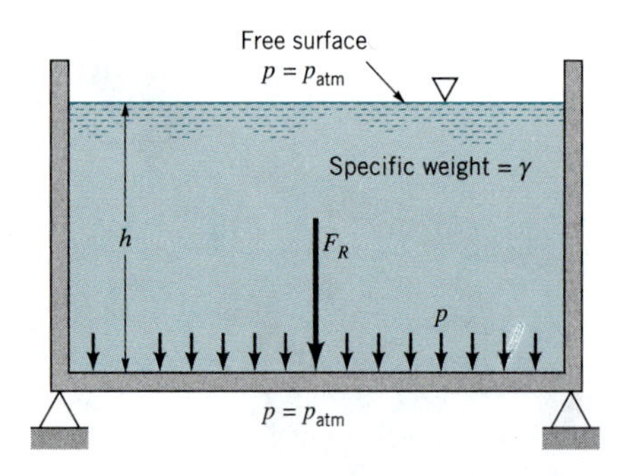

■ **FIGURE 2.16 Pressure and resultant hydrostatic force developed on the bottom of an open tank.**

The resultant force of a static fluid on a plane surface is due to the hydrostatic pressure distribution on the surface.

this surface as in Fig. 2.17. The x-y coordinate system is defined so that 0 is the origin and y is directed along the surface as shown. The area can have an arbitrary shape as shown. We wish to determine the direction, location, and magnitude of the resultant force acting on one side of this area due to the liquid in contact with the area. At any given depth, h, the force acting on dA (the differential area of Fig. 2.17) is $dF = \gamma h\, dA$ and is perpendicular to the surface. Thus, the magnitude of the resultant force can be found by summing these differential forces over the entire surface. In equation form

$$F_R = \int_A \gamma h\, dA = \int_A \gamma y \sin\theta\, dA$$

where $h = y \sin\theta$. For constant γ and θ

$$F_R = \gamma \sin\theta \int_A y\, dA \tag{2.17}$$

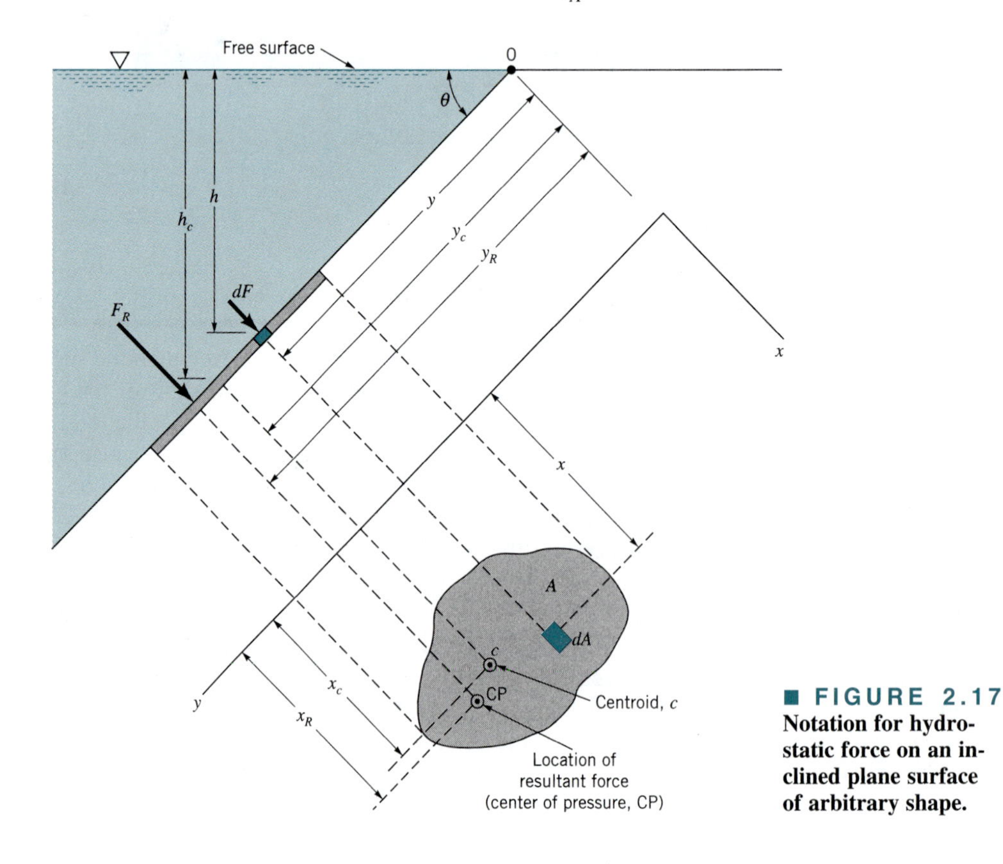

■ **FIGURE 2.17 Notation for hydrostatic force on an inclined plane surface of arbitrary shape.**

The integral appearing in Eq. 2.17 is the *first moment of the area* with respect to the x axis, so we can write

$$\int_A y\, dA = y_c A$$

where y_c is the y coordinate of the centroid measured from the x axis which passes through 0. Equation 2.17 can thus be written as

$$F_R = \gamma A y_c \sin\theta$$

The magnitude of the resultant fluid force is equal to the pressure acting at the centroid of the area multiplied by the total area.

or more simply as

$$\boxed{F_R = \gamma h_c A} \quad \textbf{(2.18)}$$

where h_c is the vertical distance from the fluid surface to the centroid of the area. Note that the magnitude of the force is independent of the angle θ and depends only on the specific weight of the fluid, the total area, and the depth of the centroid of the area below the surface. In effect, Eq. 2.18 indicates that the magnitude of the resultant force is equal to the pressure at the centroid of the area multiplied by the total area. Since all the differential forces that were summed to obtain F_R are perpendicular to the surface, the resultant F_R must also be perpendicular to the surface.

Although our intuition might suggest that the resultant force should pass through the centroid of the area, this is not actually the case. The y coordinate, y_R, of the resultant force can be determined by summation of moments around the x axis. That is, the moment of the resultant force must equal the moment of the distributed pressure force, or

$$F_R y_R = \int_A y\, dF = \int_A \gamma \sin\theta\, y^2\, dA$$

and, therefore, since $F_R = \gamma A y_c \sin\theta$

$$y_R = \frac{\int_A y^2\, dA}{y_c A}$$

The integral in the numerator is the *second moment of the area* (*moment of inertia*), I_x, with respect to an axis formed by the intersection of the plane containing the surface and the free surface (x axis). Thus, we can write

$$y_R = \frac{I_x}{y_c A}$$

Use can now be made of the parallel axis theorem to express I_x as

$$I_x = I_{xc} + A y_c^2$$

where I_{xc} is the second moment of the area with respect to an axis passing through its *centroid* and parallel to the x axis. Thus,

$$\boxed{y_R = \frac{I_{xc}}{y_c A} + y_c} \quad \textbf{(2.19)}$$

Equation 2.19 clearly shows that the resultant force does not pass through the centroid but is always *below* it, since $I_{xc}/y_c A > 0$.

The x coordinate, x_R, for the resultant force can be determined in a similar manner by summing moments about the y axis. Thus,

$$F_R x_R = \int_A \gamma \sin \theta \, xy \, dA$$

The resultant fluid force does not pass through the centroid of the area.

and, therefore,

$$x_R = \frac{\int_A xy \, dA}{y_c A} = \frac{I_{xy}}{y_c A}$$

where I_{xy} is the product of inertia with respect to the x and y axes. Again, using the parallel axis theorem,[1] we can write

$$x_R = \frac{I_{xyc}}{y_c A} + x_c \tag{2.20}$$

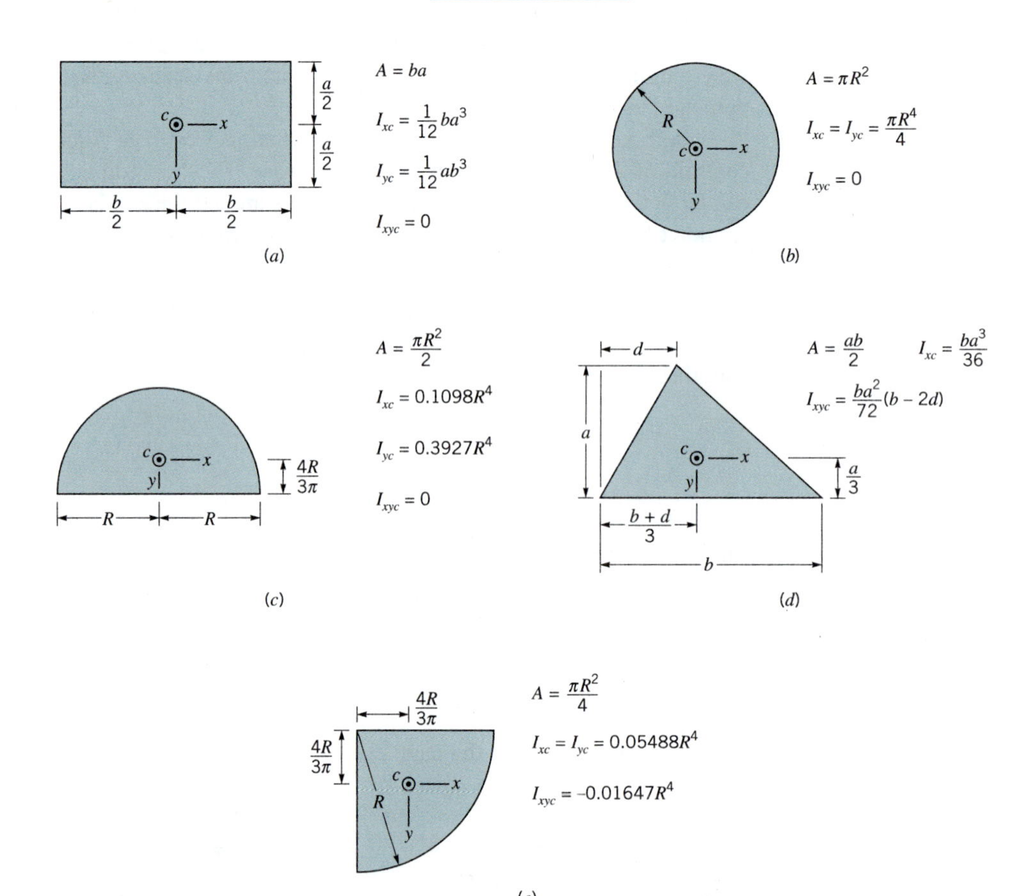

■ FIGURE 2.18 **Geometric properties of some common shapes.**

[1]Recall that the parallel axis theorem for the product of inertia of an area states that the product of inertia with respect to an orthogonal set of axes (x-y coordinate system) is equal to the product of inertia with respect to an orthogonal set of axes parallel to the original set and passing through the centroid of the area, plus the product of the area and the x and y coordinates of the centroid of the area. Thus, $I_{xy} = I_{xyc} + Ax_c y_c$.

where I_{xyc} is the product of inertia with respect to an orthogonal coordinate system passing through the *centroid* of the area and formed by a translation of the x-y coordinate system. If the submerged area is symmetrical with respect to an axis passing through the centroid and parallel to either the x or y axes, the resultant force must lie along the line $x = x_c$, since I_{xyc} is identically zero in this case. The point through which the resultant force acts is called the *center of pressure*. It is to be noted from Eqs. 2.19 and 2.20 that as y_c increases the center of pressure moves closer to the centroid of the area. Since $y_c = h_c/\sin\theta$, the distance y_c will increase if the depth of submergence, h_c, increases, or, for a given depth, the area is rotated so that the angle, θ, decreases. Centroidal coordinates and moments of inertia for some common areas are given in Fig. 2.18.

The point through which the resultant fluid force acts is called the center of pressure.

EXAMPLE 2.6

The 4-m-diameter circular gate of Fig E2.6*a* is located in the inclined wall of a large reservoir containing water ($\gamma = 9.80 \text{ kN/m}^3$). The gate is mounted on a shaft along its horizontal diameter. For a water depth of 10 m above the shaft determine: (a) the magnitude and location of the resultant force exerted on the gate by the water, and (b) the moment that would have to be applied to the shaft to open the gate.

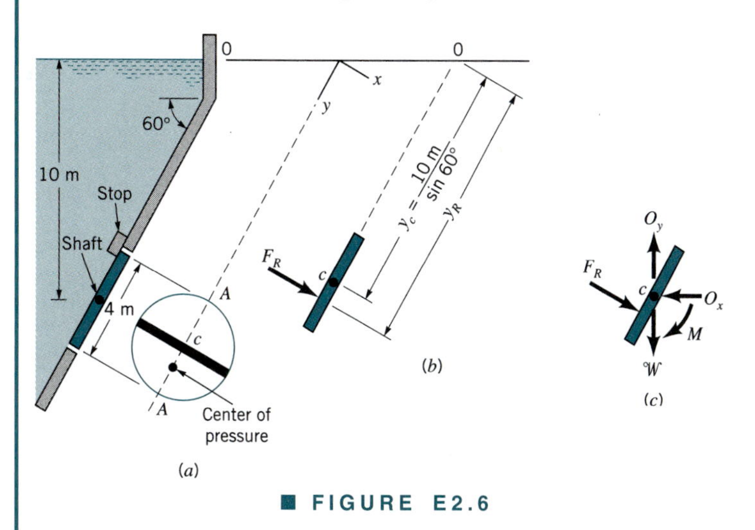

■ FIGURE E2.6

SOLUTION

(a) To find the magnitude of the force of the water we can apply Eq. 2.18,

$$F_R = \gamma h_c A$$

and since the vertical distance from the fluid surface to the centroid of the area is 10 m it follows that

$$F_R = (9.80 \times 10^3 \text{ N/m}^3)(10 \text{ m})(4\pi \text{ m}^2)$$
$$= 1230 \times 10^3 \text{ N} = 1.23 \text{ MN} \qquad \textbf{(Ans)}$$

To locate the point (center of pressure) through which F_R acts, we use Eqs. 2.19 and 2.20,

$$x_R = \frac{I_{xyc}}{y_c A} + x_c \qquad y_R = \frac{I_{xc}}{y_c A} + y_c$$

For the coordinate system shown, $x_R = 0$ since the area is symmetrical, and the center of pressure must lie along the diameter *A-A*. To obtain y_R, we have from Fig. 2.18

$$I_{xc} = \frac{\pi R^4}{4}$$

and y_c is shown in Fig. E2.6*b*. Thus,

$$y_R = \frac{(\pi/4)(2 \text{ m})^4}{(10 \text{ m}/\sin 60°)(4\pi \text{ m}^2)} + \frac{10 \text{ m}}{\sin 60°}$$

$$= 0.0866 \text{ m} + 11.55 \text{ m} = 11.6 \text{ m}$$

and the distance (along the gate) below the shaft to the center of pressure is

$$y_R - y_c = 0.0866 \text{ m} \qquad \textbf{(Ans)}$$

We can conclude from this analysis that the force on the gate due to the water has a magnitude of 1.23 MN and acts through a point along its diameter *A-A* at a distance of 0.0866 m (along the gate) below the shaft. The force is perpendicular to the gate surface as shown.

(b) The moment required to open the gate can be obtained with the aid of the free-body diagram of Fig. E2.6*c*. In this diagram $\mathcal{W}$ is the weight of the gate and O_x and O_y are the horizontal and vertical reactions of the shaft on the gate. We can now sum moments about the shaft

$$\sum M_c = 0$$

and, therefore,

$$M = F_R(y_R - y_c)$$

$$= (1230 \times 10^3 \text{ N})(0.0866 \text{ m})$$

$$= 1.07 \times 10^5 \text{ N·m} \qquad \textbf{(Ans)}$$

A large fish-holding tank contains seawater ($\gamma = 64.0 \text{ lb/ft}^3$) to a depth of 10 ft as shown in Fig. E2.7*a*. To repair some damage to one corner of the tank, a triangular section is replaced with a new section as illustrated. Determine the magnitude and location of the force of the seawater on this triangular area.

SOLUTION

The various distances needed to solve this problem are shown in Fig. E2.7*b*. Since the surface of interest lies in a vertical plane, $y_c = h_c = 9$ ft, and from Eq. 2.18 the magnitude of the force is

$$F_R = \gamma h_c A$$

$$= (64.0 \text{ lb/ft}^3)(9 \text{ ft})(9/2 \text{ ft}^2) = 2590 \text{ lb} \qquad \textbf{(Ans)}$$

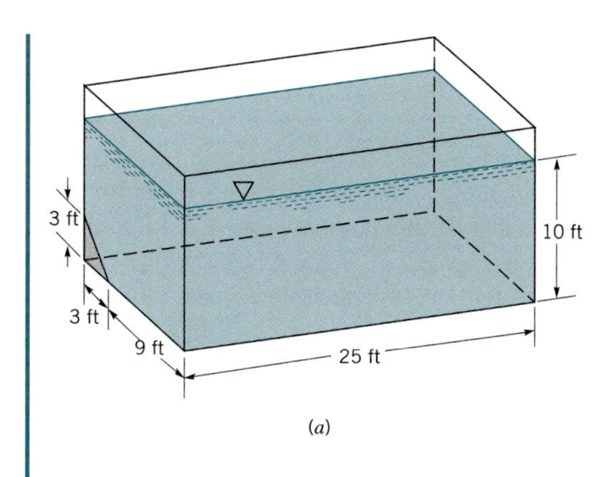

(a)

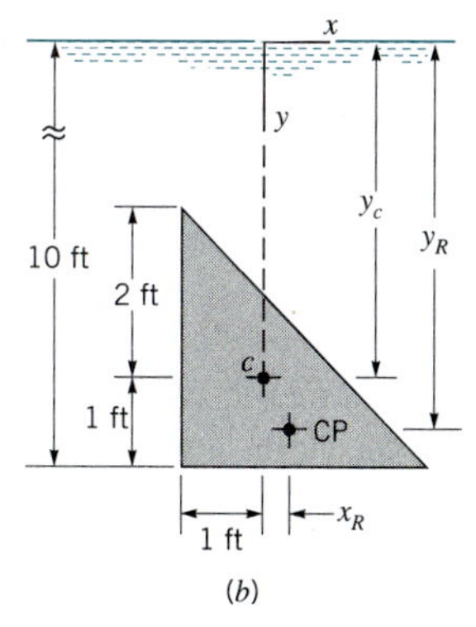

(b)

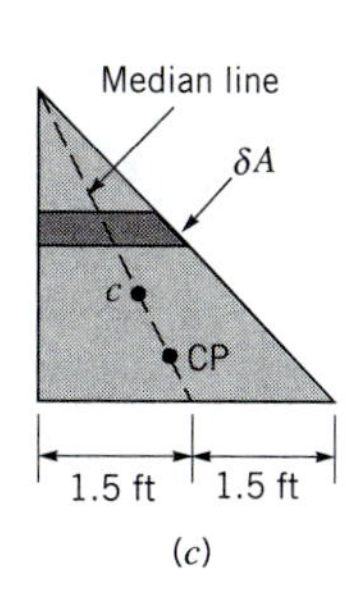

(c)

■ FIGURE E2.7

Note that this force is independent of the tank length. The result is the same if the tank is 0.25 ft, 25 ft, or 25 miles long. The y coordinate of the center of pressure (CP) is found from Eq. 2.19,

$$y_R = \frac{I_{xc}}{y_c A} + y_c$$

and from Fig. 2.18

$$I_{xc} = \frac{(3 \text{ ft})(3 \text{ ft})^3}{36} = \frac{81}{36} \text{ ft}^4$$

so that

$$y_R = \frac{81/36 \text{ ft}^4}{(9 \text{ ft})(9/2 \text{ ft}^2)} + 9 \text{ ft}$$

$$= 0.0556 \text{ ft} + 9 \text{ ft} = 9.06 \text{ ft} \qquad \textbf{(Ans)}$$

Similarly, from Eq. 2.20

$$x_R = \frac{I_{xyc}}{y_c A} + x_c$$

and from Fig. 2.18

$$I_{xyc} = \frac{(3 \text{ ft})(3 \text{ ft})^2}{72}(3 \text{ ft}) = \frac{81}{72} \text{ ft}^4$$

so that

$$x_R = \frac{81/72 \text{ ft}^4}{(9 \text{ ft})(9/2 \text{ ft}^2)} + 0 = 0.0278 \text{ ft} \qquad \textbf{(Ans)}$$

Thus, we conclude that the center of pressure is 0.0278 ft to the right of and 0.0556 ft below the centroid of the area. If this point is plotted, we find that it lies on the median line for the area as illustrated in Fig. E2.7*c*. Since we can think of the total area as consisting of a number of small rectangular strips of area δA (and the fluid force on each of these small areas acts through its center), it follows that the resultant of all these parallel forces must lie along the median.

2.9 Pressure Prism

An informative and useful graphical interpretation can be made for the force developed by a fluid acting on a plane area. Consider the pressure distribution along a vertical wall of a tank of width b, which contains a liquid having a specific weight γ. Since the pressure must vary linearly with depth, we can represent the variation as is shown in Fig. 2.19*a*, where the pressure is equal to zero at the upper surface and equal to γh at the bottom. It is apparent from this diagram that the average pressure occurs at the depth $h/2$, and therefore the resultant force acting on the rectangular area $A = bh$ is

$$F_R = p_{\text{av}} A = \gamma \left(\frac{h}{2}\right) A$$

which is the same result as obtained from Eq. 2.18. The pressure distribution shown in Fig. 2.19*a* applies across the vertical surface so we can draw the three-dimensional representation of the pressure distribution as shown in Fig. 2.19*b*. The base of this "volume" in pressure-area space is the plane surface of interest, and its altitude at each point is the pressure. This volume is called the *pressure prism*, and it is clear that the magnitude of the resultant force acting on the surface is equal to the volume of the pressure prism. Thus, for the prism of Fig. 2.19*b* the fluid force is

The pressure prism is a geometric representation of the hydrostatic force on a plane surface.

$$F_R = \text{volume} = \frac{1}{2}(\gamma h)(bh) = \gamma \left(\frac{h}{2}\right) A$$

where bh is the area of the rectangular surface, A.

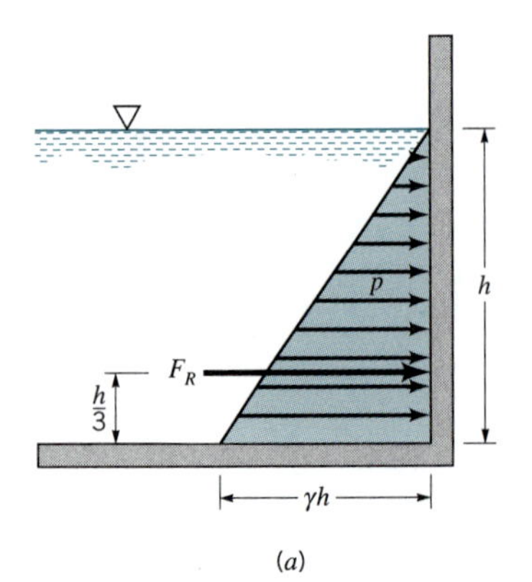

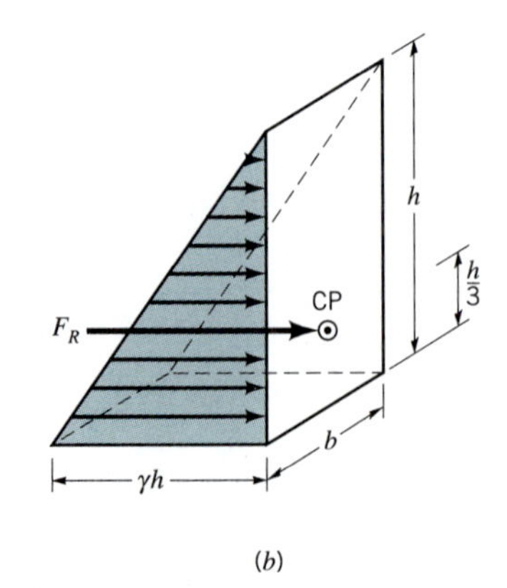

■ **FIGURE 2.19** **Pressure prism for vertical rectangular area.**

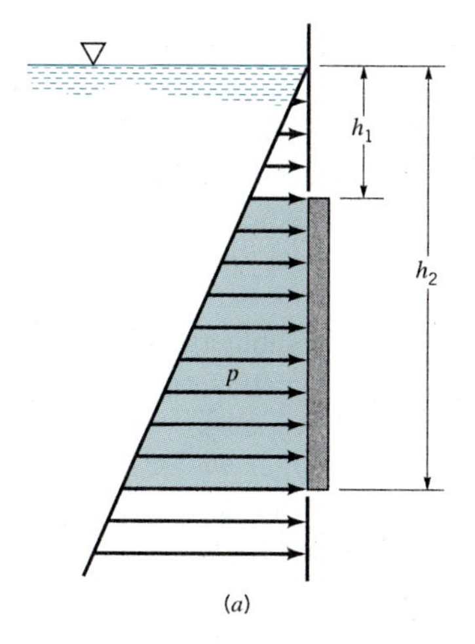

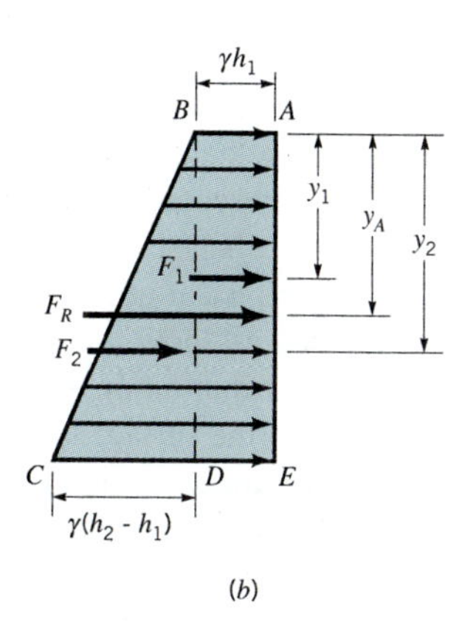

■ **FIGURE 2.20 Graphical representation of hydrostatic forces on a vertical rectangular surface.**

The magnitude of the resultant fluid force is equal to the volume of the pressure prism and passes through its centroid.

The resultant force must pass through the *centroid* of the pressure prism. For the volume under consideration the centroid is located along the vertical axis of symmetry of the surface, and at a distance of $h/3$ above the base (since the centroid of a triangle is located at $h/3$ above its base). This result can readily be shown to be consistent with that obtained from Eqs. 2.19 and 2.20.

This same graphical approach can be used for plane surfaces that do not extend up to the fluid surface as illustrated in Fig. 2.20*a*. In this instance, the cross section of the pressure prism is trapezoidal. However, the resultant force is still equal in magnitude to the volume of the pressure prism, and it passes through the centroid of the volume. Specific values can be obtained by decomposing the pressure prism into two parts, *ABDE* and *BCD*, as shown in Fig. 2.20*b*. Thus,

$$F_R = F_1 + F_2$$

where the components can readily be determined by inspection for rectangular surfaces. The location of F_R can be determined by summing moments about some convenient axis, such as one passing through A. In this instance

$$F_R y_A = F_1 y_1 + F_2 y_2$$

and y_1 and y_2 can be determined by inspection.

For inclined plane surfaces the pressure prism can still be developed, and the cross section of the prism will generally be trapezoidal as is shown in Fig. 2.21. Although it is usually convenient to measure distances along the inclined surface, the pressures developed depend on the vertical distances as illustrated.

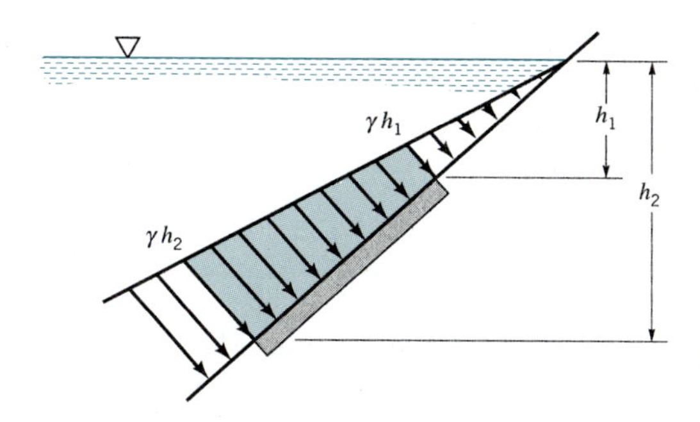

■ **FIGURE 2.21 Pressure variation along an inclined plane area.**

The use of pressure prisms for determining the force on submerged plane areas is convenient if the area is rectangular so the volume and centroid can be easily determined. However, for other nonrectangular shapes, integration would generally be needed to determine the volume and centroid. In these circumstances it is more convenient to use the equations developed in the previous section, in which the necessary integrations have been made and the results presented in a convenient and compact form that is applicable to submerged plane areas of any shape.

The effect of atmospheric pressure on a submerged area has not yet been considered, and we may ask how this pressure will influence the resultant force. If we again consider the pressure distribution on a plane vertical wall, as is shown in Fig. 2.22*a*, the pressure varies from zero at the surface to γh at the bottom. Since we are setting the surface pressure equal to zero, we are using atmospheric pressure as our datum, and thus the pressure used in the determination of the fluid force is gage pressure. If we wish to include atmospheric pressure, the pressure distribution will be as is shown in Fig. 2.22*b*. We note that in this case the force on one side of the wall now consists of F_R as a result of the hydrostatic pressure distribution, plus the contribution of the atmospheric pressure, $p_{atm}A$, where A is the area of the surface. However, if we are going to include the effect of atmospheric pressure on one side of the wall we must realize that this same pressure acts on the outside surface (assuming it is exposed to the atmosphere), so that an equal and opposite force will be developed as illustrated in the figure. Thus, we conclude that the *resultant* fluid force on the surface is that due only to the gage pressure contribution of the liquid in contact with the surface—the atmospheric pressure does not contribute to this resultant. Of course, if the surface pressure of the liquid is different from atmospheric pressure (such as might occur in a closed tank), the resultant force acting on a submerged area, A, will be changed in magnitude from that caused simply by hydrostatic pressure by an amount p_sA, where p_s is the gage pressure at the liquid surface (the outside surface is assumed to be exposed to atmospheric pressure).

The resultant fluid force acting on a submerged area is affected by the pressure at the free surface.

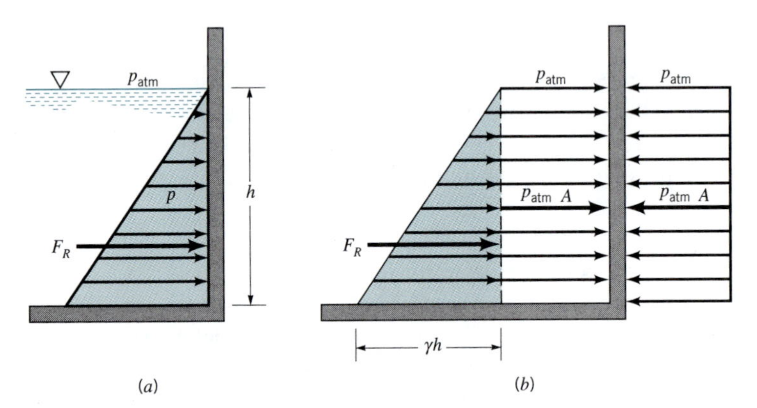

■ **FIGURE 2.22 Effect of atmospheric pressure on the resultant force acting on a plane vertical wall.**

EXAMPLE 2.8

A pressurized tank contains oil ($SG = 0.90$) and has a square, 0.6-m by 0.6-m plate bolted to its side, as is illustrated in Fig. E2.8*a*. When the pressure gage on the top of the tank reads 50 kPa, what is the magnitude and location of the resultant force on the attached plate? The outside of the tank is at atmospheric pressure.

SOLUTION

The pressure distribution acting on the inside surface of the plate is shown in Fig. E2.8*b*. The pressure at a given point on the plate is due to the air pressure, p_s, at the oil surface, and the

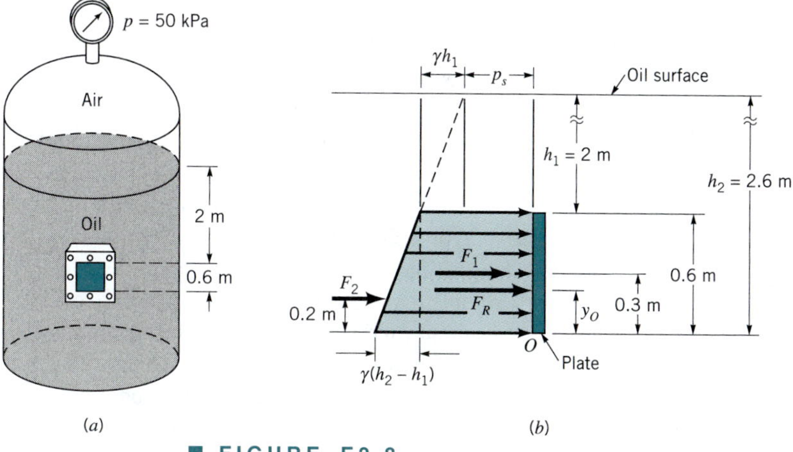

■ FIGURE E2.8

pressure due to the oil, which varies linearly with depth as is shown in the figure. The resultant force on the plate (having an area A) is due to the components, F_1 and F_2, with

$$\begin{aligned} F_1 &= (p_s + \gamma h_1)A \\ &= [50 \times 10^3 \text{ N/m}^2 + (0.90)(9.81 \times 10^3 \text{ N/m}^3)(2 \text{ m})](0.36 \text{ m}^2) \\ &= 24.4 \times 10^3 \text{ N} \end{aligned}$$

and

$$\begin{aligned} F_2 &= \gamma \left(\frac{h_2 - h_1}{2} \right) A \\ &= (0.90)(9.81 \times 10^3 \text{ N/m}^3) \left(\frac{0.6 \text{ m}}{2} \right) (0.36 \text{ m}^2) \\ &= 0.954 \times 10^3 \text{ N} \end{aligned}$$

The magnitude of the resultant force, F_R, is therefore

$$F_R = F_1 + F_2 = 25.4 \times 10^3 \text{ N} = 25.4 \text{ kN} \qquad \textbf{(Ans)}$$

The vertical location of F_R can be obtained by summing moments around an axis through point O so that

$$F_R y_O = F_1(0.3 \text{ m}) + F_2(0.2 \text{ m})$$

or

$$(25.4 \times 10^3 \text{ N})y_O = (24.4 \times 10^3 \text{ N})(0.3 \text{ m}) + (0.954 \times 10^3 \text{ N})(0.2 \text{ m})$$

$$y_O = 0.296 \text{ m} \qquad \textbf{(Ans)}$$

Thus, the force acts at a distance of 0.296 m above the bottom of the plate along the vertical axis of symmetry.

Note that the air pressure used in the calculation of the force was gage pressure. Atmospheric pressure does not affect the resultant force (magnitude or location), since it acts on both sides of the plate, thereby canceling its effect.

2.10 Hydrostatic Force on a Curved Surface

The equations developed in Section 2.8 for the magnitude and location of the resultant force acting on a submerged surface only apply to plane surfaces. However, many surfaces of interest (such as those associated with dams, pipes, and tanks) are nonplanar. Although the resultant fluid force can be determined by integration, as was done for the plane surfaces, this is generally a rather tedious process and no simple, general formulas can be developed. As an alternative approach we will consider the equilibrium of the fluid volume enclosed by the curved surface of interest and the horizontal and vertical projections of this surface.

V2.4

The development of a free-body diagram of a suitable volume of fluid can be used to determine the resultant fluid force acting on a curved surface.

For example, consider the curved section BC of the open tank of Fig. 2.23*a*. We wish to find the resultant fluid force acting on this section, which has a unit length perpendicular to the plane of the paper. We first isolate a volume of fluid that is bounded by the surface of interest, in this instance section BC, the horizontal plane surface AB, and the vertical plane surface AC. The free-body diagram for this volume is shown in Fig. 2.23*b*. The magnitude and location of forces F_1 and F_2 can be determined from the relationships for planar surfaces. The weight, $\mathcal{W}$, is simply the specific weight of the fluid times the enclosed volume and acts through the center of gravity (CG) of the mass of fluid contained within the volume. The forces F_H and F_V represent the components of the force that the tank *exerts on the fluid.*

In order for this force system to be in equilibrium, the horizontal component F_H must be equal in magnitude and collinear with F_2, and the vertical component F_V equal in magnitude and collinear with the resultant of the vertical forces F_1 and $\mathcal{W}$. This follows since the three forces acting on the fluid mass (F_2, the resultant of F_1 and $\mathcal{W}$, and the resultant force that the tank exerts on the mass) must form a *concurrent* force system. That is, from the principles of statics, it is known that when a body is held in equilibrium by three nonparallel forces they must be concurrent (their lines of action intersect at a common point), and coplanar. Thus,

$$F_H = F_2$$

$$F_V = F_1 + \mathcal{W}$$

and the magnitude of the resultant is obtained from the equation

$$F_R = \sqrt{(F_H)^2 + (F_V)^2}$$

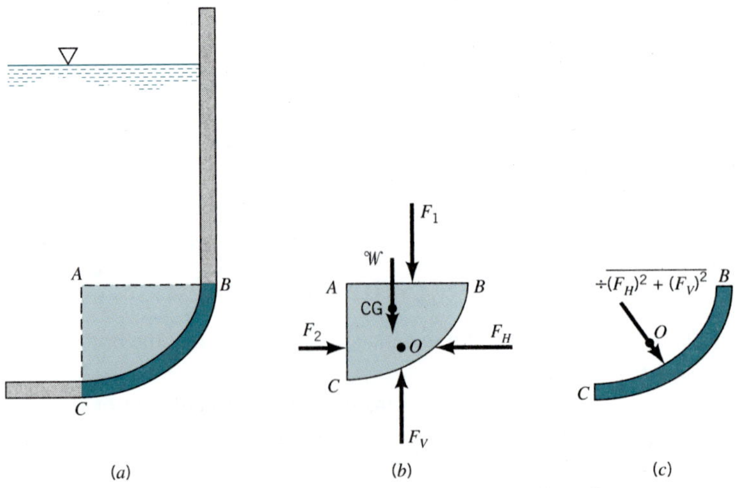

■ FIGURE 2.23 **Hydrostatic force on a curved surface.**

The resultant F_R passes through the point O, which can be located by summing moments about an appropriate axis. The resultant force of the fluid acting *on the curved surface BC* is equal and opposite in direction to that obtained from the free-body diagram of Fig. 2.23*b*. The desired fluid force is shown in Fig. 2.23*c*.

EXAMPLE 2.9

The 6-ft-diameter drainage conduit of Fig. E2.9*a* is half full of water at rest. Determine the magnitude and line of action of the resultant force that the water exerts on a 1-ft length of the curved section BC of the conduit wall.

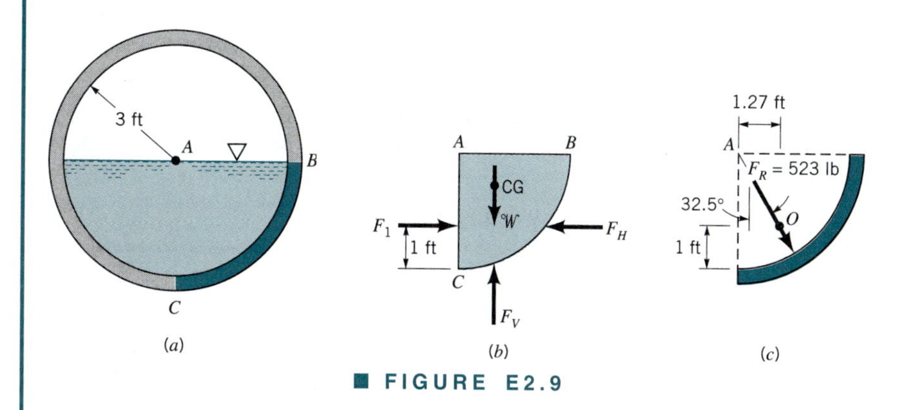

■ FIGURE E2.9

SOLUTION

We first isolate a volume of fluid bounded by the curved section BC, the horizontal surface AB, and the vertical surface AC, as shown in Fig. E2.9*b*. The volume has a length of 1 ft. The forces acting on the volume are the horizontal force, F_1, which acts on the vertical surface AC, the weight, $\mathcal{W}$, of the fluid contained within the volume, and the horizontal and vertical components of the force of the conduit wall on the fluid, F_H and F_V, respectively.

The magnitude of F_1 is found from the equation

$$F_1 = \gamma h_c A = (62.4 \text{ lb/ft}^3)(\tfrac{3}{2} \text{ ft})(3 \text{ ft}^2) = 281 \text{ lb}$$

and this force acts 1 ft above C as shown. The weight, $\mathcal{W}$, is

$$\mathcal{W} = \gamma \text{ vol} = (62.4 \text{ lb/ft}^3)(9\pi/4 \text{ ft}^2)(1 \text{ ft}) = 441 \text{ lb}$$

and acts through the center of gravity of the mass of fluid, which according to Fig. 2.18 is located 1.27 ft to the right of AC as shown. Therefore, to satisfy equilibrium

$$F_H = F_1 = 281 \text{ lb} \qquad F_V = \mathcal{W} = 441 \text{ lb}$$

and the magnitude of the resultant force is

$$F_R = \sqrt{(F_H)^2 + (F_V)^2}$$

$$= \sqrt{(281 \text{ lb})^2 + (441 \text{ lb})^2} = 523 \text{ lb} \qquad \textbf{(Ans)}$$

The force the water exerts *on* the conduit wall is equal, but *opposite in direction*, to the forces F_H and F_V shown in Fig. E2.9*b*. Thus, the resultant force *on the conduit wall* is shown in Fig. E2.9*c*. This force acts through the point O at the angle shown.

An inspection of this result will show that the line of action of the resultant force passes through the center of the conduit. In retrospect, this is not a surprising result since at each point on the curved surface of the conduit the elemental force due to the pressure is normal to the surface, and each line of action must pass through the center of the conduit. It therefore follows that the resultant of this concurrent force system must also pass through the center of concurrence of the elemental forces that make up the system.

This same general approach can also be used for determining the force on curved surfaces of pressurized, closed tanks. If these tanks contain a gas, the weight of the gas is usually negligible in comparison with the forces developed by the pressure. Thus, the forces (such as F_1 and F_2 in Fig. 2.23*b*) on horizontal and vertical projections of the curved surface of interest can simply be expressed as the internal pressure times the appropriate projected area.

2.11 Buoyancy, Flotation, and Stability

2.11.1 Archimedes' Principle

The resultant fluid force acting on a body that is completely submerged or floating in a fluid is called the buoyant force.

When a body is completely submerged in a fluid, or floating so that it is only partially submerged, the resultant fluid force acting on the body is called the *buoyant force.* A net upward vertical force results because pressure increases with depth and the pressure forces acting from below are larger than the pressure forces acting from above. This force can be determined through an approach similar to that used in the previous article for forces on curved surfaces. Consider a body of arbitrary shape, having a volume $\forall$, that is immersed in a fluid as illustrated in Fig. 2.24*a*. We enclose the body in a parallelepiped and draw a free-body diagram of the parallelepiped with the body removed as shown in Fig. 2.24*b*. Note that the forces F_1, F_2, F_3, and F_4 are simply the forces exerted on the plane surfaces of the parallelepiped (for simplicity the forces in the x direction are not shown), $\mathcal{W}$ is the weight of the shaded fluid volume (parallelepiped minus body), and F_B is the force the body is exerting *on the fluid.* The forces on the vertical surfaces, such as F_3 and F_4, are all equal and cancel, so the equilibrium equation of interest is in the z direction and can be expressed as

$$F_B = F_2 - F_1 - \mathcal{W} \tag{2.21}$$

If the specific weight of the fluid is constant, then

$$F_2 - F_1 = \gamma(h_2 - h_1)A$$

where A is the horizontal area of the upper (or lower) surface of the parallelepiped, and Eq. 2.21 can be written as

$$F_B = \gamma(h_2 - h_1)A - \gamma[(h_2 - h_1)A - \forall]$$

Simplifying, we arrive at the desired expression for the buoyant force

$$\boxed{F_B = \gamma\forall} \tag{2.22}$$

V2.5

where γ is the specific weight of the fluid and $\forall$ is the volume of the body. The direction of the buoyant force, which is the force of the fluid *on the body*, is opposite to that shown on the free-body diagram. Therefore, the buoyant force has a magnitude equal to the weight

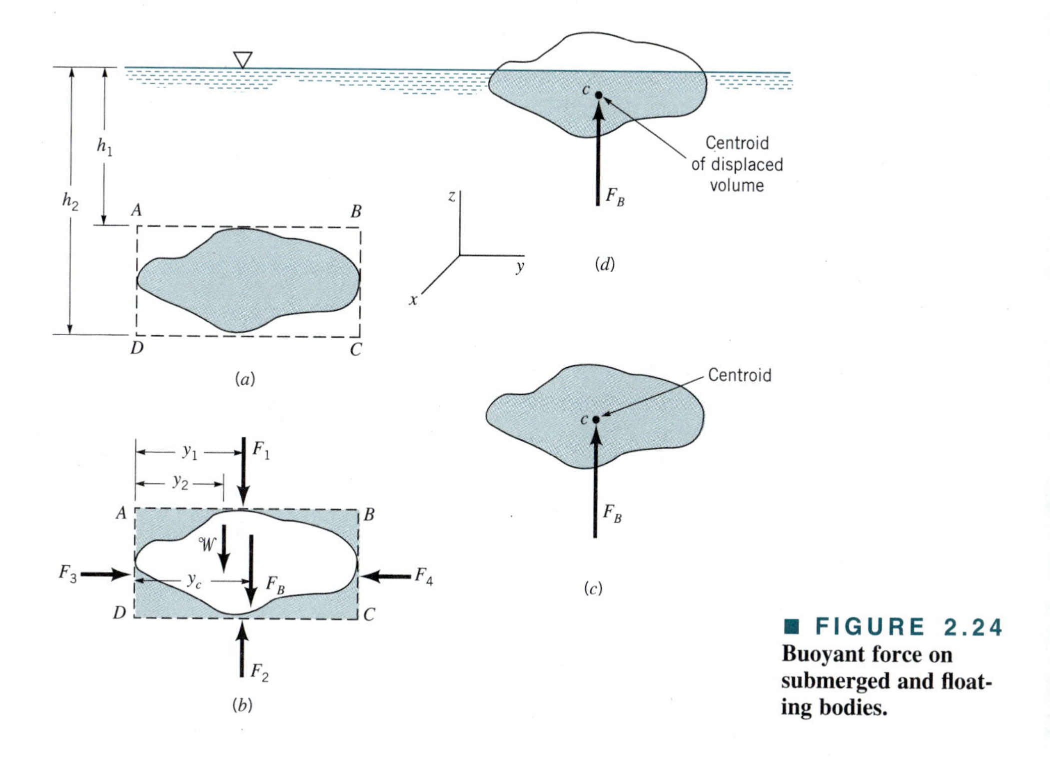

■ **FIGURE 2.24** **Buoyant force on submerged and floating bodies.**

Archimedes' principle states that the buoyant force has a magnitude equal to the weight of the fluid displaced by the body and is directed vertically upward.

of the fluid displaced by the body and is directed vertically upward. This result is commonly referred to as *Archimedes' principle* in honor of Archimedes (287–212 B.C.), a Greek mechanician and mathematician who first enunciated the basic ideas associated with hydrostatics.

The location of the line of action of the buoyant force can be determined by summing moments of the forces shown on the free-body diagram in Fig. 2.24*b* with respect to some convenient axis. For example, summing moments about an axis perpendicular to the paper through point D we have

$$F_B y_c = F_2 y_1 - F_1 y_1 - \mathcal{W} y_2$$

and on substitution for the various forces

$$\forall y_c = \forall_T y_1 - (\forall_T - \forall) y_2 \tag{2.23}$$

where $\forall_T$ is the total volume $(h_2 - h_1)A$. The right-hand side of Eq. 2.23 is the first moment of the displaced volume $\forall$ with respect to the x-z plane so that y_c is equal to the y coordinate of the centroid of the volume $\forall$. In a similar fashion it can be shown that the x coordinate of the buoyant force coincides with the x coordinate of the centroid. Thus, we conclude that the *buoyant force passes through the centroid of the displaced volume* as shown in Fig. 2.24*c*. The point through which the buoyant force acts is called the *center of buoyancy*.

V2.6

These same results apply to floating bodies which are only partially submerged, as illustrated in Fig. 2.24*d*, if the specific weight of the fluid above the liquid surface is very small compared with the liquid in which the body floats. Since the fluid above the surface is usually air, for practical purposes this condition is satisfied.

In the derivations presented above, the fluid is assumed to have a constant specific weight, γ. If a body is immersed in a fluid in which γ varies with depth, such as in a layered fluid, the magnitude of the buoyant force remains equal to the weight of the displaced fluid. However, the buoyant force does not pass through the centroid of the displaced volume, but rather, it passes through the center of gravity of the displaced volume.

EXAMPLE 2.10

A spherical buoy has a diameter of 1.5 m, weighs 8.50 kN, and is anchored to the sea floor with a cable as is shown in Fig. E2.10*a*. Although the buoy normally floats on the surface, at certain times the water depth increases so that the buoy is completely immersed as illustrated. For this condition what is the tension of the cable?

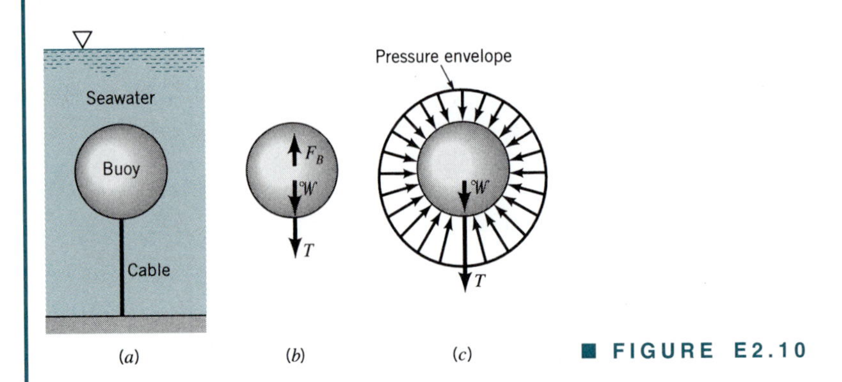

■ FIGURE E2.10

SOLUTION

We first draw a free-body diagram of the buoy as is shown in Fig. E2.10*b*, where F_B is the buoyant force acting on the buoy, $\mathcal{W}$ is the weight of the buoy, and T is the tension in the cable. For equilibrium it follows that

$$T = F_B - \mathcal{W}$$

From Eq. 2.22

$$F_B = \gamma \mathcal{V}$$

and for seawater with $\gamma = 10.1 \text{ kN/m}^3$ and $\mathcal{V} = \pi d^3/6$ then

$$F_B = (10.1 \times 10^3 \text{ N/m}^3)[(\pi/6)(1.5 \text{ m})^3] = 1.785 \times 10^4 \text{ N}$$

The tension in the cable can now be calculated as

$$T = 1.785 \times 10^4 \text{ N} - 0.850 \times 10^4 \text{ N} = 9.35 \text{ kN} \qquad \textbf{(Ans)}$$

Note that we replaced the effect of the hydrostatic pressure force on the body by the buoyant force, F_B. Another correct free-body diagram of the buoy is shown in Fig. E2.10*c*. The net effect of the pressure forces on the surface of the buoy is equal to the upward force of magnitude, F_B (the buoyant force). Do not include both the buoyant force and the hydrostatic pressure effects in your calculations—use one or the other.

2.11.2 Stability

Submerged or floating bodies can be either in a stable or unstable position.

Another interesting and important problem associated with submerged or floating bodies is concerned with the stability of the bodies. A body is said to be in a *stable equilibrium* position if, when displaced, it returns to its equilibrium position. Conversely, it is in an *unstable equilibrium* position if, when displaced (even slightly), it moves to a new equilibrium position. Stability considerations are particularly important for submerged or floating bodies since the centers of buoyancy and gravity do not necessarily coincide. A small rotation can result in either a restoring or overturning couple. For example, for the *completely* submerged body

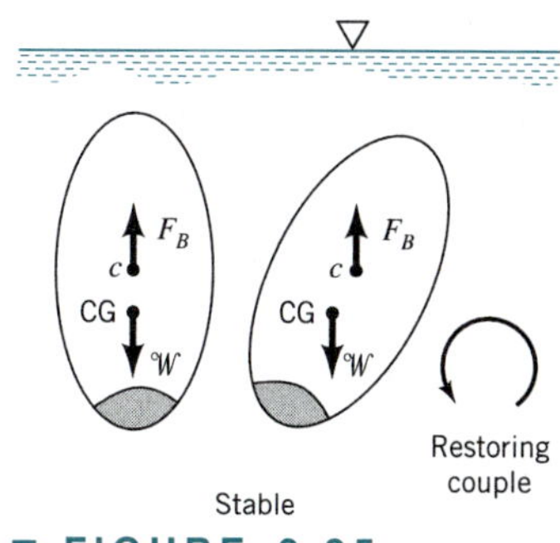

■ FIGURE 2.25
Stability of a completely immersed body—center of gravity below centroid.

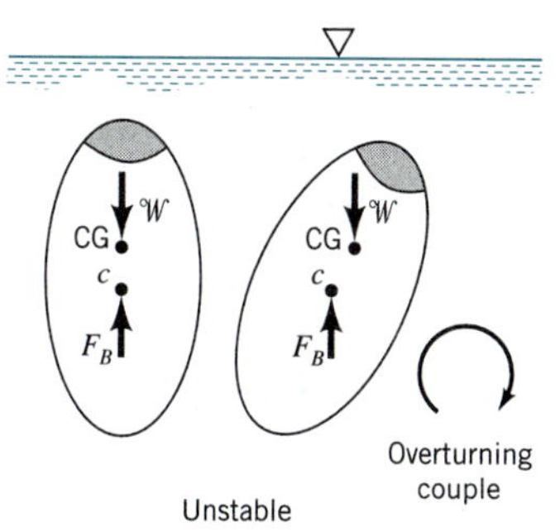

■ FIGURE 2.26
Stability of a completely immersed body—center of gravity above centroid.

The stability of a body can be determined by considering what happens when it is displaced from its equilibrium position.

shown in Fig. 2.25, which has a center of gravity below the center of buoyancy, a rotation from its equilibrium position will create a restoring couple formed by the weight, $\mathcal{W}$, and the buoyant force, F_B, which causes the body to rotate back to its original position. Thus, for this configuration the body is stable. It is to be noted that as long as the center of gravity falls *below* the center of buoyancy, this will always be true; that is, the body is in a *stable equilibrium* position with respect to small rotations. However, as is illustrated in Fig. 2.26, if the center of gravity is above the center of buoyancy, the resulting couple formed by the weight and the buoyant force will cause the body to overturn and move to a new equilibrium position. Thus, a completely submerged body with its center of gravity *above* its center of buoyancy is in an *unstable equilibrium* position.

For *floating* bodies the stability problem is more complicated, since as the body rotates the location of the center of buoyancy (which passes through the centroid of the displaced volume) may change. As is shown in Fig. 2.27, a floating body such as a barge that rides low in the water can be stable even though the center of gravity lies above the center of buoyancy. This is true since as the body rotates the buoyant force, F_B, shifts to pass through the centroid of the newly formed displaced volume and, as illustrated, combines with the weight, $\mathcal{W}$, to form a couple which will cause the body to return to its original equilibrium position. However, for the relatively tall, slender body shown in Fig. 2.28, a small rotational displacement can cause the buoyant force and the weight to form an overturning couple as illustrated.

V2.7

It is clear from these simple examples that the determination of the stability of submerged or floating bodies can be difficult since the analysis depends in a complicated fashion on the particular geometry and weight distribution of the body. The problem can be further complicated by the necessary inclusion of other types of external forces such as those induced by wind gusts or currents. Stability considerations are obviously of great importance in the design of ships, submarines, bathyscaphes, and so forth, and such considerations play a significant role in the work of naval architects (see, for example, Ref. 6).

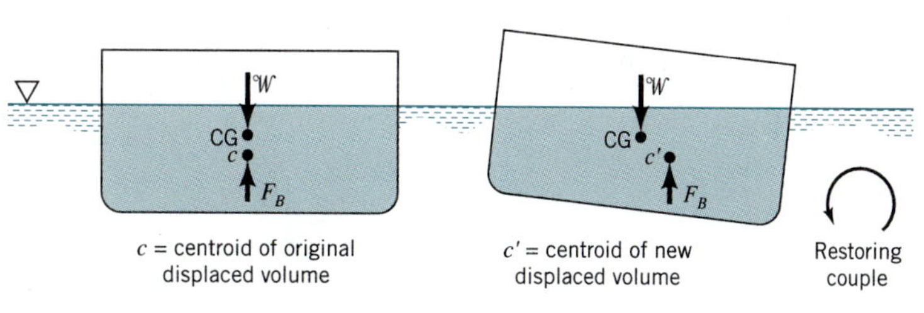

■ FIGURE 2.27
Stability of a floating body—stable configuration.

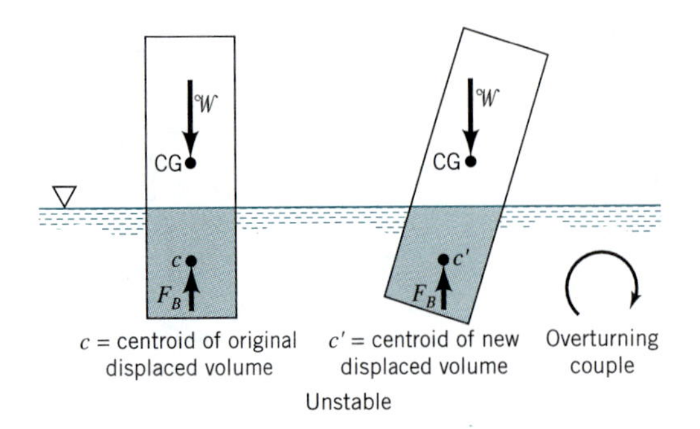

■ FIGURE 2.28 **Stability of a floating body–unstable configuration.**

2.12 Pressure Variation in a Fluid with Rigid-Body Motion

Although in this chapter we have been primarily concerned with fluids at rest, the general equation of motion (Eq. 2.2)

$$-\nabla p - \gamma \hat{\mathbf{k}} = \rho \mathbf{a}$$

was developed for both fluids at rest and fluids in motion, with the only stipulation being that there were no shearing stresses present. Equation 2.2 in component form, based on rectangular coordinates with the positive z axis being vertically upward, can be expressed as

$$-\frac{\partial p}{\partial x} = \rho a_x \qquad -\frac{\partial p}{\partial y} = \rho a_y \qquad -\frac{\partial p}{\partial z} = \gamma + \rho a_z \tag{2.24}$$

Even though a fluid may be in motion, if it moves as a rigid body there will be no shearing stresses present.

A general class of problems involving fluid motion in which there are no shearing stresses occurs when a mass of fluid undergoes rigid-body motion. For example, if a container of fluid accelerates along a straight path, the fluid will move as a rigid mass (after the initial sloshing motion has died out) with each particle having the same acceleration. Since there is no deformation, there will be no shearing stresses and, therefore, Eq. 2.2 applies. Similarly, if a fluid is contained in a tank that rotates about a fixed axis, the fluid will simply rotate with the tank as a rigid body, and again Eq. 2.2 can be applied to obtain the pressure distribution throughout the moving fluid. Specific results for these two cases (rigid-body uniform motion and rigid-body rotation) are developed in the following two sections. Although problems relating to fluids having rigid-body motion are not, strictly speaking, ''fluid statics'' problems, they are included in this chapter because, as we will see, the analysis and resulting pressure relationships are similar to those for fluids at rest.

2.12.1 Linear Motion

We first consider an open container of a liquid that is translating along a straight path with a constant acceleration $\mathbf{a}$ as illustrated in Fig. 2.29. Since $a_x = 0$ it follows from the first of Eqs. 2.24 that the pressure gradient in the x direction is zero ($\partial p/\partial x = 0$). In the y and z directions

$$\frac{\partial p}{\partial y} = -\rho a_y \tag{2.25}$$

$$\frac{\partial p}{\partial z} = -\rho(g + a_z) \tag{2.26}$$

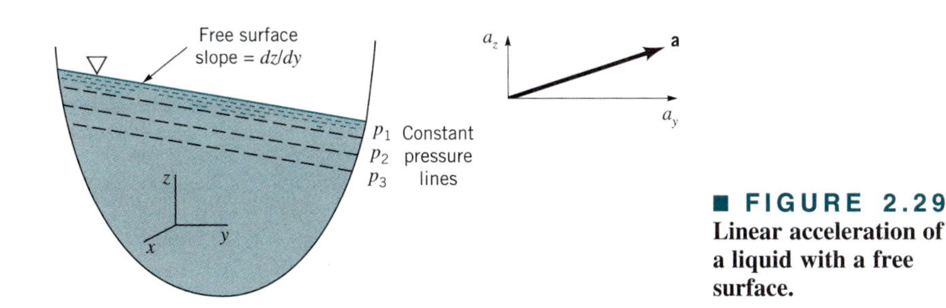

■ **FIGURE 2.29 Linear acceleration of a liquid with a free surface.**

The change in pressure between two closely spaced points located at y, z, and $y + dy$, $z + dz$ can be expressed as

$$dp = \frac{\partial p}{\partial y}\, dy + \frac{\partial p}{\partial z}\, dz$$

or in terms of the results from Eqs. 2.25 and 2.26

$$dp = -\rho a_y\, dy - \rho(g + a_z)\, dz \tag{2.27}$$

Along a line of *constant* pressure, $dp = 0$, and therefore from Eq. 2.27 it follows that the slope of this line is given by the relationship

$$\frac{dz}{dy} = -\frac{a_y}{g + a_z} \tag{2.28}$$

The pressure distribution in a fluid mass that is accelerating along a straight path is not hydrostatic.

Along a free surface the pressure is constant, so that for the accelerating mass shown in Fig. 2.29 the free surface will be inclined if $a_y \neq 0$. In addition, all lines of constant pressure will be parallel to the free surface as illustrated.

For the special circumstance in which $a_y = 0$, $a_z \neq 0$, which corresponds to the mass of fluid accelerating in the vertical direction, Eq. 2.28 indicates that the fluid surface will be horizontal. However, from Eq. 2.26 we see that the pressure distribution is not hydrostatic, but is given by the equation

$$\frac{dp}{dz} = -\rho(g + a_z)$$

For fluids of constant density this equation shows that the pressure will vary linearly with depth, but the variation is due to the combined effects of gravity and the externally induced acceleration, $\rho(g + a_z)$, rather than simply the specific weight ρg. Thus, for example, the pressure along the bottom of a liquid-filled tank which is resting on the floor of an elevator that is accelerating upward will be increased over that which exists when the tank is at rest (or moving with a constant velocity). It is to be noted that for a *freely falling* fluid mass ($a_z = -g$), the pressure gradients in all three coordinate directions are zero, which means that if the pressure surrounding the mass is zero, the pressure throughout will be zero. The pressure throughout a ''blob'' of orange juice floating in an orbiting space shuttle (a form of free fall) is zero. The only force holding the liquid together is surface tension (see Section 1.9).

EXAMPLE 2.11

The cross section for the fuel tank of an experimental vehicle is shown in Fig. E2.11. The rectangular tank is vented to the atmosphere, and a pressure transducer is located in its side as illustrated. During testing of the vehicle, the tank is subjected to a constant linear acceleration, a_y. (a) Determine an expression that relates a_y and the pressure (in lb/ft^2) at the transducer for a fuel with a $SG = 0.65$. (b) What is the maximum acceleration that can occur before the fuel level drops below the transducer?

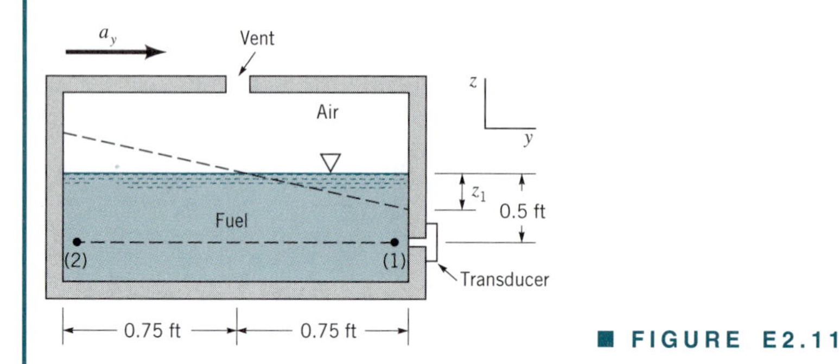

■ FIGURE E2.11

SOLUTION

(a) For a constant horizontal acceleration the fuel will move as a rigid body, and from Eq. 2.28 the slope of the fuel surface can be expressed as

$$\frac{dz}{dy} = -\frac{a_y}{g}$$

since $a_z = 0$. Thus, for some arbitrary a_y, the change in depth, z_1, of liquid on the right side of the tank can be found from the equation

$$-\frac{z_1}{0.75\text{ ft}} = -\frac{a_y}{g}$$

or

$$z_1 = (0.75\text{ ft})\left(\frac{a_y}{g}\right)$$

Since there is no acceleration in the vertical, z, direction, the pressure along the wall varies hydrostatically as shown by Eq. 2.26. Thus, the pressure at the transducer is given by the relationship

$$p = \gamma h$$

where h is the depth of fuel above the transducer, and therefore

$$p = (0.65)(62.4\text{ lb/ft}^3)[0.5\text{ ft} - (0.75\text{ ft})(a_y/g)]$$

$$= 20.3 - 30.4\frac{a_y}{g} \qquad \textbf{(Ans)}$$

for $z_1 \leq 0.5$ ft. As written, p would be given in lb/ft^2.

(b) The limiting value for a_y (when the fuel level reaches the transducer) can be found from the equation

$$0.5 \text{ ft} = (0.75 \text{ ft}) \left[\frac{(a_y)_{\text{max}}}{g} \right]$$

or

$$(a_y)_{\text{max}} = \frac{2g}{3}$$

and for standard acceleration of gravity

$$(a_y)_{\text{max}} = \tfrac{2}{3} (32.2 \text{ ft/s}^2) = 21.5 \text{ ft/s}^2 \qquad \textbf{(Ans)}$$

Note that the pressure in horizontal layers is not constant in this example since $\partial p / \partial y = -\rho a_y \neq 0$. Thus, for example, $p_1 \neq p_2$.

2.12.2 Rigid-Body Rotation

A fluid contained in a tank that is rotating with a constant angular velocity about an axis will rotate as a rigid body.

After an initial ''start-up'' transient, a fluid contained in a tank that rotates with a constant angular velocity ω about an axis as is shown in Fig. 2.30 will rotate with the tank as a rigid body. It is known from elementary particle dynamics that the acceleration of a fluid particle located at a distance r from the axis of rotation is equal in magnitude to $r\omega^2$, and the direction of the acceleration is toward the axis of rotation as is illustrated in the figure. Since the paths of the fluid particles are circular, it is convenient to use cylindrical polar coordinates r, θ, and z, defined in the insert in Fig. 2.30. It will be shown in Chapter 6 that in terms of cylindrical coordinates the pressure gradient ∇p can be expressed as

$$\nabla p = \frac{\partial p}{\partial r} \hat{\mathbf{e}}_r + \frac{1}{r} \frac{\partial p}{\partial \theta} \hat{\mathbf{e}}_\theta + \frac{\partial p}{\partial z} \hat{\mathbf{e}}_z \qquad \textbf{(2.29)}$$

Thus, in terms of this coordinate system

$$\mathbf{a}_r = -r\omega^2 \, \hat{\mathbf{e}}_r \qquad \mathbf{a}_\theta = 0 \qquad \mathbf{a}_z = 0$$

and from Eq. 2.2

$$\frac{\partial p}{\partial r} = \rho r \omega^2 \qquad \frac{\partial p}{\partial \theta} = 0 \qquad \frac{\partial p}{\partial z} = -\gamma \qquad \textbf{(2.30)}$$

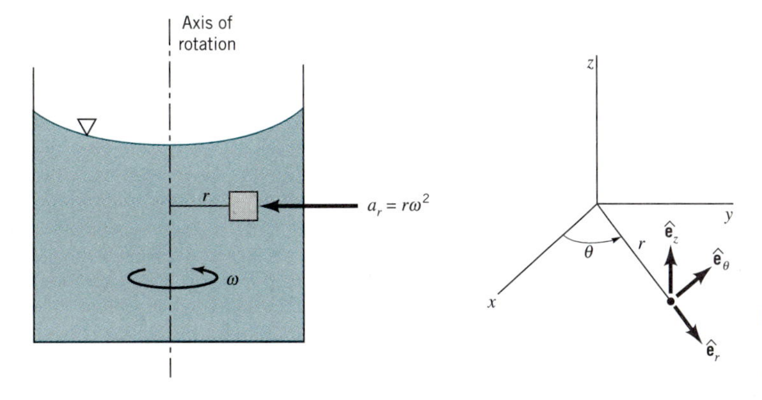

■ **FIGURE 2.30 Rigid-body rotation of a liquid in a tank.**

These results show that for this type of rigid-body rotation, the pressure is a function of two variables r and z, and therefore the differential pressure is

$$dp = \frac{\partial p}{\partial r}\,dr + \frac{\partial p}{\partial z}\,dz$$

or

$$dp = \rho r\omega^2\,dr - \gamma\,dz \tag{2.31}$$

Along a surface of constant pressure, such as the free surface, $dp = 0$, so that from Eq. 2.31 (using $\gamma = \rho g$)

$$\frac{dz}{dr} = \frac{r\omega^2}{g}$$

The free surface in a rotating liquid is curved rather than flat.

and, therefore, the equation for surfaces of constant pressure is

$$z = \frac{\omega^2 r^2}{2g} + \text{constant} \tag{2.32}$$

This equation reveals that these surfaces of constant pressure are parabolic as illustrated in Fig. 2.31.

Integration of Eq. 2.31 yields

$$\int dp = \rho\omega^2 \int r\,dr - \gamma \int dz$$

or

$$p = \frac{\rho\omega^2 r^2}{2} - \gamma z + \text{constant} \tag{2.33}$$

where the constant of integration can be expressed in terms of a specified pressure at some arbitrary point r_0, z_0. This result shows that the pressure varies with the distance from the axis of rotation, but at a fixed radius, the pressure varies hydrostatically in the vertical direction as shown in Fig. 2.31.

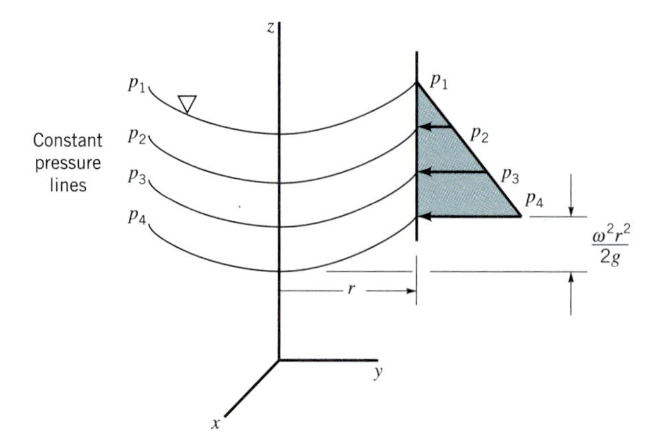

■ **FIGURE 2.31 Pressure distribution in a rotating liquid.**

EXAMPLE 2.12

It has been suggested that the angular velocity, ω, of a rotating body or shaft can be measured by attaching an open cylinder of liquid, as shown in Fig. E2.12*a*, and measuring with some type of depth gage the change in the fluid level, $H - h_0$, caused by the rotation of the fluid. Determine the relationship between this change in fluid level and the angular velocity.

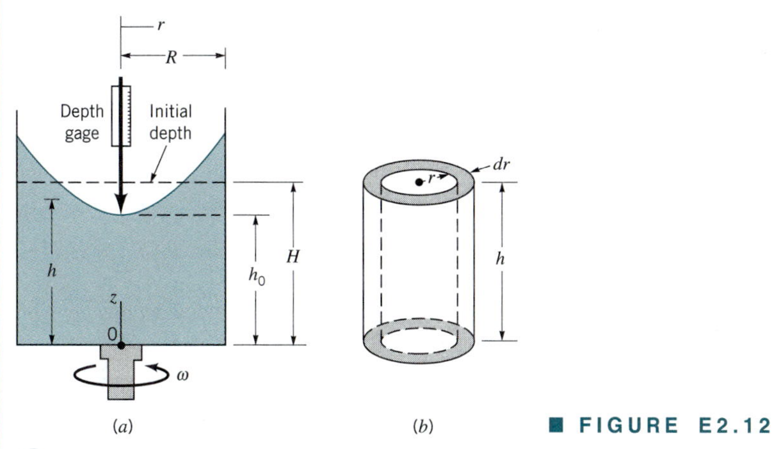

■ FIGURE E2.12

SOLUTION

The height, h, of the free surface above the tank bottom can be determined from Eq. 2.32, and it follows that

$$h = \frac{\omega^2 r^2}{2g} + h_0$$

The initial volume of fluid in the tank, $\forall_i$, is equal to

$$\forall_i = \pi R^2 H$$

The volume of the fluid with the rotating tank can be found with the aid of the differential element shown in Fig. E2.12*b*. This cylindrical shell is taken at some arbitrary radius, r, and its volume is

$$d\forall = 2\pi r h \, dr$$

The total volume is, therefore

$$\forall = 2\pi \int_0^R r\left(\frac{\omega^2 r^2}{2g} + h_0\right) dr = \frac{\pi \omega^2 R^4}{4g} + \pi R^2 h_0$$

Since the volume of the fluid in the tank must remain constant (assuming that none spills over the top), it follows that

$$\pi R^2 H = \frac{\pi \omega^2 R^4}{4g} + \pi R^2 h_0$$

or

$$H - h_0 = \frac{\omega^2 R^2}{4g} \qquad \textbf{(Ans)}$$

This is the relationship we were looking for. It shows that the change in depth could indeed be used to determine the rotational speed, although the relationship between the change in depth and speed is not a linear one.

References

1. *The U.S. Standard Atmosphere, 1962*, U.S. Government Printing Office, Washington, D.C., 1962.
2. *The U.S. Standard Atmosphere, 1976*, U.S. Government Printing Office, Washington, D.C., 1976.
3. Benedict, R. P., *Fundamentals of Temperature, Pressure, and Flow Measurements*, 3rd Ed., Wiley, New York, 1984.
4. Dally, J. W., Riley, W. F., and McConnell, K. G., *Instrumentation for Engineering Measurements*, 2nd Ed., Wiley, New York, 1993.
5. Holman, J. P., *Experimental Methods for Engineers*, 4th Ed., McGraw-Hill, New York, 1983.
6. Comstock, J. P., ed., *Principles of Naval Architecture*, Society of Naval Architects and Marine Engineers, New York, 1967.
7. Hasler, A. F., Pierce, H., Morris, K. R., and Dodge, J., ''Meteorological Data Fields 'In Perspective' '', *Bulletin of the American Meteorological Society*, Vol. 66, No. 7, July 1985.

Review Problems

Note: Problems designated with (R) are review problems. The phrases within parentheses refer to the main topics to be used in solving the problems. Complete, detailed solutions to these review problems can be found in the supplement titled *Student Solution Manual for Fundamentals of Fluid Mechanics* by Munson, Young, and Okiishi (John Wiley and Sons, New York, 1997).

2.1R (Pressure head) Compare the column heights of water, carbon tetrachloride, and mercury corresponding to a pressure of 50 kPa. Express your answer in meters.

(ANS: 5.10 m; 3.21 m; 0.376 m)

2.2R (Pressure-depth relationship) A closed tank is partially filled with glycerin. If the air pressure in the tank is 6 lb/in.^2 and the depth of glycerin is 10 ft, what is the pressure in lb/ft^2 at the bottom of the tank?

(ANS: 1650 lb/ft^2)

2.3R (Gage-absolute pressure) On the inlet side of a pump a Bourdon pressure gage reads 600 lb/ft^2 vacuum. What is the corresponding absolute pressure if the local atmospheric pressure is 14.7 psia?

(ANS: 10.5 psia)

2.4R (Manometer) A tank is constructed of a series of cylinders having diameters of 0.30, 0.25, and 0.15 m as shown in Fig. P2.4R. The tank contains oil, water, and glycerin and a mercury manometer is attached to the bottom as illustrated. Calculate the manometer reading, h.

(ANS: 0.0327 m)

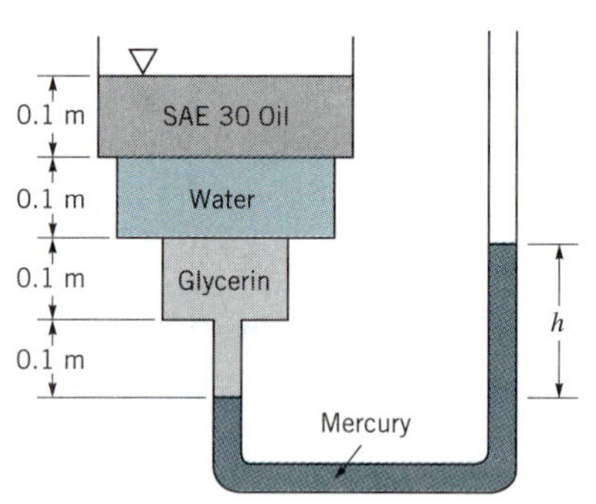

■ FIGURE P2.4R

2.5R (Manometer) A mercury manometer is used to measure the pressure difference in the two pipelines of Fig. P2.5R. Fuel oil (specific weight = 53.0 lb/ft^3) is flowing in A and SAE 30 lube oil (specific weight = 57.0 lb/ft^3) is flowing in B. An air pocket has become entrapped in the lube oil as indicated. Determine the pressure in pipe B if the pressure in A is 15.3 psi.

(ANS: 18.2 psi)

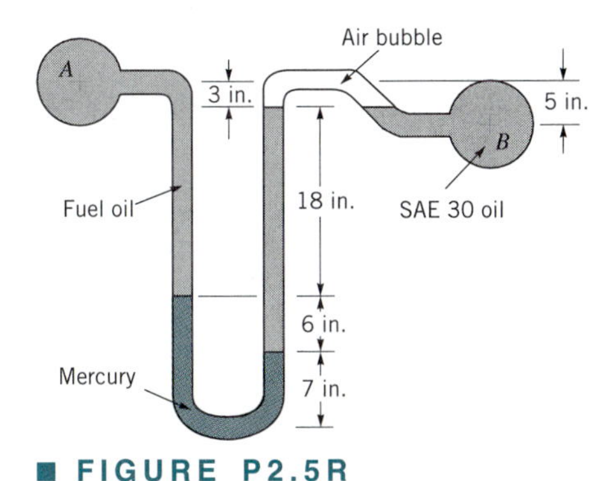

■ FIGURE P2.5R

2.6R (Manometer) Determine the angle θ of the inclined tube shown in Fig. P2.6R if the pressure at A is 1 psi greater than that at B.

(ANS: 19.3 deg)

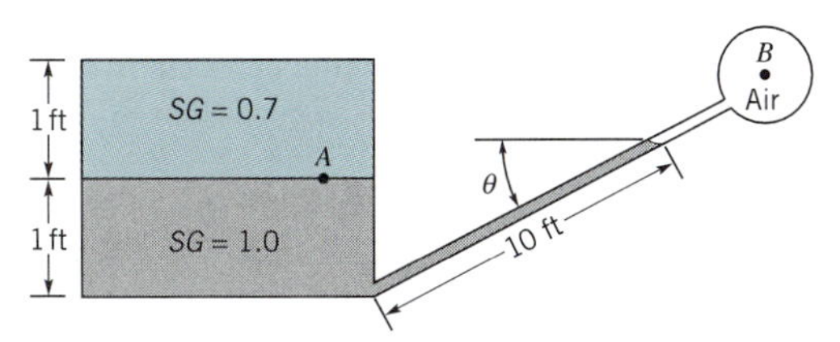

■ FIGURE P2.6R

2.7R (Force on plane surface) A swimming pool is 18 m long and 7 m wide. Determine the magnitude and location of the resultant force of the water on the vertical end of the pool where the depth is 2.5 m.

(ANS: 214 kN on centerline, 1.67 m below surface)

2.8R (Force on plane surface) The vertical cross section of a 7-m-long closed storage tank is shown in Fig. P2.8R. The tank contains ethyl alcohol and the air pressure is 40 kPa. Determine the magnitude of the resultant fluid force acting on one end of the tank.

(ANS: 847 kN)

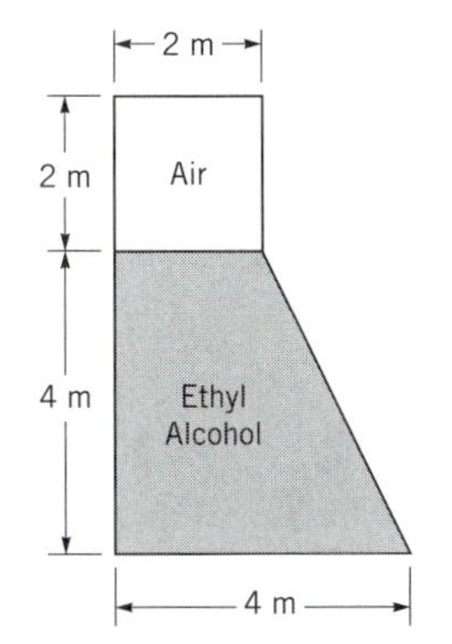

■ FIGURE P2.8R

2.9R (Center of pressure) A 3-ft-diameter circular plate is located in the vertical side of an open tank containing gasoline. The resultant force that the gasoline exerts on the plate acts 3.1 in. below the centroid of the plate. What is the depth of the liquid above the centroid?

(ANS: 2.18 ft)

2.10R (Force on plane surface) A gate having the triangular shape shown in Fig. P2.10R is located in the vertical side of an open tank. The gate is hinged about the horizontal axis AB. The force of the water on the gate creates a moment with respect to the axis AB. Determine the magnitude of this moment.

(ANS: 3890 kN·m)

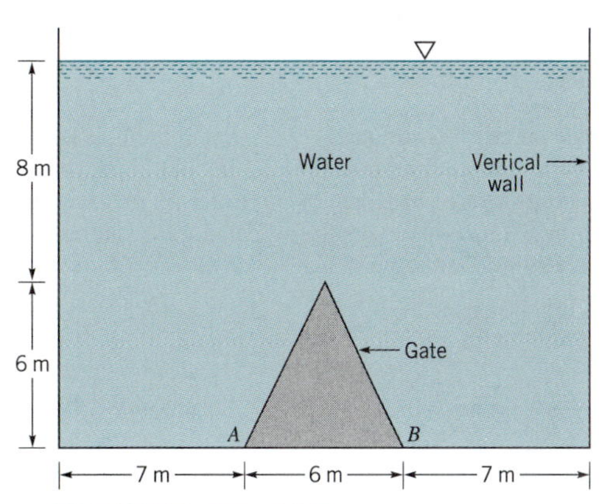

■ FIGURE P2.10R

2.11R (Force on plane surface) The rectangular gate CD of Fig P2.11R is 1.8 m wide and 2.0 m long. Assuming the material of the gate to be homogeneous and neglecting friction at the hinge C, determine the weight of the gate necessary to keep it shut until the water level rises to 2.0 m above the hinge.

(ANS: 180 kN)

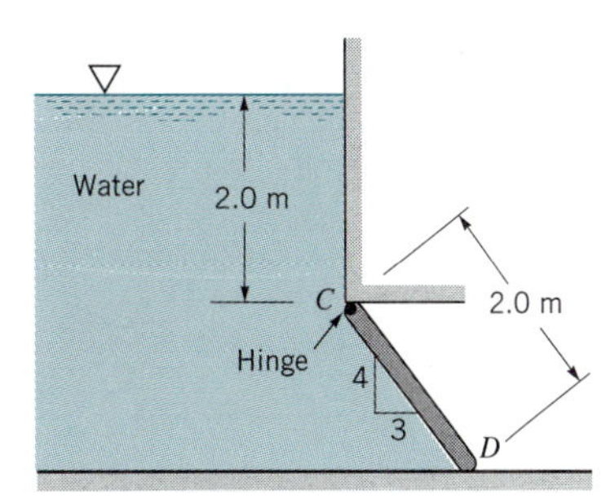

■ FIGURE P2.11R

2.12R (Force on curved surface) A gate in the form of a partial cylindrical surface (called a *Tainter gate*) holds back water on top of a dam as shown in Fig. P2.12R. The radius of the surface is 22 ft, and its length is 36 ft. The gate can pivot about point A, and the pivot point is 10 ft above the seat, C.

Determine the magnitude of the resultant water force on the gate. Will the resultant pass through the pivot? Explain.

(ANS: 118,000 lb)

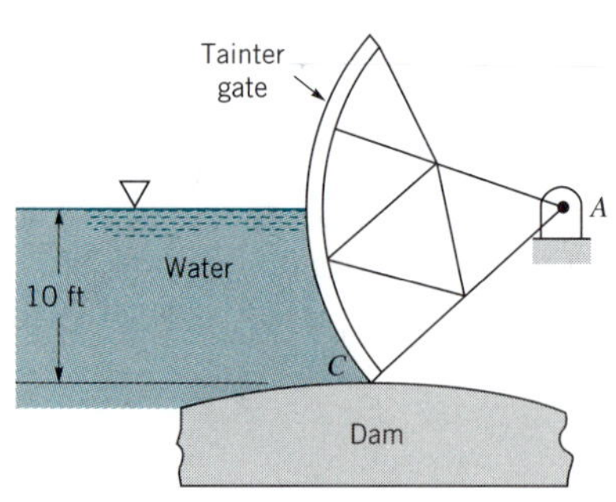

■ FIGURE P2.12R

2.13R (Force on curved surface) A conical plug is located in the side of a tank as shown in Fig. 2.13R. **(a)** Show that the horizontal component of the force of the water on the plug does not depend on h. **(b)** For the depth indicated, what is the magnitude of this component?

(ANS: 735 lb)

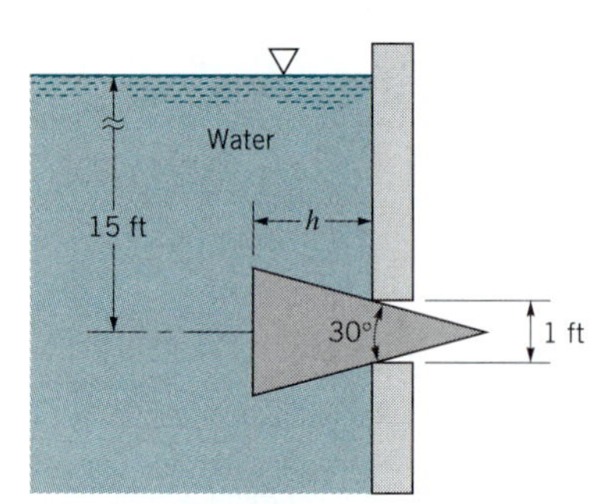

■ FIGURE P2.13R

2.14R (Force on curved surface) The 9-ft-long cylinder of Fig. P2.14R floats in oil and rests against a wall. Determine the horizontal force the cylinder exerts on the wall at the point of contact, A.

(ANS: 2300 lb)

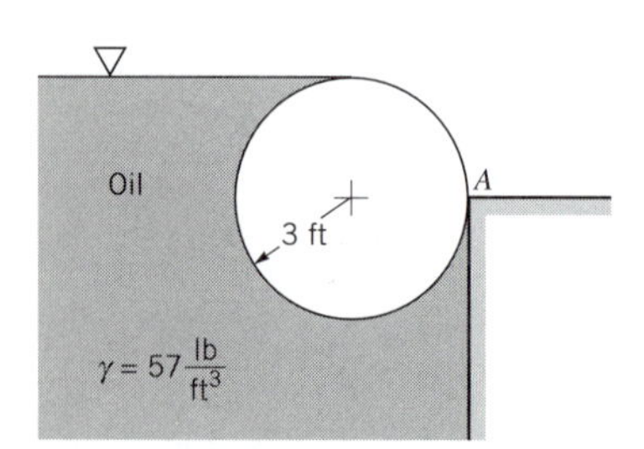

■ FIGURE P2.14R

2.15R (Buoyancy) A hot-air balloon weighs 500 lb, including the weight of the balloon, the basket, and one person. The air outside the balloon has a temperature of 80 °F, and the heated air inside the balloon has a temperature of 150 °F. Assume the inside and outside air to be at standard atmospheric pressure of 14.7 psia. Determine the required volume of the balloon to support the weight. If the balloon had a spherical shape, what would be the required diameter?

(ANS: 59,200 ft^3; 48.3 ft)

2.16R (Buoyancy) An irregularly shaped piece of a solid material weighs 8.05 lb in air and 5.26 lb when completely submerged in water. Determine the density of the material.

(ANS: 5.60 $slugs/ft^3$)

2.17R (Buoyancy, force on plane surface) A cube, 4 ft on a side, weighs 3000 lb and floats half-submerged in an open tank as shown in Fig. P2.17R. For a liquid depth of 10 ft, determine the force of the liquid on the inclined section AB of the tank wall. The width of the wall is 8 ft. Show the magnitude, direction, and location of the force on a sketch.

(ANS: 75,000 lb on centerline, 13.33 ft along wall from free surface)

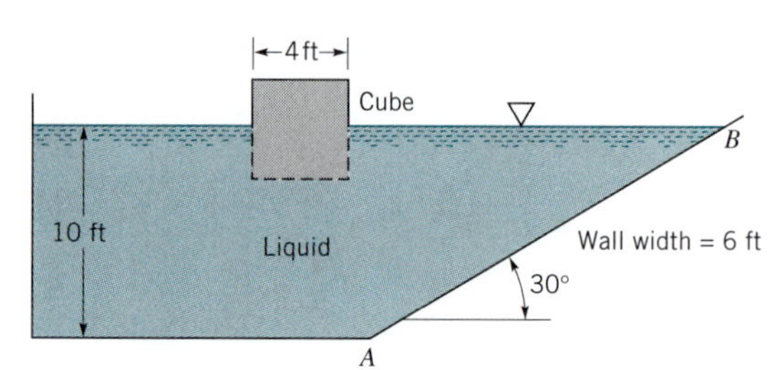

■ FIGURE P2.17R

2.18R (Rigid-body motion) A container that is partially filled with water is pulled with a constant acceleration along a plane horizontal surface. With this acceleration the water surface slopes downward at an angle of 40° with respect to the horizontal. Determine the acceleration. Express your answer in m/s^2.

(ANS: 8.23 m/s^2)

2.19R (Rigid-body motion) An open, 2-ft-diameter tank contains water to a depth of 3 ft when at rest. If the tank is rotated about its vertical axis with an angular velocity of 160 rev/min, what is the minimum height of the tank walls to prevent water from spilling over the sides?

(ANS: 5.18 ft)

Problems

Note: Unless otherwise indicated use the values of fluid properties found in the tables on the inside of the front cover. Problems designated with an (*) are intended to be solved with the aid of a programmable calculator or a computer. Problems designated with a (†) are "open-ended" problems and require critical thinking in that to work them one must make various assumptions and provide the necessary data. There is not a unique answer to these problems.

2.1 The water level in an open standpipe is 80 ft above the ground. What is the static pressure at a fire hydrant that is connected to the standpipe and located at ground level? Express your answer in psi.

2.2 How high a column of SAE 30 oil would be required to give the same pressure as 700 mm Hg?

2.3 What pressure, expressed in pascals, will a skin diver be subjected to at a depth of 40 m in seawater?

2.4 The two open tanks shown in Fig. P2.4 have the same bottom area, A, but different shapes. When the depth, h, of a liquid in the two tanks is the same, the pressure on the bottom of the two tanks will be the same in accordance with Eq. 2.7. However, the weight of the liquid in each of the tanks is different. How do you account for this apparent paradox?

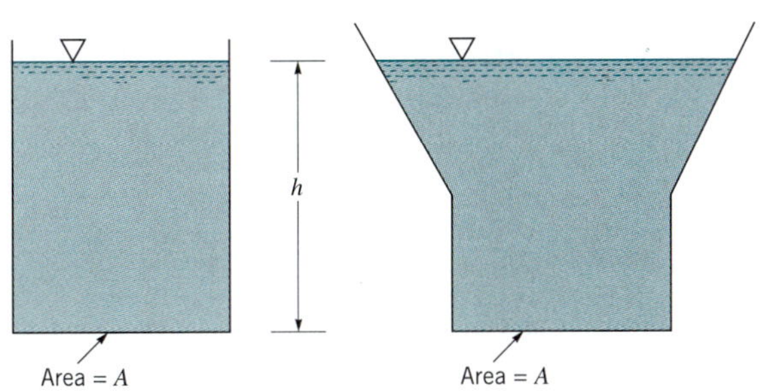

■ **FIGURE P2.4**

2.5 The closed tank of Fig. P2.5 is filled with water and is 5 ft long. The pressure gage on the tank reads 7 psi. Determine: **(a)** the height, h, in the open water column, **(b)** the gage pressure acting on the bottom tank surface AB, and **(c)** the absolute pressure of the air in the top of the tank if the local atmospheric pressure is 14.7 psia.

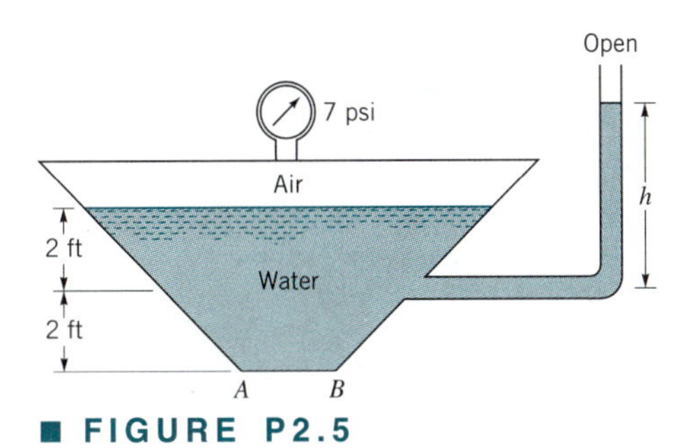

■ **FIGURE P2.5**

2.6 Bathyscaphes are capable of submerging to great depths in the ocean. What is the pressure at a depth of 6 km, assuming that seawater has a constant specific weight of 10.1 kN/m^3? Express your answer in pascals and psi.

2.7 For the great depths that may be encountered in the ocean the compressibility of seawater may become an important consideration. **(a)** Assume that the bulk modulus for seawater is constant and derive a relationship between pressure and depth which takes into account the change in fluid density with depth. **(b)** Make use of part (a) to determine the pressure at a depth of 6 km assuming seawater has a bulk modulus of 2.3×10^9 Pa and a density of 1030 kg/m^3 at the surface. Compare this result with that obtained by assuming a constant density of 1030 kg/m^3.

2.8 Blood pressure is usually given as the ratio of the maximum pressure (systolic pressure) to the minimum pressure (diastolic pressure). For example, a typical value for this ratio for a human would be 120/70, where the pressures are in mm Hg. What would these pressures be in pascals and in psi?

2.9 Two hemispherical shells are bolted together as shown in Fig. P2.9. The resulting spherical container, which weighs 400 lb, is filled with mercury and supported by a cable as shown. The container is vented at the top. If eight bolts are symmetrically located around the circumference, what is the vertical force that each bolt must carry?

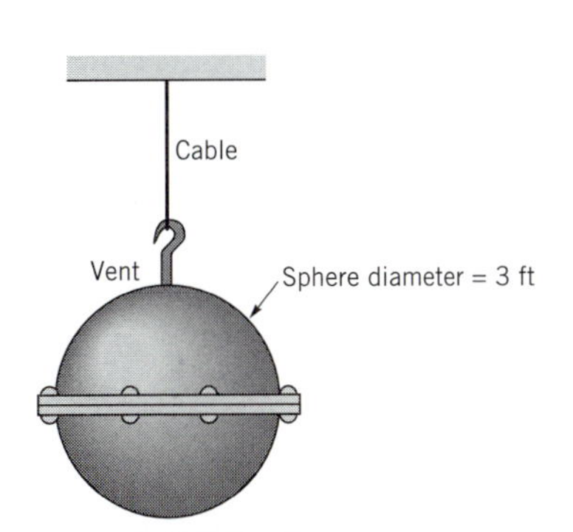

■ **FIGURE P2.9**

2.10 Develop an expression for the pressure variation in a liquid in which the specific weight increases with depth, h, as $\gamma = Kh + \gamma_0$, where K is a constant and γ_0 is the specific weight at the free surface.

*2.11 In a certain liquid at rest, measurements of the specific weight at various depths show the following variation:

h (ft)	γ (lb/ft^3)
0	70
10	76
20	84
30	91
40	97
50	102
60	107
70	110
80	112
90	114
100	115

The depth $h = 0$ corresponds to a free surface at atmospheric pressure. Determine, through numerical integration of Eq. 2.4, the corresponding variation in pressure and show the results on a plot of pressure (in psf) versus depth (in feet).

2.12 The basic elements of a hydraulic press are shown in Fig. P2.12. The plunger has an area of 1 in.^{2}, and a force, F_1, can be applied to the plunger through a lever mechanism having a mechanical advantage of 8 to 1. If the large piston has an area of 150 in.^{2}, what load, F_2, can be raised by a force of 30 lb applied to the lever? Neglect the hydrostatic pressure variation.

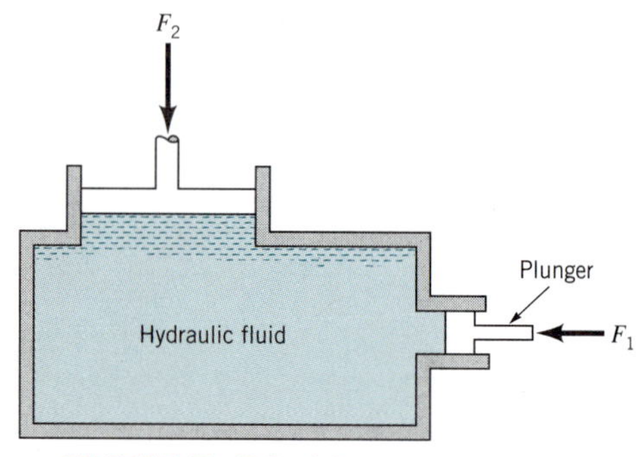

■ FIGURE P2.12

2.13 A 0.3-m-diameter pipe is connected to a 0.02-m-diameter pipe and both are rigidly held in place. Both pipes are horizontal with pistons at each end. If the space between the pistons is filled with water, what force will have to be applied to the larger piston to balance a force of 90 N applied to the smaller piston? Neglect friction.

† 2.14 Because of elevation differences, the water pressure in the second floor of your house is lower than it is in the first floor. For tall buildings this pressure difference can become unacceptable. Discuss possible ways to design the water distribution system in very tall buildings so that the hydrostatic pressure difference is within acceptable limits.

2.15 What would be the barometric pressure reading, in mm Hg, at an elevation of 4 km in the U.S. standard atmosphere? (Refer to Table C.2 in Appendix C.)

2.16 An absolute pressure of 7 psia corresponds to what gage pressure for standard atmospheric pressure of 14.7 psia?

2.17 A Bourdon pressure gage attached to the outside of a tank containing air reads 77.0 psi when the local atmospheric pressure is 760 mm Hg. What will be the gage reading if the atmospheric pressure increases to 773 mm Hg?

2.18 For an atmospheric pressure of 101 kPa (abs) determine the heights of the fluid columns in barometers containing one of the following liquids: **(a)** mercury, **(b)** water, and **(c)** ethyl alcohol. Calculate the heights including the effect of vapor pressure, and compare the results with those obtained neglecting vapor pressure. Do these results support the widespread use of mercury for barometers? Why?

2.19 Aneroid barometers can be used to measure changes in altitude. If a barometer reads 30.1 in. Hg at one elevation, what has been the change in altitude in meters when the barometer reading is 28.3 in. Hg? Assume a standard atmosphere and that Eq. 2.12 is applicable over the range of altitudes of interest.

2.20 Pikes Peak near Denver, Colorado, has an elevation of 14,110 ft. **(a)** Determine the pressure at this elevation, based on Eq. 2.12. **(b)** If the air is assumed to have a constant specific weight of 0.07647 lb/ft^3, what would the pressure be at this altitude? **(c)** If the air is assumed to have a constant temperature of 59 °F, what would the pressure be at this elevation? For all three cases assume standard atmospheric conditions at sea level (see Table 2.1).

2.21 Equation 2.12 provides the relationship between pressure and elevation in the atmosphere for those regions in which the temperature varies linearly with elevation. Derive this equation and verify the value of the pressure given in Table C.2 in Appendix C for an elevation of 5 km.

2.22 As shown in Fig. 2.6 for the U.S. standard atmosphere, the troposphere extends to an altitude of 11 km where the pressure is 22.6 kPa (abs). In the next layer, called the stratosphere, the temperature remains constant at -56.5 °C. Determine the pressure and density in this layer at an altitude of 15 km. Assume $g = 9.77$ m/s^2 in your calculations. Compare your results with those given in Table C.2 in Appendix C.

*2.23 Under normal conditions the temperature of the atmosphere decreases with increasing elevation. In some situations, however, a temperature inversion may exist so that the air temperature increases with elevation. A series of temperature probes on a mountain give the elevation–temperature data shown in the table below. If the barometric pressure at the base of the mountain is 12.1 psia, determine by means of numerical integration the pressure at the top of the mountain.

Elevation (ft)	Temperature (°F)
5000	50.1 (base)
5500	55.2
6000	60.3
6400	62.6
7100	67.0
7400	68.4
8200	70.0
8600	69.5
9200	68.0
9900	67.1 (top)

2.24 A U-tube manometer is connected to a closed tank containing air and water as shown in Fig. P2.24. At the closed end of the manometer the air pressure is 16 psia. Determine the reading on the pressure gage for a differential reading of 4 ft on the manometer. Express your answer in psi (gage). Assume standard atmospheric pressure and neglect the weight of the air columns in the manometer.

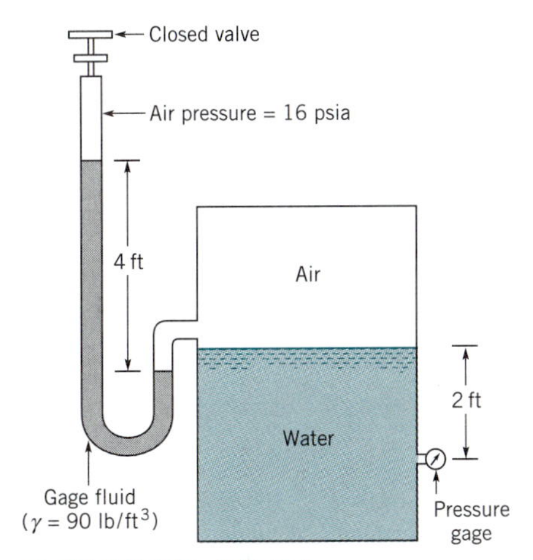

■ **FIGURE P2.24**

2.25 A closed cylindrical tank filled with water has a hemispherical dome and is connected to an inverted piping system as shown in Fig. P2.25. The liquid in the top part of the piping system has a specific gravity of 0.8, and the remaining parts of the system are filled with water. If the pressure gage reading at A is 60 kPa, determine: **(a)** the pressure in pipe B, and **(b)** the pressure head, in millimeters of mercury, at the top of the dome (point C).

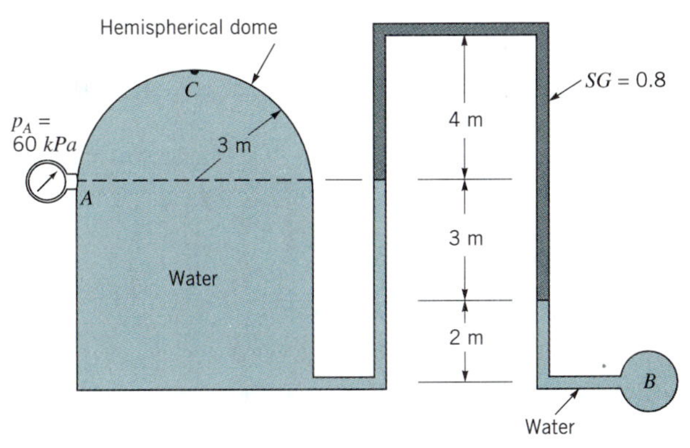

■ **FIGURE P2.25**

2.26 In Fig. P2.26 pipe A contains carbon tetrachloride (SG = 1.60) and the closed storage tank B contains a salt brine (SG = 1.15). Determine the air pressure in tank B if the pressure in pipe A is 25 psi.

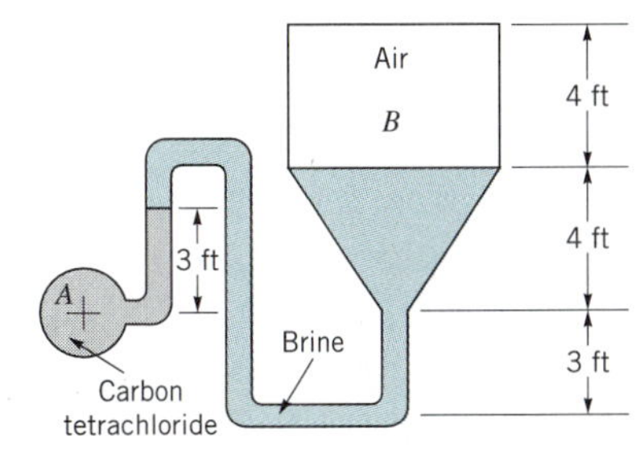

■ **FIGURE P2.26**

2.27 A U-tube mercury manometer is connected to a closed pressurized tank as illustrated in Fig. P2.27. If the air pressure is 2 psi, determine the differential reading, h. The specific weight of the air is negligible.

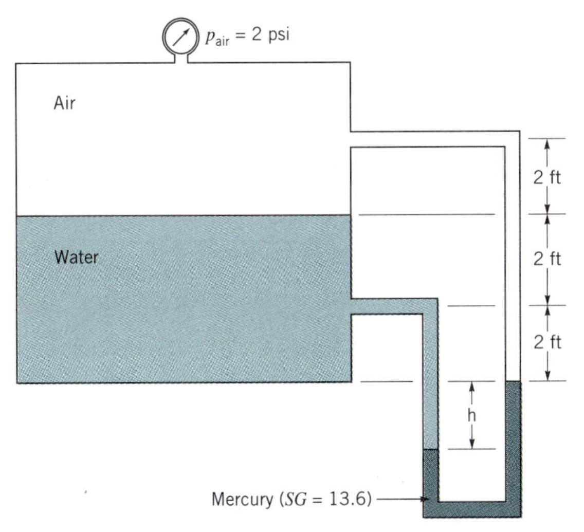

■ **FIGURE P2.27**

2.28 An inverted open tank is held in place by a force R as shown in Fig. P2.28. If the specific gravity of the manometer fluid is 2.5, determine the value of h.

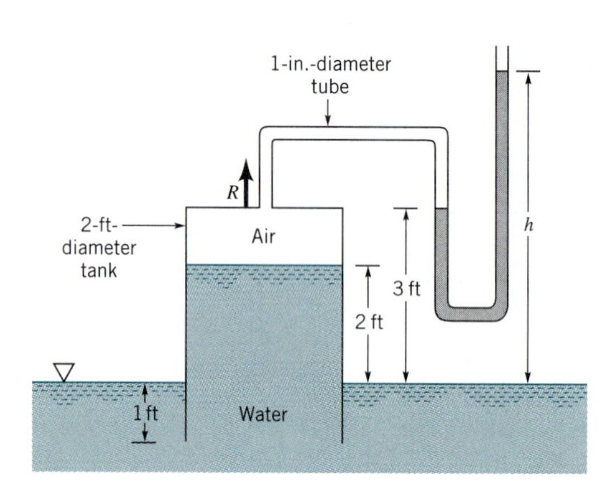

■ **FIGURE P2.28**

2.29 Water, oil, and an unknown fluid are contained in the vertical tubes shown in Fig. P2.29. Determine the density of the unknown fluid.

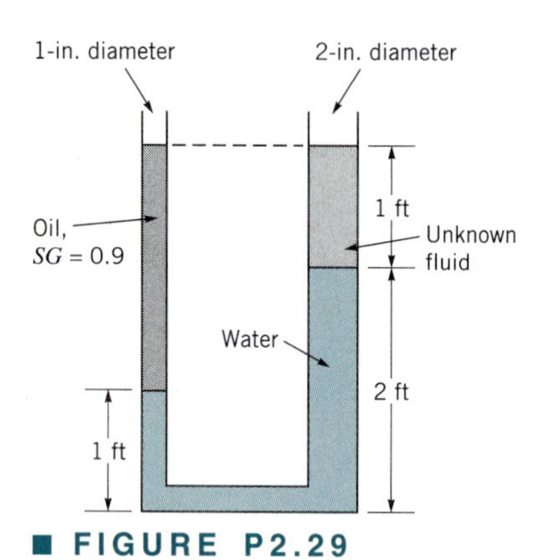

■ **FIGURE P2.29**

† **2.30** Although it is difficult to compress water, the density of water at the bottom of the ocean is greater than that at the surface because of the higher pressure at depth. Estimate how much higher the ocean's surface would be if the density of seawater were instantly changed to a uniform density equal to that at the surface.

2.31 The mercury manometer of Fig. P2.31 indicates a differential reading of 0.30 m when the pressure in pipe A is 30-mm Hg vacuum. Determine the pressure in pipe B.

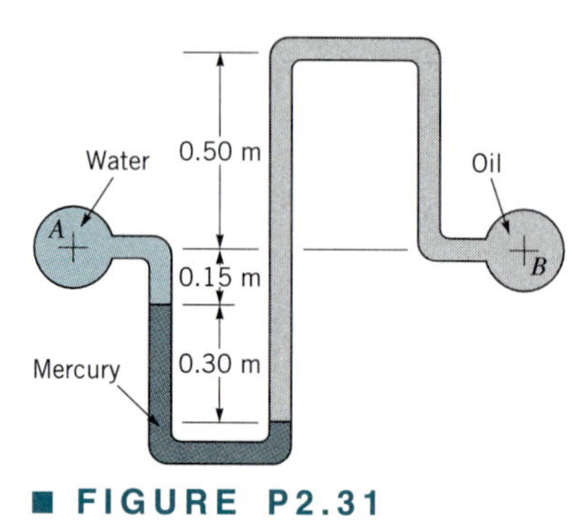

■ **FIGURE P2.31**

2.32 For the inclined-tube manometer of Fig. P2.32 the pressure in pipe A is 0.8 psi. The fluid in both pipes A and B is water, and the gage fluid in the manometer has a specific gravity of 2.6. What is the pressure in pipe B corresponding to the differential reading shown?

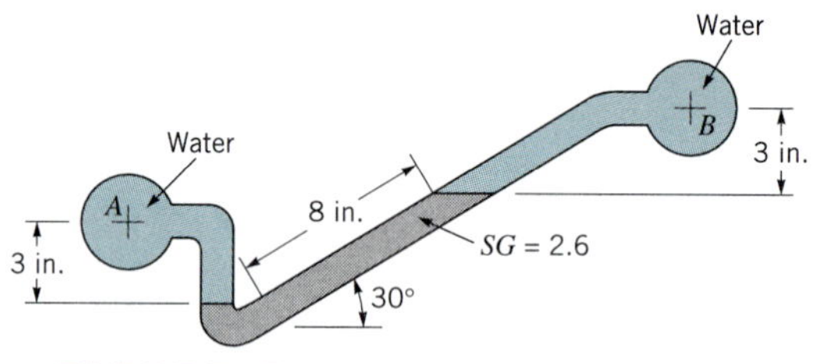

■ **FIGURE P2.32**

2.33 Compartments A and B of the tank shown in Fig. P2.33 are closed and filled with air and a liquid with a specific gravity equal to 0.6. Determine the manometer reading, h, if the barometric pressure is 14.7 psia and the pressure gage reads 0.5 psi. The effect of the weight of the air is negligible.

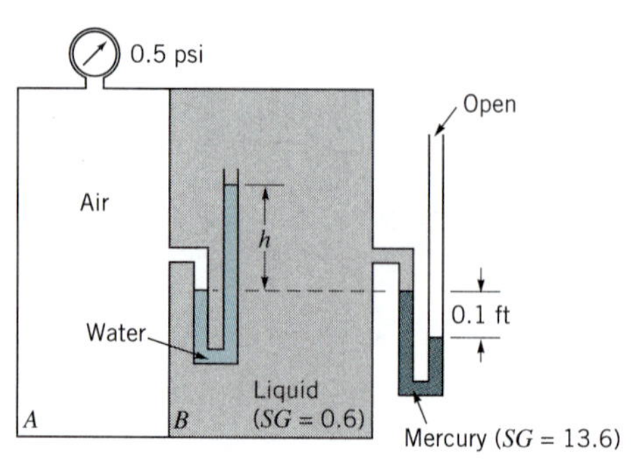

■ **FIGURE P2.33**

2.34 Small differences in gas pressures are commonly measured with a *micromanometer* of the type illustrated in Fig. P2.34. This device consists of two large reservoirs each having a cross-sectional area A_r which are filled with a liquid having a specific weight γ_1 and connected by a U-tube of cross-sectional area A_t containing a liquid of specific weight γ_2. When a differential gas pressure, $p_1 - p_2$, is applied, a differential reading, h, develops. It is desired to have this reading sufficiently large (so that it can be easily read) for small pressure differentials. Determine the relationship between h and $p_1 - p_2$ when the area ratio A_t/A_r is small, and show that the differential reading, h, can be magnified by making the difference in specific weights, $\gamma_2 - \gamma_1$, small. Assume that initially (with $p_1 = p_2$) the fluid levels in the two reservoirs are equal.

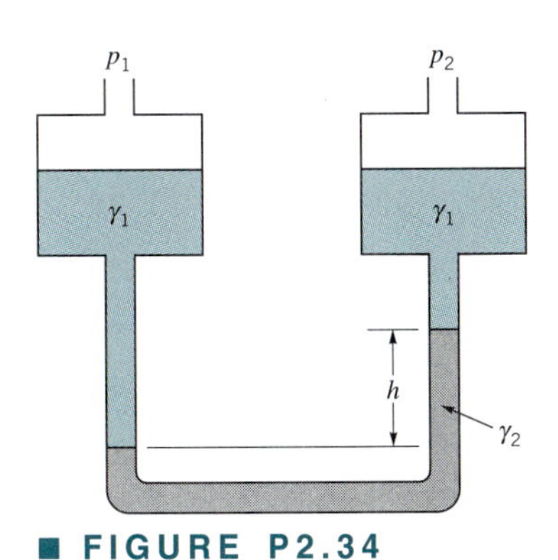

■ **FIGURE P2.34**

2.35 An inverted U-tube manometer containing oil ($SG = 0.8$) is located between two reservoirs as shown in Fig. P2.35. The reservoir on the left, which contains carbon tetrachloride, is closed and pressurized to 9 psi. The reservoir on the right contains water and is open to the atmosphere. With the given data, determine the depth of water, h, in the right reservoir.

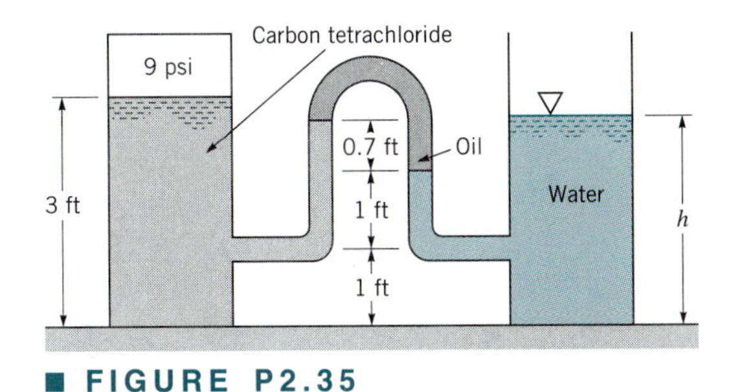

■ **FIGURE P2.35**

2.36 Determine the elevation difference, Δh, between the water levels in the two open tanks shown in Fig. P2.36.

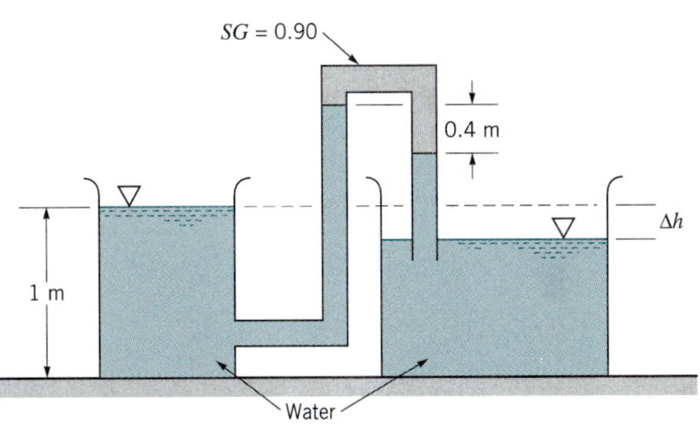

■ **FIGURE P2.36**

2.37 Water, oil, and salt water fill a tube as shown in Fig. P2.37. Determine the pressure at point 1 (inside the closed tube).

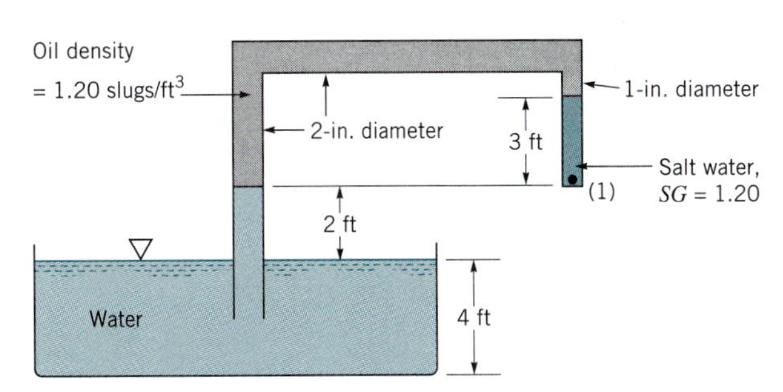

■ **FIGURE P2.37**

2.38 An air-filled, hemispherical shell is attached to the ocean floor at a depth of 10 m as shown in Fig. P2.38. A mercury barometer located inside the shell reads 765 mm Hg, and a mercury U-tube manometer designed to give the outside water pressure indicates a differential reading of 735 mm Hg as illustrated. Based on these data what is the atmospheric pressure at the ocean surface?

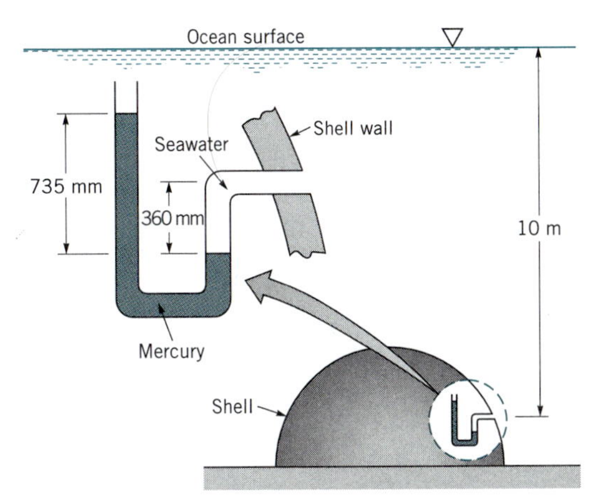

■ **FIGURE P2.38**

*2.39 Both ends of the U-tube mercury manometer of Fig. P2.39 are initially open to the atmosphere and under standard atmospheric pressure. When the valve at the top of the right leg is open, the level of mercury below the valve is h_i. After the valve is closed, air pressure is applied to the left leg. Determine the relationship between the differential reading on the manometer and the applied gage pressure, p_g. Show on a plot how the differential reading varies with p_g for h_i = 25, 50, 75, and 100 mm over the range $0 \le p_g \le 300$ kPa. Assume that the temperature of the trapped air remains constant.

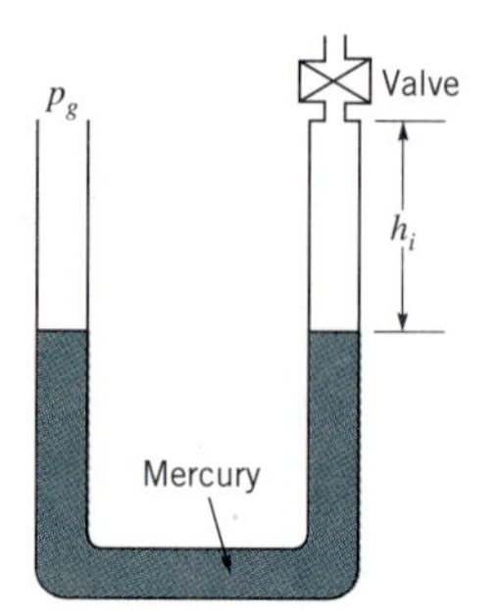

■ FIGURE P2.39

2.40 Three different liquids with properties as indicated fill the tank and manometer tubes as shown in Fig. P2.40. Determine the specific gravity of Fluid 3.

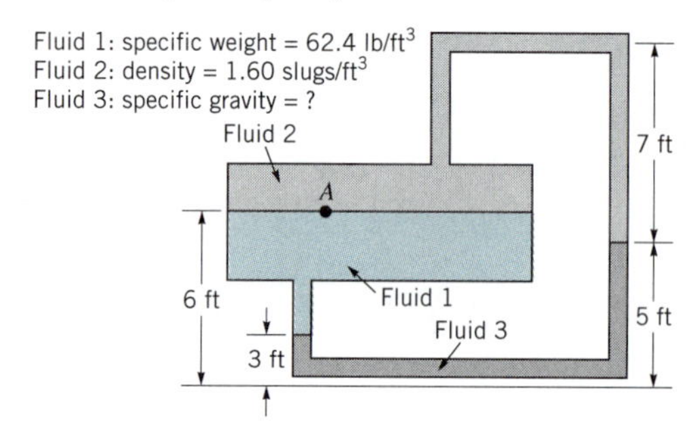

■ FIGURE P2.40

2.41 A 6-in.-diameter piston is located within a cylinder which is connected to a $\frac{1}{2}$-in.-diameter inclined-tube manometer as shown in Fig. P2.41. The fluid in the cylinder and the manometer is oil (specific weight = 59 lb/ft^3). When a weight $\mathcal{W}$ is placed on the top of the cylinder, the fluid level in the manometer tube rises from point (1) to (2). How heavy is the weight? Assume that the change in position of the piston is negligible.

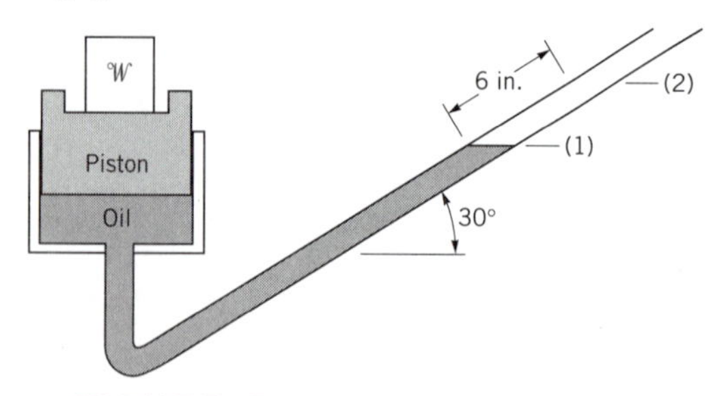

■ FIGURE P2.41

2.42 The manometer fluid in the manometer of Fig. P2.42 has a specific gravity of 3.46. Pipes A and B both contain water. If the pressure in pipe A is decreased by 1.3 psi and the pressure in pipe B increases by 0.9 psi, determine the new differential reading of the manometer.

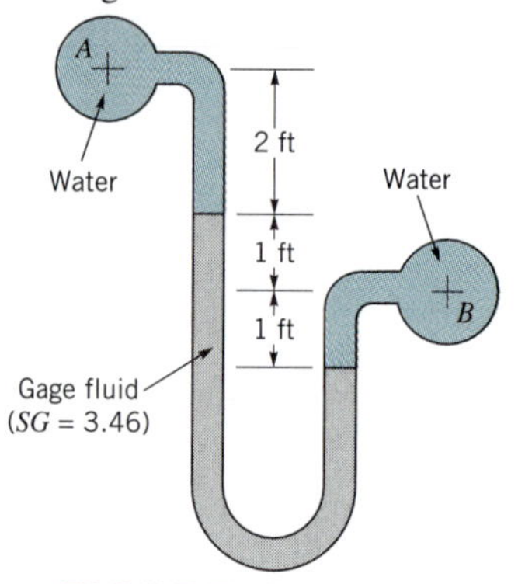

■ FIGURE P2.42

2.43 Determine the ratio of areas, A_1/A_2, of the two manometer legs of Fig. P2.43 if a change in pressure in pipe B of 0.5 psi gives a corresponding change of 1 in. in the level of the mercury in the right leg. The pressure in pipe A does not change.

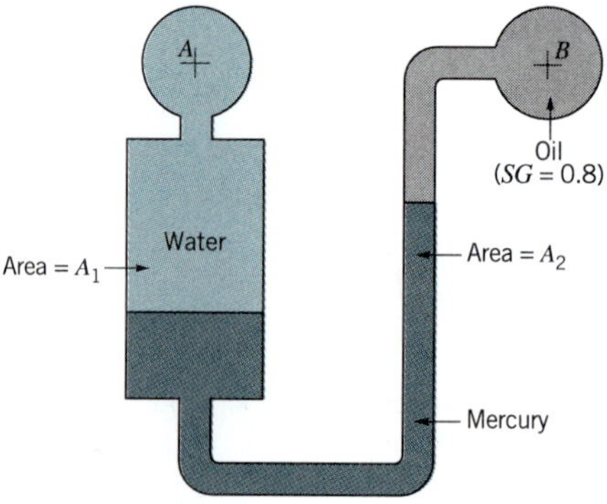

■ FIGURE P2.43

2.44 The inclined differential manometer of Fig. P2.44 contains carbon tetrachloride. Initially the pressure differential between pipes A and B, which contain a brine (SG = 1.1), is zero as illustrated in the figure. It is desired that the manometer give a differential reading of 12 in. (measured along the inclined tube) for a pressure differential of 0.1 psi. Determine the required angle of inclination, θ.

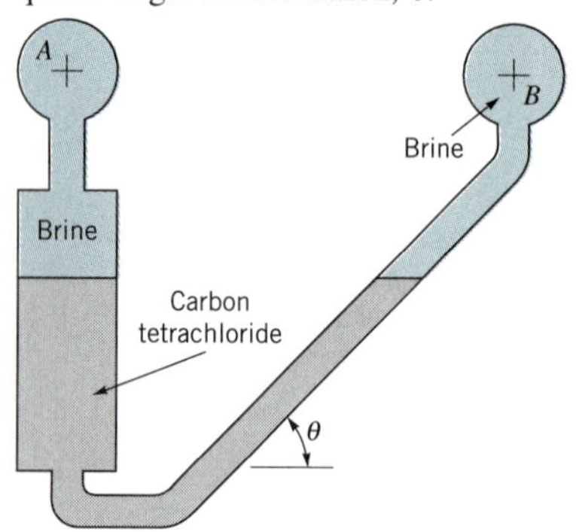

■ FIGURE P2.44

2.45 Determine the new differential reading along the inclined leg of the mercury manometer of Fig. P2.45, if the pressure in pipe A is decreased 10 kPa and the pressure in pipe B remains unchanged. The fluid in A has a specific gravity of 0.9 and the fluid in B is water.

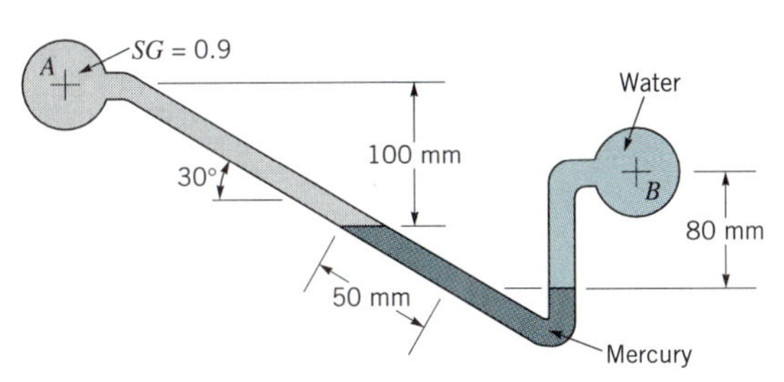

■ **FIGURE P2.45**

2.46 Determine the change in the elevation of the mercury in the left leg of the manometer of Fig. P2.46 as a result of an increase in pressure of 5 psi in pipe A while the pressure in pipe B remains constant.

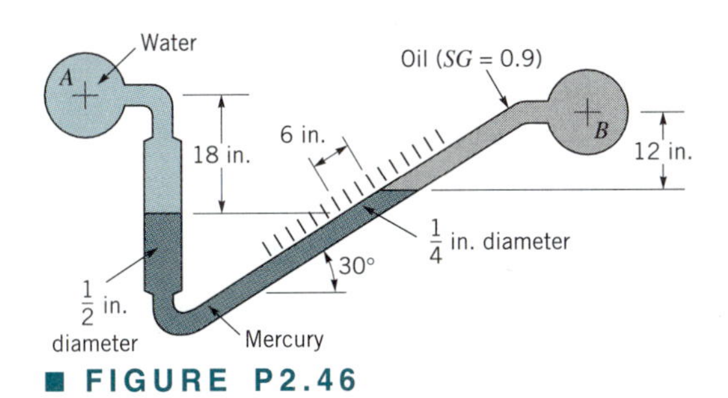

■ **FIGURE P2.46**

***2.47** Water initially fills the funnel and its connecting tube as shown in Fig. P2.47. Oil ($SG = 0.85$) is poured into the funnel until it reaches a level $h > H/2$ as indicated. Determine and plot the value of the rise in the water level in the tube, ℓ, as a function of h for $H/2 \leq h \leq H$, with $H = D = 2$ ft and $d = 0.1$ ft.

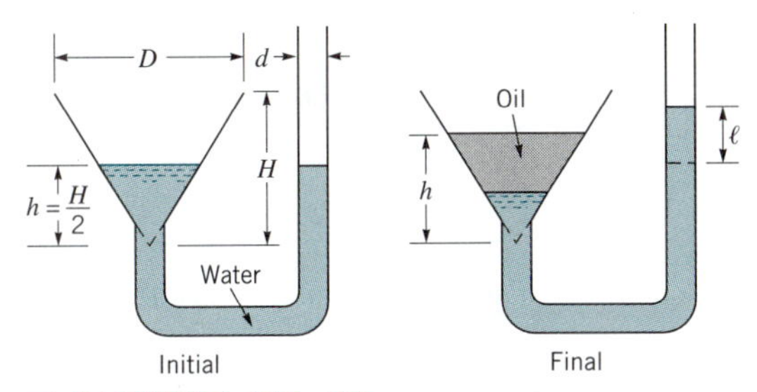

■ **FIGURE P2.47**

2.48 Concrete is poured into the forms as shown in Fig. P2.48 to produce a set of steps. Determine the weight of the sandbag needed to keep the bottomless forms from lifting off the ground. The weight of the forms is 85 lb, and the specific weight of the concrete is 150 lb/ft^3.

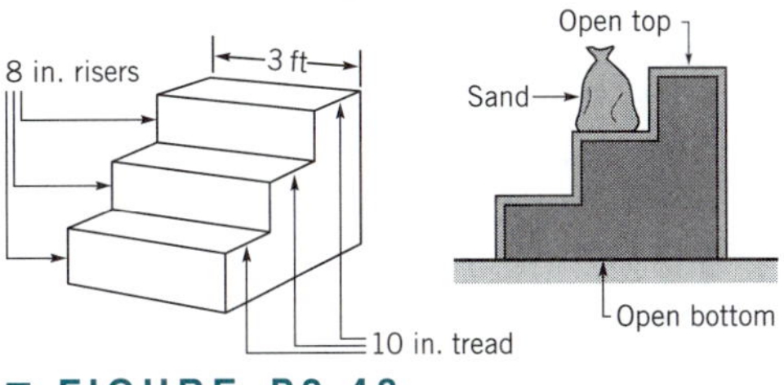

■ **FIGURE P2.48**

2.49 A square gate (4 m by 4 m) is located on the 45° face of a dam. The top edge of the gate lies 8 m below the water surface. Determine the force of the water on the gate and the point through which it acts.

2.50 An inverted 0.1-m-diameter circular cylinder is partially filled with water and held in place as shown in Fig. P2.50. A force of 20 N is needed to pull the flat plate from the cylinder. Determine the air pressure within the cylinder. The plate is not fastened to the cylinder and has negligible mass.

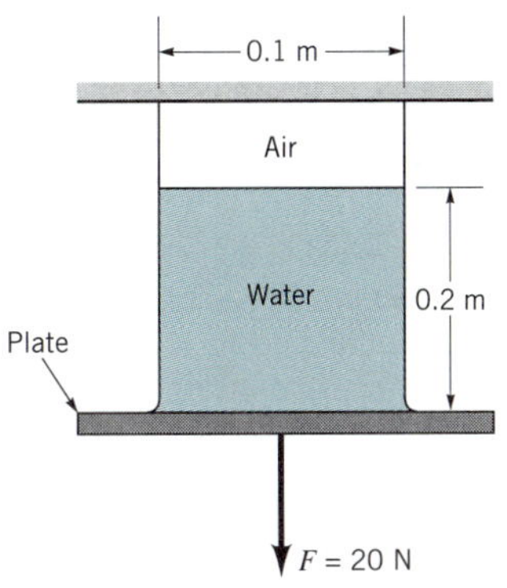

■ **FIGURE P2.50**

2.51 A large, open tank contains water and is connected to a 6-ft-diameter conduit as shown in Fig. P2.51. A circular plug is used to seal the conduit. Determine the magnitude, direction, and location of the force of the water on the plug.

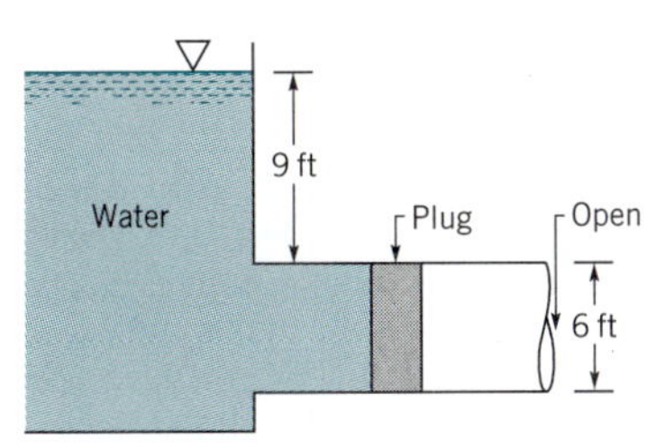

■ **FIGURE P2.51**

2.52 A homogeneous, 4-ft-wide, 8-ft-long rectangular gate weighing 800 lb is held in place by a horizontal flexible cable as shown in Fig. P2.52. Water acts against the gate which is hinged at point *A*. Friction in the hinge is negligible. Determine the tension in the cable.

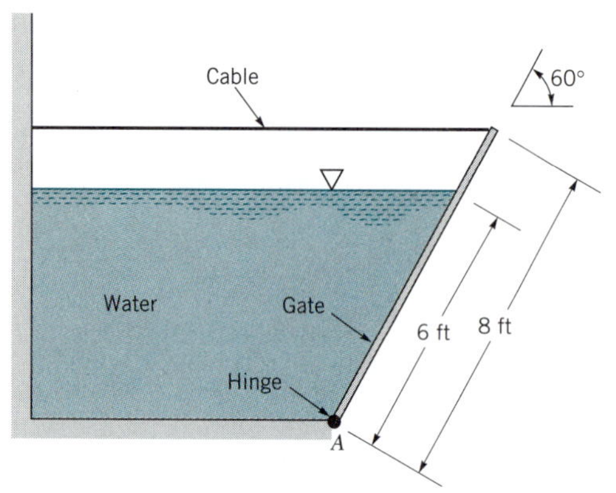

■ **FIGURE P2.52**

† **2.53** Sometimes it is difficult to open an exterior door of a building because the air distribution system maintains a pressure difference between the inside and outside of the building. Estimate how big this pressure difference can be if it is ''not too difficult'' for an average person to open the door.

2.54 An area in the form of an isosceles triangle with a base width of 6 ft and an altitude of 8 ft lies in the plane forming one wall of a tank which contains a liquid having a specific weight of 79.8 lb/ft^3. The side slopes upward making an angle of 60° with the horizontal. The base of the triangle is horizontal and the vertex is above the base. Determine the resultant force the fluid exerts on the area when the fluid depth is 20 ft above the base of the triangular area. Show, with the aid of a sketch, where the center of pressure is located.

2.55 Solve Problem 2.54 if the isosceles triangle is replaced with a right triangle having the same base width and altitude.

2.56 A horizontal 2-m-diameter conduit is half-filled with a liquid (SG = 1.6) and is capped at both ends with plane vertical surfaces. The air pressure in the conduit above the liquid surface is 150 kPa. Determine the resultant force of the fluid acting on one of the end caps, and locate this force relative to the bottom of the conduit.

2.57 Two square gates close two openings in a conduit connected to an open tank of water as shown in Fig. P2.57. When the water depth, *h*, reaches 5 m it is desired that both gates open at the same time. Determine the weight of the homogeneous horizontal gate and the horizontal force, *R*, acting on the vertical gate that is required to keep the gates closed until this depth is reached. The weight of the vertical gate is negligible, and both gates are hinged at one end as shown. Friction in the hinges is negligible.

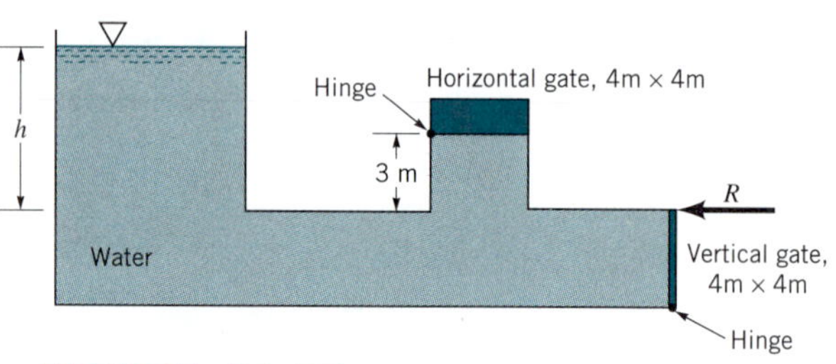

■ **FIGURE P2.57**

2.58 The rigid gate, *OAB*, of Fig. P2.58 is hinged at *O* and rests against a rigid support at *B*. What minimum horizontal force, *P*, is required to hold the gate closed if its width is 3 m? Neglect the weight of the gate and friction in the hinge. The back of the gate is exposed to the atmosphere.

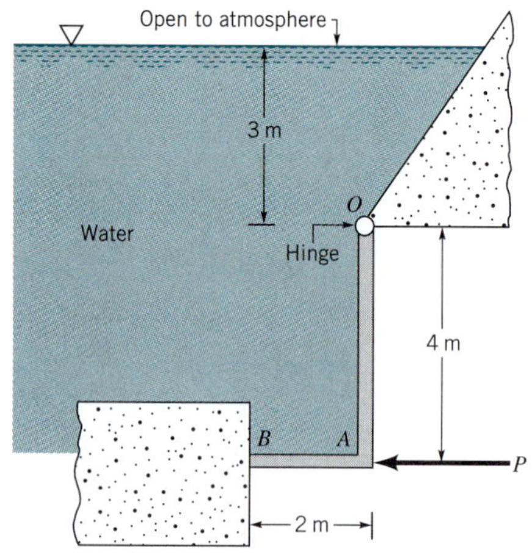

■ **FIGURE P2.58**

2.59 The massless, 4-ft-wide gate shown in Fig. P2.59 pivots about the frictionless hinge O. It is held in place by the 2000 lb counterweight, *W*. Determine the water depth, *h*.

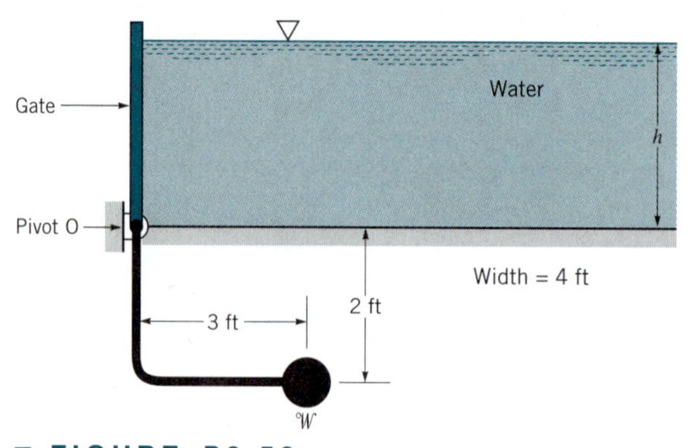

■ **FIGURE P2.59**

*2.60 A 200-lb homogeneous gate of 10-ft width and 5-ft length is hinged at point A and held in place by a 12-ft-long brace as shown in Fig. P2.60. As the bottom of the brace is moved to the right, the water level remains at the top of the gate. The line of action of the force that the brace exerts on the gate is along the brace. **(a)** Plot the magnitude of the force exerted on the gate by the brace as a function of the angle of the gate, θ, for $0 \leq \theta \leq 90°$. **(b)** Repeat the calculations for the case in which the weight of the gate is negligible. Comment on the results as $\theta \rightarrow 0$.

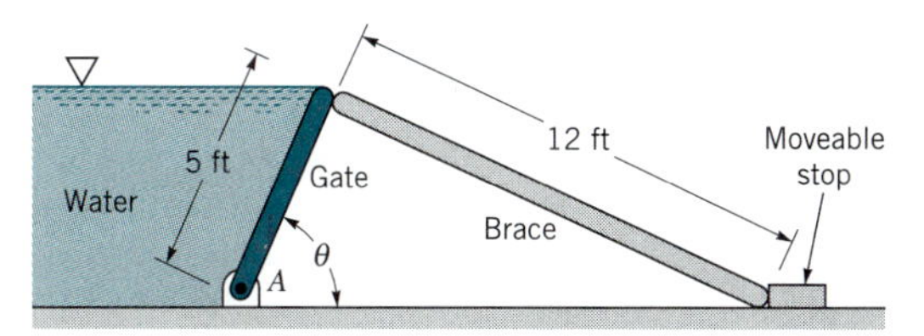

■ FIGURE P2.60

2.61 A rectangular gate 6 ft tall and 5 ft wide in the side of an open tank is held in place by the force F as indicated in Fig. P2.61. The weight of the gate is negligible, and the hinge at O is frictionless. **(a)** Determine the water depth, h, if the resultant hydrostatic force of the water acts 2.5 ft above the bottom of the gate, i.e., it is collinear with the applied force F. **(b)** For the depth of part (a), determine the magnitude of the resultant hydrostatic force. **(c)** Determine the force that the hinge puts on the gate under the above conditions.

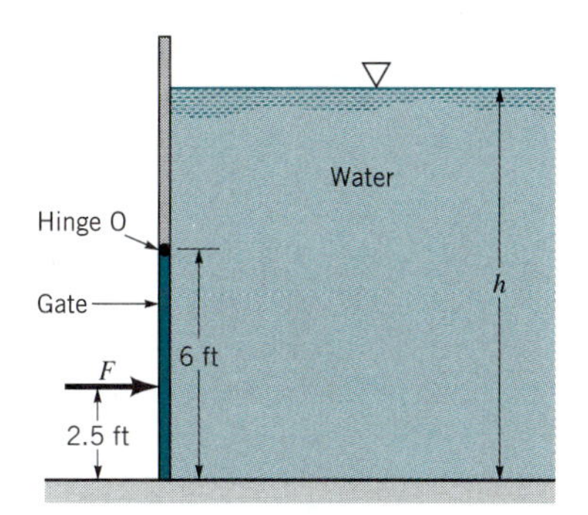

■ FIGURE P2.61

2.62 A gate having the shape shown in Fig. P2.62 is located in the vertical side of an open tank containing water. The gate is mounted on a horizontal shaft. **(a)** When the water level is at the top of the gate, determine the magnitude of the fluid force on the rectangular portion of the gate above the shaft and the magnitude of the fluid force on the semicircular portion of the gate below the shaft. **(b)** For this same fluid depth determine the moment of the force acting on the semicircular portion of the gate with respect to an axis which coincides with the shaft.

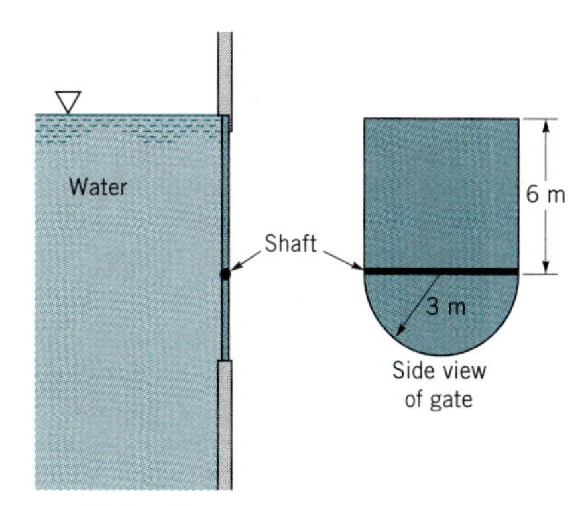

■ FIGURE P2.62

2.63 A gate having the cross section shown in Fig. P2.63 is 4 ft wide and is hinged at C. The gate weighs 18,000 lb, and its mass center is 1.67 ft to the right of the plane BC. Determine the vertical reaction at A on the gate when the water level is 3 ft above the base. All contact surfaces are smooth.

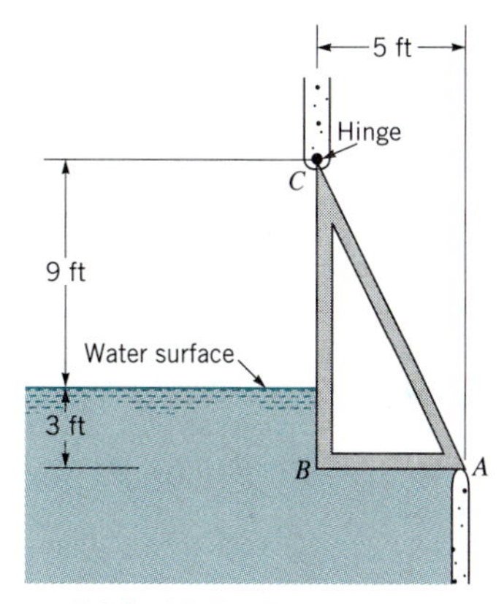

■ FIGURE P2.63

2.64 A structure is attached to the ocean floor as shown in Fig. P2.64. A 2-m-diameter hatch is located in an inclined wall and hinged on one edge. Determine the minimum air pressure, p_1, within the container to open the hatch. Neglect the weight of the hatch and friction in the hinge.

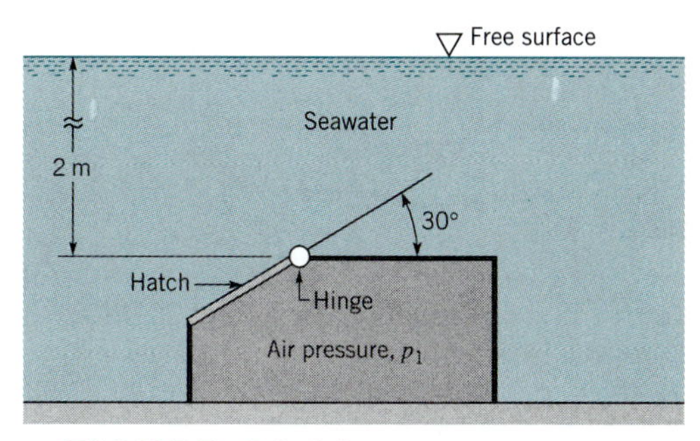

■ FIGURE P2.64

2.65 An open rectangular container contains a liquid that has a specific weight that varies according to the equation $\gamma = c_1 + c_2 h$, where c_1 and c_2 are constants and h is a vertical coordinate measured downward from the free surface. Derive an equation for the magnitude of the liquid force exerted on one wall of the container having a width b and height H and an equation that gives the vertical coordinate of this force.

***2.66** An open rectangular settling tank contains a liquid suspension that at a given time has a specific weight that varies approximately with depth according to the following data:

h (m)	γ (kN/m^3)
0	10.0
0.4	10.1
0.8	10.2
1.2	10.6
1.6	11.3
2.0	12.3
2.4	12.7
2.8	12.9
3.2	13.0
3.6	13.1

The depth $h = 0$ corresponds to the free surface. Determine, by means of numerical integration, the magnitude and location of the resultant force that the liquid suspension exerts on a vertical wall of the tank that is 6 m wide. The depth of fluid in the tank is 3.6 m.

2.67 The inclined face AD of the tank of Fig. P2.67 is a plane surface containing a gate ABC, which is hinged along line BC. The shape of the gate is shown in the plan view. If the tank contains water, determine the magnitude of the force that the water exerts on the gate.

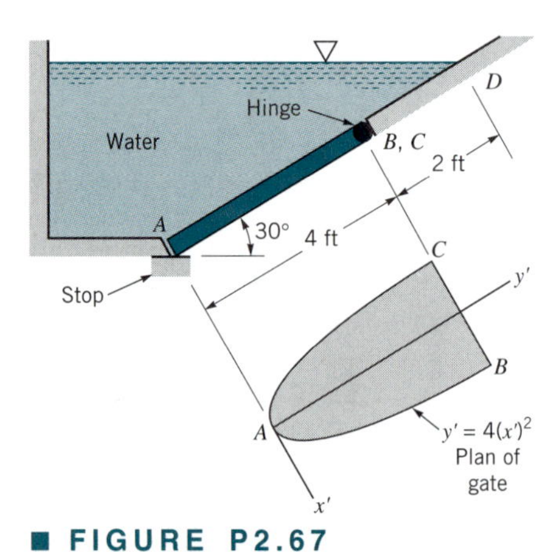

■ **FIGURE P2.67**

2.68 The concrete dam of Fig. P2.68 weighs 23.6 kN/m^3 and rests on a solid foundation. Determine the minimum coefficient of friction between the dam and the foundation required to keep the dam from sliding at the water depth shown. Assume no fluid uplift pressure along the base. Base your analysis on a unit length of the dam.

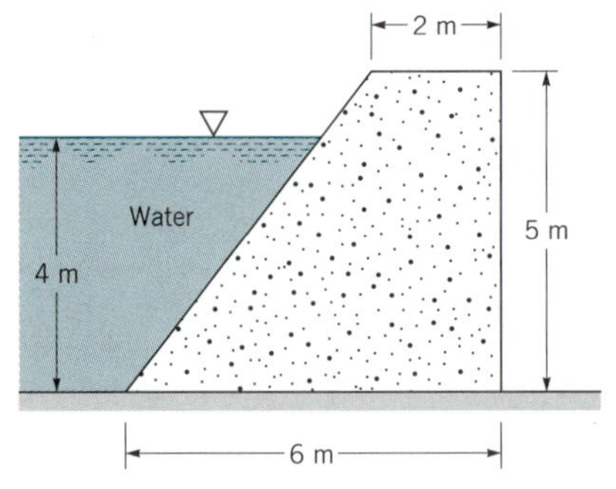

■ **FIGURE P2.68**

***2.69** Water backs up behind a concrete dam as shown in Fig. P2.69. Leakage under the foundation gives a pressure distribution under the dam as indicated. If the water depth, h, is too great, the dam will topple over about its toe (point A). For the dimensions given, determine the maximum water depth for the following widths of the dam: $\ell = 20, 30, 40, 50$, and 60 ft. Base your analysis on a unit length of the dam. The specific weight of the concrete is 150 lb/ft^3.

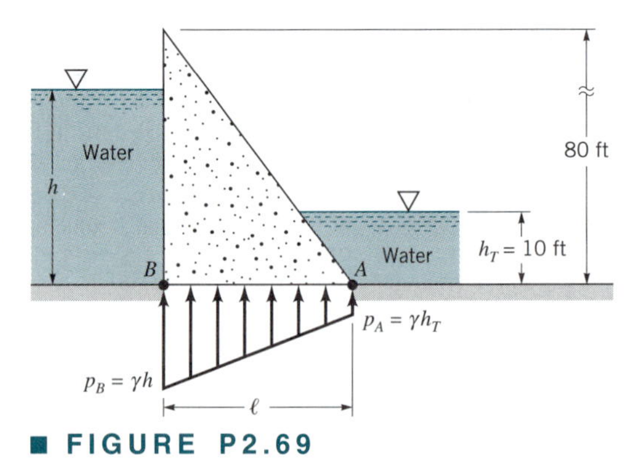

■ **FIGURE P2.69**

2.70 A 4-m-long curved gate is located in the side of a reservoir containing water as shown in Fig. P2.70. Determine the magnitude of the horizontal and vertical components of the force of the water on the gate. Will this force pass through point A? Explain.

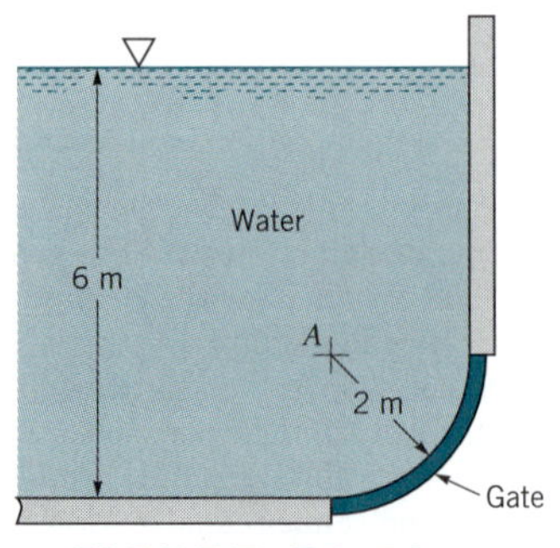

■ **FIGURE P2.70**

2.71 A 3-m-diameter open cylindrical tank contains water and has a hemispherical bottom as shown in Fig. P2.71. Determine the magnitude, line of action, and direction of the force of the water on the curved bottom.

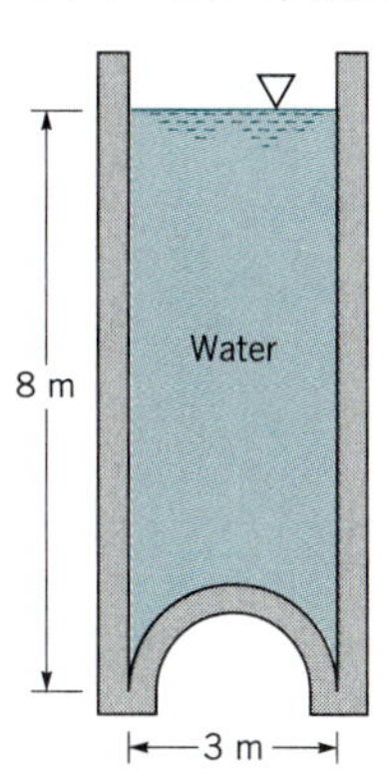

■ **FIGURE P2.71**

2.72 The 20-ft-long gate of Fig. P2.72 is a quarter circle and is hinged at H. Determine the horizontal force, P, required to hold the gate in place. Neglect friction at the hinge and the weight of the gate.

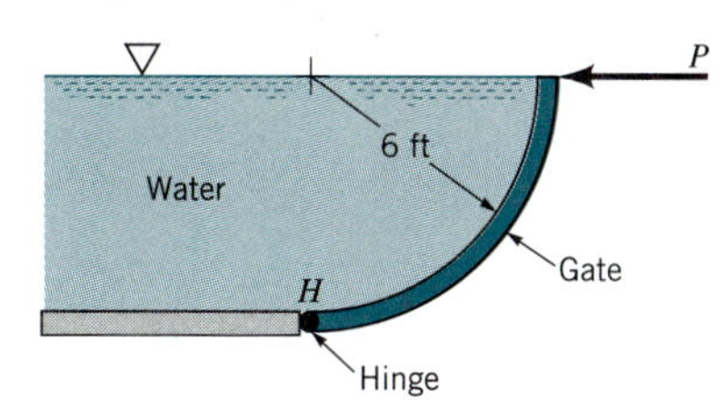

■ **FIGURE P2.72**

2.73 A plug in the bottom of a pressurized tank is conical in shape as shown in Fig. P2.73. The air pressure is 50 kPa and the liquid in the tank has a specific weight of 27 kN/m^3. Determine the magnitude, direction, and line of action of the force exerted on the curved surface of the cone within the tank due to the 50-kPa pressure and the liquid.

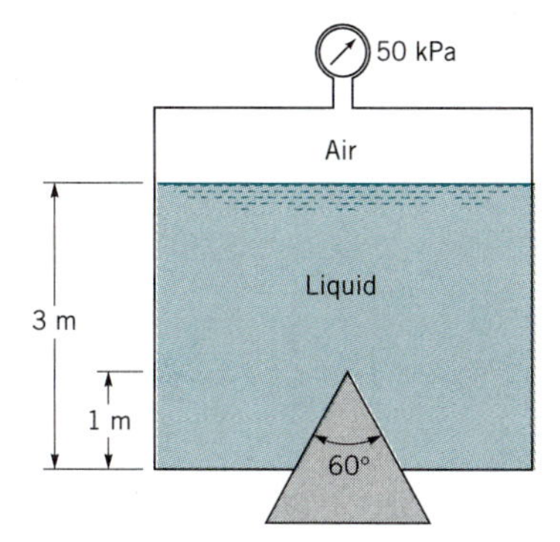

■ **FIGURE P2.73**

2.74 A 12-in.-diameter pipe contains a gas under a pressure of 140 psi. If the pipe wall thickness is $\frac{1}{4}$-in., what is the average circumferential stress developed in the pipe wall?

2.75 The concrete (specific weight $= 150 \text{ lb/ft}^3$) seawall of Fig. P2.75 has a curved surface and restrains seawater at a depth of 24 ft. The trace of the surface is a parabola as illustrated. Determine the moment of the fluid force (per unit length) with respect to an axis through the toe (point A).

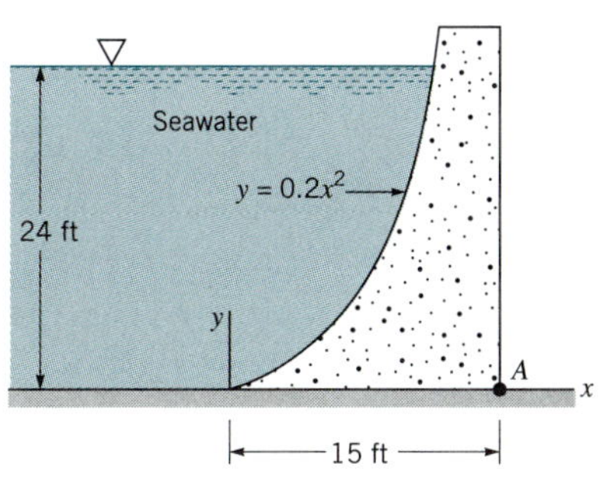

■ **FIGURE P2.75**

2.76 A cylindrical tank with its axis horizontal has a diameter of 2.0 m and a length of 4.0 m. The ends of the tank are vertical planes. A vertical, 0.1-m-diameter pipe is connected to the top of the tank. The tank and the pipe are filled with ethyl alcohol to a level of 1.5 m above the top of the tank. Determine the resultant force of the alcohol on one end of the tank and show where it acts.

2.77 If the tank ends in Problem 2.76 are hemispherical, what is the magnitude of the resultant horizontal force of the alcohol on one of the curved ends?

2.78 Imagine the tank of Problem 2.76 split by a horizontal plane. Determine the magnitude of the resultant force of the alcohol on the bottom half of the tank.

2.79 A closed tank is filled with water and has a 4-ft-diameter hemispherical dome as shown in Fig. P2.79. A U-tube manometer is connected to the tank. Determine the vertical force of the water on the dome if the differential manometer reading is 7 ft and the air pressure at the upper end of the manometer is 12.6 psi.

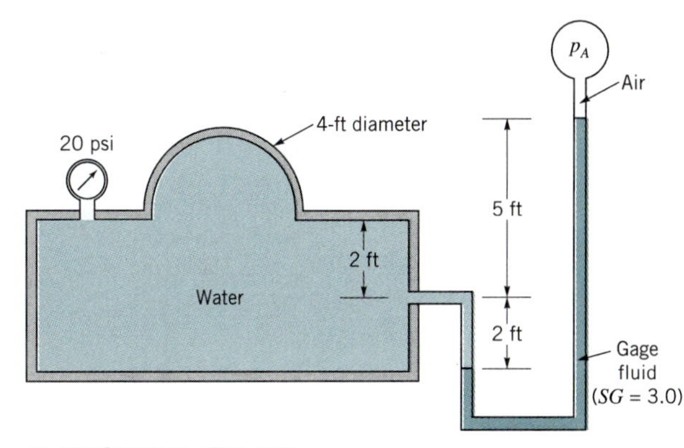

■ **FIGURE P2.79**

2.80 A tank wall has the shape shown in Fig. P2.80. Determine the horizontal and vertical components of the force of the water on a 4-ft length of the curved section AB.

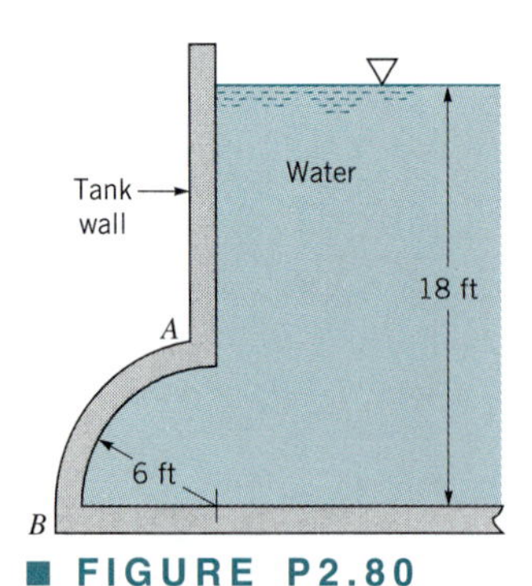

■ FIGURE P2.80

2.81 Three gates of negligible weight are used to hold back water in a channel of width b as shown in Fig. P2.81. The force of the gate against the block for gate (b) is R. Determine (in terms of R) the force against the blocks for the other two gates.

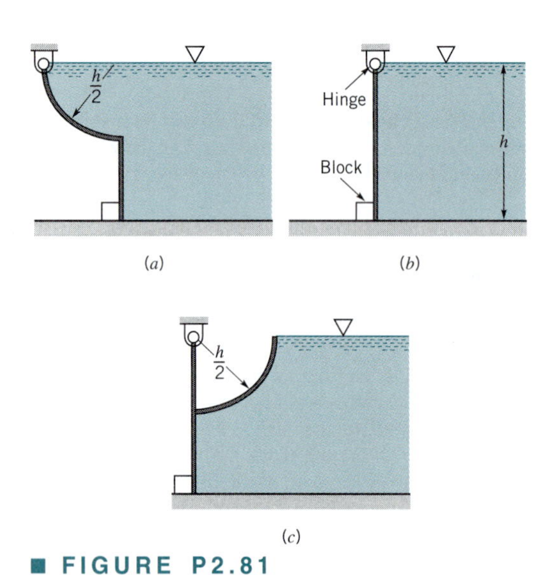

■ FIGURE P2.81

2.82 A 3 ft × 3 ft × 3 ft wooden cube (specific weight = 37 lb/ft^3) floats in a tank of water. How much of the cube extends above the water surface? If the tank were pressurized so that the air pressure at the water surface was increased to 1.0 psi, how much of the cube would extend above the water surface? Explain how you arrived at your answer.

2.83 The homogeneous timber AB of Fig. P2.83 is 0.15 m by 0.35 m in cross section. Determine the specific weight of the timber and the tension in the rope.

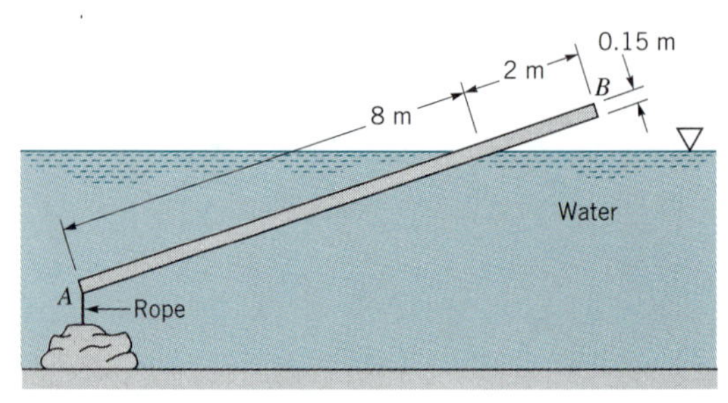

■ FIGURE P2.83

2.84 A 2-ft-thick block constructed of wood (SG = 0.6) is submerged in oil (SG = 0.9) and has a 2-ft-thick aluminum (specific weight = 168 lb/ft^3) plate attached to the bottom as indicated in Fig. P2.84. Determine completely the force required to hold the block in the position shown. Locate the force with respect to point A.

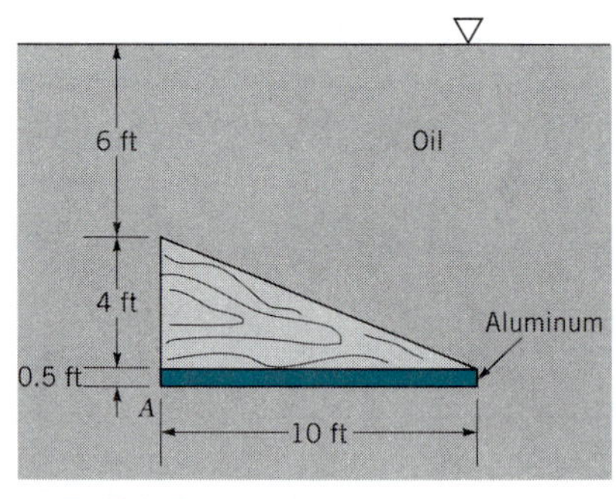

■ FIGURE P2.84

† **2.85** Estimate the minimum water depth needed to float a canoe carrying two people and their camping gear. List all assumptions and show all calculations.

2.86 An inverted test tube partially filled with air floats in a plastic water-filled soft drink bottle as shown in Fig. P2.86. The amount of air in the tube has been adjusted so that it just floats. The bottle cap is securely fastened. A slight squeezing of the plastic bottle will cause the test tube to sink to the bottom of the bottle. Explain this phenomenon.

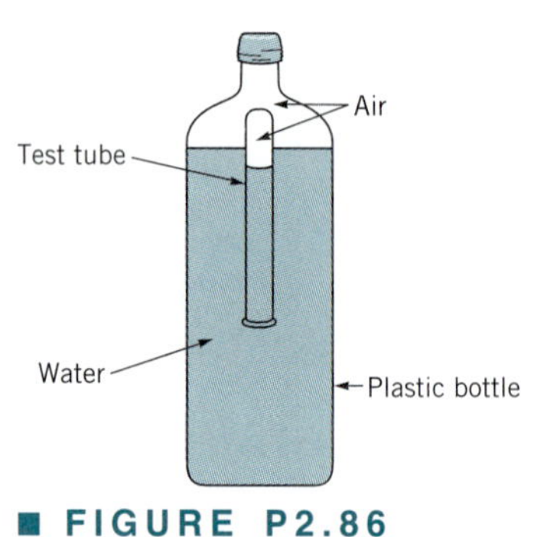

■ FIGURE P2.86

2.87 The hydrometer shown in Fig. P2.87 has a mass of 0.045 kg and the cross-sectional area of its stem is 290 mm^2. Determine the distance between graduations (on the stem) for specific gravities of 1.00 and 0.90.

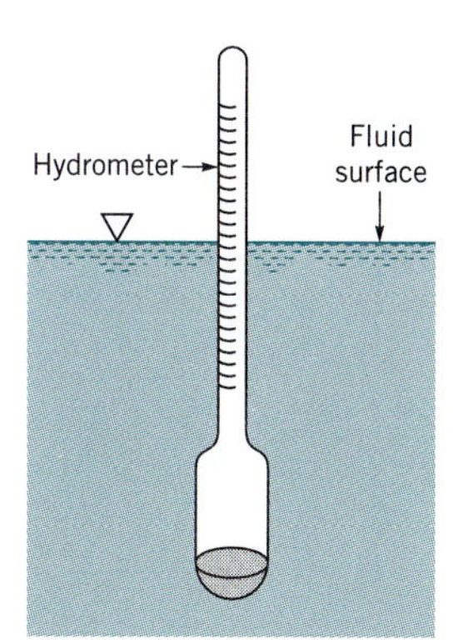

■ FIGURE P2.87

2.88 A 1-m-diameter cylindrical mass, M, is connected to a 2-m-wide rectangular gate as shown in Fig. P2.88. The gate is to open when the water level, h, drops below 2.5 m. Determine the required value for M. Neglect friction at the gate hinge and the pulley.

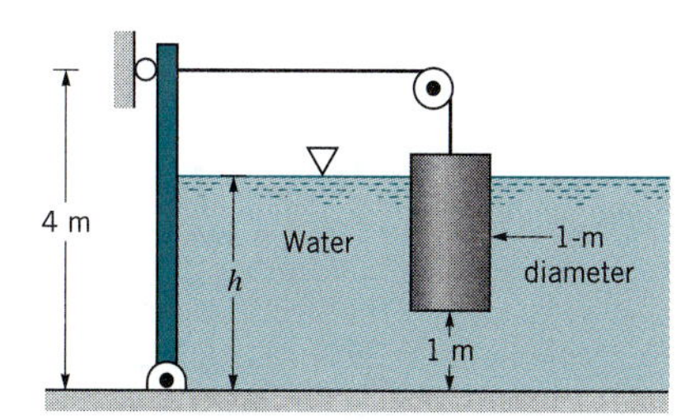

■ FIGURE P2.88

2.89 A 1-ft-diameter, 2-ft-long cylinder floats in an open tank containing a liquid having a specific weight γ. A U-tube manometer is connected to the tank as shown in Fig. P2.89. When the pressure in pipe A is 0.1 psi below atmospheric pressure, the various fluid levels are as shown. Determine the weight of the cylinder. Note that the top of the cylinder is flush with the fluid surface.

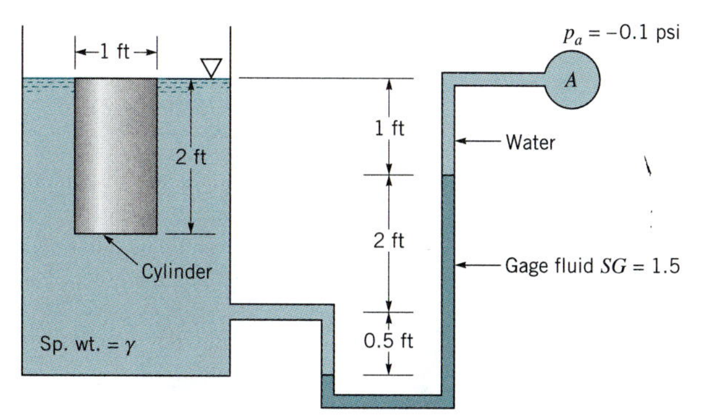

■ FIGURE P2.89

2.90 The thin-walled, 1-m-diameter tank of Fig. P2.90 is closed at one end and has a mass of 90 kg. The open end of the tank is lowered into the water and held in the position shown by a steel block having a density of 7840 kg/m^3. Assume that the air that is trapped in the tank is compressed at a constant temperature. Determine: **(a)** the reading on the pressure gage at the top of the tank, and **(b)** the volume of the steel block.

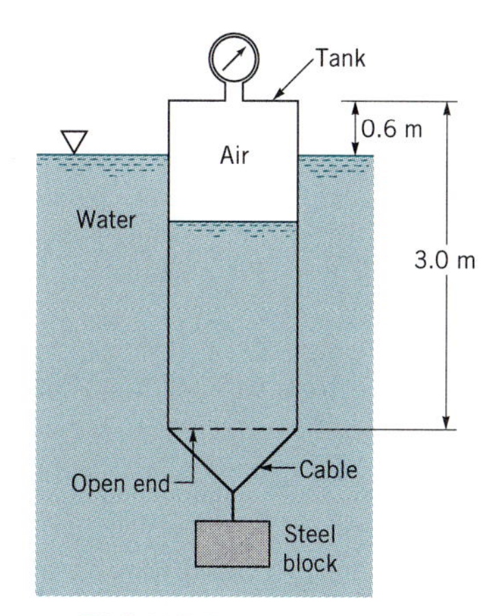

■ FIGURE P2.90

***2.91** An inverted hollow cone is pushed into the water as is shown in Fig. P2.91. Determine the distance, ℓ, that the water rises in the cone as a function of the depth, d, of the lower edge of the cone. Plot the results for $0 \le d \le H$, when H is equal to 1 m. Assume the temperature of the air within the cone remains constant.

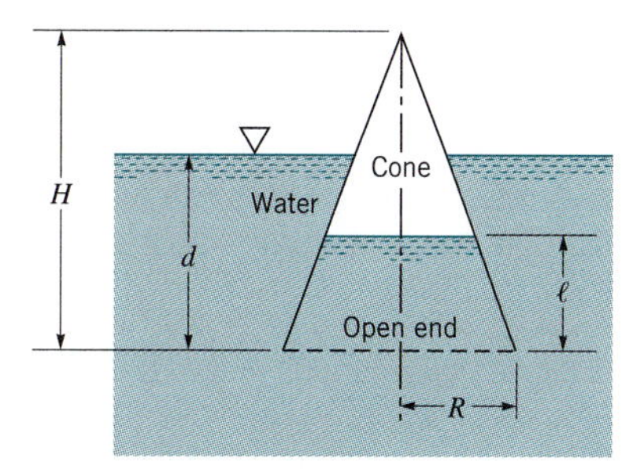

■ FIGURE P2.91

2.92 An open container of oil rests on the flatbed of a truck that is traveling along a horizontal road at 55 mi/hr. As the truck slows uniformly to a complete stop in 5 s, what will be the slope of the oil surface during the period of constant deceleration?

2.93 A 5-gal, cylindrical open container with a bottom area of 120 $in.^2$ is filled with glycerin and rests on the floor of an elevator. **(a)** Determine the fluid pressure at the bottom of the container when the elevator has an upward acceleration of 3 ft/s^2. **(b)** What resultant force does the container exert on the floor of the elevator during this acceleration? The weight of the container is negligible. (Note: 1 gal = 231 $in.^3$)

2.94 An open rectangular tank 1 m wide and 2 m long contains gasoline to a depth of 1 m. If the height of the tank sides is 1.5 m, what is the maximum horizontal acceleration (along the long axis of the tank) that can develop before the gasoline would begin to spill?

2.95 If the tank of Problem 2.94 slides down a frictionless plane that is inclined at 30° with the horizontal, determine the angle the free surface makes with the horizontal.

2.96 A closed cylindrical tank that is 8 ft in diameter and 24 ft long is completely filled with gasoline. The tank, with its long axis horizontal, is pulled by a truck along a horizontal surface. Determine the pressure difference between the ends (along the long axis of the tank) when the truck undergoes an acceleration of 5 ft/s^2.

2.97 The open U-tube of Fig. P2.97 is partially filled with a liquid. When this device is accelerated with a horizontal acceleration a, a differential reading h develops between the manometer legs which are spaced a distance ℓ apart. Determine the relationship between a, ℓ, and h.

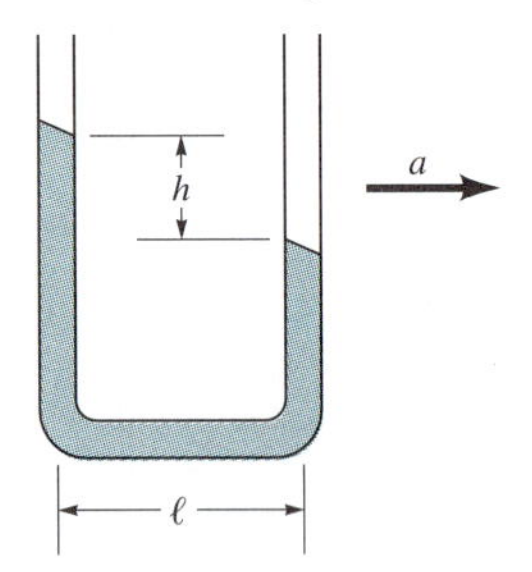

■ FIGURE P2.97

2.98 An open 1-m-diameter tank contains water at a depth of 0.7 m when at rest. As the tank is rotated about its vertical axis the center of the fluid surface is depressed. At what angular velocity will the bottom of the tank first be exposed? No water is spilled from the tank.

2.99 The U-tube of Fig. P2.99 is partially filled with water and rotates around the axis a–a. Determine the angular velocity that will cause the water to start to vaporize at the bottom of the tube (point A).

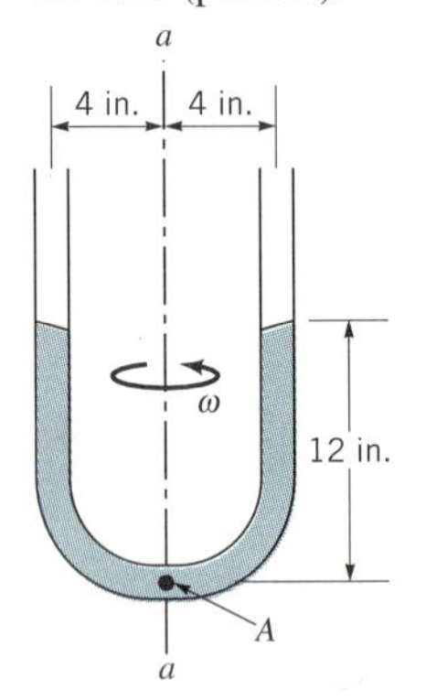

■ FIGURE P2.99

2.100 The U-tube of Fig. P2.100 contains mercury and rotates about the off-center axis a–a. At rest, the depth of mercury in each leg is 150 mm as illustrated. Determine the angular velocity for which the difference in heights between the two legs is 75 mm.

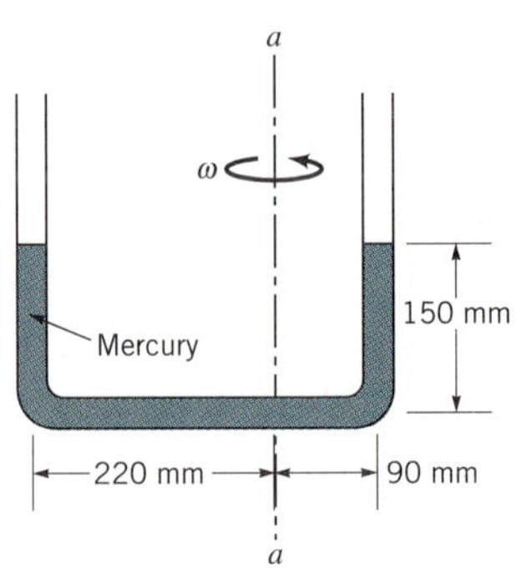

■ FIGURE P2.100

2.101 A closed, 0.4-m-diameter cylindrical tank is completely filled with oil ($SG = 0.9$) and rotates about its vertical longitudinal axis with an angular velocity of 40 rad/s. Determine the difference in pressure just under the vessel cover between a point on the circumference and a point on the axis.

2.102 The device shown in Fig. P2.102 is used to investigate the hydrostatic force on a plane rectangular surface. With no water in the tank the balance beam is horizontal. A weight, $\mathcal{W}$, is attached to the beam as shown, and the water depth, h, to the bottom of the rectangular surface is adjusted until the beam is again horizontal.

Values of $\mathcal{W}$ and h obtained experimentally are shown in the table below. Use these results to plot a graph of weight as a function of water depth. On the same graph plot the theoretical curve obtained by equating the moment that the weight produces about the pivot point to the moment produced by the hydrostatic force on the rectangular end of the block. Note that the pressure forces on the circular curved sides of the block do not produce a moment about the pivot because the line of action of these forces is through the pivot.

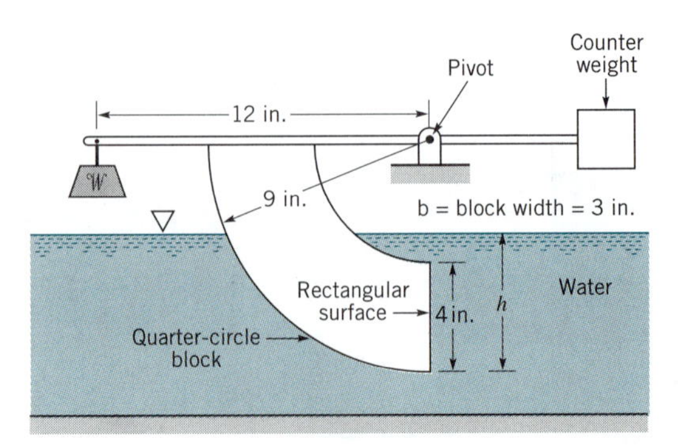

■ FIGURE P2.102

Compare the experimental and theoretical results and discuss some possible reasons for any differences between them.

$\mathcal{W}$ (lb)	h (in.)
0	0
0.044	1.11
0.132	1.92
0.264	2.76
0.352	3.20
0.440	3.60
0.573	4.17
0.663	4.51
0.882	5.39
1.101	6.27
1.211	6.70

2.103 A bottomless tank with vertical sides and slanted ends sits on a flat surface as shown in Fig. P2.103. The hydrostatic pressure force on the slanted ends has an upward vertical component. If the water depth is large enough, the vertical component of the pressure force becomes greater than the combined weight of the tank $\mathcal{W}_{tank}$ and the applied load $\mathcal{W}$. The tank will then lift slightly off the surface upon which it sits, and some of the water will drain from the tank. This happens when the depth is h.

Values of $\mathcal{W}$ and h determined experimentally are shown below. Use these results to plot a graph of water depth as a function of the applied load. On the same graph plot two theoretical curves obtained as follows. First assume that the combined weight of the tank and the applied load is supported by the vertical component of the hydrostatic forces on the slanted ends. For the second theoretical curve, assume that there is an additional upward force caused by the pressure between the edges of the box and the surface upon which it rests. Assume this pressure is equal to $\gamma h/2$ since it varies from $p_1 = \gamma h$ at the inside of the edge to $p_2 = 0$ at the outside (see the figure above).

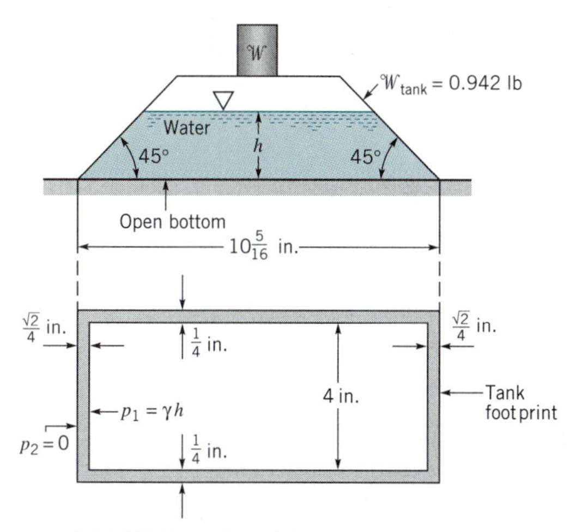

■ **FIGURE P2.103**

Compare the experimental and theoretical results and discuss some possible reasons for any differences between them.

$\mathcal{W}$ (lb)	h (in.)
0	2.06
0.221	2.44
0.441	2.67
0.662	2.94
0.881	3.16

2.104 The device shown in Fig. P2.104 is used to investigate the hydrostatic force on a plane rectangular surface. The tank is filled with water to a depth h and the force R (applied at the location indicated) needed to open the rectangular gate is measured. Values of R and h obtained experimentally are shown in the table below. Use these results to plot a graph of the force R as a function of the water depth. On the same graph plot the theoretical curve obtained by equating the moment that the applied force produces about the hinge to the moment that the resultant water force produces.

Compare the experimental and theoretical results and discuss some possible reasons for any differences between them.

R (lb)	h (in.)
9.8	22.0
8.7	19.5
7.7	17.0
6.4	15.5
6.0	14.0
4.8	12.5
4.2	11.0
2.6	8.0

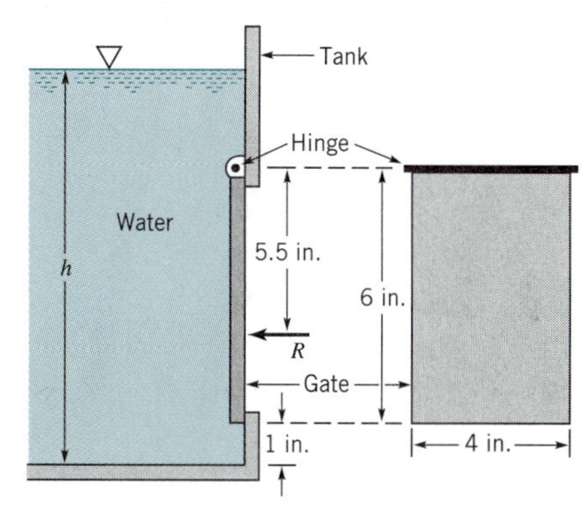

■ **FIGURE P2.104**

Flow past a blunt body: On any object placed in a moving fluid there is a stagnation point on the front of the object where the velocity is zero. This location has a relatively large pressure and divides the flow field into two portions—one flowing over the body, and one flowing under the body. (Dye in water.) (Photograph by B. R. Munson.)

3 Elementary Fluid Dynamics— The Bernoulli Equation

As was discussed in the previous chapter, there are many situations involving fluids in which the fluid can be considered as stationary. In general, however, the use of fluids involves motion of some type. In fact, a dictionary definition of the word ''fluid'' is ''free to change in form.'' In this chapter we investigate some typical fluid motions (fluid dynamics) in an elementary way.

To understand the interesting phenomena associated with fluid motion, one must consider the fundamental laws that govern the motion of fluid particles. Such considerations include the concepts of force and acceleration. We will discuss in some detail the use of Newton's second law ($\mathbf{F} = m\mathbf{a}$) as it is applied to fluid particle motion that is ''ideal'' in some sense. We will obtain the celebrated Bernoulli equation and apply it to various flows. Although this equation is one of the oldest in fluid mechanics and the assumptions involved in its derivation are numerous, it can be effectively used to predict and analyze a variety of flow situations. However, if the equation is applied without proper respect for its restrictions, serious errors can arise. Indeed, the Bernoulli equation is appropriately called ''the most used and the most abused equation in fluid mechanics.''

The Bernoulli equation may be the most used and abused equation in fluid mechanics.

A thorough understanding of the elementary approach to fluid dynamics involved in this chapter will be useful on its own. It also provides a good foundation for the material in the following chapters where some of the present restrictions are removed and ''more nearly exact'' results are presented.

3.1 Newton's Second Law

As a fluid particle moves from one location to another, it usually experiences an acceleration or deceleration. According to Newton's second law of motion, the net force acting on the fluid particle under consideration must equal its mass times its acceleration,

$$\mathbf{F} = m\mathbf{a}$$

In this chapter we consider the motion of inviscid fluids. That is, the fluid is assumed to have zero viscosity. If the viscosity is zero, then the thermal conductivity of the fluid is also zero and there can be no heat transfer (except by radiation).

In practice there are no inviscid fluids, since every fluid supports shear stresses when it is subjected to a rate of strain displacement. For many flow situations the viscous effects are relatively small compared with other effects. As a first approximation for such cases it is often possible to ignore viscous effects. For example, often the viscous forces developed in flowing water may be several orders of magnitude smaller than forces due to other influences, such as gravity or pressure differences. For other water flow situations, however, the viscous effects may be the dominant ones. Similarly, the viscous effects associated with the flow of a gas are often negligible, although in some circumstances they are very important.

Inviscid fluid flow in governed by pressure and gravity forces.

We assume that the fluid motion is governed by pressure and gravity forces only and examine Newton's second law as it applies to a fluid particle in the form:

$$\text{(Net pressure force on a particle)} + \text{(net gravity force on particle)} =$$

$$\text{(particle mass)} \times \text{(particle acceleration)}$$

The results of the interaction between the pressure, gravity, and acceleration provide numerous useful applications in fluid mechanics.

To apply Newton's second law to a fluid (or any other object), we must define an appropriate coordinate system in which to describe the motion. In general the motion will be three-dimensional and unsteady so that three space coordinates and time are needed to describe it. There are numerous coordinate systems available, including the most often used rectangular (x, y, z) and cylindrical (r, θ, z) systems. Usually the specific flow geometry dictates which system would be most appropriate.

In this chapter we will be concerned with two-dimensional motion like that confined to the x–z plane as is shown in Fig. 3.1*a*. Clearly we could choose to describe the flow in terms of the components of acceleration and forces in the x and z coordinate directions. The resulting equations are frequently referred to as a two-dimensional form of the *Euler equations* of motion in rectangular Cartesian coordinates. This approach will be discussed in Chapter 6.

As is done in the study of dynamics (Ref. 1), the motion of each fluid particle is described in terms of its velocity vector, $\mathbf{V}$, which is defined as the time rate of change of the position of the particle. The particle's velocity is a vector quantity with a magnitude (the speed, $V = |\mathbf{V}|$) and direction. As the particle moves about, it follows a particular path, the shape of which is governed by the velocity of the particle. The location of the particle along the path is a function of where the particle started at the initial time and its velocity along the path. If it is *steady flow* (i.e., nothing changes with time at a given location in the flow field), each successive particle that passes through a given point [such as point (1) in Fig. 3.1*a*] will follow the same path. For such cases the path is a fixed line in the x–z plane. Neighboring particles that pass on either side of point (1) follow their own paths, which may

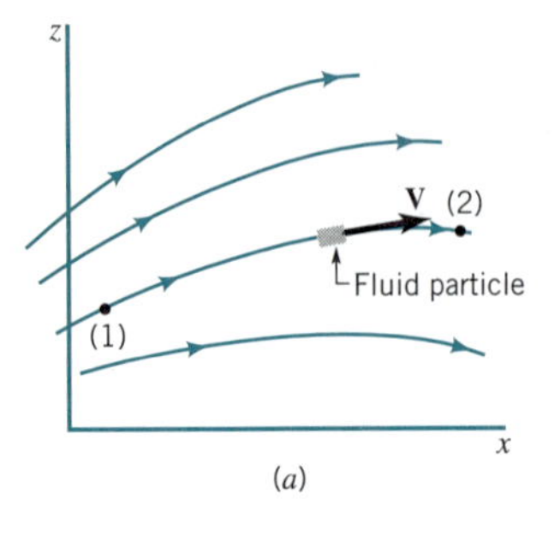

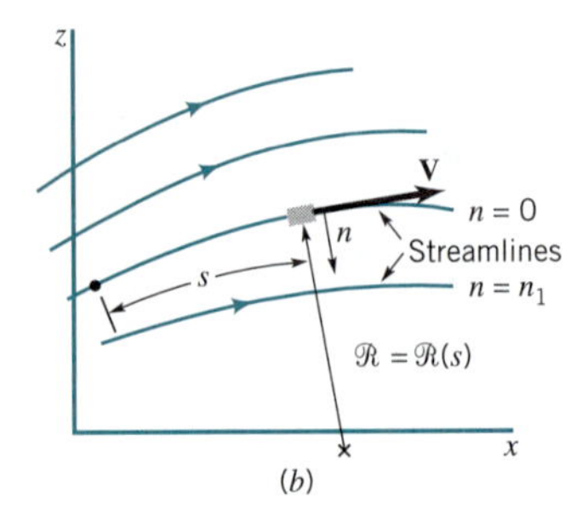

■ **FIGURE 3.1** **(*a*) Flow in the x–z plane. (*b*) Flow in terms of streamline and normal coordinates.**

be of different shape than the one passing through (1). The entire x–z plane is filled with such paths.

For steady flows each particle slides along its path, and its velocity vector is everywhere tangent to the path. The lines that are tangent to the velocity vectors throughout the flow field are called *streamlines*. For many situations it is easiest to describe the flow in terms of the "streamline" coordinates based on the streamlines as are illustrated in Fig. 3.1b The particle motion is described in terms of its distance, $s = s(t)$, along the streamline from some convenient origin and the local radius of curvature of the streamline, $\mathcal{R} = \mathcal{R}(s)$. The distance along the streamline is related to the particle's speed by $V = ds/dt$, and the radius of curvature is related to shape of the streamline. In addition to the coordinate along the streamline, s, the coordinate normal to the streamline, n, as is shown in Fig. 3.1b, will be of use.

To apply Newton's second law to a particle flowing along its streamline, we must write the particle acceleration in terms of the streamline coordinates. By definition, the acceleration is the time rate of change of the velocity of the particle, $\mathbf{a} = d\mathbf{V}/dt$. For two-dimensional flow in the x–z plane, the acceleration has two components—one along the streamline, a_s, the streamwise acceleration, and one normal to the streamline, a_n, the normal acceleration.

Fluid particles accelerate normal to and along streamlines.

The streamwise acceleration results from the fact that the speed of the particle generally varies along the streamline, $V = V(s)$. For example, in Fig. 3.1a the speed may be 100 ft/s at point (1) and 50 ft/s at point (2). Thus, by use of the chain rule of differentiation, the s component of the acceleration is given by $a_s = dV/dt = (\partial V/\partial s)(ds/dt) = (\partial V/\partial s)V$. We have used the fact that $V = ds/dt$. The normal component of acceleration, the centrifugal acceleration, is given in terms of the particle speed and the radius of curvature of its path. Thus, $a_n = V^2/\mathcal{R}$, where both V and $\mathcal{R}$ may vary along the streamline. These equations for the acceleration should be familiar from the study of particle motion in physics (Ref. 2) or dynamics (Ref. 1). A more complete derivation and discussion of these topics can be found in Chapter 4.

Thus, the components of acceleration in the s and n directions, a_s and a_n, are given by

$$a_s = V\frac{\partial V}{\partial s}, \qquad a_n = \frac{V^2}{\mathcal{R}} \tag{3.1}$$

where $\mathcal{R}$ is the local radius of curvature of the streamline, and s is the distance measured along the streamline from some arbitrary initial point. In general there is acceleration along the streamline (because the particle speed changes along its path, $\partial V/\partial s \neq 0$) and acceleration normal to the streamline (because the particle does not flow in a straight line, $\mathcal{R} \neq \infty$). To produce this acceleration there must be a net, nonzero force on the fluid particle.

To determine the forces necessary to produce a given flow (or conversely, what flow results from a given set of forces), we consider the free-body diagram of a small fluid particle as is shown in Fig. 3.2. The particle of interest is removed from its surroundings, and the reactions of the surroundings on the particle are indicated by the appropriate forces present,

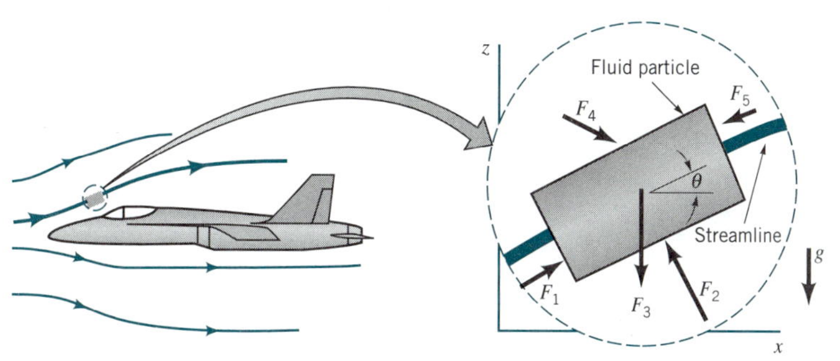

■ **FIGURE 3.2** **Isolation of a small fluid particle in a flow field.**

$\mathbf{F}_1$, $\mathbf{F}_2$, and so forth. For the present case, the important forces are assumed to be gravity and pressure. Other forces, such as viscous forces and surface tension effects, are assumed negligible. The acceleration of gravity, g, is assumed to be constant and acts vertically, in the negative z direction, at an angle θ relative to the normal to the streamline.

3.2 F = *ma* Along a Streamline

Consider the small fluid particle of size δs by δn in the plane of the figure and δy normal to the figure as shown in the free-body diagram of Fig. 3.3. Unit vectors along and normal to the streamline are denoted by $\hat{\mathbf{s}}$ and $\hat{\mathbf{n}}$, respectively. For steady flow, the component of Newton's second law along the streamline direction, s, can be written as

$$\sum \delta F_s = \delta m \, a_s = \delta m \, V \frac{\partial V}{\partial s} = \rho \, \delta \forall \, V \frac{\partial V}{\partial s} \tag{3.2}$$

where $\Sigma \, \delta F_s$ represents the sum of the s components of all the forces acting on the particle, which has mass $\delta m = \rho \, \delta \forall$, and $V \, \partial V / \partial s$ is the acceleration in the s direction. Here, $\delta \forall = \delta s \, \delta n \, \delta y$ is the particle volume. Equation 3.2 is valid for both compressible and incompressible fluids. That is, the density need not be constant throughout the flow field.

The component of weight along a streamline depends on the streamline angle.

The gravity force (weight) on the particle can be written as $\delta \mathcal{W} = \gamma \, \delta \forall$, where $\gamma = \rho g$ is the specific weight of the fluid (lb/ft^3 or N/m^3). Hence, the component of the weight force in the direction of the streamline is

$$\delta \mathcal{W}_s = -\delta \mathcal{W} \sin \theta = -\gamma \, \delta \forall \sin \theta$$

If the streamline is horizontal at the point of interest, then $\theta = 0$, and there is no component of particle weight along the streamline to contribute to its acceleration in that direction.

As is indicated in Chapter 2, the pressure is not constant throughout a stationary fluid ($\nabla p \neq 0$) because of the fluid weight. Likewise, in a flowing fluid the pressure is usually not constant. In general, for steady flow, $p = p(s, n)$. If the pressure at the center of the particle shown in Fig. 3.3 is denoted as p, then its average value on the two end faces that are perpendicular to the streamline are $p + \delta p_s$ and $p - \delta p_s$. Since the particle is "small," we

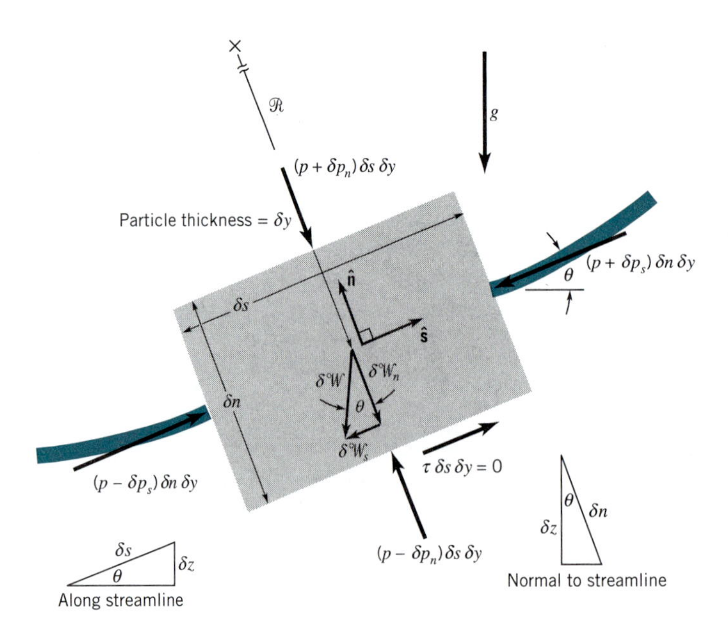

■ **FIGURE 3.3** **Free-body diagram of a fluid particle for which the important forces are those due to pressure and gravity.**

can use a one-term Taylor series expansion for the pressure field (as was done in Chapter 2 for the pressure forces in static fluids) to obtain

$$\delta p_s \approx \frac{\partial p}{\partial s} \frac{\delta s}{2}$$

Thus, if δF_{ps} is the net pressure force on the particle in the streamline direction, it follows that

The net pressure force on a particle is determined by the pressure gradient.

$$\delta F_{ps} = (p - \delta p_s)\, \delta n\, \delta y - (p + \delta p_s)\, \delta n\, \delta y = -2\, \delta p_s\, \delta n\, \delta y$$
$$= -\frac{\partial p}{\partial s}\, \delta s\, \delta n\, \delta y = -\frac{\partial p}{\partial s}\, \delta \forall$$

Note that the actual level of the pressure, p, is not important. What produces a net pressure force is the fact that the pressure is not constant throughout the fluid. The nonzero pressure gradient, $\nabla p = \partial p/\partial s\, \hat{\mathbf{s}} + \partial p/\partial n\, \hat{\mathbf{n}}$, is what provides a net pressure force on the particle. Viscous forces, represented by $\tau\, \delta s\, \delta y$, are zero, since the fluid is inviscid.

Thus, the net force acting in the streamline direction on the particle shown in Fig. 3.3 is given by

$$\sum \delta F_s = \delta \mathcal{W}_s + \delta F_{ps} = \left(-\gamma \sin \theta - \frac{\partial p}{\partial s} \right) \delta \forall \tag{3.3}$$

By combining Eqs. 3.2 and 3.3 we obtain the following equation of motion along the streamline direction:

$$-\gamma \sin \theta - \frac{\partial p}{\partial s} = \rho V \frac{\partial V}{\partial s} = \rho a_s \tag{3.4}$$

We have divided out the common particle volume factor, $\delta \forall$, that appears in both the force and the acceleration portions of the equation. This is a representation of the fact that it is the fluid density (mass per unit volume), not the mass, per se, of the fluid particle that is important.

The physical interpretation of Eq. 3.4 is that a change in fluid particle speed is accomplished by the appropriate combination of pressure gradient and particle weight along the streamline. For fluid static situations this balance between pressure and gravity forces is such that no change in particle speed is produced—the right-hand side of Eq. 3.4 is zero, and the particle remains stationary. In a flowing fluid the pressure and weight forces do not necessarily balance—the force unbalance provides the appropriate acceleration and, hence, particle motion.

Consider the inviscid, incompressible, steady flow along the horizontal streamline A–B in front of the sphere of radius a as shown in Fig. E3.1a. From a more advanced theory of flow past a sphere, the fluid velocity along this streamline is

$$V = V_0 \left(1 + \frac{a^3}{x^3} \right)$$

Determine the pressure variation along the streamline from point A far in front of the sphere ($x_A = -\infty$ and $V_A = V_0$) to point B on the sphere ($x_B = -a$ and $V_B = 0$).

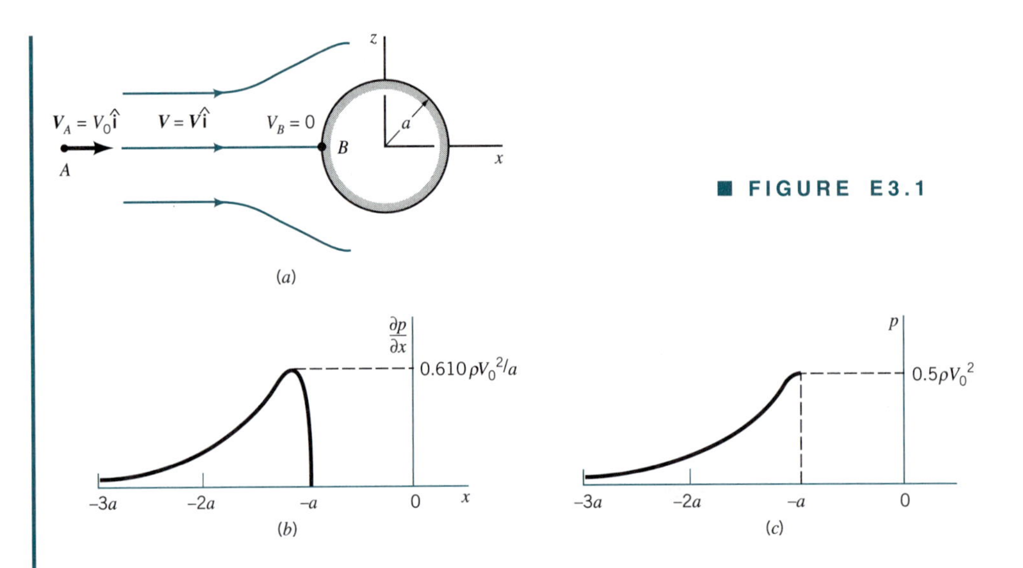

■ FIGURE E3.1

SOLUTION

Since the flow is steady and inviscid, Eq. 3.4 is valid. In addition, since the streamline is horizontal, $\sin \theta = \sin 0° = 0$ and the equation of motion along the streamline reduces to

$$\frac{\partial p}{\partial s} = -\rho V \frac{\partial V}{\partial s} \tag{1}$$

With the given velocity variation along the streamline, the acceleration term is

$$V\frac{\partial V}{\partial s} = V\frac{\partial V}{\partial x} = V_0\left(1 + \frac{a^3}{x^3}\right)\left(-\frac{3V_0 a^3}{x^4}\right) = -3V_0^2\left(1 + \frac{a^3}{x^3}\right)\frac{a^3}{x^4}$$

where we have replaced s by x since the two coordinates are identical (within an additive constant) along streamline A–B. It follows that $V\,\partial V/\partial s < 0$ along the streamline. The fluid slows down from V_0 far ahead of the sphere to zero velocity on the ''nose'' of the sphere $(x = -a)$.

Thus, according to Eq. 1, to produce the given motion the pressure gradient along the streamline is

$$\frac{\partial p}{\partial x} = \frac{3\rho a^3 V_0^2(1 + a^3/x^3)}{x^4} \tag{2}$$

This variation is indicated in Fig. E3.1b. It is seen that the pressure increases in the direction of flow $(\partial p/\partial x > 0)$ from point A to point B. The maximum pressure gradient $(0.610\ \rho V_0^2/a)$ occurs just slightly ahead of the sphere $(x = -1.205a)$. It is the pressure gradient that slows the fluid down from $V_A = V_0$ to $V_B = 0$.

The pressure distribution along the streamline can be obtained by integrating Eq. 2 from $p = 0$ (gage) at $x = -\infty$ to pressure p at location x. The result, plotted in Fig. E3.1c, is

$$p = -\rho V_0^2\left[\left(\frac{a}{x}\right)^3 + \frac{(a/x)^6}{2}\right] \tag{Ans}$$

The pressure at B, a stagnation point since $V_B = 0$, is the highest pressure along the streamline ($p_B = \rho V_0^2/2$). As shown in Chapter 9, this excess pressure on the front of the sphere (i.e., $p_B > 0$) contributes to the net drag force on the sphere. Note that the pressure gradient and pressure are directly proportional to the density of the fluid, a representation of the fact that the fluid inertia is proportional to its mass.

Equation 3.4 can be rearranged and integrated as follows. First, we note from Fig. 3.3 that along the streamline $\sin\theta = dz/ds$. Also, we can write $V\,dV/ds = \frac{1}{2}d(V^2)/ds$. Finally, along the streamline the value of n is constant ($dn = 0$) so that $dp = (\partial p/\partial s)\,ds + (\partial p/\partial n)\,dn = (\partial p/\partial s)\,ds$. Hence, along the streamline $\partial p/\partial s = dp/ds$. These ideas combined with Eq. 3.4 give the following result valid along a streamline

$$-\gamma\frac{dz}{ds} - \frac{dp}{ds} = \frac{1}{2}\rho\frac{d(V^2)}{ds}$$

This simplifies to

$$dp + \frac{1}{2}\rho d(V^2) + \gamma\,dz = 0 \qquad \text{(along a streamline)} \tag{3.5}$$

which can be integrated to give

$$\int\frac{dp}{\rho} + \frac{1}{2}V^2 + gz = C \qquad \text{(along a streamline)} \tag{3.6}$$

where C is a constant of integration to be determined by the conditions at some point on the streamline.

In general it is not possible to integrate the pressure term because the density may not be constant and, therefore, cannot be removed from under the integral sign. To carry out this integration we must know specifically how the density varies with pressure. This is not always easily determined. For example, for a perfect gas the density, pressure, and temperature are related according to $p = \rho RT$, where R is the gas constant. To know how the density varies with pressure, we must also know the temperature variation. For now we will assume that the density is constant (incompressible flow). The justification for this assumption and the consequences of compressibility will be considered further in Section 3.8.1 and more fully in Chapter 11.

With the additional assumption that the density remains constant (a very good assumption for liquids and also for gases if the speed is "not too high"), Eq. 3.6 assumes the following simple representation for steady, inviscid, incompressible flow.

The Bernoulli equation can be obtained by integrating F = ma along a streamline.

$$\boxed{p + \tfrac{1}{2}\rho V^2 + \gamma z = \text{constant along streamline}} \tag{3.7}$$

This is the celebrated *Bernoulli equation*—a very powerful tool in fluid mechanics. In 1738 Daniel Bernoulli (1700–1782) published his *Hydrodynamics* in which an equivalent of this famous equation first appeared. To use it correctly we must constantly remember the basic assumptions used in its derivation: (1) viscous effects are assumed negligible, (2) the flow is assumed to be steady, (3) the flow is assumed to be incompressible, (4) the equation is applicable along a streamline. In the derivation of Eq. 3.7, we assume that the flow takes place in a plane (the x–z plane). In general, this equation is valid for both planar and nonplanar (three-dimensional) flows, provided it is applied along the streamline.

V3.1

We will provide many examples to illustrate the correct use of the Bernoulli equation and will show how a violation of the basic assumptions used in the derivation of this equation can lead to erroneous conclusions. The constant of integration in the Bernoulli equation can be evaluated if sufficient information about the flow is known at one location along the streamline.

EXAMPLE 3.2

Consider the flow of air around a bicyclist moving through still air with velocity V_0, as is shown in Fig. E3.2. Determine the difference in the pressure between points (1) and (2).

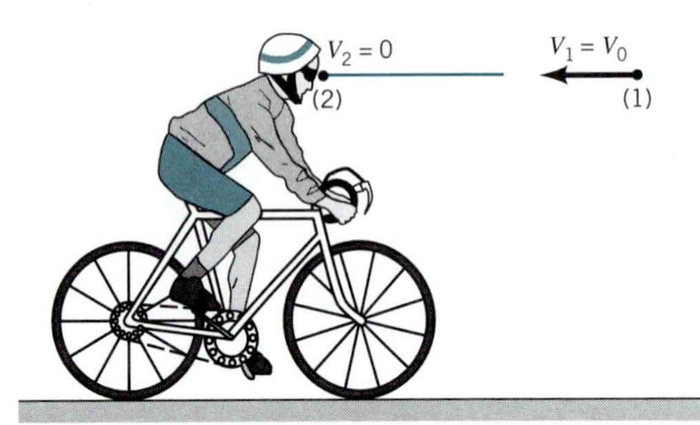

■ FIGURE E3.2

SOLUTION

In a coordinate system fixed to the bike, it appears as though the air is flowing steadily toward the bicyclist with speed V_0. If the assumptions of Bernoulli's equation are valid (steady, incompressible, inviscid flow), Eq. 3.7 can be applied as follows along the streamline that passes through (1) and (2)

$$p_1 + \tfrac{1}{2}\rho V_1^2 + \gamma z_1 = p_2 + \tfrac{1}{2}\rho V_2^2 + \gamma z_2$$

We consider (1) to be in the free stream so that $V_1 = V_0$ and (2) to be at the tip of the bicyclist's nose and assume that $z_1 = z_2$ and $V_2 = 0$ (both of which, as is discussed in Section 3.4, are reasonable assumptions). It follows that the pressure at (2) is greater than that at (1) by an amount

$$p_2 - p_1 = \tfrac{1}{2}\rho V_1^2 = \tfrac{1}{2}\rho V_0^2 \qquad \textbf{(Ans)}$$

A similar result was obtained in Example 3.1 by integrating the pressure gradient, which was known because the velocity distribution along the streamline, $V(s)$, was known. The Bernoulli equation is a general integration of $\mathbf{F} = m\mathbf{a}$. To determine $p_2 - p_1$, knowledge of the detailed velocity distribution is not needed—only the ''boundary conditions'' at (1) and (2) are required. Of course, knowledge of the value of V along the streamline is needed to determine the pressure at points between (1) and (2). Note that if we measure $p_2 - p_1$ we can determine the speed, V_0. As discussed in Section 3.5, this is the principle upon which many velocity measuring devices are based.

If the bicyclist were accelerating or decelerating, the flow would be unsteady (i.e., $V_0 \neq$ constant) and the above analysis would be incorrect since Eq. 3.7 is restricted to steady flow.

The difference in fluid velocity between two points in a flow field, V_1 and V_2, can often be controlled by appropriate geometric constraints of the fluid. For example, a garden hose nozzle is designed to give a much higher velocity at the exit of the nozzle than at its entrance where it is attached to the hose. As is shown by the Bernoulli equation, the pressure within the hose must be larger than that at the exit (for constant elevation, an increase in velocity requires a decrease in pressure if Eq. 3.7 is valid). It is this pressure drop that accelerates the water through the nozzle. Similarly, an airfoil is designed so that the fluid velocity over its upper surface is greater (on the average) than that along its lower surface. From the Bernoulli equation, therefore, the average pressure on the lower surface is greater than on the upper surface. A net upward force, the lift, results.

3.3 F = *m*a Normal to a Streamline

In this section we will consider application of Newton's second law in a direction normal to the streamline. In many flows the streamlines are relatively straight, the flow is essentially one-dimensional, and variations in parameters across streamlines (in the normal direction) can often be neglected when compared to the variations along the streamline. However, in numerous other situations valuable information can be obtained from considering $\mathbf{F} = m\mathbf{a}$ normal to the streamlines. For example, the devastating low-pressure region at the center of a tornado can be explained by applying Newton's second law across the nearly circular streamlines of the tornado.

We again consider the force balance on the fluid particle shown in Fig. 3.3. This time, however, we consider components in the normal direction, $\hat{\mathbf{n}}$, and write Newton's second law in this direction as

$$\sum \delta F_n = \frac{\delta m\, V^2}{\mathcal{R}} = \frac{\rho\, \delta \forall\, V^2}{\mathcal{R}} \tag{3.8}$$

where $\Sigma\, \delta F_n$ represents the sum of n components of all the forces acting on the particle. We assume the flow is steady with a normal acceleration $a_n = V^2/\mathcal{R}$, where $\mathcal{R}$ is the local radius of curvature of the streamlines. This acceleration is produced by the change in direction of the particle's velocity as it moves along a curved path.

We again assume that the only forces of importance are pressure and gravity. The component of the weight (gravity force) in the normal direction is

$$\delta \mathcal{W}_n = -\delta \mathcal{W} \cos\theta = -\gamma\, \delta \forall \cos\theta$$

To apply F = ma normal to streamlines, the normal components of force are needed.

If the streamline is vertical at the point of interest, $\theta = 90°$, and there is no component of the particle weight normal to the direction of flow to contribute to its acceleration in that direction.

If the pressure at the center of the particle is p, then its values on the top and bottom of the particle are $p + \delta p_n$ and $p - \delta p_n$, where $\delta p_n = (\partial p/\partial n)(\delta n/2)$. Thus, if δF_{pn} is the net pressure force on the particle in the normal direction, it follows that

$$\begin{aligned} \delta F_{pn} &= (p - \delta p_n)\, \delta s\, \delta y - (p + \delta p_n)\, \delta s\, \delta y = -2\, \delta p_n\, \delta s\, \delta y \\ &= -\frac{\partial p}{\partial n}\, \delta s\, \delta n\, \delta y = -\frac{\partial p}{\partial n}\, \delta \forall \end{aligned}$$

Hence, the net force acting in the normal direction on the particle shown in Fig 3.3 is given by

$$\sum \delta F_n = \delta \mathcal{W}_n + \delta F_{pn} = \left(-\gamma \cos\theta - \frac{\partial p}{\partial n}\right) \delta \forall \tag{3.9}$$

By combining Eqs. 3.8 and 3.9 and using the fact that along a line normal to the streamline

$\cos \theta = dz/dn$ (see Fig. 3.3), we obtain the following equation of motion along the normal direction

$$-\gamma \frac{dz}{dn} - \frac{\partial p}{\partial n} = \frac{\rho V^2}{\mathcal{R}} \tag{3.10}$$

Weight and/or pressure can produce curved streamlines.

The physical interpretation of Eq. 3.10 is that a change in the direction of flow of a fluid particle (i.e., a curved path, $\mathcal{R} < \infty$) is accomplished by the appropriate combination of pressure gradient and particle weight normal to the streamline. A larger speed or density or a smaller radius of curvature of the motion requires a larger force unbalance to produce the motion. For example, if gravity is neglected (as is commonly done for gas flows) or if the flow is in a horizontal ($dz/dn = 0$) plane, Eq. 3.10 becomes

$$\frac{\partial p}{\partial n} = -\frac{\rho V^2}{\mathcal{R}}$$

This indicates that the pressure increases with distance away from the center of curvature ($\partial p/\partial n$ is negative since $\rho V^2/\mathcal{R}$ is positive—the positive n direction points toward the ''inside'' of the curved streamline). Thus, the pressure outside a tornado (typical atmospheric pressure) is larger than it is near the center of the tornado (where an often dangerously low partial vacuum may occur). This pressure difference is needed to balance the centrifugal acceleration associated with the curved streamlines of the fluid motion.

EXAMPLE 3.3

Shown in Figs. E3.3*a*,*b* are two flow fields with circular streamlines. The velocity distributions are

$$V(r) = C_1 r \qquad \text{for case } (a)$$

and

$$V(r) = \frac{C_2}{r} \qquad \text{for case } (b)$$

where C_1 and C_2 are constant. Determine the pressure distributions, $p = p(r)$, for each, given that $p = p_0$ at $r = r_0$.

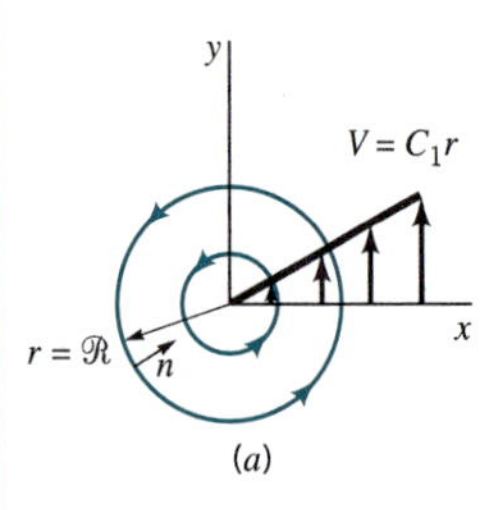

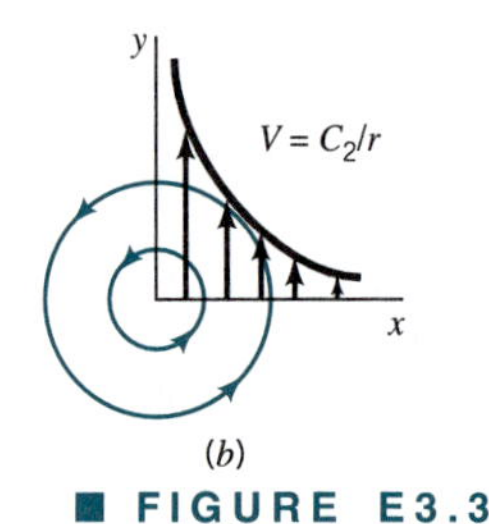

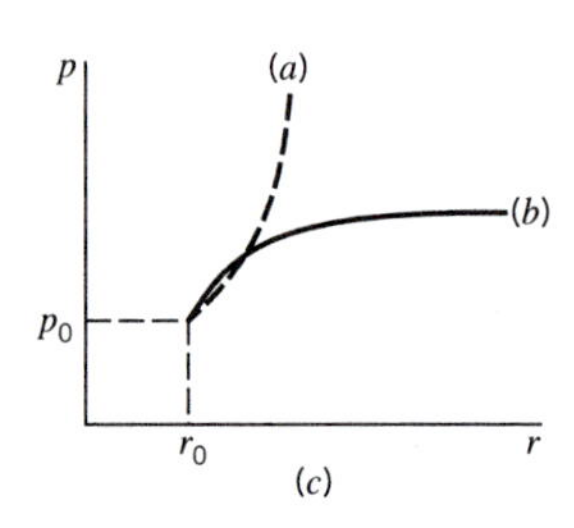

■ FIGURE E3.3

SOLUTION

We assume the flows are steady, inviscid, and incompressible with streamlines in the horizontal plane ($dz/dn = 0$). Since the streamlines are circles, the coordinate n points in a direction opposite to that of the radial coordinate, $\partial/\partial n = -\partial/\partial r$, and the radius of curvature is given by $\mathcal{R} = r$. Hence, Eq. 3.10 becomes

$$\frac{\partial p}{\partial r} = \frac{\rho V^2}{r}$$

For case (a) this gives

$$\frac{\partial p}{\partial r} = \rho C_1^2 r$$

while for case (b) it gives

$$\frac{\partial p}{\partial r} = \frac{\rho C_2^2}{r^3}$$

For either case the pressure increases as r increases since $\partial p/\partial r > 0$. Integration of these equations with respect to r, starting with a known pressure $p = p_0$ at $r = r_0$, gives

$$p = \frac{1}{2} \rho C_1^2 (r^2 - r_0^2) + p_0 \qquad \textbf{(Ans)}$$

for case (a) and

$$p = \frac{1}{2} \rho C_2^2 \left(\frac{1}{r_0^2} - \frac{1}{r^2} \right) + p_0 \qquad \textbf{(Ans)}$$

for case (b). These pressure distributions are sketched in Fig. E3.3c. The pressure distributions needed to balance the centrifugal accelerations in cases (a) and (b) are not the same because the velocity distributions are different. In fact for case (a) the pressure increases without bound as $r \rightarrow \infty$, while for case (b) the pressure approaches a finite value as $r \rightarrow \infty$. The streamline patterns are the same for each case, however.

V3.2

Physically, case (a) represents rigid body rotation (as obtained in a can of water on a turntable after it has been ''spun up'') and case (b) represents a free vortex (an approximation of a tornado or the swirl of water in a drain, the ''bathtub vortex'').

The sum of pressure, elevation, and velocity effects is constant across streamlines.

If we multiply Eq. 3.10 by dn, use the fact that $\partial p/\partial n = dp/dn$ if s is constant, and integrate across the streamline (in the n direction) we obtain

$$\int \frac{dp}{\rho} + \int \frac{V^2}{\mathcal{R}}\, dn + gz = \text{constant across the streamline} \qquad \textbf{(3.11)}$$

To complete the indicated integrations we must know how the density varies with pressure and how the fluid speed and radius of curvature vary with n. For incompressible flow the density is constant and the integration involving the pressure term gives simply p/ρ. We are still left, however, with the integration of the second term in Eq. 3.11. Without knowing the n dependence in $V = V(s, n)$ and $\mathcal{R} = \mathcal{R}(s, n)$ this integration cannot be completed.

Thus, the final form of Newton's second law applied across the streamlines for steady, inviscid, incompressible flow is

$$\boxed{p + \rho \int \frac{V^2}{\mathcal{R}}\, dn + \gamma z = \text{constant across the streamline}} \qquad \textbf{(3.12)}$$

As with the Bernoulli equation, we must be careful that the assumptions involved in the derivation of this equation are not violated when it is used.

3.4 Physical Interpretation

In the previous two sections we developed the basic equations governing fluid motion under a fairly stringent set of restrictions. In spite of the numerous assumptions imposed on these flows, a variety of flows can be readily analyzed with them. A physical interpretation of the equations will be of help in understanding the processes involved. To this end, we rewrite Eqs. 3.7 and 3.12 here and interpret them physically. Application of $\mathbf{F} = m\mathbf{a}$ along and normal to the streamline results in

$$p + \tfrac{1}{2}\rho V^2 + \gamma z = \text{constant along the streamline} \tag{3.13}$$

and

$$p + \rho \int \frac{V^2}{\mathcal{R}}\, dn + \gamma z = \text{constant across the streamline} \tag{3.14}$$

The following basic assumptions were made to obtain these equations: The flow is steady and the fluid is inviscid and incompressible. In practice none of these assumptions is exactly true.

A violation of one or more of the above assumptions is a common cause for obtaining an incorrect match between the "real world" and solutions obtained by use of the Bernoulli equation. Fortunately, many "real world" situations are adequately modeled by the use of Eqs. 3.13 and 3.14 because the flow is nearly steady and incompressible and the fluid behaves as if it were nearly inviscid.

The Bernoulli equation was obtained by integration of the equation of motion along the "natural" coordinate direction of the streamline. To produce an acceleration, there must be an unbalance of the resultant forces, of which only pressure and gravity were considered to be important. Thus, there are three processes involved in the flow—mass times acceleration (the $\rho V^2/2$ term), pressure (the p term), and weight (the γz term).

Integration of the equation of motion to give Eq. 3.13 actually corresponds to the work-energy principle often used in the study of dynamics [see any standard dynamics text (Ref. 1)]. This principle results from a general integration of the equations of motion for an object in a way very similar to that done for the fluid particle in Section 3.2. With certain assumptions, a statement of the work-energy principle may be written as follows:

> The work done on a particle by all forces acting on the particle is equal to the change of the kinetic energy of the particle.

The Bernoulli equation is a mathematical statement of this principle.

As the fluid particle moves, both gravity and pressure forces do work on the particle. Recall that the work done by a force is equal to the product of the distance the particle travels times the component of force in the direction of travel (i.e., work $= \mathbf{F} \cdot \mathbf{d}$). The terms γz and p in Eq. 3.13 are related to the work done by the weight and pressure forces, respectively. The remaining term, $\rho V^2/2$, is obviously related to the kinetic energy of the particle. In fact, an alternate method of deriving the Bernoulli equation is to use the first and second laws of thermodynamics (the energy and entropy equations), rather than Newton's second law. With the appropriate restrictions, the general energy equation reduces to the Bernoulli equation. This approach is discussed in Section 5.4.

An alternate but equivalent form of the Bernoulli equation is obtained by dividing each term of Eq. 3.7 by the specific weight, γ, to obtain

The Bernoulli equation can be written in terms of heights called heads.

$$\frac{p}{\gamma} + \frac{V^2}{2g} + z = \text{constant on a streamline}$$

Each of the terms in this equation has the units of energy per weight ($LF/F = L$) or length (feet, meters) and represents a certain type of head.

The elevation term, z, is related to the potential energy of the particle and is called the *elevation head*. The pressure term, p/γ, is called the *pressure head* and represents the height of a column of the fluid that is needed to produce the pressure p. The velocity term, $V^2/2g$, is the *velocity head* and represents the vertical distance needed for the fluid to fall freely (neglecting friction) if it is to reach velocity V from rest. The Bernoulli equation states that the sum of the pressure head, the velocity head, and the elevation head is constant along a streamline.

EXAMPLE 3.4

Consider the flow of water from the syringe shown in Fig. E3.4. A force applied to the plunger will produce a pressure greater than atmospheric at point (1) within the syringe. The water flows from the needle, point (2), with relatively high velocity and coasts up to point (3) at the top of its trajectory. Discuss the energy of the fluid at points (1), (2), and (3) by using the Bernoulli equation.

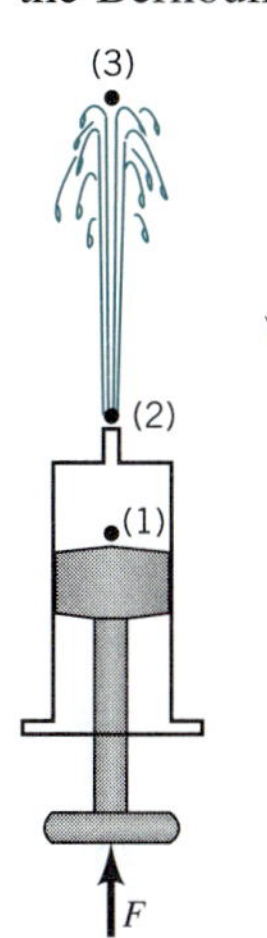

■ FIGURE E3.4

SOLUTION

If the assumptions (steady, inviscid, incompressible flow) of the Bernoulli equation are approximately valid, it then follows that the flow can be explained in terms of the partition of the total energy of the water. According to Eq. 3.13 the sum of the three types of energy (kinetic, potential, and pressure) or heads (velocity, elevation, and pressure) must remain constant. The following table indicates the relative magnitude of each of these energies at the three points shown in the figure.

	Energy Type		
Point	**Kinetic** $\rho V^2/2$	**Potential** γz	**Pressure** p
1	Small	Zero	Large
2	Large	Small	Zero
3	Zero	Large	Zero

The motion results in (or is due to) a change in the magnitude of each type of energy as the fluid flows from one location to another. An alternate way to consider this flow is as follows. The pressure gradient between (1) and (2) produces an acceleration to eject the water from the needle. Gravity acting on the particle between (2) and (3) produces a deceleration to cause the water to come to a momentary stop at the top of its flight.

If friction (viscous) effects were important, there would be an energy loss between (1) and (3) and for the given p_1 the water would not be able to reach the height indicated in the figure. Such friction may arise in the needle (see Chapter 8 on pipe flow) or between the water stream and the surrounding air (see Chapter 9 on external flow).

A net force is required to accelerate any mass. For steady flow the acceleration can be interpreted as arising from two distinct occurrences—a change in speed along the streamline and a change in direction if the streamline is not straight. Integration of the equation of motion along the streamline accounts for the change in speed (kinetic energy change) and results in the Bernoulli equation. Integration of the equation of motion normal to the streamline accounts for the centrifugal acceleration ($V^2/\mathfrak{R}$) and results in Eq. 3.14.

The pressure variation across straight streamlines is hydrostatic.

When a fluid particle travels along a curved path, a net force directed toward the center of curvature is required. Under the assumptions valid for Eq. 3.14, this force may be either gravity or pressure, or a combination of both. In many instances the streamlines are nearly straight ($\mathcal{R} = \infty$) so that centrifugal effects are negligible and the pressure variation across the streamlines is merely hydrostatic (because of gravity alone), even though the fluid is in motion.

EXAMPLE 3.5

Consider the inviscid, incompressible, steady flow shown in Fig. E3.5. From section A to B the streamlines are straight, while from C to D they follow circular paths. Describe the pressure variation between points (1) and (2) and points (3) and (4).

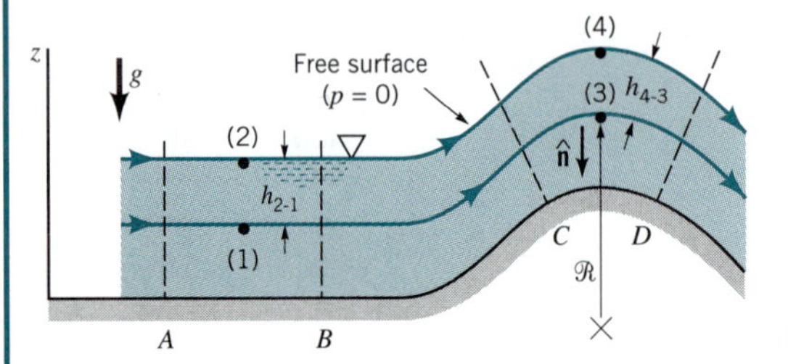

■ FIGURE E3.5

SOLUTION

With the above assumptions and the fact that $\mathcal{R} = \infty$ for the portion from A to B, Eq. 3.14 becomes

$$p + \gamma z = \text{constant}$$

The constant can be determined by evaluating the known variables at the two locations using $p_2 = 0$ (gage), $z_1 = 0$, and $z_2 = h_{2-1}$ to give

$$p_1 = p_2 + \gamma(z_2 - z_1) = p_2 + \gamma h_{2-1} \qquad \textbf{(Ans)}$$

Note that since the radius of curvature of the streamline is infinite, the pressure variation in the vertical direction is the same as if the fluid were stationary.

However, if we apply Eq. 3.14 between points (3) and (4) we obtain (using $dn = -dz$)

$$p_4 + \rho \int_{z_3}^{z_4} \frac{V^2}{\mathcal{R}} (-dz) + \gamma z_4 = p_3 + \gamma z_3$$

With $p_4 = 0$ and $z_4 - z_3 = h_{4-3}$ this becomes

$$p_3 = \gamma h_{4-3} - \rho \int_{z_3}^{z_4} \frac{V^2}{\mathcal{R}} dz \qquad \textbf{(Ans)}$$

To evaluate the integral we must know the variation of V and $\mathcal{R}$ with z. Even without this detailed information we note that the integral has a positive value. Thus, the pressure at (3) is less than the hydrostatic value, γh_{4-3}, by an amount equal to $\rho \int_{z_3}^{z_4} (V^2/\mathcal{R})\, dz$. This lower pressure, caused by the curved streamline, is necessary to accelerate the fluid around the curved path.

Note that we did not apply the Bernoulli equation (Eq. 3.13) across the streamlines from (1) to (2) or (3) to (4). Rather we used Eq. 3.14. As is discussed in Section 3.8, application of the Bernoulli equation across streamlines (rather than along them) may lead to serious errors.

3.5 Static, Stagnation, Dynamic, and Total Pressure

Each term in the Bernoulli equation can be interpreted as a form of pressure.

A useful concept associated with the Bernoulli equation deals with the stagnation and dynamic pressures. These pressures arise from the conversion of kinetic energy in a flowing fluid into a "pressure rise" as the fluid is brought to rest (as in Example 3.2). In this section we explore various results of this process. Each term of the Bernoulli equation, Eq. 3.13, has the dimensions of force per unit area—psi, lb/ft^2, N/m^2. The first term, p, is the actual thermodynamic pressure of the fluid as it flows. To measure its value, one could move along with the fluid, thus being "static" relative to the moving fluid. Hence, it is normally termed the *static pressure*. Another way to measure the static pressure would be to drill a hole in a flat surface and fasten a piezometer tube as indicated by the location of point (3) in Fig. 3.4. As we saw in Example 3.5, the pressure in the flowing fluid at (1) is $p_1 = \gamma h_{3-1} + p_3$, the same as if the fluid were static. From the manometer considerations of Chapter 2, we know that $p_3 = \gamma h_{4-3}$. Thus, since $h_{3-1} + h_{4-3} = h$ it follows that $p_1 = \gamma h$.

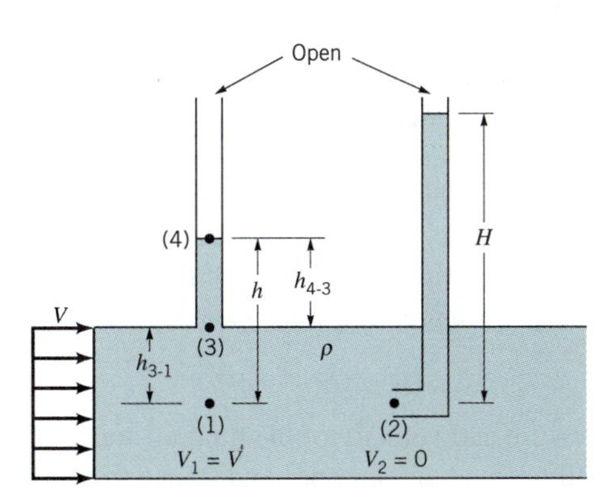

■ **FIGURE 3.4 Measurement of static and stagnation pressures.**

The third term in Eq. 3.13, γz, is termed the *hydrostatic pressure*, in obvious regard to the hydrostatic pressure variation discussed in Chapter 2. It is not actually a pressure but does represent the change in pressure possible due to potential energy variations of the fluid as a result of elevation changes.

The second term in the Bernoulli equation, $\rho V^2/2$, is termed the *dynamic pressure*. Its interpretation can be seen in Fig. 3.4 by considering the pressure at the end of a small tube inserted into the flow and pointing upstream. After the initial transient motion has died out, the liquid will fill the tube to a height of H as shown. The fluid in the tube, including that at its tip, (2), will be stationary. That is, $V_2 = 0$, or point (2) is a *stagnation point*.

If we apply the Bernoulli equation between points (1) and (2), using $V_2 = 0$ and assuming that $z_1 = z_2$, we find that

$$p_2 = p_1 + \tfrac{1}{2}\rho V_1^2$$

Hence, the pressure at the stagnation point is greater than the static pressure, p_1, by an amount $\rho V_1^2/2$, the dynamic pressure.

It can be shown that there is a stagnation point on any stationary body that is placed into a flowing fluid. Some of the fluid flows "over" and some "under" the object. The dividing line (or surface for two-dimensional flows) is termed the *stagnation streamline* and terminates at the stagnation point on the body (See the photograph on page 102.) For symmetrical objects (such as a sphere) the stagnation point is clearly at the tip or front of the object as shown in Fig. 3.5*a*. For nonsymmetrical objects such as the airplane shown in Fig. 3.5*b*, the location of the stagnation point is not always obvious.

V3.3

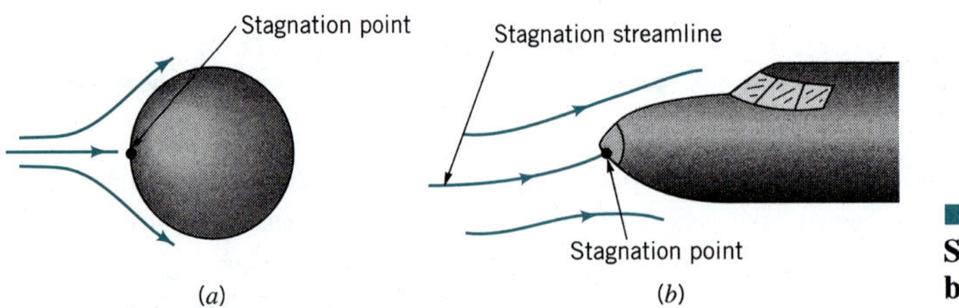

■ **FIGURE 3.5** **Stagnation points on bodies in flowing fluids.**

If elevation effects are neglected, the *stagnation pressure*, $p + \rho V^2/2$, is the largest pressure obtainable along a given streamline. It represents the conversion of all of the kinetic energy into a pressure rise. The sum of the static pressure, hydrostatic pressure, and dynamic pressure is termed the *total pressure*, p_T. The Bernoulli equation is a statement that the total pressure remains constant along a streamline. That is,

$$p + \tfrac{1}{2}\rho V^2 + \gamma z = p_T = \text{constant along a streamline} \quad \textbf{(3.15)}$$

Again, we must be careful that the assumptions used in the derivation of this equation are appropriate for the flow being considered.

Knowledge of the values of the static and stagnation pressures in a fluid implies that the fluid speed can be calculated. This is the principle on which the *Pitot-static tube* is based. As shown in Fig. 3.6, two concentric tubes are attached to two pressure gages (or a differential gage) so that the values of p_3 and p_4 (or the difference $p_3 - p_4$) can be determined. The center tube measures the stagnation pressure at its open tip. If elevation changes are negligible,

$$p_3 = p + \tfrac{1}{2}\rho V^2$$

where p and V are the pressure and velocity of the fluid upstream of point (2). The outer tube is made with several small holes at an appropriate distance from the tip so that they measure the static pressure. If the elevation difference between (1) and (4) is negligible, then

V3.4

Pitot-static tubes measure fluid velocity by converting velocity into pressure.

$$p_4 = p_1 = p$$

By combining these two equations we see that

$$p_3 - p_4 = \tfrac{1}{2}\rho V^2$$

which can be rearranged to give

$$V = \sqrt{2(p_3 - p_4)/\rho} \quad \textbf{(3.16)}$$

The actual shape and size of Pitot-static tubes vary considerably. Some of the more common types are shown in Fig. 3.7.

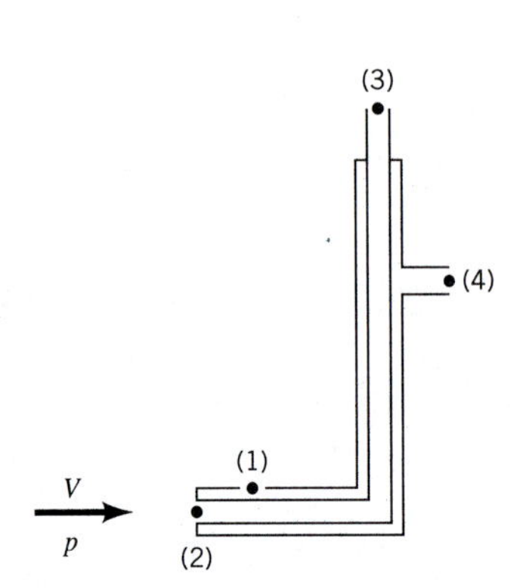

■ **FIGURE 3.6** **The Pitot-static tube.**

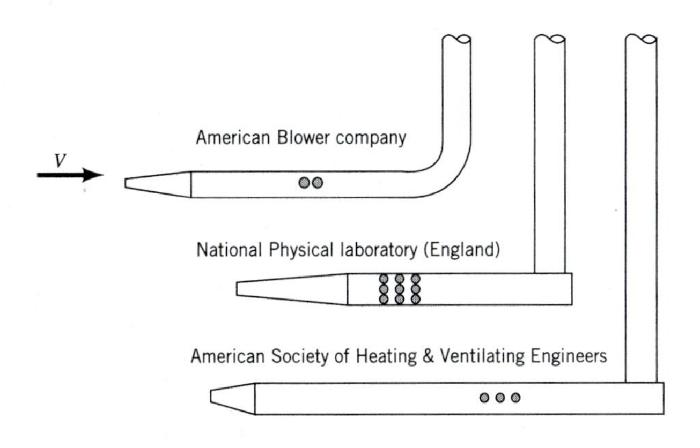

■ **FIGURE 3.7 Typical Pitot-static tube designs.**

EXAMPLE 3.6

An airplane flies 100 mi/hr at an elevation of 10,000 ft in a standard atmosphere as shown in Fig. E3.6. Determine the pressure at point (1) far ahead of the airplane, the pressure at the stagnation point on the nose of the airplane, point (2), and the pressure difference indicated by a Pitot-static probe attached to the fuselage.

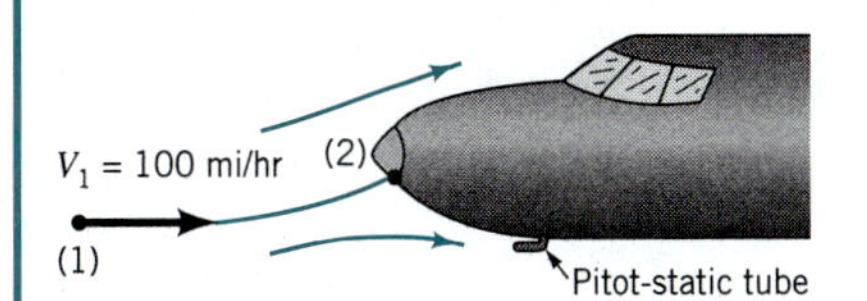

■ **FIGURE E3.6**

SOLUTION

From Table C.1 we find that the static pressure at the altitude given is

$$p_1 = 1456 \text{ lb/ft}^2 \text{ (abs)} = 10.11 \text{ psia} \qquad \textbf{(Ans)}$$

Also, the density is $\rho = 0.001756 \text{ slug/ft}^3$.

If the flow is steady, inviscid, and incompressible and elevation changes are neglected, Eq. 3.13 becomes

$$p_2 = p_1 + \frac{\rho V_1^2}{2}$$

With $V_1 = 100$ mi/hr $= 146.7$ ft/s and $V_2 = 0$ (since the coordinate system is fixed to the airplane) we obtain

$$p_2 = 1456 \text{ lb/ft}^2 + (0.001756 \text{ slugs/ft}^3)(146.7^2 \text{ ft}^2\text{/s}^2)/2$$

$$= (1456 + 18.9) \text{ lb/ft}^2 \text{ (abs)}$$

Hence, in terms of gage pressure

$$p_2 = 18.9 \text{ lb/ft}^2 = 0.1313 \text{ psi} \qquad \textbf{(Ans)}$$

Thus, the pressure difference indicated by the Pitot-static tube is

$$p_2 - p_1 = \frac{\rho V_1^2}{2} = 0.1313 \text{ psi} \qquad \textbf{(Ans)}$$

Note that it is very easy to obtain incorrect results by using improper units. Do not add lb/in.^{2} and lb/ft^2. Recall that (slug/ft^3)(ft^2/s^2) = (slug·ft/s^2)/(ft^2) = lb/ft^2.

It was assumed that the flow is incompressible—the density remains constant from (1) to (2). However, since $\rho = p/RT$, a change in pressure (or temperature) will cause a change in density. For this relatively low speed, the ratio of the absolute pressures is nearly unity [i.e., $p_1/p_2 = (10.11 \text{ psia})/(10.11 + 0.1313 \text{ psia}) = 0.987$], so that the density change is negligible. However, at high speed it is necessary to use compressible flow concepts to obtain accurate results. (See Section 3.8.1 and Chapter 11.)

The Pitot-static tube provides a simple, relatively inexpensive way to measure fluid speed. Its use depends on the ability to measure the static and stagnation pressures. Care is needed to obtain these values accurately. For example, an accurate measurement of static pressure requires that none of the fluid's kinetic energy be converted into a pressure rise at the point of measurement. This requires a smooth hole with no burrs or imperfections. As indicated in Fig. 3.8, such imperfections can cause the measured pressure to be greater or less than the actual static pressure.

Accurate measurement of static pressure requires great care.

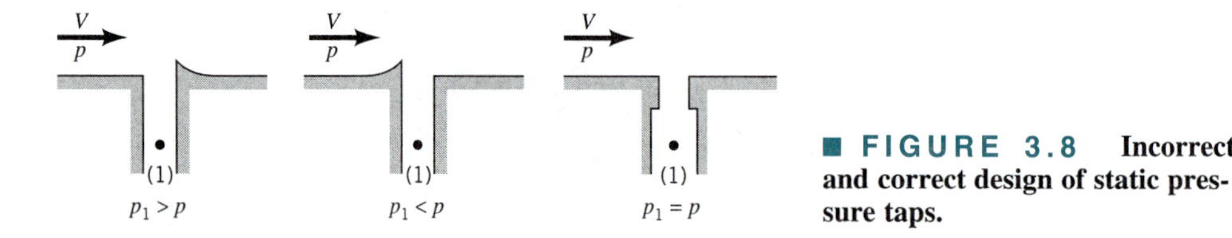

■ FIGURE 3.8 **Incorrect and correct design of static pressure taps.**

Also, the pressure along the surface of an object varies from the stagnation pressure at its stagnation point to values that may be less than the free stream static pressure. A typical pressure variation for a Pitot-static tube is indicated in Fig. 3.9. Clearly it is important that the pressure taps be properly located to ensure that the pressure measured is actually the static pressure.

In practice it is often difficult to align the Pitot-static tube directly into the flow direction. Any misalignment will produce a nonsymmetrical flow field that may introduce errors. Typically, yaw angles up to 12 to 20° (depending on the particular probe design) give results that are less than 1% in error from the perfectly aligned results. Generally it is more difficult to measure static pressure than stagnation pressure.

One method of determining the flow direction and its speed (thus the velocity) is to use a directional-finding Pitot tube as is illustrated in Fig. 3.10. Three pressure taps are drilled into a small circular cylinder, fitted with small tubes, and connected to three pressure transducers. The cylinder is rotated until the pressures in the two side holes are equal, thus indicating that the center hole points directly upstream. The center tap then measures the stag-

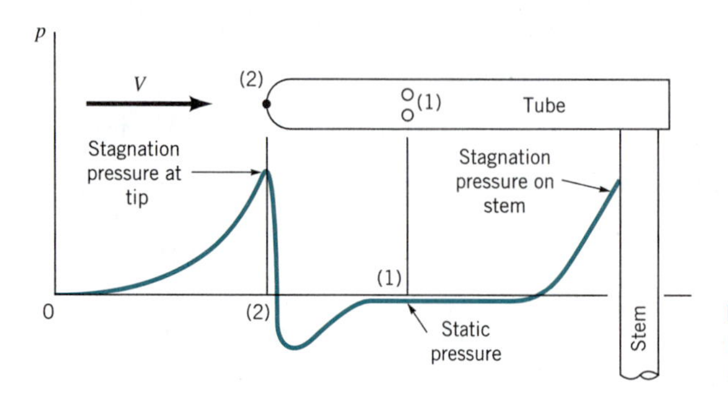

■ FIGURE 3.9 **Typical pressure distribution along a Pitot-static tube.**

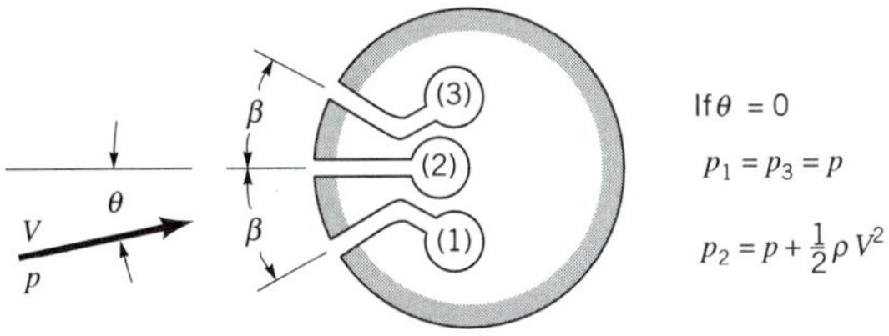

(3)
β
(2)
θ
V
p
β
(1)

If $\theta = 0$

$p_1 = p_3 = p$

$p_2 = p + \frac{1}{2}\rho V^2$

■ **FIGURE 3.10** **Cross section of a directional-finding Pitot-static tube.**

Many velocity measuring devices use Pitot-static tube principles.

nation pressure. The two side holes are located at a specific angle ($\beta = 29.5°$) so that they measure the static pressure. The speed is then obtained from $V = [2(p_2 - p_1)/\rho]^{1/2}$.

The above discussion is valid for incompressible flows. At high speeds, compressibility becomes important (the density is not constant) and other phenomena occur. Some of these ideas are discussed in Section 3.8, while others (such as shockwaves for supersonic Pitot-tube applications) are discussed in Chapter 11.

The concepts of static, dynamic, stagnation, and total pressure are useful in a variety of flow problems. These ideas will be used more fully in the remainder of the book.

3.6 Examples of Use of the Bernoulli Equation

In this section we illustrate various additional applications of the Bernoulli equation. Between any two points, (1) and (2), on a streamline in steady, inviscid, incompressible flow the Bernoulli equation can be applied in the form

$$p_1 + \tfrac{1}{2}\rho V_1^2 + \gamma z_1 = p_2 + \tfrac{1}{2}\rho V_2^2 + \gamma z_2 \tag{3.17}$$

Obviously if five of the six variables are known, the remaining one can be determined. In many instances it is necessary to introduce other equations, such as the conservation of mass. Such considerations will be discussed briefly in this section and in more detail in Chapter 5.

3.6.1 Free Jets

One of the oldest equations in fluid mechanics deals with the flow of a liquid from a large reservoir as is shown in Fig. 3.11. A jet of liquid of diameter d flows from the nozzle with velocity V as shown. (A nozzle is a device shaped to accelerate a fluid.) Application of Eq. 3.17 between points (1) and (2) on the streamline shown gives

$$\gamma h = \tfrac{1}{2}\rho V^2$$

V3.5

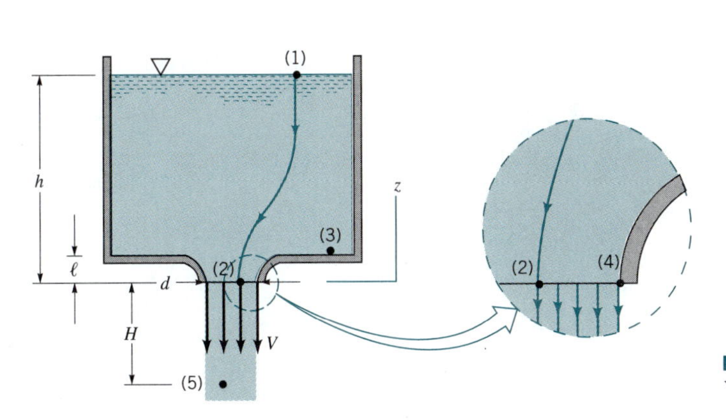

■ **FIGURE 3.11** **Vertical flow from a tank.**

We have used the facts that $z_1 = h$, $z_2 = 0$, the reservoir is large ($V_1 \cong 0$), open to the atmosphere ($p_1 = 0$ gage), and the fluid leaves as a "free jet" ($p_2 = 0$). Thus, we obtain

$$V = \sqrt{2\frac{\gamma h}{\rho}} = \sqrt{2gh} \tag{3.18}$$

which is the modern version of a result obtained in 1643 by Torricelli (1608–1647), an Italian physicist.

The exit pressure for an incompressible fluid jet is equal to the surrounding pressure.

The fact that the exit pressure equals the surrounding pressure ($p_2 = 0$) can be seen by applying $\mathbf{F} = m\mathbf{a}$, as given by Eq. 3.14, across the streamlines between (2) and (4). If the streamlines at the tip of the nozzle are straight ($\mathcal{R} = \infty$), it follows that $p_2 = p_4$. Since (4) is on the surface of the jet, in contact with the atmosphere, we have $p_4 = 0$. Thus, $p_2 = 0$ also. Since (2) is an arbitrary point in the exit plane of the nozzle, it follows that the pressure is atmospheric across this plane. Physically, since there is no component of the weight force or acceleration in the normal (horizontal) direction, the pressure is constant in that direction.

Once outside the nozzle, the stream continues to fall as a free jet with zero pressure throughout ($p_5 = 0$) and as seen by applying Eq. 3.17 between points (1) and (5), the speed increases according to

$$V = \sqrt{2g(h + H)}$$

where H is the distance the fluid has fallen outside the nozzle.

Equation 3.18 could also be obtained by writing the Bernoulli equation between points (3) and (4) using the fact that $z_4 = 0$, $z_3 = \ell$. Also, $V_3 = 0$ since it is far from the nozzle, and from hydrostatics, $p_3 = \gamma(h - \ell)$.

Recall from physics or dynamics that any object dropped from rest through a distance h in a vacuum will obtain the speed $V = \sqrt{2gh}$, the same as the liquid leaving the nozzle. This is consistent with the fact that all of the particle's potential energy is converted to kinetic energy, provided viscous (friction) effects are negligible. In terms of heads, the elevation head at point (1) is converted into the velocity head at point (2). Recall that for the case shown in Fig. 3.11 the pressure is the same (atmospheric) at points (1) and (2).

For the horizontal nozzle of Fig. 3.12, the velocity of the fluid at the centerline, V_2, will be slightly greater than that at the top, V_1, and slightly less than that at the bottom, V_3, due to the differences in elevation. In general $d \ll h$ and we can safely use the centerline velocity as a reasonable "average velocity."

If the exit is not a smooth, well-contoured nozzle, but rather a flat plate as shown in Fig. 3.13, the diameter of the jet, d_j, will be less than the diameter of the hole, d_h. This phenomenon, called a *vena contracta* effect, is a result of the inability of the fluid to turn the sharp 90° corner indicated by the dotted lines in the figure.

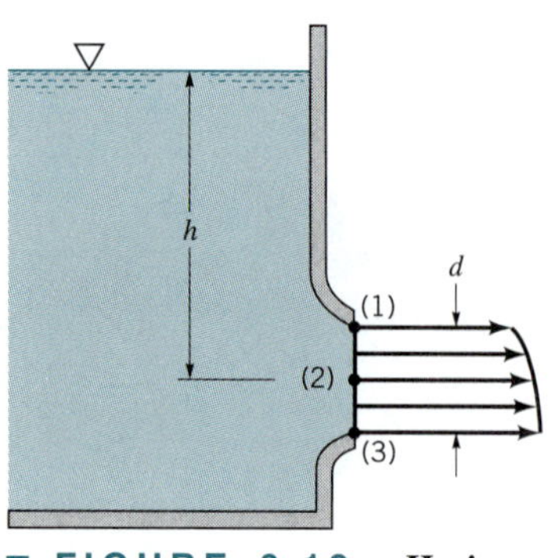

■ **FIGURE 3.12 Horizontal flow from a tank.**

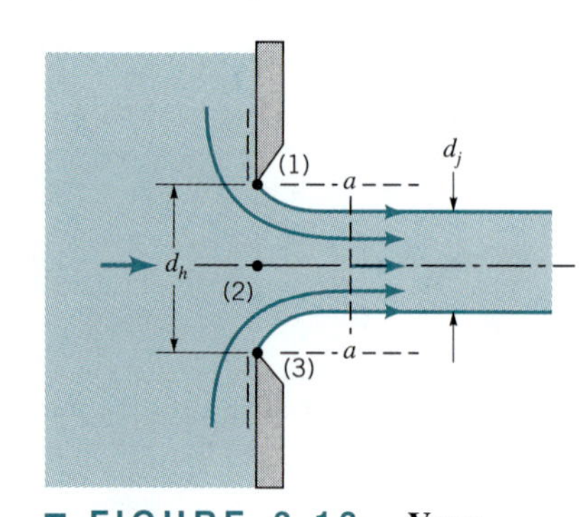

■ **FIGURE 3.13 Vena contracta effect for a sharp-edged orifice.**

Since the streamlines in the exit plane are curved ($\mathcal{R} < \infty$), the pressure across them is not constant. It would take an infinite pressure gradient across the streamlines to cause the fluid to turn a ''sharp'' corner ($\mathcal{R} = 0$). The highest pressure occurs along the centerline at (2) and the lowest pressure, $p_1 = p_3 = 0$, is at the edge of the jet. Thus, the assumption of uniform velocity with straight streamlines and constant pressure is not valid at the exit plane. It is valid, however, in the plane of the vena contracta, section a–a. The uniform velocity assumption is valid at this section provided $d_j \ll h$, as is discussed for the flow from the nozzle shown in Fig. 3.12.

The diameter of a fluid jet is often smaller than that of the hole from which it flows.

The vena contracta effect is a function of the geometry of the outlet. Some typical configurations are shown in Fig. 3.14 along with typical values of the experimentally obtained *contraction coefficient*, $C_c = A_j/A_h$, where A_j and A_h are the areas of the jet at the vena contracta and the area of the hole, respectively.

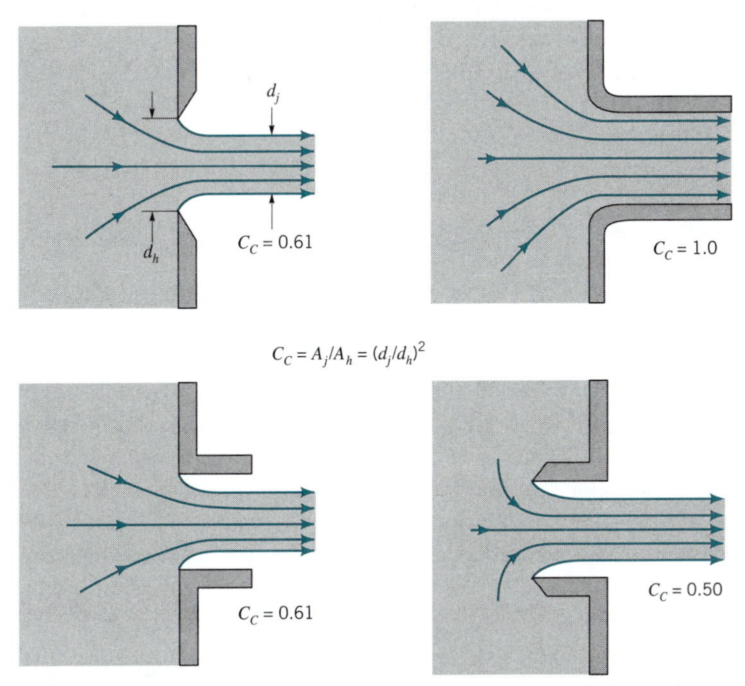

■ **FIGURE 3.14 Typical flow patterns and contraction coefficients for various round exit configurations.**

3.6.2 Confined Flows

In many cases the fluid is physically constrained within a device so that its pressure cannot be prescribed a priori as was done for the free jet examples above. Such cases include nozzles and pipes of variable diameter for which the fluid velocity changes because the flow area is different from one section to another. For these situations it is necessary to use the concept of conservation of mass (the continuity equation) along with the Bernoulli equation. The derivation and use of this equation are discussed in detail in Chapters 4 and 5. For the needs of this chapter we can use a simplified form of the continuity equation obtained from the following intuitive arguments. Consider a fluid flowing through a fixed volume (such as a tank) that has one inlet and one outlet as shown in Fig. 3.15. If the flow is steady so that there is no additional accumulation of fluid within the volume, the rate at which the fluid flows into the volume must equal the rate at which it flows out of the volume (otherwise mass would not be conserved).

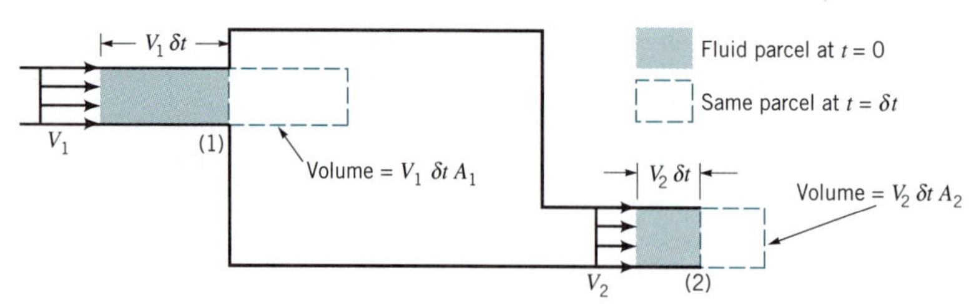

■ FIGURE 3.15 **Steady flow into and out of a tank.**

The *mass flowrate* from an outlet, $\dot{m}$ (slugs/s or kg/s), is given by $\dot{m} = \rho Q$, where Q (ft^3/s or m^3/s) is the *volume flowrate*. If the outlet area is A and the fluid flows across this area (normal to the area) with an average velocity V, then the volume of the fluid crossing this area in a time interval δt is $VA\ \delta t$, equal to that in a volume of length $V\ \delta t$ and cross-sectional area A (see Fig. 3.15). Hence, the volume flowrate (volume per unit time) is $Q = VA$. Thus, $\dot{m} = \rho VA$. To conserve mass, the inflow rate must equal the outflow rate. If the inlet is designated as (1) and the outlet as (2), it follows that $\dot{m}_1 = \dot{m}_2$. Thus, conservation of mass requires

The continuity equation states that mass cannot be created or destroyed.

$$\rho_1 A_1 V_1 = \rho_2 A_2 V_2$$

If the density remains constant, then $\rho_1 = \rho_2$, and the above becomes the *continuity equation* for incompressible flow

$$\boxed{A_1 V_1 = A_2 V_2, \text{ or } Q_1 = Q_2} \qquad \textbf{(3.19)}$$

For example, if the outlet flow area is one-half the size of the inlet flow area, it follows that the outlet velocity is twice that of the inlet velocity, since $V_2 = A_1 V_1 / A_2 = 2V_1$. The use of the Bernoulli equation and the flowrate equation (continuity equation) is demonstrated by Example 3.7.

EXAMPLE 3.7

A stream of water of diameter $d = 0.1$ m flows steadily from a tank of diameter $D = 1.0$ m as shown in Fig. E3.7*a*. Determine the flowrate, Q, needed from the inflow pipe if the water depth remains constant, $h = 2.0$ m.

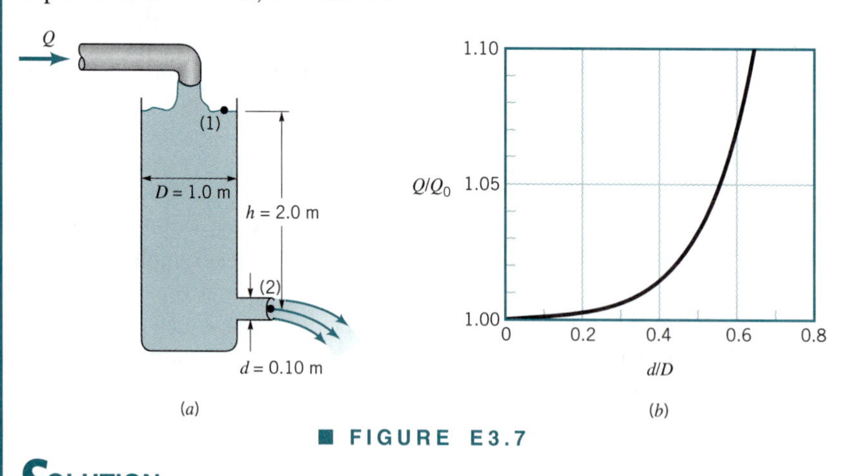

■ FIGURE E3.7

SOLUTION

For steady, inviscid, incompressible flow the Bernoulli equation applied between points (1) and (2) is

$$p_1 + \tfrac{1}{2}\rho V_1^2 + \gamma z_1 = p_2 + \tfrac{1}{2}\rho V_2^2 + \gamma z_2 \tag{1}$$

With the assumptions that $p_1 = p_2 = 0$, $z_1 = h$, and $z_2 = 0$, Eq. 1 becomes

$$\tfrac{1}{2}V_1^2 + gh = \tfrac{1}{2}V_2^2 \tag{2}$$

Although the water level remains constant (h = constant), there is an average velocity, V_1, across section (1) because of the flow from the tank. From Eq. 3.19 for steady incompressible flow, conservation of mass requires $Q_1 = Q_2$, where $Q = AV$. Thus, $A_1 V_1 = A_2 V_2$, or

$$\frac{\pi}{4} D^2 V_1 = \frac{\pi}{4} d^2 V_2$$

Hence,

$$V_1 = \left(\frac{d}{D}\right)^2 V_2 \tag{3}$$

Equations 1 and 3 can be combined to give

$$V_2 = \sqrt{\frac{2gh}{1 - (d/D)^4}} = \sqrt{\frac{2(9.81 \text{ m/s}^2)(2.0 \text{ m})}{1 - (0.1\text{m}/1\text{m})^4}} = 6.26 \text{ m/s}$$

Thus,

$$Q = A_1 V_1 = A_2 V_2 = \frac{\pi}{4}(0.1 \text{ m})^2(6.26 \text{ m/s}) = 0.0492 \text{ m}^3/\text{s} \qquad \textbf{(Ans)}$$

In this example we have not neglected the kinetic energy of the water in the tank ($V_1 \neq 0$). If the tank diameter is large compared to the jet diameter ($D \gg d$), Eq. 3 indicates that $V_1 \ll V_2$ and the assumption that $V_1 \approx 0$ would be reasonable. The error associated with this assumption can be seen by calculating the ratio of the flowrate assuming $V_1 \neq 0$, denoted Q, to that assuming $V_1 = 0$, denoted Q_0. This ratio, written as

$$\frac{Q}{Q_0} = \frac{V_2}{V_2|_{D=\infty}} = \frac{\sqrt{2gh/[1 - (d/D)^4]}}{\sqrt{2gh}} = \frac{1}{\sqrt{1 - (d/D)^4}}$$

is plotted in Fig. E3.7*b*. With $0 < d/D < 0.4$ it follows that $1 < Q/Q_0 \lesssim 1.01$, and the error in assuming $V_1 = 0$ is less than 1%. Thus, it is often reasonable to assume $V_1 = 0$.

The fact that a kinetic energy change is often accompanied by a change in pressure is shown by Example 3.8.

EXAMPLE 3.8

Air flows steadily from a tank, through a hose of diameter $D = 0.03$ m and exits to the atmosphere from a nozzle of diameter $d = 0.01$ m as shown in Fig. E3.8. The pressure in the tank remains constant at 3.0 kPa (gage) and the atmospheric conditions are standard temperature and pressure. Determine the flowrate and the pressure in the hose.

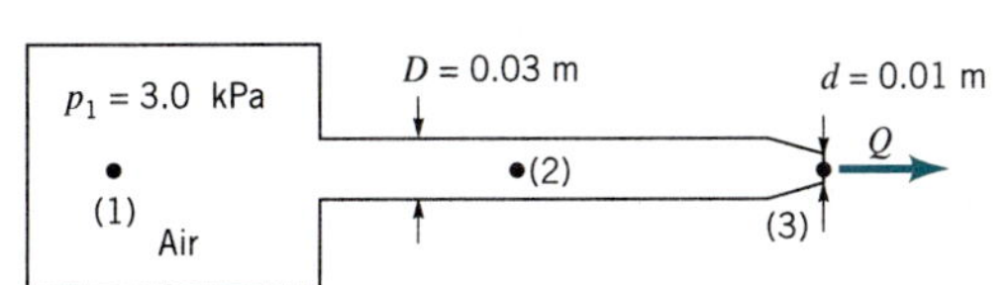

■ FIGURE E3.8

SOLUTION

If the flow is assumed steady, inviscid, and incompressible, we can apply the Bernoulli equation along the streamline shown as

$$p_1 + \tfrac{1}{2}\rho V_1^2 + \gamma z_1 = p_2 + \tfrac{1}{2}\rho V_2^2 + \gamma z_2$$
$$= p_3 + \tfrac{1}{2}\rho V_3^2 + \gamma z_3$$

With the assumption that $z_1 = z_2 = z_3$ (horizontal hose), $V_1 = 0$ (large tank), and $p_3 = 0$ (free jet) this becomes

$$V_3 = \sqrt{\frac{2p_1}{\rho}}$$

and

$$p_2 = p_1 - \tfrac{1}{2}\rho V_2^2 \qquad \textbf{(1)}$$

The density of the air in the tank is obtained from the perfect gas law, using standard absolute pressure and temperature, as

$$\rho = \frac{p_1}{RT_1}$$
$$= [(3.0 + 101)\ \text{kN/m}^2] \times \frac{10^3\ \text{N/kN}}{(286.9\ \text{N}\cdot\text{m/kg}\cdot\text{K})(15 + 273)\text{K}}$$
$$= 1.26\ \text{kg/m}^3$$

Thus, we find that

$$V_3 = \sqrt{\frac{2(3.0 \times 10^3\ \text{N/m}^2)}{1.26\ \text{kg/m}^3}} = 69.0\ \text{m/s}$$

or

$$Q = A_3V_3 = \frac{\pi}{4}d^2V_3 = \frac{\pi}{4}(0.01\ \text{m})^2(69.0\ \text{m/s})$$
$$= 0.00542\ \text{m}^3/\text{s} \qquad \textbf{(Ans)}$$

Note that the value of V_3 is determined strictly by the value of p_1 (and the assumptions involved in the Bernoulli equation), independent of the ''shape'' of the nozzle. The pressure head within the tank, $p_1/\gamma = (3.0\ \text{kPa})/(9.81\ \text{m/s}^2)(1.26\ \text{kg/m}^3) = 243$ m, is converted to the velocity head at the exit, $V_2^2/2g = (69.0\ \text{m/s})^2/(2 \times 9.81\ \text{m/s}^2) = 243$ m. Although we used gage pressure in the Bernoulli equation ($p_3 = 0$), we had to use absolute pressure in the perfect gas law when calculating the density.

The pressure within the hose can be obtained from Eq. 1 and the continuity equation (Eq. 3.19)

$$A_2V_2 = A_3V_3$$

Hence,

$$V_2 = A_3V_3/A_2 = \left(\frac{d}{D}\right)^2 V_3 = \left(\frac{0.01\ \text{m}}{0.03\ \text{m}}\right)^2 (69.0\ \text{m/s}) = 7.67\ \text{m/s}$$

and from Eq. 1

$$p_2 = 3.0 \times 10^3 \text{ N/m}^2 - \tfrac{1}{2}(1.26 \text{ kg/m}^3)(7.67 \text{ m/s})^2$$
$$= (3000 - 37.1)\text{N/m}^2 = 2963 \text{ N/m}^2 \qquad \textbf{(Ans)}$$

In the absence of viscous effects the pressure throughout the hose is constant and equal to p_2. Physically, the decreases in pressure from p_1 to p_2 to p_3 accelerate the air and increase its kinetic energy from zero in the tank to an intermediate value in the hose and finally to its maximum value at the nozzle exit. Since the air velocity in the nozzle exit is nine times that in the hose, most of the pressure drop occurs across the nozzle ($p_1 = 3000 \text{ N/m}^2$, $p_2 = 2963 \text{ N/m}^2$ and $p_3 = 0$).

Since the pressure change from (1) to (3) is not too great [that is, in terms of absolute pressure $(p_1 - p_3)/p_1 = 3.0/101 = 0.03$], it follows from the perfect gas law that the density change is also not significant. Hence, the incompressibility assumption is reasonable for this problem. If the tank pressure were considerably larger or if viscous effects were important, the above results would be incorrect.

In many situations the combined effects of kinetic energy, pressure, and gravity are important. Example 3.9 illustrates this.

EXAMPLE 3.9

Water flows through a pipe reducer as is shown in Fig. E3.9. The static pressures at (1) and (2) are measured by the inverted U-tube manometer containing oil of specific gravity, *SG*, less than one. Determine the manometer reading, *h*.

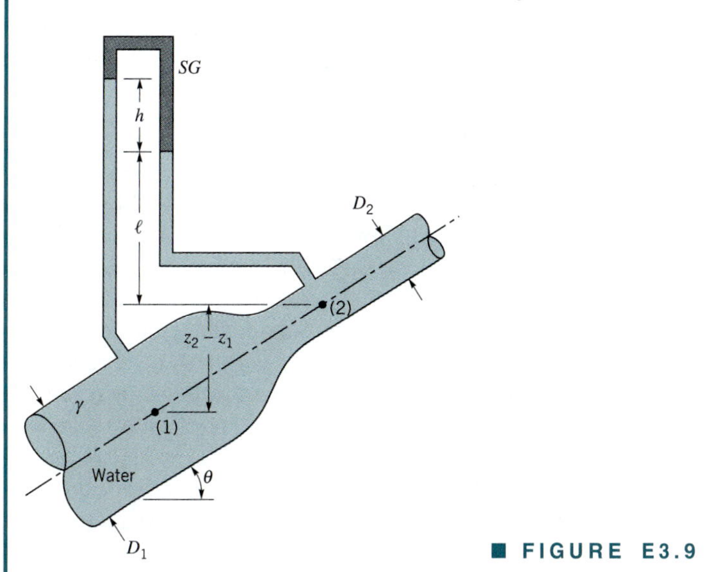

■ FIGURE E3.9

SOLUTION

With the assumptions of steady, inviscid, incompressible flow, the Bernoulli equation can be written as

$$p_1 + \tfrac{1}{2}\rho V_1^2 + \gamma z_1 = p_2 + \tfrac{1}{2}\rho V_2^2 + \gamma z_2$$

The continuity equation (Eq. 3.19) provides a second relationship between V_1 and V_2 if we assume the velocity profiles are uniform at those two locations and the fluid incompressible:

$$Q = A_1V_1 = A_2V_2$$

By combining these two equations we obtain

$$p_1 - p_2 = \gamma(z_2 - z_1) + \tfrac{1}{2}\rho V_2^2[1 - (A_2/A_1)^2] \qquad \textbf{(1)}$$

This pressure difference is measured by the manometer and can be determined by using the pressure-depth ideas developed in Chapter 2. Thus,

$$p_1 - \gamma(z_2 - z_1) - \gamma\ell - \gamma h + SG\ \gamma h + \gamma\ell = p_2$$

or

$$p_1 - p_2 = \gamma(z_2 - z_1) + (1 - SG)\gamma h \qquad \textbf{(2)}$$

As discussed in Chapter 2, this pressure difference is neither merely γh nor $\gamma(h + z_1 - z_2)$.

Equations 1 and 2 can be combined to give the desired result as follows

$$(1 - SG)\gamma h = \frac{1}{2}\rho V_2^2\left[1 - \left(\frac{A_2}{A_1}\right)^2\right]$$

or since $V_2 = Q/A_2$

$$h = (Q/A_2)^2\frac{1 - (A_2/A_1)^2}{2g(1 - SG)} \qquad \textbf{(Ans)}$$

The difference in elevation, $z_1 - z_2$, was not needed because the change in elevation term in the Bernoulli equation exactly cancels the elevation term in the manometer equation. However, the pressure difference, $p_1 - p_2$, depends on the angle θ, because of the elevation, $z_1 - z_2$, in Eq. 1. Thus, for a given flowrate, the pressure difference, $p_1 - p_2$, as measured by a pressure gage would vary with θ, but the manometer reading, h, would be independent of θ.

V3.6

Cavitation occurs when the pressure is reduced to the vapor pressure.

In general, an increase in velocity is accompanied by a decrease in pressure. For example, the velocity of the air flowing over the top surface of an airplane wing is, on the average, faster than that flowing under the bottom surface. Thus, the net pressure force is greater on the bottom than on the top—the wing generates a lift.

If the differences in velocity are considerable, the differences in pressure can also be considerable. For flows of gases, this may introduce compressibility effects as discussed in Section 3.8 and Chapter 11. For flows of liquids, this may result in cavitation, a potentially dangerous situation that results when the liquid pressure is reduced to the vapor pressure and the liquid "boils."

As discussed in Chapter 1, the vapor pressure, p_v, is the pressure at which vapor bubbles form in a liquid. It is the pressure at which the liquid starts to boil. Obviously this pressure depends on the type of liquid and its temperature. For example, water, which boils at 212 °F at standard atmospheric pressure, 14.7 psia, boils at 80 °F if the pressure is 0.507 psia. That is, $p_v = 0.507$ psia at 80 °F and $p_v = 14.7$ psia at 212 °F. (See Tables B.1 and B.2.)

One way to produce cavitation in a flowing liquid is noted from the Bernoulli equation. If the fluid velocity is increased (for example, by a reduction in flow area as shown in Fig. 3.16) the pressure will decrease. This pressure decrease (needed to accelerate the fluid through the constriction) can be large enough so that the pressure in the liquid is reduced to its vapor pressure. A simple example of cavitation can be demonstrated with an ordinary garden hose. If the hose is "kinked," a restriction in the flow area in some ways analogous to that shown

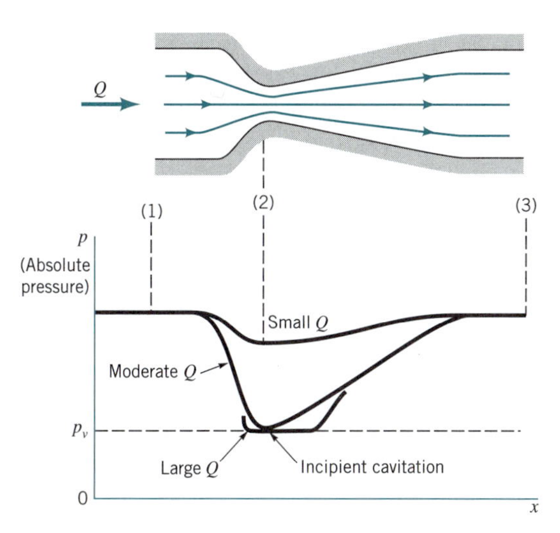

■ **FIGURE 3.16 Pressure variation and cavitation in a variable area pipe.**

in Fig. 3.16 will result. The water velocity through this restriction will be relatively large. With a sufficient amount of restriction the sound of the flowing water will change—a definite ''hissing'' sound is produced. This sound is a result of cavitation.

Cavitation can cause damage to equipment.

In such situations boiling occurs (though the temperature need not be high), vapor bubbles form, and then they collapse as the fluid moves into a region of higher pressure (lower velocity). This process can produce dynamic effects (imploding) that cause very large pressure transients in the vicinity of the bubbles. Pressures as large as 100,000 psi (690 MPa) are believed to occur. If the bubbles collapse close to a physical boundary they can, over a period of time, cause damage to the surface in the cavitation area. Tip cavitation from a propeller is shown in Fig. 3.17. In this case the high-speed rotation of the propeller produced a corresponding low pressure on the propeller. Obviously, proper design and use of equipment is needed to eliminate cavitation damage.

■ **FIGURE 3.17 Tip cavitation from a propeller. (Photograph courtesy of Garfield Thomas Water Tunnel, Pennsylvania State University.)**

EXAMPLE 3.10

Water at 60 °F is siphoned from a large tank through a constant diameter hose as shown in Fig. E3.10. Determine the maximum height of the hill, H, over which the water can be siphoned without cavitation occurring. The end of the siphon is 5 ft below the bottom of the tank. Atmospheric pressure is 14.7 psia.

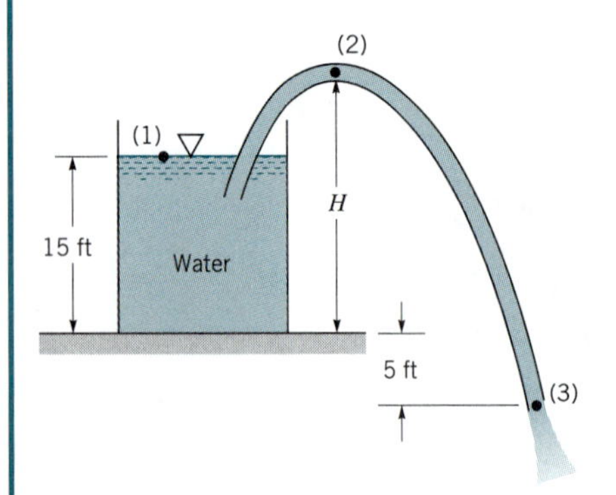

■ FIGURE E3.10

SOLUTION

If the flow is steady, inviscid, and incompressible we can apply the Bernoulli equation along the streamline from (1) to (2) to (3) as follows:

$$p_1 + \tfrac{1}{2}\rho V_1^2 + \gamma z_1 = p_2 + \tfrac{1}{2}\rho V_2^2 + \gamma z_2 = p_3 + \tfrac{1}{2}\rho V_3^2 + \gamma z_3 \tag{1}$$

With the tank bottom as the datum, we have $z_1 = 15$ ft, $z_2 = H$, and $z_3 = -5$ ft. Also, $V_1 = 0$ (large tank), $p_1 = 0$ (open tank), $p_3 = 0$ (free jet), and from the continuity equation $A_2V_2 = A_3V_3$, or because the hose is constant diameter, $V_2 = V_3$. Thus, the speed of the fluid in the hose is determined from Eq. 1 to be

$$V_3 = \sqrt{2g(z_1 - z_3)} = \sqrt{2(32.2 \text{ ft/s}^2)[15 - (-5)] \text{ ft}}$$
$$= 35.9 \text{ ft/s} = V_2$$

Use of Eq. 1 between points (1) and (2) then gives the pressure p_2 at the top of the hill as

$$p_2 = p_1 + \tfrac{1}{2}\rho V_1^2 + \gamma z_1 - \tfrac{1}{2}\rho V_2^2 - \gamma z_2 = \gamma(z_1 - z_2) - \tfrac{1}{2}\rho V_2^2 \tag{2}$$

From Table B.1, the vapor pressure of water at 60 °F is 0.256 psia. Hence, for incipient cavitation the lowest pressure in the system will be $p = 0.256$ psia. Careful consideration of Eq. 2 and Fig. E3.10 will show that this lowest pressure will occur at the top of the hill. Since we have used gage pressure at point (1) ($p_1 = 0$), we must use gage pressure at point (2) also. Thus, $p_2 = 0.256 - 14.7 = -14.4$ psi and Eq. 2 gives

$$(-14.4 \text{ lb/in.}^2)(144 \text{ in.}^2/\text{ft}^2) = (62.4 \text{ lb/ft}^3)(15 - H)\text{ft} - \tfrac{1}{2}(1.94 \text{ slugs/ft}^3)(35.9 \text{ ft/s})^2$$

or

$$H = 28.2 \text{ ft} \qquad \textbf{(Ans)}$$

For larger values of H, vapor bubbles will form at point (2) and the siphon action may stop.

Note that we could have used absolute pressure throughout ($p_2 = 0.256$ psia and $p_1 = 14.7$ psia) and obtained the same result. The lower the elevation of point (3), the larger the flowrate and, therefore, the smaller the value of H allowed.

We could also have used the Bernoulli equation between (2) and (3), with $V_2 = V_3$, to obtain the same value of H. In this case it would not have been necessary to determine V_2 by use of the Bernoulli equation between (1) and (3).

The above results are independent of the diameter and length of the hose (provided viscous effects are not important). Proper design of the hose (or pipe) is needed to ensure that it will not collapse due to the large pressure difference (vacuum) between the inside and outside of the hose.

3.6.3 Flowrate Measurement

Many types of devices using principles involved in the Bernoulli equation have been developed to measure fluid velocities and flowrates. The Pitot-static tube discussed in Section 3.5 is an example. Other examples discussed below include devices to measure flowrates in pipes and conduits and devices to measure flowrates in open channels. In this chapter we will consider ''ideal'' flow meters—those devoid of viscous, compressibility, and other ''real world'' effects. Corrections for these effects are discussed in Chapters 8 and 10. Our goal here is to understand the basic operating principles of these simple flow meters.

Various flow meters are governed by the Bernoulli and continuity equations.

An effective way to measure the flowrate through a pipe is to place some type of restriction within the pipe as shown in Fig. 3.18 and to measure the pressure difference between the low-velocity, high-pressure upstream section (1), and the high-velocity, low-pressure downstream section (2). Three commonly used types of flow meters are illustrated: the *orifice meter*, the *nozzle meter*, and the *Venturi meter*. The operation of each is based on the same physical principles—an increase in velocity causes a decrease in pressure. The difference between them is a matter of cost, accuracy, and how closely their actual operation obeys the idealized flow assumptions.

We assume the flow is horizontal ($z_1 = z_2$), steady, inviscid, and incompressible between points (1) and (2). The Bernoulli equation becomes

$$p_1 + \tfrac{1}{2}\rho V_1^2 = p_2 + \tfrac{1}{2}\rho V_2^2$$

(The effect of nonhorizontal flow can be incorporated easily by including the change in elevation, $z_1 - z_2$, in the Bernoulli equation.)

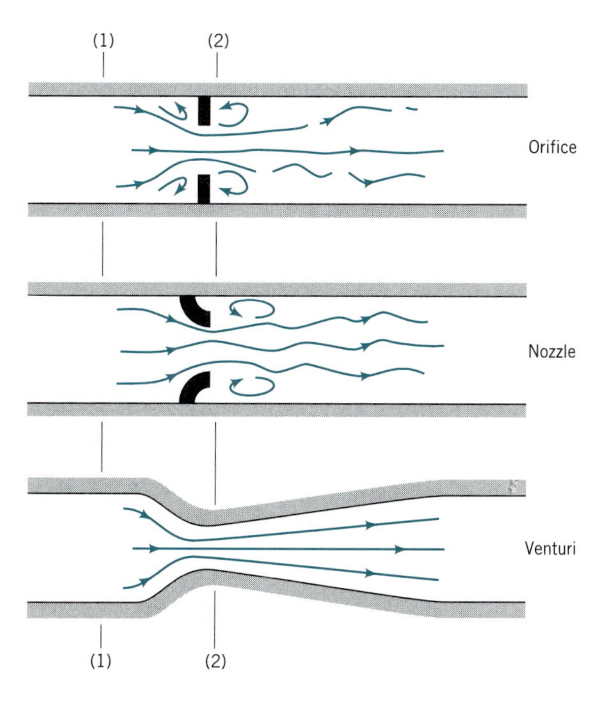

■ **FIGURE 3.18 Typical devices for measuring flowrate in pipes.**

If we assume the velocity profiles are uniform at sections (1) and (2), the continuity equation (Eq. 3.19) can be written as

$$Q = A_1V_1 = A_2V_2$$

where A_2 is the small ($A_2 < A_1$) flow area at section (2). Combination of these two equations results in the following theoretical flowrate

The flowrate is a function of the pressure difference across the flow meter.

$$Q = A_2 \sqrt{\frac{2(p_1 - p_2)}{\rho[1 - (A_2/A_1)^2]}} \tag{3.20}$$

Thus, for a given flow geometry (A_1 and A_2) the flowrate can be determined if the pressure difference, $p_1 - p_2$, is measured. The actual measured flowrate, Q_{actual}, will be smaller than this theoretical result because of various differences between the "real world" and the assumptions used in the derivation of Eq. 3.20. These differences (which are quite consistent and may be as small as 1 to 2% or as large as 40% depending on the geometry used) are discussed in Chapter 8.

EXAMPLE 3.11

Kerosene ($SG = 0.85$) flows through the Venturi meter shown in Fig. E3.11 with flowrates between 0.005 and 0.050 m^3/s. Determine the range in pressure difference, $p_1 - p_2$, needed to measure these flowrates.

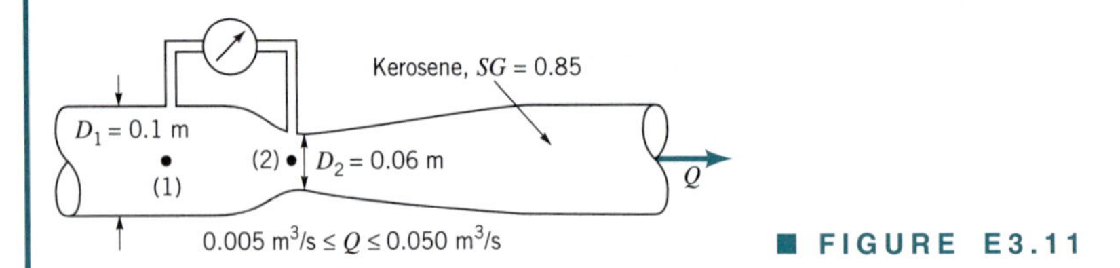

■ FIGURE E3.11

SOLUTION

If the flow is assumed to be steady, inviscid, and incompressible, the relationship between flowrate and pressure is given by Eq. 3.20. This can be rearranged to give

$$p_1 - p_2 = \frac{Q^2\rho[1 - (A_2/A_1)^2]}{2A_2^2}$$

With a density of the flowing fluid of

$$\rho = SG\ \rho_{H_2O} = 0.85(1000\ \text{kg/m}^3) = 850\ \text{kg/m}^3$$

the pressure difference for the smallest flowrate is

$$p_1 - p_2 = (0.005\ \text{m}^3/\text{s})^2(850\ \text{kg/m}^3)\,\frac{[1 - (0.06\ \text{m}/0.10\ \text{m})^4]}{2[(\pi/4)(0.06\ \text{m})^2]^2}$$

$$= 1160\ \text{N/m}^2 = 1.16\ \text{kPa}$$

Likewise, the pressure difference for the largest flowrate is

$$p_1 - p_2 = (0.05)^2(850)\,\frac{[1 - (0.06/0.10)^4]}{2[(\pi/4)(0.06)^2]^2}$$

$$= 1.16\ 10^5\ \text{N/m}^2 = 116\ \text{kPa}$$

Thus,

$$1.16 \text{ kPa} \leq p_1 - p_2 \leq 116 \text{ kPa} \qquad \textbf{(Ans)}$$

These values represent the pressure differences for inviscid, steady, incompressible conditions. The ideal results presented here are independent of the particular flow meter geometry—an orifice, nozzle, or Venturi meter (see Fig. 3.18).

It is seen from Eq. 3.20 that the flowrate varies as the square root of the pressure difference. Hence, as indicated by the numerical results, a tenfold increase in flowrate requires a one-hundredfold increase in pressure difference. This nonlinear relationship can cause difficulties when measuring flowrates over a wide range of values. Such measurements would require pressure transducers with a wide range of operation. An alternative is to use two flow meters in parallel—one for the larger and one for the smaller flowrate ranges.

Other flow meters based on the Bernoulli equation are used to measure flowrates in open channels such as flumes and irrigation ditches. Two of these devices, the *sluice gate* and the *sharp-crested weir*, are discussed below under the assumption of steady, inviscid, incompressible flow. These and other open-channel flow devices are discussed in more detail in Chapter 10.

The sluice gate as shown in Fig. 3.19 is often used to regulate and measure the flowrate in an open channel. The flowrate, Q, is a function of the water depth upstream, z_1, the width of the gate, b, and the gate opening, a. Application of the Bernoulli equation and continuity equation between points (1) and (2) can provide a good approximation to the actual flowrate obtained. We assume the velocity profiles are uniform sufficiently far upstream and downstream of the gate.

Thus, we apply the Bernoulli and continuity equations between points on the free surfaces at (1) and (2) to give

$$p_1 + \tfrac{1}{2}\rho V_1^2 + \gamma z_1 = p_2 + \tfrac{1}{2}\rho V_2^2 + \gamma z_2$$

and

$$Q = A_1 V_1 = b V_1 z_1 = A_2 V_2 = b V_2 z_2$$

The flowrate under a sluice gate depends on the water depths on either side of the gate.

With the fact that $p_1 = p_2 = 0$, these equations can be combined and rearranged to give the flowrate as

$$Q = z_2 b \sqrt{\frac{2g(z_1 - z_2)}{1 - (z_2/z_1)^2}} \tag{3.21}$$

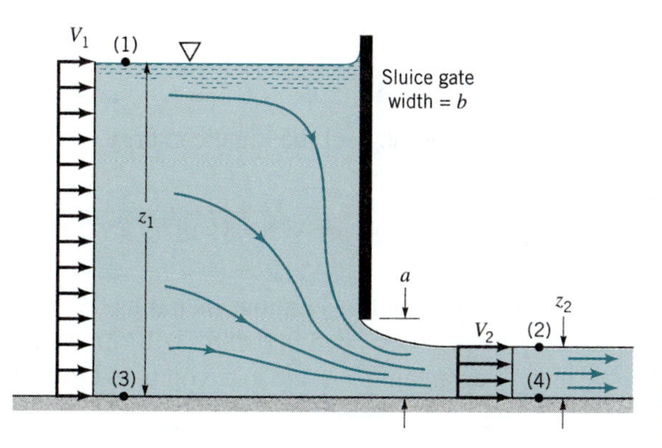

■ **FIGURE 3.19** **Sluice gate geometry.**

In the limit of $z_1 \gg z_2$ this result simply becomes

$$Q = z_2 b \sqrt{2gz_1}$$

This limiting result represents the fact that if the depth ratio, z_1/z_2, is large, the kinetic energy of the fluid upstream of the gate is negligible and the fluid velocity after it has fallen a distance $(z_1 - z_2) \approx z_1$ is approximately $V_2 = \sqrt{2gz_1}$.

The results of Eq. 3.21 could also be obtained by using the Bernoulli equation between points (3) and (4) and the fact that $p_3 = \gamma z_1$ and $p_4 = \gamma z_2$ since the streamlines at these sections are straight. In this formulation, rather than the potential energies at (1) and (2), we have the pressure contributions at (3) and (4).

A vena contracta occurs as water flows under a sluice gate.

The downstream depth, z_2, not the gate opening, a, was used to obtain the result of Eq. 3.21. As was discussed relative to flow from an orifice (Fig. 3.14), the fluid cannot turn a sharp 90° corner. A vena contracta results with a contraction coefficient, $C_c = z_2/a$, less than 1. Typically C_c is approximately 0.61 over the depth ratio range of $0 < a/z_1 < 0.2$. For larger values of a/z_1 the value of C_c increases rapidly.

EXAMPLE 3.12

Water flows under the sluice gate shown in Fig. E3.12. Determine the approximate flowrate per unit width of the channel.

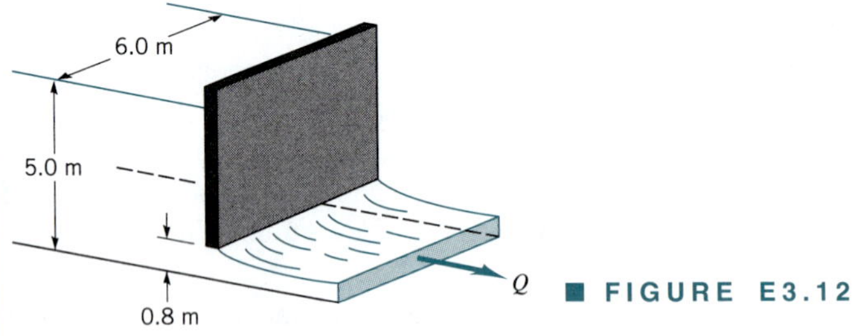

■ FIGURE E3.12

SOLUTION

Under the assumptions of steady, inviscid, incompressible flow, we can apply Eq. 3.21 to obtain Q/b, the flowrate per unit width, as

$$\frac{Q}{b} = z_2 \sqrt{\frac{2g(z_1 - z_2)}{1 - (z_2/z_1)^2}}$$

In this instance $z_1 = 5.0$ m and $a = 0.80$ m so the ratio $a/z_1 = 0.16 < 0.20$, and we can assume that the contraction coefficient is approximately $C_c = 0.61$. Thus, $z_2 = C_c a = 0.61$ (0.80 m) $= 0.488$ m and we obtain the flowrate

$$\frac{Q}{b} = (0.488 \text{ m}) \sqrt{\frac{2(9.81 \text{ m/s}^2)(5.0 \text{ m} - 0.488 \text{ m})}{1 - (0.488 \text{ m}/5.0 \text{ m})^2}}$$

$$= 4.61 \text{ m}^2/\text{s} \qquad \textbf{(Ans)}$$

If we consider $z_1 \gg z_2$ and neglect the kinetic energy of the upstream fluid, we would have

$$\frac{Q}{b} = z_2 \sqrt{2gz_1} = 0.488 \text{ m} \sqrt{2(9.81 \text{ m/s}^2)(5.0 \text{ m})} = 4.83 \text{ m}^2/\text{s}$$

In this case the difference in Q with or without including V_1 is not too significant because the depth ratio is fairly large ($z_1/z_2 = 5.0/0.488 = 10.2$). Thus, it is often reasonable to neglect the kinetic energy upstream from the gate compared to that downstream of it.

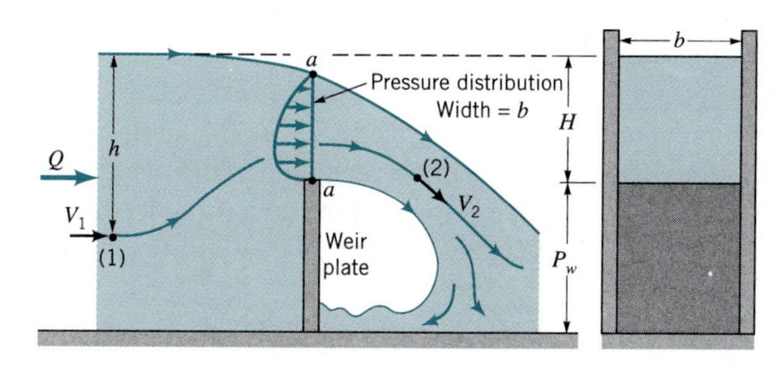

■ FIGURE 3.20 **Rectangular, sharp-crested weir geometry.**

Another device used to measure flow in an open channel is a *weir*. A typical rectangular, sharp-crested weir is shown in Fig. 3.20. For such devices the flowrate of liquid over the top of the weir plate is dependent on the weir height, P_w, the width of the channel, b, and the head, H, of the water above the top of the weir. Application of the Bernoulli equation can provide a simple approximation of the flowrate expected for these situations, even though the actual flow is quite complex.

Between points (1) and (2) the pressure and gravitational fields cause the fluid to accelerate from velocity V_1 to velocity V_2. At (1) the pressure is $p_1 = \gamma h$, while at (2) the pressure is essentially atmospheric, $p_2 = 0$. Across the curved streamlines directly above the top of the weir plate (section a–a), the pressure changes from atmospheric on the top surface to some maximum value within the fluid stream and then to atmospheric again at the bottom surface. This distribution is indicated in Fig. 3.20. Such a pressure distribution, combined with the streamline curvature and gravity, produces a rather nonuniform velocity profile across this section. This velocity distribution can be obtained from experiments or a more advanced theory.

For now, we will take a very simple approach and assume that the weir flow is similar in many respects to an orifice-type flow with a free streamline. In this instance we would expect the average velocity across the top of the weir to be proportional to $\sqrt{2gH}$ and the flow area for this rectangular weir to be proportional to Hb. Hence, it follows that

$$Q = C_1 Hb \sqrt{2gH} = C_1 b \sqrt{2g}\, H^{3/2}$$

where C_1 is a constant to be determined.

Simple use of the Bernoulli equation has provided a method to analyze the relatively complex flow over a weir. The correct functional dependence of Q on H has been obtained ($Q \sim H^{3/2}$), but the value of the coefficient C_1 is unknown. Even a more advanced analysis cannot predict its value accurately. As is discussed in Chapter 10, experiments are used to determine the value of C_1.

EXAMPLE 3.13

Water flows over a triangular weir as is shown in Fig. E3.13. Based on a simple analysis using the Bernoulli equation, determine the dependence of the flowrate on the depth H. If the flowrate is Q_0 when $H = H_0$, estimate the flowrate when the depth is increased to $H = 3H_0$.

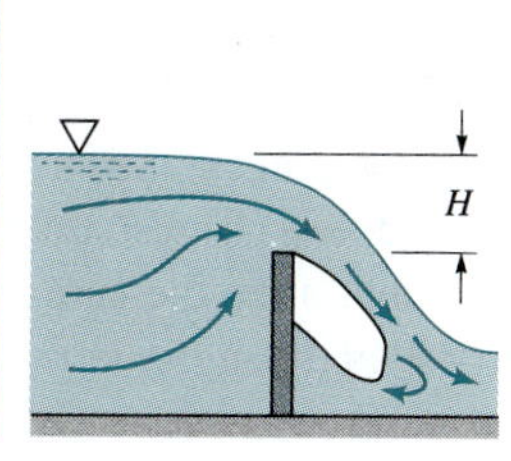

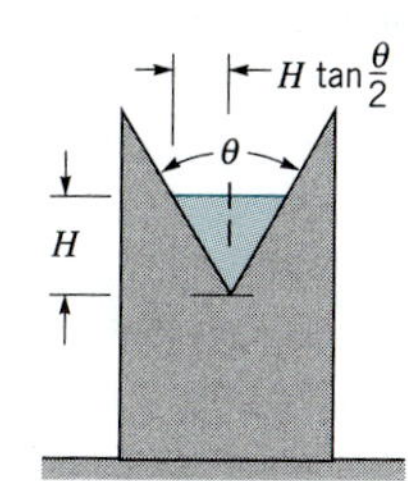

■ FIGURE E3.13

SOLUTION

With the assumption that the flow is steady, inviscid, and incompressible, it is reasonable to assume from Eq. 3.18 that the average speed of the fluid over the triangular notch in the weir plate is proportional to $\sqrt{2gH}$. Also, the flow area for a depth of H is $H\,[H \tan(\theta/2)]$. The combination of these two ideas gives

$$Q = AV = H^2 \tan\frac{\theta}{2}\,(C_2\sqrt{2gH}) = C_2 \tan\frac{\theta}{2}\sqrt{2g}\,H^{5/2} \qquad \textbf{(Ans)}$$

where C_2 is an unknown constant to be determined experimentally.

Thus, an increase in the depth by a factor of three (from H_0 to $3H_0$) results in an increase of the flowrate by a factor of

$$\frac{Q_{3H_0}}{Q_{H_0}} = \frac{C_2 \tan(\theta/2)\sqrt{2g}\,(3H_0)^{5/2}}{C_2 \tan(\theta/2)\sqrt{2g}\,(H_0)^{5/2}}$$

$$= 15.6 \qquad \textbf{(Ans)}$$

Note that for a triangular weir the flowrate is proportional to $H^{5/2}$, whereas for the rectangular weir discussed above, it is proportional to $H^{3/2}$. The triangular weir can be accurately used over a wide range of flowrates.

3.7 The Energy Line and the Hydraulic Grade Line

The hydraulic grade line and energy line are graphical forms of the Bernoulli equation.

As was discussed in Section 3.4, the Bernoulli equation is actually an energy equation representing the partitioning of energy for an inviscid, incompressible, steady flow. The sum of the various energies of the fluid remains constant as the fluid flows from one section to another. A useful interpretation of the Bernoulli equation can be obtained through the use of the concepts of the *hydraulic grade line* (HGL) and the *energy line* (EL). These ideas represent a geometrical interpretation of a flow and can often be effectively used to better grasp the fundamental processes involved.

For steady, inviscid, incompressible flow the total energy remains constant along a streamline. The concept of "head" was introduced by dividing each term in Eq. 3.7 by the specific weight, $\gamma = \rho g$, to give the Bernoulli equation in the following form

$$\frac{p}{\gamma} + \frac{V^2}{2g} + z = \text{constant on a streamline} = H \qquad \textbf{(3.22)}$$

Each of the terms in this equation has the units of length (feet or meters) and represents a certain type of head. The Bernoulli equation states that the sum of the pressure head, the velocity head, and the elevation head is constant along a streamline. This constant is called the *total head, H*.

The energy line is a line that represents the total head available to the fluid. As shown in Fig. 3.21, the elevation of the energy line can be obtained by measuring the stagnation pressure with a Pitot tube. (A Pitot tube is the portion of a Pitot-static tube that measures the stagnation pressure. See Section 3.5.) The stagnation point at the end of the Pitot tube provides a measurement of the total head (or energy) of the flow. The static pressure tap connected to the piezometer tube shown, on the other hand, measures the sum of the pressure head and the elevation head, $p/\gamma + z$. This sum is often called the *piezometric head*. The static pressure tap does not measure the velocity head.

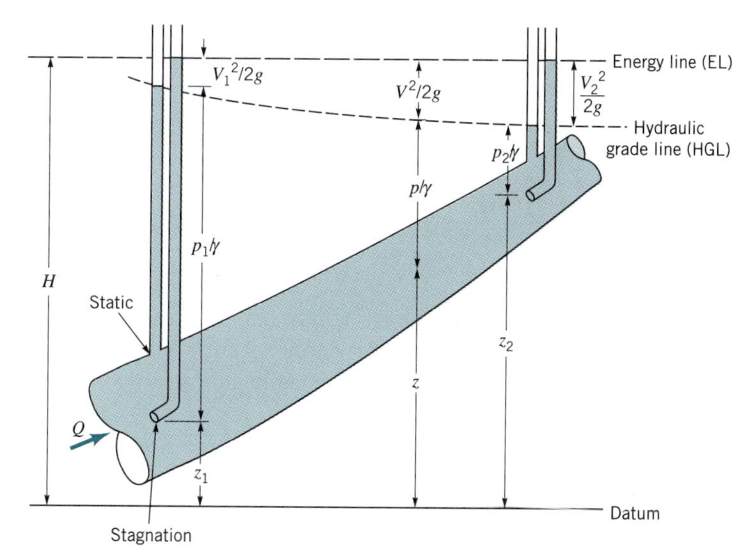

■ **FIGURE 3.21 Representation of the energy line and the hydraulic grade line.**

According to Eq. 3.22 the total head remains constant along the streamline (provided the assumptions of the Bernoulli equation are valid). Thus, a Pitot tube at any other location in the flow will measure the same total head, as is shown in the figure. The elevation head, velocity head, and pressure head may vary along the streamline, however.

The locus of elevations provided by a series of Pitot tubes is termed the energy line, EL. The locus provided by a series of piezometer taps is termed the hydraulic grade line, HGL. Under the assumptions of the Bernoulli equation, the energy line is horizontal. If the fluid velocity changes along the streamline, the hydraulic grade line will not be horizontal. If viscous effects are important (as they often are in pipe flows) the total head does not remain constant due to a loss in energy as the fluid flows along its streamline. This means that the energy line is no longer horizontal. Such viscous effects are discussed in Chapter 8.

The hydraulic grade line lies one velocity head below the energy line.

The energy line and hydraulic grade line for flow from a large tank are shown in Fig. 3.22. If the flow is steady, incompressible, and inviscid, the energy line is horizontal and at the elevation of the liquid in the tank (since the fluid velocity in the tank and the pressure on the surface are zero). The hydraulic grade line lies a distance of one velocity head, $V^2/2g$, below the energy line. Thus, a change in fluid velocity due to a change in the pipe diameter results in a change in the elevation of the hydraulic grade line. At the pipe outlet the pressure head is zero (gage) so the pipe elevation and the hydraulic grade line, coincide.

The distance from the pipe to the hydraulic grade line indicates the pressure within the pipe as is shown in Fig. 3.23. If the pipe lies below the hydraulic grade line, the pressure

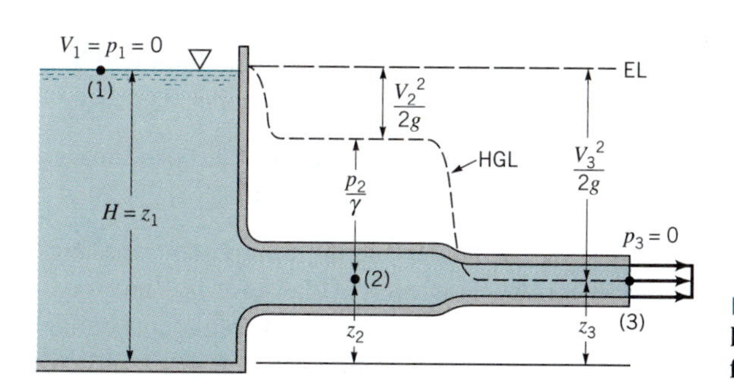

■ **FIGURE 3.22 The energy line and hydraulic grade line for flow from a tank.**

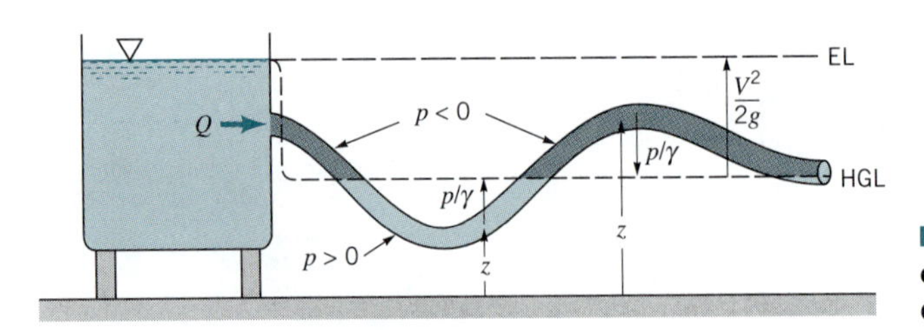

■ **FIGURE 3.23 Use of the energy line and the hydraulic grade line.**

For flow below (above) the hydraulic grade line, the pressure is positive (negative).

within the pipe is positive (above atmospheric). If the pipe lies above the hydraulic grade line, the pressure is negative (below atmospheric). Thus, a scale drawing of a pipeline and the hydraulic grade line can be used to readily indicate regions of positive or negative pressure within a pipe.

EXAMPLE 3.14

Water is siphoned from the tank shown in Fig. E3.14 through a hose of constant diameter. A small hole is found in the hose at location (1) as indicated. When the siphon is used, will water leak out of the hose, or will air leak into the hose, thereby possibly causing the siphon to malfunction?

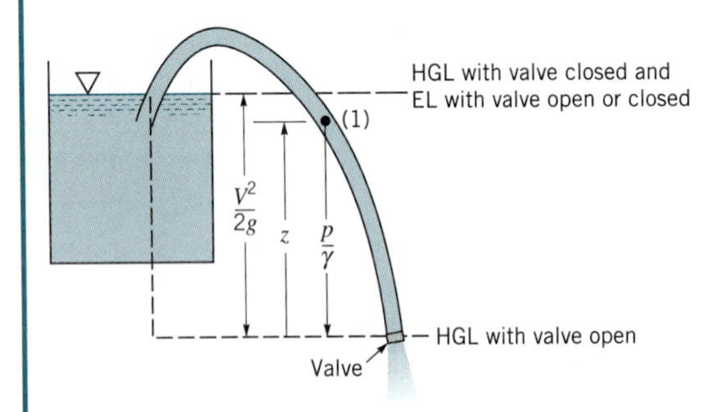

■ **FIGURE E3.14**

SOLUTION

Whether air will leak into or water will leak out of the hose depends on whether the pressure within the hose at (1) is less than or greater than atmospheric. Which happens can be easily determined by using the energy line and hydraulic grade line concepts as follows. With the assumption of steady, incompressible, inviscid flow it follows that the total head is constant—thus, the energy line is horizontal.

Since the hose diameter is constant, it follows from the continuity equation ($AV =$ constant) that the water velocity in the hose is constant throughout. Thus, the hydraulic grade line is a constant distance, $V^2/2g$, below the energy line as shown in Fig. E3.14. Since the pressure at the end of the hose is atmospheric, it follows that the hydraulic grade line is at the same elevation as the end of the hose outlet. The fluid within the hose at any point above the hydraulic grade line will be at less than atmospheric pressure.

Thus, air will leak into the hose through the hole at point (1). **(Ans)**

In practice, viscous effects may be quite important, making this simple analysis (horizontal energy line) incorrect. However, if the hose is "not too small diameter," "not too long," the fluid "not too viscous," and the flowrate "not too large," the above result may be very accurate. If any of these assumptions are relaxed, a more detailed analysis is required (see Chapter 8). If the end of the hose were closed so the flowrate were zero, the hydraulic grade line would coincide with the energy line ($V^2/2g = 0$ throughout), the pressure at (1) would be greater than atmospheric, and water would leak through the hole at (1).

The above discussion of the hydraulic grade line and the energy line is restricted to ideal situations involving inviscid, incompressible flows. Another restriction is that there are no ''sources'' or ''sinks'' of energy within the flow field. That is, there are no pumps or turbines involved. Alterations in the energy line and hydraulic grade line concepts due to these devices are discussed in Chapter 8.

3.8 Restrictions on Use of the Bernoulli Equation

Proper use of the Bernoulli equation requires close attention to the assumptions used in its derivation. In this section we review some of these assumptions and consider the consequences of incorrect use of the equation.

3.8.1 Compressibility Effects

One of the main assumptions is that the fluid is incompressible. Although this is reasonable for most liquid flows, it can, in certain instances, introduce considerable errors for gases.

In the previous section we saw that the stagnation pressure is greater than the static pressure by an amount $\rho V^2/2$, provided that the density remains constant. If this dynamic pressure is not too large compared with the static pressure, the density change between two points is not very large and the flow can be considered incompressible. However, since the dynamic pressure varies as V^2, the error associated with the assumption that a fluid is incompressible increases with the square of the velocity of the fluid. To account for compressibility effects we must return to Eq. 3.6 and properly integrate the term $\int dp/\rho$ when ρ is not constant.

The Bernoulli equation can be modified for compressible flows.

A simple, although specialized, case of compressible flow occurs when the temperature of a perfect gas remains constant along the streamline—isothermal flow. Thus, we consider $p = \rho RT$, where T is constant. (In general, p, ρ, and T will vary.) For steady, inviscid, isothermal flow, Eq. 3.6 becomes

$$RT \int \frac{dp}{p} + \frac{1}{2} V^2 + gz = \text{constant}$$

where we have used $\rho = p/RT$. The pressure term is easily integrated and the constant of integration evaluated if z_1, p_1, and V_1 are known at some location on the streamline. The result is

$$\frac{V_1^2}{2g} + z_1 + \frac{RT}{g} \ln\left(\frac{p_1}{p_2}\right) = \frac{V_2^2}{2g} + z_2 \qquad \textbf{(3.23)}$$

Equation 3.23 is the inviscid, isothermal analog of the incompressible Bernoulli equation. In the limit of small pressure difference, $p_1/p_2 = 1 + (p_1 - p_2)/p_2 = 1 + \varepsilon$, with $\varepsilon \ll 1$ and Eq. 3.23 reduces to the standard incompressible Bernoulli equation. This can be shown by use of the approximation $\ln(1 + \varepsilon) \approx \varepsilon$ for small ε. The use of Eq. 3.23 in practical applications is restricted by the inviscid flow assumption, since (as is discussed in Section 11.5) most isothermal flows are accompanied by viscous effects.

A much more common compressible flow condition is that of isentropic (constant entropy) flow of a perfect gas. Such flows are reversible adiabatic processes—''no friction or heat transfer''—and are closely approximated in many physical situations. As discussed fully in Chapter 11, for isentropic flow of a perfect gas the density and pressure are related by $p/\rho^k = C$, where k is the specific heat ratio and C is a constant. Hence, the $\int dp/\rho$ integral of Eq. 3.6 can be evaluated as follows. The density can be written in terms of the pressure as $\rho = p^{1/k}C^{-1/k}$ so that Eq. 3.6 becomes

$$C^{1/k} \int p^{-1/k}\, dp + \frac{1}{2} V^2 + gz = \text{constant}$$

The pressure term can be integrated between points (1) and (2) on the streamline and the constant C evaluated at either point ($C^{1/k} = p_1^{1/k}/\rho_1$ or $C^{1/k} = p_2^{1/k}/\rho_2$) to give the following

$$C^{1/k} \int_{p_1}^{p_2} p^{-1/k}\, dp = C^{1/k} \left(\frac{k}{k-1}\right) [p_2^{(k-1)/k} - p_1^{(k-1)/k}]$$

$$= \left(\frac{k}{k-1}\right)\left(\frac{p_2}{\rho_2} - \frac{p_1}{\rho_1}\right)$$

Thus, the final form of Eq. 3.6 for compressible, isentropic, steady flow of a perfect gas is

An equation similar to the Bernoulli equation can be obtained for isentropic flow of a perfect gas.

$$\left(\frac{k}{k-1}\right)\frac{p_1}{\rho_1} + \frac{V_1^2}{2} + gz_1 = \left(\frac{k}{k-1}\right)\frac{p_2}{\rho_2} + \frac{V_2^2}{2} + gz_2 \quad \textbf{(3.24)}$$

The similarities between the results for compressible isentropic flow (Eq. 3.24) and incompressible isentropic flow (the Bernoulli equation, Eq. 3.7) are apparent. The only differences are the factors of $[k/(k-1)]$ that multiply the pressure terms and the fact that the densities are different ($\rho_1 \neq \rho_2$). In the limit of ''low-speed flow'' the two results are exactly the same as is seen by the following.

We consider the stagnation point flow of Section 3.5 to illustrate the difference between the incompressible and compressible results. As is shown in Chapter 11, Eq. 3.24 can be written in dimensionless form as

$$\frac{p_2 - p_1}{p_1} = \left[\left(1 + \frac{k-1}{2}\text{Ma}_1^2\right)^{k/k-1} - 1\right] \text{(compressible)} \quad \textbf{(3.25)}$$

where (1) denotes the upstream conditions and (2) the stagnation conditions. We have assumed $z_1 = z_2$, $V_2 = 0$, and have denoted $\text{Ma}_1 = V_1/c_1$ as the upstream *Mach number*—the ratio of the fluid velocity to the speed of sound, $c_1 = \sqrt{kRT_1}$.

A comparison between this compressible result and the incompressible result is perhaps most easily seen if we write the incompressible flow result in terms of the pressure ratio and the Mach number. Thus, we divide each term in the Bernoulli equation, $\rho V_1^2/2 + p_1 = p_2$, by p_1 and use the perfect gas law, $p_1 = \rho RT_1$, to obtain

$$\frac{p_2}{p_1} = \frac{V_1^2}{2RT_1}$$

Since $\text{Ma}_1 = V_1/\sqrt{kRT_1}$ this can be written as

$$\frac{p_2 - p_1}{p_1} = \frac{k\text{Ma}_1^2}{2} \text{(incompressible)} \quad \textbf{(3.26)}$$

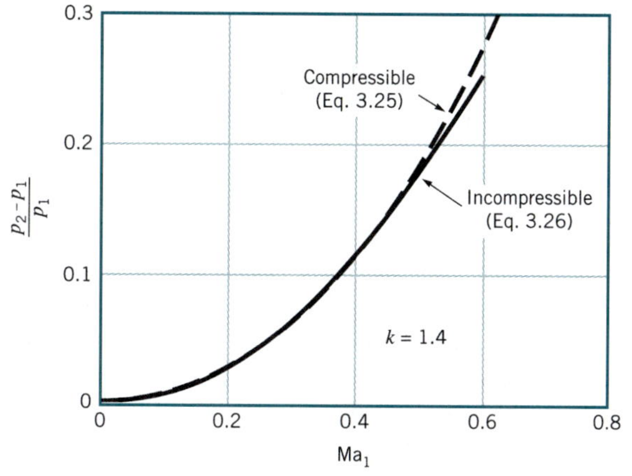

■ **FIGURE 3.24 Pressure ratio as a function of Mach number for incompressible and compressible (isentropic) flow.**

Equations 3.25 and 3.26 are plotted in Fig. 3.24. In the low-speed limit of $\text{Ma}_1 \rightarrow 0$, both of the results are the same. This can be seen by denoting $(k - 1)\text{Ma}_1^2/2 = \tilde{\varepsilon}$ and using the binomial expansion, $(1 + \tilde{\varepsilon})^n = 1 + n\tilde{\varepsilon} + n(n - 1)\,\tilde{\varepsilon}^2/2 + \cdots$, where $n = k/(k - 1)$, to write Eq. 3.25 as

$$\frac{p_2 - p_1}{p_1} = \frac{k\text{Ma}_1^2}{2}\left(1 + \frac{1}{4}\text{Ma}_1^2 + \frac{2 - k}{24}\text{Ma}_1^4 + \cdots\right) \quad \text{(compressible)}$$

For small Mach numbers the compressible and incompressible results are nearly the same.

For $\text{Ma}_1 \ll 1$ this compressible flow result agrees with Eq. 3.26. The incompressible and compressible equations agree to within about 2% up to a Mach number of approximately $\text{Ma}_1 = 0.3$. For larger Mach numbers the disagreement between the two results increases.

Thus, a ''rule of thumb'' is that the flow of a perfect gas may be considered as incompressible provided the Mach number is less than about 0.3. In standard air ($T_1 = 59°\text{F}$, $c_1 = \sqrt{kRT_1} = 1117 \text{ ft/s}$) this corresponds to a speed of $V_1 = c_1\text{Ma}_1 = 0.3\,(1117 \text{ ft/s}) = 335 \text{ ft/s} = 228 \text{ mi/hr}$. At higher speeds, compressibility may become important.

EXAMPLE 3.15

A Boeing 777 flies at Mach 0.82 at an altitude of 10 km in a standard atmosphere. Determine the stagnation pressure on the leading edge of its wing if the flow is incompressible; if the flow is compressible isentropic.

SOLUTION

From Tables 1.8 and C.2 we find that $p_1 = 26.5 \text{ kPa (abs)}$, $T_1 = -49.9\ °\text{C}$, $\rho = 0.414 \text{ kg/m}^3$, and $k = 1.4$. Thus, if we assume incompressible flow, Eq. 3.26 gives

$$\frac{p_2 - p_1}{p_1} = \frac{k\text{Ma}_1^2}{2} = 1.4\,\frac{(0.82)^2}{2} = 0.471$$

or

$$p_2 - p_1 = 0.471\,(26.5 \text{ kPa}) = 12.5 \text{ kPa} \qquad \textbf{(Ans)}$$

On the other hand, if we assume isentropic flow, Eq. 3.25 gives

$$\frac{p_2 - p_1}{p_1} = \left\{\left[1 + \frac{(1.4 - 1)}{2}(0.82)^2\right]^{1.4/(1.4-1)} - 1\right\} = 0.555$$

or

$$p_2 - p_1 = 0.555\ (26.5\ \text{kPa}) = 14.7\ \text{kPa} \qquad \textbf{(Ans)}$$

We see that at Mach 0.82 compressibility effects are of importance. The pressure (and, to a first approximation, the lift and drag on the airplane; see Chapter 9) is approximately $14.7/12.5 = 1.18$ times greater according to the compressible flow calculations. This may be very significant. As discussed in Chapter 11, for Mach numbers greater than 1 (supersonic flow) the differences between incompressible and compressible results are often not only quantitative but also qualitative.

Note that if the airplane were flying at Mach 0.30 (rather than 0.82) the corresponding values would be $p_2 - p_1 = 1.670$ kPa for incompressible flow and $p_2 - p_1 = 1.707$ kPa for compressible flow. The difference between these two results is about 2 percent.

3.8.2 Unsteady Effects

Another restriction of the Bernouilli equation (Eq. 3.7) is the assumption that the flow is steady. For such flows, on a given streamline the velocity is a function of only s, the location along the streamline. That is, along a streamline $V = V(s)$. For unsteady flows the velocity is also a function of time, so that along a streamline $V = V(s,t)$. Thus when taking the time derivative of the velocity to obtain the streamwise acceleration, we obtain $a_s = \partial V/\partial t + V\,\partial V/\partial s$ rather than just $a_s = V\,\partial V/\partial s$ as is true for steady flow. For steady flows the acceleration is due to the change in velocity resulting from a change in position of the particle (the $V\,\partial V/\partial s$ term), whereas for unsteady flow there is an additional contribution to the acceleration resulting from a change in velocity with time at a fixed location (the $\partial V/\partial t$ term). These effects are discussed in detail in Chapter 4. The net effect is that the inclusion of the unsteady term, $\partial V/\partial t$, does not allow the equation of motion to be easily integrated (as was done to obtain the Bernoulli equation) unless additional assumptions are made.

The Bernoulli equation can be modified for unsteady flows.

The Bernoulli equation was obtained by integrating the component of Newton's second law (Eq. 3.5) along the streamline. When integrated, the acceleration contribution to this equation, the $\frac{1}{2}\rho d(V^2)$ term, gave rise to the kinetic energy term in the Bernoulli equation. If the steps leading to Eq. 3.5 are repeated with the inclusion of the unsteady effect ($\partial V/\partial t \neq 0$) the following is obtained

$$\rho\frac{\partial V}{\partial t}\,ds + dp + \frac{1}{2}\rho d(V^2) + \gamma\,dz = 0 \qquad \text{(along a streamline)}$$

For incompressible flow this can be easily integrated between points (1) and (2) to give

$$p_1 + \frac{1}{2}\rho V_1^2 + \gamma z_1 = \rho\int_{s_1}^{s_2}\frac{\partial V}{\partial t}\,ds + p_2 + \frac{1}{2}\rho V_2^2 + \gamma z_2 \qquad \text{(along a streamline)} \qquad \textbf{(3.27)}$$

Equation 3.27 is an unsteady form of the Bernoulli equation valid for unsteady, incompressible, inviscid flow. Except for the integral involving the local acceleration, $\partial V/\partial t$, it is identical to the steady Bernoulli equation. In general, it is not easy to evaluate this integral because the variation of $\partial V/\partial t$ along the streamline is not known. In some situations the concepts of "irrotational flow" and the "velocity potential" can be used to simplify this integral. These topics are discussed in Chapter 6.

EXAMPLE 3.16

An incompressible, inviscid liquid is placed in a vertical, constant-diameter U-tube as indicated in Fig. E3.16. When released from the nonequilibrium position shown, the liquid column will oscillate at a specific frequency. Determine this frequency.

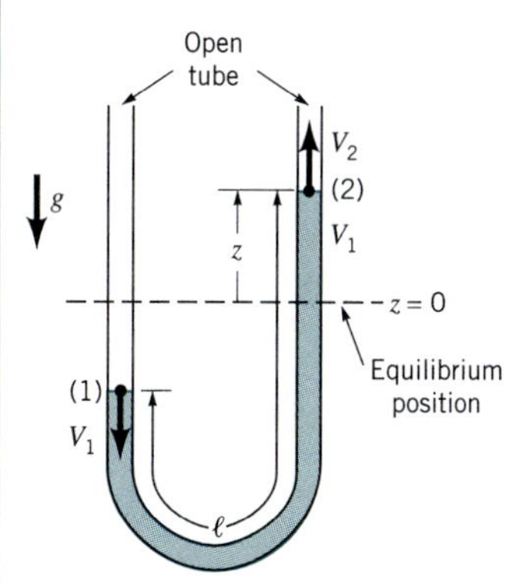

■ FIGURE E3.16

SOLUTION

The frequency of oscillation can be calculated by use of Eq. 3.27 as follows. Let points (1) and (2) be at the air-water interfaces of the two columns of the tube and $z = 0$ correspond to the equilibrium position of these interfaces. Hence, $p_1 = p_2 = 0$ and if $z_2 = z$, then $z_1 = -z$. In general, z is a function of time, $z = z(t)$. For a constant diameter tube, at any instant in time the fluid speed is constant throughout the tube, $V_1 = V_2 = V$, and the integral representing the unsteady effect in Eq. 3.27 can be written as

$$\int_{s_1}^{s_2} \frac{\partial V}{\partial t}\, ds = \frac{dV}{dt} \int_{s_1}^{s_2} ds = \ell \frac{dV}{dt}$$

where ℓ is the total length of the liquid column as shown in the figure. Thus, Eq. 3.27 can be written as

$$\gamma(-z) = \rho \ell \frac{dV}{dt} + \gamma z$$

Since $V = dz/dt$ and $\gamma = \rho g$, this can be written as the second-order differential equation describing simple harmonic motion

$$\frac{d^2 z}{dt^2} + \frac{2g}{\ell} z = 0$$

which has the solution $z(t) = C_1 \sin(\sqrt{2g/\ell}\; t) + C_2 \cos(\sqrt{2g/\ell}\; t)$. The values of the constants C_1 and C_2 depend on the initial state (velocity and position) of the liquid at $t = 0$. Thus, the liquid oscillates in the tube with a frequency

$$\omega = \sqrt{2g/\ell} \qquad \textbf{(Ans)}$$

This frequency depends on the length of the column and the acceleration of gravity (in a manner very similar to the oscillation of a pendulum). The period of this oscillation (the time required to complete an oscillation) is $t_0 = 2\pi\sqrt{\ell/2g}$.

In a few unsteady flow cases the flow can be made steady by an appropriate selection of the coordinate system. Example 3.17 illustrates this.

EXAMPLE 3.17

A submarine moves through the seawater ($SG = 1.03$) at a depth of 50 m with velocity $V_0 = 5.0$ m/s as shown in Fig. E3.17. Determine the pressure at the stagnation point (2).

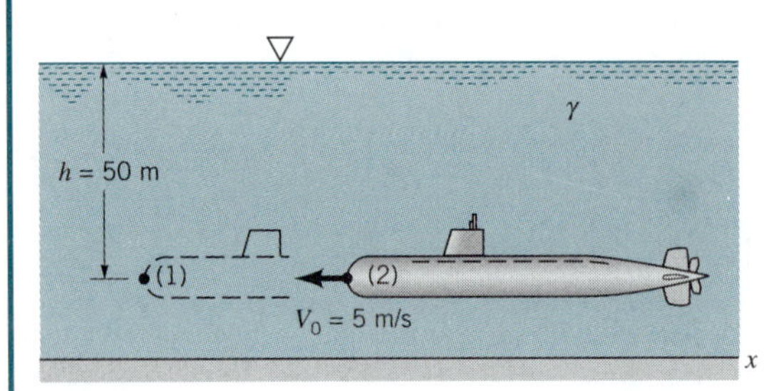

■ FIGURE E3.17

SOLUTION

In a coordinate system fixed to the ground, the flow is unsteady. For example, the water velocity at (1) is zero with the submarine in its initial position, but at the instant when the nose, (2), reaches point (1) the velocity there becomes $\mathbf{V}_1 = -V_0\,\hat{\mathbf{i}}$. Thus, $\partial \mathbf{V}_1/\partial t \neq 0$ and the flow is unsteady. Application of the steady Bernoulli equation between (1) and (2) would give the incorrect result that "$p_1 = p_2 + \rho V_0^2/2$." According to this result the static pressure is greater than the stagnation pressure—an incorrect use of the Bernoulli equation.

We can either use an unsteady analysis for the flow (which is outside the scope of this text) or redefine the coordinate system so that it is fixed on the submarine, giving steady flow with respect to this system. The correct method would be

$$p_2 = \frac{\rho V_1^2}{2} + \gamma h = [(1.03)(1000)\text{ kg/m}^3]\,(5.0\text{ m/s})^2/2$$

$$+ (9.80 \times 10^3\text{ N/m}^3)(1.03)(50\text{ m})$$

$$= (12{,}900 + 505{,}000)\text{ N/m}^2 = 518\text{ kPa} \qquad \textbf{(Ans)}$$

similar to that discussed in Example 3.2.

If the submarine were accelerating, $\partial V_0/\partial t \neq 0$, the flow would be unsteady in either of the above coordinate systems and we would be forced to use an unsteady form of the Bernoulli equation.

Some unsteady flows may be treated as "quasisteady" and solved approximately by using the steady Bernoulli equation. In these cases the unsteadiness is "not too great" (in some sense), and the steady flow results can be applied at each instant in time as though the flow were steady. The slow draining of a tank filled with liquid provides an example of this type of flow.

3.8.3 Rotational Effects

Care must be used in applying the Bernoulli equation across streamlines.

Another of the restrictions of the Bernoulli equation is that it is applicable along the streamline. Application of the Bernoulli equation across streamlines (i.e., from a point on one streamline to a point on another streamline) can lead to considerable errors, depending on the particular flow conditions involved. In general, the Bernoulli constant varies from streamline to streamline. However, under certain restrictions this constant is the same throughout the entire flow field. Example 3.18 illustrates this fact.

EXAMPLE 3.18

Consider the uniform flow in the channel shown in Fig. E3.18*a*. Discuss the use of the Bernoulli equation between points (1) and (2), points (3) and (4), and points (4) and (5). The liquid in the vertical piezometer tube is stationary.

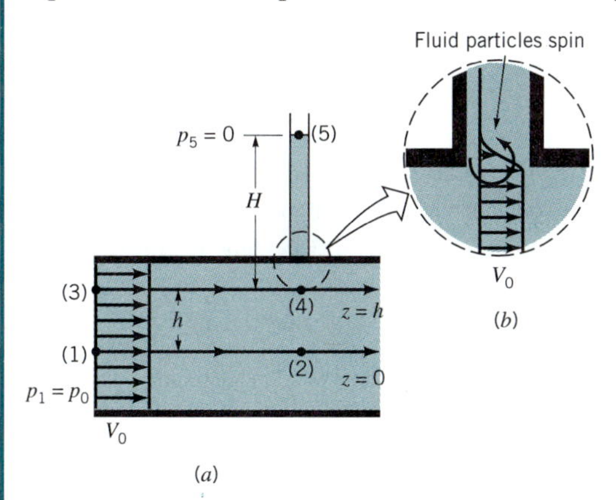

■ **FIGURE E3.18**

SOLUTION

If the flow is steady, inviscid, and incompressible, Eq. 3.7 written between points (1) and (2) gives

$$p_1 + \tfrac{1}{2}\rho V_1^2 + \gamma z_1 = p_2 + \tfrac{1}{2}\rho V_2^2 + \gamma z_2 = \text{constant} = C_{12}$$

Since $V_1 = V_2 = V_0$ and $z_1 = z_2 = 0$, it follows that $p_1 = p_2 = p_0$ and the Bernoulli constant for this streamline, C_{12}, is given by

$$C_{12} = \tfrac{1}{2}\rho V_0^2 + p_0$$

Along the streamline from (3) to (4) we note that $V_3 = V_4 = V_0$ and $z_3 = z_4 = h$. As was shown in Example 3.5, application of $\mathbf{F} = m\mathbf{a}$ across the streamline (Eq. 3.12) gives $p_3 = p_1 - \gamma h$ because the streamlines are straight and horizontal. The above facts combined with the Bernoulli equation applied between (3) and (4) show that $p_3 = p_4$ and that the Bernoulli constant along this streamline is the same as that along the streamline between (1) and (2). That is, $C_{34} = C_{12}$, or

$$p_3 + \tfrac{1}{2}\rho V_3^2 + \gamma z_3 = p_4 + \tfrac{1}{2}\rho V_4^2 + \gamma z_4 = C_{34} = C_{12}$$

Similar reasoning shows that the Bernoulli constant is the same for any streamline in Fig. E3.18. Hence,

$$p + \tfrac{1}{2}\rho V^2 + \gamma z = \text{constant throughout the flow in the channel}$$

Again from Example 3.5 we recall that

$$p_4 = p_5 + \gamma H = \gamma H$$

If we apply the Bernoulli equation across streamlines from (4) to (5) we obtain the incorrect result "$H = p_4/\gamma + V_4^2/2g$." The correct result is $H = p_4/\gamma$.

From the above we see that we can apply the Bernoulli equation across streamlines (1)–(2) and (3)–(4) (that is, $C_{12} = C_{34}$) but not across streamlines from (4) to (5). The reason for this is that while the flow in the channel is "irrotational," it is "rotational" between the flowing fluid in the channel and the stationary fluid in the piezometer tube. Because of the uniform velocity profile across the channel, it is seen that the fluid particles do not rotate or "spin" as they move. The flow is "irrotational." However, as seen in Fig. E3.18*b*, there is a very thin shear layer between (4) and (5) in which adjacent fluid particles interact and rotate or ''spin''. This produces a ''rotational'' flow. A more complete analysis would show that the Bernoulli equation cannot be applied across streamlines if the flow is ''rotational'' (see Chapter 6).

As is suggested by Example 3.18, if the flow is ''irrotational'' (that is, the fluid particles do not ''spin'' as they move), it is appropriate to use the Bernoulli equation across streamlines. However, if the flow is ''rotational'' (fluid particles ''spin''), use of the Bernoulli equation is restricted to flow along a streamline. The distinction between irrotational and rotational flow is often a very subtle and confusing one. These topics are discussed in more detail in Chapter 6. A thorough discussion can be found in more advanced texts (Ref. 3).

3.8.4 Other Restrictions

Another restriction on the Bernoulli equation is that the flow is inviscid. As is discussed in Section 3.4, the Bernoulli equation is actually a first integral of Newton's second law along a streamline. This general integration was possible because, in the absence of viscous effects, the fluid system considered was a conservative system. The total energy of the system remains constant. If viscous effects are important the system is nonconservative and energy losses occur. A more detailed analysis is needed for these cases. Such material is presented in Chapter 8.

The Bernoulli equation is not valid for flows that involve pumps or turbines.

The final basic restriction on use of the Bernoulli equation is that there are no mechanical devices (pumps or turbines) in the system between the two points along the streamline for which the equation is applied. These devices represent sources or sinks of energy. Since the Bernoulli equation is actually one form of the energy equation, it must be altered to include pumps or turbines, if these are present. The inclusion of pumps and turbines is covered in Chapters 5 and 12.

In this chapter we have spent considerable time investigating fluid dynamic situations governed by a relatively simple analysis for steady, inviscid, incompressible flows. Many flows can be adequately analyzed by use of these ideas. However, because of the rather severe restrictions imposed, many others cannot. An understanding of these basic ideas will provide a firm foundation for the remainder of the topics in this book.

References

1. Riley, W. F., and Sturges, L. D., *Engineering Mechanics: Dynamics*, 2nd Ed., Wiley, New York, 1996.
2. Tipler, P. A., *Physics*, Worth, New York, 1982.
3. Panton, R. L., *Incompressible Flow*, Wiley, New York, 1984.

Review Problems

Note: Problems designated with (R) are review problems. The phrases within parentheses refer to the main topics to be used in solving the problems. Complete, detailed solutions to these review problems can be found in the supplement titled *Student Solution Manual for Fundamentals of Fluid Mechanics*, by Munson, Young, and Okiishi (John Wiley and Sons, New York, 1997).

3.1R (F = *ma* along streamline) What pressure gradient along the streamline, dp/ds, is required to accelerate air at standard temperature and pressure in a horizontal pipe at a rate of 300 ft/s^2?

(ANS: -0.714 lb/ft^3)

3.2R (F = *ma* normal to streamline) An incompressible, inviscid fluid flows steadily with circular streamlines around a horizontal bend as shown in Fig. P3.2R. The radial variation of the velocity profile is given by $rV = r_0V_0$, where V_0 is the velocity at the inside of the bend which has radius $r = r_0$. Determine the pressure variation across the bend in terms of V_0, r_0, ρ, r, and p_0, where p_0 is the pressure at $r = r_0$. Plot the pressure distribution, $p = p(r)$, if $r_0 = 1.2$ m, $r_1 = 1.8$ m, $V_0 = 12$ m/s, $p_0 = 20$ kN/m^2, and the fluid is water. Neglect gravity.

(ANS: $p_0 + 0.5\rho V_0^2[1 - (r_0/r)^2]$)

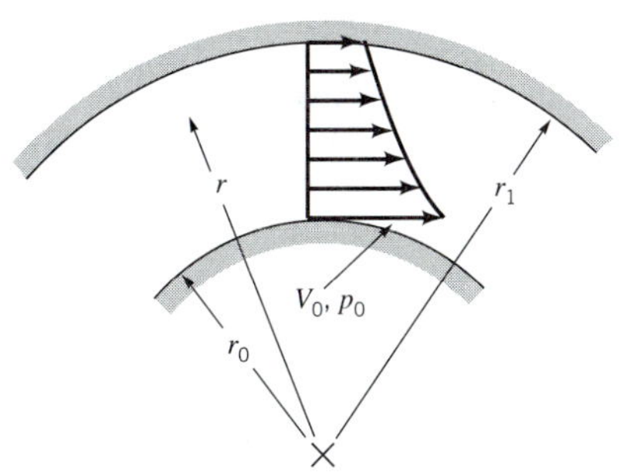

■ FIGURE P3.2R

3.3R (Stagnation pressure) A hang glider soars through standard sea level air with an airspeed of 10 m/s. What is the gage pressure at a stagnation point on the structure?

(ANS: 61.5 Pa)

3.4R (Bernoulli equation) The pressure in domestic water pipes is typically 60 psi above atmospheric. If viscous effects are neglected, determine the height reached by a jet of water through a small hole in the top of the pipe.

(ANS: 138 ft)

3.5R (Heads) A 4-in.-diameter pipe carries 300 gal/min of water at a pressure of 30 psi. Determine **(a)** the pressure head in feet of water, **(b)** the velocity head, and **(c)** the total head with reference to a datum plane 20 ft below the pipe.

(ANS: 69.2 ft; 0.909 ft; 90.1 ft)

3.6R (Free jet) Water flows from a nozzle of triangular cross section as shown in Fig. P3.6R. After it has fallen a distance of 2.7 ft, its cross section is circular (because of surface tension effects) with a diameter $D = 0.11$ ft. Determine the flowrate, Q.

(ANS: 0.158 ft^3/s)

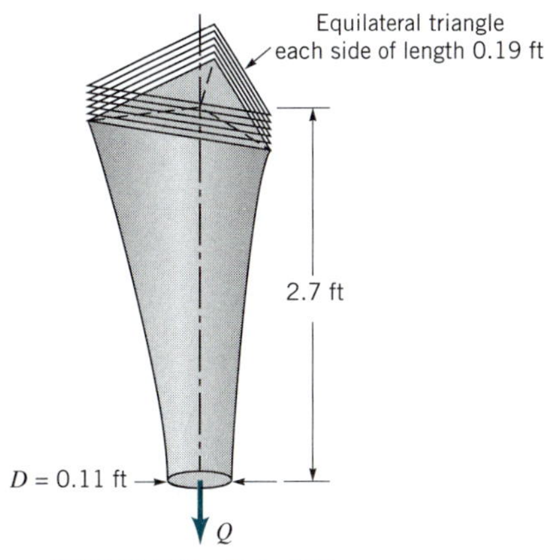

■ FIGURE P3.6R

3.7R (Bernoulli/continuity) Water flows into a large tank at a rate of 0.011 m^3/s as shown in Fig. P3.7R. The water leaves the tank through 20 holes in the bottom of the tank, each of which produces a stream of 10-mm diameter. Determine the equilibrium height, h, for steady state operation.

(ANS: 2.50 m)

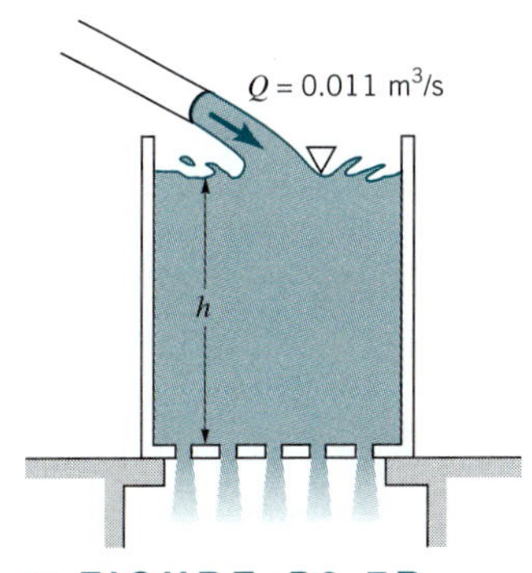

■ FIGURE P3.7R

3.8R (Bernoulli/continuity) Gasoline flows from a 0.3-m-diameter pipe in which the pressure is 300 kPa into a 0.15-m-diameter pipe in which the pressure is 120 kPa. If the pipes are horizontal and viscous effects are negligible, determine the flowrate.

(ANS: 0.420 m^3/s)

3.9R (Bernoulli/continuity) Water flows steadily through the pipe shown in Fig. P3.9R such that the pressures at sections (1) and (2) are 300 kPa and 100 kPa, respectively. Determine the diameter of the pipe at section (2), D_2, if the velocity at section 1 is 20 m/s and viscous effects are negligible.

(ANS: 0.0688 m)

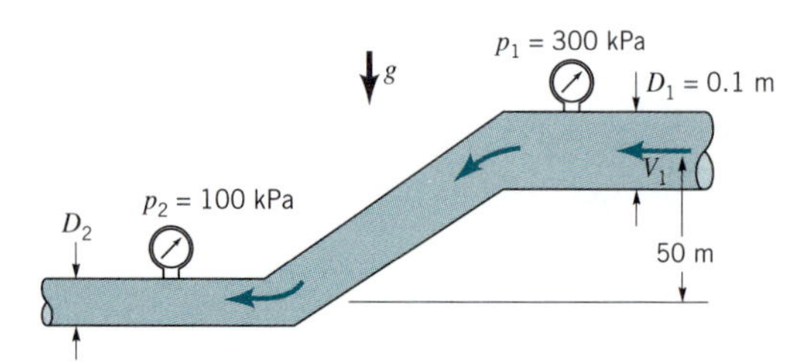

■ FIGURE P3.9R

3.10R (Bernoulli/continuity) Water flows steadily through a diverging tube as shown in Fig. P3.10R. Determine the velocity, V, at the exit of the tube if frictional effects are negligible.

(ANS: 1.04 ft/s)

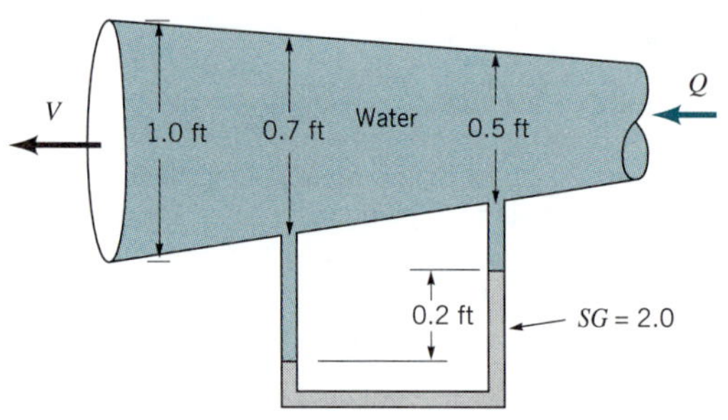

■ FIGURE P3.10R

3.11R (Bernoulli/continuity/Pitot tube) Two Pitot tubes and two static pressure taps are placed in the pipe contraction shown in Fig. P3.11R. The flowing fluid is water, and viscous effects are negligible. Determine the two manometer readings, h and H.

(ANS: 0; 0.252 ft)

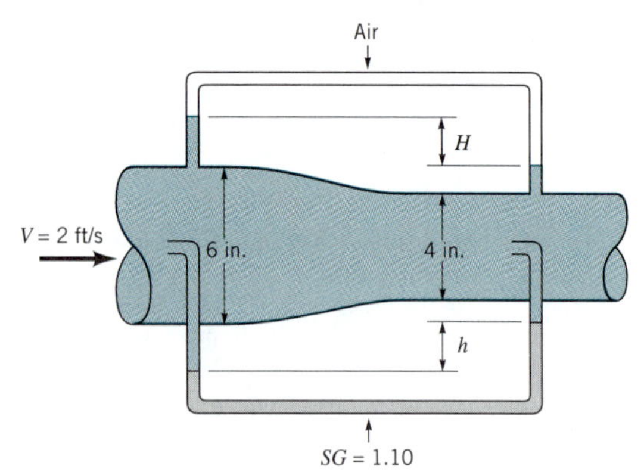

■ FIGURE P3.11R

3.12R (Bernoulli/continuity) Water collects in the bottom of a rectangular oil tank as shown in Fig. P3.12R. How long will it take for the water to drain from the tank through a 0.02-m-diameter drain hole in the bottom of the tank? Assume quasi-steady flow.

(ANS: 2.45 hr)

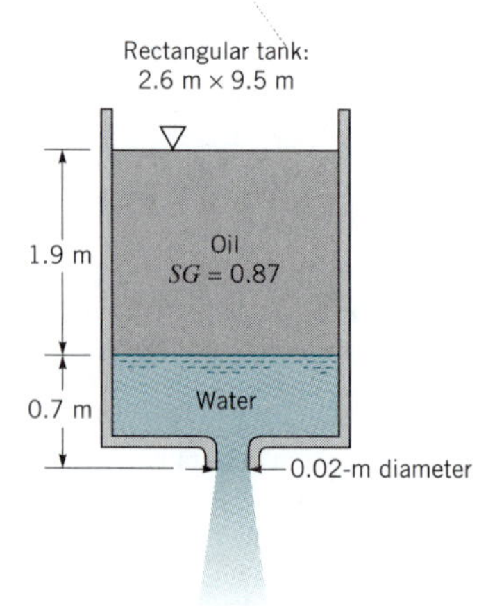

■ FIGURE P3.12R

3.13R (Cavitation) Water flows past the hydrofoil shown in Fig. P3.13R with an upstream velocity of V_0. A more advanced analysis indicates that the maximum velocity of the water in the entire flow field occurs at point B and is equal to $1.1V_0$. Calculate the velocity, V_0, at which cavitation will begin if the atmospheric pressure is 101 kPa (abs) and the vapor pressure of the water is 3.2 kPa (abs).

(ANS: 31.4 m/s)

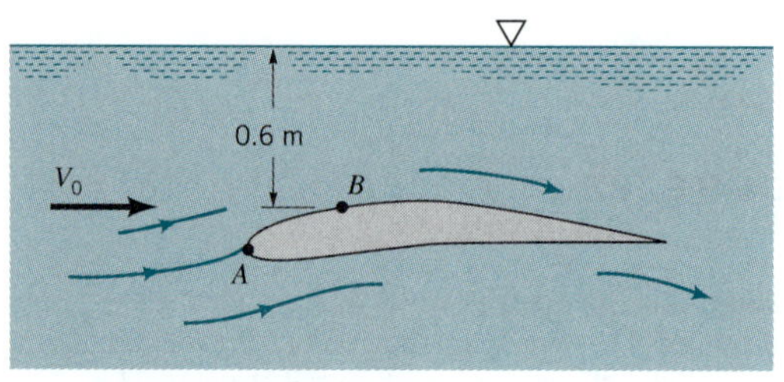

■ FIGURE P3.13R

3.14R (Flowrate) Water flows through the pipe contraction shown in Fig. P3.14R. For the given 0.2-m difference in manometer level, determine the flowrate as a function of the diameter of the small pipe, D.

(ANS: 0.0156 m³/s)

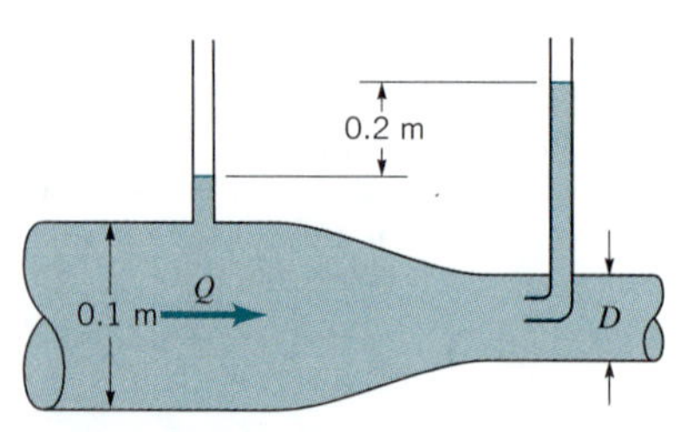

■ FIGURE P3.14R

3.15R (Channel flow) Water flows down the ramp shown in the channel of Fig. P3.15R. The channel width decreases from 15 ft at section (1) to 9 ft at section (2). For the conditions shown, determine the flowrate.

(ANS: 509 ft³/s)

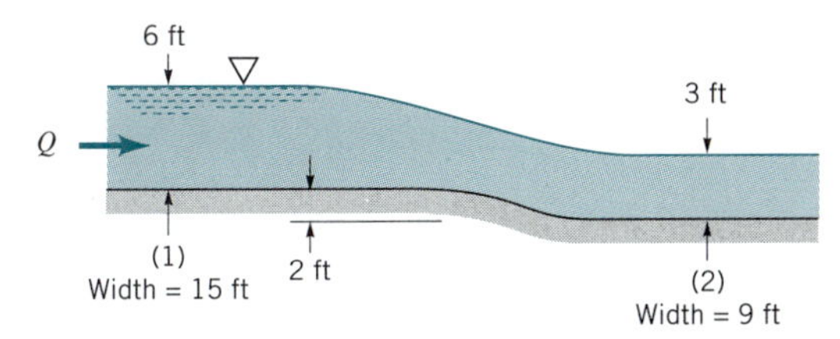

■ FIGURE P3.15R

3.16R (Channel flow) Water flows over the spillway shown in Fig. P3.16R. If the velocity is uniform at sections (1) and (2) and viscous effects are negligible, determine the flowrate per unit width of the spillway.

(ANS: 7.44 m^2/s)

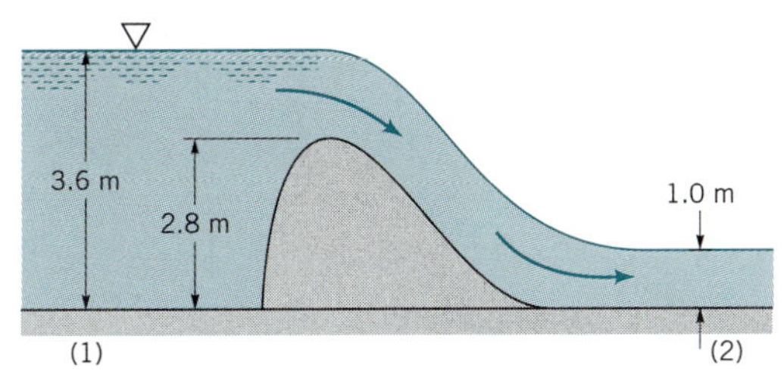

■ FIGURE P3.16R

3.17R (Energy line/hydraulic grade line) Draw the energy line and hydraulic grade line for the flow shown in Problem 3.43.

3.18R (Restrictions on Bernoulli equation) A 0.3-m-diameter soccer ball, pressurized to 20 kPa, develops a small leak with an area equivalent to 0.006 mm^2. If viscous effects are neglected and the air is assumed to be incompressible, determine the flowrate through the hole. Would the ball become noticeably softer during a 1-hr soccer game? Explain. Is it reasonable to assume incompressible flow for this situation? Explain.

(ANS: 9.96×10^{-7} m^3/s; yes; no, Ma > 0.3)

3.19R (Restrictions on Bernoulli equation) Niagara Falls is approximately 167 ft high. If the water flows over the crest of the falls with a velocity of 8 ft/s and viscous effects are neglected, with what velocity does the water strike the rocks at the bottom of the falls? What is the maximum pressure of the water on the rocks? Repeat the calculations for the 1430-ft-high Upper Yosemite Falls in Yosemite National Park. Is it reasonable to neglect viscous effects for these falls? Explain.

(ANS: 104 ft/s, 72.8 psi; 304 ft/s, 620 psi; no)

Problems

Note: Unless otherwise indicated use the values of fluid properties found in the tables on the inside of the front cover. Problems designated with an (*) are intended to be solved with the aid of a programmable calculator or a computer. Problems designated with a (†) are ''open-ended'' problems and require critical thinking in that to work them one must make various assumptions and provide the necessary data. There is not a unique answer to these problems.

3.1 Water flows steadily through the variable area horizontal pipe shown in Fig. P3.1. The centerline velocity is given by $\mathbf{V} = 10(1 + x)\hat{\mathbf{i}}$ ft/s, where x is in feet. Viscous effects are neglected. **(a)** Determine the pressure gradient, $\partial p/\partial x$, (as a function of x) needed to produce this flow. **(b)** If the pressure at section (1) is 50 psi, determine the pressure at (2) by (i) integration of the pressure gradient obtained in **(a)**, (ii) application of the Bernoulli equation.

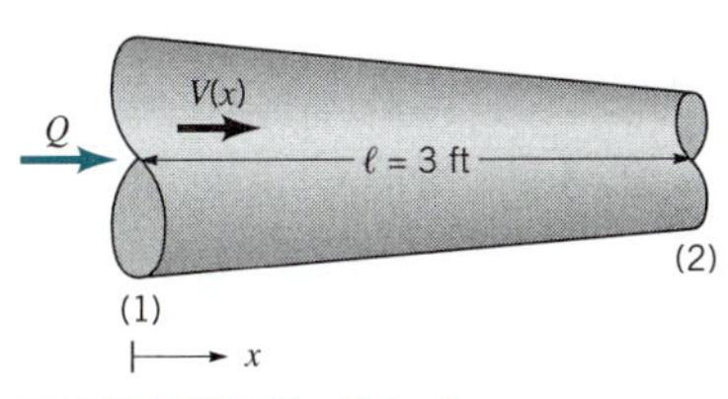

■ FIGURE P3.1

3.2 Repeat Problem 3.1 if the pipe is vertical with the flow down.

3.3 An incompressible fluid flows steadily past a circular cylinder as shown in Fig. P3.3. The fluid velocity along the dividing streamline ($-\infty \leq x \leq -a$) is found to be $V = V_0(1 - a^2/x^2)$, where a is the radius of the cylinder and V_0 is the upstream velocity. **(a)** Determine the pressure gradient along this streamline. **(b)** If the upstream pressure is p_0, integrate the pressure gradient to obtain the pressure $p(x)$ for $-\infty \leq x \leq -a$. **(c)** Show from the result of part **(b)** that the pressure at the stagnation point ($x = -a$) is $p_0 + \rho V_0^2/2$, as expected from the Bernoulli equation.

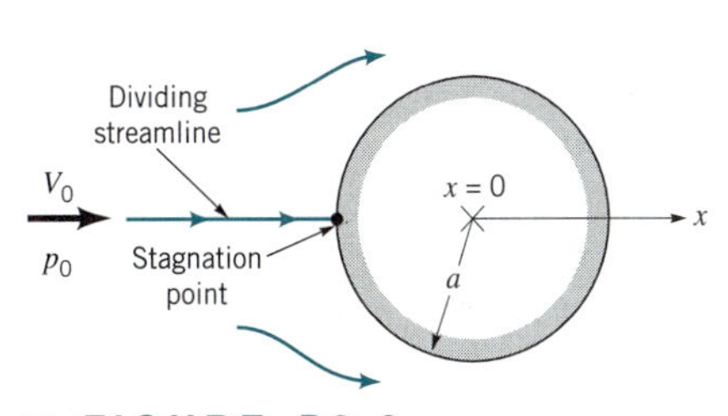

■ FIGURE P3.3

3.4 What pressure gradient along the streamline, dp/ds, is required to accelerate water in a horizontal pipe at a rate of 30 m/s^2?

3.5 At a given location the air speed is 20 m/s and the pressure gradient along the streamline is 100 N/m^3. Estimate the air speed at a point 0.5 m farther along the streamline.

3.6 What pressure gradient along the streamline, dp/ds, is required to accelerate water upward in a vertical pipe at a rate of 30 ft/s^2? What is the answer if the flow is downward?

3.7 Consider a compressible fluid for which the pressure and density are related by $p/\rho^n = C_0$, where n and C_0 are constants. Integrate the equation of motion along the streamline, Eq. 3.6, to obtain the "Bernoulli equation" for this compressible flow as $[n/(n - 1)]p/\rho + V^2/2 + gz = \text{constant}$.

3.8 The Bernoulli equation is valid for steady, inviscid, incompressible flows with constant acceleration of gravity. Consider flow on a planet where the acceleration of gravity varies with height so that $g = g_0 - cz$, where g_0 and c are constants. Integrate "$\mathbf{F} = m\mathbf{a}$" along a streamline to obtain the equivalent of the Bernoulli equation for this flow.

3.9 Consider a compressible liquid that has a constant bulk modulus. Integrate "$\mathbf{F} = m\mathbf{a}$" along a streamline to obtain the equivalent of the Bernoulli equation for this flow. Assume steady, inviscid flow.

3.10 Water flows around the vertical two-dimensional bend with circular streamlines and constant velocity as shown in Fig. P3.10. If the pressure is 40 kPa at point (1), determine the pressures at points (2) and (3). Assume that the velocity profile is uniform as indicated.

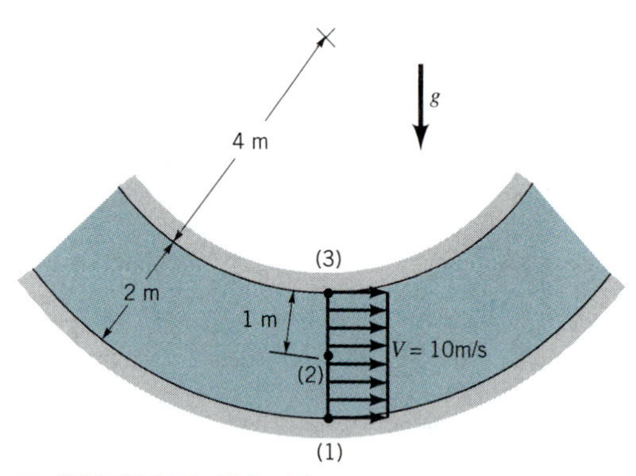

■ **FIGURE P3.10**

† **3.11** Air flows smoothly past your face as you ride your bike, but bugs and particles of dust pelt your face and get into your eyes. Explain why this is so.

***3.12** Water flows around a vertical two-dimensional bend with circular streamlines as is shown in Fig. P3.12. The pressure at point (1) is measured to be $p_1 = 25$ psi and the velocity across section a–a is as indicated in the table. Calculate and plot the pressure across section a–a of the channel [$p = p(z)$ for $0 \le z \le 2$ ft].

z (ft)	V (ft/s)
0	0
0.2	8.0
0.4	14.3
0.6	20.0
0.8	19.5
1.0	15.6
1.2	8.3
1.4	6.2
1.6	3.7
1.8	2.0
2.0	0

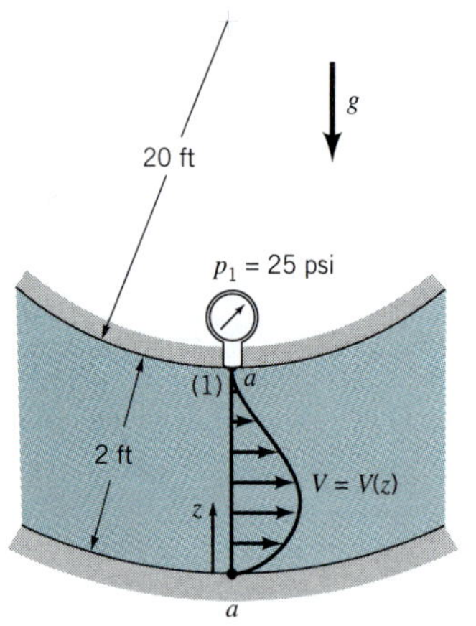

■ **FIGURE P3.12**

3.13 Some animals have learned to take advantage of the Bernoulli effect without having read a fluid mechanics book. For example, a typical prairie dog burrow contains two entrances—a flat front door, and a mounded back door as shown in Fig. P3.13. When the wind blows with velocity V_0 across the front door, the average velocity across the back door is greater than V_0 because of the mound. Assume the air velocity across the back door is $1.07V_0$. For a wind velocity of 6 m/s, what pressure difference, $p_1 - p_2$, is generated to provide a fresh air flow within the burrow?

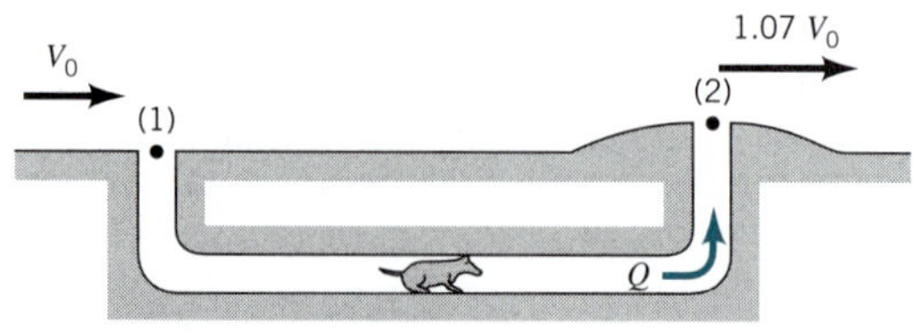

■ **FIGURE P3.13**

3.14 Water flows from the faucet on the first floor of the building shown in Fig. P3.14 with a maximum velocity of 20 ft/s. For steady inviscid flow, determine the maximum water velocity from the basement faucet and from the faucet on the second floor (assume each floor is 12 ft tall).

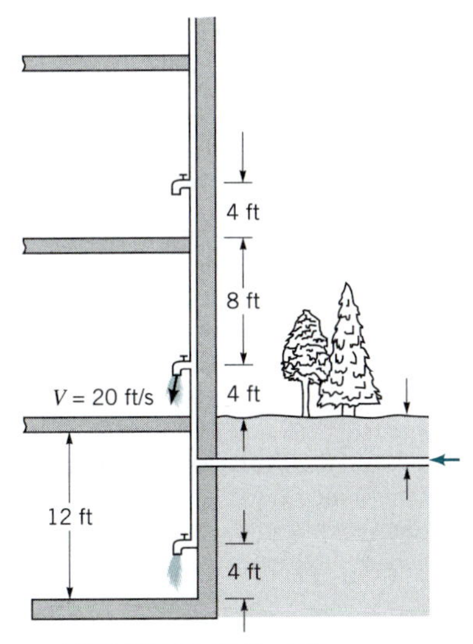

■ FIGURE P3.14

† **3.15** A drop of water in a zero-g environment (as in the Space Shuttle) will assume a spherical shape as shown in Fig. P3.15*a*. A raindrop in the cartoons is typically drawn as in Fig. P3.15*b*. The shape of an actual raindrop is more nearly like that shown in Fig. 3.15*c*. Discuss why these shapes are as indicated.

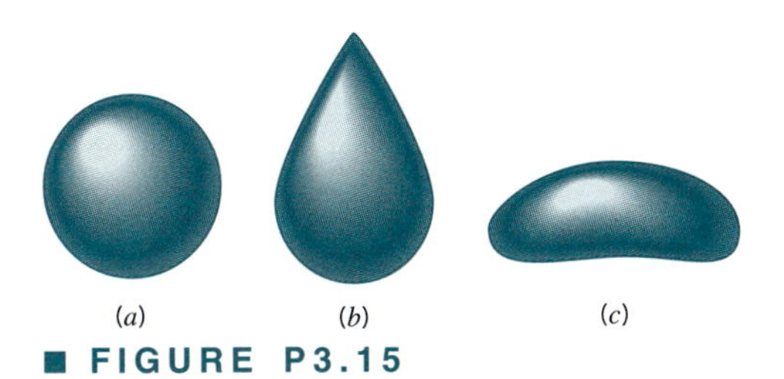

■ FIGURE P3.15

3.16 Observations show that it is not possible to blow the table tennis ball from the funnel shown in Fig. P3.16*a*. In fact, the ball can be kept in an inverted funnel, Fig. P3.16*b*, by blowing through it. The harder one blows through the funnel, the harder the ball is held within the funnel. Explain this phenomenon.

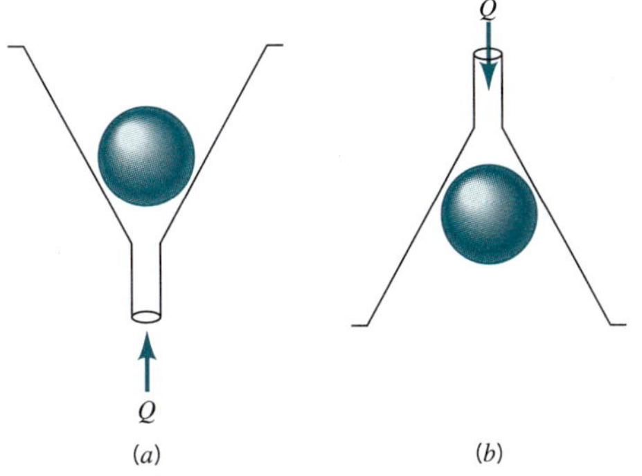

■ FIGURE P3.16

† **3.17** Estimate the pressure needed at the pumper truck in order to shoot water from the street level onto a fire on the roof of a five-story building. List all assumptions and show all calculations.

3.18 A fire hose nozzle has a diameter of $1\frac{1}{8}$ in. According to some fire codes, the nozzle must be capable of delivering at least 250 gal/min. If the nozzle is attached to a 3-in.-diameter hose, what pressure must be maintained just upstream of the nozzle to deliver this flowrate?

3.19 Water flows from a garden hose nozzle with a velocity of 15 m/s. What is the maximum height that it can reach above the nozzle?

3.20 A jet of water flows from a nozzle of diameter d_0 with speed V_0 as shown in Fig. P3.20*a*. If viscous effects are negligible, determine the jet diameter as a function of elevation, $d(z)$. Repeat your analysis for a situation in which the flowing fluid is the same as that into which it flows. For example, consider a jet of air injected into surrounding air as shown in Fig. P3.20*b* and show that the diameter remains constant, independent of z.

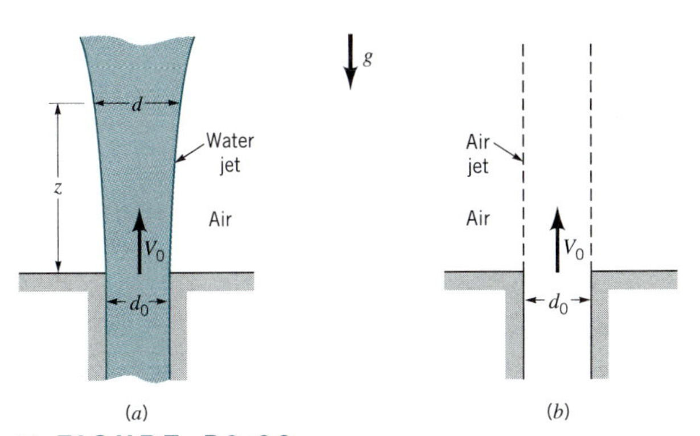

■ FIGURE P3.20

***3.21** Water flows from a pipe of diameter 20 mm with a flowrate Q as shown in Fig. P3.21. Plot the diameter of the water stream, d, as a function of distance below the faucet, h, for values of $0 \le h \le 1$ m and $0 \le Q \le 0.004\ \text{m}^3/\text{s}$. Discuss the validity of the one-dimensional assumption used to calculate $d = d(h)$, noting, in particular, the conditions of small h and small Q.

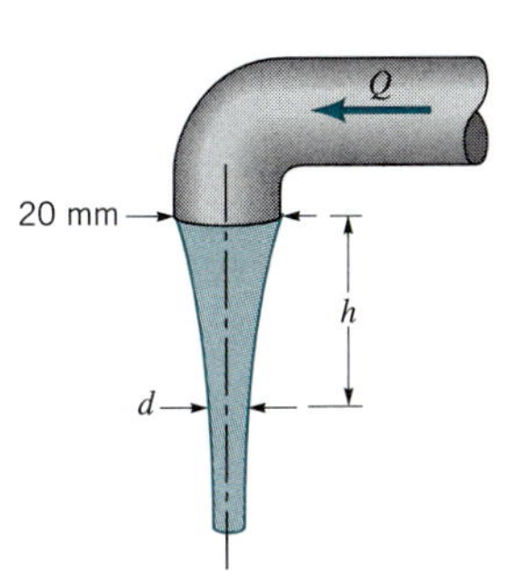

■ FIGURE P3.21

3.22 A person holds her hand out of an open car window while the car drives through still air at 65 mph. Under standard atmospheric conditions, what is the maximum pressure on her hand? What would be the maximum pressure if the ''car'' were an Indy 500 racer traveling 220 mph?

† **3.23** Estimate the pressure on your hand when you hold it in the stream of air coming from the air hose at a filling station. List all assumptions and show calculations. Warning: Do not try this experiment; it can be dangerous!

3.24 A 40-mph wind blowing past your house speeds up as it flows up and over the roof. If elevation effects are negligible, determine **(a)** the pressure at the point on the roof where the speed is 60 mph if the pressure in the free stream blowing toward your house is 14.7 psia. Would this effect tend to push the roof down against the house, or would it tend to lift the roof? **(b)** Determine the pressure on a window facing the wind if the window is assumed to be a stagnation point.

† **3.25** Estimate the maximum pressure on the surface of your car when you wash it using a garden hose connected to your outside faucet. List all assumptions and show calculations.

3.26 Small-diameter, high-pressure liquid jets can be used to cut various materials as shown in Fig. P3.26. If viscous effects are negligible, estimate the pressure needed to produce a 0.10-mm-diameter water jet with a speed of 700 m/s. Determine the flowrate.

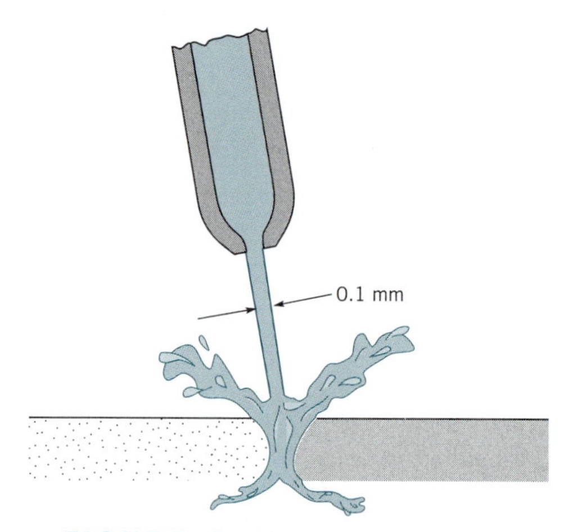

■ **FIGURE P3.26**

3.27 Air is drawn into a small open-circuit wind tunnel as shown in Fig. P3.27. Atmospheric pressure is 98.7 kPa (abs) and the temperature is 27 °C. If viscous effects are negligible, determine the pressure at the stagnation point on the nose of the airplane. Also determine the manometer reading, h, for the manometer attached to the static pressure tap within the test section of the wind tunnel if the air velocity within the test section is 50 m/s.

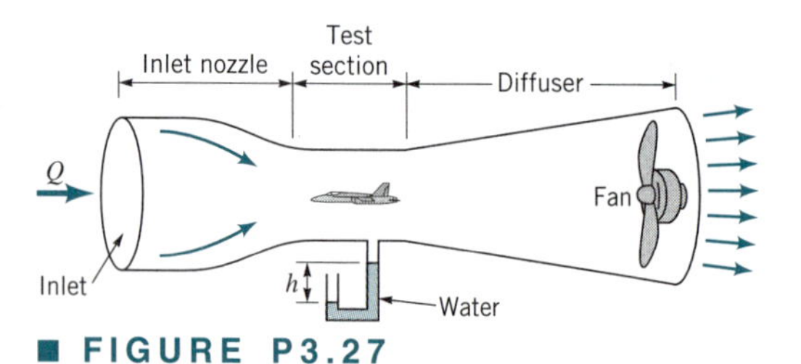

■ **FIGURE P3.27**

3.28 A loon is a diving bird equally at home ''flying'' in the air or water. What swimming velocity under water will produce a dynamic pressure equal to that when it flies in the air at 40 mph?

3.29 Water (assumed frictionless and incompressible) flows steadily from a large tank and exits through a vertical, constant diameter pipe as shown in Fig. P3.29. The air in the tank is pressurized to 50 kN/m². Determine **(a)** the height, h, to which the water rises, **(b)** the water velocity in the pipe, and **(c)** the pressure in the horizontal part of the pipe.

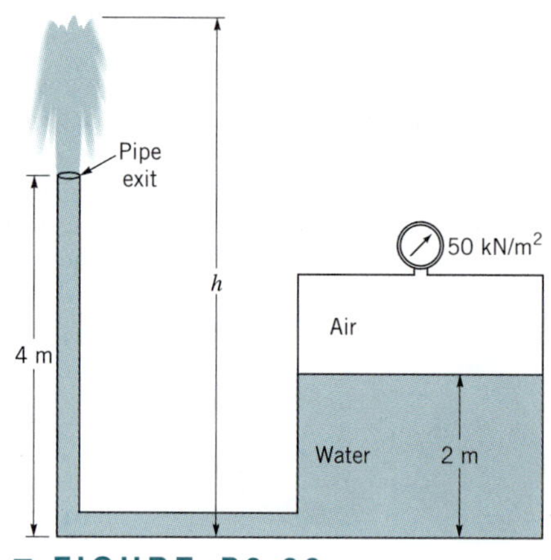

■ **FIGURE P3.29**

3.30 Water flows through the pipe contraction shown in Fig. P3.30. For the given 0.2-m difference in manometer level, determine the flowrate as a function of the diameter of the small pipe, D.

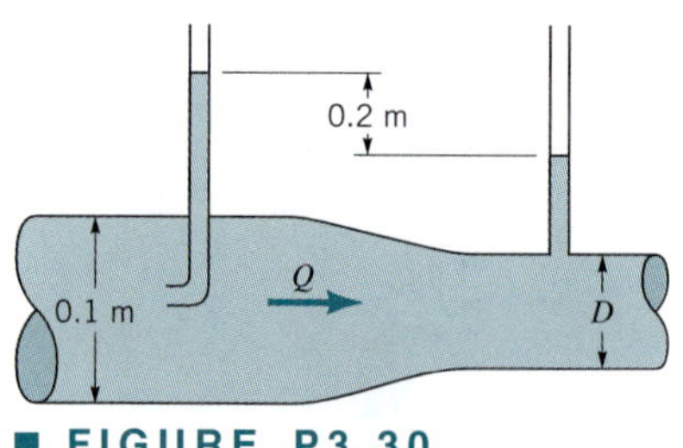

■ **FIGURE P3.30**

3.31 Water flows through the pipe contraction shown in Fig. P3.31. For the given 0.2-m difference in the manometer level, determine the flowrate as a function of the diameter of the small pipe, D.

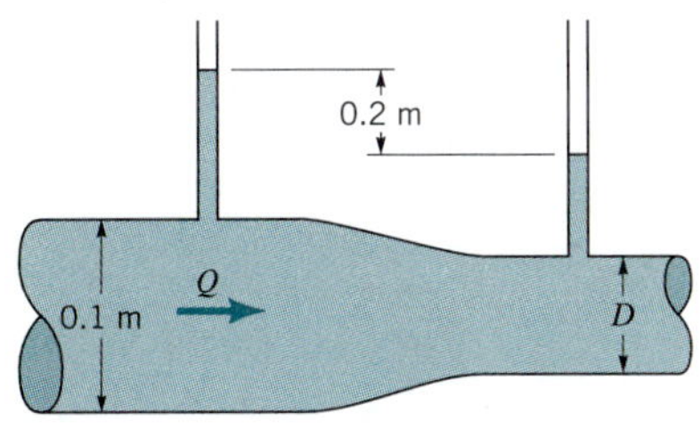

■ FIGURE P3.31

3.32 Water flows through the pipe contraction shown in Fig. P3.32. Determine the difference in manometer level, h, for a flowrate of $0.10 \text{ m}^3/\text{s}$.

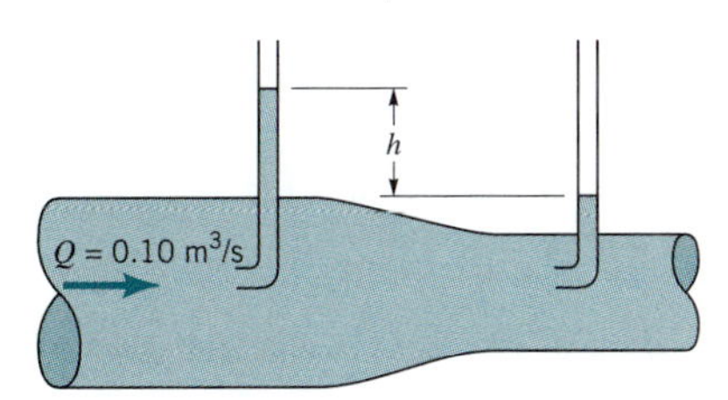

■ FIGURE P3.32

3.33 A Bourdon-type pressure gage is used to measure the pressure from a Pitot tube attached to the leading edge of an airplane wing. The gage is calibrated to read in miles per hour at standard sea level conditions (rather than psi). If the airspeed meter indicates 150 mph when flying at an altitude of 10,000 ft, what is the true airspeed?

3.34 Streams of water from two tanks impinge upon each other as shown in Fig. P3.34. If viscous effects are negligible and point A is a stagnation point, determine the height h.

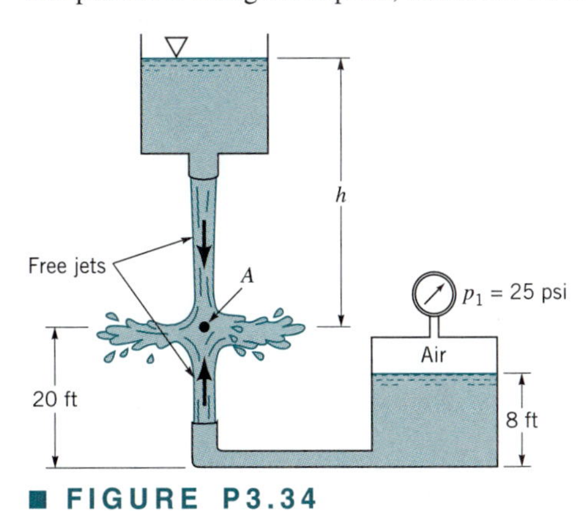

■ FIGURE P3.34

3.35 A 0.15-m-diameter pipe discharges into a 0.10-m-diameter pipe. Determine the velocity head in each pipe if they are carrying $0.12 \text{ m}^3/\text{s}$ of kerosene.

3.36 Water flows upward through a variable area pipe with a constant flowrate, Q, as shown in Fig. P3.36. If viscous effects are negligible, determine the diameter, $D(z)$, in terms of D_1 if the pressure is to remain constant throughout the pipe. That is, $p(z) = p_1$.

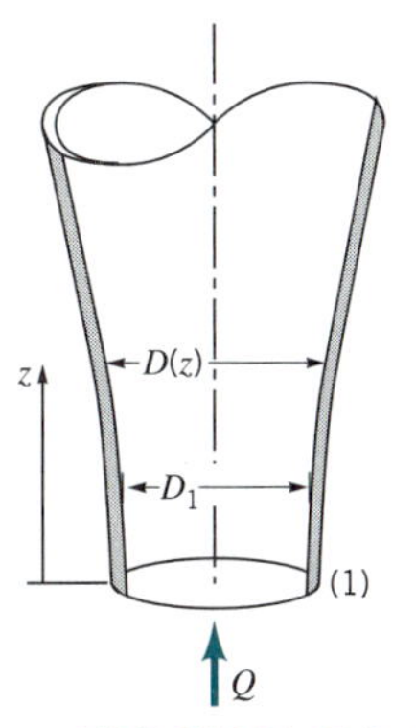

■ FIGURE P3.36

3.37 Water flows steadily from the pipe shown in Fig. P3.37 with negligible viscous effects. Determine the maximum flowrate if the water is not to flow from the open vertical tube at A.

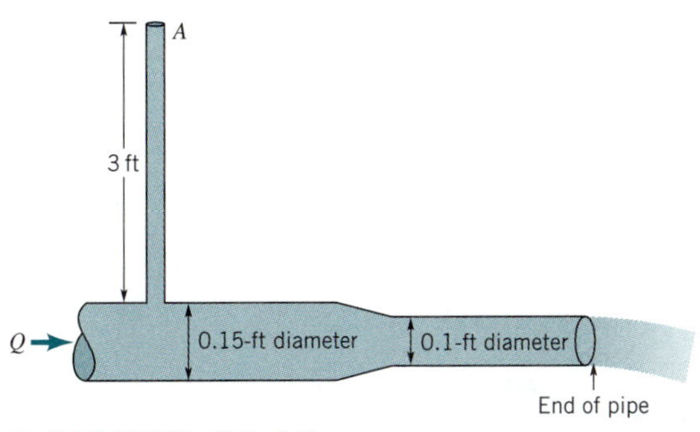

■ FIGURE P3.37

3.38 The circular stream of water from a faucet is observed to taper from a diameter of 20 mm to 10 mm in a distance of 50 cm. Determine the flowrate.

3.39 Water is siphoned from the tank shown in Fig. P3.39. The water barometer indicates a reading of 30.2 ft. Determine the maximum value of h allowed without cavitation occurring. Note that the pressure of the vapor in the closed end of the barometer equals the vapor pressure.

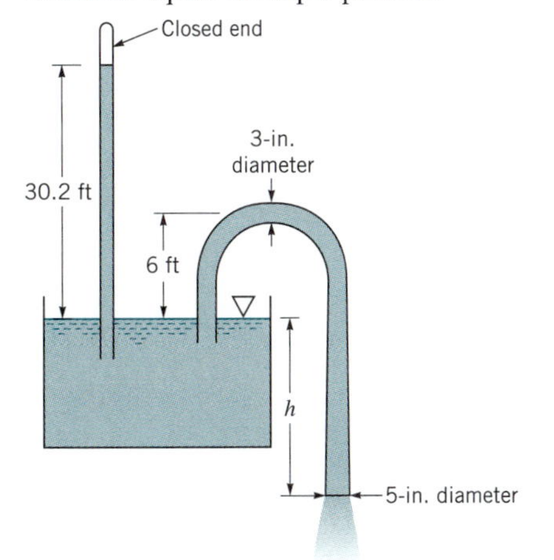

■ FIGURE P3.39

3.40 Water is siphoned from a tank as shown in Fig. P3.40. Determine the flowrate and the pressure at point A, a stagnation point.

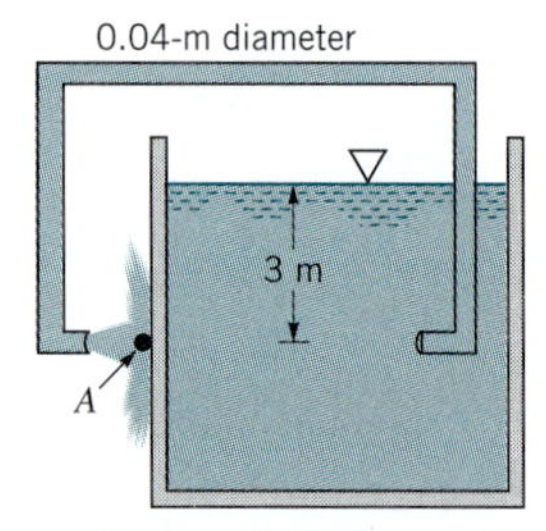

■ **FIGURE P3.40**

† **3.41** Estimate the force of a hurricane strength wind against the side of your house. List any assumptions and show all calculations.

3.42 Water from a faucet fills a 16-oz glass (volume = 28.9 in.^{3}) in 10 s. If the diameter of the jet leaving the faucet is 0.60 in., what is the diameter of the jet when it strikes the water surface in the glass which is positioned 14 in. below the faucet?

3.43 A smooth plastic, 10-m-long garden hose with an inside diameter of 20 mm is used to drain a wading pool as is shown in Fig. P3.43. If viscous effects are neglected, what is the flowrate from the pool?

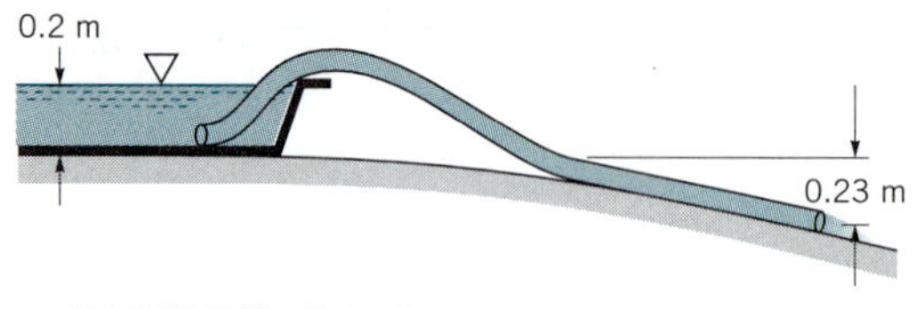

■ **FIGURE P3.43**

3.44 Carbon dioxide flows at a rate of 1.5 ft^3/s from a 3-in. pipe in which the pressure and temperature are 20 psi (gage) and 120 °F into a 1.5-in. pipe. If viscous effects are neglected and incompressible conditions are assumed, determine the pressure in the smaller pipe.

3.45 Oil of specific gravity 0.83 flows in the pipe shown in Fig. P3.45. If viscous effects are neglected, what is the flowrate?

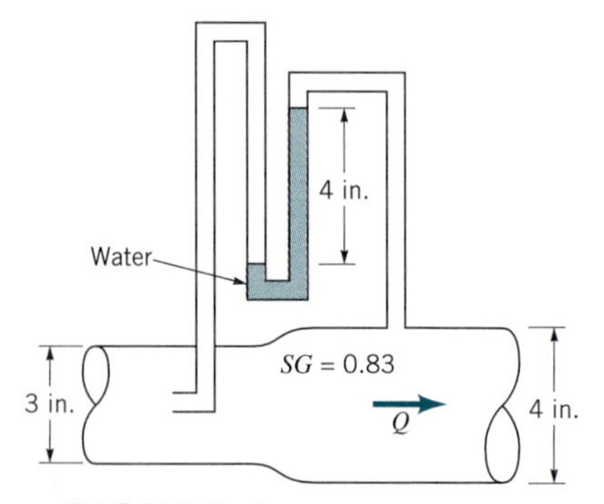

■ **FIGURE P3.45**

3.46 Water flows steadily from a large open tank and discharges into the atmosphere through a 3-in.-diameter pipe as shown in Fig. P3.46. Determine the diameter, d, in the narrowed section of the pipe at A if the pressure gages at A and B indicate the same pressure.

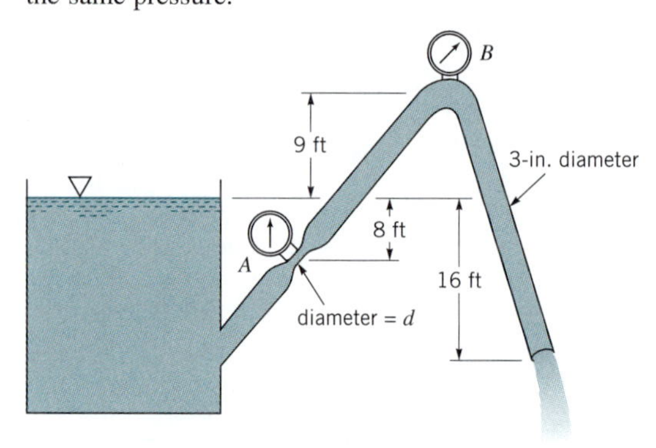

■ **FIGURE P3.46**

3.47 Determine the flowrate through the pipe in Fig. P3.47.

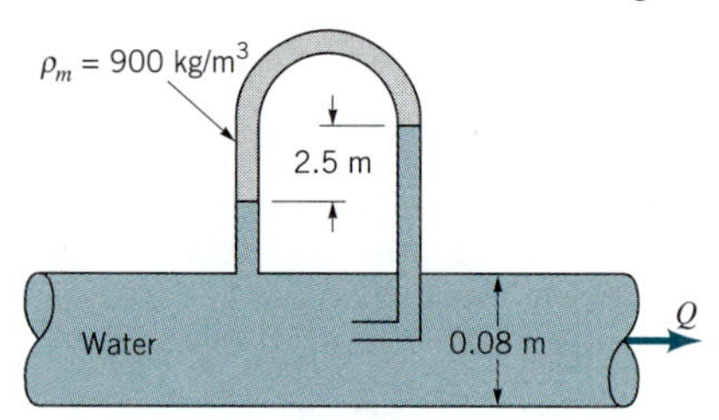

■ **FIGURE P3.47**

3.48 A plastic tube of 50-mm diameter is used to siphon water from the large tank shown in Fig. P3.48. If the pressure on the outside of the tube is more than 30 kPa greater than the pressure within the tube, the tube will collapse and the siphon will stop. If viscous effects are negligible, determine the minimum value of h allowed without the siphon stopping.

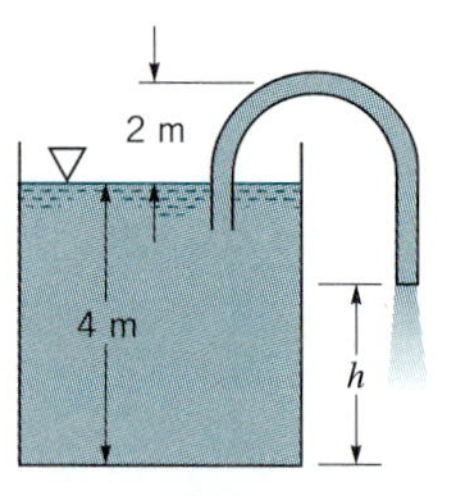

■ **FIGURE P3.48**

3.49 For the pipe enlargement shown in Fig. P3.49, the pressures at sections (1) and (2) are 56.3 and 58.2 psi, respectively. Determine the weight flowrate (lb/s) of the gasoline in the pipe.

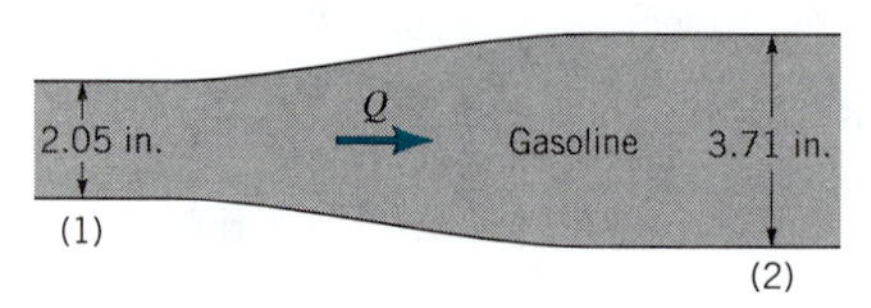

■ **FIGURE P3.49**

3.50 Water is pumped from a lake through an 8-in. pipe at a rate of 10 ft^3/s. If viscous effects are negligible, what is the pressure in the suction pipe (the pipe between the lake and the pump) at an elevation 6 ft above the lake?

3.51 Air flows through the device shown in Fig. P3.51. If the flowrate is large enough, the pressure within the constriction will be low enough to draw the water up into the tube. Determine the flowrate, Q, and the pressure needed at section (1) to draw the water into section (2). Neglect compressibility and viscous effects.

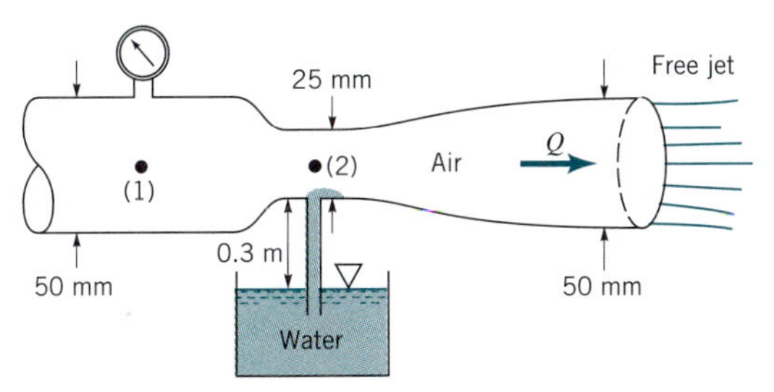

■ **FIGURE P3.51**

3.52 Natural gas (methane) flows from a 3-in.-diameter gas main, through a 1-in.-diameter pipe, and into the burner of a furnace at a rate of 100 $ft^3/hour$. Determine the pressure in the gas main if the pressure in the 1-in. pipe is to be 6 in. of water greater than atmospheric pressure. Neglect viscous effects.

3.53 Air flows through a flowmeter and out a nozzle as shown in Fig. P3.53. A Pitot tube connected to a water-filled manometer is used to measure the air speed at the exit of the 1.17-in.-diameter nozzle. Another water-filled manometer is used to measure the pressure difference across the flowmeter. The basic flowmeter equation can be written as $Q = Kh^{1/2}$, with Q in ft^3/s and h in inches. For standard air flowing through the system, determine K (including units) if $h = 5.4$ in. when $H = 2.5$ in.

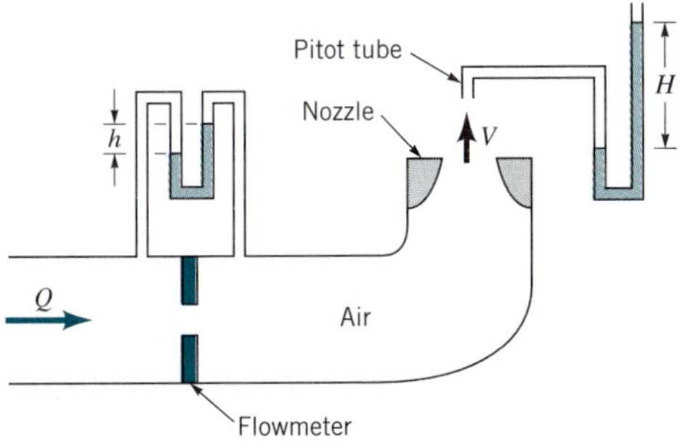

■ **FIGURE P3.53**

3.54 The center pivot irrigation system shown in Fig. P3.54 is to provide uniform watering of the entire circular field. Water flows through the common supply pipe and out through 10 evenly spaced nozzles. Water from each nozzle is to cover a strip 30 feet wide as indicated. If viscous effects are negligible, determine the diameter of each nozzle, d_i, $i = 1$ to 10, in terms of the diameter, d_{10}, of the nozzle at the outer end of the arm.

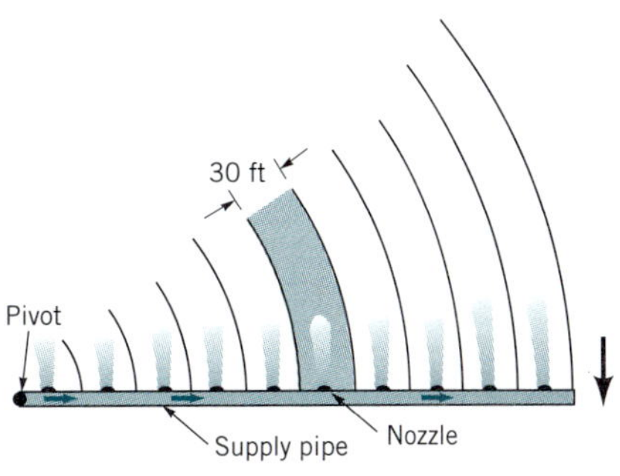

■ **FIGURE P3.54**

***3.55** Water flows from a large tank and through a pipe of variable area as shown in Fig. P3.55. The area of the pipe is given by $A = A_0\,[1 - x(1 - x/\ell)/2\ell]$, where A_0 is the area at the beginning ($x = 0$) and end ($x = \ell$) of the pipe. Plot graphs of the pressure within the pipe as a function of distance along the pipe for water depths of $h = 1$, 4, 10, and 25 m.

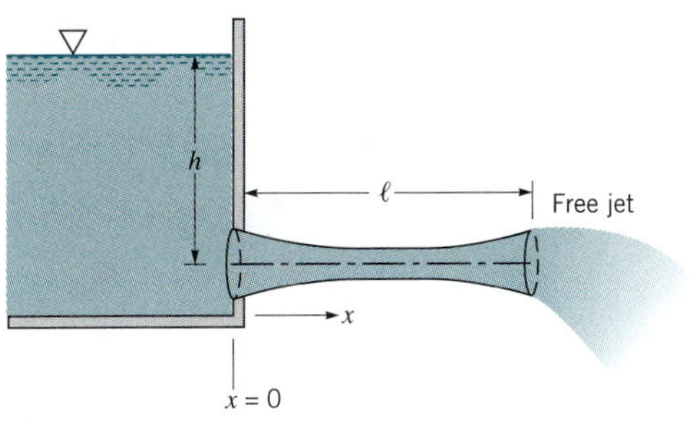

■ **FIGURE P3.55**

***3.56** Air flows through a horizontal pipe of variable diameter, $D = D(x)$, at a rate of 1.5 ft^3/s. The static pressure distribution obtained from a set of 12 static pressure taps along the pipe wall is as shown below. Plot the pipe shape, $D(x)$, if the diameter at $x = 0$ is 1, 2, or 3 in. Neglect viscous and compressibility effects.

x (in.)	p (in. H_2O)	x (in.)	p (in. H_2O)
0	1.00		
1	0.72	7	0.44
2	0.16	8	0.51
3	−0.96	9	0.65
4	−0.31	10	0.78
5	0.27	11	0.90
6	0.39	12	1.00

3.57 The vent on the tank shown in Fig. P3.57 is closed and the tank pressurized to increase the flowrate. What pressure, p_1, is needed to produce twice the flowrate of that when the vent is open?

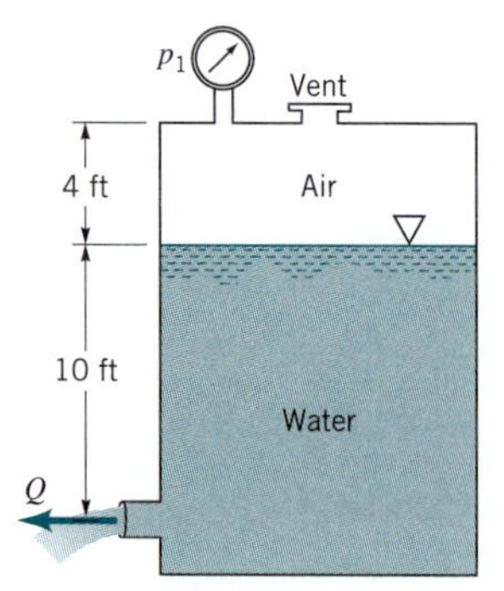

■ **FIGURE P3.57**

3.58 Water flows steadily through the large tanks shown in Fig. P3.58. Determine the water depth, h_A.

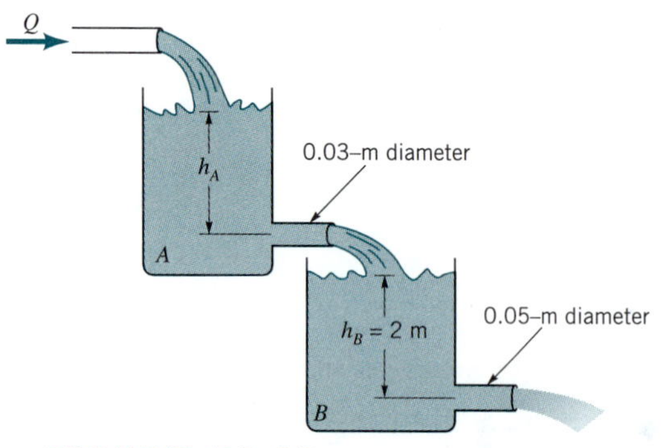

■ **FIGURE P3.58**

3.59 Air at 80 °F and 14.7 psia flows into the tank shown in Fig. P3.59. Determine the flowrate in ft^3/s, lb/s, and slugs/s. Assume incompressible flow.

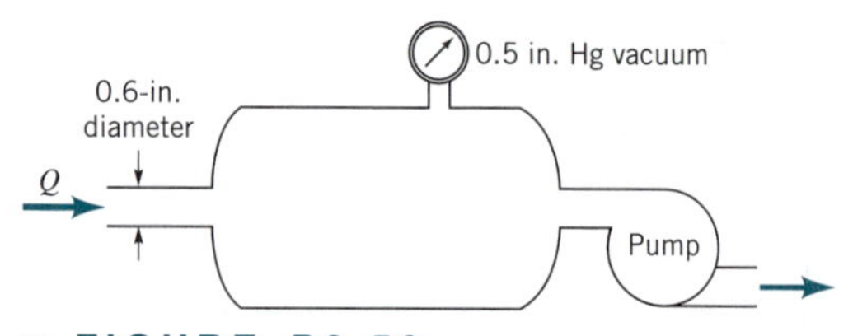

■ **FIGURE P3.59**

3.60 Water flows from a large tank as shown in Fig. P3.60. Atmospheric pressure is 14.5 psia and the vapor pressure is 1.60 psia. If viscous effects are neglected, at what height, h, will cavitation begin? To avoid cavitation, should the value of D_1 be increased or decreased? To avoid cavitation, should the value of D_2 be increased or decreased? Explain.

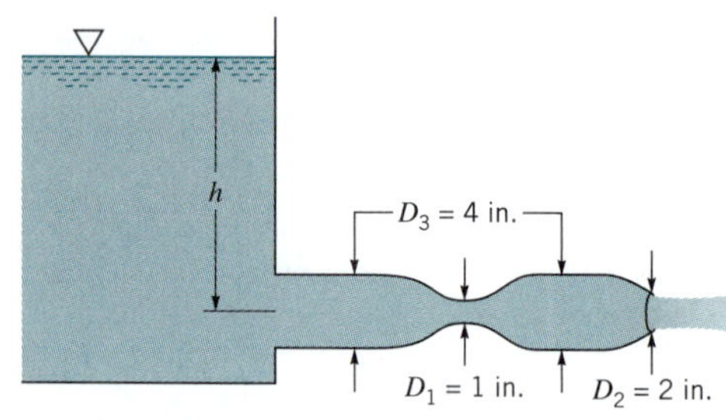

■ **FIGURE P3.60**

3.61 Water flows into the sink shown in Fig. P3.61 at a rate of 2 gal/min. If the drain is closed, the water will eventually flow through the overflow drain holes rather than over the edge of the sink. How many 0.4-in.-diameter drain holes are needed to ensure that the water does not overflow the sink? Neglect viscous effects.

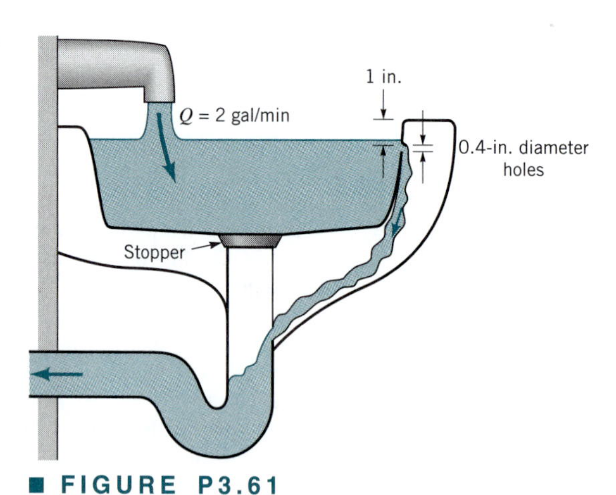

■ **FIGURE P3.61**

3.62 What pressure, p_1, is needed to produce a flowrate of 0.09 ft^3/s from the tank shown in Fig. P3.62?

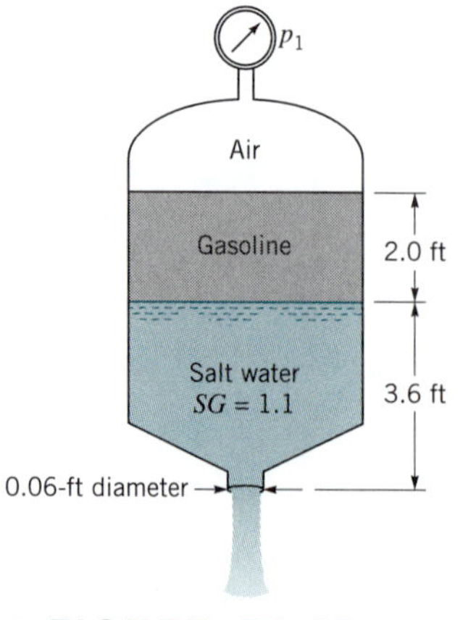

■ **FIGURE P3.62**

3.63 Laboratories containing dangerous materials are often kept at a pressure slightly less than ambient pressure so that contaminants can be filtered through an exhaust system rather than leaked through cracks around doors, etc. If the pressure in such a room is 0.1 in. of water below that of the surrounding rooms, with what velocity will air enter the room through an opening? Assume viscous effects are negligible.

3.64 Water is siphoned from the tank shown in Fig. P3.64. Determine the flowrate from the tank and the pressures at points (1), (2), and (3) if viscous effects are negligible.

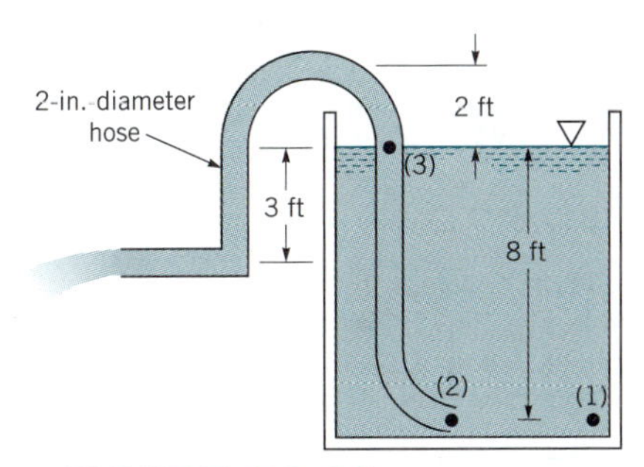

■ **FIGURE P3.64**

3.65 Redo Problem 3.64 if a 1-in.-diameter nozzle is placed at the end of the tube.

3.66 Determine the manometer reading, h, for the flow shown in Fig. P3.66.

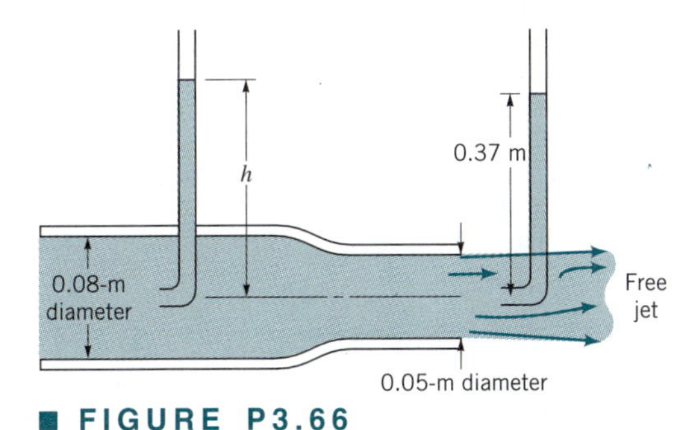

■ **FIGURE P3.66**

3.67 The specific gravity of the manometer fluid shown in Fig. P3.67 is 1.07. Determine the volume flowrate, Q, if the flow is inviscid and incompressible and the flowing fluid is **(a)** water, **(b)** gasoline, or **(c)** air at standard conditions.

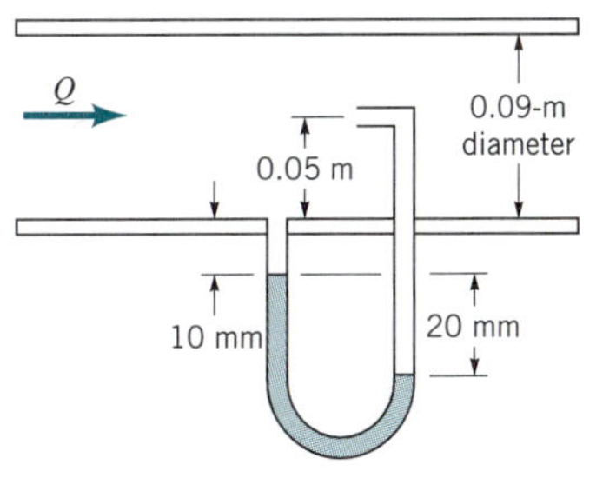

■ **FIGURE P3.67**

3.68 JP-4 fuel ($SG = 0.77$) flows through the Venturi meter shown in Fig. P3.68 with a velocity of 15 ft/s in the 6-in. pipe. If viscous effects are negligible, determine the elevation, h, of the fuel in the open tube connected to the throat of the Venturi meter.

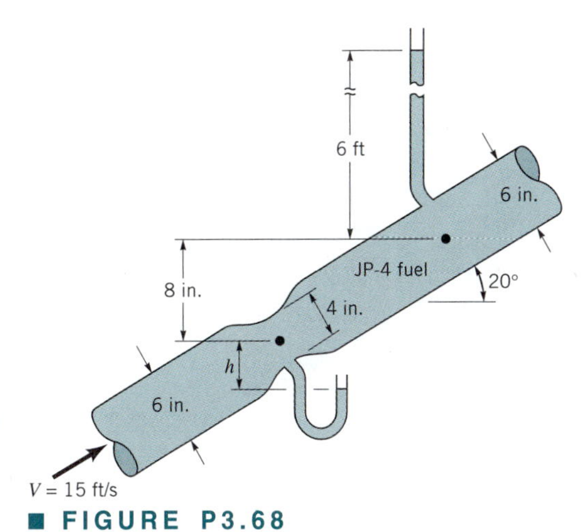

■ **FIGURE P3.68**

3.69 Repeat Problem 3.68 if the flowing fluid is water rather than JP-4 fuel.

3.70 Air at standard conditions flows through the cylindrical drying stack shown in Fig. P3.70. If viscous effects are negligible and the inclined water-filled manometer reading is 20 mm as indicated, determine the flowrate.

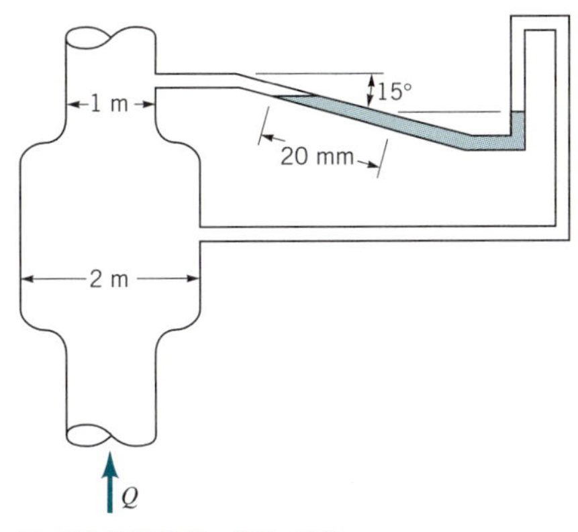

■ **FIGURE P3.70**

3.71 Oil flows through the system shown in Fig. P3.71 with negligible losses. Determine the flowrate.

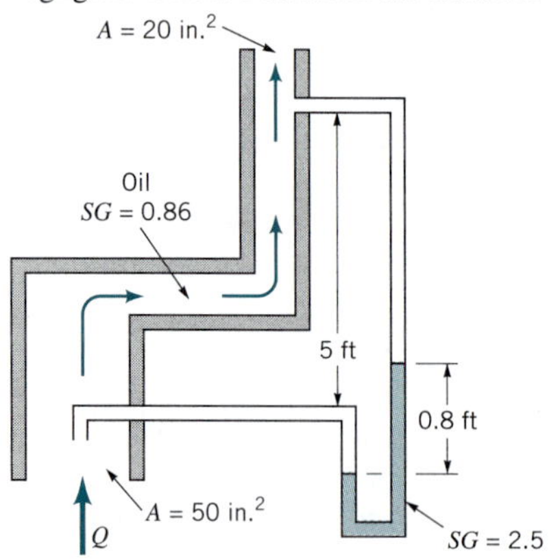

■ **FIGURE P3.71**

3.72 Determine the flowrate through the submerged orifice shown in Fig. P3.72 if the contraction coefficient is $C_c = 0.63$.

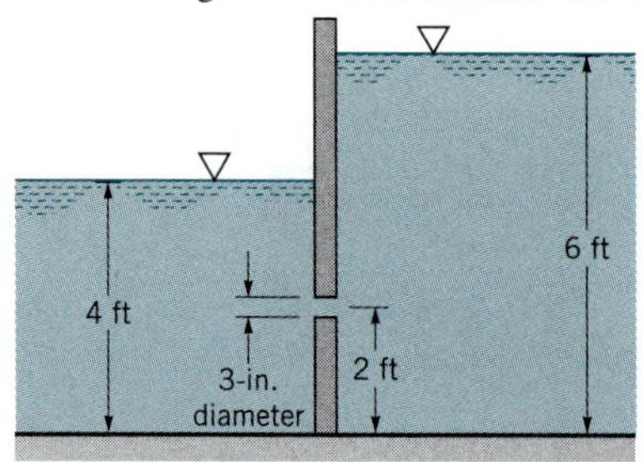

■ **FIGURE P3.72**

3.73 Determine the flowrate through the Venturi meter shown in Fig. P3.73 if ideal conditions exist.

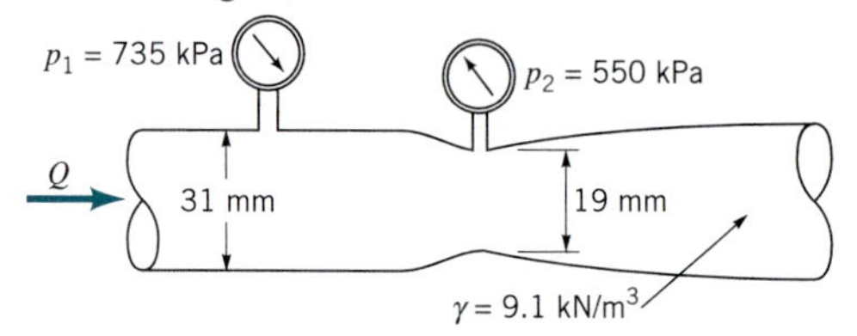

■ **FIGURE P3.73**

3.74 For what flowrate through the Venturi meter of Prob. 3.73 will cavitation begin if $p_1 = 275$ kPa gage, atmospheric pressure is 101 kPa (abs), and the vapor pressure is 3.6 kPa (abs)?

3.75 What diameter orifice hole, d, is needed if under ideal conditions the flowrate through the orifice meter of Fig. P3.75 is to be 30 gal/min of seawater with $p_1 - p_2 = 2.37$ lb/in.^{2}? The contraction coefficient is assumed to be 0.63.

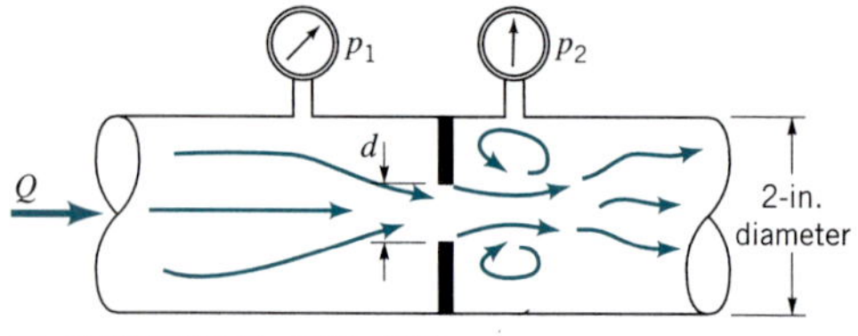

■ **FIGURE P3.75**

3.76 An ancient device for measuring time is shown in Fig. P3.76. The axisymmetric vessel is shaped so that the water level falls at a constant rate. Determine the shape of the vessel, $R = R(z)$, if the water level is to decrease at a rate of 0.10 m/hr and the drain hole is 5.0 mm in diameter. The device is to operate for 12 hr without needing refilling. Make a scale drawing of the shape of the vessel.

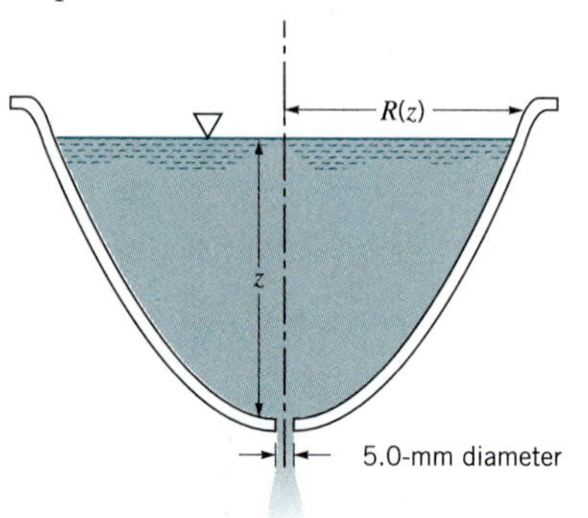

■ **FIGURE P3.76**

† **3.77** A small hole develops in the bottom of the stationary rowboat shown in Fig. P3.77. Estimate the amount of time it will take for the boat to sink. List all assumptions and show all calculations.

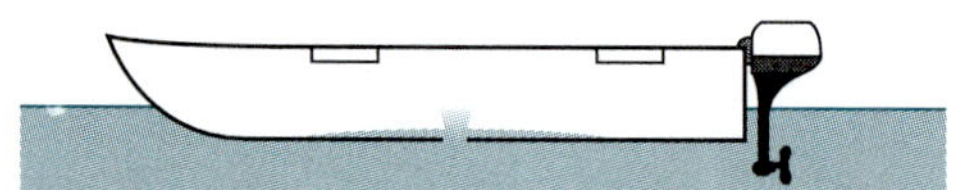

■ **FIGURE P3.77**

***3.78** A spherical tank of diameter D has a drain hole of diameter d at its bottom. A vent at the top of the tank maintains atmospheric pressure at the liquid surface within the tank. The flow is quasisteady and inviscid and the tank is full of water initially. Determine the water depth as a function of time, $h = h(t)$, and plot graphs of $h(t)$ for tank diameters of 1, 5, 10, and 20 ft if $d = 1$ in.

***3.79** An inexpensive timer is to be made from a funnel as indicated in Fig. P3.79. The funnel is filled to the top with water and the plug is removed at time $t = 0$ to allow the water to run out. Marks are to be placed on the wall of the funnel indicating the time in 15-s intervals, from 0 to 3 min (at which time the funnel becomes empty). If the funnel outlet has a diameter of $d = 0.1$ in., draw to scale the funnel with the timing marks for funnels with angles of $\theta = 30$, 45, and 60°. Repeat the problem if the diameter is changed to 0.05 in.

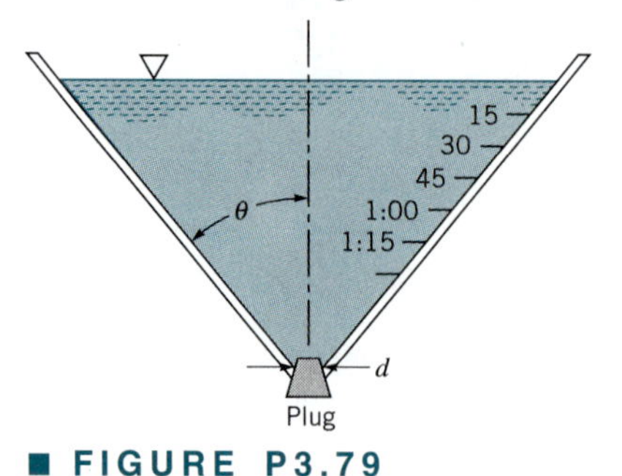

■ **FIGURE P3.79**

*3.80 The surface area, A, of the pond shown in Fig. P3.80 varies with the water depth, h, as shown in the table. At time $t = 0$ a valve is opened and the pond is allowed to drain through a pipe of diameter D. If viscous effects are negligible and quasisteady conditions are assumed, plot the water depth as a function of time from when the valve is opened ($t = 0$) until the pond is drained for pipe diameters of $D = 0.5, 1.0, 1.5, 2.0, 2.5$, and 3.0 ft. Assume $h = 18$ ft at $t = 0$.

h (ft)	**A [acres (1 acre = 43,560 ft^2)]**
0	0
2	0.3
4	0.5
6	0.8
8	0.9
10	1.1
12	1.5
14	1.8
16	2.4
18	2.8

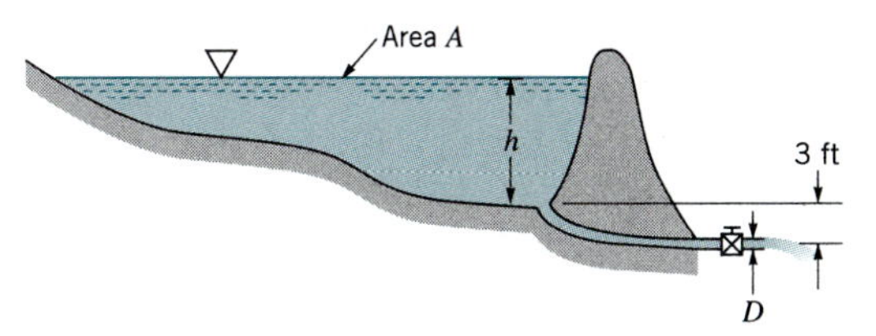

■ FIGURE P3.80

3.81 Water flows through the branching pipe shown in Fig. P3.81. If viscous effects are negligible, determine the pressure at section (2) and the pressure at section (3).

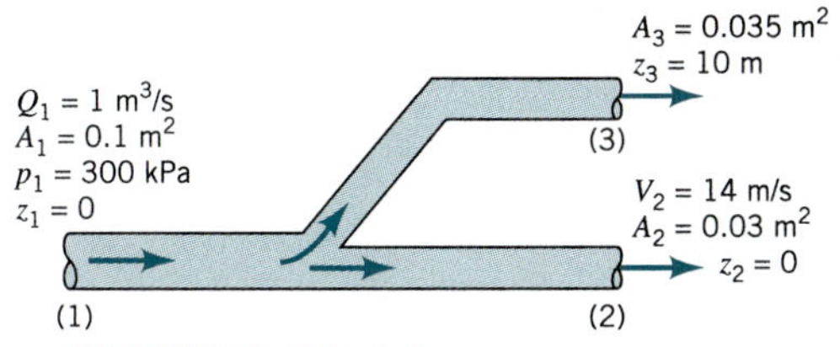

■ FIGURE P3.81

3.82 Water flows through the horizontal branching pipe shown in Fig. P3.82 at a rate of 10 ft^3/s. If viscous effects are negligible, determine the water speed at section (2), the pressure at section (3), and the flowrate at section (4).

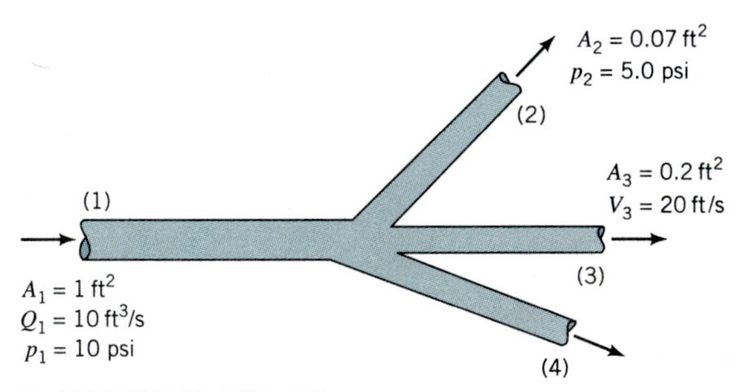

■ FIGURE P3.82

3.83 Water flows from a large tank through a large pipe that splits into two smaller pipes as shown in Fig. P3.83. If viscous effects are negligible, determine the flowrate from the tank and the pressure at point (1).

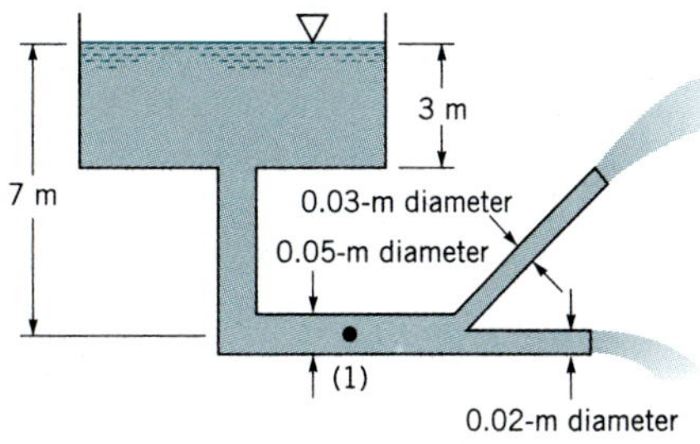

■ FIGURE P3.83

3.84 Water flows through the horizontal Y-fitting shown in Fig. P3.84. If the flowrate and pressure in pipe (1) are $Q_1 = 2.3$ ft^3/s and $p_1 = 50$ lb/in.^{2}, determine the pressures, p_2 and p_3, in pipes (2) and (3) under the assumption that the flowrate divides evenly between pipes (2) and (3).

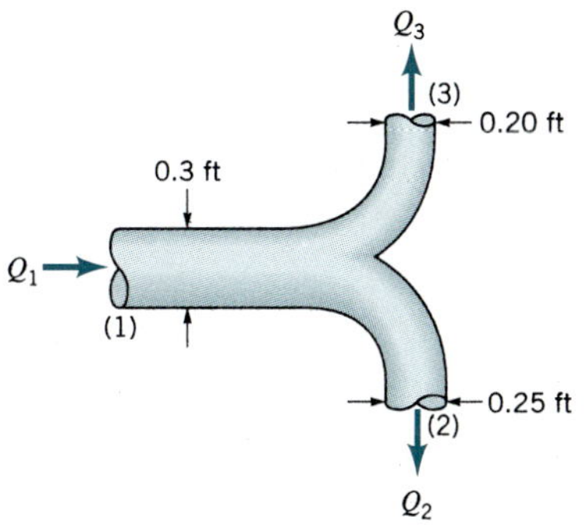

■ FIGURE P3.84

3.85 Water flows from the pipe shown in Fig. P3.85 as a free jet and strikes a circular flat plate. The flow geometry shown is axisymmetrical. Determine the flowrate and the manometer reading, H.

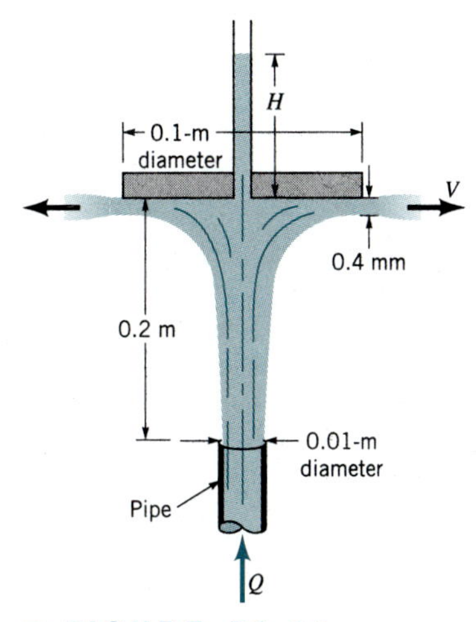

■ FIGURE P3.85

3.86 Air, assumed incompressible and inviscid, flows into the outdoor cooking grill through nine holes of 0.40-in. diameter as shown in Fig. P3.86. If a flowrate of 40 in.^{3}/s into the grill is required to maintain the correct cooking conditions, determine the pressure within the grill near the holes.

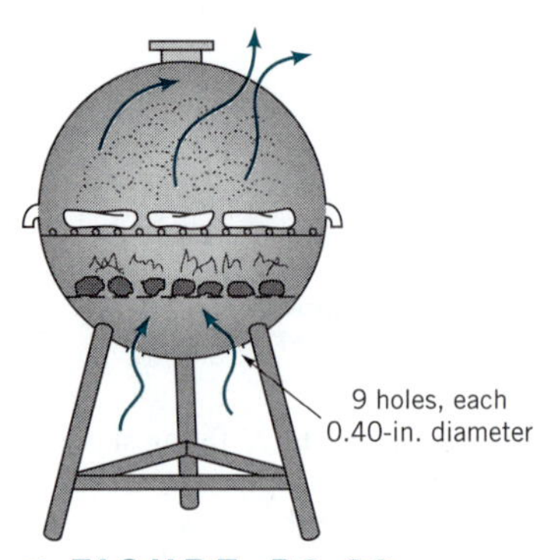

■ FIGURE P3.86

3.87 A conical plug is used to regulate the air flow from the pipe shown in Fig. P3.87. The air leaves the edge of the cone with a uniform thickness of 0.02 m. If viscous effects are negligible and the flowrate is 0.50 m^3/s, determine the pressure within the pipe.

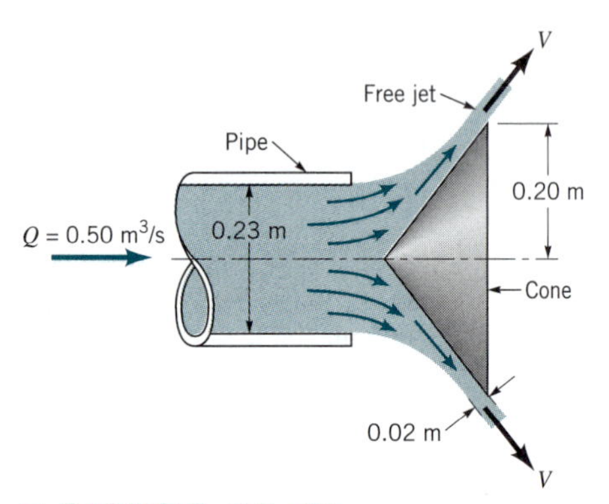

■ FIGURE P3.87

3.88 An air cushion vehicle is supported by forcing air into the chamber created by a skirt around the periphery of the vehicle as shown in Fig. P3.88. The air escapes through the 3-in. clearance between the lower end of the skirt and the ground (or water). Assume the vehicle weighs 10,000 lb and is essentially rectangular in shape, 30 by 50 ft. The volume of the chamber is large enough so that the kinetic energy of the air within the chamber is negligible. Determine the flowrate, Q, needed to support the vehicle. If the ground clearance were reduced to 2 in., what flowrate would be needed? If the vehicle weight were reduced to 5000 lb and the ground clearance maintained at 3 in., what flowrate would be needed?

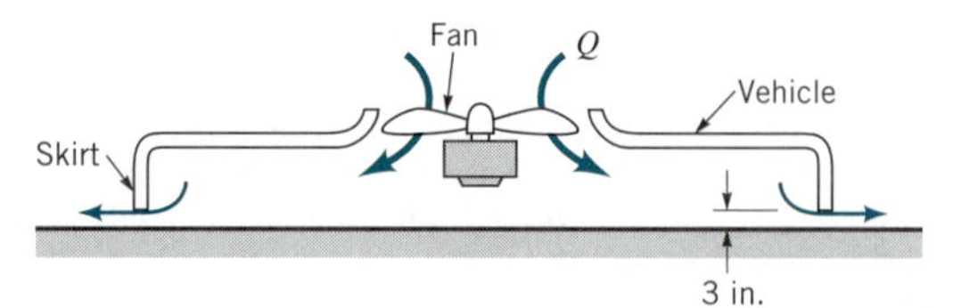

■ FIGURE P3.88

3.89 A small card is placed on top of a spool as shown in Fig. P3.89. It is not possible to blow the card off the spool by blowing air through the hole in the center of the spool. The harder one blows, the harder the card ''sticks'' to the spool. In fact, by blowing hard enough it is possible to keep the card against the spool with the spool turned upside down. (*Note:* It may be necessary to use a thumb tack to prevent the card from sliding from the spool.) Explain this phenomenon.

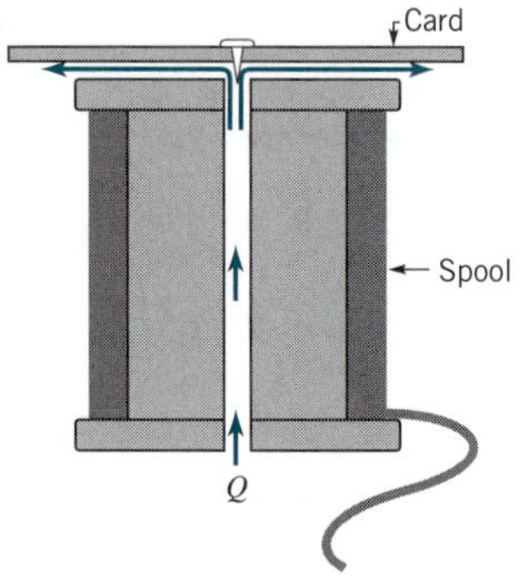

■ FIGURE P3.89

3.90 Water flows over a weir plate which has a parabolic opening as shown in Fig. P3.90. That is, the opening in the weir plate has a width $CH^{1/2}$, where C is a constant. Determine the functional dependence of the flowrate on the head, $Q = Q(H)$.

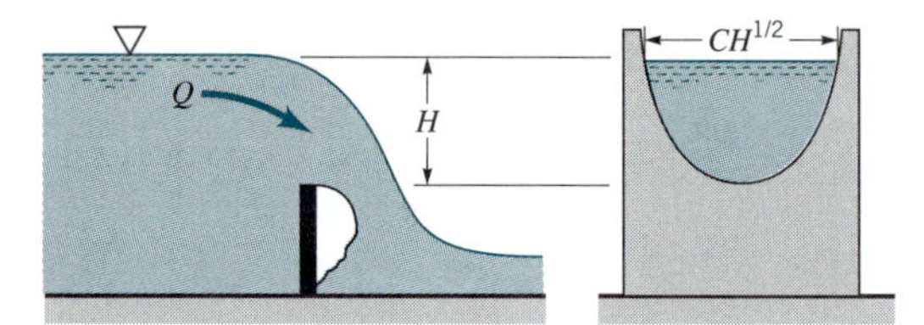

■ FIGURE P3.90

3.91 A weir of trapezoidal cross section is used to measure the flowrate in a channel as shown in Fig. P3.91. If the flowrate is Q_0 when $H = \ell/2$, what flowrate is expected when $H = \ell$?

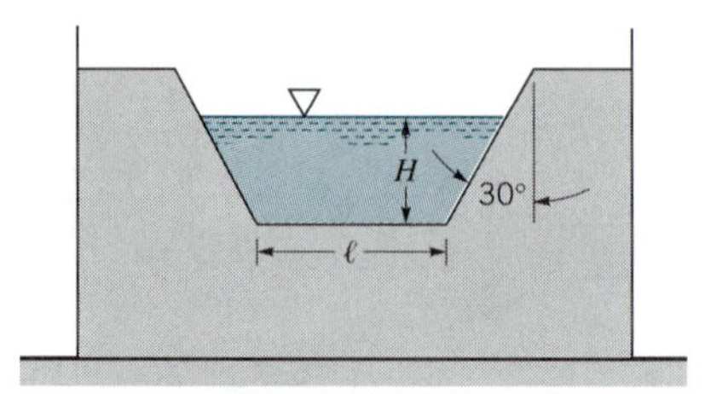

■ FIGURE P3.91

3.92 Water flows down the sloping ramp shown in Fig. P3.92 with negligible viscous effects. The flow is uniform at sections (1) and (2). For the conditions given, show that three solutions for the downstream depth, h_2, are obtained by use of the Bernoulli and continuity equations. However, show that only two of these solutions are realistic. Determine these values.

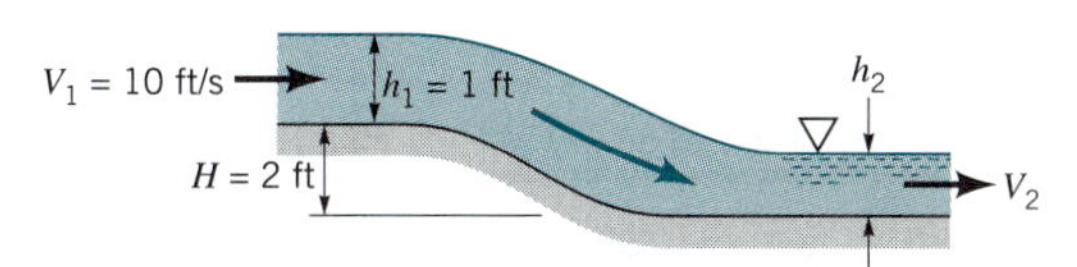

■ **FIGURE P3.92**

3.93 The flowrate in a water channel is sometimes determined by use of a device called a Venturi flume. As shown in Fig. P3.93, this device consists simply of a hump on the bottom of the channel. If the water surface dips a distance of 0.07 m for the conditions shown, what is the flowrate per width of the channel? Assume the velocity is uniform and viscous effects are negligible.

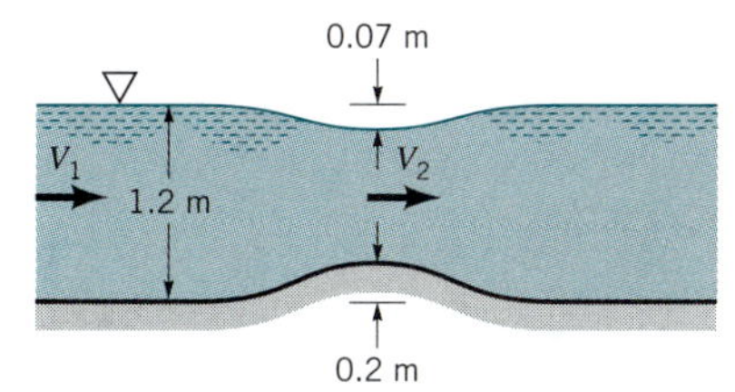

■ **FIGURE P3.93**

3.94 Water flows in a rectangular channel that is 2.0 m wide as shown in Fig. P3.94. The upstream depth is 70 mm. The water surface rises 40 mm as it passes over a portion where the channel bottom rises 10 mm. If viscous effects are negligible, what is the flowrate?

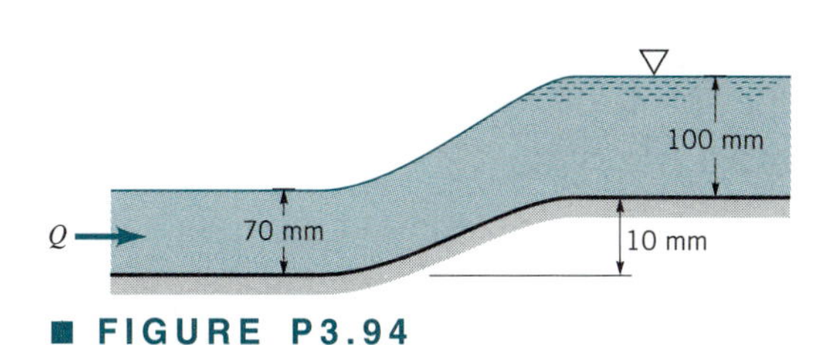

■ **FIGURE P3.94**

3.95 Water flows under the inclined sluice gate shown in Fig. P3.95. Determine the flowrate if the gate is 8 ft wide.

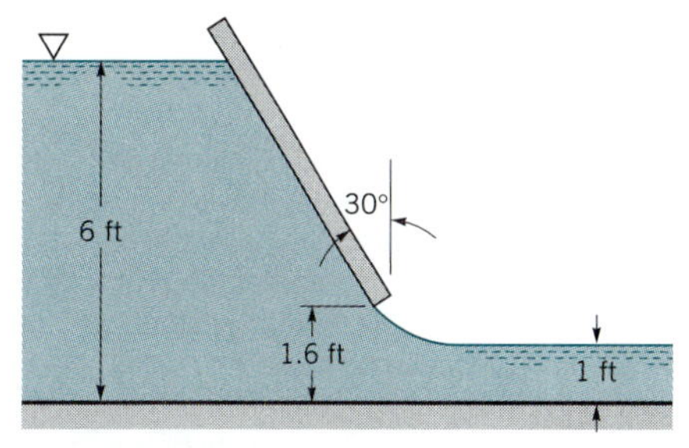

■ **FIGURE P3.95**

3.96 Water flows in a vertical pipe of 0.15-m diameter at a rate of 0.2 m^3/s and a pressure of 200 kPa at an elevation of 25 m. Determine the velocity head and pressure head at elevations of 20 and 55 m.

3.97 Draw the energy line and the hydraulic grade line for the flow shown in Problem 3.64.

3.98 Draw the energy line and the hydraulic grade line for the flow of Problem 3.60.

3.99 Draw the energy line and hydraulic grade line for the flow shown in Problem 3.65.

***3.100** Water flows up the ramp shown in Fig. P3.100 with negligible viscous losses. The upstream depth and velocity are maintained at $h_1 = 0.3$ m and $V_1 = 6$ m/s. Plot a graph of the downstream depth, h_2, as a function of the ramp height, H, for $0 \leq H \leq 2$ m. Note that for each value of H there are three solutions, not all of which are realistic.

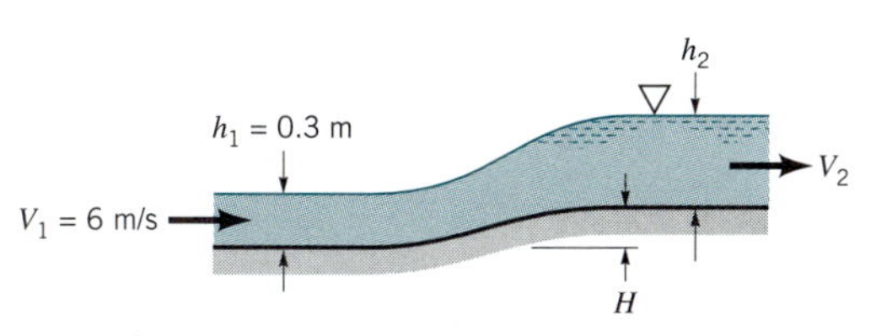

■ **FIGURE P3.100**

3.101 The device shown in Fig. P3.101 is used to investigate the flow in a radial diffuser—radial flow between two parallel circular disks. Air at a temperature of 83 °F and an absolute pressure of 29.09 in. of mercury flows at a rate of $Q = 0.837$ cfs through the inlet tube and radially out in the gap between the parallel disks as shown. The static pressure, p, as a function of radial location, r, is determined from the manometer reading, h. Since the velocity, V, decreases as r increases, it follows from the Bernoulli equation that the pressure increases in the radial direction. At the edge of the disks (the exit) the gage pressure is zero.

Experimental values of the manometer reading as a function of the radial distance are given in the table below. Use these results to plot a graph of the pressure head (feet of air) as a function of the radial location. On the same graph plot a theoretical curve obtained by using the Bernoulli equation and the given flowrate to calculate the pressure head.

Compare the experimental and theoretical results and discuss some possible reasons for any differences between them.

h (in.)	*r* (ft)
−2.79	0
−1.75	0.026
0.50	0.036
9.05	0.061
6.02	0.083
2.02	0.125
0.96	0.167
0.48	0.208
0.24	0.250
0.13	0.292
0.03	0.333
0.01	0.375
0.00	0.417

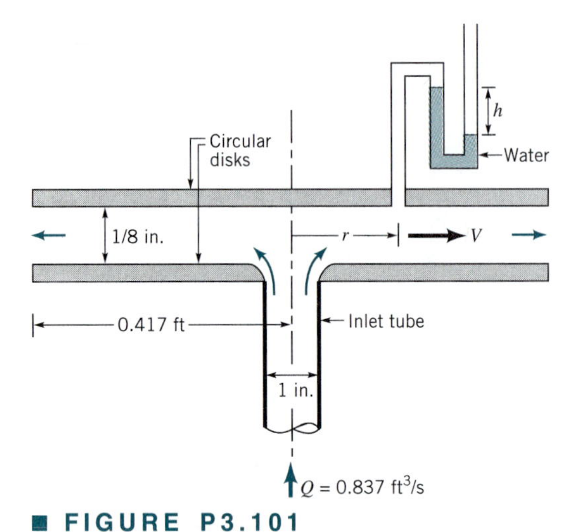

■ FIGURE P3.101

3.102 The device shown in Fig. P3.102 is used to calibrate a flowmeter. For the flowmeter shown it is known that the volume flowrate, Q, is proportional to the square root of the pressure difference across the meter. This pressure difference is given in terms of the reading, h, of a water-filled manometer. That is, $Q = K\sqrt{h}$, where K is an unknown constant dependent upon the detailed design of the flowmeter.

Air at a temperature of 75 °F and an absolute pressure of 29.0 inches of mercury flows through the flowmeter and exits into the room through a nozzle designed to give a uniform exit velocity. The velocity at the exit plane, V, can be determined from a Pitot tube attached to a water-filled manometer as indicated. The Pitot tube manometer reading is H. The flowrate can be determined by $Q = VA$, where A is the area of the nozzle exit plane.

Experimentally determined values of h and H corresponding to different flowrates are given in the table below. Use these results to plot a graph on log-log paper of the flowrate as a function of the flow meter manometer reading. Note that if $Q = K\sqrt{h}$, the resulting graph should be a straight line with a slope of $\frac{1}{2}$. Determine the value of K.

Discuss some possible sources of error in the results.

h (in.)	*H* (in.)
11.6	5.6
11.1	5.4
10.7	5.2
10.1	4.9
9.6	4.7
8.8	4.3
7.9	3.9
7.2	3.5
6.1	3.1
5.4	2.7
4.5	2.3
3.8	2.0
2.9	1.6

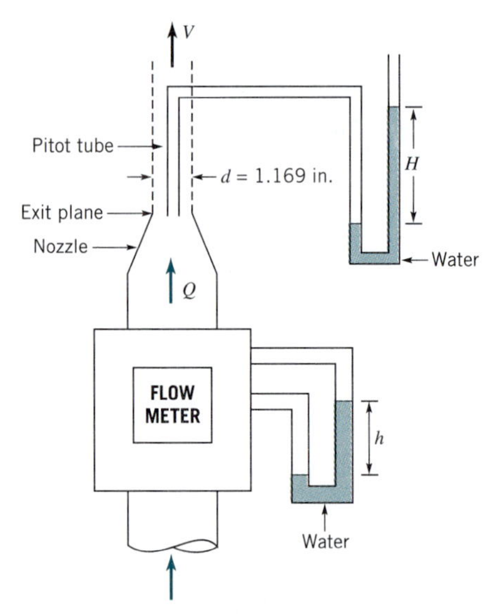

■ FIGURE P3.102

3.103 The device shown in Fig. P3.103 is used to investigate the pressure variation in a channel of variable cross section. The constant depth channel has one straight side and one curved side as indicated. The channel width, b, varies along the length of the channel. Air at a temperature of 71 °F and an absolute pressure of 28.96 inches of mercury flows through the channel at a rate of 1.32 cfs. The static pressure, p, is measured at various locations, y, along the channel (on both the flat and the curved sides) by means of a water-filled manometer whose reading is h. The pressure at the exit plane ($y = 1.81$ ft) is zero gage.

Experimentally determined values of y, b, and h are shown in the table below. Use these results to plot a graph of the pressure head in ft of air as a function of the location along the channel. (There should be two separate curves—one for the curved side and one for the straight side.) On the same graph plot a theoretical curve obtained by using the Bernoulli equation and the given flowrate to calculate the pressure head.

Compare the experimental and theoretical results and discuss some possible reasons for any differences between them.

y (ft)	b (in.)	h (curved) (in.)	h (straight) (in.)
0.06	2.00	−0.31	−0.28
0.21	2.00	−0.37	−0.18
0.33	1.28	−0.32	0.42
0.39	1.05	1.63	0.77
0.45	1.05	1.05	1.00
0.68	1.29	0.62	0.63
0.90	1.54	0.31	0.32
1.10	1.77	0.15	0.15
1.32	2.00	0.05	0.00
1.81	2.00	0.00	0.00

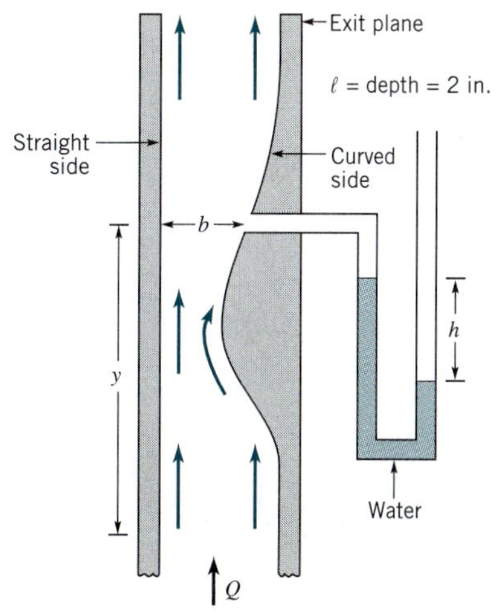

■ FIGURE P3.103

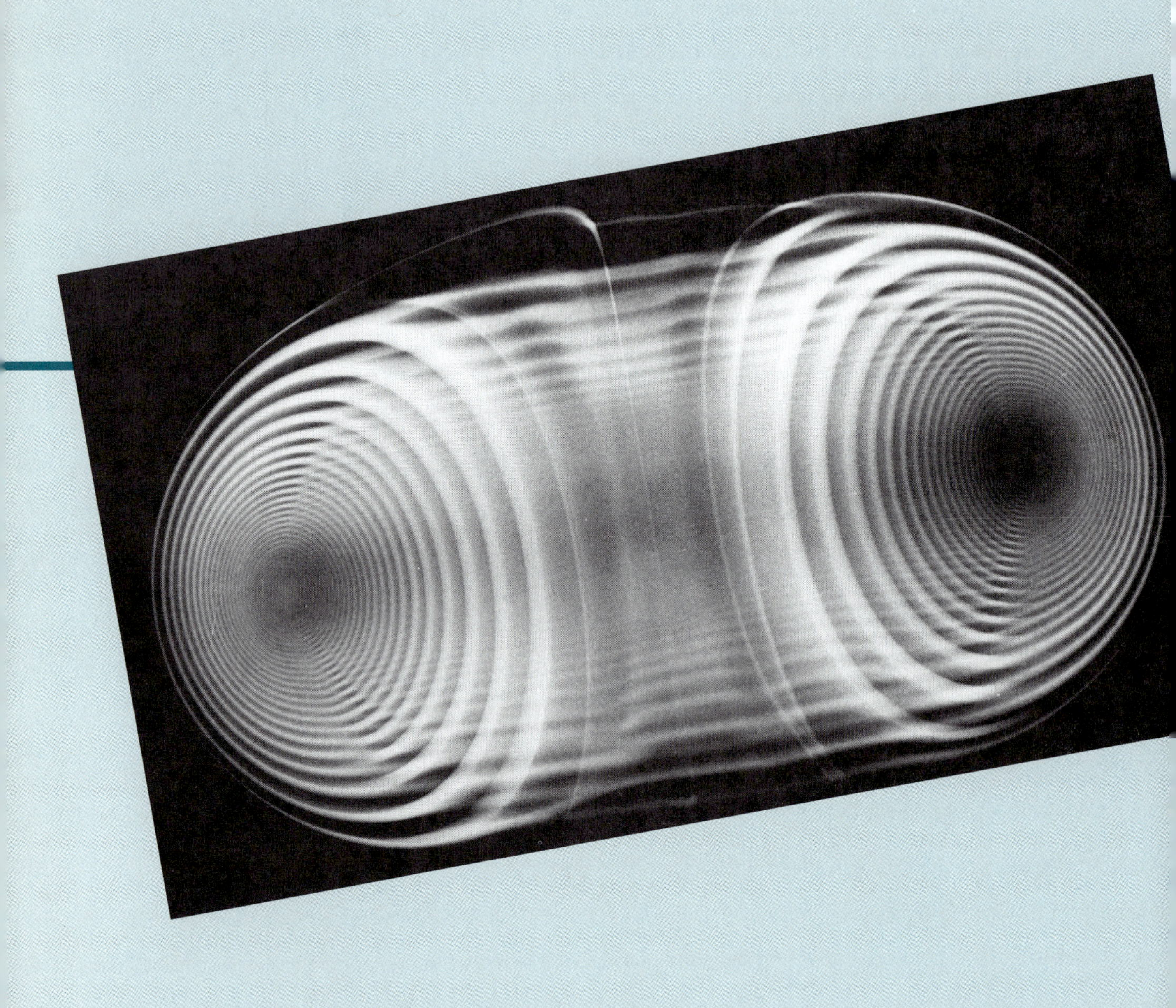

A vortex ring: The complex, three-dimensional structure of a smoke ring is indicated in this cross-sectional view. (Smoke in air.) (Photograph courtesy of R. H. Magarvey and C. S. MacLatchy, Ref. 4.)

4 Fluid Kinematics

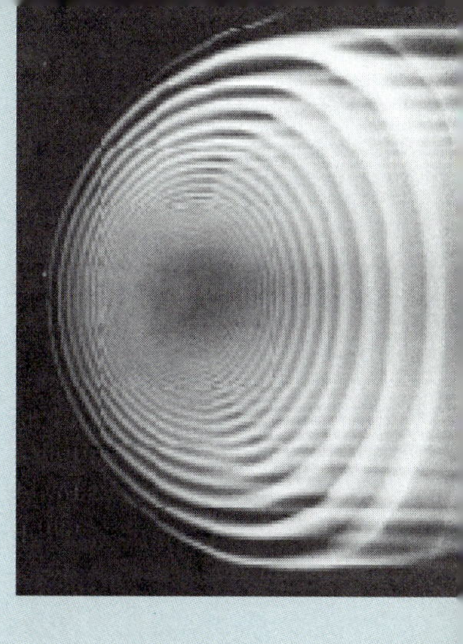

In the previous three chapters we have defined some basic properties of fluids and have considered various situations involving fluids that are either at rest or are moving in a rather elementary manner. In general, fluids have a well-known tendency to move or flow. It is very difficult to ''tie down'' a fluid and restrain it from moving. The slightest of shear stresses will cause the fluid to move. Similarly, an appropriate imbalance of normal stresses (pressure) will cause fluid motion.

Kinematics involves position, velocity, and acceleration, not force.

In this chapter we will discuss various aspects of fluid motion without being concerned with the actual forces necessary to produce the motion. That is, we will consider the *kinematics* of the motion—the velocity and acceleration of the fluid, and the description and visualization of its motion. The analysis of the specific forces necessary to produce the motion (the *dynamics* of the motion) will be discussed in detail in the following chapters. A wide variety of useful information can be gained from a thorough understanding of fluid kinematics. Such an understanding of how to describe and observe fluid motion is an essential step to the complete understanding of fluid dynamics.

We have all observed fascinating fluid motions like those associated with the smoke emerging from a chimney or the flow of the atmosphere as indicated by the motion of clouds. The motion of waves on a lake or the mixing of paint in a bucket provide other common, although quite different, examples of flow visualization. Considerable insight into these fluid motions can be gained by considering the kinematics of such flows without being concerned with the specific force that drives them.

4.1 The Velocity Field

In general, fluids flow. That is, there is a net motion of molecules from one point in space to another point as a function of time. As is discussed in Chapter 1, a typical portion of fluid contains so many molecules that it becomes totally unrealistic (except in special cases) for us to attempt to account for the motion of individual molecules. Rather, we employ the continuum hypothesis and consider fluids to be made up of fluid particles that interact with each other and with their surroundings. Each particle contains numerous molecules. Thus,

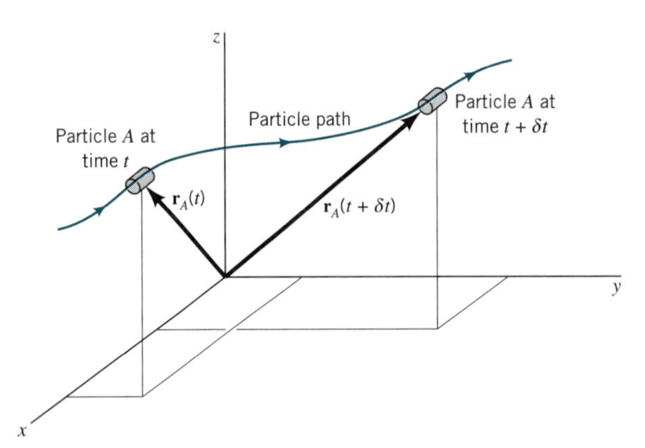

■ **FIGURE 4.1 Particle location in terms of its position vector.**

we can describe the flow of a fluid in terms of the motion of fluid particles rather than individual molecules. This motion can be described in terms of the velocity and acceleration of the fluid particles.

The infinitesimal particles of a fluid are tightly packed together (as is implied by the continuum assumption). Thus, at a given instant in time, a description of any fluid property (such as density, pressure, velocity, and acceleration) may be given as a function of the fluid's location. This representation of fluid parameters as functions of the spatial coordinates is termed a *field representation* of the flow. Of course, the specific field representation may be different at different times, so that to describe a fluid flow we must determine the various parameters not only as a function of the spatial coordinates (x, y, z, for example) but also as a function of time, t. Thus, to completely specify the temperature, T, in a room we must specify the temperature field, $T = T(x, y, z, t)$, throughout the room (from floor to ceiling and wall to wall) at any time of the day or night.

Fluid parameters can be described by a field representation.

One of the most important fluid variables is the *velocity field*,

$$\mathbf{V} = u(x, y, z, t)\hat{\mathbf{i}} + v(x, y, z, t)\hat{\mathbf{j}} + w(x, y, z, t)\hat{\mathbf{k}}$$

V4.1

where u, v, and w are the x, y, and z components of the velocity vector. By definition, the velocity of a particle is the time rate of change of the position vector for that particle. As is illustrated in Fig. 4.1, the position of particle A relative to the coordinate system is given by its *position vector*, $\mathbf{r}_A$, which (if the particle is moving) is a function of time. The time derivative of this position gives the *velocity* of the particle, $d\mathbf{r}_A/dt = \mathbf{V}_A$. By writing the velocity for all of the particles we can obtain the field description of the velocity vector $\mathbf{V} = \mathbf{V}(x, y, z, t)$.

Since the velocity is a vector, it has both a direction and a magnitude. The magnitude of $\mathbf{V}$, denoted $V = |\mathbf{V}| = (u^2 + v^2 + w^2)^{1/2}$, is the speed of the fluid. (It is very common in practical situations to call V velocity rather than speed, i.e., ''the velocity of the fluid is 12 m/s.'') As is discussed in the next section, a change in velocity results in an acceleration. This acceleration may be due to a change in speed and/or direction.

EXAMPLE 4.1

A velocity field is given by $\mathbf{V} = (V_0/\ell)(x\hat{\mathbf{i}} - y\hat{\mathbf{j}})$ where V_0 and ℓ are constants. At what location in the flow field is the speed equal to V_0? Make a sketch of the velocity field in the first quadrant ($x \geq 0$, $y \geq 0$) by drawing arrows representing the fluid velocity at representative locations.

SOLUTION

The x, y, and z components of the velocity are given by $u = V_0x/\ell$, $v = -V_0y/\ell$, and $w = 0$ so that the fluid speed, V, is

$$V = (u^2 + v^2 + w^2)^{1/2} = \frac{V_0}{\ell}(x^2 + y^2)^{1/2} \tag{1}$$

The speed is $V = V_0$ at any location on the circle of radius ℓ centered at the origin $[(x^2 + y^2)^{1/2} = \ell]$ as shown in Fig. E4.1*a*. **(Ans)**

The direction of the fluid velocity relative to the x axis is given in terms of $\theta = \arctan(v/u)$ as shown in Fig. E4.1*b*. For this flow

$$\tan\theta = \frac{v}{u} = \frac{-V_0 y/\ell}{V_0 x/\ell} = \frac{-y}{x}$$

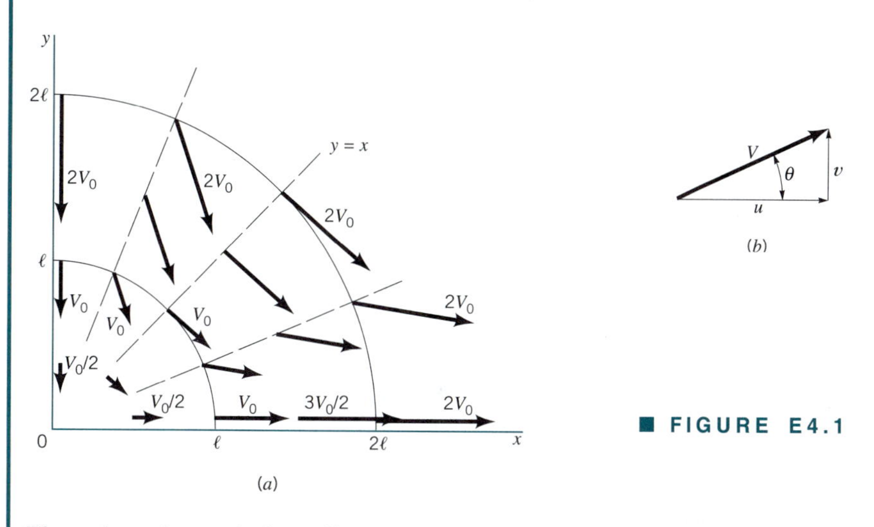

■ **FIGURE E4.1**

Thus, along the x axis ($y = 0$) we see that $\tan\theta = 0$, so that $\theta = 0°$ or $\theta = 180°$. Similarly, along the y axis ($x = 0$) we obtain $\tan\theta = \pm\infty$ so that $\theta = 90°$ or $\theta = 270°$. Also, for $y = 0$ we find $\mathbf{V} = (V_0 x/\ell)\hat{\mathbf{i}}$, while for $x = 0$ we have $\mathbf{V} = (-V_0 y/\ell)\hat{\mathbf{j}}$, indicating (if $V_0 > 0$) that the flow is directed toward the origin along the y axis and away from the origin along the x axis as shown in Fig. E4.1*a*.

By determining $\mathbf{V}$ and θ for other locations in the $x-y$ plane, the velocity field can be sketched as shown in the figure. For example, on the line $y = x$ the velocity is at a $-45°$ angle relative to the x axis ($\tan\theta = v/u = -y/x = -1$). At the origin $x = y = 0$ so that $\mathbf{V} = 0$. This point is a stagnation point. The farther from the origin the fluid is, the faster it is flowing (as seen from Eq. 1). By careful consideration of the velocity field it is possible to determine considerable information about the flow.

4.1.1 Eulerian and Lagrangian Flow Descriptions

Either Eulerian or Lagrangian methods can be used to describe flow fields.

There are two general approaches in analyzing fluid mechanics problems (or problems in other branches of the physical sciences, for that matter). The first method, called the *Eulerian method*, uses the field concept introduced above. In this case, the fluid motion is given by completely prescribing the necessary properties (pressure, density, velocity, etc.) as functions of space and time. From this method we obtain information about the flow in terms of what happens at fixed points in space as the fluid flows past those points.

The second method, called the *Lagrangian method*, involves following individual fluid particles as they move about and determining how the fluid properties associated with these particles change as a function of time. That is, the fluid particles are "tagged" or identified, and their properties determined as they move.

The difference between the two methods of analyzing fluid flow problems can be seen in the example of smoke discharging from a chimney, as is shown in Fig. 4.2. In the Eulerian method one may attach a temperature-measuring device to the top of the chimney (point 0) and record the temperature at that point as a function of time. At different times there are different fluid particles passing by the stationary device. Thus, one would obtain the temperature, T, for that location ($x = x_0$, $y = y_0$, and $z = z_0$) as a function of time. That is, $T = T(x_0, y_0, z_0, t)$. The use of numerous temperature-measuring devices fixed at various locations would provide the temperature field, $T = T(x, y, z, t)$. The temperature of a particle as a function of time would not be known unless the location of the particle were known as a function of time.

In the Lagrangian method, one would attach the temperature-measuring device to a particular fluid particle (particle A) and record that particle's temperature as it moves about. Thus, one would obtain that particle's temperature as a function of time, $T_A = T_A(t)$. The use of many such measuring devices moving with various fluid particles would provide the temperature of these fluid particles as a function of time. The temperature would not be known as a function of position unless the location of each particle were known as a function of time. If enough information in Eulerian form is available, Lagrangian information can be derived from the Eulerian data—and vice versa.

Example 4.1 provides an Eulerian description of the flow. For a Lagrangian description we would need to determine the velocity as a function of time for each particle as it flows along from one point to another.

Most fluid mechanics considerations involve the Eulerian method.

In fluid mechanics it is usually easier to use the Eulerian method to describe a flow—in either experimental or analytical investigations. There are, however, certain instances in which the Lagrangian method is more convenient. For example, some numerical fluid mechanics calculations are based on determining the motion of individual fluid particles (based on the appropriate interactions among the particles), thereby describing the motion in Lagrangian terms. Similarly, in some experiments individual fluid particles are "tagged" and are followed throughout their motion, providing a Lagrangian description. Oceanographic measurements obtained from devices that flow with the ocean currents provide this information. Similarly, by using X-ray opaque dyes it is possible to trace blood flow in arteries and to obtain a Lagrangian description of the fluid motion. A Lagrangian description may also be useful in describing fluid machinery (such as pumps and turbines) in which fluid particles gain or lose energy as they move along their flow paths.

Another illustration of the difference between the Eulerian and Lagrangian descriptions can be seen in the following biological example. Each year thousands of birds migrate between their summer and winter habitats. Ornithologists study these migrations to obtain various types of important information. One set of data obtained is the rate at which birds pass

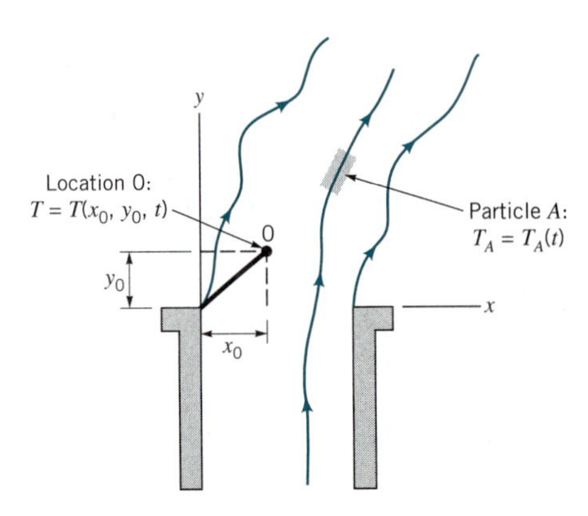

■ **FIGURE 4.2 Eulerian and Lagrangian descriptions of temperature of a flowing fluid.**

a certain location on their migration route (birds per hour). This corresponds to an Eulerian description—"flowrate" at a given location as a function of time. Individual birds need not be followed to obtain this information. Another type of information is obtained by "tagging" certain birds with radio transmitters and following their motion along the migration route. This corresponds to a Lagrangian description—"position" of a given particle as a function of time.

4.1.2 One-, Two-, and Three-Dimensional Flows

Generally, a fluid flow is a rather complex three-dimensional, time-dependent phenomenon—$\mathbf{V} = \mathbf{V}(x, y, z, t) = u\hat{\mathbf{i}} + v\hat{\mathbf{j}} + w\hat{\mathbf{k}}$. In many situations, however, it is possible to make simplifying assumptions that allow a much easier understanding of the problem without sacrificing needed accuracy. One of these simplifications involves approximating a real flow as a simpler one- or two-dimensional flow.

Most flow fields are actually three-dimensional.

In almost any flow situation, the velocity field actually contains all three velocity components (u, v, and w, for example). In many situations the *three-dimensional flow* characteristics are important in terms of the physical effects they produce. For these situations it is necessary to analyze the flow in its complete three-dimensional character. Neglect of one or two of the velocity components in these cases would lead to considerable misrepresentation of the effects produced by the actual flow.

V4.2

The flow of air past an airplane wing provides an example of a complex three-dimensional flow. A feel for the three-dimensional structure of such flows can be obtained by studying Fig. 4.3, which is a photograph of the flow past a model airfoil; the flow has been made visible by using a flow visualization technique.

In many situations one of the velocity components may be small (in some sense) relative to the two other components. In situations of this kind it may be reasonable to neglect the smaller component and assume *two-dimensional flow*. That is, $\mathbf{V} = u\hat{\mathbf{i}} + v\hat{\mathbf{j}}$, where u and v are functions of x and y (and possibly time, t).

It is sometimes possible to further simplify a flow analysis by assuming that two of the velocity components are negligible, leaving the velocity field to be approximated as a *one-dimensional flow* field. That is, $\mathbf{V} = u\hat{\mathbf{i}}$. As we will learn from examples throughout the remainder of the book, although there are very few, if any, flows that are truly one-dimensional, there are many flow fields for which the one-dimensional flow assumption provides a reasonable approximation. There are also many flow situations for which use of a one-dimensional flow field assumption will give completely erroneous results.

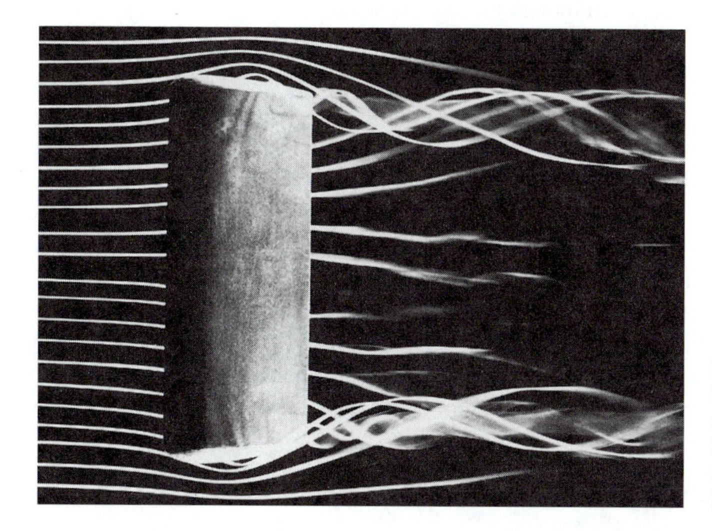

■ **FIGURE 4.3** **Flow visualization of the complex three-dimensional flow past a model airfoil. (Photograph by M. R. Head.)**

4.1.3 Steady and Unsteady Flows

In the above discussion we have assumed *steady flow*—the velocity at a given point in space does not vary with time, $\partial \mathbf{V}/\partial t = 0$. In reality, almost all flows are unsteady in some sense. That is, the velocity does vary with time. It is not difficult to believe that *unsteady flows* are usually more difficult to analyze (and to investigate experimentally) than are steady flows. Hence, considerable simplicity often results if one can make the assumption of steady flow without compromising the usefulness of the results. Among the various types of unsteady flows are nonperiodic flow, periodic flow, and truly random flow. Whether or not unsteadiness of one or more of these types must be included in an analysis is not always immediately obvious.

An example of a nonperiodic, unsteady flow is that produced by turning off a faucet to stop the flow of water. Usually this unsteady flow process is quite mundane and the forces developed as a result of the unsteady effects need not be considered. However, if the water is turned off suddenly (as with an electrically operated valve in a dishwasher), the unsteady effects can become important [as in the "water hammer" effects made apparent by the loud banging of the pipes under such conditions (Ref. 1)].

In other flows the unsteady effects may be periodic, occurring time after time in basically the same manner. The periodic injection of the air-gasoline mixture into the cylinder of an automobile engine is such an example. The unsteady effects are quite regular and repeatable in a regular sequence. They are very important in the operation of the engine.

V4.3

In many situations the unsteady character of a flow is quite random. That is, there is no repeatable sequence or regular variation to the unsteadiness. This behavior occurs in *turbulent flow* and is absent from *laminar flow*. The "smooth" flow of highly viscous syrup onto a pancake represents a "deterministic" laminar flow. It is quite different from the turbulent flow observed in the "irregular" splashing of water from a faucet onto the sink below it. The "irregular" gustiness of the wind represents another random turbulent flow. The differences between these types of flows will be discussed in considerable detail in Chapters 8 and 9.

It must be understood that the definition of steady or unsteady flow pertains to the behavior of a fluid property as observed at a fixed point in space. For steady flow, the values of all fluid properties (velocity, temperature, density, etc.) at any fixed point are independent of time. However, the value of those properties for a given fluid particle may change with time as the particle flows along, even in steady flow. Thus, the temperature of the exhaust at the exit of a car's exhaust pipe may be constant for several hours, but the temperature of a fluid particle that left the exhaust pipe five minutes ago is lower now than it was when it left the pipe, even though the flow is steady.

V4.4

4.1.4 Streamlines, Streaklines, and Pathlines

Although fluid motion can be quite complicated, there are various concepts that can be used to help in the visualization and analysis of flow fields. To this end we discuss the use of streamlines, streaklines, and pathlines in flow analysis. The streamline is often used in analytical work while the streakline and pathline are often used in experimental work.

Streamlines are lines tangent to the velocity field.

A *streamline* is a line that is everywhere tangent to the velocity field. If the flow is steady, nothing at a fixed point (including the velocity direction) changes with time, so the streamlines are fixed lines in space. For unsteady flows the streamlines may change shape with time. Streamlines are obtained analytically by integrating the equations defining lines tangent to the velocity field. For two-dimensional flows the slope of the streamline, dy/dx, must be equal to the tangent of the angle that the velocity vector makes with the x axis or

$$\frac{dy}{dx} = \frac{v}{u} \tag{4.1}$$

If the velocity field is known as a function of x and y (and t if the flow is unsteady), this equation can be integrated to give the equation of the streamlines.

For unsteady flow there is no easy way to produce streamlines experimentally in the laboratory. As discussed below, the observation of dye, smoke, or some other tracer injected into a flow can provide useful information, but for unsteady flows it is not necessarily information about the streamlines.

EXAMPLE 4.2

Determine the streamlines for the two-dimensional steady flow discussed in Example 4.1, $\mathbf{V} = (V_0/\ell)(x\hat{\mathbf{i}} - y\hat{\mathbf{j}})$.

SOLUTION

Since $u = (V_0/\ell)x$ and $\mathrm{v} = -(V_0/\ell)y$ it follows that streamlines are given by solution of the equation

$$\frac{dy}{dx} = \frac{\mathrm{v}}{u} = \frac{-(V_0/\ell)y}{(V_0/\ell)x} = -\frac{y}{x}$$

in which variables can be separated and the equation integrated to give

$$\int \frac{dy}{y} = -\int \frac{dx}{x}$$

or

$$\ln y = -\ln x + \text{constant}$$

Thus, along the streamline

$$xy = C, \qquad \text{where } C \text{ is a constant} \qquad \textbf{(Ans)}$$

By using different values of the constant C, we can plot various lines in the x–y plane—the streamlines. The usual notation for a streamline is ψ = constant on a streamline. Thus, the equation for the streamlines of this flow are

$$\psi = xy$$

As is discussed more fully in Chapter 6, the function $\psi = \psi(x, y)$ is called the *stream function*. The streamlines in the first quadrant are plotted in Fig. E4.2. A comparison of this figure with Fig. E4.1*a* illustrates the fact that streamlines are lines parallel to the velocity field.

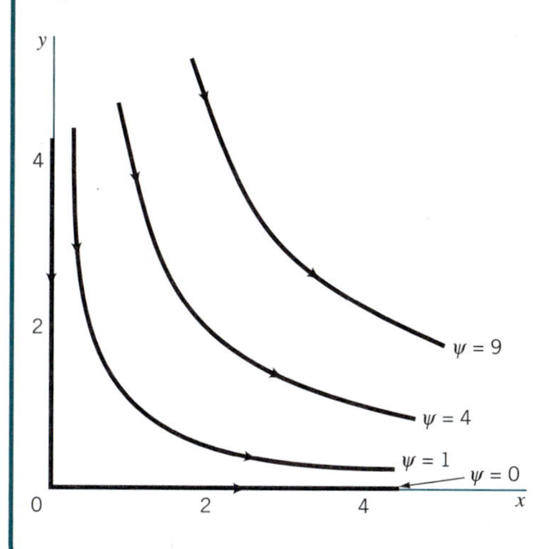

■ FIGURE E4.2

A *streakline* consists of all particles in a flow that have previously passed through a common point. Streaklines are more of a laboratory tool than an analytical tool. They can be obtained by taking instantaneous photographs of marked particles that all passed through a given location in the flow field at some earlier time. Such a line can be produced by continuously injecting marked fluid (neutrally buoyant smoke in air, or dye in water) at a given location (Ref. 2). If the flow is steady, each successively injected particle follows precisely behind the previous one, forming a steady streakline that is exactly the same as the streamline through the injection point.

For unsteady flows, particles injected at the same point at different times need not follow the same path. An instantaneous photograph of the marked fluid would show the streakline at that instant, but it would not necessarily coincide with the streamline through the point of injection at that particular time nor with the streamline through the same injection point at a different time (see Example 4.3).

V4.5

The third method used for visualizing and describing flows involves the use of *pathlines*. A pathline is the line traced out by a given particle as it flows from one point to another. The pathline is a Lagrangian concept that can be produced in the laboratory by marking a fluid particle (dying a small fluid element) and taking a time exposure photograph of its motion.

For steady flow, streamlines, streaklines, and pathlines are the same.

If the flow is steady, the path taken by a marked particle (a pathline) will be the same as the line formed by all other particles that previously passed through the point of injection (a streakline). For such cases these lines are tangent to the velocity field. Hence, pathlines, streamlines, and streaklines are the same for steady flows. For unsteady flows none of these three types of lines need be the same (Ref. 3). Often one sees pictures of ''streamlines'' made visible by the injection of smoke or dye into a flow as is shown in Fig. 4.3. Actually, such pictures show streaklines rather than streamlines. However, for steady flows the two are identical; only the nomenclature is incorrectly used.

Water flowing from the oscillating slit shown in Fig. E4.3*a* produces a velocity field given by $\mathbf{V} = u_0 \sin[\omega(t - y/v_0)]\hat{\mathbf{i}} + v_0\hat{\mathbf{j}}$, where u_0, v_0, and ω are constants. Thus, the y component of velocity remains constant ($v = v_0$) and the x component of velocity at $y = 0$ coincides with the velocity of the oscillating sprinkler head [$u = u_0 \sin(\omega t)$ at $y = 0$].

(a) Determine the streamline that passes through the origin at $t = 0$; at $t = \pi/2\omega$. (b) Determine the pathline of the particle that was at the origin at $t = 0$; at $t = \pi/2\omega$. (c) Discuss the shape of the streakline that passes through the origin.

SOLUTION

(a) Since $u = u_0 \sin[\omega(t - y/v_0)]$ and $v = v_0$ it follows from Eq. 4.1 that streamlines are given by the solution of

$$\frac{dy}{dx} = \frac{v}{u} = \frac{v_0}{u_0 \sin[\omega(t - y/v_0)]}$$

in which the variables can be separated and the equation integrated (for any given time t) to give

$$u_0 \int \sin\left[\omega\left(t - \frac{y}{v_0}\right)\right] dy = v_0 \int dx,$$

or

$$u_0(v_0/\omega) \cos\left[\omega\left(t - \frac{y}{v_0}\right)\right] = v_0 x + C \qquad \textbf{(1)}$$

where C is a constant. For the streamline at $t = 0$ that passes through the origin ($x = y = 0$), the value of C is obtained from Eq. 1 as $C = u_0 v_0/\omega$. Hence, the equation for this streamline is

$$x = \frac{u_0}{\omega}\left[\cos\left(\frac{\omega y}{v_0}\right) - 1\right] \qquad \textbf{(2)} \quad \textbf{(Ans)}$$

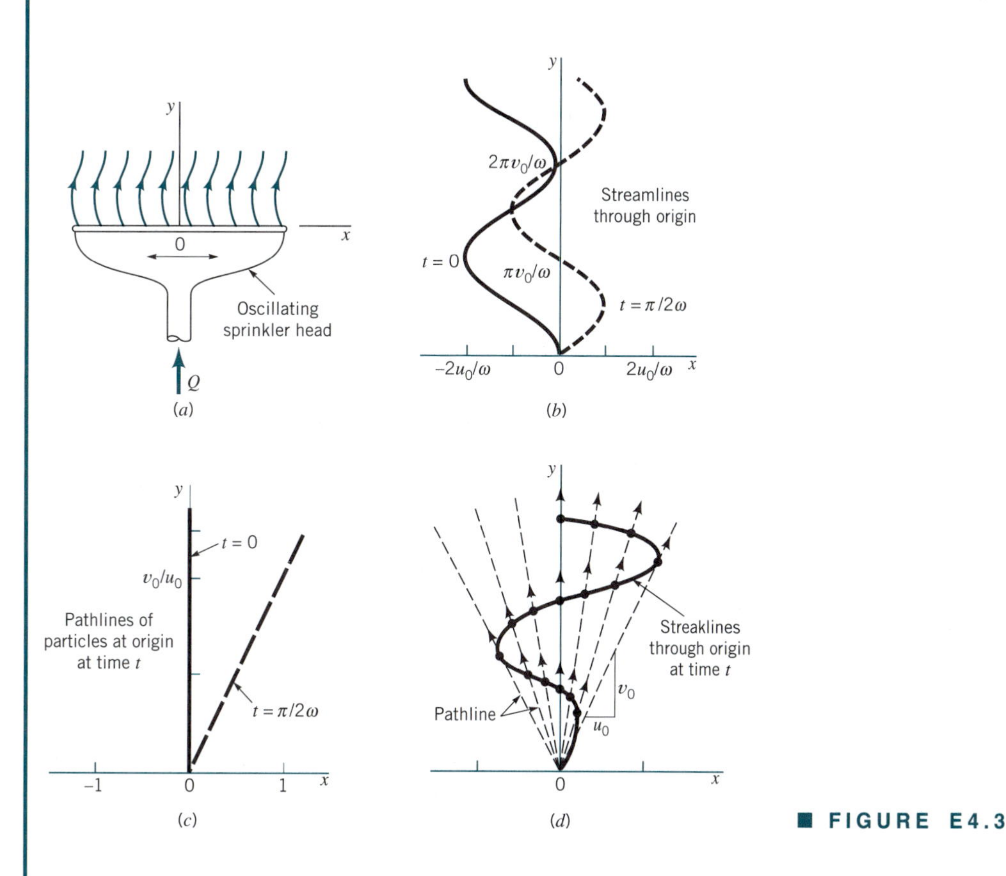

■ **FIGURE E4.3**

Similarly, for the streamline at $t = \pi/2\omega$ that passes through the origin, Eq. 1 gives $C = 0$. Thus, the equation for this streamline is

$$x = \frac{u_0}{\omega}\cos\left[\omega\left(\frac{\pi}{2\omega} - \frac{y}{v_0}\right)\right] = \frac{u_0}{\omega}\cos\left(\frac{\pi}{2} - \frac{\omega y}{v_0}\right)$$

or

$$x = \frac{u_0}{\omega}\sin\left(\frac{\omega y}{v_0}\right) \qquad \textbf{(3)} \quad \textbf{(Ans)}$$

These two streamlines, plotted in Fig. E4.3*b*, are not the same because the flow is unsteady. For example, at the origin ($x = y = 0$) the velocity is $\mathbf{V} = v_0\hat{\mathbf{j}}$ at $t = 0$ and $\mathbf{V} = u_0\hat{\mathbf{i}} + v_0\hat{\mathbf{j}}$ at $t = \pi/2\omega$. Thus, the angle of the streamline passing through the origin changes with time. Similarly, the shape of the entire streamline is a function of time.

(b) The pathline of a particle (the location of the particle as a function of time) can be obtained from the velocity field and the definition of the velocity. Since $u = dx/dt$ and $v = dy/dt$ we obtain

$$\frac{dx}{dt} = u_0 \sin\left[\omega\left(t - \frac{y}{v_0}\right)\right] \quad \text{and} \quad \frac{dy}{dt} = v_0$$

The y equation can be integrated (since $v_0 =$ constant) to give the y coordinate of the pathline as

$$y = v_0 t + C_1 \tag{4}$$

where C_1 is a constant. With this known $y = y(t)$ dependence, the x equation for the pathline becomes

$$\frac{dx}{dt} = u_0 \sin\left[\omega\left(t - \frac{v_0 t + C_1}{v_0}\right)\right] = -u_0 \sin\left(\frac{C_1\omega}{v_0}\right)$$

This can be integrated to give the x component of the pathline as

$$x = -\left[u_0 \sin\left(\frac{C_1\omega}{v_0}\right)\right] t + C_2 \tag{5}$$

where C_2 is a constant. For the particle that was at the origin ($x = y = 0$) at time $t = 0$, Eqs. 4 and 5 give $C_1 = C_2 = 0$. Thus, the pathline is

$$x = 0 \quad \text{and} \quad y = v_0 t \tag{6}$$ **(Ans)**

Similarly, for the particle that was at the origin at $t = \pi/2\omega$, Eqs. 4 and 5 give $C_1 = -\pi v_0/2\omega$ and $C_2 = -\pi u_0/2\omega$. Thus, the pathline for this particle is

$$x = u_0\left(t - \frac{\pi}{2\omega}\right) \quad \text{and} \quad y = v_0\left(t - \frac{\pi}{2\omega}\right) \tag{7}$$

The pathline can be drawn by plotting the locus of $x(t)$, $y(t)$ values for $t \geq 0$ or by eliminating the parameter t from Eq. 7 to give

$$y = \frac{v_0}{u_0} x \tag{8}$$ **(Ans)**

The pathlines given by Eqs. 6 and 8, shown in Fig. E4.3*c*, are straight lines from the origin (rays). The pathlines and streamlines do not coincide because the flow is unsteady.

V4.6

(c) The streakline through the origin at time $t = 0$ is the locus of particles at $t = 0$ that previously ($t < 0$) passed through the origin. The general shape of the streaklines can be seen as follows. Each particle that flows through the origin travels in a straight line (pathlines are rays from the origin), the slope of which lies between $\pm v_0/u_0$ as shown in Fig. E4.3*d*. Particles passing through the origin at different times are located on different rays from the origin and at different distances from the origin. The net result is that a stream of dye continually injected at the origin (a streakline) would have the shape shown in Fig. E4.3*d*. Because of the unsteadiness, the streakline will vary with time, although it will always have the oscillating, sinuous character shown. Similar streaklines are given by the stream of water from a garden hose nozzle that oscillates back and forth in a direction normal to the axis of the nozzle.

In this example neither the streamlines, pathlines, nor streaklines coincide. If the flow were steady all of these lines would be the same.

4.2 The Acceleration Field

As indicated in the previous section, we can describe fluid motion by either (1) following individual particles (Lagrangian description) or (2) remaining fixed in space and observing different particles as they pass by (Eulerian description). In either case, to apply Newton's second law ($\mathbf{F} = m\mathbf{a}$) we must be able to describe the particle acceleration in an appropriate fashion. For the infrequently used Lagrangian method, we describe the fluid acceleration just as is done in solid body dynamics—$\mathbf{a} = \mathbf{a}(t)$ for each particle. For the Eulerian description we describe the *acceleration field* as a function of position and time without actually following any particular particle. This is analogous to describing the flow in terms of the velocity field, $\mathbf{V} = \mathbf{V}(x, y, z, t)$, rather than the velocity for particular particles. In this section we will discuss how to obtain the acceleration field if the velocity field is known.

The acceleration of a particle is the time rate of change of its velocity. For unsteady flows the velocity at a given point in space (occupied by different particles) may vary with time, giving rise to a portion of the fluid acceleration. In addition, a fluid particle may experience an acceleration because its velocity changes as it flows from one point to another in space. For example, water flowing through a garden hose nozzle under steady conditions (constant number of gallons per minute from the hose) will experience an acceleration as it changes from its relatively low velocity in the hose to its relatively high velocity at the tip of the nozzle.

4.2.1 The Material Derivative

Consider a fluid particle moving along its pathline as is shown in Fig. 4.4. In general, the particle's velocity, denoted $\mathbf{V}_A$ for particle A, is a function of its location and the time. That is,

$$\mathbf{V}_A = \mathbf{V}_A(\mathbf{r}_A, t) = \mathbf{V}_A[x_A(t), y_A(t), z_A(t), t]$$

where $x_A = x_A(t)$, $y_A = y_A(t)$, and $z_A = z_A(t)$ define the location of the moving particle. By definition, the acceleration of a particle is the time rate of change of its velocity. Since the velocity may be a function of both position and time, its value may change because of the change in time as well as a change in the particle's position. Thus, we use the chain rule of differentiation to obtain the acceleration of particle A, denoted $\mathbf{a}_A$, as

Acceleration is the time rate of change of velocity for a given particle.

$$\mathbf{a}_A(t) = \frac{d\mathbf{V}_A}{dt} = \frac{\partial \mathbf{V}_A}{\partial t} + \frac{\partial \mathbf{V}_A}{\partial x}\frac{dx_A}{dt} + \frac{\partial \mathbf{V}_A}{\partial y}\frac{dy_A}{dt} + \frac{\partial \mathbf{V}_A}{\partial z}\frac{dz_A}{dt} \tag{4.2}$$

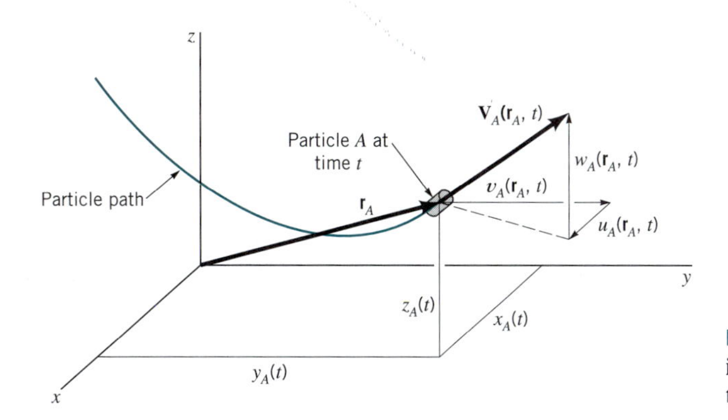

■ **FIGURE 4.4 Velocity and position of particle *A* at time *t*.**

Using the fact that the particle velocity components are given by $u_A = dx_A/dt$, $v_A = dy_A/dt$, and $w_A = dz_A/dt$, Eq. 4.2 becomes

$$\mathbf{a}_A = \frac{\partial \mathbf{V}_A}{\partial t} + u_A \frac{\partial \mathbf{V}_A}{\partial x} + v_A \frac{\partial \mathbf{V}_A}{\partial y} + w_A \frac{\partial \mathbf{V}_A}{\partial z}$$

Since the above is valid for any particle, we can drop the reference to particle A and obtain the acceleration field from the velocity field as

$$\mathbf{a} = \frac{\partial \mathbf{V}}{\partial t} + u \frac{\partial \mathbf{V}}{\partial x} + v \frac{\partial \mathbf{V}}{\partial y} + w \frac{\partial \mathbf{V}}{\partial z} \tag{4.3}$$

This is a vector result whose scalar components can be written as

$$a_x = \frac{\partial u}{\partial t} + u \frac{\partial u}{\partial x} + v \frac{\partial u}{\partial y} + w \frac{\partial u}{\partial z}$$

$$a_y = \frac{\partial v}{\partial t} + u \frac{\partial v}{\partial x} + v \frac{\partial v}{\partial y} + w \frac{\partial v}{\partial z} \tag{4.4}$$

and

$$a_z = \frac{\partial w}{\partial t} + u \frac{\partial w}{\partial x} + v \frac{\partial w}{\partial y} + w \frac{\partial w}{\partial z}$$

where a_x, a_y, and a_z are the x, y, and z components of the acceleration.

The above result is often written in shorthand notation as

$$\mathbf{a} = \frac{D\mathbf{V}}{Dt}$$

where the operator

$$\frac{D(\ )}{Dt} \equiv \frac{\partial(\ )}{\partial t} + u \frac{\partial(\ )}{\partial x} + v \frac{\partial(\ )}{\partial y} + w \frac{\partial(\ )}{\partial z} \tag{4.5}$$

The material derivative is used to describe time rates of change for a given particle.

is termed the *material derivative* or *substantial derivative*. An often-used shorthand notation for the material derivative operator is

$$\frac{D(\ )}{Dt} = \frac{\partial(\ )}{\partial t} + (\mathbf{V} \cdot \nabla)(\ ) \tag{4.6}$$

The dot product of the velocity vector, $\mathbf{V}$, and the gradient operator, $\nabla(\ ) = \partial(\ )/\partial x\,\hat{\mathbf{i}} + \partial(\ )/\partial y\,\hat{\mathbf{j}} + \partial(\ )/\partial z\,\hat{\mathbf{k}}$ (a vector operator) provides a convenient notation for the spatial derivative terms appearing in the Cartesian coordinate representation of the material derivative. Note that the notation $\mathbf{V} \cdot \nabla$ represents the operator $\mathbf{V} \cdot \nabla(\ ) = u\partial(\ )/\partial x + v\partial(\ )/\partial y + w\partial(\ )/\partial z$.

The material derivative concept is very useful in analysis involving various fluid parameters, not just the acceleration. The material derivative of any variable is the rate at which that variable changes with time for a given particle (as seen by one moving along with the fluid—the Lagrangian description). For example, consider a temperature field $T = T(x, y, z, t)$ associated with a given flow, like that shown in Fig. 4.2. It may be of interest to determine the time rate of change of temperature of a fluid particle (particle A) as it moves through this temperature field. If the velocity, $\mathbf{V} = \mathbf{V}(x, y, z, t)$, is known, we can apply the chain rule to determine the rate of change of temperature as

$$\frac{dT_A}{dt} = \frac{\partial T_A}{\partial t} + \frac{\partial T_A}{\partial x}\frac{dx_A}{dt} + \frac{\partial T_A}{\partial y}\frac{dy_A}{dt} + \frac{\partial T_A}{\partial z}\frac{dz_A}{dt}$$

This can be written as

$$\frac{DT}{Dt} = \frac{\partial T}{\partial t} + u\frac{\partial T}{\partial x} + v\frac{\partial T}{\partial y} + w\frac{\partial T}{\partial z} = \frac{\partial T}{\partial t} + \mathbf{V}\cdot\nabla T$$

As in the determination of the acceleration, the material derivative operator, $D(\ )/Dt$, appears.

EXAMPLE 4.4

An incompressible, inviscid fluid flows steadily past a sphere of radius a, as shown in Fig. E4.4*a*. According to a more advanced analysis of the flow, the fluid velocity along streamline A–B is given by

$$\mathbf{V} = u(x)\hat{\mathbf{i}} = V_0\left(1 + \frac{a^3}{x^3}\right)\hat{\mathbf{i}}$$

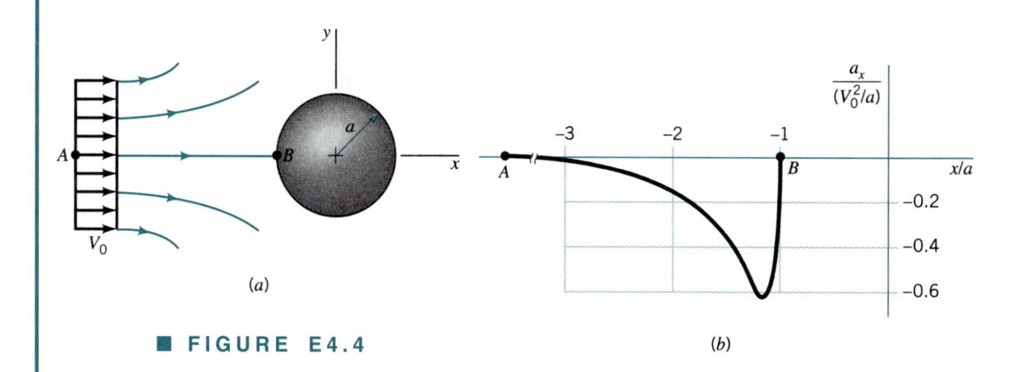

■ FIGURE E4.4

where V_0 is the upstream velocity far ahead of the sphere. Determine the acceleration experienced by fluid particles as they flow along this streamline.

SOLUTION

Along streamline A–B there is only one component of velocity ($v = w = 0$) so that from Eq. 4.3

$$\mathbf{a} = \frac{\partial \mathbf{V}}{\partial t} + u\frac{\partial \mathbf{V}}{\partial x} = \left(\frac{\partial u}{\partial t} + u\frac{\partial u}{\partial x}\right)\hat{\mathbf{i}}$$

or

$$a_x = \frac{\partial u}{\partial t} + u\frac{\partial u}{\partial x}, \qquad a_y = 0, \qquad a_z = 0$$

Since the flow is steady the velocity at a given point in space does not change with time. Thus, $\partial u/\partial t = 0$. With the given velocity distribution along the streamline, the acceleration becomes

$$a_x = u\frac{\partial u}{\partial x} = V_0\left(1 + \frac{a^3}{x^3}\right)V_0[a^3(-3x^{-4})]$$

or

$$a_x = -3(V_0^2/a)\frac{1 + (a/x)^3}{(x/a)^4} \qquad \textbf{(Ans)}$$

Along streamline A–B ($-\infty \leq x \leq -a$ and $y = 0$) the acceleration has only an x component and it is negative (a deceleration). Thus, the fluid slows down from its upstream velocity of $\mathbf{V} = V_0\hat{\mathbf{i}}$ at $x = -\infty$ to its stagnation point velocity of $\mathbf{V} = 0$ at $x = -a$, the "nose" of the sphere. The variation of a_x along streamline A–B is shown in Fig. E4.4*b*. It is the same result as is obtained in Example 3.1 by using the streamwise component of the acceleration, $a_x = V\,\partial V/\partial s$. The maximum deceleration occurs at $x = -1.205a$ and has a value of $a_x = -0.610V_0^2/a$.

In general, for fluid particles on streamlines other than A–B, all three components of the acceleration (a_x, a_y, and a_z) will be nonzero.

Fairly large accelerations (or decelerations) often occur in fluid flows. Consider air flowing past a baseball of radius $a = 0.14$ ft with a velocity of $V_0 = 100$ mi/hr $= 147$ ft/s. According to the results of Example 4.4, the maximum deceleration of an air particle approaching the stagnation point along the streamline in front of the ball is

$$|a_x|_{\max} = |a_x|_{x=-0.168 \text{ ft}} = \frac{0.610(147 \text{ ft/s})^2}{0.14 \text{ ft}} = 94.2 \times 10^3 \text{ ft/s}^2$$

This is a deceleration of approximately 3000 times that of gravity. In some situations the acceleration or deceleration experienced by fluid particles may be very large. An extreme case involves flow through shock waves that can occur in supersonic flow past objects (see Chapter 11). In such circumstances the fluid particles may experience decelerations hundreds of thousands of times greater than gravity. Large forces are obviously needed to produce such accelerations.

4.2.2 Unsteady Effects

As is seen from Eq. 4.5, the material derivative formula contains two types of terms—those involving the time derivative $[\partial(\ )/\partial t]$ and those involving spatial derivatives $[\partial(\ )/\partial x$, $\partial(\ )/\partial y$, and $\partial(\ )/\partial z]$. The time derivative portions are denoted as the *local derivative*. They represent effects of the unsteadiness of the flow. If the parameter involved is the acceleration, that portion given by $\partial\mathbf{V}/\partial t$ is termed the *local acceleration*. For steady flow the time derivative is zero throughout the flow field $[\partial(\ )/\partial t \equiv 0]$, and the local effect vanishes. Physically, there is no change in flow parameters at a fixed point in space if the flow is steady. There may be a change of those parameters for a fluid particle as it moves about, however.

The local derivative is a result of the unsteadiness of the flow.

If a flow is unsteady, its parameter values (velocity, temperature, density, etc.) at any location may change with time. For example, an unstirred ($\mathbf{V} = 0$) cup of coffee will cool down in time because of heat transfer to its surroundings. That is, $DT/Dt = \partial T/\partial t + \mathbf{V}\cdot\nabla T = \partial T/\partial t < 0$. Similarly, a fluid particle may have nonzero acceleration as a result of the unsteady effect of the flow. Consider flow in a constant diameter pipe as is shown in Fig. 4.5. The flow is assumed to be spatially uniform throughout the pipe. That is, $\mathbf{V} = V_0(t)\,\hat{\mathbf{i}}$ at all points in the pipe. The value of the acceleration depends on whether V_0 is being increased, $\partial V_0/\partial t > 0$, or decreased, $\partial V_0/\partial t < 0$. Unless V_0 is independent of time ($V_0 \equiv$ constant) there will be an acceleration, the local acceleration term. Thus, the acceleration field, $\mathbf{a} = \partial V_0/\partial t\,\hat{\mathbf{i}}$, is uniform throughout the entire flow, although it may vary with time ($\partial V_0/\partial t$ need not be constant). The acceleration due to the spatial variations of velocity ($u\,\partial u/\partial x$, $v\,\partial v/\partial y$, etc.) vanishes automatically for this flow, since $\partial u/\partial x = 0$ and $v = w = 0$. That is,

$$\mathbf{a} = \frac{\partial\mathbf{V}}{\partial t} + u\frac{\partial\mathbf{V}}{\partial x} + v\frac{\partial\mathbf{V}}{\partial y} + w\frac{\partial\mathbf{V}}{\partial z} = \frac{\partial\mathbf{V}}{\partial t} = \frac{\partial V_0}{\partial t}\hat{\mathbf{i}}$$

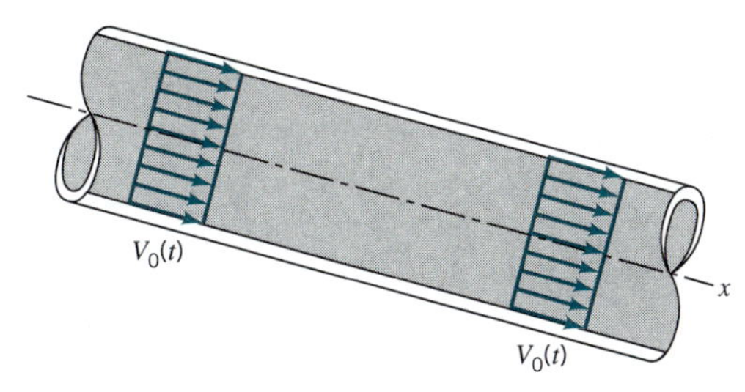

■ FIGURE 4.5 **Uniform, unsteady flow in a constant diameter pipe.**

4.2.3 Convective Effects

The convective derivative is a result of the spatial variation of the flow.

The portion of the material derivative (Eq. 4.5) represented by the spatial derivatives is termed the *convective derivative*. It represents the fact that a flow property associated with a fluid particle may vary because of the motion of the particle from one point in space where the parameter has one value to another point in space where its value is different. This contribution to the time rate of change of the parameter for the particle can occur whether the flow is steady or unsteady. It is due to the convection, or motion, of the particle through space in which there is a gradient $[\nabla(\ ) = \partial(\ )/\partial x\, \hat{\mathbf{i}} + \partial(\ )/\partial y\, \hat{\mathbf{j}} + \partial(\ )/dz\, \hat{\mathbf{k}}]$ in the parameter value. That portion of the acceleration given by the term $(\mathbf{V} \cdot \nabla)\mathbf{V}$ is termed the *convective acceleration*.

As is illustrated in Fig. 4.6, the temperature of a water particle changes as it flows through a water heater. The water entering the heater is always the same cold temperature and the water leaving the heater is always the same hot temperature. The flow is steady. However, the temperature, T, of each water particle increases as it passes through the heater—$T_{\text{out}} > T_{\text{in}}$. Thus, $DT/Dt \neq 0$ because of the convective term in the total derivative of the temperature. That is, $\partial T/\partial t = 0$, but $u\, \partial T/\partial x \neq 0$ (where x is directed along the streamline), since there is a nonzero temperature gradient along the streamline. A fluid particle traveling along this nonconstant temperature path ($\partial T/\partial x \neq 0$) at a specified speed ($u$) will have its temperature change with time at a rate of $DT/Dt = u\, \partial T/\partial x$ even though the flow is steady ($\partial T/\partial t = 0$).

The same types of processes are involved with fluid accelerations. Consider flow in a variable area pipe as shown in Fig. 4.7. It is assumed that the flow is steady and one-dimensional with velocity that increases and decreases in the flow direction as indicated. As the fluid flows from section (1) to section (2), its velocity increases from V_1 to V_2. Thus, even though $\partial \mathbf{V}/\partial t = 0$, fluid particles experience an acceleration given by $a_x = u\, \partial u/\partial x$. For $x_1 < x < x_2$, it is seen that $\partial u/\partial x > 0$ so that $a_x > 0$—the fluid accelerates. For $x_2 < x < x_3$, it is seen that $\partial u/\partial x < 0$ so that $a_x < 0$—the fluid decelerates. If $V_1 = V_3$, the amount of acceleration precisely balances the amount of deceleration even though the distances between x_2 and x_1 and x_3 and x_2 are not the same.

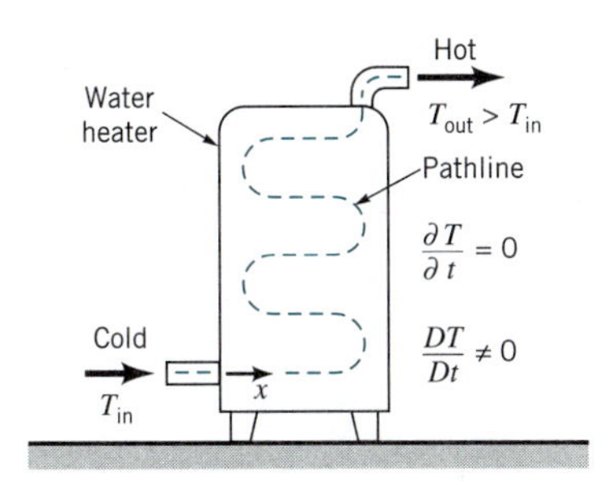

■ FIGURE 4.6 **Steady-state operation of a water heater.**

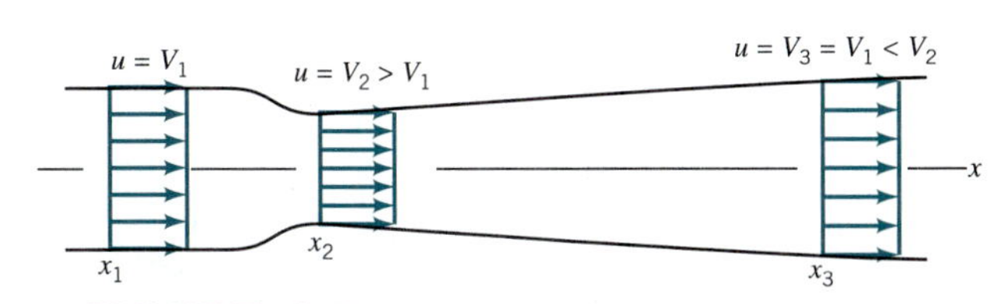

■ FIGURE 4.7 **Uniform, steady flow in a variable area pipe.**

EXAMPLE 4.5

Consider the steady, two-dimensional flow field discussed in Example 4.2. Determine the acceleration field for this flow.

SOLUTION

In general, the acceleration is given by

$$\mathbf{a} = \frac{D\mathbf{V}}{Dt} = \frac{\partial \mathbf{V}}{\partial t} + (\mathbf{V} \cdot \nabla)(\mathbf{V}) = \frac{\partial \mathbf{V}}{\partial t} + u\frac{\partial \mathbf{V}}{\partial x} + v\frac{\partial \mathbf{V}}{\partial y} + w\frac{\partial \mathbf{V}}{\partial z} \tag{1}$$

where the velocity is given by $\mathbf{V} = (V_0/\ell)(x\hat{\mathbf{i}} - y\hat{\mathbf{j}})$ so that $u = (V_0/\ell)x$ and $v = -(V_0/\ell)y$. For steady $[\partial(\ )/\partial t = 0]$, two-dimensional $[w = 0$ and $\partial(\ )/\partial z = 0]$ flow, Eq. 1 becomes

$$\mathbf{a} = u\frac{\partial \mathbf{V}}{\partial x} + v\frac{\partial \mathbf{V}}{\partial y} = \left(u\frac{\partial u}{\partial x} + v\frac{\partial u}{\partial y}\right)\hat{\mathbf{i}} + \left(u\frac{\partial v}{\partial x} + v\frac{\partial v}{\partial y}\right)\hat{\mathbf{j}}$$

Hence, for this flow the acceleration is given by

$$\mathbf{a} = \left[\left(\frac{V_0}{\ell}\right)(x)\left(\frac{V_0}{\ell}\right) + \left(\frac{V_0}{\ell}\right)(y)(0)\right]\hat{\mathbf{i}} + \left[\left(\frac{V_0}{\ell}\right)(x)(0) + \left(\frac{-V_0}{\ell}\right)(y)\left(\frac{-V_0}{\ell}\right)\right]\hat{\mathbf{j}}$$

or

$$a_x = \frac{V_0^2 x}{\ell^2}, \qquad a_y = \frac{V_0^2 y}{\ell^2} \tag{Ans}$$

The fluid experiences an acceleration in both the x and y directions. Since the flow is steady, there is no local acceleration—the fluid velocity at any given point is constant in time. However, there is a convective acceleration due to the change in velocity from one point on the particle's pathline to another. Recall that the velocity is a vector—it has both a magnitude and a direction. In this flow both the fluid speed (magnitude) and flow direction change with location (see Fig. E4.1*a*).

For this flow the magnitude of the acceleration is constant on circles centered at the origin, as is seen from the fact that

$$|\mathbf{a}| = (a_x^2 + a_y^2 + a_z^2)^{1/2} = \left(\frac{V_0}{\ell}\right)^2 (x^2 + y^2)^{1/2} \tag{2}$$

Also, the acceleration vector is oriented at an angle θ from the x axis, where

$$\tan \theta = \frac{a_y}{a_x} = \frac{y}{x}$$

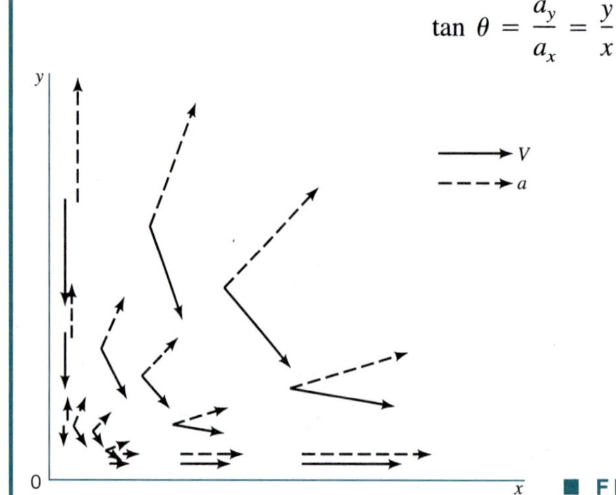

■ FIGURE E4.5

This is the same angle as that formed by a ray from the origin to point (x, y). Thus, the acceleration is directed along rays from the origin and has a magnitude proportional to the distance from the origin. Typical acceleration vectors (from Eq. 2) and velocity vectors (from Example 4.1) are shown in Fig. E4.5 for the flow in the first quadrant. Note that **a** and **V** are not parallel except along the x and y axes (a fact that is responsible for the curved pathlines of the flow), and that both the acceleration and velocity are zero at the origin $(x = y = 0)$. An infinitesimal fluid particle placed precisely at the origin will remain there, but its neighbors (no matter how close they are to the origin) will drift away.

The concept of the material derivative can be used to determine the time rate of change of any parameter associated with a particle as it moves about. Its use is not restricted to fluid mechanics alone. The basic ingredients needed to use the material derivative concept are the field description of the parameter, $P = P(x, y, z, t)$, and the rate at which the particle moves through that field, $\mathbf{V} = \mathbf{V}(x, y, z, t)$.

EXAMPLE 4.6

A manufacturer produces a perishable product in a factory located at $x = 0$ and sells the product along the distribution route $x > 0$. The selling price of the product, P, is a function of the length of time after it was produced, t, and the location at which it is sold, x. That is, $P = P(x, t)$. At a given location the price of the product decreases in time (it is perishable) according to $\partial P/\partial t = -C_1$, where C_1 is a positive constant (dollars per hour). In addition, because of shipping costs the price increases with distance from the factory according to $\partial P/\partial x = C_2$, where C_2 is a positive constant (dollars per mile). If the manufacturer wishes to sell the product for the same price anywhere along the distribution route, determine how fast he must travel along the route.

SOLUTION

For a given batch of the product (Lagrangian description), the time rate of change of the price can be obtained by using the material derivative

$$\frac{DP}{Dt} = \frac{\partial P}{\partial t} + \mathbf{V} \cdot \nabla P = \frac{\partial P}{\partial t} + u\frac{\partial P}{\partial x} + v\frac{\partial P}{\partial y} + w\frac{\partial P}{\partial z} = \frac{\partial P}{\partial t} + u\frac{\partial P}{\partial x}$$

We have used the fact that the motion is one-dimensional with $\mathbf{V} = u\hat{\mathbf{i}}$, where u is the speed at which the product is convected along its route. If the price is to remain constant as the product moves along the distribution route, then

$$\frac{DP}{Dt} = 0 \quad \text{or} \quad \frac{\partial P}{\partial t} + u\frac{\partial P}{\partial x} = 0$$

Thus, the correct delivery speed is

$$u = \frac{-\partial P/\partial t}{\partial P/\partial x} = \frac{C_1}{C_2} \qquad \textbf{(Ans)}$$

With this speed, the decrease in price because of the local effect $(\partial P/\partial t)$ is exactly balanced by the increase in price due to the convective effect $(u\, \partial P/\partial x)$. A faster delivery speed will cause the price of the given batch of the product to increase in time ($DP/Dt > 0$; it is rushed to distant markets before it spoils), while a slower delivery speed will cause its price to decrease ($DP/Dt < 0$; the increased costs due to distance from the factory is more than offset by reduced costs due to spoilage.)

4.2.4 Streamline Coordinates

In many flow situations it is convenient to use a coordinate system defined in terms of the streamlines of the flow. An example for steady, two-dimensional flows is illustrated in Fig. 4.8. Such flows can be described either in terms of the usual x, y Cartesian coordinate system (or some other system such as the r, θ polar coordinate system) or the streamline coordinate system. In the streamline coordinate system the flow is described in terms of one coordinate along the streamlines, denoted s, and the second coordinate normal to the streamlines, denoted n. Unit vectors in these two directions are denoted by $\hat{\mathbf{s}}$ and $\hat{\mathbf{n}}$, as shown in the figure. Care is needed not to confuse the coordinate distance s (a scalar) with the unit vector along the streamline direction, $\hat{\mathbf{s}}$.

Streamline coordinates provide a natural coordinate system for a flow.

The flow plane is therefore covered by an orthogonal curved net of coordinate lines. At any point the s and n directions are perpendicular, but the lines of constant s or constant n are not necessarily straight. Without knowing the actual velocity field (hence, the streamlines) it is not possible to construct this flow net. In many situations appropriate simplifying assumptions can be made so that this lack of information does not present an insurmountable difficulty. One of the major advantages of using the streamline coordinate system is that the velocity is always tangent to the s direction. That is,

$$\mathbf{V} = V\hat{\mathbf{s}}$$

This allows simplifications in describing the fluid particle acceleration and in solving the equations governing the flow.

For steady, two-dimensional flow we can determine the acceleration as

$$\mathbf{a} = \frac{D\mathbf{V}}{Dt} = a_s\hat{\mathbf{s}} + a_n\hat{\mathbf{n}}$$

where a_s and a_n are the streamline and normal components of acceleration, respectively. We use the material derivative because by definition the acceleration is the time rate of change of the velocity of a given particle as it moves about. If the streamlines are curved, both the speed of the particle and its direction of flow may change from one point to another. In general, for steady flow both the speed and the flow direction are a function of location—$V = V(s, n)$ and $\hat{\mathbf{s}} = \hat{\mathbf{s}}(s, n)$. For a given particle, the value of s changes with time, but the value of n remains fixed because the particle flows along a streamline defined by n = constant. (Recall that streamlines and pathlines coincide in steady flow.) Thus, application of the chain rule gives

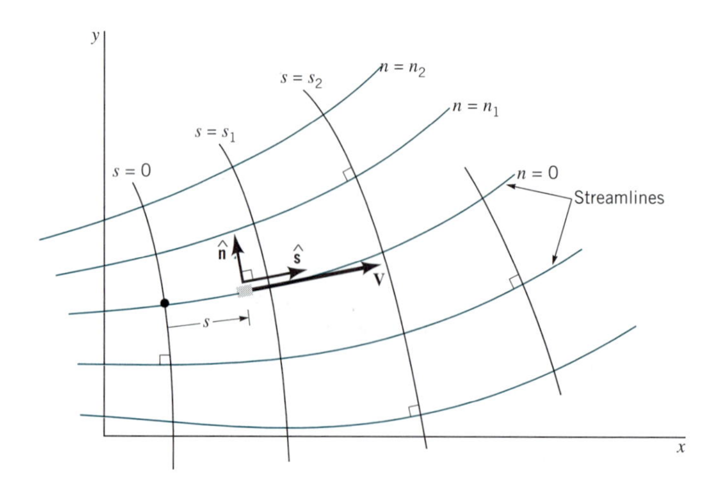

■ **FIGURE 4.8 Streamline coordinate system for two-dimensional flow.**

$$\mathbf{a} = \frac{D(V\hat{\mathbf{s}})}{Dt} = \frac{DV}{Dt}\hat{\mathbf{s}} + V\frac{D\hat{\mathbf{s}}}{Dt}$$

or

$$\mathbf{a} = \left(\frac{\partial V}{\partial t} + \frac{\partial V}{\partial s}\frac{ds}{dt} + \frac{\partial V}{\partial n}\frac{dn}{dt}\right)\hat{\mathbf{s}} + V\left(\frac{\partial \hat{\mathbf{s}}}{\partial t} + \frac{\partial \hat{\mathbf{s}}}{\partial s}\frac{ds}{dt} + \frac{\partial \hat{\mathbf{s}}}{\partial n}\frac{dn}{dt}\right)$$

This can be simplified by using the fact that for steady flow nothing changes with time at a given point so that both $\partial V/\partial t$ and $\partial \hat{\mathbf{s}}/\partial t$ are zero. Also, the velocity along the streamline is $V = ds/dt$ and the particle remains on its streamline (n = constant) so that $dn/dt = 0$. Hence,

$$\mathbf{a} = \left(V\frac{\partial V}{\partial s}\right)\hat{\mathbf{s}} + V\left(V\frac{\partial \hat{\mathbf{s}}}{\partial s}\right)$$

The quantity $\partial \hat{\mathbf{s}}/\partial s$ represents the limit as $\delta s \rightarrow 0$ of the change in the unit vector along the streamline, $\delta \hat{\mathbf{s}}$, per change in distance along the streamline, δs. The magnitude of $\hat{\mathbf{s}}$ is constant ($|\hat{\mathbf{s}}| = 1$; it is a unit vector), but its direction is variable if the streamlines are curved. From Fig. 4.9 it is seen that the magnitude of $\partial \hat{\mathbf{s}}/\partial s$ is equal to the inverse of the radius of curvature of the streamline, $\mathcal{R}$, at the point in question. This follows because the two triangles shown (AOB and $A'O'B'$) are similar triangles so that $\delta s/\mathcal{R} = |\delta \hat{\mathbf{s}}|/|\hat{\mathbf{s}}| = |\delta \hat{\mathbf{s}}|$, or $|\delta \hat{\mathbf{s}}/\delta s| = 1/\mathcal{R}$. Similarly, in the limit $\delta s \rightarrow 0$, the direction of $\delta \hat{\mathbf{s}}/\delta s$ is seen to be normal to the streamline. That is,

$$\frac{\partial \hat{\mathbf{s}}}{\partial s} = \lim_{\delta s \rightarrow 0} \frac{\delta \hat{\mathbf{s}}}{\delta s} = \frac{\hat{\mathbf{n}}}{\mathcal{R}}$$

Streamline and normal components of acceleration occur even in steady flows.

Hence, the acceleration for steady, two-dimensional flow can be written in terms of its streamwise and normal components in the form

$$\mathbf{a} = V\frac{\partial V}{\partial s}\hat{\mathbf{s}} + \frac{V^2}{\mathcal{R}}\hat{\mathbf{n}} \quad \text{or} \quad a_s = V\frac{\partial V}{\partial s}, \qquad a_n = \frac{V^2}{\mathcal{R}} \tag{4.7}$$

The first term, $a_s = V\,\partial V/\partial s$, represents the convective acceleration along the streamline and the second term, $a_n = V^2/\mathcal{R}$, represents centrifugal acceleration (one type of convective acceleration) normal to the fluid motion. These components can be noted in Fig. E4.5 by resolving the acceleration vector into its components along and normal to the velocity vector. Note that the unit vector $\hat{\mathbf{n}}$ is directed from the streamline toward the center of curvature. These forms of the acceleration are probably familiar from previous dynamics or physics considerations.

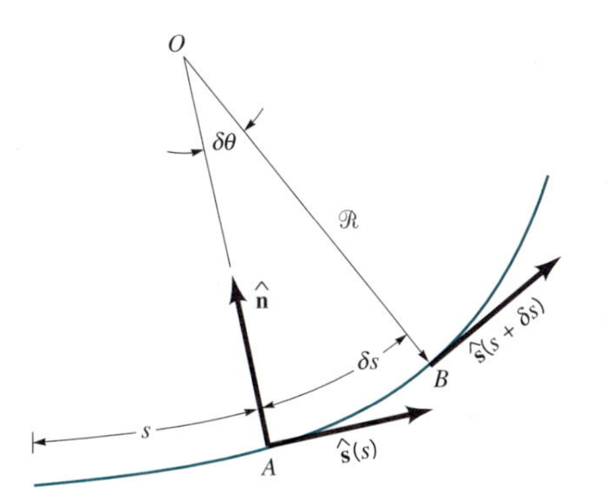

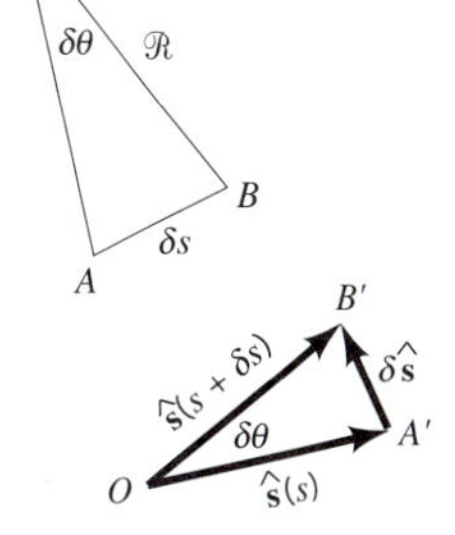

■ **FIGURE 4.9 Relationship between the unit vector along the streamline, ŝ, and the radius of curvature of the streamline, $\mathcal{R}$.**

4.3 Control Volume and System Representations

As is discussed in Chapter 1, a fluid is a type of matter that is relatively free to move and interact with its surroundings. As with any matter, a fluid's behavior is governed by a set of fundamental physical laws which are approximated by an appropriate set of equations. The application of laws such as the conservation of mass, Newton's laws of motion, and the laws of thermodynamics form the foundation of fluid mechanics analyses. There are various ways that these governing laws can be applied to a fluid, including the system approach and the control volume approach. By definition, a *system* is a collection of matter of fixed identity (always the same atoms or fluid particles), which may move, flow, and interact with its surroundings. A *control volume*, on the other hand, is a volume in space (a geometric entity, independent of mass) through which fluid may flow.

Both control volume and system concepts can be used to describe fluid flow.

A system is a specific, identifiable quantity of matter. It may consist of a relatively large amount of mass (such as all of the air in the earth's atmosphere), or it may be an infinitesimal size (such as a single fluid particle). In any case, the molecules making up the system are "tagged" in some fashion (dyed red, either actually or only in your mind) so that they can be continually identified as they move about. The system may interact with its surroundings by various means (by the transfer of heat or the exertion of a pressure force, for example). It may continually change size and shape, but it always contains the same mass.

A mass of air drawn into an air compressor can be considered as a system. It changes shape and size (it is compressed), its temperature may change, and it is eventually expelled through the outlet of the compressor. The matter associated with the original air drawn into the compressor remains as a system, however. The behavior of this material could be investigated by applying the appropriate governing equations to this system.

One of the important concepts used in the study of statics and dynamics is that of the free-body diagram. That is, we identify an object, isolate it from its surroundings, replace its surroundings by the equivalent actions that they put on the object, and apply Newton's laws of motion. The body in such cases is our system—an identified portion of matter that we follow during its interactions with its surroundings. In fluid mechanics, it is often quite difficult to identify and keep track of a specific quantity of matter. A finite portion of a fluid contains an uncountable number of fluid particles that move about quite freely, unlike a solid that may deform but usually remains relatively easy to identify. For example, we cannot as easily follow a specific portion of water flowing in a river as we can follow a branch floating on its surface.

We may often be more interested in determining the forces put on a fan, airplane, or automobile by air flowing past the object than we are in the information obtained by following a given portion of the air (a system) as it flows along. For these situations we often use the control volume approach. We identify a specific volume in space (a volume associated with the fan, airplane, or automobile, for example) and analyze the fluid flow within, through, or around that volume. In general, the control volume can be a moving volume, although for most situations considered in this book we will use only fixed, nondeformable control volumes. The matter within a control volume may change with time as the fluid flows through it. Similarly, the amount of mass within the volume may change with time. The control volume itself is a specific geometric entity, independent of the flowing fluid.

Examples of control volumes and *control surfaces* (the surface of the control volume) are shown in Fig. 4.10. For case (*a*), fluid flows through a pipe. The fixed control surface consists of the inside surface of the pipe, the outlet end at section (2), and a section across the pipe at (1). One portion of the control surface is a physical surface (the pipe), while the remainder is simply a surface in space (across the pipe). Fluid flows across part of the control surface, but not across all of it.

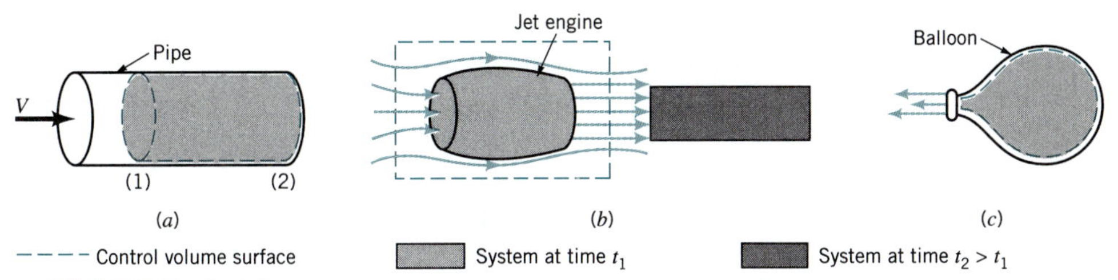

■ **FIGURE 4.10 Typical control volumes: (*a*) fixed control volume, (*b*) fixed or moving control volume, (*c*) deforming control volume.**

Another control volume is the rectangular volume surrounding the jet engine shown in Fig. 4.10*b*. If the airplane to which the engine is attached is sitting still on the runway, air flows through this control volume because of the action of the engine within it. The air that was within the engine itself at time $t = t_1$ (a system) has passed through the engine and is outside of the control volume at a later time $t = t_2$ as indicated. At this later time other air (a different system) is within the engine. If the airplane is moving, the control volume is fixed relative to an observer on the airplane, but it is a moving control volume relative to an observer on the ground. In either situation air flows through and around the engine as indicated.

The deflating balloon shown in Fig. 4.10*c* provides an example of a deforming control volume. As time increases, the control volume (whose surface is the inner surface of the balloon) decreases in size. If we do not hold onto the balloon, it becomes a moving, deforming control volume as it darts about the room. The majority of the problems we will analyze can be solved by using a fixed, nondeforming control volume. In some instances, however, it will be advantageous, in fact necessary, to use a moving, deforming control volume.

In many ways the relationship between a system and a control volume is similar to the relationship between the Lagrangian and Eulerian flow description introduced in Section 4.1.1. In the system or Lagrangian description we follow the fluid and observe its behavior as it moves about. In the control volume or Eulerian description we remain stationary and observe the fluid's behavior at a fixed location. (If a moving control volume is used, it virtually never moves with the system—the system flows through the control volume.) These ideas are discussed in more detail in the next section.

The governing laws of fluid motion are stated in terms of fluid systems, not control volumes.

All of the laws governing the motion of a fluid are stated in their basic form in terms of a system approach. For example, "the mass of a system remains constant," or "the time rate of change of momentum of a system is equal to the sum of all the forces acting on the system." Note the word system, not control volume, in these statements. To use the governing equations in a control volume approach to problem solving, we must rephrase the laws in an appropriate manner. To this end we introduce the Reynolds transport theorem in the following section.

4.4 The Reynolds Transport Theorem

We are sometimes interested in what happens to a particular part of the fluid as it moves about. Other times we may be interested in what effect the fluid has on a particular object or volume in space as fluid interacts with it. Thus, we need to describe the laws governing fluid motion using both system concepts (consider a given mass of the fluid) and control volume concepts (consider a given volume). To do this we need an analytical tool to shift from one representation to the other. The *Reynolds transport theorem* provides this tool.

All physical laws are stated in terms of various physical parameters. Velocity, acceleration, mass, temperature, and momentum are but a few of the more common parameters. Let B represent any of these (or other) fluid parameters and b represent the amount of that parameter per unit mass. That is,

$$B = mb$$

where m is the mass of the portion of fluid of interest. For example, if $B = m$, the mass, it follows that $b = 1$. (The mass per unit mass is unity.) If $B = mV^2/2$, the kinetic energy of the mass, then $b = V^2/2$, the kinetic energy per unit mass. The parameters B and b may be scalars or vectors. Thus, if $\mathbf{B} = m\mathbf{V}$, the momentum of the mass, then $\mathbf{b} = \mathbf{V}$. (The momentum per unit mass is the velocity.)

The parameter B is termed an *extensive property* and the parameter b is termed an *intensive property*. The value of B is directly proportional to the amount of the mass being considered, whereas the value of b is independent of the amount of mass. The amount of an extensive property that a system possesses at a given instant, B_{sys}, can be determined by adding up the amount associated with each fluid particle in the system. For infinitesimal fluid particles of size $\delta \mathcal{V}$ and mass $\rho\, \delta \mathcal{V}$, this summation (in the limit of $\delta \mathcal{V} \rightarrow 0$) takes the form of an integration over all the particles in the system and can be written as

$$B_{\text{sys}} = \lim_{\delta \mathcal{V} \rightarrow 0} \sum_i b_i (\rho_i\, \delta \mathcal{V}_i) = \int_{\text{sys}} \rho b \, d\mathcal{V}$$

The limits of integration cover the entire system—a (usually) moving volume. We have used the fact that the amount of B in a fluid particle of mass $\rho\, \delta \mathcal{V}$ is given in terms of b by $\delta B = b\rho\, \delta \mathcal{V}$.

Most of the laws governing fluid motion involve the time rate of change of an extensive property of a fluid system—the rate at which the momentum of a system changes with time, the rate at which the mass of a system changes with time, and so on. Thus, we often encounter terms such as

$$\frac{dB_{\text{sys}}}{dt} = \frac{d\left(\int_{\text{sys}} \rho b \, d\mathcal{V}\right)}{dt} \tag{4.8}$$

To formulate the laws into a control volume approach, we must obtain an expression for the time rate of change of an extensive property within a control volume, B_{cv}, not within a system. This can be written as

$$\frac{dB_{\text{cv}}}{dt} = \frac{d\left(\int_{\text{cv}} \rho b \, d\mathcal{V}\right)}{dt} \tag{4.9}$$

Differences between control volume and system concepts are subtle but very important.

where the limits of integration, denoted by cv, cover the control volume of interest. Although Eqs. 4.8 and 4.9 may look very similar, the physical interpretation of each is quite different. Mathematically, the difference is represented by the difference in the limits of integration. Recall that the control volume is a volume in space (in most cases stationary, although if it moves it need not move with the system). On the other hand, the system is an identifiable collection of mass that moves with the fluid (indeed it is a specified portion of the fluid). We will learn that even for those instances when the control volume and the system momentarily occupy the same volume in space, the two quantities dB_{sys}/dt and dB_{cv}/dt need not be the same. The Reynolds transport theorem provides the relationship between the time rate of change of an extensive property for a system and that for a control volume—the relationship between Eqs. 4.8 and 4.9.

EXAMPLE 4.7

Fluid flows from the fire extinguisher tank shown in Fig. E4.7. Discuss the differences between dB_{sys}/dt and dB_{cv}/dt if B represents mass.

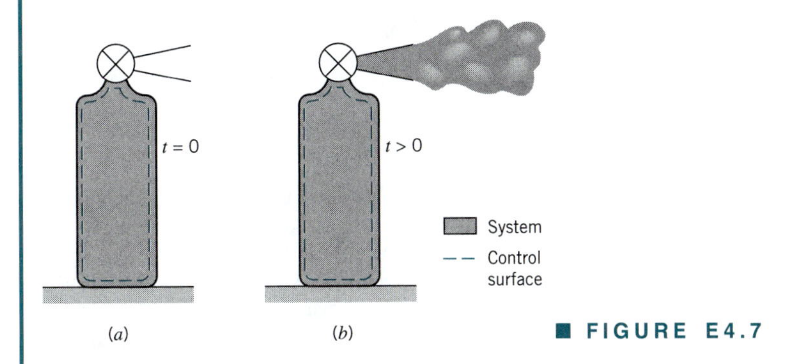

■ FIGURE E4.7

SOLUTION

With $B = m$, the system mass, it follows that $b = 1$ and Eqs. 4.8 and 4.9 can be written as

$$\frac{dB_{sys}}{dt} \equiv \frac{dm_{sys}}{dt} = \frac{d\left(\int_{sys} \rho \, d\mathcal{V}\right)}{dt}$$

and

$$\frac{dB_{cv}}{dt} \equiv \frac{dm_{cv}}{dt} = \frac{d\left(\int_{cv} \rho \, d\mathcal{V}\right)}{dt}$$

Physically these represent the time rate of change of mass within the system and the time rate of change of mass within the control volume, respectively. We choose our system to be the fluid within the tank at the time the valve was opened ($t = 0$) and the control volume to be the tank itself. A short time after the valve is opened, part of the system has moved outside of the control volume as is shown in Fig. E4.7*b*. The control volume remains fixed. The limits of integration are fixed for the control volume; they are a function of time for the system.

Clearly, if mass is to be conserved (one of the basic laws governing fluid motion), the mass of the fluid in the system is constant, so that

$$\frac{d\left(\int_{sys} \rho \, d\mathcal{V}\right)}{dt} = 0$$

On the other hand, it is equally clear that some of the fluid has left the control volume through the nozzle on the tank. Hence, the amount of mass within the tank (the control volume) decreases with time, or

$$\frac{d\left(\int_{cv} \rho \, d\mathcal{V}\right)}{dt} < 0$$

The actual numerical value of the rate at which the mass in the control volume decreases will depend on the rate at which the fluid flows through the nozzle (that is, the size of the nozzle and the speed and density of the fluid). Clearly the meanings of dB_{sys}/dt and dB_{cv}/dt are different. For this example, $dB_{\text{cv}}/dt < dB_{\text{sys}}/dt$. Other situations may have $dB_{\text{cv}}/dt \geq dB_{\text{sys}}/dt$.

4.4.1 Derivation of the Reynolds Transport Theorem

The moving system flows through the fixed control volume.

A simple version of the Reynolds transport theorem relating system concepts to control volume concepts can be obtained easily for the one-dimensional flow through a fixed control volume as is shown in Fig. 4.11*a*. We consider the control volume to be that stationary volume within the pipe or duct between sections (1) and (2) as indicated. The system that we consider is that fluid occupying the control volume at some initial time t. A short time later, at time $t + \delta t$, the system has moved slightly to the right. The fluid particles that coincided with section (2) of the control surface at time t have moved a distance $\delta \ell_2 = V_2\, \delta t$ to the right, where V_2 is the velocity of the fluid as it passes section (2). Similarly, the fluid initially at section (1) has moved a distance $\delta \ell_1 = V_1\, \delta t$, where V_1 is the fluid velocity at section (1). We assume the fluid flows across sections (1) and (2) in a direction normal to these surfaces and that V_1 and V_2 are constant across sections (1) and (2).

As is shown in Fig. 4.11*b*, the outflow from the control volume from time t to $t + \delta t$ is denoted as volume II, the inflow as volume I, and the control volume itself as CV. Thus, the system at time t consists of the fluid in section CV ("SYS = CV" at time t), while at time $t + \delta t$ the system consists of the same fluid that now occupies sections (CV − I) + II. That is, "SYS = CV − I + II" at time $t + \delta t$. The control volume remains as section CV for all time.

If B is an extensive parameter of the system, then the value of it for the system at time t is

$$B_{\text{sys}}(t) = B_{\text{cv}}(t)$$

since the system and the fluid within the control volume coincide at this time. Its value at time $t + \delta t$ is

$$B_{\text{sys}}(t + \delta t) = B_{\text{cv}}(t + \delta t) - B_{\text{I}}(t + \delta t) + B_{\text{II}}(t + \delta t)$$

Thus, the change in the amount of B in the system in the time interval δt divided by this time

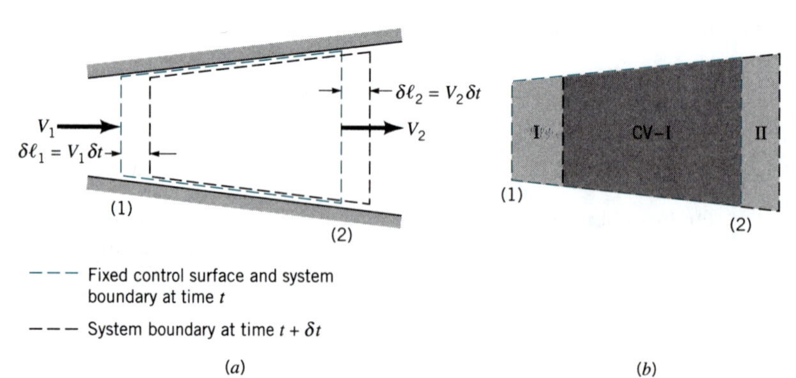

■ **FIGURE 4.11 Control volume and system for flow through a variable area pipe.**

interval is given by

$$\frac{\delta B_{\text{sys}}}{\delta t} = \frac{B_{\text{sys}}(t + \delta t) - B_{\text{sys}}(t)}{\delta t} = \frac{B_{\text{cv}}(t + \delta t) - B_{\text{I}}(t + \delta t) + B_{\text{II}}(t + \delta t) - B_{\text{sys}}(t)}{\delta t}$$

By using the fact that at the initial time t we have $B_{\text{sys}}(t) = B_{\text{cv}}(t)$, this ungainly expression may be rearranged as follows.

$$\frac{\delta B_{\text{sys}}}{\delta t} = \frac{B_{\text{cv}}(t + \delta t) - B_{\text{cv}}(t)}{\delta t} - \frac{B_{\text{I}}(t + \delta t)}{\delta t} + \frac{B_{\text{II}}(t + \delta t)}{\delta t} \tag{4.10}$$

In the limit $\delta t \to 0$, the left-hand side of Eq. 4.10 is equal to the time rate of change of B for the system and is denoted as DB_{sys}/Dt. We use the material derivative notation, $D(\ )/Dt$, to denote this time rate of change to emphasize the Lagrangian character of this term. (Recall from Section 4.2.1 that the material derivative, DP/Dt, of any quantity P represents the time rate of change of that quantity associated with a given fluid particle as it moves along.) Similarly, the quantity DB_{sys}/Dt represents the time rate of change of property B associated with a system (a given portion of fluid) as it moves along.

In the limit $\delta t \to 0$, the first term on the right-hand side of Eq. 4.10 is seen to be the time rate of change of the amount of B within the control volume

$$\lim_{\delta t \to 0} \frac{B_{\text{cv}}(t + \delta t) - B_{\text{cv}}(t)}{\delta t} = \frac{\partial B_{\text{cv}}}{\partial t} = \frac{\partial \left(\int_{\text{cv}} \rho b \, d\forall \right)}{\partial t} \tag{4.11}$$

The third term on the right-hand side of Eq. 4.10 represents the rate at which the extensive parameter B flows from the control volume, across the control surface. This can be seen from the fact that the amount of B within region II, the outflow region, is its amount per unit volume, ρb, times the volume $\delta \forall_{\text{II}} = A_2 \, \delta \ell_2 = A_2(V_2 \, \delta t)$. Hence,

$$B_{\text{II}}(t + \delta t) = (\rho_2 b_2)(\delta \forall_{\text{II}}) = \rho_2 b_2 A_2 V_2 \, \delta t$$

where b_2 and ρ_2 are the constant values of b and ρ across section (2). Thus, the rate at which this property flows from the control volume, $\dot{B}_{\text{out}}$, is given by

$$\dot{B}_{\text{out}} = \lim_{\delta t \to 0} \frac{B_{\text{II}}(t + \delta t)}{\delta t} = \rho_2 A_2 V_2 b_2 \tag{4.12}$$

Similarly, the inflow of B into the control volume across section (1) during the time interval δt corresponds to that in region I and is given by the amount per unit volume times the volume, $\delta \forall_{\text{I}} = A_1 \, \delta \ell_1 = A_1(V_1 \, \delta t)$. Hence,

$$B_{\text{I}}(t + \delta t) = (\rho_1 b_1)(\delta \forall_{\text{I}}) = \rho_1 b_1 A_1 V_1 \, \delta t$$

where b_1 and ρ_1 are the constant values of b and ρ across section (1). Thus, the rate on inflow of the property B into the control volume, $\dot{B}_{\text{in}}$, is given by

$$\dot{B}_{\text{in}} = \lim_{\delta t \to 0} \frac{B_{\text{I}}(t + \delta t)}{\delta t} = \rho_1 A_1 V_1 b_1 \tag{4.13}$$

The time derivative associated with a system may be different from that for a control volume.

If we combine Eqs. 4.10, 4.11, 4.12, and 4.13 we see that the relationship between the time rate of change of B for the system and that for the control volume is given by

$$\frac{DB_{\text{sys}}}{Dt} = \frac{\partial B_{\text{cv}}}{\partial t} + \dot{B}_{\text{out}} - \dot{B}_{\text{in}} \tag{4.14}$$

or

$$\frac{DB_{\text{sys}}}{Dt} = \frac{\partial B_{\text{cv}}}{\partial t} + \rho_2 A_2 V_2 b_2 - \rho_1 A_1 V_1 b_1 \tag{4.15}$$

The Reynolds transport theorem involves time derivatives and flow rates.

This is a version of the Reynolds transport theorem valid under the restrictive assumptions associated with the flow shown in Fig. 4.11—fixed control volume with one inlet and one outlet having uniform properties (density, velocity, and the parameter b) across the inlet and outlet with the velocity normal to sections (1) and (2). Note that the time rate of change of B for the system (the left-hand side of Eq. 4.15 or the quantity in Eq. 4.8) is not necessarily the same as the rate of change of B within the control volume (the first term on the right-hand side of Eq. 4.15 or the quantity in Eq. 4.9). This is true because the inflow rate ($b_1\rho_1V_1A_1$) and the outflow rate ($b_2\rho_2V_2A_2$) of the property B for the control volume need not be the same.

Consider again the flow from the fire extinguisher shown in Fig. E4.7. Let the extensive property of interest be the system mass ($B = m$, the system mass, or $b = 1$) and write the appropriate form of the Reynolds transport theorem for this flow.

SOLUTION

Again we take the control volume to be the fire extinguisher, and the system to be the fluid within it at time $t = 0$. For this case there is no inlet, section (1), across which the fluid flows into the control volume ($A_1 = 0$). There is, however, an outlet, section (2). Thus, the Reynolds transport theorem, Eq. 4.15, along with Eq. 4.9 with $b = 1$ can be written as

$$\frac{Dm_{\text{sys}}}{Dt} = \frac{\partial \left(\int_{\text{cv}} \rho \, d\mathcal{V} \right)}{\partial t} + \rho_2 A_2 V_2 \tag{1}$$

(Ans)

If we proceed one step further and use the basic law of conservation of mass, we may set the left-hand side of this equation equal to zero (the amount of mass in a system is constant) and rewrite Eq. 1 in the form:

$$\frac{\partial \left(\int_{\text{cv}} \rho \, d\mathcal{V} \right)}{\partial t} = -\rho_2 A_2 V_2 \tag{2}$$

The physical interpretation of this result is that the rate at which the mass in the tank decreases in time is equal in magnitude but opposite to the rate of flow of mass from the exit, $\rho_2 A_2 V_2$. Note the units for the two terms of Eq. 2 (kg/s or slugs/s). Note that if there were both an inlet and an outlet to the control volume shown in Fig. E4.7, Eq. 2 would become

$$\frac{\partial \left(\int_{\text{cv}} \rho \, d\mathcal{V} \right)}{\partial t} = \rho_1 A_1 V_1 - \rho_2 A_2 V_2 \tag{3}$$

In addition, if the flow were steady, the left-hand side of Eq. 3 would be zero (the amount of mass in the control would be constant in time) and Eq. 3 would become

$$\rho_1 A_1 V_1 = \rho_2 A_2 V_2$$

This is one form of the conservation of mass principle—the mass flowrates into and out of the control volume are equal. Other more general forms are discussed in Chapter 5.

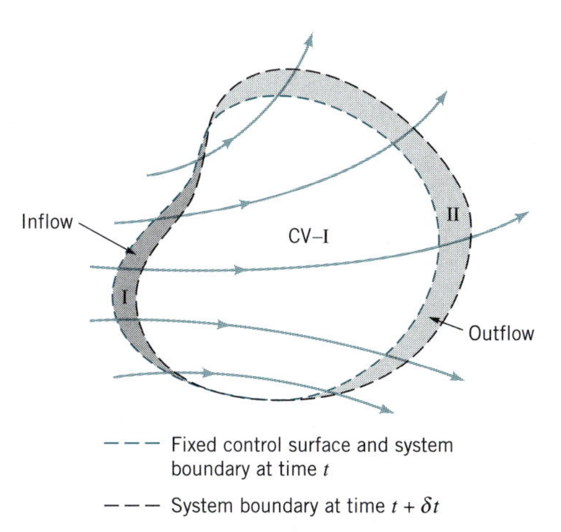

■ **FIGURE 4.12 Control volume and system for flow through an arbitrary, fixed control volume.**

Equation 4.15 is a simplified version of the Reynolds transport theorem. We will now derive it for much more general conditions. A general, fixed control volume with fluid flowing through it is shown in Fig. 4.12. The flow field may be quite simple (as in the above one-dimensional flow considerations), or it may involve a quite complex, unsteady, three-dimensional situation. In any case we again consider the system to be the fluid within the control volume at the initial time t. A short time later a portion of the fluid (region II) has exited from the control volume and additional fluid (region I, not part of the original system) has entered the control volume.

The simplified Reynolds transport theorem can be easily generalized.

We consider an extensive fluid property B and seek to determine how the rate of change of B associated with the system is related to the rate of change of B within the control volume at any instant. By repeating the exact steps that we did for the simplified control volume shown in Fig. 4.11, we see that Eq. 4.14 is valid for the general case also, provided that we give the correct interpretation to the terms $\dot{B}_{\text{out}}$ and $\dot{B}_{\text{in}}$. In general, the control volume may contain more (or less) than one inlet and one outlet. A typical pipe system may contain several inlets and outlets as are shown in Fig. 4.13. In such instances we think of all inlets grouped together ($\text{I} = \text{I}_a + \text{I}_b + \text{I}_c + \cdots$) and all outlets grouped together ($\text{II} = \text{II}_a + \text{II}_b + \text{II}_c + \cdots$), at least conceptually.

The term $\dot{B}_{\text{out}}$ represents the net flowrate of the property B from the control volume. Its value can be thought of as arising from the addition (integration) of the contributions through each infinitesimal area element of size δA on the portion of the control surface

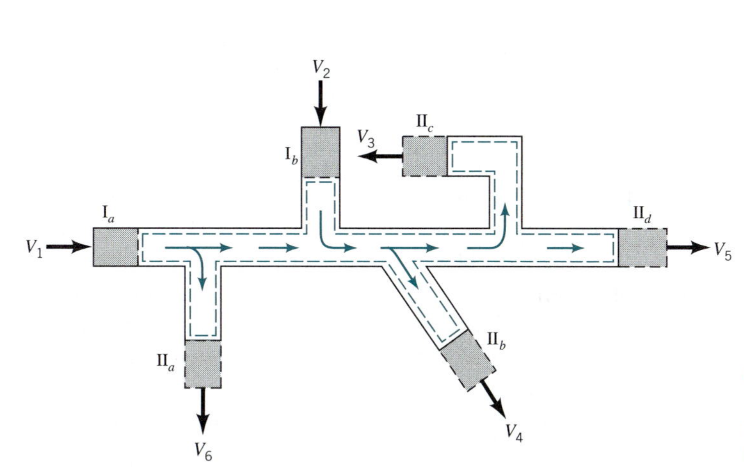

■ **FIGURE 4.13 Typical control volume with more than one inlet and outlet.**

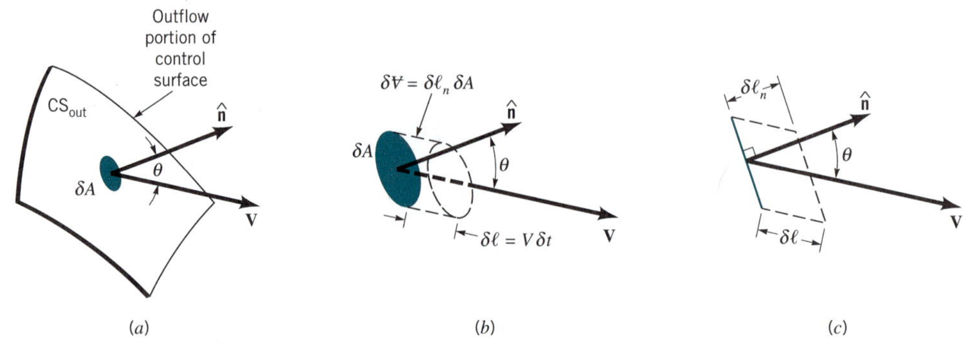

■ **FIGURE 4.14 Outflow across a typical portion of the control surface.**

dividing region II and the control volume. This surface is denoted CS_{out}. As is indicated in Fig. 4.14, in time δt the volume of fluid that passes across each area element is given by $\delta \forall = \delta \ell_n \, \delta A$, where $\delta \ell_n = \delta \ell \cos \theta$ is the height (normal to the base, δA) of the small volume element, and θ is the angle between the velocity vector and the outward pointing normal to the surface, $\hat{\mathbf{n}}$. Thus, since $\delta \ell = V \, \delta t$, the amount of the property B carried across the area element δA in the time interval δt is given by

$$\delta B = b\rho \, \delta \forall = b\rho(V \cos \theta \, \delta t) \, \delta A$$

The rate at which B is carried out of the control volume across the small area element δA, denoted $\delta \dot{B}_{out}$, is

$$\delta \dot{B}_{out} = \lim_{\delta t \to 0} \frac{\rho b \, \delta \forall}{\delta t} = \lim_{\delta t \to 0} \frac{(\rho b V \cos \theta \, \delta t) \, \delta A}{\delta t} = \rho b V \cos \theta \, \delta A$$

By integrating over the entire outflow portion of the control surface, CS_{out}, we obtain

$$\dot{B}_{out} = \int_{cs_{out}} d\dot{B}_{out} = \int_{cs_{out}} \rho b V \cos \theta \, dA$$

The quantity $V \cos \theta$ is the component of the velocity normal to the area element δA. From the definition of the dot product, this can be written as $V \cos \theta = \mathbf{V} \cdot \hat{\mathbf{n}}$. Hence, an alternate form of the outflow rate is

The flowrate of a parameter across the control surface is written in terms of a surface integral.

$$\dot{B}_{out} = \int_{cs_{out}} \rho b \mathbf{V} \cdot \hat{\mathbf{n}} \, dA \qquad \textbf{(4.16)}$$

In a similar fashion, by considering the inflow portion of the control surface, CS_{in}, as shown in Fig. 4.15, we find that the inflow rate of B into the control volume is

$$\dot{B}_{in} = -\int_{cs_{in}} \rho b V \cos \theta \, dA = -\int_{cs_{in}} \rho b \mathbf{V} \cdot \hat{\mathbf{n}} \, dA \qquad \textbf{(4.17)}$$

We use the standard notation that the unit normal vector to the control surface, $\hat{\mathbf{n}}$, points out from the control volume. Thus, as is shown in Fig. 4.16, $-90° < \theta < 90°$ for outflow regions (the normal component of $\mathbf{V}$ is positive; $\mathbf{V} \cdot \hat{\mathbf{n}} > 0$). For inflow regions $90° < \theta < 270°$ (the normal component of $\mathbf{V}$ is negative; $\mathbf{V} \cdot \hat{\mathbf{n}} < 0$). The value of $\cos \theta$ is, therefore, positive on the CV_{out} portions of the control surface and negative on the CV_{in} portions. Over the remainder of the control surface there is no inflow or outflow, leading to $\mathbf{V} \cdot \hat{\mathbf{n}} = V \cos \theta = 0$ on those portions. On such portions either $V = 0$ (the fluid "sticks" to the surface) or $\cos \theta = 0$ (the fluid "slides" along the surface without crossing it) (see Fig. 4.16). Therefore,

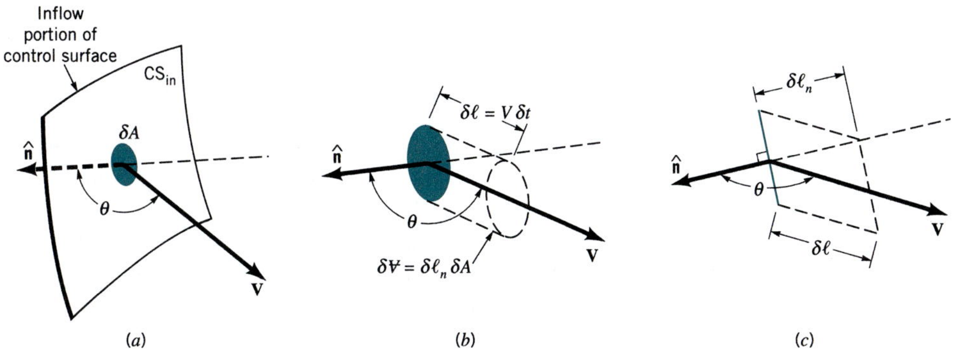

■ **FIGURE 4.15 Inflow across a typical portion of the control surface.**

the net flux (flowrate) of parameter B across the entire control surface is

$$\dot{B}_{\text{out}} - \dot{B}_{\text{in}} = \int_{\text{cs}_{\text{out}}} \rho b \mathbf{V} \cdot \hat{\mathbf{n}}\, dA - \left(-\int_{\text{cs}_{\text{in}}} \rho b \mathbf{V} \cdot \hat{\mathbf{n}}\, dA\right)$$
$$= \int_{\text{cs}} \rho b \mathbf{V} \cdot \hat{\mathbf{n}}\, dA \tag{4.18}$$

where the integration is over the entire control surface.

By combining Eqs. 4.14 and 4.18 we obtain

$$\frac{DB_{\text{sys}}}{Dt} = \frac{\partial B_{\text{cv}}}{\partial t} + \int_{\text{cs}} \rho b \mathbf{V} \cdot \hat{\mathbf{n}}\, dA$$

This can be written in a slightly different form by using $B_{\text{cv}} = \int_{\text{cv}} \rho b \, d\forall$ so that

The general Reynolds transport theorem involves volume and surface integrals.

$$\frac{DB_{\text{sys}}}{Dt} = \frac{\partial}{\partial t}\int_{\text{cv}} \rho b \, d\forall + \int_{\text{cs}} \rho b \, \mathbf{V} \cdot \hat{\mathbf{n}}\, dA \tag{4.19}$$

Equation 4.19 is the general form of the Reynolds transport theorem for a fixed, nondeforming control volume. Its interpretation and use are discussed in the following sections.

4.4.2 Physical Interpretation

The Reynolds transport theorem as given in Eq. 4.19 is widely used in fluid mechanics (and other areas as well). At first it appears to be a rather formidable mathematical expression—perhaps one to be steered clear of if possible. However, a physical understanding of the concepts involved will show that it is a rather straightforward, relatively easy-to-use tool. Its purpose is to provide a link between control volume ideas and system ideas.

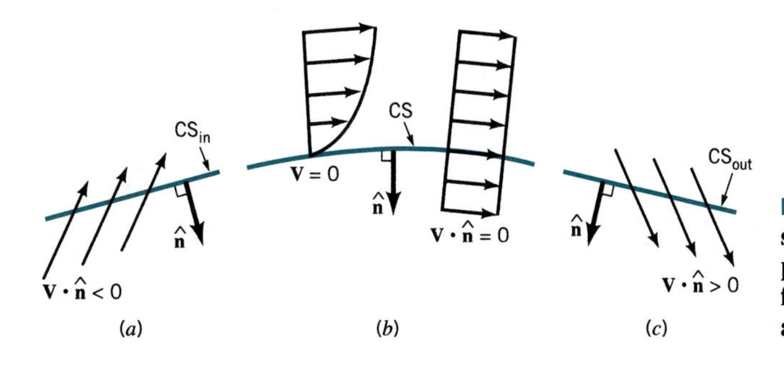

■ **FIGURE 4.16 Possible velocity configurations on portions of the control surface: (*a*) inflow, (*b*) no flow across the surface, (*c*) outflow.**

The left side of Eq. 4.19 is the time rate of change of an arbitrary extensive parameter of a system. This may represent the rate of change of mass, momentum, energy, or angular momentum of the system, depending on the choice of the parameter B.

Because the system is moving and the control volume is stationary, the time rate of change of the amount of B within the control volume is not necessarily equal to that of the system. The first term on the right side of Eq. 4.19 represents the rate of change of B within the control volume as the fluid flows through it. Recall that b is the amount of B per unit mass, so that $\rho b\, d\mathcal{V}$ is the amount of B in a small volume $d\mathcal{V}$. Thus, the time derivative of the integral of ρb throughout the control volume is the time rate of change of B within the control volume at a given time.

The last term in Eq. 4.19 (an integral over the control surface) represents the net flowrate of the parameter B across the entire control surface. Over a portion of the control surface this property is being carried out of the control volume ($\mathbf{V} \cdot \hat{\mathbf{n}} > 0$); over other portions it is being carried into the control volume ($\mathbf{V} \cdot \hat{\mathbf{n}} < 0$). Over the remainder of the control surface there is no transport of B across the surface since $b\mathbf{V} \cdot \hat{\mathbf{n}} = 0$, because either $b = 0$, $\mathbf{V} = 0$, or $\mathbf{V}$ is parallel to the surface at those locations. The mass flowrate through area element δA, given by $\rho \mathbf{V} \cdot \hat{\mathbf{n}}\, \delta A$, is positive for outflow (efflux) and negative for inflow (influx). Each fluid particle or fluid mass carries a certain amount of B with it, as given by the product of B per unit mass, b, and the mass. The rate at which this B is carried across the control surface is given by the area integral term of Eq. 4.19. This net rate across the entire control surface may be negative, zero, or positive depending on the particular situation involved.

4.4.3 Relationship to Material Derivative

In Section 4.2.1 we discussed the concept of the material derivative $D(\ )/Dt = \partial(\ )/\partial t + \mathbf{V} \cdot \nabla(\ ) = \partial(\ )\partial t + u\, \partial(\ )\partial x + v\, \partial(\ )/\partial y + w\, \partial(\ )/\partial z$. The physical interpretation of this derivative is that it provides the time rate of change of a fluid property (temperature, velocity, etc.) associated with a particular fluid particle as it flows. The value of that parameter for that particle may change because of unsteady effects [the $\partial(\ )/\partial t$ term] or because of effects associated with the particle's motion [the $\mathbf{V} \cdot \nabla(\ )$ term].

Careful consideration of Eq. 4.19 indicates the same type of physical interpretation for the Reynolds transport theorem. The term involving the time derivative of the control volume integral represents unsteady effects associated with the fact that values of the parameter within the control volume may change with time. For steady flow this effect vanishes—fluid flows through the control volume but the amount of any property, B, within the control volume is constant in time. The term involving the control surface integral represents the convective effects associated with the flow of the system across the fixed control surface. The sum of these two terms gives the rate of change of the parameter B for the system. This corresponds to the interpretation of the material derivative, $D(\ )/Dt = \partial(\ )/\partial t + \mathbf{V} \cdot \nabla(\ )$, in which the sum of the unsteady effect and the convective effect gives the rate of change of a parameter for a fluid particle. As is discussed in Section 4.2, the material derivative operator may be applied to scalars (such as temperature) or vectors (such as velocity). This is also true for the Reynolds transport theorem. The particular parameters of interest, B and b, may be scalars or vectors.

The Reynolds transport theorem is the integral counterpart of the material derivative.

Thus, both the material derivative and the Reynolds transport theorem equations represent ways to transfer from the Lagrangian viewpoint (follow a particle or follow a system) to the Eulerian viewpoint (observe the fluid at a given location in space or observe what happens in the fixed control volume). The material derivative (Eq. 4.5) is essentially the infinitesimal (or derivative) equivalent of the finite size (or integral) Reynolds transport theorem (Eq. 4.19).

4.4.4 Steady Effects

Consider a steady flow $[\partial(\ )/\partial t \equiv 0]$ so that Eq. 4.19 reduces to

$$\frac{DB_{\text{sys}}}{Dt} = \int_{\text{cs}} \rho b \mathbf{V} \cdot \hat{\mathbf{n}}\, dA \tag{4.20}$$

In such cases if there is to be a change in the amount of B associated with the system (nonzero left-hand side), there must be a net difference in the rate that B flows into the control volume compared with the rate that it flows out of the control volume. That is, the integral of $\rho b \mathbf{V} \cdot \hat{\mathbf{n}}$ over the inflow portions of the control surface would not be equal and opposite to that over the outflow portions of the surface.

Consider steady flow through the ''black box'' control volume that is shown in Fig. 4.17. If the parameter B is the mass of the system, the left-hand side of Eq. 4.20 is zero (conservation of mass for the system as discussed in detail in Section 5.1). Hence, the flowrate of mass into the box must be the same as the flowrate of mass out of the box because the right-hand side of Eq. 4.20 represents the net flowrate through the control surface. On the other hand, assume the parameter B is the momentum of the system. The momentum of the system need not be constant. In fact, according to Newton's second law the time rate of change of the system momentum equals the net force, $\mathbf{F}$, acting on the system. In general, the left-hand side of Eq. 4.20 will therefore be nonzero. Thus, the right-hand side, which then represents the net flux of momentum across the control surface, will be nonzero. The flowrate of momentum into the control volume need not be the same as the flux of momentum from the control volume. We will investigate these concepts much more fully in Chapter 5. They are the basic principles describing the operation of such devices as jet or rocket engines.

The Reynolds transport theorem involves both steady and unsteady effects.

For steady flows the amount of the property B within the control volume does not change with time. The amount of the property associated with the system may or may not change with time, depending on the particular property considered and the flow situation involved. The difference between that associated with the control volume and that associated with the system is determined by the rate at which B is carried across the control surface—the term $\int_{\text{cs}} \rho b \mathbf{V} \cdot \hat{\mathbf{n}}\, dA$.

4.4.5 Unsteady Effects

Consider unsteady flow $[\partial(\ )/\partial t \neq 0]$ so that all terms in Eq. 4.19 must be retained. When they are viewed from a control volume standpoint, the amount of parameter B within the system may change because the amount of B within the fixed control volume may change with time [the $\partial(\int_{\text{cv}} \rho b\, d\mathcal{V})/\partial t$ term] and because there may be a net nonzero flow of that parameter across the control surface (the $\int_{\text{cs}} \rho b \mathbf{V} \cdot \hat{\mathbf{n}}\, dA$ term).

For the special unsteady situations in which the rate of inflow of parameter B is exactly balanced by its rate of outflow, it follows that $\int_{\text{cs}} \rho b \mathbf{V} \cdot \hat{\mathbf{n}}\, dA = 0$, and Eq. 4.19 reduces to

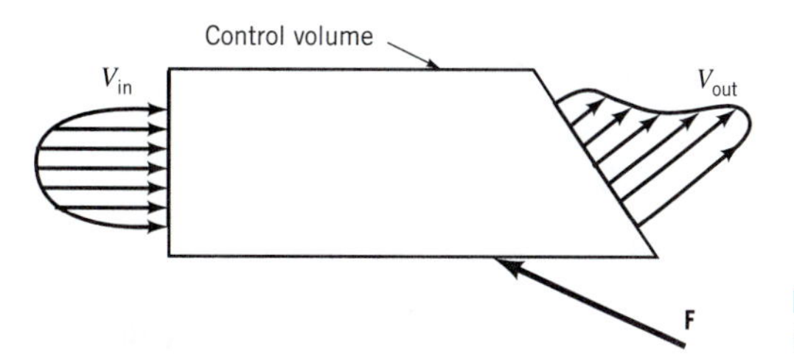

■ **FIGURE 4.17** **Steady flow through a control volume.**

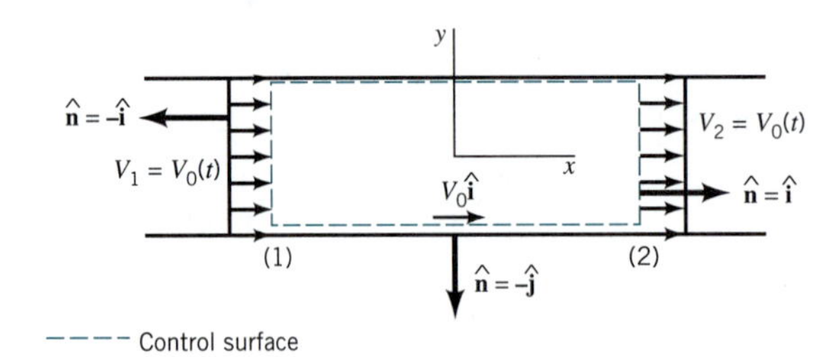

■ FIGURE 4.18 **Unsteady flow through a constant diameter pipe.**

$$\frac{DB_{\text{sys}}}{Dt} = \frac{\partial}{\partial t}\int_{\text{cv}} \rho b \, d\forall \quad \textbf{(4.21)}$$

For such cases, any rate of change in the amount of B associated with the system is equal to the rate of change of B within the control volume. This can be illustrated by considering flow through a constant diameter pipe as is shown in Fig. 4.18. The control volume is as shown, and the system is the fluid within this volume at time t_0. We assume the flow is one-dimensional with $\mathbf{V} = V_0\hat{\mathbf{i}}$, where $V_0(t)$ is a function of time, and that the density is constant. At any instant in time, all particles in the system have the same velocity. We let $\mathbf{B}$ = system momentum = $m\mathbf{V} = mV_0\hat{\mathbf{i}}$, where m is the system mass, so that $\mathbf{b} = \mathbf{B}/m = \mathbf{V} = V_0\hat{\mathbf{i}}$, the fluid velocity. The magnitude of the momentum efflux across the outlet [section (2)] is the same as the magnitude of the momentum influx across the inlet [section (1)]. However, the sign of the efflux is opposite to that of the influx since $\mathbf{V} \cdot \hat{\mathbf{n}} > 0$ for the outflow and $\mathbf{V} \cdot \hat{\mathbf{n}} < 0$ for the inflow. Note that $\mathbf{V} \cdot \hat{\mathbf{n}} = 0$ along the sides of the control volume. Thus, with $\mathbf{V} \cdot \hat{\mathbf{n}} = -V_0$ on section (1), $\mathbf{V} \cdot \hat{\mathbf{n}} = V_0$ on section (2), and $A_1 = A_2$ we obtain

In many situations the integrals involved in the Reynolds transport theorem reduce to simple algebra.

$$\begin{aligned}\int_{\text{cs}} \rho b \mathbf{V} \cdot \hat{\mathbf{n}} \, dA &= \int_{\text{cs}} \rho(V_0\hat{\mathbf{i}})(\mathbf{V} \cdot \hat{\mathbf{n}}) \, dA \\ &= \int_{(1)} \rho(V_0\hat{\mathbf{i}})(-V_0) \, dA + \int_{(2)} \rho(V_0\hat{\mathbf{i}})(V_0) \, dA \\ &= -\rho V_0^2 A_1 \hat{\mathbf{i}} + \rho V_0^2 A_2 \hat{\mathbf{i}} = 0\end{aligned}$$

It is seen that for this special case Eq. 4.21 is valid. The rate at which the momentum of the system changes with time is the same as the rate of change of momentum within the control volume. If V_0 is constant in time, there is no rate of change of momentum of the system and for this special case each of the terms in the Reynolds transport theorem is zero by itself.

Consider the flow through a variable area pipe shown in Fig. 4.19. In such cases the fluid velocity is not the same at section (1) as it is at (2). Hence, the efflux of momentum from the control volume is not equal to the influx of momentum, so that the convective term in Eq. 4.20 [the integral of $\rho\mathbf{V}(\mathbf{V} \cdot \hat{\mathbf{n}})$ over the control surface] is not zero. These topics will be discussed in considerably more detail in Chapter 5.

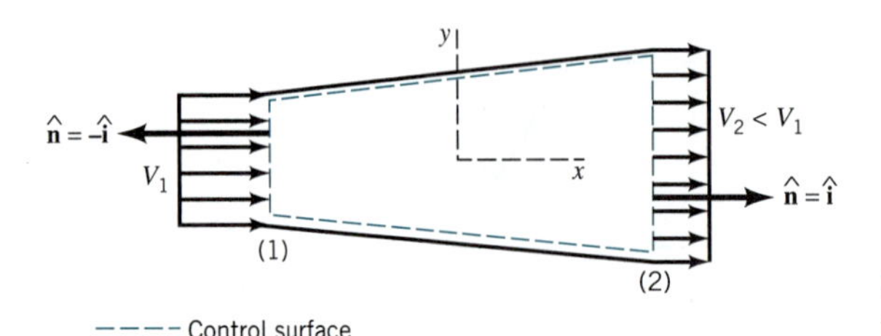

■ FIGURE 4.19 **Flow through a variable area pipe.**

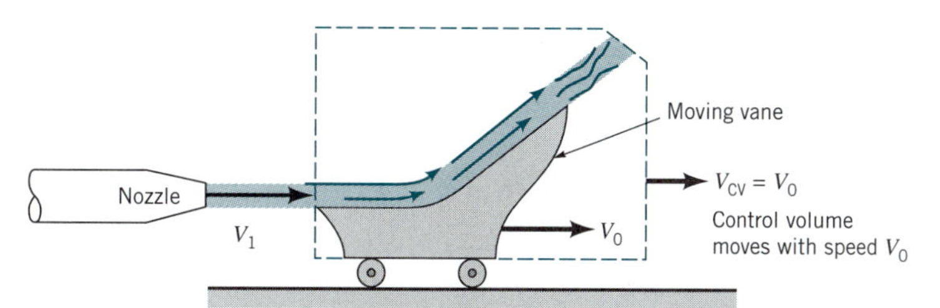

■ FIGURE 4.20 **Example of a moving control volume.**

4.4.6 Moving Control Volumes

Some problems are most easily solved by using a moving control volume.

For most problems in fluid mechanics, the control volume may be considered as a fixed volume through which the fluid flows. There are, however, situations for which the analysis is simplified if the control volume is allowed to move or deform. The most general situation would involve a control volume that moves, accelerates, and deforms. As one might expect, the use of these control volumes can become fairly complex.

A number of important problems can be most easily analyzed by using a nondeforming control volume that moves with a constant velocity. Such an example is shown in Fig. 4.20 in which a stream of water with velocity $\mathbf{V}_1$ strikes a vane that is moving with constant velocity $\mathbf{V}_0$. It may be of interest to determine the force, $\mathbf{F}$, that the water puts on the vane. Such problems frequently occur in turbines where a stream of fluid (water or steam, for example) strikes a series of blades that move past the nozzle. To analyze such problems it is advantageous to use a moving control volume. We will obtain the Reynolds transport theorem for such control volumes.

We consider a control volume that moves with a constant velocity as is shown in Fig. 4.21. The shape, size, and orientation of the control volume do not change with time. The control volume merely translates with a constant velocity, $\mathbf{V}_{cv}$, as shown. In general, the velocity of the control volume and the fluid are not the same, so that there is a flow of fluid through the moving control volume just as in the stationary control volume cases discussed in Section 4.4.2. The main difference between the fixed and the moving control volume cases is that it is the *relative velocity*, $\mathbf{W}$, that carries fluid across the moving control surface, whereas it is the *absolute velocity*, $\mathbf{V}$, that carries the fluid across the fixed control surface. The relative velocity is the fluid velocity relative to the moving control volume—the fluid velocity seen by an observer riding along on the control volume. The absolute velocity is the fluid velocity as seen by a stationary observer in a fixed coordinate system.

The difference between the absolute and relative velocities is the velocity of the control volume, $\mathbf{V}_{cv} = \mathbf{V} - \mathbf{W}$, or

$$\mathbf{V} = \mathbf{W} + \mathbf{V}_{cv} \tag{4.22}$$

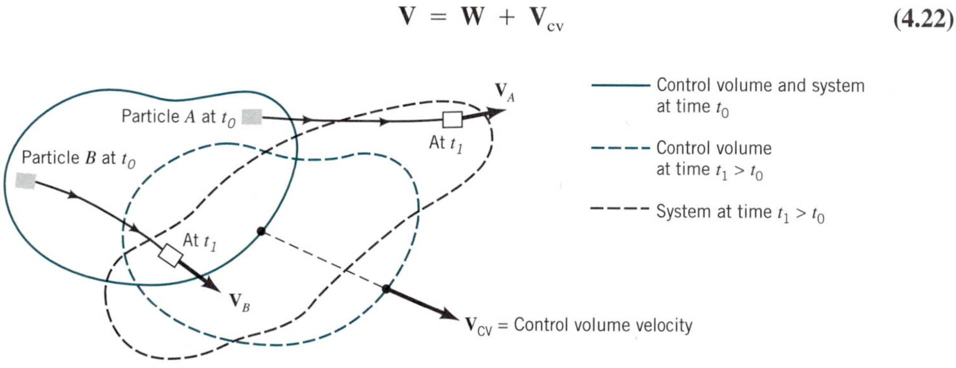

■ FIGURE 4.21 **Typical moving control volume and system.**

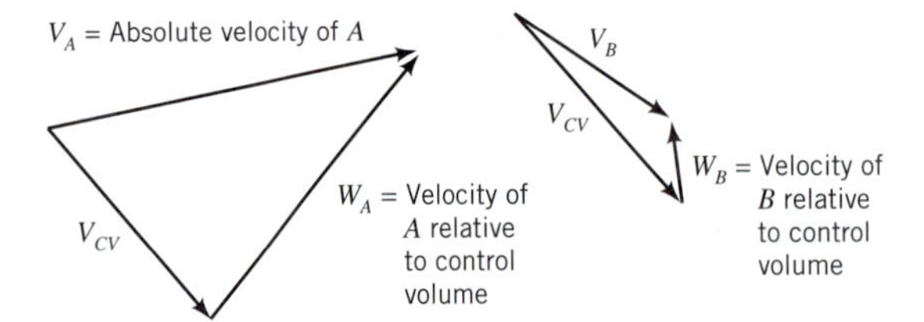

■ **FIGURE 4.22 Relationship between absolute and relative velocities.**

Since the velocity is a vector, we must use vector addition as is shown in Fig. 4.22 to obtain the relative velocity if we know the absolute velocity and the velocity of the control volume. Thus, if the water leaves the nozzle in Fig. 4.20 with a velocity of $\mathbf{V}_1 = 100\hat{\mathbf{i}}$ ft/s and the vane has a velocity of $\mathbf{V}_0 = 20\hat{\mathbf{i}}$ ft/s (the same as the control volume), it appears to an observer riding on the vane that the water approaches the vane with a velocity of $\mathbf{W} = \mathbf{V} - \mathbf{V}_{\text{cv}} = 80\hat{\mathbf{i}}$ ft/s. In general, the absolute velocity, $\mathbf{V}$, and the control volume velocity, $\mathbf{V}_{\text{cv}}$, will not be in the same direction so that the relative and absolute velocities will have different directions (see Fig. 4.22).

The Reynolds transport theorem for a moving, nondeforming control volume can be derived in the same manner that it was obtained for a fixed control volume. As is indicated in Fig. 4.23, the only difference that needs be considered is the fact that relative to the moving control volume the fluid velocity observed is the relative velocity, not the absolute velocity. An observer fixed to the moving control volume may or may not even know that he or she is moving relative to some fixed coordinate system. If we follow the derivation that led to Eq. 4.19 (the Reynolds transport theorem for a fixed control volume), we note that the corresponding result for a moving control volume can be obtained by simply replacing the absolute velocity, $\mathbf{V}$, in that equation by the relative velocity, $\mathbf{W}$. Thus, the Reynolds transport theorem for a control volume moving with constant velocity is given by

The Reynolds transport theorem for a moving control volume involves the relative velocity.

$$\frac{DB_{\text{sys}}}{Dt} = \frac{\partial}{\partial t}\int_{\text{cv}} \rho b \, d\mathcal{V} + \int_{\text{cs}} \rho b \, \mathbf{W} \cdot \hat{\mathbf{n}} \, dA \tag{4.23}$$

where the relative velocity is given by Eq. 4.22.

4.4.7 Selection of a Control Volume

Any volume in space can be considered as a control volume. It may be of finite size or it may be infinitesimal in size, depending on the type of analysis to be carried out. In most of our cases, the control volume will be a fixed, nondeforming volume. In some situations we will consider control volumes that move with constant velocity. In either case it is important that considerable thought go into the selection of the specific control volume to be used.

The selection of an appropriate control volume in fluid mechanics is very similar to the

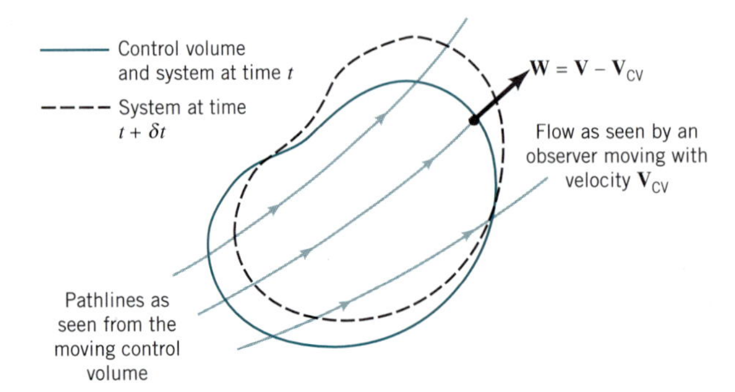

■ **FIGURE 4.23 Control volume and system as seen by an observer moving with the control volume.**

Control surface
(1)
V
(a)

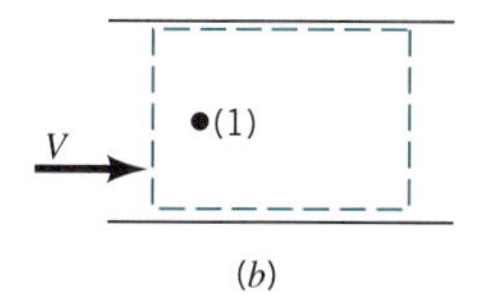

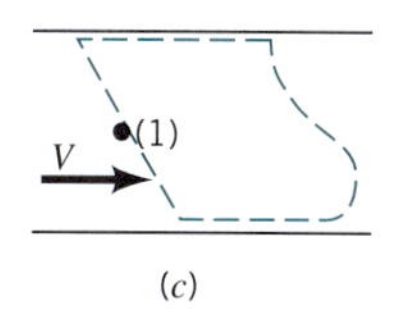

■ FIGURE 4.24 Various control volumes for flow through a pipe.

selection of an appropriate free-body diagram in dynamics or statics. In dynamics we select the body in which we are interested, represent the object in a free-body diagram, and then apply the appropriate governing laws to that body. The ease of solving a given dynamics problem is often very dependent upon the specific object that we select for use in our free-body diagram. Similarly, the ease of solving a given fluid mechanics problem is often very dependent upon the choice of the control volume used. Only by practice can we develop skill at selecting the ''best'' control volume. None are ''wrong,'' but some are ''much better'' than others.

For any problem there is an infinite variety of control volumes that can be used.

Solution of a typical problem will involve determining parameters such as velocity, pressure, and force at some point in the flow field. It is usually best to ensure that this point is located on the control surface, not ''buried'' within the control volume. The unknown will then appear in the convective term (the surface integral) of the Reynolds transport theorem. If possible, the control surface should be normal to the fluid velocity so that the angle θ ($\mathbf{V} \cdot \hat{\mathbf{n}} = V \cos \theta$) in the flux terms of Eq. 4.19 will be 0 or 180°. This will usually simplify the solution process.

Figure 4.24 illustrates three possible control volumes associated with flow through a pipe. If the problem is to determine the pressure at point (1), the selection of the control volume (*a*) is better than that of (*b*) because point (1) lies on the control surface. Similarly, control volume (*a*) is better than (*c*) because the flow is normal to the inlet and exit portions of the control volume. None of these control volumes are wrong—(*a*) will be easier to use. Proper control volume selection will become much clearer in Chapter 5 where the Reynolds transport theorem is used to transform the governing equations from the system formulation into the control volume formulation, and numerous examples using control volume ideas are discussed.

References

1. Streeter, V. L., and Wylie, E. B., *Fluid Mechanics*, 8th Ed., McGraw-Hill, New York, 1985.
2. Goldstein, R. J., *Fluid Mechanics Measurements*, Hemisphere, New York, 1983.
3. Kline, S. J., *Flow Visualization*, National Committee for Fluid Mechanics Films, distributed by Encyclopaedia Britannica Educational Corp., Chicago, 1972.
4. Magarvey, R. H., and MacLatchy, C. S., The Formation and Structure of Vortex Rings, *Canadian Journal of Physics*, Vol. 42, 1964.

Review Problems

Note: Problems designated with (R) are review problems. The phrases within parentheses refer to the main topics to be used in solving the problems. Complete, detailed solutions to these review problems can be found in the supplement titled *Student Solution Manual for Fundamentals of Fluid Mechanics*, by Munson, Young, and Okiishi (John Wiley and Sons, New York, 1997).

4.1R (Streamlines) The velocity field in a flow is given by $\mathbf{V} = x^2y\hat{\mathbf{i}} + x^2t\hat{\mathbf{j}}$. **(a)** Plot the streamline through the origin at times $t = 0$, $t = 1$, and $t = 2$. **(b)** Do the streamlines plotted in part (a) coincide with the path of particles through the origin? Explain.

(ANS: $y^2/2 = tx + C$; no)

4.2R (Streamlines) A velocity field is given by $u = y - 1$ and $v = y - 2$, where u and v are in m/s and x and y are in meters. Plot the streamline that passes through the point $(x, y) = (4, 3)$. Compare this streamline with the streakline through the point $(x, y) = (4, 3)$.

(ANS: $x = y + \ln(y - 2) + 1$)

4.3R (Material derivative) The pressure in the pipe near the discharge of a reciprocating pump fluctuates according to $p = [200 + 40 \sin(8t)]$ kPa, where t is in seconds. If the fluid speed in the pipe is 5 m/s, determine the maximum rate of change of pressure experienced by a fluid particle.

(ANS: 320 kPa/s)

4.4R (Acceleration) A shock wave is a very thin layer (thickness $= \ell$) in a high-speed (supersonic) gas flow across which the flow properties (velocity, density, pressure, etc.) change from state (1) to state (2) as shown in Fig. P4.4R. If $V_1 = 1800$ fps, $V_2 = 700$ fps, and $\ell = 10^{-4}$ in., estimate the average deceleration of the gas as it flows across the shock wave. How many g's deceleration does this represent?

(ANS: -1.65×10^{11} ft/s^2; -5.12×10^9)

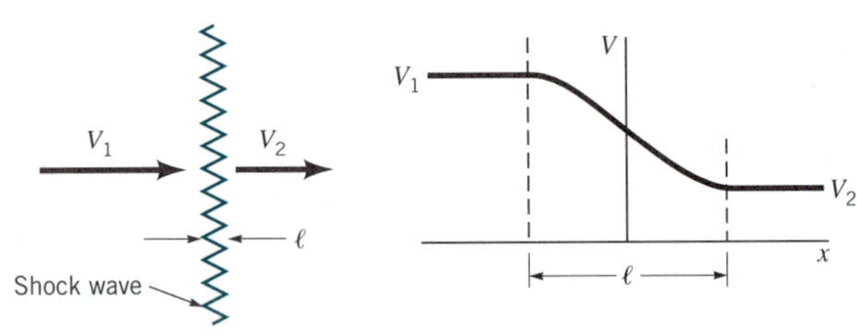

■ FIGURE P4.4R

4.5R (Acceleration) Air flows through a pipe with a uniform velocity of $\mathbf{V} = 5\,t^2\hat{\mathbf{i}}$ ft/s, where t is in seconds. Determine the acceleration at time $t = -1$, 0, and 1 s.

(ANS: $-10\,\hat{\mathbf{i}}$ ft/s^2; 0; $10\,\hat{\mathbf{i}}$ ft/s^2)

4.6R (Acceleration) A fluid flows steadily along the streamline as shown in Fig. P4.6R. Determine the acceleration at point A. At point A what is the angle between the acceleration and the x axis? At point A what is the angle between the acceleration and the streamline?

(ANS: $10\,\hat{\mathbf{n}} + 30\,\hat{\mathbf{s}}$ ft/s^2; 48.5 deg; 18.5 deg)

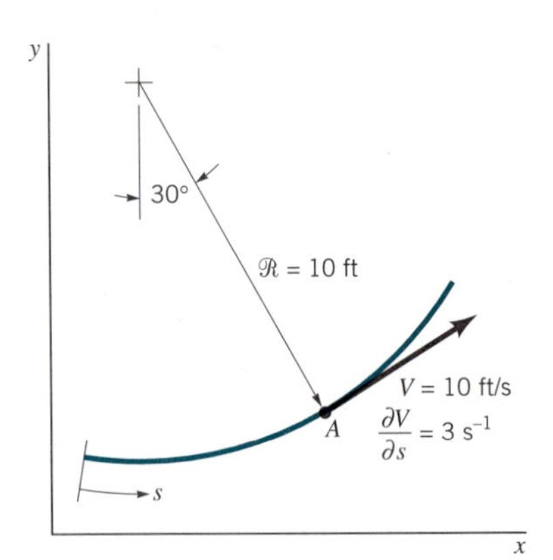

■ FIGURE P4.6R

4.7R (Acceleration) In the conical nozzle shown in Fig. P4.7R the streamlines are essentially radial lines emanating from point A and the fluid velocity is given approximately by $V = C/r^2$, where C is a constant. The fluid velocity is 2 m/s along the centerline at the beginning of the nozzle ($x = 0$). Determine the acceleration along the nozzle centerline as a function of x. What is the value of the acceleration at $x = 0$ and $x = 0.3$ m?

(ANS: $1.037/(0.6 - x)^5\,\hat{\mathbf{i}}$ m/s^2; $13.3\,\hat{\mathbf{i}}$ m/s^2; $427\,\hat{\mathbf{i}}$ m/s^2)

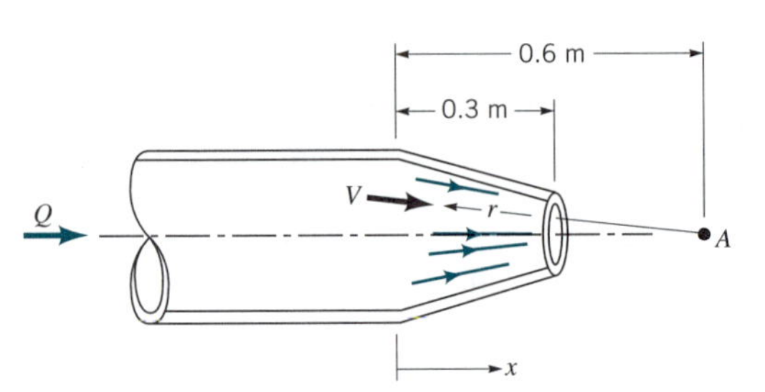

■ FIGURE P4.7R

4.8R (Reynolds transport theorem) A sanding operation injects 10^5 particles/s into the air in a room as shown in Fig. P4.8R. The amount of dust in the room is maintained at a constant level by a ventilating fan that draws clean air into the room at section (1) and expels dusty air at section (2). Consider a control volume whose surface is the interior surface of the room (excluding the sander) and a system consisting of the material within the control volume at time $t = 0$. **(a)** If N is the number of particles, discuss the physical meaning of and evaluate the terms DN_{sys}/Dt and $\partial N_{cv}/\partial t$. **(b)** Use the Reynolds transport theorem to determine the concentration of particles (particles/m^3) in the exhaust air for steady state conditions.

(ANS: 5×10^5 particles/m^3)

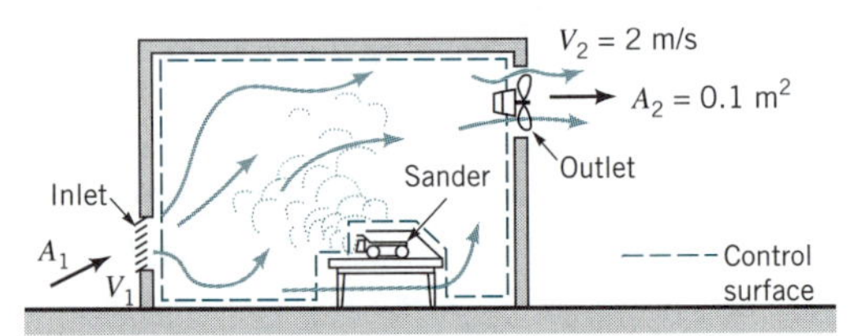

■ FIGURE P4.8R

4.9R (Flowrate) Water flows through the rectangular channel shown in Fig. P4.9R with a uniform velocity. Directly integrate Eqs. 4.16 and 4.17 with $b = 1$ to determine the mass flowrate (kg/s) across and A–B of the control volume. Repeat for C–D.

(ANS: 18,000 kg/s; 18,000 kg/s)

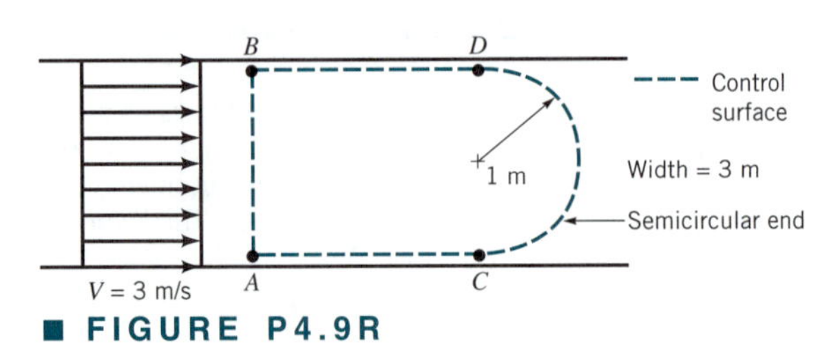

■ FIGURE P4.9R

4.10R (Flowrate) Air blows through two windows as indicated in Fig. P4.10R. Use Eq. 4.16 with $b = 1/\rho$ to determine the volume flowrate (ft^3/s) through each window. Explain the relationship between the two results you obtained.

(ANS: 80 ft^3/s; 160 ft^3/s)

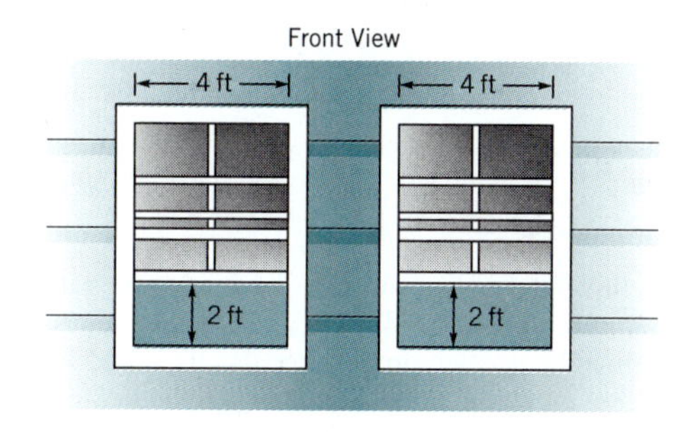

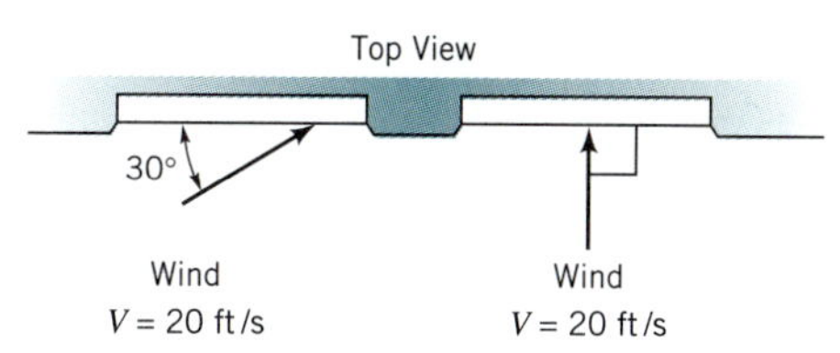

■ FIGURE P4.10R

4.11R (Control volume/system) Air flows over a flat plate with a velocity profile given by $\mathbf{V} = u(y)\hat{\mathbf{i}}$, where $u = 2y$ ft/s for $0 \leq y \leq 0.5$ ft and $u = 1$ ft/s for $y > 0.5$ ft as shown in Fig. P4.11R. The fixed rectangular control volume *ABCD* coincides with the system at time $t = 0$. Make a sketch to indicate **(a)** the boundary of the system at time $t = 0.1$ s, **(b)** the fluid that moved out of the control volume in the interval $0 \leq t \leq 0.1$ s, and **(c)** the fluid that moved into the control volume during that time interval.

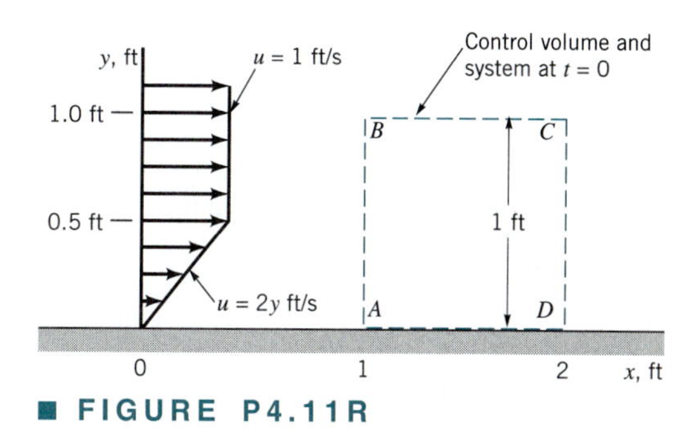

■ FIGURE P4.11R

Problems

Note: Unless otherwise indicated use the values of fluid properties found in the tables on the inside of the front cover. Problems designated with an (*) are intended to be solved with the aid of a programmable calculator or a computer. Problems designated with a (†) are "open-ended" problems and require critical thinking in that to work them one must make various assumptions and provide the necessary data. There is not a unique answer to these problems.

4.1 The velocity field of a flow is given by $\mathbf{V} = (3y + 2)\hat{\mathbf{i}} + (x - 8)\hat{\mathbf{j}} + 5z\hat{\mathbf{k}}$ ft/s, where x, y, and z are in feet. Determine the fluid speed at the origin ($x = y = z = 0$) and on the y axis ($x = z = 0$).

4.2 The velocity field of a flow is given by $\mathbf{V} = 2x^2t\hat{\mathbf{i}} + [4y(t - 1) + 2x^2t]\hat{\mathbf{j}}$ m/s, where x and y are in meters and t is in seconds. For fluid particles on the x axis, determine the speed and direction of flow.

4.3 The velocity field of a flow is given by $\mathbf{V} = 20y/(x^2 + y^2)^{1/2}\hat{\mathbf{i}} - 20x/(x^2 + y^2)^{1/2}\hat{\mathbf{j}}$ ft/s, where x and y are in feet. Determine the fluid speed at points along the x axis; along the y axis. What is the angle between the velocity vector and the x axis at points $(x, y) = (5, 0)$, $(5, 5)$, and $(0, 5)$?

4.4 The x and y components of a velocity field are given by $u = x - y$ and $v = x^2y - 8$. Determine the location of any stagnation points in the flow field. That is, at what point(s) is the velocity zero?

4.5 The x and y components of velocity for a two-dimensional flow are $u = 6y$ ft/s and $v = 4$ ft/s, where y is in feet. Determine the equation for the streamlines and sketch representative streamlines in the upper half plane.

4.6 Show that the streamlines for a flow whose velocity components are $u = c(x^2 - y^2)$ and $v = -2cxy$, where c is a constant, are given by the equation $x^2y - y^3/3 =$ constant. At which point (points) is the flow parallel to the y axis? At which point (points) is the fluid stationary?

4.7 The velocity field of a flow is given by $u = -V_0y/(x^2 + y^2)^{1/2}$ and $v = V_0x/(x^2 + y^2)^{1/2}$, where V_0 is a constant. Where in the flow field is the speed equal to V_0? Determine the equation of the streamlines and discuss the various characteristics of this flow.

† **4.8** Pathlines and streaklines provide ways to visualize flows. Another technique would be to instantly inject a line of dye across streamlines and observe how this line moves as time increases. For example, consider the initially straight dye line injected in front of the circular cylinder shown in Fig. P4.8. Discuss how this dye line would appear at later times. How would you calculate the location of this line as a function of time?

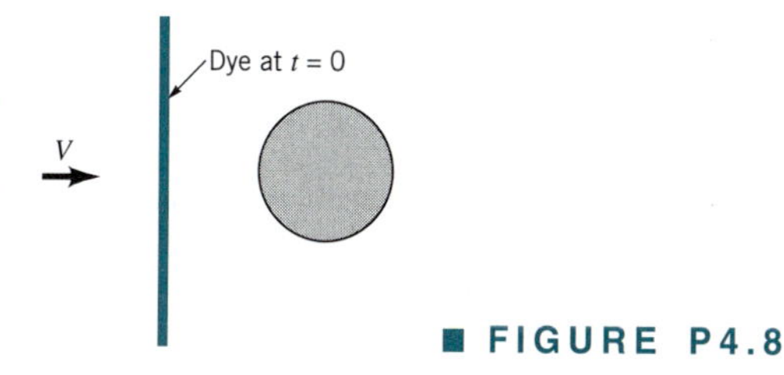

■ FIGURE P4.8

4.9 From time $t = 0$ to $t = 5$ hr, radioactive steam is released from a nuclear power plant accident located at $x = -1$ mile and $y = 3$ miles. The following wind conditions are expected: $\mathbf{V} = 10\hat{\mathbf{i}} - 5\hat{\mathbf{j}}$ mph for $0 < t < 3$ hr, $\mathbf{V} = 15\hat{\mathbf{i}} + 8\hat{\mathbf{j}}$ mph for $3 < t < 10$ hr, and $\mathbf{V} = 5\hat{\mathbf{i}}$ mph for $t > 10$ hr. Draw to scale the expected streakline of the steam for $t = 3$, 10, and 15 hr.

4.10 The x and y components of a velocity field are given by $u = x^2y$ and $v = -xy^2$. Determine the equation for the streamlines of this flow and compare with those in Example 4.2. Is the flow in this problem the same as that in Example 4.2? Explain.

† **4.11** For any steady flow the streamlines and streaklines are the same. For most unsteady flows this is not true. However, there are unsteady flows for which the streamlines and streaklines are the same. Describe a flow field for which this is true.

4.12 In addition to the customary horizontal velocity components of the air in the atmosphere (the "wind"), there often are vertical air currents (thermals) caused by buoyant effects due to uneven heating of the air as indicated in Fig. P4.12. Assume that the velocity field in a certain region is approximated by $u = u_0$, $v = v_0(1 - y/h)$ for $0 < y < h$, and $u = u_0$, $v = 0$ for $y > h$. Plot the shape of the streamline that passes through the origin for values of $u_0/v_0 = 0.5$, 1, and 2.

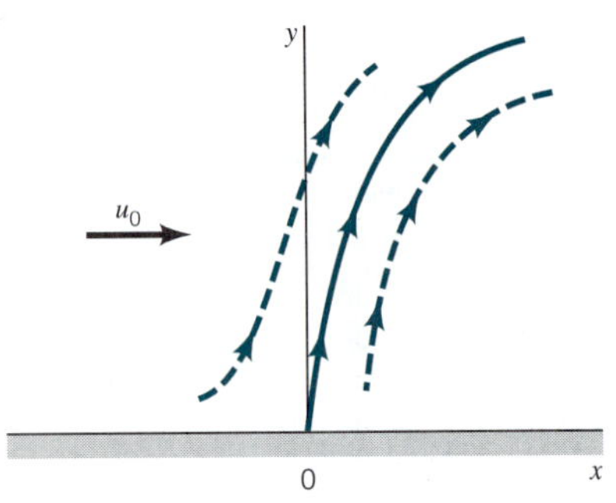

■ **FIGURE P4.12**

*__4.13__ Repeat Problem 4.12 using the same information except that $u = u_0y/h$ for $0 \le y \le h$ rather than $u = u_0$. Use values of $u_0/v_0 = 0$, 0.1, 0.2, 0.4, 0.6, 0.8, and 1.0.

4.14 A velocity field is given by $u = cx^2$ and $v = cy^2$, where c is a constant. Determine the x and y components of the acceleration. At what point (points) in the flow field is the acceleration zero?

4.15 A three-dimensional velocity field is given by $u = 2x$, $v = -y$, and $w = z$. Determine the acceleration vector.

† **4.16** Estimate the average acceleration of water as it travels through the nozzle on your garden hose. List all assumptions and show all calculations.

4.17 The velocity of air in the diverging pipe shown in Fig. P4.17 is given by $V_1 = 4t$ ft/s and $V_2 = 2t$ ft/s, where t is in seconds. **(a)** Determine the local acceleration at points (1) and (2). **(b)** Is the average convective acceleration between these two points negative, zero, or positive? Explain.

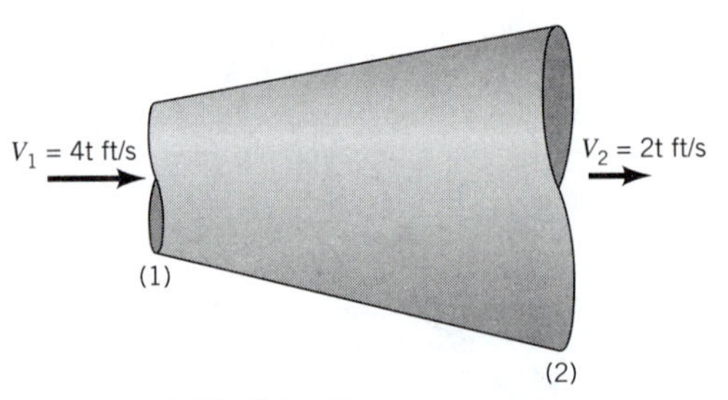

■ **FIGURE P4.17**

4.18 Water flows through a constant diameter pipe with a uniform velocity given by $\mathbf{V} = (8/t + 5)\hat{\mathbf{j}}$ m/s, where t is in seconds. Determine the acceleration at time $t = 1$, 2, and 10 s.

4.19 When a valve is opened, the velocity of water in a certain pipe is given by $u = 10(1 - e^{-t})$, $v = 0$, and $w = 0$, where u is in ft/s and t is in seconds. Determine the maximum velocity and maximum acceleration of the water.

*__4.20__ Water flows through a pipe with $\mathbf{V} = u(t)\hat{\mathbf{i}}$ where the approximate measured values of $u(t)$ are shown in the table. Plot the acceleration as a function of time for $0 \le t \le 20$ s. Plot the acceleration as a function of time if all of the values of $u(t)$ are increased by a factor of 2; by a factor of 5.

t (s)	u (ft/s)	t (s)	u (ft/s)
0	0	11.2	8.1
1.8	1.7	12.3	8.4
3.1	3.2	13.9	8.3
4.0	3.8	15.0	8.1
5.5	4.6	16.4	7.9
6.9	5.8	17.5	7.0
8.1	6.3	18.4	6.6
10.0	7.1	20.0	5.7

4.21 The fluid velocity along the x axis shown in Fig. P4.21 changes from 6 m/s at point A to 18 m/s at point B. It is also known that the velocity is a linear function of distance along the streamline. Determine the acceleration at points A, B, and C. Assume steady flow.

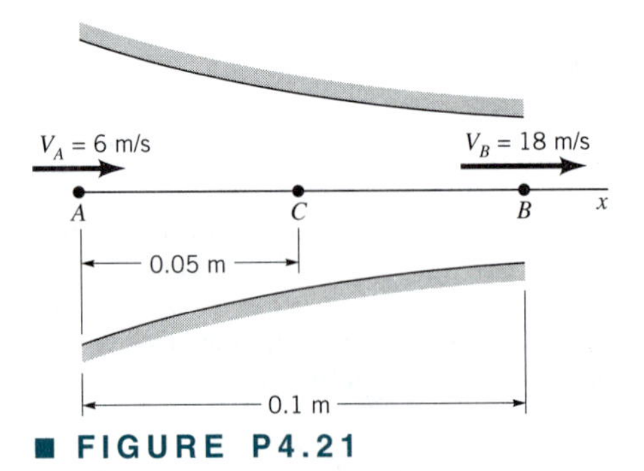

■ **FIGURE P4.21**

4.22 When a fluid flows into a round pipe as shown in Fig. P4.22, viscous effects may cause the velocity profile to change from a uniform profile ($\mathbf{V} = V_0\hat{\mathbf{i}}$) at the entrance of the pipe to a parabolic profile $\{\mathbf{V} = 2V_0 [1 - (r/R)^2]\hat{\mathbf{i}}\}$ at $x = \ell$. Velocity profiles for various values of x are as indicated in the figure. Use this graph to show that a fluid particle moving along the centerline ($r = 0$) experiences an acceleration, but a particle close to the edge of the pipe ($r \approx R$) experiences a deceleration. Does a particle traveling along the line $r = 0.5$ R experience an acceleration or deceleration, or both? Explain.

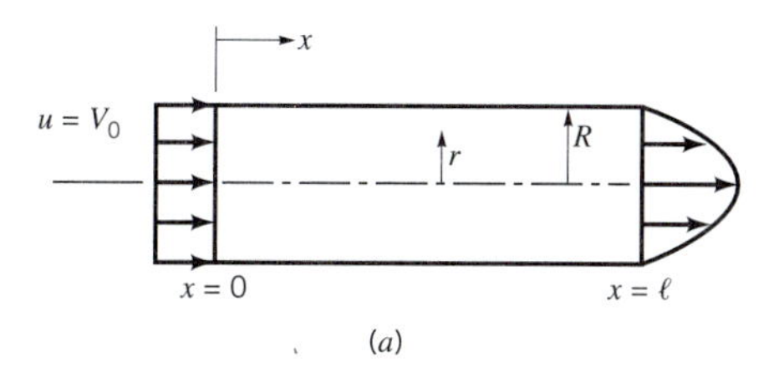

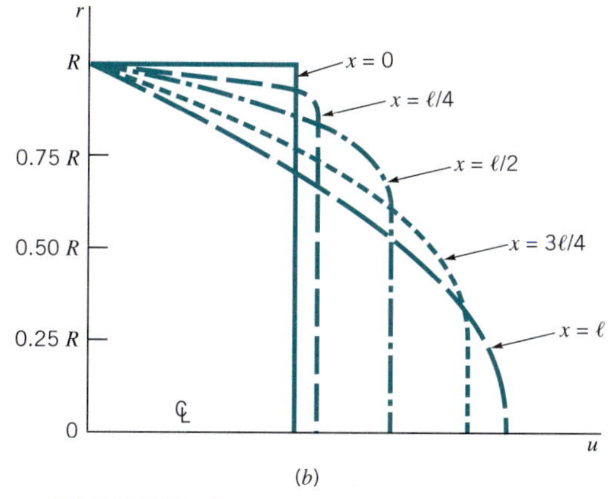

■ **FIGURE P4.22**

4.23 As a valve is opened, water flows through the diffuser shown in Fig. P4.23 at an increasing flowrate so that the velocity along the centerline is given by $\mathbf{V} = u\hat{\mathbf{i}} = V_0(1 - e^{-ct})(1 - x/\ell)\hat{\mathbf{i}}$, where u_0, c, and ℓ are constants. Determine the acceleration as a function of x and t. If $V_0 = 10$ ft/s and $\ell = 5$ ft, what value of c (other than $c = 0$) is needed to make the acceleration zero for any x at $t = 1$ s? Explain how the acceleration can be zero if the flowrate is increasing with time.

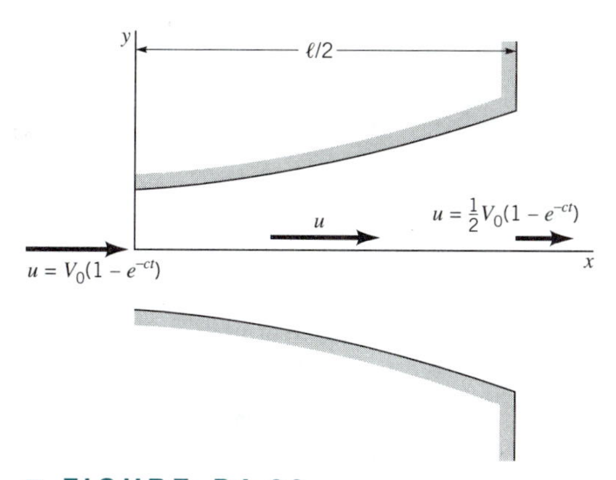

■ **FIGURE P4.23**

4.24 A fluid flows along the x axis with a velocity given by $\mathbf{V} = (x/t)\hat{\mathbf{i}}$, where x is in feet and t in seconds. **(a)** Plot the speed for $0 \leq x \leq 10$ ft and $t = 3$ s. **(b)** Plot the speed for $x = 7$ ft and $2 \leq t \leq 4$ s. **(c)** Determine the local and convective acceleration. **(d)** Show that the acceleration of any fluid particle in the flow is zero. **(e)** Explain physically how the velocity of a particle in this unsteady flow remains constant throughout its motion.

4.25 A hydraulic jump is a rather sudden change in depth of a liquid layer as it flows in an open channel as shown in Fig. P4.25. In a relatively short distance (thickness $= \ell$) the liquid depth changes from z_1 to z_2, with a corresponding change in velocity from V_1 to V_2. If $V_1 = 5$ m/s, $V_2 = 1$ m/s, and $\ell = 0.2$ m, estimate the average deceleration of the liquid as it flows across the hydraulic jump. How many g's deceleration does this represent?

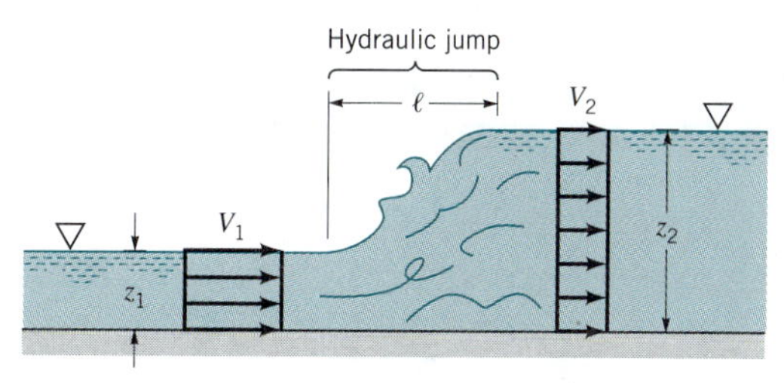

■ **FIGURE P4.25**

***4.26** Assume that the velocity of a fluid flowing past the body shown in Fig. P4.26 is given by $u = (V_0/2)(1 + e^{-x^2/\ell^2})$ for $x > 0$, where V_0 and ℓ are constants. **(a)** Determine the acceleration for $x = 0$ and $x = \infty$. **(b)** Plot the acceleration in terms of V_0 and ℓ for $0 \leq x \leq \infty$ and determine its maximum value.

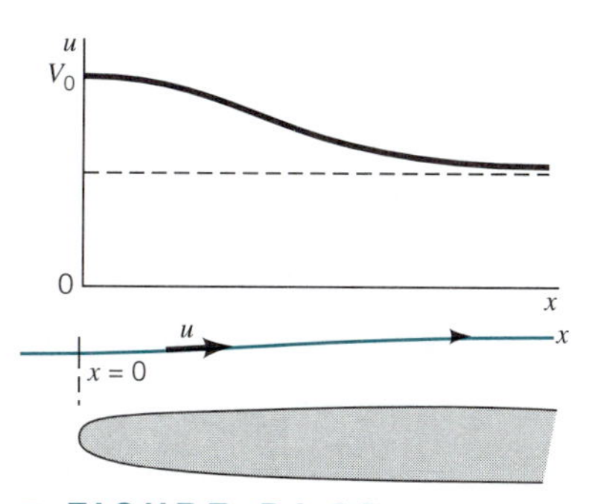

■ **FIGURE P4.26**

4.27 A nozzle is designed to accelerate the fluid from V_1 to V_2 in a linear fashion. That is, $V = ax + b$, where a and b are constants. If the flow is constant with $V_1 = 10$ m/s at $x_1 = 0$ and $V_2 = 25$ m/s at $x_2 = 1$ m, determine the local acceleration, the convective acceleration, and the acceleration of the fluid at points (1) and (2).

† **4.28** A stream of water from the faucet strikes the bottom of the sink. Estimate the maximum acceleration experienced by the water particles. List all assumptions and show calculations.

4.29 Repeat Problem 4.27 with the assumption that the flow is not steady, but at the time when $V_1 = 10$ m/s and $V_2 = 25$ m/s, it is known that $\partial V_1/\partial t = 20$ m/s^2 and $\partial V_2/\partial t = 60$ m/s^2.

4.30 An incompressible fluid flows through the converging duct shown in Fig. P4.30*a* with velocity V_0 at the entrance. Measurements indicate that the actual velocity of the fluid near the wall of the duct along streamline A–F is as shown in Fig. P4.30*b*. Sketch the component of acceleration along this streamline, a, as a function of s. Discuss the important characteristics of your result.

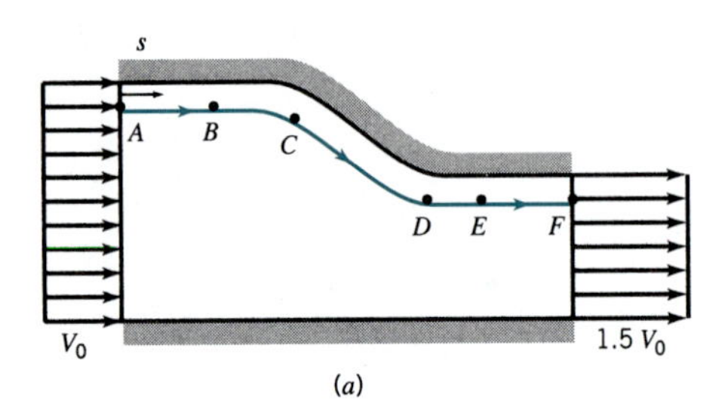

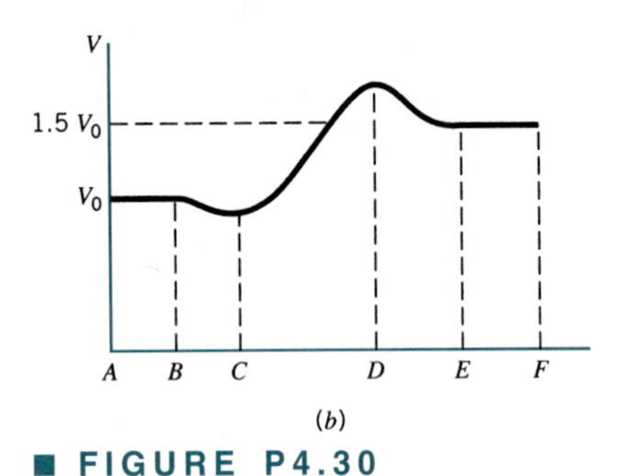

■ FIGURE P4.30

*4.31 Air flows steadily through a variable area pipe with a velocity of $\mathbf{V} = u(x)\hat{\mathbf{i}}$ ft/s, where the approximate measured values of $u(x)$ are given in the table. Plot the acceleration as a function of x for $0 \leq x \leq 12$ in. Plot the acceleration if the flowrate is increased by a factor of N (i.e., the values of u are increased by a factor of N) for $N = 2, 4, 10$.

x (in.)	u (ft/s)	x (in.)	u (ft/s)
0	10.0	7	20.1
1	10.2	8	17.4
2	13.0	9	13.5
3	20.1	10	11.9
4	28.3	11	10.3
5	28.4	12	10.0
6	25.8	13	10.0

4.32 Assume the temperature of the exhaust in an exhaust pipe can be approximated by $T = T_0(1 + ae^{-bx})[1 + c\cos(\omega t)]$, where $T_0 = 100$ °C, $a = 3$, $b = 0.03$ m^{-1}, $c = 0.05$, and $\omega = 100$ rad/s. If the exhaust speed is a constant 2 m/s, determine the time rate of change of temperature of the fluid particles at $x = 0$ and $x = 4$ m when $t = 0$.

*4.33 As is indicated in Fig. P4.33, the speed of exhaust in a car's exhaust pipe varies in time and distance because of the periodic nature of the engine's operation and the damping effect with distance from the engine. Assume that the speed is given by $V = V_0[1 + ae^{-bx}\sin(\omega t)]$, where $V_0 = 8$ fps, $a = 0.05$, $b = 0.2$ ft^{-1}, and $\omega = 50$ rad/s. Calculate and plot the fluid acceleration at $x = 0, 1, 2, 3, 4,$ and 5 ft for $0 \leq t \leq \pi/25$ s.

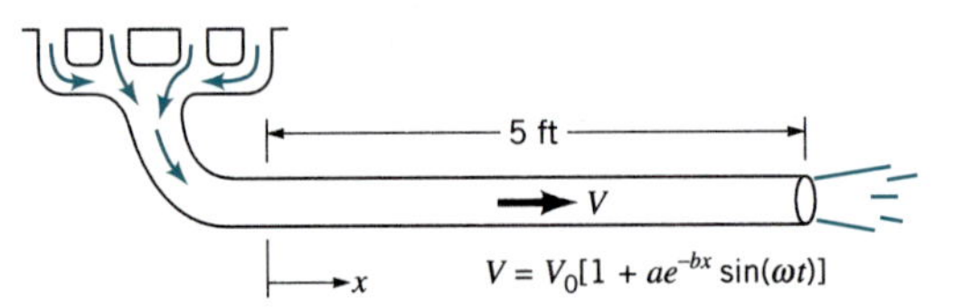

■ FIGURE P4.33

4.34 A gas flows along the x-axis with a speed of $V = 5x$ m/s and a pressure of $p = 10x^2$ N/m^2, where x is in meters. **(a)** Determine the time rate of change of pressure at the fixed location $x = 1$. **(b)** Determine the time rate of change of pressure for a fluid particle flowing past $x = 1$. **(c)** Explain without using any equations why the answers to parts (a) and (b) are different.

4.35 The temperature distribution in a fluid is given by $T = 10x + 5y$, where x and y are the horizontal and vertical coordinates in meters and T is in degrees centigrade. Determine the time rate of change of temperature of a fluid particle traveling (a) horizontally with $u = 20$ m/s, $v = 0$ or (b) vertically with $u = 0$, $v = 20$ m/s.

4.36 Water flows under the sluice gate shown in Fig. P4.36. If $V_1 = 3$ m/s, what is the normal acceleration at point (1)?

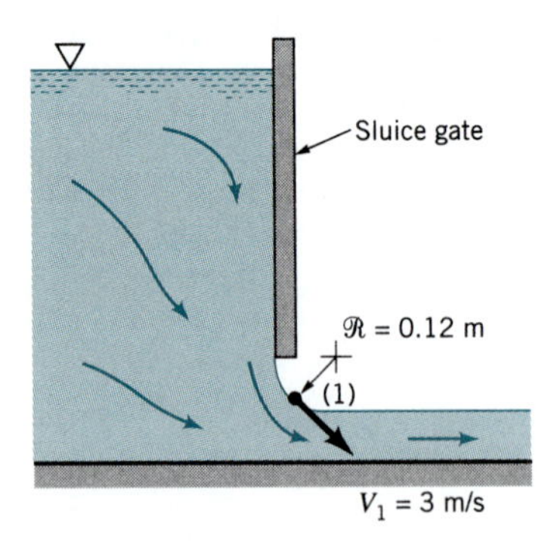

■ FIGURE P4.36

4.37 Water flows down the face of the dam shown in Fig. P4.37. The face of the dam consists of two circular arcs with radii of 10 and 20 ft as shown. If the speed of the water along streamline A–B is approximately $V = (2gh)^{1/2}$, where the distance h is as indicated, plot the normal acceleration as a function of distance along the streamline, $a_n = a_n(s)$.

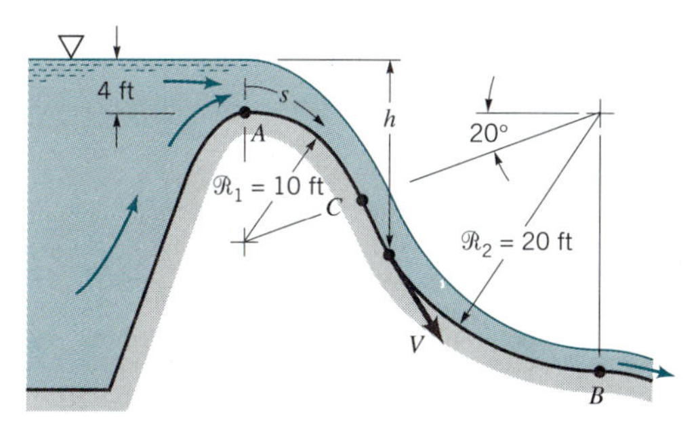

■ FIGURE P4.37

4.38 For the flow given in Problem 4.37, plot the streamwise acceleration, a_s, as a function of distance, s, along the surface of the dam from A to C.

4.39 A fluid flows past a sphere with an upstream velocity of $V_0 = 40$ m/s as shown in Fig. P4.39. From a more advanced theory it is found that the speed of the fluid along the front part of the sphere is $V = \frac{3}{2}V_0 \sin \theta$. Determine the streamwise and normal components of acceleration at point A if the radius of the sphere is $a = 0.20$ m.

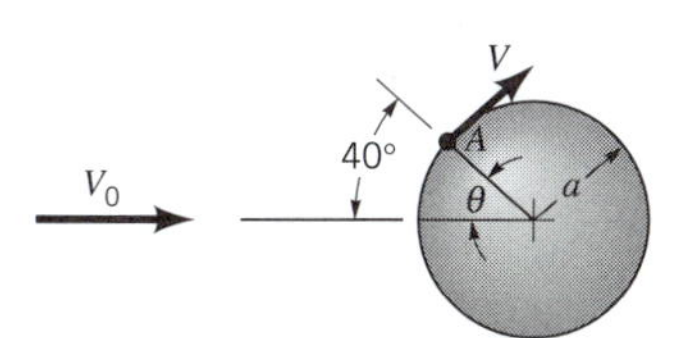

■ FIGURE P4.39

***4.40** For flow past a sphere as discussed in Problem 4.39, plot a graph of the streamwise acceleration, a_s, the normal acceleration, a_n, and the magnitude of the acceleration as a function of θ for $0 \leq \theta \leq 90°$ with $V_0 = 50$ ft/s and $a = 0.1$, 1.0, and 10 ft. Repeat for $V_0 = 5$ ft/s. At what point is the acceleration a maximum; a minimum?

4.41 A fluid flows past a circular cylinder of radius a with an upstream speed of V_0 as shown in Fig. P4.41. A more advanced theory indicates that if viscous effects are negligible, the velocity of the fluid along the surface of the cylinder is given by $V = 2V_0 \sin \theta$. Determine the streamline and normal components of acceleration on the surface of the cylinder as a function of V_0, a, and θ.

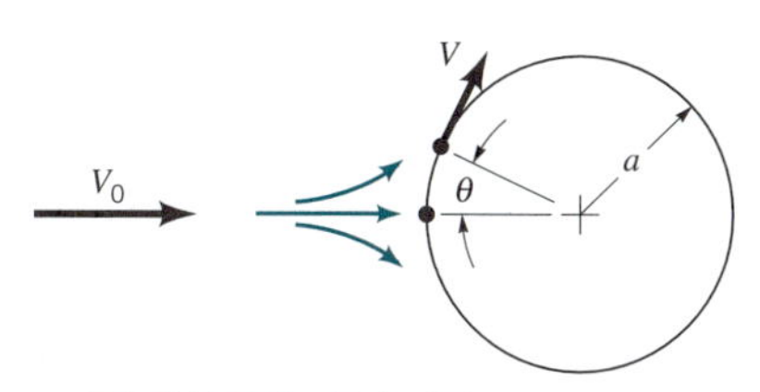

■ FIGURE P4.41

***4.42** Use the results of Problem 4.41 to plot graphs of a_s and a_n for $0 \leq \theta \leq 90°$ with $V_0 = 10$ m/s and $a = 0.01$, 0.10, 1.0, and 10.0 m.

4.43 Determine the x and y components of acceleration for the flow given in Problem 4.6. If $c > 0$, is the particle at point $x = x_0 > 0$ and $y = 0$ accelerating or decelerating? Explain. Repeat if $x_0 < 0$.

4.44 Water flows through the curved hose shown in Fig. P4.44 with an increasing speed of $V = 10t$ ft/s, where t is in seconds. For $t = 2$ s determine **(a)** the component of acceleration along the streamline, **(b)** the component of acceleration normal to the streamline, and **(c)** the net acceleration (magnitude and direction).

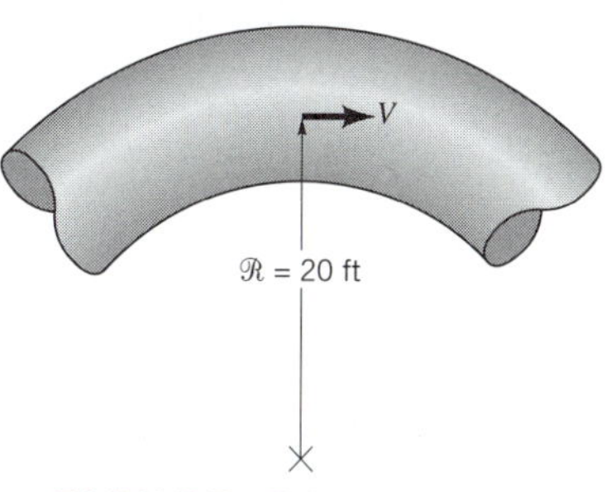

■ FIGURE P4.44

4.45 Water flows steadily through the funnel shown in Fig. P4.45. Throughout most of the funnel the flow is approximately radial (along rays from O) with a velocity of $V = c/r^2$, where r is the radial coordinate and c is a constant. If the velocity is 0.4 m/s when $r = 0.1$ m, determine the acceleration at points A and B.

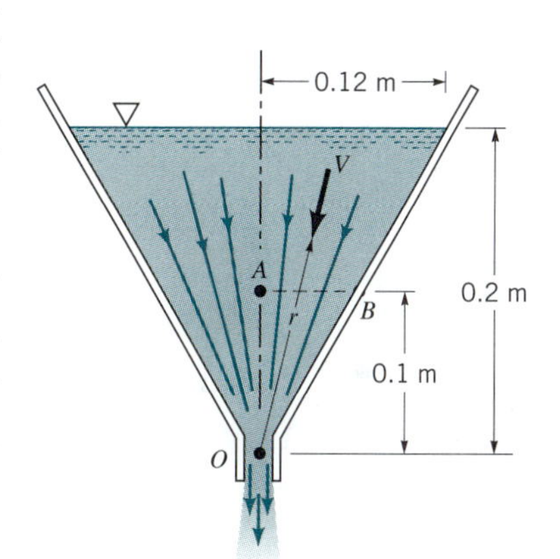

■ FIGURE P4.45

4.46 Water flows through the slit at the bottom of a two-dimensional water trough as shown in Fig. P4.46. Throughout most of the trough the flow is approximately radial (along rays from O) with a velocity of $V = c/r$, where r is the radial coordinate and c is a constant. If the velocity is 0.04 m/s when $r = 0.1$ m, determine the acceleration at points A and B.

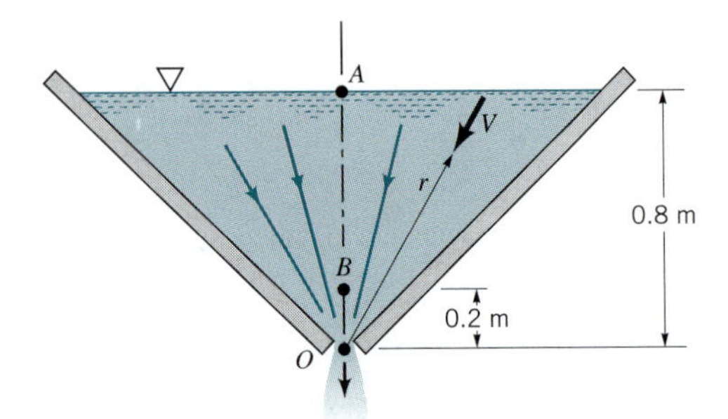

■ FIGURE P4.46

4.47 Air flows from a pipe into the region between two parallel circular disks as shown in Fig. P4.47. The fluid velocity in the gap between the disks is closely approximated by $V = V_0R/r$, where R is the radius of the disk, r is the radial coordinate, and V_0 is the fluid velocity at the edge of the disk. Determine the acceleration for $r = 1$, 2, or 3 ft if $V_0 = 5$ ft/s and $R = 3$ ft.

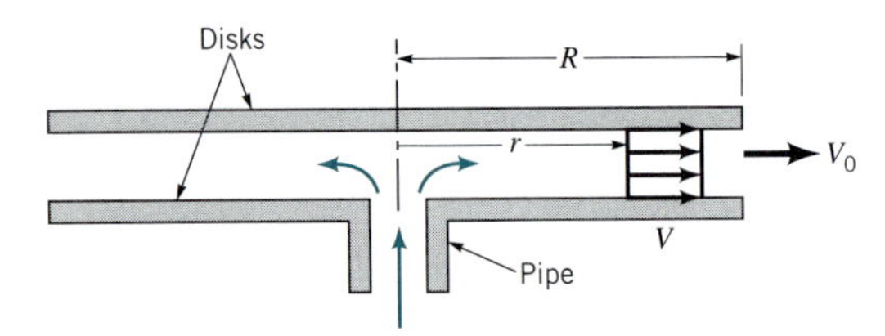

■ FIGURE P4.47

4.48 Air flows from a pipe into the region between a circular disk and a cone as shown in Fig. P4.48. The fluid velocity in the gap between the disk and the cone is closely approximated by $V = V_0R^2/r^2$, where R is the radius of the disk, r is the radial coordinate, and V_0 is the fluid velocity at the edge of the disk. Determine the acceleration for $r = 0.5$ and 2 ft if $V_0 = 5$ ft/s and $R = 2$ ft.

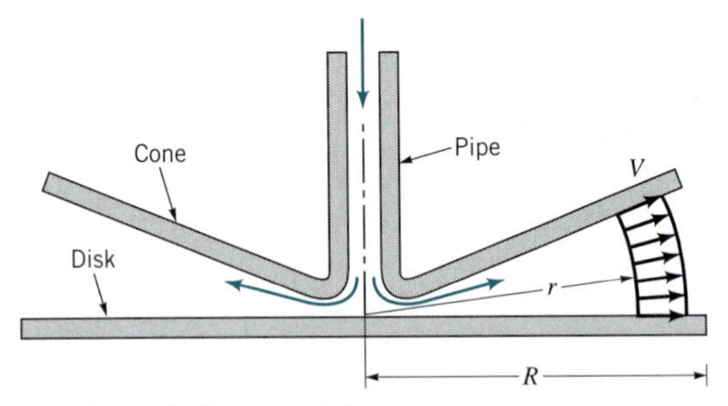

■ FIGURE P4.48

4.49 Water flows through a duct of square cross section as shown in Fig. P4.49 with a constant, uniform velocity of $V = 20$ m/s. Consider fluid particles that lie along line A–B at time $t = 0$. Determine the position of these particles, denoted by line A'–B', when $t = 0.20$ s. Use the volume of fluid in the region between lines A–B and A'–B' to determine the flowrate in the duct. Repeat the problem for fluid particles originally along line C–D; along line E–F. Compare your three answers.

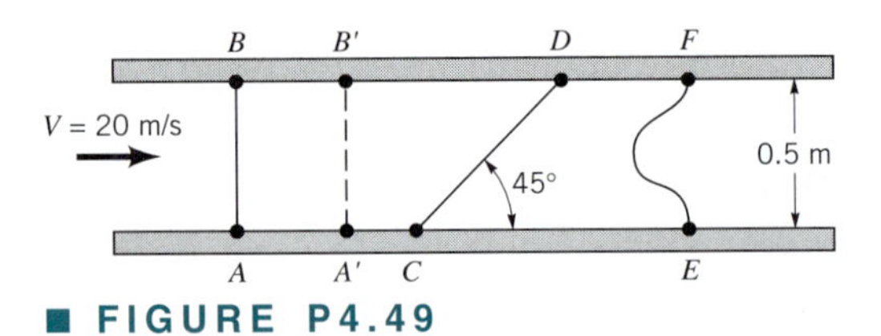

■ FIGURE P4.49

4.50 Repeat Problem 4.49 if the velocity profile is linear from 10 to 20 m/s across the duct as shown in Fig. P4.50.

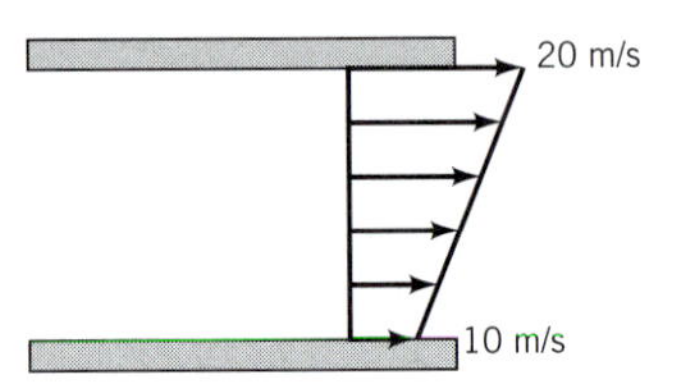

■ FIGURE P4.50

4.51 In the region just downstream of a sluice gate, the water may develop a reverse flow region as is indicated in Fig. P4.51. The velocity profile is assumed to consist of two uniform regions, one with velocity $V_a = 10$ fps and the other with $V_b = 3$ fps. Determine the net flowrate of water across the portion of the control surface at section (2) if the channel is 20 ft wide.

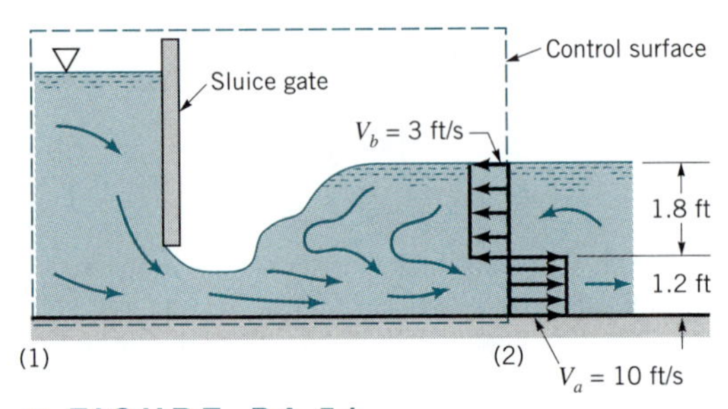

■ FIGURE P4.51

4.52 At time $t = 0$ the valve on an initially empty (perfect vacuum, $\rho = 0$) tank is opened and air rushes in. If the tank has a volume of $\mathcal{V}_0$ and the density of air within the tank increases as $\rho = \rho_\infty(1 - e^{-bt})$, where b is a constant, determine the time rate of change of mass within the tank.

† **4.53** From calculus, one obtains the following formula (Leibnitz rule) for the time derivative of an integral that contains time in both the integrand and the limits of the integration:

$$\frac{d}{dt}\int_{x_1(t)}^{x_2(t)} f(x, t)dx = \int_{x_1}^{x_2} \frac{\partial f}{\partial t}\, dx + f(x_2, t)\frac{dx_2}{dt} - f(x_1, t)\frac{dx_1}{dt}$$

Discuss how this formula is related to the time derivative of the total amount of a property in a system and to the Reynolds transport theorem.

4.54 Air enters an elbow with a uniform speed of 10 m/s as shown in Fig. P4.54. At the exit of the elbow the velocity profile is not uniform. In fact, there is a region of separation or reverse flow. The fixed control volume $ABCD$ coincides with the system at time $t = 0$. Make a sketch to indicate **(a)** the system at time $t = 0.01$ s and **(b)** the fluid that has entered and exited the control volume in that time period.

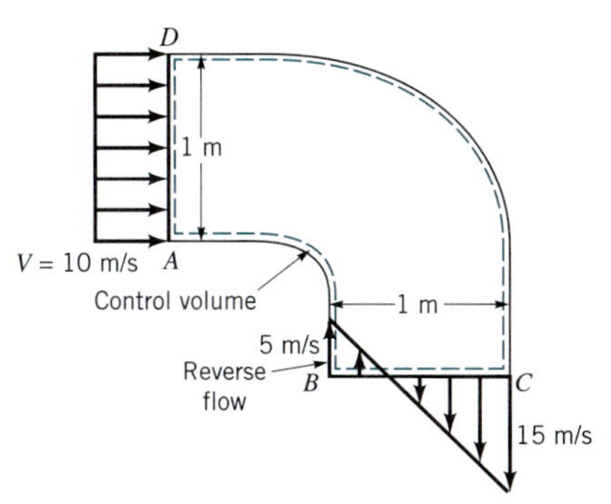

■ **FIGURE P4.54**

4.55 A layer of oil flows down a vertical plate as shown in Fig. P4.55 with a velocity of $\mathbf{V} = (V_0/h^2)(2hx - x^2)\,\hat{\mathbf{j}}$ where V_0 and h are constants. **(a)** Show that the fluid sticks to the plate and that the shear stress at the edge of the layer ($x = h$) is zero. **(b)** Determine the flowrate across surface AB. Assume the width of the plate is b.

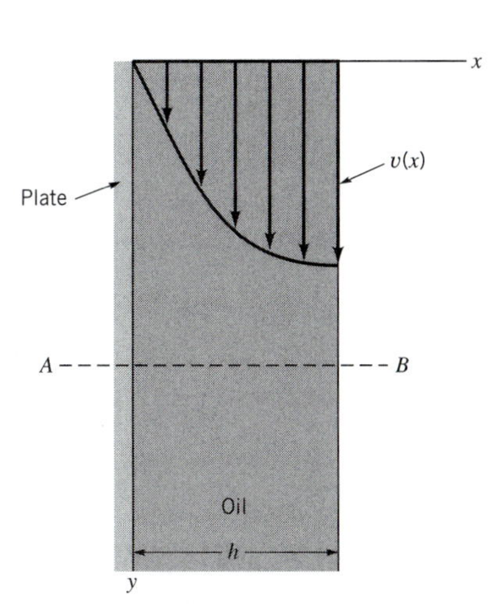

■ **FIGURE P4.55**

4.56 Water flows in the branching pipe shown in Fig. P4.56 with uniform velocity at each inlet and outlet. The fixed control volume indicated coincides with the system at time $t = 20$ s. Make a sketch to indicate **(a)** the boundary of the system at time $t = 20.2$ s, **(b)** the fluid that left the control volume during that 0.2-s interval, and **(c)** the fluid that entered the control volume during that time interval.

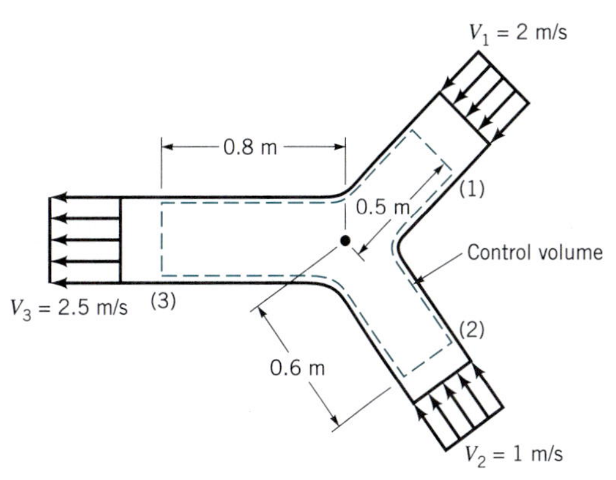

■ **FIGURE P4.56**

4.57 Two plates are pulled in opposite directions with speeds of 1.0 ft/s as shown in Fig. P4.57. The oil between the plates moves with a velocity given by $\mathbf{V} = 10\,y\hat{\mathbf{i}}$ ft/s, where y is in feet. The fixed control volume $ABCD$ coincides with the system at time $t = 0$. Make a sketch to indicate **(a)** the system at time $t = 0.2$ s and **(b)** the fluid that has entered and exited the control volume in that time period.

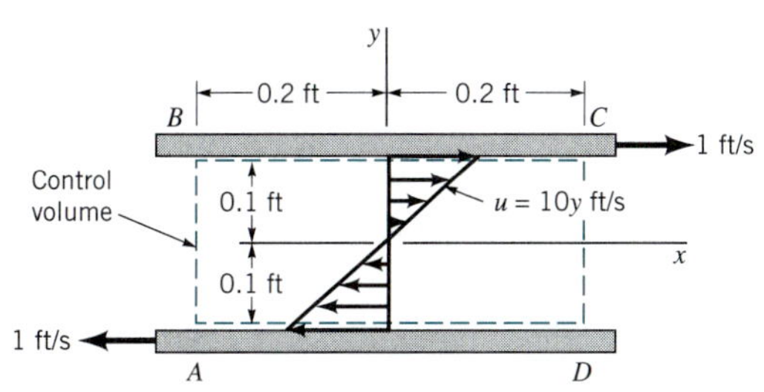

■ **FIGURE P4.57**

4.58 Water is squirted from a syringe with a speed of $V = 5$ m/s by pushing in the plunger with a speed of $V_p = 0.03$ m/s as shown in Fig. P4.58. The surface of the deforming control volume consists of the sides and end of the cylinder and the end of the plunger. The system consists of the water in the syringe at $t = 0$ when the plunger is at section (1) as shown. Make a sketch to indicate the control surface and the system when $t = 0.5$ s.

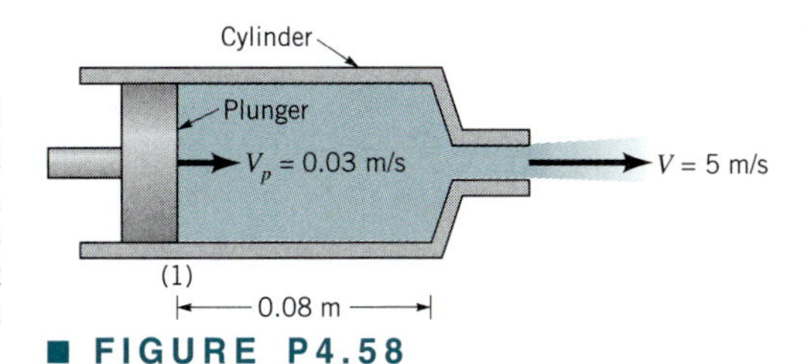

■ **FIGURE P4.58**

4.59 Water flows from a nozzle with a speed of $V = 10$ m/s and is collected in a container that moves toward the nozzle with a speed of $V_{cv} = 2$ m/s as shown in Fig. P4.59. The moving control surface consists of the inner surface of the container. The system consists of the water in the container at time $t = 0$ and the water between the nozzle and the tank in the constant diameter stream at $t = 0$. At time $t = 0.1$ s what volume of the system remains outside of the control volume? How much water has entered the control volume during this time period? Repeat the problem for $t = 0.3$ s.

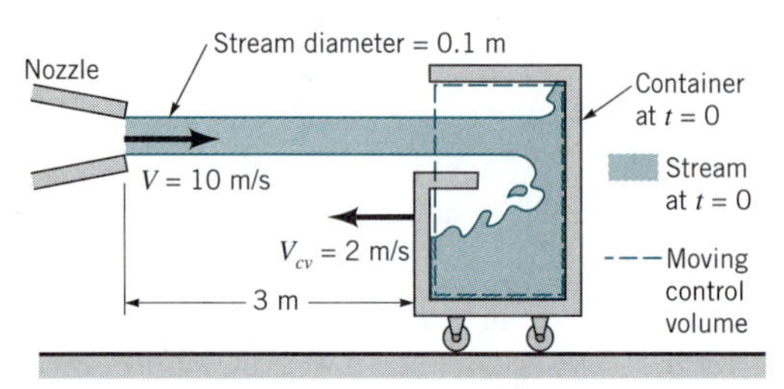

■ **FIGURE P4.59**

4.60 Water flows through the 2-m-wide rectangular channel shown in Fig. P4.60 with a uniform velocity of 3 m/s. **(a)** Directly integrate Eq. 4.16 with $b = 1$ to determine the mass flowrate (kg/s) across section CD of the control volume. **(b)** Repeat part (a) with $b = 1/\rho$, where ρ is the density. Explain the physical interpretation of the answer to part (b).

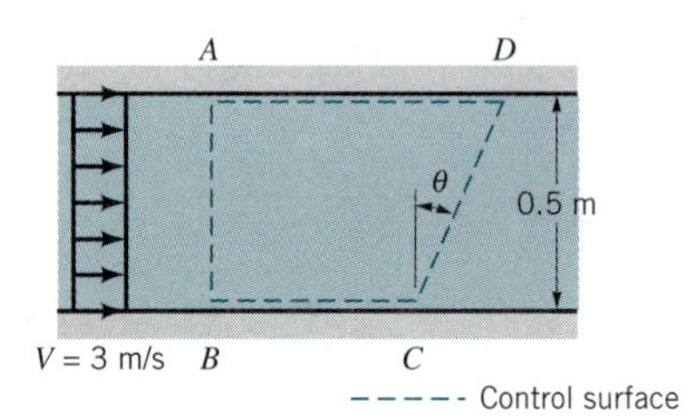

■ **FIGURE P4.60**

4.61 The wind blows across a field with an approximate velocity profile as shown in Fig. P4.61. Use Eq. 4.16 with the parameter b equal to the velocity to determine the momentum flowrate across the vertical surface A–B, which is of unit depth into the paper.

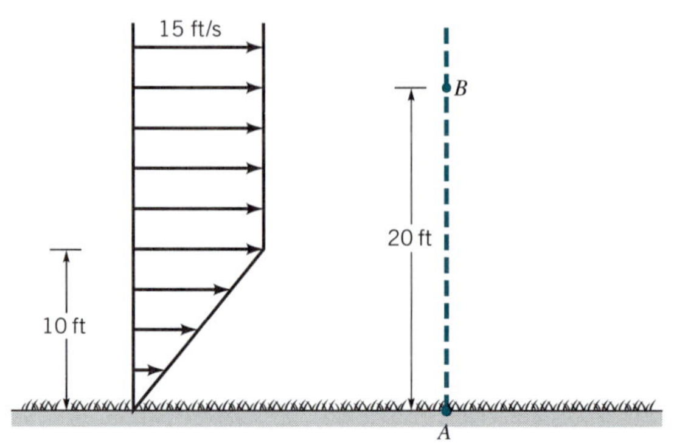

■ **FIGURE P4.61**

A jet of water injected into stationary water: Upon emerging from the slit at the left, the jet of fluid loses some of its momentum to the surrounding fluid. This causes the jet to slow down and its width to increase (air bubbles in water). (Photograph courtesy of ONERA, France.)

5 Finite Control Volume Analysis

Many fluid mechanics problems can be solved by using control volume analysis.

Many practical problems in fluid mechanics require analysis of the behavior of the contents of a finite region in space (a control volume). For example, we may be asked to calculate the anchoring force required to hold a jet engine in place during a test. Or, we could be called on to determine the amount of time to allow for complete filling of a large storage tank. An estimate of how much power it would take to move water from one location to another at a higher elevation and several miles away may be sought. As you will learn by studying the material in this chapter, these and many other important questions can be readily answered with finite control volume analyses. The bases of this analysis method are some fundamental principles of physics, namely, conservation of mass, Newton's second law of motion, and the first and second[1] laws of thermodynamics. Thus, as one might expect, the resultant techniques are powerful and applicable to a wide variety of fluid mechanical circumstances that require engineering judgment. Furthermore, the finite control volume formulas are easy to interpret physically and thus are not difficult to use.

The control volume formulas are derived from the equations representing basic laws applied to a collection of mass (a system). The system statements are probably familiar to you presently. However, in fluid mechanics, the control volume or Eulerian view is generally less complicated and, therefore, more convenient to use than the system or Lagrangian view. The concept of a control volume and system occupying the same region of space at an instant (coincident condition) and use of the Reynolds transport theorem (Eqs. 4.19 and 4.23) are key elements in the derivation of the control volume equations.

Integrals are used throughout the chapter for generality. Volume integrals can accommodate spatial variations of the material properties of the contents of a control volume. Control surface area integrals allow for surface distributions of flow variables. However, in this chapter, for simplicity we often assume that flow variables are uniformly distributed over cross-sectional areas where fluid enters or leaves the control volume. This uniform flow is called one-dimensional flow. In Chapters 8 and 9, when we discuss velocity profiles and other flow variable distributions, the effects of nonuniformities will be covered in more detail.

[1]The section (Section 5.4) on the second law of thermodynamics may be omitted without loss of continuity in the text material.

Mostly steady flows are considered. However, some simple examples of unsteady flow analyses are introduced.

Although fixed, nondeforming control volumes are emphasized in this chapter, a few examples of moving, nondeforming control volumes and deforming control volumes are also included.

5.1 Conservation of Mass—The Continuity Equation

5.1.1 Derivation of the Continuity Equation

A system is defined as a collection of unchanging contents, so the conservation of mass principle for a system is simply stated as

time rate of change of the system mass = 0

or

The amount of mass in a system is constant.

$$\frac{DM_{\text{sys}}}{Dt} = 0 \tag{5.1}$$

where the system mass, M_{sys}, is more generally expressed as

$$M_{\text{sys}} = \int_{\text{sys}} \rho \, d\forall \tag{5.2}$$

and the integration is over the volume of the system. In words, Eq. 5.2 states that the system mass is equal to the sum of all the density-volume element products for the contents of the system.

For a system and a fixed, nondeforming control volume that are coincident at an instant of time, as illustrated in Fig. 5.1, the Reynolds transport theorem (Eq. 4.19) with $B =$ mass and $b = 1$ allows us to state that

$$\frac{D}{Dt}\int_{\text{sys}} \rho \, d\forall = \frac{\partial}{\partial t}\int_{\text{cv}} \rho \, d\forall + \int_{\text{cs}} \rho \mathbf{V} \cdot \hat{\mathbf{n}} \, dA \tag{5.3}$$

or

time rate of change of the mass of the coincident system	=	time rate of change of the mass of the contents of the coincident control volume	+	net rate of flow of mass through the control surface

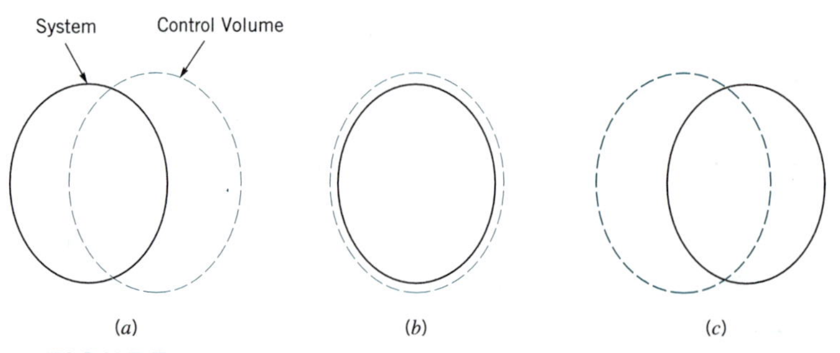

■ **FIGURE 5.1 System and control volume at three different instances of time. (*a*) System and control volume at time $t - \delta t$. (*b*) System and control volume at time t, coincident condition. (*c*) System and control volume at time $t + \delta t$.**

In Eq. 5.3, we express the time rate of change of the system mass as the sum of two control volume quantities, the time rate of change of the mass of the contents of the control volume,

$$\frac{\partial}{\partial t} \int_{\text{cv}} \rho \, d\maltese$$

and the net rate of mass flow through the control surface,

$$\int_{\text{cs}} \rho \mathbf{V} \cdot \hat{\mathbf{n}} \, dA$$

When a flow is steady, all field properties (i.e., properties at any specified point) including density remain constant with time and the time rate of change of the mass of the contents of the control volume is zero. That is,

$$\frac{\partial}{\partial t} \int_{\text{cv}} \rho \, d\maltese = 0$$

The integrand, $\mathbf{V} \cdot \hat{\mathbf{n}} \, dA$, in the mass flowrate integral represents the product of the component of velocity, $\mathbf{V}$, perpendicular to the small portion of control surface and the differential area, dA. Thus, $\mathbf{V} \cdot \hat{\mathbf{n}} \, dA$ is the volume flowrate through dA and $\rho\mathbf{V} \cdot \hat{\mathbf{n}} \, dA$ is the mass flowrate through dA. Furthermore, the sign of the dot product $\mathbf{V} \cdot \hat{\mathbf{n}}$ is "+" for flow *out* of the control volume and "−" for flow *into* the control volume since $\hat{\mathbf{n}}$ is considered positive when it points out of the control volume. When all of the differential quantities, $\rho\mathbf{V} \cdot \hat{\mathbf{n}} \, dA$, are summed over the entire control surface, as indicated by the integral

$$\int_{\text{cs}} \rho \mathbf{V} \cdot \hat{\mathbf{n}} \, dA$$

the result is the net mass flowrate through the control surface, or

$$\int_{\text{cs}} \rho \mathbf{V} \cdot \hat{\mathbf{n}} \, dA = \sum \dot{m}_{\text{out}} - \sum \dot{m}_{\text{in}} \qquad \textbf{(5.4)}$$

where $\dot{m}$ is the mass flowrate (slugs/s or kg/s). If the integral in Eq. 5.4 is positive, the net flow is out of the control volume; if the integral is negative, the net flow is into the control volume.

The continuity equation is a statement that mass is conserved.

The control volume expression for *conservation of mass*, which is commonly called the *continuity equation*, for a fixed, nondeforming control volume is obtained by combining Eqs. 5.1, 5.2, and 5.3 to obtain

$$\boxed{\frac{\partial}{\partial t} \int_{\text{cv}} \rho \, d\maltese + \int_{\text{cs}} \rho \mathbf{V} \cdot \hat{\mathbf{n}} \, dA = 0} \qquad \textbf{(5.5)}$$

In words, Eq. 5.5 states that to conserve mass the time rate of change of the mass of the contents of the control volume plus the net rate of mass flow through the control surface must equal zero. Actually, the same result could have been obtained more directly by equating the rates of mass flow into and out of the control volume to the rates of accumulation and depletion of mass within the control volume (See Section 3.6.2). It is reassuring, however, to see that the Reynolds transport theorem works for this simple-to-understand case. This confidence will serve us well as we develop control volume expressions for other important principles.

An often-used expression for mass flowrate, $\dot{m}$, through a section of control surface having area A is

Mass flowrate equals the product of density and volume flowrate.

$$\boxed{\dot{m} = \rho Q = \rho A V} \tag{5.6}$$

where ρ is the fluid density, Q is the volume flowrate (ft^3/s or m^3/s), and V is the component of fluid velocity perpendicular to area A. Since

$$\dot{m} = \int_A \rho \mathbf{V} \cdot \hat{\mathbf{n}} \, dA$$

application of Eq. 5.6 involves the use of *representative* or average values of fluid density, ρ, and fluid velocity, V. For incompressible flows, ρ is uniformly distributed over area A. For compressible flows, we will normally consider a uniformly distributed fluid density at each section of flow and allow density changes to occur only from section to section. The appropriate fluid velocity to use in Eq. 5.6 is the average value of the component of velocity normal to the section area involved. This average value, $\overline{V}$, is defined as

$$\boxed{\overline{V} = \frac{\int_A \rho \mathbf{V} \cdot \hat{\mathbf{n}} \, dA}{\rho A}} \tag{5.7}$$

If the velocity is considered uniformly distributed (one-dimensional flow) over the section area, A, then

$$\overline{V} = \frac{\int_A \rho \mathbf{V} \cdot \hat{\mathbf{n}} \, dA}{\rho A} = V \tag{5.8}$$

and the bar notation is not necessary (as in Example 5.1). When the flow is not uniformly distributed over the flow cross-sectional area, the bar notation reminds us that an average velocity is being used (as in Examples 5.2 and 5.4).

5.1.2 Fixed, Nondeforming Control Volume

V5.1

In many applications of fluid mechanics, an appropriate control volume to use is fixed and nondeforming. Several example problems that involve the continuity equation for fixed, nondeforming control volumes (Eq. 5.5) follow.

Seawater flows steadily through a simple conical-shaped nozzle at the end of a fire hose as illustrated in Fig. E5.1. If the nozzle exit velocity must be at least 20 m/s, determine the minimum pumping capacity required in m^3/s.

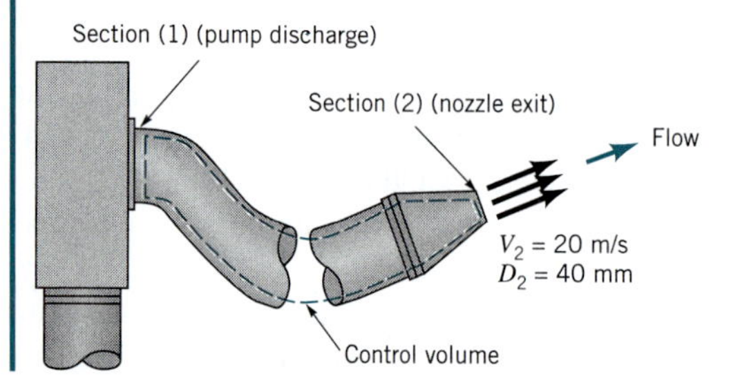

■ FIGURE E5.1

SOLUTION

The pumping capacity sought is the volume flowrate delivered by the fire pump to the hose and nozzle. Since we desire knowledge about the pump discharge flowrate and we have information about the nozzle exit flowrate, we link these two flowrates with the control volume designated with the dashed line in Fig. E5.1. This control volume contains, at any instant, seawater that is within the hose and nozzle from the pump discharge to the nozzle exit plane.

Equation 5.5 is applied to the contents of this control volume to give

$$\cancelto{0 \text{ (flow is steady)}}{\frac{\partial}{\partial t}\int_{\text{cv}}} \rho \, d\forall + \int_{\text{cs}} \rho \mathbf{V} \cdot \hat{\mathbf{n}} \, dA = 0 \qquad \textbf{(1)}$$

The time rate of change of the mass of the contents of this control volume is zero because the flow is steady. From Eq. 5.4, we see that the control surface integral in Eq. 1 involves mass flowrates at the pump discharge, section (1), and at the nozzle exit, section (2), or

$$\int_{\text{cs}} \rho \mathbf{V} \cdot \hat{\mathbf{n}} \, dA = \dot{m}_2 - \dot{m}_1 = 0$$

so that

$$\dot{m}_2 = \dot{m}_1 \qquad \textbf{(2)}$$

Since the mass flowrate is equal to the product of fluid density, ρ, and volume flowrate, Q, (see Eq. 5.6), we obtain from Eq. 2

$$\rho_2 Q_2 = \rho_1 Q_1 \qquad \textbf{(3)}$$

Liquid flow at low speeds, as in this example, may be considered incompressible. Therefore $\rho_2 = \rho_1$ and from Eq. 3

$$Q_2 = Q_1 \qquad \textbf{(4)}$$

The pumping capacity is equal to the volume flowrate at the nozzle exit. If, for simplicity, the velocity distribution at the nozzle exit plane, section (2), is considered uniform (one-dimensional), then from Eqs. 4, 5.6, and 5.8

$$Q_1 = Q_2 = V_2 A_2$$

$$= V_2 \frac{\pi}{4} D_2^2 = (20 \text{ m/s}) \frac{\pi}{4} \left(\frac{40 \text{ mm}}{1000 \text{ mm/m}} \right)^2 = 0.0251 \text{ m}^3/\text{s} \qquad \textbf{(Ans)}$$

Air flows steadily between two sections in a long, straight portion of 4-in. inside diameter pipe as indicated in Fig. E5.2. The uniformly distributed temperature and pressure at each section are given. If the average air velocity (nonuniform velocity distribution) at section (2) is 1000 ft/s, calculate the average air velocity at section (1).

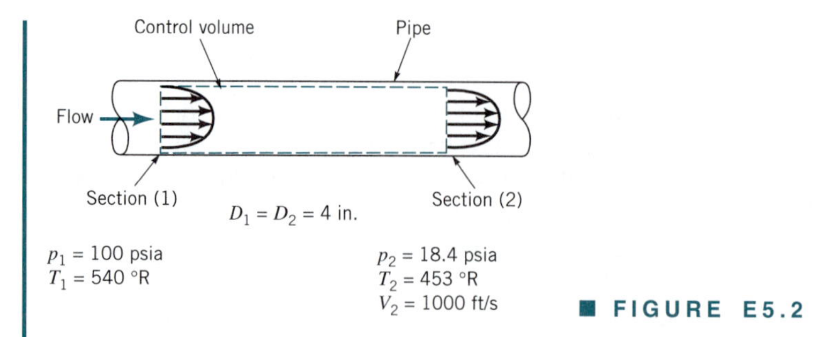

■ FIGURE E5.2

SOLUTION

The average fluid velocity at any section is that velocity which yields the section mass flowrate when multiplied by the section average fluid density and section area (Eq. 5.7). We relate the flows at sections (1) and (2) with the control volume designated with a dashed line in Fig. E5.2.

Equation 5.5 is applied to the contents of this control volume to obtain

$$\overbrace{\frac{\partial}{\partial t}\int_{\text{cv}} \rho \, d\forall}^{0 \text{ (flow is steady)}} + \int_{\text{cs}} \rho \mathbf{V} \cdot \hat{\mathbf{n}} \, dA = 0$$

The time rate of change of the mass of the contents of this control volume is zero because the flow is steady. The control surface integral involves mass flowrates at sections (1) and (2) so that from Eq. 5.4 we get

$$\int_{\text{cs}} \rho \mathbf{V} \cdot \hat{\mathbf{n}} \, dA = \dot{m}_2 - \dot{m}_1 = 0$$

or

$$\dot{m}_1 = \dot{m}_2 \tag{1}$$

and from Eqs. 1, 5.6, and 5.7 we obtain

$$\rho_1 A_1 \overline{V}_1 = \rho_2 A_2 \overline{V}_2 \tag{2}$$

or since $A_1 = A_2$

$$\overline{V}_1 = \frac{\rho_2}{\rho_1} \overline{V}_2 \tag{3}$$

Air at the pressures and temperatures involved in this example problem behaves like an ideal gas. The ideal gas equation of state (Eq. 1.8) is

$$\rho = \frac{p}{RT} \tag{4}$$

Thus, combining Eqs. 3 and 4 we obtain

$$\overline{V}_1 = \frac{p_2 T_1 \overline{V}_2}{p_1 T_2}$$

$$= \frac{(18.4 \text{ psia})(540 \text{ °R})(1000 \text{ ft/s})}{(100 \text{ psia})(453 \text{ °R})} = 219 \text{ ft/s} \qquad \textbf{(Ans)}$$

We learn from this example that the continuity equation (Eq. 5.5) is valid for compressible as well as incompressible flows. Also, nonuniform velocity distributions can be handled with the average velocity concept.

EXAMPLE 5.3

Moist air (a mixture of dry air and water vapor) enters a dehumidifier at the rate of 22 slugs/hr. Liquid water drains out of the dehumidifier at a rate of 0.5 slugs/hr. Determine the mass flowrate of the dry air and the water vapor leaving the dehumidifier. A simplified sketch of the process is provided in Fig. E5.3.

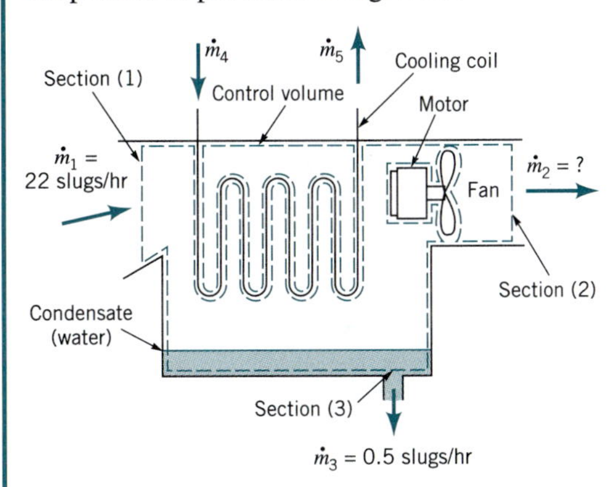

■ FIGURE E5.3

SOLUTION

The unknown mass flowrate at section (2) is linked with the known flowrates at sections (1) and (3) with the control volume designated with a dashed line in Fig. E5.3. The contents of the control volume are the air and water vapor mixture and the condensate (liquid water) in the dehumidifier at any instant.

Not included in the control volume are the fan and its motor, and the condenser coils and refrigerant. Even though the flow in the vicinity of the fan blade is unsteady, it is unsteady in a cyclical way. Thus, the flowrates at sections (1), (2), and (3) appear steady and the time rate of change of the mass of the contents of the control volume may be considered equal to zero on a time-average basis. The application of Eqs. 5.4 and 5.5 to the control volume contents results in

$$\int_{\text{cs}} \rho \mathbf{V} \cdot \hat{\mathbf{n}}\, dA = -\dot{m}_1 + \dot{m}_2 + \dot{m}_3 = 0$$

or

$$\dot{m}_2 = \dot{m}_1 - \dot{m}_3 = 22 \text{ slugs/hr} - 0.5 \text{ slugs/hr} = 21.5 \text{ slugs/hr} \qquad \textbf{(Ans)}$$

Note that the continuity equation (Eq. 5.5) can be used when there is more than one stream of fluid flowing through the control volume.

The answer is the same regardless of which control volume is chosen. For example, if we select the same control volume as before except that we include the cooling coils to be within the control volume, the continuity equation becomes

$$\dot{m}_2 = \dot{m}_1 - \dot{m}_3 + \dot{m}_4 - \dot{m}_5 \qquad \textbf{(1)}$$

where $\dot{m}_4$ is the mass flowrate of the cooling fluid flowing into the control volume, and $\dot{m}_5$ is the flowrate out of the control volume through the cooling coil. Since the flow through the coils is steady, it follows that $\dot{m}_4 = \dot{m}_5$. Hence, Eq. 1 gives the same answer as obtained with the original control volume.

EXAMPLE 5.4

Incompressible, laminar water flow develops in a straight pipe having radius R as indicated in Fig. E5.4. At section (1), the velocity profile is uniform; the velocity is equal to a constant value U and is parallel to the pipe axis everywhere. At section (2), the velocity profile is axisymmetric and parabolic, with zero velocity at the pipe wall and a maximum value of u_{max} at the centerline. How are U and u_{max} related? How are the average velocity at section (2), $\overline{V}_2$, and u_{max} related?

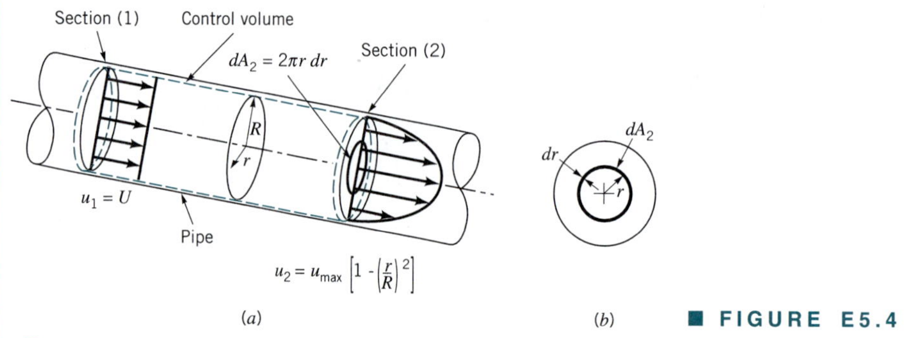

■ FIGURE E5.4

SOLUTION

An appropriate control volume is sketched (dashed lines) in Fig. E5.4*a*. The application of Eq. 5.5 to the contents of this control volume yields

$$\int_{cs} \rho \mathbf{V} \cdot \hat{\mathbf{n}}\, dA = 0$$

The surface integral is evaluated at sections (1) and (2) to give

$$-\rho_1 A_1 U + \int_{A_2} \rho \mathbf{V} \cdot \hat{\mathbf{n}}\, dA_2 = 0 \tag{1}$$

or, since the component of velocity, $\mathbf{V}$, perpendicular to the area at section (2) is u_2, and the element cross-sectional area, dA_2, is equal to $2\pi r\, dr$ (see shaded area element in Fig. E5.4*b*), Eq. 1 becomes

$$-\rho_1 A_1 U + \rho_2 \int_0^R u_2 2\pi r\, dr = 0 \tag{2}$$

Since the flow is considered incompressible, $\rho_1 = \rho_2$. The parabolic velocity relationship for flow through section (2) is used in Eq. 2 to yield

$$-A_1 U + 2\pi u_{max} \int_0^R \left[1 - \left(\frac{r}{R}\right)^2\right] r\, dr = 0 \tag{3}$$

Integrating, we get from Eq. 3

$$-\pi R^2 U + 2\pi u_{max} \left(\frac{r^2}{2} - \frac{r^4}{4R^2}\right)_0^R = 0$$

or

$$u_{max} = 2U \qquad \textbf{(Ans)}$$

Since this flow is incompressible, we conclude from Eq. 5.8 that U is the average velocity at all sections of the control volume. Thus, the average velocity at section (2), $\overline{V}_2$, is one-half the maximum velocity, u_{max}, there, or

$$\overline{V}_2 = \frac{u_{max}}{2} \qquad \textbf{(Ans)}$$

EXAMPLE 5.5

A bathtub is being filled with water from a faucet. The rate of flow from the faucet is steady at 9 gal/min. The tub volume is approximated by a rectangular space as indicated in Fig. E5.5. Estimate the time rate of change of the depth of water in the tub, $\partial h/\partial t$, in in./min at any instant.

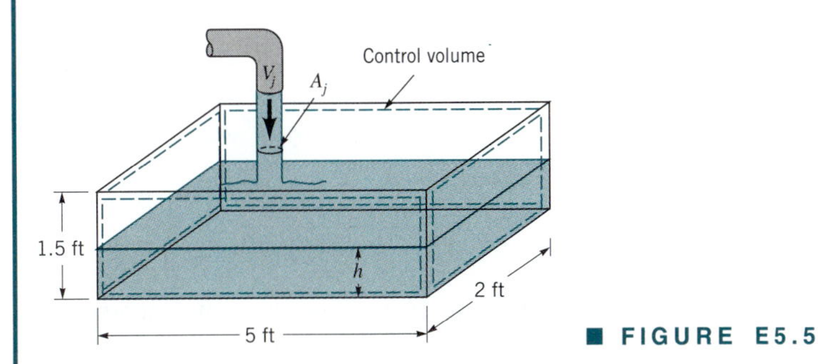

■ FIGURE E5.5

SOLUTION

We will see later (Example 5.9) that this problem can also be solved with a deforming control volume that includes only the water in the tub at any instant. We presently use the fixed, nondeforming control volume outlined with a dashed line in Fig. E5.5. This control volume includes in it, at any instant, the water accumulated in the tub, some of the water flowing from the faucet into the tub, and some air. Application of Eqs. 5.4 and 5.5 to these contents of the control volume results in

$$\frac{\partial}{\partial t}\int_{\substack{\text{air}\\\text{volume}}} \rho_{\text{air}}\, d\forall_{\text{air}} + \frac{\partial}{\partial t}\int_{\substack{\text{water}\\\text{volume}}} \rho_{\text{water}}\, d\forall_{\text{water}} - \dot{m}_{\text{water}} + \dot{m}_{\text{air}} = 0$$

Note that the time rate of change of air mass and water mass are each not zero. Recognizing, however, that the air mass must be conserved, we know that the time rate of change of the mass of air in the control volume must be equal to the rate of air mass flow out of the control volume. For simplicity, we disregard any water evaporation that occurs. Thus, applying Eqs. 5.4 and 5.5 to the air only and to the water only, we obtain

$$\frac{\partial}{\partial t}\int_{\substack{\text{air}\\\text{volume}}} \rho_{\text{air}}\, d\forall_{\text{air}} + \dot{m}_{\text{air}} = 0$$

for air, and

$$\frac{\partial}{\partial t}\int_{\substack{\text{water}\\\text{volume}}} \rho_{\text{water}}\, d\forall_{\text{water}} = \dot{m}_{\text{water}} \tag{1}$$

for water. For the water,

$$\int_{\substack{\text{water}\\\text{volume}}} \rho_{\text{water}}\, d\forall_{\text{water}} = \rho_{\text{water}}\,[h(2\text{ ft})(5\text{ ft}) + (1.5\text{ ft} - h)A_j] \tag{2}$$

where A_j is the cross-sectional area of the jet of water flowing from the faucet into the tub. Combining Eqs. 1 and 2, we obtain

$$\rho_{\text{water}}\,(10\text{ ft}^2 - A_j)\,\frac{\partial h}{\partial t} = \dot{m}_{\text{water}}$$

and, thus

$$\frac{\partial h}{\partial t} = \frac{Q_{\text{water}}}{(10 \text{ ft}^2 - A_j)}$$

For $A_j \ll 10 \text{ ft}^2$ we can conclude that

$$\frac{\partial h}{\partial t} = \frac{Q_{\text{water}}}{(10 \text{ ft}^2)} = \frac{(9 \text{ gal/min})(12 \text{ in./ft})}{(7.48 \text{ gal/ft}^3)(10 \text{ ft}^2)} = 1.44 \text{ in./min} \quad \textbf{(Ans)}$$

The preceding example problems illustrate some important results of applying the conservation of mass principle to the contents of a fixed, nondeforming control volume. The dot product $\mathbf{V} \cdot \hat{\mathbf{n}}$ is considered "+" for flow out of the control volume and "−" for flow into the control volume. Thus, mass flowrate out of the control volume is "+" and mass flowrate in is "−". When the flow is steady, the time rate of change of the mass of the contents of the control volume,

The appropriate sign convention must be followed.

$$\frac{\partial}{\partial t} \int_{\text{cv}} \rho \, d\forall$$

is zero and the net amount of mass flowrate, $\dot{m}$, through the control surface is therefore also zero

$$\sum \dot{m}_{\text{out}} - \sum \dot{m}_{\text{in}} = 0 \quad \textbf{(5.9)}$$

If the steady flow is also incompressible, the net amount of volume flowrate, Q, through the control surface is also zero

V5.2

$$\sum Q_{\text{out}} - \sum Q_{\text{in}} = 0 \quad \textbf{(5.10)}$$

An unsteady, but cyclical flow can be considered steady on a time-average basis. When the flow is unsteady, the instantaneous time rate of change of the mass of the contents of the control volume is not necessarily zero and can be an important variable. When the value of

$$\frac{\partial}{\partial t} \int_{\text{cv}} \rho \, d\forall$$

is "+," the mass of the contents of the control volume is increasing. When it is "−," the mass of the contents of the control volume is decreasing.

When the flow is uniformly distributed over the opening in the control surface (one-dimensional flow),

$$\dot{m} = \rho A V$$

where V is the uniform value of velocity component normal to the section area A. When the velocity is nonuniformly distributed over the opening in the control surface,

$$\dot{m} = \rho A \overline{V} \quad \textbf{(5.11)}$$

where $\overline{V}$ is the average value of the component of velocity normal to the section area A as defined by Eq. 5.7.

For steady flow involving only one stream of a specific fluid flowing through the control volume at sections (1) and (2),

$$\dot{m} = \rho_1 A_1 \overline{V}_1 = \rho_2 A_2 \overline{V}_2 \tag{5.12}$$

and for incompressible flow,

$$Q = A_1 \overline{V}_1 = A_2 \overline{V}_2 \tag{5.13}$$

For steady flow involving more than one stream of a specific fluid or more than one specific fluid flowing through the control volume,

$$\sum \dot{m}_{\text{in}} = \sum \dot{m}_{\text{out}}$$

The variety of example problems solved above should give the correct impression that the fixed, nondeforming control volume is versatile and useful.

5.1.3 Moving, Nondeforming Control Volume

It is sometimes necessary to use a nondeforming control volume attached to a moving reference frame. Examples include control volumes containing a gas turbine engine on an aircraft in flight, the exhaust stack of a ship at sea, and the gasoline tank of an automobile passing by.

Some problems are most easily solved by using a moving control volume.

As discussed in Section 4.4.6, when a moving control volume is used, the fluid velocity relative to the moving control volume (relative velocity) is an important flow field variable. The relative velocity, **W**, is the fluid velocity seen by an observer moving with the control volume. The control volume velocity, $\mathbf{V}_{\text{cv}}$, is the velocity of the control volume as seen from a fixed coordinate system. The absolute velocity, **V**, is the fluid velocity seen by a stationary observer in a fixed coordinate system. These velocities are related to each other by the vector equation

$$\mathbf{V} = \mathbf{W} + \mathbf{V}_{\text{cv}} \tag{5.14}$$

which is the same as Eq. 4.22, introduced earlier.

For a system and a moving, nondeforming control volume that are coincident at an instant of time, the Reynolds transport theorem (Eq. 4.23) for a moving control volume leads to

$$\frac{DM_{\text{sys}}}{Dt} = \frac{\partial}{\partial t} \int_{\text{cv}} \rho \, d\forall + \int_{\text{cs}} \rho \mathbf{W} \cdot \hat{\mathbf{n}} \, dA \tag{5.15}$$

From Eqs. 5.1 and 5.15, we can get the control volume expression for conservation of mass (the continuity equation) for a moving, nondeforming control volume, namely,

$$\boxed{\frac{\partial}{\partial t} \int_{\text{cv}} \rho \, d\forall + \int_{\text{cs}} \rho \mathbf{W} \cdot \hat{\mathbf{n}} \, dA = 0} \tag{5.16}$$

Some examples of the application of Eq. 5.16 follow.

An airplane moves forward at a speed of 971 km/hr as shown in Fig. E5.6. The frontal intake area of the jet engine is 0.80 m^2 and the entering air density is 0.736 kg/m^3. A stationary observer determines that relative to the earth, the jet engine exhaust gases move away from the engine with a speed of 1050 km/hr. The engine exhaust area is 0.558 m^2 and the exhaust gas density is 0.515 kg/m^3. Estimate the mass flowrate of fuel into the engine in kg/hr.

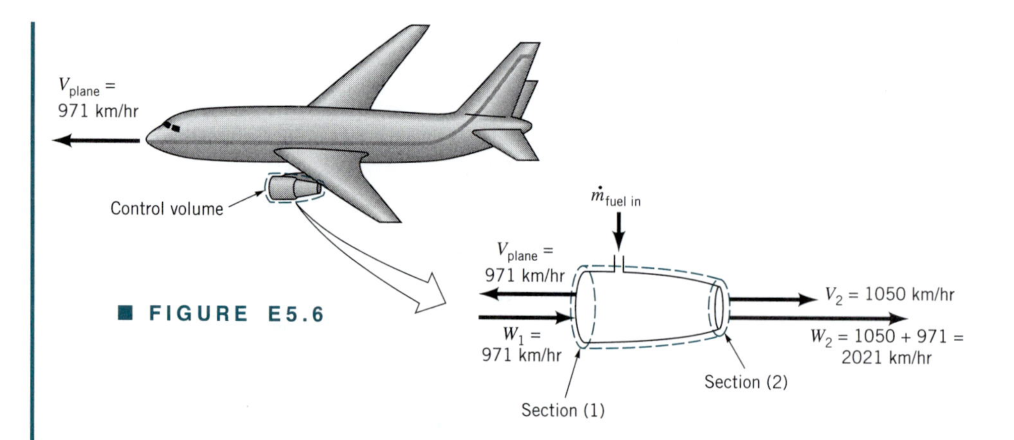

■ FIGURE E5.6

SOLUTION

The control volume, which moves with the airplane (see Fig. E5.6), surrounds the engine and its contents and includes all fluids involved at an instant. The application of Eq. 5.16 to these contents of the control volume yields

$$\cancelto{0}{\frac{\partial}{\partial t}\int_{\text{cv}} \rho \, d\V} + \int_{\text{cs}} \rho \mathbf{W} \cdot \hat{\mathbf{n}} \, dA = 0 \tag{1}$$

0 (flow relative to moving control volume is considered steady on a time-average basis)

Assuming one-dimensional flow, we evaluate the surface integral in Eq. 1 and get

$$-\dot{m}_{\substack{\text{fuel}\\\text{in}}} - \rho_1 A_1 W_1 + \rho_2 A_2 W_2 = 0$$

or

$$\dot{m}_{\substack{\text{fuel}\\\text{in}}} = \rho_2 A_2 W_2 - \rho_1 A_1 W_1 \tag{2}$$

We consider the intake velocity, W_1 relative to the moving control volume, as being equal in magnitude to the speed of the airplane, 971 km/hr. The exhaust velocity, W_2, also needs to be measured relative to the moving control volume. Since a fixed observer noted that the exhaust gases were moving away from the engine at a speed of 1050 km/hr, the speed of the exhaust gases relative to the moving control volume, W_2, is determined as follows by using Eq. 5.14

$$V_2 = W_2 + V_{\text{plane}}$$

or

$$W_2 = V_2 - V_{\text{plane}} = 1050 \text{ km/hr} + 971 \text{ km/hr} = 2021 \text{ km/hr}$$

and is shown in Fig. E5.6*b*.

From Eq. 2,

$$\dot{m}_{\substack{\text{fuel}\\\text{in}}} = (0.515 \text{ kg/m}^3)(0.558 \text{ m}^2)(2021 \text{ km/hr})(1000 \text{ m/km})$$

$$- (0.736 \text{ kg/m}^3)(0.80 \text{ m}^2)(971 \text{ km/hr})(1000 \text{ m/km})$$

$$= (580{,}800 - 571{,}700) \text{ kg/hr}$$

$$\dot{m}_{\substack{\text{fuel}\\\text{in}}} = 9100 \text{ kg/hr} \qquad \textbf{(Ans)}$$

Note that the fuel flowrate was obtained as the difference of two large, nearly equal numbers. Precise values of W_2 and W_1 are needed to obtain a modestly accurate value of $\dot{m}_{\substack{\text{fuel}\\\text{in}}}$.

EXAMPLE 5.7

Water enters a rotating lawn sprinkler through its base at the steady rate of 1000 ml/s as sketched in Fig. E5.7. If the exit area of each of the two nozzles is 30 mm^2, determine the average speed of the water leaving each nozzle, relative to the nozzle, if **(a)** the rotary sprinkler head is stationary, **(b)** the sprinkler head rotates at 600 rpm, **(c)** the sprinkler head accelerates from 0 to 600 rpm.

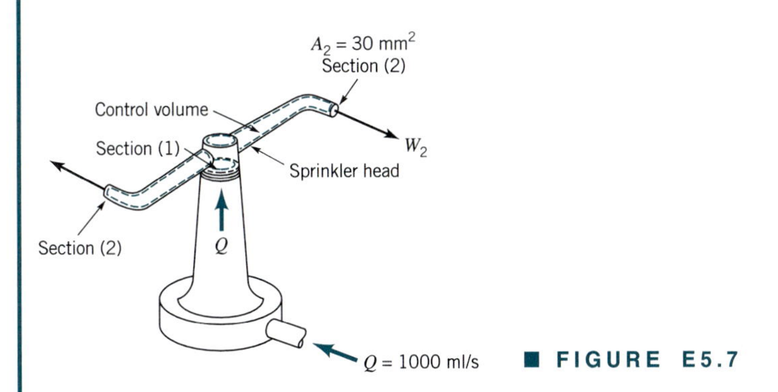

■ **FIGURE E5.7**

SOLUTION

We specify a control volume that contains the water in the rotary sprinkler head at any instant. This control volume is nondeforming, but it moves (rotates) with the sprinkler head. The application of Eq. 5.16 to the contents of this control volume for situation **(a)**, **(b)**, or **(c)** of the problem results in the same expression, namely,

0 flow is steady or the control volume is filled with an incompressible fluid

$$\frac{\partial}{\partial t}\int_{\text{cv}} \rho \, d\forall + \int_{\text{cs}} \rho \, \mathbf{W} \cdot \hat{\mathbf{n}} \, dA = 0$$

The flow is steady in the control volume reference frame when the control volume is stationary [part (a)] and when it moves [parts (b) and (c)]. Also, the control volume is filled with water. Thus, the time rate of change of the mass of the water in the control volume is zero. The control surface integral has a nonzero value only where water enters and leaves the control volume; thus,

$$\int_{\text{cv}} \rho \mathbf{W} \cdot \mathbf{n} \, dA = -\dot{m}_{\text{in}} + \dot{m}_{\text{out}} = 0$$

or

$$\dot{m}_{\text{out}} = \dot{m}_{\text{in}} \qquad \textbf{(1)}$$

Since $\dot{m}_{\text{out}} = 2\rho A_2 \overline{W}_2$ and $\dot{m}_{\text{in}} = \rho Q$
it follows from Eq. 1 that

$$\overline{W}_2 = \frac{Q}{2A_2}$$

or

$$\overline{W}_2 = \frac{(1000 \text{ ml/s})(0.001 \text{ m}^3/\text{liter})(10^6 \text{ mm}^2/\text{m}^2)}{(1000 \text{ ml/liter})(2)(30 \text{ mm}^2)} = 16.7 \text{ m/s} = W_2 \qquad \textbf{(Ans)}$$

The value of W_2 is independent of the speed of rotation of the sprinkler head and represents the average velocity of the water exiting from each nozzle with respect to the nozzle for cases **(a)**, **(b)**, **(c)**. The velocity of water discharging from each nozzle, when viewed from a stationary reference (i.e., V_2), will vary as the rotation speed of the sprinkler head varies since from Eq. 5.14,

$$V_2 = W_2 - U$$

where $U = \omega R$ is the speed of the nozzle and ω and R are the angular velocity and radius of the sprinkler head, respectively.

When a moving, nondeforming control volume is used, the dot product sign convention used earlier for fixed, nondeforming control volume applications is still valid. Also, if the flow within the moving control volume is steady, or steady on a time-average basis, the time rate of change of the mass of the contents of the control volume is zero. Velocities seen from the control volume reference frame (relative velocities) must be used in the continuity equation. Relative and absolute velocities are related by a vector equation (Eq. 5.14), which also involves the control volume velocity.

5.1.4 Deforming Control Volume

Some problems are most easily solved by using a deforming control volume.

Occasionally, a deforming control volume can simplify the solution of a problem. A deforming control volume involves changing volume size and control surface movement. Thus, the Reynolds transport theorem for a moving control volume can be used for this case, and Eqs. 4.23 and 5.1 lead to

$$\frac{DM_{\text{sys}}}{Dt} = \frac{\partial}{\partial t} \int_{\text{cv}} \rho \, d\forall + \int_{\text{cs}} \rho \mathbf{W} \cdot \hat{\mathbf{n}} \, dA = 0 \qquad \textbf{(5.17)}$$

The time rate of change term in Eq. 5.17,

$$\frac{\partial}{\partial t}\int_{\text{cv}} \rho \, d\forall$$

is usually nonzero and must be carefully evaluated because the extent of the control volume varies with time. The mass flowrate term in Eq. 5.17,

$$\int_{\text{cs}} \rho \mathbf{W} \cdot \hat{\mathbf{n}} \, dA$$

must be determined with the relative velocity, **W**, the velocity referenced to the control surface. Since the control volume is deforming, the control surface velocity is not necessarily uniform and identical to the control volume velocity, $\mathbf{V}_{\text{cv}}$, as was true for moving, nondeforming control volumes. For the deforming control volume,

The velocity of the surface of a deforming control volume is not the same at all points on the surface.

$$\mathbf{V} = \mathbf{W} + \mathbf{V}_{\text{cs}} \tag{5.18}$$

where $\mathbf{V}_{\text{cs}}$ is the velocity of the control surface as seen by a fixed observer. The relative velocity, **W**, must be ascertained with care wherever fluid crosses the control surface. Two example problems that illustrate the use of the continuity equation for a deforming control volume, Eq. 5.17, follow.

EXAMPLE 5.8

A syringe (Fig. E5.8) is used to inoculate a cow. The plunger has a face area of 500 mm². If the liquid in the syringe is to be injected steadily at a rate of 300 cm³/min, at what speed should the plunger be advanced? The leakage rate past the plunger is 0.10 times the volume flowrate out of the needle.

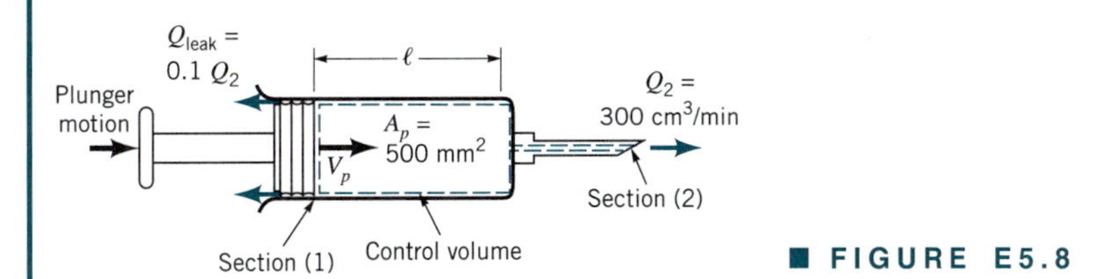

■ FIGURE E5.8

SOLUTION

The control volume selected for solving this problem is the deforming one illustrated in Fig. E5.8. Section (1) of the control surface moves with the plunger. The surface area of section (1), A_1, is considered equal to the circular area of the face of the plunger, A_p although this is not strictly true, since leakage occurs. The difference is small, however. Thus,

$$A_1 = A_p \tag{1}$$

Liquid also leaves the needle through section (2), which involves fixed area A_2. The application of Eq. 5.17 to the contents of this control volume gives

$$\frac{\partial}{\partial t}\int_{\text{cv}} \rho \, d\forall + \dot{m}_2 + \rho Q_{\text{leak}} = 0 \tag{2}$$

Even though Q_{leak} and the flow through section area A_2 are steady, the time rate of change of the mass of liquid in the shrinking control volume is not zero because the control volume is getting smaller. To evaluate the first term of Eq. 2, we note that

$$\int_{cv} \rho \, d\forall = \rho(\ell A_1 + \forall_{\text{needle}}) \quad \textbf{(3)}$$

where ℓ is the changing length of the control volume (see Fig. E5.8) and $\forall_{\text{needle}}$ is the volume of the needle. From Eq. 3, we obtain

$$\frac{\partial}{\partial t} \int_{cv} \rho \, d\forall = \rho A_1 \frac{\partial \ell}{\partial t} \quad \textbf{(4)}$$

Note that

$$-\frac{\partial \ell}{\partial t} = V_p \quad \textbf{(5)}$$

where V_p is the speed of the plunger sought in the problem statement. Combining Eqs. 2, 4, and 5 we obtain

$$-\rho A_1 V_p + \dot{m}_2 + \rho Q_{\text{leak}} = 0 \quad \textbf{(6)}$$

However, from Eq. 5.6, we see that

$$\dot{m}_2 = \rho Q_2 \quad \textbf{(7)}$$

and Eq. 6 becomes

$$-\rho A_1 V_p + \rho Q_2 + \rho Q_{\text{leak}} = 0 \quad \textbf{(8)}$$

Solving Eq. 8 for V_p yields

$$V_p = \frac{Q_2 + Q_{\text{leak}}}{A_1} \quad \textbf{(9)}$$

Since $Q_{\text{leak}} = 0.1Q_2$, Eq. 9 becomes

$$V_p = \frac{Q_2 + 0.1Q_2}{A_1} = \frac{1.1Q_2}{A_1}$$

and

$$V_p = \frac{(1.1)(300 \text{ cm}^3/\text{min})}{(500 \text{ mm}^2)} \left(\frac{1000 \text{ mm}^3}{\text{cm}^3} \right) = 660 \text{ mm/min} \quad \textbf{(Ans)}$$

Solve the problem of Example 5.5 using a deforming control volume that includes only the water accumulating in the bathtub.

SOLUTION

For this deforming control volume, Eq. 5.17 leads to

$$\frac{\partial}{\partial t} \int_{\substack{\text{water} \\ \text{volume}}} \rho \, d\forall + \int_{cs} \rho \mathbf{W} \cdot \hat{\mathbf{n}} \, dA = 0 \quad \textbf{(1)}$$

The first term of Eq. 1 can be evaluated as

$$\frac{\partial}{\partial t}\int_{\substack{\text{water}\\\text{volume}}} \rho \, d\forall = \frac{\partial}{\partial t}[\rho h(2 \text{ ft})(5 \text{ ft})]$$

$$= \rho\,(10 \text{ ft}^2)\frac{\partial h}{\partial t} \tag{2}$$

The second term of Eq. 1 can be evaluated as

$$\int_{\text{cs}} \rho \, \mathbf{W} \cdot \hat{\mathbf{n}} \, dA = -\rho\left(V_j + \frac{\partial h}{\partial t}\right) A_j \tag{3}$$

where A_j and V_j are the cross-sectional area and velocity of the water flowing from the faucet into the tube. Thus, from Eqs. 1, 2, and 3 we obtain

$$\frac{\partial h}{\partial t} = \frac{V_j A_j}{(10 \text{ ft}^2 - A_j)} = \frac{Q_{\text{water}}}{(10 \text{ ft}^2 - A_j)}$$

or for $A_j \ll 10 \text{ ft}^2$

$$\frac{\partial h}{\partial t} = \frac{9(\text{gal/min})(12 \text{ in./ft})}{(7.48 \text{ gal/ft}^3)(10 \text{ ft}^2)} = 1.44 \text{ in./min} \qquad \text{(Ans)}$$

Note that these results using a deforming control volume are the same as that obtained in Example 5.5 with a fixed control volume.

The conservation of mass principle is easily applied to the contents of a control volume. The appropriate selection of a specific kind of control volume (for example, fixed and nondeforming, moving and nondeforming, or deforming) can make the solution of a particular problem less complicated. In general, where fluid flows through the control surface, it is advisable to make the control surface perpendicular to the flow. In the sections ahead we learn that the conservation of mass principle is primarily used in combination with other important laws to solve problems.

5.2 Newton's Second Law—The Linear Momentum and Moment-of-Momentum Equations

V5.3

5.2.1 Derivation of the Linear Momentum Equation

Newton's second law deals with system momentum and forces.

Newton's second law of motion for a system is

time rate of change of the linear momentum of the system = sum of external forces acting on the system

Since momentum is mass times velocity, the momentum of a small particle of mass $\rho d\forall$ is $\mathbf{V}\rho d\forall$. Thus, the momentum of the entire system is $\int_{\text{sys}} \mathbf{V}\rho d\forall$ and Newton's law becomes

$$\frac{D}{Dt}\int_{\text{sys}} \mathbf{V}\rho \, d\forall = \sum \mathbf{F}_{\text{sys}} \tag{5.19}$$

Any reference or coordinate system for which this statement is true is called *inertial.* A fixed coordinate system is inertial. A coordinate system that moves in a straight line with constant velocity and is thus without acceleration is also inertial. We proceed to develop the control volume formula for this important law. When a control volume is coincident with a system at an instant of time, the forces acting on the system and the forces acting on the contents of the coincident control volume (see Fig. 5.2) are instantaneously identical, that is,

$$\sum \mathbf{F}_{\text{sys}} = \sum \mathbf{F}_{\substack{\text{contents of the} \\ \text{coincident control volume}}} \tag{5.20}$$

Furthermore, for a system and the contents of a coincident control volume that is fixed and nondeforming, the Reynolds transport theorem (Eq. 4.19 with b set equal to the velocity, and B_{sys} being the system momentum) allows us to conclude that

$$\frac{D}{Dt}\int_{\text{sys}} \mathbf{V}\rho \, d\forall = \frac{\partial}{\partial t}\int_{\text{cv}} \mathbf{V}\, \rho \, d\forall + \int_{\text{cs}} \mathbf{V}\rho\mathbf{V} \cdot \hat{\mathbf{n}} \, dA \tag{5.21}$$

or

time rate of change of the linear momentum of the system	=	time rate of change of the linear momentum of the contents of the control volume	+	net rate of flow of linear momentum through the control surface

Equation 5.21 states that the time rate of change of system linear momentum is expressed as the sum of the two control volume quantities: the time rate of change of the *linear momentum of the contents of the control volume*, and the net rate of *linear momentum flow through the control surface*. As particles of mass move into or out of a control volume through the control surface, they carry linear momentum in or out. Thus, linear momentum flow should seem no more unusual than mass flow.

For a control volume that is fixed (and thus inertial) and nondeforming, Eqs. 5.19, 5.20, and 5.21 suggest that an appropriate mathematical statement of Newton's second law of motion is

The linear momentum equation is a statement of Newton's second law.

$$\boxed{\frac{\partial}{\partial t}\int_{\text{cv}} \mathbf{V}\rho \, d\forall + \int_{\text{cs}} \mathbf{V}\rho\mathbf{V} \cdot \hat{\mathbf{n}} \, dA = \sum \mathbf{F}_{\substack{\text{contents of the} \\ \text{control volume}}}} \tag{5.22}$$

We call Eq. 5.22 the *linear momentum equation.*

In our application of the linear momentum equation, we initially confine ourselves to fixed, nondeforming control volumes for simplicity. Subsequently, we discuss the use of a moving but inertial, nondeforming control volume. We do not consider deforming control volumes and accelerating (noninertial) control volumes. If a control volume is noninertial,

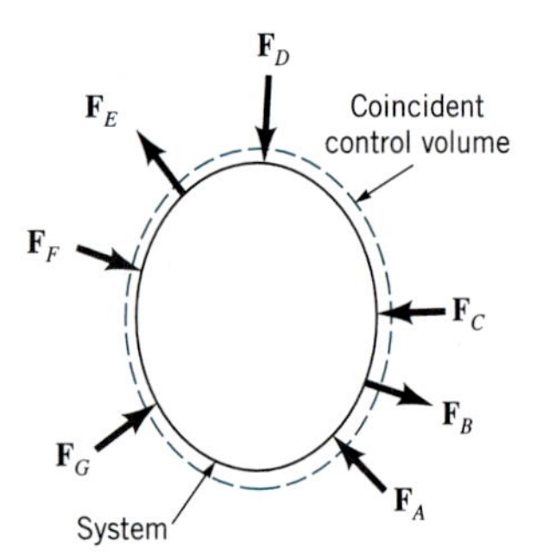

■ **FIGURE 5.2 External forces acting on system and coincident control volume.**

the acceleration components involved (for example, translation acceleration, Coriolis acceleration, and centrifugal acceleration) require consideration.

Both surface and body forces act on the contents of the control volume.

The forces involved in Eq. 5.22 are body and surface forces that act on what is contained in the control volume. The only body force we consider in this chapter is the one associated with the action of gravity. We experience this body force as weight. The surface forces are basically exerted on the contents of the control volume by material just outside the control volume in contact with material just inside the control volume. For example, a wall in contact with fluid can exert a reaction surface force on the fluid it bounds. Similarly, fluid just outside the control volume can push on fluid just inside the control volume at a common interface, usually an opening in the control surface through which fluid flow occurs. An immersed object can resist fluid motion with surface forces.

The linear momentum terms in the momentum equation deserve careful explanation. We clarify their physical significance in the following sections.

5.2.2 Application of the Linear Momentum Equation

The linear momentum equation for an inertial control volume is a vector equation (Eq. 5.22). In engineering applications, components of this vector equation resolved along orthogonal coordinates, for example, x, y, and z (rectangular coordinate system) or r, θ, and x (cylindrical coordinate system), will normally be used. A simple example involving steady, incompressible flow is considered first.

EXAMPLE 5.10

As shown in Fig. E5.10a, a horizontal jet of water exits a nozzle with a uniform speed of $V_1 = 10$ ft/s, strikes a vane, and is turned through an angle θ. Determine the anchoring force needed to hold the vane stationary. Neglect gravity and viscous effects.

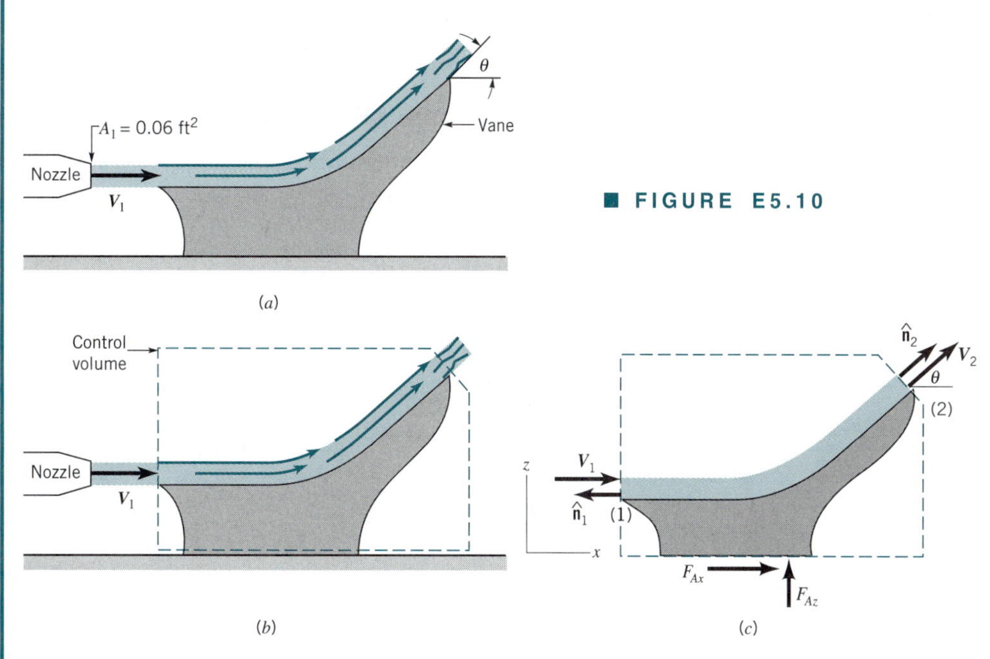

■ FIGURE E5.10

SOLUTION

We select a control volume that includes the vane and a portion of the water (see Figs. E5.10b,c) and apply the linear momentum equation to this fixed control volume. The x and z components of Eq. 5.17 become

$$\frac{\partial}{\partial t}\int_{\text{cv}} u\,\rho\, d\forall + \int_{\text{cs}} u\,\rho\,\mathbf{V}\cdot\hat{\mathbf{n}}\, dA = \sum F_x \tag{1}$$

(the first term → 0 (flow is steady))

and

$$\frac{\partial}{\partial t}\int_{\text{cv}} w\,\rho\, d\forall + \int_{\text{cs}} w\,\rho\,\mathbf{V}\cdot\hat{\mathbf{n}}\, dA = \sum F_z \tag{2}$$

(the first term → 0 (flow is steady))

where $\mathbf{V} = u\,\hat{\mathbf{i}} + w\,\hat{\mathbf{k}}$, and $\Sigma\,\mathbf{F}_x$ and $\Sigma\,F_z$ are the net x and z components of force acting on the contents of the control volume.

The water enters and leaves the control volume as a free jet at atmospheric pressure. Hence, there is atmospheric pressure surrounding the entire control volume, and the net pressure force on the control volume surface is zero. If we neglect the weight of the water and vane, the only forces applied to the control volume contents are the horizontal and vertical components of the anchoring force, F_{Ax} and F_{Az}, respectively.

The only portions of the control surface across which fluid flows are section (1) (the entrance) where $\mathbf{V}\cdot\hat{\mathbf{n}} = -V_1$ and section (2) (the exit) where $\mathbf{V}\cdot\hat{\mathbf{n}} = +V_2$. (Recall that the unit normal vector is directed out from the control surface.) Also, with negligible gravity and viscous effects, and since $p_1 = p_2$, the speed of the fluid remains constant, so that $V_1 = V_2 = 10$ ft/s (see the Bernoulli equation, Eq. 3.6). Hence, at section (1), $u = V_1$, $w = 0$, and at section (2), $u = V_1 \cos\theta$, $w = V_1 \sin\theta$.

By using the above information, Eqs. 1 and 2 can be written as

$$V_1\,\rho(-V_1)A_1 + V_1\cos\theta\,\rho(V_1)A_2 = F_{Ax} \tag{3}$$

and

$$(0)\rho(-V_1)A_1 + V_1\sin\theta\,\rho(V_1)A_2 = F_{Az} \tag{4}$$

Note that since the flow is uniform across the inlet and exit, the integrals simply reduce to multiplications. Equations 3 and 4 can be simplified by using conservation of mass which states that for this incompressible flow $A_1V_1 = A_2V_2$, or $A_1 = A_2$ since $V_1 = V_2$. Thus

$$F_{Ax} = -\rho A_1 V^2 + \rho A_1 V_1^2\cos\theta = -\rho A_1 V_1^2(1-\cos\theta) \tag{5}$$

and

$$F_{Az} = \rho A_1 V_1^2 \sin\theta \tag{6}$$

With the given data we obtain

$$\begin{aligned} F_{Ax} &= -(1.94\ \text{slugs/ft}^3)(0.06\ \text{ft}^2)(10\ \text{ft/s})^2(1-\cos\theta) \\ &= -11.64(1-\cos\theta)\ \text{slugs}\cdot\text{ft/s}^2 = -11.64(1-\cos\theta)\ \text{lb} \end{aligned}$$ **(Ans)**

and

$$\begin{aligned} F_{Az} &= (1.94\ \text{slugs/ft}^3)(0.06\ \text{ft}^2)(10\ \text{ft/s})^2\sin\theta \\ &= 11.64\sin\theta\ \text{lb} \end{aligned}$$ **(Ans)**

Note that if $\theta = 0$ (i.e., the vane does not turn the water), the anchoring force is zero. The inviscid fluid merely slides along the vane without putting any force on it. If $\theta = 90°$, then $F_{Ax} = -11.64$ lb and $F_{Az} = 11.64$ lb. It is necessary to push on the vane (and, hence, for the vane to push on the water) to the left (F_{Ax} is negative) and up in order to change the direction of flow of the water from horizontal to vertical. A momentum change requires a force. If $\theta = 180°$, the water jet is turned back on itself. This requires no vertical force

($F_{Az} = 0$), but the horizontal force ($F_{Ax} = -23.3$ lb) is two times that required if $\theta = 90°$. This force must eliminate the incoming fluid momentum and create the outgoing momentum.

Note that the anchoring force (Eqs. 5, 6) can be written in terms of the mass flowrate, $\dot{m} = \rho A_1 V_1$, as

$$F_{Ax} = -\dot{m}V_1 \,(1 - \cos\theta)$$

and

$$F_{Az} = \dot{m}V_1 \sin\theta$$

In this example the anchoring force is needed to produce the nonzero net momentum flowrate (mass flowrate times the change in x or z component of velocity) across the control surface.

EXAMPLE 5.11

Determine the anchoring force required to hold in place a conical nozzle attached to the end of a laboratory sink faucet (see Fig. E5.11*a*) when the water flowrate is 0.6 liter/s. The nozzle mass is 0.1 kg. The nozzle inlet and exit diameters are 16 mm and 5 mm, respectively. The nozzle axis is vertical and the axial distance between sections (1) and (2) is 30 mm. The pressure at section (1) is 464 kPa.

SOLUTION

The anchoring force sought is the reaction force between the faucet and nozzle threads. To evaluate this force we select a control volume that includes the entire nozzle and the water contained in the nozzle at an instant, as is indicated in Figs. E5.11*a* and E5.11*b*. All of the

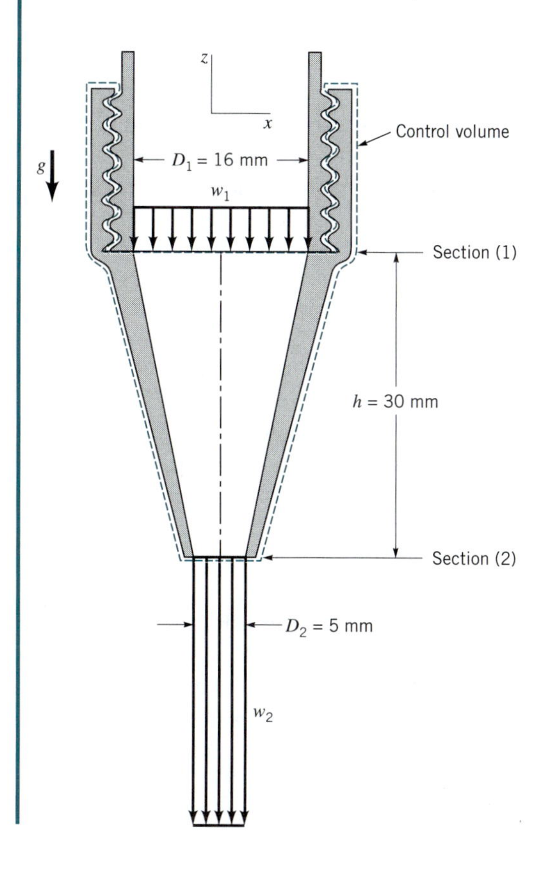

■ FIGURE E5.11*a*

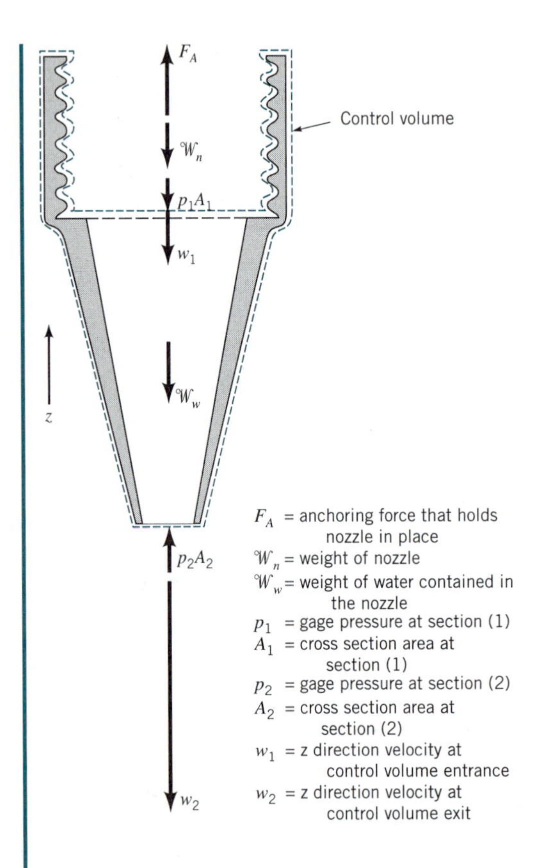

■ **FIGURE E5.11*b***

vertical forces acting on the contents of this control volume are identified in Fig. E5.11*b*. The action of atmospheric pressure cancels out in every direction and is not shown. Gage pressure forces do not cancel out in the vertical direction and are shown. Application of the vertical or z direction component of Eq. 5.22 to the contents of this control volume leads to

$$\overbrace{\frac{\partial}{\partial t}\int_{\text{cv}} w\rho\, d\mathcal{V}}^{0 \text{ (flow is steady)}} + \int_{\text{cs}} w\rho \mathbf{V} \cdot \hat{\mathbf{n}}\, dA = F_A - \mathcal{W}_n - p_1A_1 - \mathcal{W}_w + p_2A_2 \tag{1}$$

where w is the z direction component of fluid velocity, and the various parameters are identified in the figure.

Note that the positive direction is considered ''up'' for the forces. We will use this same sign convention for the fluid velocity, w, in Eq. 1. In Eq. 1, the dot product, $\mathbf{V} \cdot \hat{\mathbf{n}}$, is ''+'' for flow out of the control volume and ''−'' for flow into the control volume. For this particular example

$$\mathbf{V} \cdot \hat{\mathbf{n}}\, dA = \pm |w|\, dA \tag{2}$$

with the ''+'' used for flow out of the control volume and ''−'' used for flow in. To evaluate the control surface integral in Eq. 1, we need to assume a distribution for fluid velocity, w, and fluid density, ρ. For simplicity, we assume that w is uniformly distributed or constant, with magnitudes of w_1 and w_2 over cross-sectional areas A_1 and A_2. Also, this flow is incompressible so the fluid density, ρ, is constant throughout. Proceeding further we obtain for Eq. 1

$$(-\dot{m}_1)(-w_1) + \dot{m}_2(-w_2)$$
$$= F_A - \mathcal{W}_n - p_1A_1 - \mathcal{W}_w + p_2A_2 \qquad \textbf{(3)}$$

where $\dot{m} = \rho AV$ is the mass flowrate.

Note that $-w_1$ and $-w_2$ are used because both of these velocities are "down." Also, $-\dot{m}_1$ is used because it is associated with flow into the control volume. Similarly, $+\dot{m}_2$ is used because it is associated with flow out of the control volume. Solving Eq. 3 for the anchoring force, F_A, we obtain

$$F_A = \dot{m}_1w_1 - \dot{m}_2w_2 + \mathcal{W}_n + p_1A_1 + \mathcal{W}_w - p_2A_2 \qquad \textbf{(4)}$$

From the conservation of mass equation, Eq. 5.12, we obtain

$$\dot{m}_1 = \dot{m}_2 = \dot{m} \qquad \textbf{(5)}$$

which when combined with Eq. 4 gives

$$F_A = \dot{m}(w_1 - w_2) + \mathcal{W}_n + p_1A_1 + \mathcal{W}_w - p_2A_2 \qquad \textbf{(6)}$$

It is instructive to note how the anchoring force is affected by the different actions involved. As expected, the nozzle weight, $\mathcal{W}_n$, the water weight, $\mathcal{W}_w$, and gage pressure force at section (1), p_1A_1, all increase the anchoring force, while the gage pressure force at section (2), p_2A_2, acts to decrease the anchoring force. The change in the vertical momentum flowrate, $\dot{m}(w_1 - w_2)$, will, in this instance, decrease the anchoring force because this change is negative ($w_2 > w_1$).

To complete this example we use quantities given in the problem statement to quantify the terms on the right-hand side of Eq. 6.

From Eq. 5.6,

$$\dot{m} = \rho w_1 A_1 = \rho Q = (999\ \text{kg/m}^3)(0.6\ \text{liter/s})(10^{-3}\ \text{m}^3/\text{liter}) = 0.599\ \text{kg/s} \qquad \textbf{(7)}$$

and

$$w_1 = \frac{Q}{A_1} = \frac{Q}{\pi(D_1^2/4)} = \frac{(0.6\ \text{liter/s})(10^{-3}\ \text{m}^3/\text{liter})}{\pi(16\ \text{mm})^2/4(1000^2\ \text{mm}^2/\text{m}^2)} = 2.98\ \text{m/s} \qquad \textbf{(8)}$$

Also from Eq. 5.6,

$$w_2 = \frac{Q}{A_2} = \frac{Q}{\pi(D_2^2/4)} = \frac{(0.6\ \text{liter/s})(10^{-3}\ \text{m}^3/\text{liter})}{\pi(5\ \text{mm})^2/4(1000^2\ \text{mm}^2/\text{m}^2)} = 30.6\ \text{m/s} \qquad \textbf{(9)}$$

The weight of the nozzle, $\mathcal{W}_n$, can be obtained from the nozzle mass, m_n, with

$$\mathcal{W}_n = m_n g = (0.1\ \text{kg})(9.81\ \text{m/s}^2) = 0.981\ \text{N} \qquad \textbf{(10)}$$

The weight of the water in the control volume, $\mathcal{W}_w$, can be obtained from the water density, ρ, and the volume of water, $\mathcal{V}_w$, in the truncated cone of height h. That is,

$$\mathcal{W}_w = \rho \mathcal{V}_w g = \rho \tfrac{1}{12}\pi h(D_1^2 + D_2^2 + D_1D_2)g$$

Thus,

$$\mathcal{W}_w = (999\ \text{kg/m}^3)\frac{1}{12}\pi\frac{(30\ \text{mm})}{(1000\ \text{mm/m})}$$
$$\times\left[\frac{(16\ \text{mm})^2 + (5\ \text{mm})^2 + (16\ \text{mm})(5\ \text{mm})}{(1000^2\ \text{mm}^2/\text{m}^2)}\right](9.81\ \text{m/s}^2) = 0.0278\ \text{N} \qquad \textbf{(11)}$$

The gage pressure at section (2), p_2, is zero since, as discussed in Section 3.6.1, when a subsonic flow discharges to the atmosphere as in the present situation, the discharge pressure

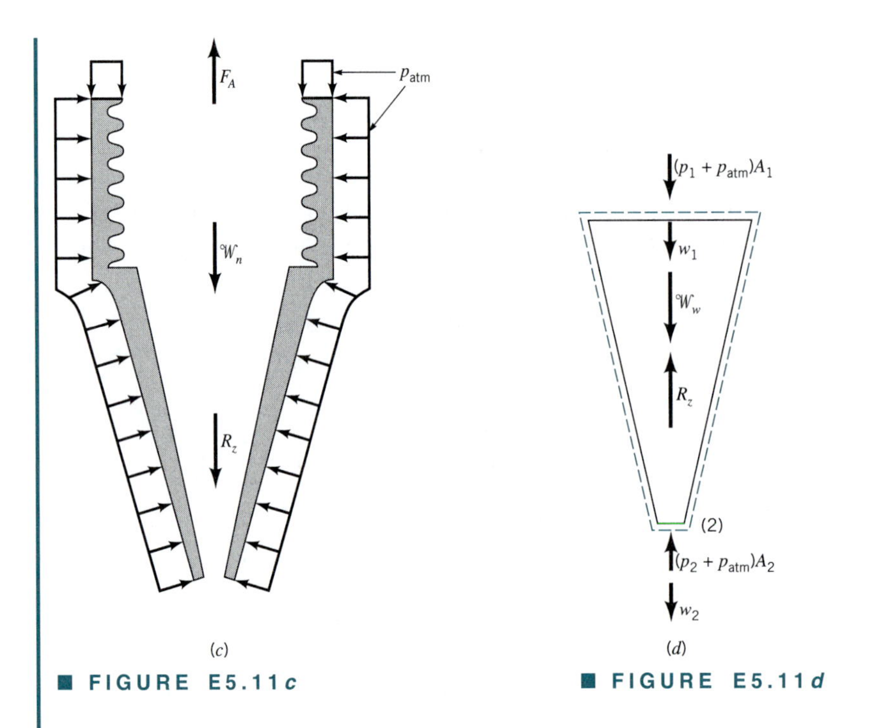

■ **FIGURE E5.11c** ■ **FIGURE E5.11d**

is essentially atmospheric. The anchoring force, F_A, can now be determined from Eqs. 6 through 11 with

$$F_A = (0.599 \text{ kg/s})(2.98 \text{ m/s} - 30.6 \text{ m/s}) + 0.981 \text{ N} + (464 \text{ kPa})(1000 \text{ Pa/kPa}) \frac{\pi(16 \text{ mm})^2}{4(1000^2 \text{ mm}^2/\text{m}^2)} + 0.0278 \text{ N} - 0$$

or

$$F_A = -16.5 \text{ N} + 0.981 \text{ N} + 93.3 \text{ N} + 0.0278 \text{ N} = 77.8 \text{ N} \qquad \textbf{(Ans)}$$

Since the anchoring force, F_A, is positive, it acts upward in the z direction. The nozzle would be pushed off the pipe if it were not fastened securely.

The control volume selected above to solve problems such as these is not unique. The following is an alternate solution that involves two other control volumes—one containing only the nozzle and the other containing only the water in the nozzle. These control volumes are shown in Figs. E5.11*c* and E5.11*d* along with the vertical forces acting on the contents of each control volume. The new force involved, R_z, represents the interaction between the water and the conical inside surface of the nozzle. It includes the net pressure and viscous forces at this interface.

Application of Eq. 5.22 to the contents of the control volume of Fig. E5.11*c* leads to

$$F_A = \mathcal{W}_n + R_z - p_{atm}(A_1 - A_2) \qquad \textbf{(12)}$$

The term $p_{atm}(A_1 - A_2)$ is the resultant force from the atmospheric pressure acting upon the exterior surface of the nozzle (i.e., that portion of the surface of the nozzle that is not in contact with the water). Recall that the pressure force on a curved surface (such as the exterior surface of the nozzle) is equal to the pressure times the projection of the surface area on a plane perpendicular to the axis of the nozzle. The projection of this area on a plane perpendicular to the z direction is $A_1 - A_2$. The effect of the atmospheric pressure on the internal area (between the nozzle and the water) is already included in R_z which represents the net force on this area.

Similarly, for the control volume of Fig. E5.11*d* we obtain

$$R_z = \dot{m}(w_1 - w_2) + \mathcal{W}_w + (p_1 + p_{atm})A_1 - (p_2 + p_{atm})A_2 \quad \textbf{(13)}$$

where p_1 and p_2 are gage pressures. From Eq. 13 it is clear that the value of R_z depends on the value of the atmospheric pressure, p_{atm}, since $A_1 \neq A_2$. That is, we must use absolute pressure, not gage pressure, to obtain the correct value of R_z.

By combining Eqs. 12 and 13 we obtain the same result for F_A as before (Eq. 6)

$$F_A = \dot{m}(w_1 - w_2) + \mathcal{W}_n + p_1A_1 + W_w - p_2A_2$$

Note that although the force between the fluid and the nozzle wall, R_z, is a function of p_{atm}, the anchoring force, F_A, is not. That is, we were correct in using gage pressure when solving for F_A by means of the original control volume shown in Fig. E5.11*b*.

Several important generalities about the application of the linear momentum equation (Eq. 5.22) are apparent in the example just considered.

Linear momentum is a vector quantity.

1. When the flow is uniformly distributed over a section of the control surface where flow into or out of the control volume occurs, the integral operations are simplified. Thus, one-dimensional flows are easier to work with than flows involving nonuniform velocity distributions.
2. Linear momentum is directional; it can have components in as many as three orthogonal coordinate directions. Furthermore, along any one coordinate, the linear momentum of a fluid particle can be in the positive or negative direction and thus be considered as a positive or a negative quantity. In Example 5.11, only the linear momentum in the z direction was considered (all of it was in the negative z direction and was hence treated as being negative).
3. The flow of positive or negative linear momentum *into* a control volume involves a negative $\mathbf{V} \cdot \hat{\mathbf{n}}$ product. Momentum flow *out* of the control volume involves a positive $\mathbf{V} \cdot \hat{\mathbf{n}}$ product. The correct algebraic sign ($+$ or $-$) to assign to momentum flow ($\mathbf{V}\rho\mathbf{V} \cdot \hat{\mathbf{n}}\, dA$) will depend on the sense of the velocity ($+$ in positive coordinate direction, $-$ in negative coordinate direction) and the $\mathbf{V} \cdot \hat{\mathbf{n}}$ product ($+$ for flow out of the control volume, $-$ for flow into the control volume). In Example 5.11, the momentum flow into the control volume past section (1) was a positive ($+$) quantity while the momentum flow out of the control volume at section (2) was a negative ($-$) quantity.
4. The time rate of change of the linear momentum of the contents of a nondeforming control volume (i.e., $\partial/\partial t \int_{cv} \mathbf{V}\rho\, d\mathcal{V}$) is zero for steady flow. The momentum problems considered in this text all involve steady flow.
5. If the control surface is selected so that it is perpendicular to the flow where fluid enters or leaves the control volume, the surface force exerted at these locations by fluid outside the control volume on fluid inside will be due to pressure. Furthermore, when subsonic flow exits from a control volume into the atmosphere, atmospheric pressure prevails at the exit cross section. In Example 5.11, the flow was subsonic and so we set the exit flow pressure at the atmospheric level. The continuity equation (Eq. 5.12) allowed us to evaluate the fluid flow velocities w_1 and w_2 at sections (1) and (2).
6. The forces due to atmospheric pressure acting on the control surface may need consideration as indicated by Eq. 13 for the reaction force between the nozzle and the fluid.

When calculating the anchoring force, F_A, the forces due to atmospheric pressure on the control surface cancel each other (for example, after combining Eqs. 12 and 13 the atmospheric pressure forces are no longer involved) and gage pressures may be used.

7. The external forces have an algebraic sign, positive if the force is in the assigned positive coordinate direction and negative otherwise.

A control volume diagram is similar to a free body diagram.

8. Only external forces acting on the contents of the control volume are considered in the linear momentum equation (Eq. 5.22). If the fluid alone is included in a control volume, reaction forces between the fluid and the surface or surfaces in contact with the fluid [wetted surface(s)] will need to be in Eq. 5.22. If the fluid and the wetted surface or surfaces are within the control volume, the reaction forces between fluid and wetted surface(s) do not appear in the linear momentum equation (Eq. 5.22) because they are internal, not external forces. The anchoring force that holds the wetted surface(s) in place is an external force, however, and must therefore be in Eq. 5.22.
9. The force required to anchor an object will generally exist in response to surface pressure and/or shear forces acting on the control surface, to a change in linear momentum flow through the control volume containing the object, and to the weight of the object and the fluid contained in the control volume. In Example 5.11 the nozzle anchoring force was required mainly because of pressure forces and partly because of a change in linear momentum flow associated with accelerating the fluid in the nozzle. The weight of the water and the nozzle contained in the control volume influenced the size of the anchoring force only slightly.

To further demonstrate the use of the linear momentum equation (Eq. 5.22), we consider another one-dimensional flow example before moving on to other facets of this important equation.

EXAMPLE 5.12

Water flows through a horizontal, 180° pipe bend as illustrated in Fig. E5.12*a*. The flow cross-sectional area is constant at a value of 0.1 ft^2 through the bend. The flow velocity everywhere in the bend is axial and 50 ft/s. The absolute pressures at the entrance and exit of the bend are 30 psia and 24 psia, respectively. Calculate the horizontal (x and y) components of the anchoring force required to hold the bend in place.

SOLUTION

Since we want to evaluate components of the anchoring force to hold the pipe bend in place, an appropriate control volume (see dashed line in Fig. E5.12*a*) contains the bend and the water in the bend at an instant. The horizontal forces acting on the contents of this control volume are identified in Fig. E5.12*b*. Note that the weight of the water is vertical (in the negative z direction) and does not contribute to the x and y components of the anchoring force. All of the horizontal normal and tangential forces exerted on the fluid and the pipe bend are resolved and combined into the two resultant components, F_{Ax} and F_{Ay}. These two forces act on the control volume contents, and thus for the x direction, Eq. 5.22 leads to

$$\int_{\text{cs}} u\rho \mathbf{V} \cdot \hat{\mathbf{n}}\, dA = F_{Ax} \qquad \textbf{(1)}$$

At sections (1) and (2), the flow is in the y direction and therefore $u = 0$ at both cross sections. There is no x direction momentum flow into or out of the control volume and we conclude from Eq. 1 that

$$F_{Ax} = 0. \qquad \textbf{(Ans)}$$

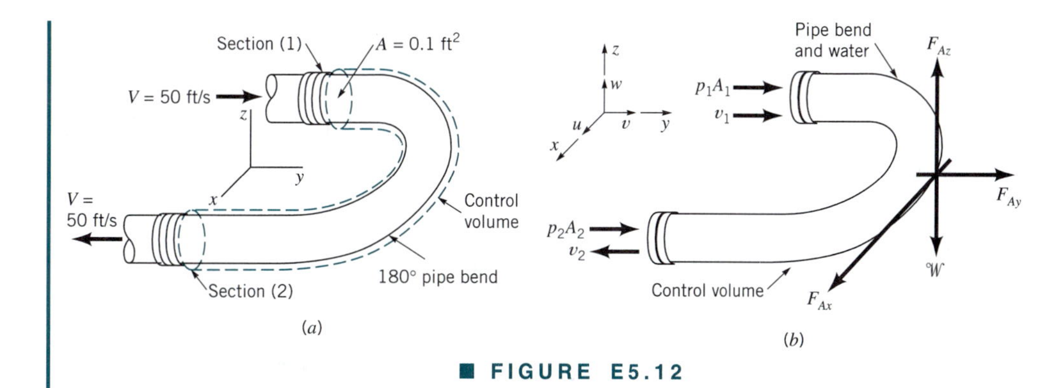

■ FIGURE E5.12

For the y direction, we get from Eq. 5.22

$$\int_{\mathrm{cs}} v\rho \mathbf{V} \cdot \hat{\mathbf{n}}\, dA = F_{Ay} + p_1 A_1 + p_2 A_2 \tag{2}$$

For one-dimensional flow, the surface integral in Eq. 2 is easy to evaluate and Eq. 2 becomes

$$(+v_1)(-\dot{m}_1) + (-v_2)(+\dot{m}_2) = F_{Ay} + p_1 A_1 + p_2 A_2 \tag{3}$$

Note that the y component of velocity is positive at section (1) but is negative at section (2). Also, the mass flowrate term is negative at section (1) (flow in) and is positive at section (2) (flow out). From the continuity equation (Eq. 5.12), we get

$$\dot{m} = \dot{m}_1 = \dot{m}_2 \tag{4}$$

and thus Eq. 3 can be written as

$$-\dot{m}(v_1 + v_2) = F_{Ay} + p_1 A_1 + p_2 A_2 \tag{5}$$

Solving Eq. 5 for F_{Ay} we obtain

$$F_{Ay} = -\dot{m}(v_1 + v_2) - p_1 A_1 - p_2 A_2 \tag{6}$$

From the given data we can calculate $\dot{m}$ from Eq. 5.6 as

$$\dot{m} = \rho_1 A_1 v_1 = (1.94 \text{ slugs/ft}^3)(0.1 \text{ ft}^2)(50 \text{ ft/s}) = 9.70 \text{ slugs/s}$$

For determining the anchoring force, F_{Ay}, the effects of atmospheric pressure cancel and thus gage pressures for p_1 and p_2 are appropriate. By substituting numerical values of variables into Eq. 6, we get

$$\begin{aligned} F_{Ay} &= -(9.70 \text{ slugs/s})(50 \text{ ft/s} + 50 \text{ ft/s})\{1 \text{ lb/[slug·(ft/s}^2)]\} \\ &\quad - (30 \text{ psia} - 14.7 \text{ psia})(144 \text{ in.}^2/\text{ft}^2)(0.1 \text{ ft}^2) \\ &\quad - (24 \text{ psia} - 14.7 \text{ psia})(144 \text{ in.}^2/\text{ft}^2)(0.1 \text{ ft}^2) \end{aligned}$$

$$F_{Ay} = -970 \text{ lb} - 220 \text{ lb} - 134 \text{ lb} = -1324 \text{ lb} \qquad \textbf{(Ans)}$$

The negative sign for F_{Ay} is interpreted as meaning that the y component of the anchoring force is actually in the negative y direction, not the positive y direction as originally indicated in Fig. E5.12*b*.

As with Example 5.11, the anchoring force for the pipe bend is independent of the atmospheric pressure. However, the force that the bend puts on the fluid inside of it, R_y, depends on the atmospheric pressure. We can see this by using a control volume which

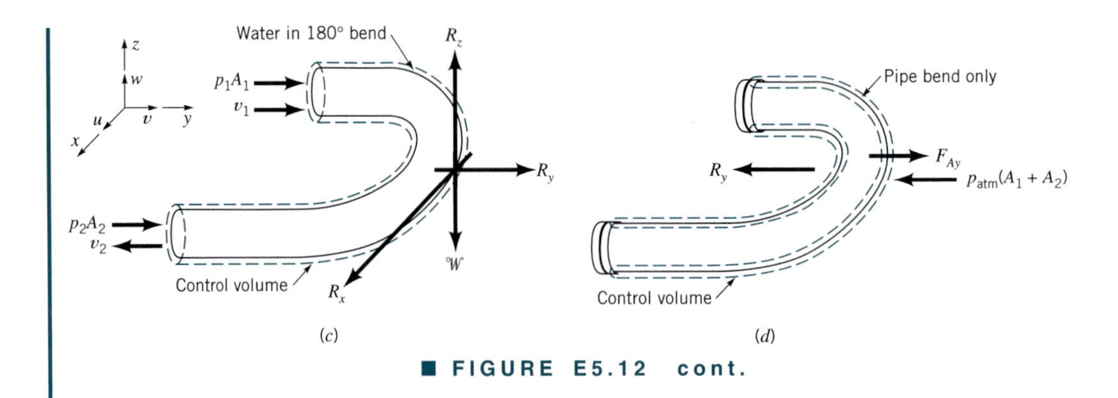

■ FIGURE E5.12 cont.

surrounds only the fluid within the bend as shown in Fig. E5.12*c*. Application of the momentum equation to this situation gives

$$R_y = -\dot{m}(v_1 + v_2) - p_1A_1 - p_2A_2$$

where p_1 and p_2 must be in terms of absolute pressure because the force between the fluid and the pipe wall, R_y, is the complete pressure effect (i.e., absolute pressure).

Thus, we obtain

$$\begin{aligned} R_y &= -(9.70 \text{ slugs/s})(50 \text{ ft/s} + 50 \text{ ft/s}) - (30 \text{ psia})(144 \text{ in.}^2/\text{ft}^2)(0.1 \text{ ft}^2) \\ &\quad - (24 \text{ psia})(144 \text{ in.}^2/\text{ft}^2)(0.1 \text{ ft}^2) \\ &= -1748 \text{ lb} \end{aligned} \quad \textbf{(7)}$$

We can use the control volume that includes just the pipe bend (without the fluid inside it) as shown in Fig. E5.12*d* to determine F_{Ay}, the anchoring force component in the *y* direction necessary to hold the bend stationary. The *y* component of the momentum equation applied to this control volume gives

$$F_{Ay} = R_y + p_{\text{atm}} (A_1 + A_2) \quad \textbf{(8)}$$

where R_y is given by Eq. 7. The $p_{\text{atm}} (A_1 + A_2)$ term represents the net pressure force on the outside portion of the control volume. Recall that the pressure force on the inside of the bend is accounted for by R_y. By combining Eqs. 7 and 8 we obtain

$$F_{Ay} = -1748 \text{ lb} + 14.7 \text{ lb/in.}^2 (0.1 \text{ ft}^2 + 0.1 \text{ ft}^2)(144 \text{ in.}^2/\text{ft}^2) = -1324 \text{ lb}$$

in agreement with the original answer obtained using the control volume of Fig. E5.12*b*.

In Example 5.12, the direction of flow entering the control volume was different from the direction of flow leaving the control volume by 180°. This change in flow direction only (the flow speed remained constant) resulted in a large portion of the reaction force exerted by the pipe on the water. This is in contrast to the small contribution of fluid acceleration to the anchoring force of Example 5.11. From Examples 5.11 and 5.12 we see that changes in flow speed and/or direction result in a reaction force. Other types of problems that can be solved with the linear momentum equation (Eq. 5.22) are illustrated in the following examples.

EXAMPLE 5.13

Air flows steadily between two cross sections in a long, straight portion of 4-in. inside diameter pipe as indicated in Fig. E5.13, where the uniformly distributed temperature and pressure at each cross section are given. If the average air velocity at section (2) is 1000 ft/s, we found in Example 5.2 that the average air velocity at section (1) must be 219 ft/s. Assuming uniform velocity distributions at sections (1) and (2), determine the frictional force exerted by the pipe wall on the air flow between sections (1) and (2).

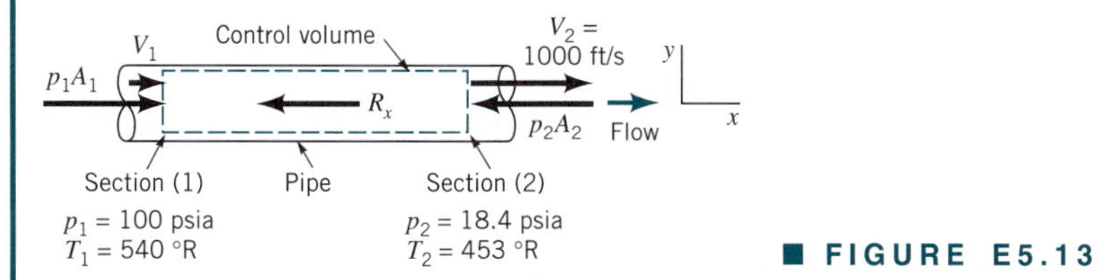

■ FIGURE E5.13

SOLUTION

The control volume of Example 5.2 is appropriate for this problem. The forces acting on the air between sections (1) and (2) are identified in Fig. E5.13. The weight of air is considered negligibly small. The reaction force between the wetted wall of the pipe and the flowing air, R_x, is the frictional force sought. Application of the axial component of Eq. 5.22 to this control volume yields

$$\int_{\text{cs}} u\rho \mathbf{V} \cdot \hat{\mathbf{n}}\, dA = -R_x + p_1A_1 - p_2A_2 \tag{1}$$

The positive x direction is set as being to the right. Furthermore, for uniform velocity distributions (one-dimensional flow), Eq. 1 becomes

$$(+u_1)(-\dot{m}_1) + (+u_2)(+\dot{m}_2) = -R_x + p_1A_1 - p_2A_2 \tag{2}$$

From conservation of mass (Eq. 5.12) we get

$$\dot{m} = \dot{m}_1 = \dot{m}_2 \tag{3}$$

so that Eq. 2 becomes

$$\dot{m}(u_2 - u_1) = -R_x + A_2(p_1 - p_2) \tag{4}$$

Solving Eq. 4 for R_x, we get

$$R_x = A_2(p_1 - p_2) - \dot{m}(u_2 - u_1) \tag{5}$$

The equation of state gives

$$\rho_2 = \frac{p_2}{RT_2} \tag{6}$$

and the equation for area A_2 is

$$A_2 = \frac{\pi D_2^2}{4} \tag{7}$$

Thus, from Eqs. 3, 6, and 7

$$\dot{m} = \left(\frac{p_2}{RT_2}\right)\left(\frac{\pi D_2^2}{4}\right)u_2 = \frac{(18.4 \text{ psia})(144 \text{ in.}^2/\text{ft}^2)}{(1716 \text{ ft}\cdot\text{lb/slug}\cdot{}^\circ\text{R})(453\ {}^\circ\text{R})}$$

$$\times \frac{\pi(4 \text{ in.})^2}{4(144 \text{ in.}^2/\text{ft}^2)}(1000 \text{ ft/s}) = 0.297 \text{ slugs/s} \tag{8}$$

Thus, from Eqs. 5 and 8

$$R_x = \frac{\pi(4 \text{ in.})^2}{4}(100 \text{ psia} - 18.4 \text{ psia})$$
$$- (0.297 \text{ slugs/s})(1000 \text{ ft/s} - 219 \text{ ft/s})[1 \text{ lb/(slug}\cdot\text{ft/s}^2)]$$

or

$$R_x = 1025 \text{ lb} - 232 \text{ lb} = 793 \text{ lb} \qquad \textbf{(Ans)}$$

Note that both the pressure and momentum contribute to the friction force, R_x. If the fluid flow were incompressible, then $u_1 = u_2$ and there would be no momentum contribution to R_x.

EXAMPLE 5.14

If the flow of Example 5.4 is vertically upward, develop an expression for the fluid pressure drop that occurs between section (1) and section (2).

SOLUTION

A control volume (see dashed lines in Fig. E5.4) that includes only fluid from section (1) to section (2) is selected. The forces acting on the fluid in this control volume are identified in Fig. E5.14. The application of the axial component of Eq. 5.22 to the fluid in this control volume results in

$$\int_{\text{cs}} w\rho \mathbf{V} \cdot \hat{\mathbf{n}}\, dA = p_1A_1 - R_z - \mathcal{W} - p_2A_2 \qquad \textbf{(1)}$$

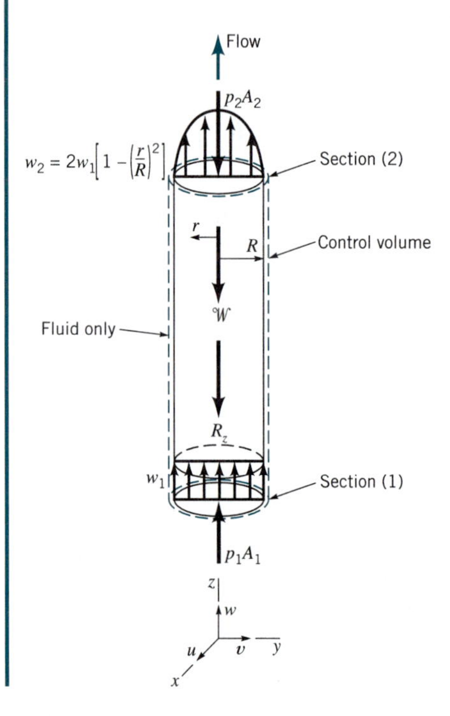

■ FIGURE E5.14

where R_z is the resultant force of the wetted pipe wall on the fluid. Further, for uniform flow at section (1), and because the flow at section (2) is out of the control volume, Eq. 1 becomes

$$(+w_1)(-\dot{m}_1) + \int_{A_2} (+w_2)\rho(+w_2\, dA_2) = p_1A_1 - R_z - \mathcal{W} - p_2A_2 \qquad \textbf{(2)}$$

The positive direction is considered up. The surface integral over the cross-sectional area at section (2), A_2, is evaluated by using the parabolic velocity profile obtained in Example 5.4, $w_2 = 2w_1[1 - (r/R)^2]$, as

$$\int_{A_2} w_2\rho w_2\, dA_2 = \rho \int_0^R w_2^2\, 2\, \pi\, r\, dr = 2\pi\rho \int_0^R (2w_1)^2 \left[1 - \left(\frac{r}{R}\right)^2\right]^2 r\, dr$$

or

$$\int_{A_2} w_2\rho w_2\, dA_2 = 4\, \pi\rho w_1^2 \frac{R^2}{3} \qquad \textbf{(3)}$$

Combining Eqs. 2 and 3 we obtain

$$-w_1^2\rho\pi R^2 + \tfrac{4}{3}w_1^2\rho\pi R^2 = p_1A_1 - R_z - \mathcal{W} - p_2A_2 \qquad \textbf{(4)}$$

Solving Eq. 4 for the pressure drop from section (1) to section (2), $p_1 - p_2$, we obtain

$$p_1 - p_2 = \frac{\rho w_1^2}{3} + \frac{R_z}{A_1} + \frac{\mathcal{W}}{A_1} \qquad \textbf{(Ans)}$$

We see that the drop in pressure from section (1) to section (2) occurs because of the following:

1. The change in momentum flow between the two sections associated with going from a uniform velocity profile to a parabolic velocity profile.
2. Pipe wall friction.
3. The weight of the water column; a hydrostatic pressure effect.

If the velocity profiles had been identically parabolic at sections (1) and (2), the momentum flowrate at each section would have been identical, a condition we call "fully developed" flow. Then, the pressure drop, $p_1 - p_2$, would be due only to pipe wall friction and the weight of the water column. If in addition to being fully developed, the flow involved negligible weight effects (for example, horizontal flow of liquids or the flow of gases in any direction) the drop in pressure between any two sections, $p_1 - p_2$, would be a result of pipe wall friction only.

Note that although the average velocity is the same at section (1) as it is at section (2) ($\overline{V}_1 = \overline{V}_2 = w_1$), the momentum flux across section (1) is not the same as it is across section (2). If it were, the left-hand side of Eq. (4) would be zero. For this nonuniform flow the momentum flux can be written in terms of the average velocity, $\overline{V}$, and the *momentum coefficient*, β, as

$$\beta = \frac{\int w\rho \mathbf{V} \cdot \hat{\mathbf{n}}\, dA}{\rho \overline{V}^2 A}$$

Hence the momentum flux can be written as

$$\int_{\text{cs}} w\rho \mathbf{V} \cdot \hat{\mathbf{n}}\, dA = -\beta_1 w_1^2 \rho\pi R^2 + \beta_2 w_1^2 \rho\pi R^2$$

where $\beta_1 = 1$ ($\beta = 1$ for uniform flow) and $\beta_2 = 4/3$ ($\beta > 1$ for nonuniform flow).

EXAMPLE 5.15

A static thrust stand as sketched in Fig. E5.15 is to be designed for testing a jet engine. The following conditions are known for a typical test:

Intake air velocity = 200 m/s; exhaust gas velocity = 500 m/s; intake cross-sectional area = 1 m^2; intake static pressure = −22.5 kPa = 78.5 kPa (abs); intake static temperature = 268 K; exhaust static pressure = 0 kPa = 101 kPa (abs). Estimate the nominal thrust for which to design.

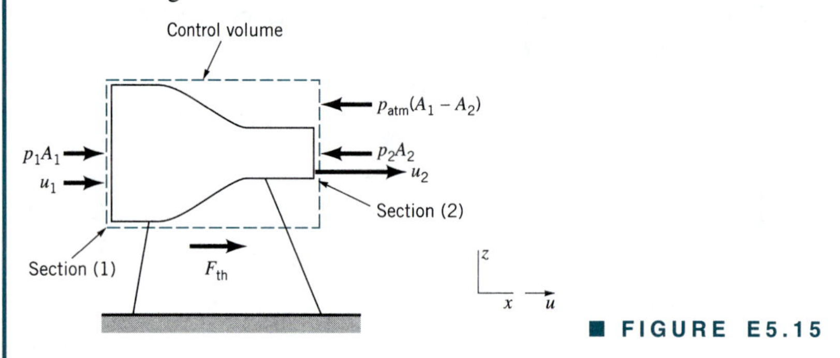

■ FIGURE E5.15

SOLUTION

The cylindrical control volume outlined with a dashed line in Fig. E5.15 is selected. The external forces acting in the axial direction are also shown. Application of the momentum equation (Eq. 5.22) to the contents of this control volume yields

$$\int_{cs} u\rho \mathbf{V} \cdot \hat{\mathbf{n}}\, dA = p_1A_1 + F_{th} - p_2A_2 - p_{atm}(A_1 - A_2) \quad (1)$$

where the pressures are absolute. Thus, for one-dimensional flow, Eq. 1 becomes

$$(+u_1)(-\dot{m}_1) + (+u_2)(+\dot{m}_2) = (p_1 - p_{atm})A_1 - (p_2 - p_{atm})A_2 + F_{th} \quad (2)$$

The positive direction is to the right. The conservation of mass equation (Eq. 5.12) leads to

$$\dot{m} = \dot{m}_1 = \rho_1A_1u_1 = \dot{m}_2 = \rho_2A_2u_2 \quad (3)$$

Combining Eqs. 2 and 3 and using gage pressure we obtain

$$\dot{m}(u_2 - u_1) = p_1A_1 - p_2A_2 + F_{th} \quad (4)$$

Solving Eq. 4 for the thrust force, F_{th}, we obtain

$$F_{th} = -p_1A_1 + p_2A_2 + \dot{m}(u_2 - u_1) \quad (5)$$

We need to determine the mass flowrate, $\dot{m}$, to calculate F_{th}, and to calculate $\dot{m} = \rho_1A_1u_1$, we need ρ_1. From the ideal gas equation of state

$$\rho_1 = \frac{p_1}{RT_1} = \frac{(78.5\text{ kPa})(1000\text{ Pa/kPa})[1(\text{N/m}^2)/\text{Pa}]}{(286.9\text{ J/kg}\cdot\text{K})(268\text{ K})(1\text{ N}\cdot\text{m/J})} = 1.02\text{ kg/m}^3$$

Thus,

$$\dot{m} = \rho_1A_1u_1 = (1.02\text{ kg/m}^3)(1\text{ m}^2)(200\text{ m/s}) = 204\text{ kg/s} \quad (6)$$

Finally, combining Eqs. 5 and 6 and substituting given data with $p_2 = 0$, we obtain

$$F_{th} = -(1\text{ m}^2)(-22.5\text{ kPa})(1000\text{ Pa/kPa})[1(\text{N/m}^2)/\text{Pa}]$$
$$+ (204\text{ kg/s})(500\text{ m/s} - 200\text{ m/s})[1\text{ N/(kg}\cdot\text{m/s}^2)]$$

and

$$F_{\text{th}} = 22{,}500 \text{ N} + 61{,}200 \text{ N} = 83{,}700 \text{ N} \qquad \textbf{(Ans)}$$

The force of the thrust stand on the engine is directed toward the right. Conversely, the engine pushes to the left on the thrust stand (or aircraft).

EXAMPLE 5.16

A sluice gate across a channel of width b is shown in the closed and open positions in Figs. E5.16*a* and E5.16*b*. Is the anchoring force required to hold the gate in place larger when the gate is closed or when it is open?

SOLUTION

We will answer this question by comparing expressions for the horizontal reaction force, R_x, between the gate and the water when the gate is closed and when the gate is open. The control volume used in each case is indicated with dashed lines in Figs. E5.16*a* and E5.16*b*.

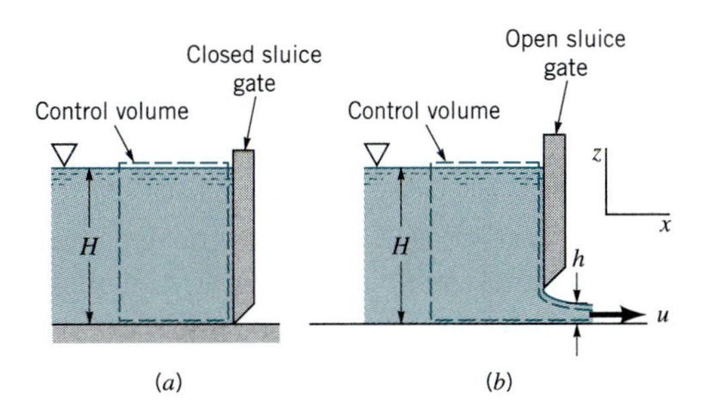

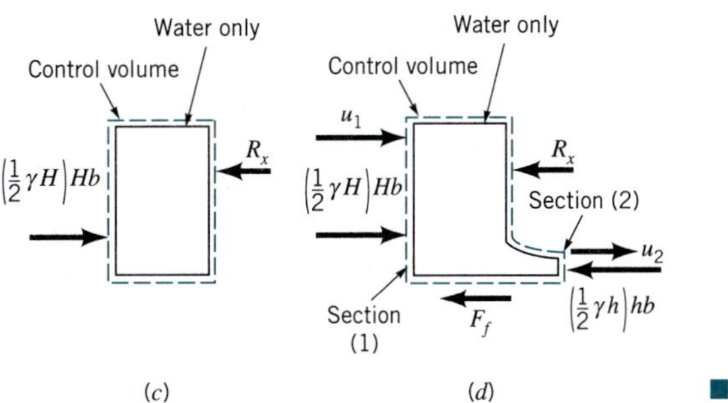

■ FIGURE E5.16

When the gate is closed, the horizontal forces acting on the contents of the control volume are identified in Fig. E5.16*c*. Application of Eq. 5.22 to the contents of this control volume yields

$$\int_{\text{cs}} u\rho \overset{0 \text{ (no flow)}}{\cancel{\mathbf{V}}} \cdot \hat{\mathbf{n}}\, dA = \tfrac{1}{2}\gamma H^2 b - R_x \qquad \textbf{(1)}$$

Note that the hydrostatic pressure force, $\gamma H^2 b/2$, is used. From Eq. 1, the force exerted on the water by the gate (which is equal to the force necessary to hold the gate stationary) is

$$R_x = \tfrac{1}{2}\gamma H^2 b \qquad \textbf{(2)}$$

which is equal in magnitude to the hydrostatic force exerted on the gate by the water.

When the gate is open, the horizontal forces acting on the contents of the control volume are shown in Fig. E5.16*d*. Application of Eq. 5.22 to the contents of this control volume leads to

$$\int_{\text{cs}} u\rho \mathbf{V} \cdot \hat{\mathbf{n}}\, dA = \tfrac{1}{2}\gamma H^2 b - R_x - \tfrac{1}{2}\gamma h^2 b - F_f \tag{3}$$

Note that we have assumed that the pressure distribution is hydrostatic in the water at sections (1) and (2) (see Section 3.4). Also, the frictional force between the channel bottom and the water is specified as F_f. The surface integral in Eq. 3 is nonzero only where there is flow across the control surface. With the assumption of uniform velocity distributions

$$\int_{\text{cs}} u\rho \mathbf{V} \cdot \hat{\mathbf{n}}\, dA = (u_1)\rho(-u_1)Hb + (+u_2)\rho(+u_2)hb \tag{4}$$

Thus, Eqs. 3 and 4 combine to form

$$-\rho u_1^2 Hb + \rho u_2^2 hb = \tfrac{1}{2}\gamma H^2 b - R_x - \tfrac{1}{2}\gamma h^2 b - F_f \tag{5}$$

If $H \gg h$, the upstream velocity, u_1, is much less than u_2 so that the contribution of the incoming momentum flow to the control surface integral can be neglected and from Eq. 5 we obtain

$$R_x = \tfrac{1}{2}\gamma H^2 b - \tfrac{1}{2}\gamma h^2 b - F_f - \rho u_2^2 hb \tag{6}$$

Comparing the expressions for R_x (Eqs. 2 and 6) we conclude that the reaction force between the gate and the water (and therefore the anchoring force required to hold the gate in place) is smaller when the gate is open than when it is closed. **(Ans)**

The linear momentum equation can be written for a moving control volume.

All of the linear momentum examples considered thus far have involved stationary and nondeforming control volumes which are thus inertial because there is no acceleration. A nondeforming control volume translating in a straight line at constant speed is also inertial because there is no acceleration. For a system and an inertial, moving, nondeforming control volume that are both coincident at an instant of time, the Reynolds transport theorem (Eq. 4.23) leads to

$$\frac{D}{Dt}\int_{\text{sys}} \mathbf{V}\rho\, d\forall = \frac{\partial}{\partial t}\int_{\text{cv}} \mathbf{V}\rho\, d\forall + \int_{\text{cs}} \mathbf{V}\rho \mathbf{W} \cdot \hat{\mathbf{n}}\, dA \tag{5.23}$$

When we combine Eq. 5.23 with Eqs. 5.19 and 5.20, we get

$$\frac{\partial}{\partial t}\int_{\text{cv}} \mathbf{V}\rho\, d\forall + \int_{\text{cs}} \mathbf{V}\rho \mathbf{W} \cdot \hat{\mathbf{n}}\, dA = \sum \mathbf{F}_{\substack{\text{contents of the}\\ \text{control volume}}} \tag{5.24}$$

When the equation relating absolute, relative, and control volume velocities (Eq. 5.14) is used with Eq. 5.24, the result is

$$\frac{\partial}{\partial t}\int_{\text{cv}} (\mathbf{W} + \mathbf{V}_{\text{cv}})\rho\, d\forall + \int_{\text{cs}} (\mathbf{W} + \mathbf{V}_{\text{cv}})\rho \mathbf{W} \cdot \hat{\mathbf{n}}\, dA = \sum \mathbf{F}_{\substack{\text{contents of the}\\ \text{control volume}}} \tag{5.25}$$

For a constant control volume velocity, $\mathbf{V}_{\text{cv}}$, and steady flow in the control volume reference frame,

$$\frac{\partial}{\partial t}\int_{\text{cv}} (\mathbf{W} + \mathbf{V}_{\text{cv}})\rho \, d\forall = 0 \qquad (5.26)$$

Also, for this inertial, nondeforming control volume

$$\int_{\text{cs}} (\mathbf{W} + \mathbf{V}_{\text{cv}})\rho\mathbf{W} \cdot \hat{\mathbf{n}} \, dA = \int_{\text{cs}} \mathbf{W}\rho\mathbf{W} \cdot \hat{\mathbf{n}} \, dA + \mathbf{V}_{\text{cv}} \int_{\text{cs}} \rho\mathbf{W} \cdot \hat{\mathbf{n}} \, dA \qquad (5.27)$$

For steady flow (on an instantaneous or time-average basis), Eq. 5.15 gives

$$\int_{\text{cs}} \rho\mathbf{W} \cdot \hat{\mathbf{n}} \, dA = 0 \qquad (5.28)$$

Combining Eqs. 5.25, 5.26, 5.27, and 5.28, we conclude that the linear momentum equation for an inertial, moving, nondeforming control volume that involves steady (instantaneous or time-average) flow is

The linear momentum equation for a moving control volume involves the relative velocity.

$$\boxed{\int_{\text{cs}} \mathbf{W}\rho\mathbf{W} \cdot \hat{\mathbf{n}} \, dA = \sum \mathbf{F}_{\substack{\text{contents of the} \\ \text{control volume}}}} \qquad (5.29)$$

Example 5.17 illustrates the use of Eq. 5.29.

EXAMPLE 5.17

A vane on wheels moves with constant velocity $\mathbf{V}_0$ when a stream of water having a nozzle exit velocity of $\mathbf{V}_1$ is turned 45° by the vane as indicated in Fig. E5.17*a*. Note that this is the same moving vane considered in Section 4.4.6 earlier. Determine the magnitude and direction of the force, $\mathbf{F}$, exerted by the stream of water on the vane surface. The speed of the water jet leaving the nozzle is 100 ft/s and the vane is moving to the right with a constant speed of 20 ft/s.

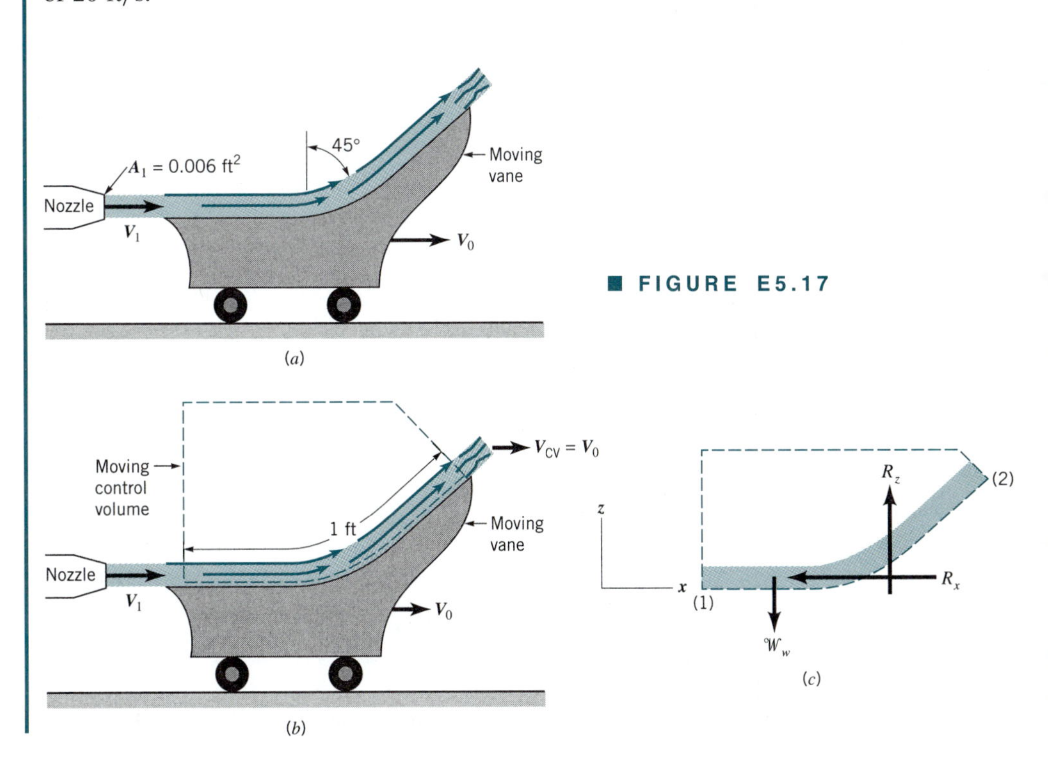

■ FIGURE E5.17

SOLUTION

To determine the magnitude and direction of the force, **F**, exerted by the water on the vane, we apply Eq. 5.29 to the contents of the moving control volume shown in Fig. E5.17*b*. The forces acting on the contents of this control volume are indicated in Fig. E5.17*c*. Note that since the ambient pressure is atmospheric, all pressure forces cancel each other out. Equation 5.29 is applied to the contents of the moving control volume in component directions. For the x direction (positive to the right), we get

$$\int_{\text{cs}} W_x \rho \, \mathbf{W} \cdot \hat{\mathbf{n}} \, dA = -R_x$$

or

$$(+W_1)(-\dot{m}_1) + (+W_2 \cos 45°)(+\dot{m}_2) = -R_x \quad \textbf{(1)}$$

where

$$\dot{m}_1 = \rho_1 W_1 A_1 \qquad \text{and} \qquad \dot{m}_2 = \rho_2 W_2 A_2.$$

For the vertical or z direction (positive up) we get

$$\int_{\text{cs}} W_z \rho \mathbf{W} \cdot \hat{\mathbf{n}} \, dA = R_z - \mathcal{W}_w$$

or

$$(+W_2 \sin 45°)(+\dot{m}_2) = R_z - \mathcal{W}_w \quad \textbf{(2)}$$

We assume for simplicity, that the water flow is frictionless and that the change in water elevation across the vane is negligible. Thus, from the Bernoulli equation (Eq. 3.7) we conclude that the speed of the water relative to the moving control volume, W, is constant or

$$W_1 = W_2$$

The relative speed of the stream of water entering the control volume, W_1, is

$$W_1 = V_1 - V_0 = 100 \text{ ft/s} - 20 \text{ ft/s} = 80 \text{ ft/s} = W_2$$

The water density is constant so that

$$\rho_1 = \rho_2 = 1.94 \text{ slugs/ft}^3$$

Application of the conservation of mass principle to the contents of the moving control volume (Eq. 5.16) leads to

$$\dot{m}_1 = \rho_1 W_1 A_1 = \rho_2 W_2 A_2 = \dot{m}_2$$

Combining results we get

$$R_x = \rho W_1^2 A_1 (1 - \cos 45°)$$

or

$$R_x = (1.94 \text{ slugs/ft}^3)(80 \text{ ft/s})^2(0.006 \text{ ft}^2)(1 - \cos 45°)$$

$$R_x = 21.8 \text{ lb}$$

Also,

$$R_z = \rho W_1^2 (\sin 45°) A_1 + \mathcal{W}_w$$

where

$$\mathcal{W}_w = \rho g A_1 \ell$$

Thus,

$$R_z = (1.94 \text{ slugs/ft}^3)(80 \text{ ft/s})^2(\sin 45°)(0.006 \text{ ft}^2) + (62.4 \text{ lb/ft}^3)(0.006 \text{ ft}^2)(1 \text{ ft}) = 52.6 \text{ lb} + 0.37 \text{ lb} = 53 \text{ lb}$$

Combining the components we get

$$R = \sqrt{R_x^2 + R_z^2} = [(21.8 \text{ lb})^2 + (53 \text{ lb})^2]^{1/2} = 57.3 \text{ lb}$$

The angle of **R** from the x direction, α, is

$$\alpha = \tan^{-1} \frac{R_z}{R_x} = \tan^{-1} (53 \text{ lb}/21.8 \text{ lb}) = 67.6°$$

The force of the water on the vane is equal in magnitude but opposite in direction from **R**; thus it points to the right and down at an angle of 67.6° from the x direction and is equal in magnitude to 57.3 lb. **(Ans)**

It should be clear from the preceding examples that fluid flows can lead to a reaction force in the following ways:

1. Linear momentum flow variation in direction and/or magnitude.
2. Fluid pressure forces.
3. Fluid friction forces.
4. Fluid weight.

The selection of a control volume is an important matter. An appropriate control volume can make a problem solution straightforward.

5.2.3 Derivation of the Moment-of-Momentum Equation[2]

The moment-of-momentum equation involves torques and angular momentum.

In many engineering problems, the moment of a force with respect to an axis, namely, *torque*, is important. Newton's second law of motion has already led to a useful relationship between forces and linear momentum flow. The linear momentum equation can also be used to solve problems involving torques. However, by forming the moment of the linear momentum and the resultant force associated with each particle of fluid with respect to a point in an inertial coordinate system, we will develop a *moment-of-momentum equation* that relates *torques* and *angular momentum flow* for the contents of a control volume. When torques are important, the moment-of-momentum equation is often more convenient to use than the linear momentum equation.

Application of Newton's second law of motion to a particle of fluid yields

$$\frac{D}{Dt}(\mathbf{V}\rho\, \delta \mathcal{V}) = \delta \mathbf{F}_{\text{particle}} \tag{5.30}$$

[2]This section may be omitted, along with Sections 5.2.4 and 5.3.5, without loss of continuity in the text material. However, these sections are recommended for those interested in Chapter 12.

where **V** is the particle velocity measured in an inertial reference system, ρ is the particle density, $\delta \forall$ is the infinitesimally small particle volume, and $\delta \mathbf{F}_{\text{particle}}$ is the resultant external force acting on the particle. If we form the moment of each side of Eq. 5.30 with respect to the origin of an inertial coordinate system, we obtain

The angular momentum equation is derived from Newton's second law.

$$\mathbf{r} \times \frac{D}{Dt}(\mathbf{V}\rho\, \delta \forall) = \mathbf{r} \times \delta \mathbf{F}_{\text{particle}} \tag{5.31}$$

where **r** is the position vector from the origin of the inertial coordinate system to the fluid particle (Fig. 5.3). We note that

$$\frac{D}{Dt}[(\mathbf{r} \times \mathbf{V})\rho\, \delta \forall] = \frac{D\mathbf{r}}{Dt} \times \mathbf{V}\rho\, \delta \forall + \mathbf{r} \times \frac{D(\mathbf{V}\rho\, \delta \forall)}{Dt} \tag{5.32}$$

and

$$\frac{D\mathbf{r}}{Dt} = \mathbf{V} \tag{5.33}$$

Thus, since

$$\mathbf{V} \times \mathbf{V} = 0 \tag{5.34}$$

by combining Eqs. 5.31, 5.32, 5.33, and 5.34, we obtain the expression

$$\frac{D}{Dt}[(\mathbf{r} \times \mathbf{V})\rho\, \delta \forall] = \mathbf{r} \times \delta \mathbf{F}_{\text{particle}} \tag{5.35}$$

Equation 5.35 is valid for every particle of a system. For a system (collection of fluid particles), we need to use the sum of both sides of Eq. 5.35 to obtain

$$\int_{\text{sys}} \frac{D}{Dt}[(\mathbf{r} \times \mathbf{V})\rho\, d\forall] = \sum (\mathbf{r} \times \mathbf{F})_{\text{sys}} \tag{5.36}$$

where

$$\sum \mathbf{r} \times \delta \mathbf{F}_{\text{particle}} = \sum (\mathbf{r} \times \mathbf{F})_{\text{sys}} \tag{5.37}$$

We note that

$$\frac{D}{Dt}\int_{\text{sys}} (\mathbf{r} \times \mathbf{V})\rho\, d\forall = \int_{\text{sys}} \frac{D}{Dt}[(\mathbf{r} \times \mathbf{V})\rho\, d\forall] \tag{5.38}$$

since the sequential order of differentiation and integration can be reversed without consequence. (Recall that the material derivative, $D(\ )/Dt$, denotes the time derivative following a given system; see Section 4.2.1.) Thus, from Eqs. 5.36 and 5.38 we get

$$\frac{D}{Dt}\int_{\text{sys}} (\mathbf{r} \times \mathbf{V})\rho\, d\forall = \sum (\mathbf{r} \times \mathbf{F})_{\text{sys}} \tag{5.39}$$

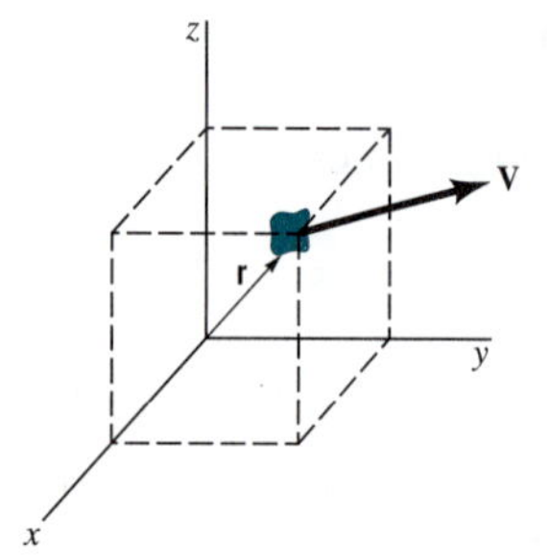

■ **FIGURE 5.3** **Inertial coordinate system.**

or

the time rate of change of the moment-of-momentum of the system	=	sum of the external torques acting on the system

For a control volume that is instantaneously coincident with the system, the torques acting on the system and on the control volume contents will be identical:

$$\sum (\mathbf{r} \times \mathbf{F})_{\text{sys}} = \sum (\mathbf{r} \times \mathbf{F})_{\text{cv}} \tag{5.40}$$

Further, for the system and the contents of the coincident control volume that is fixed and nondeforming, the Reynolds transport theorem (Eq. 4.19) leads to

$$\frac{D}{Dt}\int_{\text{sys}} (\mathbf{r} \times \mathbf{V})\rho \, d\forall = \frac{\partial}{\partial t}\int_{\text{cv}} (\mathbf{r} \times \mathbf{V})\, \rho \, d\forall + \int_{\text{cs}} (\mathbf{r} \times \mathbf{V})\rho \mathbf{V} \cdot \hat{\mathbf{n}} \, dA \tag{5.41}$$

or

For a system, the rate of change of moment-of-momentum equals the net torque.

time rate of change of the moment-of-momentum of the system	=	time rate of change of the moment-of-momentum of the contents of the control volume	+	net rate of flow of the moment-of-momentum through the control surface

For a control volume that is fixed (and therefore inertial) and nondeforming, we combine Eqs. 5.39, 5.40, and 5.41 to obtain the moment-of-momentum equation:

$$\frac{\partial}{\partial t}\int_{\text{cv}} (\mathbf{r} \times \mathbf{V})\rho \, d\forall + \int_{\text{cs}} (\mathbf{r} \times \mathbf{V})\rho \mathbf{V} \cdot \hat{\mathbf{n}} \, dA = \sum (\mathbf{r} \times \mathbf{F})_{\substack{\text{contents of the} \\ \text{control volume}}} \tag{5.42}$$

An important category of fluid mechanical problems that is readily solved with the help of the moment-of-momentum equation (Eq. 5.42) involves machines that rotate or tend to rotate around a single axis. Examples of these machines include rotary lawn sprinklers, ceiling fans, lawn mower blades, wind turbines, turbochargers, and gas turbine engines. As a class, these devices are often called turbomachines.

5.2.4 Application of the Moment-of-Momentum Equation[3]

We simplify our use of Eq. 5.42 in several ways:

1. We assume that flows considered are one-dimensional (uniform distributions of average velocity at any section).
2. We confine ourselves to steady or steady-in-the-mean cyclical flows. Thus,

$$\frac{\partial}{\partial t}\int_{\text{cv}} (\mathbf{r} \times \mathbf{V})\rho \, d\forall = 0$$

   at any instant of time for steady flows or on a time-average basis for cyclical unsteady flows.
3. We work only with the component of Eq. 5.42 resolved along the axis of rotation.

Consider the rotating sprinkler sketched in Fig. 5.4. Because the direction and magnitude of the flow through the sprinkler from the inlet [section (1)] to the outlet [section (2)] of the arm

[3]This section may be omitted, along with Sections 5.2.3 and 5.3.5, without loss of continuity in the text material. However, these sections are recommended for those interested in Chapter 12.

changes, the water exerts a torque on the sprinkler head causing it to tend to rotate or to actually rotate in the direction shown, much like a turbine rotor. In applying the moment-of-momentum equation (Eq. 5.42) to this flow situation, we elect to use the fixed and nondeforming control volume shown in Fig. 5.4. This disk-shaped control volume contains within its boundaries the spinning or stationary sprinkler head and the portion of the water flowing through the sprinkler contained in the control volume at an instant. The control surface cuts through the sprinkler head's solid material so that the shaft torque that resists motion can be clearly identified. When the sprinkler is rotating, the flow field in the stationary control volume is cyclical and unsteady, but steady in the mean. We proceed to use the axial component of the moment-of-momentum equation (Eq. 5.42) to analyze this flow.

The integrand of the moment-of-momentum flow term in Eq. 5.42,

$$\int_{\text{cs}} (\mathbf{r} \times \mathbf{V})\rho \mathbf{V} \cdot \hat{\mathbf{n}} \, dA$$

The component of angular momentum along the axis of rotation is often the most important.

can be nonzero only where fluid is crossing the control surface. Everywhere else on the control surface this term will be zero because $\mathbf{V} \cdot \hat{\mathbf{n}} = 0$. Water enters the control volume axially through the hollow stem of the sprinkler at section (1). At this portion of the control surface, the component of $\mathbf{r} \times \mathbf{V}$ resolved along the axis of rotation is zero because $\mathbf{r} \times \mathbf{V}$ and the axis of rotation are perpendicular. Thus, there is no axial moment-of-momentum flow in at section (1). Water leaves the control volume through each of the two nozzle openings at section (2). For the exiting flow, the magnitude of the axial component of $\mathbf{r} \times \mathbf{V}$ is $r_2 V_{\theta 2}$, where r_2 is the radius from the axis of rotation to the nozzle centerline and $V_{\theta 2}$ is the value of the tangential component of the velocity of the flow exiting each nozzle as observed from a frame of reference attached to the fixed and nondeforming control volume. The fluid velocity measured relative to a fixed control surface is an absolute velocity, $\mathbf{V}$. The velocity of the nozzle exit flow as viewed from the nozzle is called the relative velocity, $\mathbf{W}$. The absolute and relative velocities, $\mathbf{V}$ and $\mathbf{W}$, are related by the vector relationship

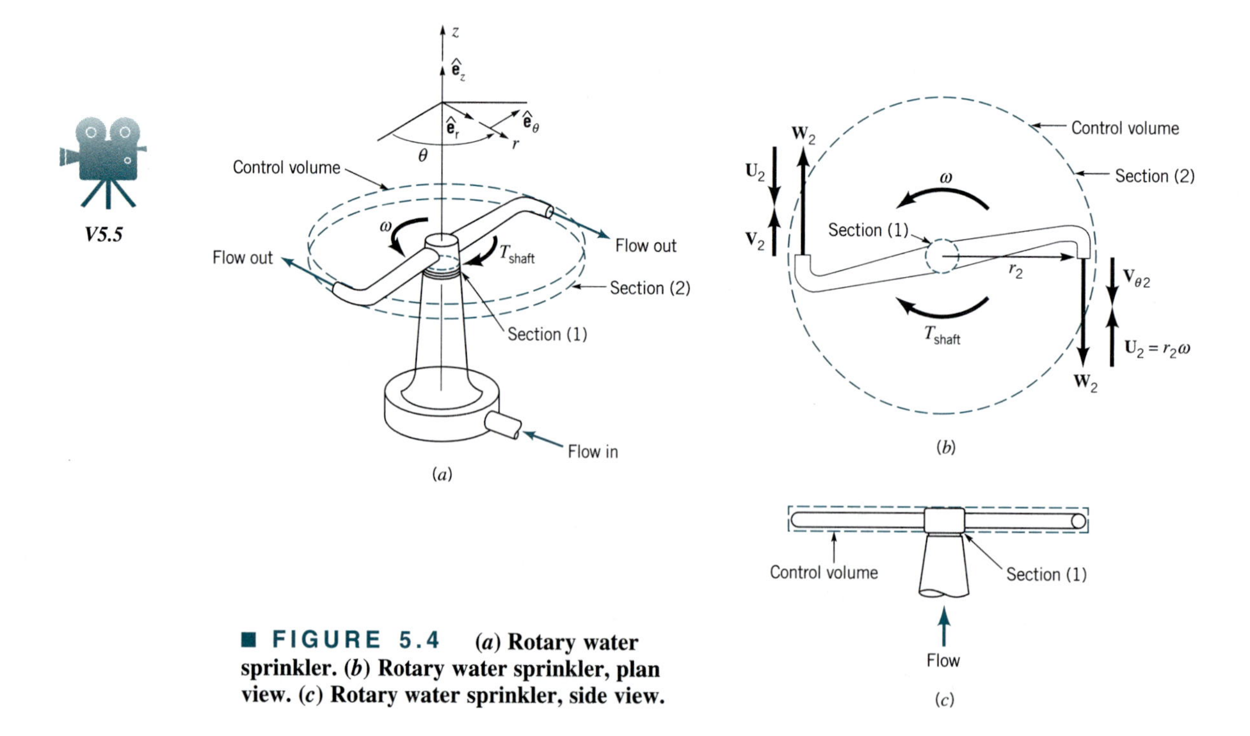

■ **FIGURE 5.4** ***(a)* Rotary water sprinkler. *(b)* Rotary water sprinkler, plan view. *(c)* Rotary water sprinkler, side view.**

$$\mathbf{V} = \mathbf{W} + \mathbf{U} \tag{5.43}$$

where $\mathbf{U}$ is the velocity of the moving nozzle as measured relative to the fixed control surface.

The cross product and the dot product involved in the moment-of-momentum flow term of Eq. 5.42,

$$\int_{\text{cs}} (\mathbf{r} \times \mathbf{V})\rho \mathbf{V} \cdot \hat{\mathbf{n}}\, dA$$

can each result in a positive or negative value. For flow into the control volume, $\mathbf{V} \cdot \hat{\mathbf{n}}$ is negative. For flow out, $\mathbf{V} \cdot \hat{\mathbf{n}}$ is positive. The correct algebraic sign to assign the axis component of $\mathbf{r} \times \mathbf{V}$ can be ascertained by using the right-hand rule. The positive direction along the axis of rotation is the direction the thumb of the right hand points when it is extended and the remaining fingers are curled around the rotation axis in the positive direction of rotation as illustrated in Fig. 5.5. The direction of the axial component of $\mathbf{r} \times \mathbf{V}$ is similarly ascertained by noting the direction of the cross product of the radius from the axis of rotation, $r\hat{\mathbf{e}}_r$, and the tangential component of absolute velocity, $V_\theta \hat{\mathbf{e}}_\theta$. Thus, for the sprinkler of Fig. 5.4, we can state that

The algebraic sign is obtained by the right-hand rule.

$$\left[\int_{\text{cs}} (\mathbf{r} \times \mathbf{V})\rho \mathbf{V} \cdot \hat{\mathbf{n}}\, dA\right]_{\text{axial}} = (-r_2 V_{\theta 2})(+\dot{m}) \tag{5.44}$$

where, because of mass conservation, $\dot{m}$ is the total mass flowrate through both nozzles. As was demonstrated in Example 5.7, the mass flowrate is the same whether the sprinkler rotates or not. The correct algebraic sign of the axial component of $\mathbf{r} \times \mathbf{V}$ can be easily remembered in the following way: if $\mathbf{V}_\theta$ and $\mathbf{U}$ are in the same direction, use $+$; if $\mathbf{V}_\theta$ and $\mathbf{U}$ are in opposite directions, use $-$.

The torque term $[\sum(\mathbf{r} \times \mathbf{F})_{\text{contents of the control volume}}]$ of the moment-of-momentum equation (Eq. 5.42) is analyzed next. Confining ourselves to torques acting with respect to the axis of rotation only, we conclude that the shaft torque is important. The net torque with respect to the axis of rotation associated with normal forces exerted on the contents of the control volume will be very small if not zero. The net axial torque due to fluid tangential forces is also negligibly small for the control volume of Fig. 5.4. Thus, for the sprinkler of Fig. 5.4

$$\sum\left[(\mathbf{r} \times \mathbf{F})_{\substack{\text{contents of the}\\ \text{control volume}}}\right]_{\text{axial}} = T_{\text{shaft}} \tag{5.45}$$

Note that we have entered T_{shaft} as a positive quantity in Eq. 5.45. This is equivalent to assuming that T_{shaft} is in the same direction as rotation.

For the sprinkler of Fig. 5.4, the axial component of the moment-of-momentum equation (Eq. 5.42) is, from Eqs. 5.44 and 5.45

$$-r_2 V_{\theta 2} \dot{m} = T_{\text{shaft}} \tag{5.46}$$

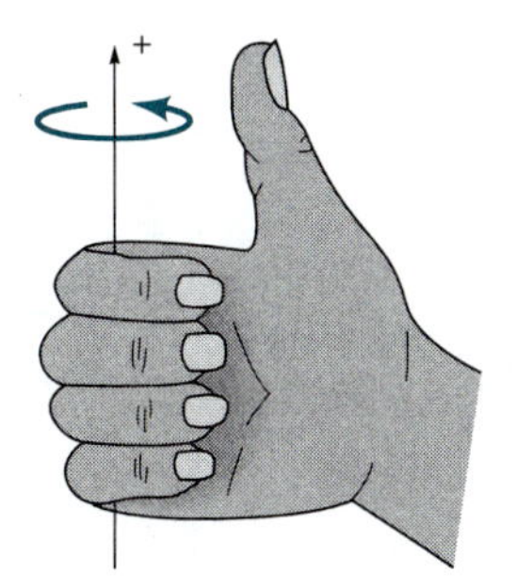

■ FIGURE 5.5 Right-hand rule convention.

We interpret T_{shaft} being a negative quantity from Eq. 5.46 to mean that the shaft torque actually opposes the rotation of the sprinkler arms as shown in Fig. 5.4. The shaft torque, T_{shaft}, opposes rotation in all turbine devices.

Power is equal to angular velocity times torque.

We could evaluate the shaft power, $\dot{W}_{\text{shaft}}$, associated with shaft torque, T_{shaft}, by forming the product of T_{shaft} and the rotational speed of the shaft, ω. (We use the notation that W = work, $(\dot{\ }) = d(\ )/dt$, and thus $\dot{W}$ = power.) Thus, from Eq. 5.46 we get

$$\dot{W}_{\text{shaft}} = T_{\text{shaft}}\omega = -r_2 V_{\theta 2}\dot{m}\omega \qquad \textbf{(5.47)}$$

Since $r_2\omega$ is the speed of each sprinkler nozzle, U, we can also state Eq. 5.47 in the form

$$\dot{W}_{\text{shaft}} = -U_2 V_{\theta 2}\dot{m} \qquad \textbf{(5.48)}$$

Shaft work per unit mass, w_{shaft}, is equal to $\dot{W}_{\text{shaft}}/\dot{m}$. Dividing Eq. 5.48 by the mass flowrate, $\dot{m}$, we obtain

$$w_{\text{shaft}} = -U_2 V_{\theta 2} \qquad \textbf{(5.49)}$$

Negative shaft work as in Eqs. 5.47, 5.48, and 5.49 is work out of the control volume, i.e., work done by the fluid on the rotor and thus its shaft.

V5.6

The principles associated with this sprinkler example can be extended to handle most simplified turbomachine flows. The fundamental technique is not difficult. However, the geometry of some turbomachine flows is quite complicated.

Example 5.18 further illustrates how the axial component of the moment-of-momentum equation (Eq. 5.46) can be used.

EXAMPLE 5.18

Water enters a rotating lawn sprinkler through its base at the steady rate of 1000 ml/s as sketched in Fig. E5.18. The exit area of each of the two nozzles is 30 mm^2 and the flow leaving each nozzle is in the tangential direction. The radius from the axis of rotation to the centerline of each nozzle is 200 mm.

(a) Determine the resisting torque required to hold the sprinkler head stationary.
(b) Determine the resisting torque associated with the sprinkler rotating with a constant speed of 500 rev/min.
(c) Determine the speed of the sprinkler if no resisting torque is applied.

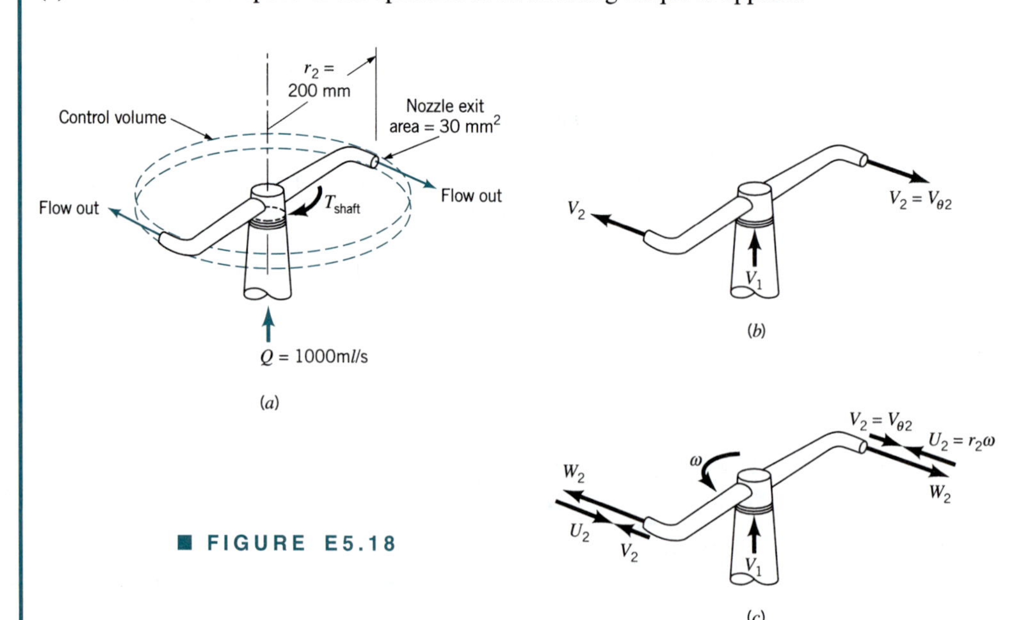

■ FIGURE E5.18

SOLUTION

To solve parts (a), (b), and (c) of this example we can use the same fixed and nondeforming, disk shaped control volume illustrated in Fig. 5.4. As is indicated in Fig. E.5.18*a*, the only axial torque considered is the one resisting motion, T_{shaft}.

When the sprinkler head is held stationary as specified in part (a) of this example problem, the velocities of the fluid entering and leaving the control volume are as shown in Fig. E.18*b*. Equation 5.46 applies to the contents of this control volume. Thus

$$T_{\text{shaft}} = -r_2 V_{\theta 2} \dot{m} \tag{1}$$

Since the control volume is fixed and nondeforming and the flow exiting from each nozzle is tangential,

$$V_{\theta 2} = V_2 \tag{2}$$

Equations 1 and 2 give

$$T_{\text{shaft}} = -r_2 V_2 \dot{m} \tag{3}$$

In Example 5.7, we ascertained that $V_2 = 16.7 \text{ m/s}$. Thus, from Eq. 3,

$$T_{\text{shaft}} = -\frac{(200 \text{ mm})(16.7 \text{ m/s})(1000 \text{ ml/s})(10^{-3} \text{ m}^3/\text{liter})(999 \text{ kg/m}^3)[1 \text{ (N/kg)/(m/s}^2)]}{(1000 \text{ mm/m})(1000 \text{ ml/liter})}$$

or

$$T_{\text{shaft}} = -3.34 \text{ N·m} \qquad \textbf{(Ans)}$$

When the sprinkler is rotating at a constant speed of 500 rpm, the flow field in the control volume is unsteady but cyclical. Thus, the flow field is steady in the mean. The velocities of the flow entering and leaving the control volume are as indicated in Fig. E5.18*c*. The absolute velocity of the fluid leaving each nozzle, V_2, is, from Eq. 5.43,

$$V_2 = W_2 - U_2 \tag{4}$$

where $W_2 = 16.7 \text{ m/s}$ as determined in Example 5.7. The speed of the nozzle, U_2, is obtained from

$$U_2 = r_2 \omega \tag{5}$$

Application of the axial component of the moment-of-momentum equation (Eq. 5.46) leads again to Eq. 3. From Eqs. 4 and 5,

$$V_2 = 16.7 \text{ m/s} - r_2 \omega = 16.7 \text{ m/s} - \frac{(200 \text{ mm})(500 \text{ rev/min})(2\pi \text{ rad/rev})}{(1000 \text{ mm/m})(60 \text{ s/min})}$$

or

$$V_2 = 16.7 \text{ m/s} - 10.5 \text{ m/s} = 6.2 \text{ m/s}$$

Thus, using Eq. 3, we get

$$T_{\text{shaft}} = -\frac{(200 \text{ mm})(6.2 \text{ m/s})(1000 \text{ ml/s})(10^{-3} \text{ m}^3/\text{liter})(999 \text{ kg/m}^3)[1 \text{ (N/kg)/(m/s}^2)]}{(1000 \text{ mm/m})(1000 \text{ ml/liter})}$$

or

$$T_{\text{shaft}} = -1.24 \text{ N·m} \qquad \textbf{(Ans)}$$

Note that the resisting torque associated with sprinkler head rotation is much less than the resisting torque that is required to hold the sprinkler stationary.

When no resisting torque is applied to the rotating sprinkler head, a maximum constant speed of rotation will occur as demonstrated below. Application of Eqs. 3, 4, and 5 to the contents of the control volume results in

$$T_{\text{shaft}} = -r_2(W_2 - r_2\omega)\dot{m} \tag{6}$$

For no resisting torque ($T_{\text{shaft}} = 0$), Eq. 6 yields

$$\omega = \frac{W_2}{r_2} \tag{7}$$

In Example 5.7, we learned that the relative velocity of the fluid leaving each nozzle, W_2, is the same regardless of the speed of rotation of the sprinkler head, ω, as long as the mass flowrate of the fluid, $\dot{m}$, remains constant. Thus, by using Eq. 7 we obtain

$$\omega = \frac{W_2}{r_2} = \frac{(16.7 \text{ m/s})(1000 \text{ mm/m})}{(200 \text{ mm})} = 83.5 \text{ rad/s}$$

or

$$\omega = \frac{(83.5 \text{ rad/s})(60 \text{ s/min})}{2\ \pi \text{ rad/rev}} = 797 \text{ rpm} \qquad \textbf{(Ans)}$$

For this condition ($T_{\text{shaft}} = 0$), the water both enters and leaves the control volume with zero angular momentum.

In summary, we observe that the resisting torque associated with rotation is less than the torque required to hold a rotor stationary. Even in the absence of a resisting torque, the rotor maximum speed is finite.

When the moment-of-momentum equation (Eq. 5.42) is applied to a more general, one-dimensional flow through a rotating machine, we obtain

$$\boxed{T_{\text{shaft}} = (-\dot{m}_{\text{in}})(\pm r_{\text{in}} V_{\theta\text{in}}) + \dot{m}_{\text{out}}(\pm r_{\text{out}} V_{\theta\text{out}})} \tag{5.50}$$

by applying the same kind of analysis used with the sprinkler of Fig. 5.4. The "$-$" is used with mass flowrate into the control volume, $\dot{m}_{\text{in}}$, and the "$+$" is used with mass flowrate out of the control volume, $\dot{m}_{\text{out}}$, to account for the sign of the dot product, $\mathbf{V} \cdot \hat{\mathbf{n}}$, involved. Whether "$+$" or "$-$" is used with the rV_θ product depends on the direction of $(\mathbf{r} \times \mathbf{V})_{\text{axial}}$. A simple way to determine the sign of the rV_θ product is to compare the direction of V_θ and the blade speed, U. If V_θ and U are in the same direction, then the rV_θ product is positive. If V_θ and U are in opposite directions, the rV_θ product is negative. The sign of the shaft torque is "$+$" if T_{shaft} is in the same direction along the axis of rotation as ω, and "$-$" otherwise.

The angular momentum equation is used to obtain torque and power for rotating machines.

The shaft power, $\dot{W}_{\text{shaft}}$, is related to shaft torque, T_{shaft}, by

$$\dot{W}_{\text{shaft}} = T_{\text{shaft}}\omega \tag{5.51}$$

Thus, using Eq. 5.50 and 5.51 with a "$+$" sign for T_{shaft} in Eq. 5.50, we obtain

$$\dot{W}_{\text{shaft}} = (-\dot{m}_{\text{in}})(\pm r_{\text{in}}\omega V_{\theta\text{in}}) + \dot{m}_{\text{out}}(\pm r_{\text{out}}\omega V_{\theta\text{out}}) \tag{5.52}$$

or since $r\omega = U$

$$\boxed{\dot{W}_{\text{shaft}} = (-\dot{m}_{\text{in}})(\pm U_{\text{in}} V_{\theta\text{in}}) + \dot{m}_{\text{out}}(\pm U_{\text{out}} V_{\theta\text{out}})} \tag{5.53}$$

The "+" is used for the UV_θ product when U and V_θ are in the same direction; the "−" is used when U and V_θ are in opposite directions. Also, since $+T_{\text{shaft}}$ was used to obtain Eq. 5.53, when $\dot{W}_{\text{shaft}}$ is positive, power is into the control volume (e.g., pump), and when $\dot{W}_{\text{shaft}}$ is negative, power is out of the control volume (e.g., turbine).

The shaft work per unit mass, w_{shaft}, can be obtained from the shaft power, $\dot{W}_{\text{shaft}}$, by dividing Eq. 5.53 by the mass flowrate, $\dot{m}$. By conservation of mass,

$$\dot{m} = \dot{m}_{\text{in}} = \dot{m}_{\text{out}}$$

The shaft work per unit mass involves only the blade velocity and the tangential fluid velocity.

From Eq. 5.53, we obtain

$$w_{\text{shaft}} = -(\pm U_{\text{in}} V_{\theta\text{in}}) + (\pm U_{\text{out}} V_{\theta\text{out}}) \tag{5.54}$$

The application of Eqs. 5.50, 5.53, and 5.54 is demonstrated in Example 5.19. More examples of the application of Eqs. 5.50, 5.53, and 5.54 are included in Chapter 12.

An air fan has a bladed rotor of 12-in. outside diameter and 10-in. inside diameter as illustrated in Fig. E5.19*a*. The height of each rotor blade is constant at 1 in. from blade inlet to outlet. The flowrate is steady, on a time-average basis, at 230 ft^3/min and the absolute velocity of the air at blade inlet, $\mathbf{V}_1$, is radial. The blade discharge angle is 30° from the tangential direction. If the rotor rotates at a constant speed of 1725 rpm, estimate the power required to run the fan.

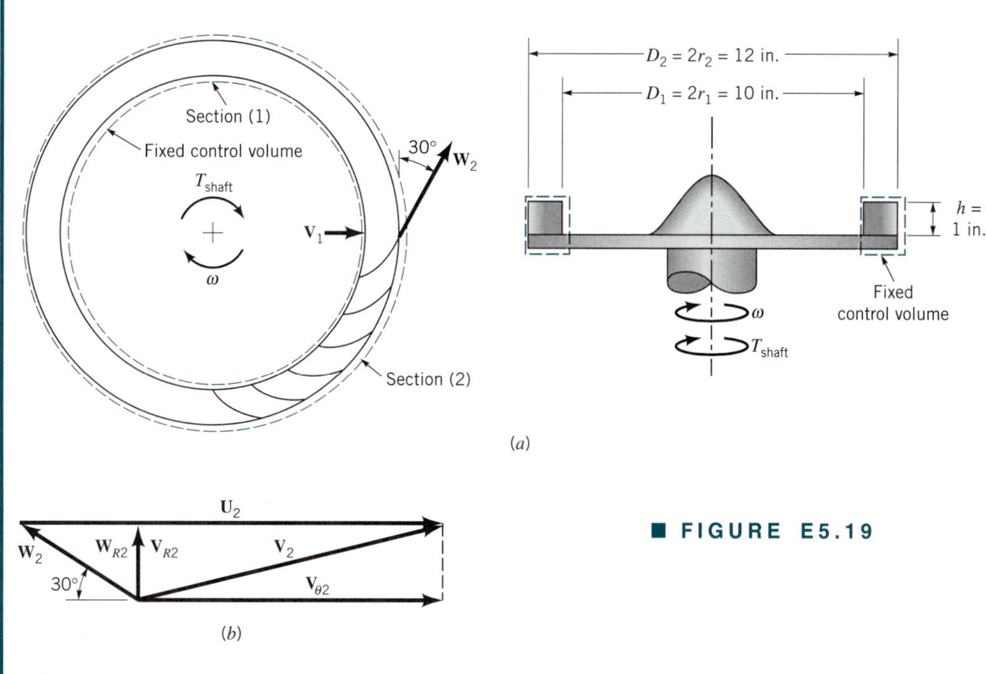

■ **FIGURE E5.19**

SOLUTION

We select a fixed and nondeforming control volume that includes the rotating blades and the fluid within the blade row at an instant, as shown with a dashed line in Fig. E5.19*a*. The flow within this control volume is cyclical, but steady in the mean. The only torque we consider is the driving shaft torque, T_{shaft}. This torque is provided by a motor. We assume that the entering and leaving flows are each represented by uniformly distributed velocities and flow properties. Since shaft power is sought, Eq. 5.53 is appropriate. Application of Eq. 5.53 to the contents of the control volume in Fig. E5.19 gives

$$\dot{W}_{\text{shaft}} = (-\dot{m}_1)(\pm \overset{0\ (\mathbf{V}_1 \text{ is radial})}{\cancel{U_1 V_{\theta 1}}}) + \dot{m}_2(\pm U_2 V_{\theta 2}) \tag{1}$$

From Eq. 1 we see that to calculate fan power, we need mass flowrate, $\dot{m}$, rotor exit blade velocity, U_2, and fluid tangential velocity at blade exit, $V_{\theta 2}$. The mass flowrate, $\dot{m}$, is easily obtained from Eq. 5.6 as

$$\dot{m} = \rho Q = \frac{(2.38 \times 10^{-3} \text{ slug/ft}^3)(230 \text{ ft}^3\text{/min})}{(60 \text{ s/min})} = 0.00912 \text{ slug/s} \tag{2}$$

The rotor exit blade speed, U_2, is

$$U_2 = r_2\omega = \frac{(6 \text{ in.})(1725 \text{ rpm})(2\pi \text{ rad/rev})}{(12 \text{ in./ft})(60 \text{ s/min})} = 90.3 \text{ ft/s} \tag{3}$$

To determine the fluid tangential speed at the fan rotor exit, $V_{\theta 2}$, we use Eq. 5.43 to get

$$\mathbf{V}_2 = \mathbf{W}_2 + \mathbf{U}_2 \tag{4}$$

The vector addition of Eq. 4 is shown in the form of a "velocity triangle" in Fig. E5.19*b*. From Fig. E5.19*b*, we can see that

$$V_{\theta 2} = U_2 - W_2 \cos 30° \tag{5}$$

To solve Eq. 5 for $V_{\theta 2}$ we need a value of W_2, in addition to the value of U_2 already determined (Eq. 3). To get W_2, we recognize that

$$W_2 \sin 30° = V_{R2} \tag{6}$$

where V_{R2} is the radial component of either $\mathbf{W}_2$ or $\mathbf{V}_2$. Also, using Eq. 5.6, we obtain

$$\dot{m} = \rho A_2 V_{R2} \tag{7}$$

or since

$$A_2 = 2\ \pi r_2 h \tag{8}$$

where h is the blade height, Eqs. 7 and 8 combine to form

$$\dot{m} = \rho 2\pi r_2 h V_{R2} \tag{9}$$

Taking Eqs. 6 and 9 together we get

$$W_2 = \frac{\dot{m}}{\rho 2\pi r_2 h \sin 30°} \tag{10}$$

Substituting known values into Eq. 10, we obtain

$$W_2 = \frac{(0.00912 \text{ slugs/s})(12 \text{ in./ft})(12 \text{ in./ft})}{(2.38 \times 10^{-3} \text{ slugs/ft}^3)2\pi(6 \text{ in.})(1 \text{ in.}) \sin 30°} = 29.3 \text{ ft/s}$$

By using this value of W_2 in Eq. 5 we get

$$V_{\theta 2} = U_2 - W_2 \cos 30°$$
$$= 90.3 \text{ ft/s} - (29.3 \text{ ft/s})(0.866) = 64.9 \text{ ft/s}$$

Equation 1 can now be used to obtain

$$\dot{W}_{\text{shaft}} = \dot{m}U_2V_{\theta 2} = \frac{(0.00912 \text{ slug/s})(90.3 \text{ ft/s})(64.9 \text{ ft/s})}{[1 \text{ (slug·ft/s}^2\text{)/lb}][550 \text{ (ft·lb)/(hp·s)}]}$$

or

$$\dot{W}_{\text{shaft}} = 0.0972 \text{ hp} \quad \textbf{(Ans)}$$

Note that the "+" was used with the $U_2V_{\theta 2}$ product because U_2 and $V_{\theta 2}$ are in the same direction. This result, 0.0972 hp, is the power that needs to be delivered through the fan shaft for the given conditions. Ideally, all of this power would go into the flowing air. However, because of fluid friction, only some of this power will produce a useful effect (e.g., pressure rise) in the air. How much useful effect depends on the efficiency of the energy transfer between the fan blades and the fluid. This important aspect of energy transfer is covered in Section 5.3.5.

5.3 First Law of Thermodynamics—The Energy Equation

5.3.1 Derivation of the Energy Equation

The first law of thermodynamics for a system is, in words

time rate of increase of the total stored energy of the system	=	net time rate of energy addition by heat transfer into the system	+	net time rate of energy addition by work transfer into the system

The energy equation is a statement of the first law of thermodynamics.

In symbolic form, this statement is

$$\frac{D}{Dt}\int_{\text{sys}} e\rho \, d\mathcal{V} = \left(\sum \dot{Q}_{\text{in}} - \sum \dot{Q}_{\text{out}}\right)_{\text{sys}} + \left(\sum \dot{W}_{\text{in}} - \sum \dot{W}_{\text{out}}\right)_{\text{sys}}$$

or

$$\frac{D}{Dt}\int_{\text{sys}} e\rho \, d\mathcal{V} = (\dot{Q}_{\substack{\text{net}\\ \text{in}}} + \dot{W}_{\substack{\text{net}\\ \text{in}}})_{\text{sys}} \tag{5.55}$$

Some of these variables deserve a brief explanation before proceeding further. The total stored energy per unit mass for each particle in the system, e, is related to the internal energy per unit mass, $\check{u}$, the kinetic energy per unit mass, $V^2/2$, and the potential energy per unit mass, gz, by the equation

$$e = \check{u} + \frac{V^2}{2} + gz \tag{5.56}$$

The net rate of heat transfer into the system is denoted with $\dot{Q}_{\text{net in}}$, and the net rate of work transfer into the system is labeled $\dot{W}_{\text{net in}}$. Heat transfer and work transfer are considered "+" going into the system and "−" coming out.

Equation 5.55 is valid for inertial and noninertial reference systems. We proceed to develop the control volume statement of the first law of thermodynamics. For the control volume that is coincident with the system at an instant of time

$$(\dot{Q}_{\substack{\text{net}\\ \text{in}}} + \dot{W}_{\substack{\text{net}\\ \text{in}}})_{\text{sys}} = (\dot{Q}_{\substack{\text{net}\\ \text{in}}} + \dot{W}_{\substack{\text{net}\\ \text{in}}})_{\substack{\text{coincident}\\ \text{control}\\ \text{volume}}} \tag{5.57}$$

Furthermore, for the system and the contents of the coincident control volume that is fixed and nondeforming, the Reynolds transport theorem (Eq. 4.19 with the parameter b set equal to e) allows us to conclude that

$$\frac{D}{Dt}\int_{\text{sys}} e\rho \, d\mathcal{V} = \frac{\partial}{\partial t}\int_{\text{cv}} e\rho \, d\mathcal{V} + \int_{\text{cs}} e\rho \mathbf{V} \cdot \hat{\mathbf{n}} \, dA \tag{5.58}$$

or in words,

the time rate of increase of the total stored energy of the system	=	the time rate of increase of the total stored energy of the contents of the control volume	+	the net rate of flow of the total stored energy out of the control volume through the control surface

Combining Eqs. 5.55, 5.57, and 5.58 we get the control volume formula for the first law of thermodynamics:

The energy equation involves energy, heat transfer, and work.

$$\frac{\partial}{\partial t}\int_{\text{cv}} e\rho\, d\mathcal{V} + \int_{\text{cs}} e\rho\mathbf{V}\cdot\hat{\mathbf{n}}\, dA = (\dot{Q}_{\substack{\text{net}\\\text{in}}} + \dot{W}_{\substack{\text{net}\\\text{in}}})_{\text{cv}} \tag{5.59}$$

The total stored energy per unit mass, e, in Eq. 5.59 is for fluid particles entering, leaving, and within the control volume. Further explanation of the heat transfer and work transfer involved in this equation follows.

The heat transfer rate, $\dot{Q}$, represents all of the ways in which energy is exchanged between the control volume contents and surroundings because of a temperature difference. Thus, radiation, conduction, and/or convection are possible. Heat transfer into the control volume is considered positive, heat transfer out is negative. In many engineering applications, the process is *adiabatic*; the heat transfer rate, $\dot{Q}$, is zero. The net heat transfer rate, $\dot{Q}_{\text{net in}}$, can also be zero when $\Sigma\dot{Q}_{\text{in}} - \Sigma\dot{Q}_{\text{out}} = 0$.

The work transfer rate, $\dot{W}$, also called *power*, is positive when work is done on the contents of the control volume by the surroundings. Otherwise, it is considered negative. Work can be transferred across the control surface in several ways. In the following paragraphs, we consider some important forms of work transfer.

In many instances, work is transferred across the control surface by a moving shaft. In rotary devices such as turbines, fans, and propellers, a rotating shaft transfers work across that portion of the control surface that slices through the shaft. Even in reciprocating machines like positive displacement internal combustion engines and compressors that utilize piston-in-cylinder arrangements, a rotating crankshaft is used. Since work is the dot product of force and related displacement, rate of work (or power) is the dot product of force and related displacement per unit time. For a rotating shaft, the power transfer, $\dot{W}_{\text{shaft}}$, is related to the shaft torque that causes the rotation, T_{shaft}, and the angular velocity of the shaft, ω, by the relationship

$$\dot{W}_{\text{shaft}} = T_{\text{shaft}}\omega$$

When the control surface cuts through the shaft material, the shaft torque is exerted by shaft material at the control surface. To allow for consideration of problems involving more than one shaft we use the notation

$$\dot{W}_{\substack{\text{shaft}\\\text{net in}}} = \sum \dot{W}_{\substack{\text{shaft}\\\text{in}}} - \sum \dot{W}_{\substack{\text{shaft}\\\text{out}}} \tag{5.60}$$

Work transfer can also occur at the control surface when a force associated with fluid normal stress acts over a distance. Consider the simple pipe flow illustrated in Fig. 5.6 and the control volume shown. For this situation, the fluid normal stress, σ, is simply equal to the negative of fluid pressure, p, in all directions; that is,

$$\sigma = -p \tag{5.61}$$

This relationship can be used with varying amounts of approximation for many engineering problems (see Chapter 6).

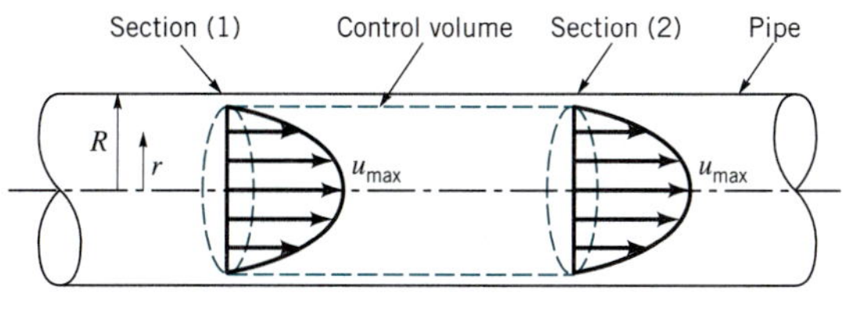

■ **FIGURE 5.6 Simple, fully developed pipe flow.**

The power transfer associated with normal stresses acting on a single fluid particle, $\delta\dot{W}_{\text{normal stress}}$, can be evaluated as the dot product of the normal stress force, $\delta\mathbf{F}_{\text{normal stress}}$, and the fluid particle velocity, $\mathbf{V}$, as

$$\delta\dot{W}_{\text{normal stress}} = \delta\mathbf{F}_{\text{normal stress}} \cdot \mathbf{V}$$

If the normal stress force is expressed as the product of local normal stress, $\sigma = -p$, and fluid particle surface area, $\hat{\mathbf{n}}\,\delta A$, the result is

$$\delta\dot{W}_{\text{normal stress}} = \sigma\hat{\mathbf{n}}\,\delta A \cdot \mathbf{V} = -p\hat{\mathbf{n}}\,\delta A \cdot \mathbf{V} = -p\mathbf{V} \cdot \hat{\mathbf{n}}\,\delta A$$

Work is done by rotating shafts, normal stresses, and tangential stresses.

For all fluid particles on the control surface of Fig. 5.6 at the instant considered, power transfer due to fluid normal stress, $\dot{W}_{\text{normal stress}}$, is

$$\dot{W}_{\substack{\text{normal}\\ \text{stress}}} = \int_{\text{cs}} \sigma\mathbf{V} \cdot \hat{\mathbf{n}}\,dA = \int_{\text{cs}} -p\mathbf{V} \cdot \hat{\mathbf{n}}\,dA \tag{5.62}$$

Note that the value of $\dot{W}_{\text{normal stress}}$ for particles on the wetted inside surface of the pipe is zero because $\mathbf{V} \cdot \hat{\mathbf{n}}$ is zero there. Thus, $\dot{W}_{\text{normal stress}}$ can be nonzero only where fluid enters and leaves the control volume. Although only a simple pipe flow was considered, Eq. 5.62 is quite general and the control volume used in this example can serve as a general model for other cases.

Work transfer can also occur at the control surface because of tangential stress forces. Rotating shaft work is transferred by tangential stresses in the shaft material. For a fluid particle, shear stress force power, $\delta\dot{W}_{\text{tangential stress}}$, can be evaluated as the dot product of tangential stress force, $\delta\mathbf{F}_{\text{tangential stress}}$, and the fluid particle velocity, $\mathbf{V}$. That is,

$$\delta\dot{W}_{\text{tangential stress}} = \delta\mathbf{F}_{\text{tangential stress}} \cdot \mathbf{V}$$

For the control volume of Fig. 5.6, the fluid particle velocity is zero everywhere on the wetted inside surface of the pipe. Thus, no tangential stress work is transferred across that portion of the control surface. Furthermore, where fluid crosses the control surface, the tangential stress force is perpendicular to the fluid particle velocity and therefore tangential stress work transfer is also zero there. In general, we select control volumes like the one of Fig. 5.6 and consider fluid tangential stress power transfer to be negligibly small.

Using the information we have developed about power, we can express the first law of thermodynamics for the contents of a control volume by combining Eqs. 5.59, 5.60, and 5.62 to obtain

$$\frac{\partial}{\partial t}\int_{\text{cv}} e\rho\,dV\!\!\!/ + \int_{\text{cs}} e\rho\mathbf{V} \cdot \hat{\mathbf{n}}\,dA = \dot{Q}_{\substack{\text{net}\\ \text{in}}} + \dot{W}_{\substack{\text{shaft}\\ \text{net in}}} - \int_{\text{cs}} p\mathbf{V} \cdot \hat{\mathbf{n}}\,dA \tag{5.63}$$

When the equation for total stored energy (Eq. 5.56) is considered with Eq. 5.63, we obtain the energy equation:

$$\frac{\partial}{\partial t}\int_{\text{cv}} e\rho\,dV\!\!\!/ + \int_{\text{cs}} \left(\check{u} + \frac{p}{\rho} + \frac{V^2}{2} + gz\right)\rho\mathbf{V} \cdot \hat{\mathbf{n}}\,dA = \dot{Q}_{\substack{\text{net}\\ \text{in}}} + \dot{W}_{\substack{\text{shaft}\\ \text{net in}}} \tag{5.64}$$

5.3.2 Application of the Energy Equation

In Eq. 5.64, the term $\partial/\partial t \int_{\text{cv}} e\rho \, d\forall$ represents the time rate of change of the total stored energy, e, of the contents of the control volume. This term is zero when the flow is steady. This term is also zero in the mean when the flow is steady in the mean (cyclical).

In Eq. 5.64, the integrand of

$$\int_{\text{cs}} \left(\check{u} + \frac{p}{\rho} + \frac{V^2}{2} + gz \right) \rho \mathbf{V} \cdot \hat{\mathbf{n}} \, dA$$

can be nonzero only where fluid crosses the control surface ($\mathbf{V} \cdot \hat{\mathbf{n}} \neq 0$). Otherwise, $\mathbf{V} \cdot \hat{\mathbf{n}}$ is zero and the integrand is zero for that portion of the control surface. If the properties within parentheses, $\check{u}$, p/ρ, $V^2/2$, and gz, are all assumed to be uniformly distributed over the flow cross-sectional areas involved, the integration becomes simple and gives

$$\int_{\text{cs}} \left(\check{u} + \frac{p}{\rho} + \frac{V^2}{2} + gz \right) \rho \mathbf{V} \cdot \hat{\mathbf{n}} \, dA = \sum_{\substack{\text{flow} \\ \text{out}}} \left(\check{u} + \frac{p}{\rho} + \frac{V^2}{2} + gz \right) \dot{m} - \sum_{\substack{\text{flow} \\ \text{in}}} \left(\check{u} + \frac{p}{\rho} + \frac{V^2}{2} + gz \right) \dot{m} \qquad \textbf{(5.65)}$$

Furthermore, if there is only one stream entering and leaving the control volume, then

$$\int_{\text{cs}} \left(\check{u} + \frac{p}{\rho} + \frac{V^2}{2} + gz \right) \rho \mathbf{V} \cdot \hat{\mathbf{n}} \, dA = \left(\check{u} + \frac{p}{\rho} + \frac{V^2}{2} + gz \right)_{\text{out}} \dot{m}_{\text{out}} - \left(\check{u} + \frac{p}{\rho} + \frac{V^2}{2} + gz \right)_{\text{in}} \dot{m}_{\text{in}} \qquad \textbf{(5.66)}$$

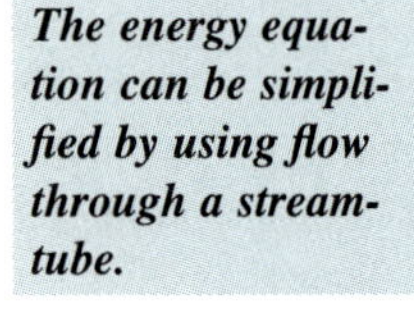

Uniform flow as described above will occur in an infinitesimally small diameter streamtube as illustrated in Fig. 5.7. This kind of streamtube flow is representative of the steady flow of a particle of fluid along a pathline. We can also idealize actual conditions by disregarding nonuniformities in a finite cross section of flow. We call this one-dimensional flow and although such uniform flow rarely occurs in reality, the simplicity achieved with the one-dimensional approximation often justifies its use. More details about the effects of nonuniform distributions of velocities and other fluid flow variables are considered in Section 5.3.4 and in Chapters 7, 8, 9, and 10.

If shaft work is involved, the flow must be unsteady, at least locally (see Refs. 1 and 2). The flow in any fluid machine that involves shaft work is unsteady within that machine. For example, the velocity and pressure at a fixed location near the rotating blades of a fan are unsteady. However, upstream and downstream of the machine, the flow may be steady. Most often shaft work is associated with flow that is unsteady in a recurring or cyclical way. On a time-average basis for flow that is one-dimensional, cyclical, and involves only one

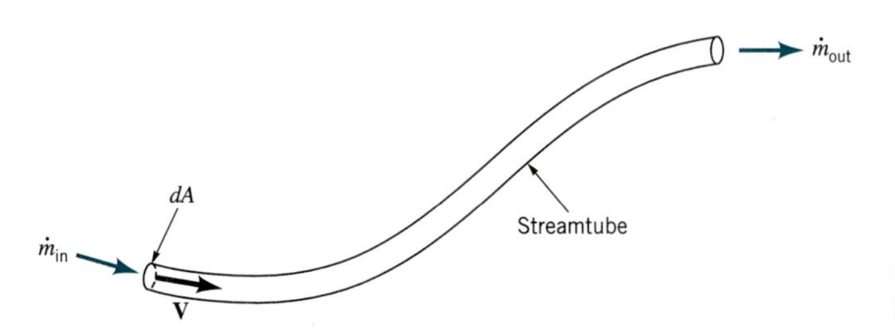

■ **FIGURE 5.7**
Streamtube flow.

stream of fluid entering and leaving the control volume, Eq. 5.64 can be simplified with the help of Eqs. 5.9 and 5.66 to form

$$\dot{m}\left[\check{u}_{\text{out}} - \check{u}_{\text{in}} + \left(\frac{p}{\rho}\right)_{\text{out}} - \left(\frac{p}{\rho}\right)_{\text{in}} + \frac{V_{\text{out}}^2 - V_{\text{in}}^2}{2} + g(z_{\text{out}} - z_{\text{in}})\right] = \dot{Q}_{\substack{\text{net}\\ \text{in}}} + \dot{W}_{\substack{\text{shaft}\\ \text{net in}}} \quad \textbf{(5.67)}$$

We call Eq. 5.67 the *one-dimensional energy equation for steady-in-the-mean flow*. Note that Eq. 5.67 is valid for incompressible and compressible flows. Often, the fluid property called *enthalpy*, $\check{h}$, where

$$\check{h} = \check{u} + \frac{p}{\rho} \quad \textbf{(5.68)}$$

The energy equation is sometimes written in terms of enthalpy.

is used in Eq. 5.67. With enthalpy, the one-dimensional energy equation for steady-in-the-mean flow (Eq. 5.67) is

$$\dot{m}\left[\check{h}_{\text{out}} - \check{h}_{\text{in}} + \frac{V_{\text{out}}^2 - V_{\text{in}}^2}{2} + g(z_{\text{out}} - z_{\text{in}})\right] = \dot{Q}_{\substack{\text{net}\\ \text{in}}} + \dot{W}_{\substack{\text{shaft}\\ \text{net in}}} \quad \textbf{(5.69)}$$

Equation 5.69 is often used for solving compressible flow problems. Examples 5.20 and 5.21 illustrate how Eqs. 5.67 and 5.69 can be used.

EXAMPLE 5.20

A pump delivers water at a steady rate of 300 gal/min as shown in Fig. E5.20. Just upstream of the pump [section (1)] where the pipe diameter is 3.5 in., the pressure is 18 psi. Just downstream of the pump [section (2)] where the pipe diameter is 1 in., the pressure is 60 psi. The change in water elevation across the pump is zero. The rise in internal energy of water, $\check{u}_2 - \check{u}_1$, associated with a temperature rise across the pump is 3000 ft·lb/slug. If the pumping process is considered to be adiabatic, determine the power (hp) required by the pump.

SOLUTION

We include in our control volume the water contained in the pump between its entrance and exit sections. Application of Eq. 5.67 to the contents of this control volume on a time-average basis yields

$$\dot{m}\left[\check{u}_2 - \check{u}_1 + \left(\frac{p}{\rho}\right)_2 - \left(\frac{p}{\rho}\right)_1 + \frac{V_2^2 - V_1^2}{2} + \underbrace{g(z_2 - z_1)}_{0 \text{ (no elevation change)}}\right] = \underbrace{\dot{Q}_{\substack{\text{net}\\ \text{in}}}}_{0 \text{ (adiabatic flow)}} + \dot{W}_{\substack{\text{shaft}\\ \text{net in}}} \quad (1)$$

We can solve directly for the power required by the pump, $\dot{W}_{\text{shaft net in}}$, from Eq. 1, after we first determine the mass flowrate, $\dot{m}$, the speed of flow into the pump, V_1, and the speed of the flow out of the pump, V_2. All other quantities in Eq. 1 are given in the problem statement.

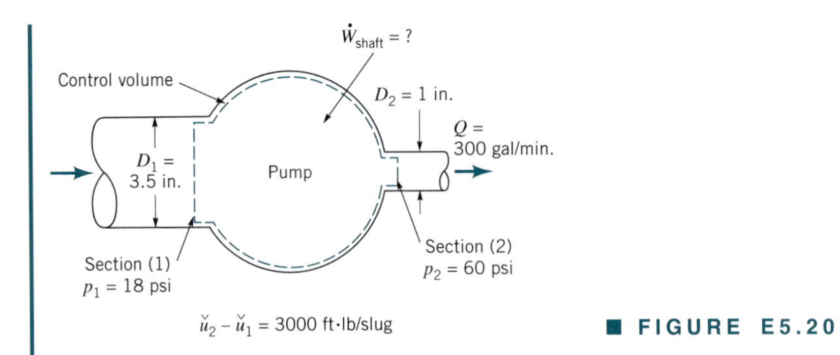

■ FIGURE E5.20

From Eq. 5.6, we get

$$\dot{m} = \rho Q = \frac{(1.94 \text{ slugs/ft}^3)(300 \text{ gal/min})}{(7.48 \text{ gal/ft}^3)(60 \text{ s/min})} = 1.30 \text{ slugs/s} \tag{2}$$

Also from Eq. 5.6,

$$V = \frac{Q}{A} = \frac{Q}{\pi D^2/4}$$

so

$$V_1 = \frac{Q}{A_1} = \frac{(300 \text{ gal/min})4\ (12 \text{ in./ft})^2}{(7.48 \text{ gal/ft}^3)(60 \text{ s/min})\pi(3.5 \text{ in.})^2} = 10.0 \text{ ft/s} \tag{3}$$

and

$$V_2 = \frac{Q}{A_2} = \frac{(300 \text{ gal/min})4\ (12 \text{ in./ft})^2}{(7.48 \text{ gal/ft}^3)(60 \text{ s/min})\pi(1 \text{ in.})^2} = 123 \text{ ft/s} \tag{4}$$

Substituting the values of Eqs. 2, 3, and 4 and values from the problem statement into Eq. 1 we obtain

$$\dot{W}_{\substack{\text{shaft}\\ \text{net in}}} = (1.30 \text{ slugs/s}) \Bigg[(3000 \text{ ft·lb/slug}) + \frac{(60 \text{ psi})(144 \text{ in.}^2/\text{ft}^2)}{(1.94 \text{ slugs/ft}^3)} - \frac{(18 \text{ psi})(144 \text{ in.}^2/\text{ft}^2)}{(1.94 \text{ slugs/ft}^3)} + \frac{(123 \text{ ft/s})^2 - (10.0 \text{ ft/s})^2}{2[1\ (\text{slug·ft})/(\text{lb·s}^2)]} \Bigg] \times \frac{1}{[550(\text{ft·lb/s})/\text{hp}]} = 32.2 \text{ hp}$$

(Ans)

Of the total 32.2 hp, internal energy change accounts for 7.09 hp, the pressure rise accounts for 7.37 hp, and the kinetic energy increase accounts for 17.8 hp.

Steam enters a turbine with a velocity of 30 m/s and enthalpy, $\check{h}_1$, of 3348 kJ/kg (see Fig. E5.21). The steam leaves the turbine as a mixture of vapor and liquid having a velocity of 60 m/s and an enthalpy of 2550 kJ/kg. If the flow through the turbine is adiabatic and changes in elevation are negligible, determine the work output involved per unit mass of steam through-flow.

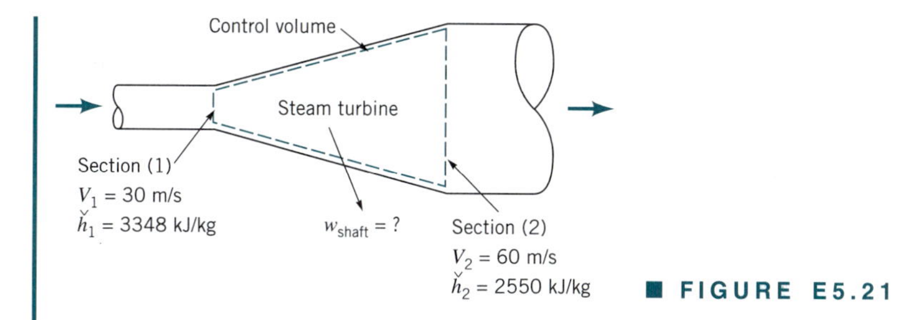

■ FIGURE E5.21

SOLUTION

We use a control volume that includes the steam in the turbine from the entrance to the exit as shown in Fig. E5.21. Applying Eq. 5.69 to the steam in this control volume we get

$$\dot{m}\left[\check{h}_2 - \check{h}_1 + \frac{V_2^2 - V_1^2}{2} + g(z_2 - z_1)\right] = \dot{Q}_{\substack{\text{net}\\ \text{in}}} + \dot{W}_{\substack{\text{shaft}\\ \text{net in}}} \tag{1}$$

where $g(z_2 - z_1) = 0$ (elevation change is negligible) and $\dot{Q}_{\text{net in}} = 0$ (adiabatic flow).

The work output per unit mass of steam through-flow, $w_{\substack{\text{shaft}\\ \text{net in}}}$, can be obtained by dividing Eq. 1 by the mass flow rate, $\dot{m}$, to obtain

$$w_{\substack{\text{shaft}\\ \text{net in}}} = \frac{\dot{W}_{\substack{\text{shaft}\\ \text{net in}}}}{\dot{m}} = \check{h}_2 - \check{h}_1 + \frac{V_2^2 - V_1^2}{2} \tag{2}$$

Since $w_{\text{shaft net out}} = -w_{\text{shaft net in}}$, we obtain

$$w_{\substack{\text{shaft}\\ \text{net out}}} = \check{h}_1 - \check{h}_2 + \frac{V_1^2 - V_2^2}{2}$$

or

$$w_{\substack{\text{shaft}\\ \text{net out}}} = 3348 \text{ kJ/kg} - 2550 \text{ kJ/kg} + \frac{[(30 \text{ m/s})^2 - (60 \text{ m/s})^2][1 \text{ J/(N}\cdot\text{m)}]}{2[1 \text{ (kg}\cdot\text{m)/(N}\cdot\text{s}^2)](1000 \text{ J/kJ})}$$

Thus

$$w_{\substack{\text{shaft}\\ \text{net out}}} = 3348 \text{ kJ/kg} - 2550 \text{ kJ/kg} - 1.35 \text{ kJ/kg} = 797 \text{ kJ/kg} \qquad \textbf{(Ans)}$$

Note that in this particular example, the change in kinetic energy is small in comparison to the difference in enthalpy involved. This is often true in applications involving steam turbines. To determine the power output, $\dot{W}_{\text{shaft}}$, we must know the mass flowrate, $\dot{m}$.

If the flow is steady throughout, one-dimensional, and only one fluid stream is involved, then the shaft work is zero and the energy equation is

$$\dot{m}\left[\check{u}_{\text{out}} - \check{u}_{\text{in}} + \left(\frac{p}{\rho}\right)_{\text{out}} - \left(\frac{p}{\rho}\right)_{\text{in}} + \frac{V_{\text{out}}^2 - V_{\text{in}}^2}{2} + g(z_{\text{out}} - z_{\text{in}})\right] = \dot{Q}_{\substack{\text{net}\\ \text{in}}} \tag{5.70}$$

We call Eq. 5.70 the *one-dimensional, steady flow energy equation.* This equation is valid for incompressible and compressible flows. For compressible flows, enthalpy is most often used in the one-dimensional, steady flow energy equation and, thus, we have

$$\dot{m}\left[\check{h}_{\text{out}} - \check{h}_{\text{in}} + \frac{V_{\text{out}}^2 - V_{\text{in}}^2}{2} + g(z_{\text{out}} - z_{\text{in}})\right] = \dot{Q}_{\substack{\text{net}\\ \text{in}}} \tag{5.71}$$

An example of the application of Eq. 5.70 follows.

EXAMPLE 5.22

A 500-ft waterfall involves steady flow from one large body of water to another. Determine the temperature change associated with this flow.

SOLUTION

To solve this problem we consider a control volume consisting of a small cross section streamtube from the nearly motionless surface of the upper body of water to the nearly motionless surface of the lower body of water as is sketched in Fig. E5.22. We need to determine $T_2 - T_1$. This temperature change is related to the change of internal energy of the water, $\check{u}_2 - \check{u}_1$, by the relationship

$$T_2 - T_1 = \frac{\check{u}_2 - \check{u}_1}{\check{c}} \tag{1}$$

where $\check{c} = 1 \text{ Btu/(lbm}\cdot{}^\circ\text{R)}$ is the specific heat of water. The application of Eq. 5.70 to the contents of this control volume leads to

$$\dot{m}\left[\check{u}_2 - \check{u}_1 + \left(\frac{p}{\rho}\right)_2 - \left(\frac{p}{\rho}\right)_1 + \frac{V_2^2 - V_1^2}{2} + g(z_2 - z_1)\right] = \dot{Q}_{\substack{\text{net}\\ \text{in}}} \tag{2}$$

We assume that the flow is adiabatic. Thus $\dot{Q}_{\text{net in}} = 0$. Also,

$$\left(\frac{p}{\rho}\right)_1 = \left(\frac{p}{\rho}\right)_2 \tag{3}$$

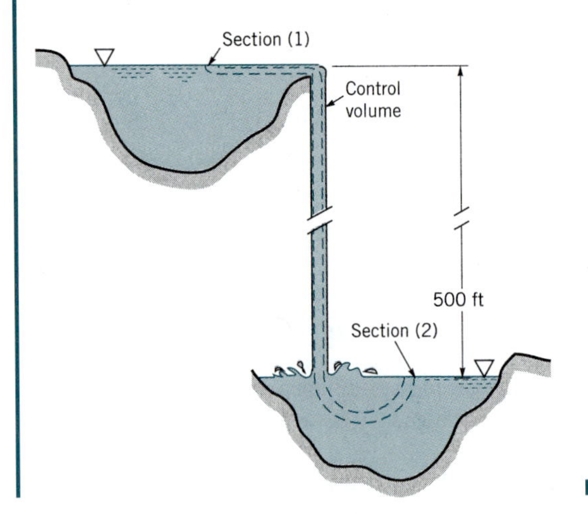

■ FIGURE E5.22

because the flow is incompressible and atmospheric pressure prevails at sections (1) and (2). Furthermore,

$$V_1 = V_2 = 0 \tag{4}$$

because the surface of each large body of water is considered motionless. Thus, Eqs. 1 through 4 combine to yield

$$T_2 - T_1 = \frac{g(z_1 - z_2)}{\check{c}}$$

or

$$T_2 - T_1 = \frac{(32.2 \text{ ft/s}^2)(500 \text{ ft})}{[1 \text{ Btu/(lbm}\cdot{}^\circ\text{R)}][32.2 \text{ (lbm}\cdot\text{ft)/(lb}\cdot\text{s}^2\text{)}](778 \text{ ft}\cdot\text{lb/Btu})} = 0.643\ {}^\circ\text{R} \quad \textbf{(Ans)}$$

Note that it takes a considerable change of potential energy to produce even a small increase in temperature.

A form of the energy equation that is most often used to solve incompressible flow problems is developed in the next section.

5.3.3 Comparison of the Energy Equation with the Bernoulli Equation

When the one-dimensional energy equation for steady-in-the-mean flow, Eq. 5.67, is applied to a flow that is steady, Eq. 5.67 becomes the one-dimensional, steady-flow energy equation, Eq. 5.70. The only difference between Eq. 5.67 and Eq. 5.70 is that shaft power, $\dot{W}_{\text{shaft net in}}$, is zero if the flow is steady throughout the control volume (fluid machines involve locally unsteady flow). If in addition to being steady, the flow is incompressible, we get from Eq. 5.70

$$\dot{m}\left[\check{u}_{\text{out}} - \check{u}_{\text{in}} + \frac{p_{\text{out}}}{\rho} - \frac{p_{\text{in}}}{\rho} + \frac{V_{\text{out}}^2 - V_{\text{in}}^2}{2} + g(z_{\text{out}} - z_{\text{in}})\right] = \dot{Q}_{\substack{\text{net}\\\text{in}}} \tag{5.72}$$

Dividing Eq. 5.72 by the mass flowrate, $\dot{m}$, and rearranging terms we obtain

With certain assumptions, the energy equation and the Bernoulli equation are similar.

$$\frac{p_{\text{out}}}{\rho} + \frac{V_{\text{out}}^2}{2} + gz_{\text{out}} = \frac{p_{\text{in}}}{\rho} + \frac{V_{\text{in}}^2}{2} + gz_{\text{in}} - (\check{u}_{\text{out}} - \check{u}_{\text{in}} - q_{\substack{\text{net}\\\text{in}}}) \tag{5.73}$$

where

$$q_{\substack{\text{net}\\\text{in}}} = \frac{\dot{Q}_{\text{net in}}}{\dot{m}}$$

is the heat transfer rate per mass flowrate, or heat transfer per unit mass. Note that Eq. 5.73 involves energy per unit mass and is applicable to one-dimensional flow of a single stream of fluid between two sections or flow along a streamline between two sections.

If the steady, incompressible flow we are considering also involves negligible viscous effects (frictionless flow), then the Bernoulli equation, Eq. 3.7, can be used to describe what happens between two sections in the flow as

$$p_{\text{out}} + \frac{\rho V_{\text{out}}^2}{2} + \gamma z_{\text{out}} = p_{\text{in}} + \frac{\rho V_{\text{in}}^2}{2} + \gamma z_{\text{in}} \tag{5.74}$$

where $\gamma = \rho g$ is the specific weight of the fluid. To get Eq. 5.74 in terms of energy per unit mass, so that it can be compared directly with Eq. 5.73, we divide Eq. 5.74 by density, ρ, and obtain

$$\frac{p_{\text{out}}}{\rho} + \frac{V_{\text{out}}^2}{2} + gz_{\text{out}} = \frac{p_{\text{in}}}{\rho} + \frac{V_{\text{in}}^2}{2} + gz_{\text{in}} \tag{5.75}$$

A comparison of Eqs. 5.73 and 5.75 prompts us to conclude that

$$\check{u}_{\text{out}} - \check{u}_{\text{in}} - q_{\underset{\text{in}}{\text{net}}} = 0 \tag{5.76}$$

when the steady incompressible flow is frictionless. For steady incompressible flow with friction, we learn from experience that

$$\check{u}_{\text{out}} - \check{u}_{\text{in}} - q_{\underset{\text{in}}{\text{net}}} > 0 \tag{5.77}$$

In Eqs. 5.73 and 5.75, we can consider the combination of variables

$$\frac{p}{\rho} + \frac{V^2}{2} + gz$$

A loss of useful energy occurs because of friction.

as equal to *useful* or *available energy*. Thus, from inspection of Eqs. 5.73 and 5.75, we can conclude that $\check{u}_{\text{out}} - \check{u}_{\text{in}} - q_{\text{net in}}$ represents the *loss* of useful or available energy that occurs in an incompressible fluid flow because of friction. In equation form we have

$$\check{u}_{\text{out}} - \check{u}_{\text{in}} - q_{\underset{\text{in}}{\text{net}}} = \text{loss} \tag{5.78}$$

For a frictionless flow, Eqs. 5.73 and 5.75 tell us that loss equals zero.

It is often convenient to express Eq. 5.73 in terms of loss as

$$\frac{p_{\text{out}}}{\rho} + \frac{V_{\text{out}}^2}{2} + gz_{\text{out}} = \frac{p_{\text{in}}}{\rho} + \frac{V_{\text{in}}^2}{2} + gz_{\text{in}} - \text{loss} \tag{5.79}$$

An example of the application of Eq. 5.79 follows.

Compare the volume flowrates associated with two different vent configurations, a cylindrical hole in the wall having a diameter of 120 mm and the same diameter cylindrical hole in the wall but with a well-rounded entrance (see Fig. E5.23). The room pressure is held constant at 1.0 kPa above atmospheric pressure. Both vents exhaust into the atmosphere. As discussed in Section 8.4.2, the loss in available energy associated with flow through the cylindrical vent from the room to the vent exit is $0.5V_2^2/2$ where V_2 is the uniformly distributed exit velocity of air. The loss in available energy associated with flow through the rounded entrance vent from the room to the vent exit is $0.05V_2^2/2$, where V_2 is the uniformly distributed exit velocity of air.

SOLUTION

We use the control volume for each vent sketched in Fig. E5.23. What is sought is the flowrate, $Q = A_2V_2$, where A_2 is the vent exit cross-sectional area, and V_2 is the uniformly distributed exit velocity. For both vents, application of Eq. 5.79 leads to

0 (no elevation change)

$$\frac{p_2}{\rho} + \frac{V_2^2}{2} + gz_2 = \frac{p_1}{\rho} + \frac{V_1^2}{2} + gz_1 - {}_1\text{loss}_2 \tag{1}$$

$0 \; (V_1 \approx 0)$

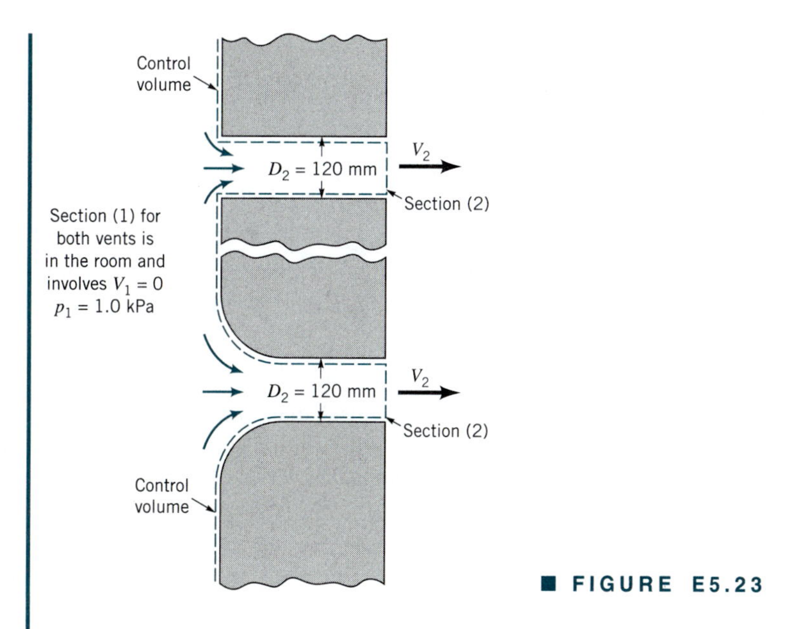

■ FIGURE E5.23

where $_1\text{loss}_2$ is the loss between sections (1) and (2). Solving Eq. 1 for V_2 we get

$$V_2 = \sqrt{2\left[\left(\frac{p_1 - p_2}{\rho}\right) - {}_1\text{loss}_2\right]} \quad \textbf{(2)}$$

Since

$$_1\text{loss}_2 = K_L \frac{V_2^2}{2} \quad \textbf{(3)}$$

where K_L is the loss coefficient ($K_L = 0.5$ and 0.05 for the two vent configurations involved), we can combine Eqs. 2 and 3 to get

$$V_2 = \sqrt{2\left[\left(\frac{p_1 - p_2}{\rho}\right) - K_L \frac{V_2^2}{2}\right]} \quad \textbf{(4)}$$

Solving Eq. 4 for V_2 we obtain

$$V_2 = \sqrt{\frac{p_1 - p_2}{\rho[(1 + K_L)/2]}} \quad \textbf{(5)}$$

Therefore, for flowrate, Q, we obtain

$$Q = A_2 V_2 = \frac{\pi D_2^2}{4} \sqrt{\frac{p_1 - p_2}{\rho[(1 + K_L)/2]}} \quad \textbf{(6)}$$

For the rounded entrance cylindrical vent, Eq. 6 gives

$$Q = \frac{\pi(120 \text{ mm})^2}{4(1000 \text{ mm/m})^2} \sqrt{\frac{(1.0 \text{ kPa})(1000 \text{ Pa/kPa})[1(\text{N/m}^2)/(\text{Pa})]}{(1.23 \text{ kg/m}^3)[(1 + 0.05)/2][1(\text{N}\cdot\text{s}^2)/(\text{kg}\cdot\text{m})]}}$$

or

$$Q = 0.445 \text{ m}^3/\text{s} \quad \textbf{(Ans)}$$

For the cylindrical vent, Eq. 6 gives us

$$Q = \frac{\pi(120 \text{ mm})^2}{4(1000 \text{ mm/m})^2} \sqrt{\frac{(1.0 \text{ kPa})(1000 \text{ Pa/kPa})[1(\text{N/m}^2)/(\text{Pa})]}{(1.23 \text{ kg/m}^3)[(1 + 0.5)/2][1(\text{N}\cdot\text{s}^2)/(\text{kg}\cdot\text{m})]}}$$

or

$$Q = 0.372 \text{ m}^3/\text{s} \qquad \textbf{(Ans)}$$

Note that the rounded entrance vent passes more air than the cylindrical vent because the loss associated with the rounded entrance vent is less than with the cylindrical vent.

An important group of fluid mechanics problems involves one-dimensional, incompressible, steady-in-the-mean flow with friction and shaft work. Included in this category are constant density flows through pumps, blowers, fans, and turbines. For this kind of flow, Eq. 5.67 becomes

$$\dot{m}\left[\check{u}_{\text{out}} - \check{u}_{\text{in}} + \frac{p_{\text{out}}}{\rho} - \frac{p_{\text{in}}}{\rho} + \frac{V_{\text{out}}^2 - V_{\text{in}}^2}{2} + g(z_{\text{out}} - z_{\text{in}})\right] = \dot{Q}_{\substack{\text{net}\\\text{in}}} + \dot{W}_{\substack{\text{shaft}\\\text{net in}}} \qquad \textbf{(5.80)}$$

Dividing Eq. 5.80 by mass flowrate and using the work per unit mass, $w_{\substack{\text{shaft}\\\text{net in}}} = \dot{W}_{\substack{\text{shaft}\\\text{net in}}}/\dot{m}$, we obtain

$$\frac{p_{\text{out}}}{\rho} + \frac{V_{\text{out}}^2}{2} + gz_{\text{out}} = \frac{p_{\text{in}}}{\rho} + \frac{V_{\text{in}}^2}{2} + gz_{\text{in}} + w_{\substack{\text{shaft}\\\text{net in}}} - (\check{u}_{\text{out}} - \check{u}_{\text{in}} - q_{\substack{\text{net}\\\text{in}}}) \qquad \textbf{(5.81)}$$

If the flow is steady throughout, Eq. 5.81 becomes identical to Eq. 5.73, and the previous observation that $\check{u}_{\text{out}} - \check{u}_{\text{in}} - q_{\text{net in}}$ equals the loss of available energy is valid. Thus, we conclude that Eq. 5.81 can be expressed as

The mechanical energy equation can be written in terms of energy per unit mass.

$$\boxed{\frac{p_{\text{out}}}{\rho} + \frac{V_{\text{out}}^2}{2} + gz_{\text{out}} = \frac{p_{\text{in}}}{\rho} + \frac{V_{\text{in}}^2}{2} + gz_{\text{in}} + w_{\substack{\text{shaft}\\\text{net in}}} - \text{loss}} \qquad \textbf{(5.82)}$$

V5.7

This is a form of the energy equation for steady-in-the-mean flow that is often used for incompressible flow problems. It is sometimes called the *mechanical energy* equation or the *extended Bernoulli* equation. Note that Eq. 5.82 involves energy per unit mass (ft·lb/slug = ft^2/s^2 or N·m/kg = m^2/s^2).

According to Eq. 5.82, when the shaft work is into the control volume, as for example with a pump, a larger amount of loss will result in more shaft work being required for the same rise in available energy. Similarly, when the shaft work is out of the control volume (e.g., a turbine), a larger loss will result in less shaft work out for the same drop in available energy. Designers spend a great deal of effort on minimizing losses in fluid flow components. The following examples demonstrate why losses should be kept as small as possible in fluid systems.

EXAMPLE 5.24

An axial-flow ventilating fan driven by a motor that delivers 0.4 kW of power to the fan blades produces a 0.6-m-diameter axial stream of air having a speed of 12 m/s. The flow upstream of the fan involves negligible speed. Determine how much of the work to the air actually produces a useful effect, that is, a rise in available energy and estimate the fluid mechanical efficiency of this fan.

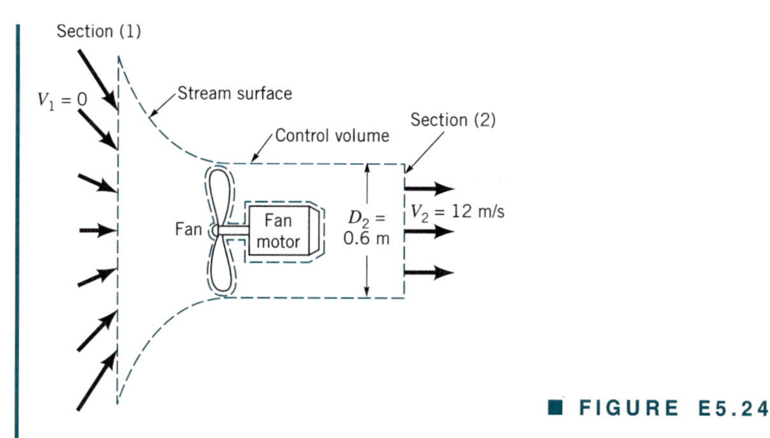

■ FIGURE E5.24

SOLUTION

We select a fixed and nondeforming control volume as is illustrated in Fig. E5.24. The application of Eq. 5.82 to the contents of this control volume leads to

$$w_{\substack{\text{shaft}\\\text{net in}}} - \text{loss} = \left(\frac{p_2}{\rho} + \frac{V_2^2}{2} + gz_2\right) - \left(\frac{p_1}{\rho} + \frac{V_1^2}{2} + gz_1\right) \tag{1}$$

0 (atmospheric pressures cancel); 0 ($V_1 \approx 0$); 0 (no elevation change)

where $w_{\text{shaft net in}}$ − loss is the amount of work added to the air that produces a useful effect. Equation 1 leads to

$$w_{\substack{\text{shaft}\\\text{net in}}} - \text{loss} = \frac{V_2^2}{2} = \frac{(12 \text{ m/s})^2}{2[1(\text{kg}\cdot\text{m})/(\text{N}\cdot\text{s}^2)]} = 72.0 \text{ N}\cdot\text{m/kg} \tag{2}$$

(Ans)

A reasonable estimate of *efficiency*, η, would be the ratio of amount of work that produces a useful effect, Eq. 2, to the amount of work delivered to the fan blades. That is

$$\eta = \frac{w_{\substack{\text{shaft}\\\text{net in}}} - \text{loss}}{w_{\substack{\text{shaft}\\\text{net in}}}} \tag{3}$$

To calculate the efficiency we need a value of $w_{\text{shaft net in}}$ which is related to the power delivered to the blades, $\dot{W}_{\text{shaft net in}}$. We note that

$$w_{\substack{\text{shaft}\\\text{net in}}} = \frac{\dot{W}_{\substack{\text{shaft}\\\text{net in}}}}{\dot{m}} \tag{4}$$

where the mass flowrate, $\dot{m}$, is (from Eq. 5.6)

$$\dot{m} = \rho A V = \rho \frac{\pi D_2^2}{4} V_2 \tag{5}$$

For fluid density, ρ, we use 1.23 kg/m^3 (standard air) and thus from Eqs. 4 and 5 we obtain

$$w_{\substack{\text{shaft}\\\text{net in}}} = \frac{\dot{W}_{\substack{\text{shaft}\\\text{net in}}}}{(\rho\pi D_2^2/4)V_2} = \frac{(0.4 \text{ kW})[1000 \text{ (N}\cdot\text{m)/(s}\cdot\text{kW)}]}{(1.23 \text{ kg/m}^3)[(\pi)(0.6 \text{ m})^2/4](12 \text{ m/s})}$$

or

$$w_{\substack{\text{shaft}\\ \text{net in}}} = 95.8 \text{ N·m/kg} \tag{6}$$

From Eqs. 2, 3, and 6 we obtain

$$\eta = \frac{72.0 \text{ N·m/kg}}{95.8 \text{ N·m/kg}} = 0.752 \qquad \textbf{(Ans)}$$

Note that only 75% of the power that was delivered to the air resulted in a useful effect, and thus 25% of the shaft power is lost to air friction.

If Eq. 5.82, which involves energy per unit mass, is multiplied by fluid density, ρ, we obtain

$$p_{\text{out}} + \frac{\rho V_{\text{out}}^2}{2} + \gamma z_{\text{out}} = p_{\text{in}} + \frac{\rho V_{\text{in}}^2}{2} + \gamma z_{\text{in}} + \rho w_{\substack{\text{shaft}\\ \text{net in}}} - \rho(\text{loss}) \tag{5.83}$$

where $\gamma = \rho g$ is the specific weight of the fluid. Equation 5.83 involves *energy per unit volume* and the units involved are identical with those used for pressure ($\text{ft·lb/ft}^3 = \text{lb/ft}^2$ or $\text{N·m/m}^3 = \text{N/m}^2$).

V5.8

If Eq. 5.82 is divided by the acceleration of gravity, g, we get

$$\boxed{\frac{p_{\text{out}}}{\gamma} + \frac{V_{\text{out}}^2}{2g} + z_{\text{out}} = \frac{p_{\text{in}}}{\gamma} + \frac{V_{\text{in}}^2}{2g} + z_{\text{in}} + h_s - h_L} \tag{5.84}$$

where

The energy equation written in terms of energy per unit weight involves heads.

$$h_s = w_{\text{shaft net in}}/g = \frac{\dot{W}_{\substack{\text{shaft}\\ \text{net in}}}}{\dot{m}g} = \frac{\dot{W}_{\substack{\text{shaft}\\ \text{net in}}}}{\gamma Q} \tag{5.85}$$

and $h_L = \text{loss}/g$. Equation 5.84 involves *energy per unit weight* (ft·lb/lb = ft or N·m/N = m). In Section 3.7, we introduced the notion of "head," which is energy per unit weight. Units of length (e.g., ft, m) are used to quantify the amount of head involved. If a turbine is in the control volume, the notation $h_s = -h_T$ (with $h_T > 0$) is sometimes used, particularly in the field of hydraulics. For a pump in the control volume, $h_s = h_P$. The quantity h_T is termed the *turbine head* and h_P is the *pump head*. The loss term, h_L is often referred to as *head loss*. The turbine head is often written as

$$h_T = -(h_s + h_L)_T$$

where the subscript T refers to the turbine component of the contents of the control volume only. The quantity h_T is the actual head drop across the turbine and is the sum of the shaft work head out of the turbine and the head loss within the turbine. When a pump is in the control volume,

$$h_P = (h_s - h_L)_P$$

is often used where h_P is the actual head rise across the pump and is equal to the difference between the shaft work head into the pump and the head loss within the pump. Notice that the h_L used for the turbine and the pump is the head loss within that component only. When h_s is used in Eq. 5.84, h_L involves all losses including those within the turbine or compressor. When h_T or h_P is used for h_s, then h_L includes all losses except those associated with the turbine or pump flows.

EXAMPLE 5.25

The pump shown in Fig. E5.25 adds 10 horsepower to the water as it pumps 2 ft^3/s from the lower lake to the upper lake. The elevation difference between the lake surfaces is 30 ft. Determine the head loss and power loss associated with this flow.

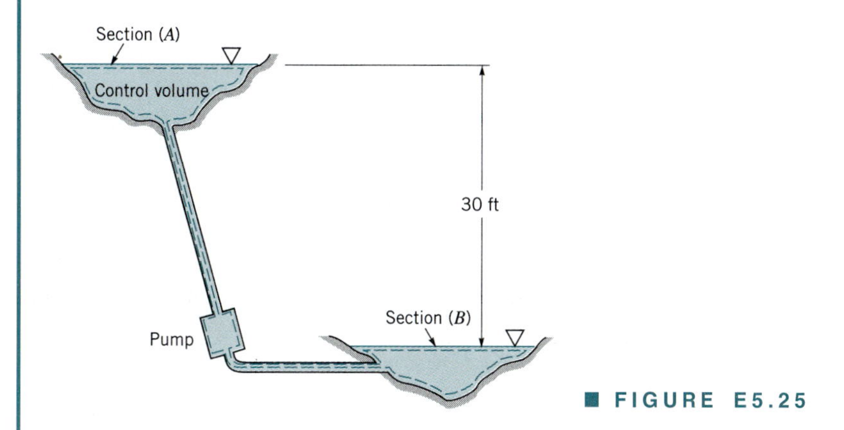

■ FIGURE E5.25

SOLUTION

The energy equation (Eq. 5.84) for this flow is

$$\frac{p_A}{\gamma} + \frac{V_A^2}{2g} + z_A = \frac{p_B}{\gamma} + \frac{V_B^2}{2g} + z_B + h_s - h_L \tag{1}$$

where points A and B (corresponding to "out" and "in" in Eq. 5.84) are located on the lake surfaces. Thus, $p_A = p_B = 0$ and $V_A = V_B = 0$ so that Eq. 1 becomes

$$h_L = h_s + z_B - z_A \tag{2}$$

where $z_B = 0$ and $z_A = 30$ ft. The pump head is obtained from Eq. 5.85 as

$$\begin{aligned} h_s &= \dot{W}_{\text{shaft net in}}/\gamma\, Q \\ &= (10 \text{ hp})(550 \text{ ft}\cdot\text{lb/s/hp})/(62.4 \text{ lb/ft}^3)(2 \text{ ft}^3/\text{s}) \\ &= 44.1 \text{ ft} \end{aligned}$$

Hence, from Eq. 2,

$$h_L = 44.1 \text{ ft} - 30 \text{ ft} = 14.1 \text{ ft} \qquad \textbf{(Ans)}$$

Note that in this example the purpose of the pump is to lift the water (a 30 ft head) and overcome the head loss (a 14.1 ft head); it does not, overall, alter the water's pressure or velocity.

The power lost due to friction can be obtained from Eq. 5.85 as

$$\begin{aligned} \dot{W}_{\text{loss}} &= \gamma\, Q h_L = (62.4 \text{ lb/ft}^3)(2 \text{ ft}^3/\text{s})(14.1 \text{ ft}) \\ &= 1760 \text{ ft}\cdot\text{lb/s } (1 \text{ hp}/550 \text{ ft}\cdot\text{lb/s}) \qquad \textbf{(Ans)} \\ &= 3.20 \text{ hp} \end{aligned}$$

The remaining 10 hp − 3.20 hp = 6.80 hp that the pump adds to the water is used to lift the water from the lower to the upper lake. This energy is not "lost," but it is stored as potential energy.

A comparison of the energy equation and the Bernoulli equation has led to the concept of loss of available energy in incompressible fluid flows with friction. In Chapter 8, we discuss in detail some methods for estimating loss in incompressible flows with friction. In Section 5.4 and Chapter 11, we demonstrate that loss of available energy is also an important factor to consider in compressible flows with friction.

5.3.4 Application of the Energy Equation to Nonuniform Flows

The forms of the energy equation discussed in Sections 5.3.2 and 5.3.3 are applicable to one-dimensional flows, flows that are approximated with uniform velocity distributions where fluid crosses the control surface.

If the velocity profile at any section where flow crosses the control surface is not uniform, inspection of the energy equation for a control volume, Eq. 5.64, suggests that the integral

$$\int_{cs} \frac{V^2}{2} \rho \mathbf{V} \cdot \hat{\mathbf{n}}\, dA$$

will require special attention. The other terms of Eq. 5.64 can be accounted for as already discussed in Sections 5.3.2 and 5.3.3.

For one stream of fluid entering and leaving the control volume, we can define the relationship

$$\int_{cs} \frac{V^2}{2} \rho \mathbf{V} \cdot \hat{\mathbf{n}}\, dA = \dot{m}\left(\frac{\alpha_{out}\overline{V}_{out}^2}{2} - \frac{\alpha_{in}\overline{V}_{in}^2}{2}\right)$$

The kinetic energy coefficient is used to account for nonuniform flows.

where α is the *kinetic energy coefficient* and $\overline{V}$ is the average velocity defined earlier in Eq. 5.7. From the above we can conclude that

$$\frac{\dot{m}\alpha\overline{V}^2}{2} = \int_A \frac{V^2}{2} \rho \mathbf{V} \cdot \hat{\mathbf{n}}\, dA$$

for flow through surface area A of the control surface. Thus,

$$\alpha = \frac{\int_A (V^2/2)\rho \mathbf{V} \cdot \hat{\mathbf{n}}\, dA}{\dot{m}\overline{V}^2/2} \tag{5.86}$$

It can be shown that for any velocity profile, $\alpha \geq 1$, with $\alpha = 1$ only for uniform flow. For nonuniform velocity profiles, the energy equation on an energy per unit mass basis for the incompressible flow of one stream of fluid through a control volume that is steady in the mean is

$$\frac{p_{out}}{\rho} + \frac{\alpha_{out}\overline{V}_{out}^2}{2} + gz_{out} = \frac{p_{in}}{\rho} + \frac{\alpha_{in}\overline{V}_{in}^2}{2} + gz_{in} + w_{\substack{shaft\\ net\ in}} - \text{loss} \tag{5.87}$$

On an energy per unit volume basis we have

$$p_{out} + \frac{\rho\alpha_{out}\overline{V}_{out}^2}{2} + \gamma z_{out} = p_{in} + \frac{\rho\alpha_{in}\overline{V}_{in}^2}{2} + \gamma z_{in} + \rho w_{\substack{shaft\\ net\ in}} - \rho(\text{loss}) \tag{5.88}$$

and on an energy per unit weight or head basis we have

$$\frac{p_{\text{out}}}{\gamma} + \frac{\alpha_{\text{out}}\overline{V}_{\text{out}}^2}{2g} + z_{\text{out}} = \frac{p_{\text{in}}}{\gamma} + \frac{\alpha_{\text{in}}\overline{V}_{\text{in}}^2}{2g} + z_{\text{in}} + \frac{w_{\substack{\text{shaft}\\\text{net in}}}}{g} - h_L \tag{5.89}$$

The following examples illustrate the use of the kinetic energy coefficient.

EXAMPLE 5.26

The small fan shown in Fig. E5.26 moves air at a mass flowrate of 0.1 kg/min. Upstream of the fan, the pipe diameter is 60 mm, the flow is laminar, the velocity distribution is parabolic, and the kinetic energy coefficient, α_1, is equal to 2.0. Downstream of the fan, the pipe diameter is 30 mm, the flow is turbulent, the velocity profile is quite uniform, and the kinetic energy coefficient, α_2, is equal to 1.08. If the rise in static pressure across the fan is 0.1 kPa and the fan motor draws 0.14 W, compare the value of loss calculated: (a) assuming uniform velocity distributions, (b) considering actual velocity distributions.

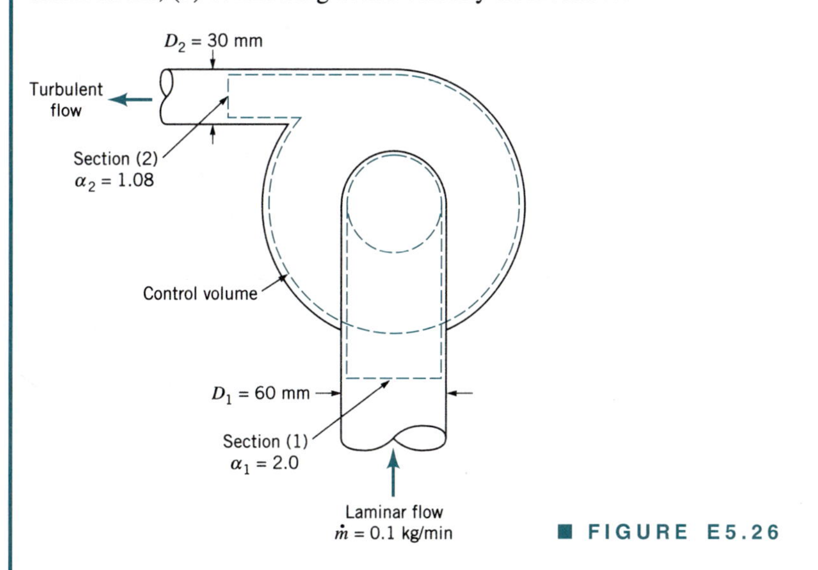

■ FIGURE E5.26

SOLUTION

Application of Eq. 5.87 to the contents of the control volume shown in Fig. E5.26 leads to

$$\frac{p_2}{\rho} + \frac{\alpha_2 \overline{V}_2^2}{2} + \underbrace{gz_2 = \frac{p_1}{\rho} + \frac{\alpha_1 \overline{V}_1^2}{2} + gz_1}_{0 \text{ (change in } gz \text{ is negligible)}} - \text{loss} + w_{\substack{\text{shaft}\\\text{net in}}} \tag{1}$$

or solving Eq. 1 for loss we get

$$\text{loss} = w_{\substack{\text{shaft}\\\text{net in}}} - \left(\frac{p_2 - p_1}{\rho}\right) + \frac{\alpha_1 \overline{V}_1^2}{2} - \frac{\alpha_2 \overline{V}_2^2}{2} \tag{2}$$

To proceed further, we need values of $w_{\text{shaft net in}}$, $\overline{V}_1$, and $\overline{V}_2$. These quantities can be obtained as follows. For shaft work

$$w_{\substack{\text{shaft}\\ \text{net in}}} = \frac{\text{power to fan motor}}{\dot{m}}$$

or

$$w_{\substack{\text{shaft}\\ \text{net in}}} = \frac{(0.14 \text{ W})([1 \text{ (N·m/s)/W}]}{0.1 \text{ kg/min}}(60 \text{ s/min})[1 \text{ (kg·m)/(N·s}^2)] = 84.0 \text{ N·m/kg} \quad \textbf{(3)}$$

For the average velocity at section (1), $\overline{V}_1$, from Eq. 5.11 we obtain

$$\begin{aligned}\overline{V}_1 &= \frac{\dot{m}}{\rho A_1}\\ &= \frac{\dot{m}}{\rho(\pi D_1^2/4)} \quad \textbf{(4)}\\ &= \frac{(0.1 \text{ kg/min})}{(1.23 \text{ kg/m}^3)[\pi(60 \text{ mm})^2/4][(60 \text{ s/min})/(1000 \text{ mm/m})^2]}\\ &= 0.479 \text{ m/s}\end{aligned}$$

For the average velocity at section (2), $\overline{V}_2$,

$$\overline{V}_2 = \frac{(0.1 \text{ kg/min})}{(1.23 \text{ kg/m}^3)[\pi(30 \text{ mm})^2/4][(60 \text{ s/min})/(1000 \text{ mm/m})^2]} = 1.92 \text{ m/s} \quad \textbf{(5)}$$

(a) For the assumed uniform velocity profiles ($\alpha_1 = \alpha_2 = 1.0$), Eq. 2 yields

$$\text{loss} = w_{\substack{\text{shaft}\\ \text{net in}}} - \left(\frac{p_2 - p_1}{\rho}\right) + \frac{\overline{V}_1^2}{2} - \frac{\overline{V}_2^2}{2} \quad \textbf{(6)}$$

Using Eqs. 3, 4, and 5 and the pressure rise given in the problem statement, Eq. 6 gives

$$\text{loss} = 84.0\,\frac{\text{N·m}}{\text{kg}} - \frac{(0.1 \text{ kPa})(1000 \text{ Pa/kPa})(1 \text{ N/m}^2/\text{Pa})}{1.23 \text{ kg/m}^3} + \frac{(0.479 \text{ m/s})^2}{2[1 \text{ (kg·m)/(N·s}^2)]} - \frac{(1.92 \text{ m/s})^2}{2[1 \text{ (kg·m)/(N·s}^2)]}$$

or

$$\begin{aligned}\text{loss} &= 84.0 \text{ N·m/kg} - 81.3 \text{ N·m/kg} + 0.115 \text{ N·m/kg} - 1.84 \text{ N·m/kg}\\ &= 0.975 \text{ N·m/kg}\end{aligned}$$ **(Ans)**

(b) For the actual velocity profiles ($\alpha_1 = 2$, $\alpha_2 = 1.08$), Eq. 1 gives

$$\text{loss} = w_{\substack{\text{shaft}\\ \text{net in}}} - \left(\frac{p_2 - p_1}{\rho}\right) + \alpha_1\frac{\overline{V}_1^2}{2} - \alpha_2\frac{\overline{V}_2^2}{2} \quad \textbf{(7)}$$

If we use Eqs. 3, 4, and 5 and the given pressure rise, Eq. 7 yields

$$\text{loss} = 84 \text{ N·m/kg} - \frac{(0.1 \text{ kPa})(1000 \text{ Pa/kPa})(1 \text{ N/m}^2/\text{Pa})}{1.23 \text{ kg/m}^3} + \frac{2(0.479 \text{ m/s})^2}{2[1 \text{ (kg·m)/(N·s}^2)]} - \frac{1.08(1.92 \text{ m/s})^2}{2[1 \text{ (kg·m)/(N·s}^2)]}$$

or

$$\begin{aligned}\text{loss} &= 84.0 \text{ N·m/kg} - 81.3 \text{ N·m/kg} + 0.230 \text{ N·m/kg} - 1.99 \text{ N·m/kg}\\ &= 0.940 \text{ N·m/kg}\end{aligned}$$ **(Ans)**

The difference in loss calculated assuming uniform velocity profiles and actual velocity profiles is not large compared to $w_{\text{shaft net in}}$ for this fluid flow situation.

EXAMPLE 5.27

Apply Eq. 5.87 to the flow situation of Example 5.14 and develop an expression for the fluid pressure drop that occurs between sections (1) and (2). By comparing the equation for pressure drop obtained presently with the result of Example 5.14, obtain an expression for loss between sections (1) and (2).

SOLUTION

Application of Eq. 5.87 to the flow of Example 5.14 (see Fig. E5.14) leads to

$$\frac{p_2}{\rho} + \frac{\alpha_2 \overline{w}_2^2}{2} + gz_2 = \frac{p_1}{\rho} + \frac{\alpha_1 \overline{w}_1^2}{2} + gz_1 - \text{loss} + \underbrace{w_{\substack{\text{shaft}\\ \text{net in}}}}_{0 \text{ (no shaft work)}} \tag{1}$$

Solving Eq. 1 for the pressure drop, $p_1 - p_2$, we obtain

$$p_1 - p_2 = \rho\left[\frac{\alpha_2 \overline{w}_2^2}{2} - \frac{\alpha_1 \overline{w}_1^2}{2} + g(z_2 - z_1) + \text{loss}\right] \tag{2}$$

Since the fluid velocity at section (1), w_1, is uniformly distributed over cross-sectional area A_1, the corresponding kinetic energy coefficient, α_1, is equal to 1.0. The kinetic energy coefficient at section (2), α_2, needs to be determined from the velocity profile distribution given in Example 5.4. Using Eq. 5.86 we get

$$\alpha_2 = \frac{\int_{A_2} \rho w_2^3 \, dA_2}{\dot{m}\overline{w}_2^2} \tag{3}$$

Substituting the parabolic velocity profile equation into Eq. 3 we obtain

$$\alpha_2 = \frac{\rho \int_0^R (2w_1)^3[1 - (r/R)^2]^3 2\pi r \, dr}{(\rho A_2 \overline{w}_2)\overline{w}_2^2}$$

From conservation of mass, since $A_1 = A_2$

$$w_1 = \overline{w}_2 \tag{4}$$

Then, substituting Eq. 4 into Eq. 3, we obtain

$$\alpha_2 = \frac{\rho 8 \overline{w}_2^3 2\pi \int_0^R [1 - (r/R)^2]^3 r \, dr}{\rho \pi R^2 \overline{w}_2^3}$$

or

$$\alpha_2 = \frac{16}{R^2}\int_0^R [1 - 3(r/R)^2 + 3(r/R)^4 - (r/R)^6] r \, dr = 2 \tag{5}$$

Now we combine Eqs. 2 and 5 to get

$$p_1 - p_2 = \rho \left[\frac{2.0\overline{w}_2^2}{2} - \frac{1.0\overline{w}_1^2}{2} + g(z_2 - z_1) + \text{loss} \right] \tag{6}$$

However, from conservation of mass $\overline{w}_2 = \overline{w}_1 = \overline{w}$ so that Eq. 6 becomes

$$p_1 - p_2 = \frac{\rho \overline{w}^2}{2} + \rho g(z_2 - z_1) + \rho(\text{loss}) \tag{7}$$

The term associated with change in elevation, $\rho g(z_2 - z_1)$, is equal to the weight per unit cross-sectional area, $\mathcal{W}/A$, of the water contained between sections (1) and (2) at any instant,

$$\rho g(z_2 - z_1) = \frac{\mathcal{W}}{A} \tag{8}$$

Thus, combining Eqs. 7 and 8 we get

$$p_1 - p_2 = \frac{\rho \overline{w}^2}{2} + \frac{\mathcal{W}}{A} + \rho(\text{loss}) \tag{9}$$

The pressure drop between sections (1) and (2) is due to:

1. The change in kinetic energy between sections (1) and (2) associated with going from a uniform velocity profile to a parabolic velocity profile.
2. The weight of the water column; that is, hydrostatic pressure effect.
3. Viscous loss.

Comparing Eq. 9 for pressure drop with the one obtained in Example 5.14 (i.e., the answer of Example 5.14) we obtain

$$\frac{\rho \overline{w}^2}{2} + \frac{\mathcal{W}}{A} + \rho(\text{loss}) = \frac{\rho \overline{w}^2}{3} + \frac{R_z}{A} + \frac{\mathcal{W}}{A} \tag{10}$$

or

$$\text{loss} = \frac{R_z}{\rho A} - \frac{\overline{w}^2}{6} \tag{Ans}$$

We conclude that while some of the pipe wall friction force, R_z, resulted in loss of available energy, a portion of this friction, $\rho A \overline{w}^2/6$, led to the velocity profile change.

5.3.5 Combination of the Energy Equation and the Moment-of-Momentum Equation[4]

If Eq. 5.82 is used for one-dimensional incompressible flow through a turbomachine, we can use Eq. 5.54, developed in Section 5.2.4 from the moment-of-momentum equation (Eq. 5.42), to evaluate shaft work. This application of both Eqs. 5.54 and 5.82 allows us to ascertain the

[4]This section may be omitted without loss of continuity in the text material. This section should not be considered without prior study of Sections 5.2.3 and 5.2.4. All of these sections are recommended for those interested in Chapter 12.

amount of loss that occurs in incompressible turbomachine flows as is demonstrated in Example 5.28.

For the fan of Example 5.19, show that only some of the shaft power into the air is converted into a useful effect. Develop a meaningful efficiency equation and a practical means for estimating lost shaft energy.

SOLUTION

We use the same control volume used in Example 5.19. Application of Eq. 5.82 to the contents of this control volume yields

$$\frac{p_2}{\rho} + \frac{V_2^2}{2} + gz_2 = \frac{p_1}{\rho} + \frac{V_1^2}{2} + gz_1 + w_{\substack{\text{shaft}\\ \text{net in}}} - \text{loss} \tag{1}$$

As in Example 5.26, we can see with Eq. 1 that a "useful effect" in this fan can be defined as

$$\text{useful effect} = w_{\substack{\text{shaft}\\ \text{net in}}} - \text{loss} = \left(\frac{p_2}{\rho} + \frac{V_2^2}{2} + gz_2\right) - \left(\frac{p_1}{\rho} + \frac{V_1^2}{2} + gz_1\right) \tag{2}$$ **(Ans)**

In other words, only a portion of the shaft work delivered to the air by the fan blades is used to increase the available energy of the air; the rest is lost because of fluid friction.

A meaningful efficiency equation would involve the ratio of shaft work converted into a useful effect (Eq. 2) to shaft work into the air, $w_{\text{shaft net in}}$. Thus, we can express efficiency, η, as

$$\eta = \frac{w_{\substack{\text{shaft}\\ \text{net in}}} - \text{loss}}{w_{\substack{\text{shaft}\\ \text{net in}}}} \tag{3}$$

However, when Eq. 5.54, which was developed from the moment-of-momentum equation (Eq. 5.42), is applied to the contents of the control volume of Fig. E5.19, we obtain

$$w_{\substack{\text{shaft}\\ \text{net in}}} = +U_2 V_{\theta 2} \tag{4}$$

Combining Eqs. 2, 3, and 4, we obtain

$$\eta = \frac{[(p_2/\rho) + (V_2^2/2) + gz_2] - [(p_1/\rho) + (V_1^2/2) + gz_1]}{U_2 V_{\theta 2}} \tag{5}$$ **(Ans)**

Equation 5 provides us with a practical means to evaluate the efficiency of the fan of Example 5.19.

Combining Eqs. 2 and 4, we obtain

$$\text{loss} = U_2 V_{\theta 2} - \left[\left(\frac{p_2}{\rho} + \frac{V_2^2}{2} + gz_2\right) - \left(\frac{p_1}{\rho} + \frac{V_1^2}{2} + gz_1\right)\right] \tag{6}$$ **(Ans)**

Equation 6 provides us with a useful method of evaluating the loss due to fluid friction in the fan of Example 5.19 in terms of fluid mechanical variables that can be measured.

5.4 Second Law of Thermodynamics—Irreversible Flow[5]

The second law of thermodynamics affords us with a means to formalize the inequality

$$\check{u}_2 - \check{u}_1 - q_{\substack{\text{net}\\\text{in}}} \geq 0 \tag{5.90}$$

The second law of thermodynamics formalizes the notion of loss.

for steady, incompressible, one-dimensional flow with friction (see Eq. 5.73). In this section we continue to develop the notion of loss of useful or available energy for flow with friction. Minimization of loss of available energy in any flow situation is of obvious engineering importance.

5.4.1 Semi-infinitesimal Control Volume Statement of the Energy Equation

If we apply the one-dimensional, steady flow energy equation, Eq. 5.70, to the contents of a control volume that is infinitesimally thin as illustrated in Fig 5.8, the result is

$$\dot{m}\left[d\check{u} + d\left(\frac{p}{\rho}\right) + d\left(\frac{V^2}{2}\right) + g\,(dz)\right] = \delta\dot{Q}_{\substack{\text{net}\\\text{in}}} \tag{5.91}$$

For all pure substances including common engineering working fluids, such as air, water, oil, and gasoline, the following relationship is valid (see, for example, Ref. 3).

$$T\,ds = d\check{u} + pd\left(\frac{1}{\rho}\right) \tag{5.92}$$

where T is the absolute temperature and s is the *entropy* per unit mass.

Combining Eqs. 5.91 and 5.92 we get

$$\dot{m}\left[T\,ds - pd\left(\frac{1}{\rho}\right) + d\left(\frac{p}{\rho}\right) + d\left(\frac{V^2}{2}\right) + g\,dz\right] = \delta\dot{Q}_{\substack{\text{net}\\\text{in}}}$$

or, dividing through by $\dot{m}$ and letting $\delta q_{\substack{\text{net}\\\text{in}}} = \delta\dot{Q}_{\substack{\text{net}\\\text{in}}}/\dot{m}$, we obtain

$$\frac{dp}{\rho} + d\left(\frac{V^2}{2}\right) + g\,dz = -(T\,ds - \delta q_{\substack{\text{net}\\\text{in}}}) \tag{5.93}$$

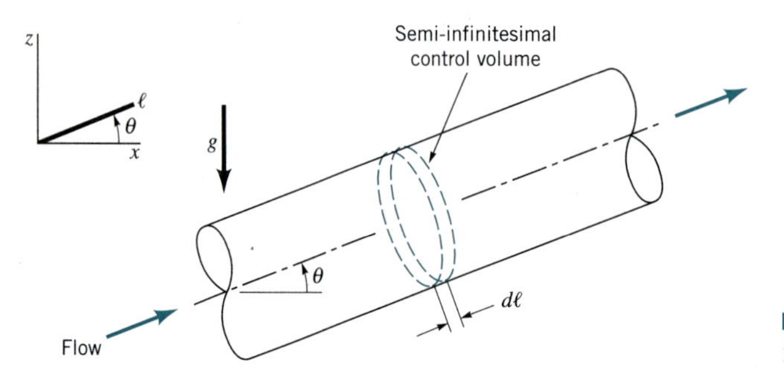

■ **FIGURE 5.8** **Semi-infinitesimal control volume.**

[5]This entire section may be omitted without loss of continuity in the text material.

5.4.2 Semi-infinitesimal Control Volume Statement of the Second Law of Thermodynamics

The second law of thermodynamics involves entropy, heat transfer, and temperature.

A general statement of the second law of thermodynamics is

$$\frac{D}{Dt}\int_{\text{sys}} s\rho \, d\mathcal{V} \geq \sum \left(\frac{\delta \dot{Q}_{\substack{\text{net}\\ \text{in}}}}{T}\right)_{\text{sys}} \tag{5.94}$$

or in words,

the time rate of increase of the entropy of a system	$\geq$	sum of the ratio of net heat transfer rate into system to absolute temperature for each particle of mass in the system receiving heat from surroundings

The right-hand side of Eq. 5.94 is identical for the system and control volume at the instant when system and control volume are coincident; thus,

$$\sum \left(\frac{\delta \dot{Q}_{\substack{\text{net}\\ \text{in}}}}{T}\right)_{\text{sys}} = \sum \left(\frac{\delta \dot{Q}_{\substack{\text{net}\\ \text{in}}}}{T}\right)_{\text{cv}} \tag{5.95}$$

With the help of the Reynolds transport theorem (Eq. 4.19) the system time derivative can be expressed for the contents of the coincident control volume that is fixed and nondeforming. Using Eq. 4.19, we obtain

$$\frac{D}{Dt}\int_{\text{sys}} s\rho \, d\mathcal{V} = \frac{\partial}{\partial t}\int_{\text{cv}} s\rho \, d\mathcal{V} + \int_{\text{cs}} s\rho \mathbf{V} \cdot \hat{\mathbf{n}} \, dA \tag{5.96}$$

For a fixed, nondeforming control volume, Eqs. 5.94, 5.95, and 5.96 combine to give

$$\frac{\partial}{\partial t}\int_{\text{cv}} s\rho \, d\mathcal{V} + \int_{\text{cs}} s\rho \mathbf{V} \cdot \hat{\mathbf{n}} \, dA \geq \sum \left(\frac{\delta \dot{Q}_{\substack{\text{net}\\ \text{in}}}}{T}\right)_{\text{cv}} \tag{5.97}$$

At any instant for steady flow

$$\frac{\partial}{\partial t}\int_{\text{cv}} s\rho \, d\mathcal{V} = 0 \tag{5.98}$$

If the flow consists of only one stream through the control volume and if the properties are uniformly distributed (one-dimensional flow), Eqs. 5.97 and 5.98 lead to

$$\dot{m}(s_{\text{out}} - s_{\text{in}}) \geq \sum \frac{\delta \dot{Q}_{\substack{\text{net}\\ \text{in}}}}{T} \tag{5.99}$$

For the infinitesimally thin control volume of Fig. 5.8, Eq. 5.99 yields

$$\dot{m} \, ds \geq \sum \frac{\delta \dot{Q}_{\substack{\text{net}\\ \text{in}}}}{T} \tag{5.100}$$

If all of the fluid in the infinitesimally thin control volume is considered as being at a uniform temperature, T, then from Eq. 5.100 we get

$$T \, ds \geq \delta q_{\substack{\text{net}\\ \text{in}}}$$

or

$$T\,ds - \delta q_{\substack{\text{net}\\\text{in}}} \geq 0 \tag{5.101}$$

The equality is for any reversible (frictionless) process; the inequality is for all irreversible (friction) processes.

5.4.3 Combination of the Equations of the First and Second Laws of Thermodynamics

Combining Eqs. 5.93 and 5.101, we conclude that

$$-\left[\frac{dp}{\rho} + d\left(\frac{V^2}{2}\right) + g\,dz\right] \geq 0 \tag{5.102}$$

The equality is for any steady, reversible (frictionless) flow, an important example being flow for which the Bernoulli equation (Eq. 3.7) is applicable. The inequality is for all steady, irreversible (friction) flows. The actual amount of the inequality has physical significance. It represents the extent of loss of useful or available energy which occurs because of irreversible flow phenomena including viscous effects. Thus, Eq. 5.102 can be expressed as

$$-\left[\frac{dp}{\rho} + d\left(\frac{V^2}{2}\right) + g\,dz\right] = \delta(\text{loss}) = (T\,ds - \delta q_{\substack{\text{net}\\\text{in}}}) \tag{5.103}$$

The irreversible flow loss is zero for frictionless flow.

The irreversible flow loss is zero for a frictionless flow and greater than zero for a flow with frictional effects. Note that when the flow is frictionless, Eq. 5.103 multiplied by density, ρ, is identical to Eq. 3.5. Thus, for steady frictionless flow, Newton's second law of motion (see Section 3.1) and the first and second laws of thermodynamics lead to the same differential equation,

$$\frac{dp}{\rho} + d\left(\frac{V^2}{2}\right) + g\,dz = 0 \tag{5.104}$$

If some shaft work is involved, then the flow must be at least locally unsteady in a cyclical way and the appropriate form of the energy equation for the contents of an infinitesimally thin control volume can be developed starting with Eq. 5.67. The resulting equation is

$$-\left[\frac{dp}{\rho} + d\left(\frac{V^2}{2}\right) + g\,dz\right] = \delta(\text{loss}) - \delta w_{\substack{\text{shaft}\\\text{net in}}} \tag{5.105}$$

Equations 5.103 and 5.105 are valid for incompressible and compressible flows. If we combine Eqs. 5.92 and 5.103, we obtain

$$d\check{u} + p\,d\left(\frac{1}{\rho}\right) - \delta q_{\substack{\text{net}\\\text{in}}} = \delta(\text{loss}) \tag{5.106}$$

For incompressible flow, $d(1/\rho) = 0$ and, thus, from Eq. 5.106,

$$d\check{u} - \delta q_{\substack{\text{net}\\\text{in}}} = \delta(\text{loss}) \tag{5.107}$$

Applying Eq. 5.107 to a finite control volume, we obtain

$$\check{u}_{\text{out}} - \check{u}_{\text{in}} - q_{\substack{\text{net} \\ \text{in}}} = \text{loss}$$

which is the same conclusion we reached earlier (see Eq. 5.78) for incompressible flows.

For compressible flow, $d(1/\rho) \neq 0$, and thus when we apply Eq. 5.106 to a finite control volume we obtain

$$\check{u}_{\text{out}} - \check{u}_{\text{in}} + \int_{\text{in}}^{\text{out}} p\, d\left(\frac{1}{\rho}\right) - q_{\substack{\text{net} \\ \text{in}}} = \text{loss} \tag{5.108}$$

indicating that $u_{\text{out}} - u_{\text{in}} - q_{\text{net in}}$ is not equal to loss.

5.4.4 Application of the Loss Form of the Energy Equation

Zero loss is associated with the Bernoulli equation.

Steady flow along a pathline in an incompressible and frictionless flow field provides a simple application of the loss form of the energy equation (Eq. 5.105). We start with Eq. 5.105 and integrate it term by term from one location on the pathline, section (1), to another one downstream, section (2). Note that because the flow is frictionless, loss $= 0$. Also, because the flow is steady throughout, $w_{\text{shaft net in}} = 0$. Since the flow is incompressible, the density is constant. The control volume in this case is an infinitesimally small diameter streamtube (Fig. 5.7). The resultant equation is

$$\frac{p_2}{\rho} + \frac{V_2^2}{2} + gz_2 = \frac{p_1}{\rho} + \frac{V_1^2}{2} + gz_1 \tag{5.109}$$

which is identical to the Bernoulli equation (Eq. 3.7) already discussed in Chapter 3.

If the frictionless and steady pathline flow of the fluid particle considered above was compressible, application of Eq. 5.105 would yield

$$\int_1^2 \frac{dp}{\rho} + \frac{V_2^2}{2} + gz_2 = \frac{V_1^2}{2} + gz_1 \tag{5.110}$$

To carry out the integration required, $\int_1^2 (dp/\rho)$, a relationship between fluid density, ρ, and pressure, p, must be known. If the frictionless compressible flow we are considering is adiabatic and involves the flow of an ideal gas, it is shown in Section 11.1 that

$$\frac{p}{\rho^k} = \text{constant} \tag{5.111}$$

where $k = c_p/c_v$ is the ratio of gas specific heats, c_p and c_v, which are properties of the fluid. Using Eq. 5.111 we get

$$\int_1^2 \frac{dp}{\rho} = \frac{k}{k-1}\left(\frac{p_2}{\rho_2} - \frac{p_1}{\rho_1}\right) \tag{5.112}$$

Thus, Eqs. 5.110 and 5.112 lead to

$$\frac{k}{k-1}\frac{p_2}{\rho_2} + \frac{V_2^2}{2} + gz_2 = \frac{k}{k-1}\frac{p_1}{\rho_1} + \frac{V_1^2}{2} + gz_1 \tag{5.113}$$

Note that this equation is identical to Eq. 3.24. An example application of Eqs. 5.109 and 5.113 follows.

EXAMPLE 5.29

Air steadily expands adiabatically and without friction from stagnation conditions of 100 psia and 520 °R to 14.7 psia. Determine the velocity of the expanded air assuming (a) incompressible flow, (b) compressible flow.

SOLUTION

(a) If the flow is considered incompressible, the Bernoulli equation, Eq. 5.109, can be applied to flow through an infinitesimal cross-sectional streamtube, like the one in Fig. 5.7, from the stagnation state (1) to the expanded state (2). From Eq. 109 we get

0 (1 is the stagnation state)

$$\frac{p_2}{\rho} + \frac{V_2^2}{2} + \cancel{gz_2} = \frac{p_1}{\rho} + \cancel{\frac{V_1^2}{2}} + \cancel{gz_1} \tag{1}$$

0 (changes in gz are negligible for air flow)

or

$$V_2 = \sqrt{2\left(\frac{p_1 - p_2}{\rho}\right)}$$

We can calculate the density at state (1) by assuming that air behaves like an ideal gas,

$$\rho = \frac{p_1}{RT_1} = \frac{(100 \text{ psia})(144 \text{ in.}^2/\text{ft}^2)}{(1716 \text{ ft·lb/slug·°R})(520 \text{ °R})} = 0.0161 \text{ slug/ft}^3 \tag{2}$$

Thus,

$$V_2 = \sqrt{\frac{2(100 \text{ psia} - 14.7 \text{ psia})(144 \text{ in.}^2/\text{ft}^2)}{(0.016 \text{ slug/ft}^3)[1 \text{ (lb·s}^2)/(\text{slug·ft})]}} = 1240 \text{ ft/s} \qquad \textbf{(Ans)}$$

The assumption of incompressible flow is not valid in this case since for air a change from 100 psia to 14.7 psia would undoubtedly result in a significant density change.

(b) If the flow is considered compressible, Eq. 5.113 can be applied to the flow through an infinitesimal cross-sectional control volume, like the one in Fig. 5.7, from the stagnation state (1) to the expanded state (2). We obtain

0 (1 is the stagnation state)

$$\frac{k}{k-1}\frac{p_2}{\rho_2} + \frac{V_2^2}{2} + \cancel{gz_2} = \frac{k}{k-1}\frac{p_1}{\rho_1} + \cancel{\frac{V_1^2}{2}} + \cancel{gz_1} \tag{3}$$

0 (changes in gz are negligible for air flow)

or

$$V_2 = \sqrt{\frac{2k}{k-1}\left(\frac{p_1}{\rho_1} - \frac{p_2}{\rho_2}\right)} \tag{4}$$

Given in the problem statement are values of p_1 and p_2. A value of ρ_1 was calculated earlier (Eq. 2). To determine ρ_2 we need to make use of a property relationship for reversible (frictionless) and adiabatic flow of an ideal gas that is derived in Chapter 11; namely,

$$\frac{p}{\rho^k} = \text{constant} \tag{5}$$

where $k = 1.4$ for air. Solving Eq. 5 for ρ_2 we get

$$\rho_2 = \rho_1 \left(\frac{p_2}{p_1}\right)^{1/k}$$

or

$$\rho_2 = (0.0161 \text{ slug/ft}^3) \left[\frac{14.7 \text{ psia}}{100 \text{ psia}}\right]^{1/1.4} = 0.00409 \text{ slug/ft}^3$$

Then, from Eq. 4,

$$V_2 = \sqrt{\frac{(2)(1.4)}{1.4 - 1}\left(\frac{100 \text{ psia}}{0.0161 \text{ slug/ft}^3} - \frac{14.7 \text{ psia}}{0.00409 \text{ slug/ft}^3}\right)\frac{(144 \text{ in.}^2/\text{ft}^2)}{1 \text{ (slug·ft)/(lb·s}^2)}}$$

or

$$V_2 = 1620 \text{ ft/s} \qquad \textbf{(Ans)}$$

A considerable difference exists between the air velocities calculated assuming incompressible and compressible flow.

References

1. Eck, B., *Technische Stromungslehre*, Springer-Verlag, Berlin, Germany, 1957.
2. Dean, R. C., ''On the Necessity of Unsteady Flow in Fluid Machines,'' *ASME Journal of Basic Engineering* 81D; 24–28, March 1959.
3. Moran, M. J., and Shapiro, H. N., *Fundamentals of Engineering Thermodynamics*, 2nd Ed., Wiley, New York, 1992.

Review Problems

Note: Problems designated with (R) are review problems. The phrases within parentheses refer to the main topics to be used in solving the problems. Complete, detailed solutions to these review problems can be found in the supplement titled *Student Solution Manual for Fundamentals of Fluid Mechanics*, by Munson, Young, and Okiishi (John Wiley and Sons, New York, 1997).

5.1R (Continuity equation) Water flows steadily through a 2-in.-inside-diameter pipe at the rate of 200 gal/min. The 2-in. pipe branches into two 1-in.-inside-diameter pipes. If the average velocity in one of the 1-in. pipes is 30 ft/s, what is the average velocity in the other 1-in. pipe?

(ANS: 51.7 ft/s)

5.2R (Continuity equation) Air (assumed incompressible) flows steadily into the square inlet of an air scoop with the nonuniform velocity profile indicated in Fig. P5.2R. The air exits as a uniform flow through a round pipe 1 ft in diameter. **(a)** Determine the average velocity at the exit plane. **(b)** In one minute, how many pounds of air pass through the scoop?

(ANS: 191 ft/s; 688 lb/min)

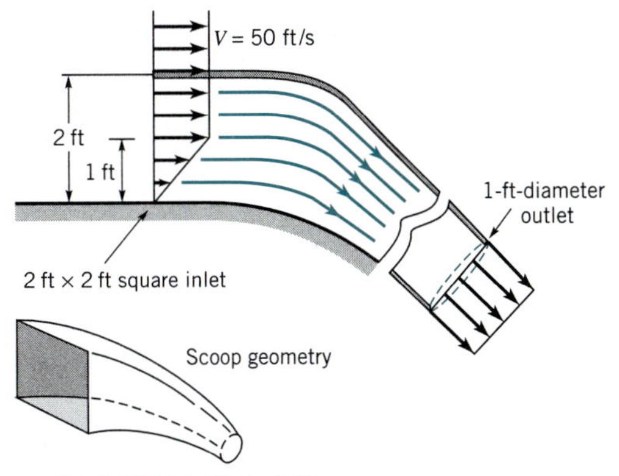

■ FIGURE P5.2R

5.3R (Continuity equation) Water at 0.1 m^3/s and alcohol ($SG = 0.8$) at 0.3 m^3/s are mixed in a y-duct as shown in Fig. P5.3R. What is the average density of the mixture of alcohol and water?

(ANS: 849 kg/m^3)

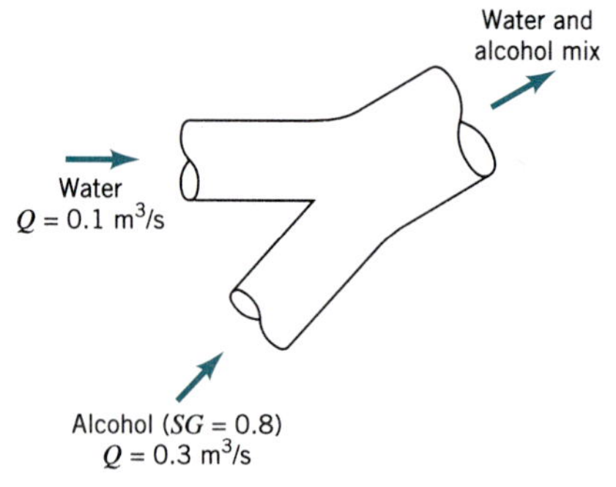

■ FIGURE P5.3R

5.4R (Average velocity) The flow in an open channel has a velocity distribution

$$\mathbf{V} = U(y/h)^{1/5}\hat{\mathbf{i}}\ \text{ft/s}$$

where U = free-surface velocity, y = perpendicular distance from the channel bottom in feet, and h = depth of the channel in feet. Determine the average velocity of the channel stream as a fraction of U.

(ANS: 0.833)

5.5R (Linear momentum) Water flows through a right angle valve at the rate of 1000 lbm/s as is shown in Fig. P5.5R. The pressure just upstream of the valve is 90 psi and the pressure drop across the valve is 50 psi. The inside diameters of the valve inlet and exit pipes are 12 and 24 in. If the flow through the valve occurs in a horizontal plane, determine the x and y components of the force exerted by the valve on the water.

(ANS: 18,200 lb; 10,800 lb)

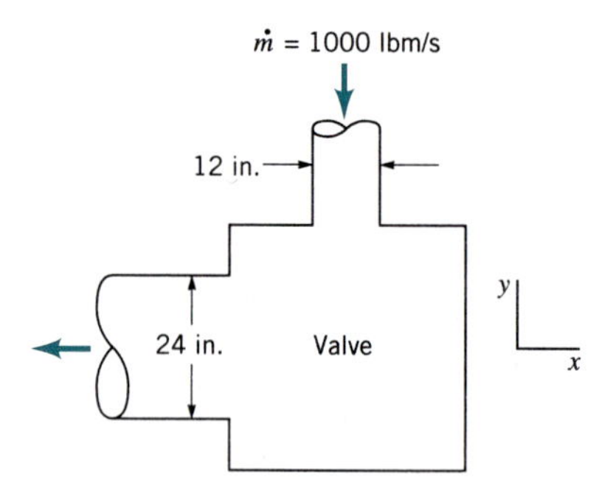

■ FIGURE P5.5R

5.6R (Linear momentum) A horizontal circular jet of air strikes a stationary flat plate as indicated in Fig. P5.6R. The jet velocity is 40 m/s and the jet diameter is 30 mm. If the air velocity magnitude remains constant as the air flows over the plate surface in the directions shown, determine: **(a)** the magnitude of F_A, the anchoring force required to hold the plate stationary, **(b)** the fraction of mass flow along the plate surface in each of the two directions shown, **(c)** the magnitude of F_A, the anchoring force required to allow the plate to move to the right at a constant speed of 10 m/s.

(ANS: 0.696 N; 0.933 and 0.0670; 0.391 N)

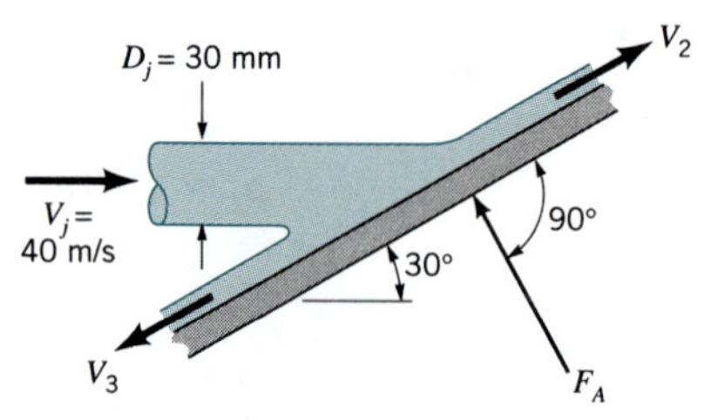

■ FIGURE P5.6R

5.7R (Linear momentum) An axisymmetric device is used to partially "plug" the end of the round pipe shown in Fig. P5.7R. The air leaves in a radial direction with a speed of 50 ft/s as indicated. Gravity and viscous forces are negligible. Determine the **(a)** flowrate through the pipe, **(b)** gage pressure at point (1), **(c)** gage pressure at the tip of the plug, point (2), **(d)** force, F, needed to hold the plug in place.

(ANS: 23.6 ft³/s; 1.90 lb/ft²; 2.97 lb/ft²; 3.18 lb)

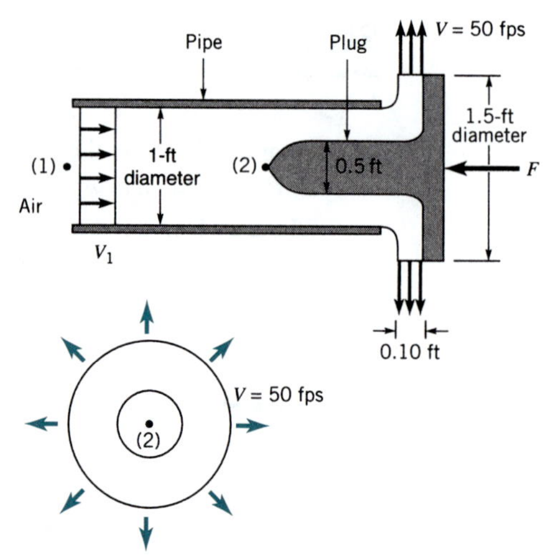

■ FIGURE P5.7R

5.8R (Linear momentum) A nozzle is attached to an 80-mm inside-diameter flexible hose. The nozzle area is 500 mm². If the delivery pressure of water at the nozzle inlet is 700 kPa, could you hold the hose and nozzle stationary? Explain.

(ANS: yes, 707 N or 159 lb)

5.9R (Linear momentum) A horizontal air jet having a velocity of 50 m/s and a diameter of 20 mm strikes the inside surface of a hollow hemisphere as indicated in Fig. P5.9R. How large is the horizontal anchoring force needed to hold the hemisphere in place? The magnitude of velocity of the air remains constant.

(ANS: 1.93 N)

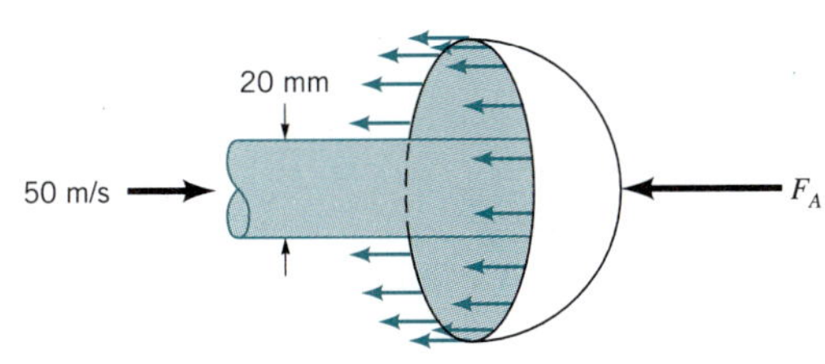

■ FIGURE P5.9R

5.10R (Linear momentum) Determine the magnitude of the horizontal component of the anchoring force required to hold in place the 10-foot-wide sluice gate shown in Fig. P5.10R. Compare this result with the size of the horizontal component of the anchoring force required to hold in place the sluice gate when it is closed and the depth of water upstream is 6 ft.

(ANS: 5310 lb; 11,200 lb)

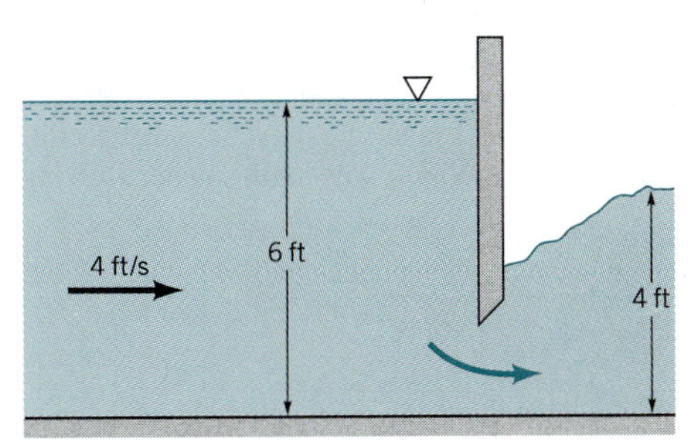

■ FIGURE P5.10R

5.11R (Linear momentum) Two jets of liquid, one with specific gravity 1.0 and the other with specific gravity 1.3, collide and form one homogeneous jet as shown in Fig. P5.11R. Determine the speed, V, and the direction, θ, of the combined jet. Gravity is negligible.

(ANS: 6.97 ft/s; 70.3 deg)

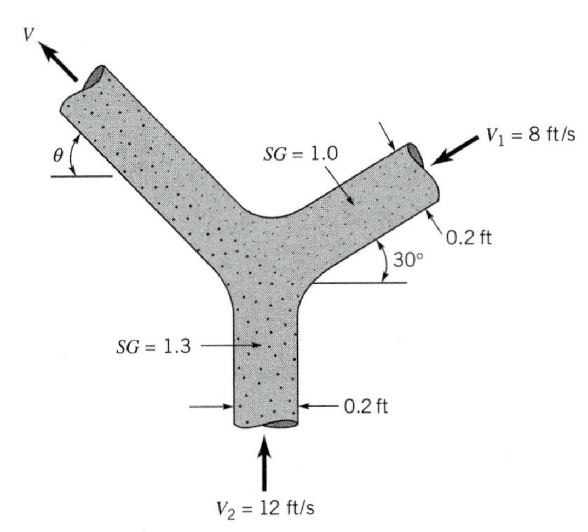

■ FIGURE P5.11R

5.12R (Linear momentum) Water flows vertically upward in a circular cross-sectional pipe as shown in Fig. P5.12R. At section (1), the velocity profile over the cross-sectional area is uniform. At section (2), the velocity profile is

$$\mathbf{V} = w_c\left(\frac{R - r}{R}\right)\hat{\mathbf{k}}$$

where $\mathbf{V}$ = local velocity vector, w_c = centerline velocity in the axial direction, R = pipe radius, and r = radius from pipe axis. Develop an expression for the fluid pressure drop that occurs between sections (1) and (2).

(ANS: $p_1 - p_2 = R_z/\pi R^2 + 0.50\,\rho w_1^2 + g\rho h$, where R_z = friction force)

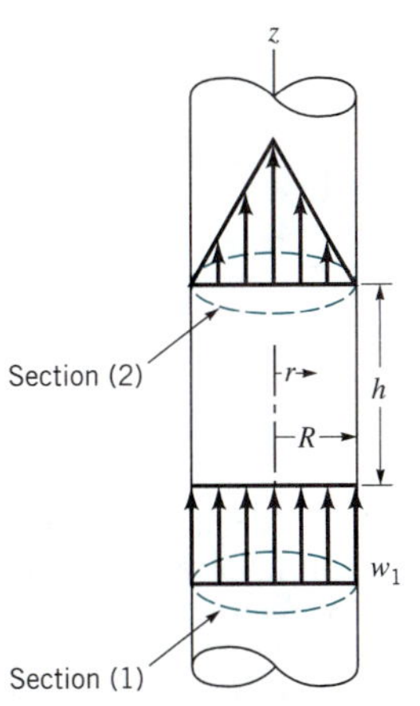

■ FIGURE P5.12R

5.13R (Moment-of-momentum) A lawn sprinkler is constructed from pipe with $\frac{1}{4}$-in. inside diameter as indicated in Fig. P5.13R. Each arm is 6 in. in length. Water flows through the sprinkler at the rate of 1.5 lb/s. A force of 3 lb positioned halfway along one arm holds the sprinkler stationary. Compute the angle, θ, which the exiting water stream makes with the tangential direction. The flow leaves the nozzles in the horizontal plane.

(ANS: 23.9 deg)

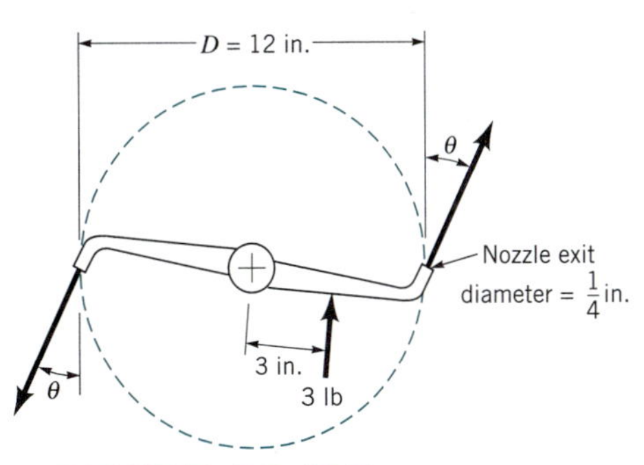

■ FIGURE P5.13R

5.14R (Moment-of-momentum) A water turbine with radial flow has the dimensions shown in Fig. P5.14R. The absolute entering velocity is 15 m/s, and it makes an angle of 30°

with the tangent to the rotor. The absolute exit velocity is directed radially inward. The angular speed of the rotor is 30 rpm. Find the power delivered to the shaft of the turbine.

(ANS: −7.68 MW)

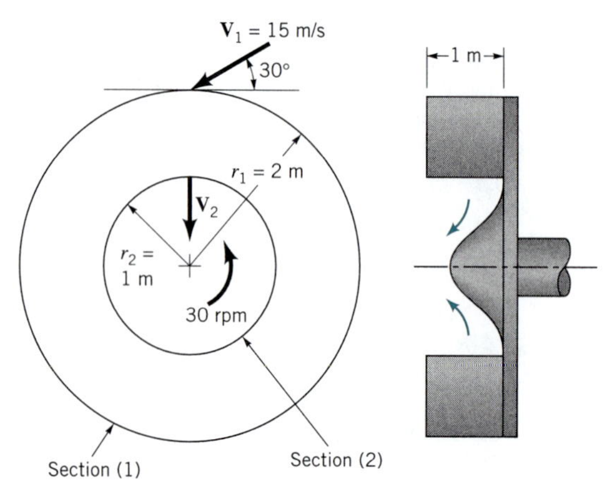

■ FIGURE P5.14R

5.15R (Moment-of-momentum) The single stage, axial-flow turbomachine shown in Fig. P5.15R involves water flow at a volumetric flowrate of 11 m^3/s. The rotor revolves at 600 rpm. The inner and outer radii of the annular flow path through the stage are 0.46 and 0.61 m, and $\beta_2 = 30°$. The flow entering the rotor row and leaving the stator row is axial viewed from the stationary casing. Is this device a turbine or a pump? Estimate the amount of power transferred to or from the fluid.

(ANS: pump; 7760 kW)

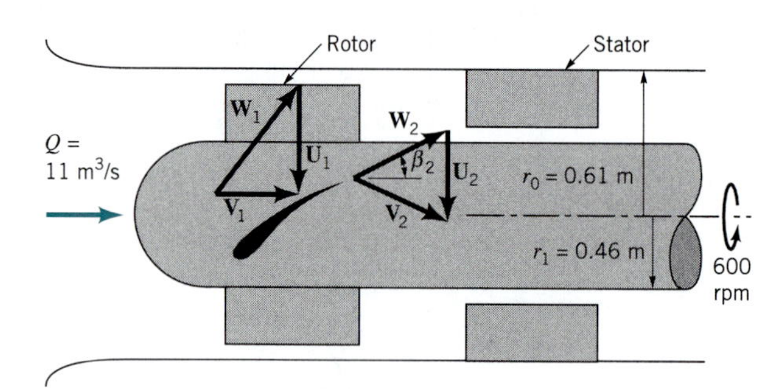

■ FIGURE P5.15R

5.16R (Moment-of-momentum) A small water turbine is designed as shown in Fig. P5.16R. If the flowrate through the turbine is 0.0030 slugs/s, and the rotor speed is 300 rpm, estimate the shaft torque and shaft power involved. Each nozzle exit cross-sectional area is 3.5×10^{-5} ft^2.

(ANS: −0.0107 ft·lb; −0.336 ft·lb/s)

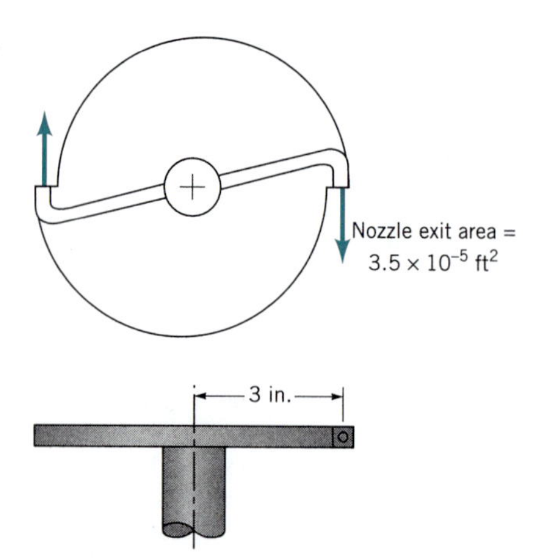

■ FIGURE P5.16R

5.17R (Energy equation) Water flows steadily from one location to another in the inclined pipe shown in Fig. P5.17R. At one section, the static pressure is 12 psi. At the other section, the static pressure is 5 psi. Which way is the water flowing? Explain.

(ANS: from A to B)

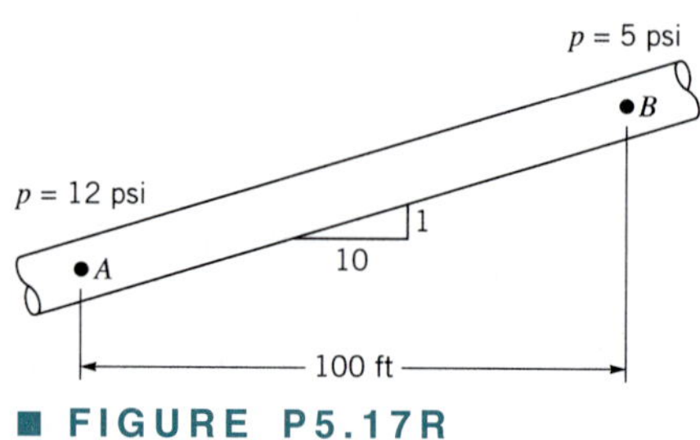

■ FIGURE P5.17R

5.18R (Energy equation) The pump shown in Fig. P5.18R adds 20 kW of power to the flowing water. The only loss is that which occurs across the filter at the inlet of the pump. Determine the head loss for this filter.

(ANS: 7.69 m)

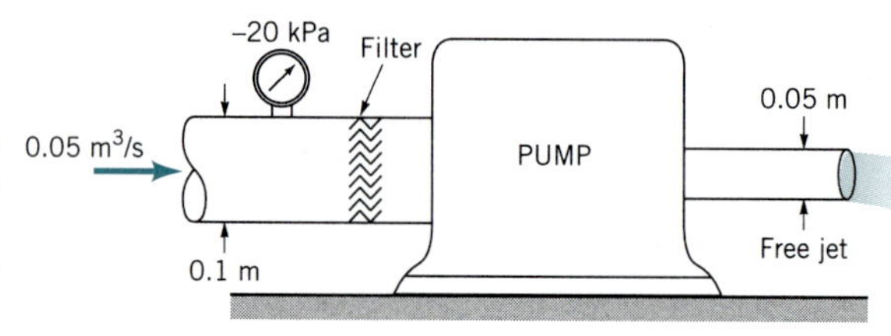

■ FIGURE P5.18R

5.19R (Linear momentum/energy) Eleven equally spaced turning vanes are used in the horizontal plane 90° bend as indicated in Fig. P5.19R. The depth of the rectangular cross-sectional bend remains constant at 3 in. The velocity distributions upstream and downstream of the vanes may be considered uniform. The loss in available energy across the vanes is $0.2V_1^2/2$. The required velocity and pressure downstream of the vanes, section (2), are 180 ft/s and 15 psia. What is the average magnitude of the force exerted by the air flow on each vane? Assume the force of the air on the duct walls is equivalent to the force of the air on one vane.

(ANS: 4.61 lb)

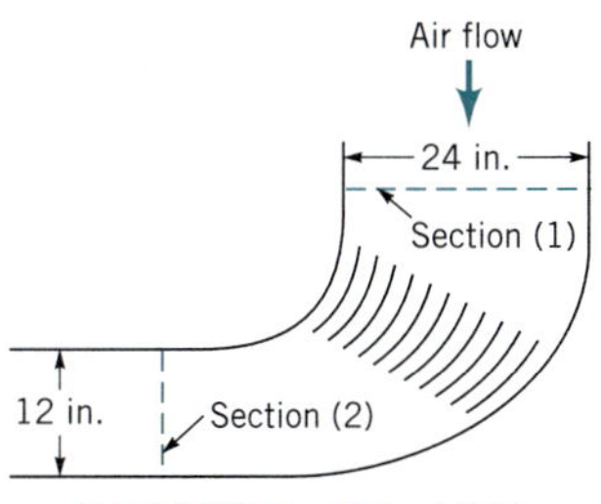

■ FIGURE P5.19R

5.20R (Energy equation) A hydroelectric power plant operates under the conditions illustrated in Fig. P5.20R. The head loss associated with flow from the water level upstream of the dam, section (1), to the turbine discharge at atmospheric pressure, section (2), is 20 m. How much power is transferred from the water to the turbine blades?

(ANS: 23.5 MW)

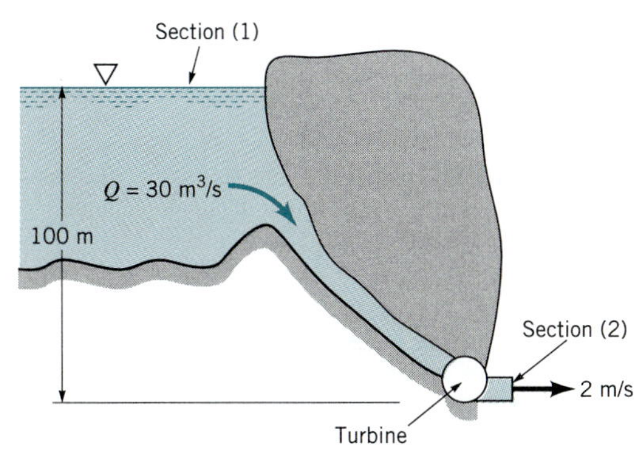

■ FIGURE P5.20R

5.21R (Energy equation) A pump transfers water from one large reservoir to another as shown in Fig. P5.21R*a*. The difference in elevation between the two reservoirs is 100 ft. The friction head loss in the piping is given by $K_L\overline{V}^2/2g$, where $\overline{V}$ is the average fluid velocity in the pipe and K_L is the loss coefficient, which is considered constant. The relation between the total head rise, H, across the pump and the flowrate, Q, through the pump is given in Fig. 5.21R*b*. If $K_L = 40$, and the pipe diameter is 4 in., what is the flowrate through the pump?

(ANS: 0.653 ft^3/s)

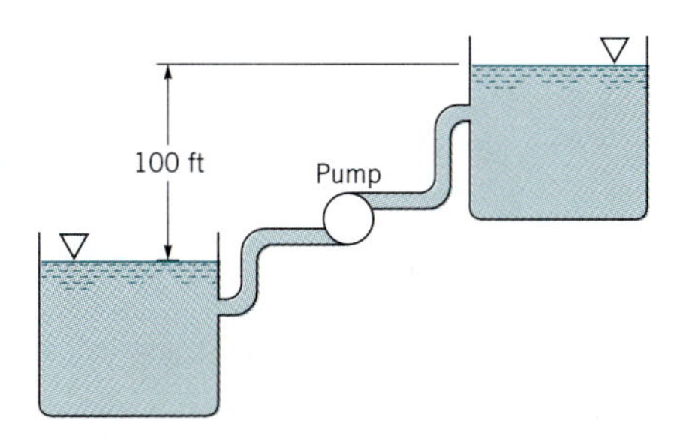

(*a*)

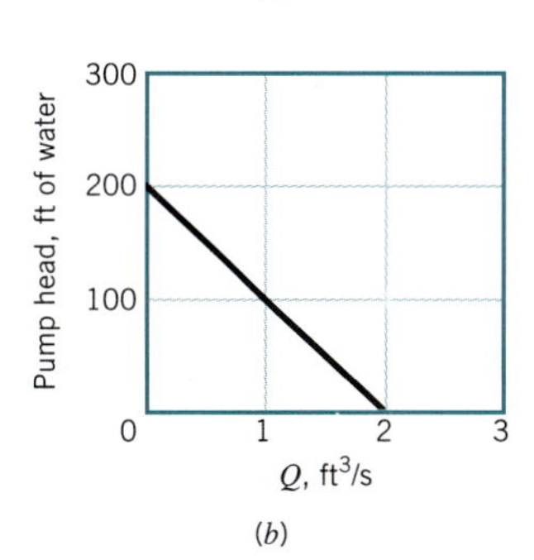

(*b*)

■ FIGURE P5.21R

5.22R (Energy equation) The pump shown in Fig. P5.22R adds 1.6 horsepower to the water when the flowrate is 0.6 ft^3/s. Determine the head loss bewteen the free surface in the large, open tank and the top of the fountain (where the velocity is zero).

(ANS: 7.50 ft)

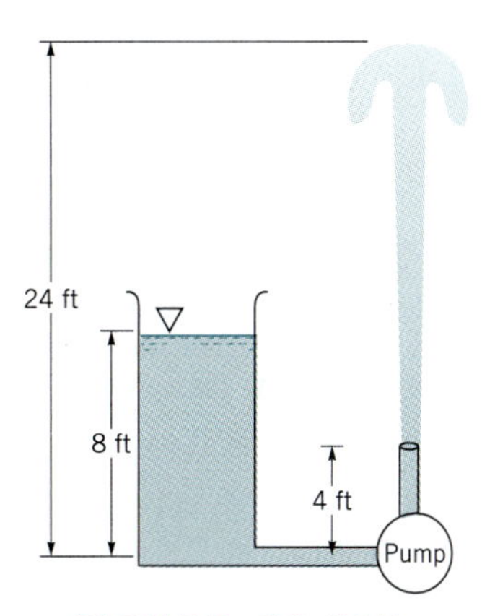

■ FIGURE P5.22R

Problems

Note: Unless otherwise indicated, use the values of fluid properties found in the tables on the inside of the front cover. Problems designated with an (*) are intended to be solved with the aid of a programmable calculator or a computer. Problems designated with a (†) are "open-ended" problems and require critical thinking in that to work them one must make various assumptions and provide the necessary data. There is not a unique answer to these problems.

5.1 Use the Reynolds transport theorem (Eq. 4.19) with $B =$ volume and, therefore, $b =$ volume/mass $= 1/\text{density}$ to obtain the continuity equation for steady or unsteady incompressible flow through a fixed control volume: $\int_{\text{cv}} \mathbf{V} \cdot \hat{\mathbf{n}}\, dA = 0$.

5.2 An incompressible flow velocity field (water) is given as

$$\mathbf{V} = -\frac{1}{r}\,\hat{\mathbf{e}}_r + \frac{1}{r}\,\hat{\mathbf{e}}_\theta \text{ m/s}$$

where r is in meters. **(a)** Calculate the mass flowrate through the cylindrical surface at $r = 1$ m from $z = 0$ to $z = 1$ m as shown in Fig. P5.2*a*. **(b)** Show that mass is conserved in the annular control volume from $r = 1$ m to $r = 2$ m and $z = 0$ to $z = 1$ m as shown in Fig. P5.2*b*.

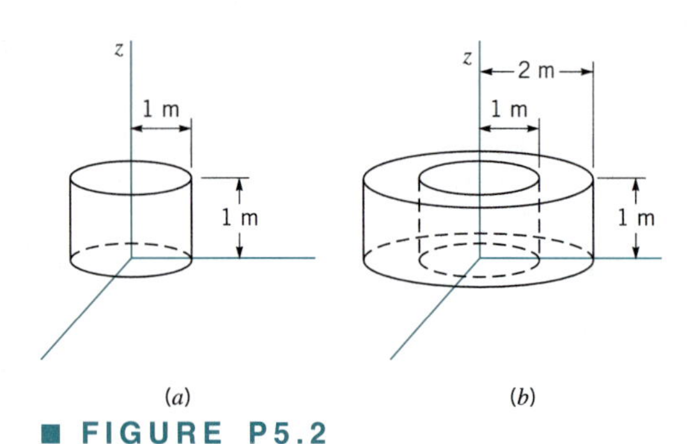

■ FIGURE P5.2

5.3 Water flows steadily through the horizontal piping system shown in Fig. P5.3. The velocity is uniform at section (1), the mass flowrate is 10 slugs/s at section (2), and the velocity is nonuniform at section (3). **(a)** Determine the value of the quantity $\frac{D}{Dt}\int_{\text{sys}} \rho\, d\mathcal{V}$, where the system is the water contained in the pipe bounded by sections (1), (2), and (3). **(b)** Determine the mean velocity at section (2). **(c)** Determine, if possible, the value of the integral $\int_{(3)} \rho \mathbf{V} \cdot \hat{\mathbf{n}}\, dA$ over section (3). If it is not possible, explain what additional information is needed to do so.

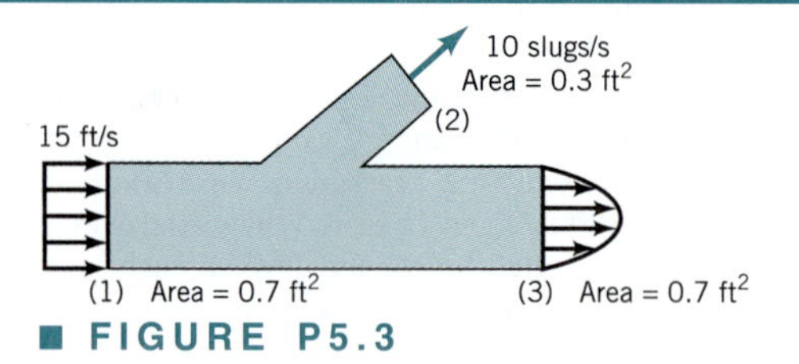

■ FIGURE P5.3

5.4 Air flows steadily between two cross sections in a long, straight section of 0.25-m inside-diameter pipe. The static temperature and pressure at each section are indicated in Fig. P5.4. If the average air velocity at section (2) is 320 m/s, determine the average air velocity at section (1).

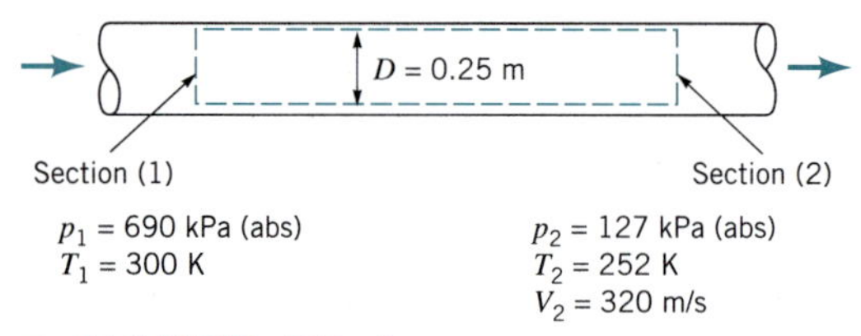

■ FIGURE P5.4

5.5 The wind blows through a 7 ft × 10 ft garage door with a speed of 5 ft/s as shown in Fig. P5.5. Determine the average speed, V, of the air through the two 3 ft × 4 ft windows.

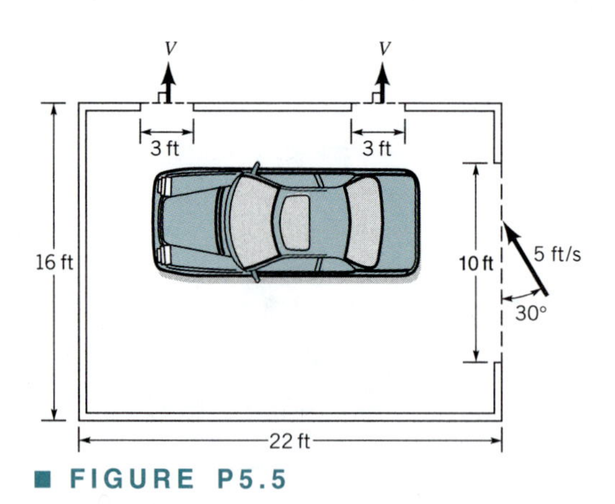

■ FIGURE P5.5

5.6 A hydroelectric turbine passes 2 million gal/min through its blades. If the average velocity of the flow in the circular cross-section conduit leading to the turbine is not to exceed 30 ft/s, determine the minimum allowable diameter of the conduit.

5.7 Water flows out through a set of thin, closely spaced blades as shown in Fig. P5.7 with a speed of $V = 10$ ft/s around the entire circumference of the outlet. Determine the mass flowrate through the inlet pipe.

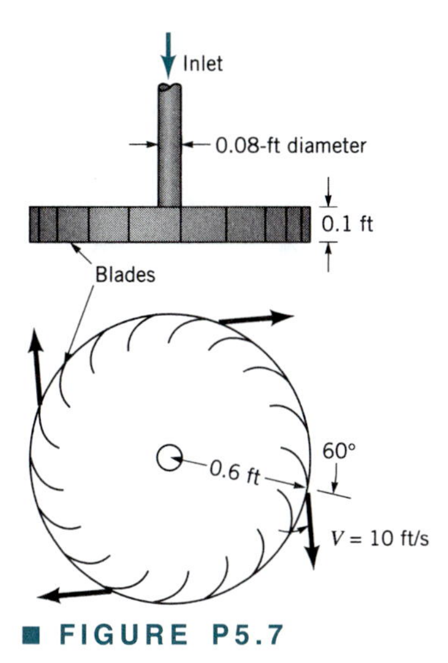

■ FIGURE P5.7

5.8 A hydraulic jump is in place downstream from a spillway as indicated in Fig. P5.8. Upstream of the jump, the depth of the stream is 0.6 ft and the average stream velocity is 18 ft/s. Just downstream of the jump, the average stream velocity is 3.4 ft/s. Calculate the depth of the stream, h, just downstream of the jump.

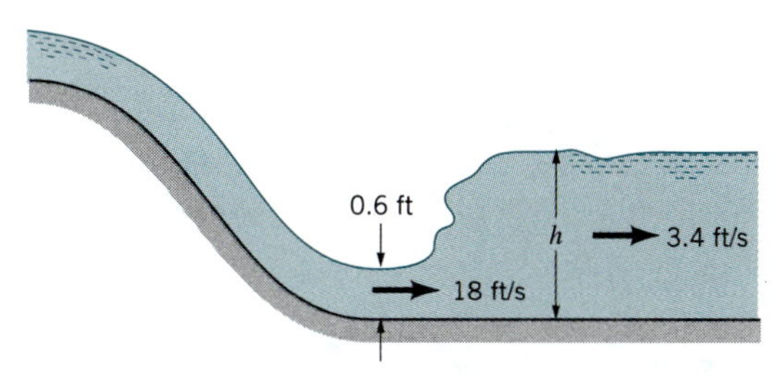

■ FIGURE P5.8

5.9 A water jet pump (see Fig. P5.9) involves a jet cross-section area of 0.01 m^2, and a jet velocity of 30 m/s. The jet is surrounded by entrained water. The total cross-sectional area associated with the jet and entrained streams is 0.075 m^2. These two fluid streams leave the pump thoroughly mixed with an average velocity of 6 m/s through a cross-sectional area of 0.075 m^2. Determine the pumping rate (i.e., the entrained fluid flowrate) involved in liters/s.

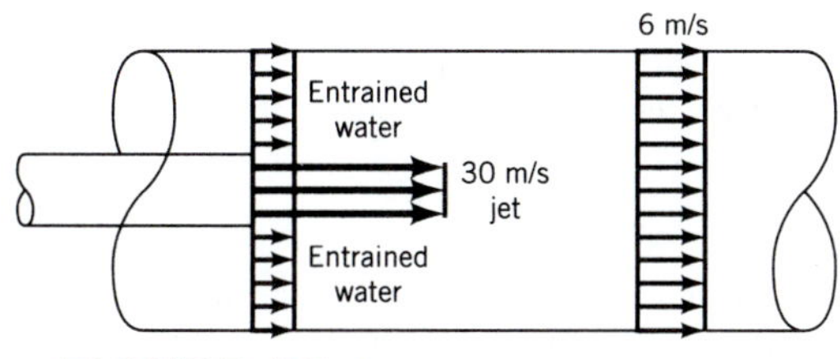

■ FIGURE P5.9

5.10 Water enters a cylindrical tank through two pipes at rates of 250 and 100 gal/min (see Fig. P5.10). If the level of the water in the tank remains constant, calculate the average velocity of the flow leaving the tank through an 8-in. inside-diameter pipe.

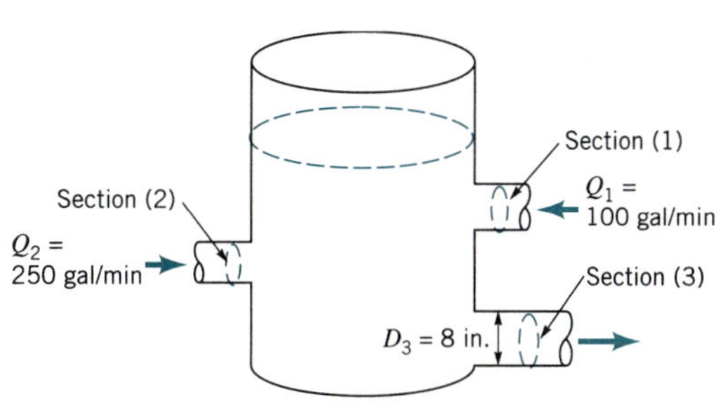

■ FIGURE P5.10

5.11 At cruise conditions, air flows into a jet engine at a steady rate of 65 lbm/s. Fuel enters the engine at a steady rate of 0.60 lbm/s. The average velocity of the exhaust gases is 1500 ft/s relative to the engine. If the engine exhaust effective cross-sectional area is 3.5 ft^2, estimate the density of the exhaust gases in lbm/ft^3.

5.12 Air at standard atmospheric conditions is drawn into a compressor at the steady rate of 30 m^3/min. The compressor pressure ratio, p_{exit}/p_{inlet}, is 10 to 1. Through the compressor p/ρ^n remains constant with $n = 1.4$. If the average velocity in the compressor discharge pipe is not to exceed 30 m/s, calculate the minimum discharge pipe diameter required.

5.13 Two rivers merge to form a larger river as shown in Fig. P5.13. At a location downstream from the junction (before the two streams completely merge), the nonuniform velocity profile is as shown and the depth is 6 ft. Determine the value of V.

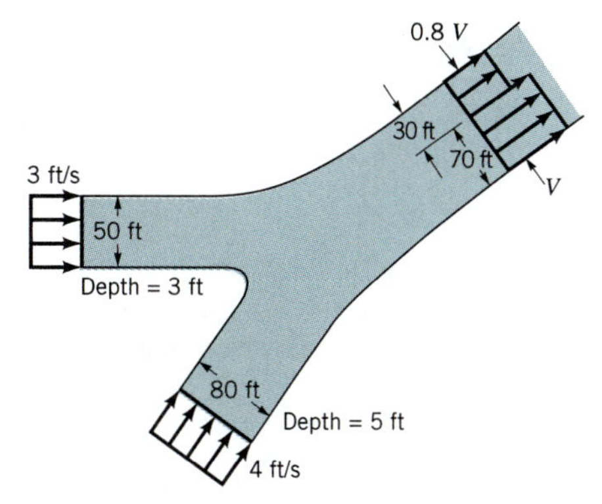

■ FIGURE P5.13

5.14 Oil having a specific gravity of 0.9 is pumped as illustrated in Fig. P5.14 with a water jet pump. The water volume flowrate is 1 m^3/s. The water and oil mixture has an average specific gravity of 0.95. Calculate the rate, in m^3/s, at which the pump moves oil.

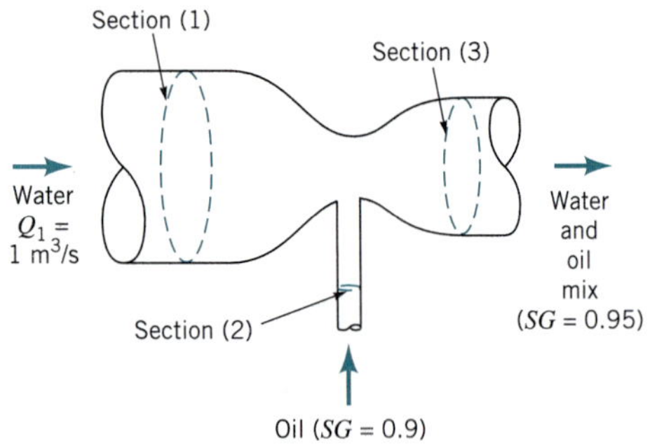

■ FIGURE P5.14

5.15 Air at standard conditions enters the compressor shown in Fig. P5.15 at a rate of 10 ft^3/s. It leaves the tank through a 1.2-in.-diameter pipe with a density of 0.0035 $slugs/ft^3$ and a uniform speed of 700 ft/s. **(a)** Determine the rate (slugs/s) at which the mass of air in the tank is increasing or decreasing. **(b)** Determine the average time rate of change of air density within the tank.

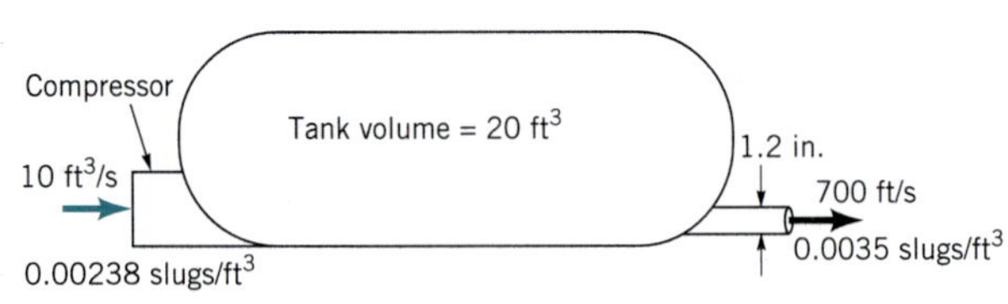

■ FIGURE P5.15

5.16 An appropriate turbulent pipe flow velocity profile is

$$\mathbf{V} = u_c \left(\frac{R - r}{R} \right)^{1/n} \hat{\mathbf{i}}$$

where u_c = centerline velocity, r = local radius, R = pipe radius, and $\hat{\mathbf{i}}$ = unit vector along pipe centerline. Determine the ratio of average velocity, $\bar{u}$, to centerline velocity, u_c, for **(a)** $n = 4$, **(b)** $n = 6$, **(c)** $n = 8$, **(d)** $n = 10$.

5.17 The velocity and temperature profiles for one circular cross section in laminar pipe flow of air with heat transfer are

$$\mathbf{V} = u_c \left[1 - \left(\frac{r}{R} \right)^2 \right] \hat{\mathbf{i}}$$

where the unit vector $\hat{\mathbf{i}}$ is along the pipe axis, and

$$T = T_c \left[1 + \frac{1}{2} \left(\frac{r}{R} \right)^2 - \frac{1}{4} \left(\frac{r}{R} \right)^4 \right]$$

The subscript c refers to centerline value, r = local radius, R = pipe radius, and T = local temperature. Show how you would evaluate the mass flowrate through this cross-sectional area.

***5.18** To measure the mass flowrate of air through a 6-in. inside-diameter pipe, local velocity data are collected at different radii from the pipe axis (see Table). Determine the mass flowrate corresponding to the data listed below.

r (in.)	Axial Velocity (ft/s)
0	30
0.2	29.71
0.4	29.39
0.6	29.06
0.8	28.70
1.0	28.31
1.2	27.89
1.4	27.42
1.6	26.90
1.8	26.32
2.0	25.64
2.2	24.84
2.4	23.84
2.6	22.50
2.8	20.38
2.9	18.45
2.95	16.71
2.98	14.66
3.00	0

5.19 As shown in Fig. P5.19, at the entrance to a 3-ft-wide channel the velocity distribution is uniform with a velocity V. Further downstream the velocity profile is given by $u = 4y - 2y^2$, where u is in ft/s and y is in ft. Determine the value of V.

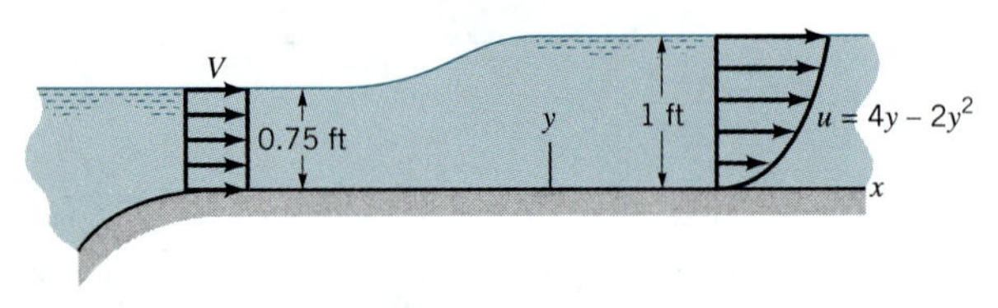

■ FIGURE P5.19

5.20 Flow of a viscous fluid over a flat plate surface results in the development of a region of reduced velocity adjacent to the wetted surface as depicted in Fig. P5.20. This region of reduced flow is called a boundary layer. At the leading edge of the plate, the velocity profile may be considered uniformly distributed with a value U. All along the outer edge of the boundary layer, the fluid velocity component parallel to the plate surface is also U. If the x direction velocity profile at section (2) is

$$\frac{u}{U} = \left(\frac{y}{\delta} \right)^{1/7}$$

develop an expression for the volume flowrate through the edge of the boundary layer from the leading edge to a location downstream at x where the boundary layer thickness is δ.

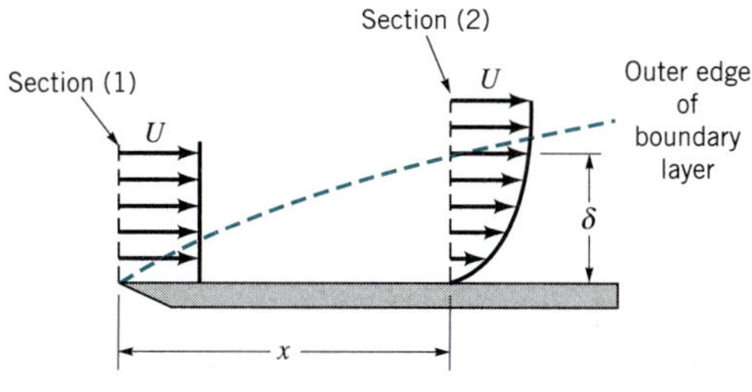

■ FIGURE P5.20

† **5.21** Estimate the rate (in gallons per hour) that your car uses gasoline when it is being driven on an interstate highway. Determine how long it would take to empty a 12-oz soft drink container at this flowrate. List all assumptions and show calculations.

5.22 How long would it take to fill a cylindrical-shaped swimming pool having a diameter of 8 m to a depth of 1.5 m with water from a garden hose if the flowrate is 1.0 liter/s?

5.23 To determine the mass flowrate of cooling water through a test engine, the amount of water collected in a large container over a time interval is measured. Using an appropriate control volume, comment on the validity of this procedure.

5.24 Storm sewer backup causes your basement to flood at the steady rate of 1 in. of depth per hour. The basement floor area is 1500 ft^2. What capacity (gal/min) pump would you rent to **(a)** keep the water accumulated in your basement at a constant level until the storm sewer is blocked off, **(b)** reduce the water accumulation in your basement at a rate of 3 in./hr even while the backup problem exists?

5.25 A hypodermic syringe (see Fig. P5.25) is used to apply a vaccine. If the plunger is moved forward at the steady rate of 20 mm/s and if vaccine leaks past the plunger at 0.1 of the volume flowrate out the needle opening, calculate the average velocity of the needle exit flow. The inside diameters of the syringe and the needle are 20 mm and 0.7 mm.

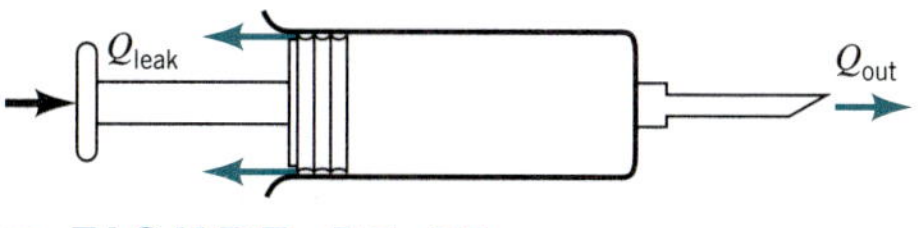

■ FIGURE P5.25

† **5.26** Estimate the maximum flowrate of rainwater (during a heavy rain) that you would expect from the downspout connected to the gutters of your house. List all assumptions and show all calculations.

5.27 Water enters a rigid, sealed, cylindrical tank at a steady rate of 100 liters/hr and forces gasoline ($SG = 0.68$) out as is indicated in Fig. P5.27. What is the time rate of change of mass of gasoline contained in the tank?

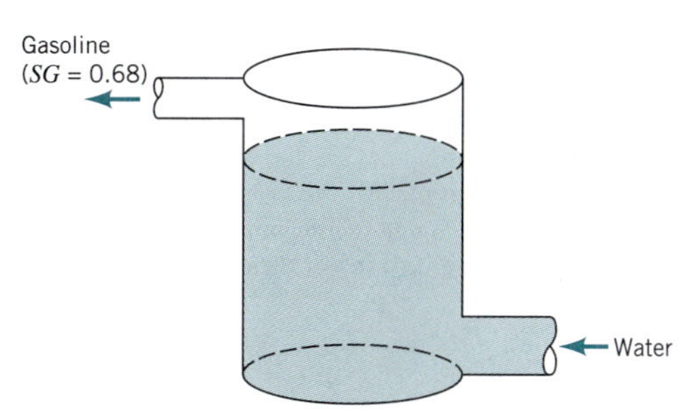

■ FIGURE P5.27

5.28 A gas flows steadily through a duct of varying cross-sectional area. If the gas density is assumed to be uniformly distributed at any cross section, show that the conservation of mass principle leads to

$$\frac{d\rho}{\rho} + \frac{d\overline{V}}{\overline{V}} + \frac{dA}{A} = 0$$

where ρ = gas density, $\overline{V}$ = average speed of gas, and A = cross-sectional area.

5.29 Water flows steadily from a tank mounted on a cart as shown in Fig. P5.29. After the water jet leaves the nozzle of the tank, it falls and strikes a vane attached to another cart. The cart's wheels are frictionless, and the fluid is inviscid. **(a)** Determine the speed of the water leaving the tank, V_1, and the water speed leaving the cart, V_2. **(b)** Determine the tension in rope A. **(c)** Determine the tension in rope B.

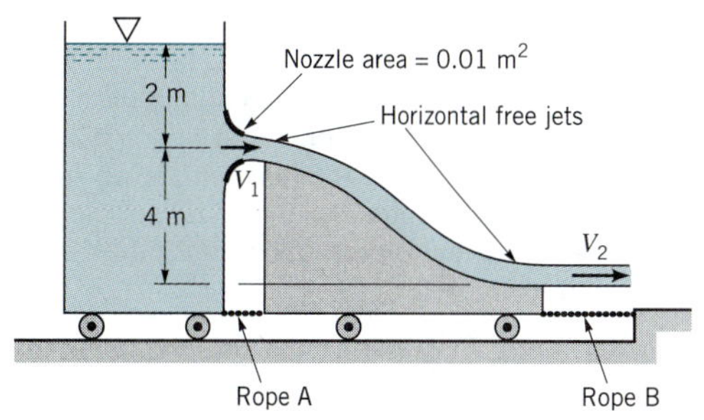

■ FIGURE P5.29

5.30 Water enters the horizontal, circular cross-sectional, sudden contraction nozzle sketched in Fig. P5.30 at section (1) with a uniformly distributed velocity of 25 ft/s and a pressure of 75 psi. The water exits from the nozzle into the atmosphere at section (2) where the uniformly distributed velocity is 100 ft/s. Determine the axial component of the anchoring force required to hold the contraction in place.

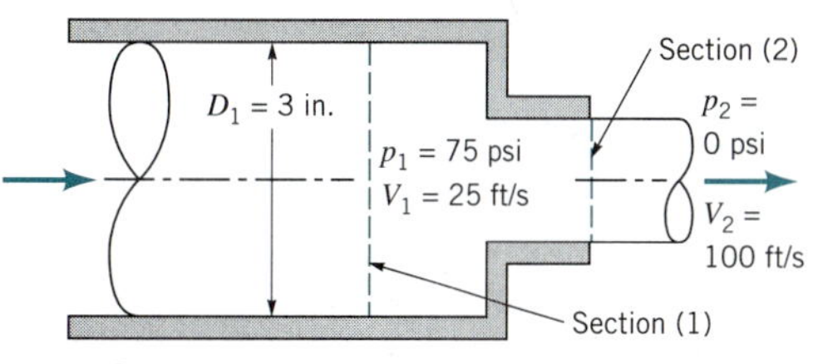

■ FIGURE P5.30

5.31 A nozzle is attached to a vertical pipe and discharges water into the atmosphere as shown in Fig. P5.31. When the discharge is 0.1 m^3/s, the gage pressure at the flange is 40 kPa. Determine the vertical component of the anchoring force required to hold the nozzle in place. The nozzle has a weight of 200 N, and the volume of water in the nozzle is 0.012 m^3. Is the anchoring force directed upward or downward?

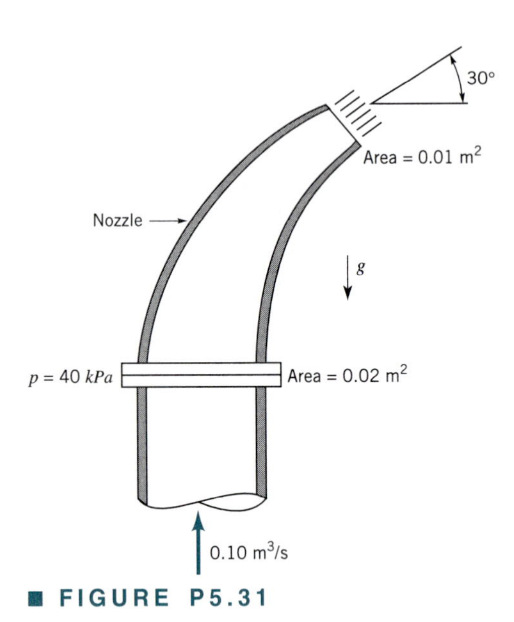

■ FIGURE P5.31

5.32 Determine the magnitude and direction of the x and y components of the anchoring force required to hold in place the horizontal 180° elbow and nozzle combination shown in Fig. P5.32. Also determine the magnitude and direction of the x and y components of the reaction force exerted by the 180° elbow and nozzle on the flowing water.

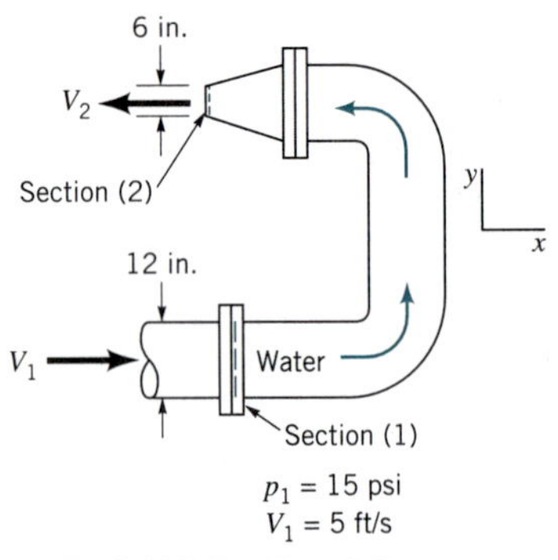

■ FIGURE P5.32

5.33 Water flows as two free jets from the tee attached to the pipe shown in Fig. P5.33. The exit speed is 15 m/s. If viscous effects and gravity are negligible, determine the x and y components of the force that the pipe exerts on the tee.

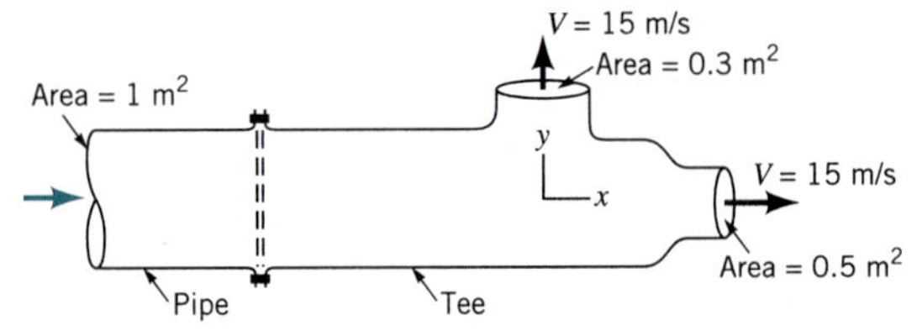

■ FIGURE P5.33

5.34 Water flows through a horizontal bend and discharges into the atmosphere as shown in Fig. P5.34. When the pressure gage reads 10 psi, the resultant x direction anchoring force, F_{Ax}, in the horizontal plane required to hold the bend in place is shown on the figure. Determine the flowrate through the bend and the y direction anchoring force, F_{Ay}, required to hold the bend in place. The flow is not frictionless.

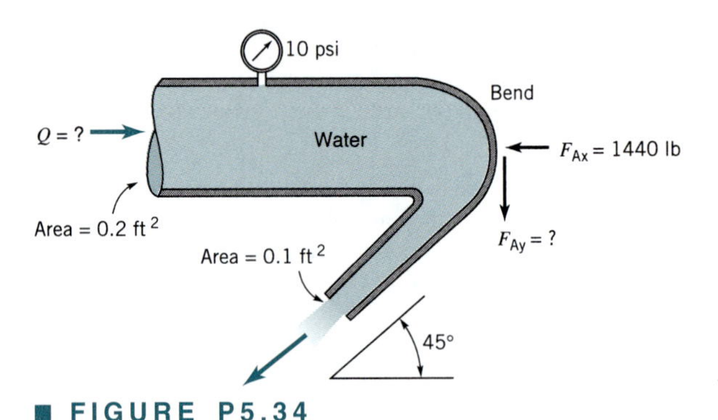

■ FIGURE P5.34

5.35 Thrust vector control is a new technique that can be used to greatly improve the maneuverability of military fighter aircraft. It consists of using a set of vanes in the exit of a jet engine to deflect the exhaust gases as shown in Fig. P5.35. **(a)** Determine the pitching moment (the moment tending to rotate the nose of the aircraft up) about the aircraft's mass center (cg) for the conditions indicated in the figure. **(b)** By how much is the thrust (force along the centerline of the aircraft) reduced for the case indicated compared to normal flight when the exhaust is parallel to the centerline?

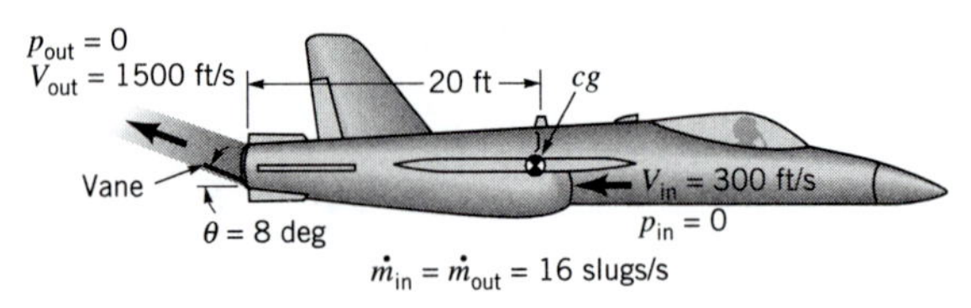

■ FIGURE P5.35

5.36 For the conditions of Problem 5.4, determine the frictional force exerted by the pipe wall on the air flowing between sections (1) and (2). Assume uniform velocity distributions at each section.

5.37 Water is sprayed radially outward over 180° as indicated in Fig. P5.37. The jet sheet is in the horizontal plane. If the jet velocity at the nozzle exit is 20 ft/s, determine the direction and magnitude of the resultant horizontal anchoring force required to hold the nozzle in place.

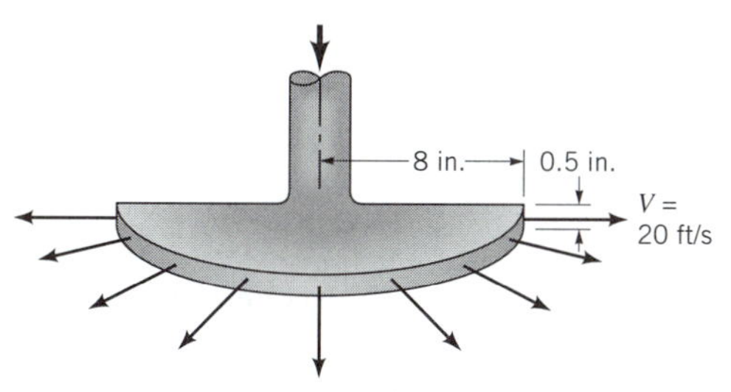

■ FIGURE P5.37

5.38 A circular plate having a diameter of 300 mm is held perpendicular to an axisymmetric horizontal jet of air having a velocity of 40 m/s and a diameter of 80 mm as shown in Fig. P5.38. A hole at the center of the plate results in a discharge jet of air having a velocity of 40 m/s and a diameter of 20 mm. Determine the horizontal component of force required to hold the plate stationary.

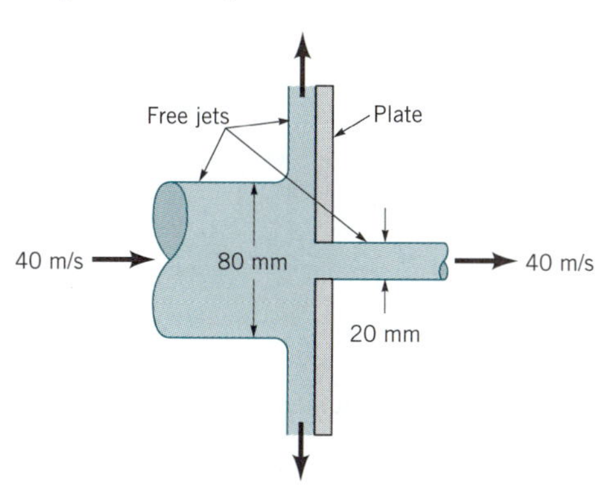

■ FIGURE P5.38

5.39 A sheet of water of uniform thickness ($h = 0.01$ m) flows from the device shown in Fig. P5.39. The water enters vertically through the inlet pipe and exits horizontally with a speed that varies linearly from 0 to 10 m/s along the 0.2-m length of the slit. Determine the y component of anchoring force necessary to hold this device stationary.

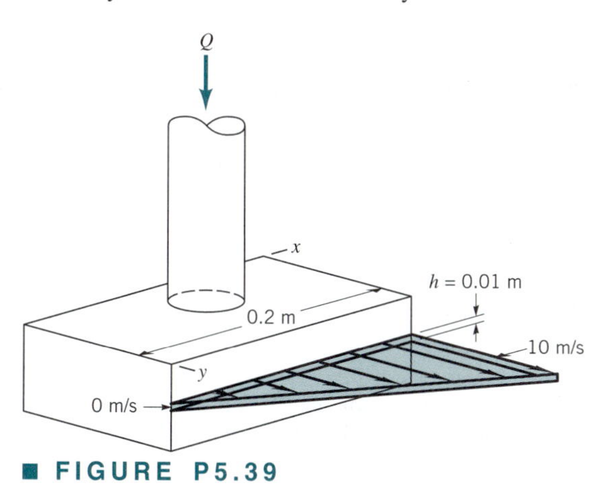

■ FIGURE P5.39

5.40 The results of a wind tunnel test to determine the drag on a body (see Fig. P5.40) are summarized below. The upstream [section (1)] velocity is uniform at 100 ft/s. The static pressures are given by $p_1 = p_2 = 14.7$ psia. The downstream velocity distribution, which is symmetrical about the centerline, is given by

$$u = 100 - 30\left(1 - \frac{|y|}{3}\right) \qquad |y| \leq 3 \text{ ft}$$

$$u = 100 \qquad |y| > 3 \text{ ft}$$

where u is the velocity in ft/s and y is the distance on either side of the centerline in feet (see Fig. P5.40). Assume that the body shape does not change in the direction normal to the paper. Calculate the drag force (reaction force in x direction) exerted on the air by the body per unit length normal to the plane of the sketch.

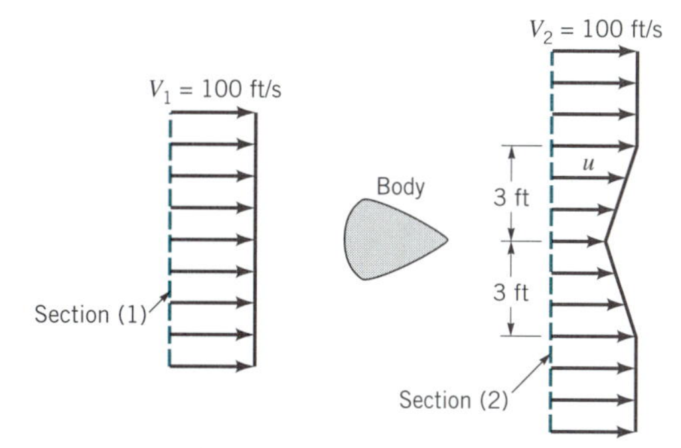

■ FIGURE P5.40

5.41 The exhaust gas from the rocket shown in Fig. P5.41*a* leaves the nozzle with a uniform velocity parallel to the x axis. The gas is assumed to be discharged from the nozzle as a free jet. **(a)** Show that the thrust that is developed is equal to ρAV^2, where $A = \pi D^2/4$. **(b)** The exhaust gas from the rocket nozzle shown in Fig. P5.41*b* is also uniform, but rather than being directed along the x axis, it is directed along rays from point 0 as indicated. Determine the thrust for this rocket.

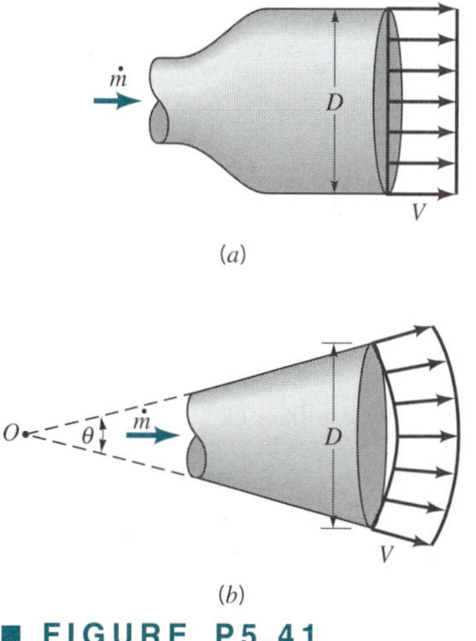

■ FIGURE P5.41

5.42 Water flows vertically upward in a circular cross-sectional pipe as shown in Fig. P5.42. At section (1), the velocity profile over the cross-sectional area is uniform. At section (2), the velocity profile is

$$\mathbf{V} = w_c\left(\frac{R - r}{R}\right)^{1/7} \hat{\mathbf{k}}$$

where $\mathbf{V}$ = local velocity vector, w_c = centerline velocity in the axial direction, R = pipe radius, and r = radius from pipe axis. Develop an expression for the fluid pressure drop that occurs between sections (1) and (2).

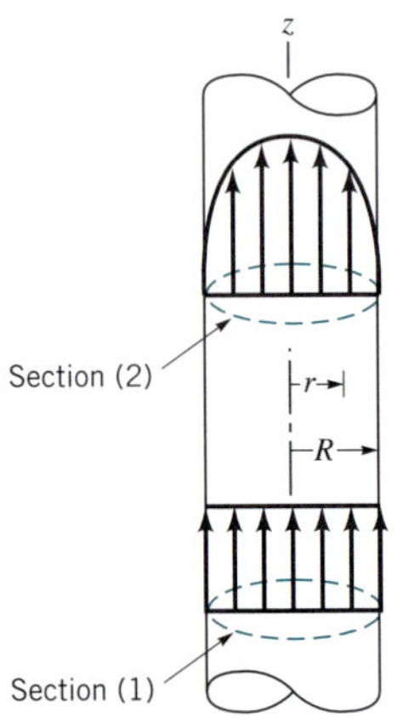

■ **FIGURE P5.42**

5.43 In a laminar pipe flow that is fully developed, the axial velocity profile is parabolic. That is,

$$u = u_c\left[1 - \left(\frac{r}{R}\right)^2\right]$$

as is illustrated in Fig. P5.43. Compare the axial direction momentum flowrate calculated with the average velocity, $\bar{u}$, with the axial direction momentum flowrate calculated with the nonuniform velocity distribution taken into account.

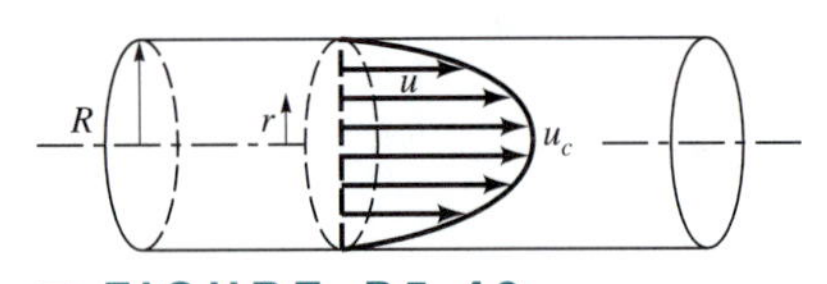

■ **FIGURE P5.43**

***5.44** For the pipe (6-in. inside diameter) air flow data of Problem 5.18, calculate the rate of flow of axial direction momentum. How large would the error be if the average axial velocity were used to calculate axial direction momentum flow?

5.45 Consider unsteady flow in the constant diameter, horizontal pipe shown in Fig. P5.45. The velocity is uniform throughout the entire pipe, but it is a function of time: $\mathbf{V} = u(t)\,\hat{\mathbf{i}}$. Use the x component of the unsteady momentum equation to determine the pressure difference $p_1 - p_2$. Discuss how this result is related to $F_x = ma_x$.

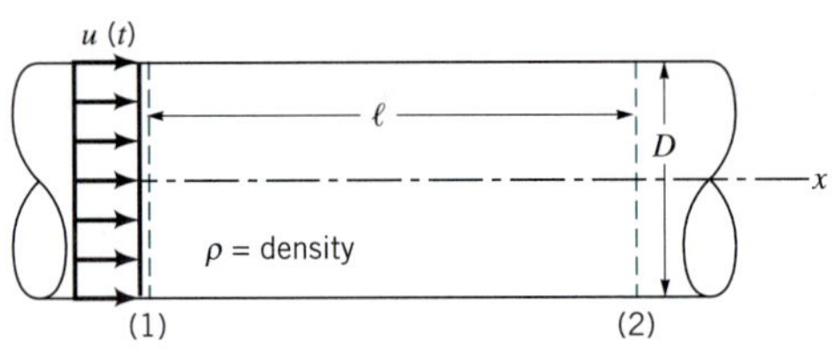

■ **FIGURE P5.45**

† **5.46** If a valve in a pipe is suddenly closed, a large pressure surge may develop. For example, when the electrically operated shutoff valve in a dishwasher closes quickly, the pipes supplying the dishwasher may rattle or ''bang'' because of this large pressure pulse. Explain the physical mechanism for this ''water hammer'' phenomenon. How could this phenomenon be analyzed?

5.47 A free jet of fluid strikes a wedge as shown in Fig. P5.47. Of the total flow, a portion is deflected 30°; the remainder is not deflected. The horizontal and vertical components of force needed to hold the wedge stationary are F_H and F_V, respectively. Gravity is negligible, and the fluid speed remains constant. Determine the force ratio, F_H/F_V.

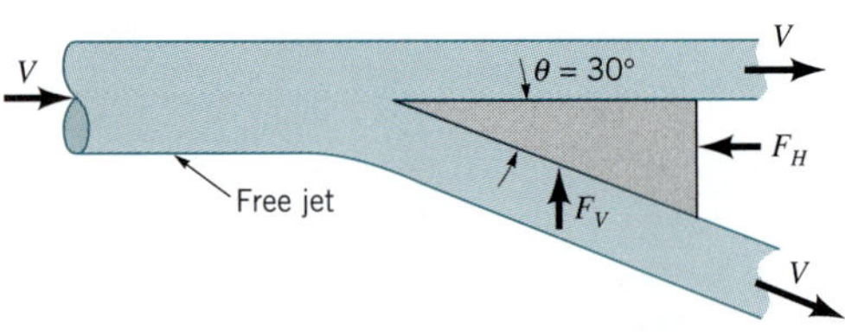

■ **FIGURE P5.47**

5.48 Water flows from a two-dimensional open channel and is diverted by an inclined plate as illustrated in Fig. P5.48. When the velocity at section (1) is 10 ft/s, what horizontal force (per unit width) is required to hold the plate in position? At section (1) the pressure distribution is hydrostatic, and the fluid acts as a free jet at section (2). Neglect friction.

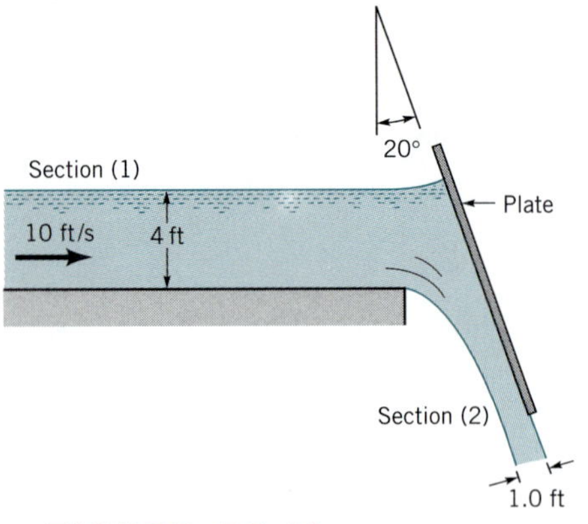

■ **FIGURE P5.48**

† **5.49** When a baseball player catches a ball, the force of the ball on her glove is as shown as a function of time in Fig. P5.49. Describe how this situation is similar to the force gen-

erated by the deflection of a jet of water by a vane. Note: Consider many baseballs being caught in quick succession.

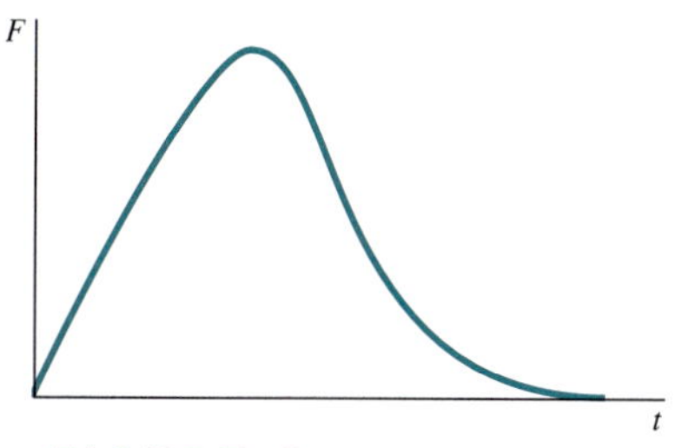

■ FIGURE P5.49

5.50 A vertical, circular cross-sectional jet of air strikes a conical deflector as indicated in Fig. P5.50. A vertical anchoring force of 0.1 N is required to hold the deflector in place. Determine the mass (kg) of the deflector. The magnitude of velocity of the air remains constant.

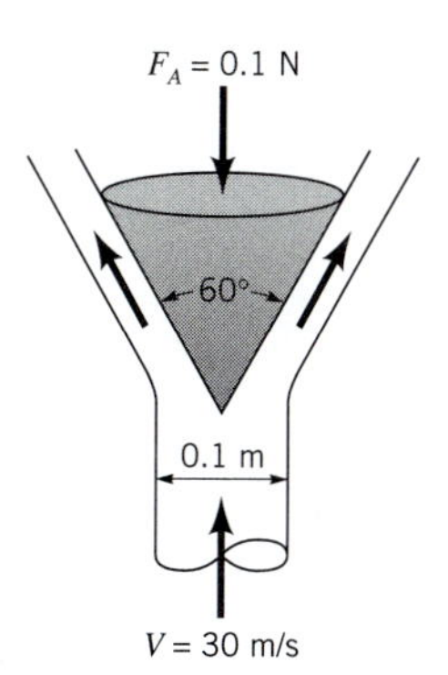

■ FIGURE P5.50

5.51 Water flows from a large tank into a dish as shown in Fig. P5.51. **(a)** If at the instant shown the tank and the water in it weigh W_1 lb, what is the tension, T_1, in the cable supporting the tank? **(b)** If at the instant shown the dish and the water in it weigh W_2 lb, what is the force, F_2, needed to support the dish?

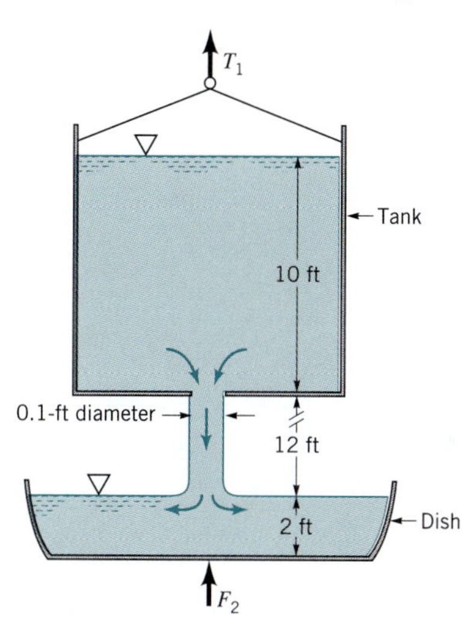

■ FIGURE P5.51

5.52 Air flows into the atmosphere from a nozzle and strikes a vertical plate as shown in Fig. P5.52. A horizontal force of 12 N is required to hold the plate in place. Determine the reading on the pressure gage. Assume the flow to be incompressible and frictionless.

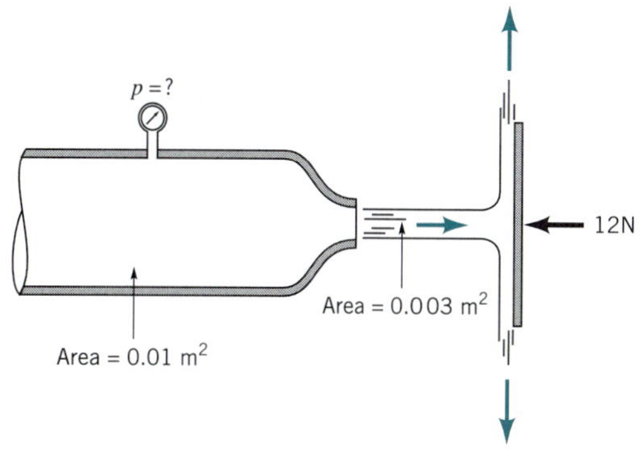

■ FIGURE P5.52

† **5.53** A beachball "balances" on a jet of air as shown in Fig. P5.53. Explain why the ball sits at the height it does and why the ball does not "roll off" the jet.

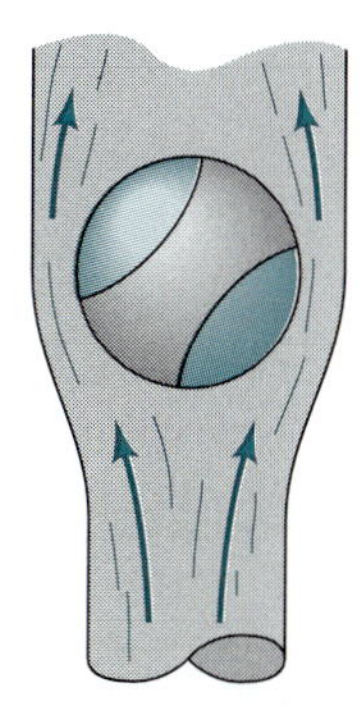

■ FIGURE P5.53

5.54 Two water jets of equal size and speed strike each other as shown in Fig. P5.54. Determine the speed, V, and direction, θ, of the resulting combined jet. Gravity is negligible.

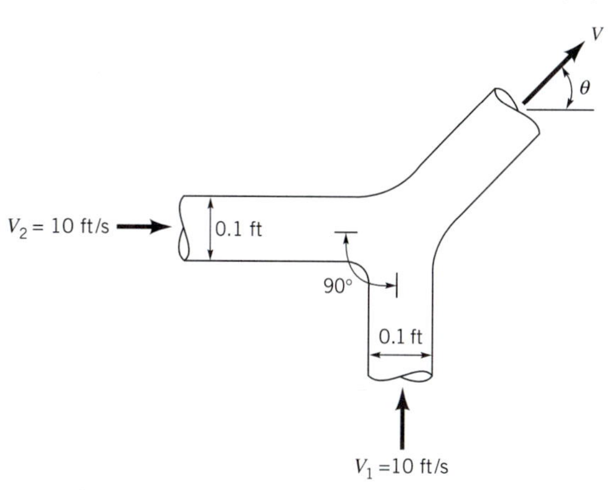

■ FIGURE P5.54

5.55 Assuming frictionless, incompressible, one-dimensional flow of water through the horizontal tee connection sketched in Fig. P5.55, estimate values of the x and y components of the force exerted by the tee on the water. Each pipe has an inside diameter of 1 m.

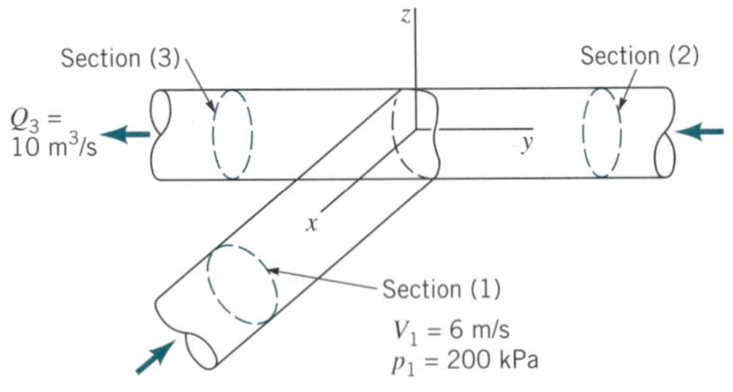

■ FIGURE P5.55

5.56 Water is added to the tank shown in Fig. P5.56 through a vertical pipe to maintain a constant (water) level. The tank is placed on a horizontal plane which has a frictionless surface. Determine the horizontal force, F, required to hold the tank stationary. Neglect all losses.

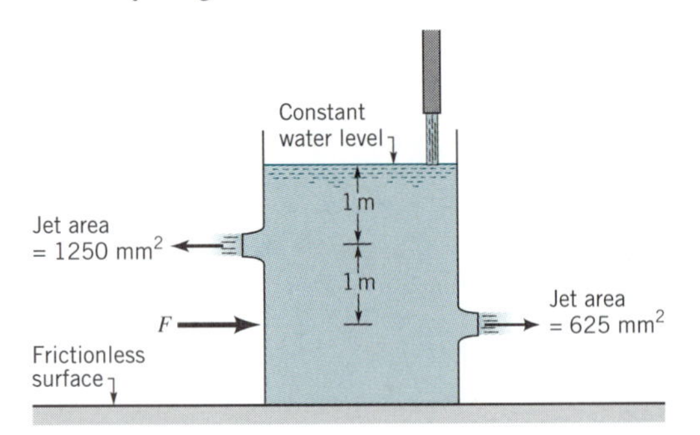

■ FIGURE P5.56

5.57 Water flows steadily into and out of a tank that sits on frictionless wheels as shown in Fig. P5.57. Determine the diameter D so that the tank remains motionless if $F = 0$.

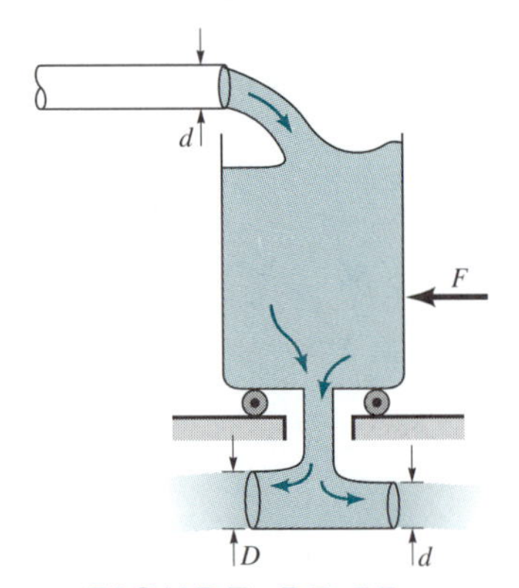

■ FIGURE P5.57

5.58 The four devices shown in Fig. P5.58 rest on frictionless wheels, are restricted to move in the x direction only and are initially held stationary. The pressure at the inlets and outlets of each is atmospheric, and the flow is incompressible. The contents of each device is not known. When released, which devices will move to the right and which to the left? Explain.

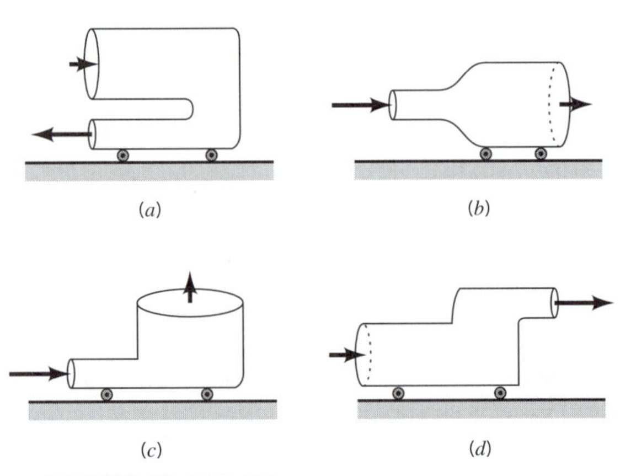

■ FIGURE P5.58

5.59 Water discharges into the atmosphere through the device shown in Fig. P5.59. Determine the x component of force at the flange required to hold the device in place. Neglect the effect of gravity and friction.

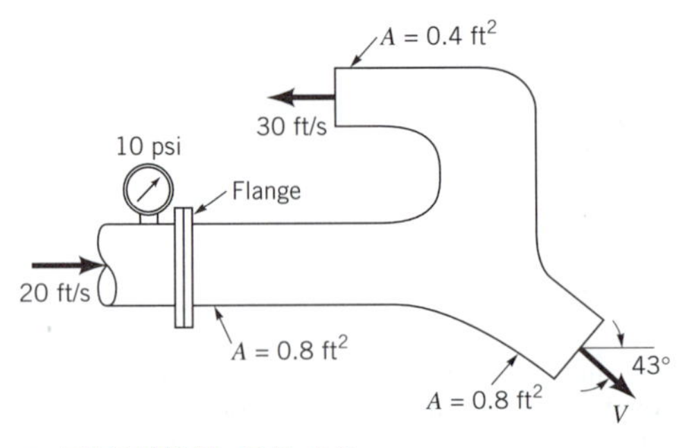

■ FIGURE P5.59

5.60 A vertical jet of water leaves a nozzle at a speed of 10 m/s and a diameter of 20 mm. It suspends a plate having a mass of 1.5 kg as indicated in Fig. P5.60. What is the vertical distance h?

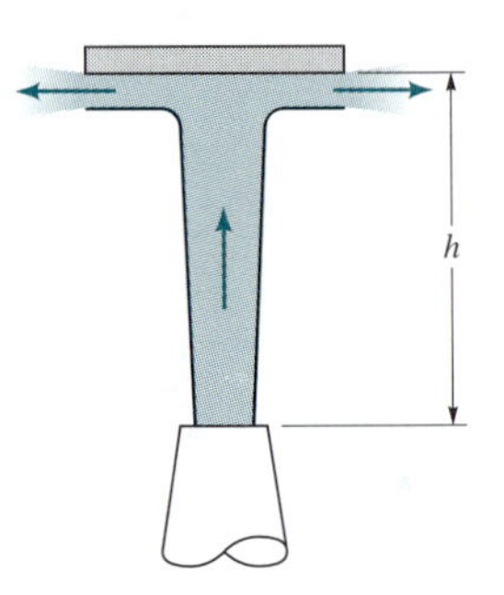

■ FIGURE P5.60

5.61 A vertical jet of water having a nozzle exit velocity of 15 ft/s with a diameter of 1 in. suspends a hollow hemisphere as indicated in Fig. P5.61. If the hemisphere is stationary at an elevation of 12 in., determine its weight.

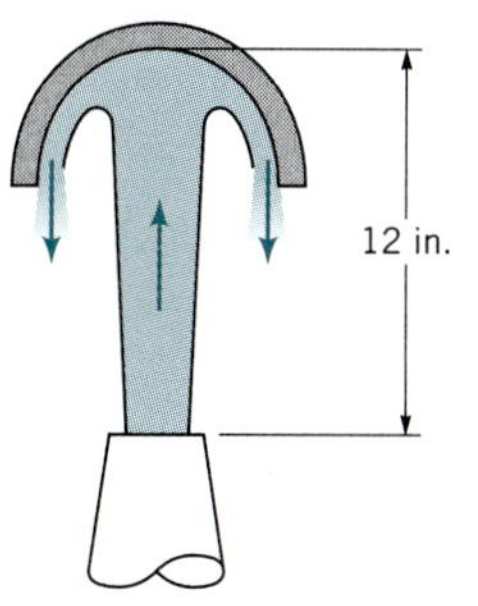

■ FIGURE P5.61

5.62 Air discharges from a 2-in.-diameter nozzle and strikes a curved vane, which is in a vertical plane as shown in Fig. P5.62. A stagnation tube connected to a water U-tube manometer is located in the free air jet. Determine the horizontal component of the force that the air jet exerts on the vane. Neglect the weight of the air and all friction.

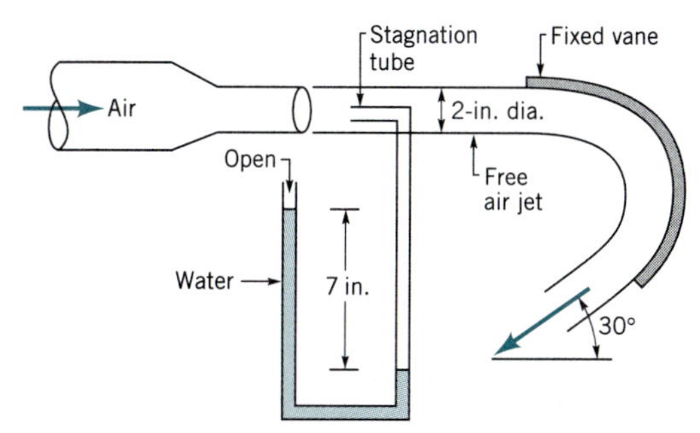

■ FIGURE P5.62

† **5.63** Water from a garden hose is sprayed against your car to rinse dirt from it. Estimate the force that the water exerts on the car. List all assumptions and show calculations.

† **5.64** A truck carrying chickens is too heavy for a bridge that it needs to cross. The empty truck is within the weight limits; with the chickens it is overweight. It is suggested that if one could get the chickens to fly around the truck (i.e., by banging on the truck side) it would be safe to cross the bridge. Do you agree? Explain.

5.65 A 3-in.-diameter horizontal jet of water strikes a flat plate as indicated in Fig. P5.65. Determine the jet velocity if a 10-lb horizontal force is required to **(a)** hold the plate stationary, **(b)** allow the plate to move at a constant speed of 10 ft/s to the right.

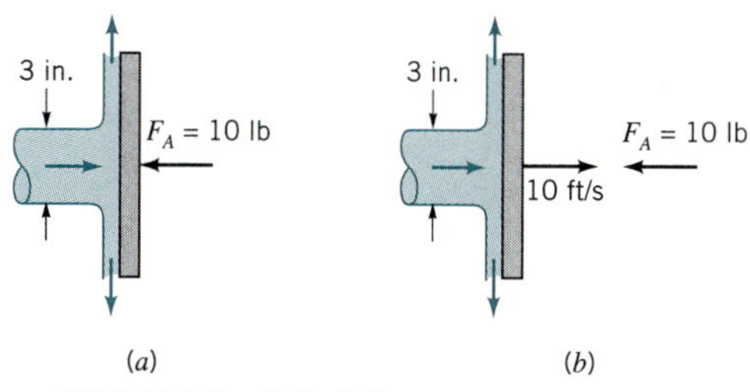

■ FIGURE P5.65

5.66 A vane directs a horizontal, circular cross-sectional jet of water symmetrically as indicated in Fig. P5.66. The jet leaves the nozzle with a velocity of 100 ft/s. Determine the x direction component of anchoring force required to **(a)** hold the vane stationary, **(b)** confine the speed of the vane to a value of 10 ft/s to the right. The fluid speed magnitude remains constant along the vane surface.

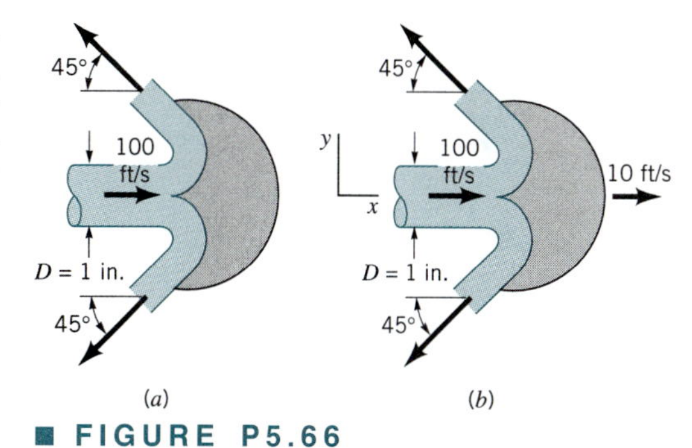

■ FIGURE P5.66

5.67 How much power is transferred to the moving vane of Problem 5.67?

5.68 Water enters a rotating lawn sprinkler through its base at the steady rate of 16 gal/min as shown in Fig. P5.68. The exit cross-sectional area of each of the two nozzles is 0.04 in.^{2}, and the flow leaving each nozzle is tangential. The radius from the axis of rotation to the centerline of each nozzle is 8 in. **(a)** Determine the resisting torque required to hold the sprinkler head stationary. **(b)** Determine the resisting torque associated with the sprinkler rotating with a constant speed of 500 rev/min. **(c)** Determine the angular velocity of the sprinkler if no resisting torque is applied.

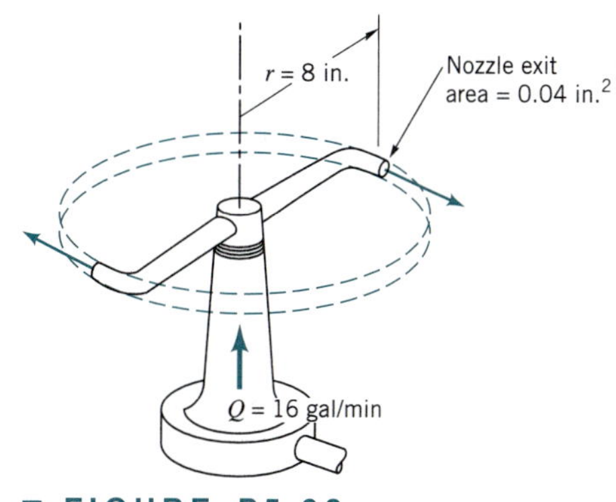

■ FIGURE P5.68

5.69 Five liters/s of water enters the rotor shown in Fig. P5.69 along the axis of rotation. The cross-sectional area of each of the three nozzle exits normal to the relative velocity is 18 mm^2. How large is the resisting torque required to hold the rotor stationary? How fast will the rotor spin steadily if the resisting torque is reduced to zero and **(a)** $\theta = 0°$, **(b)** $\theta = 30°$, **(c)** $\theta = 60°$?

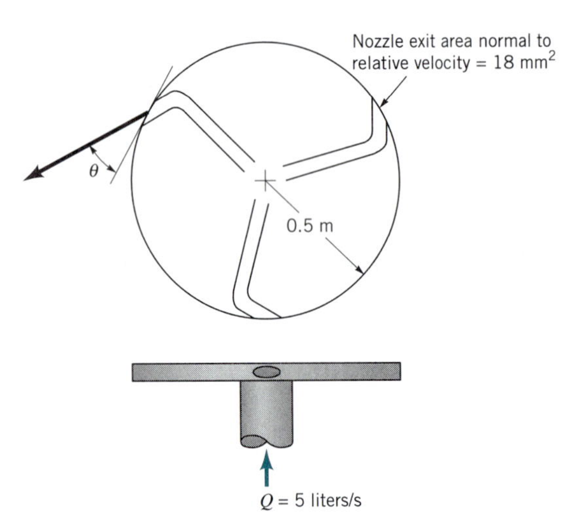

■ FIGURE P5.69

5.70 An inward flow radial turbine (see Fig. P5.70) involves a nozzle angle, α_1, of 60° and an inlet rotor tip speed, U_1, of 6 m/s. The ratio of rotor inlet to outlet diameters is 1.8. The absolute velocity leaving the rotor at section (2) is radial with a magnitude of 12 m/s. Determine the energy transfer per unit mass of fluid flowing through this turbine if the fluid is **(a)** air, **(b)** water.

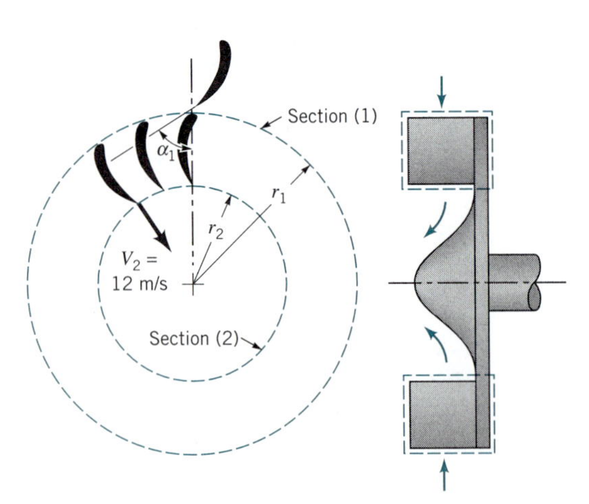

■ FIGURE P5.70

5.71 A water turbine wheel rotates at the rate of 50 rpm in the direction shown in Fig. P5.71. The inner radius, r_2, of the blade row is 2 ft, and the outer radius, r_1, is 4 ft. The absolute velocity vector at the turbine rotor entrance makes an angle of 20° with the tangential direction. The inlet blade angle is 60° relative to the tangential direction. The blade outlet angle is 120°. The flowrate is 20 ft^3/s. For the flow tangent to the rotor blade surface at inlet and outlet, determine an appropriate constant blade height, b, and the corresponding power available at the rotor shaft.

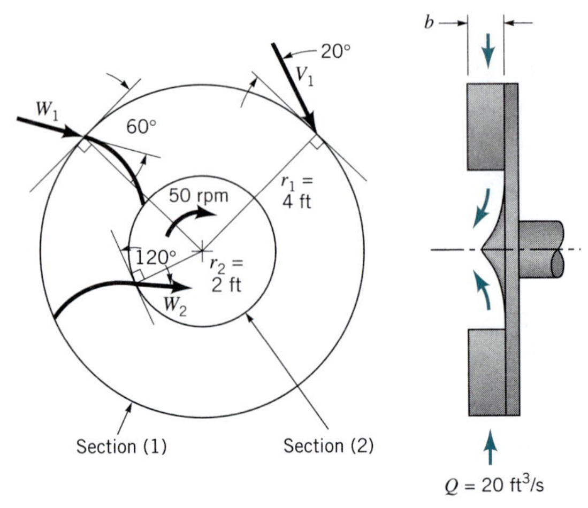

■ FIGURE P5.71

5.72 An incompressible fluid flows outward through a blower as indicated in Fig. P5.72. The shaft torque involved, T_{shaft}, is estimated with the following relationship:

$$T_{shaft} = \dot{m} r_2 V_{\theta 2}$$

where $\dot{m}$ = mass flowrate through the blower, r_2 = outer radius of blower, and $V_{\theta 2}$ = tangential component of absolute fluid velocity leaving the blower. State the flow conditions that make this formula valid.

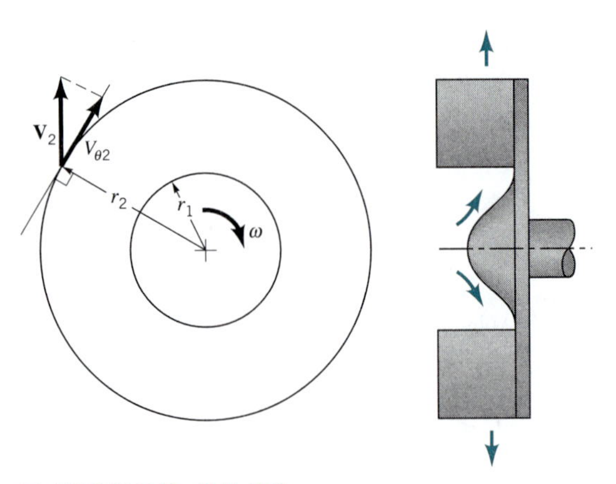

■ FIGURE P5.72

5.73 The radial component of velocity of water leaving the centrifugal pump sketched in Fig. P5.73 is 30 ft/s. The magnitude of the absolute velocity at the pump exit is 60 ft/s. The fluid enters the pump rotor radially. Calculate the shaft work required per unit mass flowing through the pump.

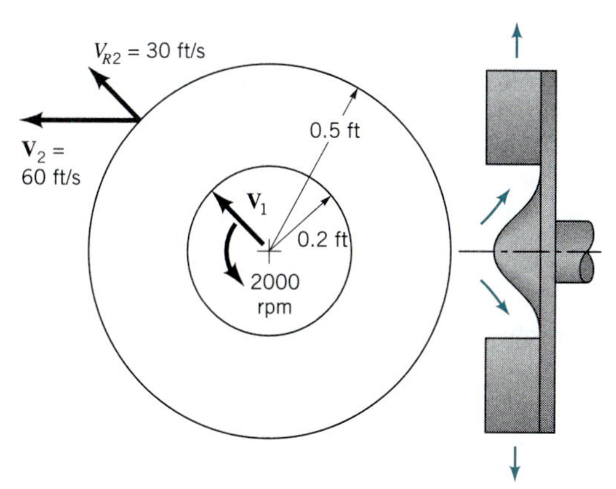

■ FIGURE P5.73

5.74 A fan (see Fig. P5.74) has a bladed rotor of 12-in. outside diameter and 5-in. inside diameter and runs at 1725 rpm. The width of each rotor blade is 1 in. from blade inlet to outlet. The volume flowrate is steady at 230 ft^3/min, and the absolute velocity of the air at blade inlet, V_1, is purely radial. The blade discharge angle is 30° measured with respect to the tangential direction at the outside diameter of the rotor. **(a)** What would be a reasonable blade inlet angle (measured with respect to the tangential direction at the inside diameter of the rotor)? **(b)** Find the power required to run the fan.

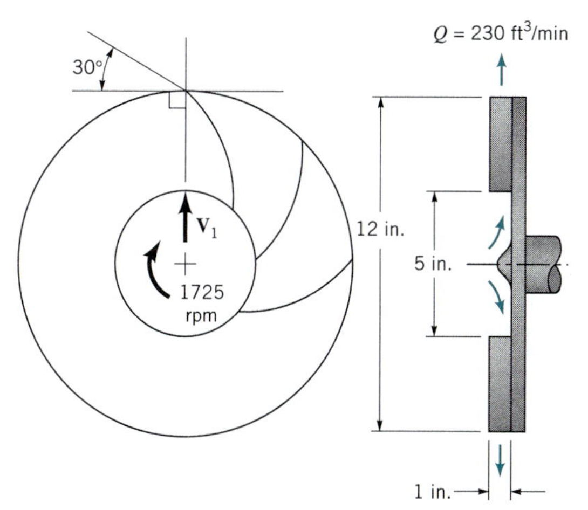

■ FIGURE P5.74

5.75 An axial flow gasoline pump (see Fig. P5.75) consists of a rotating row of blades (rotor) followed downstream by a stationary row of blades (stator). The gasoline enters the rotor axially (without any angular momentum) with an absolute velocity of 3 m/s. The rotor blade inlet and exit angles are 60° and 45° from the axial direction. The pump annulus passage cross-sectional area is constant. Consider the flow as being tangent to the blades involved. Sketch velocity triangles for flow just upstream and downstream of the rotor and just downstream of the stator where the flow is axial. How much energy is added to each kilogram of gasoline?

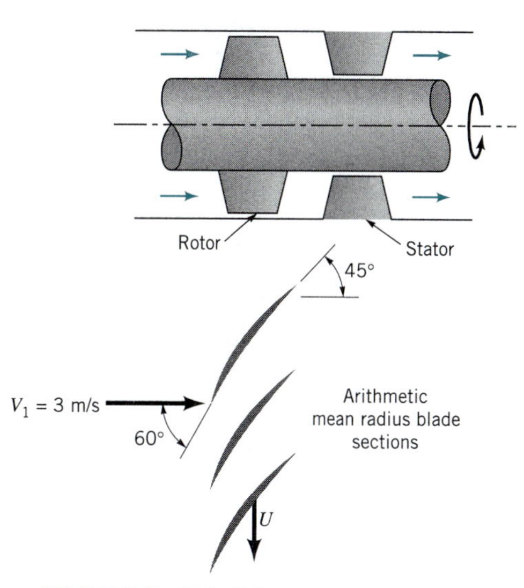

■ FIGURE P5.75

5.76 A sketch of the arithmetic mean radius blade sections of an axial-flow water turbine stage is shown in Fig. P5.76. The rotor speed is 1000 rpm. **(a)** Sketch and label velocity triangles for the flow entering and leaving the rotor row. Use **V** for absolute velocity, **W** for relative velocity, and **U** for blade velocity. Assume flow enters and leaves each blade row at the blade angles shown. **(b)** Calculate the work per unit mass delivered at the shaft.

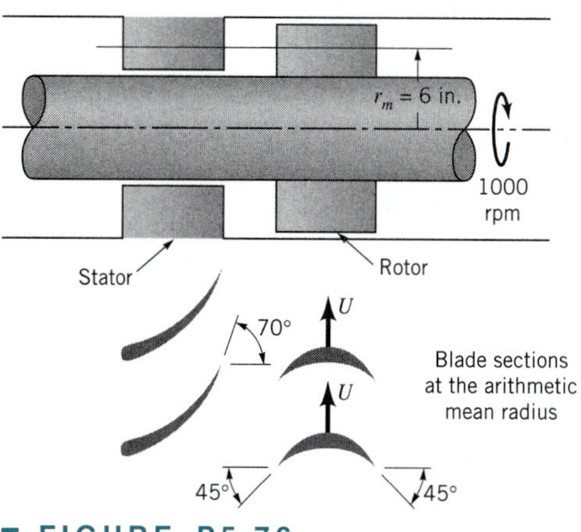

■ FIGURE P5.76

5.77 An axial-flow turbomachine rotor involves the upstream (1) and downstream (2) velocity triangles shown in Fig. P5.77. Is this turbomachine a turbine or a fan? Sketch an appropriate blade section and determine the energy transferred per unit mass of fluid.

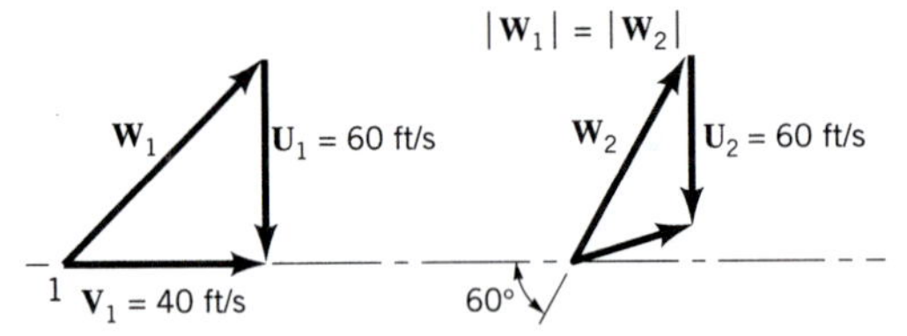

■ FIGURE P5.77

5.78 By using velocity triangles for flow upstream (1) and downstream (2) of a turbomachine rotor, prove that the shaft work in per unit mass flowing through the rotor is

$$w_{\substack{\text{shaft}\\\text{net in}}} = \frac{V_2^2 - V_1^2 + U_2^2 - U_1^2 + W_1^2 - W_2^2}{2}$$

where V = absolute flow velocity magnitude, W = relative flow velocity magnitude, and U = blade speed.

***5.79** Summarized below are air flow data for flow across a low-speed axial-flow fan. Calculate the change in rate of flow of axial direction angular momentum across this rotor and evaluate the shaft power input involved. The inner and outer radii of the fan annulus are 142 and 203 mm. The rotor speed is 2400 rpm.

	Upstream of Rotor		Downstream of Rotor	
Radius (mm)	Axial Velocity (m/s)	Absolute Tangential Velocity (m/s)	Axial Velocity (m/s)	Absolute Tangential Velocity (m/s)
142	0	0	0	0
148	32.03	0	32.28	12.64
169	32.03	0	32.37	12.24
173	32.04	0	31.78	11.91
185	32.03	0	31.50	11.35
197	31.09	0	29.64	11.66
203	0	0	0	0

5.80 Air enters a radial blower with zero angular momentum. It leaves with an absolute tangential velocity, V_θ, of 100 ft/s. The rotor blade speed at rotor exit is 70 ft/s. If the stagnation pressure rise across the rotor is 0.1 psi, calculate the loss of available energy across the rotor.

5.81 Water enters a pump impeller radially. It leaves the impeller with a tangential component of absolute velocity of 10 m/s. The impeller exit diameter is 60 mm and the impeller speed is 1800 rpm. If the stagnation pressure rise across the impeller is 45 kPa, determine the loss of available energy across the impeller and the hydraulic efficiency of the pump.

5.82 Water enters an axial-flow turbine rotor with an absolute velocity tangential component, V_θ, of 15 ft/s. The corresponding blade velocity, U, is 50 ft/s. The water leaves the rotor blade row with no angular momentum. If the stagnation pressure drop across the turbine is 12 psi, determine the hydraulic efficiency of the turbine.

5.83 An inward flow radial turbine (see Fig. P5.83) involves a nozzle angle, α_1, of 60° and an inlet rotor tip speed, U_1, of 30 ft/s. The ratio of rotor inlet to outlet diameters is 2.0. The radial component of velocity remains constant at 20 ft/s through the rotor and the flow leaving the rotor at section (2) is without angular momentum. If the flowing fluid is water and the stagnation pressure drop across the rotor is 16 psi, determine the loss of available energy across the rotor and the hydraulic efficiency involved.

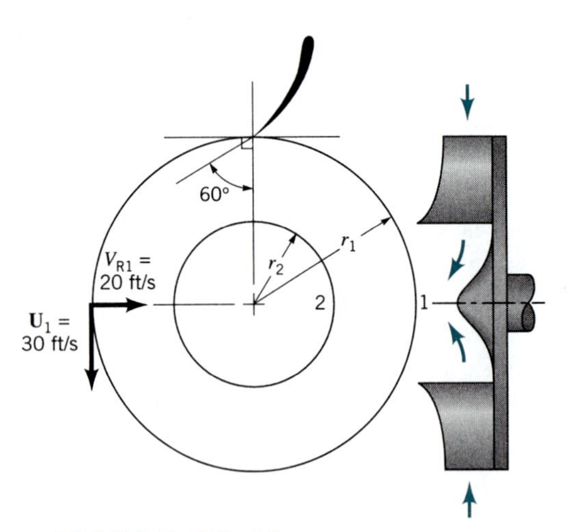

■ FIGURE P5.83

5.84 An inward flow radial turbine (see Fig. P5.83) involves a nozzle angle, α_1, of 60° and an inlet rotor tip speed of 30 ft/s. The ratio of rotor inlet to outlet diameters is 2.0. The radial component of velocity remains constant at 20 ft/s through the rotor and the flow leaving the rotor at section (2) is without angular momentum. If the flowing fluid is air and the static pressure drop across the rotor is 0.01 psi, determine the loss of available energy across the rotor and the rotor aerodynamic efficiency.

5.85 For adiabatic flow through a turbomachine rotor, prove that the quantity

$$\check{h} + \frac{W^2}{2} - \frac{U^2}{2}$$

remains constant, where

$\check{h}$ = enthalpy
W = relative flow velocity
U = blade speed

5.86 What is the size of the head loss that is needed to raise the temperature of water by 1°F?

† **5.87** Based on flowrate and pressure rise information, estimate the power output of a human heart.

5.88 Air steadily expands adiabatically and without friction from stagnation conditions of 650 kPa (abs) and 290 K to a static pressure of 101 kPa (abs). Determine the velocity of the expanded air assuming **(a)** incompressible flow, **(b)** compressible flow.

5.89 Air flows past an object in a pipe of 2-m diameter and exits as a free jet as shown in Fig. P5.89. The velocity and pressure upstream are uniform at 10 m/s and 50 N/m^2, respectively. At the pipe exit the velocity is nonuniform as indicated. The shear stress along the pipe wall is negligible. **(a)** Determine the head loss associated with a particle as it flows from the uniform velocity upstream of the object to a location in the wake at the exit plane of the pipe. **(b)** Determine the force that the air puts on the object.

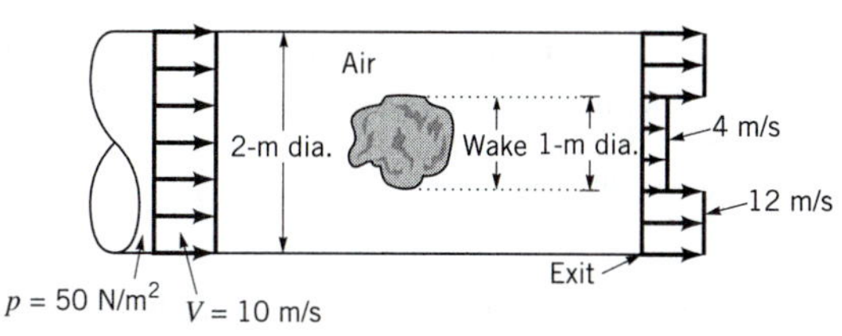

■ FIGURE P5.89

5.90 Oil ($SG = 0.9$) flows downward through a vertical pipe contraction as shown in Fig. P5.90. If the mercury manometer reading, h, is 100 mm, determine the volume flowrate for frictionless flow. Is the actual flowrate more or less than the frictionless value? Explain.

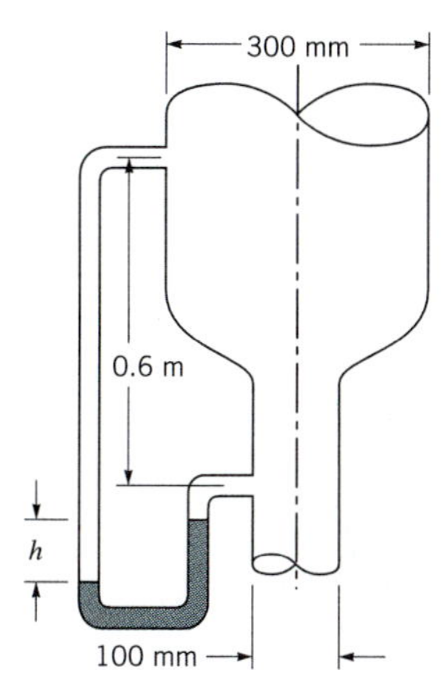

■ FIGURE P5.90

5.91 An incompressible liquid flows steadily along the pipe shown in Fig. P5.91. Determine the direction of flow and the head loss over the 6-m length of pipe.

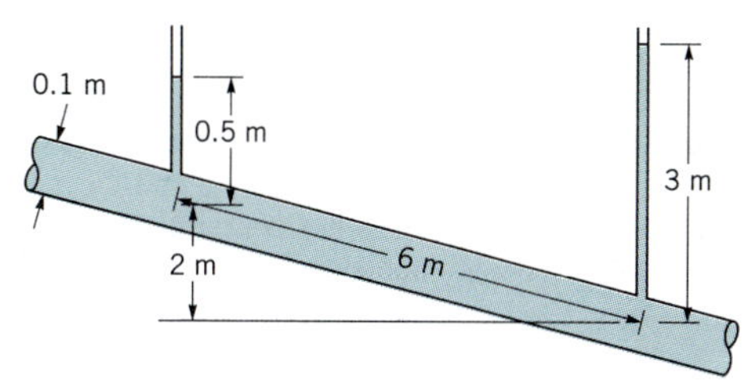

■ FIGURE P5.91

5.92 A siphon is used to draw water at 70 °F from a large container as indicated in Fig. P5.92. The inside diameter of the siphon line is 1 in. and the pipe centerline rises 3 ft above the essentially constant water level in the tank. Show that by varying the length of the siphon below the water level, h, the rate of flow through the siphon can be changed. Assuming frictionless flow, determine the maximum flowrate possible through the siphon. The limiting condition is the occurrence of cavitation in the siphon. Will the actual maximum flow be more or less than the frictionless value? Explain.

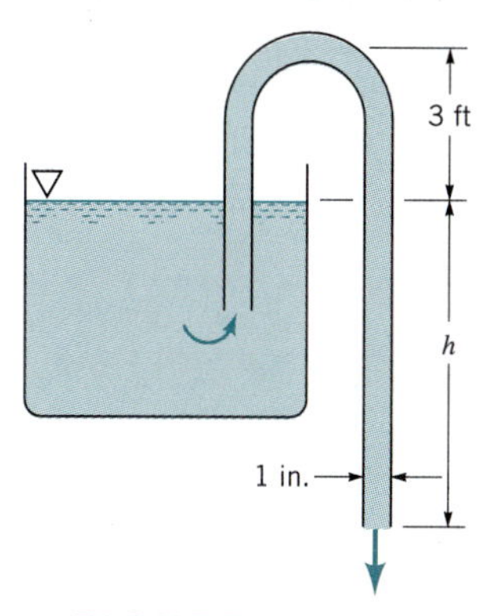

■ FIGURE P5.92

5.93 A water siphon having a constant inside diameter of 3 in. is arranged as shown in Fig. P5.93. If the friction loss between A and B is $0.8V^2/2$, where V is the velocity of flow in the siphon, determine the flowrate involved.

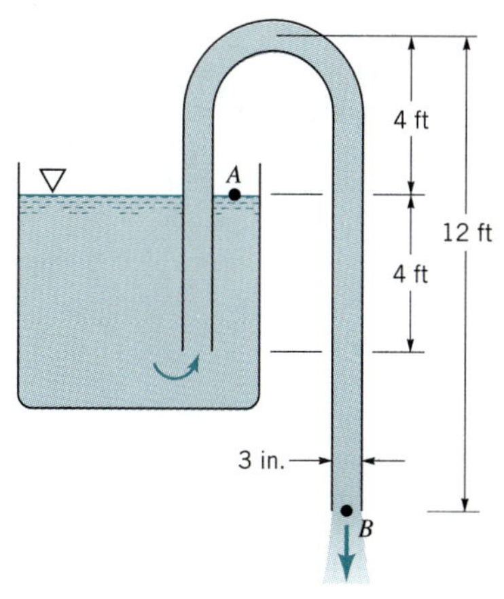

■ FIGURE P5.93

5.94 Water flows through a valve (see Fig. P5.94) at the rate of 1000 lbm/s. The pressure just upstream of the valve is 90 psi and the pressure drop across the valve is 5 psi. The inside diameters of the valve inlet and exit pipes are 12 and 24 in. If the flow through the valve occurs in a horizontal plane, determine the loss in available energy across the valve.

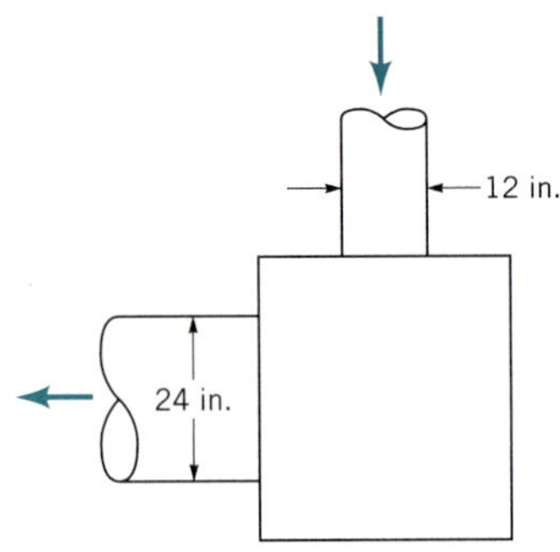

■ FIGURE P5.94

5.95 Water flows through a vertical pipe as is indicated in Fig. P5.95. Is the flow up or down in the pipe? Explain.

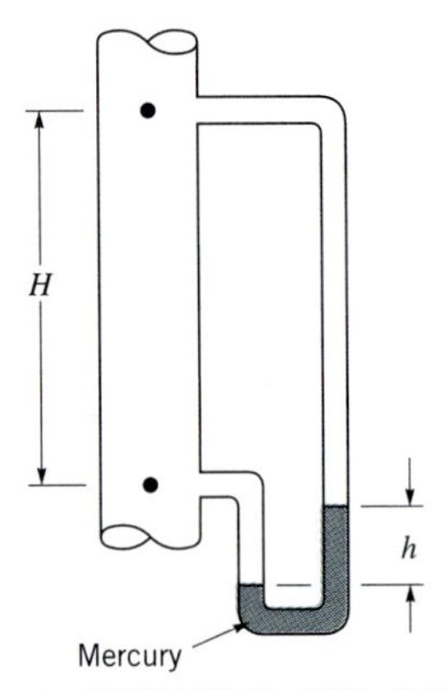

■ FIGURE P5.95

5.96 A fire hose nozzle is designed to deliver water that will rise 40 m vertically. Calculate the stagnation pressure required at the nozzle inlet if **(a)** no loss is assumed, **(b)** a loss of 30 N·m/kg is assumed.

5.97 For the 180° elbow and nozzle flow shown in Fig. P5.97, determine the loss in available energy from section (1) to section (2). How much additional available energy is lost from section (2) to where the water comes to rest?

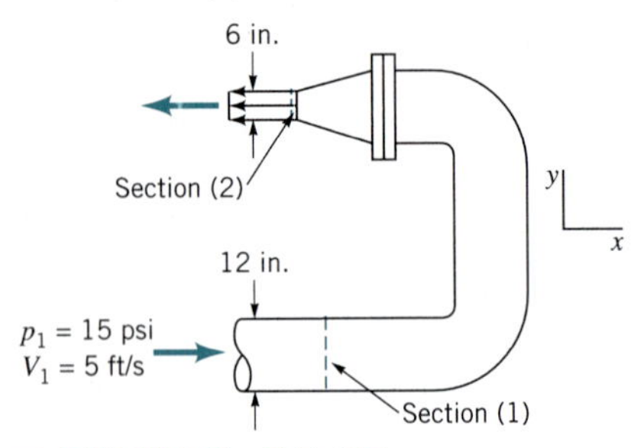

■ FIGURE P5.97

5.98 An automobile engine will work best when the back pressure at the interface of the exhaust manifold and the engine block is minimized. Show how reduction of losses in the exhaust manifold, piping, and muffler will also reduce the back pressure. How could losses in the exhaust system be reduced? What primarily limits the minimization of exhaust system losses?

5.99 Water flows vertically upward in a circular cross-sectional pipe. At section (1), the velocity profile over the cross-sectional area is uniform. At section (2), the velocity profile is

$$\mathbf{V} = w_c \left(\frac{R - r}{R}\right)^{1/7} \hat{\mathbf{k}}$$

where $\mathbf{V}$ = local velocity vector, w_c = centerline velocity in the axial direction, R = pipe inside radius, and, r = radius from pipe axis. Develop an expression for the loss in available energy between sections (1) and (2).

5.100 Discuss the causes of loss of available energy in a fluid flow.

5.101 Consider the flow shown in Fig. P5.91. If the flowing fluid is water, determine the axial (along the pipe) and normal (perpendicular to the pipe) components of force that the pipe puts on the fluid in the 6-m section shown.

5.102 Water flows steadily down the inclined pipe as indicated in Fig. P5.102. Determine the following: **(a)** the difference in pressure $p_1 - p_2$, **(b)** the loss between sections (1) and (2), **(c)** the net axial force exerted by the pipe wall on the flowing water between sections (1) and (2).

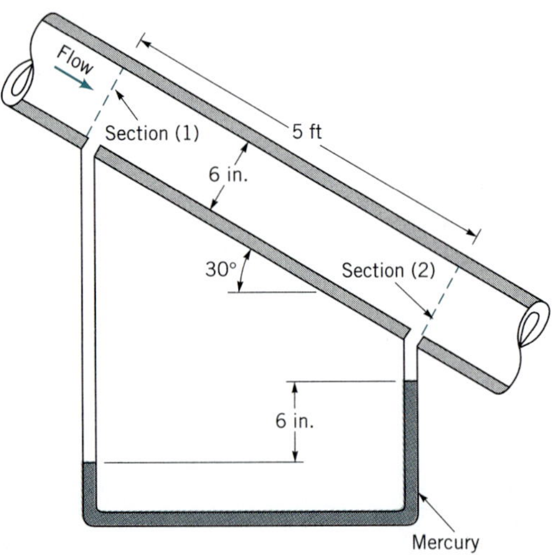

■ FIGURE P5.102

5.103 Water flows through a 2-ft-diameter pipe arranged horizontally in a circular arc as shown in Fig. P5.103. If the pipe discharges to the atmosphere (p = 14.7 psia), determine the x and y components of the resultant force exerted by the water on the piping between sections (1) and (2). The steady flowrate is 3000 ft^3/min. The loss in pressure due to fluid friction between sections (1) and (2) is 25 psi.

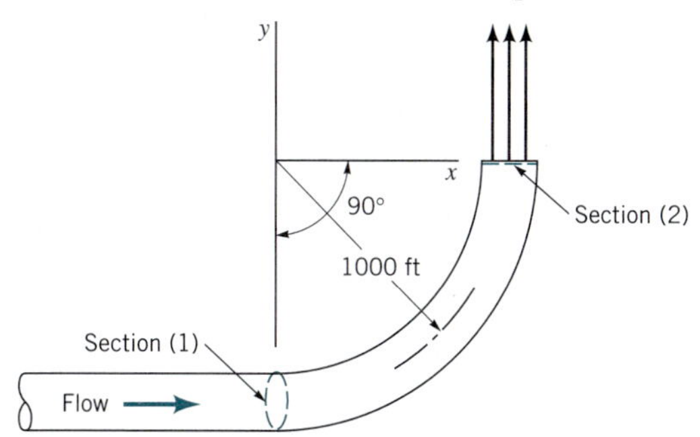

■ **FIGURE P5.103**

5.104 When fluid flows through an abrupt expansion as indicated in Fig. P5.104, the loss in available energy across the expansion, loss_{ex}, is often expressed as

$$\text{loss}_{ex} = \left(1 - \frac{A_1}{A_2}\right)^2 \frac{V_1^2}{2}$$

where A_1 = cross-sectional area upstream of expansion, A_2 = cross-sectional area downstream of expansion, and V_1 = velocity of flow upstream of expansion. Derive this relationship.

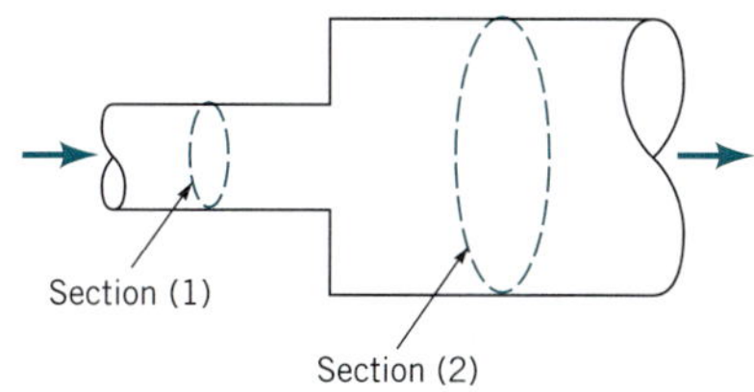

■ **FIGURE P5.104**

5.105 Near the downstream end of a river spillway, a hydraulic jump often forms, as illustrated in Fig. P5.105. The velocity of the channel flow is reduced abruptly across the jump. Using the conservation of mass and linear momentum principles, derive the following expression for h_2,

$$h_2 = -\frac{h_1}{2} + \sqrt{\left(\frac{h_1}{2}\right)^2 + \frac{2V_1^2 h_1}{g}}$$

The loss of available energy across the jump can also be determined if energy conservation is considered. Derive the loss expression

$$\text{jump loss} = \frac{g(h_2 - h_1)^3}{4h_1h_2}$$

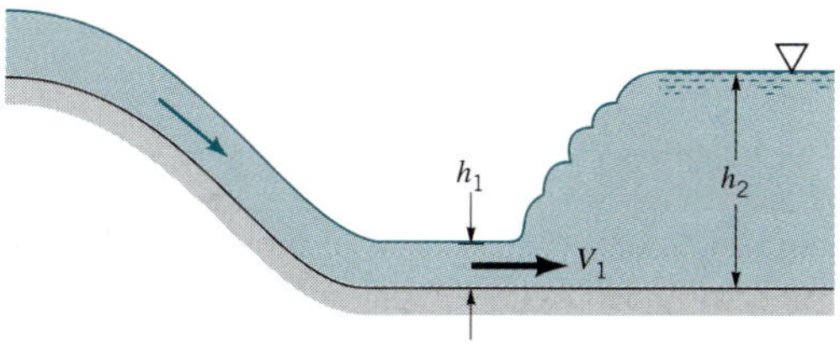

■ **FIGURE P5.105**

5.106 Two water jets collide and form one homogeneous jet as shown in Fig. P5.106. **(a)** Determine the speed, V, and direction, θ, of the combined jet. **(b)** Determine the loss for a fluid particle flowing from (1) to (3), from (2) to (3). Gravity is negligible.

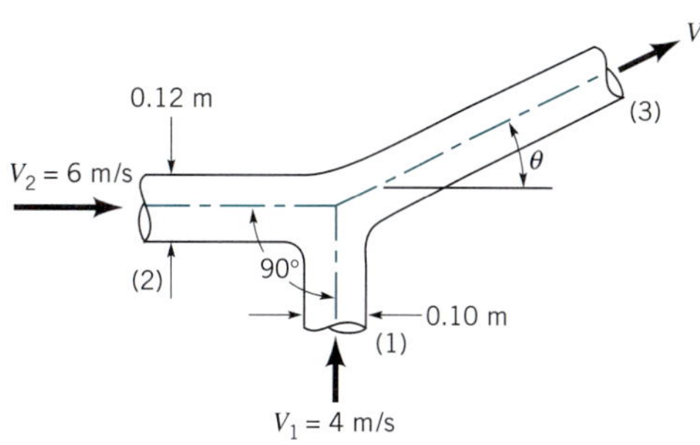

■ **FIGURE P5.106**

5.107 A gas expands through a nozzle from a pressure of 300 psia to a pressure of 5 psia. The enthalpy change involved, $\check{h}_1 - \check{h}_2$, is 150 Btu/lbm. If the expansion is adiabatic but with frictional effects and the inlet gas speed is negligibly small, determine the exit gas velocity.

5.108 What is the maximum possible power output of the hydroelectric turbine shown in Fig. P5.108?

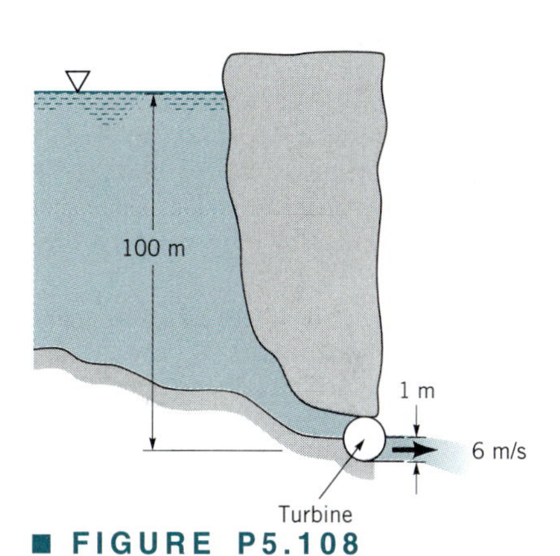

■ **FIGURE P5.108**

5.109 A hydraulic turbine is provided with 4.25 m^3/s of water at 415 kPa. A vacuum gage in the turbine discharge 3 m below the turbine inlet centerline reads 250-mm Hg vacuum. If the turbine shaft output power is 1100 kW, calculate the power loss through the turbine. The supply and discharge pipe inside diameters are identically 800 mm.

5.110 Water is supplied at 150 ft^3/s and 60 psi to a hydraulic turbine through a 3-ft inside-diameter inlet pipe as indicated in Fig. P5.110. The turbine discharge pipe has a 4-ft inside diameter. The static pressure at section (2), 10 ft below the turbine inlet, is 10-in. Hg vacuum. If the turbine develops 2500 hp, determine the power lost between sections (1) and (2).

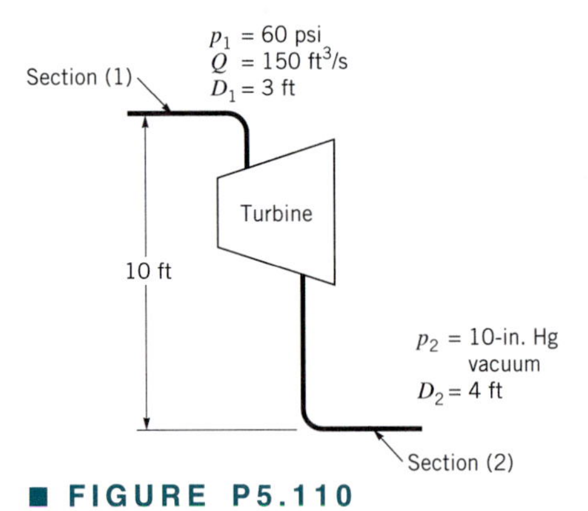

■ **FIGURE P5.110**

5.111 A steam turbine receives steam having a static pressure, p_1, of 400 psia, an enthalpy, $\check{h}_1$, of 1407 Btu/lbm, and a velocity, V_1, of 100 ft/s. The steam leaves the turbine as a mixture of vapor and liquid having an enthalpy, $\check{h}_2$, of 1098 Btu/lbm, a pressure, p_2, of 2 psia, and a velocity, V_2, of 200 ft/s. If the flow through the turbine is essentially adiabatic and the change in elevation of the steam is negligible, calculate **(a)** the actual work output per unit mass of steam, **(b)** the efficiency of the turbine if the ideal work output is 467 Btu/lbm.

5.112 A centrifugal air compressor stage operates between an inlet stagnation pressure of 14.7 psia and an exit stagnation pressure of 60 psia. The inlet stagnation temperature is 80 °F. If the loss of total pressure through the compressor stage associated with irreversible flow phenomena is 10 psi, estimate the actual and ideal stagnation temperature rise through the compressor. Estimate the ratio of ideal to actual temperature rise to obtain an approximate value of the efficiency.

5.113 Explain how, in terms of the loss of available energy involved, a home water faucet valve works to vary the flow through a lawn hose from the shutoff condition to maximum flow.

***5.114** Total head-rise values measured for air flowing across a fan are listed below as a function of volume flowrate.

Q (m³/s)	Total Head Rise (mm H_2O)
0	79
0.14	79
0.28	76
0.42	67
0.57	65
0.71	70
0.85	76
0.99	79
1.13	75
1.27	64

Determine the flowrate that will result when this fan is connected to a piping system whose loss in total head is described by loss $= K_L Q^2$ when: **(a)** K_L = 49 mm $H_2O/(m^3/s)^2$; **(b)** K_L = 91 mm $H_2O/(m^3/s)^2$; **(c)** K_L = 140 mm $H_2O(m^3/s)^2$.

5.115 Water is pumped from the tank shown in Fig. P5.115*a*. The head loss is known to be 1.2 $V^2/2g$, where V is the average velocity in the pipe. According to the pump manufacturer, the relationship between the pump head and the flowrate is as shown in Fig. P5.115*b*: $h_p = 20 - 2000\,Q^2$, where h_p is in meters and Q is in m³/s. Determine the flowrate, Q.

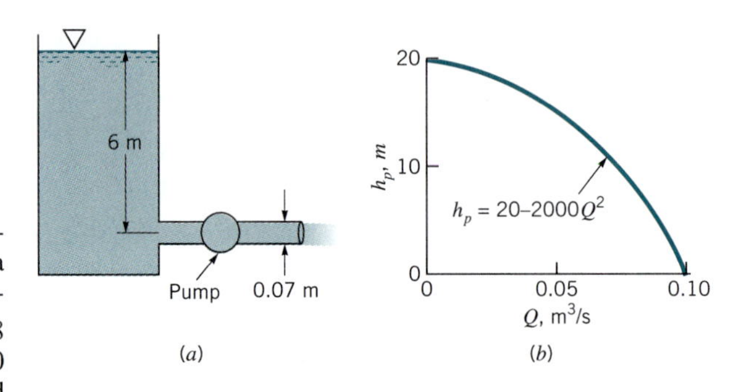

■ **FIGURE P5.115**

5.116 Water flows by gravity from one lake to another as sketched in Fig. P5.116 at the steady rate of 80 gpm. What is the loss in available energy associated with this flow? If this same amount of loss is associated with pumping the fluid from the lower lake to the higher one at the same flowrate, estimate the amount of pumping power required.

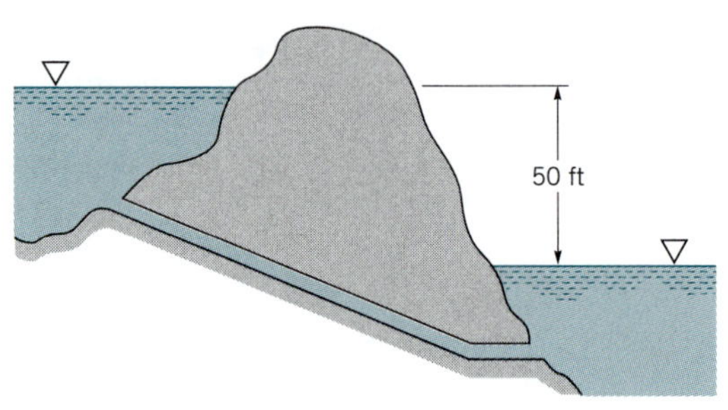

■ **FIGURE P5.116**

5.117 A $\frac{3}{4}$-hp motor is required by an air ventilating fan to produce a 24-in.-diameter stream of air having a uniform speed of 40 ft/s. Determine the aerodynamic efficiency of the fan.

5.118 A pump moves water horizontally at a rate of 0.02 m³/s. Upstream of the pump where the pipe diameter is 90 mm, the pressure is 120 kPa. Downstream of the pump where the pipe diameter is 30 mm, the pressure is 400 kPa. If the loss in energy across the pump due to fluid friction effects is 170 N·m/kg, determine the hydraulic efficiency of the pump.

5.119 The turbine shown in Fig. P5.119 develops 100 hp when the flowrate of water is 20 ft^3/s. If all losses are negligible, determine **(a)** the elevation h, **(b)** the pressure difference across the turbine, and **(c)** the flowrate expected if the turbine were removed.

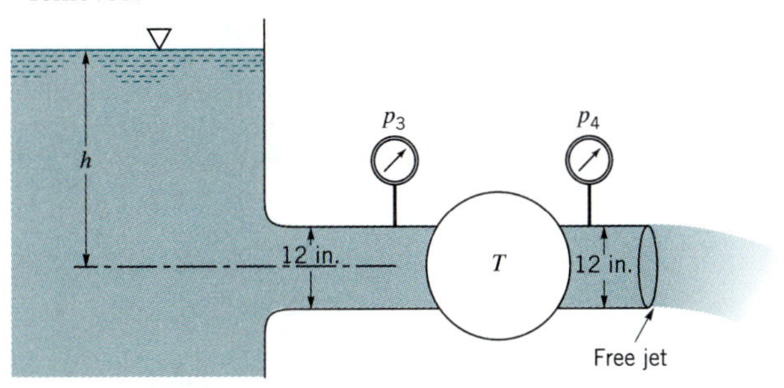

■ **FIGURE P5.119**

5.120 A liquid enters a fluid machine at section (1) and leaves at sections (2) and (3) as shown in Fig. P5.120. The density of the fluid is constant at 2 slugs/ft^3. All of the flow occurs in a horizontal plane and is frictionless and adiabatic. For the above-mentioned and additional conditions indicated in Fig. 5.120, determine the amount of shaft power involved.

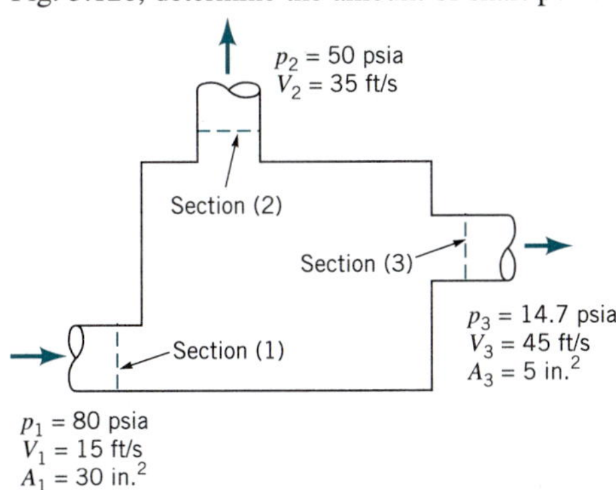

■ **FIGURE P5.120**

5.121 Water is to be moved from one large reservoir to another at a higher elevation as indicated in Fig. P5.121. The loss available energy associated with 2.5 ft^3/s being pumped from sections (1) to (2) is $61\overline{V}^2/2\ \text{ft}^2/\text{s}^2$, where $\overline{V}$ is the average velocity of water in the 8-in. inside-diameter piping involved. Determine the amount of shaft power required.

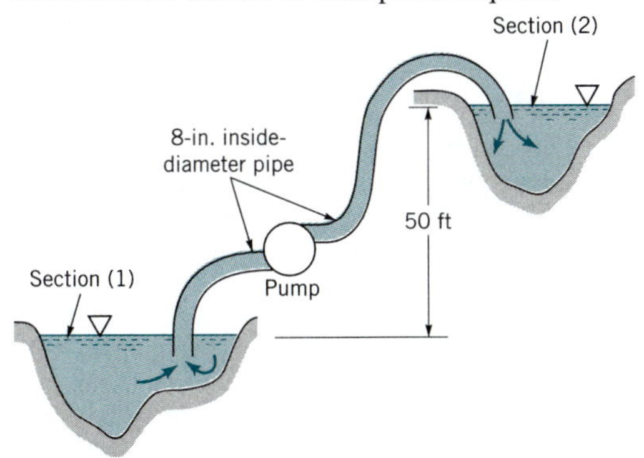

■ **FIGURE P5.121**

5.122 Oil ($SG = 0.88$) flows in an inclined pipe at a rate of 5 ft^3/s as shown in Fig. P5.122. If the differential reading in the mercury manometer is 3 ft, calculate the power that the pump supplies to the oil if head losses are negligible.

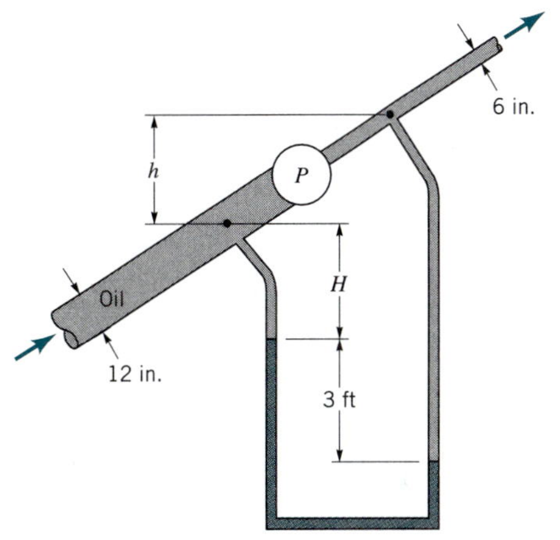

■ **FIGURE P5.122**

5.123 The distribution of axial direction velocity, u, in a pipe flow is linear from zero at the wall to maximum of u_c at the centerline. Determine the average velocity, $\overline{u}$, and the kinetic energy coefficient, α.

5.124 The velocity profile in a turbulent pipe flow may be approximated with the expression

$$\frac{u}{u_c} = \left(\frac{R - r}{R}\right)^{1/n}$$

where u = local velocity in the axial direction, u_c = centerline velocity in the axial direction, R = pipe inner radius from pipe axis, r = local radius from pipe axis, and n = constant. Determine the kinetic energy coefficient, α, for **(a)** $n = 5$, **(b)** $n = 6$, **(c)** $n = 7$, **(d)** $n = 8$, **(e)** $n = 9$, **(f)** $n = 10$.

5.125 A small fan moves air at a mass flowrate of 0.004 lbm/s. Upstream of the fan, the pipe diameter is 2.5 in., the flow is laminar, the velocity distribution is parabolic, and the kinetic energy coefficient, α_1, is equal to 2.0. Downstream of the fan, the pipe diameter is 1 in., the flow is turbulent, the velocity profile is quite flat, and the kinetic energy coefficient, α_2, is equal to 1.08. If the rise in static pressure across the fan is 0.015 psi and the fan shaft draws 0.00024 hp, compare the value of loss calculated: **(a)** assuming uniform velocity distributions, **(b)** considering actual velocity distributions.

5.126 The device shown in Fig. P5.126 is used to determine the force put on a flat plate by a jet of air that is deflected by it. Air at a temperature of 77 °F and an absolute pressure of 29.25 inches of mercury flows from the nozzle at a measured rate of $Q = 1.40 \text{ ft}^3/\text{s}$. The air jet strikes a flat plate and is deflected through a 90 degree angle as shown. A water-filled manometer is used to measure the pressure, p, on the plate as a function of the radial distance, r, from the center of the plate. The manometer reading is h. Since the flow is axisymmetric, the net force, R, of the air against the plate is

$$R = 2\pi \int_0^{D/2} p\, r\, dr$$

where D is the diameter of the plate.

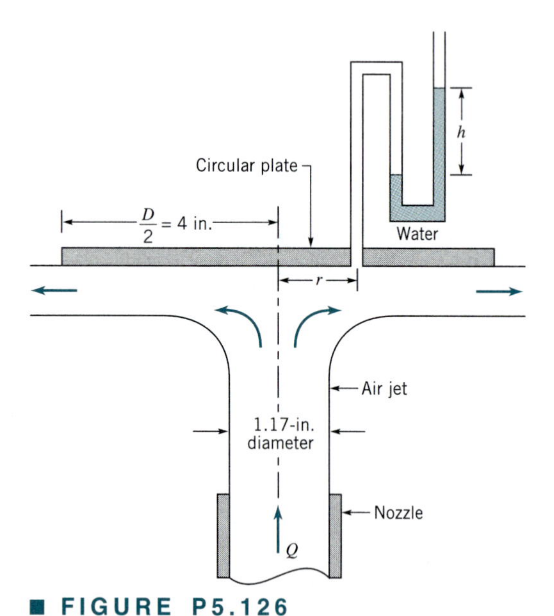

■ FIGURE P5.126

Experimentally determined values of h and r are shown in the table below. Use these results to determine the value of R. Note: Some type of numerical or graphical integration of the experimental data is needed.

Compare your experimentally determined value of R with that obtained from the momentum equation, $R = \rho V^2 A$. Discuss some possible reasons for any difference between the two values.

h (in.)	r (in.)
2.95	0
2.80	0.399
1.83	0.791
1.25	1.242
0.60	1.599
0.33	2.04
0.15	2.41
0.07	2.85
0.03	3.23
0.02	3.67
0.00	4.00

5.127 The device shown in Fig. P5.127 is used to investigate the force needed to deflect a stream of air. A fan forces air at a temperature of 73 °F and an absolute pressure of 29.07 in. of mercury through a circular galvanized iron duct. The average speed, V, is determined by use of a Pitot-static tube. A long-radius elbow at the end of the duct deflects the air through a 90-degree bend as indicated. The elbow rests on a scale, and the horizontal pipe is attached to the fan by means of a flexible bellows about which the pipe is free to pivot. The scale reading, R, is adjusted to zero when the fan is turned off.

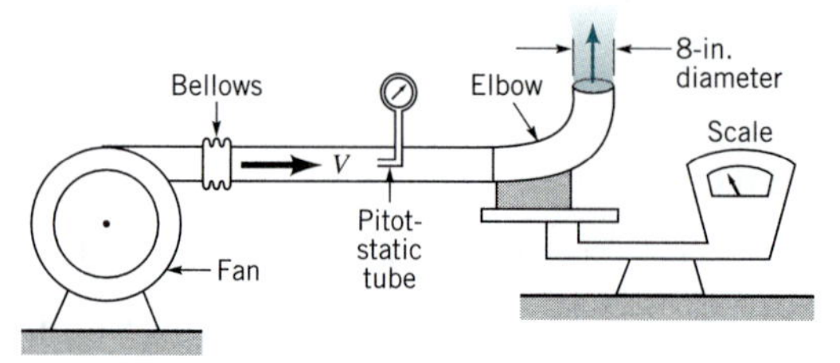

■ FIGURE P5.127

Experimentally determined values of V and R are shown in the table below. Use these results to plot a graph of the force that the air puts on the elbow as a function of the air speed. On the same graph, plot the theoretical curve as obtained from the momentum equation.

Compare the experimental and theoretical results and discuss some possible reasons for any differences between them.

V (ft/min)	R (lb)
1200	0.38
1420	0.53
1800	0.79
2160	1.05
2440	1.38
2700	1.65
2900	1.90
3100	2.19
3520	2.83
3750	3.12
3950	3.38

5.128 The device shown in Fig. P5.128 is used to investigate the force needed to deflect a jet of water. Water pumped from a tank at a given flowrate, Q, issues from a nozzle with speed V and is deflected through a known angle, θ, by means of a vane as shown. The vane is constrained from moving horizontally. The tension in the spring attached to the vane is adjusted to give a null reading on the position indicator when the flowrate is zero and $\mathcal{W} = 0$. With a known weight, $\mathcal{W}$, on the balance pan, the flowrate is adjusted to return the position indicator to its null position. The flowrate is determined by measuring the weight of water, $\mathcal{W}_w$, that is pumped from the tank in a given time, t. That is, Q = $\mathcal{W}_w/(\gamma t)$, where γ is the specific weight of water in the table below for two values of θ. Use these results to plot a graph of the force that the water puts on the vane as a function of the speed of the water from the nozzle. There will be two curves, one for each value of θ. On the same graph plot the theoretical curves for the two cases tested.

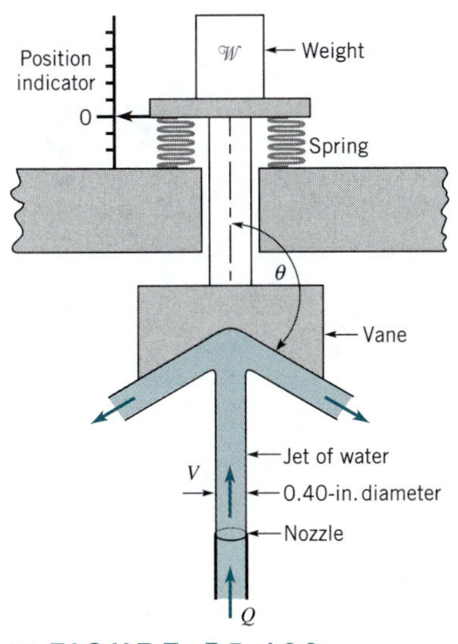

■ FIGURE P5.128

Compare the experimental and theoretical results and discuss some possible reasons for any differences between them.

For $\theta = 90°$:

$\mathcal{W}$ (lb)	$\mathcal{W}_w$ (lb)	t (s)
0.044	7.71	26.8
0.154	8.66	18.2
0.264	8.92	12.6
0.374	8.78	10.0
0.485	9.96	10.6

For $\theta = 180°$:

$\mathcal{W}$ (lb)	$\mathcal{W}_w$ (lb)	t (s)
0.110	6.81	24.5
0.220	9.00	20.8
0.551	7.88	10.9
0.771	7.97	9.5
0.880	6.37	7.6

Flow past an inclined plate: The streamlines of a viscous fluid flowing slowly past a two-dimensional object placed between two closely spaced plates (a Hele-Shaw cell) approximate inviscid, irrotational (potential) flow. (Dye in water between glass plates spaced 1 mm apart.) (Photography courtesy of D. H. Peregrine.)

6 Differential Analysis of Fluid Flow

In the previous chapter attention is focused on the use of finite control volumes for the solution of a variety of fluid mechanics problems. This approach is very practical and useful, since it does not generally require a detailed knowledge of the pressure and velocity variations within the control volume. Typically, we found that only conditions on the surface of the control volume entered the problem, and thus problems could be solved without a detailed knowledge of the flow field. Unfortunately, there are many situations that arise in which the details of the flow are important and the finite control volume approach will not yield the desired information. For example, we may need to know how the velocity varies over the cross section of a pipe, or how the pressure and shear stress vary along the surface of an airplane wing. In these circumstances we need to develop relationships that apply at a point, or at least in a very small region (infinitesimal volume) within a given flow field. This approach, which involves an *infinitesimal control volume*, as distinguished from a finite control volume, is commonly referred to as *differential analysis*, since (as we will soon discover) the governing equations are differential equations.

Differential analysis provides very detailed knowledge of a flow field.

In this chapter we will provide an introduction to the differential equations that describe (in detail) the motion of fluids. Unfortunately, we will also find that these equations are rather complicated, partial differential equations that cannot be solved exactly except in a few cases, at least without making some simplifying assumptions. Thus, although differential analysis has the potential for supplying very detailed information about flow fields, this information is not easily extracted. Nevertheless, this approach provides a fundamental basis for the study of fluid mechanics. We do not want to be too discouraging at this point, since there are some exact solutions for laminar flow that can be obtained, and these have proved to be very useful. A few of these are included in this chapter. In addition, by making some simplifying assumptions many other analytical solutions can be obtained. For example, in some circumstances it may be reasonable to assume that the effect of viscosity is small and can be neglected. This rather drastic assumption greatly simplifies the analysis and provides the opportunity to obtain detailed solutions to a variety of complex flow problems. Some examples of these so-called *inviscid flow* solutions are also described in this chapter.

It is known that for certain types of flows the flow field can be conceptually divided into two regions—a very thin region near the boundaries of the system in which viscous effects are important, and a region away from the boundaries in which the flow is essentially inviscid. By making certain assumptions about the behavior of the fluid in the thin layer near the boundaries, and using the assumption of inviscid flow outside this layer, a large class of problems can be solved using differential analysis. These boundary layer problems are discussed in Chapter 9. Finally, it is to be noted that with the availability of powerful digital computers it is feasible to attempt to solve the differential equations using the techniques of numerical analysis. Although it is beyond the scope of this book to delve into this approach, which is generally referred to as *computational fluid dynamics* (CFD), the reader should be aware of this approach to complex flow problems. A few additional comments about CFD and other aspects of differential analysis are given in the last section of this chapter.

We begin our introduction to differential analysis by reviewing and extending some of the ideas associated with fluid kinematics that were introduced in Chapter 4. With this background the remainder of the chapter will be devoted to the derivation of the basic differential equations (which will be based on the principle of conservation of mass and Newton's second law of motion) and to some applications.

6.1 Fluid Element Kinematics

In this section we will be concerned with the mathematical description of the motion of fluid elements moving in a flow field. A small fluid element in the shape of a cube which is initially in one position will move to another position during a short time interval δt as illustrated in Fig. 6.1. Because of the generally complex velocity variation within the field, we expect the element not only to translate from one position but also to have its volume changed (linear deformation), to rotate, and to undergo a change in shape (angular deformation). Although these movements and deformations occur simultaneously, we can consider each one separately as illustrated in Fig. 6.1. Since element motion and deformation are intimately related to the velocity and variation of velocity throughout the flow field, we will briefly review the manner in which velocity and acceleration fields can be described.

6.1.1 Velocity and Acceleration Fields Revisited

Point velocities are conveniently described using the Eulerian method.

As discussed in detail in Section 4.1, the velocity field can be described by specifying the velocity **V** at all points, and at all times, within the flow field of interest. Thus, in terms of rectangular coordinates, the notation **V** (x, y, z, t) means that the velocity of a fluid particle depends on where it is located within the flow field (as determined by its coordinates, x, y, and z) and when it occupies the particular point (as determined by the time, t). As is pointed out in Section 4.1.1, this method of describing the fluid motion is called the Eulerian method.

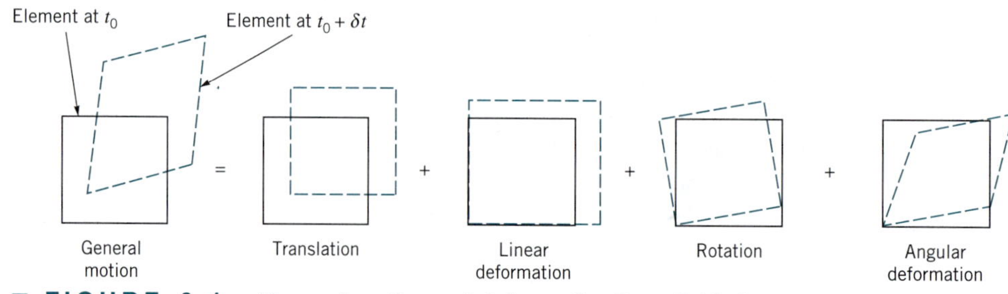

■ FIGURE 6.1 **Types of motion and deformation for a fluid element.**

It is also convenient to express the velocity in terms of three rectangular components so that

$$\mathbf{V} = u\hat{\mathbf{i}} + v\hat{\mathbf{j}} + w\hat{\mathbf{k}} \tag{6.1}$$

where u, v, and w are the velocity components in the x, y, and z directions, respectively, and $\hat{\mathbf{i}}$, $\hat{\mathbf{j}}$, and $\hat{\mathbf{k}}$ are the corresponding unit vectors. Of course, each of these components will, in general, be a function of x, y, z, and t. One of the goals of differential analysis is to determine how these velocity components specifically depend on x, y, z, and t for a particular problem.

With this description of the velocity field it was also shown in Section 4.2.1 that the acceleration of a fluid particle can be expressed as

$$\mathbf{a} = \frac{\partial \mathbf{V}}{\partial t} + u\frac{\partial \mathbf{V}}{\partial x} + v\frac{\partial \mathbf{V}}{\partial y} + w\frac{\partial \mathbf{V}}{\partial z} \tag{6.2}$$

and in component form:

$$a_x = \frac{\partial u}{\partial t} + u\frac{\partial u}{\partial x} + v\frac{\partial u}{\partial y} + w\frac{\partial u}{\partial z} \tag{6.3a}$$

$$a_y = \frac{\partial v}{\partial t} + u\frac{\partial v}{\partial x} + v\frac{\partial v}{\partial y} + w\frac{\partial v}{\partial z} \tag{6.3b}$$

$$a_z = \frac{\partial w}{\partial t} + u\frac{\partial w}{\partial x} + v\frac{\partial w}{\partial y} + w\frac{\partial w}{\partial z} \tag{6.3c}$$

The acceleration of a fluid particle is described using the concept of the material derivative.

The acceleration is also concisely expressed as

$$\mathbf{a} = \frac{D\mathbf{V}}{Dt} \tag{6.4}$$

where the operator

$$\frac{D(\;)}{Dt} = \frac{\partial(\;)}{\partial t} + u\frac{\partial(\;)}{\partial x} + v\frac{\partial(\;)}{\partial y} + w\frac{\partial(\;)}{\partial z} \tag{6.5}$$

is termed the *material derivative*, or *substantial derivative*. In vector notation

$$\frac{D(\;)}{Dt} = \frac{\partial(\;)}{\partial t} + (\mathbf{V}\cdot\boldsymbol{\nabla})(\;) \tag{6.6}$$

where the gradient operator, $\boldsymbol{\nabla}(\;)$, is

$$\boldsymbol{\nabla}(\;) = \frac{\partial(\;)}{\partial x}\hat{\mathbf{i}} + \frac{\partial(\;)}{\partial y}\hat{\mathbf{j}} + \frac{\partial(\;)}{\partial z}\hat{\mathbf{k}} \tag{6.7}$$

which was introduced in Chapter 2. As we will see in the following sections, the motion and deformation of a fluid element depend on the velocity field. The relationship between the motion and the forces causing the motion depends on the acceleration field.

6.1.2 Linear Motion and Deformation

The simplest type of motion that a fluid element can undergo is translation as illustrated in Fig. 6.2. In a small time interval δt a particle located at point O will move to point O' as is illustrated in the figure. If all points in the element have the same velocity (which is only true if there are no velocity gradients), then the element will simply translate from one position to another. However, because of the presence of velocity gradients, the element will generally be deformed and rotated as it moves. For example, consider the effect of a single velocity

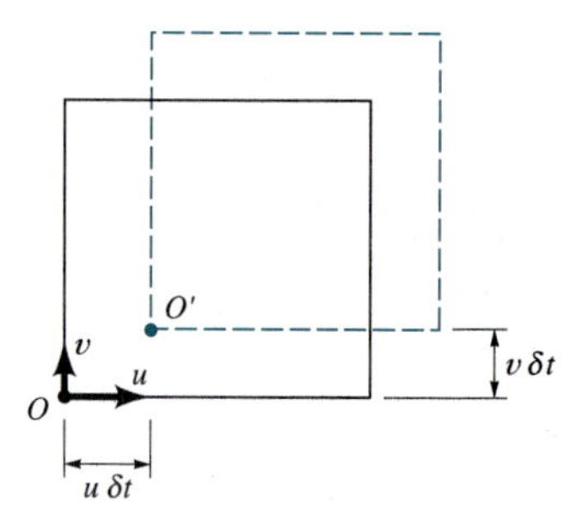

■ FIGURE 6.2 **Translation of a fluid element.**

gradient, $\partial u/\partial x$, on a small cube having sides δx, δy, and δz. As is shown in Fig. 6.3*a*, if the x component of velocity of O and B is u, then at nearby points A and C the x component of the velocity can be expressed as $u + (\partial u/\partial x)\,\delta x$. This difference in velocity causes a ''stretching'' of the volume element by an amount $(\partial u/\partial x)(\delta x)(\delta t)$ during the short time interval δt in which line OA stretches to OA' and BC to BC' (Fig. 6.3*b*). The corresponding change in the original volume, $\delta \forall = \delta x\,\delta y\,\delta z$, would be

$$\text{Change in } \delta \forall = \left(\frac{\partial u}{\partial x}\,\delta x\right)(\delta y\,\delta z)(\delta t)$$

and the *rate* at which the volume $\delta \forall$ is changing *per unit volume* due to the gradient $\partial u/\partial x$ is

$$\frac{1}{\delta \forall}\frac{d(\delta \forall)}{dt} = \lim_{\delta t \to 0}\left[\frac{(\partial u/\partial x)\,\delta t}{\delta t}\right] = \frac{\partial u}{\partial x} \tag{6.8}$$

If velocity gradients $\partial v/\partial y$ and $\partial w/\partial z$ are also present, then using a similar analysis it follows that, in the general case,

$$\frac{1}{\delta \forall}\frac{d(\delta \forall)}{dt} = \frac{\partial u}{\partial x} + \frac{\partial v}{\partial y} + \frac{\partial w}{\partial z} = \nabla \cdot \mathbf{V} \tag{6.9}$$

The volumetric dilatation rate is zero for an incompressible fluid.

This rate of change of the volume per unit volume is called the *volumetric dilatation rate*. Thus, we see that the volume of a fluid may change as the element moves from one location to another in the flow field. However, for an *incompressible fluid* the volumetric dilatation rate is zero, since the element volume cannot change without a change in fluid density (the element mass must be conserved). Variations in the velocity in the direction of the velocity, as represented by the derivatives $\partial u/\partial x$, $\partial v/\partial y$, and $\partial w/\partial z$, simply cause a *linear deformation* of the element in the sense that the shape of the element does not change. Cross derivatives,

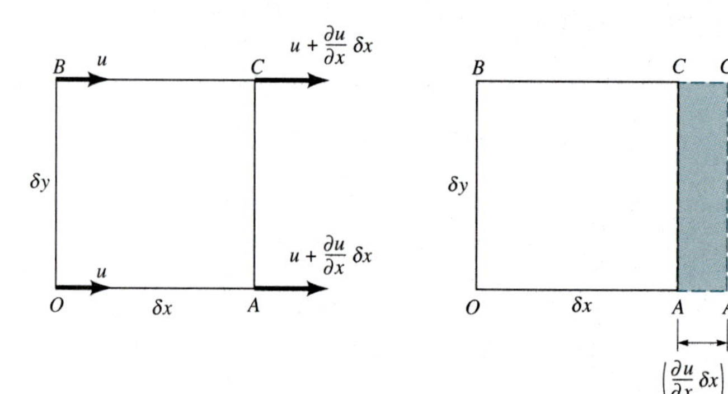

■ FIGURE 6.3 **Linear deformation of a fluid element.**

such as $\partial u/\partial y$, and $\partial v/\partial x$, will cause the element to rotate and generally to undergo an *angular deformation*, which changes the shape of the element.

6.1.3 Angular Motion and Deformation

For simplicity we will consider motion in the x–y plane, but the results can be readily extended to the more general case. The velocity variation that causes rotation and angular deformation is illustrated in Fig. 6.4a. In a short time interval δt the line segments OA and OB will rotate through the angles $\delta\alpha$ and $\delta\beta$ to the new positions OA' and OB' as is shown in Fig. 6.4b. The angular velocity of line OA, ω_{OA}, is

$$\omega_{OA} = \lim_{\delta t \to 0} \frac{\delta\alpha}{\delta t}$$

For small angles

$$\tan \delta\alpha \approx \delta\alpha = \frac{(\partial v/\partial x)\ \delta x\ \delta t}{\delta x} = \frac{\partial v}{\partial x}\ \delta t \tag{6.10}$$

Rotation of fluid particles is related to certain velocity gradients in the flow field.

so that

$$\omega_{OA} = \lim_{\delta t \to 0} \left[\frac{(\partial v/\partial x)\ \delta t}{\delta t}\right] = \frac{\partial v}{\partial x}$$

Note that if $\partial v/\partial x$ is positive, ω_{OA} will be counterclockwise. Similarly, the angular velocity of the line OB is

$$\omega_{OB} = \lim_{\delta t \to 0} \frac{\delta\beta}{\delta t}$$

and

$$\tan \delta\beta \approx \delta\beta = \frac{(\partial u/\partial y)\ \delta y\ \delta t}{\delta y} = \frac{\partial u}{\partial y}\ \delta t \tag{6.11}$$

so that

$$\omega_{OB} = \lim_{\delta t \to 0} \left[\frac{(\partial u/\partial y)\ \delta t}{\delta t}\right] = \frac{\partial u}{\partial y}$$

In this instance if $\partial u/\partial y$ is positive, ω_{OB} will be clockwise. The *rotation*, ω_z, of the element about the z axis is defined as the average of the angular velocities ω_{OA} and ω_{OB} of the two

V6.1

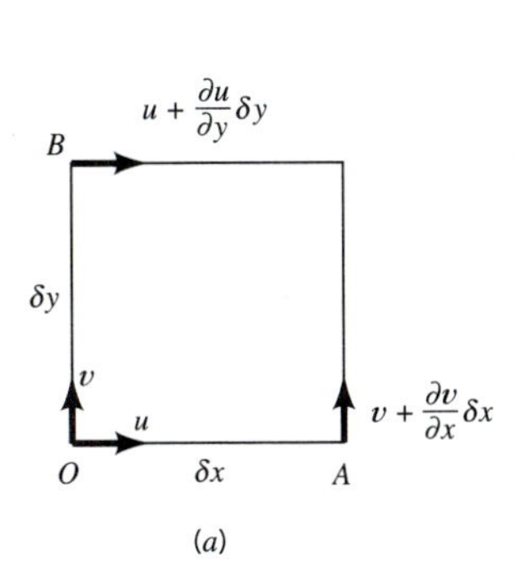

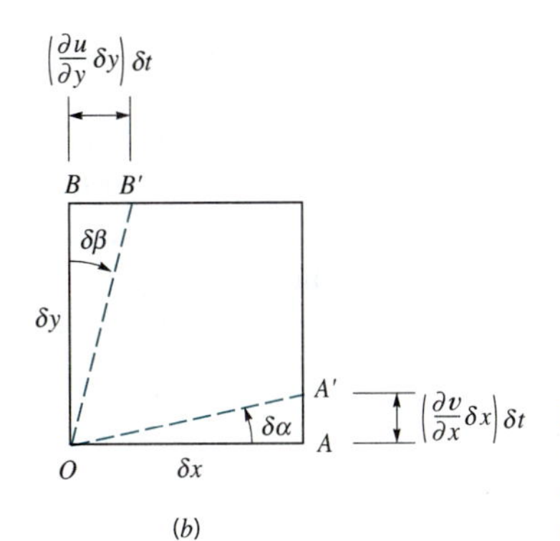

■ **FIGURE 6.4 Angular motion and deformation of a fluid element.**

mutually perpendicular lines OA and OB.[1] Thus, if counterclockwise rotation is considered to be positive, it follows that

$$\omega_z = \frac{1}{2}\left(\frac{\partial v}{\partial x} - \frac{\partial u}{\partial y}\right) \tag{6.12}$$

Rotation of the field element about the other two coordinate axes can be obtained in a similar manner with the result that for rotation about the x axis

$$\omega_x = \frac{1}{2}\left(\frac{\partial w}{\partial y} - \frac{\partial v}{\partial z}\right) \tag{6.13}$$

and for rotation about the y axis

$$\omega_y = \frac{1}{2}\left(\frac{\partial u}{\partial z} - \frac{\partial w}{\partial x}\right) \tag{6.14}$$

The three components, ω_x, ω_y, and ω_z can be combined to give the rotation vector, $\boldsymbol{\omega}$, in the form

$$\boldsymbol{\omega} = \omega_x\hat{\mathbf{i}} + \omega_y\hat{\mathbf{j}} + \omega_z\hat{\mathbf{k}} \tag{6.15}$$

An examination of this result reveals that $\boldsymbol{\omega}$ is equal to one-half the curl of the velocity vector. That is,

$$\boldsymbol{\omega} = \tfrac{1}{2}\,\text{curl}\,\mathbf{V} = \tfrac{1}{2}\,\nabla \times \mathbf{V} \tag{6.16}$$

since by definition of the vector operator $\nabla \times \mathbf{V}$

$$\frac{1}{2}\nabla \times \mathbf{V} = \frac{1}{2}\begin{vmatrix} \hat{\mathbf{i}} & \hat{\mathbf{j}} & \hat{\mathbf{k}} \\ \dfrac{\partial}{\partial x} & \dfrac{\partial}{\partial y} & \dfrac{\partial}{\partial z} \\ u & v & w \end{vmatrix}$$

$$= \frac{1}{2}\left(\frac{\partial w}{\partial y} - \frac{\partial v}{\partial z}\right)\hat{\mathbf{i}} + \frac{1}{2}\left(\frac{\partial u}{\partial z} - \frac{\partial w}{\partial x}\right)\hat{\mathbf{j}} + \frac{1}{2}\left(\frac{\partial v}{\partial x} - \frac{\partial u}{\partial y}\right)\hat{\mathbf{k}}$$

Vorticity in a flow field is related to fluid particle rotation.

The *vorticity*, $\boldsymbol{\zeta}$, is defined as a vector that is twice the rotation vector; that is,

$$\boldsymbol{\zeta} = 2\,\boldsymbol{\omega} = \nabla \times \mathbf{V} \tag{6.17}$$

The use of the vorticity to describe the rotational characteristics of the fluid simply eliminates the $(\frac{1}{2})$ factor associated with the rotation vector.

We observe from Eq. 6.12 that the fluid element will rotate about the z axis as an *undeformed* block (i.e., $\omega_{OA} = -\omega_{OB}$) only when $\partial u/\partial y = -\partial v/\partial x$. Otherwise the rotation will be associated with an angular deformation. We also note from Eq. 6.12 that when $\partial u/\partial y = \partial v/\partial x$ the rotation around the z axis is zero. More generally if $\nabla \times \mathbf{V} = 0$, then the rotation (and the vorticity) are zero, and flow fields for which this condition applies are termed *irrotational*. We will find in Section 6.4 that the condition of irrotationality often greatly simplifies the analysis of complex flow fields. However, it is probably not immediately obvious why some flow fields would be irrotational, and we will need to examine this concept more fully in Section 6.4.

[1]With this definition ω_z can also be interpreted to be the angular velocity of the bisector of the angle between the lines OA and OB.

EXAMPLE 6.1

For a certain two-dimensional flow field the velocity is given by the equation

$$\mathbf{V} = 4xy\hat{\mathbf{i}} + 2(x^2 - y^2)\hat{\mathbf{j}}$$

Is this flow irrotational?

SOLUTION

For an irrotational flow the rotation vector, $\boldsymbol{\omega}$, having the components given by Eqs. 6.12, 6.13, and 6.14 must be zero. For the prescribed velocity field

$$u = 4xy \qquad v = 2(x^2 - y^2) \qquad w = 0$$

and therefore

$$\omega_x = \frac{1}{2}\left(\frac{\partial w}{\partial y} - \frac{\partial v}{\partial z}\right) = 0$$

$$\omega_y = \frac{1}{2}\left(\frac{\partial u}{\partial z} - \frac{\partial w}{\partial x}\right) = 0$$

$$\omega_z = \frac{1}{2}\left(\frac{\partial v}{\partial x} - \frac{\partial u}{\partial y}\right) = \frac{1}{2}(4x - 4x) = 0$$

Thus, the flow is irrotational. **(Ans)**

It is to be noted that for a two-dimensional flow field (where the flow is in the x–y plane) ω_x and ω_y will always be zero, since by definition of two-dimensional flow u and v are not functions of z, and w is zero. In this instance the condition for irrotationality simply becomes $\omega_z = 0$ or $\partial v/\partial x = \partial u/\partial y$. (Lines OA and OB of Fig. 6.4 rotate with the same speed but in opposite directions so that there is no rotation of the fluid element.)

In addition to the rotation associated with the derivatives $\partial u/\partial y$ and $\partial v/\partial x$, it is observed from Fig. 6.4*b* that these derivatives can cause the fluid element to undergo an *angular deformation*, which results in a change in shape of the element. The change in the original right angle formed by the lines OA and OB is termed the shearing strain, $\delta\gamma$, and from Fig. 6.4*b*

$$\delta\gamma = \delta\alpha + \delta\beta$$

In fluid mechanics the rate of angular deformation is an important flow characteristic.

where $\delta\gamma$ is considered to be positive if the original right angle is decreasing. The rate of change of $\delta\gamma$ is called the *rate of shearing strain* or the *rate of angular deformation* and is commonly denoted with the symbol $\dot{\gamma}$. The angles $\delta\alpha$ and $\delta\beta$ are related to the velocity gradients through Eqs. 6.10 and 6.11 so that

$$\dot{\gamma} = \lim_{\delta t \to 0} \frac{\delta\gamma}{\delta t} = \lim_{\delta t \to 0} \left[\frac{(\partial v/\partial x)\,\delta t + (\partial u/\partial y)\,\delta t}{\delta t}\right]$$

and, therefore,

$$\dot{\gamma} = \frac{\partial v}{\partial x} + \frac{\partial u}{\partial y} \tag{6.18}$$

As we will learn in Section 6.7, the rate of angular deformation is related to a corresponding shearing stress which causes the fluid element to change in shape. From Eq. 6.18 we note that if $\partial u/\partial y = -\partial v/\partial x$, the rate of angular deformation is zero, and this condition corresponds to the case in which the element is simply rotating as an undeformed block (Eq. 6.12). In the remainder of this chapter we will see how the various kinematical relationships developed in this section play an important role in the development and subsequent analysis of the differential equations that govern fluid motion.

6.2 Conservation of Mass

Conservation of mass requires that the mass of a system remain constant.

As is discussed in Section 5.2, conservation of mass requires that the mass, M, of a system remain constant as the system moves through the flow field. In equation form this principle is expressed as

$$\frac{DM_{\text{sys}}}{Dt} = 0$$

We found it convenient to use the control volume approach for fluid flow problems, with the control volume representation of the conservation of mass written as

$$\frac{\partial}{\partial t}\int_{\text{cv}} \rho \, d\forall + \int_{\text{cs}} \rho \, \mathbf{V} \cdot \hat{\mathbf{n}} \, dA = 0 \tag{6.19}$$

where the equation (commonly called the continuity equation) can be applied to a finite control volume (cv), which is bounded by a control surface (cs). The first integral on the left side of Eq. 6.19 represents the rate at which the mass within the control volume is increasing, and the second integral represents the net rate at which mass is flowing out through the control surface (rate of mass outflow − rate of mass inflow). To obtain the differential form of the continuity equation, Eq. 6.19 is applied to an infinitesimal control volume.

6.2.1 Differential Form of Continuity Equation

We will take as our control volume the small, stationary cubical element shown in Fig. 6.5*a*. At the center of the element the fluid density is ρ and the velocity has components u, v, and w. Since the element is small the volume integral in Eq. 6.19 can be expressed as

$$\frac{\partial}{\partial t}\int_{\text{cv}} \rho \, d\forall \approx \frac{\partial \rho}{\partial t} \, \delta x \, \delta y \, \delta z \tag{6.20}$$

The rate of mass flow through the surfaces of the element can be obtained by considering the flow in each of the coordinate directions separately. For example, in Fig. 6.5*b* flow in the x

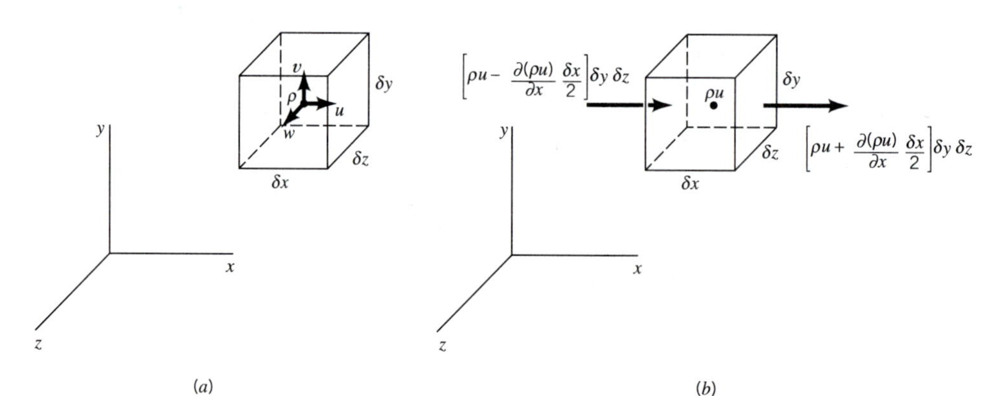

■ **FIGURE 6.5 A differential element for the development of conservation of mass equation.**

direction is depicted. If we let ρu represent the x component of the mass rate of flow per unit area at the center of the element, then on the right face

$$\rho u\big|_{x+(\delta x/2)} = \rho u + \frac{\partial(\rho u)}{\partial x}\frac{\delta x}{2} \tag{6.21}$$

and on the left face

$$\rho u\big|_{x-(\delta x/2)} = \rho u - \frac{\partial(\rho u)}{\partial x}\frac{\delta x}{2} \tag{6.22}$$

Note that we are really using a Taylor series expansion of ρu and neglecting higher order terms such as $(\delta x)^2$, $(\delta x)^3$, and so on. When the right-hand sides of Eqs. 6.21 and 6.22 are multiplied by the area $\delta y\ \delta z$, the rate at which mass is crossing the right and left sides of the element are obtained as is illustrated in Fig. 6.5*b*. When these two expressions are combined, the net rate of mass flowing from the element through the two surfaces can be expressed as

$$\begin{aligned}\text{Net rate of mass outflow in } x \text{ direction} &= \left[\rho u + \frac{\partial(\rho u)}{\partial x}\frac{\delta x}{2}\right]\delta y\ \delta z \\ &\quad - \left[\rho u - \frac{\partial(\rho u)}{\partial x}\frac{\delta x}{2}\right]\delta y\ \delta z = \frac{\partial(\rho u)}{\partial x}\delta x\ \delta y\ \delta z\end{aligned} \tag{6.23}$$

For simplicity, only flow in the x direction has been considered in Fig. 6.5*b*, but, in general, there will also be flow in the y and z directions. An analysis similar to the one used for flow in the x direction shows that

$$\text{Net rate of mass outflow in } y \text{ direction} = \frac{\partial(\rho v)}{\partial y}\delta x\ \delta y\ \delta z \tag{6.24}$$

and

$$\text{Net rate of mass outflow in } z \text{ direction} = \frac{\partial(\rho w)}{\partial z}\delta x\ \delta y\ \delta z \tag{6.25}$$

Thus,

$$\text{Net rate of mass outflow} = \left[\frac{\partial(\rho u)}{\partial x} + \frac{\partial(\rho v)}{\partial y} + \frac{\partial(\rho w)}{\partial z}\right]\delta x\ \delta y\ \delta z \tag{6.26}$$

From Eqs. 6.19, 6.20, and 6.26 it now follows that the differential equation for conservation of mass is

The continuity equation is one of the fundamental equations of fluid mechanics.

$$\boxed{\frac{\partial \rho}{\partial t} + \frac{\partial(\rho u)}{\partial x} + \frac{\partial(\rho v)}{\partial y} + \frac{\partial(\rho w)}{\partial z} = 0} \tag{6.27}$$

As previously mentioned, this equation is also commonly referred to as the continuity equation.

The continuity equation is one of the fundamental equations of fluid mechanics and, as expressed in Eq. 6.27, is valid for steady or unsteady flow, and compressible or incompressible fluids. In vector notation Eq. 6.27 can be written as

$$\frac{\partial \rho}{\partial t} + \nabla \cdot \rho \mathbf{V} = 0 \tag{6.28}$$

Two special cases are of particular interest. For *steady* flow of *compressible* fluids

$$\nabla \cdot \rho \mathbf{V} = 0$$

or

$$\frac{\partial(\rho u)}{\partial x} + \frac{\partial(\rho v)}{\partial y} + \frac{\partial(\rho w)}{\partial z} = 0 \tag{6.29}$$

For incompressible fluids the continuity equation reduces to a simple relationship involving certain velocity gradients.

This follows since by definition ρ is not a function of time for steady flow, but could be a function of position. For *incompressible* fluids the fluid density, ρ, is a constant throughout the flow field so that Eq. 6.28 becomes

$$\nabla \cdot \mathbf{V} = 0 \tag{6.30}$$

or

$$\frac{\partial u}{\partial x} + \frac{\partial v}{\partial y} + \frac{\partial w}{\partial z} = 0 \tag{6.31}$$

Equation 6.31 applies to both steady and unsteady flow of incompressible fluids. Note that Eq. 6.31 is the same as that obtained by setting the volumetric dilatation rate (Eq. 6.9) equal to zero. This result should not be surprising since both relationships are based on conservation of mass for incompressible fluids. However, the expression for the volumetric dilation rate was developed from a system approach, whereas Eq. 6.31 was developed from a control volume approach. In the former case the deformation of a particular differential mass of fluid was studied, and in the latter case mass flow through a fixed differential volume was studied.

EXAMPLE 6.2

The velocity components for a certain incompressible, steady flow field are

$$u = x^2 + y^2 + z^2$$
$$v = xy + yz + z$$
$$w = ?$$

Determine the form of the z component, w, required to satisfy the continuity equation.

SOLUTION

Any physically possible velocity distribution must for an incompressible fluid satisfy conservation of mass as expressed by the continuity equation

$$\frac{\partial u}{\partial x} + \frac{\partial v}{\partial y} + \frac{\partial w}{\partial z} = 0$$

For the given velocity distribution

$$\frac{\partial u}{\partial x} = 2x \quad \text{and} \quad \frac{\partial v}{\partial y} = x + z$$

so that the required expression for $\partial w/\partial z$ is

$$\frac{\partial w}{\partial z} = -2x - (x + z) = -3x - z$$

Integration with respect to z yields

$$w = -3xz - \frac{z^2}{2} + f(x, y) \quad \textbf{(Ans)}$$

The third velocity component cannot be explicitly determined since the function $f(x, y)$ can have any form and conservation of mass will still be satisfied. The specific form of this function will be governed by the flow field described by these velocity components—that is, some additional information is needed to completely determine w.

6.2.2 Cylindrical Polar Coordinates

For some problems, velocity components expressed in cylindrical polar coordinates will be convenient.

For some problems it is more convenient to express the various differential relationships in cylindrical polar coordinates rather than Cartesian coordinates. As is shown in Fig. 6.6, with cylindrical coordinates a point is located by specifying the coordinates r, θ, and z. The coordinate r is the radial distance from the z axis, θ is the angle measured from a line parallel to the x axis (with counterclockwise taken as positive), and z is the coordinate along the z axis. The velocity components, as sketched in Fig. 6.6, are the radial velocity, v_r, the tangential velocity, v_θ, and the axial velocity, v_z. Thus, the velocity at some arbitrary point P can be expressed as

$$\mathbf{V} = v_r\hat{\mathbf{e}}_r + v_\theta\hat{\mathbf{e}}_\theta + v_z\hat{\mathbf{e}}_z \quad \textbf{(6.32)}$$

where $\hat{\mathbf{e}}_r$, $\hat{\mathbf{e}}_\theta$, and $\hat{\mathbf{e}}_z$ are the unit vectors in the r, θ, and z directions, respectively, as are illustrated in Fig. 6.6. The use of cylindrical coordinates is particularly convenient when the boundaries of the flow system are cylindrical. Several examples illustrating the use of cylindrical coordinates will be given in succeeding sections in this chapter.

The differential form of the continuity equation in cylindrical coordinates is

$$\boxed{\frac{\partial \rho}{\partial t} + \frac{1}{r}\frac{\partial(r\rho v_r)}{\partial r} + \frac{1}{r}\frac{\partial(\rho v_\theta)}{\partial \theta} + \frac{\partial(\rho v_z)}{\partial z} = 0} \quad \textbf{(6.33)}$$

This equation can be derived by following the same procedure used in the preceding section (see Problem 6.17). For steady, compressible flow

$$\frac{1}{r}\frac{\partial(r\rho v_r)}{\partial r} + \frac{1}{r}\frac{\partial(\rho v_\theta)}{\partial \theta} + \frac{\partial(\rho v_z)}{\partial z} = 0 \quad \textbf{(6.34)}$$

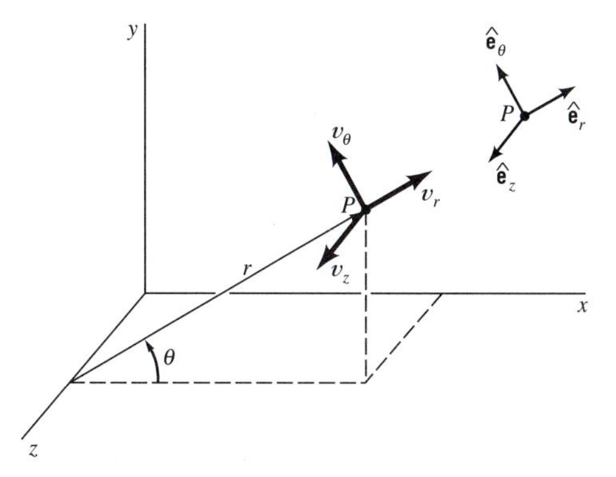

■ **FIGURE 6.6 The representation of velocity components in cylindrical polar coordinates.**

For incompressible fluids (for steady or unsteady flow)

$$\frac{1}{r}\frac{\partial(rv_r)}{\partial r} + \frac{1}{r}\frac{\partial v_\theta}{\partial \theta} + \frac{\partial v_z}{\partial z} = 0 \tag{6.35}$$

6.2.3 The Stream Function

Steady, incompressible, plane, two-dimensional flow represents one of the simplest types of flow of practical importance. By plane, two-dimensional flow we mean that there are only two velocity components, such as u and v, when the flow is considered to be in the x–y plane. For this flow the continuity equation, Eq. 6.31, reduces to

$$\frac{\partial u}{\partial x} + \frac{\partial v}{\partial y} = 0 \tag{6.36}$$

Velocity components in a two-dimensional flow field can be expressed in terms of a stream function.

We still have two variables, u and v, to deal with, but they must be related in a special way as indicated by Eq. 6.36. This equation suggests that if we define a function $\psi(x, y)$, called the *stream function*, which relates the velocities as

$$u = \frac{\partial \psi}{\partial y} \qquad v = -\frac{\partial \psi}{\partial x} \tag{6.37}$$

then the continuity equation is identically satisfied. This conclusion can be verified by simply substituting the expressions for u and v into Eq. 6.36 so that

$$\frac{\partial}{\partial x}\left(\frac{\partial \psi}{\partial y}\right) + \frac{\partial}{\partial y}\left(-\frac{\partial \psi}{\partial x}\right) = \frac{\partial^2 \psi}{\partial x\, \partial y} - \frac{\partial^2 \psi}{\partial y\, \partial x} = 0$$

Thus, whenever the velocity components are defined in terms of the stream function we know that conservation of mass will be satisfied. Of course, we still do not know what $\psi(x, y)$ is for a particular problem, but at least we have simplified the analysis by having to determine only one unknown function, $\psi(x, y)$, rather than the two functions, $u(x, y)$ and $v(x, y)$.

Another particular advantage of using the stream function is related to the fact that *lines along which ψ is constant are streamlines.* Recall from Section 4.1.4 that streamlines are lines in the flow field that are everywhere tangent to the velocities, as is illustrated in Fig. 6.7. It follows from the definition of the streamline that the slope at any point along a streamline is given by

$$\frac{dy}{dx} = \frac{v}{u}$$

The change in the value of ψ as we move from one point (x, y) to a nearby point $(x + dx, y + dy)$ is given by the relationship:

$$d\psi = \frac{\partial \psi}{\partial x}\,dx + \frac{\partial \psi}{\partial y}\,dy = -v\,dx + u\,dy$$

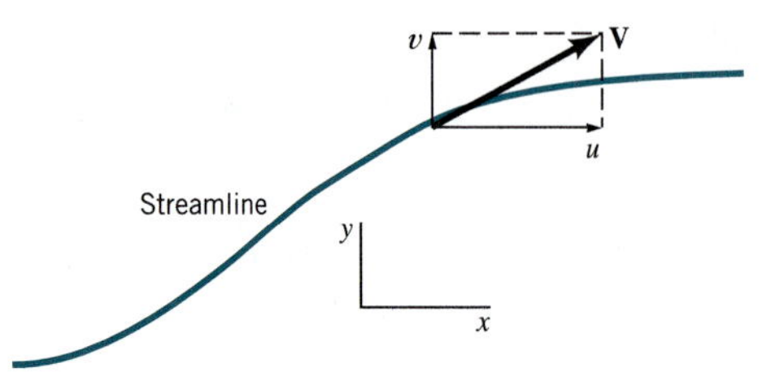

■ **FIGURE 6.7** **Velocity and velocity components along a streamline.**

Along a line of constant ψ we have $d\psi = 0$ so that

$$-v\,dx + u\,dy = 0$$

and, therefore, along a line of constant ψ

$$\frac{dy}{dx} = \frac{v}{u}$$

which is the defining equation for a streamline. Thus, if we know the function $\psi(x, y)$ we can plot lines of constant ψ to provide the family of streamlines that are helpful in visualizing the pattern of flow. There are an infinite number of streamlines that make up a particular flow field, since for each constant value assigned to ψ a streamline can be drawn.

The change in the value of the stream function is related to the volume rate of flow.

The actual numerical value associated with a particular streamline is not of particular significance, but the change in the value of ψ is related to the volume rate of flow. Consider two closely spaced streamlines, as are shown in Fig. 6.8*a*. The lower streamline is designated ψ and the upper one $\psi + d\psi$. Let dq represent the volume rate of flow (per unit width perpendicular to the x–y plane) passing between the two streamlines. Note that flow never crosses streamlines, since by definition the velocity is tangent to the streamline. From conservation of mass we know that the inflow, dq, crossing the arbitrary surface AC of Fig. 6.8*a* must equal the net outflow through surfaces AB and BC. Thus,

$$dq = u\,dy - v\,dx$$

or in terms of the stream function

$$dq = \frac{\partial \psi}{\partial y}\,dy + \frac{\partial \psi}{\partial x}\,dx \tag{6.38}$$

The right-hand side of Eq. 6.38 is equal to $d\psi$ so that

$$dq = d\psi \tag{6.39}$$

Thus, the volume rate of flow, q, between two streamlines such as ψ_1 and ψ_2 of Fig. 6.8*b* can be determined by integrating Eq. 6.39 to yield

$$q = \int_{\psi_1}^{\psi_2} d\psi = \psi_2 - \psi_1 \tag{6.40}$$

If the upper streamline, ψ_2, has a value greater than the lower streamline, ψ_1, then q is positive, which indicates that the flow is from left to right. For $\psi_1 > \psi_2$ the flow is from right to left.

In cylindrical coordinates the continuity equation (Eq. 6.35) for incompressible, plane, two-dimensional flow reduces to

$$\frac{1}{r}\frac{\partial(rv_r)}{\partial r} + \frac{1}{r}\frac{\partial v_\theta}{\partial \theta} = 0 \tag{6.41}$$

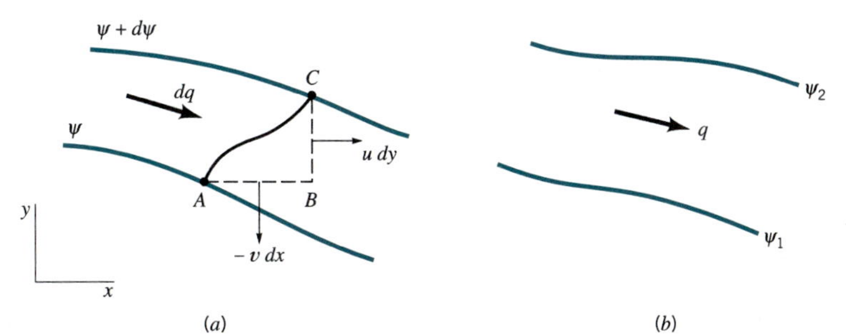

■ **FIGURE 6.8** **The flow between two streamlines.**

and the velocity components, v_r and v_θ, can be related to the stream function, $\psi(r, \theta)$, through the equations

$$v_r = \frac{1}{r}\frac{\partial \psi}{\partial \theta} \qquad v_\theta = -\frac{\partial \psi}{\partial r} \tag{6.42}$$

The concept of the stream function is not applicable to general three-dimensional flows.

Substitution of these expressions for the velocity components into Eq. 6.41 shows that the continuity equation is identically satisfied. The stream function concept can be extended to axisymmetric flows, such as flow in pipes or flow around bodies of revolution, and to two-dimensional compressible flows. However, the concept is not applicable to general three-dimensional flows.

EXAMPLE 6.3

The velocity components in a steady, incompressible, two-dimensional flow field are

$$u = 2y$$
$$v = 4x$$

Determine the corresponding stream function and show on a sketch several streamlines. Indicate the direction of flow along the streamlines.

SOLUTION

From the definition of the stream function (Eqs. 6.37)

$$u = \frac{\partial \psi}{\partial y} = 2y$$

and

$$v = -\frac{\partial \psi}{\partial x} = 4x$$

The first of these equations can be integrated to give

$$\psi = y^2 + f_1(x)$$

where $f_1(x)$ is an arbitrary function of x. Similarly from the second equation

$$\psi = -2x^2 + f_2(y)$$

where $f_2(y)$ is an arbitrary function of y. It now follows that in order to satisfy both expressions for the stream function

$$\psi = -2x^2 + y^2 + C \qquad \textbf{(Ans)}$$

where C is an arbitrary constant.

Since the velocities are related to the derivatives of the stream function, an arbitrary constant can always be added to the function, and the value of the constant is actually of no consequence. Usually, for simplicity, we set $C = 0$ so that for this particular example the simplest form for the stream function is

$$\psi = -2x^2 + y^2 \qquad \textbf{(1)} \quad \textbf{(Ans)}$$

Either answer indicated would be acceptable.

Streamlines can now be determined by setting $\psi =$ constant and plotting the resulting curve. With the above expression for ψ (with $C = 0$) the value of ψ at the origin is zero so

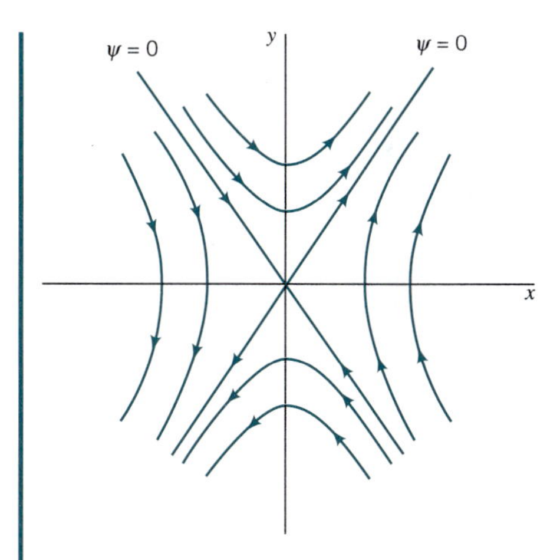

■ FIGURE E6.3

that the equation of the streamline passing through the origin (the $\psi = 0$ streamline) is

$$0 = -2x^2 + y^2$$

or

$$y = \pm\sqrt{2}x$$

Other streamlines can be obtained by setting ψ equal to various constants. It follows from Eq. 1 that the equations of these streamlines (for $\psi \neq 0$) can be expressed in the form

$$\frac{y^2}{\psi} - \frac{x^2}{\psi/2} = 1$$

which we recognize as the equation of a hyperbola. Thus, the streamlines are a family of hyperbolas with the $\psi = 0$ streamlines as asymptotes. Several of the streamlines are plotted in Fig. E6.3. Since the velocities can be calculated at any point, the direction of flow along a given streamline can be easily deduced. For example $v = -\partial\psi/\partial x = 4x$ so that $v > 0$ if $x > 0$ and $v < 0$ if $x < 0$. The direction of flow is indicated on the figure.

6.3 Conservation of Linear Momentum

The resultant force acting on a fluid mass is equal to the time rate of change of the linear momentum of the mass.

To develop the differential, linear momentum equations we can start with the linear momentum equation

$$\mathbf{F} = \left.\frac{D\mathbf{P}}{Dt}\right|_{\text{sys}} \tag{6.43}$$

where $\mathbf{F}$ is the resultant force acting on a fluid mass, $\mathbf{P}$ is the linear momentum defined as

$$\mathbf{P} = \int_{\text{sys}} \mathbf{V}\, dm$$

and the operator $D(\ )/Dt$ is the material derivative (see Section 4.2.1). In the last chapter it was demonstrated how Eq. 6.43 in the form

$$\sum \mathbf{F}_{\substack{\text{contents of the}\\ \text{control volume}}} = \frac{\partial}{\partial t}\int_{\text{cv}} \mathbf{V}\rho\, d\maltese + \int_{\text{cs}} \mathbf{V}\rho\mathbf{V}\cdot\hat{\mathbf{n}}\, dA \tag{6.44}$$

could be applied to a finite control volume to solve a variety of flow problems. To obtain the differential form of the linear momentum equation, we can either apply Eq 6.43 to a differential system, consisting of a mass, δm, or apply Eq. 6.44 to an infinitesimal control volume, $\delta \forall$, which initially bounds the mass δm. It is probably simpler to use the system approach since application of Eq. 6.43 to the differential mass, δm, yields

$$\delta \mathbf{F} = \frac{D(\mathbf{V}\ \delta m)}{Dt}$$

where $\delta \mathbf{F}$ is the resultant force acting on δm. Using this system approach δm can be treated as a constant so that

$$\delta \mathbf{F} = \delta m \frac{D\mathbf{V}}{Dt}$$

But $D\mathbf{V}/Dt$ is the acceleration, **a**, of the element. Thus,

$$\delta \mathbf{F} = \delta m\ \mathbf{a} \tag{6.45}$$

which is simply Newton's second law applied to the mass δm. This is the same result that would be obtained by applying Eq. 6.44 to an infinitesimal control volume (see Ref. 1). Before we can proceed, it is necessary to examine how the force $\delta \mathbf{F}$ can be most conveniently expressed.

6.3.1 Description of Forces Acting on the Differential Element

Both surface forces and body forces generally act on fluid particles.

In general, two types of forces need to be considered: *surface forces*, which act on the surface of the differential element, and *body forces*, which are distributed throughout the element. For our purpose, the only body force, $\delta \mathbf{F}_b$, of interest is the weight of the element, which can be expressed as

$$\delta \mathbf{F}_b = \delta m\ \mathbf{g} \tag{6.46}$$

where **g** is the vector representation of the acceleration of gravity. In component form

$$\delta F_{bx} = \delta m\ g_x \tag{6.47a}$$

$$\delta F_{by} = \delta m\ g_y \tag{6.47b}$$

$$\delta F_{bz} = \delta m\ g_z \tag{6.47c}$$

where g_x, g_y, and g_z are the components of the acceleration of gravity vector in the x, y, and z directions, respectively.

Surface forces act on the element as a result of its interaction with its surroundings. At any arbitrary location within a fluid mass, the force acting on a small area, δA, which lies in an arbitrary surface, can be represented by $\delta \mathbf{F}_s$, as is shown in Fig. 6.9. In general, $\delta \mathbf{F}_s$ will be inclined with respect to the surface. The force $\delta \mathbf{F}_s$ can be resolved into three components,

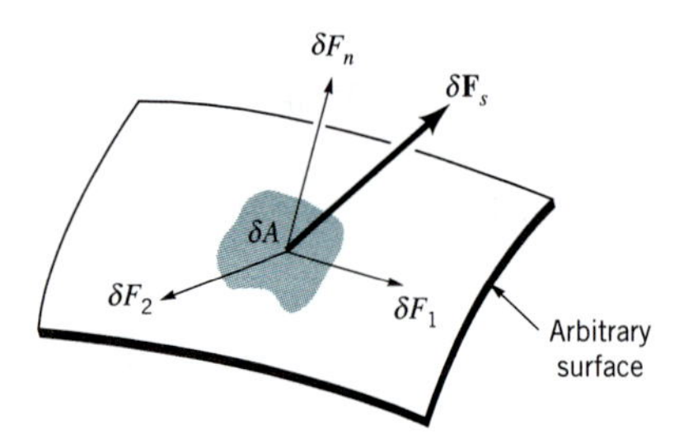

■ **FIGURE 6.9 Component of force acting on an arbitrary differential area.**

Surface forces acting on a fluid element can be described in terms of normal and shearing stresses.

δF_n, δF_1, and δF_2, where δF_n is normal to the area, δA, and δF_1 and δF_2 are parallel to the area and orthogonal to each other. The *normal stress*, σ_n is defined as

$$\sigma_n = \lim_{\delta A \to 0} \frac{\delta F_n}{\delta A}$$

and the *shearing stresses* are defined as

$$\tau_1 = \lim_{\delta A \to 0} \frac{\delta F_1}{\delta A}$$

and

$$\tau_2 = \lim_{\delta A \to 0} \frac{\delta F_2}{\delta A}$$

We will use σ for normal stresses and τ for shearing stresses. The intensity of the force per unit area at a point in a body can thus be characterized by a normal stress and two shearing stresses, if the orientation of the area is specified. For purposes of analysis it is usually convenient to reference the area to the coordinate system. For example, for the rectangular coordinate system shown in Fig. 6.10 we choose to consider the stresses acting on planes parallel to the coordinate planes. On the plane $ABCD$ of Fig. 6.10*a*, which is parallel to the y–z plane, the normal stress is denoted σ_{xx} and the shearing stresses are denoted as τ_{xy} and τ_{xz}. To easily identify the particular stress component we use a double subscript notation. The first subscript indicates the direction of the *normal* to the plane on which the stress acts, and the second subscript indicates the direction of the stress. Thus, normal stresses have repeated subscripts, whereas the subscripts for the shearing stresses are always different.

It is also necessary to establish a sign convention for the stresses. We define the positive direction for the stress as the positive coordinate direction on the surfaces for which the outward normal is in the positive coordinate direction. This is the case illustrated in Fig. 6.10*a* where the outward normal to the area $ABCD$ is in the positive x direction. The positive directions for σ_{xx}, τ_{xy}, and τ_{xz} are as shown in Fig. 6.10*a*. If the outward normal points in the negative coordinate direction, as in Fig. 6.10*b* for the area $A'B'C'D'$, then the stresses are considered positive if directed in the negative coordinate directions. Thus, the stresses shown in Fig. 6.10*b* are considered to be positive when directed as shown. Note that positive normal stresses are tensile stresses; that is, they tend to ''stretch'' the material.

It should be emphasized that the state of stress at a point in a material is not completely defined by simply three components of a ''stress vector.'' This follows, since any particular stress vector depends on the orientation of the plane passing through the point. However, it can be shown that the normal and shearing stresses acting on *any* plane passing through a

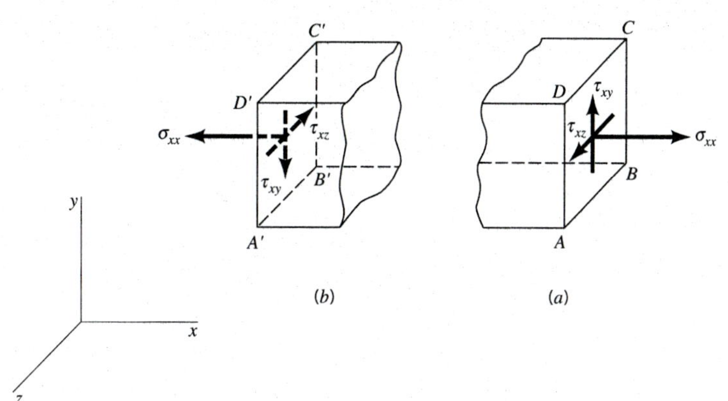

■ **FIGURE 6.10 Double subscript notation for stresses.**

point can be expressed in terms of the stresses acting on three orthogonal planes passing through the point (Ref. 2).

We now can express the surface forces acting on a small cubical element of fluid in terms of the stresses acting on the faces of the element as shown in Fig. 6.11. It is expected that in general the stresses will vary from point to point within the flow field. Thus, we will express the stresses on the various faces in terms of the corresponding stresses at the center of the element of Fig. 6.11 and their gradients in the coordinate directions. For simplicity only the forces in the x direction are shown. Note that the stresses must be multiplied by the area on which they act to obtain the force. Summing all these forces in the x direction yields

$$\delta F_{sx} = \left(\frac{\partial \sigma_{xx}}{\partial x} + \frac{\partial \tau_{yx}}{\partial y} + \frac{\partial \tau_{zx}}{\partial z} \right) \delta x \; \delta y \; \delta z \tag{6.48a}$$

for the resultant surface force in the x direction. In a similar manner the resultant surface forces in the y and z directions can be obtained and expressed as

$$\delta F_{sy} = \left(\frac{\partial \tau_{xy}}{\partial x} + \frac{\partial \sigma_{yy}}{\partial y} + \frac{\partial \tau_{zy}}{\partial z} \right) \delta x \; \delta y \; \delta z \tag{6.48b}$$

and

$$\delta F_{sz} = \left(\frac{\partial \tau_{xz}}{\partial x} + \frac{\partial \tau_{yz}}{\partial y} + \frac{\partial \sigma_{zz}}{\partial z} \right) \delta x \; \delta y \; \delta z \tag{6.48c}$$

The resultant surface force can now be expressed as

$$\delta \mathbf{F}_s = \delta F_{sx} \hat{\mathbf{i}} + \delta F_{sy} \hat{\mathbf{j}} + \delta F_{sz} \hat{\mathbf{k}} \tag{6.49}$$

and this force combined with the body force, $\delta \mathbf{F}_b$, yields the resultant force, $\delta \mathbf{F}$, acting on the differential mass, δm. That is, $\delta \mathbf{F} = \delta \mathbf{F}_s + \delta \mathbf{F}_b$.

6.3.2 Equations of Motion

The expressions for the body and surface forces can now be used in conjunction with Eq. 6.45 to develop the equations of motion. In component form Eq. 6.45 can be written as

The resultant force acting on a fluid element must equal the mass times the acceleration of the element.

$$\delta F_x = \delta m \, a_x$$

$$\delta F_y = \delta m \, a_y$$

$$\delta F_z = \delta m \, a_z$$

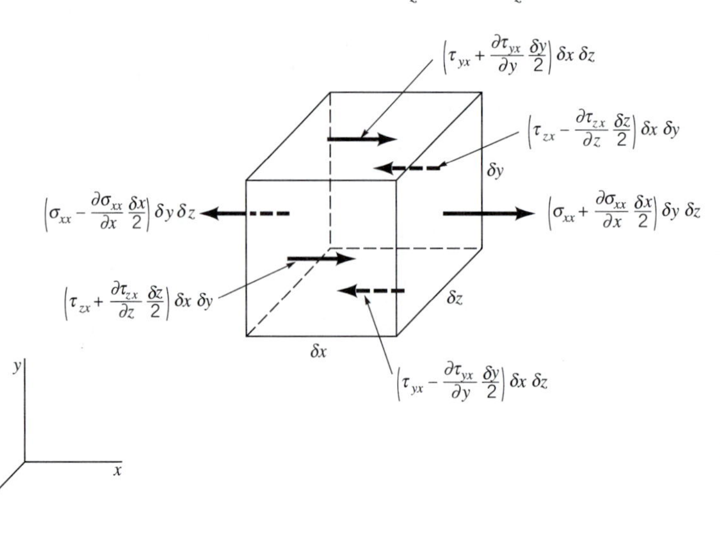

■ **FIGURE 6.11** **Surface forces in the *x* direction acting on a fluid element.**

where $\delta m = \rho\, \delta x\, \delta y\, \delta z$, and the acceleration components are given by Eq. 6.3. It now follows (using Eqs. 6.47 and 6.48 for the forces on the element) that

$$\rho g_x + \frac{\partial \sigma_{xx}}{\partial x} + \frac{\partial \tau_{yx}}{\partial y} + \frac{\partial \tau_{zx}}{\partial z} = \rho \left(\frac{\partial u}{\partial t} + u \frac{\partial u}{\partial x} + v \frac{\partial u}{\partial y} + w \frac{\partial u}{\partial z} \right) \quad \textbf{(6.50a)}$$

$$\rho g_y + \frac{\partial \tau_{xy}}{\partial x} + \frac{\partial \sigma_{yy}}{\partial y} + \frac{\partial \tau_{zy}}{\partial z} = \rho \left(\frac{\partial v}{\partial t} + u \frac{\partial v}{\partial x} + v \frac{\partial v}{\partial y} + w \frac{\partial v}{\partial z} \right) \quad \textbf{(6.50b)}$$

$$\rho g_z + \frac{\partial \tau_{xz}}{\partial x} + \frac{\partial \tau_{yz}}{\partial y} + \frac{\partial \sigma_{zz}}{\partial z} = \rho \left(\frac{\partial w}{\partial t} + u \frac{\partial w}{\partial x} + v \frac{\partial w}{\partial y} + w \frac{\partial w}{\partial z} \right) \quad \textbf{(6.50c)}$$

where the element volume $\delta x\, \delta y\, \delta z$ cancels out.

Equations 6.50 are the general differential equations of motion for a fluid. In fact, they are applicable to any continuum (solid or fluid) in motion or at rest. However, before we can use the equations to solve specific problems, some additional information about the stresses must be obtained. Otherwise, we will have more unknowns (all of the stresses and velocities and the density) than equations. It should not be too surprising that the differential analysis of fluid motion is complicated. We are attempting to describe, in detail, complex fluid motion.

6.4 Inviscid Flow

Flow fields in which the shearing stresses are zero are said to be inviscid, nonviscous, or frictionless.

As is discussed in Section 1.6, shearing stresses develop in a moving fluid because of the viscosity of the fluid. We know that for some common fluids, such as air and water, the viscosity is small, and therefore it seems reasonable to assume that under some circumstances we may be able to simply neglect the effect of viscosity (and thus shearing stresses). Flow fields in which the shearing stresses are assumed to be negligible are said to be *inviscid*, *nonviscous*, or *frictionless*. These terms are used interchangeably. As is discussed in Section 2.1, for fluids in which there are no shearing stresses the normal stress at a point is independent of direction—that is, $\sigma_{xx} = \sigma_{yy} = \sigma_{zz}$. In this instance we define the pressure, p, as the negative of the normal stress so that

$$-p = \sigma_{xx} = \sigma_{yy} = \sigma_{zz}$$

The negative sign is used so that a *compressive* normal stress (which is what we expect in a fluid) will give a *positive* value for p.

In Chapter 3 the inviscid flow concept was used in the development of the Bernoulli equation, and numerous applications of this important equation were considered. In this section we will again consider the Bernoulli equation and will show how it can be derived from the general equations of motion for inviscid flow.

6.4.1 Euler's Equations of Motion

For an inviscid flow in which all the shearing stresses are zero, and the normal stresses are replaced by $-p$, the general equations of motion (Eqs. 6.50) reduce to

$$\rho g_x - \frac{\partial p}{\partial x} = \rho \left(\frac{\partial u}{\partial t} + u \frac{\partial u}{\partial x} + v \frac{\partial u}{\partial y} + w \frac{\partial u}{\partial z} \right) \quad \textbf{(6.51a)}$$

$$\rho g_y - \frac{\partial p}{\partial y} = \rho \left(\frac{\partial v}{\partial t} + u \frac{\partial v}{\partial x} + v \frac{\partial v}{\partial y} + w \frac{\partial v}{\partial z} \right) \quad \textbf{(6.51b)}$$

$$\rho g_z - \frac{\partial p}{\partial z} = \rho \left(\frac{\partial w}{\partial t} + u \frac{\partial w}{\partial x} + v \frac{\partial w}{\partial y} + w \frac{\partial w}{\partial z} \right) \quad \textbf{(6.51c)}$$

These equations are commonly referred to as *Euler's equations of motion*, named in honor of Leonhard Euler (1707–1783), a famous Swiss mathematician who pioneered work on the relationship between pressure and flow. In vector notation Euler's equations can be expressed as

Euler's equations of motion apply to an inviscid flow field.

$$\rho \mathbf{g} - \nabla p = \rho \left[\frac{\partial \mathbf{V}}{\partial t} + (\mathbf{V} \cdot \nabla)\mathbf{V} \right] \tag{6.52}$$

Although Eqs. 6.51 are considerably simpler than the general equations of motion, they are still not amenable to a general analytical solution that would allow us to determine the pressure and velocity at all points within an inviscid flow field. The main difficulty arises from the nonlinear velocity terms (such as $u\ \partial u/\partial x$, $v\ \partial u/\partial y$, etc.), which appear in the convective acceleration. Because of these terms, Euler's equations are nonlinear partial differential equations for which we do not have a general method of solving. However, under some circumstances we can use them to obtain useful information about inviscid flow fields. For example, as shown in the following section we can integrate Eq. 6.52 to obtain a relationship (the Bernoulli equation) between elevation, pressure, and velocity along a streamline.

6.4.2 The Bernoulli Equation

In Section 3.2 the Bernoulli equation was derived by a direct application of Newton's second law to a fluid particle moving along a streamline. In this section we will again derive this important equation, starting from Euler's equations. Of course, we should obtain the same result since Euler's equations simply represent a statement of Newton's second law expressed in a general form that is useful for flow problems. We will restrict our attention to steady flow so Euler's equation in vector form becomes

$$\rho \mathbf{g} - \nabla p = \rho(\mathbf{V} \cdot \nabla)\mathbf{V} \tag{6.53}$$

We wish to integrate this differential equation along some arbitrary streamline (Fig. 6.12) and select the coordinate system with the z axis vertical (with "up" being positive) so that the acceleration of gravity vector can be expressed as

$$\mathbf{g} = -g\ \nabla z$$

where g is the magnitude of the acceleration of gravity vector. Also, it will be convenient to use the vector identity

$$(\mathbf{V} \cdot \nabla)\mathbf{V} = \tfrac{1}{2}\nabla(\mathbf{V} \cdot \mathbf{V}) - \mathbf{V} \times (\nabla \times \mathbf{V})$$

Equation 6.53 can now be written in the form

$$-\rho g\ \nabla z - \nabla p = \frac{\rho}{2}\nabla(\mathbf{V} \cdot \mathbf{V}) - \rho\mathbf{V} \times (\nabla \times \mathbf{V})$$

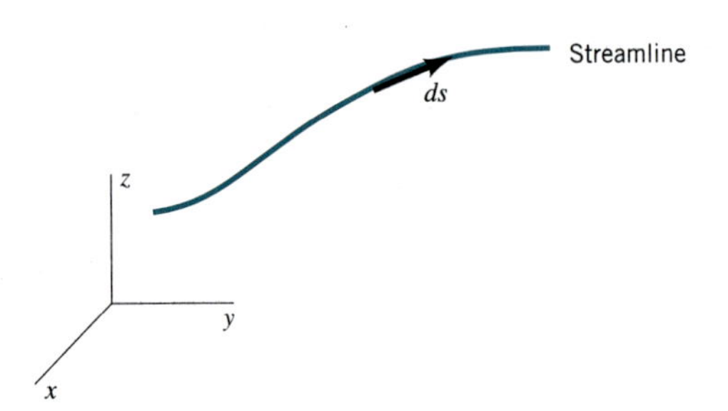

■ **FIGURE 6.12** **The notation for differential length along a streamline.**

and this equation can be rearranged to yield

$$\frac{\boldsymbol{\nabla} p}{\rho} + \frac{1}{2}\boldsymbol{\nabla}(V^2) + g\boldsymbol{\nabla} z = \mathbf{V} \times (\boldsymbol{\nabla} \times \mathbf{V})$$

We next take the dot product of each term with a differential length $d\mathbf{s}$ along a streamline (Fig. 6.12). Thus,

$$\frac{\boldsymbol{\nabla} p}{\rho} \cdot d\mathbf{s} + \frac{1}{2}\boldsymbol{\nabla}(V^2) \cdot d\mathbf{s} + g\,\boldsymbol{\nabla} z \cdot d\mathbf{s} = [\mathbf{V} \times (\boldsymbol{\nabla} \times \mathbf{V})] \cdot d\mathbf{s} \tag{6.54}$$

Since $d\mathbf{s}$ has a direction along the streamline, the vectors $d\mathbf{s}$ and $\mathbf{V}$ are parallel. However, the vector $\mathbf{V} \times (\boldsymbol{\nabla} \times \mathbf{V})$ is perpendicular to $\mathbf{V}$ (why?), so it follows that

$$[\mathbf{V} \times (\boldsymbol{\nabla} \times \mathbf{V})] \cdot d\mathbf{s} = 0$$

Recall also that the dot product of the gradient of a scalar and a differential length gives the differential change in the scalar in the direction of the differential length. That is, with $d\mathbf{s} = dx\,\hat{\mathbf{i}} + dy\,\hat{\mathbf{j}} + dz\,\hat{\mathbf{k}}$ we can write $\boldsymbol{\nabla} p \cdot d\mathbf{s} = (\partial p/\partial x)\,dx + (\partial p/\partial y)\,dy + (\partial p/\partial z)\,dz = dp$. Thus, Eq. 6.54 becomes

$$\frac{dp}{\rho} + \frac{1}{2}d(V^2) + g\,dz = 0 \tag{6.55}$$

where the change in p, V, and z is along the streamline. Equation 6.55 can now be integrated to give

The Bernoulli equation applies along a streamline for inviscid fluids.

$$\int \frac{dp}{\rho} + \frac{V^2}{2} + gz = \text{constant} \tag{6.56}$$

which indicates that the sum of the three terms on the left side of the equation must remain a constant along a given streamline. Equation 6.56 is valid for both compressible and incompressible inviscid flows, but for compressible fluids the variation in ρ with p must be specified before the first term in Eq. 6.56 can be evaluated.

For inviscid, incompressible fluids (commonly called *ideal fluids*) Eq. 6.56 can be written as

$$\boxed{\frac{p}{\rho} + \frac{V^2}{2} + gz = \text{constant}} \tag{6.57}$$

and this equation is the Bernoulli equation used extensively in Chapter 3. It is often convenient to write Eq. 6.57 between two points (1) and (2) along a streamline and to express the equation in the ''head'' form by dividing each term by g so that

$$\boxed{\frac{p_1}{\gamma} + \frac{V_1^2}{2g} + z_1 = \frac{p_2}{\gamma} + \frac{V_2^2}{2g} + z_2} \tag{6.58}$$

It should be again emphasized that the Bernoulli equation, as expressed by Eqs. 6.57 and 6.58, is restricted to the following:

- inviscid flow
- steady flow
- incompressible flow
- flow along a streamline

You may want to go back and review some of the examples in Chapter 3 that illustrate the use of the Bernoulli equation.

6.4.3 Irrotational Flow

The vorticity is zero in an irrotational flow field.

If we make one additional assumption—that the flow is *irrotational*—the analysis of inviscid flow problems is further simplified. Recall from Section 6.1.3 that the rotation of a fluid element is equal to $\frac{1}{2}(\nabla \times \mathbf{V})$, and an irrotational flow field is one for which $\nabla \times \mathbf{V} = 0$. Since the vorticity, ζ, is defined as $\nabla \times \mathbf{V}$, it also follows that in an irrotational flow field the vorticity is zero. The concept of irrotationality may seem to be a rather strange condition for a flow field. Why would a flow field be irrotational? To answer this question we note that if $\frac{1}{2}(\nabla \times \mathbf{V}) = 0$, then each of the components of this vector, as are given by Eqs. 6.12, 6.13, and 6.14, must be equal to zero. Since these components include the various velocity gradients in the flow field, the condition of irrotationality imposes specific relationships among these velocity gradients. For example, for rotation about the z axis to be zero, it follows from Eq. 6.12 that

$$\omega_z = \frac{1}{2}\left(\frac{\partial v}{\partial x} - \frac{\partial u}{\partial y}\right) = 0$$

and, therefore,

$$\frac{\partial v}{\partial x} = \frac{\partial u}{\partial y} \tag{6.59}$$

Similarly from Eqs. 6.13 and 6.14

$$\frac{\partial w}{\partial y} = \frac{\partial v}{\partial z} \tag{6.60}$$

and

$$\frac{\partial u}{\partial z} = \frac{\partial w}{\partial x} \tag{6.61}$$

A general flow field would not satisfy these three equations. However, a uniform flow as is illustrated in Fig. 6.13 does. Since $u = U$ (a constant), $v = 0$, and $w = 0$, it follows that Eqs. 6.59, 6.60, and 6.61 are all satisfied. Therefore, a uniform flow field (in which there are no velocity gradients) is certainly an example of an irrotational flow.

Uniform flows by themselves are not very interesting. However, many interesting and important flow problems include uniform flow in some part of the flow field. Two examples are shown in Fig. 6.14. In Fig. 6.14*a* a solid body is placed in a uniform stream of fluid. Far away from the body the flow remains uniform, and in this far region the flow is irrotational. In Fig. 6.14*b*, flow from a large reservoir enters a pipe through a streamlined entrance where the velocity distribution is essentially uniform. Thus, at the entrance the flow is irrotational.

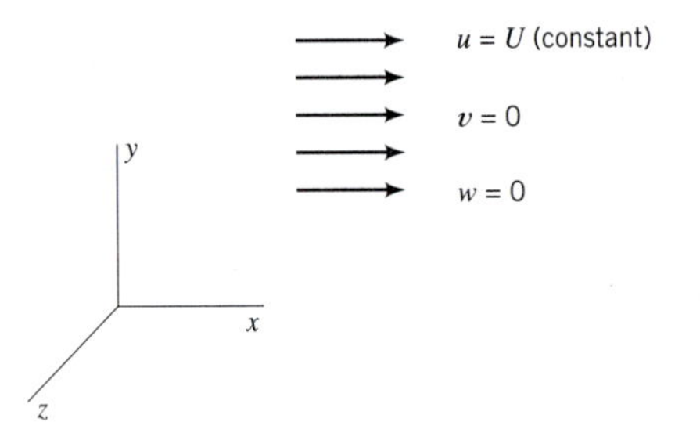

■ **FIGURE 6.13** **Uniform flow in the x direction.**

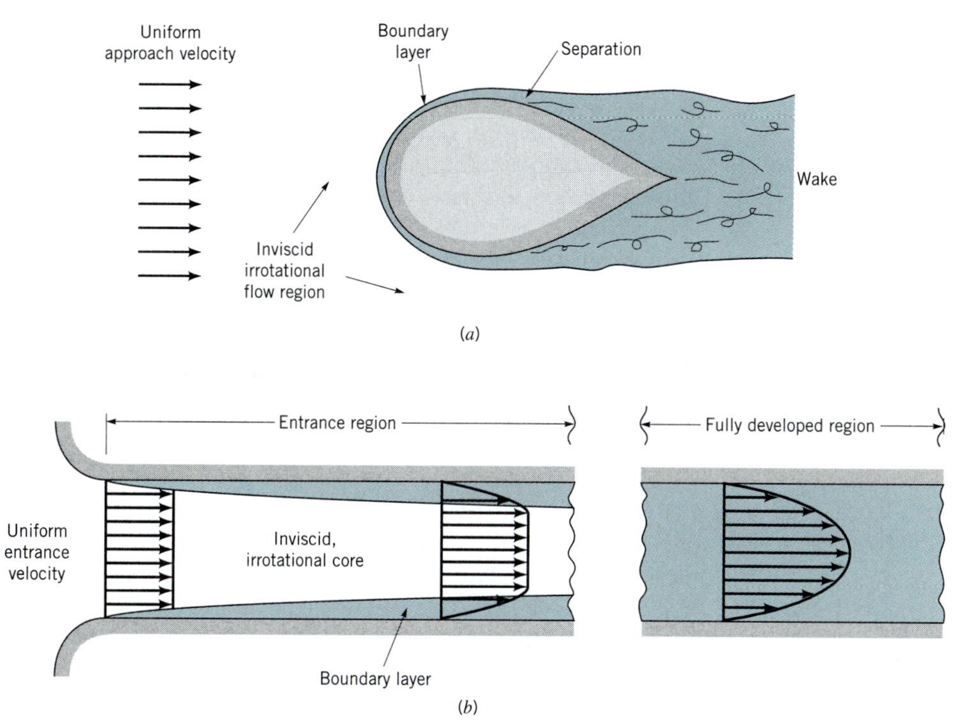

■ **FIGURE 6.14** **Various regions of flow: (*a*) around bodies; (*b*) through channels.**

For an inviscid fluid there are no shearing stresses—the only forces acting on a fluid element are its weight and pressure forces. Since the weight acts through the element center of gravity, and the pressure acts in a direction normal to the element surface, neither of these forces can cause the element to rotate. Therefore, for an inviscid fluid, if some part of the flow field is irrotational, the fluid elements emanating from this region will not take on any rotation as they progress through the flow field. This phenomenon is illustrated in Fig. 6.14*a* in which fluid elements flowing far away from the body have irrotational motion, and as they flow around the body the motion remains irrotational except very near the boundary. Near the boundary the velocity changes rapidly from zero at the boundary (no-slip condition) to some relatively large value in a short distance from the boundary. This rapid change in velocity gives rise to a large velocity gradient normal to the boundary and produces significant shearing stresses, even though the viscosity is small. Of course if we had a truly inviscid fluid, the fluid would simply "slide" past the boundary and the flow would be irrotational everywhere. But this is not the case for real fluids, so we will typically have a layer (usually very thin) near any fixed surface in a moving stream in which shearing stresses are not negligible. This layer is called the *boundary layer*. Outside the boundary layer the flow can be treated as an irrotational flow. Another possible consequence of the boundary layer is that the main stream may "separate" from the surface and form a *wake* downstream from the body. The wake would include a region of slow, randomly moving fluid. To completely analyze this type of problem it is necessary to consider both the inviscid, irrotational flow outside the boundary layer, and the viscous, rotational flow within the boundary layer and to somehow "match" these two regions. This type of analysis is considered in Chapter 9.

Flow fields involving real fluids often include both regions of negligible shearing stresses and regions of significant shearing stresses.

As is illustrated in Fig. 6.14*b*, the flow in the entrance to a pipe may be uniform (if the entrance is streamlined), and thus will be irrotational. In the central core of the pipe the flow remains irrotational for some distance. However, a boundary layer will develop along the wall and grow in thickness until it fills the pipe. Thus, for this type of internal flow there will be an *entrance region* in which there is a central irrotational core, followed by a so-called *fully developed region* in which viscous forces are dominant. The concept of irrotationality is completely invalid in the fully developed region. This type of internal flow problem is considered in detail in Chapter 8.

The two preceding examples are intended to illustrate the possible applicability of irrotational flow to some ''real fluid'' flow problems and to indicate some limitations of the irrotationality concept. We proceed to develop some useful equations based on the assumptions of inviscid, incompressible, irrotational flow, with the admonition to use caution when applying the equations.

6.4.4 The Bernoulli Equation for Irrotational Flow

In the development of the Bernoulli equation in Section 6.4.2, Eq. 6.54 was integrated along a streamline. This restriction was imposed so the right side of the equation could be set equal to zero; that is

$$[\mathbf{V} \times (\nabla \times \mathbf{V})] \cdot d\mathbf{s} = 0$$

(since $d\mathbf{s}$ is parallel to $\mathbf{V}$). However, for irrotational flow, $\nabla \times \mathbf{V} = 0$, so the right side of Eq. 6.54 is zero regardless of the direction of $d\mathbf{s}$. We can now follow the same procedure used to obtain Eq. 6.55, where the differential changes dp, $d(V^2)$, and dz can be taken in any direction. Integration of Eq. 6.55 again yields

$$\int \frac{dp}{\rho} + \frac{V^2}{2} + gz = \text{constant} \tag{6.62}$$

but the constant is the same throughout the flow field. Thus, for incompressible, irrotational flow the Bernoulli equation can be written as

$$\boxed{\frac{p_1}{\gamma} + \frac{V_1^2}{2g} + z_1 = \frac{p_2}{\gamma} + \frac{V_2^2}{2g} + z_2} \tag{6.63}$$

The Bernoulli equation can be applied between any two points in an irrotational flow field.

between *any two points in the flow field*. Equation 6.63 is exactly the same form as Eq. 6.58 but is not limited to application along a streamline. However, Eq. 6.63 is restricted to

- inviscid flow
- steady flow
- incompressible flow
- irrotational flow

It may be worthwhile to review the use and misuse of the Bernoulli equation for rotational flow as is illustrated in Example 3.19.

6.4.5 The Velocity Potential

For an irrotational flow the velocity gradients are related through Eqs. 6.59, 6.60, and 6.61. It follows that in this case the velocity components can be expressed in terms of a scalar function $\phi(x, y, z, t)$ as

$$u = \frac{\partial \phi}{\partial x} \qquad v = \frac{\partial \phi}{\partial y} \qquad w = \frac{\partial \phi}{\partial z} \tag{6.64}$$

where ϕ is called the *velocity potential*. Direct substitution of these expressions for the velocity components into Eqs. 6.59, 6.60, and 6.61 will verify that a velocity field defined by

Eqs. 6.64 is indeed irrotational. In vector form, Eqs. 6.64 can be written as

$$\boxed{\mathbf{V} = \nabla\phi} \tag{6.65}$$

so that for an irrotational flow the velocity is expressible as the gradient of a scalar function ϕ.

The velocity potential is a consequence of the irrotationality of the flow field, whereas the stream function is a consequence of conservation of mass. It is to be noted, however, that the velocity potential can be defined for a general three-dimensional flow, whereas the stream function is restricted to two-dimensional flows.

For an incompressible fluid we know from conservation of mass that

$$\nabla \cdot \mathbf{V} = 0$$

and therefore for incompressible, irrotational flow (with $\mathbf{V} = \nabla\phi$) it follows that

$$\nabla^2\phi = 0 \tag{6.66}$$

where $\nabla^2(\ ) = \nabla \cdot \nabla(\ )$ is the *Laplacian operator*. In Cartesian coordinates

$$\frac{\partial^2\phi}{\partial x^2} + \frac{\partial^2\phi}{\partial y^2} + \frac{\partial^2\phi}{\partial z^2} = 0$$

Inviscid, incompressible, irrotational flow fields are governed by Laplace's equation and are called potential flows.

This differential equation arises in many different areas of engineering and physics and is called *Laplace's equation*. Thus, inviscid, incompressible, irrotational flow fields are governed by Laplace's equation. This type of flow is commonly called a *potential flow*. To complete the mathematical formulation of a given problem, boundary conditions have to be specified. These are usually velocities specified on the boundaries of the flow field of interest. It follows that if the potential function can be determined, then the velocity at all points in the flow field can be determined from Eq. 6.64, and the pressure at all points can be determined from the Bernoulli equation (Eq. 6.63). Although the concept of the velocity potential is applicable to both steady and unsteady flow, we will confine our attention to steady flow.

For some problems it will be convenient to use cylindrical coordinates, r, θ, and z. In this coordinate system the gradient operator is

$$\nabla(\ ) = \frac{\partial(\ )}{\partial r}\,\hat{\mathbf{e}}_r + \frac{1}{r}\frac{\partial(\ )}{\partial\theta}\,\hat{\mathbf{e}}_\theta + \frac{\partial(\ )}{\partial z}\,\hat{\mathbf{e}}_z \tag{6.67}$$

so that

$$\nabla\phi = \frac{\partial\phi}{\partial r}\,\hat{\mathbf{e}}_r + \frac{1}{r}\frac{\partial\phi}{\partial\theta}\,\hat{\mathbf{e}}_\theta + \frac{\partial\phi}{\partial z}\,\hat{\mathbf{e}}_z \tag{6.68}$$

where $\phi = \phi(r, \theta, z)$. Since

$$\mathbf{V} = v_r\hat{\mathbf{e}}_r + v_\theta\hat{\mathbf{e}}_\theta + v_z\hat{\mathbf{e}}_z \tag{6.69}$$

it follows for an irrotational flow (with $\mathbf{V} = \nabla\phi$)

$$v_r = \frac{\partial\phi}{\partial r} \qquad v_\theta = \frac{1}{r}\frac{\partial\phi}{\partial\theta} \qquad v_z = \frac{\partial\phi}{\partial z} \tag{6.70}$$

Also, Laplace's equation in cylindrical coordinates is

$$\frac{1}{r}\frac{\partial}{\partial r}\left(r\frac{\partial\phi}{\partial r}\right) + \frac{1}{r^2}\frac{\partial^2\phi}{\partial\theta^2} + \frac{\partial^2\phi}{\partial z^2} = 0 \tag{6.71}$$

EXAMPLE 6.4

The two-dimensional flow of a nonviscous, incompressible fluid in the vicinity of the 90° corner of Fig. E6.4*a* is described by the stream function

$$\psi = 2r^2 \sin 2\theta$$

where ψ has units of m^2/s when r is in meters. (a) Determine, if possible, the corresponding velocity potential. (b) If the pressure at point (1) on the wall is 30 kPa, what is the pressure at point (2)? Assume the fluid density is $10^3\ kg/m^3$ and the x–y plane is horizontal—that is, there is no difference in elevation between points (1) and (2).

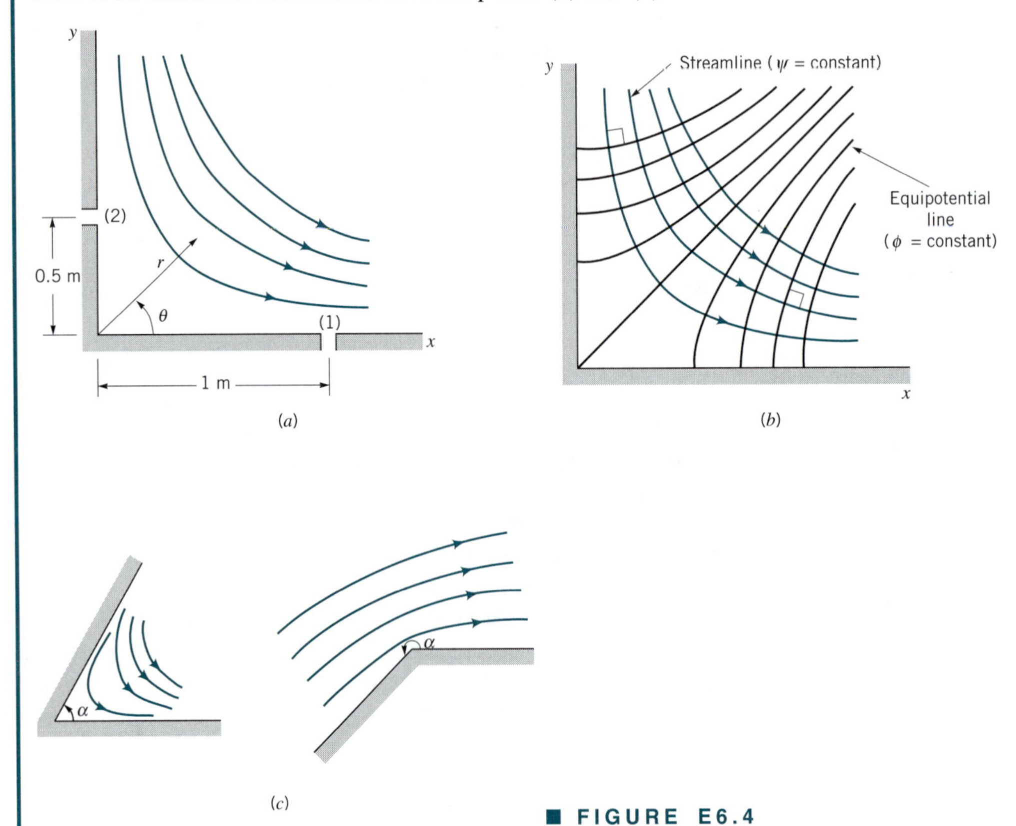

■ FIGURE E6.4

SOLUTION

(a) The radial and tangential velocity components can be obtained from the stream function as (see Eq. 6.42)

$$v_r = \frac{1}{r}\frac{\partial \psi}{\partial \theta} = 4r \cos 2\theta$$

and

$$v_\theta = -\frac{\partial \psi}{\partial r} = -4r \sin 2\theta$$

Since

$$v_r = \frac{\partial \phi}{\partial r}$$

it follows that

$$\frac{\partial \phi}{\partial r} = 4r \cos 2\theta$$

and therefore by integration

$$\phi = 2r^2 \cos 2\theta + f_1(\theta) \quad \textbf{(1)}$$

where $f_1(\theta)$ is an arbitrary function of θ. Similarly

$$v_\theta = \frac{1}{r}\frac{\partial \phi}{\partial \theta} = -4r \sin 2\theta$$

and integration yields

$$\phi = 2r^2 \cos 2\theta + f_2(r) \quad \textbf{(2)}$$

where $f_2(r)$ is an arbitrary function of r. To satisfy both Eqs. 1 and 2, the velocity potential must have the form

$$\phi = 2r^2 \cos 2\theta + C \quad \textbf{(Ans)}$$

where C is an arbitrary constant. As is the case for stream functions, the specific value of C is not important, and it is customary to let $C = 0$ so that the velocity potential for this corner flow is

$$\phi = 2r^2 \cos 2\theta \quad \textbf{(Ans)}$$

In the statement of this problem it was implied by the wording ''if possible'' that we might not be able to find a corresponding velocity potential. The reason for this concern is that we can always define a stream function for two-dimensional flow, but the flow must be *irrotational* if there is a corresponding velocity potential. Thus, the fact that we were able to determine a velocity potential means that the flow is irrotational. Several streamlines and lines of constant ϕ are plotted in Fig. E6.4*b*. These two sets of lines are *orthogonal*. The reason why streamlines and lines of constant ϕ are always orthogonal is explained in Section 6.5.

(b) Since we have an irrotational flow of a nonviscous, incompressible fluid, the Bernoulli equation can be applied between any two points. Thus, between points (1) and (2) with no elevation change

$$\frac{p_1}{\gamma} + \frac{V_1^2}{2g} = \frac{p_2}{\gamma} + \frac{V_2^2}{2g}$$

or

$$p_2 = p_1 + \frac{\rho}{2}(V_1^2 - V_2^2) \quad \textbf{(3)}$$

Since

$$V^2 = v_r^2 + v_\theta^2$$

it follows that for any point within the flow field

$$\begin{aligned} V^2 &= (4r \cos 2\theta)^2 + (-4r \sin 2\theta)^2 \\ &= 16r^2(\cos^2 2\theta + \sin^2 2\theta) \\ &= 16r^2 \end{aligned}$$

This result indicates that the square of the velocity at any point depends only on the radial distance, r, to the point. Note that the constant, 16, has units of s^{-2}. Thus,

$$V_1^2 = (16 \text{ s}^{-2})(1 \text{ m})^2 = 16 \text{ m}^2/\text{s}^2$$

and

$$V_2^2 = (16 \text{ s}^{-2})(0.5 \text{ m})^2 = 4 \text{ m}^2/\text{s}^2$$

Substitution of these velocities into Eq. 3 gives

$$p_2 = 30 \times 10^3 \text{ N/m}^2 + \frac{10^3 \text{ kg/m}^3}{2}(16 \text{ m}^2/\text{s}^2 - 4 \text{ m}^2/\text{s}^2) = 36 \text{ kPa} \quad \textbf{(Ans)}$$

The stream function used in this example could also be expressed in Cartesian coordinates as

$$\psi = 2r^2 \sin 2\theta = 4r^2 \sin \theta \cos \theta$$

or

$$\psi = 4xy$$

since $x = r \cos \theta$ and $y = r \sin \theta$. However, in the cylindrical polar form the results can be generalized to describe flow in the vicinity of a corner of angle α (see Fig. E6.4*c*) with the equations

$$\psi = Ar^{\pi/\alpha} \sin \frac{\pi\theta}{\alpha}$$

and

$$\phi = Ar^{\pi/\alpha} \cos \frac{\pi\theta}{\alpha}$$

where A is a constant.

6.5 Some Basic, Plane Potential Flows

For potential flow, basic solutions can be simply added to obtain more complicated solutions.

A major advantage of Laplace's equation is that it is a linear partial differential equation. Since it is linear, various solutions can be added to obtain other solutions—that is, if $\phi_1(x, y, z)$ and $\phi_2(x, y, z)$ are two solutions to Laplace's equation, then $\phi_3 = \phi_1 + \phi_2$ is also a solution. The practical implication of this result is that if we have certain basic solutions we can combine them to obtain more complicated and interesting solutions. In this section several basic velocity potentials, which describe some relatively simple flows, will be determined. In the next section these basic potentials will be combined to represent more complicated flows.

For simplicity, only plane (two-dimensional) flows will be considered. In this case, by using Cartesian coordinates

$$u = \frac{\partial \phi}{\partial x} \qquad v = \frac{\partial \phi}{\partial y} \tag{6.72}$$

or by using cylindrical coordinates

$$v_r = \frac{\partial \phi}{\partial r} \qquad v_\theta = \frac{1}{r}\frac{\partial \phi}{\partial \theta} \tag{6.73}$$

Since we can define a stream function for plane flow, we can also let

$$u = \frac{\partial \psi}{\partial y} \qquad v = -\frac{\partial \psi}{\partial x} \tag{6.74}$$

or

$$v_r = \frac{1}{r}\frac{\partial \psi}{\partial \theta} \qquad v_\theta = -\frac{\partial \psi}{\partial r} \tag{6.75}$$

where the stream function was previously defined in Eqs. 6.37 and 6.42. We know that by defining the velocities in terms of the stream function, conservation of mass is identically satisfied. If we now impose the condition of irrotationality, it follows from Eq. 6.59 that

$$\frac{\partial u}{\partial y} = \frac{\partial v}{\partial x}$$

and in terms of the stream function

$$\frac{\partial}{\partial y}\left(\frac{\partial \psi}{\partial y}\right) = \frac{\partial}{\partial x}\left(-\frac{\partial \psi}{\partial x}\right)$$

or

$$\frac{\partial^2 \psi}{\partial x^2} + \frac{\partial^2 \psi}{\partial y^2} = 0$$

Thus, for a plane irrotational flow we can use either the velocity potential or the stream function—both must satisfy Laplace's equation in two dimensions. It is apparent from these results that the velocity potential and the stream function are somehow related. We have previously shown that lines of constant ψ are streamlines; that is,

$$\left.\frac{dy}{dx}\right|_{\text{along } \psi=\text{constant}} = \frac{v}{u} \tag{6.76}$$

The change in ϕ as we move from one point (x, y) to a nearby point $(x + dx, y + dy)$ is given by the relationship:

$$d\phi = \frac{\partial \phi}{\partial x} dx + \frac{\partial \phi}{\partial y} dy = u\,dx + v\,dy$$

Along a line of constant ϕ we have $d\phi = 0$ so that

$$\left.\frac{dy}{dx}\right|_{\text{along } \phi=\text{constant}} = -\frac{u}{v} \tag{6.77}$$

A comparison of Eqs. 6.76 and 6.77 shows that lines of constant ϕ (called *equipotential lines*) are orthogonal to lines of constant ψ (streamlines) at all points where they intersect. (Recall that two lines are orthogonal if the product of their slopes is minus one.) For any potential flow field a "*flow net*" can be drawn that consists of a family of streamlines and equipotential lines. The flow net is useful in visualizing flow patterns and can be used to obtain graphical solutions by sketching in streamlines and equipotential lines and adjusting the lines until the lines are approximately orthogonal at all points where they intersect. An example of a flow net is shown in Fig. 6.15. Velocities can be estimated from the flow net, since the velocity is inversely proportional to the streamline spacing. Thus, for example, from Fig. 6.15 we can see that the velocity near the inside corner will be higher than the velocity along the outer part of the bend.

A flow net consists of a family of streamlines and equipotential lines.

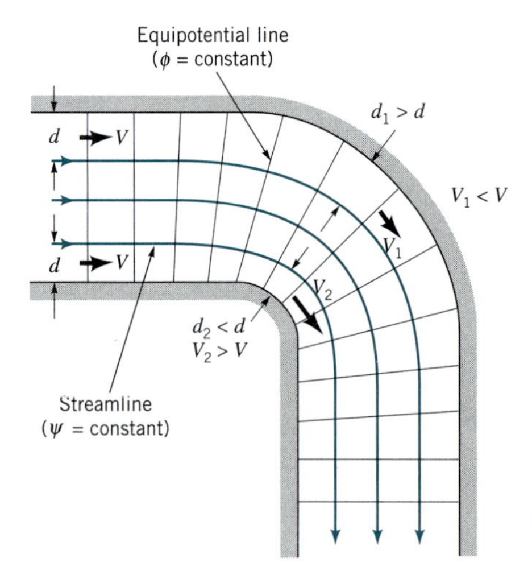

■ FIGURE 6.15 **Flow net for a 90° bend. (From Ref. 3, used by permission.)**

6.5.1 Uniform Flow

The simplest plane flow is one for which the streamlines are all straight and parallel, and the magnitude of the velocity is constant. This type of flow is called a *uniform flow*. For example, consider a uniform flow in the positive x direction as is illustrated in Fig. 6.16a. In this instance, $u = U$ and $v = 0$, and in terms of the velocity potential

$$\frac{\partial \phi}{\partial x} = U \qquad \frac{\partial \phi}{\partial y} = 0$$

These two equations can be integrated to yield

$$\phi = Ux + C$$

where C is an arbitrary constant, which can be set equal to zero. Thus, for a uniform flow in the positive x direction

$$\phi = Ux \tag{6.78}$$

Uniform flow can be simply described by either a stream function or a velocity potential.

The corresponding stream function can be obtained in a similar manner, since

$$\frac{\partial \psi}{\partial y} = U \qquad \frac{\partial \psi}{\partial x} = 0$$

and, therefore,

$$\psi = Uy \tag{6.79}$$

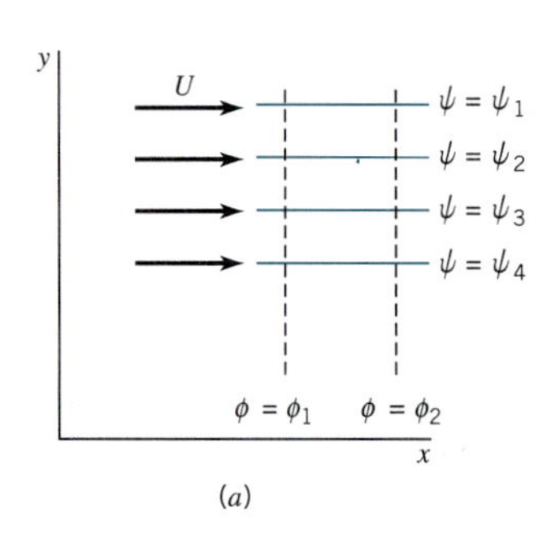

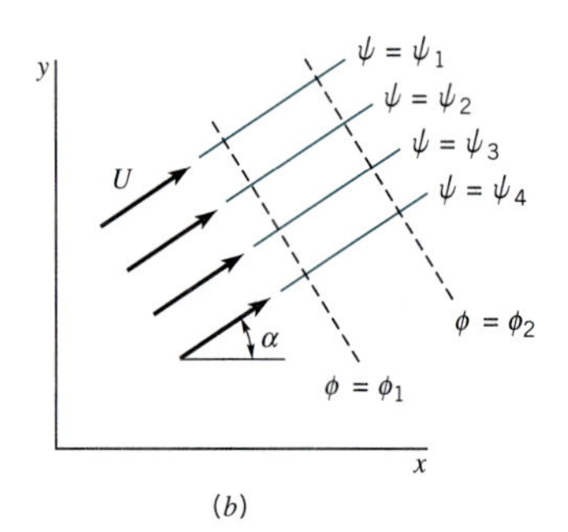

■ FIGURE 6.16 **Uniform flow: (*a*) in the *x* direction; (*b*) in an arbitrary direction, α.**

These results can be generalized to provide the velocity potential and stream function for a uniform flow at an angle α with the x axis, as in Fig. 6.16*b*. For this case

$$\phi = U(x \cos \alpha + y \sin \alpha) \tag{6.80}$$

and

$$\psi = U(y \cos \alpha - x \sin \alpha) \tag{6.81}$$

6.5.2 Source and Sink

Consider a fluid flowing radially outward from a line through the origin perpendicular to the x–y plane as is shown in Fig. 6.17. Let m be the volume rate of flow emanating from the line (per unit length), and therefore to satisfy conservation of mass

$$(2\pi r)v_r = m$$

or

$$v_r = \frac{m}{2\pi r}$$

A source or sink represents a purely radial flow.

Also, since the flow is a purely radial flow, $v_\theta = 0$, the corresponding velocity potential can be obtained by integrating the equations

$$\frac{\partial \phi}{\partial r} = \frac{m}{2\pi r} \qquad \frac{1}{r}\frac{\partial \phi}{\partial \theta} = 0$$

It follows that

$$\phi = \frac{m}{2\pi} \ln r \tag{6.82}$$

If m is positive, the flow is radially outward, and the flow is considered to be a *source* flow. If m is negative, the flow is toward the origin, and the flow is considered to be a *sink* flow. The flowrate, m, is the *strength* of the source or sink.

We note that at the origin where $r = 0$ the velocity becomes infinite, which is of course physically impossible. Thus, sources and sinks do not really exist in real flow fields, and the line representing the source or sink is a mathematical *singularity* in the flow field. However, some real flows can be approximated at points away from the origin by using sources or sinks. Also, the velocity potential representing this hypothetical flow can be combined with other basic velocity potentials to describe approximately some real flow fields. This idea is further discussed in Section 6.6.

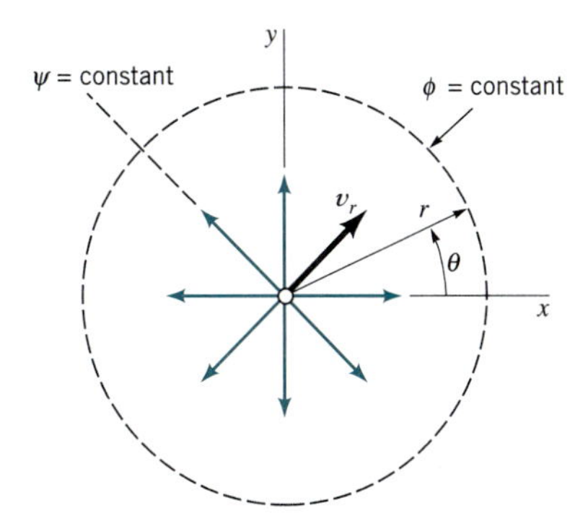

■ **FIGURE 6.17** **The streamline pattern for a source.**

The stream function for the source can be obtained by integrating the relationships

$$\frac{1}{r}\frac{\partial \psi}{\partial \theta} = \frac{m}{2\pi r} \qquad \frac{\partial \psi}{\partial r} = 0$$

to yield

$$\psi = \frac{m}{2\pi}\,\theta \tag{6.83}$$

It is apparent from Eq. 6.83 that the streamlines (lines of ψ = constant) are radial lines, and from Eq. 6.82 the equipotential lines (lines of ϕ = constant) are concentric circles centered at the origin.

A nonviscous, incompressible fluid flows between wedge-shaped walls into a small opening as shown in Fig. E6.5. The velocity potential (in ft^2/s), which approximately describes this flow is

$$\phi = -2 \ln r$$

Determine the volume rate of flow (per unit length) into the opening.

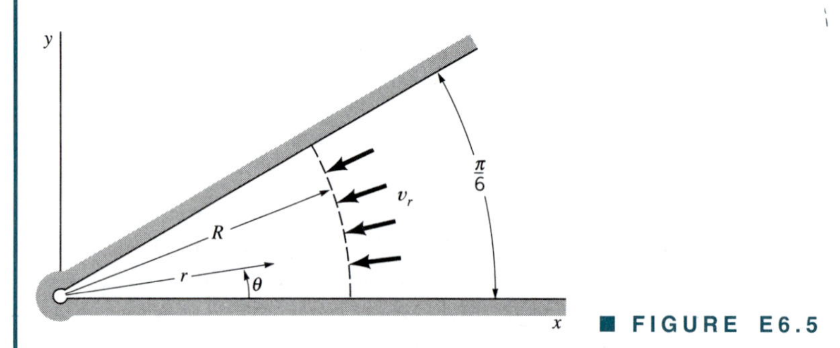

■ FIGURE E6.5

SOLUTION

The components of velocity are

$$v_r = \frac{\partial \phi}{\partial r} = -\frac{2}{r} \qquad v_\theta = \frac{1}{r}\frac{\partial \phi}{\partial \theta} = 0$$

which indicates we have a purely radial flow. The flowrate per unit width, q, crossing the arc of length $R\pi/6$ can thus be obtained by integrating the expression

$$q = \int_0^{\pi/6} v_r R\, d\theta = -\int_0^{\pi/6} \left(\frac{2}{R}\right) R\, d\theta = -\frac{\pi}{3} = -1.05 \text{ ft}^2/\text{s} \qquad \textbf{(Ans)}$$

Note that the radius R is arbitrary since the flowrate crossing any curve between the two walls must be the same. The negative sign indicates that the flow is toward the opening, that is, in the negative radial direction.

6.5.3 Vortex

A vortex represents a flow in which the streamlines are concentric circles.

We next consider a flow field in which the streamlines are concentric circles—that is, we interchange the velocity potential and stream function for the source. Thus, let

$$\phi = K\theta \tag{6.84}$$

and

$$\psi = -K \ln r \tag{6.85}$$

where K is a constant. In this case the streamlines are concentric circles as are illustrated in Fig. 6.18, with $v_r = 0$ and

$$v_\theta = \frac{1}{r}\frac{\partial \phi}{\partial \theta} = -\frac{\partial \psi}{\partial r} = \frac{K}{r} \tag{6.86}$$

This result indicates that the tangential velocity varies inversely with the distance from the origin, with a singularity occurring at $r = 0$ (where the velocity becomes infinite).

It may seem strange that this *vortex* motion is irrotational (and it is since the flow field is described by a velocity potential). However, it must be recalled that rotation refers to the orientation of a fluid element and not the path followed by the element. Thus, for an irrotational vortex, if a pair of small sticks were placed in the flow field at location A, as indicated in Fig. 6.19a, the sticks would rotate as they move to location B. One of the sticks, the one that is aligned along the streamline, would follow a circular path and rotate in a counterclock-

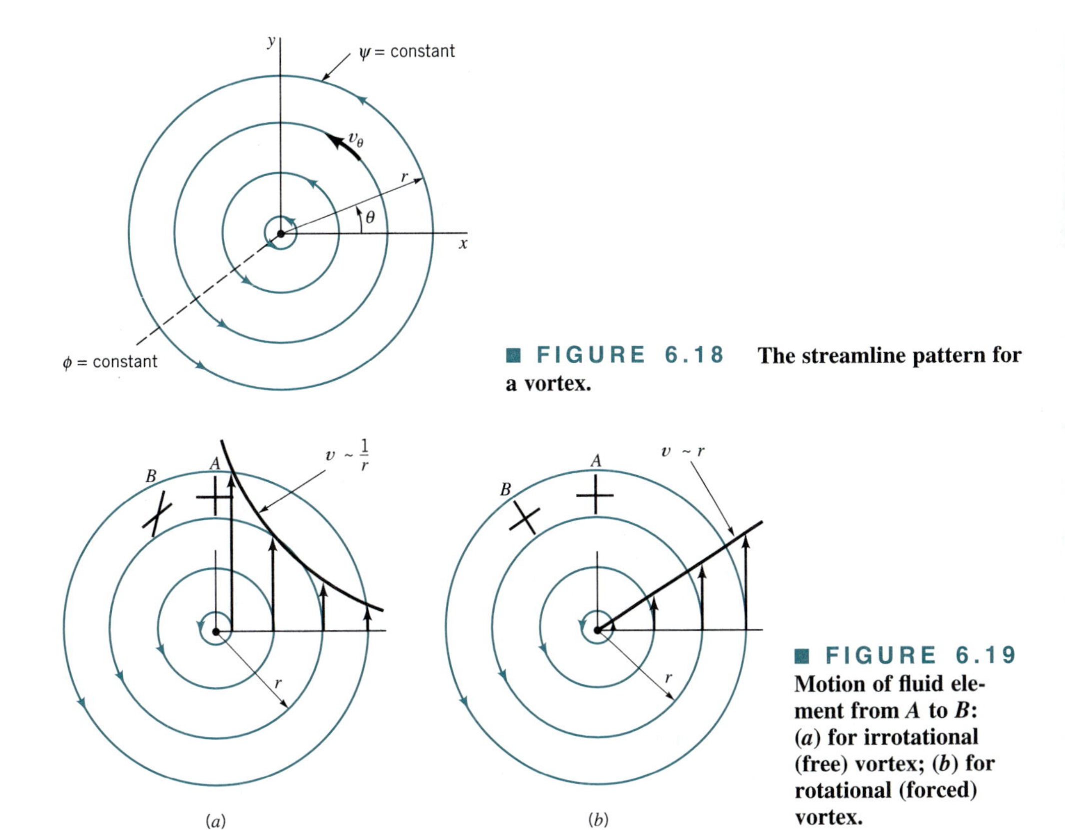

■ FIGURE 6.18 The streamline pattern for a vortex.

■ FIGURE 6.19 Motion of fluid element from A to B: (a) for irrotational (free) vortex; (b) for rotational (forced) vortex.

wise direction. The other stick would rotate in a clockwise direction due to the nature of the flow field—that is, the part of the stick nearest the origin moves faster than the opposite end. Although both sticks are rotating, the average angular velocity of the two sticks is zero since the flow is irrotational.

Vortex motion can be either rotational or irrotational.

If the fluid were rotating as a rigid body, such that $v_\theta = K_1 r$ where K_1 is a constant, then sticks similarly placed in the flow field would rotate as is illustrated in Fig. 6.19*b*. This type of vortex motion is *rotational* and cannot be described with a velocity potential. The rotational vortex is commonly called a *forced vortex*, whereas the irrotational vortex is usually called a *free vortex*. The swirling motion of the water as it drains from a bathtub is similar to that of a free vortex, whereas the motion of a liquid contained in a tank that is rotated about its axis with angular velocity ω corresponds to a forced vortex.

A *combined vortex* is one with a forced vortex as a central core and a velocity distribution corresponding to that of a free vortex outside the core. Thus, for a combined vortex

$$v_\theta = \omega r \qquad r \leq r_0 \tag{6.87}$$

and

$$v_\theta = \frac{K}{r} \qquad r > r_0 \tag{6.88}$$

where K and ω are constants and r_0 corresponds to the radius of the central core. The pressure distribution in both the free and forced vortex was previously considered in Example 3.3.

A mathematical concept commonly associated with vortex motion is that of *circulation*. The circulation, Γ, is defined as the line integral of the tangential component of the velocity taken around a closed curve in the flow field. In equation form, Γ can be expressed as

$$\Gamma = \oint_C \mathbf{V} \cdot d\mathbf{s} \tag{6.89}$$

where the integral sign means that the integration is taken around a closed curve, C, in the counterclockwise direction, and $d\mathbf{s}$ is a differential length along the curve as is illustrated in Fig. 6.20. For an irrotational flow, $\mathbf{V} = \nabla\phi$ so that $\mathbf{V} \cdot d\mathbf{s} = \nabla\phi \cdot d\mathbf{s} = d\phi$ and, therefore,

$$\Gamma = \oint_C d\phi = 0$$

This result indicates that for an irrotational flow the circulation will generally be zero. However, if there are singularities enclosed within the curve the circulation may not be zero. For example, for the free vortex with $v_\theta = K/r$ the circulation around the circular path of radius r shown in Fig. 6.21 is

$$\Gamma = \int_0^{2\pi} \frac{K}{r}(r\,d\theta) = 2\pi K$$

which shows that the circulation is nonzero and the constant $K = \Gamma/2\pi$. However, the circulation around any path that does not include the singular point at the origin will be zero.

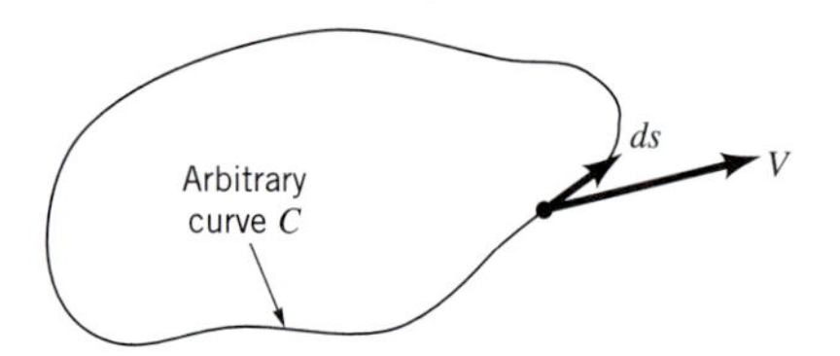

■ **FIGURE 6.20 The notation for determining circulation around closed curve *C*.**

■ **FIGURE 6.21** **Circulation around various paths in a free vortex.**

The numerical value of the circulation may depend on the particular closed path considered.

This can be easily confirmed by evaluating the circulation around a closed path such as *ABCD* of Fig. 6.21, which does not include the origin.

The velocity potential and stream function for the free vortex are commonly expressed in terms of the circulation as

$$\phi = \frac{\Gamma}{2\pi}\,\theta \qquad \textbf{(6.90)}$$

and

$$\psi = -\frac{\Gamma}{2\pi}\ln r \qquad \textbf{(6.91)}$$

The concept of circulation is often useful when evaluating the forces developed on bodies immersed in moving fluids. This application will be considered in Section 6.6.3.

A liquid drains from a large tank through a small opening as illustrated in Fig. E6.6. A vortex forms whose velocity distribution away from the tank opening can be approximated as that of a free vortex having a velocity potential

$$\phi = \frac{\Gamma}{2\pi}\,\theta$$

Determine an expression relating the surface shape to the strength of the vortex as specified by the circulation Γ.

V6.2

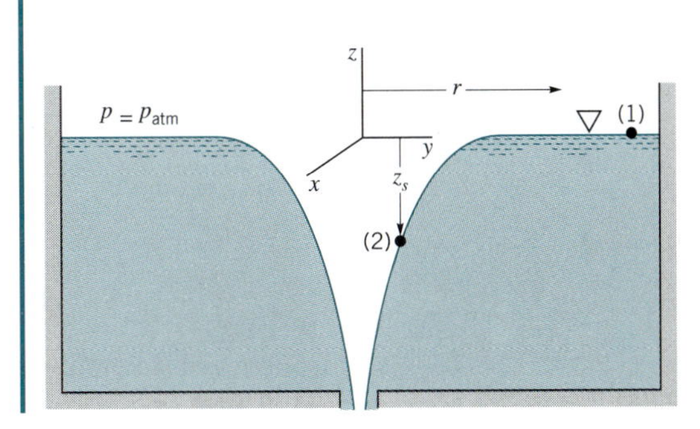

■ **FIGURE E6.6**

SOLUTION

Since the free vortex represents an irrotational flow field, the Bernoulli equation

$$\frac{p_1}{\gamma} + \frac{V_1^2}{2g} + z_1 = \frac{p_2}{\gamma} + \frac{V_2^2}{2g} + z_2$$

can be written between any two points. If the points are selected at the free surface, $p_1 = p_2 = 0$, so that

$$\frac{V_1^2}{2g} = z_s + \frac{V_2^2}{2g} \quad \textbf{(1)}$$

where the free surface elevation, z_s, is measured relative to a datum passing through point (1).

The velocity is given by the equation

$$v_\theta = \frac{1}{r}\frac{\partial \phi}{\partial \theta} = \frac{\Gamma}{2\pi r}$$

We note that far from the origin at point (1), $V_1 = v_\theta \approx 0$ so that Eq. 1 becomes

$$z_s = -\frac{\Gamma^2}{8\pi^2 r^2 g} \quad \textbf{(Ans)}$$

which is the desired equation for the surface profile. The negative sign indicates that the surface falls as the origin is approached as shown in Fig. E6.6. This solution is not valid very near the origin since the predicted velocity becomes excessively large as the origin is approached.

6.5.4 Doublet

A doublet is formed by an appropriate source-sink pair.

The final, basic potential flow to be considered is one that is formed by combining a source and sink in a special way. Consider the equal strength, source-sink pair of Fig. 6.22. The combined stream function for the pair is

$$\psi = -\frac{m}{2\pi}(\theta_1 - \theta_2)$$

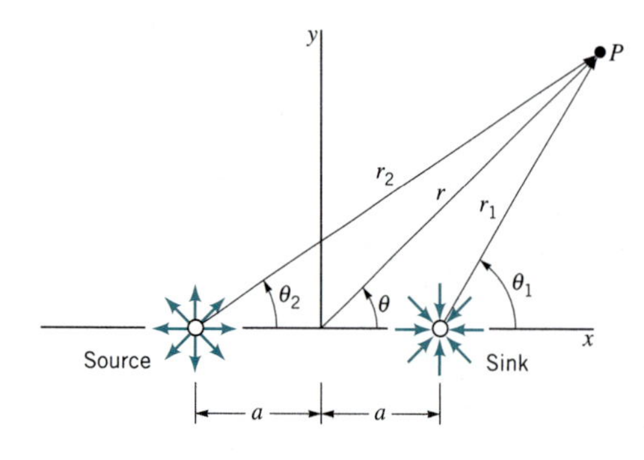

■ **FIGURE 6.22** **The combination of a source and sink of equal strength located along the x axis.**

which can be rewritten as

$$\tan\left(-\frac{2\pi\psi}{m}\right) = \tan(\theta_1 - \theta_2) = \frac{\tan\theta_1 - \tan\theta_2}{1 + \tan\theta_1 \tan\theta_2} \quad \textbf{(6.92)}$$

From Fig. 6.22 it follows that

$$\tan\theta_1 = \frac{r\sin\theta}{r\cos\theta - a}$$

and

$$\tan\theta_2 = \frac{r\sin\theta}{r\cos\theta + a}$$

These results substituted into Eq. 6.92 give

$$\tan\left(-\frac{2\pi\psi}{m}\right) = \frac{2ar\sin\theta}{r^2 - a^2}$$

so that

$$\psi = -\frac{m}{2\pi}\tan^{-1}\left(\frac{2ar\sin\theta}{r^2 - a^2}\right) \quad \textbf{(6.93)}$$

For small values of a

$$\psi = -\frac{m}{2\pi}\frac{2ar\sin\theta}{r^2 - a^2} = -\frac{mar\sin\theta}{\pi(r^2 - a^2)} \quad \textbf{(6.94)}$$

since the tangent of an angle approaches the value of the angle for small angles.

A doublet is formed by letting a source and sink approach one another.

The so-called *doublet* is formed by letting the source and sink approach one another ($a \to 0$) while increasing the strength m ($m \to \infty$) so that the product ma/π remains constant. In this case, since $r/(r^2 - a^2) \to 1/r$, Eq. 6.94 reduces to

$$\psi = -\frac{K\sin\theta}{r} \quad \textbf{(6.95)}$$

where K, a constant equal to ma/π, is called the *strength* of the doublet. The corresponding velocity potential for the doublet is

$$\phi = \frac{K\cos\theta}{r} \quad \textbf{(6.96)}$$

Plots of lines of constant ψ reveal that the streamlines for a doublet are circles through the origin tangent to the x axis as shown in Fig. 6.23. Just as sources and sinks are not physically realistic entities, neither are doublets. However, the doublet when combined with other basic potential flows provides a useful representation of some flow fields of practical interest. For example, we will determine in Section 6.6.3 that the combination of a uniform flow and a doublet can be used to represent the flow around a circular cylinder. Table 6.1 provides a summary of the pertinent equations for the basic, plane potential flows considered in the preceding sections.

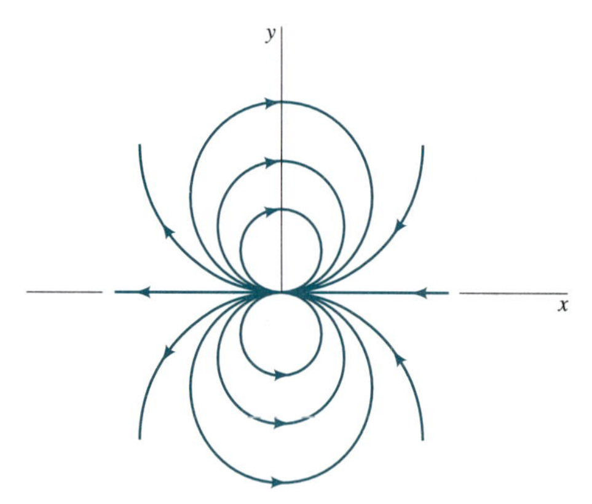

■ FIGURE 6.23 **Streamlines for a doublet.**

■ TABLE 6.1
Summary of Basic, Plane Potential Flows.

Description of Flow Field	Velocity Potential	Stream Function	Velocity Components[a]
Uniform flow at angle α with the x axis (see Fig. 6.16*b*)	$\phi = U(x \cos \alpha + y \sin \alpha)$	$\psi = U(y \cos \alpha - x \sin \alpha)$	$u = U \cos \alpha$ $v = U \sin \alpha$
Source or sink (see Fig. 6.17) $m > 0$ source $m < 0$ sink	$\phi = \dfrac{m}{2\pi} \ln r$	$\psi = \dfrac{m}{2\pi} \theta$	$v_r = \dfrac{m}{2\pi r}$ $v_\theta = 0$
Free vortex (see Fig. 6.18) $\Gamma > 0$ counterclockwise motion $\Gamma < 0$ clockwise motion	$\phi = \dfrac{\Gamma}{2\pi} \theta$	$\psi = -\dfrac{\Gamma}{2\pi} \ln r$	$v_r = 0$ $v_\theta = \dfrac{\Gamma}{2\pi r}$
Doublet (see Fig. 6.23)	$\phi = \dfrac{K \cos \theta}{r}$	$\psi = -\dfrac{K \sin \theta}{r}$	$v_r = -\dfrac{K \cos \theta}{r^2}$ $v_\theta = -\dfrac{K \sin \theta}{r^2}$

[a]Velocity components are related to the velocity potential and stream function through the relationships:
$$u = \frac{\partial \phi}{\partial x} = \frac{\partial \psi}{\partial y} \quad v = \frac{\partial \phi}{\partial y} = -\frac{\partial \psi}{\partial x} \quad v_r = \frac{\partial \phi}{\partial r} = \frac{1}{r}\frac{\partial \psi}{\partial \theta} \quad v_\theta = \frac{1}{r}\frac{\partial \phi}{\partial \theta} = -\frac{\partial \psi}{\partial r}$$

Several basic velocity potentials and stream functions can be used to describe simple plane potential flows.

6.6 Superposition of Basic, Plane Potential Flows

As was discussed in the previous section, potential flows are governed by Laplace's equation, which is a linear partial differential equation. It therefore follows that the various basic velocity potentials and stream functions can be combined to form new potentials and stream functions. (Why is this true?) Whether such combinations yield useful results remains to be seen. It is to be noted that *any streamline in an inviscid flow field can be considered as a solid boundary*, since the conditions along a solid boundary and a streamline are the same—

that is, there is no flow through the boundary or the streamline. Thus, if we can combine some of the basic velocity potentials or stream functions to yield a streamline that corresponds to a particular body shape of interest, that combination can be used to describe in detail the flow around that body. This method of solving some interesting flow problems, commonly called the *method of superposition*, is illustrated in the following three sections.

6.6.1 Source in a Uniform Stream—Half-Body

Flow around a half-body is obtained by the addition of a source to a uniform flow.

Consider the superposition of a source and a uniform flow as shown in Fig. 6.24*a*. The resulting stream function is

$$\psi = \psi_{\text{uniform flow}} + \psi_{\text{source}}$$
$$= Ur \sin \theta + \frac{m}{2\pi} \theta \tag{6.97}$$

and the corresponding velocity potential is

$$\phi = Ur \cos \theta + \frac{m}{2\pi} \ln r \tag{6.98}$$

It is clear that at some point along the negative x axis the velocity due to the source will just cancel that due to the uniform flow and a stagnation point will be created. For the source alone

V6.3

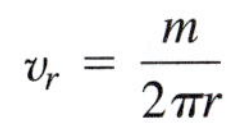

$$v_r = \frac{m}{2\pi r}$$

so that the stagnation point will occur at $x = -b$ where

$$U = \frac{m}{2\pi b}$$

or

$$b = \frac{m}{2\pi U} \tag{6.99}$$

The value of the stream function at the stagnation point can be obtained by evaluating ψ at $r = b$ and $\theta = \pi$, which yields from Eq. 6.97

$$\psi_{\text{stagnation}} = \frac{m}{2}$$

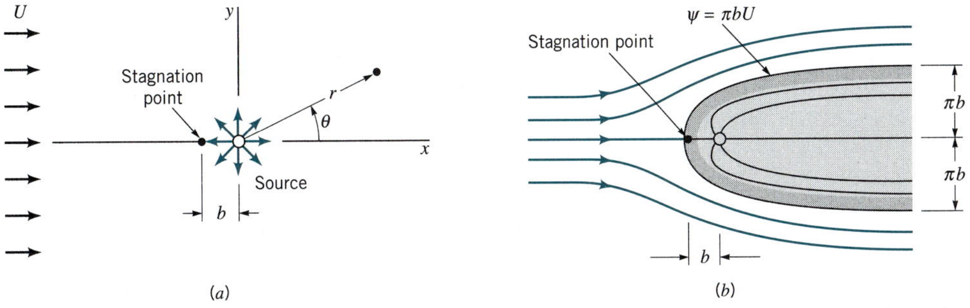

■ **FIGURE 6.24** **The flow around a half-body: (*a*) superposition of a source and a uniform flow; (*b*) replacement of streamline $\psi = \pi bU$ with solid boundary to form half-body.**

Since $m/2 = \pi bU$ (from Eq. 6.99) it follows that the equation of the streamline passing through the stagnation point is

$$\pi bU = Ur \sin \theta + bU\theta$$

or

$$r = \frac{b(\pi - \theta)}{\sin \theta} \tag{6.100}$$

For inviscid flow, a streamline can be replaced by a solid boundary.

where θ can vary between 0 and 2π. A plot of this streamline is shown in Fig. 6.24*b*. If we replace this streamline with a solid boundary, as indicated in the figure, then it is clear that this combination of a uniform flow and a source can be used to describe the flow around a streamlined body placed in a uniform stream. The body is open at the downstream end, and thus is called a *half-body*. Other streamlines in the flow field can be obtained by setting $\psi =$ constant in Eq. 6.97 and plotting the resulting equation. A number of these streamlines are shown in Fig. 6.24*b*. Although the streamlines inside the body are shown, they are actually of no interest in this case, since we are concerned with the flow field outside the body. It should be noted that the singularity in the flow field (the source) occurs inside the body, and there are no singularities in the flow field of interest (outside the body).

The width of the half-body asymptotically approaches $2\pi b$. This follows from Eq. 6.100, which can be written as

$$y = b(\pi - \theta)$$

so that as $\theta \to 0$ or $\theta \to 2\pi$ the half-width approaches $\pm b\pi$. With the stream function (or velocity potential) known, the velocity components at any point can be obtained. For the half-body, using the stream function given by Eq. 6.97,

$$v_r = \frac{1}{r}\frac{\partial \psi}{\partial \theta} = U \cos \theta + \frac{m}{2\pi r}$$

and

$$v_\theta = -\frac{\partial \psi}{\partial r} = -U \sin \theta$$

Thus, the square of the magnitude of the velocity, V, at any point is

$$V^2 = v_r^2 + v_\theta^2 = U^2 + \frac{Um \cos \theta}{\pi r} + \left(\frac{m}{2\pi r}\right)^2$$

and since $b = m/2\pi U$

$$V^2 = U^2 \left(1 + 2\frac{b}{r} \cos \theta + \frac{b^2}{r^2}\right) \tag{6.101}$$

With the velocity known, the pressure at any point can be determined from the Bernoulli equation, which can be written between any two points in the flow field since the flow is irrotational. Thus, applying the Bernoulli equation between a point far from the body, where the pressure is p_0 and the velocity is U, and some arbitrary point with pressure p and velocity V, it follows that

$$p_0 + \tfrac{1}{2}\rho U^2 = p + \tfrac{1}{2}\rho V^2 \tag{6.102}$$

where elevation changes have been neglected. Equation 6.101 can now be substituted into Eq. 6.102 to obtain the pressure at any point in terms of the reference pressure, p_0, and the upstream velocity, U.

For a potential flow the fluid is allowed to slip past a fixed solid boundary.

This relatively simple potential flow provides some useful information about the flow around the front part of a streamlined body, such as a bridge pier or strut placed in a uniform stream. An important point to be noted is that the velocity tangent to the surface of the body is not zero; that is, the fluid ''slips'' by the boundary. This result is a consequence of neglecting viscosity, the fluid property that causes real fluids to stick to the boundary, thus creating a ''no-slip'' condition. All potential flows differ from the flow of real fluids in this respect and do not accurately represent the velocity very near the boundary. However, outside this very thin boundary layer the velocity distribution will generally correspond to that predicted by potential flow theory if flow separation does not occur. Also, the pressure distribution along the surface will closely approximate that predicted from the potential flow theory, since the boundary layer is thin and there is little opportunity for the pressure to vary through the thin layer. In fact, as will be discussed in more detail in Chapter 9, the pressure distribution obtained from potential flow theory is used in conjunction with viscous flow theory to determine the nature of flow within the boundary layer.

EXAMPLE 6.7

The shape of a hill arising from a plain can be approximated with the top section of a half-body as is illustrated in Fig. E6.7. The height of the hill approaches 200 ft as shown. (a) When a 40 mi/hr wind blows toward the hill, what is the magnitude of the air velocity at a point on the hill directly above the origin [point (2)]? (b) What is the elevation of point (2) above the plain and what is the difference in pressure between point (1) on the plain far from the hill and point (2)? Assume an air density of 0.00238 slugs/ft^3.

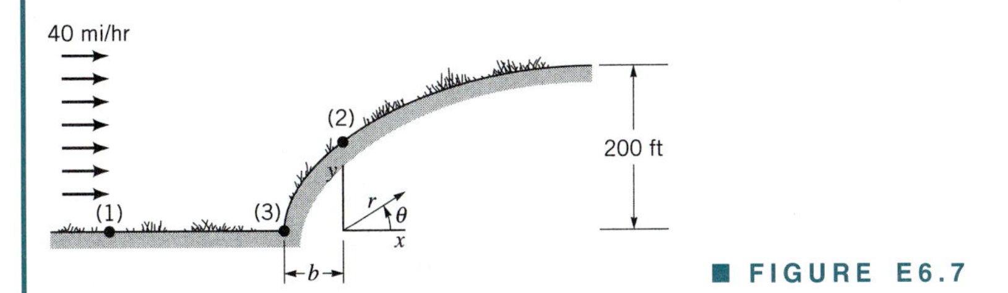

■ FIGURE E6.7

SOLUTION

(a) The velocity is given by Eq. 6.101 as

$$V^2 = U^2\left(1 + 2\frac{b}{r}\cos\theta + \frac{b^2}{r^2}\right)$$

At point (2), $\theta = \pi/2$, and since this point is on the surface (Eq. 6.100)

$$r = \frac{b(\pi - \theta)}{\sin\theta} = \frac{\pi b}{2} \tag{1}$$

Thus,

$$V_2^2 = U^2\left[1 + \frac{b^2}{(\pi b/2)^2}\right]$$

$$= U^2\left(1 + \frac{4}{\pi^2}\right)$$

and the magnitude of the velocity at (2) for a 40 mi/hr approaching wind is

$$V_2 = \left(1 + \frac{4}{\pi^2}\right)^{1/2} (40 \text{ mi/hr}) = 47.4 \text{ mi/hr} \qquad \textbf{(Ans)}$$

(b) The elevation at (2) above the plain is given by Eq. 1 as

$$y_2 = \frac{\pi b}{2}$$

Since the height of the hill approaches 200 ft and this height is equal to πb, it follows that

$$y_2 = \frac{200 \text{ ft}}{2} = 100 \text{ ft} \qquad \textbf{(Ans)}$$

From the Bernoulli equation (with the y axis the vertical axis)

$$\frac{p_1}{\gamma} + \frac{V_1^2}{2g} + y_1 = \frac{p_2}{\gamma} + \frac{V_2^2}{2g} + y_2$$

so that

$$p_1 - p_2 = \frac{\rho}{2}(V_2^2 - V_1^2) + \gamma(y_2 - y_1)$$

and with

$$V_1 = (40 \text{ mi/hr})\left(\frac{5280 \text{ ft/mi}}{3600 \text{ s/hr}}\right) = 58.7 \text{ ft/s}$$

and

$$V_2 = (47.4 \text{ mi/hr})\left(\frac{5280 \text{ ft/mi}}{3600 \text{ s/hr}}\right) = 69.5 \text{ ft/s}$$

it follows that

$$\begin{aligned} p_1 - p_2 &= \frac{(0.00238 \text{ slugs/ft}^3)}{2}[(69.5 \text{ ft/s})^2 - (58.7 \text{ ft/s})^2] \\ &\quad + (0.00238 \text{ slugs/ft}^3)(32.2 \text{ ft/s}^2)(100 \text{ ft} - 0 \text{ ft}) \\ &= 9.31 \text{ lb/ft}^2 = 0.0647 \text{ psi} \end{aligned} \qquad \textbf{(Ans)}$$

This result indicates that the pressure on the hill at point (2) is slightly lower than the pressure on the plain at some distance from the base of the hill with a 0.0533 psi difference due to the elevation increase and a 0.0114 psi difference due to the velocity increase.

The maximum velocity along the hill surface does not occur at point (2) but farther up the hill at $\theta = 63°$. At this point $V_{\text{surface}} = 1.26U$ (Problem 6.55). The minimum velocity ($V = 0$) and maximum pressure occur at point (3), the stagnation point.

6.6.2 Rankine Ovals

The half-body described in the previous section is a body that is ''open'' at one end. To study the flow around a closed body a source and a sink of equal strength can be combined with a uniform flow as shown in Fig. 6.25*a*. The stream function for this combination is

$$\psi = Ur \sin \theta - \frac{m}{2\pi}(\theta_1 - \theta_2) \tag{6.103}$$

and the velocity potential is

$$\phi = Ur \cos \theta - \frac{m}{2\pi}(\ln r_1 - \ln r_2) \tag{6.104}$$

As discussed in Section 6.5.4, the stream function for the source-sink pair can be expressed as in Eq. 6.93 and, therefore, Eq. 6.103 can also be written as

$$\psi = Ur \sin \theta - \frac{m}{2\pi} \tan^{-1}\left(\frac{2ar \sin \theta}{r^2 - a^2}\right)$$

or

$$\psi = Uy - \frac{m}{2\pi} \tan^{-1}\left(\frac{2ay}{x^2 + y^2 - a^2}\right) \tag{6.105}$$

Rankine ovals are formed by combining a source and sink with a uniform flow.

The corresponding streamlines for this flow field are obtained by setting ψ = constant. If several of these streamlines are plotted, it will be discovered that the streamline $\psi = 0$ forms a closed body as is illustrated in Fig. 6.25*b*. We can think of this streamline as forming the surface of a body of length 2ℓ and width $2h$ placed in a uniform stream. The streamlines inside the body are of no practical interest and are not shown. Note that since the body is closed, all of the flow emanating from the source flows into the sink. These bodies have an oval shape and are termed *Rankine ovals*.

Stagnation points occur at the upstream and downstream ends of the body as are indicated in Fig. 6.25*b*. These points can be located by determining where along the x axis the velocity is zero. The stagnation points correspond to the points where the uniform velocity, the source velocity, and the sink velocity all combine to give a zero velocity. The location of the stagnation points depend on the value of a, m, and U. The body half-length, ℓ (the value of $|x|$ that gives $\mathbf{V} = 0$ when $y = 0$), can be expressed as

$$\ell = \left(\frac{ma}{\pi U} + a^2\right)^{1/2} \tag{6.106}$$

or

$$\frac{\ell}{a} = \left(\frac{m}{\pi U a} + 1\right)^{1/2} \tag{6.107}$$

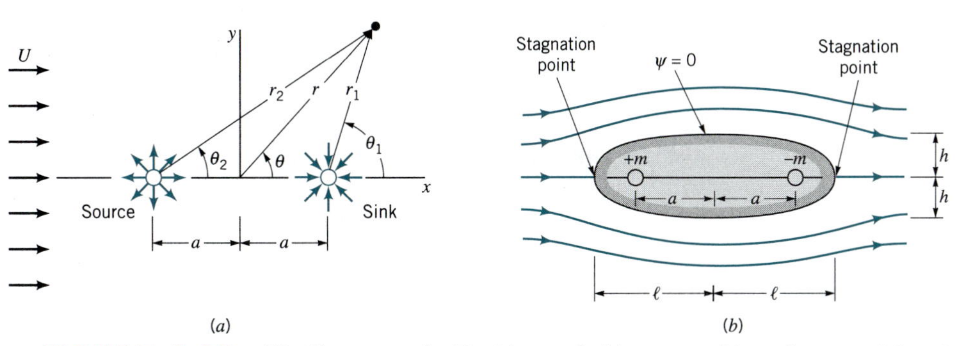

■ **FIGURE 6.25 The flow around a Rankine oval: (*a*) superposition of source-sink pair and a uniform flow; (*b*) replacement of streamline $\psi = 0$ with solid boundary to form Rankine oval.**

The body half-width, h, can be obtained by determining the value of y where the y axis intersects the $\psi = 0$ streamline. Thus, from Eq. 6.105 with $\psi = 0$, $x = 0$, and $y = h$, it follows that

$$h = \frac{h^2 - a^2}{2a} \tan \frac{2\pi Uh}{m} \qquad \textbf{(6.108)}$$

or

$$\frac{h}{a} = \frac{1}{2}\left[\left(\frac{h}{a}\right)^2 - 1\right] \tan\left[2\left(\frac{\pi Ua}{m}\right)\frac{h}{a}\right] \qquad \textbf{(6.109)}$$

Equations 6.107 and 6.109 show that both ℓ/a and h/a are functions of the dimensionless parameter, $\pi Ua/m$. Although for a given value of Ua/m the corresponding value of ℓ/a can be determined directly from Eq. 6.107, h/a must be determined by a trial and error solution of 6.109.

A large variety of body shapes with different length to width ratios can be obtained by using different values of Ua/m. As this parameter becomes large, flow around a long slender body is described, whereas for small values of the parameter, flow around a more blunt shape is obtained. Downstream from the point of maximum body width the surface pressure increases with distance along the surface. This condition (called an adverse pressure gradient) typically leads to separation of the flow from the surface, resulting in a large low pressure wake on the downstream side of the body. Separation is not predicted by potential theory (which simply indicates a symmetrical flow) and, therefore, the potential solution for the Rankine ovals will give a reasonable approximation of the velocity outside the thin, viscous boundary layer and the pressure distribution on the front part of the body only.

6.6.3 Flow Around a Circular Cylinder

A doublet combined with a uniform flow can be used to represent flow around a circular cylinder.

As was noted in the previous section, when the distance between the source-sink pair approaches zero, the shape of the Rankine oval becomes more blunt and in fact approaches a circular shape. Since the doublet described in Section 6.5.4 was developed by letting a source-sink pair approach one another, it might be expected that a uniform flow in the positive x direction combined with a doublet could be used to represent flow around a circular cylinder. This combination gives for the stream function

$$\psi = Ur \sin\theta - \frac{K \sin\theta}{r} \qquad \textbf{(6.110)}$$

and for the velocity potential

$$\phi = Ur \cos\theta + \frac{K \cos\theta}{r} \qquad \textbf{(6.111)}$$

In order for the stream function to represent flow around a circular cylinder it is necessary that $\psi =$ constant for $r = a$, where a is the radius of the cylinder. Since Eq. 6.110 can be written as

$$\psi = \left(U - \frac{K}{r^2}\right) r \sin\theta$$

it follows that $\psi = 0$ for $r = a$ if

$$U - \frac{K}{a^2} = 0$$

which indicates that the doublet strength, K, must be equal to Ua^2. Thus, the stream function for flow around a circular cylinder can be expressed as

$$\psi = Ur\left(1 - \frac{a^2}{r^2}\right)\sin\theta \tag{6.112}$$

and the corresponding velocity potential is

$$\phi = Ur\left(1 + \frac{a^2}{r^2}\right)\cos\theta \tag{6.113}$$

A sketch of the streamlines for this flow field is shown in Fig. 6.26.

The velocity components can be obtained from either Eq. 6.112 or 6.113 as

$$v_r = \frac{\partial\phi}{\partial r} = \frac{1}{r}\frac{\partial\psi}{\partial\theta} = U\left(1 - \frac{a^2}{r^2}\right)\cos\theta \tag{6.114}$$

and

$$v_\theta = \frac{1}{r}\frac{\partial\phi}{\partial\theta} = -\frac{\partial\psi}{\partial r} = -U\left(1 + \frac{a^2}{r^2}\right)\sin\theta \tag{6.115}$$

On the surface of the cylinder ($r = a$) it follows from Eq. 6.114 and 6.115 that $v_r = 0$ and

$$v_{\theta s} = -2U\sin\theta$$

We observe from this result that the maximum velocity occurs at the top and bottom of the cylinder ($\theta = \pm\pi/2$) and has a magnitude of twice the upstream velocity, U. As we move away from the cylinder along the ray $\theta = \pi/2$ the velocity varies as is illustrated in Fig. 6.26.

The pressure distribution on the cylinder surface is obtained from the Bernoulli equation.

The pressure distribution on the cylinder surface is obtained from the Bernoulli equation written from a point far from the cylinder where the pressure is p_0 and the velocity is U so that

$$p_0 + \tfrac{1}{2}\rho U^2 = p_s + \tfrac{1}{2}\rho v_{\theta s}^2$$

where p_s is the surface pressure. Elevation changes are neglected. Since $v_{\theta s} = -2U\sin\theta$, the surface pressure can be expressed as

$$p_s = p_0 + \tfrac{1}{2}\rho U^2(1 - 4\sin^2\theta) \tag{6.116}$$

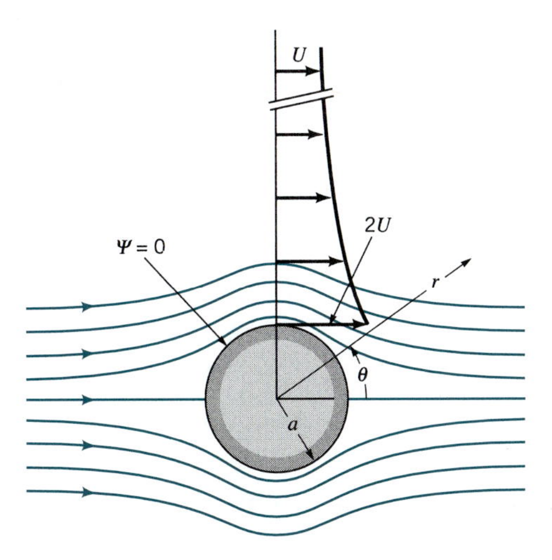

■ FIGURE 6.26 **The flow around a circular cylinder.**

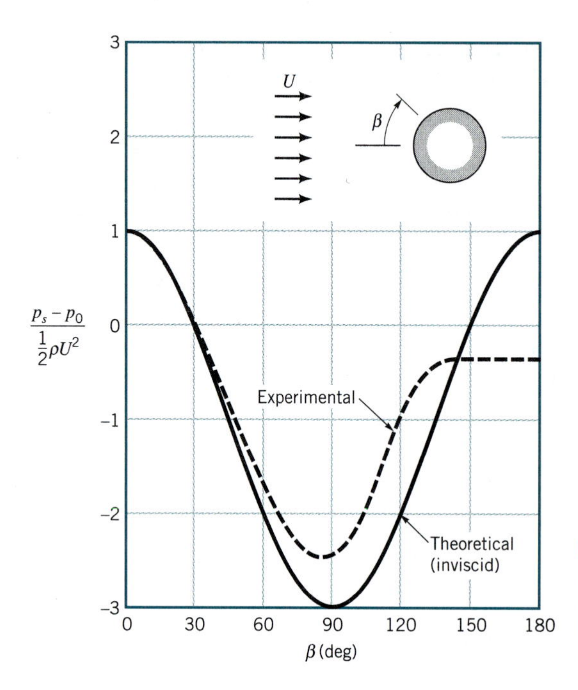

■ **FIGURE 6.27 A comparison of theoretical (inviscid) pressure distribution on the surface of a circular cylinder with typical experimental distribution.**

A comparison of this theoretical, symmetrical pressure distribution expressed in dimensionless form with a typical measured distribution is shown in Fig. 6.27. This figure clearly reveals that only on the upstream part of the cylinder is there approximate agreement between the potential flow and the experimental results. Because of the viscous boundary layer that develops on the cylinder, the main flow separates from the surface of the cylinder, leading to the large difference between the theoretical, frictionless fluid solution and the experimental results on the downstream side of the cylinder (see Chapter 9).

The resultant fluid force can be obtained from a knowledge of the pressure distribution on the surface.

The resultant force (per unit length) developed on the cylinder can be determined by integrating the pressure over the surface. From Fig. 6.28 it can be seen that

$$F_x = -\int_0^{2\pi} p_s \cos\theta \, a \, d\theta \qquad \textbf{(6.117)}$$

and

$$F_y = -\int_0^{2\pi} p_s \sin\theta \, a \, d\theta \qquad \textbf{(6.118)}$$

where F_x is the *drag* (force parallel to direction of the uniform flow) and F_y is the *lift* (force perpendicular to the direction of the uniform flow). Substitution for p_s from Eq. 6.116 into these two equations, and subsequent integration, reveals that $F_x = 0$ and $F_y = 0$ (Problem 6.64). These results indicate that both the drag and left as predicted by potential theory for a

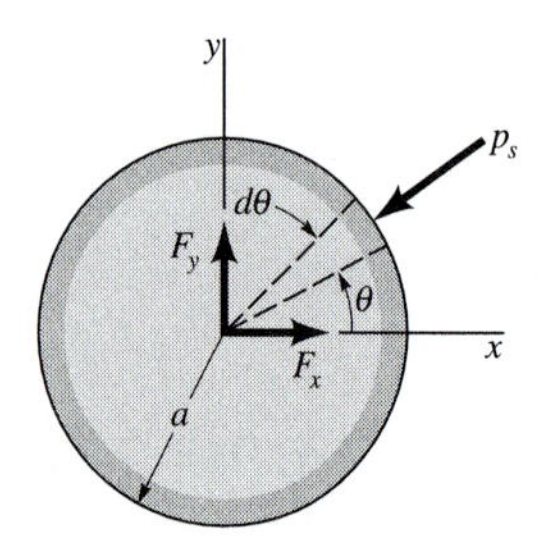

■ **FIGURE 6.28 The notation for determining lift and drag on a circular cylinder.**

Potential theory incorrectly predicts that the drag on a cylinder is zero.

fixed cylinder in a uniform stream are zero. Since the pressure distribution is symmetrical around the cylinder, this is not really a surprising result. However, we know from experience that there is a significant drag developed on a cylinder when it is placed in a moving fluid. This discrepancy is known as d'Alembert's paradox. The paradox is named after Jean le Rond d'Alembert (1717–1783), a French mathematician and philosopher, who first showed that the drag on bodies immersed in inviscid fluids is zero. It was not until the latter part of the nineteenth century and the early part of the twentieth century that the role viscosity plays in the steady fluid motion was understood and d'Alembert's paradox explained (see Section 9.1).

EXAMPLE 6.8

When a circular cylinder is placed in a uniform stream, a stagnation point is created on the cylinder as is shown in Fig. E6.8*a*. If a small hole is located at this point, the stagnation pressure, p_{stag}, can be measured and used to determine the approach velocity, U. (a) Show how p_{stag} and U are related. (b) If the cylinder is misaligned by an angle α (Fig. E6.8*b*), but the measured pressure still interpreted as the stagnation pressure, determine an expression for the ratio of the true velocity, U, to the predicted velocity, U'. Plot this ratio as a function of α for the range $-20° \leq \alpha \leq 20°$.

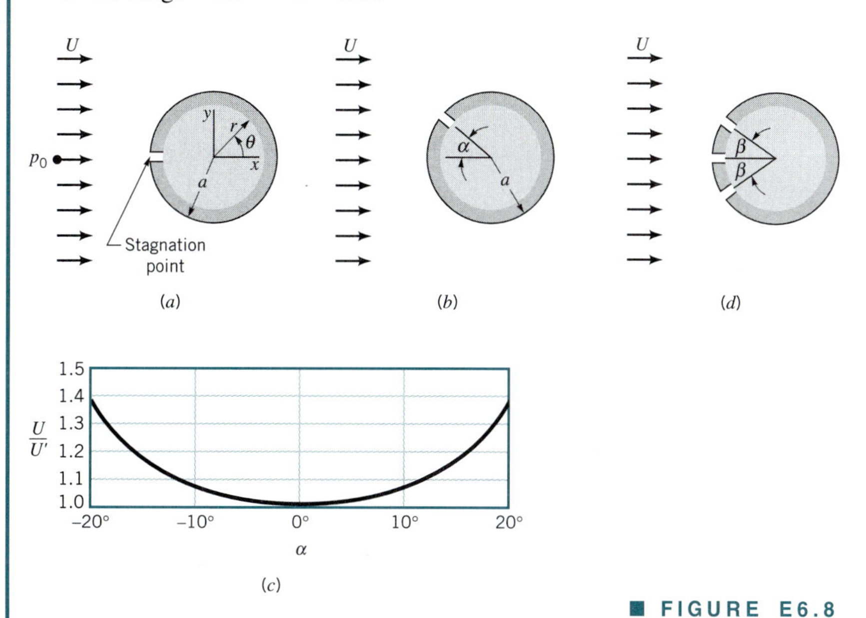

■ FIGURE E6.8

SOLUTION

(a) The velocity at the stagnation point is zero so the Bernoulli equation written between a point on the stagnation streamline upstream from the cylinder and the stagnation point gives

$$\frac{p_0}{\gamma} + \frac{U^2}{2g} = \frac{p_{stag}}{\gamma}$$

Thus,

$$U = \left[\frac{2}{\rho}(p_{stag} - p_0)\right]^{1/2} \qquad \textbf{(Ans)}$$

A measurement of the difference between the pressure at the stagnation point and the upstream pressure can be used to measure the approach velocity. This is, of course, the same result that was obtained in Section 3.5 for Pitot-static tubes.

(b) If the direction of the fluid approaching the cylinder is not known precisely, it is possible that the cylinder is misaligned by some angle, α. In this instance the pressure actually measured, p_α, will be different from the stagnation pressure, but if the misalignment is not recognized the predicted approach velocity, U', would still be calculated as

$$U' = \left[\frac{2}{\rho}(p_\alpha - p_0)\right]^{1/2}$$

Thus,

$$\frac{U(\text{true})}{U'(\text{predicted})} = \left(\frac{p_{\text{stag}} - p_0}{p_\alpha - p_0}\right)^{1/2} \tag{1}$$

The velocity on the surface of the cylinder, v_θ, where $r = a$, is obtained from Eq. 6.115 as

$$v_\theta = -2U \sin\theta$$

If we now write the Bernoulli equation between a point upstream of the cylinder and the point on the cylinder where $r = a$, $\theta = \alpha$, it follows that

$$p_0 + \frac{1}{2}\rho U^2 = p_\alpha + \frac{1}{2}\rho(-2U\sin\alpha)^2$$

and, therefore,

$$p_\alpha - p_0 = \tfrac{1}{2}\rho U^2(1 - 4\sin^2\alpha) \tag{2}$$

Since $p_{\text{stag}} - p_0 = \frac{1}{2}\rho U^2$ it follows from Eqs. 1 and 2 that

$$\frac{U(\text{true})}{U'(\text{predicted})} = (1 - 4\sin^2\alpha)^{-1/2} \qquad \textbf{(Ans)}$$

This velocity ratio is plotted as a function of the misalignment angle α in Fig. E6.8*c*.

It is clear from these results that significant errors can arise if the stagnation pressure tap is not aligned with the stagnation streamline. As is discussed in Section 3.5, if two additional, symmetrically located holes are drilled on the cylinder, as are illustrated in Fig. E6.8*d*, the correct orientation of the cylinder can be determined. The cylinder is rotated until the pressure in the two symmetrically placed holes are equal, thus indicating that the center hole coincides with the stagnation streamline. For $\beta = 30°$ the pressure at the two holes theoretically corresponds to the upstream pressure, p_0. With this orientation a measurement of the difference in pressure between the center hole and the side holes can be used to determine U.

An additional, interesting potential flow can be developed by adding a free vortex to the stream function or velocity potential for the flow around a cylinder. In this case

$$\psi = Ur\left(1 - \frac{a^2}{r^2}\right)\sin\theta - \frac{\Gamma}{2\pi}\ln r \tag{6.119}$$

and

$$\phi = Ur\left(1 + \frac{a^2}{r^2}\right)\cos\theta + \frac{\Gamma}{2\pi}\theta \tag{6.120}$$

where Γ is the circulation. We note that the circle $r = a$ will still be a streamline (and thus can be replaced with a solid cylinder), since the streamlines for the added free vortex are all circular. However, the tangential velocity, v_θ, on the surface of the cylinder ($r = a$) now becomes

$$v_{\theta s} = -\left.\frac{\partial \psi}{\partial r}\right|_{r=a} = -2U\sin\theta + \frac{\Gamma}{2\pi a} \tag{6.121}$$

Flow around a rotating cylinder is approximated by the addition of a free vortex.

This type of flow field could be approximately created by placing a rotating cylinder in a uniform stream. Because of the presence of viscosity in any real fluid, the fluid in contact with the rotating cylinder would rotate with the same velocity as the cylinder, and the resulting flow field would resemble that developed by the combination of a uniform flow past a cylinder and a free vortex.

A variety of streamline patterns can be developed, depending on the vortex strength, Γ. For example, from Eq. 6.121 we can determine the location of stagnation points on the surface of the cylinder. These points will occur at $\theta = \theta_{\text{stag}}$ where $v_\theta = 0$ and therefore from Eq. 6.121

$$\sin\theta_{\text{stag}} = \frac{\Gamma}{4\pi U a} \tag{6.122}$$

If $\Gamma = 0$, then $\theta_{\text{stag}} = 0$ or π—that is, the stagnation points occur at the front and rear of the cylinder as are shown in Fig. 6.29*a*. However, for $-1 \leq \Gamma/4\pi Ua \leq 1$, the stagnation points will occur at some other location on the surface as illustrated in Figs. 6.29*b*, *c*. If the absolute value of the parameter $\Gamma/4\pi Ua$ exceeds 1, Eq. 6.122 cannot be satisfied, and the stagnation point is located away from the cylinder as shown in Fig. 6.29*d*.

The force per unit length developed on the cylinder can again be obtained by integrating the differential pressure forces around the circumference as in Eqs. 6.117 and 6.118. For the cylinder with circulation, the surface pressure, p_s, is obtained from the Bernoulli equation (with the surface velocity given by Eq. 6.121)

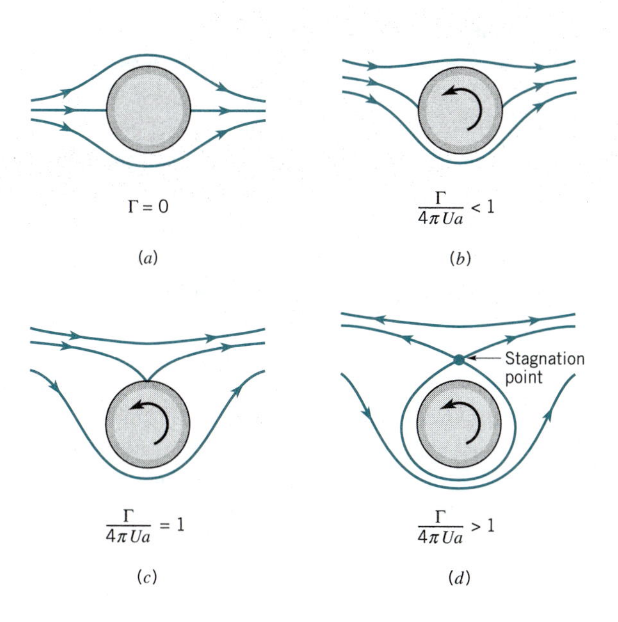

■ **FIGURE 6.29** **The location of stagnation points on a circular cylinder: (*a*) without circulation; (*b, c, d*) with circulation.**

$$p_0 + \frac{1}{2}\rho U^2 = p_s + \frac{1}{2}\rho\left(-2U\sin\theta + \frac{\Gamma}{2\pi a}\right)^2$$

or

$$p_s = p_0 + \frac{1}{2}\rho U^2\left(1 - 4\sin^2\theta + \frac{2\Gamma\sin\theta}{\pi a U} - \frac{\Gamma^2}{4\pi^2 a^2 U^2}\right) \quad \textbf{(6.123)}$$

Equation 6.123 substituted into Eq. 6.117 for the drag, and integrated, again yields (Problem 6.65)

$$F_x = 0$$

That is, even for the rotating cylinder no force in the direction of the uniform flow is developed. However, use of Eq. 6.123 with the equation for the lift, F_y (Eq. 6.118), yields (Problem 6.65)

$$F_y = -\rho U\Gamma \quad \textbf{(6.124)}$$

The development of lift on rotating bodies is called the Magnus effect.

Thus, for the cylinder with circulation, lift is developed equal to the product of the fluid density, the upstream velocity, and the circulation. The negative sign means that if U is positive (in the positive x direction) and Γ is positive (a free vortex with counterclockwise rotation), the direction of the F_y is downward. Of course, if the cylinder is rotated in the clockwise direction ($\Gamma < 0$) the direction of F_y would be upward. It is this force acting in a direction perpendicular to the direction of the approach velocity that causes baseballs and golf balls to curve when they spin as they are propelled through the air. The development of this lift on rotating bodies is called the *Magnus effect.* (See Section 9.4 for further comments.) Although Eq. 6.124 was developed for a cylinder with circulation, it gives the lift per unit length for a right cylinder of any cross-sectional shape placed in a uniform, inviscid stream. The circulation is determined around any closed curve containing the body. The generalized equation relating lift to the fluid density, velocity, and circulation is called the *Kutta–Joukowski law.* It is commonly used to determine the lift on airfoils (see Section 9.4.2 and Refs. 2–6).

6.7 Other Aspects of Potential Flow Analysis

In the preceding section the method of superposition of basic potentials has been used to obtain detailed descriptions of irrotational flow around certain body shapes immersed in a uniform stream. For the cases considered, two or more of the basic potentials were combined and the question is asked: What kind of flow does this combination represent? This approach is relatively simple and does not require the use of advanced mathematical techniques. It is, however, restrictive in its general applicability. It does not allow us to specify a priori the body shape and then determine the velocity potential or stream function that describes the flow around the particular body. Determining the velocity potential or stream function for a given body shape is a much more complicated problem.

It is possible to extend the idea of superposition by considering a *distribution* of sources and sinks, or doublets, which when combined with a uniform flow can describe the flow around bodies of arbitrary shape. Techniques are available to determine the required distribution to give a prescribed body shape. Also, for plane potential flow problems it can be shown that complex variable theory (the use of real and imaginary numbers) can be effectively used to obtain solutions to a great variety of important flow problems. There are, of course, numerical techniques that can be used to solve not only plane two-dimensional problems, but the more general three-dimensional problems. Since potential flow is governed by Laplace's equation, any procedure that is available for solving this equation can be applied to the analysis of irrotational flow of frictionless fluids. Potential flow theory is an old and well-established discipline within the general field of fluid mechanics. The interested reader can find many detailed references on this subject, including Refs. 2, 3, 4, 5, and 6 given at the end of this chapter.

Potential flow solutions are always approximate because the fluid is assumed to be frictionless.

An important point to remember is that regardless of the particular technique used to obtain a solution to a potential flow problem, the solution remains approximate because of the fundamental assumption of a frictionless fluid. Thus, "exact" solutions based on potential flow theory represent, at best, only approximate solutions to real fluid problems. The applicability of potential flow theory to real fluid problems has been alluded to in a number of examples considered in the previous section. As a rule of thumb, potential flow theory will usually provide a reasonable approximation in those circumstances when we are dealing with a low viscosity fluid moving at a relatively high velocity, in regions of the flow field in which the flow is accelerating. Under these circumstances we generally find that the effect of viscosity is confined to the thin boundary layer that develops at a solid boundary. Outside the boundary layer the velocity distribution and the pressure distribution are closely approximated by the potential flow solution. However, in those regions of the flow field in which the flow is decelerating (for example, in the rearward portion of a bluff body or in the expanding region of a conduit) the pressure near a solid boundary will increase in the direction of flow. This so-called adverse pressure gradient can lead to flow separation, a phenomenon that causes dramatic changes in the flow field which are generally not accounted for by potential theory. However, as discussed in Chapter 9, in which boundary layer theory is developed, it is found that potential flow theory is used to obtain the appropriate pressure distribution that can then be combined with the viscous flow equations to obtain solutions near the boundary (and also to predict separation). The general differential equations that describe viscous fluid behavior and some simple solutions to these equations are considered in the remaining sections of this chapter.

V6.4

6.8 Viscous Flow

To incorporate viscous effects into the differential analysis of fluid motion we must return to the previously derived general equations of motion, Eq. 6.50. Since these equations include both stresses and velocities, there are more unknowns than equations, and therefore before proceeding it is necessary to establish a relationship between the stresses and velocities.

6.8.1 Stress-Deformation Relationships

For incompressible Newtonian fluids it is known that the stresses are linearly related to the rates of deformation and can be expressed in Cartesian coordinates as (for normal stresses)

$$\sigma_{xx} = -p + 2\mu \frac{\partial u}{\partial x} \quad \textbf{(6.125a)}$$

$$\sigma_{yy} = -p + 2\mu \frac{\partial v}{\partial y} \quad \textbf{(6.125b)}$$

$$\sigma_{zz} = -p + 2\mu \frac{\partial w}{\partial z} \quad \textbf{(6.125c)}$$

(for shearing stresses)

$$\tau_{xy} = \tau_{yx} = \mu \left(\frac{\partial u}{\partial y} + \frac{\partial v}{\partial x} \right) \quad \textbf{(6.125d)}$$

$$\tau_{yz} = \tau_{zy} = \mu \left(\frac{\partial v}{\partial z} + \frac{\partial w}{\partial y} \right) \quad \textbf{(6.125e)}$$

$$\tau_{zx} = \tau_{xz} = \mu \left(\frac{\partial w}{\partial x} + \frac{\partial u}{\partial z} \right) \quad \textbf{(6.125f)}$$

where p is the pressure, the negative of the average of the three normal stresses; that is $-p = (\frac{1}{3})(\sigma_{xx} + \sigma_{yy} + \sigma_{zz})$. For viscous fluids in motion the normal stresses are not necessarily the same in different directions, thus, the need to define the pressure as the average of the three normal stresses. For fluids at rest, or frictionless fluids, the normal stresses are equal in all directions. (We have made use of this fact in the chapter on fluid statics and in developing the equations for inviscid flow.) Detailed discussions of the development of these stress–velocity gradient relationships can be found in Refs. 3, 7, and 8. An important point to note is that whereas for elastic solids the stresses are linearly related to the deformation (or strain), for Newtonian fluids the stresses are linearly related to the rate of deformation (or rate of strain).

For Newtonian fluids, stresses are linearly related to the rate of strain.

In cylindrical polar coordinates the stresses for incompressible Newtonian fluids are expressed as (for normal stresses)

$$\sigma_{rr} = -p + 2\mu \frac{\partial v_r}{\partial r} \quad \textbf{(6.126a)}$$

$$\sigma_{\theta\theta} = -p + 2\mu \left(\frac{1}{r} \frac{\partial v_\theta}{\partial \theta} + \frac{v_r}{r} \right) \quad \textbf{(6.126b)}$$

$$\sigma_{zz} = -p + 2\mu \frac{\partial v_z}{\partial z} \quad \textbf{(6.126c)}$$

(for shearing stresses)

$$\tau_{r\theta} = \tau_{\theta r} = \mu \left[r \frac{\partial}{\partial r} \left(\frac{v_\theta}{r} \right) + \frac{1}{r} \frac{\partial v_r}{\partial \theta} \right] \quad \textbf{(6.126d)}$$

$$\tau_{\theta z} = \tau_{z\theta} = \mu \left(\frac{\partial v_\theta}{\partial z} + \frac{1}{r} \frac{\partial v_z}{\partial \theta} \right) \quad \textbf{(6.126e)}$$

$$\tau_{zr} = \tau_{rz} = \mu \left(\frac{\partial v_r}{\partial z} + \frac{\partial v_z}{\partial r} \right) \quad \textbf{(6.126f)}$$

The double subscript has a meaning similar to that of stresses expressed in Cartesian coordinates—that is, the first subscript indicates the plane on which the stress acts, and the second subscript the direction. Thus, for example, σ_{rr} refers to a stress acting on a plane perpendicular to the radial direction and in the radial direction (thus a normal stress). Similarly, $\tau_{r\theta}$ refers to a stress acting on a plane perpendicular to the radial direction but in the tangential (θ direction) and is therefore a shearing stress.

6.8.2 The Navier–Stokes Equations

The stresses as defined in the preceding section can be substituted into the differential equations of motion (Eqs. 6.50) and simplified by using the continuity equation (Eq. 6.31) to obtain (x direction)

$$\rho\left(\frac{\partial u}{\partial t} + u\frac{\partial u}{\partial x} + v\frac{\partial u}{\partial y} + w\frac{\partial u}{\partial z}\right) = -\frac{\partial p}{\partial x} + \rho g_x + \mu\left(\frac{\partial^2 u}{\partial x^2} + \frac{\partial^2 u}{\partial y^2} + \frac{\partial^2 u}{\partial z^2}\right) \quad \textbf{(6.127a)}$$

(y direction)

$$\rho\left(\frac{\partial v}{\partial t} + u\frac{\partial v}{\partial x} + v\frac{\partial v}{\partial y} + w\frac{\partial v}{\partial z}\right) = -\frac{\partial p}{\partial y} + \rho g_y + \mu\left(\frac{\partial^2 v}{\partial x^2} + \frac{\partial^2 v}{\partial y^2} + \frac{\partial^2 v}{\partial z^2}\right) \quad \textbf{(6.127b)}$$

(z direction)

$$\rho\left(\frac{\partial w}{\partial t} + u\frac{\partial w}{\partial x} + v\frac{\partial w}{\partial y} + w\frac{\partial w}{\partial z}\right) = -\frac{\partial p}{\partial z} + \rho g_z + \mu\left(\frac{\partial^2 w}{\partial x^2} + \frac{\partial^2 w}{\partial y^2} + \frac{\partial^2 w}{\partial z^2}\right) \quad \textbf{(6.127c)}$$

The Navier–Stokes equations are the basic differential equations describing the flow of incompressible Newtonian fluids.

where we have rearranged the equations so the acceleration terms are on the left side and the force terms are on the right. These equations are commonly called the *Navier–Stokes* equations, named in honor of the French mathematician L. M. H. Navier (1758–1836) and the English mechanician Sir G. G. Stokes (1819–1903), who were responsible for their formulation. These three equations of motion, when combined with the conservation of mass equation (Eq. 6.31), provide a complete mathematical description of the flow of incompressible Newtonian fluids. We have four equations and four unknowns (u, v, w, and p), and therefore the problem is ''well-posed'' in mathematical terms. Unfortunately, because of the general complexity of the Navier–Stokes equations (they are nonlinear, second-order, partial differential equations), they are not amenable to exact mathematical solutions except in a few instances. However, in those few instances in which solutions have been obtained and compared with experimental results, the results have been in close agreement. Thus, the Navier–Stokes equations are considered to be the governing differential equations of motion for incompressible Newtonian fluids.

In terms of cylindrical polar coordinates (see Fig. 6.6) the Navier–Stokes equation can be written as (r direction)

$$\rho\left(\frac{\partial v_r}{\partial t} + v_r\frac{\partial v_r}{\partial r} + \frac{v_\theta}{r}\frac{\partial v_r}{\partial \theta} - \frac{v_\theta^2}{r} + v_z\frac{\partial v_r}{\partial z}\right)$$
$$= -\frac{\partial p}{\partial r} + \rho g_r + \mu\left[\frac{1}{r}\frac{\partial}{\partial r}\left(r\frac{\partial v_r}{\partial r}\right) - \frac{v_r}{r^2} + \frac{1}{r^2}\frac{\partial^2 v_r}{\partial \theta^2} - \frac{2}{r^2}\frac{\partial v_\theta}{\partial \theta} + \frac{\partial^2 v_r}{\partial z^2}\right] \quad \textbf{(6.128a)}$$

(θ direction)

$$\rho\left(\frac{\partial v_\theta}{\partial t} + v_r\frac{\partial v_\theta}{\partial r} + \frac{v_\theta}{r}\frac{\partial v_\theta}{\partial \theta} + \frac{v_r v_\theta}{r} + v_z\frac{\partial v_\theta}{\partial z}\right)$$
$$= -\frac{1}{r}\frac{\partial p}{\partial \theta} + \rho g_\theta + \mu\left[\frac{1}{r}\frac{\partial}{\partial r}\left(r\frac{\partial v_\theta}{\partial r}\right) - \frac{v_\theta}{r^2} + \frac{1}{r^2}\frac{\partial^2 v_\theta}{\partial \theta^2} + \frac{2}{r^2}\frac{\partial v_r}{\partial \theta} + \frac{\partial^2 v_\theta}{\partial z^2}\right] \quad \textbf{(6.128b)}$$

(z direction)

$$\rho\left(\frac{\partial v_z}{\partial t} + v_r\frac{\partial v_z}{\partial r} + \frac{v_\theta}{r}\frac{\partial v_z}{\partial \theta} + v_z\frac{\partial v_z}{\partial z}\right)$$
$$= -\frac{\partial p}{\partial z} + \rho g_z + \mu\left[\frac{1}{r}\frac{\partial}{\partial r}\left(r\frac{\partial v_z}{\partial r}\right) + \frac{1}{r^2}\frac{\partial^2 v_z}{\partial \theta^2} + \frac{\partial^2 v_z}{\partial z^2}\right] \quad \textbf{(6.128c)}$$

To provide a brief introduction to the use of the Navier–Stokes equations, a few of the simplest exact solutions are developed in the next section. Although these solutions will prove to be relatively simple, this is not the case in general. In fact, only a few other exact solutions have been obtained.

6.9 Some Simple Solutions for Viscous, Incompressible Fluids

A principal difficulty in solving the Navier–Stokes equations is because of their nonlinearity arising from the convective acceleration terms (i.e., $u\,\partial u/\partial x$, $w\,\partial v/\partial z$, etc.). There are no general analytical schemes for solving nonlinear partial differential equations (e.g., superposition of solutions cannot be used), and each problem must be considered individually. For most practical flow problems, fluid particles do have accelerated motion as they move from one location to another in the flow field. Thus, the convective acceleration terms are usually important. However, there are a few special cases for which the convective acceleration vanishes because of the nature of the geometry of the flow system. In these cases exact solutions are usually possible. The Navier–Stokes equations apply to both laminar and turbulent flow, but for turbulent flow each velocity component fluctuates randomly with respect to time and this added complication makes an analytical solution intractable. Thus, the exact solutions referred to are for laminar flows in which the velocity is either independent of time (steady flow) or dependent on time (unsteady flow) in a well-defined manner.

6.9.1 Steady Laminar Flow Between Fixed Parallel Plates

An exact solution can be obtained for steady laminar flow between fixed parallel plates.

We first consider flow between the two horizontal, infinite parallel plates of Fig. 6.30*a*. For this geometry the fluid particles move in the x direction parallel to the plates, and there is no velocity in the y or z direction—that is, $v = 0$ and $w = 0$. In this case it follows from the continuity equation (Eq. 6.31) that $\partial u/\partial x = 0$. Furthermore, there would be no variation of u in the z direction for infinite plates, and for steady flow $\partial u/\partial t = 0$ so that $u = u(y)$. If these conditions are used in the Navier–Stokes equations (Eqs. 6.127), they reduce to

$$0 = -\frac{\partial p}{\partial x} + \mu\left(\frac{\partial^2 u}{\partial y^2}\right) \tag{6.129}$$

$$0 = -\frac{\partial p}{\partial y} - \rho g \tag{6.130}$$

$$0 = -\frac{\partial p}{\partial z} \tag{6.131}$$

where we have set $g_x = 0$, $g_y = -g$, and $g_z = 0$. That is, the y axis points up. We see that for this particular problem the Navier–Stokes equations reduce to some rather simple equations.

Equations 6.130 and 6.131 can be integrated to yield

$$p = -\rho g y + f_1(x) \tag{6.132}$$

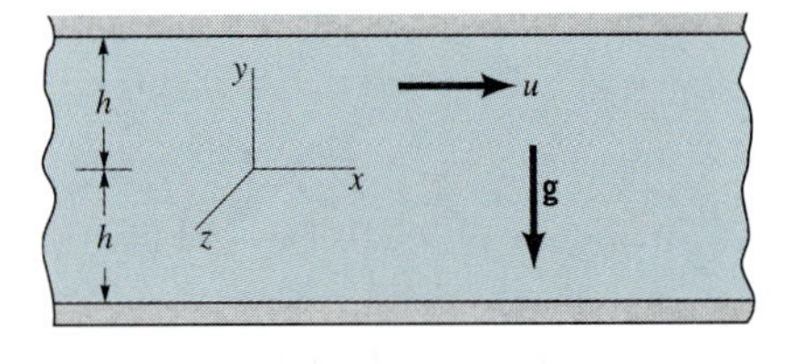

(*a*)

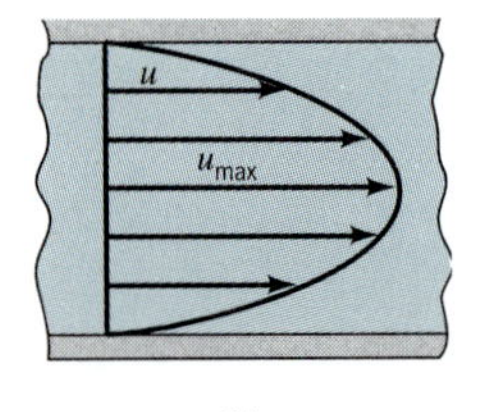

(*b*)

■ **FIGURE 6.30 The viscous flow between parallel plates: (*a*) coordinate system and notation used in analysis; (*b*) parabolic velocity distribution for flow between parallel fixed plates.**

which shows that the pressure varies hydrostatically in the y direction. Equation 6.129, rewritten as

$$\frac{d^2u}{dy^2} = \frac{1}{\mu}\frac{\partial p}{\partial x}$$

can be integrated to give

$$\frac{du}{dy} = \frac{1}{\mu}\left(\frac{\partial p}{\partial x}\right) y + c_1$$

and integrated again to yield

$$u = \frac{1}{2\mu}\left(\frac{\partial p}{\partial x}\right) y^2 + c_1 y + c_2 \tag{6.133}$$

Note that for this simple flow the pressure gradient, $\partial p/\partial x$, is treated as constant as far as the integration is concerned, since (as shown in Eq. 6.132) it is not a function of y. The two constants c_1 and c_2 must be determined from the boundary conditions. For example, if the two plates are fixed, then $u = 0$ for $y = \pm h$ (because of the no-slip condition for viscous fluids). To satisfy this condition $c_1 = 0$ and

V6.5

$$c_2 = -\frac{1}{2\mu}\left(\frac{\partial p}{\partial x}\right) h^2$$

The velocity profile between two fixed, parallel plates is parabolic.

Thus, the velocity distribution becomes

$$u = \frac{1}{2\mu}\left(\frac{\partial p}{\partial x}\right)(y^2 - h^2) \tag{6.134}$$

Equation 6.134 shows that the velocity profile between the two fixed plates is parabolic as illustrated in Fig. 6.30*b*.

The volume rate of flow, q, passing between the plates (for a unit width in the z direction) is obtained from the relationship

$$q = \int_{-h}^{h} u\, dy = \int_{-h}^{h} \frac{1}{2\mu}\left(\frac{\partial p}{\partial x}\right)(y^2 - h^2)\, dy$$

or

$$q = -\frac{2h^3}{3\mu}\left(\frac{\partial p}{\partial x}\right) \tag{6.135}$$

The pressure gradient $\partial p/\partial x$ is negative, since the pressure decreases in the direction of flow. If we let Δp represent the pressure *drop* between two points a distance ℓ apart, then

$$\frac{\Delta p}{\ell} = -\frac{\partial p}{\partial x}$$

and Eq. 6.135 can be expressed as

$$q = \frac{2h^3\,\Delta p}{3\mu\ell} \tag{6.136}$$

The flow is proportional to the pressure gradient, inversely proportional to the viscosity, and strongly dependent ($\sim h^3$) on the gap width. In terms of the mean velocity, V, where $V = q/2h$, Eq. 6.136 becomes

$$V = \frac{h^2\,\Delta p}{3\mu\ell} \tag{6.137}$$

Equations 6.136 and 6.137 provide convenient relationships for relating the pressure drop along a parallel-plate channel and the rate of flow or mean velocity. The maximum velocity, u_{max}, occurs midway ($y = 0$) between the two plates so that from Eq. 6.134

$$u_{max} = -\frac{h^2}{2\mu}\left(\frac{\partial p}{\partial x}\right)$$

or

$$u_{max} = \tfrac{3}{2}V \tag{6.138}$$

The Navier–Stokes equations provide detailed flow characteristics for laminar flow between fixed parallel plates.

The details of the steady laminar flow between infinite parallel plates are completely predicted by this solution to the Navier–Stokes equations. For example, if the pressure gradient, viscosity, and plate spacing are specified, then from Eq. 6.134 the velocity profile can be determined, and from Eqs. 6.136 and 6.137 the corresponding flowrate and mean velocity determined. In addition, from Eq. 6.132 it follows that

$$f_1(x) = \left(\frac{\partial p}{\partial x}\right)x + p_0$$

where p_0 is a reference pressure at $x = y = 0$, and the pressure variation throughout the fluid can be obtained from

$$p = -\rho g y + \left(\frac{\partial p}{\partial x}\right)x + p_0 \tag{6.139}$$

For a given fluid and reference pressure, p_0, the pressure at any point can be predicted. This relatively simple example of an exact solution illustrates the detailed information about the flow field which can be obtained. The flow will be laminar if the Reynolds number, $\text{Re} = \rho V(2h)/\mu$, remains below about 1400. For flow with larger Reynolds numbers the flow becomes turbulent and the preceding analysis is not valid since the flow field is complex, three-dimensional, and unsteady.

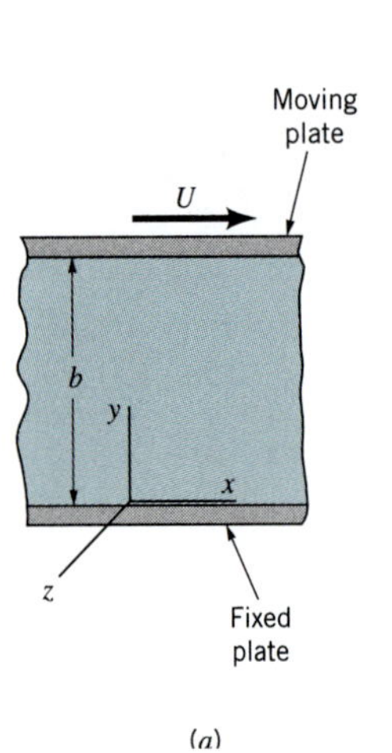

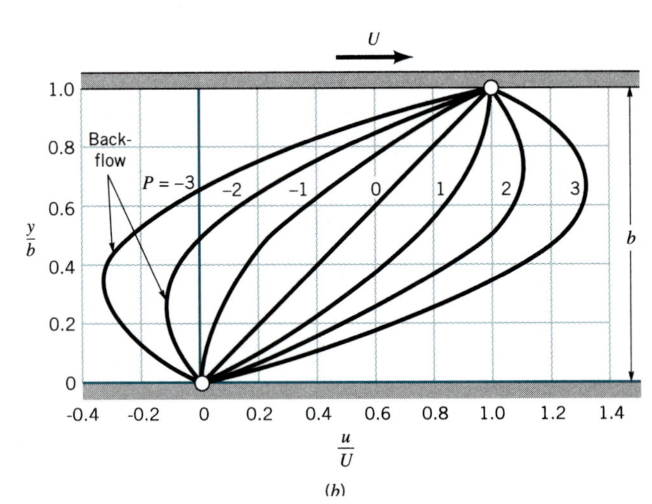

■ **FIGURE 6.31** **The viscous flow between parallel plates with bottom plate fixed and upper plate moving (Couette flow): *(a)* coordinate system and notation used in analysis; *(b)* velocity distribution as a function of parameter, *P*, where $P = -(b^2/2\mu U)\partial p/\partial x$. (From Ref. 8, used by permission.)**

6.9.2 Couette Flow

Another simple parallel-plate flow can be developed by fixing one plate and letting the other plate move with a constant velocity, U, as is illustrated in Fig. 6.31*a*. The Navier–Stokes equations reduce to the same form as those in the preceding section and the solution for the pressure and velocity distribution are still given by Eqs. 6.132 and 6.133, respectively. However, for the moving plate problem the boundary conditions for the velocity are different. For this case we locate the origin of the coordinate system at the bottom plate and designate the distance between the two plates as b (see Fig. 6.31*a*). The two constants c_1 and c_2 in Eq. 6.133 can be determined from the boundary conditions, $u = 0$ at $y = 0$ and $u = U$ at $y = b$. It follows that

$$u = U\frac{y}{b} + \frac{1}{2\mu}\left(\frac{\partial p}{\partial x}\right)(y^2 - by) \tag{6.140}$$

or, in dimensionless form

$$\frac{u}{U} = \frac{y}{b} - \frac{b^2}{2\mu U}\left(\frac{\partial p}{\partial x}\right)\left(\frac{y}{b}\right)\left(1 - \frac{y}{b}\right) \tag{6.141}$$

The actual velocity profile will depend on the dimensionless parameter

$$P = -\frac{b^2}{2\mu U}\left(\frac{\partial p}{\partial x}\right)$$

Flow between parallel plates with one plate fixed and the other moving is called Couette flow.

Several profiles are shown in Fig. 6.31*b*. This type of flow is called *Couette flow*.

The simplest type of Couette flow is one for which the pressure gradient is zero; that is, the fluid motion is caused by the fluid being dragged along by the moving boundary. In this case, with $\partial p/\partial x = 0$, Eq. 6.140 simply reduces to

$$u = U\frac{y}{b} \tag{6.142}$$

which indicates that the velocity varies linearly between the two plates as shown in Fig. 6.31*b* for $P = 0$. This situation would be approximated by the flow between closely spaced concentric cylinders in which one cylinder is fixed and the other cylinder rotates with a constant angular velocity, ω. As illustrated in Fig. 6.32 the flow in an unloaded journal bearing might be approximated by this simple Couette flow if the gap width is very small (i.e., $r_o - r_i \ll r_i$). In this case $U = r_i\,\omega$, $b = r_o - r_i$, and the shearing stress resisting the rotation of the shaft can be simply calculated as $\tau = \mu r_i\,\omega/(r_o - r_i)$. When the bearing is loaded (i.e., a force applied normal to the axis of rotation) the shaft will no longer remain concentric with the housing and the flow cannot be treated as flow between parallel boundaries. Such problems are dealt with in lubrication theory (see, for example, Ref. 9).

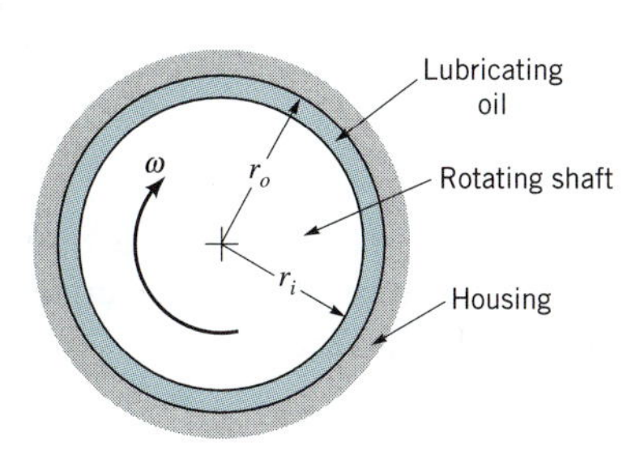

■ **FIGURE 6.32 Flow in the narrow gap of a journal bearing.**

A wide moving belt passes through a container of a viscous liquid. The belt moves vertically upward with a constant velocity, V_0, as illustrated in Fig. E6.9. Because of viscous forces the belt picks up a film of fluid of thickness h. Gravity tends to make the fluid drain down the belt. Use the Navier–Stokes equations to determine an expression for the average velocity of the fluid film as it is dragged up the belt. Assume that the flow is laminar, steady, and uniform.

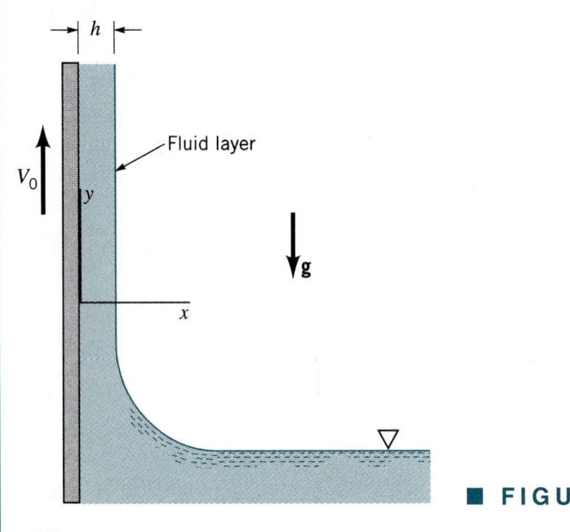

■ FIGURE E6.9

SOLUTION

Since the flow is assumed to be uniform, the only velocity component is in the y direction (the v component) so that $u = w = 0$. It follows from the continuity equation that $\partial v/\partial y = 0$, and for steady flow $\partial v/\partial t = 0$, so that $v = v(x)$. Under these conditions the Navier–Stokes equations for the x direction (Eq. 6.127a) and the z direction (perpendicular to the paper) (Eq. 6.127c) simply reduce to

$$\frac{\partial p}{\partial x} = 0 \qquad \frac{\partial p}{\partial z} = 0$$

This result indicates that the pressure does not vary over a horizontal plane, and since the pressure on the surface of the film ($x = h$) is atmospheric, the pressure throughout the film must be atmospheric (or zero gage pressure). The equation of motion in the y direction (Eq. 6.127b) thus reduces to

$$0 = -\rho g + \mu \frac{d^2 v}{dx^2}$$

or

$$\frac{d^2 v}{dx^2} = \frac{\gamma}{\mu} \tag{1}$$

Integration of Eq. 1 yields

$$\frac{dv}{dx} = \frac{\gamma}{\mu} x + c_1 \tag{2}$$

On the film surface ($x = h$) we assume the shearing stress is zero—that is, the drag of the air on the film is negligible. The shearing stress at the free surface (or any interior parallel surface) is designated as τ_{xy} where from Eq. 6.125d

$$\tau_{xy} = \mu \left(\frac{dv}{dx}\right)$$

Thus, if $\tau_{xy} = 0$ at $x = h$, it follows from Eq. 2 that

$$c_1 = -\frac{\gamma h}{\mu}$$

A second integration of Eq. 2 gives the velocity distribution in the film as

$$v = \frac{\gamma}{2\mu} x^2 - \frac{\gamma h}{\mu} x + c_2$$

At the belt ($x = 0$) the fluid velocity must match the belt velocity, V_0, so that

$$c_2 = V_0$$

and the velocity distribution is therefore

$$v = \frac{\gamma}{2\mu} x^2 - \frac{\gamma h}{\mu} x + V_0$$

With the velocity distribution known we can determine the flowrate per unit width, q, from the relationship

$$q = \int_0^h v\, dx = \int_0^h \left(\frac{\gamma}{2\mu} x^2 - \frac{\gamma h}{\mu} x + V_0 \right) dx$$

and thus

$$q = V_0 h - \frac{\gamma h^3}{3\mu}$$

The average film velocity, V (where $q = Vh$), is therefore

$$V = V_0 - \frac{\gamma h^2}{3\mu} \qquad \textbf{(Ans)}$$

It is interesting to note from this result that there will be a net upward flow of liquid (positive V) only if $V_0 > \gamma h^2/3\mu$. It takes a relatively large belt speed to lift a small viscosity fluid.

6.9.3 Steady, Laminar Flow in Circular Tubes

An exact solution can be obtained for steady, incompressible, laminar flow in circular tubes.

Probably the best known exact solution to the Navier–Stokes equations is for steady, incompressible, laminar flow through a straight circular tube of constant cross section. This type of flow is commonly called *Hagen-Poiseuille flow*, or simply *Poiseuille flow*. It is named in honor of J. L. Poiseuille (1799–1869), a French physician, and G. H. L. Hagen (1797–1884), a German hydraulic engineer. Poiseuille was interested in blood flow through capillaries and deduced experimentally the resistance laws for laminar flow through circular tubes. Hagen's investigation of flow in tubes was also experimental. It was actually after the work of Hagen and Poiseuille that the theoretical results presented in this section were determined, but their names are commonly associated with the solution of this problem.

Consider the flow through a horizontal circular tube of radius R as is shown in Fig. 6.33*a*. Because of the cylindrical geometry it is convenient to use cylindrical coordinates. We assume that the flow is parallel to the walls so that $v_r = 0$ and $v_\theta = 0$, and from the continuity equation (6.34) $\partial v_z/\partial z = 0$. Also, for steady, axisymmetric flow, v_z is not a function of t or θ so the velocity, v_z, is only a function of the radial position within the tube—that is, $v_z =$

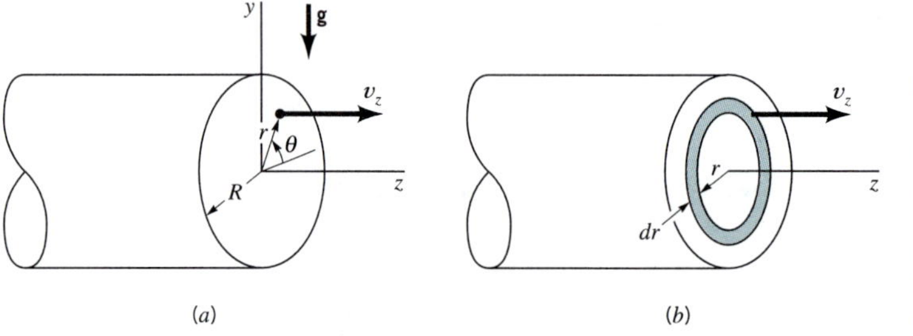

■ **FIGURE 6.33 The viscous flow in a horizontal, circular tube: (*a*) coordinate system and notation used in analysis; (*b*) flow through differential annular ring.**

$v_z(r)$. Under these conditions the Navier–Stokes equations (Eqs. 6.128) reduce to

$$0 = -\rho g \sin \theta - \frac{\partial p}{\partial r} \tag{6.143}$$

$$0 = -\rho g \cos \theta - \frac{1}{r}\frac{\partial p}{\partial \theta} \tag{6.144}$$

$$0 = -\frac{\partial p}{\partial z} + \mu \left[\frac{1}{r}\frac{\partial}{\partial r}\left(r\frac{\partial v_z}{\partial r}\right)\right] \tag{6.145}$$

where we have used the relationships $g_r = -g \sin \theta$ and $g_\theta = -g \cos \theta$ (with θ measured from the horizontal plane).

Equations 6.143 and 6.144 can be integrated to give

$$p = -\rho g(r \sin \theta) + f_1(z)$$

or

$$p = -\rho g y + f_1(z) \tag{6.146}$$

Equation 6.146 indicates that the pressure is hydrostatically distributed at any particular cross section, and the z component of the pressure gradient, $\partial p/\partial z$, is not a function of r or θ.

The equation of motion in the z direction (Eq. 6.145) can be written in the form

$$\frac{1}{r}\frac{\partial}{\partial r}\left(r\frac{\partial v_z}{\partial r}\right) = \frac{1}{\mu}\frac{\partial p}{\partial z}$$

and integrated (using the fact that $\partial p/\partial z = \text{constant}$) to give

$$r\frac{\partial v_z}{\partial r} = \frac{1}{2\mu}\left(\frac{\partial p}{\partial z}\right) r^2 + c_1$$

Integrating again we obtain

$$v_z = \frac{1}{4\mu}\left(\frac{\partial p}{\partial z}\right) r^2 + c_1 \ln r + c_2 \tag{6.147}$$

V6.6

The velocity distribution is parabolic for steady, laminar flow in circular tubes.

Since we wish v_z to be finite at the center of the tube ($r = 0$), it follows that $c_1 = 0$ [since $\ln(0) = -\infty$]. At the wall ($r = R$) the velocity must be zero so that

$$c_2 = -\frac{1}{4\mu}\left(\frac{\partial p}{\partial z}\right) R^2$$

and the velocity distribution becomes

$$v_z = \frac{1}{4\mu}\left(\frac{\partial p}{\partial z}\right)(r^2 - R^2) \tag{6.148}$$

Thus, at any cross section the velocity distribution is parabolic.

To obtain a relationship between the volume rate of flow, Q, passing through the tube and the pressure gradient, we consider the flow through the differential, washer-shaped ring of Fig. 6.33*b*. Since v_z is constant on this ring the volume rate of flow through the differential area $dA = (2\pi r)\,dr$ is

$$dQ = v_z(2\pi r)\,dr$$

and therefore

$$Q = 2\pi \int_0^R v_z r\,dr \tag{6.149}$$

Equation 6.148 for v_z can be substituted into Eq. 6.149, and the resulting equation integrated to yield

$$Q = -\frac{\pi R^4}{8\mu}\left(\frac{\partial p}{\partial z}\right) \tag{6.150}$$

This relationship can be expressed in terms of the pressure *drop*, Δp, which occurs over a length, ℓ, along the tube, since

$$\frac{\Delta p}{\ell} = -\frac{\partial p}{\partial z}$$

and therefore

$$Q = \frac{\pi R^4\,\Delta p}{8\mu\ell} \tag{6.151}$$

Poiseuille's law relates pressure drop and flowrate for steady, laminar flow in circular tubes.

For a given pressure drop per unit length, the volume rate of flow is inversely proportional to the viscosity and proportional to the tube radius to the fourth power. A doubling of the tube radius produces a sixteenfold increase in flow! Equation 6.151 is commonly called *Poiseuille's law*.

In terms of the mean velocity, V, where $V = Q/\pi R^2$, Eq. 6.151 becomes

$$V = \frac{R^2\,\Delta p}{8\mu\ell} \tag{6.152}$$

The maximum velocity v_{max} occurs at the center of the tube, where from Eq. 6.148

$$v_{\text{max}} = -\frac{R^2}{4\mu}\left(\frac{\partial p}{\partial z}\right) = \frac{R^2\,\Delta p}{4\mu\ell} \tag{6.153}$$

so that

$$v_{\text{max}} = 2V$$

The velocity distribution can be written in terms of v_{max} as

$$\frac{v_z}{v_{\text{max}}} = 1 - \left(\frac{r}{R}\right)^2 \tag{6.154}$$

As was true for the similar case of flow between parallel plates (sometimes referred to as *plane Poiseuille flow*), a very detailed description of the pressure and velocity distribution in tube flow results from this solution to the Navier–Stokes equations. Numerous experiments performed to substantiate the theoretical results show that the theory and experiment are in agreement for the laminar flow of Newtonian fluids in circular tubes or pipes. The flow remains laminar for Reynolds numbers, $\text{Re} = \rho V(2R)/\mu$, below 2100. Turbulent flow in tubes is considered in Chapter 8.

6.9.4 Steady, Axial, Laminar Flow in an Annulus

An exact solution can be obtained for axial flow in the annular space between two fixed, concentric cylinders.

The differential equations (Eqs. 6.143, 6.144, 6.145) used in the preceding section for flow in a tube also apply to the axial flow in the annular space between two fixed, concentric cylinders (Fig. 6.34). Equation 6.147 for the velocity distribution still applies, but for the stationary annulus the boundary conditions become $v_z = 0$ at $r = r_o$ and $v_z = 0$ for $r = r_i$. With these two conditions the constants c_1 and c_2 in Eq. 6.147 can be determined and the velocity distribution becomes

$$v_z = \frac{1}{4\mu}\left(\frac{\partial p}{\partial z}\right)\left[r^2 - r_o^2 + \frac{r_i^2 - r_o^2}{\ln(r_o/r_i)} \ln \frac{r}{r_o}\right] \tag{6.155}$$

The corresponding volume rate of flow is

$$Q = \int_{r_i}^{r_o} v_z(2\pi r)\, dr = -\frac{\pi}{8\mu}\left(\frac{\partial p}{\partial z}\right)\left[r_o^4 - r_i^4 - \frac{(r_o^2 - r_i^2)^2}{\ln(r_o/r_i)}\right]$$

or in terms of the pressure drop, Δp, in length ℓ of the annulus

$$Q = \frac{\pi\,\Delta p}{8\mu\ell}\left[r_o^4 - r_i^4 - \frac{(r_o^2 - r_i^2)^2}{\ln(r_o/r_i)}\right] \tag{6.156}$$

The velocity at any radial location within the annular space can be obtained from Eq. 6.155. The maximum velocity occurs at the radius $r = r_m$ where $\partial v_z/\partial r = 0$. Thus,

$$r_m = \left[\frac{r_o^2 - r_i^2}{2\ln(r_o/r_i)}\right]^{1/2} \tag{6.157}$$

An inspection of this result shows that the maximum velocity does not occur at the midpoint of the annular space, but rather it occurs nearer the inner cylinder. The specific location depends on r_o and r_i.

These results for flow through an annulus are only valid if the flow is laminar. A criterion based on the conventional Reynolds number (which is defined in terms of the tube diameter) cannot be directly applied to the annulus, since there are really "two" diameters involved. For tube cross sections other than simple circular tubes it is common practice to use an "effective" diameter, termed the *hydraulic diameter*, D_h, which is defined as

$$D_h = \frac{4 \times \text{cross-sectional area}}{\text{wetted perimeter}}$$

The wetted perimeter is the perimeter in contact with the fluid. For an annulus

$$D_h = \frac{4\pi(r_o^2 - r_i^2)}{2\pi(r_o + r_i)} = 2(r_o - r_i)$$

In terms of the hydraulic diameter the Reynolds number is $\text{Re} = \rho D_h V/\mu$ (where $V = Q/\text{cross-sectional area}$), and it is commonly assumed that if this Reynolds number remains

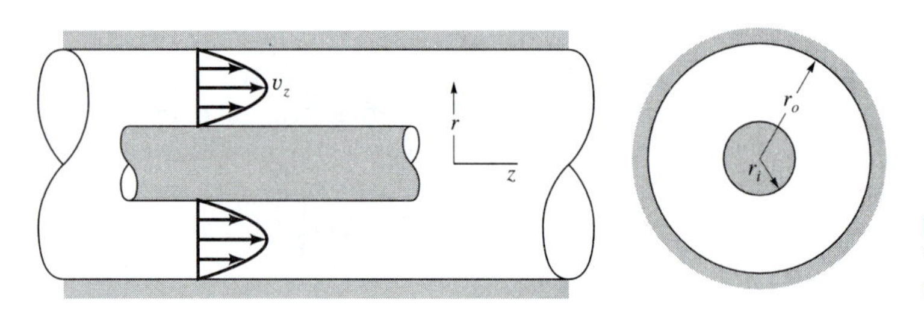

■ **FIGURE 6.34** **The viscous flow through an annulus.**

below 2100 the flow will be laminar. A further discussion of the concept of the hydraulic diameter as it applies to other noncircular cross sections is given in Section 8.4.3.

EXAMPLE 6.10

A viscous liquid ($\rho = 1.18 \times 10^3$ kg/m^3; $\mu = 0.0045$ N·s/m^2) flows at a rate of 12 ml/s through a horizontal, 4-mm-diameter tube. (a) Determine the pressure drop along a 1-m length of the tube which is far from the tube entrance so that the only component of velocity is parallel to the tube axis. (b) If a 2-mm-diameter rod is placed in the 4-mm-diameter tube to form a symmetric annulus, what is the pressure drop along a 1-m length if the flowrate remains the same as in part (a)?

SOLUTION

(a) We first calculate the Reynolds number, Re, to determine whether or not the flow is laminar. The mean velocity is

$$V = \frac{Q}{(\pi/4)D^2} = \frac{(12 \text{ ml/s})(10^{-6} \text{ m}^3/\text{ml})}{(\pi/4)(4 \text{ mm} \times 10^{-3} \text{ m/mm})^2}$$

$$= 0.955 \text{ m/s}$$

and, therefore,

$$\text{Re} = \frac{\rho V D}{\mu} = \frac{(1.18 \times 10^3 \text{ kg/m}^3)(0.955 \text{ m/s})(4 \text{ mm} \times 10^{-3} \text{ m/mm})}{0.0045 \text{ N·s/m}^2}$$

$$= 1000$$

Since the Reynolds number is well below the critical value of 2100 we can safely assume that the flow is laminar. Thus, we can apply Eq. 6.151 which gives for the pressure drop

$$\Delta p = \frac{8\mu \ell Q}{\pi R^4}$$

$$= \frac{8(0.0045 \text{ N·s/m}^2)(1 \text{ m})(12 \times 10^{-6} \text{ m}^3/\text{s})}{\pi(2 \text{ mm} \times 10^{-3} \text{ m/mm})^4}$$

$$= 8.59 \text{ kPa}$$ **(Ans)**

(b) For flow in the annulus, the mean velocity is

$$V = \frac{Q}{\pi(r_o^2 - r_i^2)} = \frac{12 \times 10^{-6} \text{ m}^3/\text{s}}{(\pi)[(2 \text{ mm} \times 10^{-3} \text{ m/mm})^2 - (1 \text{ mm} \times 10^{-3} \text{ m/mm})^2]}$$

$$= 1.27 \text{ m/s}$$

and the Reynolds number (based on the hydraulic diameter) is

$$\text{Re} = \frac{\rho 2(r_o - r_i)V}{\mu}$$

$$= \frac{(1.18 \times 10^3 \text{ kg/m}^3)(2)(2 \text{ mm} - 1 \text{ mm})(10^{-3} \text{ m/mm})(1.27 \text{ m/s})}{0.0045 \text{ N·s/m}^2}$$

$$= 666$$

This value is also well below 2100 so the flow in the annulus should also be laminar. From Eq. 6.156

$$\Delta p = \frac{8\mu\ell Q}{\pi}\left[r_o^4 - r_i^4 - \frac{(r_o^2 - r_i^2)^2}{\ln(r_o/r_i)}\right]^{-1}$$

so that

$$\Delta p = \frac{8(0.0045 \text{ N·s/m}^2)(1 \text{ m})(12 \times 10^{-6} \text{ m}^3\text{/s})}{\pi} \times \left\{(2 \times 10^{-3} \text{ m})^4 - (1 \times 10^{-3} \text{ m})^4 - \frac{[(2 \times 10^{-3} \text{ m})^2 - (1 \times 10^{-3} \text{ m})^2]^2}{\ln(2 \text{ mm}/1 \text{ mm})}\right\}^{-1}$$

$$= 68.2 \text{ kPa}$$ **(Ans)**

The pressure drop in the annulus is much larger than that of the tube. This is not a surprising result, since to maintain the same flow in the annulus as that in the open tube the average velocity must be larger and the pressure difference along the annulus must overcome the shearing stresses that develop along both an inner and an outer wall. Even an annulus with a very small inner diameter will have a pressure drop significantly higher than that of an open tube. For example, if the inner diameter is only 1/100 of the outer diameter, Δp (annulus)/Δp (tube) $= 1.28$.

6.10 Other Aspects of Differential Analysis

In this chapter the basic differential equations that govern the flow of fluids have been developed. The Navier–Stokes equations, which can be compactly expressed in vector notation as

$$\rho\left(\frac{\partial \mathbf{V}}{\partial t} + \mathbf{V} \cdot \nabla\mathbf{V}\right) = -\nabla p + \rho\mathbf{g} + \mu \nabla^2\mathbf{V} \tag{6.158}$$

along with the continuity equation

$$\nabla \cdot \mathbf{V} = 0 \tag{6.159}$$

are the general equations of motion for incompressible Newtonian fluids. Although we have restricted our attention to incompressible fluids, these equations can be readily extended to include compressible fluids. It is well beyond the scope of this introductory text to consider in depth the variety of analytical and numerical techniques that can be used to obtain both exact and approximate solutions to the Navier–Stokes equations. Students, however, should be aware of the existence of these very general equations, which are frequently used as the basis for many advanced analyses of fluid motion. A few relatively simple solutions have been obtained and discussed in this chapter to indicate the type of detailed flow information that can be obtained by using differential analysis. However, it is hoped that the relative ease with which these solutions were obtained does not give the false impression that solutions to the Navier–Stokes equations are readily available. This is certainly not true, and as previously mentioned there are actually very few practical fluid flow problems that can be solved by using an exact analytical approach. In fact, there are no known analytical solutions to Eq. 6.158 for flow past any object such as a sphere, cube, or airplane.

Very few practical fluid flow problems can be solved using an exact analytical approach.

Because of the difficulty in solving the Navier–Stokes equations, much attention has been given to various types of approximate solutions. For example, if the viscosity is set equal to zero, the Navier–Stokes equations reduce to Euler's equations. Thus, the frictionless fluid solutions discussed previously are actually approximate solutions to the Navier–Stokes equations. At the other extreme, for problems involving slowly moving fluids, viscous effects may be dominant and the nonlinear (convective) acceleration terms can be neglected. This assumption greatly simplifies the analysis, since the equations now become linear. There are numerous analytical solutions to these ''*slow flow*'' or ''*creeping flow*'' problems. Another broad class of approximate solutions is concerned with flow in the very thin boundary layer. L. Prandtl showed in 1904 how the Navier–Stokes equations could be simplified to study flow in boundary layers. Such ''boundary layer solutions'' play a very important role in the study of fluid mechanics. A further discussion of boundary layers is given in Chapter 9.

6.10.1 Numerical Methods

Numerical methods using digital computers are, of course, commonly utilized to solve a wide variety of flow problems. As discussed previously, although the differential equations that govern the flow of Newtonian fluids [the Navier-Stokes equations (6.158)] were derived many years ago, there are few known analytical solutions to them. With the advent of high-speed digital computers it has become possible to obtain approximate numerical solutions to these (and other fluid mechanics) equations for a wide variety of circumstances.

Computer-based numerical techniques are widely used to solve complex fluid flow problems.

Of the various techniques available for the numerical solution of the governing differential equations of fluid flow, the following three types are most common: (1) the finite difference method, (2) the finite element (or finite volume) method, and (3) the boundary element method. In each of these methods the continuous flow field (i.e., velocity or pressure as a function of space and time) is described in terms of discrete (rather than continuous) values at prescribed locations. By this technique the differential equations are replaced by a set of algebraic equations that can be solved on the computer.

For the finite element (or finite volume) method, the flow field is broken into a set of small fluid elements (usually triangular areas if the flow is two-dimensional, or small volume elements if the flow is three-dimensional). The conservation equations (i.e., conservation of mass, momentum, and energy) are written in an appropriate form for each element, and the set of resulting algebraic equations is solved numerically for the flow field. The number, size, and shape of the elements are dictated in part by the particular flow geometry and flow conditions for the problem at hand. As the number of elements increases (as is necessary for flows with complex boundaries), the number of simultaneous algebraic equations that must be solved increases rapidly. Problems involving 1000 to 10,000 elements and 50,000 equations are not uncommon. A mesh for calculating flow past an airfoil is shown in Fig. 6.35. Further information about this method can be found in Refs. 10 and 13.

V6.7

For the boundary element method, the boundary of the flow field (not the entire flow field as in the finite element method) is broken into discrete segments (Ref. 14), and appropriate singularities such as sources, sinks, doublets, and vortices are distributed on these

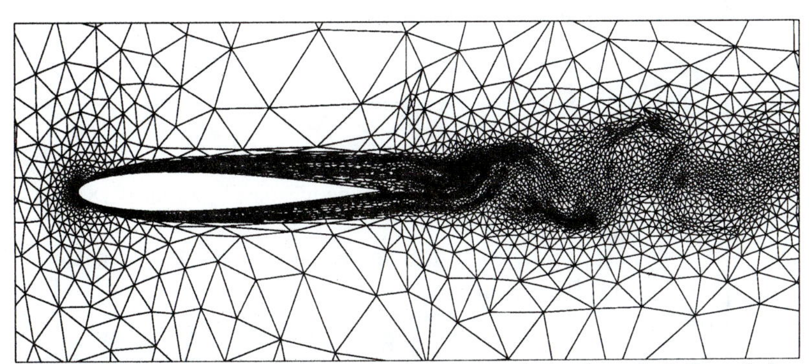

■ FIGURE 6.35 **Anisotropic adaptive mesh for the calculation of viscous flow over a NACA 0012 airfoil at a Reynolds number of 10,000, Mach number of 0.755, and angle of attack of 1.5°. (From CFD Laboratory, Concordia University, Montreal, Canada. Used by permission.)**

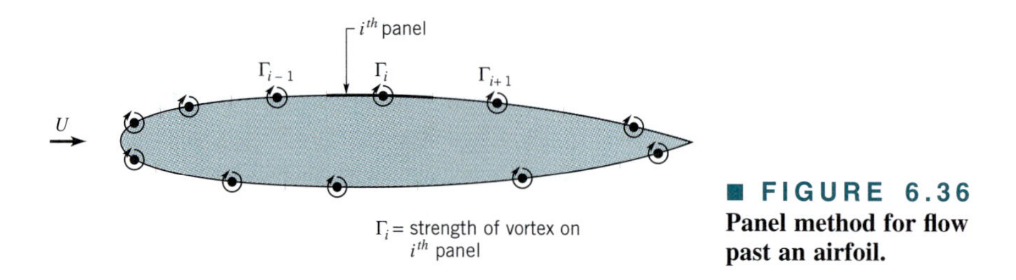

■ **FIGURE 6.36 Panel method for flow past an airfoil.**

boundary elements. The strength and type of the singularities are chosen so that the appropriate boundary conditions of the flow are obtained on the boundary elements. For points in the flow field not on the boundary, the flow is calculated by adding the contributions from the various singularities on the boundary. Although the details of this method are rather mathematically sophisticated, it may (depending on the particular problem) require less computational time and space than the finite element method.

Typical boundary elements and their associated singularities (vortices) for two-dimensional flow past an airfoil are shown in Fig. 6.36. Such use of the boundary element method in aerodynamics is often termed the *panel method* in recognition of the fact that each element plays the role of a panel on the airfoil surface (Ref. 15).

The finite-difference method is commonly used to solve, numerically, a variety of fluid flow problems.

The finite difference method for computational fluid dynamics is perhaps the most easily understood and widely used of the three methods listed above. For this method the flow field is dissected into a set of grid points and the continuous functions (velocity, pressure, etc.) are approximated by discrete values of these functions calculated at the grid points. Derivatives of the functions are approximated by using the differences between the function values at neighboring grid points divided by the grid spacing. The differential equations are thereby transferred into a set of algebraic equations, which is solved by appropriate numerical techniques. The larger the number of grid points used, the larger the number of equations that must be solved. It is usually necessary to increase the number of grid points (i.e., use a finer mesh) where large gradients are to be expected, such as in the boundary layer near a solid surface.

A very simple one-dimensional example of the finite difference technique is presented in the following example.

EXAMPLE 6.11

A viscous oil flows from a large, open tank and through a long, small-diameter pipe as shown in Fig. E6.11*a*. At time $t = 0$ the fluid depth is H. Use a finite difference technique to determine the liquid depth as a function of time, $h = h(t)$. Compare this result with the exact solution of the governing equation.

SOLUTION

Although this is an unsteady flow (i.e., the deeper the oil, the faster it flows from the tank) we assume that the flow is "quasisteady" (see Example 3.18) and apply steady flow equations as follows.

As shown by Eq. 6.152, the mean velocity, V, for steady laminar flow in a round pipe of diameter D is given by

$$V = \frac{D^2 \Delta p}{32 \mu \ell} \qquad \textbf{(1)}$$

where Δp is the pressure drop over the length ℓ. For this problem the pressure at the bottom of the tank (the inlet of the pipe) is γh and that at the pipe exit is zero. Hence, $\Delta p = \gamma h$ and

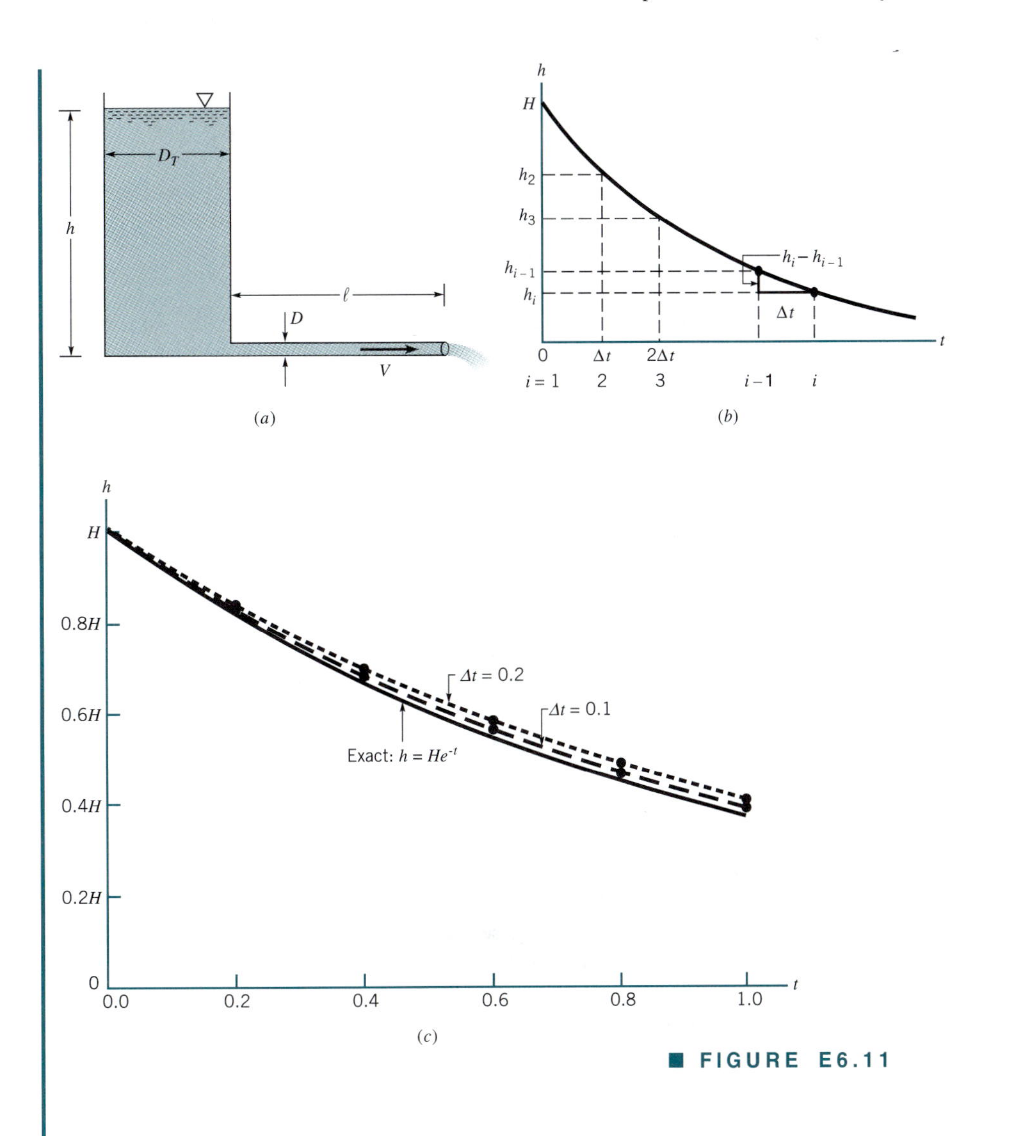

■ FIGURE E6.11

Eq. 1 becomes

$$V = \frac{D^2 \gamma h}{32 \mu \ell} \quad \textbf{(2)}$$

Conservation of mass requires that the flowrate from the tank, $Q = \pi D^2 V/4$, is related to the rate of change of depth of oil in the tank, dh/dt, by

$$Q = -\frac{\pi}{4} D_T^2 \frac{dh}{dt}$$

where D_T is the tank diameter. Thus

$$\frac{\pi}{4} D^2 V = -\frac{\pi}{4} D_T^2 \frac{dh}{dt}$$

or

$$V = -\left(\frac{D_T}{D}\right)^2 \frac{dh}{dt} \quad \textbf{(3)}$$

By combining Eqs. 2 and 3 we obtain

$$\frac{D^2 \gamma h}{32 \mu \ell} = -\left(\frac{D_T}{D}\right)^2 \frac{dh}{dt}$$

or

$$\frac{dh}{dt} = -Ch$$

where $C = \gamma D^4 / 32 \mu \ell D_T^2$ is a constant. For simplicity we assume the conditions are such that $C = 1$. Thus, we must solve

$$\frac{dh}{dt} = -h \quad \text{with} \quad h = H \text{ at } t = 0 \tag{4}$$

The exact solution to Eq. 4 is obtained by separating the variables and integrating to obtain

$$h = He^{-t} \tag{5}$$

However, assume this solution were not known. The following finite difference technique can be used to obtain an approximate solution.

As shown in Fig. E6.11*b*, we select discrete points (nodes or grid points) in time and approximate the time derivative of h by the expression

$$\left.\frac{dh}{dt}\right|_{t=t_i} \approx \frac{h_i - h_{i-1}}{\Delta t} \tag{6}$$

where Δt is the time step between the different node points on the time axis and h_i and h_{i-1} are the approximate values of h at nodes i and $i - 1$. Equation 6 is called the backward-difference approximation to dh/dt. We are free to select whatever value of Δt that we wish. (Although we do not need to space the nodes at equal distances, it is often convenient to do so.) Since the governing equation (Eq. 4) is an ordinary differential equation, the ''grid'' for the finite difference method is a one-dimensional grid as shown in Fig. E6.11*b* rather than a two-dimensional grid (which occurs for partial differential equations) as shown in Fig. E6.12*b*.

Thus, for each value of $i = 2, 3, 4, \ldots$ we can approximate the governing equation, Eq. 4, as

$$\frac{h_i - h_{i-1}}{\Delta t} = -h_i$$

or

$$h_i = \frac{h_{i-1}}{(1 + \Delta t)} \tag{7}$$

We cannot use Eq. 7 for $i = 1$ since it would involve the nonexisting h_0. Rather we use the initial condition (Eq. 4), which gives

$$h_1 = H$$

The result is the following set of N algebraic equations for the N approximate values of h at times $t_1 = 0, t_2 = \Delta t, \ldots, t_N = (\Delta t)^{N-1}$.

$$\begin{aligned} h_1 &= H \\ h_2 &= h_1/(1 + \Delta t) \\ h_3 &= h_2/(1 + \Delta t) \\ &\vdots \\ h_N &= h_{N-1}/(1 + \Delta t) \end{aligned}$$

For most problems the corresponding equations would be more complicated than those given above, and a computer would be used to solve for the h_i. For this problem the solution is simply

$$h_2 = H/(1 + \Delta t)$$
$$h_3 = H/(1 + \Delta t)^2$$
$$\vdots \qquad \vdots$$

or in general

$$h_i = H/(1 + \Delta t)^{i-1}$$

The results for $0 < t < 1$ are shown in Fig. E6.11*c*. Tabulated values of the depth for $t = 1$ are listed in the table below.

Δt	i for $t = 1$	h_i for $t = 1$
0.2	6	0.4019H
0.1	11	0.3855H
0.01	101	0.3697H
0.001	1001	0.3681H
Exact (Eq. 4)	—	0.3678H

It is seen that the approximate results compare quite favorably with the exact solution given by Eq. 5. It is expected that the finite difference results would more closely approximate the exact results as Δt is decreased since in the limit of $\Delta t \rightarrow 0$ the finite difference approximation for the derivatives (Eq. 6) approaches the actual definition of the derivative.

In general the governing equations to be solved are partial differential equations [rather than ordinary differential equations as in the above example (Eq. 4)] and the finite difference method becomes considerably more involved. The following example illustrates some of the concepts involved.

Consider steady, incompressible flow of an inviscid fluid past a circular cylinder as shown in Fig. E6.12*a*. The stream function, ψ, for this flow is governed by the Laplace equation (see Section 6.5)

$$\frac{\partial^2 \psi}{\partial x^2} + \frac{\partial^2 \psi}{\partial y^2} = 0 \tag{1}$$

The exact analytical solution is given in Section 6.6.3.

Describe a simple finite difference technique that can be used to solve this problem.

SOLUTION

The first step is to define a flow domain and set up an appropriate grid for the finite difference scheme. Since we expect the flow field to be symmetrical both above and below and in front of and behind the cylinder, we consider only one quarter of the entire flow domain as indicated in Fig. E6.12*b*. We locate the upper boundary and right-hand boundary far enough from the

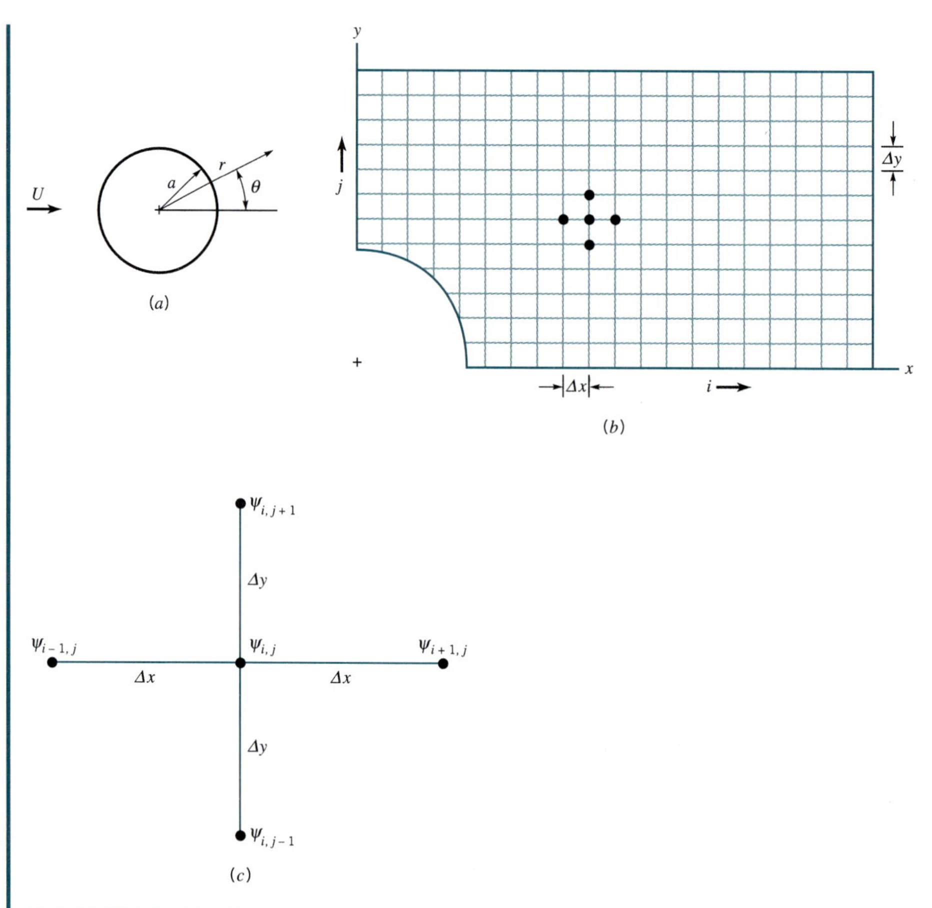

■ FIGURE E6.12

cylinder so that we expect the flow to be essentially uniform at these locations. It is not always clear how far from the object these boundaries must be located. If they are not far enough, the solution obtained will be incorrect because we have imposed artificial, uniform flow conditions at a location where the actual flow is not uniform. If these boundaries are farther than necessary from the object, the flow domain will be larger than necessary and excessive computer time and storage will be required. Experience in solving such problems is invaluable!

Once the flow domain has been selected, an appropriate grid is imposed on this domain (see Fig. E6.12*b*). Various grid structures can be used. If the grid is too coarse, the numerical solution may not be capable of capturing the fine scale structure of the actual flow field. If the grid is too fine, excessive computer time and storage may be required. Considerable work has gone into forming appropriate grids (Ref. 16). We consider a grid that is uniformly spaced in the x and y directions as shown in Fig. E6.12*b*.

As shown in Eq. 6.112, the exact solution to Eq. 1 (in terms of polar coordinates r, θ rather than Cartesian coordinates x, y) is $\psi = Ur(1 - a^2/r^2) \sin \theta$. The finite difference solution approximates these stream function values at a discrete (finite) number of locations (the grid points) as $\psi_{i,j}$, where the i and j indices refer to the corresponding x_i and y_j locations.

The derivatives of ψ can be approximated as follows

$$\frac{\partial \psi}{\partial x} \approx \frac{1}{\Delta x}\,(\psi_{i+1,j} - \psi_{i,j})$$

and

$$\frac{\partial \psi}{\partial y} \approx \frac{1}{\Delta y} (\psi_{i,j+1} - \psi_{i,j})$$

This particular approximation is called a forward-difference approximation. Other approximations are possible. By similar reasoning, it is possible to show that the second derivatives of ψ can be written as follows

$$\frac{\partial^2 \psi}{\partial x^2} \approx \frac{1}{(\Delta x)^2} (\psi_{i+1,j} - 2\psi_{i,j} + \psi_{i-1,j}) \tag{2}$$

and

$$\frac{\partial^2 \psi}{\partial y^2} \approx \frac{1}{(\Delta y)^2} (\psi_{i,j+1} - 2\psi_{i,j} + \psi_{i,j-1}) \tag{3}$$

Thus, by combining Eqs. 1, 2, and 3 we obtain

$$\frac{\partial^2 \psi}{\partial x^2} + \frac{\partial^2 \psi}{\partial y^2} \approx \frac{1}{(\Delta x)^2} (\psi_{i+1,j} + \psi_{i-1,j}) + \frac{1}{(\Delta y)^2} (\psi_{i,j+1} + \psi_{i,j-1}) - 2\left(\frac{1}{(\Delta x)^2} + \frac{1}{(\Delta y)^2}\right) \psi_{i,j} = 0 \tag{4}$$

Equation 4 can be solved for the stream function at x_i and y_j to give

$$\psi_{i,j} = \frac{1}{2[(\Delta x)^2 + (\Delta y)^2]} [(\Delta y)^2(\psi_{i+1,j} + \psi_{i-1,j}) + (\Delta x)^2(\psi_{i,j+1} + \psi_{i,j-1})] \tag{5}$$

Note that the value of $\psi_{i,j}$ depends on the values of the stream function at neighboring grid points on either side and above and below the point of interest (see Eq. 5 and Fig. E6.12*c*).

To solve the problem (either exactly or by the finite difference technique) it is necessary to specify boundary conditions for points located on the boundary of the flow domain (see Section 6.6.3). For example we may specify that $\psi = 0$ on the lower boundary of the domain (see Fig. E6.12*b*) and $\psi = C$, a constant, on the upper boundary of the domain. Appropriate boundary conditions on the two vertical ends of the flow domain can also be specified. Thus, for points interior to the boundary Eq. 5 is valid; similar equations or specified values of $\psi_{i,j}$ are valid for boundary points. The result is an equal number of equations and unknowns, $\psi_{i,j}$, one for every grid point. For this problem, these equations represent a set of linear algebraic equations for $\psi_{i,j}$, the solution of which provides the finite difference approximation for the stream function at discrete grid points in the flow field. Streamlines (lines of constant ψ) can be obtained by interpolating values of $\psi_{i,j}$ between the grid points and "connecting the dots" of $\psi =$ constant. The velocity field can be obtained from the derivatives of the stream function according to Eq. 6.74. That is

$$u = \frac{\partial \psi}{\partial y} \approx \frac{1}{\Delta y} (\psi_{i,j+1} - \psi_{i,j})$$

and

$$v = -\frac{\partial \psi}{\partial x} \approx -\frac{1}{\Delta x} (\psi_{i+1,j} - \psi_{i,j})$$

Further details of the finite difference technique are beyond the scope of this text but can be found in standard references on the topic (Refs. 11 and 12).

The above two examples are rather simple because the governing equations are not too complex. A finite difference solution of the more complicated, nonlinear Navier–Stokes equation (Eq. 6.158) requires considerably more effort and insight and larger and faster computers. A typical finite difference grid for the flow past a turbine blade is shown in Fig. 6.37. Note that the mesh is much finer in regions where large gradients are to be expected (i.e., near the leading and trailing edges of the blade) and more coarse away from the blade.

In Fig. 6.38 are shown the streamlines for viscous flow past a circular cylinder at a given instant after it was impulsively started from rest. The lower half of the figure represents the results of a finite difference calculation; the upper half of the figure is a flow visualization photograph of the same flow situation. It is clear that the numerical and experimental results agree quite well.

Any numerical technique (including those discussed above), no matter how simple in concept, contains many hidden subtleties and potential problems. For example, it may seem reasonable that a finer discretization (i.e., smaller elements of finer grid spacing) would ensure a more accurate numerical solution. While this may sometimes be so (as in Example 6.11), it is not always so; a variety of stability or convergence problems may occur. In such cases the numerical ''solution'' obtained may exhibit unreasonable ''wiggles'' or the numerical result may ''diverge'' to an unreasonable (and incorrect) result.

Computational fluid dynamics (CFD) represents an extremely important area in advanced fluid mechanics.

A great deal of care must be used in obtaining approximate numerical solutions to the governing equations of fluid motion. The process is not as simple as the often-heard ''just let the computer do it.'' The general field of computational fluid dynamics (CFD), in which computers and numerical analysis are combined to solve fluid flow problems, represents an extremely important subject area in advanced fluid mechanics. Considerable progress has been made in the past relatively few years, but much remains to be done. The reader is encouraged to consult some of the available literature.

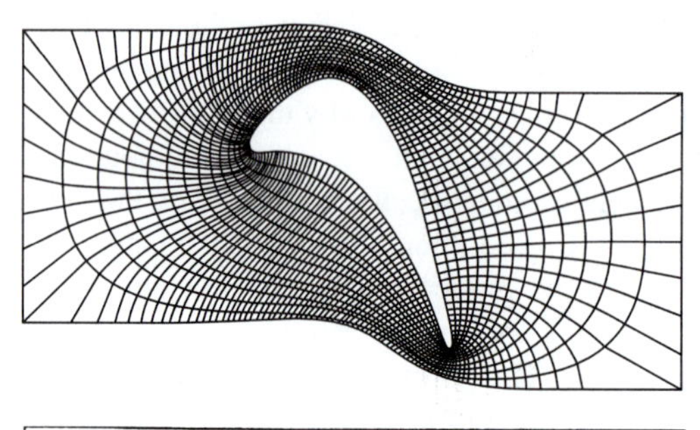

■ **FIGURE 6.37 Finite difference grid for flow past a turbine blade. (From Ref. 17, used by permission.)**

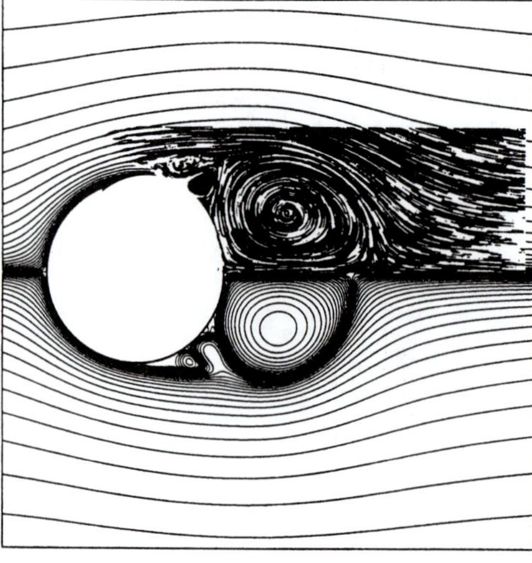

■ **FIGURE 6.38 Streamlines for flow past a circular cylinder at a short time after the flow was impulsively started. The upper half is a photograph from a flow visualization experiment. The lower half is from a finite difference calculation. (From Ref. 17, used by permission.)**

References

1. White, F. M., *Fluid Mechanics*, 2nd Ed., McGraw-Hill, New York, 1986.
2. Streeter, V. L., *Fluid Dynamics*, McGraw-Hill, New York, 1948.
3. Rouse, H., *Advanced Mechanics of Fluids*, Wiley, New York, 1959.
4. Milne-Thomson, L. M., *Theoretical Hydrodynamics*, 4th Ed., Macmillan, New York, 1960.
5. Robertson, J. M., *Hydrodynamics in Theory and Application*, Prentice-Hall, Englewood Cliffs, N.J., 1965.
6. Panton, R. L., *Incompressible Flow*, Wiley, New York, 1984.
7. Li, W. H., and Lam, S. H., *Principles of Fluid Mechanics*, Addison-Wesley, Reading, Mass., 1964.
8. Schlichting, H., *Boundary-Layer Theory*, 7th Ed., McGraw-Hill, New York, 1979.
9. Fuller, D. D., *Theory and Practice of Lubrication for Engineers*, Wiley, New York, 1984.
10. Baker, A. J., *Finite Element Computational Fluid Mechanics*, McGraw-Hill, New York, 1983.
11. Peyret, R., and Taylor, T. D., *Computational Methods for Fluid Flow*, Springer-Verlag, New York, 1983.
12. Tannehill, J. C., Anderson, D. A., and Pletcher, R. H., *Computational Fluid Mechanics and Heat Transfer*, 2nd Ed., Taylor and Francis, Washington, D.C., 1997.
13. Carey, G. F., and Oden, J. T., *Finite Elements: Fluid Mechanics*, Prentice-Hall, Englewood Cliffs, N.J., 1986.
14. Brebbia, C. A., and Dominguez, J., *Boundary Elements: An Introductory Course*, McGraw-Hill, New York, 1989.
15. Moran, J., *An Introduction to Theoretical and Computational Aerodynamics*, Wiley, New York, 1984.
16. Thompson, J. F., Warsi, Z. U. A., and Mastin, C. W., *Numerical Grid Generation: Foundations and Applications*, North-Holland, New York, 1985.
17. Hall, E. J., and Pletcher, R. H., *Simulation of Time Dependent, Compressible Viscous Flow Using Central and Upwind-Biased Finite-Difference Techniques*, Technical Report HTL-52, CFD-22, College of Engineering, Iowa State Unversity, 1990.

Review Problems

Note: Problems designated with (R) are review problems. The phrases within parentheses refer to the main topics to be used in solving the problems. Complete, detailed solutions to these review problems can be found in the supplement titled *Student Solution Manual for Fundamentals of Fluid Mechanics*, by Munson, Young, and Okiishi (John Wiley and Sons, New York, 1997).

6.1R (Acceleration) The velocity in a certain flow field is given by the equation

$$\mathbf{V} = 3yz^2\hat{\mathbf{i}} + xz\hat{\mathbf{j}} + y\hat{\mathbf{k}}$$

Determine the expressions for the three rectangular components of acceleration.

(ANS: $3xz^3 + 6y^2z$; $3yz^3 + xy$; xz)

6.2R (Vorticity) Determine an expression for the vorticity of the flow field described by

$$\mathbf{V} = x^2y\hat{\mathbf{i}} - xy^2\hat{\mathbf{j}}$$

Is the flow irrotational?

(ANS: $-(x^2 + y^2)\,\hat{\mathbf{k}}$; no)

6.3R (Conservation of mass) For a certain incompressible, two-dimensional flow field the velocity component in the y direction is given by the equation

$$v = x^2 + 2xy$$

Determine the velocity component in the x direction so that the continuity equation is satisfied.

(ANS: $-x^2 + f(y)$)

6.4R (Conservation of mass) For a certain incompressible flow field it is suggested that the velocity components are given by the equations

$$u = x^2y \qquad v = 4y^3z \qquad w = 2z$$

Is this a physically possible flow field? Explain.

(ANS: No)

6.5R (Stream function) The velocity potential for a certain flow field is

$$\phi = 4xy$$

Determine the corresponding stream function.

(ANS: $2(y^2 - x^2) + C$)

6.6R (Velocity potential) A two-dimensional flow field is formed by adding a source at the origin of the coordinate system to the velocity potential

$$\phi = r^2 \cos 2\theta$$

Locate any stagnation points in the upper half of the coordinate plane ($0 \leq \theta \leq \pi$).

(ANS: $\theta_s = \pi/2$; $r_s = (m/4\pi)^{1/2}$)

6.7R (Potential flow) The stream function for a two-dimensional, incompressible flow field is given by the equation

$$\psi = 2x - 2y$$

where the stream function has the units of ft^2/s with x and y in feet. **(a)** Sketch the streamlines for this flow field. Indicate the direction of flow along the streamlines. **(b)** Is this an irrotational flow field? **(c)** Determine the acceleration of a fluid particle at the point $x = 1$ ft, $y = 2$ ft.

(ANS: yes; no acceleration)

6.8R (Inviscid flow) In a certain steady, incompressible, inviscid, two-dimensional flow field ($w = 0$, and all variables independent of z) the x component of velocity is given by the equation:

$$u = x^2 - y$$

Will the corresponding pressure gradient in the horizontal x direction be a function only of x, only of y, or of both x and y? Justify your answer.

(ANS: only of x)

6.9R (Inviscid flow) The stream function for the flow of a nonviscous, incompressible fluid in the vicinity of a corner (Fig. P6.9R) is

$$\psi = 2r^{4/3} \sin \tfrac{4}{3}\theta$$

Determine an expression for the pressure *gradient* along the boundary $\theta = 3\pi/4$.

(ANS: $-64\ \rho/27\ r^{1/3}$)

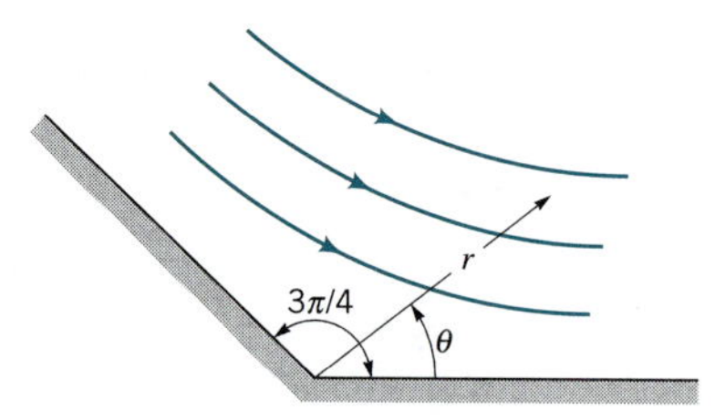

■ FIGURE P6.9R

6.10R (Potential flow) A certain body has the shape of a half-body with a thickness of 0.5 m. If this body is to be placed in an airstream moving at 20 m/s, what source strength is required to simulate flow around the body?

(ANS: $10.0\ m^2/s$)

6.11R (Potential flow) A source and a sink are located along the x axis with the source at $x = -1$ ft and the sink at $x = 1$ ft. Both the source and the sink have a strength of $10\ ft^2/s$. Determine the location of the stagnation points along the x axis when this source-sink pair is combined with a uniform velocity of 20 ft/s in the positive x direction.

(ANS: ± 1.08 ft)

6.12R (Viscous flow) In a certain viscous, incompressible flow field with zero body forces the velocity components are

$$u = ay - b(cy - y^2)$$
$$v = w = 0$$

where a, b, and c are constant. **(a)** Use the Navier–Stokes equations to determine an expression for the pressure gradient in the x direction. **(b)** For what combination of the constants a, b, and c (if any) will the shearing stress, τ_{yx}, be zero at $y = 0$ where the velocity is zero?

(ANS: $2b\mu$; $a = bc$)

6.13R (Viscous flow) A viscous fluid is contained between two infinite, horizontal parallel plates that are spaced 0.5 in. apart. The bottom plate is fixed, and the upper plate moves with a constant velocity, U. The fluid motion is caused by the movement of the upper plate, and there is no pressure gradient in the direction of flow. The flow is laminar. If the velocity of the upper plate is 2 ft/s and the fluid has a viscosity of $0.03\ lb{\cdot}s/ft^2$ and a specific weight of $70\ lb/ft^3$, what is the required horizontal force per square foot on the upper plate to maintain the 2 ft/s velocity? What is the pressure differential in the fluid between the top and bottom plates?

(ANS: $1.44\ lb/ft^2$; $2.92\ lb/ft^2$)

6.14R (Viscous flow) A viscous liquid ($\mu = 0.016$ lb·s/ft^2, $\rho = 1.79$ slugs/ft^3) flows through the annular space between two horizontal, fixed, concentric cylinders. If the radius of the inner cylinder is 1.5 in. and the radius of the outer cylinder is 2.5 in., what is the volume flowrate when the pressure drop along the axis of the annulus is 100 lb/ft^2 per ft?

(ANS: 0.317 ft^3/s)

6.15R (Viscous flow) Consider the steady, laminar flow of an incompressible fluid through the horizontal rectangular channel of Fig. P6.15R. Assume that the velocity components in the x and y directions are zero and the only body force is the weight. Start with the Navier–Stokes equations. **(a)** Determine the appropriate set of differential equations and boundary conditions for this problem. You need not solve the equations. **(b)** Show that the pressure distribution is hydrostatic at any particular cross section.

(ANS: $\partial p/\partial x = 0$; $\partial p/\partial y = -\rho g$; $\partial p/\partial z = \mu(\partial^2 w/\partial x^2 + \partial^2 w/\partial y^2)$ with $w = 0$ for $x = \pm b/2$ and $y = \pm a/2$)

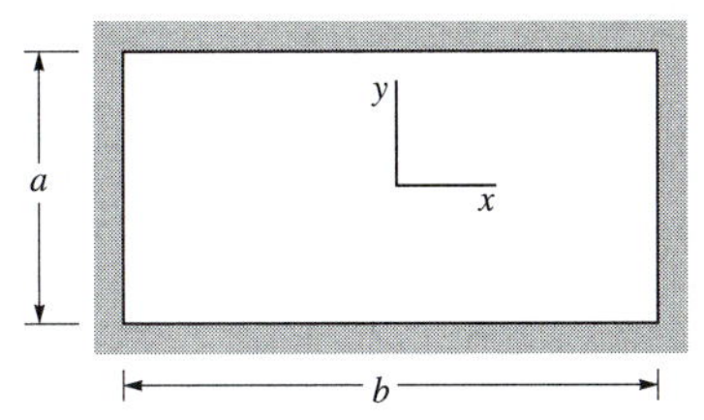

■ FIGURE P6.15R

6.16R (Viscous flow) A viscous liquid, having a viscosity of 10^{-4} lb·s/ft^2 and a specific weight of 50 lb/ft^3, flows steadily through the 2-in.-diameter, horizontal, smooth pipe shown in Fig. P6.16R. The mean velocity in the pipe is 0.5 ft/s. Determine the differential reading, Δh, on the inclined-tube manometer.

(ANS: 0.0640 ft)

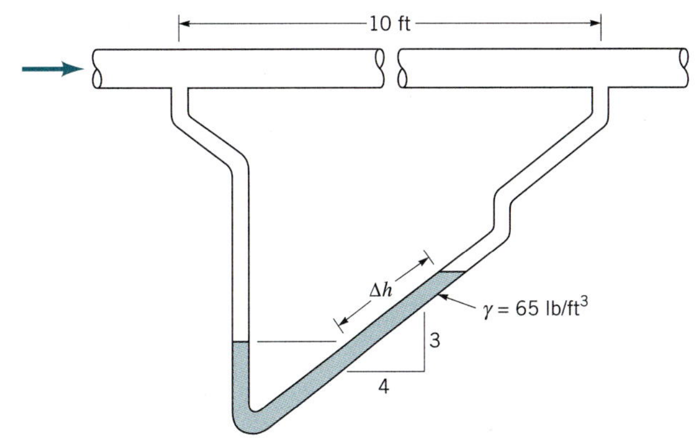

■ FIGURE P6.16R

Problems

Note: Unless otherwise indicated, use the values of fluid properties found in the tables on the inside of the front cover. Problems designated with an (*) are intended to be solved with the aid of a programmable calculator or a computer. Problems designated with a (†) are "open-ended" problems and require critical thinking in that to work them one must make various assumptions and provide the necessary data. There is not a unique answer to these problems.

6.1 The velocity in a certain two-dimensional flow field is given by the equation

$$\mathbf{V} = 2xt\hat{\mathbf{i}} - 2yt\hat{\mathbf{j}}$$

where the velocity is in ft/s when x, y, and t are in feet and seconds, respectively. Determine expressions for the local and convective components of acceleration in the x and y directions. What is the magnitude and direction of the velocity and the acceleration at the point $x = y = 2$ ft at the time $t = 0$?

6.2 Repeat Problem 6.1 if the flow field is described by the equation

$$\mathbf{V} = 3(x^2 - y^2)\hat{\mathbf{i}} - 6xy\hat{\mathbf{j}}$$

where the velocity is in ft/s when x and y are in feet.

6.3 The velocity in a certain flow field is given by the equation

$$\mathbf{V} = yz\hat{\mathbf{i}} + x^2z\hat{\mathbf{j}} + x\hat{\mathbf{k}}$$

Determine the expressions for the three rectangular components of acceleration.

6.4 The three components of velocity in a flow field are given by

$$u = x^2 + y^2 + z^2$$
$$v = xy + yz + z^2$$
$$w = -3xz - z^2/2 + 4$$

(a) Determine the volumetric dilatation rate and interpret the results. **(b)** Determine an expression for the rotation vector. Is this an irrotational flow field?

6.5 Determine an expression for the vorticity of the flow field described by

$$\mathbf{V} = -4xy^3\,\hat{\mathbf{i}} + y^4\hat{\mathbf{j}}$$

Is the flow irrotational?

6.6 A one-dimensional flow is described by the velocity field

$$u = ay + by^2$$

$$v = w = 0$$

where a and b are constants. Is the flow irrotational? For what combination of constants (if any) will the rate of angular deformation as given by Eq. 6.18 be zero?

6.7 For incompressible fluids the volumetric dilatation rate must be zero; that is, $\nabla \cdot \mathbf{V} = 0$. For what combination of constants, a, b, c, and e can the velocity components

$$u = ax + by$$

$$v = cx + ey$$

$$w = 0$$

be used to describe an incompressible flow field?

6.8 An incompressible viscous fluid is placed between two large parallel plates as shown in Fig. P6.8. The bottom plate is fixed and the upper plate moves with a constant velocity, U. For these conditions the velocity distribution between the plates is linear and can be expressed as

$$u = U\frac{y}{b}$$

Determine: **(a)** the volumetric dilatation rate, **(b)** the rotation vector, **(c)** the vorticity, and **(d)** the rate of angular deformation.

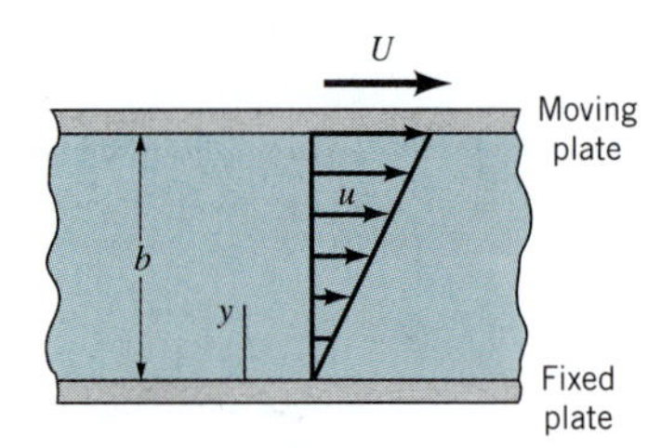

■ **FIGURE P6.8**

6.9 A two-dimensional flow field described by

$$\mathbf{V} = (2x^2y + x)\hat{\mathbf{i}} + (2xy^2 + y + 1)\hat{\mathbf{j}}$$

where the velocity is in m/s when x and y are in meters. Determine the angular rotation of a fluid element located at $x = 0.5$ m, $y = 1.0$ m.

6.10 Some velocity measurements in a three-dimensional incompressible flow field indicate that $u = 6xy^2$ and $v = -4y^2z$. There is some conflicting data for the velocity component in the z direction. One set of data indicates that $w = 4yz^2$ and the other set indicates that $w = 4yz^2 - 6y^2z$. Which set do you think is correct? Explain.

6.11 The velocity components of an incompressible, two-dimensional velocity field are given by the equations

$$u = 2xy$$

$$v = x^2 - y^2$$

Show that the flow is irrotational and satisfies conservation of mass.

6.12 The velocity components in an incompressible, two-dimensional flow field are given by the equations

$$u = x^2$$

$$v = -2xy + x$$

Determine, if possible, the corresponding stream function.

6.13 The stream function for a certain incompressible flow field is given by the equation

$$\psi = 2x^2y - \tfrac{2}{3}y^3$$

Show that the velocity field represented by this stream function satisfies the continuity equation.

6.14 The stream function for an incompressible, two-dimensional flow field is

$$\psi = ay^2 - bx$$

where a and b are constants. Is this an irrotational flow? Explain.

6.15 The velocity components for an incompressible, plane flow are

$$v_r = Ar^{-1} + Br^{-2}\cos\theta$$

$$v_\theta = Br^{-2}\sin\theta$$

where A and B are constants. Determine the corresponding stream function.

6.16 For a certain two-dimensional flow field

$$u = 0$$

$$v = V$$

(a) What are the corresponding radial and tangential velocity components? **(b)** Determine the corresponding stream function expressed in Cartesian coordinates and in cylindrical polar coordinates.

6.17 Make use of the control volume shown in Fig. P6.17 to derive the continuity equation in cylindrical coordinates (Eq. 6.33 in text).

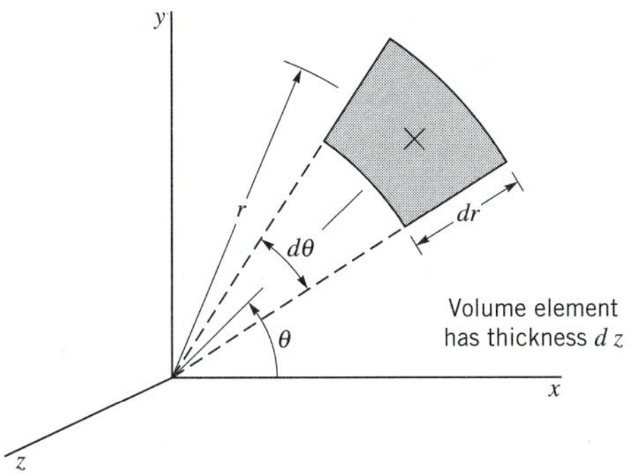

■ FIGURE P6.17

6.18 It is proposed that a two-dimensional, incompressible flow field be described by the velocity components

$$u = Ay$$
$$v = Bx$$

where A and B are both positive constants. **(a)** Will the continuity equation be satisfied? **(b)** Is the flow irrotational? **(c)** Determine the equation for the streamlines and show a sketch of the streamline that passes through the origin. Indicate the direction of flow along this streamline.

6.19 In a certain steady, two-dimensional flow field the fluid density varies linearly with respect to the coordinate x; that is, $\rho = Ax$ where A is a constant. If the x component of velocity u is given by the equation $u = y$, determine an expression for v.

6.20 In a two-dimensional, incompressible flow field, the x component of velocity is given by the equation $u = 2x$. **(a)** Determine the corresponding equation for the y component of velocity if $v = 0$ along the x axis. **(b)** For this flow field, what is the magnitude of the average velocity of the fluid crossing the surface OA of Fig. P6.20? Assume that the velocities are in feet per second when x and y are in feet.

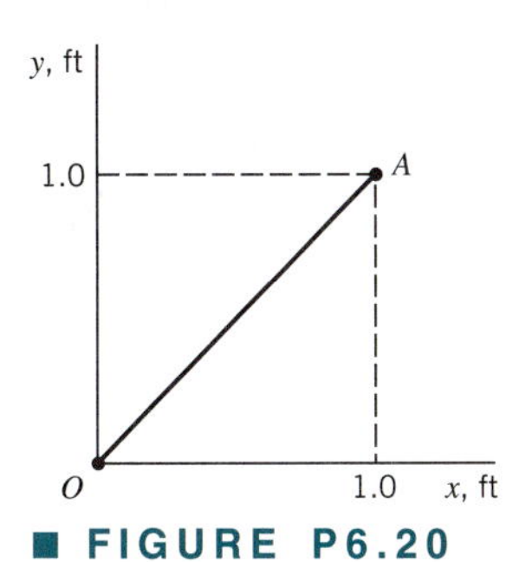

■ FIGURE P6.20

6.21 The radial velocity component in an incompressible, two-dimensional flow field ($v_z = 0$) is

$$v_r = 2r + 3r^2 \sin \theta$$

Determine the corresponding tangential velocity component, v_θ, required to satisfy conservation of mass.

6.22 The stream function for an incompressible flow field is given by the equation

$$\psi = 3x^2y - y^3$$

where the stream function has the units of m^2/s with x and y in meters. **(a)** Sketch the streamline(s) passing through the origin. **(b)** Determine the rate of flow across the straight path AB shown in Fig. P6.22.

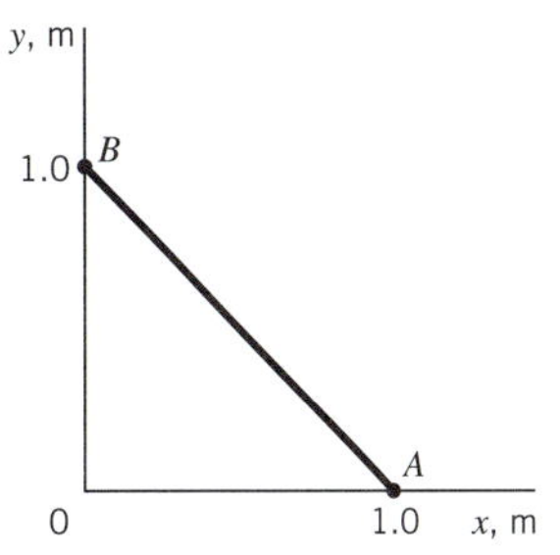

■ FIGURE P6.22

6.23 The streamlines in a certain incompressible, two-dimensional flow field are all concentric circles so that $v_r = 0$. Determine the stream function for **(a)** $v_\theta = Ar$ and for **(b)** $v_\theta = Ar^{-1}$, where A is a constant.

***6.24** The stream function for an incompressible, two-dimensional flow field is

$$\psi = 3x^2y + y$$

For this flow field, plot several streamlines.

***6.25** The stream function for an incompressible, two-dimensional flow field is

$$\psi = 2r^3 \sin 3\theta$$

For this flow field, plot several streamlines for $0 \leq \theta \leq \pi/3$.

6.26 A two-dimensional flow field for a nonviscous, incompressible fluid is described by the velocity components

$$u = U_0 + 2y$$
$$v = 0$$

where U_0 is a constant. If the pressure at the origin (Fig. P6.26) is p_0, determine an expression for the pressure at **(a)** point A, and **(b)** point B. Explain clearly how you obtained your answer. Assume the units are consistent and body forces may be neglected.

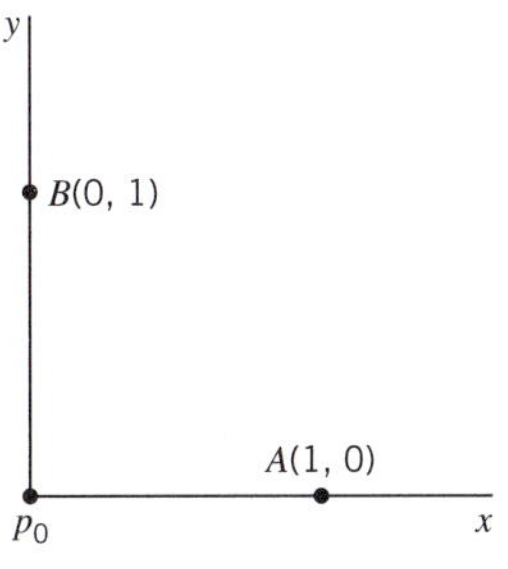

■ FIGURE P6.26

6.27 In a certain two-dimensional flow field the velocity is constant with components $u = -4$ ft/s and $v = -2$ ft/s. Determine the corresponding stream function and velocity potential for this flow field. Sketch the equipotential line $\phi = 0$ which passes through the origin of the coordinate system.

6.28 The velocity potential for a given two-dimensional flow field is

$$\phi = (\tfrac{5}{3})x^3 - 5xy^2$$

Show that the continuity equation is satisfied and determine the corresponding stream function.

6.29 Determine the stream function corresponding to the velocity potential

$$\phi = x^3 - 3xy^2$$

Sketch the streamline $\psi = 0$, which passes through the origin.

6.30 A certain flow field is described by the velocity potential

$$\phi = A \ln r + Br \cos \theta$$

where A and B are positive constants. Determine the corresponding stream function and locate any stagnation points in this flow field.

6.31 It is known that the velocity distribution for two-dimensional flow of a viscous fluid between wide parallel plates (Fig. P6.31) is parabolic; that is

$$u = U_c\left[1 - \left(\frac{y}{h}\right)^2\right]$$

with $v = 0$. Determine, if possible, the corresponding stream function and velocity potential.

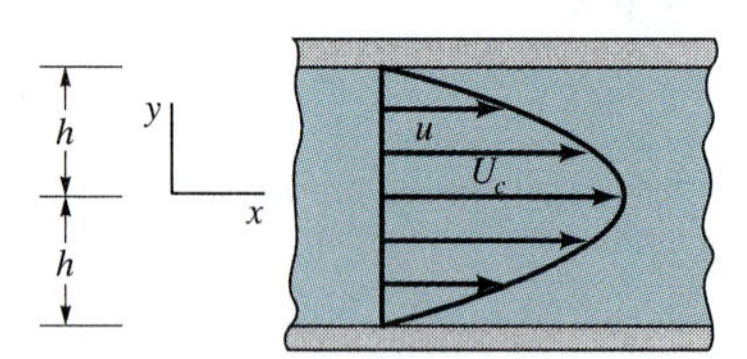

■ **FIGURE P6.31**

6.32 The velocity potential for a certain inviscid flow field is

$$\phi = -(3x^2y - y^3)$$

where ϕ has the units of ft²/s when x and y are in feet. Determine the pressure difference (in psi) between the points (1, 2) and (4, 4), where the coordinates are in feet, if the fluid is water and elevation changes are negligible.

6.33 Consider the incompressible, two-dimensional flow of a nonviscous fluid between the boundaries shown in Fig. P6.33. The velocity potential for this flow field is

$$\phi = x^2 - y^2$$

(a) Determine the corresponding stream function. **(b)** What is the relationship between the discharge, q (per unit width normal to plane of paper), passing between the walls and the coordinates x_i, y_i of any point on the curved wall? Neglect body forces.

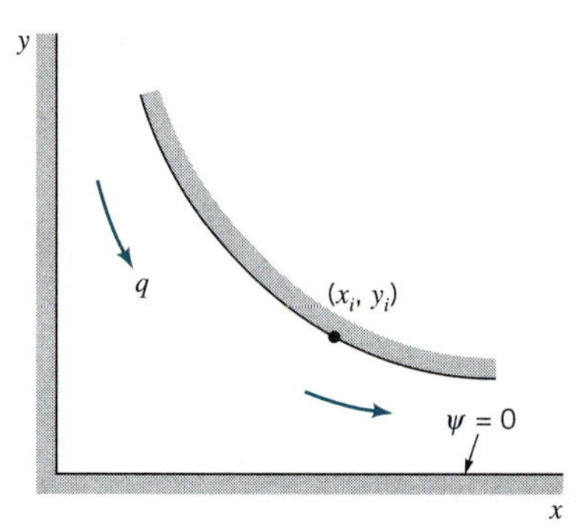

■ **FIGURE P6.33**

6.34 The stream function for a two-dimensional, nonviscous, incompressible flow field is given by the expression

$$\psi = -2(x - y)$$

where the stream function has the units of ft²/s with x and y in feet. **(a)** Is the continuity equation satisfied? **(b)** Is the flow field irrotational? If so, determine the corresponding velocity potential. **(c)** Determine the pressure gradient in the horizontal x direction at the point $x = 2$ ft, $y = 2$ ft.

6.35 In a certain steady, two-dimensional flow field the fluid may be assumed to be ideal and the weight of the fluid (specific weight $= 50$ lb/ft³) is the only body force. The x component of velocity is known to be $u = 6x$ which gives the velocity in ft/s when x is measured in feet, and the y component of velocity is known to be a function of only y. The y axis is vertical, and at the origin the velocity is zero. **(a)** Determine the y component of velocity so that the continuity equation is satisfied. **(b)** Can the difference in pressures between the points $x = 1$ ft, $y = 1$ ft and $x = 1$ ft, $y = 4$ ft be determined from the Bernoulli equation? If so, determine the value in lb/ft². If not, explain why not.

6.36 The velocity potential for a certain inviscid, incompressible flow field is given by the equation

$$\phi = 2x^2y - (\tfrac{2}{3})y^3$$

where ϕ has the units of m²/s when x and y are in meters. Determine the pressure at the point $x = 2$ m, $y = 2$ m if the pressure at $x = 1$ m, $y = 1$ m is 200 kPa. Elevation changes can be neglected and the fluid is water.

6.37 The velocity components in an ideal, two-dimensional velocity field are given by the equations

$$u = 3(x^2 - y^2)$$
$$v = -6xy$$

All body forces are negligible. **(a)** Does this velocity field satisfy the continuity equation? **(b)** Determine the equation for the pressure gradient in the y direction at any point in the field.

6.38 The streamlines for an incompressible, inviscid, two-dimensional flow field are all concentric circles and the velocity varies directly with the distance from the common center of the streamlines; that is

$$v_\theta = Kr$$

where K is a constant. **(a)** For this *rotational* flow, determine, if possible, the stream function. **(b)** Can the pressure difference between the origin and any other point be determined from the Bernoulli equation? Explain.

6.39 The velocity potential

$$\phi = -k(x^2 - y^2) \qquad (k = \text{constant})$$

may be used to represent the flow against an infinite plane boundary as illustrated in Fig. P6.39. For flow in the vicinity of a stagnation point, it is frequently assumed that the pressure gradient along the surface is of the form

$$\frac{\partial p}{\partial x} = Ax$$

where A is a constant. Use the given velocity potential to show that this is true.

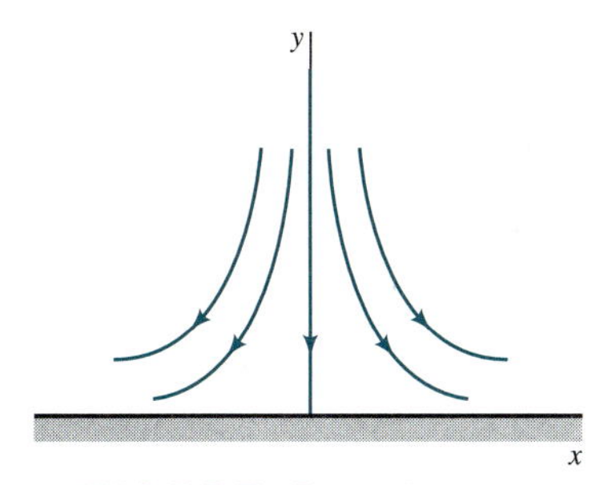

■ **FIGURE P6.39**

6.40 Water flows through a two-dimensional diffuser having a 20° expansion angle as shown in Fig. P6.40. Assume that the flow in the diffuser can be treated as a radial flow emanating from a source at the origin O. **(a)** If the velocity at the entrance is 20 m/s, determine an expression for the pressure gradient along the diffuser walls. **(b)** What is the pressure rise between the entrance and exit?

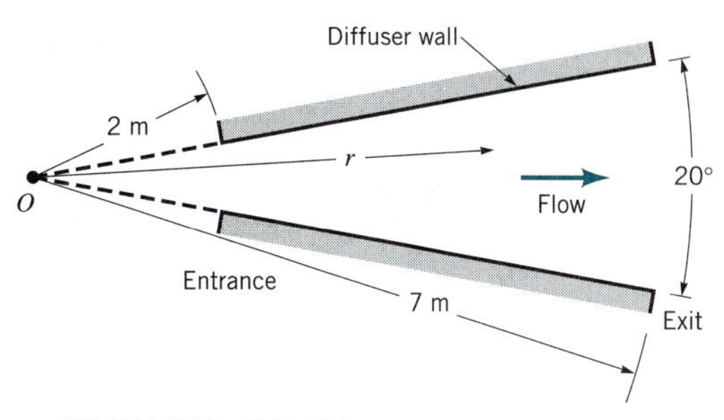

■ **FIGURE P6.40**

6.41 An ideal fluid flows between the inclined walls of a two-dimensional channel into a sink located at the origin (Fig. P6.41). The velocity potential for this flow field is

$$\phi = \frac{m}{2\pi} \ln r$$

where m is a constant. **(a)** Determine the corresponding stream function. Note that the value of the stream function along the wall OA is zero. **(b)** Determine the equation of the streamline passing through the point B, located at $x = 1$, $y = 4$.

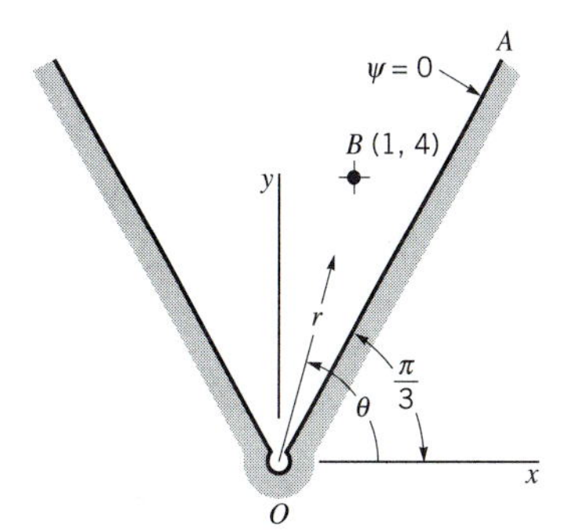

■ **FIGURE P6.41**

6.42 It is suggested that the velocity potential for the flow of an incompressible, nonviscous, two-dimensional flow along the wall shown in Fig. P6.42 is

$$\phi = r^{4/3} \cos \tfrac{4}{3}\theta$$

Is this a suitable velocity potential for flow along the wall? Explain.

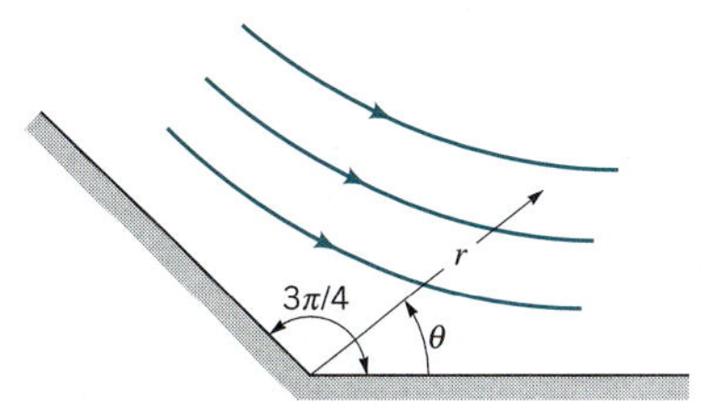

■ **FIGURE P6.42**

6.43 As illustrated in Fig. P6.43 a tornado can be approximated by a free vortex of strength Γ for $r > R_c$, where R_c is the radius of the core. Velocity measurements at points A and B indicate that $V_A = 125$ ft/s and $V_B = 75$ ft/s. Determine the distance from point A to the center of the tornado. Why can the free vortex model not be used to approximate the tornado throughout the flow field ($r \geq 0$)?

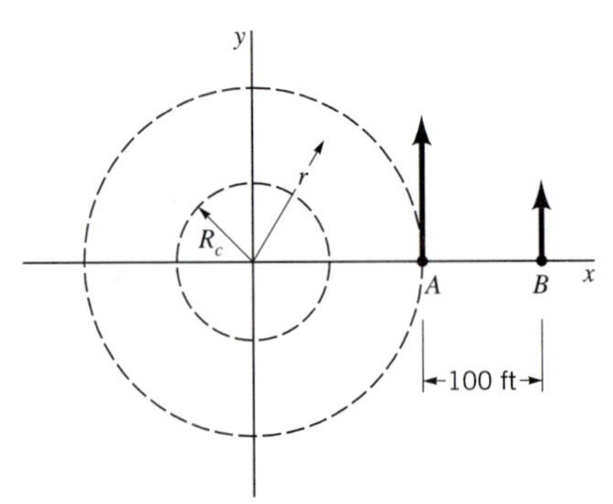

■ FIGURE P6.43

6.44 The velocity distribution in a horizontal, two-dimensional bend through which an ideal fluid flows can be approximated with a free vortex as shown in Fig. P6.44. Show how the discharge (per unit width normal to plane of paper) through the channel can be expressed as

$$q = C\sqrt{\frac{\Delta p}{\rho}}$$

where $\Delta p = p_B - p_A$. Determine the value of the constant C for the bend dimensions given.

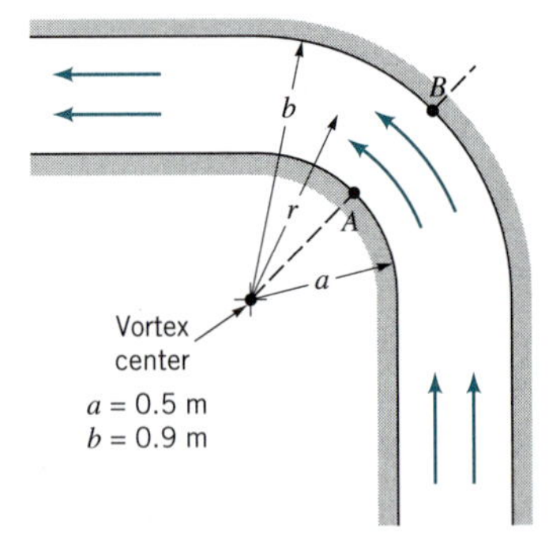

■ FIGURE P6.44

6.45 Water discharges from the large tank of Fig. P6.45 through an opening at A and a vortex forms above A. The velocity distribution around this vortex is approximately the same as that for a free vortex. At the same time that the fluid is being discharged from A, it is necessary to discharge a small quantity of water through the pipe B. As the discharge through A is increased, the strength of the vortex, as indicated by its circulation, is increased. Determine the maximum strength that the vortex can have in order that no air is sucked in at B. Express your answer in terms of the circulation and assume that the fluid level in the tank at a large distance from the opening at A remains constant, and viscous effects are negligible.

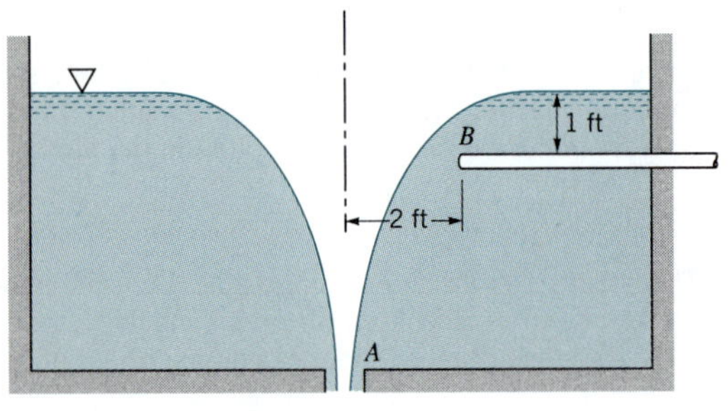

■ FIGURE P6.45

6.46 The streamlines in a particular two-dimensional flow field are all concentric circles, as shown in Fig. P6.46. The velocity is given by the equation $v_\theta = \omega r$ where ω is the angular velocity of the rotating mass of fluid. Determine the circulation around the path $ABCD$.

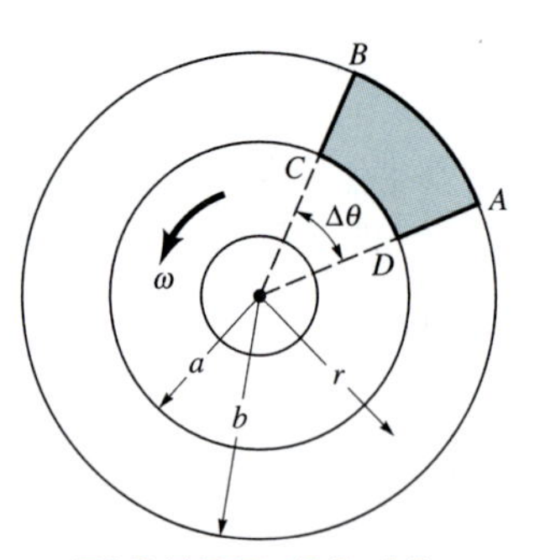

■ FIGURE P6.46

6.47 Water flows over a flat surface at 5 ft/s as shown in Fig. P6.47. A pump draws off water through a narrow slit at a volume rate of 0.1 ft^3/s per foot length of the slit. Assume that the fluid is incompressible and inviscid and can be represented by the combination of a uniform flow and a sink. Locate the stagnation point on the wall (point A) and determine the equation for the stagnation streamline. How far above the surface, H, must the fluid be so that it does not get sucked into the slit?

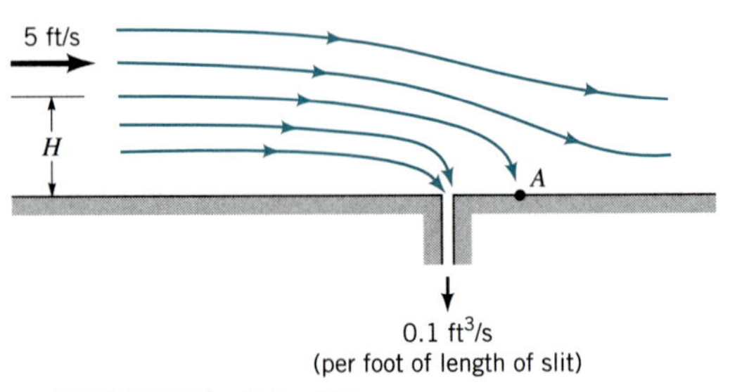

■ FIGURE P6.47

6.48 Consider two sources having equal strengths located along the x axis at $x = 0$ and $x = 2$ m, and a sink located on the y axis at $y = 2$ m. Determine the magnitude and direction of the fluid velocity at $x = 5$ m and $y = 0$ due to this combination if the flowrate from each of the sources is $0.5\ m^3/s$ per m and the flowrate into the sink is $1.0\ m^3/s$ per m.

6.49 The velocity potential for a spiral vortex flow is given by $\phi = (\Gamma/2\pi)\,\theta - (m/2\pi)\ln r$, where Γ and m are constants. Show that the angle, α, between the velocity vector and the radial direction is constant throughout the flow field (see Fig. P6.49).

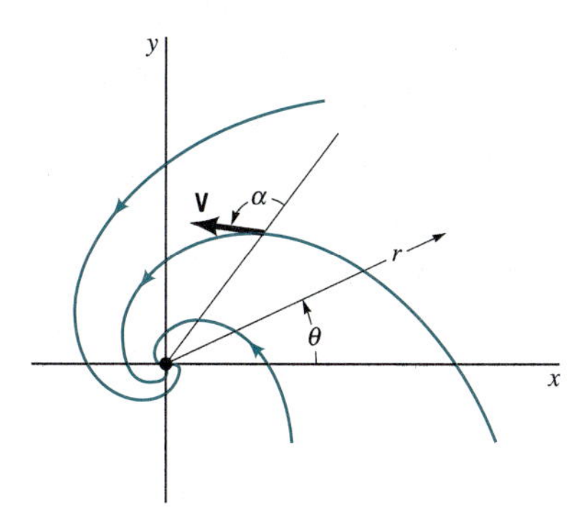

■ FIGURE P6.49

6.50 Consider a uniform flow in the positive x direction combined with a free vortex located at the origin of the coordinate system. The streamline $\psi = 0$ passes through the point $x = 4$, $y = 0$. Determine the equation of this streamline.

6.51 Potential flow against a flat plate (Fig. P6.51a) can be described with the stream function

$$\psi = Axy$$

where A is a constant. This type of flow is commonly called a ''stagnation point'' flow since it can be used to describe the flow in the vicinity of the stagnation point at O. By adding a source of strength m at O, stagnation point flow against a flat plate with a ''bump'' is obtained as illustrated in Fig. P6.51b. Determine the relationship between the bump height, h, the constant, A, and the source strength, m.

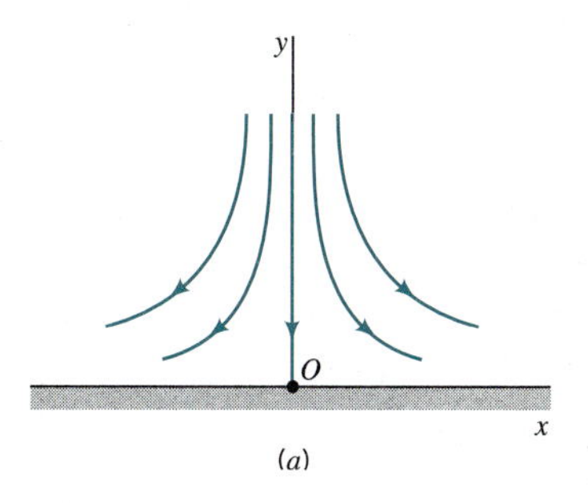

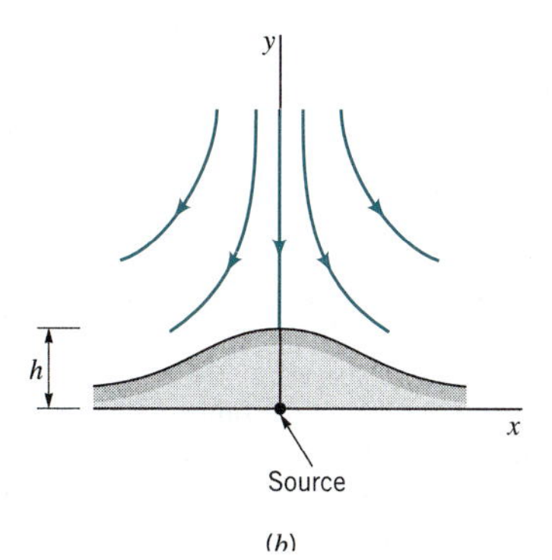

■ FIGURE P6.51

***6.52** Show on a plot several ''bumps'' that can be generated by combining stagnation point flow against a flat plate and a source as described in Problem 6.51.

6.53 A body having the general shape of a half-body is placed in a stream of fluid. At a great distance upstream the velocity is U as shown in Fig. P6.53. Show how a measurement of the differential pressure between the stagnation point and point A can be used to predict the free-stream velocity, U. Express the pressure differential in terms of U and fluid density. Neglect body forces and assume that the fluid is nonviscous and incompressible.

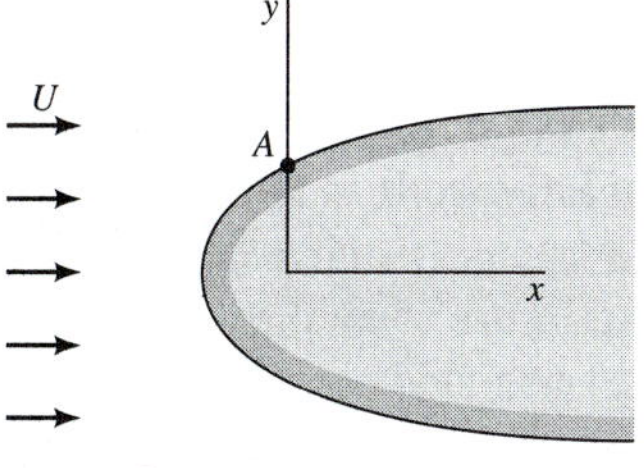

■ FIGURE P6.53

6.54 One end of a pond has a shoreline that resembles a half-body as shown in Fig. P6.54. A vertical porous pipe is located near the end of the pond so that water can be pumped out. When water is pumped at the rate of $0.08\ m^3/s$ through a 3-m-long pipe, what will be the velocity at point A? *Hint:* Consider the flow *inside* a half-body.

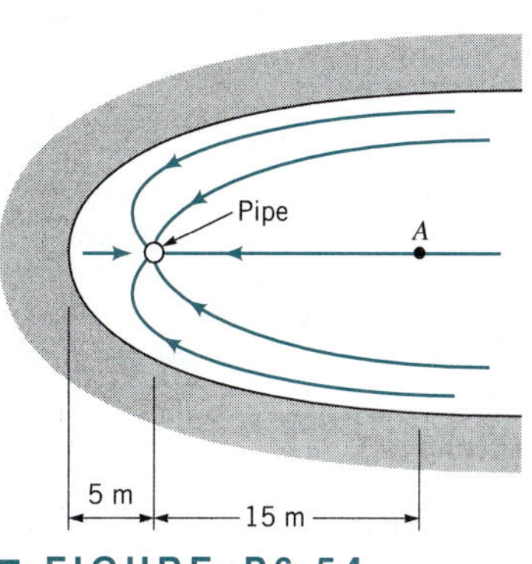

■ FIGURE P6.54

*6.55 For the half-body described in Section 6.6.1, show on a plot how the magnitude of the velocity on the surface, V_s, varies as a function of the distance, s (measured along the surface), from the stagnation point. Use the dimensionless variables V_s/U and s/b where U and b are defined in Fig. 6.24.

*6.56 Consider a uniform flow with velocity U in the positive x-direction combined with two free vortices of equal strength located along the y-axis. Let one vortex located at $y = a$ be a clockwise vortex ($\psi = K \ln r$) and the other at $y = -a$ be a counterclockwise vortex, where K is a positive constant. It can be shown by plotting streamlines that for $Ua/K < 2$ the streamline $\psi = 0$ forms a closed contour, as shown in Fig. P6.56. Thus, this combination can be used to represent flow around a family of bodies (called *Kelvin ovals*). Show, with the aid of a graph, how the dimensionless height, H/a, varies with the parameter Ua/K in the range $0.3 < Ua/K < 1.75$.

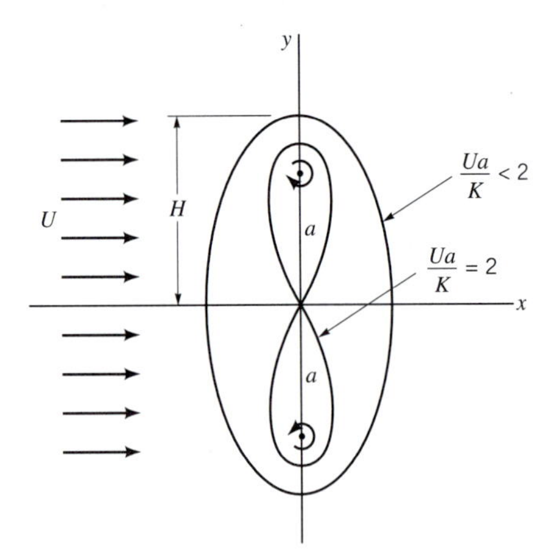

■ FIGURE P6.56

6.57 A Rankine oval is formed by combining a source-sink pair, each having a strength of 36 ft²/s and separated by a distance of 12 ft along the x axis, with a uniform velocity of 10 ft/s (in the positive x direction). Determine the length and thickness of the oval.

*6.58 Make use of Eqs. 6.107 and 6.109 to construct a table showing how ℓ/a, h/a, and ℓ/h for Rankine ovals depend on the parameter $\pi Ua/m$. Plot ℓ/h versus $\pi Ua/m$ and describe how this plot could be used to obtain the required values of m and a for a Rankine oval having a specific value of ℓ and h when placed in a uniform fluid stream of velocity, U.

6.59 Assume that the flow around the long circular cylinder of Fig. P6.59 is nonviscous and incompressible. Two pressures, p_1 and p_2, are measured on the surface of the cylinder, as illustrated. It is proposed that the free-stream velocity, U, can be related to the pressure difference $\Delta p = p_1 - p_2$ by the equation

$$U = C\sqrt{\frac{\Delta p}{\rho}}$$

where ρ is the fluid density. Determine the value of the constant C. Neglect body forces.

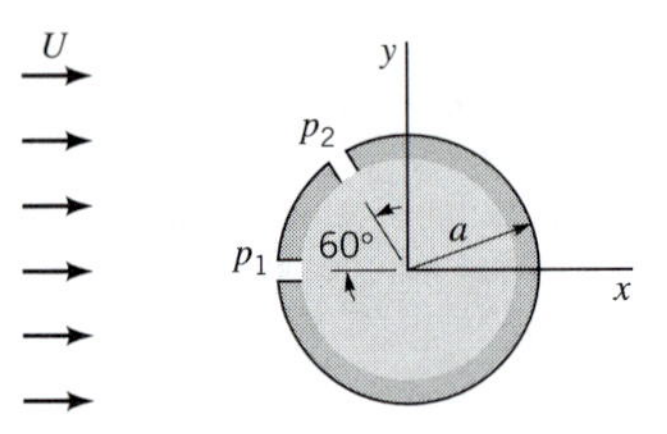

■ FIGURE P6.59

6.60 An ideal fluid flows past an infinitely long semicircular "hump" located along a plane boundary as shown in Fig. P6.60. Far from the hump the velocity field is uniform, and the pressure is p_0. **(a)** Determine expressions for the maximum and minimum values of the pressure along the hump, and indicate where these points are located. Express your answer in terms of ρ, U, and p_0. **(b)** If the solid surface is the $\psi = 0$ streamline, determine the equation of the streamline passing through the point $\theta = \pi/2$, $r = 2a$.

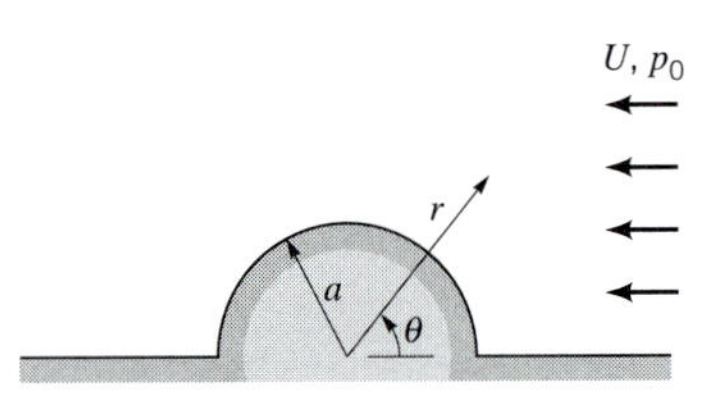

■ FIGURE P6.60

6.61 Water flows around a 6-ft-diameter bridge pier with a velocity of 12 ft/s. Estimate the force (per unit length) that the flowing water exerts on the front half of the pier. Assume that the flow can be approximated as ideal fluid around a circular cylinder.

*6.62 Consider the steady potential flow around the circular cylinder shown in Fig. 6.26. On a plot show the variation of the magnitude of the dimensionless fluid velocity, V/U, along the positive y axis. At what distance, y/a (along the y axis), is the velocity within 1% of the free-stream velocity?

6.63 The velocity potential for a cylinder (Fig. P6.63) rotating in a uniform stream of fluid is

$$\phi = Ur\left(1 + \frac{a^2}{r^2}\right)\cos\theta + \frac{\Gamma}{2\pi}\theta$$

where Γ is the circulation. For what value of the circulation will the stagnation point be located at: **(a)** point A, **(b)** point B?

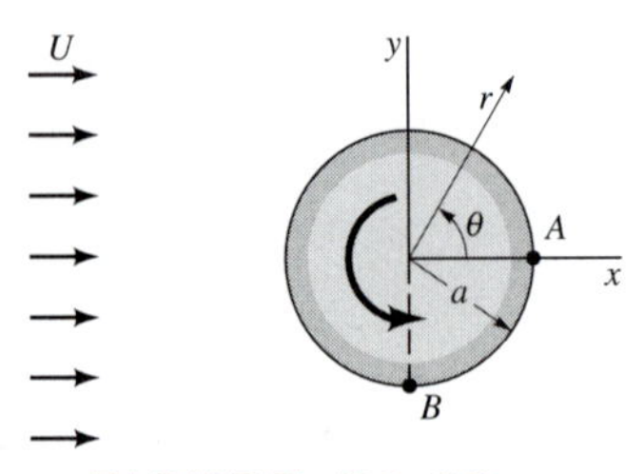

■ FIGURE P6.63

6.64 A fixed circular cylinder of infinite length is placed in a steady, uniform stream of an incompressible, nonviscous fluid. Assume that the flow is irrotational. Prove that the drag on the cylinder is zero. Neglect body forces.

6.65 Repeat Problem 6.64 for a rotating cylinder for which the stream function and velocity potential are given by Eqs. 6.119 and 6.120, respectively. Verify that the lift is not zero and can be expressed by Eq. 6.124.

6.66 A source of strength m is located a distance ℓ from a vertical solid wall as shown in Fig. P6.66. The velocity potential for this incompressible, irrotational flow is given by

$$\phi = \frac{m}{4\pi}\{\ln[(x-\ell)^2+y^2] + \ln[(x+\ell)^2+y^2]\}$$

(a) Show that there is no flow through the wall. **(b)** Determine the velocity distribution along the wall. **(c)** Determine the pressure distribution along the wall, assuming $p = p_0$ far from the source. Neglect the effect of the fluid weight on the pressure.

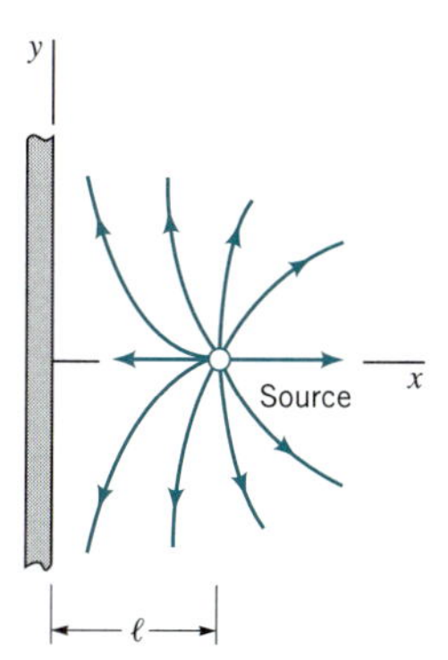

■ **FIGURE P6.66**

6.67 A long porous pipe runs parallel to a horizontal plane surface as shown in Fig. P6.67. The longitudinal axis of the pipe is perpendicular to the plane of the paper. Water flows radially from the pipe at a rate of $0.5\ \pi\ \text{ft}^3/\text{s}$ per foot of pipe. Determine the difference in pressure (in lb/ft^2) between point B and point A. The flow from the pipe may be approximated by a two-dimensional source. *Hint:* To develop the stream function or velocity potential for this type of flow, place (symmetrically) another equal source on the other side of the wall. With this combination there is no flow across the x axis, and this axis can be replaced with a solid boundary. This technique is called the *method of images*.

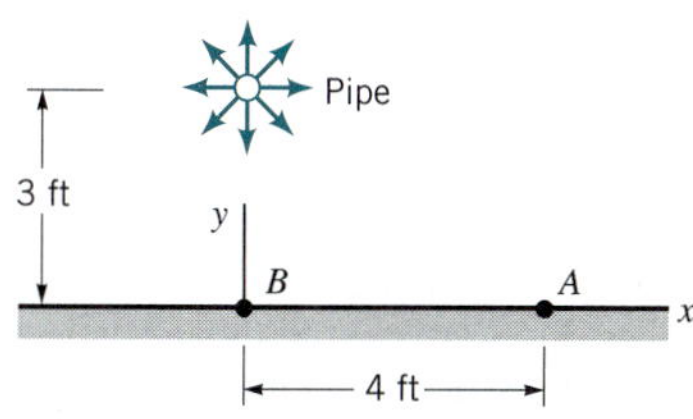

■ **FIGURE P6.67**

6.68 A plane flow field is developed by the addition of a free vortex and a uniform stream in the positive x direction. If the vortex is located at the origin, determine the pressure variation in this flow field. Neglect the effect of the fluid weight. Express your answer in terms of the uniform velocity, U, the strength of the vortex, Γ, and the pressure, p_0, far from the origin.

6.69 The two-dimensional velocity field for an incompressible Newtonian fluid is described by the relationship

$$\mathbf{V} = (12xy^2 - 6x^3)\hat{\mathbf{i}} + (18x^2y - 4y^3)\hat{\mathbf{j}}$$

where the velocity has units of m/s when x and y are in meters. Determine the stresses σ_{xx}, σ_{yy}, and τ_{xy} at the point $x = 0.5$ m, $y = 1.0$ m if pressure at this point is 6 kPa and the fluid is glycerin at 20 °C. Show these stresses on a sketch.

6.70 The stream function for a certain incompressible, two-dimensional flow field is

$$\psi = 3r^3 \sin 2\theta + 2\theta$$

where ψ is in ft^2/s when r is in feet and θ in radians. Determine the shearing stress, $\tau_{r\theta}$, at the point $r = 2$ ft, $\theta = \pi/3$ radians if the fluid is water.

6.71 For a two-dimensional incompressible flow in the x–y plane show that the z component of the vorticity, ζ_z, varies in accordance with the equation

$$\frac{D\zeta_z}{Dt} = \nu\, \nabla^2 \zeta_z$$

What is the physical interpretation of this equation for a nonviscous fluid? *Hint:* This *vorticity transport equation* can be derived from the Navier-Stokes equations by differentiating and eliminating the pressure between Eqs. 6.127a and 6.127b.

6.72 The velocity of a fluid particle moving along a horizontal streamline that coincides with the x axis in a plane, two-dimensional, incompressible flow field was experimentally found to be described by the equation $u = x^2$. Along this streamline determine an expression for **(a)** the rate of change of the v component of velocity with respect to y, **(b)** the acceleration of the particle, and **(c)** the pressure gradient in the x direction. The fluid is Newtonian.

6.73 The velocity field for a two-dimensional source or sink flow is a solution of the Euler equations of motion for inviscid flow. Show that this flow field is also a solution of the Navier-Stokes equations of motion for viscous flow.

6.74 Oil (SAE 30) at 15.6 °C flows steadily between fixed, horizontal, parallel plates. The pressure drop per unit length along the channel is 30 kPa/m, and the distance between the plates is 4 mm. The flow is laminar. Determine: **(a)** the volume rate of flow (per meter of width), **(b)** the magnitude and direction of the shearing stress acting on the bottom plate, and **(c)** the velocity along the centerline of the channel.

6.75 Two fixed, horizontal, parallel plates are spaced 0.2 in. apart. A viscous liquid ($\mu = 8 \times 10^{-3}$ lb·s/ft^2, $SG = 0.9$) flows between the plates with a mean velocity of 0.9 ft/s. Determine the pressure drop per unit length in the direction of flow. What is the maximum velocity in the channel?

6.76 A layer of viscous liquid of constant thickness (no velocity perpendicular to plate) flows steadily down an infinite, inclined plane. Determine, by means of the Navier-Stokes equations, the relationship between the thickness of the layer and the discharge per unit width. The flow is laminar, and assume air resistance is negligible so that the shearing stress at the free surface is zero.

6.77 A viscous, incompressible fluid flows between the two infinite, vertical, parallel plates of Fig. P6.77. Determine, by use of the Navier-Stokes equations, an expression for the pressure gradient in the direction of flow. Express your answer in terms of the mean velocity. Assume that the flow is laminar, steady, and uniform.

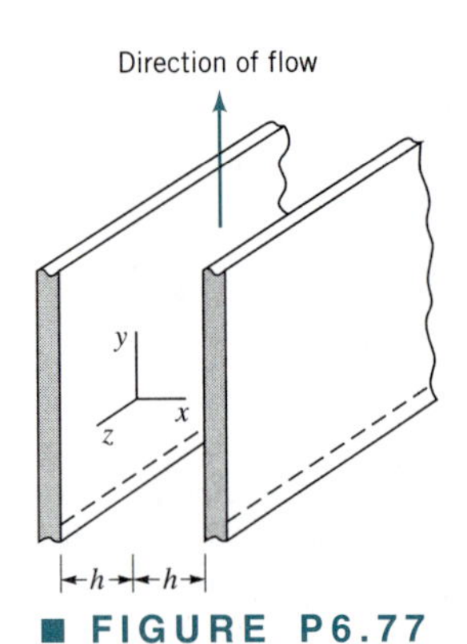

■ FIGURE P6.77

6.78 A fluid of density ρ flows steadily *downward* between the two vertical, infinite, parallel plates shown in the figure for Problem 6.77. The flow is fully developed and laminar. Make use of the Navier–Stokes equation to determine the relationship between the discharge and the other parameters involved, for the case in which the change in pressure along the channel is zero.

6.79 For the problem described in Example 6.9 determine the equation for the velocity distribution in the fluid layer when the net upward flow is zero. Show this velocity distribution on a sketch.

6.80 An incompressible, viscous fluid is placed between horizontal, infinite, parallel plates as is shown in Fig. P6.80. The two plates move in opposite directions with constant velocities, U_1 and U_2, as shown. The pressure gradient in the x direction is zero and the only body force is due to the fluid weight. Use the Navier–Stokes equations to derive an expression for the velocity distribution between the plates. Assume laminar flow.

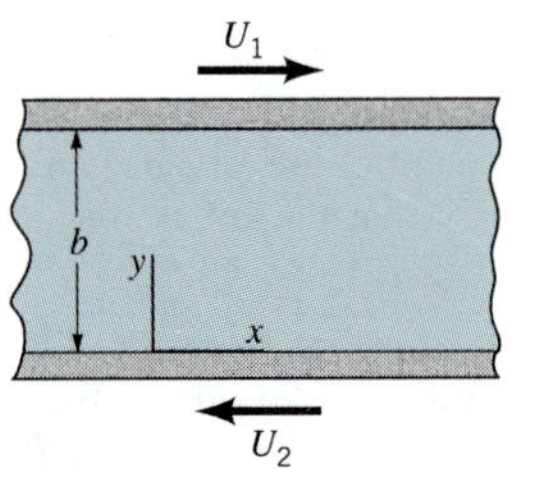

■ FIGURE P6.80

6.81 Two immiscible, incompressible, viscous fluids having the same densities but different viscosities are contained between two infinite, horizontal, parallel plates (Fig. P6.81). The bottom plate is fixed and the upper plate moves with a constant velocity U. Determine the velocity at the interface. Express your answer in terms of U, μ_1, and μ_2. The motion of the fluid is caused entirely by the movement of the upper plate; that is, there is no pressure gradient in the x direction. The fluid velocity and shearing stress are continuous across the interface between the two fluids. Assume laminar flow.

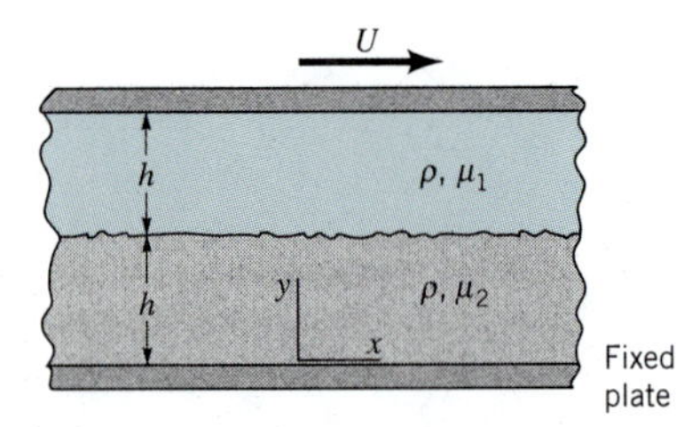

■ FIGURE P6.81

6.82 The viscous, incompressible flow between the parallel plates shown in Fig. P6.82 is caused by both the motion of the bottom plate and a pressure gradient, $\partial p/\partial x$. Determine the relationship between U and $\partial p/\partial x$ so that the shearing stress acting on the fixed plate is zero.

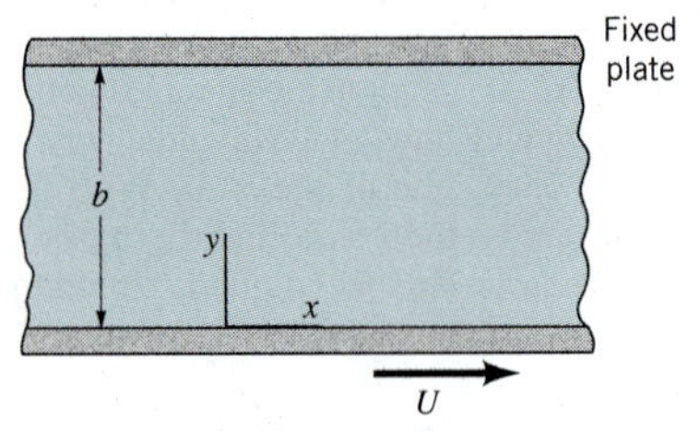

■ FIGURE P6.82

6.83 A viscous fluid (specific weight = 76 lb/ft^3; viscosity = 0.02 lb·s/ft^2) is contained between two infinite, horizontal parallel plates as shown in Fig. P6.83. The fluid moves between the plates under the action of a pressure gradient, and the upper plate moves with a velocity U while the bottom plate is fixed. A U-tube manometer connected between two points along the bottom indicates a differential reading of 0.1 in. If the upper plate moves with a velocity of 0.02 ft/s, at what distance from the bottom plate does the maximum velocity in the gap between the two plates occur? Assume laminar flow.

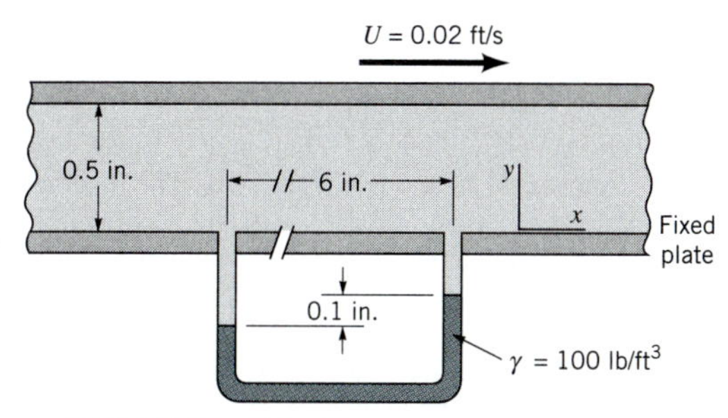

■ FIGURE P6.83

6.84 A vertical shaft passes through a bearing and is lubricated with an oil having a viscosity of 0.2 N·s/m² as shown in Fig. P6.84. Assume that the flow characteristics in the gap between the shaft and bearing are the same as those for laminar flow between infinite parallel plates with zero pressure gradient in the direction of flow. Estimate the torque required to overcome viscous resistance when the shaft is turning at 80 rev/min.

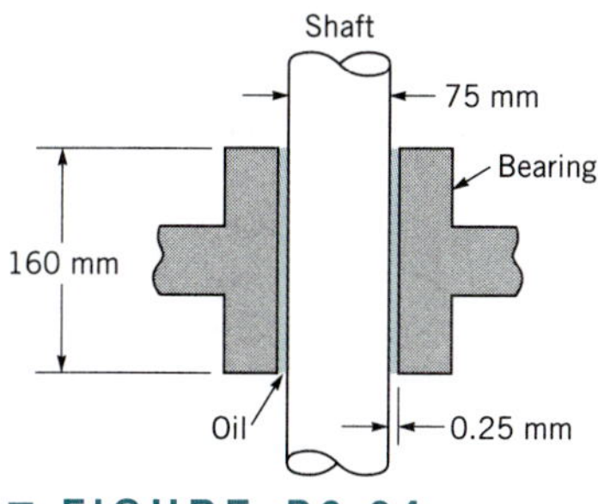

■ FIGURE P6.84

6.85 A viscous fluid is contained between two long concentric cylinders. The geometry of the system is such that the flow between the cylinders is approximately the same as the laminar flow between two infinite parallel plates. Determine an expression for the torque required to rotate the outer cylinder with an angular velocity ω. The inner cylinder is fixed. Express your answer in terms of the geometry of the system, the viscosity of the fluid, and the angular velocity.

***6.86** Oil (SAE 30) flows between parallel plates spaced 5 mm apart. The bottom plate is fixed but the upper plate moves with a velocity of 0.2 m/s in the positive x direction. The pressure gradient is 60 kPa/m and is negative. Compute the velocity at various points across the channel and show the results on a plot. Assume laminar flow.

6.87 Consider a steady, laminar flow through a straight horizontal tube having the constant elliptical cross section given by the equation:

$$\frac{x^2}{a^2} + \frac{y^2}{b^2} = 1$$

The streamlines are all straight and parallel. Investigate the possibility of using an equation for the z component of velocity of the form

$$w = A\left(1 - \frac{x^2}{a^2} - \frac{y^2}{b^2}\right)$$

as an exact solution to this problem. With this velocity distribution, what is the relationship between the pressure gradient along the tube and the volume flowrate through the tube?

6.88 A fluid is initially at rest between two horizontal, infinite, parallel plates. A constant pressure gradient in a direction parallel to the plates is suddenly applied and the fluid starts to move. Determine the appropriate differential equation(s), initial condition, and boundary conditions that govern this type of flow. You need not solve the equation(s).

6.89 Ethyl alcohol flows through a horizontal tube having a diameter of 10 mm. If the mean velocity is 0.15 m/s, what is the pressure drop per unit length along the tube? What is the velocity at a distance of 2 mm from the tube axis?

6.90 A simple flow system to be used for steady flow tests consists of a constant head tank connected to a length of 4-mm-diameter tubing as shown in Fig. P6.90. The liquid has a viscosity of 0.015 N·s/m², a density of 1200 kg/m³, and discharges into the atmosphere with a mean velocity of 1 m/s. **(a)** Verify that the flow will be laminar. **(b)** The flow is fully developed in the last 3 m of the tube. What is the pressure at the pressure gage? **(c)** What is the magnitude of the wall shearing stress, τ_{rz}, in the fully developed region?

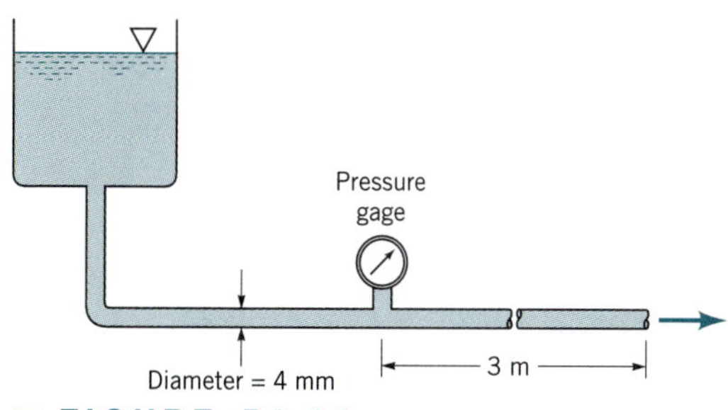

■ FIGURE P6.90

6.91 A liquid (viscosity = 0.002 N·s/m²; density = 1000 kg/m³) is forced through the circular tube shown in Fig. P6.91.

A differential manometer is connected to the tube as shown to measure the pressure drop along the tube. When the differential reading, Δh, is 7 mm, what is the mean velocity in the tube?

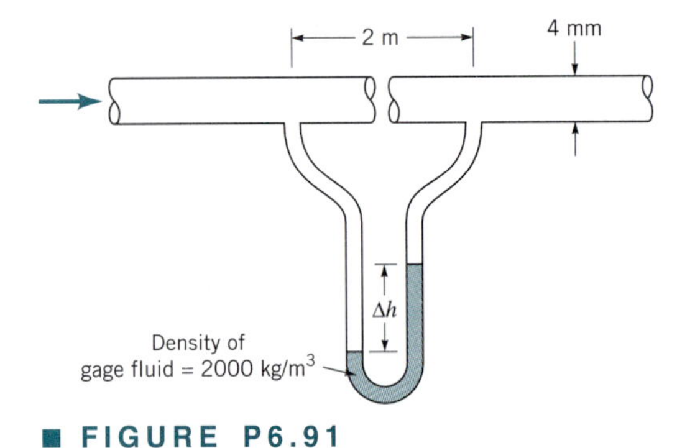

■ FIGURE P6.91

6.92 **(a)** Show that for Poiseuille flow in a tube of radius R the magnitude of the wall shearing stress, τ_{rz}, can be obtained from the relationship

$$|(\tau_{rz})_{\text{wall}}| = \frac{4\mu Q}{\pi R^3}$$

for a Newtonian fluid of viscosity μ. The volume rate of flow is Q. **(b)** Determine the magnitude of the wall shearing stress for a fluid having a viscosity of 0.003 N·s/m^2 flowing with an average velocity of 100 mm/s in a 2-mm-diameter tube.

6.93 An incompressible Newtonian fluid flows steadily between two infinitely long, concentric cylinders as shown in Fig. P6.93. The outer cylinder is fixed, but the inner cylinder moves with a longitudinal velocity V_0 as shown. For what value of V_0 will the drag on the inner cylinder be zero? Assume that the flow is laminar, axisymmetric, and fully developed.

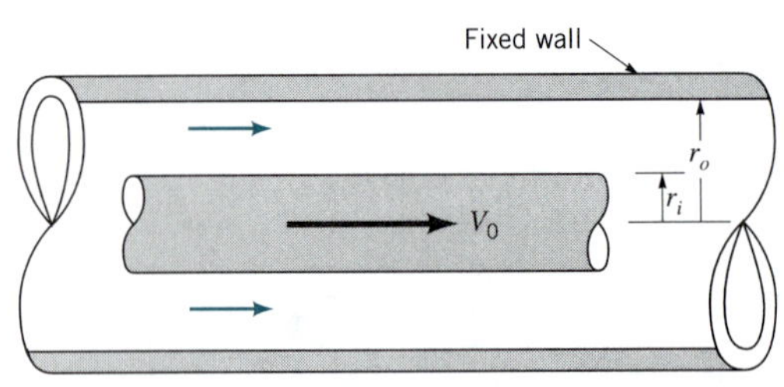

■ FIGURE P6.93

6.94 An infinitely long, solid, vertical cylinder of radius R is located in an infinite mass of an incompressible fluid. Start with the Navier-Stokes equation in the θ direction and derive an expression for the velocity distribution for the steady flow case in which the cylinder is rotating about a fixed axis with a constant angular velocity ω. You need not consider body forces. Assume that the flow is axisymmetric and the fluid is at rest at infinity.

6.95 A viscous fluid is contained between two infinitely long, vertical, concentric cylinders. The outer cylinder has a radius r_o and rotates with an angular velocity ω. The inner cylinder is fixed and has a radius r_i. Make use of the Navier-Stokes equations to obtain an exact solution for the velocity distribution in the gap. Assume that the flow in the gap is axisymmetric (neither velocity nor pressure are functions of angular position θ within the gap) and that there are no velocity components other than the tangential component. The only body force is the weight.

6.96 For flow between concentric cylinders, with the outer cylinder rotating at an angular velocity ω and the inner cylinder fixed, it is commonly assumed that the tangential velocity (v_θ) distribution in the gap between the cylinders is linear. Based on the exact solution to this problem (see Problem 6.95) the velocity distribution in the gap is not linear. For an outer cylinder with radius r_o = 2.00 in. and an inner cylinder with radius r_i = 1.80 in., show, with the aid of a plot, how the dimensionless velocity distribution, $v_\theta / r_o\omega$, varies with the dimensionless radial position, r/r_o, for the exact and approximate solutions.

6.97 A viscous liquid (μ = 0.016 lb·s/ft^2, ρ = 1.79 slugs/ft^3) flows through the annular space between two horizontal, fixed, concentric cylinders. If the radius of the inner cylinder is 1.5 in. and the radius of the outer cylinder is 2.5 in., what is the pressure drop along the axis of the annulus per foot when the volume flowrate is 0.09 ft^3/s?

***6.98** Plot the velocity profile for the fluid flowing in the annular space described in Problem P6.97. Determine from the plot the radius at which the maximum velocity occurs and compare with the value predicted from Eq. 6.157.

***6.99** As is shown by Eq. 6.150 the pressure gradient for laminar flow through a tube of constant radius is given by the expression:

$$\frac{\partial p}{\partial z} = -\frac{8\mu Q}{\pi R^4}$$

For a tube whose radius is changing very gradually, such as the one illustrated in Fig. P6.99, it is expected that this equation can be used to approximate the pressure change along the tube if the actual radius, $R(z)$, is used at each cross section. The following measurements were obtained along a particular tube.

z/ℓ	0	0.1	0.2	0.3	0.4	0.5	0.6	0.7	0.8	0.9	1.0
$R(z)/R_o$	1.00	0.73	0.67	0.65	0.67	0.80	0.80	0.71	0.73	0.77	1.00

Compare the pressure drop over the length ℓ for this nonuniform tube with one having the constant radius R_o. *Hint:* To solve this problem you will need to numerically integrate the equation for the pressure gradient given above.

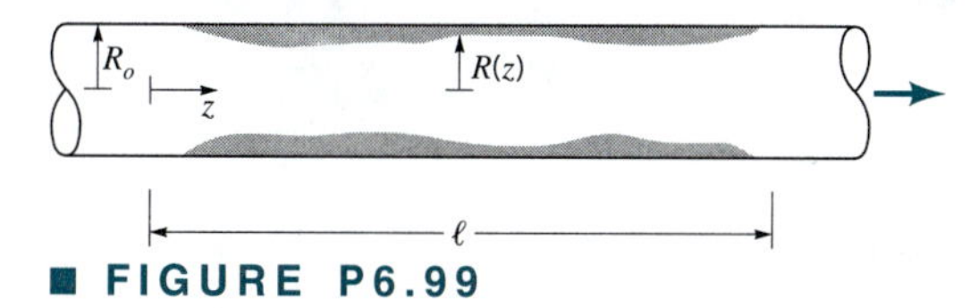

■ FIGURE P6.99

6.100 Show how Eq. 6.155 is obtained.

6.101 A wire of diameter d is stretched along the centerline of a pipe of diameter D. For a given pressure drop per unit length of pipe, by how much does the presence of the wire reduce the flowrate if **(a)** $d/D = 0.1$; **(b)** $d/D = 0.01$?

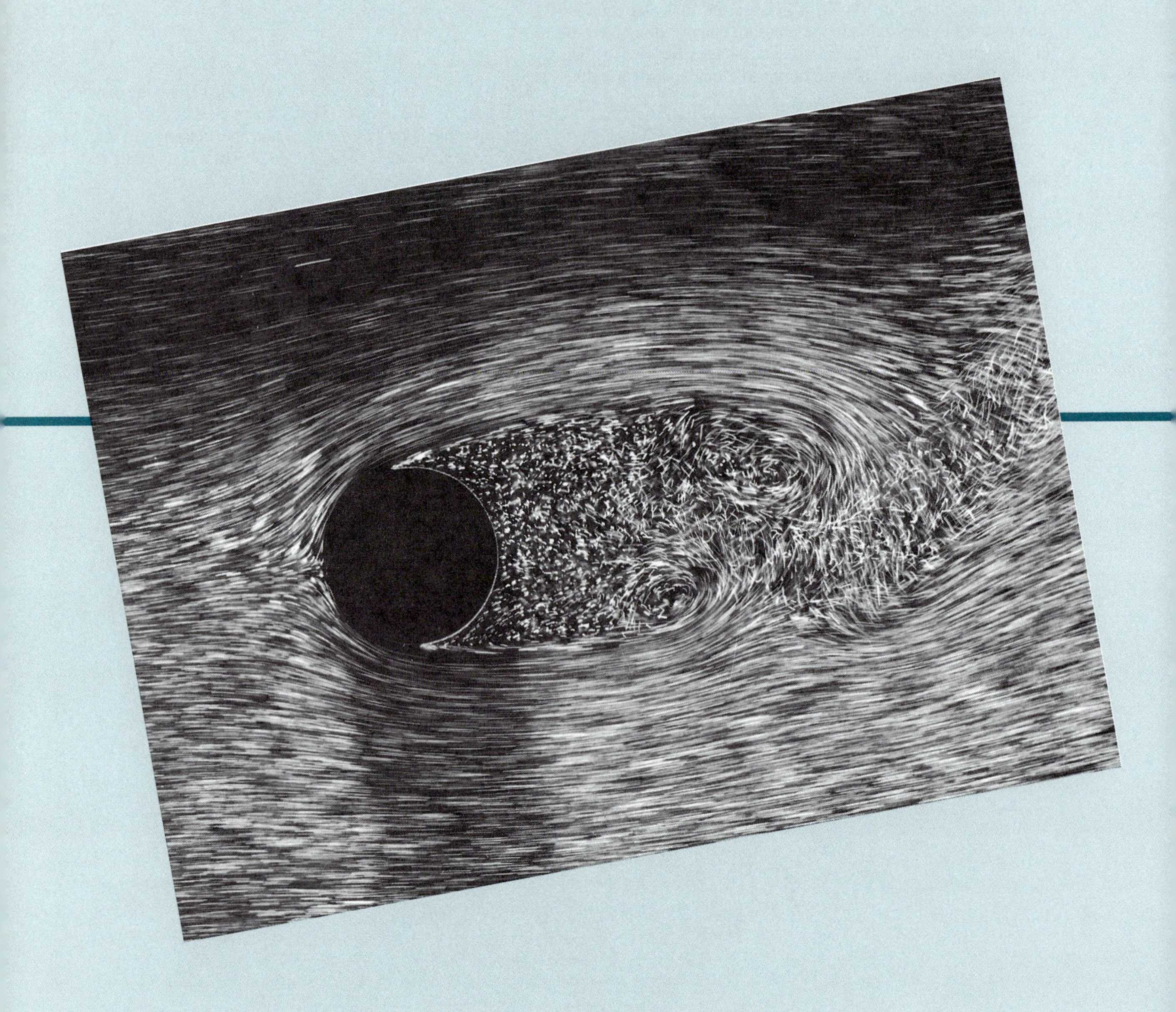

Flow past a circular cylinder with Re = 2000: The streaklines of flow past any circular cylinder (regardless of size, velocity, or fluid) are as shown provided that the dimensionless parameter called the Reynolds number, Re, is equal to 2000. For other values of Re the flow pattern will be different (air bubbles in water). (Photograph courtesy of ONERA, France.)

7 Similitude, Dimensional Analysis, and Modeling

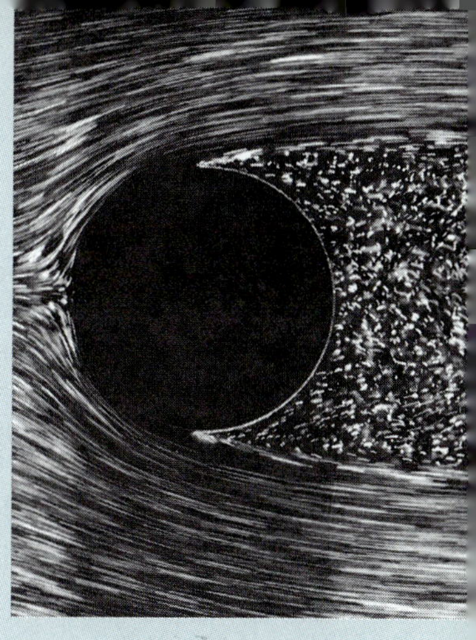

Experimentation and modeling are widely used techniques in fluid mechanics.

Although many practical engineering problems involving fluid mechanics can be solved by using the equations and analytical procedures described in the preceding chapters, there remain a large number of problems that rely on experimentally obtained data for their solution. In fact, it is probably fair to say that very few problems involving real fluids can be solved by analysis alone. The solution to many problems is achieved through the use of a combination of analysis and experimental data. Thus, engineers working on fluid mechanics problems should be familiar with the experimental approach to these problems so that they can interpret and make use of data obtained by others, such as might appear in handbooks, or be able to plan and execute the necessary experiments in their own laboratories. In this chapter we consider some techniques and ideas that are important in the planning and execution of experiments, as well as in understanding and correlating data that may have been obtained by other experimenters.

An obvious goal of any experiment is to make the results as widely applicable as possible. To achieve this end, the concept of *similitude* is often used so that measurements made on one system (for example, in the laboratory) can be used to describe the behavior of other similar systems (outside the laboratory). The laboratory systems are usually thought of as *models* and are used to study the phenomenon of interest under carefully controlled conditions. From these model studies, empirical formulations can be developed, or specific predictions of one or more characteristics of some other similar system can be made. To do this, it is necessary to establish the relationship between the laboratory model and the ''other'' system. In the following sections, we find out how this can be accomplished in a systematic manner.

7.1 Dimensional Analysis

To illustrate a typical fluid mechanics problem in which experimentation is required, consider the steady flow of an incompressible Newtonian fluid through a long, smooth-walled, horizontal, circular pipe. An important characteristic of this system, which would be of interest to an engineer designing a pipeline, is the pressure drop per unit length that develops along the pipe as a result of friction. Although this would appear to be a relatively simple flow problem, it cannot generally be solved analytically (even with the aid of large computers) without the use of experimental data.

The first step in the planning of an experiment to study this problem would be to decide on the factors, or variables, that will have an effect on the pressure drop per unit length, Δp_ℓ. We expect the list to include the pipe diameter, D, the fluid density, ρ, fluid viscosity, μ, and the mean velocity, V, at which the fluid is flowing through the pipe. Thus, we can express this relationship as

$$\Delta p_\ell = f(D, \rho, \mu, V) \tag{7.1}$$

which simply indicates mathematically that we expect the pressure drop per unit length to be some function of the factors contained within the parentheses. At this point the nature of the function is unknown and the objective of the experiments to be performed is to determine the nature of this function.

It is important to develop a meaningful and systematic way to perform an experiment.

To perform the experiments in a meaningful and systematic manner, it would be necessary to change one of the variables, such as the velocity, while holding all others constant, and measure the corresponding pressure drop. This series of tests would yield data that could be represented graphically as is illustrated in Fig. 7.1*a*. It is to be noted that this plot would only be valid for the specific pipe and for the specific fluid used in the tests; this certainly does not give us the general formulation we are looking for. We could repeat the process by varying each of the other variables in turn, as is illustrated in Figs. 7.1*b*, 7.1*c*, and 7.1*d*. This approach to determining the functional relationship between the pressure drop and the various factors that influence it, although logical in concept, is fraught with difficulties. Some of the experiments would be hard to carry out—for example, to obtain the data illustrated in Fig. 7.1*c* it would be necessary to vary fluid density while holding viscosity constant. How would you do this? Finally, once we obtained the various curves shown in Figs. 7.1*a*, 7.1*b*, 7.1*c*, and 7.1*d*, how could we combine these data to obtain the desired general functional relationship between Δp_ℓ, D, ρ, μ, and V which would be valid for any similar pipe system?

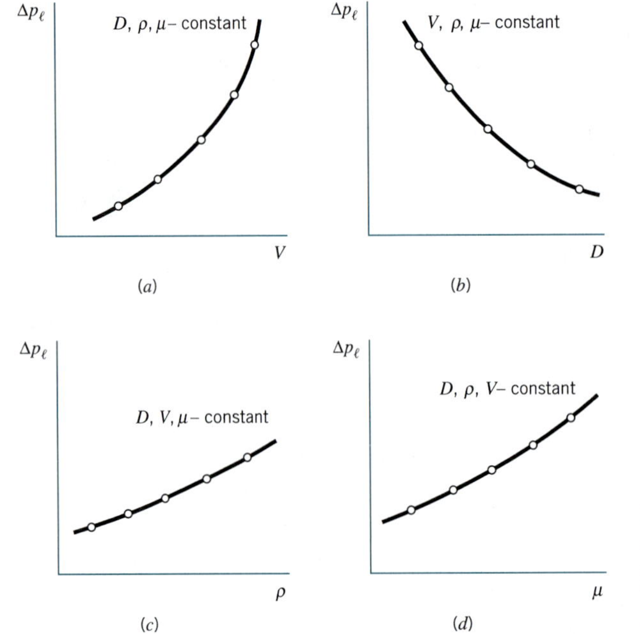

■ **FIGURE 7.1 Illustrative plots showing how the pressure drop in a pipe may be affected by several different factors.**

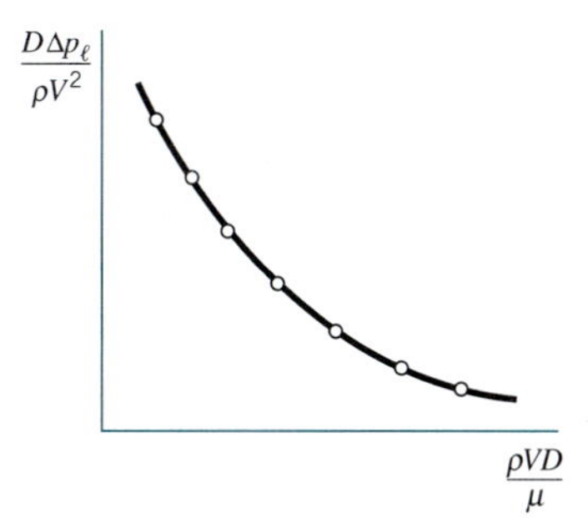

■ **FIGURE 7.2** **An illustrative plot of pressure drop data using dimensionless parameters.**

Dimensionless products are important and useful in the planning, execution, and interpretation of experiments.

Fortunately, there is a much simpler approach to this problem that will eliminate the difficulties described above. In the following sections we will show that rather than working with the original list of variables, as described in Eq. 7.1, we can collect these into two nondimensional combinations of variables (called *dimensionless products* or *dimensionless groups*) so that

$$\frac{D\,\Delta p_\ell}{\rho V^2} = \phi\left(\frac{\rho VD}{\mu}\right) \tag{7.2}$$

Thus, instead of having to work with five variables, we now have only two. The necessary experiment would simply consist of varying the dimensionless product $\rho VD/\mu$ and determining the corresponding value of $D\,\Delta p_\ell/\rho V^2$. The results of the experiment could then be represented by a single, universal curve as is illustrated in Fig. 7.2. This curve would be valid for any combination of smooth-walled pipe and incompressible Newtonian fluid. To obtain this curve we could choose a pipe of convenient size and a fluid that is easy to work with. Note that we wouldn't have to use different pipe sizes or even different fluids. It is clear that the experiment would be much simpler, easier to do, and less expensive (which would certainly make an impression on your boss).

The basis for this simplification lies in a consideration of the dimensions of the variables involved. As was discussed in Chapter 1, a qualitative description of physical quantities can be given in terms of basic dimensions such as mass, M, length, L, and time, T.[1] Alternatively, we could use force, F, L, and T as basic dimensions, since from Newton's second law

$$F \doteq MLT^{-2}$$

(Recall from Chapter 1 that the notation $\doteq$ is used to indicate dimensional equality.) The dimensions of the variables in the pipe flow example are $\Delta p_\ell \doteq FL^{-3}$, $D \doteq L$, $\rho \doteq FL^{-4}T^2$, $\mu \doteq FL^{-2}T$, and $V \doteq LT^{-1}$. A quick check of the dimensions of the two groups that appear in Eq. 7.2 shows that they are in fact *dimensionless* products; that is,

$$\frac{D\,\Delta p_\ell}{\rho V^2} \doteq \frac{L(F/L^3)}{(FL^{-4}T^2)(LT^{-1})^2} \doteq F^0L^0T^0$$

and

$$\frac{\rho VD}{\mu} \doteq \frac{(FL^{-4}T^2)(LT^{-1})(L)}{(FL^{-2}T)} \doteq F^0L^0T^0$$

Not only have we reduced the numbers of variables from five to two, but the new groups are dimensionless combinations of variables, which means that the results presented in the form

[1]As noted in Chapter 1, we will use T to represent the basic dimension of time, although T is also used for temperature in thermodynamic relationships (such as the ideal gas law).

of Fig. 7.2 will be independent of the system of units we choose to use. This type of analysis is called *dimensional analysis*, and the basis for its application to a wide variety of problems is found in the *Buckingham pi theorem* described in the following section.

7.2 Buckingham Pi Theorem

A fundamental question we must answer is how many dimensionless products are required to replace the original list of variables? The answer to this question is supplied by the basic theorem of dimensional analysis that states the following:

> If an equation involving k variables is dimensionally homogeneous, it can be reduced to a relationship among $k - r$ independent dimensionless products, where r is the minimum number of reference dimensions required to describe the variables.

Dimensional analysis is based on the Buckingham pi theorem.

The dimensionless products are frequently referred to as "pi terms," and the theorem is called the Buckingham pi theorem.[2] Buckingham used the symbol Π to represent a dimensionless product, and this notation is commonly used. Although the pi theorem is a simple one, its proof is not so simple and we will not include it here. Many entire books have been devoted to the subject of similitude and dimensional analysis, and a number of these are listed at the end of this chapter (Refs. 1–15). Students interested in pursuing the subject in more depth (including the proof of the pi theorem) can refer to one of these books.

The pi theorem is based on the idea of dimensional homogeneity which was introduced in Chapter 1. Essentially we assume that for any physically meaningful equation involving k variables, such as

$$u_1 = f(u_2, u_3, \ldots, u_k)$$

the dimensions of the variable on the left side of the equal sign must be equal to the dimensions of any term that stands by itself on the right side of the equal sign. It then follows that we can rearrange the equation into a set of dimensionless products (pi terms) so that

$$\Pi_1 = \phi(\Pi_2, \Pi_3, \ldots, \Pi_{k-r})$$

The required number of pi terms is fewer than the number of original variables by r, where r is determined by the minimum number of reference dimensions required to describe the original list of variables. Usually the reference dimensions required to describe the variables will be the basic dimensions M, L, and T or F, L, and T. However, in some instances perhaps only two dimensions, such as L and T, are required, or maybe just one, such as L. Also, in a few rare cases the variables may be described by some combination of basic dimensions, such as M/T^2, and L, and in this case r would be equal to two rather than three. Although the use of the pi theorem may appear to be a little mysterious and complicated, we can actually develop a simple, systematic procedure for developing the pi terms for a given problem.

7.3 Determination of Pi Terms

Several methods can be used to form the dimensionless products, or pi terms, that arise in a dimensional analysis. Essentially we are looking for a method that will allow us to syste-

[2]Although several early investigators, including Lord Rayleigh (1842–1919) in the nineteenth century, contributed to the development of dimensional analysis, Edgar Buckingham's (1867–1940) name is usually associated with the basic theorem. He stimulated interest in the subject in the United States through his publications during the early part of the twentieth century. See, for example, E. Buckingham, On Physically Similar Systems: Illustrations of the Use of Dimensional Equations, *Phys. Rev.*, 4 (1914), 345–376.

matically form the pi terms so that we are sure that they are dimensionless and independent, and that we have the right number. The method we will describe in detail in this section is called the *method of repeating variables.*

A dimensional analysis can be performed using a series of distinct steps.

It will be helpful to break the repeating variable method down into a series of distinct steps that can be followed for any given problem. With a little practice you will be able to readily complete a dimensional analysis for your problem.

Step 1. **List all the variables that are involved in the problem.** This step is the most difficult one and it is, of course, vitally important that all pertinent variables be included. Otherwise the dimensional analysis will not be correct! We are using the term ''variable'' to include any quantity, including dimensional and nondimensional constants, which play a role in the phenomenon under investigation. All such quantities should be included in the list of ''variables'' to be considered for the dimensional analysis. The determination of the variables must be accomplished by the experimenter's knowledge of the problem and the physical laws that govern the phenomenon. Typically the variables will include those that are necessary to describe the *geometry* of the system (such as a pipe diameter), to define any *fluid properties* (such as a fluid viscosity), and to indicate *external effects* that influence the system (such as a driving pressure drop per unit length). These general classes of variables are intended as broad categories that should be helpful in identifying variables. It is likely, however, that there will be variables that do not fit easily into one of these categories, and each problem needs to be carefully analyzed.

Since we wish to keep the number of variables to a minimum, so that we can minimize the amount of laboratory work, it is important that all variables be independent. For example, if in a certain problem the cross-sectional area of a pipe is an important variable, either the area or the pipe diameter could be used, but not both, since they are obviously not independent. Similarly, if both fluid density, ρ, and specific weight, γ, are important variables, we could list ρ and γ, or ρ and g (acceleration of gravity), or γ and g. However, it would be incorrect to use all three since $\gamma = \rho g$; that is, ρ, γ, and g are not independent. Note that although g would normally be constant in a given experiment, that fact is irrelevant as far as a dimensional analysis is concerned.

Step 2. **Express each of the variables in terms of basic dimensions.** For the typical fluid mechanics problem the basic dimensions will be either M, L, and T or F, L, and T. Dimensionally these two sets are related through Newton's second law ($\mathbf{F} = m\mathbf{a}$) so that $F \doteq MLT^{-2}$. For example, $\rho \doteq ML^{-3}$ or $\rho \doteq FL^{-4}T^2$. Thus, either set can be used. The basic dimensions for typical variables found in fluid mechanics problems are listed in Table 1.1 in Chapter 1.

Step 3. **Determine the required number of pi terms.** This can be accomplished by means of the Buckingham pi theorem, which indicates that the number of pi terms is equal to $k - r$, where k is the number of variables in the problem (which is determined from Step 1) and r is the number of reference dimensions required to describe these variables (which is determined from Step 2). The reference dimensions usually correspond to the basic dimensions and can be determined by an inspection of the dimensions of the variables obtained in Step 2. As previously noted, there may be occasions (usually rare) in which the basic dimensions appear in combinations so that the number of reference dimensions is less than the number of basic dimensions. This possibility is illustrated in Example 7.2.

Step 4. **Select a number of repeating variables, where the number required is equal to the number of reference dimensions.** Essentially what we are doing here is selecting from the original list of variables several of which can be combined with each of the remaining variables to form a pi term. All of the required reference dimensions must be included within the group of repeating variables, and each repeating variable must be dimensionally independent of the others (i.e., the dimensions of one repeating variable cannot be reproduced by some combination of products of powers of the remaining repeating variables). This means that the repeating variables cannot themselves be combined to form a dimensionless product.

For any given problem we usually are interested in determining how one particular variable is influenced by the other variables. We would consider this variable to be the dependent variable, and we would want this to appear in only one pi term. Thus, do *not* choose the dependent variable as one of the repeating variables, since the repeating variables will generally appear in more than one pi term.

Step 5. **Form a pi term by multiplying one of the nonrepeating variables by the product of the repeating variables, each raised to an exponent that will make the combination dimensionless.** Essentially each pi term will be of the form $u_i u_1^{a_i} u_2^{b_i} u_3^{c_i}$ where u_i is one of the nonrepeating variables; u_1, u_2, and u_3 are the repeating variables; and the exponents a_i, b_i, and c_i are determined so that the combination is dimensionless.

Step 6. **Repeat Step 5 for each of the remaining nonrepeating variables.** The resulting set of pi terms will correspond to the required number obtained from Step 3. If not, check your work—you have made a mistake!

Step 7. **Check all the resulting pi terms to make sure they are dimensionless.** It is easy to make a mistake in forming the pi terms. However, this can be checked by simply substituting the dimensions of the variables into the pi terms to confirm that they are all dimensionless. One good way to do this is to express the variables in terms of M, L, and T if the basic dimensions F, L, and T were used initially, or vice versa, and then check to make sure the pi terms are dimensionless.

Step 8. **Express the final form as a relationship among the pi terms, and think about what it means.** Typically the final form can be written as

$$\Pi_1 = \phi(\Pi_2, \Pi_3, \ldots, \Pi_{k-r})$$

where Π_1 would contain the dependent variable in the numerator. It should be emphasized that if you started out with the correct list of variables (and the other steps were completed correctly), then the relationship in terms of the pi terms can be used to describe the problem. You need only work with the pi terms—not with the individual variables. However, it should be clearly noted that this is as far as we can go with the dimensional analysis; that is, the actual functional relationship among the pi terms must be determined by experiment.

By using dimensional analysis, the original problem is simplified and defined with pi terms.

To illustrate these various steps we will again consider the problem discussed earlier in this chapter which was concerned with the steady flow of an incompressible Newtonian fluid through a long, smooth-walled, horizontal circular pipe. We are interested in the pressure drop per unit length, Δp_ℓ, along the pipe. According to Step 1 we must list all of the pertinent variables that are involved based on the experimenter's knowledge of the problem. In this problem we assume that

$$\Delta p_\ell = f(D, \rho, \mu, V)$$

where D is the pipe diameter, ρ and μ are the fluid density and viscosity, respectively, and V is the mean velocity.

Next (Step 2) we express all the variables in terms of basic dimensions. Using F, L, and T as basic dimensions it follows that

$$\Delta p_\ell \doteq FL^{-3}$$
$$D \doteq L$$
$$\rho \doteq FL^{-4}T^2$$
$$\mu \doteq FL^{-2}T$$
$$V \doteq LT^{-1}$$

We could also use M, L, and T as basic dimensions if desired—the final result will be the same. Note that for density, which is a mass per unit volume (ML^{-3}), we have used the relationship $F \doteq MLT^{-2}$ to express the density in terms of F, L, and T. Do not mix the basic dimensions; that is, use either F, L, and T *or* M, L, and T.

We can now apply the pi theorem to determine the required number of pi terms (Step 3). An inspection of the dimensions of the variables from Step 2 reveals that all three basic dimensions are required to describe the variables. Since there are five ($k = 5$) variables (do not forget to count the dependent variable, Δp_ℓ) and three required reference dimensions ($r = 3$), then according to the pi theorem there will be $(5 - 3)$, or two pi terms required.

Special attention should be given to the selection of repeating variables as detailed in Step 4.

The repeating variables to be used to form the pi terms (Step 4) need to be selected from the list D, ρ, μ, and V. Remember, we do not want to use the dependent variable as one of the repeating variables. Since three reference dimensions are required, we will need to select three repeating variables. Generally, we would try to select from repeating variables those that are the simplest, dimensionally. For example, if one of the variables has the dimension of a length, choose it as one of the repeating variables. In this example we will use D, V, and ρ as repeating variables. Note that these are dimensionally independent, since D is a length, V involves both length and time, and ρ involves force, length, and time. This means that we cannot form a dimensionless product from this set.

We are now ready to form the two pi terms (Step 5). Typically, we would start with the dependent variable and combine it with the repeating variables to form the first pi term; that is,

$$\Pi_1 = \Delta p_\ell D^a V^b \rho^c$$

Since this combination is to be dimensionless, it follows that

$$(FL^{-3})(L)^a(LT^{-1})^b(FL^{-4}T^2)^c \doteq F^0L^0T^0$$

The exponents, a, b, and c must be determined such that the resulting exponent for each of the basic dimensions—F, L, and T—must be zero (so that the resulting combination is dimensionless). Thus, we can write

$$1 + c = 0 \qquad \text{(for } F\text{)}$$
$$-3 + a + b - 4c = 0 \qquad \text{(for } L\text{)}$$
$$-b + 2c = 0 \qquad \text{(for } T\text{)}$$

The solution of this system of algebraic equations gives the desired values for a, b, and c. It follows that $a = 1$, $b = -2$, $c = -1$ and, therefore,

$$\Pi_1 = \frac{\Delta p_\ell D}{\rho V^2}$$

The process is now repeated for the remaining nonrepeating variables (Step 6). In this example there is only one additional variable (μ) so that

$$\Pi_2 = \mu D^a V^b \rho^c$$

or

$$(FL^{-2}T)(L)^a(LT^{-1})^b(FL^{-4}T^2)^c \doteq F^0L^0T^0$$

and, therefore,

$$\begin{aligned} 1 + c &= 0 \qquad \text{(for } F) \\ -2 + a + b - 4c &= 0 \qquad \text{(for } L) \\ 1 - b + 2c &= 0 \qquad \text{(for } T) \end{aligned}$$

Solving these equations simultaneously it follows that $a = -1, b = -1, c = -1$ so that

$$\Pi_2 = \frac{\mu}{DV\rho}$$

Always check the pi terms to make sure they are dimensionless.

Note that we end up with the correct number of pi terms as determined from Step 3.

At this point stop and check to make sure the pi terms are actually dimensionless (Step 7). We will check using both FLT and MLT dimensions. Thus,

$$\Pi_1 = \frac{\Delta p_\ell D}{\rho V^2} \doteq \frac{(FL^{-3})(L)}{(FL^{-4}T^2)(LT^{-1})^2} \doteq F^0L^0T^0$$

$$\Pi_2 = \frac{\mu}{DV\rho} \doteq \frac{(FL^{-2}T)}{(L)(LT^{-1})(FL^{-4}T^2)} \doteq F^0L^0T^0$$

or alternatively,

$$\Pi_1 = \frac{\Delta p_\ell D}{\rho V^2} \doteq \frac{(ML^{-2}T^{-2})(L)}{(ML^{-3})(LT^{-1})^2} \doteq M^0L^0T^0$$

$$\Pi_2 = \frac{\mu}{DV\rho} \doteq \frac{(ML^{-1}T^{-1})}{(L)(LT^{-1})(ML^{-3})} \doteq M^0L^0T^0$$

Finally (Step 8), we can express the result of the dimensional analysis as

$$\frac{\Delta p_\ell D}{\rho V^2} = \tilde{\phi}\left(\frac{\mu}{DV\rho}\right)$$

This result indicates that this problem can be studied in terms of these two pi terms, rather than the original five variables we started with. However, dimensional analysis will *not* provide the form of the function $\tilde{\phi}$. This can only be obtained from a suitable set of experiments. If desired, the pi terms can be rearranged; that is, the reciprocal of $\mu/DV\rho$ could be used, and of course the order in which we write the variables can be changed. Thus, for example, Π_2 could be expressed as

$$\Pi_2 = \frac{\rho VD}{\mu}$$

and the relationship between Π_1 and Π_2 as

$$\frac{D\,\Delta p_\ell}{\rho V^2} = \phi\left(\frac{\rho VD}{\mu}\right)$$

This is the form we previously used in our initial discussion of this problem (Eq. 7.2). The dimensionless product, $\rho VD/\mu$, is a very famous one in fluid mechanics—the Reynolds number. This number has been briefly alluded to in Chapters 1 and 6 and will be further discussed in Section 7.6.

To summarize, the steps to be followed in performing a dimensional analysis using the method of repeating variables are as follows:

The method of repeating variables can be most easily carried out by following a step-by-step procedure.

Step 1. List all the variables that are involved in the problem.

Step 2. Express each of the variables in terms of basic dimensions.

Step 3. Determine the required number of pi terms.

Step 4. Select a number of repeating variables, where the number required is equal to the number of reference dimensions (usually the same as the number of basic dimensions).

Step 5. Form a pi term by multiplying one of the nonrepeating variables by the product of repeating variables each raised to an exponent that will make the combination dimensionless.

Step 6. Repeat Step 5 for each of the remaining repeating variables.

Step 7. Check all the resulting pi terms to make sure they are dimensionless.

Step 8. Express the final form as a relationship among the pi terms and think about what it means.

EXAMPLE 7.1

A thin rectangular plate having a width w and a height h is located so that it is normal to a moving stream of fluid. Assume the drag, $\mathcal{D}$, that the fluid exerts on the plate is a function of w and h, the fluid viscosity and density, μ and ρ, respectively, and the velocity V of the fluid approaching the plate. Determine a suitable set of pi terms to study this problem experimentally.

SOLUTION

From the statement of the problem we can write

$$\mathcal{D} = f(w, h, \mu, \rho, V)$$

where this equation expresses the general functional relationship between the drag and the several variables that will affect it. The dimensions of the variables (using the MLT system) are

$$\mathcal{D} \doteq MLT^{-2}$$
$$w \doteq L$$
$$h \doteq L$$
$$\mu \doteq ML^{-1}T^{-1}$$
$$\rho \doteq ML^{-3}$$
$$V \doteq LT^{-1}$$

We see that all three basic dimensions are required to define the six variables so that the Buckingham pi theorem tells us that three pi terms will be needed (six variables minus three reference dimensions, $k - r = 6 - 3$).

We will next select three repeating variables such as w, V, and ρ. A quick inspection of these three reveals that they are dimensionally independent, since each one contains a basic dimension not included in the others. Note that it would be incorrect to use both w and h as repeating variables since they have the same dimensions.

Starting with the dependent variable, $\mathcal{D}$, the first pi term can be formed by combining $\mathcal{D}$ with the repeating variables such that

$$\Pi_1 = \mathcal{D} w^a V^b \rho^c$$

and in terms of dimensions

$$(MLT^{-2})(L)^a(LT^{-1})^b(ML^{-3})^c \doteq M^0L^0T^0$$

Thus, for Π_1 to be dimensionless it follows that

$$1 + c = 0 \qquad \text{(for } M\text{)}$$

$$1 + a + b - 3c = 0 \qquad \text{(for } L\text{)}$$

$$-2 - b = 0 \qquad \text{(for } T\text{)}$$

and, therefore, $a = -2$, $b = -2$, and $c = -1$. The pi term then becomes

$$\Pi_1 = \frac{\mathcal{D}}{w^2V^2\rho}$$

Next the procedure is repeated with the second nonrepeating variable, h, so that

$$\Pi_2 = h w^a V^b \rho^c$$

It follows that

$$(L)(L)^a(LT^{-1})^b(ML^{-3})^c \doteq M^0L^0T^0$$

and

$$c = 0 \qquad \text{(for } M\text{)}$$

$$1 + a + b - 3c = 0 \qquad \text{(for } L\text{)}$$

$$b = 0 \qquad \text{(for } T\text{)}$$

so that $a = -1$, $b = 0$, $c = 0$, and therefore

$$\Pi_2 = \frac{h}{w}$$

The remaining nonrepeating variable is μ so that

$$\Pi_3 = \mu w^a V^b \rho^c$$

with

$$(ML^{-1}T^{-1})(L)^a(LT^{-1})^b(ML^{-3})^c \doteq M^0L^0T^0$$

and, therefore,

$$1 + c = 0 \qquad \text{(for } M\text{)}$$

$$-1 + a + b - 3c = 0 \qquad \text{(for } L\text{)}$$

$$-1 - b = 0 \qquad \text{(for } T\text{)}$$

Solving for the exponents, we obtain $a = -1$, $b = -1$, $c = -1$ so that

$$\Pi_3 = \frac{\mu}{wV\rho}$$

Now that we have the three required pi terms we should check to make sure they are dimensionless. To make this check we use F, L, and T, which will also verify the correctness of the original dimensions used for the variables. Thus,

$$\Pi_1 = \frac{\mathscr{D}}{w^2V^2\rho} \doteq \frac{(F)}{(L)^2(LT^{-1})^2(FL^{-4}T^2)} \doteq F^0L^0T^0$$

$$\Pi_2 = \frac{h}{w} \doteq \frac{(L)}{(L)} \doteq F^0L^0T^0$$

$$\Pi_3 = \frac{\mu}{wV\rho} \doteq \frac{(FL^{-2}T)}{(L)(LT^{-1})(FL^{-4}T^2)} \doteq F^0L^0T^0$$

If these do not check, go back to the original list of variables and make sure you have the correct dimensions for each of the variables and then check the algebra you used to obtain the exponents a, b, and c.

Finally, we can express the results of the dimensional analysis in the form

$$\frac{\mathscr{D}}{w^2V^2\rho} = \tilde{\phi}\left(\frac{h}{w}, \frac{\mu}{wV\rho}\right) \qquad \textbf{(Ans)}$$

Since at this stage in the analysis the nature of the function $\tilde{\phi}$ is unknown, we could rearrange the pi terms if we so desire. For example, we could express the final result in the form

$$\frac{\mathscr{D}}{w^2\rho V^2} = \phi\left(\frac{w}{h}, \frac{\rho V w}{\mu}\right) \qquad \textbf{(Ans)}$$

which would be more conventional, since the ratio of the plate width to height, w/h, is called the *aspect ratio*, and $\rho Vw/\mu$ is the Reynolds number. To proceed, it would be necessary to perform a set of experiments to determine the nature of the function ϕ, as discussed in Section 7.7.

7.4 Some Additional Comments About Dimensional Analysis

The preceding section provides a systematic procedure for performing a dimensional analysis. Other methods could be used, although we think the method of repeating variables is the easiest for the beginning student to use. Pi terms can also be formed by inspection, as is discussed in Section 7.5. Regardless of the specific method used for the dimensional analysis, there are certain aspects of this important engineering tool that must seem a little baffling and mysterious to the student (and sometimes to the experienced investigator as well). In this section we will attempt to elaborate on some of the more subtle points that, based on our experience, can prove to be puzzling to students.

7.4.1 Selection of Variables

One of the most important and difficult steps in dimensional analysis is the selection of variables.

One of the most important, and difficult, steps in applying dimensional analysis to any given problem is the selection of the variables that are involved. As noted previously, for convenience we will use the term variable to indicate any quantity involved, including dimensional and nondimensional constants. There is no simple procedure whereby the variables can be easily identified. Generally, one must rely on a good understanding of the phenomenon involved and the governing physical laws. If extraneous variables are included, then too many pi terms appear in the final solution, and it may be difficult, time consuming, and expensive to eliminate these experimentally. If important variables are omitted, then an incorrect result will be obtained; and again, this may prove to be costly and difficult to ascertain. It is, therefore, imperative that sufficient time and attention be given to this first step in which the variables are determined.

Most engineering problems involve certain simplifying assumptions that have an influence on the variables to be considered. Usually we wish to keep the problem as simple as possible, perhaps even if some accuracy is sacrificed. A suitable balance between simplicity and accuracy is a desirable goal. How "accurate" the solution must be depends on the objective of the study; that is, we may be only concerned with general trends and, therefore, some variables that are thought to have only a minor influence in the problem may be neglected for simplicity.

It is often helpful to classify variables into three groups—geometry, material properties, and external effects.

For most engineering problems (including areas outside of fluid mechanics), pertinent variables can be classified into three general groups—geometry, material properties, and external effects.

Geometry. The geometric characteristics can usually be described by a series of lengths and angles. In most problems the geometry of the system plays an important role, and a sufficient number of geometric variables must be included to describe the system. These variables can usually be readily identified.

Material Properties. Since the response of a system to applied external effects such as forces, pressures, and changes in temperature is dependent on the nature of the materials involved in the system, the material properties that relate the external effects and the responses must be included as variables. For example, for Newtonian fluids the viscosity of the fluid is the property that relates the applied forces to the rates of deformation of the fluid. As the material behavior becomes more complex, such as would be true for non-Newtonian fluids, the determination of material properties becomes difficult, and this class of variables can be troublesome to identify.

External Effects. This terminology is used to denote any variable that produces, or tends to produce, a change in the system. For example, in structural mechanics, forces (either concentrated or distributed) applied to a system tend to change its geometry, and such forces would need to be considered as pertinent variables. For fluid mechanics, variables in this class would be related to pressures, velocities, or gravity.

The above general classes of variables are intended as broad categories that should be helpful in identifying variables. It is likely, however, that there will be important variables that do not fit easily into one of the above categories and each problem needs to be carefully analyzed.

Since we wish to keep the number of variables to a minimum, it is important that all variables are independent. For example, if in a given problem we know that the moment of inertia of the area of a circular plate is an important variable, we could list either the moment of inertia or the plate diameter as the pertinent variable. However, it would be unnecessary to include both moment of inertia and diameter, assuming that the diameter enters the problem only through the moment of inertia. In more general terms, if we have a problem in which the variables are

$$f(p, q, r, \ldots, u, v, w, \ldots) = 0 \tag{7.3}$$

and it is known that there is an additional relationship among some of the variables, for example,

$$q = f_1(u, v, w, \ldots) \tag{7.4}$$

then q is not required and can be omitted. Conversely, if it is known that the only way the variables $u, v, w, \ldots$ enter the problem is through the relationship expressed by Eq. 7.4, then the variables $u, v, w, \ldots$ can be replaced by the single variable q, therefore reducing the number of variables.

In summary, the following points should be considered in the selection of variables:

1. Clearly define the problem. What is the main variable of interest (the dependent variable)?
2. Consider the basic laws that govern the phenomenon. Even a crude theory that describes the essential aspects of the system may be helpful.
3. Start the variable selection process by grouping the variables into three broad classes: geometry, material properties, and external effects.
4. Consider other variables that may not fall into one of the above categories. For example, time will be an important variable if any of the variables are time dependent.
5. Be sure to include all quantities that enter the problem even though some of them may be held constant (e.g., the acceleration of gravity, g). For a dimensional analysis it is the dimensions of the quantities that are important—not specific values!
6. Make sure that all variables are independent. Look for relationships among subsets of the variables.

7.4.2 Determination of Reference Dimensions

For any given problem it is obviously desirable to reduce the number of pi terms to a minimum and, therefore, we wish to reduce the number of variables to a minimum; that is, we certainly do not want to include extraneous variables. It is also important to know how many reference dimensions are required to describe the variables. As we have seen in the preceding examples, F, L, and T appear to be a convenient set of basic dimensions for characterizing fluid-mechanical quantities. There is, however, really nothing ''fundamental'' about this set, and as previously noted M, L, and T would also be suitable. Actually any set of measurable quantities could be used as basic dimensions provided that the selected combination can be used to describe all secondary quantities. However, the use of FLT or MLT as basic dimensions is the simplest, and these dimensions can be used to describe fluid-mechanical phenomena. Of course, in some problems only one or two of these are required. In addition, we occasionally find that the number of reference dimensions needed to describe all variables is smaller than the number of basic dimensions. This point is illustrated in Example 7.2. Interesting discussions, both practical and philosophical, relative to the concept of basic dimensions can be found in the books by Huntley (Ref. 4) and by Isaacson and Isaacson (Ref. 12).

Typically, in fluid mechanics, the required number of reference dimensions is three, but in some problems only one or two are required.

An open, cylindrical tank having a diameter D is supported around its bottom circumference and is filled to a depth h with a liquid having a specific weight γ. The vertical deflection, δ, of the center of the bottom is a function of D, h, d, γ, and E, where d is the thickness of the bottom and E is the modulus of elasticity of the bottom material. Perform a dimensional analysis of this problem.

SOLUTION

From the statement of the problem

$$\delta = f(D, h, d, \gamma, E)$$

and the dimensions of the variables are

$$\delta \doteq L$$
$$D \doteq L$$
$$h \doteq L$$
$$d \doteq L$$
$$\gamma \doteq FL^{-3} \doteq ML^{-2}T^{-2}$$
$$E \doteq FL^{-2} \doteq ML^{-1}T^{-2}$$

where the dimensions have been expressed in terms of both the *FLT* and *MLT* systems.

We now apply the pi theorem to determine the required number of pi terms. First, let us use *F*, *L*, and *T* as our system of basic dimensions. There are six variables and two reference dimensions (*F* and *L*) required so that four pi terms are needed. For repeating variables, we can select D and γ so that

$$\Pi_1 = \delta\, D^a \gamma^b$$
$$(L)(L)^a(FL^{-3})^b \doteq F^0L^0$$

and

$$1 + a - 3b = 0 \qquad \text{(for } L)$$
$$b = 0 \qquad \text{(for } F)$$

Therefore, $a = -1$, $b = 0$, and

$$\Pi_1 = \frac{\delta}{D}$$

Similarly,

$$\Pi_2 = h\, D^a \gamma^b$$

and following the same procedure as above, $a = -1$, $b = 0$ so that

$$\Pi_2 = \frac{h}{D}$$

The remaining two pi terms can be found using the same procedure, with the result

$$\Pi_3 = \frac{d}{D} \qquad \Pi_4 = \frac{E}{D\gamma}$$

Thus, this problem can be studied by using the relationship

$$\frac{\delta}{D} = \phi\left(\frac{h}{D}, \frac{d}{D}, \frac{E}{D\gamma}\right) \qquad \textbf{(Ans)}$$

Let us now solve the same problem using the *MLT* system. Although the number of variables is obviously the same, it would seem that there are three reference dimensions required, rather than two. If this were indeed true it would certainly be fortuitous, since we would reduce the number of required pi terms from four to three. Does this seem right? How can we reduce the number of required pi terms by simply using the *MLT* system of basic dimensions? The answer is that we cannot, and a closer look at the dimensions of the variables listed above reveals that actually only two reference dimensions, MT^{-2} and L, are required.

This is an example of the situation in which the number of reference dimensions differs from the number of basic dimensions. It does not happen very often and can be detected by looking at the dimensions of the variables (regardless of the systems used) and making sure how many reference dimensions are actually required to describe the variables. Once the number of reference dimensions has been determined, we can proceed as before. Since the number of repeating variables must equal the number of reference dimensions, it follows that two reference dimensions are still required and we could again use D and γ as repeating variables. The pi terms would be determined in the same manner. For example, the pi term containing E would be developed as

$$\Pi_4 = ED^a\gamma^b$$

$$(ML^{-1}T^{-2})(L)^a(ML^{-2}T^{-2})^b \doteq (MT^{-2})^0L^0$$

$$1 + b = 0 \qquad \text{(for } MT^{-2}\text{)}$$

$$-1 + a - 2b = 0 \qquad \text{(for } L\text{)}$$

and, therefore, $a = -1$, $b = -1$ so that

$$\Pi_4 = \frac{E}{D\gamma}$$

which is the same as Π_4 obtained using the FLT system. The other pi terms would be the same, and the final result is the same; that is,

$$\frac{\delta}{D} = \phi\left(\frac{h}{D}, \frac{d}{D}, \frac{E}{D\gamma}\right) \qquad \textbf{(Ans)}$$

This will always be true—you cannot affect the required number of pi terms by using M, L, and T instead of F, L, and T, or vice versa.

7.4.3 Uniqueness of Pi Terms

There is not a unique set of pi terms for a given problem.

A little reflection on the process used to determine pi terms by the method of repeating variables reveals that the specific pi terms obtained depend on the somewhat arbitrary selection of repeating variables. For example, in the problem of studying the pressure drop in a pipe, we selected D, V, and ρ as repeating variables. This led to the formulation of the problem in terms of pi terms as

$$\frac{\Delta p_\ell D}{\rho V^2} = \phi\left(\frac{\rho VD}{\mu}\right) \qquad \textbf{(7.5)}$$

What if we had selected D, V, and μ as repeating variables? A quick check will reveal that the pi term involving Δp_ℓ becomes

$$\frac{\Delta p_\ell D^2}{V\mu}$$

and the second pi term remains the same. Thus, we can express the final result as

$$\frac{\Delta p_\ell D^2}{V\mu} = \phi_1\left(\frac{\rho VD}{\mu}\right) \qquad \textbf{(7.6)}$$

Both results are correct, and both would lead to the same final equation for Δp_ℓ. Note, however, that the functions ϕ and ϕ_1 in Eqs. 7.5 and 7.6 will be different because the dependent pi terms are different for the two relationships.

We can conclude from this illustration that there is *not* a unique set of pi terms which arises from a dimensional analysis. However, the required *number* of pi terms is fixed, and once a correct set is determined all other possible sets can be developed from this set by combinations of products of powers of the original set. Thus, if we have a problem involving, say, three pi terms,

Once a correct set of pi terms is obtained, any other set can be obtained by manipulation of the original set.

$$\Pi_1 = \phi(\Pi_2, \Pi_3)$$

we could always form a new set from this one by combining the pi terms. For example, we could form a new pi term, Π_2', by letting

$$\Pi_2' = \Pi_2^a \, \Pi_3^b$$

where a and b are arbitrary exponents. Then the relationship could be expressed as

$$\Pi_1 = \phi_1(\Pi_2', \Pi_3)$$

or

$$\Pi_1 = \phi_2(\Pi_2, \Pi_2')$$

All of these would be correct. It should be emphasized, however, that the required number of pi terms cannot be reduced by this manipulation; only the form of the pi terms is altered. By using this technique we see that the pi terms in Eq. 7.6 could be obtained from those in Eq. 7.5; that is, we multiply Π_1 in Eq. 7.5 by Π_2 so that

$$\left(\frac{\Delta p_\ell D}{\rho V^2}\right)\left(\frac{\rho V D}{\mu}\right) = \frac{\Delta p_\ell D^2}{V \mu}$$

which is the Π_1 of Eq. 7.6.

There is no simple answer to the question: Which form for the pi terms is best? Usually our only guideline is to keep the pi terms as simple as possible. Also, it may be that certain pi terms will be easier to work with in actually performing experiments. The final choice remains an arbitrary one and generally will depend on the background and experience of the investigator. It should again be emphasized, however, that although there is no unique set of pi terms for a given problem, the *number* required is fixed in accordance with the pi theorem.

7.5 Determination of Pi Terms by Inspection

One method for forming pi terms has been presented in Section 7.3. This method provides a step-by-step procedure that if executed properly will provide a correct and complete set of pi terms. Although this method is simple and straightforward, it is rather tedious, particularly for problems in which large numbers of variables are involved. Since the only restrictions placed on the pi terms are that they be (1) correct in number, (2) dimensionless, and (3) independent, it is possible to simply form the pi terms by inspection, without resorting to the more formal procedure.

To illustrate this approach, we again consider the pressure drop per unit length along a smooth pipe. Regardless of the technique to be used, the starting point remains the same—determine the variables, which in this case are

$$\Delta p_\ell = f(D, \rho, \mu, V)$$

Next, the dimensions of the variables are listed:

$$\Delta p_\ell \doteq FL^{-3}$$
$$D \doteq L$$
$$\rho \doteq FL^{-4}T^2$$
$$\mu \doteq FL^{-2}T$$
$$V \doteq LT^{-1}$$

and subsequently the number of reference dimensions determined. The application of the pi theorem then tells us how many pi terms are required. In this problem, since there are five variables and three reference dimensions, two pi terms are needed. Thus, the required number of pi terms can be easily determined, and the determination of this number should always be done at the beginning of the analysis.

Pi terms can be formed by inspection by simply making use of the fact that each pi term must be dimensionless.

Once the number of pi terms is known, we can form each pi term by inspection, simply making use of the fact that each pi term must be dimensionless. We will always let Π_1 contain the dependent variable, which in this example is Δp_ℓ. Since this variable has the dimensions FL^{-3}, we need to combine it with other variables so that a nondimensional product will result. One possibility is to first divide Δp_ℓ by ρ so that

$$\frac{\Delta p_\ell}{\rho} \doteq \frac{(FL^{-3})}{(FL^{-4}T^2)} \doteq \frac{L}{T^2}$$

The dependence on F has been eliminated, but $\Delta p_\ell/\rho$ is obviously not dimensionless. To eliminate the dependence on T, we can divide by V^2 so that

$$\left(\frac{\Delta p_\ell}{\rho}\right)\frac{1}{V^2} \doteq \left(\frac{L}{T^2}\right)\frac{1}{(LT^{-1})^2} \doteq \frac{1}{L}$$

Finally, to make the combination dimensionless we multiply by D so that

$$\left(\frac{\Delta p_\ell}{\rho V^2}\right) D \doteq \left(\frac{1}{L}\right)(L) \doteq L^0$$

Thus,

$$\Pi_1 = \frac{\Delta p_\ell D}{\rho V^2}$$

Next, we will form the second pi term by selecting the variable that was not used in Π_1, which in this case is μ. We simply combine μ with the other variables to make the combination dimensionless (but do not use Δp_ℓ in Π_2, since we want the dependent variable to appear only in Π_1). For example, divide μ by ρ (to eliminate F), then by V (to eliminate T), and finally by D (to eliminate L). Thus,

$$\Pi_2 = \frac{\mu}{\rho V D} \doteq \frac{(FL^{-2}T)}{(FL^{-4}T^2)(LT^{-1})(L)} \doteq F^0L^0T^0$$

and, therefore,

$$\frac{\Delta p_\ell D}{\rho V^2} = \phi\left(\frac{\mu}{\rho V D}\right)$$

which is, of course, the same result we obtained by using the method of repeating variables.

An additional concern, when one is forming pi terms by inspection, is to make certain that they are all independent. In the pipe flow example, Π_2 contains μ, which does not appear in Π_1, and therefore these two pi terms are obviously independent. In a more general case a pi term would not be independent of the others in a given problem if it can be formed by some combination of the others. For example, if Π_2 can be formed by a combination of say Π_3, Π_4, and Π_5 such as

$$\Pi_2 = \frac{\Pi_3^2 \, \Pi_4}{\Pi_5}$$

then Π_2 is not an independent pi term. We can ensure that each pi term is independent of those preceding it by incorporating a new variable in each pi term.

Although forming pi terms by inspection is essentially equivalent to the repeating variable method, it is less structured. With a little practice the pi terms can be readily formed by inspection, and this method offers an alternative to the more formal procedures.

7.6 Common Dimensionless Groups in Fluid Mechanics

At the top of Table 7.1 is a list of variables that commonly arise in fluid mechanics problems. The list is obviously not exhaustive but does indicate a broad range of variables likely to be found in a typical problem. Fortunately, not all of these variables would be encountered in all problems. However, when combinations of these variables are present, it is standard practice to combine them into some of the common dimensionless groups (pi terms) given in Table 7.1. These combinations appear so frequently that special names are associated with them as indicated in the table.

A useful physical interpretation can often be given to dimensionless groups.

It is also often possible to provide a physical interpretation to the dimensionless groups which can be helpful in assessing their influence in a particular application. For example, the Froude number is an index of the ratio of the force due to the acceleration of a fluid particle to the force due to gravity (weight). This can be demonstrated by considering a fluid particle moving along a streamline (Fig. 7.3). The magnitude of the component of inertia force F_I along the streamline can be expressed as $F_I = a_s m$, where a_s is the magnitude of the acceleration along the streamline for a particle having a mass m. From our study of particle motion along a curved path (see Section 3.1) we know that

$$a_s = \frac{dV_s}{dt} = V_s \frac{dV_s}{ds}$$

where s is measured along the streamline. If we write the velocity, V_s, and length, s, in dimensionless form, that is,

$$V_s^* = \frac{V_s}{V} \qquad s^* = \frac{s}{\ell}$$

where V and ℓ represent some characteristic velocity and length, respectively, then

$$a_s = \frac{V^2}{\ell} V_s^* \frac{dV_s^*}{ds^*}$$

and

$$F_I = \frac{V^2}{\ell} V_s^* \frac{dV_s^*}{ds^*} m$$

The magnitude of the weight of the particle, F_G, is $F_G = gm$, so the ratio of the inertia to the gravitational force is

$$\frac{F_I}{F_G} = \frac{V^2}{g\ell} V_s^* \frac{dV_s^*}{ds^*}$$

■ **TABLE 7.1**
Some Common Variables and Dimensionless Groups in Fluid Mechanics

Variables: Acceleration of gravity, g; Bulk modulus, E_v; Characteristic length, ℓ; Density, ρ; Frequency of oscillating flow, ω; Pressure, p (or Δp); Speed of sound, c; Surface tension, σ; Velocity, V; Viscosity, μ

Dimensionless Groups	Name	Interpretation (Index of Force Ratio Indicated)	Types of Applications
$\frac{\rho V \ell}{\mu}$	Reynolds number, Re	$\frac{\text{inertia force}}{\text{viscous force}}$	Generally of importance in all types of fluid dynamics problems
$\frac{V}{\sqrt{g\ell}}$	Froude number, Fr	$\frac{\text{inertia force}}{\text{gravitational force}}$	Flow with a free surface
$\frac{p}{\rho V^2}$	Euler number, Eu	$\frac{\text{pressure force}}{\text{inertia force}}$	Problems in which pressure, or pressure differences, are of interest
$\frac{\rho V^2}{E_v}$	Cauchy number,[a] Ca	$\frac{\text{inertia force}}{\text{compressibility force}}$	Flows in which the compressibility of the fluid is important
$\frac{V}{c}$	Mach number,[a] Ma	$\frac{\text{inertia force}}{\text{compressibility force}}$	Flows in which the compressibility of the fluid is important
$\frac{\omega \ell}{V}$	Strouhal number, St	$\frac{\text{inertia (local) force}}{\text{inertia (convective) force}}$	Unsteady flow with a characteristic frequency of oscillation
$\frac{\rho V^2 \ell}{\sigma}$	Weber number, We	$\frac{\text{inertia force}}{\text{surface tension force}}$	Problems in which surface tension is important

[a]The Cauchy number and the Mach number are related and either can be used as an index of the relative effects of inertia and compressibility. See accompanying discussion.

Special names along with physical interpretations are given to the most common dimensionless groups.

Thus, the force ratio F_I/F_G is proportional to $V^2/g\ell$, and the square root of this ratio, $V/\sqrt{g\ell}$, is called the *Froude number*. We see that a physical interpretation of the Froude number is that it is a measure of, or an index of, the relative importance of inertial forces acting on fluid particles to the weight of the particle. Note that the Froude number is not really *equal* to this force ratio, but is simply some type of average measure of the influence of these two forces. In a problem in which gravity (or weight) is not important, the Froude number would not appear as an important pi term. A similar interpretation in terms of indices of force ratios can be given to the other dimensionless groups, as indicated in Table 7.1, and

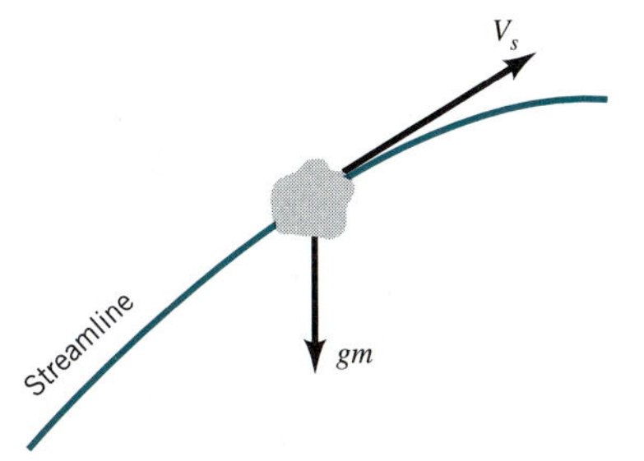

■ **FIGURE 7.3 The force of gravity acting on a fluid particle moving along a streamline.**

a further discussion of the basis for this type of interpretation is given in the last section in this chapter. Some additional details about these important dimensionless groups are given below, and the types of application or problem in which they arise are briefly noted in the last column of Table 7.1.

The Reynolds number is undoubtedly the most famous dimensionless parameter in fluid mechanics.

Reynolds Number The Reynolds number is undoubtedly the most famous dimensionless parameter in fluid mechanics. It is named in honor of Osborne Reynolds (1842–1912), a British engineer who first demonstrated that this combination of variables could be used as a criterion to distinguish between laminar and turbulent flow. In most fluid flow problems there will be a characteristic length, ℓ, and a velocity, V, as well as the fluid properties of density, ρ, and viscosity, μ, which are relevant variables in the problem. Thus, with these variables the Reynolds number

$$\text{Re} = \frac{\rho V \ell}{\mu}$$

arises naturally from the dimensional analysis. The Reynolds number is a measure of the ratio of the inertia force on an element of fluid to the viscous force on an element. When these two types of forces are important in a given problem, the Reynolds number will play an important role. However, if the Reynolds number is very small ($\text{Re} \ll 1$), this is an indication that the viscous forces are dominant in the problem, and it may be possible to neglect the inertial effects; that is, the density of the fluid will not be an important variable. Flows at very small Reynolds numbers are commonly referred to as "creeping flows" as discussed in Section 6.10. Conversely, for large Reynolds number flows, viscous effects are small relative to inertial effects and for these cases it may be possible to neglect the effect of viscosity and consider the problem as one involving a "nonviscous" fluid. This type of problem is considered in detail in Sections 6.4 through 6.7.

V7.1

Froude Number The Froude number

$$\text{Fr} = \frac{V}{\sqrt{g\ell}}$$

is distinguished from the other dimensionless groups in Table 7.1 in that it contains the acceleration of gravity, g. The acceleration of gravity becomes an important variable in a fluid dynamics problem in which the fluid weight is an important force. As discussed, the Froude number is a measure of the ratio of the inertia force on an element of fluid to the weight of the element. It will generally be important in problems involving flows with free surfaces since gravity principally affects this type of flow. Typical problems would include the study of the flow of water around ships (with the resulting wave action) or flow through rivers or open conduits. The Froude number is named in honor of William Froude (1810–1879), a British civil engineer, mathematician, and naval architect who pioneered the use of towing tanks for the study of ship design. It is to be noted that the Froude number is also commonly defined as the square of the Froude number listed in Table 7.1.

Euler Number The Euler number

$$\text{Eu} = \frac{p}{\rho V^2}$$

can be interpreted as a measure of the ratio of pressure forces to inertial forces, where p is some characteristic pressure in the flow field. Very often the Euler number is written in terms of a pressure difference, Δp, so that $\text{Eu} = \Delta p / \rho V^2$. Also, this combination expressed as $\Delta p / \frac{1}{2}\rho V^2$ is called the *pressure coefficient*. Some form of the Euler number would normally be used in problems in which pressure or the pressure difference between two points is an important variable. The Euler number is named in honor of Leonhard Euler (1707–1783), a famous Swiss mathematician who pioneered work on the relationship between pressure and flow. For problems in which cavitation is of concern, the dimensionless group $(p_r - p_v)/\frac{1}{2}\rho V^2$ is commonly used, where p_v is the vapor pressure and p_r is some reference pressure. Although this dimensionless group has the same form as the Euler number, it is generally referred to as the *cavitation number.*

Cauchy Number and Mach Number The Cauchy number

$$\text{Ca} = \frac{\rho V^2}{E_v}$$

and the Mach number

The Mach number is a commonly used dimensionless parameter in compressible flow problems.

$$\text{Ma} = \frac{V}{c}$$

are important dimensionless groups in problems in which fluid compressibility is a significant factor. Since the speed of sound, c, in a fluid is equal to $c = \sqrt{E_v/\rho}$ (see Section 1.7.3), it follows that

$$\text{Ma} = V\sqrt{\frac{\rho}{E_v}}$$

and the square of the Mach number

$$\text{Ma}^2 = \frac{\rho V^2}{E_v} = \text{Ca}$$

is equal to the Cauchy number. Thus, either number (but not both) may be used in problems in which fluid compressibility is important. Both numbers can be interpreted as representing an index of the ratio of inertial forces to compressibility forces. When the Mach number is relatively small (say, less than 0.3), the inertial forces induced by the fluid motion are not sufficiently large to cause a significant change in the fluid density, and in this case the compressibility of the fluid can be neglected. The Mach number is the more commonly used parameter in compressible flow problems, particularly in the fields of gas dynamics and aerodynamics. The Cauchy number is named in honor of Augustin Louis de Cauchy (1789–1857), a French engineer, mathematician, and hydrodynamicist. The Mach number is named in honor of Ernst Mach (1838–1916), an Austrian physicist and philosopher.

Strouhal Number The Strouhal number

$$\text{St} = \frac{\omega \ell}{V}$$

is a dimensionless parameter that is likely to be important in unsteady, oscillating flow problems in which the frequency of the oscillation is ω. It represents a measure of the ratio of inertial forces due to the unsteadiness of the flow (local acceleration) to the inertial forces due to changes in velocity from point to point in the flow field (convective acceleration). This type of unsteady flow may develop when a fluid flows past a solid body (such as a wire or cable) placed in the moving stream. For example, in a certain Reynolds number range, a periodic flow will develop downstream from a cylinder placed in a moving fluid due to a regular pattern of vortices that are shed from the body. (See the photograph at the beginning of this chapter and Fig. 9.23.) This system of vortices, called a *Kármán vortex trail* [named after Theodor von Kármán (1881–1963), a famous fluid mechanician], creates an oscillating flow at a discrete frequency, ω, such that the Strouhal number can be closely correlated with the Reynolds number. When the frequency is in the audible range, a sound can be heard and the bodies appear to "sing." In fact, the Strouhal number is named in honor of Vincenz Strouhal (1850–1922), who used this parameter in his study of "singing wires." The most dramatic evidence of this phenomenon occurred in 1940 with the collapse of the Tacoma Narrows bridge. The shedding frequency of the vortices coincided with the natural frequency of the bridge, thereby setting up a resonant condition that eventually led to the collapse of the bridge.

There are, of course, other types of oscillating flows. For example, blood flow in arteries is periodic and can be analyzed by breaking up the periodic motion into a series of harmonic components (Fourier series analysis), with each component having a frequency that is a multiple of the fundamental frequency, ω (the pulse rate). Rather than use the Strouhal number in this type of problem, a dimensionless group formed by the product of St and Re is used; that is

$$\text{St} \times \text{Re} = \frac{\rho \omega \ell^2}{\mu}$$

The square root of this dimensionless group is often referred to as the *frequency parameter.*

Weber Number The Weber number

$$\text{We} = \frac{\rho V^2 \ell}{\sigma}$$

may be important in problems in which there is an interface between two fluids. In this situation the surface tension may play an important role in the phenomenon of interest. The Weber number can be thought of as an index of the inertial force to the surface tension force acting on a fluid element. Common examples of problems in which this parameter may be important include the flow of thin films of liquid, or in the formation of droplets or bubbles. Clearly, not all problems involving flows with an interface will require the inclusion of surface tension. The flow of water in a river is not affected significantly by surface tension, since inertial and gravitational effects are dominant (We $\gg$ 1). However, as discussed in a later section, for river models (which may have small depths) caution is required so that surface tension does not become important in the model, whereas it is not important in the actual river. The Weber number is named after Moritz Weber (1871–1951), a German professor of naval mechanics who was instrumental in formalizing the general use of common dimensionless groups as a basis for similitude studies.

7.7 Correlation of Experimental Data

Dimensional analysis greatly facilitates the efficient handling, interpretation, and correlation of experimental data.

One of the most important uses of dimensional analysis is as an aid in the efficient handling, interpretation, and correlation of experimental data. Since the field of fluid mechanics relies heavily on empirical data, it is not surprising that dimensional analysis is such an important tool in this field. As noted previously, a dimensional analysis cannot provide a complete answer to any given problem, since the analysis only provides the dimensionless groups describing the phenomenon, and not the specific relationship among the groups. To determine this relationship, suitable experimental data must be obtained. The degree of difficulty involved in this process depends on the number of pi terms, and the nature of the experiments (How hard is it to obtain the measurements?). The simplest problems are obviously those involving the fewest pi terms, and the following sections indicate how the complexity of the analysis increases with the increasing number of pi terms.

7.7.1 Problems with One Pi Term

Application of the pi theorem indicates that if the number of variables minus the number of reference dimensions is equal to unity, then only *one* pi term is required to describe the

If only one pi term is involved in a problem, it must be equal to a constant.

phenomenon. The functional relationship that must exist for one pi term is

$$\Pi_1 = C$$

where C is a constant. This is one situation in which a dimensional analysis reveals the specific form of the relationship and, as is illustrated by the following example, shows how the individual variables are related. The value of the constant, however, must still be determined by experiment.

Assume that the drag, $\mathcal{D}$, acting on a spherical particle that falls very slowly through a viscous fluid is a function of the particle diameter, d, the particle velocity, V, and the fluid viscosity, μ. Determine, with the aid of dimensional analysis, how the drag depends on the particle velocity.

SOLUTION

From the information given, it follows that

$$\mathcal{D} = f(d, V, \mu)$$

and the dimensions of the variables are

$$\mathcal{D} \doteq F$$
$$d \doteq L$$
$$V \doteq LT^{-1}$$
$$\mu \doteq FL^{-2}T$$

We see that there are four variables and three reference dimensions (F, L, and T) required to describe the variables. Thus, according to the pi theorem, one pi term is required. This pi term can be easily formed by inspection and can be expressed as

$$\Pi_1 = \frac{\mathcal{D}}{\mu V d}$$

Since there is only one pi term, it follows that

$$\frac{\mathcal{D}}{\mu V d} = C$$

or

$$\mathcal{D} = C\mu V d$$

Thus, for a given particle and fluid, the drag varies directly with the velocity so that

$$\mathcal{D} \propto V \qquad \textbf{(Ans)}$$

Actually, the dimensional analysis reveals that the drag not only varies directly with the velocity, but it also varies directly with the particle diameter and the fluid viscosity. We could not, however, predict the value of the drag, since the constant, C, is unknown. An experiment would have to be performed in which the drag and the corresponding velocity are measured for a given particle and fluid. Although in principle we would only have to run a single test, we would certainly want to repeat it several times to obtain a reliable value for C. It should be emphasized that once the value of C is determined it is not necessary to run similar tests by using different spherical particles and fluids; that is, C is a universal constant so long as the drag is a function only of particle diameter, velocity, and fluid viscosity.

An approximate solution to this problem can also be obtained theoretically, from which it is found that $C = 3\pi$ so that

$$\mathscr{D} = 3\pi\mu V d$$

This equation is commonly called *Stokes law* and is used in the study of the settling of particles. Our experiments would reveal that this result is only valid for small Reynolds numbers ($\rho V d/\mu \ll 1$). This follows, since in the original list of variables, we have neglected inertial effects (fluid density is not included as a variable). The inclusion of an additional variable would lead to another pi term so that there would be two pi terms rather than one.

7.7.2 Problems with Two or More Pi Terms

If a given phenomenon can be described with two pi terms such that

$$\Pi_1 = \phi(\Pi_2)$$

For problems involving only two pi terms, results of an experiment can be conveniently presented in a simple graph.

the functional relationship among the variables can then be determined by varying Π_2 and measuring the corresponding values of Π_1. For this case the results can be conveniently presented in graphical form by plotting Π_1 versus Π_2 as is illustrated in Fig. 7.4. It should be emphasized that the curve shown in Fig. 7.4 would be a "universal" one for the particular phenomenon studied. This means that if the variables and the resulting dimensional analysis are correct, then there is only a single relationship between Π_1 and Π_2, as illustrated in Fig. 7.4. However, since this is an empirical relationship, we can only say that it is valid over the range of Π_2 covered by the experiments. It would be unwise to extrapolate beyond this range, since as illustrated with the dashed lines in the figure, the nature of the phenomenon could dramatically change as the range of Π_2 is extended. In addition to presenting the data graphically, it may be possible (and desirable) to obtain an empirical equation relating Π_1 and Π_2 by using a standard curve-fitting technique.

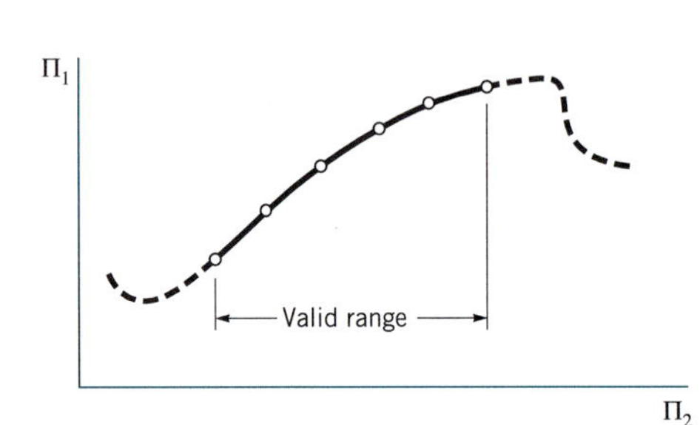

■ **FIGURE 7.4 The graphical presentation of data for problems involving two pi terms, with an illustration of the potential danger of extrapolation of data.**

The relationship between the pressure drop per unit length along a smooth-walled, horizontal pipe and the variables that affect the pressure drop is to be determined experimentally. In the laboratory the pressure drop was measured over a 5-ft length of smooth-walled pipe having an inside diameter of 0.496 in. The fluid used was water at 60 °F ($\mu = 2.34 \times 10^{-5}$ lb·s/ft^2, $\rho = 1.94$ slugs/ft^3). Tests were run in which the velocity was varied and the corresponding pressure drop measured. The results of these tests are shown below:

Velocity (ft/s)	1.17	1.95	2.91	5.84	11.13	16.92	23.34	28.73
Pressure drop (lb/ft^2) (for 5-ft length)	6.26	15.6	30.9	106	329	681	1200	1730

Make use of these data to obtain a general relationship between the pressure drop per unit length and the other variables.

SOLUTION

The first step is to perform a dimensional analysis during the planning stage *before* the experiments are actually run. As was discussed in Section 7.3, we will assume that the pressure drop per unit length, Δp_ℓ, is a function of the pipe diameter, D, fluid density, ρ, fluid viscosity, μ, and the velocity, V. Thus,

$$\Delta p_\ell = f(D, \rho, \mu, V)$$

and application of the pi theorem yields

$$\frac{D\,\Delta p_\ell}{\rho V^2} = \phi\left(\frac{\rho V D}{\mu}\right)$$

To determine the form of the relationship, we need to vary the Reynolds number, $\rho VD/\mu$, and to measure the corresponding values of $D\,\Delta p_\ell/\rho V^2$. The Reynolds number could be varied by changing any one of the variables, ρ, V, D, or μ, or any combination of them. However, the simplest way to do this is to vary the velocity, since this will allow us to use the same fluid and pipe. Based on the data given, values for the two pi terms can be computed with the result:

$D\,\Delta p_\ell/\rho V^2$	0.0195	0.0175	0.0155	0.0132	0.0113	0.0101	0.00939	0.00893
$\rho VD/\mu$	4.01×10^3	6.68×10^3	9.97×10^3	2.00×10^4	3.81×10^4	5.80×10^4	8.00×10^4	9.85×10^4

These are dimensionless groups so that their values are independent of the system of units used so long as a consistent system is used. For example, if the velocity is in ft/s, then the diameter should be in feet, not inches or meters. Note that since the Reynolds numbers are all greater than 2100, the flow in the pipe is turbulent (see Section 8.1.1).

A plot of these two pi terms can now be made with the results shown in Fig. E7.4*a*. The correlation appears to be quite good, and if it was not, this would suggest that either we had large experimental measurement errors or that we had perhaps omitted an important variable. The curve shown in Fig. E7.4*a* represents the general relationship between the pressure drop and the other factors in the range of Reynolds numbers between 4.01×10^3 and 9.85×10^4. Thus, for this range of Reynolds numbers it is *not* necessary to repeat the tests for other pipe sizes or other fluids provided the assumed independent variables (D, ρ, μ, V) are the only important ones.

Since the relationship between Π_1 and Π_2 is nonlinear, it is not immediately obvious what form of empirical equation might be used to describe the relationship. If, however, the same data are plotted on logarithmic graph paper, as is shown in Fig. E7.4*b*, the data form a straight line, suggesting that a suitable equation is of the form $\Pi_1 = A\Pi_2^n$ where A and n are empirical constants to be determined from the data by using a suitable curve-fitting technique, such as a nonlinear regression program. For the data given in this example, a good fit of the data is obtained with the equation

$$\Pi_1 = 0.150\,\Pi_2^{-0.25} \qquad \textbf{(Ans)}$$

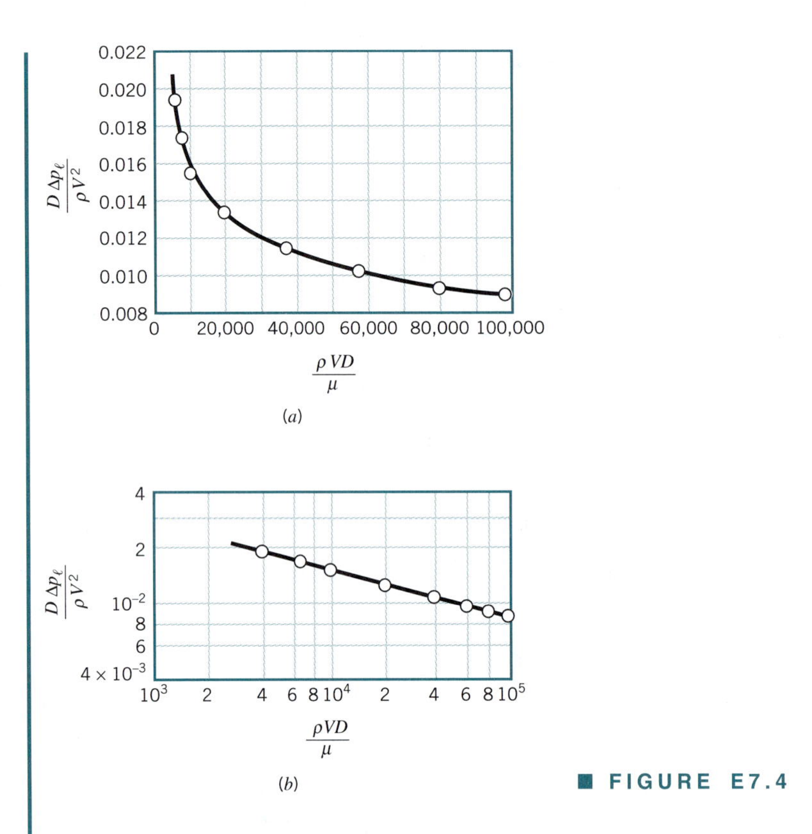

■ FIGURE E7.4

In 1911, H. Blasius (1883–1970), a German fluid mechanician, established a similar empirical equation that is used widely for predicting the pressure drop in smooth pipes in the range $4 \times 10^3 < \text{Re} < 10^5$ (Ref. 16). This equation can be expressed in the form

$$\frac{D\,\Delta p_\ell}{\rho V^2} = 0.1582\left(\frac{\rho VD}{\mu}\right)^{-1/4}$$

The so-called Blasius formula is based on numerous experimental results of the type used in this example. Flow in pipes is discussed in more detail in the next chapter, where it is shown how pipe roughness (which introduces another variable) may affect the results given in this example (which is for smooth-walled pipes).

As the number of required pi terms increases, it becomes more difficult to display the results in a convenient graphical form and to determine a specific empirical equation that describes the phenomenon. For problems involving three pi terms

$$\Pi_1 = \phi(\Pi_2, \Pi_3)$$

For problems involving more than two or three pi terms, it is often necessary to use a model to predict specific characteristics.

it is still possible to show data correlations on simple graphs by plotting families of curves as illustrated in Fig. 7.5. This is an informative and useful way of representing the data in a general way. It may also be possible to determine a suitable empirical equation relating the three pi terms. However, as the number of pi terms continues to increase, corresponding to an increase in the general complexity of the problem of interest, both the graphical presentation and the determination of a suitable empirical equation become intractable. For these more complicated problems, it is often more feasible to use models to predict specific characteristics of the system rather than to try to develop general correlations.

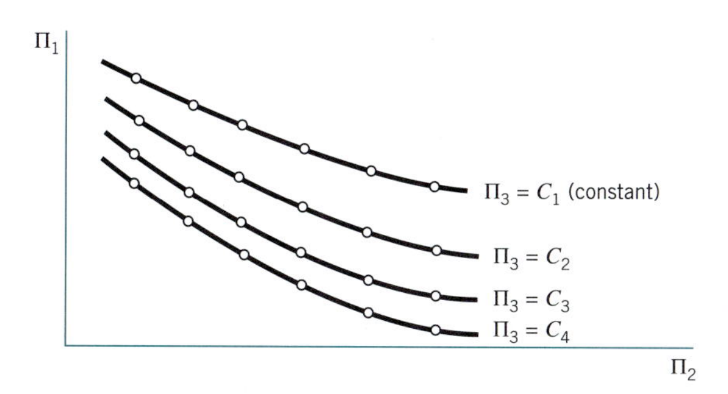

■ FIGURE 7.5 **The graphical presentation of data for problems involving three pi terms.**

7.8 Modeling and Similitude

An engineering model is used to predict the behavior of the physical system of interest.

Models are widely used in fluid mechanics. Major engineering projects involving structures, aircraft, ships, rivers, harbors, dams, air and water pollution, and so on, frequently involve the use of models. Although the term ''model'' is used in many different contexts, the ''engineering model'' generally conforms to the following definition. *A model is a representation of a physical system that may be used to predict the behavior of the system in some desired respect.* The physical system for which the predictions are to be made is called the *prototype.* Although *mathematical* or *computer* models may also conform to this definition, our interest will be in physical models, that is, models that resemble the prototype but are generally of a different size, may involve different fluids, and often operate under different conditions (pressures, velocities, etc.). Usually a model is smaller than the prototype. Therefore, it is more easily handled in the laboratory and less expensive to construct and operate than a large prototype. Occasionally, if the prototype is very small, it may be advantageous to have a model that is larger than the prototype so that it can be more easily studied. For example, large models have been used to study the motion of red blood cells, which are approximately 8 μm in diameter. With the successful development of a valid model, it is possible to predict the behavior of the prototype under a certain set of conditions. We may also wish to examine a priori the effect of possible design changes that are proposed for a hydraulic structure or fluid-flow system. There is, of course, an inherent danger in the use of models in that predictions can be made that are in error and the error not detected until the prototype is found not to perform as predicted. It is, therefore, imperative that the model be properly designed and tested and that the results be interpreted correctly. In the following sections we will develop the procedures for designing models so that the model and prototype will behave in a similar fashion.

7.8.1 Theory of Models

The theory of models can be readily developed by using the principles of dimensional analysis. It has been shown that any given problem can be described in terms of a set of pi terms as

$$\Pi_1 = \phi(\Pi_2, \Pi_3, \ldots, \Pi_n) \tag{7.7}$$

In formulating this relationship, only a knowledge of the general nature of the physical phenomenon, and the variables involved, is required. Specific values for variables (size of components, fluid properties, and so on) are not needed to perform the dimensional analysis. Thus, Eq. 7.7 applies to any system that is governed by the same variables. If Eq. 7.7 describes the behavior of a particular prototype, a similar relationship can be written for a model of this prototype; that is,

$$\Pi_{1m} = \phi(\Pi_{2m}, \Pi_{3m}, \ldots, \Pi_{nm}) \tag{7.8}$$

where the form of the function will be the same as long as the same phenomenon is involved in both the prototype and the model. Variables, or pi terms, without a subscript will refer to the prototype, whereas the subscript m will be used to designate the model variables or pi terms.

The pi terms can be developed so that Π_1 contains the variable that is to be predicted from observations made on the model. Therefore, if the model is designed and operated under the following conditions,

The similarity requirements for a model can be readily obtained with the aid of dimensional analysis.

$$\begin{aligned} \Pi_{2m} &= \Pi_2 \\ \Pi_{3m} &= \Pi_3 \\ &\vdots \\ \Pi_{nm} &= \Pi_n \end{aligned} \tag{7.9}$$

then with the presumption that the form of ϕ is the same for model and prototype, it follows that

$$\Pi_1 = \Pi_{1m} \tag{7.10}$$

Equation 7.10 is the desired *prediction equation* and indicates that the measured value of Π_{1m} obtained with the model will be equal to the corresponding Π_1 for the prototype as long as the other pi terms are equal. The conditions specified by Eqs. 7.9 provide the *model design conditions*, also called *similarity requirements* or *modeling laws.*

As an example of the procedure, consider the problem of determining the drag, $\mathscr{D}$, on a thin rectangular plate ($w \times h$ in size) placed normal to a fluid with velocity, V. The dimensional analysis of this problem was performed in Example 7.1, where it was assumed that

$$\mathscr{D} = f(w, h, \mu, \rho, V)$$

Application of the pi theorem yielded

$$\frac{\mathscr{D}}{w^2\rho V^2} = \phi\left(\frac{w}{h}, \frac{\rho V w}{\mu}\right) \tag{7.11}$$

We are now concerned with designing a model that could be used to predict the drag on a certain prototype (which presumably has a different size than the model). Since the relationship expressed by Eq. 7.11 applies to both prototype and model, Eq. 7.11 is assumed to govern the prototype, with a similar relationship

$$\frac{\mathscr{D}_m}{w_m^2\rho_m V_m^2} = \phi\left(\frac{w_m}{h_m}, \frac{\rho_m V_m w_m}{\mu_m}\right) \tag{7.12}$$

for the model. The model design conditions, or similarity requirements, are therefore

$$\frac{w_m}{h_m} = \frac{w}{h} \qquad \frac{\rho_m V_m w_m}{\mu_m} = \frac{\rho V w}{\mu}$$

The size of the model is obtained from the first requirement which indicates that

$$w_m = \frac{h_m}{h} w \tag{7.13}$$

We are free to establish the height ratio h_m/h, but then the model plate width, w_m, is fixed in accordance with Eq. 7.13.

The second similarity requirement indicates that the model and prototype must be operated at the same Reynolds number. Thus, the required velocity for the model is obtained from the relationship

$$V_m = \frac{\mu_m}{\mu} \frac{\rho}{\rho_m} \frac{w}{w_m} V \tag{7.14}$$

Note that this model design requires not only geometric scaling, as specified by Eq. 7.13, but also the correct scaling of the velocity in accordance with Eq. 7.14. This result is typical of most model designs—there is more to the design than simply scaling the geometry!

With the foregoing similarity requirements satisfied, the prediction equation for the drag is

$$\frac{\mathscr{D}}{w^2 \rho V^2} = \frac{\mathscr{D}_m}{w_m^2 \rho_m V_m^2}$$

or

$$\mathscr{D} = \left(\frac{w}{w_m}\right)^2 \left(\frac{\rho}{\rho_m}\right) \left(\frac{V}{V_m}\right)^2 \mathscr{D}_m$$

Thus, a measured drag on the model, $\mathscr{D}_m$, must be multiplied by the ratio of the square of the plate widths, the ratio of the fluid densities, and the ratio of the square of the velocities to obtain the predicted value of the prototype drag, $\mathscr{D}$.

Similarity between a model and a prototype is achieved by equating pi terms.

Generally, as is illustrated in this example, to achieve similarity between model and prototype behavior, *all the corresponding pi terms must be equated between model and prototype.* Usually, one or more of these pi terms will involve ratios of important lengths (such as w/h in the foregoing example); that is, they are purely geometrical. Thus, when we equate the pi terms involving length ratios, we are requiring that there be complete *geometric similarity* between the model and prototype. This means that the model must be a scaled version of the prototype. Geometric scaling may extend to the finest features of the system, such as surface roughness, or small protuberances on a structure, since these kinds of geometric features may significantly influence the flow. Any deviation from complete geometric similarity for a model must be carefully considered. Sometimes complete geometric scaling may be difficult to achieve, particularly when dealing with surface roughness, since roughness is difficult to characterize and control.

V7.2

Another group of typical pi terms (such as the Reynolds number in the foregoing example) involves force ratios as noted in Table 7.1. The equality of these pi terms requires the ratio of like forces in model and prototype to be the same. Thus, for flows in which the Reynolds numbers are equal, the ratio of viscous forces in model and prototype is equal to the ratio of inertia forces. If other pi terms are involved, such as the Froude number or Weber number, a similar conclusion can be drawn; that is, the equality of these pi terms requires the ratio of like forces in model and prototype to be the same. Thus, when these types of pi terms are equal in model and prototype, we have *dynamic similarity* between model and prototype. It follows that with both geometric and dynamic similarity the streamline patterns will be the same and corresponding velocity ratios (V_m/V) and acceleration ratios (a_m/a) are constant throughout the flow field. Thus, *kinematic similarity* exists between model and prototype. To have complete similarity between model and prototype, we must maintain geometric, kinematic, and dynamic similarity between the two systems. This will automatically follow if all the important variables are included in the dimensional analysis, and if all the similarity requirements based on the resulting pi terms are satisfied.

EXAMPLE 7.5

A long structural component of a bridge has the cross section shown in Fig. E7.5. It is known that when a steady wind blows past this type of bluff body, vortices may develop on the downwind side that are shed in a regular fashion at some definite frequency. Since these vortices can create harmful periodic forces acting on the structure, it is important to determine the shedding frequency. For the specific structure of interest, $D = 0.1$ m, $H = 0.3$ m, and a representative wind velocity is 50 km/hr. Standard air can be assumed. The shedding frequency is to be determined through the use of a small-scale model that is to be tested in a water tunnel. For the model $D_m = 20$ mm and the water temperature is 20 °C. Determine the model dimension, H_m, and the velocity at which the test should be performed. If the shedding frequency for the model is found to be 49.9 Hz, what is the corresponding frequency for the prototype?

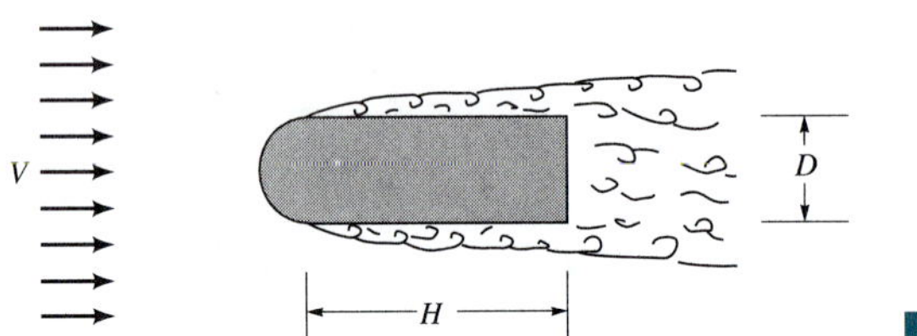

■ FIGURE E7.5

SOLUTION

We expect the shedding frequency, ω, to depend on the lengths D and H, the approach velocity, V, and the fluid density, ρ, and viscosity, μ. Thus,

$$\omega = f(D, H, V, \rho, \mu)$$

where

$$\omega \doteq T^{-1}$$
$$D \doteq L$$
$$H \doteq L$$
$$V \doteq LT^{-1}$$
$$\rho \doteq ML^{-3}$$
$$\mu \doteq ML^{-1}T^{-1}$$

Since there are six variables and three reference dimensions (MLT), three pi terms are required. Application of the pi theorem yields

$$\frac{\omega D}{V} = \phi\left(\frac{D}{H}, \frac{\rho VD}{\mu}\right)$$

We recognize the pi term on the left as the Strouhal number, and the dimensional analysis indicates that the Strouhal number is a function of the geometric parameter, D/H, and the Reynolds number. Thus, to maintain similarity between model and prototype

$$\frac{D_m}{H_m} = \frac{D}{H}$$

and

$$\frac{\rho_m V_m D_m}{\mu_m} = \frac{\rho VD}{\mu}$$

From the first similarity requirement

$$H_m = \frac{D_m}{D} H$$

$$= \frac{(20 \times 10^{-3}\ \text{m})}{(0.1\ \text{m})} (0.3\ \text{m})$$

$$H_m = 60 \times 10^{-3}\ \text{m} = 60\ \text{mm}$$ **(Ans)**

The second similarity requirement indicates that the Reynolds number must be the same for model and prototype so that the model velocity must satisfy the condition

$$V_m = \frac{\mu_m}{\mu} \frac{\rho}{\rho_m} \frac{D}{D_m} V \qquad \textbf{(1)}$$

For air at standard conditions, $\mu = 1.79 \times 10^{-5}$ kg/m·s, $\rho = 1.23$ kg/m^3, and for water at 20 °C, $\mu = 1.00 \times 10^{-3}$ kg/m·s, $\rho = 998$ kg/m^3. The fluid velocity for the prototype is

$$V = \frac{(50 \times 10^3\ \text{m/hr})}{(3600\ \text{s/hr})} = 13.9\ \text{m/s}$$

The required velocity can now be calculated from Eq. 1 as

$$V_m = \frac{[1.00 \times 10^{-3}\ \text{kg/(m·s)}](1.23\ \text{kg/m}^3)(0.1\ \text{m})}{[1.79 \times 10^{-5}\ \text{kg/(m·s)}](998\ \text{kg/m}^3)(20 \times 10^{-3}\ \text{m})} (13.9\ \text{m/s})$$

$$V_m = 4.79\ \text{m/s}$$ **(Ans)**

This is a reasonable velocity that could be readily achieved in a water tunnel.

With the two similarity requirements satisfied, it follows that the Strouhal numbers for prototype and model will be the same so that

$$\frac{\omega D}{V} = \frac{\omega_m D_m}{V_m}$$

and the predicted prototype vortex shedding frequency is

$$\omega = \frac{V}{V_m} \frac{D_m}{D} \omega_m$$

$$= \frac{(13.9\ \text{m/s})}{(4.79\ \text{m/s})} \frac{(20 \times 10^{-3}\ \text{m})}{(0.1\ \text{m})} (49.9\ \text{Hz})$$

$$\omega = 29.0\ \text{Hz}$$ **(Ans)**

This same model could also be used to predict the drag per unit length, $\mathscr{D}_\ell$ (lb/ft or N/m), on the prototype, since the drag would depend on the same variables as those used for the frequency. Thus, the similarity requirements would be the same and with these requirements satisfied it follows that the drag per unit length expressed in dimensionless form, such as $\mathscr{D}_\ell / D\rho V^2$, would be equal in model and prototype. The measured drag on the model could then be related to the corresponding drag on the prototype through the relationship

$$\mathscr{D}_\ell = \left(\frac{D}{D_m}\right)\left(\frac{\rho}{\rho_m}\right)\left(\frac{V}{V_m}\right)^2 \mathscr{D}_{\ell m}$$

7.8.2 Model Scales

It is clear from the preceding section that the ratio of like quantities for the model and prototype naturally arises from the similarity requirements. For example, if in a given problem there are two length variables ℓ_1 and ℓ_2, the resulting similarity requirement based on a pi term obtained from these two variables is

$$\frac{\ell_1}{\ell_2} = \frac{\ell_{1m}}{\ell_{2m}}$$

so that

$$\frac{\ell_{1m}}{\ell_1} = \frac{\ell_{2m}}{\ell_2}$$

The ratio of a model variable to the corresponding prototype variable is called the scale for that variable.

We define the ratio ℓ_{1m}/ℓ_1 or ℓ_{2m}/ℓ_2 as the *length scale*. For true models there will be only one length scale, and all lengths are fixed in accordance with this scale. There are, however, other scales such as the velocity scale, V_m/V, density scale, ρ_m/ρ, viscosity scale, μ_m/μ, and so on. In fact, we can define a scale for each of the variables in the problem. Thus, it is actually meaningless to talk about a "scale" of a model without specifying which scale.

We will designate the length scale as λ_ℓ, and other scales as λ_V, λ_ρ, λ_μ, and so on, where the subscript indicates the particular scale. Also, we will take the ratio of the model value to the prototype value as the scale (rather than the inverse). Length scales are often specified, for example, as 1 : 10 or as a $\frac{1}{10}$-scale model. The meaning of this specification is that the model is one-tenth the size of the prototype, and the tacit assumption is that all relevant lengths are scaled accordingly so the model is geometrically similar to the prototype.

7.8.3 Practical Aspects of Using Models

Validation of Model Design Most model studies involve simplifying assumptions with regard to the variables to be considered. Although the number of assumptions is frequently less stringent than that required for mathematical models, they nevertheless introduce some uncertainty in the model design. It is, therefore, desirable to check the design experimentally whenever possible. In some situations the purpose of the model is to predict the effects of certain proposed changes in a given prototype, and in this instance some actual prototype data may be available. The model can be designed, constructed, and tested, and the model prediction can be compared with these data. If the agreement is satisfactory, then the model can be changed in the desired manner, and the corresponding effect on the prototype can be predicted with increased confidence.

V7.3

Another useful and informative procedure is to run tests with a series of models of different sizes, where one of the models can be thought of as the prototype and the others as "models" of this prototype. With the models designed and operated on the basis of the proposed design, a necessary condition for the validity of the model design is that an accurate prediction be made between any pair of models, since one can always be considered as a model of the other. Another suitable agreement in validation tests of this type does not unequivocally indicate a correct model sign (e.g., the length scales between laboratory models may be significantly different than required for actual prototype prediction), it is certainly true that if agreement between models cannot be achieved in these tests, there is no reason to expect that the same model design can be used to predict prototype behavior correctly.

Distorted Models Although the general idea behind establishing similarity requirements for models is straightforward (we simply equate pi terms), it is not always possible to satisfy all the known requirements. If one or more of the similarity requirements are not

Models for which one or more similarity requirements are not satisfied are called distorted models.

met, for example, if $\Pi_{2m} \neq \Pi_2$, then it follows that the prediction equation $\Pi_1 = \Pi_{1m}$ is not true; that is, $\Pi_1 \neq \Pi_{1m}$. Models for which one or more of the similarity requirements are not satisfied are called *distorted models.*

Distorted models are rather commonplace, and they can arise for a variety of reasons. For example, perhaps a suitable fluid cannot be found for the model. The classic example of a distorted model occurs in the study of open channel or free-surface flows. Typically in these problems both the Reynolds number, $\rho V\ell/\mu$, and the Froude number, $V/\sqrt{g\ell}$, are involved.

Froude number similarity requires

$$\frac{V_m}{\sqrt{g_m \ell_m}} = \frac{V}{\sqrt{g\ell}}$$

If the model and prototype are operated in the same gravitational field, then the required velocity scale is

$$\frac{V_m}{V} = \sqrt{\frac{\ell_m}{\ell}} = \sqrt{\lambda_\ell}$$

Reynolds number similarity requires

$$\frac{\rho_m V_m \ell_m}{\mu_m} = \frac{\rho V \ell}{\mu}$$

and the velocity scale is

$$\frac{V_m}{V} = \frac{\mu_m}{\mu} \frac{\rho}{\rho_m} \frac{\ell}{\ell_m}$$

Since the velocity scale must be equal to the square root of the length scale, it follows that

$$\frac{\mu_m/\rho_m}{\mu/\rho} = \frac{\nu_m}{\nu} = (\lambda_\ell)^{3/2} \qquad \textbf{(7.15)}$$

where the ratio μ/ρ is the kinematic viscosity, ν. Although in principle it may be possible to satisfy this design condition, it may be quite difficult, if not impossible, to find a suitable model fluid, particularly for small length scales. For problems involving rivers, spillways, and harbors, for which the prototype fluid is water, the models are also relatively large so that the only practical model fluid is water. However, in this case (with the kinematic viscosity scale equal to unity) Eq. 7.15 will not be satisfied, and a distorted model will result. Generally, hydraulic models of this type are distorted and are designed on the basis of the Froude number, with the Reynolds number different in model and prototype.

Distorted models can be successfully used, but the interpretation of results obtained with this type of model is obviously more difficult than the interpretation of results obtained with *true models* for which all similarity requirements are met. There are no general rules for handling distorted models, and essentially each problem must be considered on its own merits. The success of using distorted models depends to a large extent on the skill and experience of the investigator responsible for the design of the model and in the interpretation of experimental data obtained from the model. Distorted models are widely used, and additional information can be found in the references at the end of the chapter. References 14 and 15 contain detailed discussions of several practical examples of distorted fluid flow and hydraulic models.

7.9 Some Typical Model Studies

Models are used to investigate many different types of fluid mechanics problems, and it is difficult to characterize in a general way all necessary similarity requirements, since each problem is unique. We can, however, broadly classify many of the problems on the basis of the general nature of the flow and subsequently develop some general characteristics of model designs in each of these classifications. In the following sections we will consider models for the study of (1) flow through closed conduits, (2) flow around immersed bodies, and (3) flow with a free surface. Turbomachine models are considered in Chapter 12.

7.9.1 Flow Through Closed Conduits

Geometric and Reynolds number similarity is usually required for models involving flow through closed conduits.

Common examples of this type of flow include pipe flow and flow through valves, fittings, and metering devices. Although the conduits are often circular, they could have other shapes as well and may contain expansions or contractions. Since there are no fluid interfaces or free surfaces, the dominant forces are inertial and viscous so that the Reynolds number is an important similarity parameter. For low Mach numbers (Ma < 0.3), compressibility effects are negligible for both the flow of liquids or gases. For this class of problems, geometric similarity between model and prototype must be maintained. Generally the geometric characteristics can be described by a series of length terms, $\ell_1, \ell_2, \ell_3, \ldots, \ell_i$, and ℓ, where ℓ is some particular length dimension for the system. Such a series of length terms leads to a set of pi terms of the form

$$\Pi_i = \frac{\ell_i}{\ell}$$

where $i = 1, 2, \ldots,$ and so on. In addition to the basic geometry of the system, the roughness of the internal surface in contact with the fluid may be important. If the average height of surface roughness elements is defined as ε, then the pi term representing roughness will be ε/ℓ. This parameter indicates that for complete geometric similarity, surface roughness would also have to be scaled. Note that this implies that for length scales less than 1, the model surfaces should be smoother than those in the prototype since $\varepsilon_m = \lambda_\ell \varepsilon$. To further complicate matters, the pattern of roughness elements in model and prototype would have to be similar. These are conditions that are virtually impossible to satisfy exactly. Fortunately, in some problems the surface roughness plays a minor role and can be neglected. However, in other problems (such as turbulent flow through pipes) roughness can be very important.

It follows from this discussion that for flow in closed conduits at low Mach numbers, any dependent pi term (the one that contains the particular variable of interest, such as pressure drop) can be expressed as

$$\text{Dependent pi term} = \phi\left(\frac{\ell_i}{\ell}, \frac{\varepsilon}{\ell}, \frac{\rho V \ell}{\mu}\right) \tag{7.16}$$

This is a general formulation for this type of problem. The first two pi terms of the right side of Eq. 7.16 lead to the requirement of geometric similarity so that

$$\frac{\ell_{im}}{\ell_m} = \frac{\ell_i}{\ell} \qquad \frac{\varepsilon_m}{\ell_m} = \frac{\varepsilon}{\ell}$$

or

$$\frac{\ell_{im}}{\ell_i} = \frac{\varepsilon_m}{\varepsilon} = \frac{\ell_m}{\ell} = \lambda_\ell$$

This result indicates that the investigator is free to choose a length scale, λ_ℓ, but once this scale is selected, all other pertinent lengths must be scaled in the same ratio.

The additional similarity requirement arises from the equality of Reynolds numbers

$$\frac{\rho_m V_m \ell_m}{\mu_m} = \frac{\rho V \ell}{\mu}$$

From this condition the velocity scale is established so that

Accurate predictions of flow behavior require the correct scaling of velocities.

$$\frac{V_m}{V} = \frac{\mu_m}{\mu}\frac{\rho}{\rho_m}\frac{\ell}{\ell_m} \tag{7.17}$$

and the actual value of the velocity scale depends on the viscosity and density scales, as well as the length scale. Different fluids can be used in model and prototype. However, if the same fluid is used (with $\mu_m = \mu$ and $\rho_m = \rho$), then

$$\frac{V_m}{V} = \frac{\ell}{\ell_m}$$

Thus, $V_m = V/\lambda_\ell$, which indicates that the fluid velocity in the model will be larger than that in the prototype for any length scale less than 1. Since length scales are typically much less than unity, Reynolds number similarity may be difficult to achieve because of the large model velocities required.

With these similarity requirements satisfied, it follows that the dependent pi term will be equal in model and prototype. For example, if the dependent variable of interest is the pressure differential,[3] Δp, between two points along a closed conduit, then the dependent pi term could be expressed as

$$\Pi_1 = \frac{\Delta p}{\rho V^2}$$

The prototype pressure drop would then be obtained from the relationship

$$\Delta p = \frac{\rho}{\rho_m}\left(\frac{V}{V_m}\right)^2 \Delta p_m$$

so that from a measured pressure differential in the model, Δp_m, the corresponding pressure differential for the prototype could be predicted. Note that in general $\Delta p \neq \Delta p_m$.

EXAMPLE 7.6

Model tests are to be performed to study the flow through a large valve having a 2-ft-diameter inlet and carrying water at a flowrate of 30 cfs. The working fluid in the model is water at the same temperature as that in the prototype. Complete geometric similarity exists between model and prototype, and the model inlet diameter is 3 in. Determine the required flowrate in the model.

SOLUTION

To ensure dynamic similarity, the model tests should be run so that

$$\text{Re}_m = \text{Re}$$

[3] In some previous examples the pressure differential *per unit length*, Δp_ℓ, was used. This is appropriate for flow in long pipes or conduits in which the pressure would vary linearly with distance. However, in the more general situation the pressure may not vary linearly with position so that it is necessary to consider the pressure differential, Δp, as the dependent variable. In this case the distance between pressure taps is an additional variable (as well as the distance of one of the taps measured from some reference point within the flow system).

or

$$\frac{V_m D_m}{\nu_m} = \frac{VD}{\nu}$$

where V and D correspond to the inlet velocity and diameter, respectively. Since the same fluid is to be used in model and prototype, $\nu = \nu_m$, and therefore

$$\frac{V_m}{V} = \frac{D}{D_m}$$

The discharge, Q, is equal to VA, where A is the inlet area, so

$$\frac{Q_m}{Q} = \frac{V_m A_m}{VA} = \left(\frac{D}{D_m}\right) \frac{[(\pi/4)D_m^2]}{[(\pi/4)D^2]}$$

$$= \frac{D_m}{D}$$

and for the data given

$$Q_m = \frac{(3/12 \text{ ft})}{(2 \text{ ft})} (30 \text{ ft}^3/\text{s})$$

$$Q_m = 3.75 \text{ cfs}$$ **(Ans)**

Although this is a large flowrate to be carried through a 3-in.-diameter pipe (the corresponding velocity is 76.4 ft/s), it could be attained in a laboratory facility. However, it is to be noted that if we tried to use a smaller model, say one with $D = 1$ in., the required model velocity is 229 ft/s, a very high velocity that would be difficult to achieve. These results are indicative of one of the difficulties encountered in maintaining Reynolds number similarity—the required model velocities may be impractical to obtain.

In some problems Reynolds number similarity may be relaxed.

Two additional points should be made with regard to modeling flows in closed conduits. First, for large Reynolds numbers, inertial forces are much larger than viscous forces, and in this case it may be possible to neglect viscous effects. The important practical consequence of this is that it would not be necessary to maintain Reynolds number similarity between model and prototype. However, *both* model and prototype would have to operate at large Reynolds numbers. Since we do not know, a priori, what is a ''large Reynolds number,'' the effect of Reynolds numbers would have to be determined from the model. This could be accomplished by varying the model Reynolds number to determine the range (if any) over which the dependent pi term ceases to be affected by changes in Reynolds number.

The second point relates to the possibility of cavitation in flow through closed conduits. For example, flow through the complex passages that may exist in valves may lead to local regions of high velocity (and thus low pressure), which can cause the fluid to cavitate. If the model is to be used to study cavitation phenomena, then the vapor pressure, p_v, becomes an important variable and an additional similarity requirement such as equality of the cavitation number $(p_r - p_v)/\frac{1}{2}\rho V^2$ is required, where p_r is some reference pressure. The use of models to study cavitation is complicated, since it is not fully understood how vapor bubbles form and grow. The initiation of bubbles seems to be influenced by the microscopic particles that exist in most liquids, and how this aspect of the problem influences model studies is not clear. Additional details can be found in Ref. 17.

7.9.2 Flow Around Immersed Bodies

Geometric and Reynolds number similarity is usually required for models involving flow around bodies.

Models have been widely used to study the flow characteristics associated with bodies that are completely immersed in a moving fluid. Examples include flow around aircraft, automobiles, golf balls, and buildings. (These types of models are usually tested in wind tunnels as is illustrated in Fig. 7.6.) Modeling laws for these problems are similar to those described in the preceding section; that is, geometric and Reynolds number similarity is required. Since there are no fluid interfaces, surface tension (and therefore the Weber number) is not important. Also, gravity will not affect the flow patterns, so the Froude number need not be considered. The Mach number will be important for high-speed flows in which compressibility becomes an important factor, but for incompressible fluids (such as liquids or for gases at relatively low speeds) the Mach number can be omitted as a similarity requirement. In this case, a general formulation for these problems is

V7.4

$$\text{Dependent pi term} = \phi\left(\frac{\ell_i}{\ell}, \frac{\varepsilon}{\ell}, \frac{\rho V \ell}{\mu}\right) \tag{7.18}$$

where ℓ is some characteristic length of the system and ℓ_i represents other pertinent lengths, ε/ℓ is the relative roughness of the surface (or surfaces), and $\rho V\ell/\mu$ is the Reynolds number.

Frequently, the dependent variable of interest for this type of problem is the drag, $\mathscr{D}$, developed on the body, and in this situation the dependent pi term would usually be expressed in the form of a *drag coefficient*, C_D, where

$$C_D = \frac{\mathscr{D}}{\frac{1}{2}\rho V^2 \ell^2}$$

The numerical factor, $\frac{1}{2}$, is arbitrary but commonly included, and ℓ^2 is usually taken as some representative area of the object. Thus, drag studies can be undertaken with the formulation

$$\frac{\mathscr{D}}{\frac{1}{2}\rho V^2 \ell^2} = C_D = \phi\left(\frac{\ell_i}{\ell}, \frac{\varepsilon}{\ell}, \frac{\rho V \ell}{\mu}\right) \tag{7.19}$$

It is clear from Eq. 7.19 that geometric similarity

$$\frac{\ell_{im}}{\ell_m} = \frac{\ell_i}{\ell} \qquad \frac{\varepsilon_m}{\ell_m} = \frac{\varepsilon}{\ell}$$

as well as Reynolds number similarity

$$\frac{\rho_m V_m \ell_m}{\mu_m} = \frac{\rho V \ell}{\mu}$$

■ **FIGURE 7.6 Model of the National Bank of Commerce, San Antonio, Texas, for measurement of peak, rms, and mean pressure distributions. The model is located in a long-test-section, meteorological wind tunnel. (Photograph courtesy of Cermak Peterka Petersen, Inc.)**

must be maintained. If these conditions are met, then

For flow around bodies, drag is often the dependent variable of interest.

$$\frac{\mathscr{D}}{\frac{1}{2}\rho V^2 \ell^2} = \frac{\mathscr{D}_m}{\frac{1}{2}\rho_m V_m^2 \ell_m^2}$$

or

$$\mathscr{D} = \frac{\rho}{\rho_m}\left(\frac{V}{V_m}\right)^2 \left(\frac{\ell}{\ell}\right)^2 \mathscr{D}_m$$

Measurements of model drag, $\mathscr{D}_m$, can then be used to predict the corresponding drag, $\mathscr{D}$, on the prototype from this relationship.

As was discussed in the previous section, one of the common difficulties with models is related to the Reynolds number similarity requirement which establishes the model velocity as

$$V_m = \frac{\mu_m}{\mu}\frac{\rho}{\rho_m}\frac{\ell}{\ell_m} V \tag{7.20}$$

or

$$V_m = \frac{\nu_m}{\nu}\frac{\ell}{\ell_m} V \tag{7.21}$$

where ν_m/ν is the ratio of kinematic viscosities. If the same fluid is used for model and prototype so that $\nu_m = \nu$, then

$$V_m = \frac{\ell}{\ell_m} V$$

and, therefore, the required model velocity will be higher than the prototype velocity for ℓ/ℓ_m greater than 1. Since this ratio is often relatively large, the required value of V_m may be large. For example, for a $\frac{1}{10}$-length scale, and a prototype velocity of 50 mph, the required model velocity is 500 mph. This is a value that is unreasonably high to achieve with liquids, and for gas flows this would be in the range where compressibility would be important in the model (but not in the prototype).

As an alternative, we see from Eq. 7.21 that V_m could be reduced by using a different fluid in the model such that $\nu_m/\nu < 1$. For example, the ratio of the kinematic viscosity of water to that of air is approximately $\frac{1}{10}$, so that if the prototype fluid were air, tests might be run on the model using water. This would reduce the required model velocity, but it still may be difficult to achieve the necessary velocity in a suitable test facility, such as a water tunnel.

Another possibility for wind tunnel tests would be to increase the air pressure in the tunnel so that $\rho_m > \rho$, thus reducing the required model velocity as specified by Eq. 7.20. Fluid viscosity is not strongly influenced by pressure. Although pressurized tunnels have been used, they are obviously more complicated and expensive.

The required model velocity can also be reduced if the length scale is modest; that is, the model is relatively large. For wind tunnel testing, this requires a large test section which greatly increases the cost of the facility. However, large wind tunnels suitable for testing very large models (or prototypes) are in use. One such tunnel, located at the NASA Ames Research Center, Moffett Field, California, has a test section that is 40 ft by 80 ft and can accommodate test speeds to 345 mph. Such a large and expensive test facility is obviously not feasible for university or industrial laboratories, so most model testing has to be accomplished with relatively small models.

The drag on an airplane cruising at 240 mph in standard air is to be determined from tests on a 1:10 scale model placed in a pressurized wind tunnel. To minimize compressibility effects, the air speed in the wind tunnel is also to be 240 mph. Determine the required air pressure in the tunnel (assuming the same air temperature for model and prototype), and the drag on the prototype corresponding to a measured force of 1 lb on the model.

SOLUTION

From Eq. 7.19 it follows that drag can be predicted from a geometrically similar model if the Reynolds numbers in model and prototype are the same. Thus,

$$\frac{\rho_m V_m \ell_m}{\mu_m} = \frac{\rho V \ell}{\mu}$$

For this example, $V_m = V$ and $\ell_m/\ell = \frac{1}{10}$ so that

$$\frac{\rho_m}{\rho} = \frac{\mu_m}{\mu} \frac{V}{V_m} \frac{\ell}{\ell_m}$$

$$= \frac{\mu_m}{\mu}(1)(10)$$

and therefore

$$\frac{\rho_m}{\rho} = 10 \frac{\mu_m}{\mu}$$

This result shows that the same fluid with $\rho_m = \rho$ and $\mu_m = \mu$ cannot be used if Reynolds number similarity is to be maintained. One possibility is to pressurize the wind tunnel to increase the density of the air. We assume that an increase in pressure does not significantly change the viscosity so that the required increase in density is given by the relationship

$$\frac{\rho_m}{\rho} = 10$$

For an ideal gas, $p = \rho RT$ so that

$$\frac{p_m}{p} = \frac{\rho_m}{\rho}$$

for constant temperature ($T = T_m$). Therefore, the wind tunnel would need to be pressurized so that

$$\frac{p_m}{p} = 10$$

Since the prototype operates at standard atmospheric pressure, the required pressure in the wind tunnel is 10 atmospheres or

$$p_m = 10 \text{ (14.7 psia)}$$

$$= 147 \text{ psia} \qquad \textbf{(Ans)}$$

Thus, we see that a high pressure would be required and this could not be easily or inexpensively achieved. However, under these conditions Reynolds similarity would be attained and the drag could be obtained from Eq. 7.19 so that

$$\frac{\mathscr{D}}{\frac{1}{2}\rho V^2 \ell^2} = \frac{\mathscr{D}_m}{\frac{1}{2}\rho_m V_m^2 \ell_m^2}$$

or

$$\mathcal{D} = \frac{\rho}{\rho_m}\left(\frac{V}{V_m}\right)^2\left(\frac{\ell}{\ell_m}\right)^2\mathcal{D}_m$$

$$= \left(\frac{1}{10}\right)(1)^2(10)^2\mathcal{D}_m$$

$$= 10\mathcal{D}_m$$

Thus, for a drag of 1 lb on the model the corresponding drag on the prototype is

$$\mathcal{D} = 10 \text{ lb}$$ **(Ans)**

V7.5

At high Reynolds numbers the drag is often essentially independent of the Reynolds number.

Fortunately, in many situations the flow characteristics are not strongly influenced by the Reynolds number over the operating range of interest. In these cases we can avoid the rather stringent similarity requirement of matching Reynolds numbers. To illustrate this point, consider the variation in the drag coefficient with the Reynolds number for a smooth sphere of diameter d placed in a uniform stream with approach velocity, V. Some typical data are shown in Fig. 7.7. We observe that for Reynolds numbers between approximately 10^3 and 2×10^5 the drag coefficient is relatively constant and does not depend on the specific value of the Reynolds number. Thus, exact Reynolds number similarity is not required in this range. For other geometric shapes we would typically find that for high Reynolds numbers, inertial forces are dominant (rather than viscous forces), and the drag is essentially independent of the Reynolds number.

Another interesting point to note from Fig. 7.7 is the rather abrupt drop in the drag coefficient near a Reynolds number of 3×10^5. As is discussed in Section 9.3.3, this is due to a change in the flow conditions near the surface of the sphere. These changes are influenced

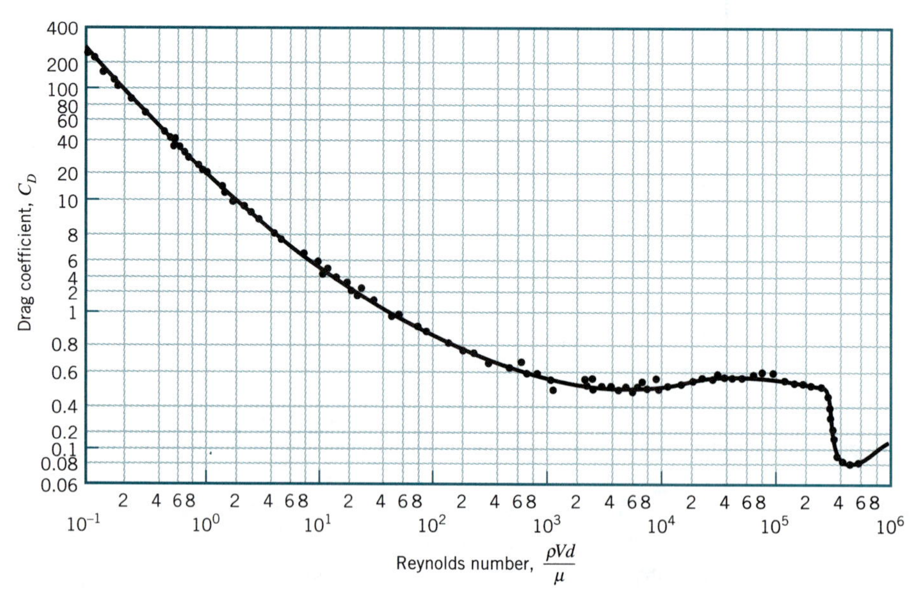

■ **FIGURE 7.7 The effect of Reynolds number on the drag coefficient, C_D, for a smooth sphere with $C_D = \mathcal{D}/\frac{1}{2}A\rho V^2$, where A is the projected area of sphere, $\pi d^2/4$. (Data from Ref. 16, used by permission.)**

by the surface roughness and, in fact, the drag coefficient for a sphere with a "rougher" surface will generally be less than that of the smooth sphere for high Reynolds number. For example, the dimples on a golf ball are used to reduce the drag over that which would occur for a smooth golf ball. Although this is undoubtedly of great interest to the avid golfer, it is also important to engineers responsible for fluid-flow models, since it does emphasize the potential importance of the surface roughness. However, for bodies that are sufficiently angular with sharp corners, the actual surface roughness is likely to play a secondary role compared with the main geometric features of the body.

For problems involving high velocities in which the Mach number is greater than about 0.3, the influence of compressibility, and therefore the Mach number (or Cauchy number), becomes significant. In this case complete similarity requires not only geometric and Reynolds number similarity but also Mach number similarity so that

$$\frac{V_m}{c_m} = \frac{V}{c} \tag{7.22}$$

This similarity requirement, when combined with that for Reynolds number similarity (Eq. 7.21), yields

$$\frac{c}{c_m} = \frac{\nu}{\nu_m}\frac{\ell_m}{\ell} \tag{7.23}$$

Clearly the same fluid with $c = c_m$ and $\nu = \nu_m$ cannot be used in model and prototype unless the length scale is unity (which means that we are running tests on the prototype). In high-speed aerodynamics the prototype fluid is usually air, and it is difficult to satisfy Eq. 7.23 for reasonable length scales. Thus, models involving high-speed flows are often distorted with respect to Reynolds number similarity, but Mach number similarity is maintained.

7.9.3 Flow with a Free Surface

Froude number similarity is usually required for models involving free-surface flows.

Flows in canals, rivers, spillways, and stilling basins, as well as flow around ships, are all examples of flow phenomena involving a free surface. For this class of problems, both gravitational and inertial forces are important and, therefore, the Froude number becomes an important similarity parameter. Also, since there is a free surface with a liquid-air interface, forces due to surface tension may be significant, and the Weber number becomes another similarity parameter that needs to be considered along with the Reynolds number. Geometric variables will obviously still be important. Thus a general formulation for problems involving flow with a free surface can be expressed as

$$\text{Dependent pi term} = \phi\left(\frac{\ell_i}{\ell}, \frac{\varepsilon}{\ell}, \frac{\rho V \ell}{\mu}, \frac{V}{\sqrt{g\ell}}, \frac{\rho V^2 \ell}{\sigma}\right) \tag{7.24}$$

As discussed previously, ℓ is some characteristic length of the system, ℓ_i represents other pertinent lengths, and ε/ℓ is the relative roughness of the various surfaces. Since gravity is the driving force in these problems, Froude number similarity is definitely required so that

V7.6

$$\frac{V_m}{\sqrt{g_m \ell_m}} = \frac{V}{\sqrt{g\ell}}$$

The model and prototype are expected to operate in the same gravitational field ($g_m = g$), and therefore it follows that

$$\frac{V_m}{V} = \sqrt{\frac{\ell_m}{\ell}} = \sqrt{\lambda_\ell} \tag{7.25}$$

Thus, when models are designed on the basis of Froude number similarity, the velocity scale is determined by the square root of the length scale. As is discussed in Section 7.8.3, to simultaneously have Reynolds and Froude number similarity it is necessary that the kinematic viscosity scale be related to the length scale as

$$\frac{\nu_m}{\nu} = (\lambda_\ell)^{3/2} \tag{7.26}$$

The working fluid for the prototype is normally either freshwater or seawater and the length scale is small. Under these circumstances it is virtually impossible to satisfy Eq. 7.26, so models involving free-surface flows are usually distorted. The problem is further complicated if an attempt is made to model surface tension effects, since this requires the equality of Weber numbers, which leads to the condition

$$\frac{\sigma_m/\rho_m}{\sigma/\rho} = (\lambda_\ell)^2 \tag{7.27}$$

for the kinematic surface tension (σ/ρ). It is again evident that the same fluid cannot be used in model and prototype if we are to have similitude with respect to surface tension effects for $\lambda_\ell \neq 1$.

Surface tension and viscous effects are often negligible in free-surface flows.

Fortunately, in many problems involving free-surface flows, both surface tension and viscous effects are small and consequently strict adherence to Weber and Reynolds number similarity is not required. Certainly, surface tension is not important in large hydraulic structures and rivers. Our only concern would be if in a model the depths were reduced to the point where surface tension becomes an important factor, whereas it is not in the prototype. This is of particular importance in the design of river models, since the length scales are typically small (so that the width of the model is reasonable), but with a small length scale the required model depth may be very small. To overcome this problem, different horizontal and vertical length scales are often used for river models. Although this approach eliminates surface tension effects in the model, it introduces geometric distortion that must be accounted for empirically, usually by increasing the model surface roughness. It is important in these circumstances that verification tests with the model be performed (if possible) in which model data are compared with available prototype river flow data. Model roughness can be adjusted to give satisfactory agreement between model and prototype, and then the model subsequently used to predict the effect of proposed changes on river characteristics (such as velocity patterns or surface elevations).

V7.7

For large hydraulic structures, such as dam spillways, the Reynolds numbers are large so that viscous forces are small in comparison to the forces due to gravity and inertia. In this case Reynolds number similarity is not maintained and models are designed on the basis of Froude number similarity. Care must be taken to ensure that the model Reynolds numbers are also large, but they are not required to be equal to those of the prototype. This type of hydraulic model is usually made as large as possible so that the Reynolds number will be

■ FIGURE 7.8 A scale hydraulic model (1:197) of the Guri Dam in Venezuela which is used to simulate the characteristics of the flow over and below the spillway and the erosion below the spillway. (Photograph courtesy of St. Anthony Falls Hydraulic Laboratory.)

large. A spillway model is shown in Fig. 7.8. Also, for relatively large models the geometric features of the prototype can be accurately scaled, as well as surface roughness. Note that $\varepsilon_m = \lambda_\ell \varepsilon$, which indicates that the model surfaces must be smoother than the corresponding prototype surfaces for $\lambda_\ell < 1$.

EXAMPLE 7.8

A certain spillway for a dam is 20 m wide and is designed to carry 125 m^3/s at flood stage. A 1 : 15 model is constructed to study the flow characteristics through the spillway. Determine the required model width and flowrate. What operating time for the model corresponds to a 24-hr period in the prototype? The effects of surface tension and viscosity are to be neglected.

SOLUTION

The width, w_m, of the model spillway is obtained from the length scale, λ_ℓ, so that

$$\frac{w_m}{w} = \lambda_\ell = \frac{1}{15}$$

and

$$w_m = \frac{20 \text{ m}}{15} = 1.33 \text{ m} \qquad \textbf{(Ans)}$$

Of course, all other geometric features (including surface roughness) of the spillway must be scaled in accordance with the same length scale.

With the neglect of surface tension and viscosity, Eq. 7.24 indicates that dynamic similarity will be achieved if the Froude numbers are equal between model and prototype. Thus,

$$\frac{V_m}{\sqrt{g_m \ell_m}} = \frac{V}{\sqrt{g\ell}}$$

and for $g_m = g$

$$\frac{V_m}{V} = \sqrt{\frac{\ell_m}{\ell}}$$

Since the flowrate is given by $Q = VA$, where A is an appropriate cross-sectional area, it follows that

$$\frac{Q_m}{Q} = \frac{V_m A_m}{VA} = \sqrt{\frac{\ell_m}{\ell}}\left(\frac{\ell_m}{\ell}\right)^2$$
$$= (\lambda_\ell)^{5/2}$$

where we have made use of the relationship $A_m/A = (\ell_m/\ell)^2$. For $\lambda_\ell = \frac{1}{15}$ and $Q =$ 125 m^3/s

$$Q_m = (\tfrac{1}{15})^{5/2}\,(125 \text{ m}^3/\text{s}) = 0.143 \text{ m}^3/\text{s} \qquad \textbf{(Ans)}$$

The time scale can be obtained from the velocity scale, since the velocity is distance divided by time ($V = \ell/t$), and therefore

$$\frac{V}{V_m} = \frac{\ell}{t}\frac{t_m}{\ell_m}$$

or

$$\frac{t_m}{t} = \frac{V}{V_m}\frac{\ell_m}{\ell} = \sqrt{\frac{\ell_m}{\ell}} = \sqrt{\lambda_\ell}$$

This result indicates that time intervals in the model will be smaller than the corresponding intervals in the prototype if $\lambda_\ell < 1$. For $\lambda_\ell = \frac{1}{15}$ and a prototype time interval of 24 hr

$$t_m = \sqrt{\tfrac{1}{15}}\,(24 \text{ hr}) = 6.20 \text{ hr} \qquad \textbf{(Ans)}$$

The ability to scale times may be very useful, since it is possible to ''speed up'' events in the model which may occur over a relatively long time in the prototype.

The drag on a ship depends on both the Reynolds and Froude numbers, which greatly complicates the use of a model.

There are, unfortunately, problems involving flow with a free surface in which viscous, inertial, and gravitational forces are all important. The drag on a ship as it moves through water is due to the viscous shearing stresses that develop along its hull, as well as a pressure-induced component of drag caused by both the shape of the hull and wave action. The shear drag is a function of the Reynolds number, whereas the pressure drag is a function of the Froude number. Since both Reynolds number and Froude number similarity cannot be simultaneously achieved by using water as the model fluid (which is the only practical fluid for ship models), some technique other than a straightforward model test must be employed. One common approach is to measure the total drag on a small, geometrically similar model as it is towed through a model basin at Froude numbers matching those of the prototype. The shear drag on the model is calculated using analytical techniques of the type described in Chapter 9. This calculated value is then subtracted from the total drag to obtain pressure drag, and using Froude number scaling the pressure drag on the prototype can then be predicted. The experimentally determined value can then be combined with a calculated value of the shear drag (again using analytical techniques) to provide the desired total drag on the ship. Ship models are widely used to study new designs, but the tests require extensive facilities (see Fig. 7.9).

It is clear from this brief discussion of various types of models involving free-surface flows that the design and use of such models requires considerable ingenuity, as well as a good understanding of the physical phenomena involved. This is generally true for most model studies. Modeling is both an art and a science. Motion picture producers make extensive use of model ships, fires, explosions, and the like. It is interesting to attempt to observe the flow differences between these distorted model flows and the real thing.

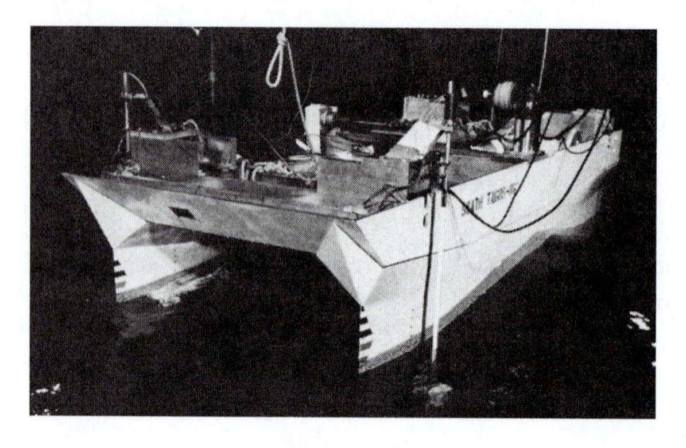

■ FIGURE 7.9 Instrumented, small-waterplane-area, twin hull (SWATH) model suspended from a towing carriage. (Photograph courtesy of the U.S. Navy's David W. Taylor Research Center.)

7.10 Similitude Based on Governing Differential Equations

Similarity laws can be directly developed from the equations governing the phenomenon of interest.

In the preceding sections of this chapter, dimensional analysis has been used to obtain similarity laws. This is a simple, straightforward approach to modeling, which is widely used. The use of dimensional analysis requires only a knowledge of the variables that influence the phenomenon of interest. Although the simplicity of this approach is attractive, it must be recognized that omission of one or more important variables may lead to serious errors in the model design. An alternative approach is available if the equations (usually differential equations) governing the phenomenon are known. In this situation similarity laws can be developed from the governing equations, even though it may not be possible to obtain analytic solutions to the equations.

To illustrate the procedure, consider the flow of an incompressible Newtonian fluid. For simplicity we will restrict our attention to two-dimensional flow, although the results are applicable to the general three-dimensional case. From Chapter 6 we know that the governing equations are the continuity equation

$$\frac{\partial u}{\partial x} + \frac{\partial v}{\partial y} = 0 \tag{7.28}$$

and the Navier–Stokes equations

$$\rho\left(\frac{\partial u}{\partial t} + u\frac{\partial u}{\partial x} + v\frac{\partial u}{\partial y}\right) = -\frac{\partial p}{\partial x} + \mu\left(\frac{\partial^2 u}{\partial x^2} + \frac{\partial^2 u}{\partial^2 y}\right) \tag{7.29}$$

$$\rho\left(\frac{\partial v}{\partial t} + u\frac{\partial v}{\partial x} + v\frac{\partial v}{\partial y}\right) = -\frac{\partial p}{\partial y} - \rho g + \mu\left(\frac{\partial^2 v}{\partial x^2} + \frac{\partial^2 v}{\partial y^2}\right) \tag{7.30}$$

where the y axis is vertical, so that the gravitational body force, ρg, only appears in the "y equation." To continue the mathematical description of the problem, boundary conditions are required. For example, velocities on all boundaries may be specified; that is, $u = u_B$ and $v = v_B$ at all boundary points $x = x_B$ and $y = y_B$. In some types of problems it may be necessary to specify the pressure over some part of the boundary. For time-dependent problems, initial conditions would also have to be provided, which means that the values of all dependent variables would be given at some time (usually taken at $t = 0$).

Once the governing equations, including boundary and initial conditions, are known, we are ready to proceed to develop similarity requirements. The next step is to define a new set of variables that are dimensionless. To do this we select a reference quantity for each type of variable. In this problem the variables are u, v, p, x, y, and t so we will need a reference velocity, V, a reference pressure, p_0, a reference length, ℓ, and a reference time, τ. These reference quantities should be parameters that appear in the problem. For example, ℓ may be a characteristic length of a body immersed in a fluid or the width of a channel through which a fluid is flowing. The velocity, V, may be the free-stream velocity or the inlet velocity. The new dimensionless (starred) variables can be expressed as

$$u^* = \frac{u}{V} \qquad v^* = \frac{v}{V} \qquad p^* = \frac{p}{p_0}$$

$$x^* = \frac{x}{\ell} \qquad y^* = \frac{y}{\ell} \qquad t^* = \frac{t}{\tau}$$

The governing equations can now be rewritten in terms of these new variables. For example,

$$\frac{\partial u}{\partial x} = \frac{\partial V u^*}{\partial x^*}\frac{\partial x^*}{\partial x} = \frac{V}{\ell}\frac{\partial u^*}{\partial x^*}$$

and

$$\frac{\partial^2 u}{\partial x^2} = \frac{V}{\ell}\frac{\partial}{\partial x^*}\left(\frac{\partial u^*}{\partial x^*}\right)\frac{\partial x^*}{\partial x} = \frac{V}{\ell^2}\frac{\partial^2 u^*}{\partial x^{*2}}$$

The other terms that appear in the equations can be expressed in a similar fashion. Thus, in terms of the new variables the governing equations become

$$\frac{\partial u^*}{\partial x^*} + \frac{\partial v^*}{\partial y^*} = 0 \quad \textbf{(7.31)}$$

and

$$\left[\frac{\rho V}{\tau}\right]\frac{\partial u^*}{\partial t^*} + \left[\frac{\rho V^2}{\ell}\right]\left(u^*\frac{\partial u^*}{\partial x^*} + v^*\frac{\partial u^*}{\partial y^*}\right) = -\left[\frac{p_0}{\ell}\right]\frac{\partial p^*}{\partial x^*} + \left[\frac{\mu V}{\ell^2}\right]\left(\frac{\partial^2 u^*}{\partial x^{*2}} + \frac{\partial^2 u^*}{\partial y^{*2}}\right) \quad \textbf{(7.32)}$$

$$\underbrace{\left[\frac{\rho V}{\tau}\right]}_{F_{I\ell}}\frac{\partial v^*}{\partial t^*} + \underbrace{\left[\frac{\rho V^2}{\ell}\right]}_{F_{Ic}}\left(u^*\frac{\partial v^*}{\partial x^*} + v^*\frac{\partial v^*}{\partial y^*}\right) = -\underbrace{\left[\frac{p_0}{\ell}\right]}_{F_P}\frac{\partial p^*}{\partial y^*} - \underbrace{[\rho g]}_{F_G} + \underbrace{\left[\frac{\mu V}{\ell^2}\right]}_{F_V}\left(\frac{\partial^2 v^*}{\partial x^{*2}} + \frac{\partial^2 v^*}{\partial y^{*2}}\right) \quad \textbf{(7.33)}$$

The terms appearing in brackets contain the reference quantities and can be interpreted as indices of the various forces (per unit volume) that are involved. Thus, as is indicated in Eq. 7.33, $F_{I\ell}$ = inertia (local) force, F_{Ic} = inertia (convective) force, F_P = pressure force, F_G = gravitational force, and F_V = viscous force. As the final step in the nondimensionalization process, we will divide each term in Eqs. 7.32 and 7.33 by one of the bracketed quantities. Although any one of these quantities could be used, it is conventional to divide by the bracketed quantity $\rho V^2/\ell$ which is the index of the convective inertia force. The final nondimensional form then becomes

Governing equations expressed in terms of dimensionless variables lead to the appropriate dimensionless groups.

$$\left[\frac{\ell}{\tau V}\right]\frac{\partial u^*}{\partial t^*} + u^*\frac{\partial u^*}{\partial x^*} + v^*\frac{\partial u^*}{\partial y^*} = -\left[\frac{p_0}{\rho V^2}\right]\frac{\partial p^*}{\partial x^*} + \left[\frac{\mu}{\rho V\ell}\right]\left(\frac{\partial^2 u^*}{\partial x^{*2}} + \frac{\partial^2 u^*}{\partial y^{*2}}\right) \quad \textbf{(7.34)}$$

$$\left[\frac{\ell}{\tau V}\right]\frac{\partial v^*}{\partial t^*} + u^*\frac{\partial v^*}{\partial x^*} + v^*\frac{\partial v^*}{\partial y^*} = -\left[\frac{p_0}{\rho V^2}\right]\frac{\partial p^*}{\partial y^*} - \left[\frac{g\ell}{V^2}\right] + \left[\frac{\mu}{\rho V\ell}\right]\left(\frac{\partial^2 v^*}{\partial x^{*2}} + \frac{\partial^2 v^*}{\partial y^{*2}}\right) \quad \textbf{(7.35)}$$

We see that bracketed terms are the standard dimensionless groups (or their reciprocals) which were developed from dimensional analysis; that is, $\ell/\tau V$ is a form of Strouhal number, $p_0/\rho V^2$ the Euler number, $g\ell/V^2$ the reciprocal of the square of the Froude number, and $\mu/\rho V\ell$ the reciprocal of the Reynolds number. From this analysis it is now clear how each of the dimensionless groups can be interpreted as the ratio of two forces, and how these groups arise naturally from the governing equations.

Although we really have not helped ourselves with regard to obtaining an analytical solution to these equations (they are still complicated and not amenable to an analytical solution), the dimensionless forms of the equations, Eqs. 7.31, 7.34, and 7.35, can be used to establish similarity requirements. From these equations it follows that if two systems are governed by these equations, then the solutions (in terms of u^*, v^*, p^*, x^*, y^*, and t^*) will be the same if the four parameters $\ell/\tau V$, $p_0/\rho V^2$, $V^2/g\ell$, and $\rho V\ell/\mu$ are equal for the two systems. The two systems will be dynamically similar. Of course, boundary and initial conditions expressed in dimensionless form must also be equal for the two systems, and this will require complete geometric similarity. These are the same similarity requirements that would be determined by a dimensional analysis if the same variables were considered. However, the advantage of working with the governing equations is that the variables appear naturally in the equations, and we do not have to worry about omitting an important one, provided the governing equations are correctly specified. We can thus use this method to deduce the conditions under which two solutions will be similar even though one of the solutions will most likely be obtained experimentally.

In the foregoing analysis we have considered a general case in which the flow may be unsteady, and both the actual pressure level, p_0, and the effect of gravity are important. A reduction in the number of similarity requirements can be achieved if one or more of these conditions is removed. For example, if the flow is steady the dimensionless group, $\ell/\tau V$, can be eliminated.

The actual pressure level will only be of importance if we are concerned with cavitation. If not, the flow patterns and the pressure differences will not depend on the pressure level. In this case, p_0 can be taken as ρV^2 (or $\frac{1}{2}\rho V^2$), and the Euler number can be eliminated as a similarity requirement. However, if we are concerned about cavitation (which will occur in the flow field if the pressure at certain points reaches the vapor pressure, p_v), then the actual pressure level is important. Usually, in this case, the characteristic pressure, p_0, is defined relative to the vapor pressure such that $p_0 = p_r - p_v$ where p_r is some reference pressure within the flow field. With p_0 defined in this manner, the similarity parameter $p_0/\rho V^2$ becomes $(p_r - p_v)/\rho V^2$. This parameter is frequently written as $(p_r - p_v)/\frac{1}{2}\rho V^2$, and in this form, as was noted previously in Section 7.6, is called the cavitation number. Thus we can conclude that if cavitation is not of concern we do not need a similarity parameter involving p_0, but if cavitation is to be modeled, then the cavitation number becomes an important similarity parameter.

The Froude number, which arises because of the inclusion of gravity, is important for problems in which there is a free surface. Examples of these types of problems include the study of rivers, flow through hydraulic structures such as spillways, and the drag on ships. In these situations the shape of the free surface is influenced by gravity, and therefore the Froude number becomes an important similarity parameter. However, if there are no free surfaces, the only effect of gravity is to superimpose a hydrostatic pressure distribution on the pressure distribution created by the fluid motion. The hydrostatic distribution can be eliminated from the governing equation (Eq. 7.30) by defining a new pressure, $p' = p - \rho g y$, and with this change the Froude number does not appear in the nondimensional governing equations.

We conclude from this discussion that for the steady flow of an incompressible fluid without free surfaces, dynamic and kinematic similarity will be achieved if (for geometrically similar systems) Reynolds number similarity exists. If free surfaces are involved, Froude number similarity must also be maintained. For free-surface flows we have tacitly assumed that surface tension is not important. We would find, however, that if surface tension is included, its effect would appear in the free-surface boundary condition, and the Weber number, $\rho V^2 \ell/\sigma$, would become an additional similarity parameter. In addition, if the governing equations for compressible fluids are considered, the Mach number, V/c, would appear as an additional similarity parameter.

The use of governing equations to obtain similarity laws provides an alternative to conventional dimensional analysis.

It is clear that all the common dimensionless groups that we previously developed by using dimensional analysis appear in the governing equations that describe fluid motion when these equations are expressed in terms of dimensionless variables. Thus, the use of the governing equations to obtain similarity laws provides an alternative to dimensional analysis.

This approach has the advantage that the variables are known and the assumptions involved are clearly identified. In addition, a physical interpretation of the various dimensionless groups can often be obtained.

References

1. Bridgman, P. W., *Dimensional Analysis*, Yale University Press, New Haven, Conn., 1922.
2. Murphy, G., *Similitude in Engineering*, Ronald Press, New York, 1950.
3. Langhaar, H. L., *Dimensional Analysis and Theory of Models*, Wiley, New York, 1951.
4. Huntley, H. E., *Dimensional Analysis*, Macdonald, London, 1952.
5. Duncan, W. J., *Physical Similarity and Dimensional Analysis: An Elementary Treatise*, Edward Arnold, London, 1953.
6. Sedov, K. I., *Similarity and Dimensional Methods in Mechanics*, Academic Press, New York, 1959.
7. Ipsen, D. C., *Units, Dimensions, and Dimensionless Numbers*, McGraw-Hill, New York, 1960.
8. Kline, S. J., *Similitude and Approximation Theory*, McGraw-Hill, New York, 1965.
9. Skoglund, V. J., *Similitude—Theory and Applications*, International Textbook, Scranton, Pa., 1967.
10. Baker, W. E., Westline, P. S., and Dodge, F. T., *Similarity Methods in Engineering Dynamics—Theory and Practice of Scale Modeling*, Hayden (Spartan Books), Rochelle Park, N.J., 1973.
11. Taylor, E. S., *Dimensional Analysis for Engineers*, Clarendon Press, Oxford, 1974.
12. Isaacson, E. de St. Q., and Isaacson, M. de St. Q., *Dimensional Methods in Engineering and Physics*, Wiley, New York, 1975.
13. Schuring, D. J., *Scale Models in Engineering*, Pergamon Press, New York, 1977.
14. Yalin, M. S., *Theory of Hydraulic Models*, Macmillan, London, 1971.
15. Sharp, J. J., *Hydraulic Modeling*, Butterworth, London, 1981.
16. Schlichting, H., *Boundary-Layer Theory*, 7th Ed., McGraw-Hill, New York, 1979.
17. Knapp, R. T., Daily, J. W., and Hammitt, F. G., *Cavitation*, McGraw-Hill, New York, 1970.

Review Problems

Note: Problems designated with (R) are review problems. The phrases within parentheses refer to the main topics to be used in solving the problems. Complete, detailed solutions to these review problems can be found in the supplement titled *Student Solution Manual for Fundamentals of Fluid Mechanics*, by Munson, Young, and Okiishi (John Wiley and Sons, New York, 1997).

7.1R (Common Pi terms) Standard air with velocity V flows past an airfoil having a chord length, b, of 6 ft. **(a)** Determine the Reynolds number, $\rho V b/\mu$, for $V = 150$ mph. **(b)** If this airfoil were attached to an airplane flying at the same speed in a standard atmosphere at an altitude of 10,000 ft, what would be the value of the Reynolds number?

(ANS: 8.40×10^6; 6.56×10^6)

7.2R (Dimensionless variables) Some common variables in fluid mechanics include: volume flowrate, Q, acceleration of gravity, g, viscosity, μ, density, ρ and a length, ℓ. Which of the following combinations of these variables are dimensionless? **(a)** $Q^2/g\ell^2$. **(b)** $\rho Q/\mu\ell$. **(c)** $g\ell^5/Q^2$. **(d)** $\rho Q\ell/\mu$.

(ANS: (b); (c))

7.3R (Determination of Pi terms) A fluid flows at a velocity V through a horizontal pipe of diameter D. An orifice plate containing a hole of diameter d is placed in the pipe. It is desired to investigate the pressure drop, Δp, across the plate. Assume that

$$\Delta p = f(D, d, \rho, V)$$

where ρ is the fluid density. Determine a suitable set of pi terms.
(ANS: $\Delta p/\rho V^2 = \phi(d/D)$)

7.4R (Determination of Pi terms) The flowrate, Q, in an open canal or channel can be measured by placing a plate with a V-notch across the channel as illustrated in Fig. P7.4R. This type of device is called a V-notch *weir*. The height, H, of the liquid above the crest can be used to determine Q. Assume that

$$Q = f(H, g, \theta)$$

where g is the acceleration of gravity. What are the significant dimensionless parameters for this problem?
(ANS: $Q/(gH^5)^{1/2} = \phi(\theta)$)

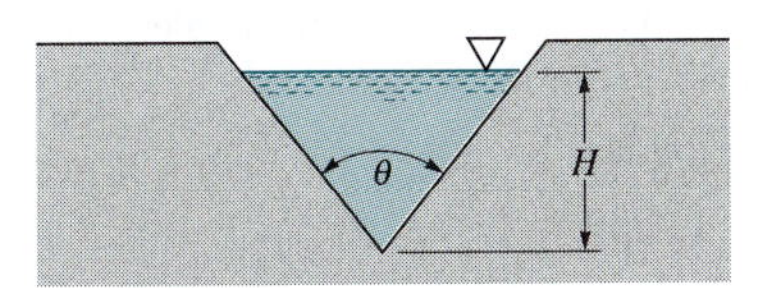

■ FIGURE P7.4R

7.5R (Determination of Pi terms) In a fuel injection system, small droplets are formed due to the breakup of the liquid jet. Assume the droplet diameter, d, is a function of the liquid density, ρ, viscosity, μ, and surface tension, σ, and the jet velocity, V, and diameter, D. Form an appropriate set of dimensionless parameters using μ, V, and D as repeating variables.
(ANS: $d/D = \phi(\rho VD/\mu, \sigma/\mu V)$)

7.6R (Determination of Pi terms) The thrust, $\mathcal{T}$, developed by a propeller of a given shape depends on its diameter, D, the fluid density, ρ, and viscosity, μ, the angular speed of rotation, ω, and the advance velocity, V. Develop a suitable set of pi terms, one of which should be $\rho D^2\omega/\mu$. Form the pi terms by inspection.
(ANS: $\mathcal{T}/\rho V^2D^2 = \phi(\rho VD/\mu, \rho D^2\omega/\mu)$)

7.7R (Modeling/similarity) The water velocity at a certain point along a 1:10 scale model of a dam spillway is 5 m/s. What is the corresponding prototype velocity if the model and prototype operate in accordance with Froude number similarity?
(ANS: 15.8 m/s)

7.8R (Modeling/similarity) The pressure drop per unit length in a 0.25-in.-diameter gasoline fuel line is to be determined from a laboratory test using the same tubing but with water as the fluid. The pressure drop at a gasoline velocity of 1.0 ft/s is of interest. **(a)** What water velocity is required? **(b)** At the properly scaled velocity from part (a), the pressure drop per unit length (using water) was found to be 0.45 psf/ft. What is the predicted pressure drop per unit length for the gasoline line?
(ANS: 2.45 ft/s; 0.0510 lb/ft^2 per ft)

7.9R (Modeling/similarity) A thin layer of an incompressible fluid flows steadily over a horizontal smooth plate as shown in Fig. P7.9R. The fluid surface is open to the atmosphere, and an obstruction having a square cross section is placed on the plate as shown. A model with a length scale of $\frac{1}{4}$ and a fluid density scale of 1.0 is to be designed to predict the depth of fluid, y, along the plate. Assume that inertial, gravitational, surface tension, and viscous effects are all important. What are the required viscosity and surface tension scales?
(ANS: 0.125; 0.0625)

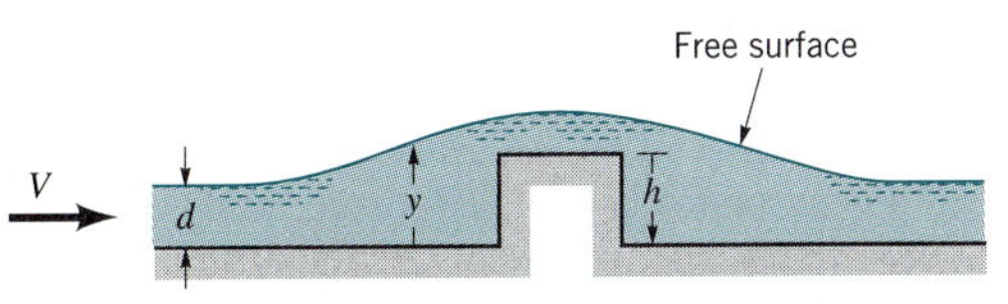

■ FIGURE P7.9R

7.10R (Correlation of experimental data) The drag on a 30-ft long, vertical, 1.25-ft diameter pole subjected to a 30 mph wind is to be determined with a model study. It is expected that the drag is a function of the pole length and diameter, the fluid density and viscosity, and the fluid velocity. Laboratory model tests were performed in a high-speed water tunnel using a model pole having a length of 2 ft and a diameter of 1 in. Some model drag data are shown in Fig. P7.10R. Based on these data, predict the drag on the full-sized pole.
(ANS: 52.2 lb)

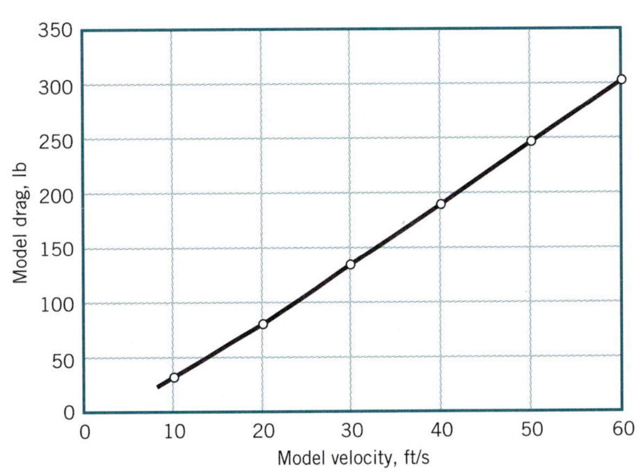

■ FIGURE P7.10R

7.11R (Correlation of experimental data) A liquid is contained in a U-tube as is shown in Fig. P7.11R. When the liquid is displaced from its equilibrium position and released, it oscillates with a period τ. Assume that τ is a function of the acceleration of gravity, g, and the column length, ℓ. Some laboratory measurements made by varying ℓ and measuring τ, with $g = 32.2 \text{ ft/s}^2$, are given in the following table.

τ (s)	0.548	0.783	0.939	1.174
ℓ (ft)	0.49	1.00	1.44	2.25

Based on these data, determine a general equation for the period. (ANS: $\tau = 4.44(\ell/g)^{1/2}$)

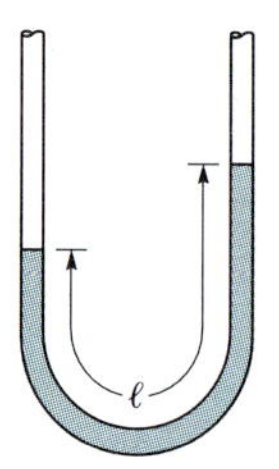

■ FIGURE P7.11R

7.12R (Dimensionless governing equations) An incompressible fluid is contained between two large parallel plates as shown in Fig. P7.12R. The upper plate is fixed. If the fluid is initially at rest and the bottom plate suddenly starts to move with a constant velocity, U, the governing differential equation describing the fluid motion is

$$\rho \frac{\partial u}{\partial t} = \mu \frac{\partial^2 u}{\partial y^2}$$

where u is the velocity in the x direction, and ρ and μ are the fluid density and viscosity, respectively. Rewrite the equation and the initial and boundary conditions in dimensionless form using h and U as reference parameters for length and velocity, and $h^2\rho/\mu$ as a reference parameter for time.

(ANS: $\partial u^*/\partial t^* = \partial^2 u^*/\partial y^{*2}$ with $u^* = 0$ at $t^* = 0$, $u^* = 1$ at $y^* = 0$, and $u^* = 0$ at $y^* = 1$)

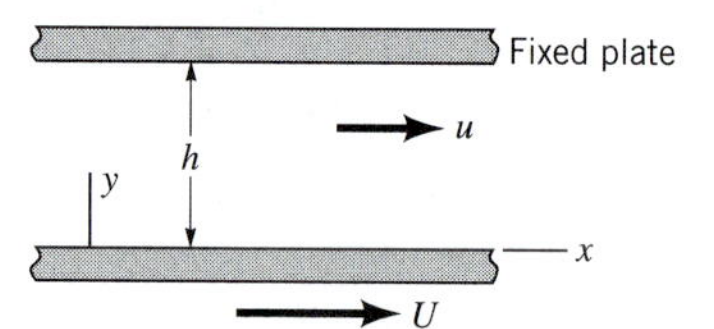

■ FIGURE P7.12R

7.13R (Dimensionless governing equations) The flow between two concentric cylinders (see Fig. P7.13R) is governed by the differential equation

$$\frac{d^2 v_\theta}{dr^2} + \frac{d}{dr}\left(\frac{v_\theta}{r}\right) = 0$$

where v_θ is the tangential velocity at any radial location, r. The inner cylinder is fixed and the outer cylinder rotates with an angular velocity ω. Express the equation in dimensionless form using R_o and ω as reference parameters.

(ANS: $d^2v_\theta^*/dr^{*2} + d(v_\theta^*/r^*)dr^* = 0$)

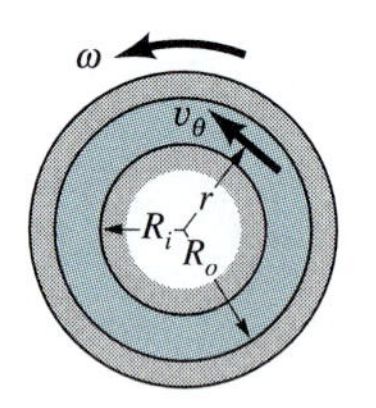

■ FIGURE P7.13R

Problems

Note: Unless otherwise indicated use the values of fluid properties found in the tables on the inside of the front cover. Problems designated with an (*) are intended to be solved with the aid of a programmable calculator or a computer. Problems designated with a (†) are "open-ended" problems and require critical thinking in that to work them one must make various assumptions and provide the necessary data. There is not a unique answer to these problems.

7.1 The Reynolds number, $\rho VD/\mu$, is a very important parameter in fluid mechanics. Verify that the Reynolds number is dimensionless, using both the *FLT* system and the *MLT* system for basic dimensions, and determine its value for water (at 70 °C) flowing at a velocity of 2 m/s through a 2-in.-diameter pipe.

7.2 What are the dimensions of density, pressure, specific weight, surface tension, and dynamic viscosity in **(a)** the *FLT* system, and **(b)** the *MLT* system? Compare your results with those given in Table 1.1 in Chapter 1.

7.3 For the flow of a thin film of a liquid with a depth h and a free surface, two important dimensionless parameters are the Froude number, $V/\sqrt{gh}$, and the Weber number, $\rho V^2 h/\sigma$. Determine the value of these two parameters for glycerin (at 20 °C) flowing with a velocity of 0.5 m/s at a depth of 2 mm.

7.4 The Mach number for a body moving through a fluid with velocity V is defined as V/c, where c is the speed of sound in the fluid. This dimensionless parameter is usually considered to be important in fluid dynamics problems when its value exceeds 0.3. What would be the velocity of a body at a Mach number of 0.3 if the fluid is **(a)** air at standard atmospheric pressure and 20 °C, and **(b)** water at the same temperature and pressure?

7.5 At a sudden contraction in a pipe the diameter changes from D_1 to D_2. The pressure drop, Δp, which develops across the contraction is a function of D_1 and D_2, as well as the velocity, V, in the larger pipe, and the fluid density, ρ, and viscosity,

μ. Use D_1, V, and μ as repeating variables to determine a suitable set of dimensionless parameters. Why would it be incorrect to include the velocity in the smaller pipe as an additional variable?

7.6 Assume that the power, $\mathcal{P}$, required to drive a fan is a function of the fan diameter, D, the fluid density, ρ, the rotational speed, ω, and the flowrate, Q. Use D, ω, and ρ as repeating variables to determine a suitable set of pi terms.

7.7 It is desired to determine the wave height when wind blows across a lake. The wave height, H, is assumed to be a function of the wind speed, V, the water density, ρ, the air density, ρ_a, the water depth, d, the distance from the shore, ℓ, and the acceleration of gravity, g, as shown in Fig. P7.7. Use d, V, and ρ as repeating variables to determine a suitable set of pi terms that could be used to describe this problem.

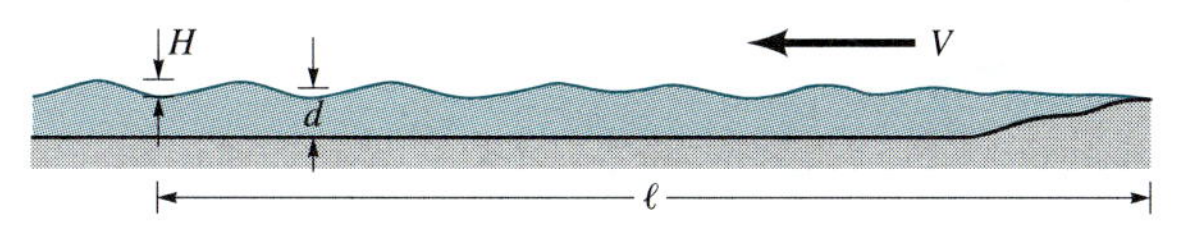

■ FIGURE P7.7

7.8 Water flows over a dam as illustrated in Fig. P7.10. Assume the flowrate, q, per unit length along the dam depends on the head, H, width, b, acceleration of gravity, g, fluid density, ρ, and fluid viscosity, μ. Develop a suitable set of dimensionless parameters for this problem using b, g, and ρ as repeating variables.

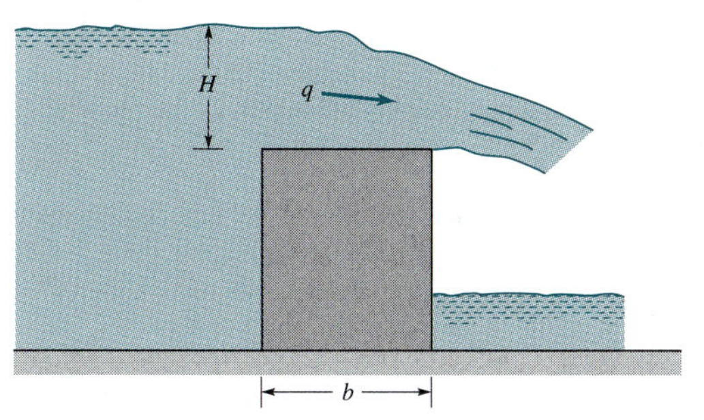

■ FIGURE P7.8

7.9 The pressure rise, Δp, across a pump can be expressed as

$$\Delta p = f(D, \rho, \omega, Q)$$

where D is the impeller diameter, ρ the fluid density, ω the rotational speed, and Q the flowrate. Determine a suitable set of dimensionless parameters.

7.10 The drag, $\mathfrak{D}$, on a washer-shaped plate placed normal to a stream of fluid can be expressed as

$$\mathfrak{D} = f(d_1, d_2, V, \mu, \rho)$$

where d_1 is the outer diameter, d_2 the inner diameter, V the fluid velocity, μ the fluid viscosity, and ρ the fluid density. Some experiments are to be performed in a wind tunnel to determine the drag. What dimensionless parameters would you use to organize these data?

7.11 A thin elastic wire is placed between rigid supports. A fluid flows past the wire, and it is desired to study the static deflection, δ, at the center of the wire due to the fluid drag. Assume that

$$\delta = f(\ell, d, \rho, \mu, V, E)$$

where ℓ is the wire length, d the wire diameter, ρ the fluid density, μ the fluid viscosity, V the fluid velocity, and E the modulus of elasticity of the wire material. Develop a suitable set of pi terms for this problem.

7.12 The flowrate, Q, of water in an open channel is assumed to be a function of the cross-sectional area of the channel, A, the height of the roughness of the channel surface, ε, the acceleration of gravity, g, and the slope, S_o, of the hill on which the channel sits. Put this relationship into dimensionless form.

7.13 Because of surface tension, it is possible, with care, to support an object heavier than water on the water surface as shown in Fig. P7.13. The maximum thickness, h, of a square of material that can be supported is assumed to be a function of the length of the side of the square, ℓ, the density of the material, ρ, the acceleration of gravity, g, and the surface tension of the liquid, σ. Develop a suitable set of dimensionless parameters for this problem.

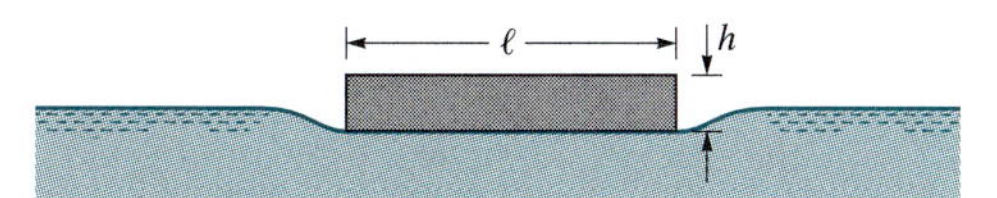

■ FIGURE P7.13

7.14 The velocity, c, at which pressure pulses travel through arteries (pulse-wave velocity) is a function of the artery diameter, D, and wall thickness, h, the density of blood, ρ, and the modulus of elasticity, E, of the arterial wall. Determine a set of nondimensional parameters that can be used to study experimentally the relationship between the pulse-wave velocity and the variables listed. Form the nondimensional parameters by inspection.

7.15 A viscous fluid is poured onto a horizontal plate as shown in Fig. P7.15. Assume that the time, t, required for the fluid to flow a certain distance, d, along the plate is a function of the volume of fluid poured, $\forall$, acceleration of gravity, g, fluid density, ρ, and fluid viscosity, μ. Determine an appropriate set of pi terms to describe this process. Form the pi terms by inspection.

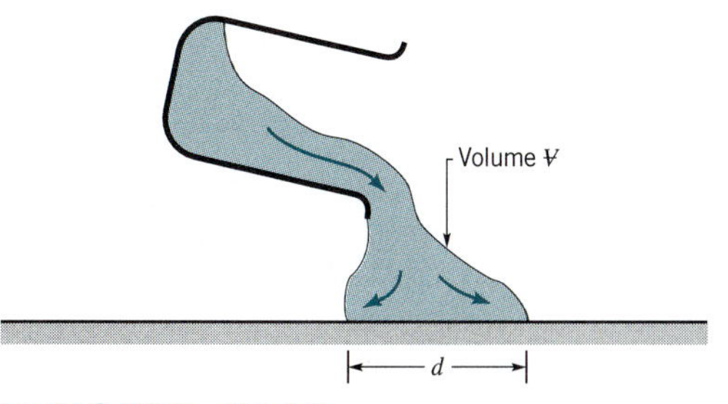

■ FIGURE P7.15

7.16 Assume that the drag, $\mathscr{D}$, on an aircraft flying at supersonic speeds is a function of its velocity, V, fluid density, ρ, speed of sound, c, and a series of lengths, $\ell_1, \ldots, \ell_i$, which describe the geometry of the aircraft. Develop a set of pi terms that could be used to investigate experimentally how the drag is affected by the various factors listed. Form the pi terms by inspection.

7.17 When a fluid flows slowly past a vertical plate of height h and width b (see Fig. P7.17), pressure develops on the face of the plate. Assume that the pressure, p, at the midpoint of the plate is a function of plate height and width, the approach velocity, V, and the fluid viscosity, μ. Make use of dimensional analysis to determine how the pressure, p, will change when the fluid velocity, V, is doubled.

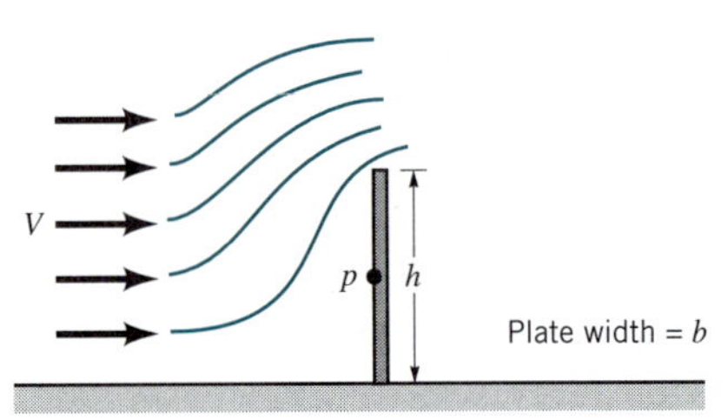

■ FIGURE P7.17

7.18 The pressure drop, Δp, along a straight pipe of diameter D has been experimentally studied, and it is observed that for laminar flow of a given fluid and pipe, the pressure drop varies directly with the distance, ℓ, between pressure taps. Assume that Δp is a function of D and ℓ, the velocity, V, and the fluid viscosity, μ. Use dimensional analysis to deduce how the pressure drop varies with pipe diameter.

7.19 The viscosity, μ, of a liquid can be measured by determining the time, t, it takes for a sphere of diameter, d, to settle slowly through a distance, ℓ, in a vertical cylinder of diameter, D, containing the liquid (see Fig. P7.19). Assume that

$$t = f(\ell, d, D, \mu, \Delta\gamma)$$

where $\Delta\gamma$ is the difference in specific weights between the sphere and the liquid. Use dimensional analysis to show how t is related to μ, and describe how such an apparatus might be used to measure viscosity.

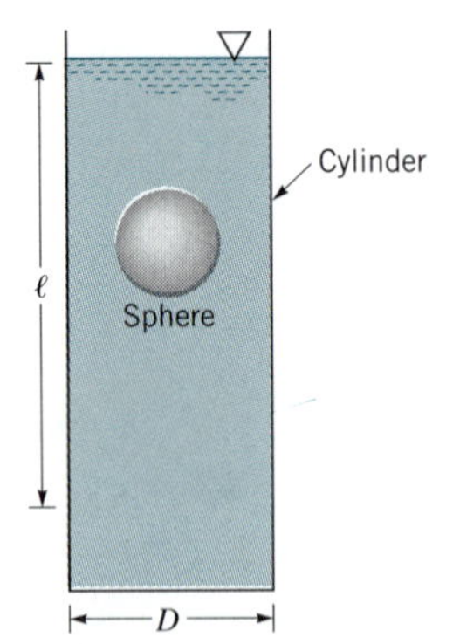

■ FIGURE P7.19

7.20 A cylinder with a diameter D floats upright in a liquid as shown in Fig. P7.20. When the cylinder is displaced slightly along its vertical axis it will oscillate about its equilibrium position with a frequency, ω. Assume that this frequency is a function of the diameter, D, the mass of the cylinder, m, and the specific weight, γ, of the liquid. Determine, with the aid of dimensional analysis, how the frequency is related to these variables. If the mass of the cylinder were increased, would the frequency increase or decrease?

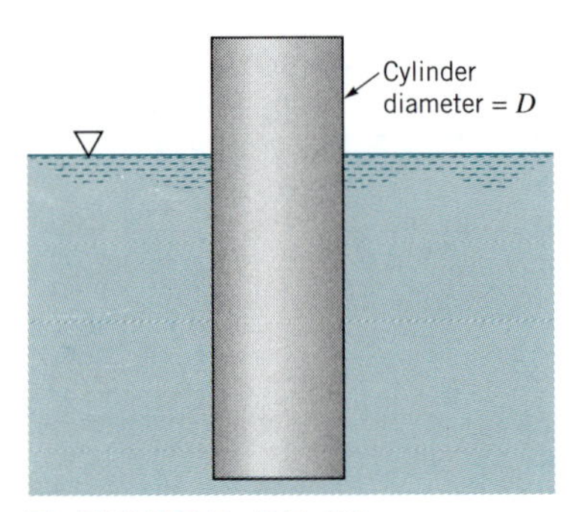

■ FIGURE P7.20

7.21 When a sphere of diameter d falls slowly in a highly viscous fluid, the settling velocity, V, is known to be a function of d, the fluid viscosity, μ, and the difference, $\Delta\gamma$, between the specific weight of the sphere and the specific weight of the fluid. Due to a tight budget situation, only one experiment can be performed, and the following data were obtained: $V = 0.42$ ft/s for $d = 0.1$ in., $\mu = 0.03$ lb·s/ft^2, and $\Delta\gamma = 10$ lb/ft^3. If possible, based on this limited amount of data, determine the general equation for the settling velocity. If you do not think it is possible, indicate what additional data would be required.

7.22 The height, h, that a liquid will rise in a capillary tube is a function of the tube diameter, D, the specific weight of the liquid, γ, and the surface tension, σ. Perform a dimensional analysis using both the FLT and MLT systems for basic dimensions. Note: The results should obviously be the same regardless of the system of dimensions used. If your analysis indicates otherwise, go back and check your work, giving particular attention to the required number of reference dimensions.

7.23 The speed of sound in a gas, c, is a function of the gas pressure, p, and density, ρ. Determine, with the aid of dimensional analysis, how the velocity is related to the pressure and density. Be careful when you decide on how many reference dimensions are required.

***7.24** The pressure rise, $\Delta p = p_2 - p_1$, across the abrupt expansion of Fig. P7.24 through which a liquid is flowing can be expressed as

$$\Delta p = f(A_1, A_2, \rho, V_1)$$

where A_1 and A_2 are the upstream and downstream cross-sectional areas, respectively, ρ is the fluid density, and V_1 is the upstream velocity. Some experimental data obtained with $A_2 = 1.25\ \text{ft}^2$, $V_1 = 5.00\ \text{ft/s}$, and using water with $\rho = 1.94\ \text{slugs/ft}^3$ are given in the following table:

A_1 (ft^2)	0.10	0.25	0.37	0.52	0.61
Δp (lb/ft^2)	3.25	7.85	10.3	11.6	12.3

Plot the results of these tests using suitable dimensionless parameters. With the aid of a standard curve fitting program determine a general equation for Δp and use this equation to predict Δp for water flowing through an abrupt expansion with an area ratio $A_1/A_2 = 0.35$ at a velocity $V_1 = 3.75\ \text{ft/s}$.

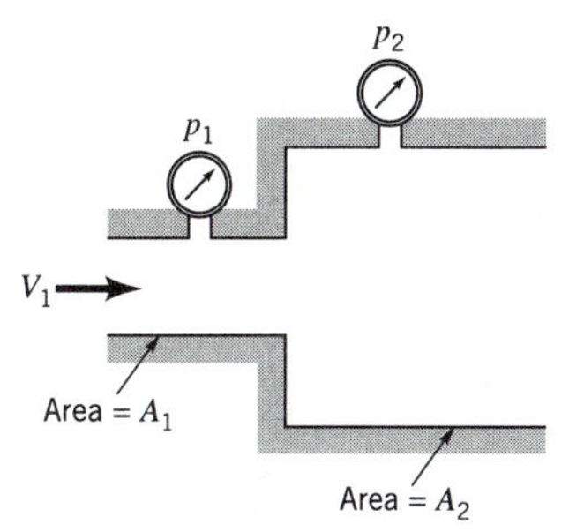

■ FIGURE P7.24

7.25 A liquid flows with a velocity V through a hole in the side of a large tank. Assume that

$$V = f(h, g, \rho, \sigma)$$

where h is the depth of fluid above the hole, g is the acceleration of gravity, ρ the fluid density, and σ the surface tension. The following data were obtained by changing h and measuring V, with a fluid having a density $= 10^3\ \text{kg/m}^3$ and surface tension $= 0.074\ \text{N/m}$.

V (m/s)	3.13	4.43	5.42	6.25	7.00
h (m)	0.50	1.00	1.50	2.00	2.50

Plot these data by using appropriate dimensionless variables. Could any of the original variables have been omitted?

***7.26** The concentric cylinder device of the type shown in Fig. P7.26 is commonly used to measure the viscosity, μ, of liquids by relating the angle of twist, θ, of the inner cylinder to the angular velocity, ω, of the outer cylinder. Assume that

$$\theta = f(\omega, \mu, K, D_1, D_2, \ell)$$

where K depends on the suspending wire properties and has the dimensions FL. The following data were obtained in a series of tests for which $\mu = 0.01\ \text{lb·s/ft}^2$, $K = 10\ \text{lb·ft}$, $\ell = 1\ \text{ft}$, and D_1 and D_2 were constant.

θ (rad)	ω (rad/s)
0.89	0.30
1.50	0.50
2.51	0.82
3.05	1.05
4.28	1.43
5.52	1.86
6.40	2.14

Determine from these data, with the aid of dimensional analysis, the relationship between θ, ω, and μ for this particular apparatus. *Hint:* Plot the data using appropriate dimensionless parameters, and determine the equation of the resulting curve using a standard curve-fitting technique. The equation should satisfy the condition that $\theta = 0$ for $\omega = 0$.

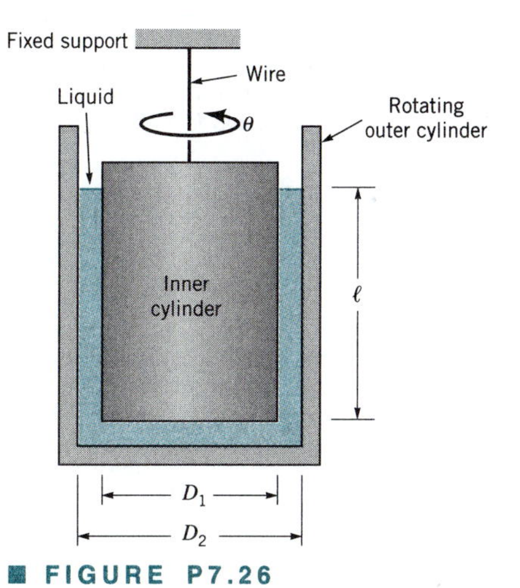

■ FIGURE P7.26

7.27 The pressure drop per unit length, Δp_ℓ, for the flow of blood through a horizontal small-diameter tube is a function of the volume rate of flow, Q, the diameter, D, and the blood viscosity, μ. For a series of tests in which $D = 2\ \text{mm}$ and $\mu = 0.004\ \text{N·s/m}^2$, the following data were obtained, where the Δp listed was measured over the length, $\ell = 300\ \text{mm}$.

Q (m^3/s)	Δp (N/m^2)
3.6×10^{-6}	1.1×10^4
4.9×10^{-6}	1.5×10^4
6.3×10^{-6}	1.9×10^4
7.9×10^{-6}	2.4×10^4
9.8×10^{-6}	3.0×10^4

Perform a dimensional analysis for this problem, and make use of the data given to determine a general relationship between Δp_ℓ and Q (one that is valid for other values of D, ℓ, and μ).

***7.28** A rectangular steel bar is rigidly attached to the floor of a channel through which a fluid is flowing with velocity V as shown in Fig. P7.28. Because of the force of the moving fluid the tip of the bar deflects a distance δ. Assume that δ is a function of the height, h, thickness, d, modulus of elasticity, E, and kinetic energy, ρV^2, of the moving fluid. (Note that the kinetic energy can be treated as a single variable.) The plate width is not an important variable. It is known from theoretical considerations that the deflection varies inversely with modulus of elasticity. In a laboratory experiment the following data were obtained when the height, h, was varied while holding the other variables constant at the values $d = 0.01\ \text{m}$, $E = 2.1 \times 10^9\ \text{N/m}^2$, $\rho = 10^3\ \text{kg/m}^3$, and $V = 2.0\ \text{m/s}$.

δ (m)	h (m)
0.19×10^{-3}	0.10
1.52×10^{-3}	0.20
5.13×10^{-3}	0.30
12.16×10^{-3}	0.40
23.75×10^{-3}	0.50

Based on these data, predict the deflection of a similar steel bar under the following conditions: $d = 0.008$ m, $E = 2.1 \times 10^9$ N/m^2, $\rho = 10^3$ kg/m^3, $V = 1.0$ m/s, and $h = 0.25$ m.

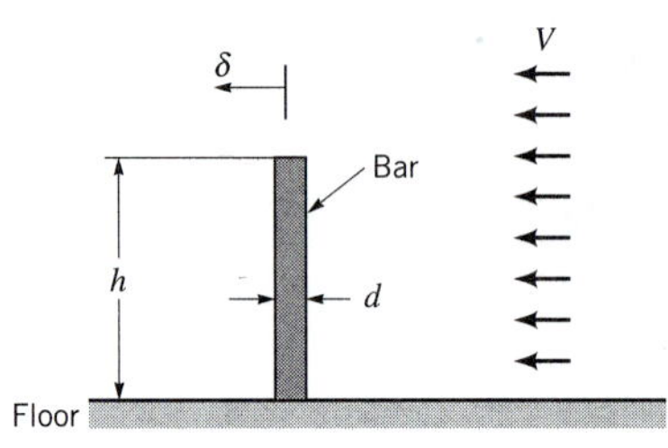

■ FIGURE P7.28

7.29 A fluid flows through the horizontal curved pipe of Fig. P7.29 with a velocity V. The pressure drop, Δp, between the entrance and the exit to the bend is thought to be a function of the velocity, bend radius, R, pipe diameter, D, and fluid density, ρ. The data shown in the following table were obtained in the laboratory. For these tests $\rho = 2.0$ slugs/ft^3, $R = 0.5$ ft, and $D = 0.1$ ft. Perform a dimensional analysis and based on the data given, determine if the variables used for this problem appear to be correct. Explain how you arrive at your answer.

V (ft/s)	2.1	3.0	3.9	5.1
Δp (lb/ft^2)	1.2	1.8	6.0	6.5

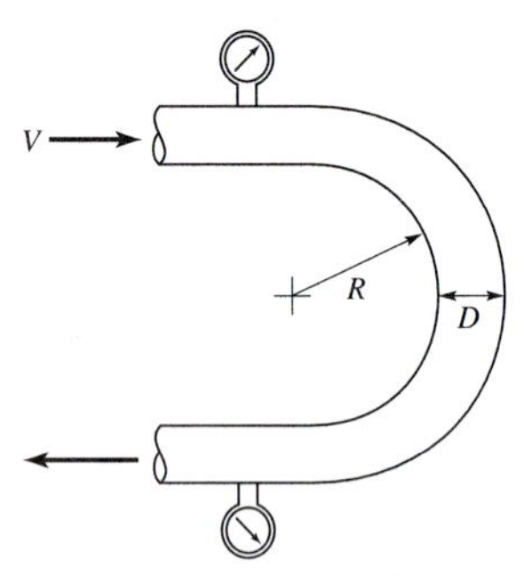

■ FIGURE P7.29

7.30 The water flowrate, Q, in an open rectangular channel can be measured by placing a plate across the channel as shown in Fig. P7.30. This type of a device is called a *weir*. The height of the water, H, above the weir crest is referred to as the head and can be used to determine the flowrate through the channel. Assume that Q is a function of the head, H, the channel width, b, and the acceleration of gravity, g. Determine a suitable set of dimensionless variables for this problem.

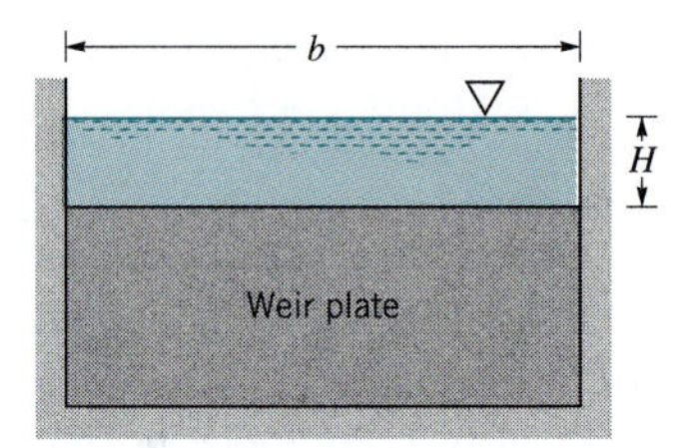

■ FIGURE P7.30

7.31 From theoretical considerations it is known that for the weir described in Problem 7.30 the flowrate, Q, must be directly proportional to the channel width, b. In some laboratory tests it was determined that if $b = 3$ ft and $H = 4$ in., then $Q = 1.96$ ft^3/s. Based on these limited data, determine a general equation for the flowrate over this type of weir.

7.32 SAE 30 oil at 60 °F is pumped through a 3-ft-diameter pipeline at a rate of 5700 gal/min. A model of this pipeline is to be designed using a 3-in.-diameter pipe and water at 60 °F as the working fluid. To maintain Reynolds number similarity between these two systems, what fluid velocity will be required in the model?

7.33 Glycerin at 20 °C flows with a velocity of 4 m/s through a 30-mm-diameter tube. A model of this system is to be developed using standard air as the model fluid. The air velocity is to be 2 m/s. What tube diameter is required for the model if dynamic similarity is to be maintained between model and prototype?

7.34 The drag characteristics of a torpedo are to be studied in a water tunnel using a 1:5 scale model. The tunnel operates with freshwater at 20 °C, whereas the prototype torpedo is to be used in seawater at 15.6 °C. To correctly simulate the behavior of the prototype moving with a velocity of 30 m/s, what velocity is required in the water tunnel?

7.35 The design of a river model is to be based on Froude number similarity, and a river depth of 3 m is to correspond to a model depth of 100 mm. Under these conditions what is the prototype velocity corresponding to a model velocity of 2 m/s?

7.36 For a certain fluid flow problem it is known that both the Froude number and the Weber number are important dimensionless parameters. If the problem is to be studied by using a 1:15 scale model, determine the required surface tension scale if the density scale is equal to 1. The model and prototype operate in the same gravitational field.

7.37 The fluid dynamic characteristics of an airplane flying 240 mph at 10,000 ft are to be investigated with the aid of a 1:20 scale model. If the model tests are to be performed in a wind tunnel using standard air, what is the required air velocity in the wind tunnel? Is this a realistic velocity?

7.38 If an airplane travels at a speed of 1120 km/hr at an altitude of 15 km, what is the required speed at an altitude of 8 km to satisfy Mach number similarity? Assume the air properties correspond to those for the U.S. standard atmosphere.

† **7.39** Describe some everyday situations involving fluid flow and estimate the Reynolds numbers for them. Based on your results, do you think fluid inertia is important in most typical flow situations? Explain.

7.40 The lift and drag developed on a hydrofoil are to be determined through wind tunnel tests using standard air. If full scale tests are to be run, what is the required wind tunnel velocity corresponding to a hydrofoil velocity in seawater at 15 mph? Assume Reynolds number similarity is required.

7.41 Water flows at a rate of 40 m^3/s through the spillway of a dam which is 65-m wide. A model spillway, having a width of 0.9 m, is to be constructed and tested in the laboratory. What is the required flowrate in the model?

7.42 A 1:40 scale model of a ship is to be tested in a towing tank. Determine the required kinematic viscosity of the model fluid so that both the Reynolds number and the Froude number are the same for model and prototype. Assume the prototype fluid to be seawater at 60 °F. Could any of the liquids with viscosities given in Fig. B.2 in Appendix B be used as the model fluid?

7.43 A solid block in the shape of a cube rests partially submerged on the bottom of a river as shown in Fig. P7.43. The drag, $\mathfrak{D}$, on the block depends on the river depth, d, the block dimension, h, the stream velocity, V, the fluid density, ρ, and the acceleration of gravity, g. **(a)** Perform a dimensional analysis for this problem. **(b)** The drag is to be determined from a model study using a length scale of 1/5. What model velocity should be used to predict the drag on the prototype located in a river with a velocity of 9 ft/s? Water is to be used for the model fluid. Determine the expected prototype drag in terms of the model drag.

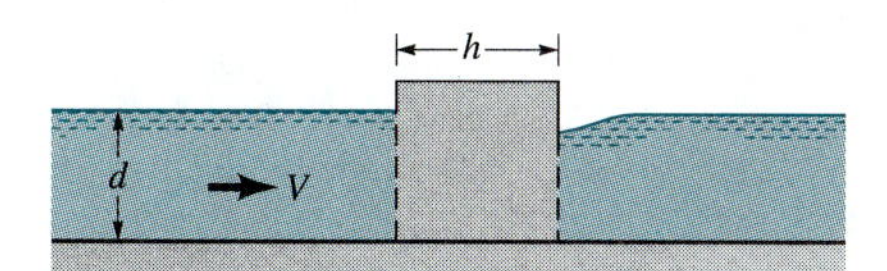

■ **FIGURE P7.43**

7.44 The drag on a 2-m-diameter satellite dish due to an 80 km/hr wind is to be determined through a wind tunnel test using a geometrically similar 0.4-m-diameter model dish. Assume standard air for both model and prototype. **(a)** At what air speed should the model test be run? **(b)** With all similarity conditions satisfied, the measured drag on the model was determined to be 170 N. What is the predicted drag on the prototype dish?

7.45 The pressure drop between the entrance and exit of a 150-mm-diameter 90° elbow, through which ethyl alcohol at 20 °C is flowing, is to be determined with a geometrically similar model. The velocity of the alcohol is 5 m/s. The model fluid is to be water at 20 °C, and the model velocity is limited to 10 m/s. **(a)** What is the required diameter of the model elbow to maintain dynamic similarity? **(b)** A measured pressure drop of 20 kPa in the model will correspond to what prototype value?

7.46 For a certain model study involving a 1:5 scale model it is known that Froude number similarity must be maintained. The possibility of cavitation is also to be investigated, and it is assumed that the cavitation number must be the same for model and prototype. The prototype fluid is water at 30 °C, and the model fluid is water at 70 °C. If the prototype operates at an ambient pressure of 101 kPa (abs), what is the required ambient pressure for the model system?

7.47 To study the sedimentation rates of small particles, an experiment is to be designed to measure the settling velocity, V, of spheres of diameter d falling slowly through a viscous fluid having a viscosity μ. It is thought that V depends on d and μ, the sphere material specific weight, γ_s, and the liquid specific weight, γ. **(a)** Develop a suitable set of dimensionless variables to study this problem. **(b)** Consider two spheres made of the same material but with different diameters settling in the same fluid. One sphere can be considered as a model of the other one. If possible, determine how the velocity scale, V_m/V, is related to the diameter scale, d_m/d, where the subscript m refers to the model sphere. If it is not possible, explain why.

7.48 A viscous fluid ($\mu = 10^{-3}$ lb·s/ft^2; $\rho = 2.0$ slugs/ft^3) flows from a large open tank and discharges into the atmosphere as shown in Fig. P7.48. A model is to be used to predict the velocity of the fluid as it discharges from the pipe. Establish the similarity requirements for the model. Assume steady flow and be sure to include the acceleration of gravity as a variable since gravity is causing the flow. If water at 70 °F is used as a model fluid, are there any restrictions on the length scale that can be used? Explain how you arrived at your answer.

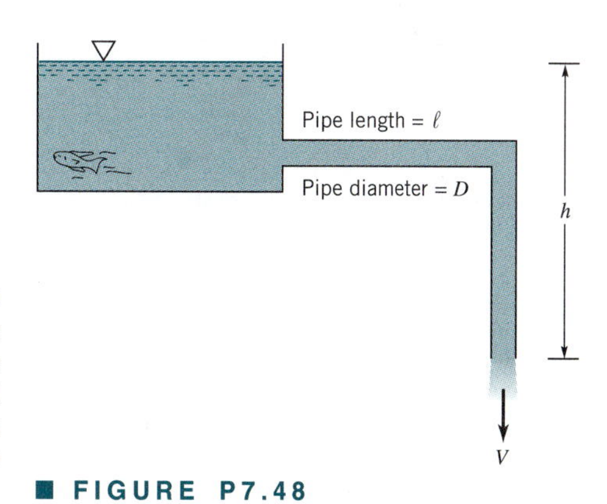

■ **FIGURE P7.48**

7.49 When small particles of diameter d are transported by a moving fluid having a velocity V, they settle to the ground at some distance ℓ after starting from a height h as shown in Fig. P7.49. The variation in ℓ with various factors is to be studied with a model having a length scale of $\frac{1}{10}$. Assume that

$$\ell = f(h, d, V, \gamma, \mu)$$

where γ is the particle specific weight and μ is the fluid viscosity. The same fluid is to be used in both the model and the prototype, but γ (model) $= 9 \times \gamma$ (prototype). **(a)** If $V = 50$ mph, at what velocity should the model tests be run? **(b)** During a certain model test it was found that ℓ (model) $= 0.8$ ft. What would be the predicted ℓ for this test?

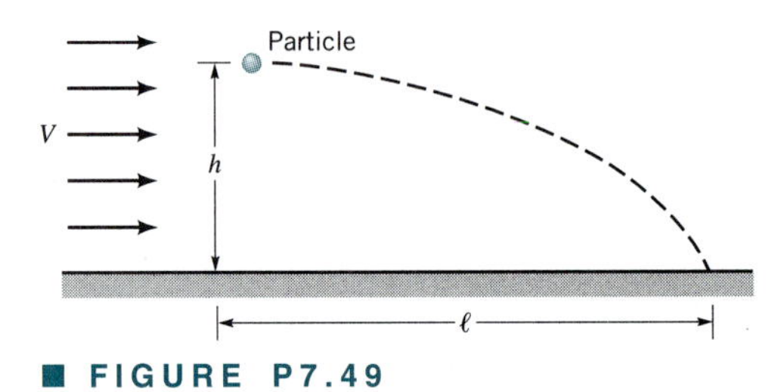

■ FIGURE P7.49

7.50 The drag, $\mathscr{D}$, on a sphere located in a pipe through which a fluid is flowing is to be determined experimentally (see Fig. P7.50). Assume that the drag is a function of the sphere diameter, d, the pipe diameter, D, the fluid velocity, V, and the fluid density, ρ. **(a)** What dimensionless parameters would you use for this problem? **(b)** Some experiments using water indicate that for $d = 0.2$ in., $D = 0.5$ in., and $V = 2$ ft/s, the drag is 1.5×10^{-3} lb. If possible, estimate the drag on a sphere located in a 2-ft-diameter pipe through which water is flowing with a velocity of 6 ft/s. The sphere diameter is such that geometric similarity is maintained. If it is not possible, explain why not.

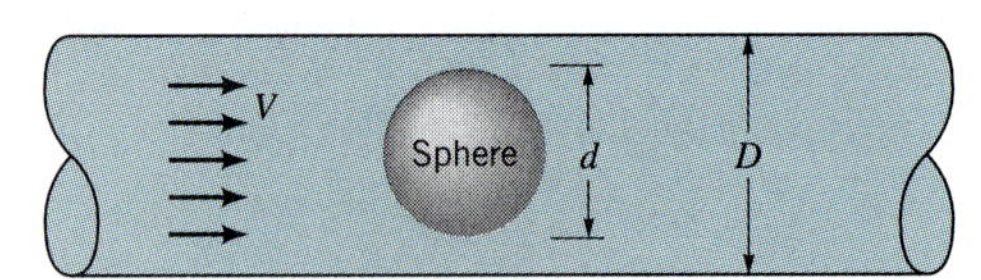

■ FIGURE P7.50

7.51 A very viscous fluid flows slowly past the submerged rectangular plate of Fig. P7.51. The drag, $\mathscr{D}$, is known to be a function of the plate height, h, plate width, b, fluid velocity, V, and fluid viscosity, μ. A model is to be used to predict the drag and during a certain model test using glycerin ($\mu_m = 0.03$ lb·s/ft^2), with $h_m = 1$ in. and $b_m = 3$ in., it was found that $\mathscr{D}_m = 0.2$ lb when $V_m = 0.5$ ft/s. If possible, predict the drag on a geometrically similar larger plate with $h = 4$ in. and $b = 12$ in. immersed in the same glycerin moving with a velocity of 2 ft/s. If it is *not* possible, explain why.

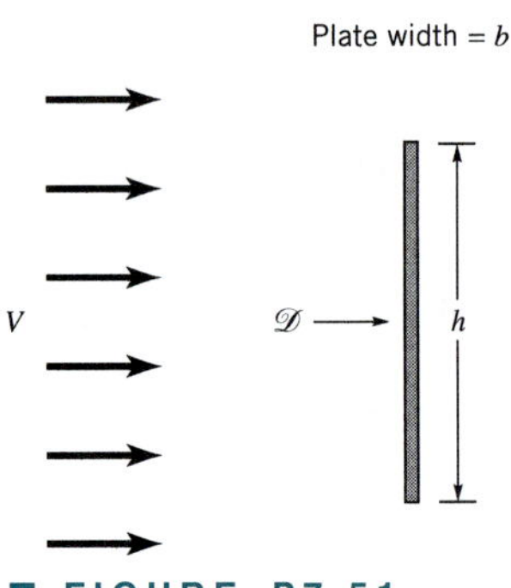

■ FIGURE P7.51

7.52 An orifice flowmeter uses a pressure drop measurement to determine the flowrate through a pipe. A particular orifice flowmeter, when tested in the laboratory, yielded a pressure drop of 8 psi for a flow of 2.9 ft^3/s through a 6-in. pipe. For a geometrically similar system using the same fluid with a 24-in. pipe, what is the required flow if similarity between the two systems is to be maintained? What is the corresponding pressure drop?

7.53 During a storm, a snow drift is formed behind a snow fence as shown in Fig. P7.53. Assume that the height of the drift, h, is a function of the number of inches of snow deposited by the storm, d, height of the fence, H, width of slats in the fence, b, wind speed, V, acceleration of gravity, g, air density, ρ, and specific weight of snow, γ_s. **(a)** If this problem is to be studied with a model, determine the similarity requirements for the model and the relationship between the drift depth for model and prototype (prediction equation). **(b)** A storm with winds of 30 mph deposits 16 in. of snow having a specific weight of 5.0 lb/ft^3. A $\frac{1}{2}$-sized scale model is to be used to investigate the effectiveness of a proposed snow fence. If the air density is the same for the model and the storm, determine the required specific weight for the model snow and required wind speed for the model.

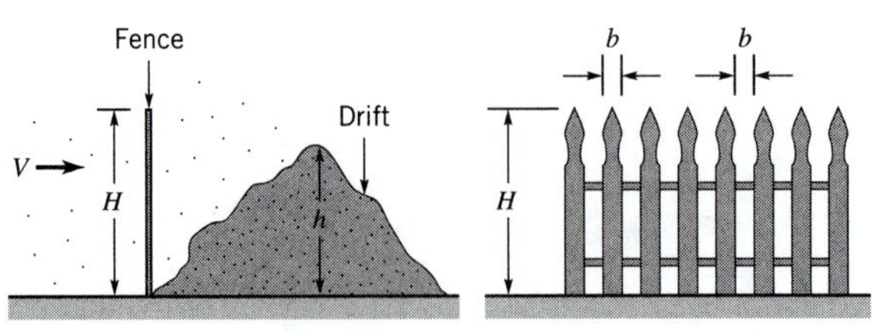

■ FIGURE P7.53

7.54 Air bubbles discharge from the end of a submerged tube as shown in Fig. P7.54. The bubble diameter, D, is assumed to be a function of the air flowrate, Q, the tube diameter, d, the acceleration of gravity, g, the density of the liquid, ρ, and the surface tension of the liquid, σ. **(a)** Determine a suitable set of dimensionless variables for this problem. **(b)** Model tests are to be run on the earth for a prototype that is to be operated on a planet where the acceleration of gravity is 10 times greater than that on earth. The model and prototype are to use the same fluid, and the prototype tube diameter is 0.25 in. Determine the tube diameter for the model and the required model flowrate if the prototype flowrate is to be 0.001 ft^3/s.

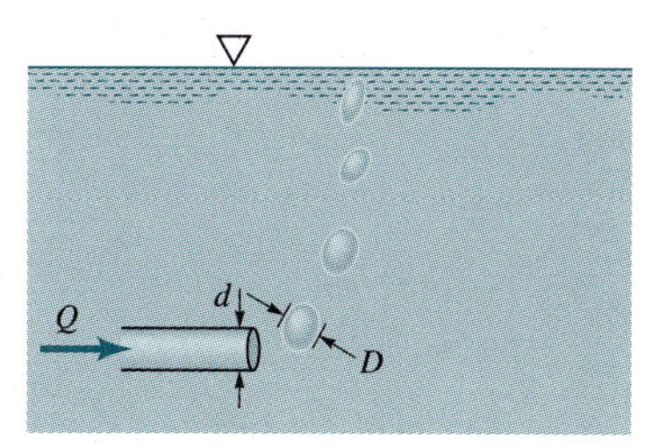

■ FIGURE P7.54

7.55 The drag on a small, completely submerged solid body having a characteristic length of 2.5 mm and moving with a velocity of 10 m/s through water is to be determined with the aid of a model. The length scale is to be 50, which indicates that the model is to be *larger* than the prototype. Investigate the possibility of using either an unpressurized wind tunnel or a water tunnel for this study. Determine the required velocity in both the wind and water tunnels and the relationship between the model drag and the prototype drag for both systems. Would either type of test facility be suitable for this study?

7.56 The drag characteristics for a newly designed automobile having a maximum characteristic length of 20 ft are to be determined through a model study. The characteristics at both low speed (approximately 20 mph) and high speed (90 mph) are of interest. For a series of projected model tests, an unpressurized wind tunnel that will accommodate a model with a maximum characteristic length of 4 ft is to be used. Determine the range of air velocities that would be required for the wind tunnel if Reynolds number similarity is desired. Are the velocities suitable? Explain.

7.57 If the unpressurized wind tunnel of Problem 7.56 were replaced with a tunnel in which the air can be pressurized isothermally to 8 atm (abs), what range of air velocities would be required to maintain Reynolds number similarity for the same prototype velocities given in Problem 7.56? For the pressurized tunnel the maximum characteristic model length that can be accommodated is 2 ft, whereas the maximum characteristic prototype length remains at 20 ft.

7.58 The drag characteristics of an airplane are to be determined by model tests in a wind tunnel operated at an absolute pressure of 1300 kPa. If the prototype is to cruise in standard air at 385 km/hr, and the corresponding speed of the model is not to differ by more than 20% from this (so that compressibility effects may be ignored), what range of length scales may be used if Reynolds number similarity is to be maintained? Assume the viscosity of air is unaffected by pressure, and the temperature of air in the tunnel is equal to the temperature of the air in which the airplane will fly.

7.59 Wind blowing past a flag causes it to "flutter in the breeze." The frequency of this fluttering, ω, is assumed to be a function of the wind speed, V, the air density, ρ, the acceleration of gravity, g, the length of the flag, ℓ, and the "area density," ρ_A, (with dimensions of ML^{-2}) of the flag material. It is desired to predict the flutter frequency of a large $\ell = 40$ ft flag in a $V = 30$ ft/s wind. To do this a model flag with $\ell = 4$ ft is to be tested in a wind tunnel. **(a)** Determine the required area density of the model flag material if the large flag has $\rho_A = 0.006$ slugs/ft^2. **(b)** What wind tunnel velocity is required for testing the model? **(c)** If the model flag flutters at 6 Hz, predict the frequency for the large flag.

7.60 An open channel with a rectangular cross section has a width of 20 ft and carries water at a depth of 4 ft at a flowrate of 60 ft^3/s. A model is to be designed, based on Froude number similarity, so that the discharge scale is 1/1024. At what depth and flowrate would the model operate?

7.61 A square parking lot of width w is bounded on all sides by a curb of height d with only one opening of width b as shown in Fig. P7.61. During a heavy rain the lot fills with water and it is of interest to determine the time, t, it takes for the water to completely drain from the lot after the rain stops. A scale model is to be used to study this problem, and it is assumed that

$$t = f(w, b, d, g, \mu, \rho)$$

where g is the acceleration of gravity, μ is the fluid viscosity, and ρ is the fluid density. **(a)** A dimensional analysis indicates that two important dimensionless parameters are b/w and d/w. What additional dimensionless parameters are required? **(b)** For a geometrically similar model having a length scale of 1/10, what is the relationship between the drain time for the model and the corresponding drain time for the actual parking lot? Assume all similarity requirements are satisfied. Can water be used as the model fluid? Explain and justify your answer.

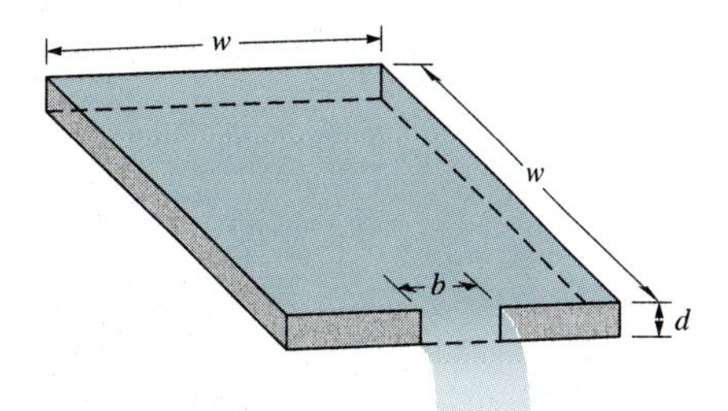

■ FIGURE P7.61

7.62 A $\frac{1}{50}$ scale model is to be used in a towing tank to determine the drag on the hull of a ship. The model is operated in accordance with the Froude number criteria for dynamic similitude. The prototype ship is designed to cruise at 18 knots.

At what velocity (in m/s) should the model be towed? Under these conditions what will be the ratio of the prototype drag to the model drag? Assume the water in the towing tank to have the same properties as those for the prototype and that shear drag is negligible.

† **7.63** If a large oil spill occurs from a tanker operating near a coastline, the time it would take for the oil to reach shore is of great concern. Design a model system that can be used to investigate this type of problem in the laboratory. Indicate all assumptions made in developing the design and discuss any difficulty that may arise in satisfying the similarity requirements arising from your model design.

7.64 A circular cylinder of diameter d is placed in a uniform stream of fluid as shown in Fig. P7.64*a*. Far from the cylinder the velocity is V and the pressure is atmospheric. The gage pressure, p, at point A on the cylinder surface is to be determined from a model study for an 18-in.-diameter prototype placed in an airstream having a speed of 8 ft/s. A 1:12 scale model is to be used with water as the working fluid. Some experimental data obtained from the model are shown in Fig. P7.64*b*. Predict the prototype pressure.

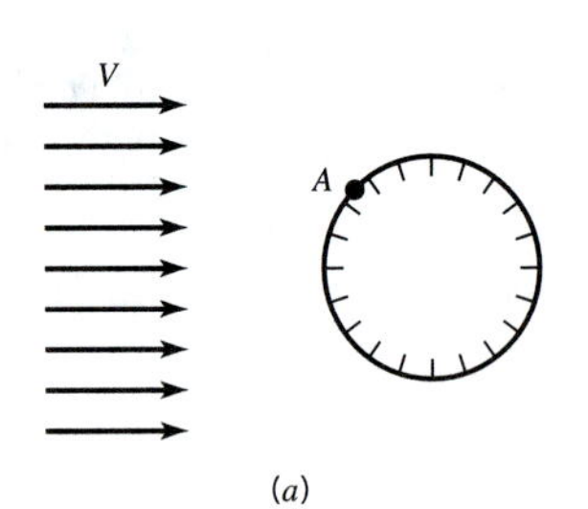

(*a*)

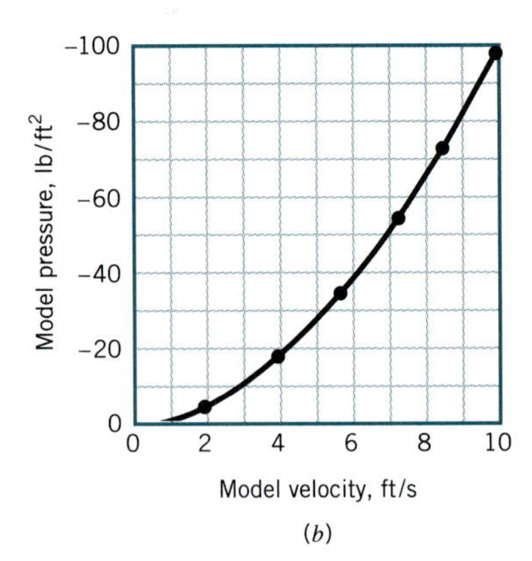

(*b*)

■ **FIGURE P7.64**

7.65 The pressure rise, Δp, across a centrifugal pump of a given shape (see Fig. P7.65*a*) can be expressed as

$$\Delta p = f(D, \omega, \rho, Q)$$

where D is the impeller diameter, ω the angular velocity of the impeller, ρ the fluid density, and Q the volume rate of flow through the pump. A model pump having a diameter of 8 in. is tested in the laboratory using water. When operated at an angular velocity of 40π rad/s the model pressure rise as a function of Q is shown in Fig. P7.65*b*. Use this curve to predict the pressure rise across a geometrically similar pump (prototype) for a prototype flowrate of 6 ft³/s. The prototype has a diameter of 12 in. and operates at an angular velocity of 60π rad/s. The prototype fluid is also water.

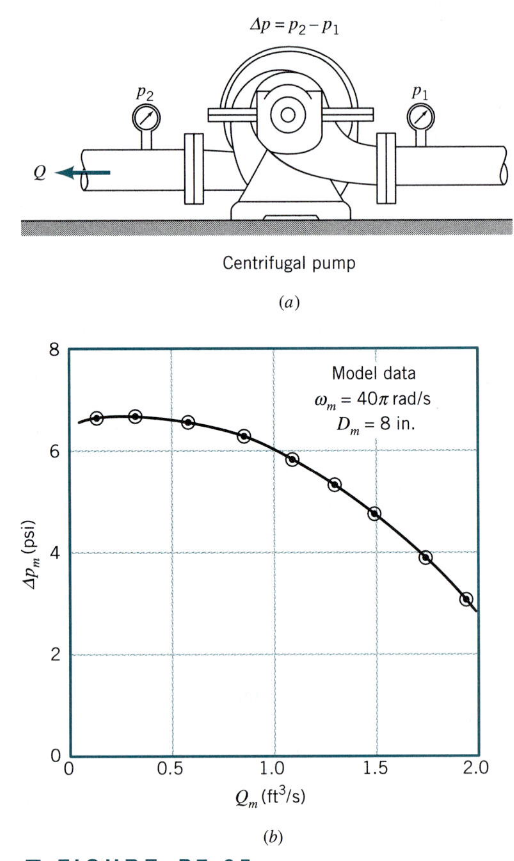

(*a*)

(*b*)

■ **FIGURE P7.65**

7.66 Start with the two-dimensional continuity equation and the Navier–Stokes equations (Eqs. 7.28, 7.29, and 7.30) and verify the nondimensional forms of these equations (Eqs. 7.31, 7.34, and 7.35).

7.67 A viscous fluid is contained between wide, parallel plates spaced a distance h apart as shown in Fig. P7.67. The upper plate is fixed, and the bottom plate oscillates harmonically with a velocity amplitude U and frequency ω. The differential equation for the velocity distribution between the plates is

$$\rho \frac{\partial u}{\partial t} = \mu \frac{\partial^2 u}{\partial y^2}$$

where u is the velocity, t is time, and ρ and μ are fluid density and viscosity, respectively. Rewrite this equation in a suitable nondimensional form using h, U, and ω as reference parameters.

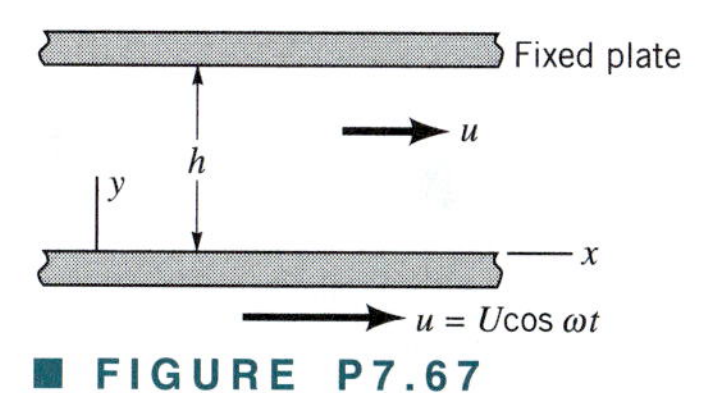

■ FIGURE P7.67

7.68 The deflection of the cantilever beam of Fig. P7.68 is governed by the differential equation

$$EI\frac{d^2y}{dx^2} = P(x - \ell)$$

where E is the modulus of elasticity and I is the moment of inertia of the beam cross section. The boundary conditions are $y = 0$ at $x = 0$ and $dy/dx = 0$ at $x = 0$. **(a)** Rewrite the equation and boundary conditions in dimensionless form using the beam length, ℓ, as the reference length. **(b)** Based on the results of part (a), what are the similarity requirements and the prediction equation for a model to predict deflections?

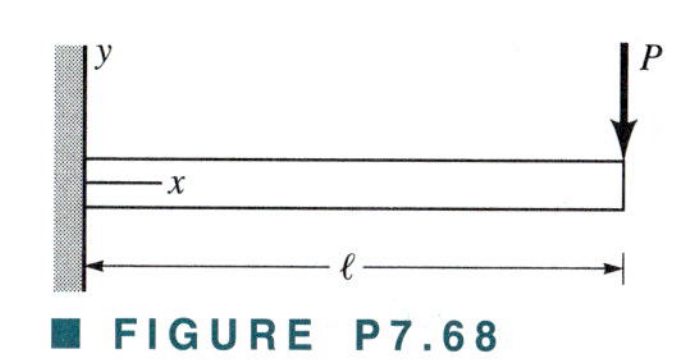

■ FIGURE P7.68

7.69 A liquid is contained in a pipe that is closed at one end as shown in Fig. P7.69. Initially the liquid is at rest, but if the end is suddenly opened the liquid starts to move. Assume the pressure p_1 remains constant. The differential equation that describes the resulting motion of the liquid is

$$\rho\frac{\partial v_z}{\partial t} = \frac{p_1}{\ell} + \mu\left(\frac{\partial^2 v_z}{\partial r^2} + \frac{1}{r}\frac{\partial v_z}{\partial r}\right)$$

where v_z is the velocity at any radial location, r, and t is time. Rewrite this equation in dimensionless form using the liquid density, ρ, the viscosity, μ, and the pipe radius, R, as reference parameters.

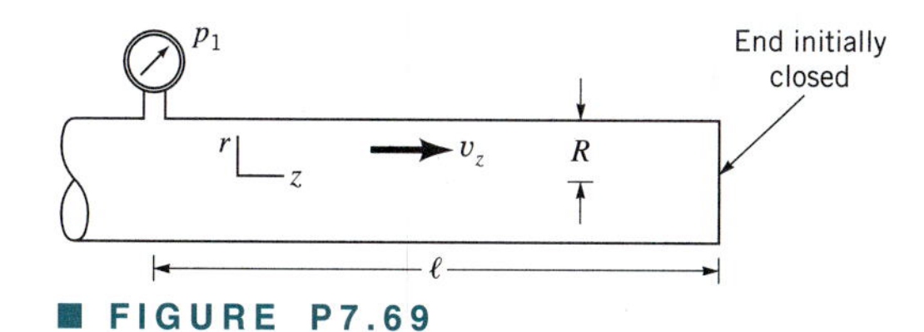

■ FIGURE P7.69

7.70 An incompressible fluid is contained between two infinite parallel plates as illustrated in Fig. P7.70. Under the influence of a harmonically varying pressure gradient in the x direction, the fluid oscillates harmonically with a frequency ω. The differential equation describing the fluid motion is

$$\rho\frac{\partial u}{\partial t} = X\cos\omega t + \mu\frac{\partial^2 u}{\partial y^2}$$

where X is the amplitude of the pressure gradient. Express this equation in nondimensional form using h and ω as reference parameters.

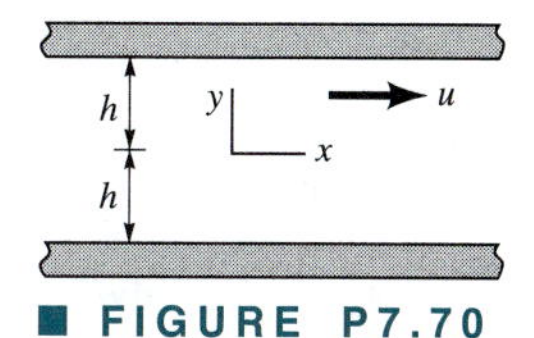

■ FIGURE P7.70

7.71 A viscous fluid flows through a vertical, square channel as shown in Fig. P7.71. The velocity w can be expressed as

$$w = f(x, y, b, \mu, \gamma, V, \partial p/\partial z)$$

where μ is the fluid viscosity, γ the fluid specific weight, V the mean velocity, and $\partial p/\partial z$ the pressure gradient in the z direction. **(a)** Use dimensional analysis to find a suitable set of dimensionless variables and parameters for this problem. **(b)** The differential equation governing the fluid motion for this problem is

$$\frac{\partial p}{\partial z} = -\gamma + \mu\left(\frac{\partial^2 w}{\partial x^2} + \frac{\partial^2 w}{\partial y^2}\right)$$

Write this equation in a suitable dimensionless form, and show that the similarity requirements obtained from this analysis are the same as those resulting from the dimensional analysis of part (a).

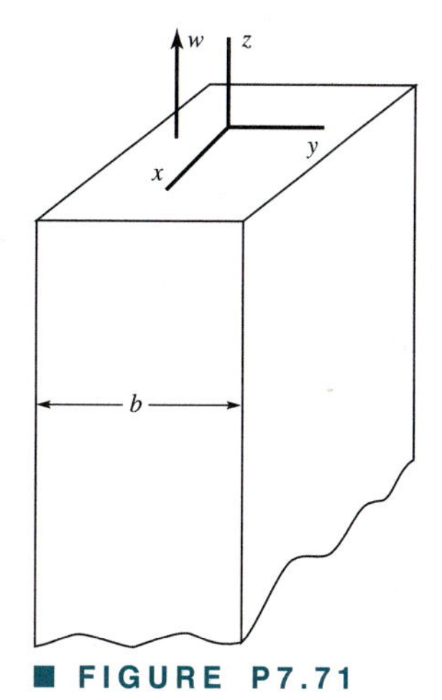

■ FIGURE P7.71

7.72 As fluid flows through a valve located in a pipe, there is a pressure drop, Δp, across the valve. An experiment is designed to investigate how Δp for a gate valve varies with the fluid velocity through the valve and the valve setting. It is assumed that

$$\Delta p = f(a, D, \lambda_i, \rho, V)$$

where a and D are defined in Fig. P7.72, λ_i represents all other lengths required to define the valve geometry, ρ is the fluid density, and V is the velocity of the fluid in the pipe in which the valve is located. Perform a dimensional analysis and show that

$$\frac{\Delta p}{(1/2)\rho V^2} = \phi_1\left(\frac{a}{D}, \frac{\lambda_i}{D}\right)$$

where the factor of $\frac{1}{2}$ has been arbitrarily added. Note that for a given valve, $\phi_1(a/D, \lambda_i/D)$ will depend only on the valve setting. Since the valve setting can also be expressed as a percentage, R, (percent open) with $R = 0\%$ for a closed valve and $R = 100\%$ for a fully open valve, it follows that

$$\frac{\Delta p}{(1/2)\rho V^2} = \phi(R)$$

As discussed in Chapter 8, $\phi(R)$ is called the minor loss coefficient.

To determine the pressure drop characteristics of a gate valve, the experimental arrangement shown in Fig. P7.72 is used with air having a specific weight of 0.0719 lb/ft³. A U-tube manometer measures the upstream pressure, and the downstream pressure is zero since the air discharges into the atmosphere. The flowrate, Q, is determined with a flow meter as illustrated in Fig. P7.72. Some experimental data obtained with this set-up are given in the following table.

For each valve setting, plot a graph of the flowrate, Q, as a function of the pressure drop, Δp. Show how these data can be collapsed into a single curve through the use of the dimensional analysis previously described. Do the plotted data support the dimensional analysis? Explain.

R (%open)	*Q* (cfs)	*H* (in.)
25	0.169	5.04
25	0.195	6.50
25	0.220	7.69
25	0.235	9.20
38	0.150	1.92
38	0.238	4.90
38	0.321	8.73
38	0.350	10.44
75	0.238	0.98
75	0.456	2.60
75	0.578	4.32
75	0.767	7.65
100	0.426	0.64
100	0.517	0.97
100	0.618	1.40
100	0.799	2.37

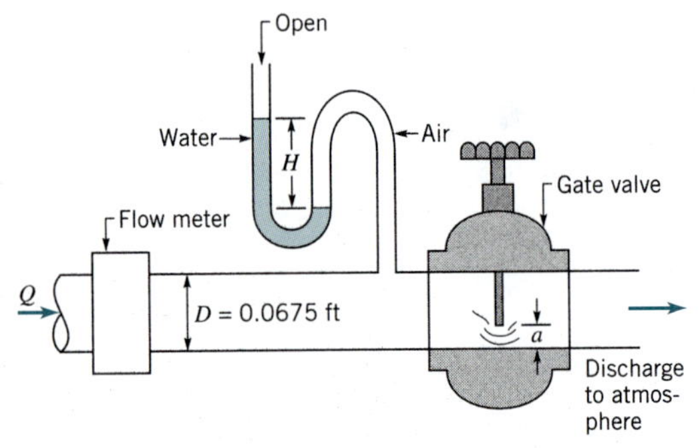

■ FIGURE P7.72

7.73 As a liquid drains from an open cylindrical tank through a small orifice in its bottom, the liquid depth, h, decreases with time (see Fig. P7.73). This change in depth as a function of time, t, is to be studied with a one-half scale model. The liquid in the prototype tank has an initial depth, H, of 16 in., a diameter, D, equal to 4.0 in., and an orifice diameter, d, of 0.25 in. The fluid is water at 20 °C. Develop a suitable set of dimensionless parameters for this problem assuming that

$$h = f(H, D, d, \gamma, \rho, t)$$

where γ and ρ are the specific weight and density of the liquid, respectively. Based on these dimensionless parameters, establish the similarity requirements for the model and the prediction equation relating the model depth to the prototype depth.

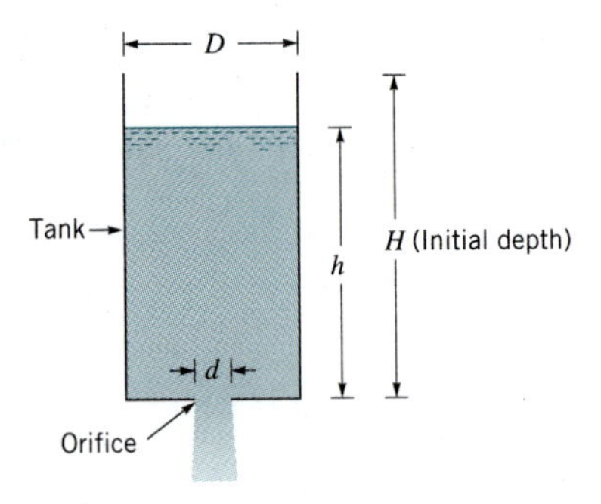

■ FIGURE P7.73

Some experimental data obtained from a geometrically similar model ($D_m = 2.0$ in., $d_m = 0.125$ in., and $H_m = 8.0$ in.) using water at 20 °C are given in the following table. On a graph, plot the values of the water depth, h_m, as a function of time, t_m. On another graph, plot these data in dimensionless form.

Some prototype data are also given in the following table. On the same graphs used for the model, plot the corresponding prototype data. Based on a comparison of the model data with the prototype data, does the model design seem correct? Explain. The effect of viscosity has been neglected in this experiment. Does this appear to be a reasonable assumption? If viscosity is included as an important variable, how would the model design be affected? Explain.

Model Data		Prototype Data	
h_m (in.)	t_m (s)	h (in.)	t (s)
8.0	0.0	16.0	0.0
7.0	3.1	14.0	4.5
6.0	6.2	12.0	8.9
5.0	9.9	10.0	14.0
4.0	13.5	8.0	20.2
3.0	18.1	6.0	25.9
2.0	24.0	4.0	32.8
1.0	32.5	2.0	45.7
0.0	43.0	0.0	59.8

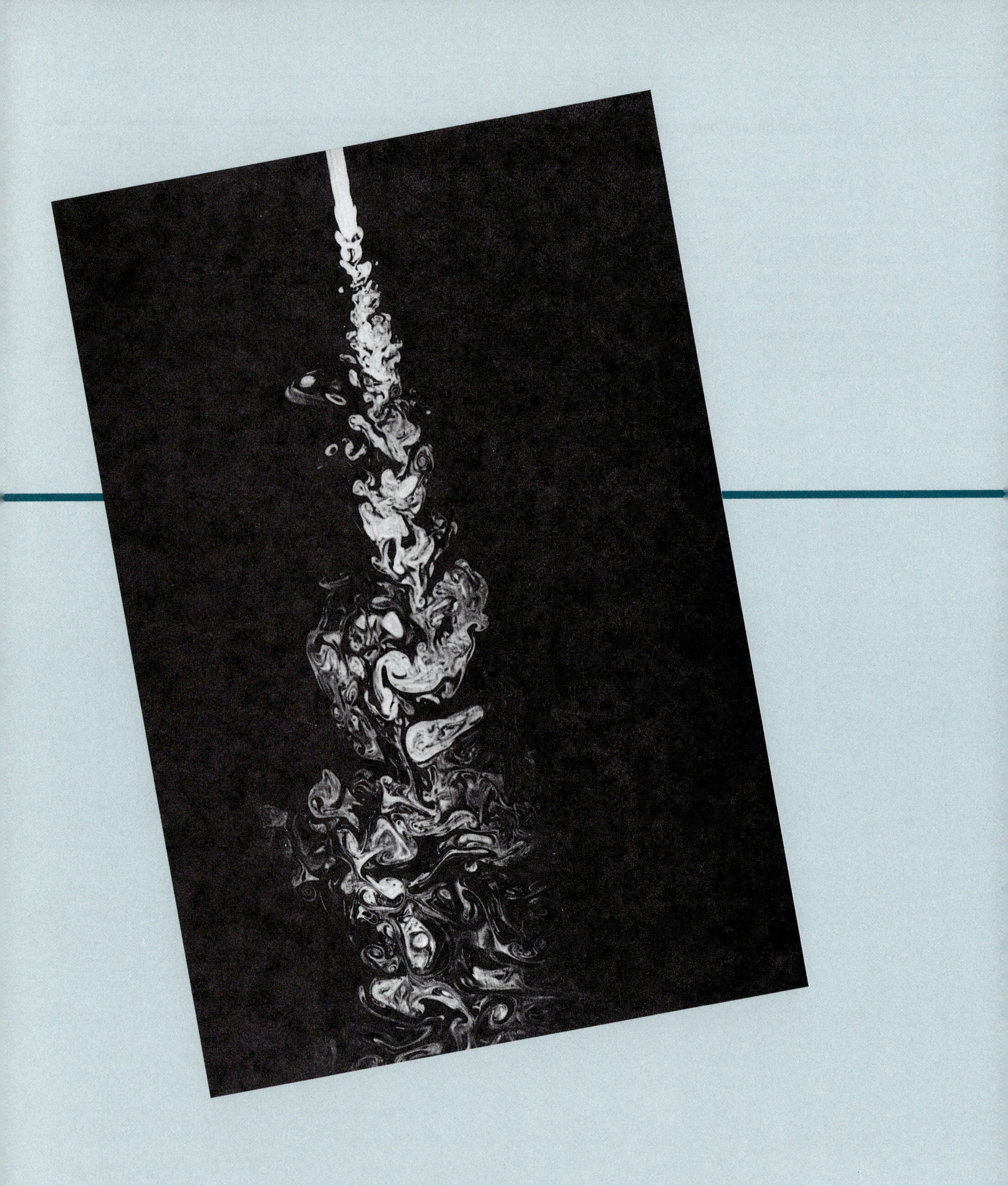

Turbulent jet: The jet of water from the pipe is turbulent. The complex, irregular, unsteady structure typical of turbulent flows is apparent. (Laser-induced fluorescence of dye in water.) (Photography by P. E. Dimotakis, R. C. Lye, and D. Z. Papantoniou.)

8
Viscous Flow in Pipes

In the previous chapters we have considered a variety of topics concerning the motion of fluids. The basic governing principles concerning mass, momentum, and energy were developed and applied, in conjunction with rather severe assumptions, to numerous flow situations. In this chapter we will apply the basic principles to a specific, important topic—the flow of viscous, incompressible fluids in pipes and ducts.

Pipe flow is very important in our daily operations.

The transport of a fluid (liquid or gas) in a closed conduit (commonly called a *pipe* if it is of round cross section or a *duct* if it is not round) is extremely important in our daily operations. A brief consideration of the world around us will indicate that there is a wide variety of applications of pipe flow. Such applications range from the large, man-made Alaskan pipeline that carries crude oil almost 800 miles across Alaska, to the more complex (and certainly not less useful) natural systems of ''pipes'' that carry blood throughout our body and air into and out of our lungs. Other examples include the water pipes in our homes and the distribution system that delivers the water from the city well to the house. Numerous hoses and pipes carry hydraulic fluid or other fluids to various components of vehicles and machines. The air quality within our buildings is maintained at comfortable levels by the distribution of conditioned (heated, cooled, humidified/dehumidified) air through a maze of pipes and ducts. Although all of these systems are different, the fluid-mechanics principles governing the fluid motions are common. The purpose of this chapter is to understand the basic processes involved in such flows.

Some of the basic components of a typical *pipe system* are shown in Fig. 8.1. They include the pipes themselves (perhaps of more than one diameter), the various fittings used to connect the individual pipes to form the desired system, the flowrate control devices (valves), and the pumps or turbines that add energy to or remove energy from the fluid. Even the most simple pipe systems are actually quite complex when they are viewed in terms of rigorous analytical considerations. We will use an ''exact'' analysis of the simplest pipe flow topics (such as laminar flow in long, straight, constant diameter pipes) and dimensional analysis considerations combined with experimental results for the other pipe flow topics. Such an approach is not unusual in fluid mechanics investigations. When ''real world'' effects are important (such as viscous effects in pipe flows), it is often difficult or ''impossible'' to

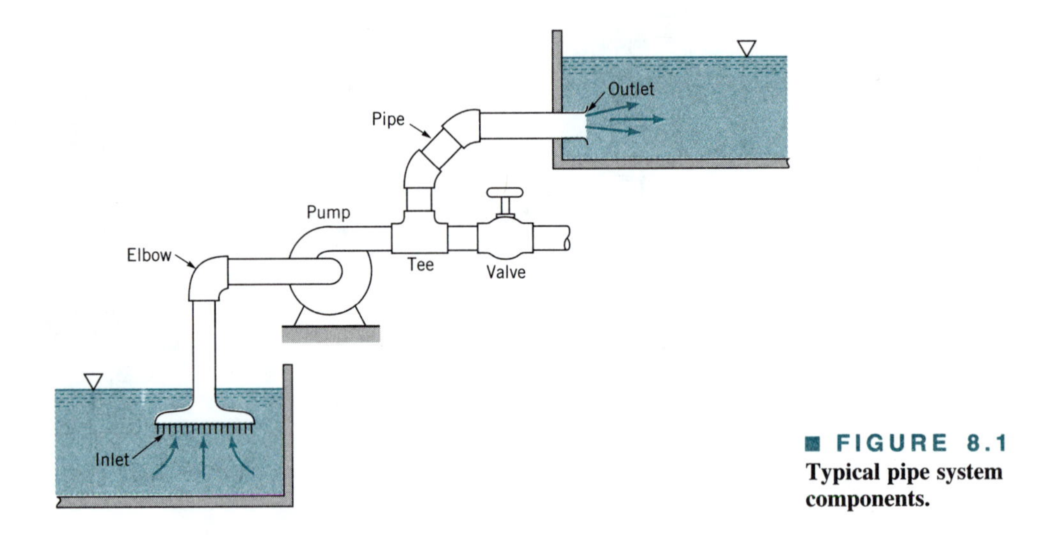

■ FIGURE 8.1 **Typical pipe system components.**

use only theoretical methods to obtain the desired results. A judicious combination of experimental data with theoretical considerations and dimensional analysis often provides the desired results. The flow in pipes discussed in this chapter is an example of such an analysis.

8.1 General Characteristics of Pipe Flow

Before we apply the various governing equations to pipe flow examples, we will discuss some of the basic concepts of pipe flow. With these ground rules established we can then proceed to formulate and solve various important flow problems.

Although not all conduits used to transport fluid from one location to another are round in cross section, most of the common ones are. These include typical water pipes, hydraulic hoses, and other conduits that are designed to withstand a considerable pressure difference across their walls without undue distortion of their shape. Typical conduits of noncircular cross section include heating and air conditioning ducts that are often of rectangular cross section. Normally the pressure difference between the inside and outside of these ducts is relatively small. Most of the basic principles involved are independent of the cross-sectional shape, although the details of the flow may be dependent on it. Unless otherwise specified, we will assume that the conduit is round, although we will show how to account for other shapes.

The pipe is assumed to be completely full of the flowing fluid.

For all flows involved in this chapter, we assume that the pipe is completely filled with the fluid being transported as is shown in Fig. 8.2*a*. Thus, we will not consider a concrete pipe through which rainwater flows without completely filling the pipe, as is shown in Fig. 8.2*b*. Such flows, called open-channel flow, are treated in Chapter 10. The difference between open-channel flow and the pipe flow of this chapter is in the fundamental mechanism that drives the flow. For open-channel flow, gravity alone is the driving force—the water flows

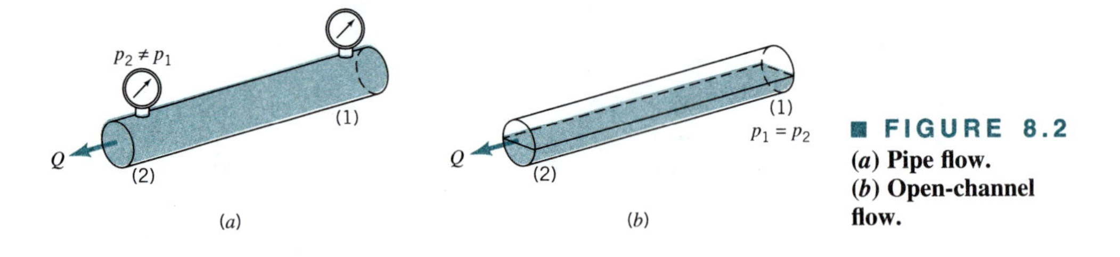

■ FIGURE 8.2 **(*a*) Pipe flow. (*b*) Open-channel flow.**

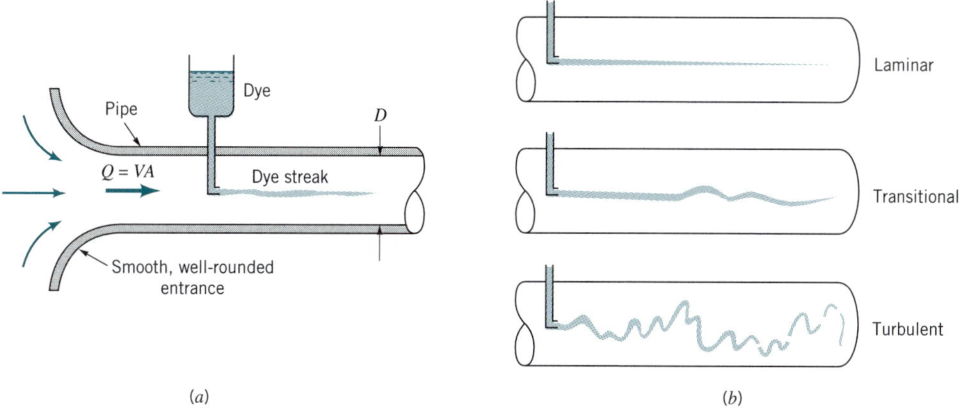

■ FIGURE 8.3 **(*a*) Experiment to illustrate type of flow. (*b*) Typical dye streaks.**

down a hill. For pipe flow, gravity may be important (the pipe need not be horizontal), but the main driving force is likely to be a pressure gradient along the pipe. If the pipe is not full, it is not possible to maintain this pressure difference, $p_1 - p_2$.

8.1.1 Laminar or Turbulent Flow

The flow of a fluid in a pipe may be laminar flow or it may be turbulent flow. Osborne Reynolds (1842–1912), a British scientist and mathematician, was the first to distinguish the difference between these two classifications of flow by using a simple apparatus as shown in Fig. 8.3*a*. If water runs through a pipe of diameter D with an average velocity V, the following characteristics are observed by injecting neutrally buoyant dye as shown. For ''small enough flowrates'' the dye streak (a streakline) will remain as a well-defined line as it flows along, with only slight blurring due to molecular diffusion of the dye into the surrounding water. For a somewhat larger ''intermediate flowrate'' the dye streak fluctuates in time and space, and intermittent bursts of irregular behavior appear along the streak. On the other hand, for ''large enough flowrates'' the dye streak almost immediately becomes blurred and spreads across the entire pipe in a random fashion. These three characteristics, denoted as *laminar*, *transitional*, and *turbulent* flow, respectively, are illustrated in Fig. 8.3*b*.

V8.1

A flow may be laminar, transitional, or turbulent.

The curves shown in Fig. 8.4 represent the x component of the velocity as a function of time at a point A in the flow. The random fluctuations of the turbulent flow (with the associated particle mixing) are what disperse the dye throughout the pipe and cause the blurred appearance illustrated in Fig. 8.3*b*. For laminar flow in a pipe there is only one component

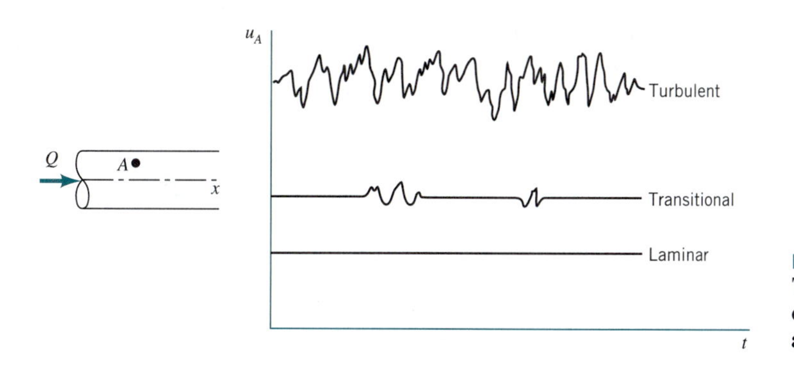

■ FIGURE 8.4 **Time dependence of fluid velocity at a point.**

of velocity, $\mathbf{V} = u\hat{\mathbf{i}}$. For turbulent flow the predominant component of velocity is also along the pipe, but it is unsteady (random) and accompanied by random components normal to the pipe axis, $\mathbf{V} = u\hat{\mathbf{i}} + v\hat{\mathbf{j}} + w\hat{\mathbf{k}}$. Such motion in a typical flow occurs too fast for our eyes to follow. Slow motion pictures of the flow can more clearly reveal the irregular, random, turbulent nature of the flow.

As was discussed in Chapter 7, we should not label dimensional quantities as being "large" or "small," such as "small enough flowrates" in the preceding paragraphs. Rather, the appropriate dimensionless quantity should be identified and the "small" or "large" character attached to it. A quantity is "large" or "small" only relative to a reference quantity. The ratio of those quantities results in a dimensionless quantity. For pipe flow the most important dimensionless parameter is the Reynolds number, Re—the ratio of the inertia to viscous effects in the flow. Hence, in the previous paragraph the term flowrate should be replaced by Reynolds number, $\text{Re} = \rho VD/\mu$, where V is the average velocity in the pipe. That is, the flow in a pipe is laminar, transitional, or turbulent provided the Reynolds number is "small enough," "intermediate," or "large enough." It is not only the fluid velocity that determines the character of the flow—its density, viscosity, and the pipe size are of equal importance. These parameters combine to produce the Reynolds number. The distinction between laminar and turbulent pipe flow and its dependence on an appropriate dimensionless quantity was first pointed out by Osborne Reynolds in 1883.

Pipe flow characteristics are dependent on the value of the Reynolds number.

The Reynolds number ranges for which laminar, transitional, or turbulent pipe flows are obtained cannot be precisely given. The actual transition from laminar to turbulent flow may take place at various Reynolds numbers, depending on how much the flow is disturbed by vibrations of the pipe, roughness of the entrance region, and the like. For general engineering purposes (i.e., without undue precautions to eliminate such disturbances), the following values are appropriate: The flow in a round pipe is laminar if the Reynolds number is less than approximately 2100. The flow in a round pipe is turbulent if the Reynolds number is greater than approximately 4000. For Reynolds numbers between these two limits, the flow may switch between laminar and turbulent conditions in an apparently random fashion (transitional flow).

EXAMPLE 8.1

Water at a temperature of 50 °F flows through a pipe of diameter $D = 0.73$ in. (a) Determine the minimum time taken to fill a 12-oz glass (volume $= 0.0125 \text{ ft}^3$) with water if the flow in the pipe is to be laminar. (b) Determine the maximum time taken to fill the glass if the flow is to be turbulent. Repeat the calculations if the water temperature is 140 °F.

SOLUTION

(a) If the flow in the pipe is to remain laminar, the minimum time to fill the glass will occur if the Reynolds number is the maximum allowed for laminar flow, typically $\text{Re} = \rho VD/\mu = 2100$. Thus, $V = 2100\ \mu/\rho D$, where from Table B.1, $\rho = 1.94 \text{ slugs/ft}^3$ and $\mu = 2.73 \times 10^{-5} \text{ lb·s/ft}^2$ at 50 °F, while $\rho = 1.91 \text{ slugs/ft}^3$ and $\mu = 0.974 \times 10^{-5} \text{ lb·s/ft}^2$ at 140 °F. Thus, the maximum average velocity for laminar flow in the pipe is

$$V = \frac{2100\mu}{\rho D} = \frac{2100(2.73 \times 10^{-5} \text{ lb·s/ft}^2)}{(1.94 \text{ slugs/ft}^3)(0.73/12 \text{ ft})} = 0.486 \text{ lb·s/slug}$$

$$= 0.486 \text{ ft/s}$$

Similarly, $V = 0.176$ ft/s at 140 °F. With $\forall$ = volume of glass and $\forall = Qt$ we obtain

$$t = \frac{\forall}{Q} = \frac{\forall}{(\pi/4)D^2V} = \frac{4(0.0125 \text{ ft}^3)}{(\pi[0.73/12]^2\text{ft}^2)(0.486 \text{ ft/s})}$$

$$= 8.85 \text{ s at } T = 50 \text{ °F} \qquad \textbf{(Ans)}$$

Similarly, $t = 24.4$ s at 140 °F. To maintain laminar flow, the less viscous hot water requires a lower flowrate than the cold water.

(b) If the flow in the pipe is to be turbulent, the maximum time to fill the glass will occur if the Reynolds number is the minimum allowed for turbulent flow, Re = 4000. Thus, $V = 4000\mu/\rho D = 0.925$ ft/s and $t = 4.65$ s at 50 °F, while $V = 0.335$ ft/s and $t =$ 12.8 s at 140 °F.

Note that because water is "not very viscous," the velocity must be "fairly small" to maintain laminar flow. In general, turbulent flows are encountered more often than laminar flows because of the relatively small viscosity of most common fluids (water, gasoline, air). If the flowing fluid had been honey with a kinematic viscosity ($\nu = \mu/\rho$) 3000 times greater than that of water, the above velocities would be increased by a factor of 3000 and the times reduced by the same factor. As we will see in the following sections, the pressure needed to force a very viscous fluid through a pipe at such a high velocity may be unreasonably large.

8.1.2 Entrance Region and Fully Developed Flow

Any fluid flowing in a pipe had to enter the pipe at some location. The region of flow near where the fluid enters the pipe is termed the *entrance region* and is illustrated in Fig. 8.5. It may be the first few feet of a pipe connected to a tank or the initial portion of a long run of a hot air duct coming from a furnace.

Flow in the entrance region of a pipe is quite complex.

As is shown in Fig. 8.5, the fluid typically enters the pipe with a nearly uniform velocity profile at section (1). As the fluid moves through the pipe, viscous effects cause it to stick to the pipe wall (the no-slip boundary condition). This is true whether the fluid is relatively inviscid air or a very viscous oil. Thus, a *boundary layer* in which viscous effects are important is produced along the pipe wall such that the initial velocity profile changes with distance along the pipe, x, until the fluid reaches the end of the entrance length, section (2), beyond which the velocity profile does not vary with x. The boundary layer has grown in thickness

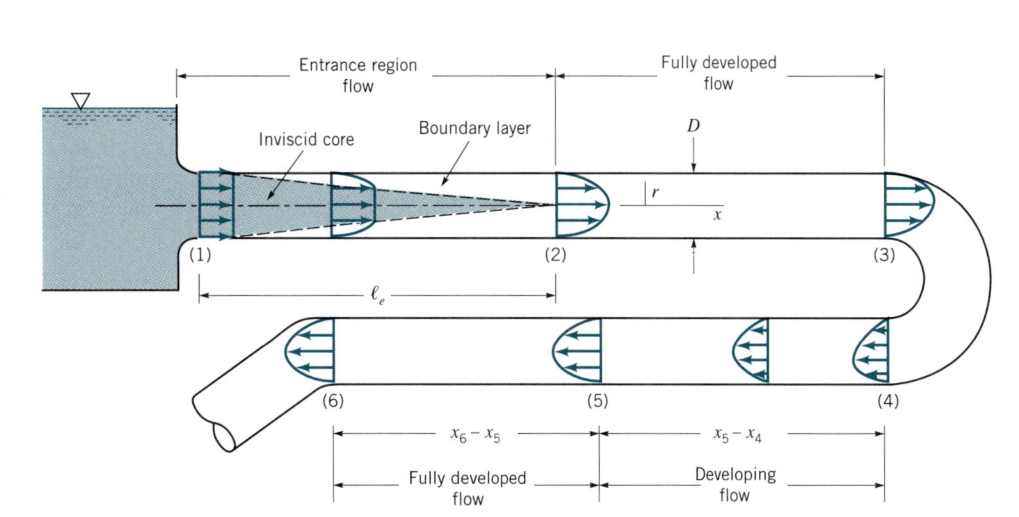

■ **FIGURE 8.5 Entrance region, developing flow, and fully developed flow in a pipe system.**

to completely fill the pipe. Viscous effects are of considerable importance within the boundary layer. For fluid outside the boundary layer [within the *inviscid core* surrounding the centerline from (1) to (2)], viscous effects are negligible.

The shape of the velocity profile in the pipe depends on whether the flow is laminar or turbulent, as does the length of the entrance region, ℓ_e. As with many other properties of pipe flow, the dimensionless *entrance length*, ℓ_e/D, correlates quite well with the Reynolds number. Typical entrance lengths are given by

$$\frac{\ell_e}{D} = 0.06 \text{ Re for laminar flow} \quad \textbf{(8.1)}$$

The entrance length is a function of the Reynolds number.

and

$$\frac{\ell_e}{D} = 4.4 \text{ (Re)}^{1/6} \text{ for turbulent flow} \quad \textbf{(8.2)}$$

For very low Reynolds number flows the entrance length can be quite short ($\ell_e = 0.6D$ if $\text{Re} = 10$), whereas for large Reynolds number flows it may take a length equal to many pipe diameters before the end of the entrance region is reached ($\ell_e = 120D$ for $\text{Re} = 2000$). For many practical engineering problems, $10^4 < \text{Re} < 10^5$ so that $20D < \ell_e < 30D$.

Calculation of the velocity profile and pressure distribution within the entrance region is quite complex. However, once the fluid reaches the end of the entrance region, section (2) of Fig. 8.5, the flow is simpler to describe because the velocity is a function of only the distance from the pipe centerline, r, and independent of x. This is true until the character of the pipe changes in some way, such as a change in diameter, or the fluid flows through a bend, valve, or some other component at section (3). The flow between (2) and (3) is termed *fully developed.* Beyond the interruption of the fully developed flow [at section (4)], the flow gradually begins its return to its fully developed character [section (5)] and continues with this profile until the next pipe system component is reached [section (6)]. In many cases the pipe is long enough so that there is a considerable length of fully developed flow compared with the developing flow length [$(x_3 - x_2) \gg \ell_e$ and $(x_6 - x_5) \gg (x_5 - x_4)$]. In other cases the distances between one component (bend, tee, valve, etc.) of the pipe system and the next component is so short that fully developed flow is never achieved.

8.1.3 Pressure and Shear Stress

Fully developed steady flow in a constant diameter pipe may be driven by gravity and/or pressure forces. For horizontal pipe flow, gravity has no effect except for a hydrostatic pressure variation across the pipe, γD, that is usually negligible. It is the pressure difference, $\Delta p = p_1 - p_2$, between one section of the horizontal pipe and another which forces the fluid through the pipe. Viscous effects provide the restraining force that exactly balances the pressure force, thereby allowing the fluid to flow through the pipe with no acceleration. If viscous effects were absent in such flows, the pressure would be constant throughout the pipe, except for the hydrostatic variation.

In non-fully developed flow regions, such as the entrance region of a pipe, the fluid accelerates or decelerates as it flows (the velocity profile changes from a uniform profile at the entrance of the pipe to its fully developed profile at the end of the entrance region). Thus, in the entrance region there is a balance between pressure, viscous, and inertia (acceleration) forces. The result is a pressure distribution along the horizontal pipe as shown in Fig. 8.6. The magnitude of the pressure gradient, $\partial p/\partial x$, is larger in the entrance region than in the fully developed region, where it is a constant, $\partial p/\partial x = -\Delta p/\ell < 0$.

The fact that there is a nonzero pressure gradient along the horizontal pipe is a result of viscous effects. As is discussed in Chapter 3, if the viscosity were zero, the pressure would not vary with x. The need for the pressure drop can be viewed from two different standpoints.

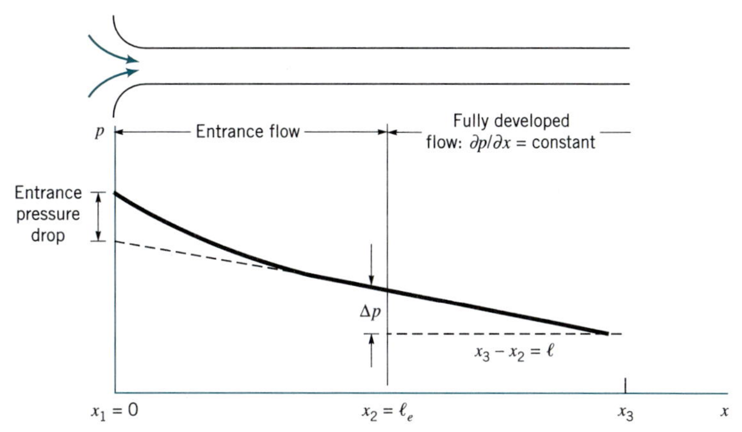

■ FIGURE 8.6 **Pressure distribution along a horizontal pipe.**

In terms of a force balance, the pressure force is needed to overcome the viscous forces generated. In terms of an energy balance, the work done by the pressure force is needed to overcome the viscous dissipation of energy throughout the fluid. If the pipe is not horizontal, the pressure gradient along it is due in part to the component of weight in that direction. As is discussed in Section 8.2.1, this contribution due to the weight either enhances or retards the flow, depending on whether the flow is downhill or uphill.

Laminar flow characteristics are different than those for turbulent flow.

The nature of the pipe flow is strongly dependent on whether the flow is laminar or turbulent. This is a direct consequence of the differences in the nature of the shear stress in laminar and turbulent flows. As is discussed in some detail in Section 8.3.3, the shear stress in laminar flow is a direct result of momentum transfer among the randomly moving molecules (a microscopic phenomenon). The shear stress in turbulent flow is largely a result of momentum transfer among the randomly moving, finite-sized bundles of fluid particles (a macroscopic phenomenon). The net result is that the physical properties of the shear stress are quite different for laminar flow than for turbulent flow.

8.2 Fully Developed Laminar Flow

As is indicated in the previous section, the flow in long, straight, constant diameter sections of a pipe becomes fully developed. That is, the velocity profile is the same at any cross section of the pipe. Although this is true whether the flow is laminar or turbulent, the details of the velocity profile (and other flow properties) are quite different for these two types of flow. As will be seen in the remainder of this chapter, knowledge of the velocity profile can lead directly to other useful information such as pressure drop, head loss, flowrate, and the like. Thus, we begin by developing the equation for the velocity profile in fully developed laminar flow. If the flow is not fully developed, a theoretical analysis becomes much more complex and is outside the scope of this text. If the flow is turbulent, a rigorous theoretical analysis is as yet not possible.

Although most flows are turbulent rather than laminar, and many pipes are not long enough to allow the attainment of fully developed flow, a theoretical treatment and full understanding of fully developed laminar flow is of considerable importance. First, it represents one of the few theoretical viscous analyses that can be carried out "exactly" (within the framework of quite general assumptions) without using other ad hoc assumptions or approximations. An understanding of the method of analysis and the results obtained provides a foundation from which to carry out more complicated analyses. Second, there are many practical situations involving the use of fully developed laminar pipe flow.

There are numerous ways to derive important results pertaining to fully developed laminar flow. Three alternatives include: (1) from $\mathbf{F} = m\mathbf{a}$ applied directly to a fluid element, (2) from the Navier-Stokes equations of motion, and (3) from dimensional analysis methods.

8.2.1 From F = *m*a Applied Directly to a Fluid Element

We consider the fluid element at time t as is shown in Fig. 8.7. It is a circular cylinder of fluid of length ℓ and radius r centered on the axis of a horizontal pipe of diameter D. Because the velocity is not uniform across the pipe, the initially flat ends of the cylinder of fluid at time t become distorted at time $t + \delta t$ when the fluid element has moved to its new location along the pipe as shown in the figure. If the flow is fully developed and steady, the distortion on each end of the fluid element is the same, and no part of the fluid experiences any acceleration as it flows. The local acceleration is zero ($\partial \mathbf{V}/\partial t = 0$) because the flow is steady, and the convective acceleration is zero ($\mathbf{V} \cdot \nabla \mathbf{V} = u\, \partial u/\partial x\, \hat{\mathbf{i}} = 0$) because the flow is fully developed. Thus, every part of the fluid merely flows along its pathline parallel to the pipe walls with constant velocity, although neighboring particles have slightly different velocities. The velocity varies from one pathline to the next. This velocity variation, combined with the fluid viscosity, produces the shear stress.

Steady, fully developed pipe flow experiences no acceleration.

If gravitational effects are neglected, the pressure is constant across any vertical cross section of the pipe, although it varies along the pipe from one section to the next. Thus, if the pressure is $p = p_1$ at section (1), it is $p_2 = p_1 - \Delta p$ at section (2). We anticipate the fact that the pressure decreases in the direction of flow so that $\Delta p > 0$. A shear stress, τ, acts on the surface of the cylinder of fluid. This viscous stress is a function of the radius of the cylinder, $\tau = \tau(r)$.

As was done in fluid statics analysis (Chapter 2), we isolate the cylinder of fluid as is shown in Fig. 8.8 and apply Newton's second law, $F_x = ma_x$. In this case even though the fluid is moving, it is not accelerating, so that $a_x = 0$. Thus, fully developed horizontal pipe flow is merely a balance between pressure and viscous forces—the pressure difference acting on the end of the cylinder of area πr^2, and the shear stress acting on the lateral surface of the cylinder of area $2\pi r\ell$. This force balance can be written as

$$(p_1)\pi r^2 - (p_1 - \Delta p)\pi r^2 - (\tau)2\pi r\ell = 0$$

which can be simplified to give

$$\frac{\Delta p}{\ell} = \frac{2\tau}{r} \qquad \textbf{(8.3)}$$

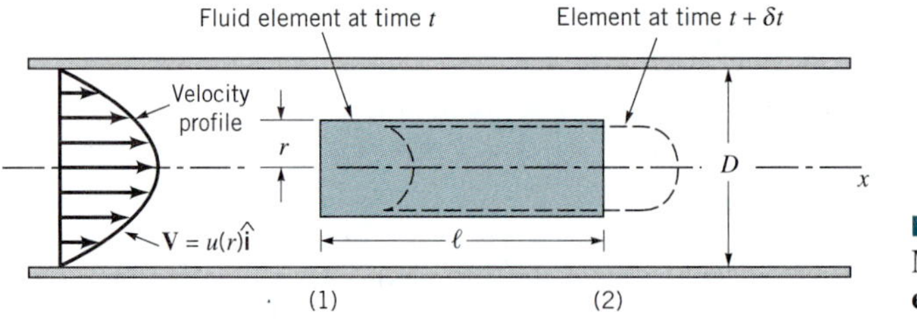

■ **FIGURE 8.7 Motion of a cylindrical fluid element within a pipe.**

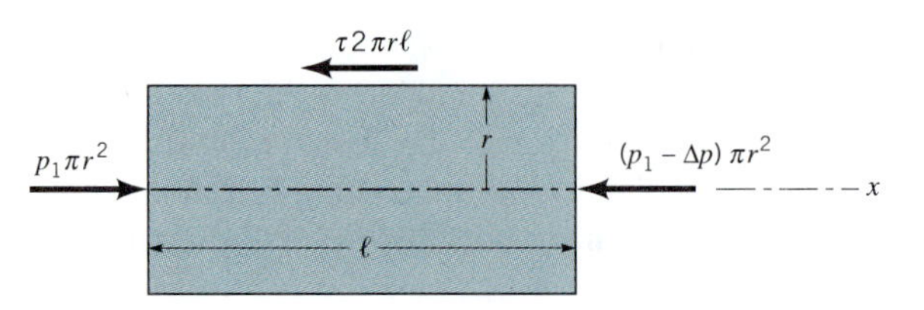

■ **FIGURE 8.8 Free-body diagram of a cylinder of fluid.**

Equation 8.3 represents the basic balance in forces needed to drive each fluid particle along the pipe with constant velocity. Since neither Δp nor ℓ are functions of the radial coordinate, r, it follows that $2\tau/r$ must also be independent of r. That is, $\tau = Cr$, where C is a constant. At $r = 0$ (the centerline of the pipe) there is no shear stress ($\tau = 0$). At $r = D/2$ (the pipe wall) the shear stress is a maximum, denoted τ_w, the *wall shear stress*. Hence, $C = 2\tau_w/D$ and the shear stress distribution throughout the pipe is a linear function of the radial coordinate

$$\tau = \frac{2\tau_w r}{D} \tag{8.4}$$

as is indicated in Fig. 8.9. The linear dependence of τ on r is a result of the pressure force being proportional to r^2 (the pressure acts on the end of the fluid cylinder; area $= \pi r^2$) and the shear force being proportional to r (the shear stress acts on the lateral sides of the cylinder; area $= 2\pi r\ell$). If the viscosity were zero there would be no shear stress, and the pressure would be constant throughout the horizontal pipe ($\Delta p = 0$). As is seen from Eqs. 8.3 and 8.4, the pressure drop and wall shear stress are related by

Basic pipe flow is governed by a balance between viscous and pressure forces.

$$\Delta p = \frac{4\ell\tau_w}{D} \tag{8.5}$$

A small shear stress can produce a large pressure difference if the pipe is relatively long ($\ell/D \gg 1$).

Although we are discussing laminar flow, a closer consideration of the assumptions involved in the derivation of Eqs. 8.3, 8.4, and 8.5 reveals that these equations are valid for both laminar and turbulent flow. To carry the analysis further we must prescribe how the shear stress is related to the velocity. This is the critical step that separates the analysis of laminar from that of turbulent flow—from being able to solve for the laminar flow properties and not being able to solve for the turbulent flow properties without additional ad hoc assumptions. As is discussed in Section 8.3, the shear stress dependence for turbulent flow is very complex. However, for laminar flow of a Newtonian fluid, the shear stress is simply proportional to the velocity gradient, "$\tau = \mu\, du/dy$" (see Section 1.6). In the notation associated with our pipe flow, this becomes

$$\tau = -\mu \frac{du}{dr} \tag{8.6}$$

The negative sign is included to give $\tau > 0$ with $du/dr < 0$ (the velocity decreases from the pipe centerline to the pipe wall).

Equations 8.3 and 8.6 represent the two governing laws for fully developed laminar flow of a Newtonian fluid within a horizontal pipe. The one is Newton's second law of motion and the other is the definition of a Newtonian fluid. By combining these two equations we obtain

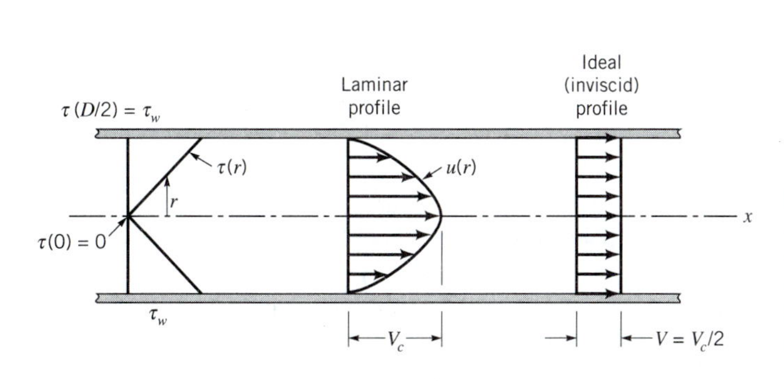

■ **FIGURE 8.9 Shear stress distribution within the fluid in a pipe (laminar or turbulent flow) and typical velocity profiles.**

$$\frac{du}{dr} = -\left(\frac{\Delta p}{2\mu \ell}\right) r$$

which can be integrated to give the velocity profile as follows:

$$\int du = -\frac{\Delta p}{2\mu \ell} \int r\, dr$$

or

$$u = -\left(\frac{\Delta p}{4\mu \ell}\right) r^2 + C_1$$

where C_1 is a constant. Because the fluid is viscous it sticks to the pipe wall so that $u = 0$ at $r = D/2$. Thus, $C_1 = (\Delta p / 16\mu\ell)D^2$. Hence, the velocity profile can be written as

$$u(r) = \left(\frac{\Delta p D^2}{16\mu \ell}\right)\left[1 - \left(\frac{2r}{D}\right)^2\right] = V_c \left[1 - \left(\frac{2r}{D}\right)^2\right] \tag{8.7}$$

__Under certain restrictions the velocity profile in a pipe is parabolic.__

where $V_c = \Delta p D^2/(16\mu\ell)$ is the centerline velocity. An alternative expression can be written by using the relationship between the wall shear stress and the pressure gradient (Eqs. 8.5 and 8.7) to give

$$u(r) = \frac{\tau_w D}{4\mu}\left[1 - \left(\frac{r}{R}\right)^2\right]$$

where $R = D/2$ is the pipe radius.

This velocity profile, plotted in Fig. 8.9, is parabolic in the radial coordinate, r, has a maximum velocity, V_c, at the pipe centerline, and a minimum velocity (zero) at the pipe wall. The volume flowrate through the pipe can be obtained by integrating the velocity profile across the pipe. Since the flow is axisymmetric about the centerline, the velocity is constant on small area elements consisting of rings of radius r and thickness dr. Thus,

$$Q = \int u\, dA = \int_{r=0}^{r=R} u(r) 2\pi r\, dr = 2\pi\, V_c \int_0^R \left[1 - \left(\frac{r}{R}\right)^2\right] r\, dr$$

or

$$Q = \frac{\pi R^2 V_c}{2}$$

By definition, the average velocity is the flowrate divided by the cross-sectional area, $V = Q/A = Q/\pi R^2$, so that for this flow

$$V = \frac{\pi R^2 V_c}{2\pi R^2} = \frac{V_c}{2} = \frac{\Delta p D^2}{32\mu \ell} \tag{8.8}$$

and

$$\boxed{Q = \frac{\pi D^4\, \Delta p}{128\mu \ell}} \tag{8.9}$$

As is indicated in Eq. 8.8, the average velocity is one-half of the maximum velocity. In general, for velocity profiles of other shapes (such as for turbulent pipe flow), the average velocity is not merely the average of the maximum (V_c) and minimum (0) velocities as it is for the laminar parabolic profile. The two velocity profiles indicated in Fig. 8.9 provide the same flowrate—one is the fictitious ideal ($\mu = 0$) profile; the other is the actual laminar flow profile.

Poiseuille's law is valid for laminar flow only.

The above results confirm the following properties of laminar pipe flow. For a horizontal pipe the flowrate is (a) directly proportional to the pressure drop, (b) inversely proportional to the viscosity, (c) inversely proportional to the pipe length, and (d) proportional to the pipe diameter to the fourth power. With all other parameters fixed, an increase in diameter by a factor of 2 will increase the flowrate by a factor of 16—the flowrate is very strongly dependent on pipe size. A 2% error in diameter gives an 8% error in flowrate ($Q \sim D^4$ or $\delta Q \sim 4D^3\,\delta D$, so that $\delta Q/Q = 4\,\delta D/D$). This flow, the properties of which were first established experimentally by two independent workers, G. Hagen (1797–1884) in 1839 and J. Poiseuille (1799–1869) in 1840, is termed *Hagen–Poiseuille flow*. Equation 8.9 is commonly referred to as Poiseuille's law. Recall that all of these results are restricted to laminar flow (those with Reynolds numbers less than approximately 2100) in a horizontal pipe.

The adjustment necessary to account for nonhorizontal pipes, as shown in Fig. 8.10, can be easily included by replacing the pressure drop, Δp, by the combined effect of pressure and gravity, $\Delta p - \gamma \ell \sin\theta$, where θ is the angle between the pipe and the horizontal. (Note that $\theta > 0$ if the flow is uphill, while $\theta < 0$ if the flow is downhill.) This can be seen from the force balance in the x direction (along the pipe axis) on the cylinder of fluid shown in Fig. 8.10*b*. The method is exactly analogous to that used to obtain the Bernoulli equation (Eq. 3.6) when the streamline is not horizontal. The net force in the x direction is a combination of the pressure force in that direction, $\Delta p \pi r^2$, and the component of weight in that direction, $-\gamma \pi r^2 \ell \sin\theta$. The result is a slightly modified form of Eq. 8.3 given by

$$\frac{\Delta p - \gamma \ell \sin\theta}{\ell} = \frac{2\tau}{r} \tag{8.10}$$

Thus, all of the results for the horizontal pipe are valid provided the pressure gradient is adjusted for the elevation term, that is, Δp is replaced by $\Delta p - \gamma \ell \sin\theta$ so that

$$V = \frac{(\Delta p - \gamma \ell \sin\theta)D^2}{32\mu\ell} \tag{8.11}$$

and

$$Q = \frac{\pi(\Delta p - \gamma \ell \sin\theta)D^4}{128\mu\ell} \tag{8.12}$$

It is seen that the driving force for pipe flow can be either a pressure drop in the flow direction, Δp, or the component of weight in the flow direction, $-\gamma \ell \sin\theta$. If the flow is downhill, gravity helps the flow (a smaller pressure drop is required; $\sin\theta < 0$). If the flow is uphill,

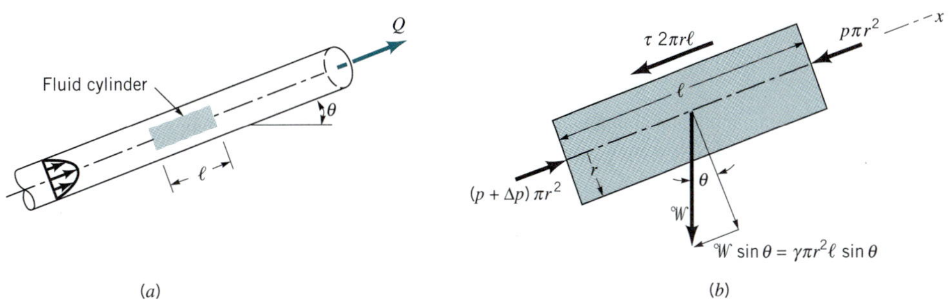

■ FIGURE 8.10 Free-body diagram of a fluid cylinder for flow in a nonhorizontal pipe.

gravity works against the flow (a larger pressure drop is required; $\sin\theta > 0$). Note that $\gamma\ell\sin\theta = \gamma\,\Delta z$ (where Δz is the change in elevation) is a hydrostatic type pressure term. If there is no flow, $V = 0$ and $\Delta p = \gamma\ell\sin\theta = \gamma\,\Delta z$, as expected for fluid statics.

EXAMPLE 8.2

An oil with a viscosity of $\mu = 0.40\ \text{N·s/m}^2$ and density $\rho = 900\ \text{kg/m}^3$ flows in a pipe of diameter $D = 0.020$ m. (a) What pressure drop, $p_1 - p_2$, is needed to produce a flowrate of $Q = 2.0 \times 10^{-5}\ \text{m}^3/\text{s}$ if the pipe is horizontal with $x_1 = 0$ and $x_2 = 10$ m? (b) How steep a hill, θ, must the pipe be on if the oil is to flow through the pipe at the same rate as in part (a), but with $p_1 = p_2$? (c) For the conditions of part (b), if $p_1 = 200$ kPa, what is the pressure at section $x_3 = 5$ m, where x is measured along the pipe?

SOLUTION

(a) If the Reynolds number is less than 2100 the flow is laminar and the equations derived in this section are valid. Since the average velocity is $V = Q/A = (2.0 \times 10^{-5}\ \text{m}^3/\text{s})/[\pi(0.020)^2\text{m}^2/4] = 0.0637\ \text{m/s}$, the Reynolds number is $\text{Re} = \rho VD/\mu = 2.87 < 2100$. Hence, the flow is laminar and from Eq. 8.9 with $\ell = x_2 - x_1 = 10$ m, the pressure drop is

$$\Delta p = p_1 - p_2 = \frac{128\mu\ell Q}{\pi D^4}$$

$$= \frac{128(0.40\ \text{N·s/m}^2)(10.0\ \text{m})(2.0 \times 10^{-5}\ \text{m}^3/\text{s})}{\pi(0.020\ \text{m})^4}$$

or

$$\Delta p = 20{,}400\ \text{N/m}^2 = 20.4\ \text{kPa} \qquad \textbf{(Ans)}$$

(b) If the pipe is on a hill of angle θ such that $\Delta p = p_1 - p_2 = 0$, Eq. 8.12 gives

$$\sin\theta = -\frac{128\mu Q}{\pi\rho g D^4} \qquad \textbf{(1)}$$

or

$$\sin\theta = \frac{-128(0.40\ \text{N·s/m}^2)(2.0 \times 10^{-5}\ \text{m}^3/\text{s})}{\pi(900\ \text{kg/m}^3)(9.81\ \text{m/s}^2)(0.020\ \text{m})^4}$$

Thus, $\theta = -13.34°$. **(Ans)**

This checks with the previous horizontal result as is seen from the fact that a change in elevation of $\Delta z = \ell\sin\theta = (10\ \text{m})\sin(-13.34°) = -2.31$ m is equivalent to a pressure change of $\Delta p = \rho g\,\Delta z = (900\ \text{kg/m}^3)(9.81\ \text{m/s}^2)(2.31\ \text{m}) = 20{,}400\ \text{N/m}^2$, which is equivalent to that needed for the horizontal pipe. For the horizontal pipe it is the work done by the pressure forces that overcomes the viscous dissipation. For the zero-pressure-drop pipe on the hill, it is the change in potential energy of the fluid "falling" down the hill that is converted to the energy lost by viscous dissipation. Note that if it is desired to increase the flowrate to $Q = 1.0 \times 10^{-4}\ \text{m}^3/\text{s}$ with $p_1 = p_2$, the value of θ given by Eq. 1 is $\sin\theta = -1.15$. Since the sine of an angle cannot be greater than 1, this flow would not be possible. The weight of the fluid would not be large enough to offset the viscous force generated for the flowrate desired. A larger diameter pipe would be needed.

(c) With $p_1 = p_2$ the length of the pipe, ℓ, does not appear in the flowrate equation (Eq. 1). This is a statement of the fact that for such cases the pressure is constant all along the pipe (provided the pipe lies on a hill of constant slope). This can be seen by substituting the values of Q and θ from case (b) into Eq. 8.12 and noting that $\Delta p = 0$ for any ℓ. For example, $\Delta p = p_1 - p_3 = 0$ if $\ell = x_3 - x_1 = 5$ m. Thus, $p_1 = p_2 = p_3$ so that

$$p_3 = 200 \text{ kPa} \qquad \textbf{(Ans)}$$

Note that if the fluid were gasoline ($\mu = 3.1 \times 10^{-4}$ N·s/m^2 and $\rho = 680$ kg/m^3), the Reynolds number would be Re $= 2790$, the flow would probably not be laminar, and a use of Eqs. 8.9 and 8.12 would give incorrect results. Also note from Eq. 1 that the kinematic viscosity, $\nu = \mu/\rho$, is the important viscous parameter. This is a statement of the fact that with constant pressure along the pipe, it is the ratio of the viscous force ($\sim\mu$) to the weight force ($\sim\gamma = \rho g$) that determines the value of θ.

8.2.2 From the Navier–Stokes Equations

Poiseuille's law can be obtained from the Navier–Stokes equations.

In the previous section we obtained results for fully developed laminar pipe flow by applying Newton's second law and the assumption of a Newtonian fluid to a specific portion of the fluid—a cylinder of fluid centered on the axis of a long, round pipe. When this governing law and assumptions are applied to a general fluid flow (not restricted to pipe flow), the result is the Navier–Stokes equations as discussed in Chapter 6. In Section 6.9.3 these equations were solved for the specific geometry of fully developed laminar flow in a round pipe. The results are the same as those given in Eq. 8.7.

We will not repeat the detailed steps used to obtain the laminar pipe flow from the Navier-Stokes equations (see Section 6.9.3) but will indicate how the various assumptions used and steps applied in the derivation correlate with the analysis used in the previous section.

General motion of an incompressible Newtonian fluid is governed by the continuity equation (conservation of mass, Eq. 6.31) and the momentum equation (Eq. 6.127), which are rewritten here for convenience:

$$\nabla \cdot \mathbf{V} = 0 \qquad \textbf{(8.13)}$$

$$\frac{\partial \mathbf{V}}{\partial t} + \mathbf{V} \cdot \nabla \mathbf{V} = -\frac{\nabla p}{\rho} + \mathbf{g} + \nu \nabla^2 \mathbf{V} \qquad \textbf{(8.14)}$$

For steady, fully developed flow in a pipe, the velocity contains only an axial component, which is a function of only the radial coordinate [$\mathbf{V} = u(r)\hat{\mathbf{i}}$]. For such conditions, the left-hand side of the Eq. 8.14 is zero. This is equivalent to saying that the fluid experiences no acceleration as it flows along. The same constraint was used in the previous section when considering $\mathbf{F} = m\mathbf{a}$ for the fluid cylinder. Thus, with $\mathbf{g} = -g\hat{\mathbf{k}}$ the Navier–Stokes equations become

$$\nabla \cdot \mathbf{V} = 0$$

$$\nabla p + \rho g \hat{\mathbf{k}} = \mu \nabla^2 \mathbf{V} \qquad \textbf{(8.15)}$$

The flow is governed by a balance of pressure, weight, and viscous forces in the flow direction, similar to that shown in Fig. 8.10 and Eq. 8.10. If the flow were not fully developed (as in an entrance region, for example), it would not be possible to simplify the Navier–Stokes equations to that form given in Eq. 8.15 (the nonlinear term $\mathbf{V} \cdot \nabla \mathbf{V}$ would not be zero), and the solution would be very difficult to obtain.

Because of the assumption that $\mathbf{V} = u(r)\hat{\mathbf{i}}$, the continuity equation, Eq. 8.13, is automatically satisfied. This conservation of mass condition was also automatically satisfied by the incompressible flow assumption in the derivation in the previous section. The fluid flows across one section of the pipe at the same rate that it flows across any other section (see Fig. 8.8).

When it is written in terms of polar coordinates (as was done in Section 6.9.3), the component of Eq. 8.15 along the pipe becomes

The governing differential equations can be simplified by appropriate assumptions.

$$\frac{\partial p}{\partial x} + \rho g \sin \theta = \mu \frac{1}{r} \frac{\partial}{\partial r}\left(r \frac{\partial u}{\partial r}\right) \tag{8.16}$$

Since the flow is fully developed, $u = u(r)$ and the right-hand side is a function of, at most, only r. The left-hand side is a function of, at most, only x. It was shown that this leads to the condition that the pressure gradient in the x direction is a constant—$\partial p/\partial x = -\Delta p/\ell$. The same condition was used in the derivation of the previous section (Eq. 8.3).

It is seen from Eq. 8.16 that the effect of a nonhorizontal pipe enters into the Navier–Stokes equations in the same manner as was discussed in the previous section. The pressure gradient in the flow direction is coupled with the effect of the weight in that direction to produce an effective pressure gradient of $-\Delta p/\ell + \rho g \sin \theta$.

The velocity profile is obtained by integration of Eq. 8.16. Since it is a second-order equation, two boundary conditions are needed—(1) the fluid sticks to the pipe wall (as was also done in Eq. 8.7) and (2) either of the equivalent forms that the velocity remains finite throughout the flow (in particular $u < \infty$ at $r = 0$), or because of symmetry, $\partial u/\partial r = 0$ at $r = 0$. In the derivation of the previous section, only one boundary condition (the no-slip condition at the wall) was needed because the equation integrated was a first-order equation. The other condition ($\partial u/\partial r = 0$ at $r = 0$) was automatically built into the analysis because of the fact that $\tau = -\mu \, du/dr$ and $\tau = 2\tau_w r/D = 0$ at $r = 0$.

The results obtained by either applying $\mathbf{F} = m\mathbf{a}$ to a fluid cylinder (Section 8.2.1) or solving the Navier–Stokes equations (Section 6.9.3) are exactly the same. Similarly, the basic assumptions regarding the flow structure are the same. This should not be surprising because the two methods are based on the same principle—Newton's second law. One is restricted to fully developed laminar pipe flow from the beginning (the drawing of the free-body diagram), and the other starts with the general governing equations (the Navier–Stokes equations) with the appropriate restrictions concerning fully developed laminar flow applied as the solution process progresses.

8.2.3 From Dimensional Analysis

Although fully developed laminar pipe flow is simple enough to allow the rather straightforward solutions discussed in the previous two sections, it may be worthwhile to consider this flow from a dimensional analysis standpoint. Thus, we assume that the pressure drop in the horizontal pipe, Δp, is a function of the average velocity of the fluid in the pipe, V, the length of the pipe, ℓ, the pipe diameter, D, and the viscosity of the fluid, μ. We have not included the density or the specific weight of the fluid as parameters because for such flows they are not important parameters. There is neither mass (density) times acceleration nor a component of weight (specific weight times volume) in the flow direction involved. Thus,

$$\Delta p = F(V, \ell, D, \mu)$$

There are five variables that can be described in terms of three reference dimensions (M, L, T). According to the results of dimensional analysis (Chapter 7), this flow can be described in terms of $k - r = 5 - 3 = 2$ dimensionless groups. One such representation is

$$\frac{D\,\Delta p}{\mu V} = \phi\left(\frac{\ell}{D}\right) \tag{8.17}$$

where $\phi(\ell/D)$ is an unknown function of the length to diameter ratio of the pipe.

Although this is as far as dimensional analysis can take us, it seems reasonable to impose a further assumption that the pressure drop is directly proportional to the pipe length. That is, it takes twice the pressure drop to force fluid through a pipe if its length is doubled. The only way that this can be true is if $\phi(\ell/D) = C\ell/D$, where C is a constant. Thus, Eq. 8.17 becomes

$$\frac{D\,\Delta p}{\mu V} = \frac{C\ell}{D}$$

which can be rewritten as

$$\frac{\Delta p}{\ell} = \frac{C\mu\, V}{D^2}$$

or

$$Q = AV = \frac{(\pi/4C)\,\Delta p D^4}{\mu\ell} \tag{8.18}$$

The basic functional dependence for laminar pipe flow given by Eq. 8.18 is the same as that obtained by the analysis of the two previous sections. The value of C must be determined by theory (as done in the previous two sections) or experiment. For a round pipe, $C = 32$. For ducts of other cross-sectional shapes, the value of C is different (see Section 8.4.3).

Dimensional analysis can be used to put pipe flow parameters into dimensionless form.

It is usually advantageous to describe a process in terms of dimensionless quantities. To this end we rewrite the pressure drop equation for laminar horizontal pipe flow, Eq. 8.8, as $\Delta p = 32\mu\ell V/D^2$ and divide both sides by the dynamic pressure, $\rho V^2/2$, to obtain the dimensionless form as

$$\frac{\Delta p}{\frac{1}{2}\rho V^2} = \frac{(32\mu\ell V/D^2)}{\frac{1}{2}\rho V^2} = 64\left(\frac{\mu}{\rho VD}\right)\left(\frac{\ell}{D}\right) = \frac{64}{\text{Re}}\left(\frac{\ell}{D}\right)$$

This is often written as

$$\Delta p = f\,\frac{\ell}{D}\,\frac{\rho V^2}{2}$$

where the dimensionless quantity

$$f = \Delta p(D/\ell)/(\rho V^2/2)$$

is termed the *friction factor*, or sometimes the *Darcy friction factor*. (This parameter should not be confused with the less-used Fanning friction factor, which is defined to be $f/4$. In this text we will use only the Darcy friction factor.) Thus, the friction factor for laminar fully developed pipe flow is simply

$$f = \frac{64}{\text{Re}} \tag{8.19}$$

By substituting the pressure drop in terms of the wall shear stress (Eq. 8.5), we obtain an alternate expression for the friction factor as a dimensionless wall shear stress

$$f = \frac{8\tau_w}{\rho V^2} \tag{8.20}$$

Knowledge of the friction factor will allow us to obtain a variety of information regarding pipe flow. For turbulent flow the dependence of the friction factor on the Reynolds number is much more complex than that given by Eq. 8.19 for laminar flow. This is discussed in detail in Section 8.4.

8.2.4 Energy Considerations

In the previous three sections we derived the basic laminar flow results from application of $\mathbf{F} = m\mathbf{a}$ or dimensional analysis considerations. It is equally important to understand the implications of energy considerations of such flows. To this end we consider the energy equation for incompressible, steady flow between two locations as is given in Eq. 5.89

$$\frac{p_1}{\gamma} + \alpha_1 \frac{V_1^2}{2g} + z_1 = \frac{p_2}{\gamma} + \alpha_2 \frac{V_2^2}{2g} + z_2 + h_L \quad \textbf{(8.21)}$$

Recall that the kinetic energy coefficients, α_1 and α_2, compensate for the fact that the velocity profile across the pipe is not uniform. For uniform velocity profiles $\alpha = 1$, whereas for any nonuniform profile, $\alpha > 1$. The head loss term, h_L, accounts for any energy loss associated with the flow. This loss is a direct consequence of the viscous dissipation that occurs throughout the fluid in the pipe. For the ideal (inviscid) cases discussed in previous chapters, $\alpha_1 = \alpha_2 = 1$, $h_L = 0$, and the energy equation reduces to the familiar Bernoulli equation discussed in Chapter 3 (Eq. 3.7).

Even though the velocity profile in viscous pipe flow is not uniform, for fully developed flow it does not change from section (1) to section (2) so that $\alpha_1 = \alpha_2$. Thus, the kinetic energy is the same at any section ($\alpha_1 V_1^2/2 = \alpha_2 V_2^2/2$) and the energy equation becomes

$$\left(\frac{p_1}{\gamma} + z_1\right) - \left(\frac{p_2}{\gamma} + z_2\right) = h_L \quad \textbf{(8.22)}$$

The energy dissipated by the viscous forces within the fluid is supplied by the excess work done by the pressure and gravity forces.

A comparison of Eqs. 8.22 and 8.10 shows that the head loss is given by

$$h_L = \frac{2\tau\ell}{\gamma r}$$

(recall $p_1 = p_2 + \Delta p$ and $z_2 - z_1 = \ell \sin\theta$), which, by use of Eq. 8.4, can be rewritten in the form

The head loss in a pipe is a result of the viscous shear stress on the wall.

$$h_L = \frac{4\ell\tau_w}{\gamma D} \quad \textbf{(8.23)}$$

It is the shear stress at the wall (which is directly related to the viscosity and the shear stress throughout the fluid) that is responsible for the head loss. A closer consideration of the assumptions involved in the derivation of Eq. 8.23 will show that it is valid for both laminar and turbulent flow.

The flowrate, Q, of corn syrup through the horizontal pipe shown in Fig. E8.3 is to be monitored by measuring the pressure difference between sections (1) and (2). It is proposed that $Q = K\,\Delta p$, where the calibration constant, K, is a function of temperature, T, because of the temperature variation of the syrup's viscosity and density. These variations are given in Table E8.3. (a) Plot $K(T)$ versus T for $60\ ^\circ\text{F} \le T \le 160\ ^\circ\text{F}$. (b) Determine the wall shear stress and the pressure drop, $\Delta p = p_1 - p_2$, for $Q = 0.5$ ft^3/s and $T = 100$ °F. (c) For the conditions of part (b), determine the net pressure force, $(\pi D^2/4)\,\Delta p$, and the net shear force, $\pi D \ell \tau_w$, on the fluid within the pipe between the sections (1) and (2).

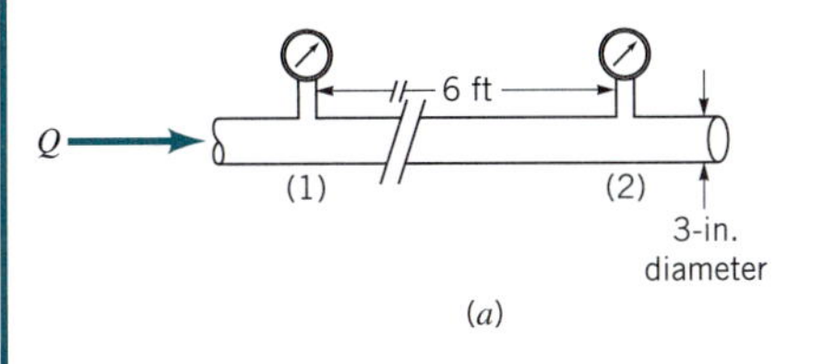

(a)

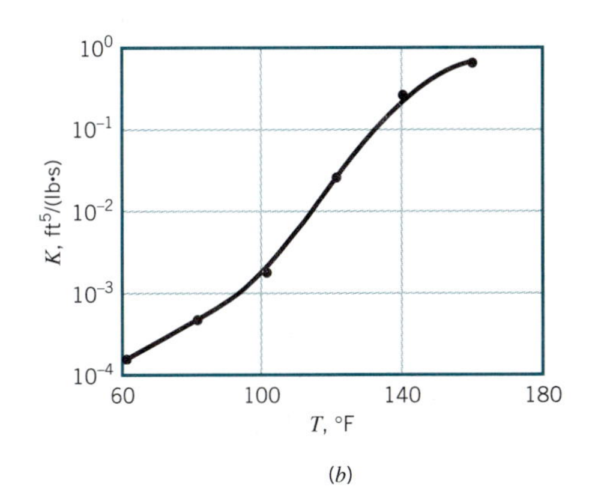

(b)

■ FIGURE E8.3

■ TABLE E8-3

T (°F)	ρ (slugs/ft^3)	μ (lb·s/ft^2)
60	2.07	4.0×10^{-2}
80	2.06	1.9×10^{-2}
100	2.05	3.8×10^{-3}
120	2.04	4.4×10^{-4}
140	2.03	9.2×10^{-5}
160	2.02	2.3×10^{-5}

SOLUTION

(a) If the flow is laminar it follows from Eq. 8.9 that

$$Q = \frac{\pi D^4\,\Delta p}{128\mu\ell} = \frac{\pi(\frac{3}{12}\text{ ft})^4\,\Delta p}{128\mu(6\text{ ft})}$$

or

$$Q = K\,\Delta p = \frac{1.60 \times 10^{-5}}{\mu}\,\Delta p \qquad \textbf{(1)}$$

where the units on Q, Δp, and μ are ft^3/s, lb/ft^2, and lb·s/ft^2, respectively. Thus

$$K = \frac{1.60 \times 10^{-5}}{\mu} \qquad \textbf{(Ans)}$$

where the units of K are ft^5/lb·s. By using values of the viscosity from Table E8.3, the calibration curve shown in Fig. E8.3*b* is obtained. This result is valid only if the flow is laminar. As shown in Section 8.5, for turbulent flow the flowrate is not linearly related to the pressure drop so it would not be possible to have $Q = K\,\Delta p$. Note also that the value of K is independent of the syrup density (ρ was not used in the calculations) since laminar pipe flow is governed by pressure and viscous effects; inertia is not important.

(b) For $T = 100$ °F, the viscosity is $\mu = 3.8 \times 10^{-3}$ lb·s/ft^2 so that with a flowrate of $Q = 0.5$ ft^3/s the pressure drop (according to Eq. 8.9) is

$$\Delta p = \frac{128\mu\ell Q}{\pi D^4} = \frac{128(3.8 \times 10^{-3} \text{ lb}\cdot\text{s/ft}^2)(6 \text{ ft})(0.5 \text{ ft}^3/\text{s})}{\pi(\frac{3}{12} \text{ ft})^4}$$

$$= 119 \text{ lb/ft}^2 \qquad \textbf{(Ans)}$$

provided the flow is laminar. For this case

$$V = \frac{Q}{A} = \frac{0.5 \text{ ft}^3/\text{s}}{\frac{\pi}{4}(\frac{3}{12} \text{ ft})^2} = 10.2 \text{ ft/s}$$

so that

$$\text{Re} = \frac{\rho VD}{\mu} = \frac{(2.05 \text{ slugs/ft}^3)(10.2 \text{ ft/s})(\frac{3}{12} \text{ ft})}{(3.8 \times 10^{-3} \text{ lb}\cdot\text{s/ft}^2)}$$

$$= 1380 < 2100$$

Hence, the flow is laminar. From Eq. 8.5 the wall shear stress is

$$\tau_w = \frac{\Delta pD}{4\ell} = \frac{(119 \text{ lb/ft}^2)(\frac{3}{12} \text{ ft})}{4(6 \text{ ft})} = 1.24 \text{ lb/ft}^2 \qquad \textbf{(Ans)}$$

(c) For the conditions of part (b), the net pressure force, F_p, on the fluid within the pipe between sections (1) and (2) is

$$F_p = \frac{\pi}{4} D^2 \Delta p = \frac{\pi}{4}\left(\frac{3}{12} \text{ ft}\right)^2 (119 \text{ lb/ft}^2) = 5.84 \text{ lb} \qquad \textbf{(Ans)}$$

Similarly, the net viscous force, F_v, on that portion of the fluid is

$$F_v = 2\pi\left(\frac{D}{2}\right)\ell\tau_w$$

$$= 2\pi\left[\frac{3}{2(12)} \text{ ft}\right](6 \text{ ft})(1.24 \text{ lb/ft}^2) = 5.84 \text{ lb} \qquad \textbf{(Ans)}$$

Note that the values of these two forces are the same. The net force is zero; there is no acceleration.

8.3 Fully Developed Turbulent Flow

Much remains to be learned about the nature of turbulent flow.

In the previous section various properties of fully developed laminar pipe flow were discussed. Since turbulent pipe flow is actually more likely to occur than laminar flow in practical situations, it is necessary to obtain similar information for turbulent pipe flow. However, turbulent flow is a very complex process. Numerous persons have devoted considerable effort in attempting to understand the variety of baffling aspects of turbulence. Although a considerable amount of knowledge about the topic has been developed, the field of turbulent flow still remains the least understood area of fluid mechanics. In this book we can provide only some of the very basic ideas concerning turbulence. The interested reader should consult some of the many books available for further reading (Refs. 1, 2, and 3).

8.3.1 Transition from Laminar to Turbulent Flow

Flows are classified as laminar or turbulent. For any flow geometry, there is one (or more) dimensionless parameter such that with this parameter value below a particular value the flow is laminar, whereas with the parameter value larger than a certain value the flow is turbulent. The important parameters involved (i.e., Reynolds number, Mach number, etc.) and their critical values depend on the specific flow situation involved. For example, flow in a pipe and flow along a flat plate (boundary layer flow, as is discussed in Section 9.2.4) can be laminar or turbulent, depending on the value of the Reynolds number involved. For pipe flow the value of the Reynolds number must be less than approximately 2100 for laminar flow and greater than approximately 4000 for turbulent flow. For flow along a flat plate the transition between laminar and turbulent flow occurs at a Reynolds number of approximately 500,000 (see Section 9.2.4), where the length term in the Reynolds number is the distance measured from the leading edge of the plate.

Consider a long section of pipe that is initially filled with a fluid at rest. As the valve is opened to start the flow, the flow velocity and, hence, the Reynolds number increase from zero (no flow) to their maximum steady-state flow values, as is shown in Fig. 8.11. Assume this transient process is slow enough so that unsteady effects are negligible (quasisteady flow). For an initial time period the Reynolds number is small enough for laminar flow to occur. At some time the Reynolds number reaches 2100, and the flow begins its transition to turbulent conditions. Intermittent spots or bursts of turbulence appear. As the Reynolds number is increased the entire flow field becomes turbulent. The flow remains turbulent as long as the Reynolds number exceeds approximately 4000.

Turbulent flows involve randomly fluctuating parameters.

A typical trace of the axial component of velocity measured at a given location in the flow, $u = u(t)$, is shown in Fig. 8.12. Its irregular, random nature is the distinguishing feature of turbulent flow. The character of many of the important properties of the flow (pressure drop, heat transfer, etc.) depends strongly on the existence and nature of the turbulent fluctuations or randomness indicated. In previous considerations involving inviscid flow, the Reynolds number is (strictly speaking) infinite (because the viscosity is zero), and the flow most surely would be turbulent. However, reasonable results were obtained by using the inviscid Bernoulli equation as the governing equation. The reason that such simplified inviscid analyses gave reasonable results is that viscous effects were not very important and the velocity used in the calculations was actually the time-averaged velocity, $\overline{u}$, indicated in Fig. 8.12. Calculation of the heat transfer, pressure drop, and many other parameters would not

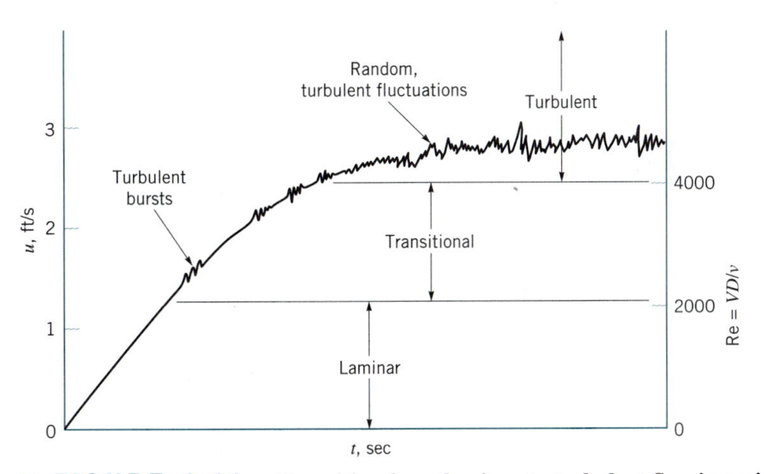

■ FIGURE 8.11 **Transition from laminar to turbulent flow in a pipe.**

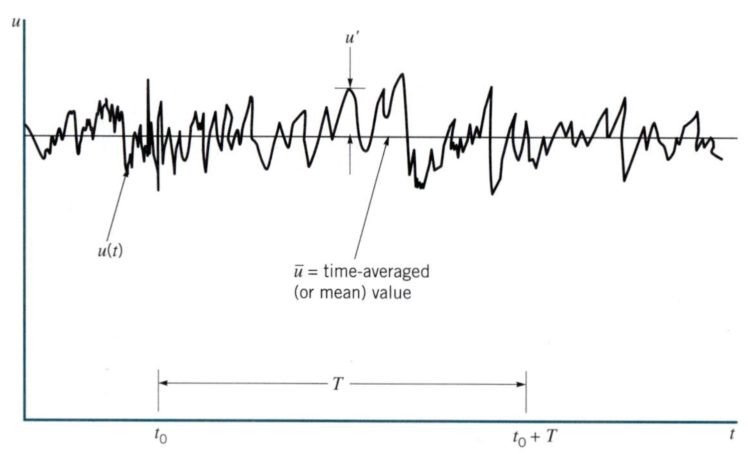

■ **FIGURE 8.12 The time-averaged, $\overline{u}$, and fluctuating, u', description of a parameter for turbulent flow.**

be possible without inclusion of the seemingly small, but very important, effects associated with the randomness of the flow.

Consider flow in a pan of water placed on a stove. With the stove turned off, the fluid is stationary. The initial sloshing has died out because of viscous dissipation within the water. With the stove turned on, a temperature gradient in the vertical direction, $\partial T/\partial z$, is produced. The water temperature is greatest near the pan bottom and decreases toward the top of the fluid layer. If the temperature difference is very small, the water will remain stationary, even though the water density is smallest near the bottom of the pan because of the decrease in density with an increase in temperature. A further increase in the temperature gradient will cause a buoyancy-driven instability that results in fluid motion—the light, warm water rises to the top, and the heavy cold water sinks to the bottom. This slow, regular ''turning over'' increases the heat transfer from the pan to the water and promotes mixing within the pan. As the temperature gradient increases still further, the fluid motion becomes more vigorous and eventually turns into a chaotic, random, turbulent flow with considerable mixing and greatly increased heat transfer rate. The flow has progressed from a stationary fluid, to laminar flow, and finally to turbulent flow.

Mixing processes and heat and mass transfer processes are considerably enhanced in turbulent flow compared to laminar flow. This is due to the macroscopic scale of the randomness in turbulent flow. We are all familiar with the ''rolling,'' vigorous eddy type motion of the water in a pan being heated on the stove (even if it is not heated to boiling). Such finite-sized random mixing is very effective in transporting energy and mass throughout the flow field, thereby increasing the various rate processes involved. Laminar flow, on the other hand, can be thought of as very small but finite-sized fluid particles flowing smoothly in layers, one over another. The only randomness and mixing take place on the molecular scale and result in relatively small heat, mass, and momentum transfer rates.

Laminar (turbulent) flow involves randomness on the molecular (macroscopic) scale.

Without turbulence it would be virtually impossible to carry out life as we now know it. In some situations turbulent flow is desirable. To transfer the required heat between a solid and an adjacent fluid (such as in the cooling coils of an air conditioner or a boiler of a power plant) would require an enormously large heat exchanger if the flow were laminar. Similarly, the required mass transfer of a liquid state to a vapor state (such as is needed in the evaporated cooling system associated with sweating) would require very large surfaces if the fluid flowing past the surface were laminar rather than turbulent.

Turbulence is also of importance in the mixing of fluids. Smoke from a stack would continue for miles as a ribbon of pollutant without rapid dispersion within the surrounding air if the flow were laminar rather than turbulent. Under certain atmospheric conditions this is observed to occur. Although there is mixing on a molecular scale (laminar flow), it is several orders of magnitude slower and less effective than the mixing on a macroscopic scale (turbulent flow). It is considerably easier to mix cream into a cup of coffee (turbulent flow) than to thoroughly mix two colors of a viscous paint (laminar flow).

In other situations laminar (rather than turbulent) flow is desirable. The pressure drop in pipes (hence, the power requirements for pumping) can be considerably lower if the flow is laminar rather than turbulent. Fortunately, the blood flow through a person's arteries is normally laminar, except in the largest arteries with high blood flowrates. The aerodynamic drag on an airplane wing can be considerably smaller with laminar flow past it than with turbulent flow.

8.3.2 Turbulent Shear Stress

The fundamental difference between laminar and turbulent flow lies in the chaotic, random behavior of the various fluid parameters. Such variations occur in the three components of velocity, the pressure, the shear stress, the temperature, and any other variable that has a field description. Turbulent flow is characterized by random, three-dimensional vorticity (i.e., fluid particle rotation or spin; see Section 6.1.3). As is indicated in Fig. 8.12, such flows can be described in terms of their mean values (denoted with an overbar) on which are superimposed the fluctuations (denoted with a prime). Thus, if $u = u(x, y, z, t)$ is the x component of instantaneous velocity, then its time mean (or *time average*) value, $\overline{u}$, is

$$\overline{u} = \frac{1}{T}\int_{t_0}^{t_0+T} u(x, y, z, t)\, dt \tag{8.24}$$

Turbulent flow parameters can be described in terms of mean and fluctuating portions.

where the time interval, T, is considerably longer than the period of the longest fluctuations, but considerably shorter than any unsteadiness of the average velocity. This is illustrated in Fig. 8.12.

The *fluctuating part* of the velocity, u', is that time-varying portion that differs from the average value

$$u = \overline{u} + u' \quad \text{or} \quad u' = u - \overline{u} \tag{8.25}$$

Clearly, the time average of the fluctuations is zero, since

$$\begin{aligned} \overline{u'} &= \frac{1}{T}\int_{t_0}^{t_0+T} (u - \overline{u})\, dt = \frac{1}{T}\left(\int_{t_0}^{t_0+T} u\, dt - \overline{u}\int_{t_0}^{t_0+T} dt\right) \\ &= \frac{1}{T}(T\overline{u} - T\overline{u}) = 0 \end{aligned}$$

The fluctuations are equally distributed on either side of the average. It is also clear, as is indicated in Fig. 8.13, that since the square of a fluctuation quantity cannot be negative $[(u')^2 \geq 0]$, its average value is positive. Thus,

$$\overline{(u')^2} = \frac{1}{T}\int_{t_0}^{t_0+T} (u')^2\, dt > 0$$

On the other hand, it may be that the average of products of the fluctuations, such as $\overline{u'v'}$, are zero or nonzero (either positive or negative).

The structure and characteristics of turbulence may vary from one flow situation to another. For example, the *turbulence intensity* (or the level of the turbulence) may be larger in a very gusty wind than it is in a relatively steady (although turbulent) wind. The turbulence

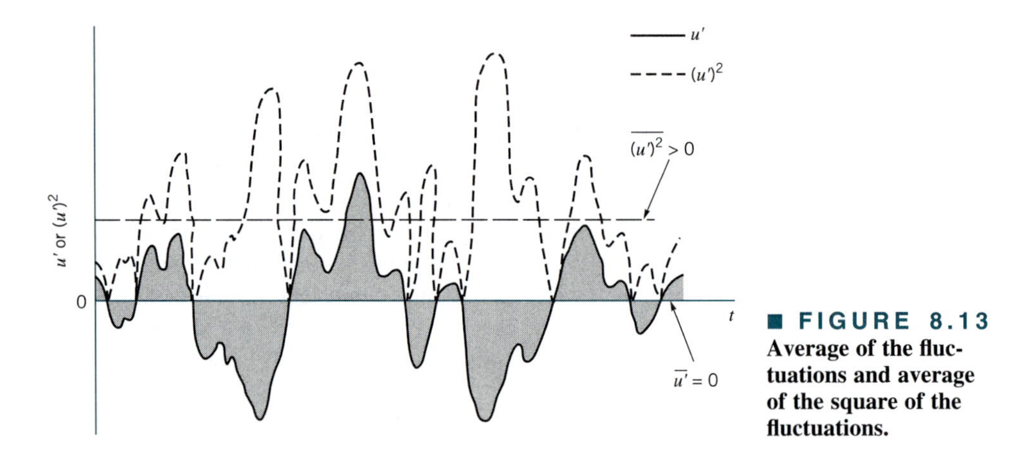

■ **FIGURE 8.13 Average of the fluctuations and average of the square of the fluctuations.**

intensity, $\mathcal{I}$, is often defined as the square root of the mean square of the fluctuating velocity divided by the time-averaged velocity, or

$$\mathcal{I} = \frac{\sqrt{\overline{(u')^2}}}{\overline{u}} = \frac{\left[\frac{1}{T}\int_{t_0}^{t_0+T} (u')^2\, dt\right]^{1/2}}{\overline{u}}$$

The larger the turbulence intensity, the larger the fluctuations of the velocity (and other flow parameters). Well-designed wind tunnels have typical values of $\mathcal{I} \approx 0.01$, although with extreme care, values as low as $\mathcal{I} = 0.0002$ have been obtained. On the other hand, values of $\mathcal{I} \gtrsim 0.1$ are found for the flow in the atmosphere and rivers.

Another turbulence parameter that is different from one flow situation to another is the period of the fluctuations—the *time scale* of the fluctuations shown in Fig. 8.12. In many flows, such as the flow of water from a faucet, typical frequencies are on the order of 10, 100, or 1000 cycles per second (cps). For other flows, such as the Gulf Stream current in the Atlantic Ocean or flow of the atmosphere of Jupiter, characteristic random oscillations may have a period on the order of hours, days, or more.

The relationship between fluid motion and shear stress is very complex for turbulent flow.

It is tempting to extend the concept of viscous shear stress for laminar flow ($\tau = \mu\, du/dy$) to that of turbulent flow by replacing u, the instantaneous velocity, by $\overline{u}$, the time-averaged velocity. However, numerous experimental and theoretical studies have shown that such an approach leads to completely incorrect results. That is, $\tau \neq \mu\, d\overline{u}/dy$. A physical explanation for this behavior can be found in the concept of what produces a shear stress.

Laminar flow is modeled as fluid particles that flow smoothly along in layers, gliding past the slightly slower or faster ones on either side. As is discussed in Chapter 1, the fluid actually consists of numerous molecules darting about in an almost random fashion as is indicated in Fig. 8.14*a*. The motion is not entirely random—a slight bias in one direction produces the flowrate we associate with the motion of fluid particles, $\overline{u}$. As the molecules dart across a given plane (plane *A–A*, for example), the ones moving upward have come from an area of smaller average x component of velocity than the ones moving downward, which have come from an area of larger velocity.

The momentum flux in the x direction across plane *A–A* gives rise to a drag (to the left) of the lower fluid on the upper fluid and an equal but opposite effect of the upper fluid on the lower fluid. The sluggish molecules moving upward across plane *A–A* must be accelerated by the fluid above this plane. The rate of change of momentum in this process produces (on the macroscopic scale) a shear force. Similarly, the more energetic molecules moving down across plane *A–A* must be slowed down by the fluid below that plane. This shear force

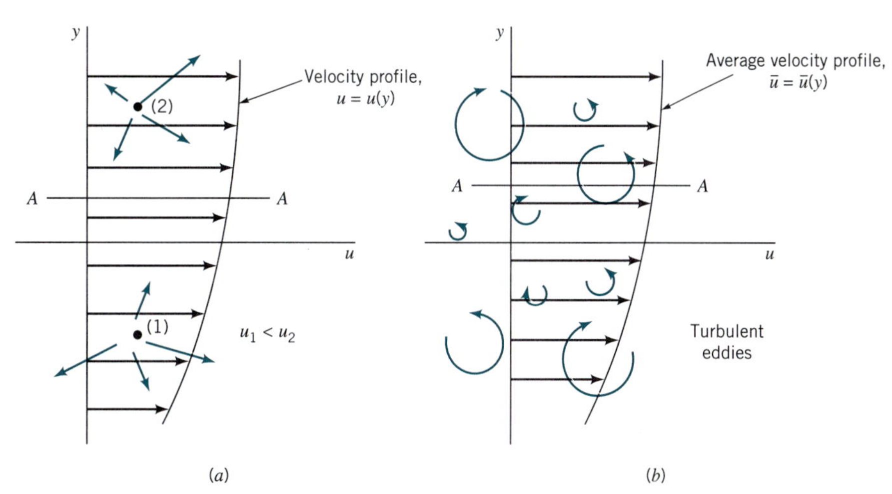

■ **FIGURE 8.14** **(*a*) Laminar flow shear stress caused by random motion of molecules. (*b*) Turbulent flow as a series of random, three-dimensional eddies.**

is present only if there is a gradient in $u = u(y)$, otherwise the average x component of velocity (and momentum) of the upward and downward molecules is exactly the same. In addition, there are attractive forces between molecules. By combining these effects we obtain the well-known Newton viscosity law: $\tau = \mu \, du/dy$, where on a molecular basis μ is related to the mass and speed (temperature) of the random motion of the molecules.

Turbulent flows involve random motions of finite-sized fluid particles.

Although the above random motion of the molecules is also present in turbulent flow, there is another factor that is generally more important. A simplistic way of thinking about turbulent flow is to consider it as consisting of a series of random, three-dimensional eddy type motions as is depicted (in one dimension only) in Fig. 8.14*b*. These eddies range in size from very small diameter (on the order of the size of a fluid particle) to fairly large diameter (on the order of the size of the object or flow geometry considered). They move about randomly, conveying mass with an average velocity $\bar{u} = \bar{u}(y)$. This eddy structure greatly promotes mixing within the fluid. It also greatly increases the transport of x momentum across plane A–A. That is, finite parcels of fluid (not merely individual molecules as in laminar flow) are randomly transported across this plane, resulting in a relatively large (when compared with laminar flow) shear force.

V8.2

The random velocity components that account for this momentum transfer (hence, the shear force) are u' (for the x component of velocity) and v' (for the rate of mass transfer crossing the plane). A more detailed consideration of the processes involved will show that the apparent shear stress on plane A–A is given by the following (Ref. 2):

$$\tau = \mu \frac{d\bar{u}}{dy} - \rho \overline{u'v'} = \tau_{\text{lam}} + \tau_{\text{turb}} \tag{8.26}$$

Note that if the flow is laminar, $u' = v' = 0$, so that $\overline{u'v'} = 0$ and Eq. 8.26 reduces to the customary random molecule-motion-induced *laminar shear stress*, $\tau_{\text{lam}} = \mu \, d\bar{u}/dy$. For turbulent flow it is found that the *turbulent shear stress*, $\tau_{\text{turb}} = -\rho\overline{u'v'}$, is positive. Hence, the shear stress is greater in turbulent flow than in laminar flow. Note the units on τ_{turb} are (density)(velocity)^2 = (slugs/ft^3)(ft/s)^2 = (slugs·ft/s^2)/ft^2 = lb/ft^2, or N/m^2, as expected. Terms of the form $-\rho\overline{u'v'}$ (or $-\rho\overline{v'w'}$, etc.) are called *Reynolds stresses* in honor of Osborne Reynolds who first discussed them in 1895.

It is seen from Eq. 8.26 that the shear stress in turbulent flow is not merely proportional to the gradient of the time-averaged velocity, $\bar{u}y)$. It also contains a contribution due to the random fluctuations of the x and y components of velocity. The density is involved because

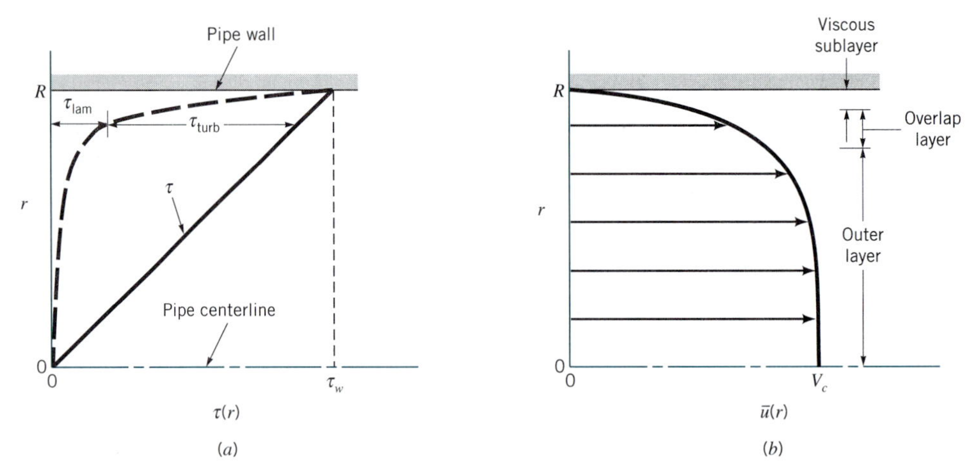

■ **FIGURE 8.15 Structure of turbulent flow in a pipe. (*a*) Shear stress. (*b*) Average velocity.**

The shear stress is the sum of a laminar portion and a turbulent portion.

of the momentum transfer of the fluid within the random eddies. Although the relative magnitude of τ_{lam} compared to τ_{turb} is a complex function dependent on the specific flow involved, typical measurements indicate the structure shown in Fig. 8.15*a*. (Recall from Eq. 8.4 that the shear stress is proportional to the distance from the centerline of the pipe). In a very narrow region near the wall (the *viscous sublayer*), the laminar shear stress is dominant. Away from the wall (in the *outer layer*) the turbulent portion of the shear stress is dominant. The transition between these two regions occurs in the *overlap layer*. The corresponding typical velocity profile is shown in Fig. 8.15*b*.

The scale of the sketches shown in Fig. 8.15 is not necessarily correct. Typically the value of τ_{turb} is 100 to 1000 times greater than τ_{lam} in the outer region, while the converse is true in the viscous sublayer. A correct modeling of turbulent flow is strongly dependent on an accurate knowledge of τ_{turb}. This, in turn, requires an accurate knowledge of the fluctuations u' and v', or $\rho\overline{u'v'}$. As yet it is not possible to solve the governing equations (the Navier–Stokes equations) for these details of the flow, although numerical techniques using the largest and fastest computers available have produced important information about some of the characteristics of turbulence. Considerable effort has gone into the study of turbulence. Much remains to be learned. Perhaps studies in the new areas of chaos and fractal geometry will provide the tools for a better understanding of turbulence (see Section 8.3.5).

The vertical scale of Fig. 8.15 is also distorted. The viscous sublayer is usually a very thin layer adjacent to the wall. For example, for water flow in a 3-in.-diameter pipe with an average velocity of 10 ft/s, the viscous sublayer is approximately 0.002 in. thick. Since the fluid motion within this thin layer is critical in terms of the overall flow (the no-slip condition and the wall shear stress occur in this layer), it is not surprising to find that turbulent pipe flow properties can be quite dependent on the roughness of the pipe wall, unlike laminar pipe flow which is independent of roughness. Small roughness elements (scratches, rust, sand or dirt particles, etc.) can easily disturb this viscous sublayer (see Section 8.4), thereby affecting the entire flow.

An alternate form for the shear stress for turbulent flow is given in terms of the *eddy viscosity*, η, where

$$\tau = \eta \frac{d\bar{u}}{dy} \tag{8.27}$$

This extension of laminar flow terminology was introduced by J. Boussinesq, a French scientist, in 1877. Although the concept of an eddy viscosity is intriguing, in practice it is not an easy parameter to use. Unlike the absolute viscosity, μ, which is a known value for a given fluid, the eddy viscosity is a function of both the fluid and the flow conditions. That is, the eddy viscosity of water cannot be looked up in handbooks—its value changes from one turbulent flow condition to another and from one point in a turbulent flow to another.

The inability to accurately determine the Reynolds stress, $\rho\overline{u'v'}$, is equivalent to not knowing the eddy viscosity. Several semiempirical theories have been proposed (Ref. 3) to determine approximate values of η. L. Prandtl (1875–1953), a German physicist and aerodynamicist, proposed that the turbulent process could be viewed as the random transport of bundles of fluid particles over a certain distance, ℓ_m, the *mixing length*, from a region of one velocity to another region of a different velocity. By the use of some ad hoc assumptions and physical reasoning, it was concluded that the eddy viscosity was given by

Various ad hoc assumptions have been used to approximate turbulent shear stresses.

$$\eta = \rho \ell_m^2 \left| \frac{d\overline{u}}{dy} \right|$$

Thus, the turbulent shear stress is

$$\tau_{\text{turb}} = \rho \ell_m^2 \left(\frac{d\overline{u}}{dy} \right)^2 \tag{8.28}$$

The problem is thus shifted to that of determining the mixing length, ℓ_m. Further considerations indicate that ℓ_m is not a constant throughout the flow field. Near a solid surface the turbulence is dependent on the distance from the surface. Thus, additional assumptions are made regarding how the mixing length varies throughout the flow.

The net result is that as yet there is no general, all-encompassing, useful model that can accurately predict the shear stress throughout a general incompressible, viscous turbulent flow. Without such information it is impossible to integrate the force balance equation to obtain the turbulent velocity profile and other useful information, as was done for laminar flow.

8.3.3 Turbulent Velocity Profile

Considerable information concerning turbulent velocity profiles has been obtained through the use of dimensional analysis, experimentation, and semiempirical theoretical efforts. As is indicated in Fig. 8.15, fully developed turbulent flow in a pipe can be broken into three regions which are characterized by their distances from the wall: the viscous sublayer very near the pipe wall, the overlap region, and the outer turbulent layer throughout the center portion of the flow. Within the viscous sublayer the viscous shear stress is dominant compared with the turbulent (or Reynolds) stress, and the random, eddying nature of the flow is essentially absent. In the outer turbulent layer the Reynolds stress is dominant, and there is considerable mixing and randomness to the flow.

The character of the flow within these two regions is entirely different. For example, within the viscous sublayer the fluid viscosity is an important parameter; the density is unimportant. In the outer layer the opposite is true. By a careful use of dimensional analysis arguments for the flow in each layer and by a matching of the results in the common overlap layer, it has been possible to obtain the following conclusions about the turbulent velocity profile in a smooth pipe (Ref. 5).

In the viscous sublayer the velocity profile can be written in dimensionless form as

$$\frac{\overline{u}}{u^*} = \frac{yu^*}{\nu} \tag{8.29}$$

where $y = R - r$ is the distance measured from the wall, $\overline{u}$ is the time-averaged x component

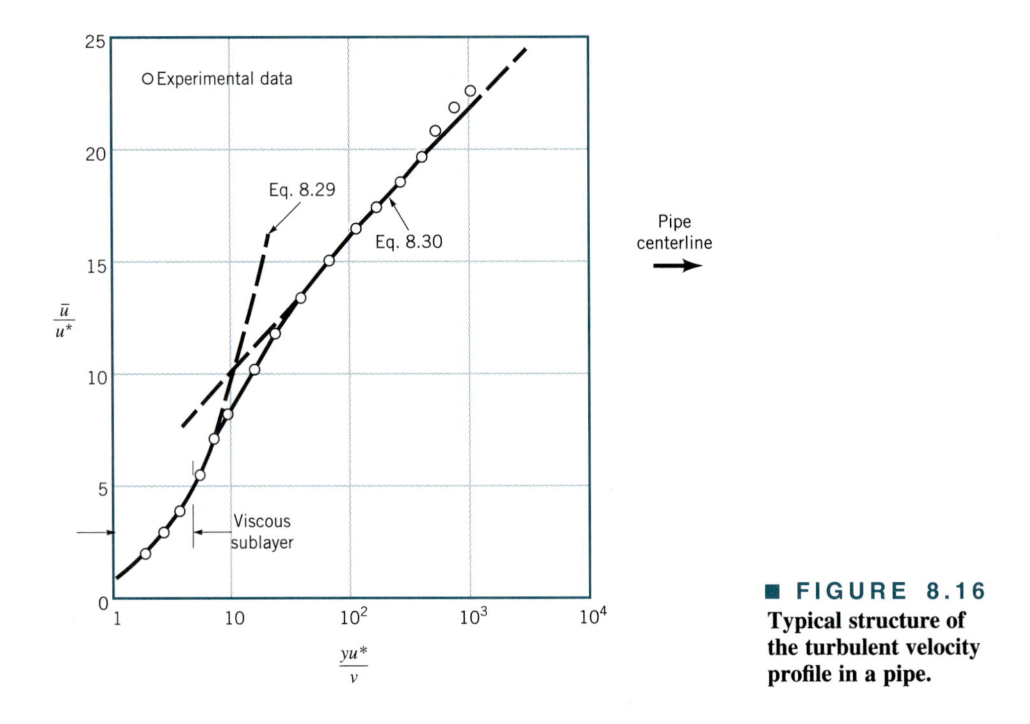

■ **FIGURE 8.16** **Typical structure of the turbulent velocity profile in a pipe.**

of velocity, and $u^* = (\tau_w/\rho)^{1/2}$ is termed the *friction velocity*. Note that u^* is not an actual velocity of the fluid—it is merely a quantity that has dimensions of velocity. As is indicated in Fig. 8.16, Eq. 8.29 (commonly called the *law of the wall*) is valid very near the smooth wall, for $0 \leq yu^*/\nu \lesssim 5$.

Dimensional analysis arguments indicate that in the overlap region the velocity should vary as the logarithm of y. Thus, the following expression has been proposed:

A turbulent flow velocity profile can be divided into various regions.

$$\frac{\bar{u}}{u^*} = 2.5 \ln\left(\frac{yu^*}{\nu}\right) + 5.0 \qquad \textbf{(8.30)}$$

where the constants 2.5 and 5.0 have been determined experimentally. As is indicated in Fig. 8.16, for regions not too close to the smooth wall, but not all the way out to the pipe center, Eq. 8.30 gives a reasonable correlation with the experimental data. Note that the horizontal scale is a logarithmic scale. This tends to exaggerate the size of the viscous sublayer relative to the remainder of the flow. As is shown in Example 8.4, the viscous sublayer is usually quite thin. Similar results can be obtained for turbulent flow past rough walls (Ref. 17).

A number of other correlations exist for the velocity profile in turbulent pipe flow. In the central region (the outer turbulent layer) the expression $(V_c - \bar{u})/u^* = 2.5 \ln(R/y)$, where V_c is the centerline velocity, is often suggested as a good correlation with experimental data. Another often-used (and relatively easy to use) correlation is the empirical *power-law velocity profile*

$$\frac{\bar{u}}{V_c} = \left(1 - \frac{r}{R}\right)^{1/n} \qquad \textbf{(8.31)}$$

In this representation, the value of n is a function of the Reynolds number, as is indicated in Fig. 8.17. The one-seventh power-law velocity profile ($n = 7$) is often used as a reasonable approximation for many practical flows. Typical turbulent velocity profiles based on this power-law representation are shown in Fig. 8.18.

A closer examination of Eq. 8.31 shows that the power-law profile cannot be valid near

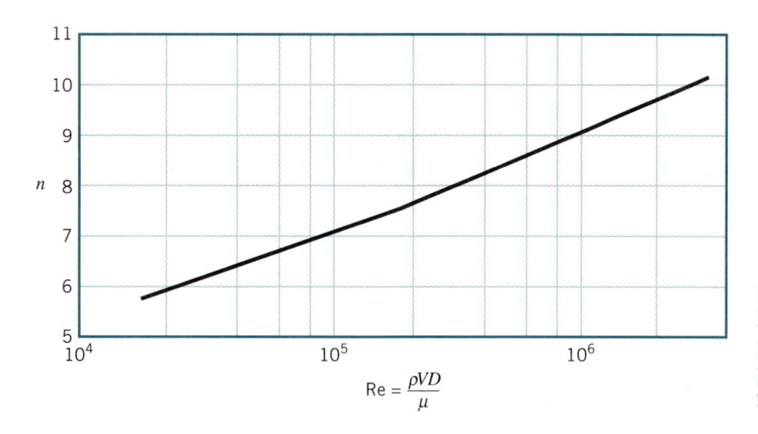

■ **FIGURE 8.17 Exponent, n, for power-law velocity profiles. (Adapted from Ref. 1.)**

the wall, since according to this equation the velocity gradient is infinite there. In addition, Eq. 8.31 cannot be precisely valid near the centerline because it does not give $d\bar{u}/dr = 0$ at $r = 0$. However, it does provide a reasonable approximation to the measured velocity profiles across most of the pipe.

A power-law velocity profile approximates the actual turbulent velocity profile.

Note from Fig. 8.18 that the turbulent profiles are much ''flatter'' than the laminar profile and that this flatness increases with Reynolds number (i.e., with n). Recall from Chapter 3 that reasonable approximate results are often obtained by using the inviscid Bernoulli equation and by assuming a fictitious uniform velocity profile. Since most flows are turbulent and turbulent flows tend to have nearly uniform velocity profiles, the usefulness of the Bernoulli equation and the uniform profile assumption is not unexpected. Of course, many properties of the flow cannot be accounted for without including viscous effects.

V8.3

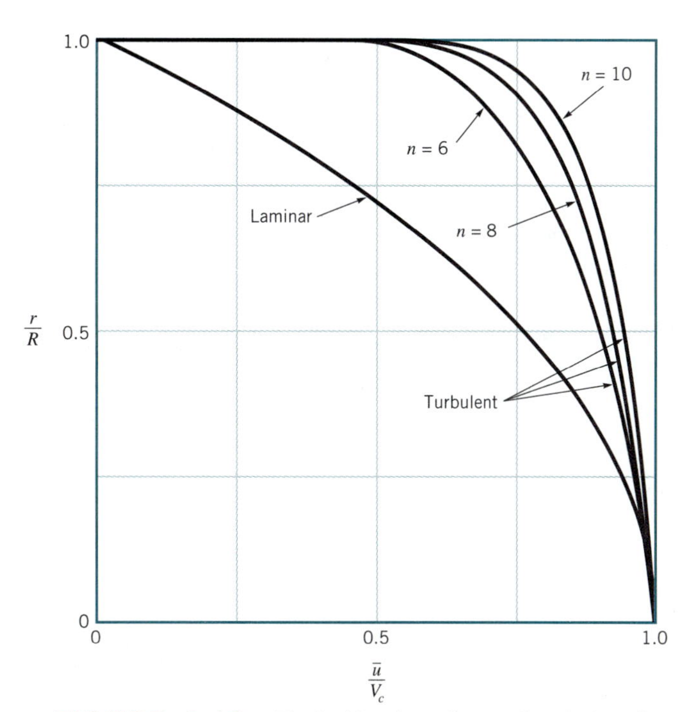

■ **FIGURE 8.18 Typical laminar flow and turbulent flow velocity profiles.**

EXAMPLE 8.4

Water at 20 °C ($\rho = 998 \text{ kg/m}^3$ and $\nu = 1.004 \times 10^{-6} \text{ m}^2/\text{s}$) flows through a horizontal pipe of 0.1-m diameter with a flowrate of $Q = 4 \times 10^{-2} \text{ m}^3/\text{s}$ and a pressure gradient of 2.59 kPa/m. (a) Determine the approximate thickness of the viscous sublayer. (b) Determine the approximate centerline velocity, V_c. (c) Determine the ratio of the turbulent to laminar shear stress, $\tau_{\text{turb}}/\tau_{\text{lam}}$, at a point midway between the centerline and the pipe wall (i.e., at $r = 0.025$ m).

SOLUTION

(a) According to Fig. 8.16, the thickness of the viscous sublayer, δ_s, is approximately

$$\frac{\delta_s u^*}{\nu} = 5$$

or

$$\delta_s = 5\,\frac{\nu}{u^*}$$

where

$$u^* = \left(\frac{\tau_w}{\rho}\right)^{1/2} \qquad \textbf{(1)}$$

The wall shear stress can be obtained from the pressure drop data and Eq. 8.5, which is valid for either laminar or turbulent flow. Thus,

$$\tau_w = \frac{D\,\Delta p}{4\ell} = \frac{(0.1 \text{ m})(2.59 \times 10^3 \text{ N/m}^2)}{4(1 \text{ m})} = 64.8 \text{ N/m}^2$$

Hence, from Eq. 1 we obtain

$$u^* = \left(\frac{64.8 \text{ N/m}^2}{998 \text{ kg/m}^3}\right)^{1/2} = 0.255 \text{ m/s}$$

so that

$$\delta_s = \frac{5(1.004 \times 10^{-6} \text{ m}^2/\text{s})}{0.255 \text{ m/s}} = 1.97 \times 10^{-5} \text{ m} \approx 0.02 \text{ mm} \qquad \textbf{(Ans)}$$

As stated previously, the viscous sublayer is very thin. Minute imperfections on the pipe wall will protrude into this sublayer and affect some of the characteristics of the flow (i.e., wall shear stress and pressure drop).

(b) The centerline velocity can be obtained from the average velocity and the assumption of a power-law velocity profile as follows. For this flow with

$$V = \frac{Q}{A} = \frac{0.04 \text{ m}^3/\text{s}}{\pi(0.1 \text{ m})^2/4} = 5.09 \text{ m/s}$$

the Reynolds number is

$$\text{Re} = \frac{VD}{\nu} = \frac{(5.09 \text{ m/s})(0.1 \text{ m})}{(1.004 \times 10^{-6} \text{ m}^2/\text{s})} = 5.07 \times 10^5$$

Thus, from Fig. 8.17, $n = 8.4$ so that

$$\frac{\bar{u}}{V_c} \approx \left(1 - \frac{r}{R}\right)^{1/8.4}$$

To determine the centerline velocity, V_c, we must know the relationship between V (the average velocity) and V_c. This can be obtained by integration of the power-law velocity profile as follows. Since the flow is axisymmetric,

$$Q = AV = \int \bar{u}\, dA = V_c \int_{r=0}^{r=R} \left(1 - \frac{r}{R}\right)^{1/n} (2\pi r)\, dr$$

which can be integrated to give

$$Q = 2\pi R^2 V_c \frac{n^2}{(n+1)(2n+1)}$$

Thus, since $Q = \pi R^2 V$, we obtain

$$\frac{V}{V_c} = \frac{2n^2}{(n+1)(2n+1)}$$

With $n = 8.4$ in the present case, this gives

$$V_c = \frac{(n+1)(2n+1)}{2n^2} V = 1.186V = 1.186\ (5.09\ \text{m/s})$$
$$= 6.04\ \text{m/s} \qquad \textbf{(Ans)}$$

Recall that $V_c = 2V$ for laminar pipe flow.

(c) From Eq. 8.4, which is valid for laminar or turbulent flow, the shear stress at $r = 0.025$ m is

$$\tau = \frac{2\tau_w r}{D} = \frac{2(64.8\ \text{N/m}^2)(0.025\ \text{m})}{(0.1\ \text{m})}$$

or

$$\tau = \tau_{\text{lam}} + \tau_{\text{turb}} = 32.4\ \text{N/m}^2$$

where $\tau_{\text{lam}} = -\mu\, d\bar{u}/dr$. From the power-law velocity profile (Eq. 8.31) we obtain the gradient of the average velocity as

$$\frac{d\bar{u}}{dr} = -\frac{V_c}{nR}\left(1 - \frac{r}{R}\right)^{(1-n)/n}$$

which gives

$$\frac{d\bar{u}}{dr} = -\frac{(6.04\ \text{m/s})}{8.4(0.05\text{m})}\left(1 - \frac{0.025\ \text{m}}{0.05\ \text{m}}\right)^{(1-8.4)/8.4} = -26.5/\text{s}$$

Thus,

$$\tau_{\text{lam}} = -\mu \frac{d\bar{u}}{dr} = -(\nu\rho)\frac{d\bar{u}}{dr}$$
$$= -(1.004 \times 10^{-6}\ \text{m}^2/\text{s})(998\ \text{kg/m}^3)(-26.5/\text{s})$$
$$= 0.0266\ \text{N/m}^2$$

Thus, the ratio of turbulent to laminar shear stress is given by

$$\frac{\tau_{\text{turb}}}{\tau_{\text{lam}}} = \frac{\tau - \tau_{\text{lam}}}{\tau_{\text{lam}}} = \frac{32.4 - 0.0266}{0.0266} = 1220 \qquad \textbf{(Ans)}$$

As expected, most of the shear stress at this location in the turbulent flow is due to the turbulent shear stress.

The turbulent flow characteristics discussed in this section are not unique to turbulent flow in round pipes. Many of the characteristics introduced (i.e., the Reynolds stress, the viscous sublayer, the overlap layer, the outer layer, the general characteristics of the velocity profile, etc.) are found in other turbulent flows. In particular, turbulent pipe flow and turbulent flow past a solid wall (boundary layer flow) share many of these common traits. Such ideas are discussed more fully in Chapter 9.

8.3.4 Turbulence Modeling

Although it is not yet possible to theoretically predict the random, irregular details of turbulent flows, it would be useful to be able to predict the time-averaged flow fields (pressure, velocity, etc.) directly from the basic governing equations. To this end one can time average the governing Navier-Stokes equations (Eqs. 6.31 and 6.127) to obtain equations for the average velocity and pressure. However, because the Navier-Stokes equations are nonlinear, the resulting time-averaged differential equations contain not only the desired average pressure and velocity as variables, but also averages of products of the fluctuations—terms of the type that one tried to eliminate by averaging the equations! For example, the Reynolds stress $-\rho\overline{u'v'}$ (see Eq. 8.26) occurs in the time-averaged momentum equation.

Thus, it is not possible to merely average the basic differential equations and obtain governing equations involving only the desired averaged quantities. This is the reason for the variety of ad hoc assumptions that have been proposed to provide "closure" to the equations governing the average flow. That is, the set of governing equations must be a complete or closed set of equations—the same number of equation as unknowns.

Various attempts have been made to solve this closure problem (Refs. 1, 32). Such schemes involving the introduction of an eddy viscosity or the mixing length (as introduced in Section 8.3.2) are termed algebraic or zero-equation models. Other methods, which are beyond the scope of this book, include the one-equation model and the two-equation model. These turbulence models are based on the equation for the turbulence kinetic energy and require significant computer usage.

Turbulence modeling is an important and extremely difficult topic. Although considerable progress has been made, much remains to be done in this area.

8.3.5 Chaos and Turbulence

Chaos theory may eventually provide a deeper understanding of turbulence.

Chaos theory is a relatively new branch of mathematical physics that may provide insight into the complex nature of turbulence. This method combines mathematics and numerical (computer) techniques to provide a new way to analyze certain problems. Chaos theory, which is quite complex and is only now being developed, involves the behavior of nonlinear dynamical systems and their response to initial and boundary conditions. The flow of a viscous fluid, which is governed by the nonlinear Navier-Stokes equations (Eq. 6.127), may be such a system.

To solve the Navier-Stokes equations for the velocity and pressure fields in a viscous flow, one must specify the particular flow geometry being considered (the boundary conditions) and the condition of the flow at some particular time (the initial conditions). If, as some researchers predict, the Navier-Stokes equations allow chaotic behavior, then the state of the flow at times after the initial time may be very, very sensitive to the initial conditions. A slight variation to the initial flow conditions may cause the flow at later times to be quite different than it would have been with the original, only slightly different initial conditions. When carried to the extreme, the flow may be "chaotic," "random," or perhaps (in current terminology), "turbulent."

The occurrence of such behavior would depend on the value of the Reynolds number.

For example, it may be found that for sufficiently small Reynolds numbers the flow is not chaotic (i.e., it is laminar), while for large Reynolds numbers it is chaotic with turbulent characteristics.

Thus, with the advancement of chaos theory it may be found that the numerous ad hoc turbulence ideas mentioned in previous sections (i.e., eddy viscosity, mixing length, law of the wall, etc.) may not be needed. It may be that chaos theory can provide the turbulence properties and structure directly from the governing equations. As of now we must wait until this exciting topic is developed further. The interested reader is encouraged to consult Ref. 33 for a general introduction to chaos or Ref. 34 for additional material.

8.4 Dimensional Analysis of Pipe Flow

Most turbulent pipe flow information is based on experimental data.

As was discussed in previous sections, turbulent flow can be a very complex, difficult topic—one that as yet has defied a rigorous theoretical treatment. Thus, most turbulent pipe flow analyses are based on experimental data and semiempirical formulas, even if the flow is fully developed. These results are given in dimensionless form and cover a very wide range of flow parameters, including arbitrary fluids, pipes, and flowrates. In addition to these fully developed flow considerations, a variety of useful data are available regarding flow through pipe fittings, such as elbows, tees, valves, and the like. These data are conveniently expressed in dimensionless form.

8.4.1 The Moody Chart

A dimensional analysis treatment of pipe flow provides the most convenient base from which to consider turbulent, fully developed pipe flow. An introduction to this topic was given in Section 8.3. As is discussed in Sections 8.2.1 and 8.2.4, the pressure drop and head loss in a pipe are dependent on the wall shear stress, τ_w, between the fluid and pipe surface. A fundamental difference between laminar and turbulent flow is that the shear stress for turbulent flow is a function of the density of the fluid, ρ. For laminar flow, the shear stress is independent of the density, leaving the viscosity, μ, as the only important fluid property.

Thus, the pressure drop, Δp, for steady, incompressible turbulent flow in a horizontal round pipe of diameter D can be written in functional form as

$$\Delta p = F(V, D, \ell, \varepsilon, \mu, \rho) \tag{8.32}$$

where V is the average velocity, ℓ is the pipe length, and ε is a measure of the roughness of the pipe wall. It is clear that Δp should be a function of V, D, and ℓ. The dependence of Δp on the fluid properties μ and ρ is expected because of the dependence of τ on these parameters.

Although the pressure drop for laminar pipe flow is found to be independent of the roughness of the pipe, it is necessary to include this parameter when considering turbulent flow. As is discussed in Section 8.3.3 and illustrated in Fig. 8.19, for turbulent flow there is a relatively thin viscous sublayer formed in the fluid near the pipe wall. In many instances this layer is very thin; $\delta_s/D \ll 1$, where δ_s is the sublayer thickness. If a typical wall roughness element protrudes sufficiently far into (or even through) this layer, the structure and properties of the viscous sublayer (along with Δp and τ_w) will be different than if the wall were smooth. Thus, for turbulent flow the pressure drop is expected to be a function of the wall roughness. For laminar flow there is no thin viscous layer–viscous effects are important across the entire pipe. Thus, relatively small roughness elements have completely negligible effects on laminar pipe flow. Of course, for pipes with very large wall "roughness," ($\varepsilon/D \gtrsim 0.1$), such as that in corrugated pipes, the flowrate may be a function of the "roughness." We will consider only typical constant diameter pipes with relative roughnesses in the range $0 \leq \varepsilon/D \lesssim 0.05$.

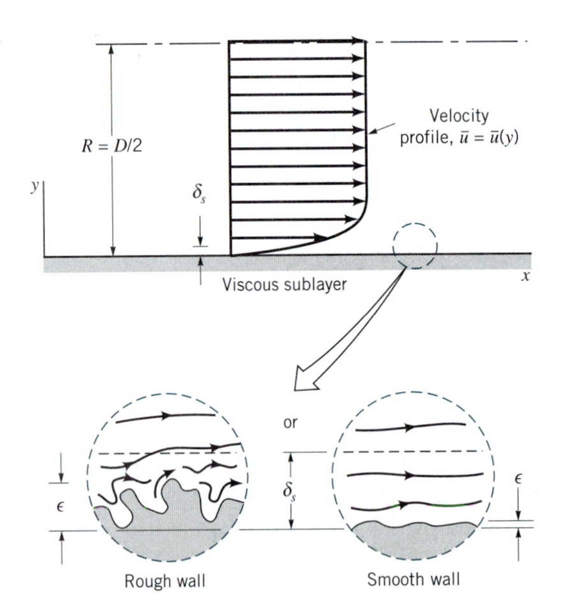

■ **FIGURE 8.19** **Flow in the viscous sublayer near rough and smooth walls.**

Analysis of flow in corrugated pipes does not fit into the standard constant diameter pipe category, although experimental results for such pipes are available (Ref. 30).

The list of parameters given in Eq. 8.32 is apparently a complete one. That is, experiments have shown that other parameters (such as surface tension, vapor pressure, etc.) do not affect the pressure drop for the conditions stated (steady, incompressible flow; round, horizontal pipe). Since there are seven variables ($k = 7$) which can be written in terms of the three reference dimensions MLT ($r = 3$), Eq. 8.32 can be written in dimensionless form in terms of $k - r = 4$ dimensionless groups. As was discussed in Section 7.9.1, one such representation is

Turbulent pipe flow properties depend on the fluid density and the pipe roughness.

$$\frac{\Delta p}{\frac{1}{2}\rho V^2} = \tilde{\phi}\left(\frac{\rho V D}{\mu}, \frac{\ell}{D}, \frac{\varepsilon}{D}\right)$$

This result differs from that used for laminar flow (see Eq. 8.17) in two ways. First, we have chosen to make the pressure dimensionless by dividing by the dynamic pressure, $\rho V^2/2$, rather than a characteristic viscous shear stress, $\mu V/D$. This convention was chosen in recognition of the fact that the shear stress for turbulent flow is normally dominated by τ_{turb}, which is a stronger function of the density than it is of viscosity. Second, we have introduced two additional dimensionless parameters, the Reynolds number, $\text{Re} = \rho VD/\mu$, and the *relative roughness*, ε/D, which are not present in the laminar formulation because the two parameters ρ and ε are not important in fully developed laminar pipe flow.

As was done for laminar flow, the functional representation can be simplified by imposing the reasonable assumption that the pressure drop should be proportional to the pipe length. (Such a step is not within the realm of dimensional analysis. It is merely a logical assumption supported by experiments.) The only way that this can be true is if the ℓ/D dependence is factored out as

$$\frac{\Delta p}{\frac{1}{2}\rho V^2} = \frac{\ell}{D}\,\phi\left(\text{Re}, \frac{\varepsilon}{D}\right)$$

As was discussed in Section 8.2.3, the quantity $\Delta p D/(\ell \rho V^2/2)$ is termed the friction factor, f. Thus, for a horizontal pipe

$$\Delta p = f \frac{\ell}{D} \frac{\rho V^2}{2} \tag{8.33}$$

where

$$f = \phi\left(\text{Re}, \frac{\varepsilon}{D}\right)$$

For laminar fully developed flow, the value of f is simply $f = 64/\text{Re}$, independent of ε/D. For turbulent flow, the functional dependence of the friction factor on the Reynolds number and the relative roughness, $f = \phi(\text{Re}, \varepsilon/D)$, is a rather complex one that cannot, as yet, be obtained from a theoretical analysis. The results are obtained from an exhaustive set of experiments and usually presented in terms of a curve-fitting formula or the equivalent graphical form.

From Eq. 5.89 the energy equation for steady incompressible flow is

$$\frac{p_1}{\gamma} + \alpha_1 \frac{V_1^2}{2g} + z_1 = \frac{p_2}{\gamma} + \alpha_2 \frac{V_2^2}{2g} + z_2 + h_L$$

where h_L is the head loss between sections (1) and (2). With the assumption of a constant diameter ($D_1 = D_2$ so that $V_1 = V_2$), horizontal ($z_1 = z_2$) pipe with fully developed flow ($\alpha_1 = \alpha_2$), this becomes $\Delta p = p_1 - p_2 = \gamma h_L$, which can be combined with Eq. 8.33 to give

The head loss in pipe flow is given in terms of the friction factor.

$$\boxed{h_L = f \frac{\ell}{D} \frac{V^2}{2g}} \tag{8.34}$$

Equation 8.34, called the *Darcy–Weisbach equation*, is valid for any fully developed, steady, incompressible pipe flow—whether the pipe is horizontal or on a hill. On the other hand, Eq. 8.33 is valid only for horizontal pipes. In general, with $V_1 = V_2$ the energy equation gives

$$p_1 - p_2 = \gamma(z_2 - z_1) + \gamma h_L = \gamma(z_2 - z_1) + f \frac{\ell}{D} \frac{\rho V^2}{2}$$

Part of the pressure change is due to the elevation change and part is due to the head loss associated with frictional effects, which are given in terms of the friction factor, f.

It is not easy to determine the functional dependence of the friction factor on the Reynolds number and relative roughness. Much of this information is a result of experiments conducted by J. Nikuradse in 1933 (Ref. 6) and amplified by many others since then. One difficulty lies in the determination of the roughness of the pipe. Nikuradse used artificially roughened pipes produced by gluing sand grains of known size onto pipe walls to produce pipes with sandpaper-type surfaces. The pressure drop needed to produce a desired flowrate was measured and the data were converted into the friction factor for the corresponding Reynolds number and relative roughness. The tests were repeated numerous times for a wide range of Re and ε/D to determine the $f = \phi(\text{Re}, \varepsilon/D)$ dependence.

In commercially available pipes the roughness is not as uniform and well defined as in the artificially roughened pipes used by Nikuradse. However, it is possible to obtain a measure of the effective relative roughness of typical pipes and thus to obtain the friction factor. Typical roughness values for various pipe surfaces are given in Table 8.1. Figure 8.20 shows the functional dependence of f on Re and ε/D and is called the *Moody chart* in honor of L. F. Moody, who, along with C. F. Colebrook, correlated the original data of Nikuradse in terms of the relative roughness of commercially available pipe materials. It should be noted that the values of ε/D do not necessarily correspond to the actual values obtained by a

■ **TABLE 8.1**

Equivalent Roughness for New Pipes [From Moody (Ref. 7) and Colebrook (Ref. 8)]

	Equivalent Roughness, ε	
Pipe	**Feet**	**Millimeters**
Riveted steel	0.003–0.03	0.9–9.0
Concrete	0.001–0.01	0.3–3.0
Wood stave	0.0006–0.003	0.18–0.9
Cast iron	0.00085	0.26
Galvanized iron	0.0005	0.15
Commercial steel or wrought iron	0.00015	0.045
Drawn tubing	0.000005	0.0015
Plastic, glass	0.0 (smooth)	0.0 (smooth)

microscopic determination of the average height of the roughness of the surface. They do, however, provide the correct correlation for $f = \phi(\text{Re}, \varepsilon/D)$.

It is important to observe that the values of relative roughness given pertain to new, clean pipes. After considerable use, most pipes (because of a buildup of corrosion or scale) may have a relative roughness that is considerably larger (perhaps by an order of magnitude) than that given. Very old pipes may have enough scale buildup to not only alter the value of ε but also to change their effective diameter by a considerable amount.

The Moody chart gives the friction factor in terms of the Reynolds number and relative roughness.

The following characteristics are observed from the data of Fig. 8.20. For laminar flow, $f = 64/\text{Re}$, which is independent of relative roughness. For very large Reynolds numbers, $f = \phi(\varepsilon/D)$, which is independent of the Reynolds number. For such flows, commonly termed *completely turbulent flow* (or *wholly turbulent flow*), the laminar sublayer is so thin (its thickness decreases with increasing Re) that the surface roughness completely dominates the character of the flow near the wall. Hence, the pressure drop required is a result of an inertia-dominated turbulent shear stress rather than the viscosity-dominated laminar shear stress normally found in the viscous sublayer. For flows with moderate values of Re, the friction factor is indeed dependent on both the Reynolds number and relative roughness—$f = \phi(\text{Re}, \varepsilon/D)$. The gap in the figure for which no values of f are given (the $2100 < \text{Re} < 4000$ range) is a result of the fact that the flow in this transition range may be laminar or turbulent (or an unsteady mix of both) depending on the specific circumstances involved.

Note that even for smooth pipes ($\varepsilon = 0$) the friction factor is not zero. That is, there is a head loss in any pipe, no matter how smooth the surface is made. This is a result of the no-slip boundary condition that requires any fluid to stick to any solid surface it flows over. There is always some microscopic surface roughness that produces the no-slip behavior (and thus $f \neq 0$) on the molecular level, even when the roughness is considerably less than the viscous sublayer thickness. Such pipes are called *hydraulically smooth.*

Various investigators have attempted to obtain an analytical expression for $f = \phi(\text{Re}, \varepsilon/D)$. Note that the Moody chart covers an extremely wide range in flow parameters. The nonlaminar region covers more than four orders of magnitude in Reynolds number—from $\text{Re} = 4 \times 10^3$ to $\text{Re} = 10^8$. Obviously, for a given pipe and fluid, typical values of the average velocity do not cover this range. However, because of the large variety in pipes (D), fluids (ρ and μ), and velocities (V), such a wide range in Re is needed to accommodate nearly all applications of pipe flow. In many cases the particular pipe flow of interest is confined to a relatively small region of the Moody chart, and simple semiempirical expres-

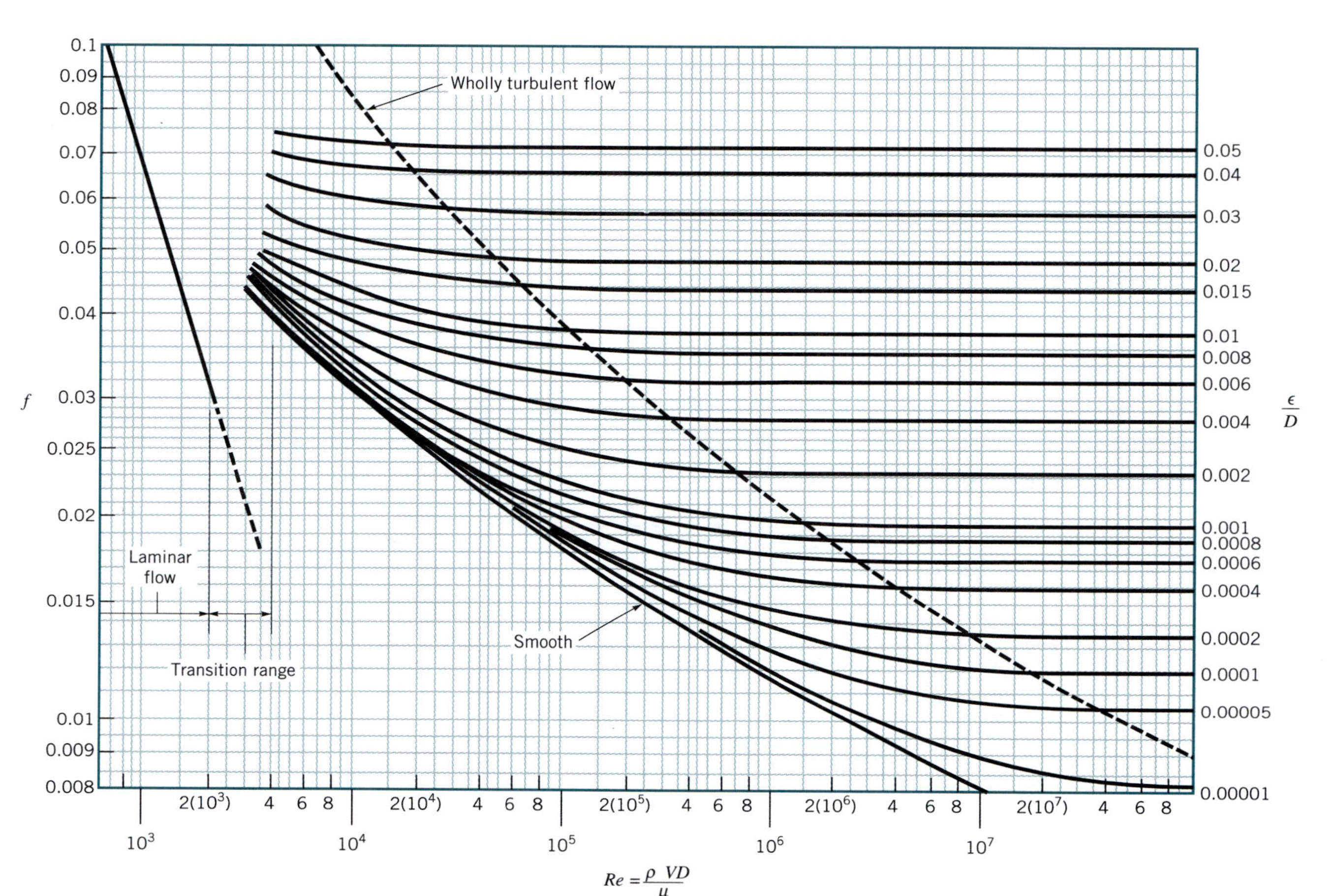

■ **FIGURE 8.20** **Friction factor as a function of Reynolds number and relative roughness for round pipes—the Moody chart. (Data from Ref. 7 with permission.)**

sions can be developed for those conditions. For example, a company that manufactures cast-iron water pipes with diameters between 2 and 12 in. may use a simple equation valid for their conditions only. The Moody chart, on the other hand, is universally valid for all steady, fully developed, incompressible pipe flows.

The following equation from Colebrook is valid for the entire nonlaminar range of the Moody chart

The turbulent portion of the Moody chart is represented by the Colebrook formula.

$$\frac{1}{\sqrt{f}} = -2.0 \log \left(\frac{\varepsilon/D}{3.7} + \frac{2.51}{\text{Re}\sqrt{f}} \right) \tag{8.35}$$

In fact, the Moody chart is a graphical representation of this equation, which is an empirical fit of the pipe flow pressure drop data. Equation 8.35 is called the *Colebrook formula*. A difficulty with its use is that it is implicit in the dependence of f. That is, for given conditions (Re and ε/D), it is not possible to solve for f without some sort of iterative scheme. With the use of modern computers and calculators, such calculations are not difficult. (As shown in Problem 8.37 at the end of this chapter, it is possible to obtain an equation that adequately approximates the Colebrook/Moody chart relationship but does not require an iterative scheme.) A word of caution is in order concerning the use of the Moody chart or the equivalent Colebrook formula. Because of various inherent inaccuracies involved (uncertainty in the relative roughness, uncertainty in the experimental data used to produce the Moody chart, etc.), the use of several place accuracy in pipe flow problems is usually not justified. As a rule of thumb, a 10% accuracy is the best expected.

Air under standard conditions flows through a 4.0-mm-diameter drawn tubing with an average velocity of $V = 50$ m/s. For such conditions the flow would normally be turbulent. However, if precautions are taken to eliminate disturbances to the flow (the entrance to the tube is very smooth, the air is dust free, the tube does not vibrate, etc.), it may be possible to maintain laminar flow. (a) Determine the pressure drop in a 0.1-m section of the tube if the flow is laminar. (b) Repeat the calculations if the flow is turbulent.

SOLUTION

Under standard temperature and pressure conditions the density and viscosity are $\rho = 1.23$ kg/m^3 and $\mu = 1.79 \times 10^{-5}$ N·s/m^2. Thus, the Reynolds number is

$$\text{Re} = \frac{\rho V D}{\mu} = \frac{(1.23 \text{ kg/m}^3)(50 \text{ m/s})(0.004 \text{ m})}{1.79 \times 10^{-5} \text{ N·s/m}^2} = 13{,}700$$

which would normally indicate turbulent flow.

(a) If the flow were laminar, then $f = 64/\text{Re} = 64/13{,}700 = 0.00467$ and the pressure drop in a 0.1-m-long horizontal section of the pipe would be

$$\Delta p = f \frac{\ell}{D} \frac{1}{2} \rho V^2 = (0.00467) \frac{(0.1 \text{ m})}{(0.004 \text{ m})} \frac{1}{2} (1.23 \text{ kg/m}^3)(50 \text{ m/s})^2$$

or

$$\Delta p = 0.179 \text{ kPa} \qquad \textbf{(Ans)}$$

Note that the same result is obtained from Eq. 8.8.

$$\Delta p = \frac{32\mu\ell}{D^2} V = \frac{32(1.79 \times 10^{-5} \text{ N·s/m}^2)(0.1 \text{ m})(50 \text{ m/s})}{(0.004 \text{ m})^2} = 179 \text{ N/m}^2$$

(b) If the flow were turbulent, then $f = \phi(\text{Re}, \varepsilon/D)$, where from Table 8.1, $\varepsilon = 0.0015$ mm so that $\varepsilon/D = 0.0015 \text{ mm}/4.0 \text{ mm} = 0.000375$. From the Moody chart with $\text{Re} = 1.37 \times 10^4$ and $\varepsilon/D = 0.000375$ we obtain $f = 0.028$. Thus, the pressure drop in this case would be approximately

$$\Delta p = f \frac{\ell}{D} \frac{1}{2} \rho V^2 = (0.028) \frac{(0.1 \text{ m})}{(0.004 \text{ m})} \frac{1}{2} (1.23 \text{ kg/m}^3)(50 \text{ m/s})^2$$

or

$$\Delta p = 1.076 \text{ kPa} \qquad \textbf{(Ans)}$$

A considerable savings in effort to force the fluid through the pipe could be realized (0.179 kPa rather than 1.076 kPa) if the flow could be maintained as laminar flow at this Reynolds number. In general this is very difficult to do, although laminar flow in pipes has been maintained up to $\text{Re} \approx 100{,}000$ in rare instances.

An alternate method to determine the friction factor for the turbulent flow would be to use the Colebrook formula, Eq. 8.35. Thus,

$$\frac{1}{\sqrt{f}} = -2.0 \log \left(\frac{\varepsilon/D}{3.7} + \frac{2.51}{\text{Re}\sqrt{f}} \right) = -2.0 \log \left(\frac{0.000375}{3.7} + \frac{2.51}{1.37 \times 10^4 \sqrt{f}} \right)$$

or

$$\frac{1}{\sqrt{f}} = -2.0 \log \left(1.01 \times 10^{-4} + \frac{1.83 \times 10^{-4}}{\sqrt{f}} \right) \qquad \textbf{(1)}$$

An iterative procedure to obtain f can be done as follows. We assume a value of f ($f = 0.02$, for example), substitute it into the right-hand side of Eq. 1, and calculate a new f ($f = 0.0307$ in this case). Since the two values do not agree, the assumed value is not the solution. Hence, we try again. This time we assume $f = 0.0307$ (the last value calculated) and calculate the new value as $f = 0.0289$. Again this is still not the solution. Two more iterations show that the assumed and calculated values converge to the solution $f = 0.0291$, in agreement (within the accuracy of reading the graph) with the Moody chart method of $f = 0.028$.

Numerous other empirical formulas can be found in the literature (Ref. 5) for portions of the Moody chart. For example, an often-used equation, commonly referred to as the Blasius formula, for turbulent flow in smooth pipes ($\varepsilon/D = 0$) with $\text{Re} < 10^5$ is

$$f = \frac{0.316}{\text{Re}^{1/4}}$$

For our case this gives

$$f = 0.316(13{,}700)^{-0.25} = 0.0292$$

which is in agreement with the previous results. Note that the value of f is relatively insensitive to ε/D for this particular situation. Whether the tube was smooth glass ($\varepsilon/D = 0$) or the drawn tubing ($\varepsilon/D = 0.000375$) would not make much difference in the pressure drop. For this flow, an increase in relative roughness by a factor of 30 to $\varepsilon/D = 0.0113$ (equivalent to a commercial steel surface; see Table 8.1) would give $f = 0.043$. This would represent an increase in pressure drop and head loss by a factor of $0.043/0.0291 = 1.48$ compared with that for the original drawn tubing.

The pressure drop of 1.076 kPa in a length of 0.1 m of pipe corresponds to a change in absolute pressure [assuming $p = 101$ kPa (abs) at $x = 0$] of approximately

1.076/101 = 0.0107, or about 1%. Thus, the incompressible flow assumption on which the above calculations (and all of the formulas in this chapter) are based is reasonable. However, if the pipe were 2-m long the pressure drop would be 21.5 kPa, approximately 20% of the original pressure. In this case the density would not be approximately constant along the pipe, and a compressible flow analysis would be needed. Such considerations are discussed in Chapter 11.

8.4.2 Minor Losses

Losses occur in straight pipes (major losses) and pipe system components (minor losses).

As discussed in the previous section, the head loss in long, straight sections of pipe can be calculated by use of the friction factor obtained from either the Moody chart or the Colebrook equation. Most pipe systems, however, consist of considerably more than straight pipes. These additional components (valves, bends, tees, and the like) add to the overall head loss of the system. Such losses are generally termed *minor losses*, with the apparent implication being that the majority of the system loss is associated with the friction in the straight portions of the pipes, the *major losses*. In many cases this is true. In other cases the minor losses are greater than the major losses. In this section we indicate how to determine the various minor losses that commonly occur in pipe systems.

The head loss associated with flow through a valve is a common minor loss. The purpose of a valve is to provide a means to regulate the flowrate. This is accomplished by changing the geometry of the system (i.e., closing or opening the valve alters the flow pattern through the valve), which in turn alters the losses associated with the flow through the valve. The flow resistance or head loss through the valve may be a significant portion of the resistance in the system. In fact, with the valve closed, the resistance to the flow is infinite—the fluid cannot flow. Such minor losses may be very important indeed. With the valve wide open the extra resistance due to the presence of the valve may or may not be negligible.

The flow pattern through a typical component such as a valve is shown in Fig. 8.21. It is not difficult to realize that a theoretical analysis to predict the details of such flows to obtain the head loss for these components is not, as yet, possible. Thus, the head loss information for essentially all components is given in dimensionless form and based on experimental data.

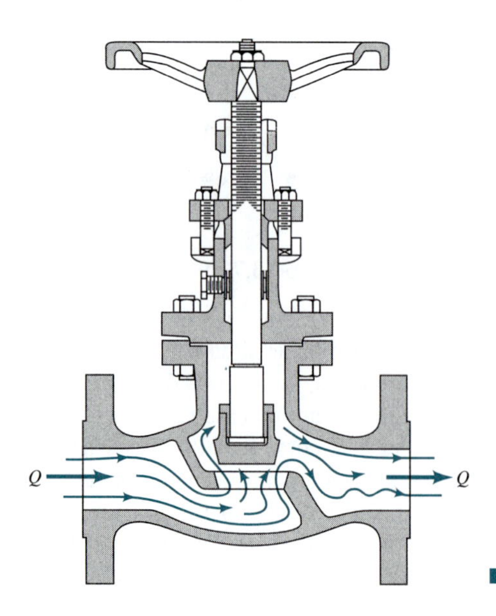

■ FIGURE 8.21 Flow through a valve.

Losses due to pipe system components are given in terms of loss coefficients.

The most common method used to determine these head losses or pressure drops is to specify the *loss coefficient*, K_L, which is defined as

$$K_L = \frac{h_L}{(V^2/2g)} = \frac{\Delta p}{\frac{1}{2}\rho V^2}$$

so that

$$\Delta p = K_L \tfrac{1}{2}\rho V^2$$

or

$$\boxed{h_L = K_L \frac{V^2}{2g}} \qquad \textbf{(8.36)}$$

The pressure drop across a component that has a loss coefficient of $K_L = 1$ is equal to the dynamic pressure, $\rho V^2/2$.

The actual value of K_L is strongly dependent on the geometry of the component considered. It may also be dependent on the fluid properties. That is,

$$K_L = \phi(\text{geometry, Re})$$

where $\text{Re} = \rho VD/\mu$ is the pipe Reynolds number. For many practical applications the Reynolds number is large enough so that the flow through the component is dominated by inertia effects, with viscous effects being of secondary importance. This is true because of the relatively large accelerations and decelerations experienced by the fluid as it flows along a rather curved, variable-area (perhaps even torturous) path through the component (see Fig. 8.21). In a flow that is dominated by inertia effects rather than viscous effects, it is usually found that pressure drops and head losses correlate directly with the dynamic pressure. This is the reason why the friction factor for very large Reynolds number, fully developed pipe flow is independent of the Reynolds number. The same condition is found to be true for flow through pipe components. Thus, in most cases of practical interest the loss coefficients for components are a function of geometry only, $K_L = \phi(\text{geometry})$.

Minor losses are sometimes given in terms of an *equivalent length*, ℓ_{eq}. In this terminology, the head loss through a component is given in terms of the equivalent length of pipe that would produce the same head loss as the component. That is,

$$h_L = K_L \frac{V^2}{2g} = f \frac{\ell_{eq}}{D} \frac{V^2}{2g}$$

or

$$\ell_{eq} = \frac{K_L D}{f}$$

where D and f are based on the pipe containing the component. The head loss of the pipe system is the same as that produced in a straight pipe whose length is equal to the pipes of the original system plus the sum of the additional equivalent lengths of all of the components of the system. Most pipe flow analyses, including those in this book, use the loss coefficient method rather than the equivalent length method to determine the minor losses.

Many pipe systems contain various transition sections in which the pipe diameter changes from one size to another. Such changes may occur abruptly or rather smoothly through some type of area change section. Any change in flow area contributes losses that are not accounted for in the fully developed head loss calculation (the friction factor). The extreme cases involve flow into a pipe from a reservoir (an entrance) or out of a pipe into a reservoir (an exit).

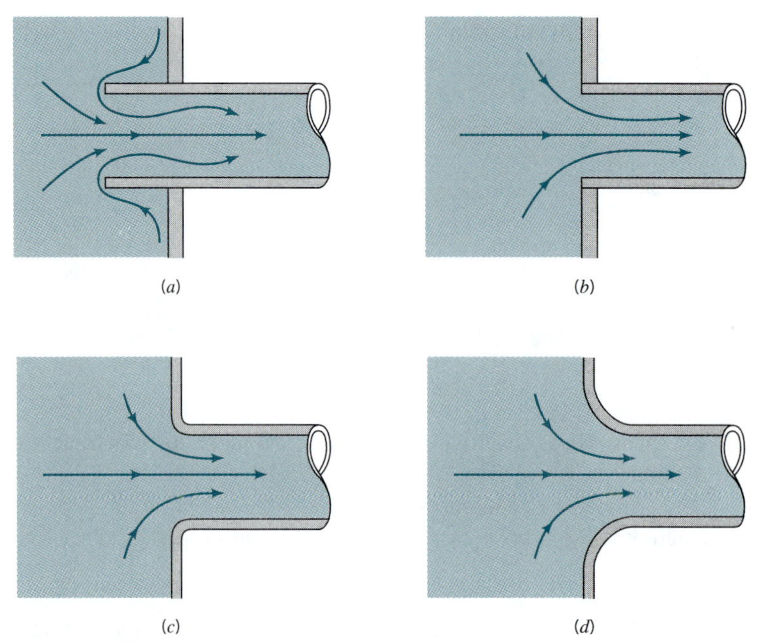

■ **FIGURE 8.22** **Entrance flow conditions and loss coefficient (Refs. 28, 29). (*a*) Reentrant, $K_L = 0.8$, (*b*) sharp-edged, $K_L = 0.5$, (*c*) slightly rounded, $K_L = 0.2$ (see Fig. 8.24), (*d*) well-rounded, $K_L = 0.04$ (see Fig. 8.24).**

A vena contracta region is often developed at the entrance to a pipe.

A fluid may flow from a reservoir into a pipe through any number of different shaped entrance regions as are sketched in Fig. 8.22. Each geometry has an associated loss coefficient. A typical flow pattern for flow entering a pipe through a square-edged entrance is sketched in Fig. 8.23. As was discussed in Chapter 3, a vena contracta region may result because the

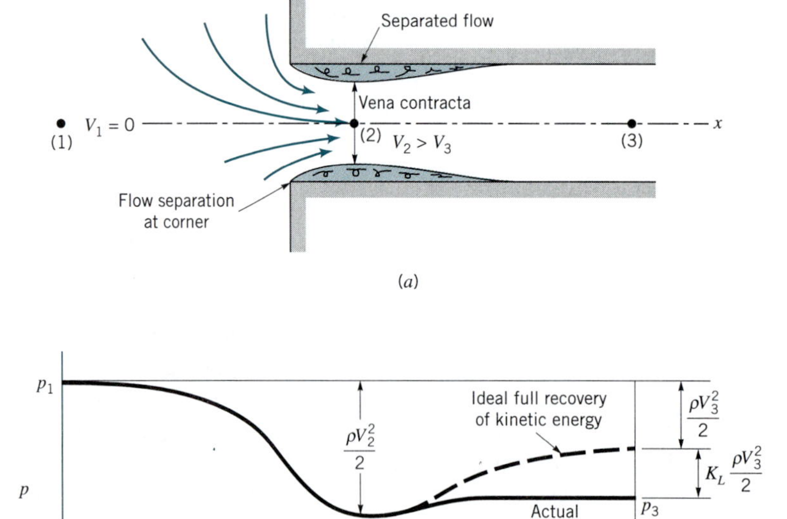

■ **FIGURE 8.23** **Flow pattern and pressure distribution for a sharp-edged entrance.**

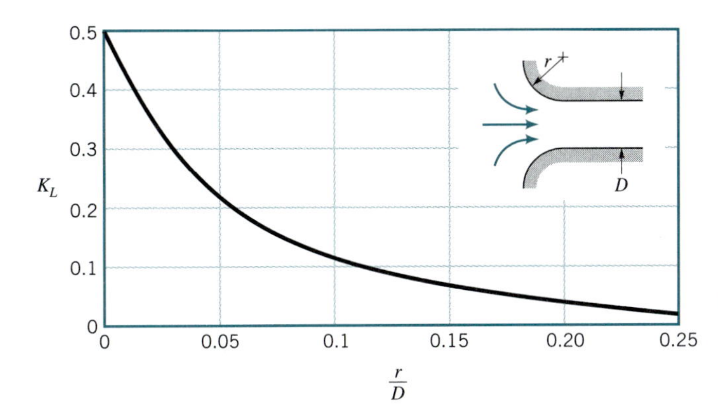

■ **FIGURE 8.24** **Entrance loss coefficient as a function of rounding of the inlet edge (Ref. 9).**

fluid cannot turn a sharp right-angle corner. The flow is said to separate from the sharp corner. The maximum velocity at section (2) is greater than that in the pipe at section (3), and the pressure there is lower. If this high-speed fluid could slow down efficiently, the kinetic energy could be converted into pressure (the Bernoulli effect), and the ideal pressure distribution indicated in Fig. 8.23 would result. The head loss for the entrance would be essentially zero.

Minor head losses are often a result of the dissipation of kinetic energy.

Such is not the case. Although a fluid may be accelerated very efficiently, it is very difficult to slow down (decelerate) a fluid efficiently. Thus, the extra kinetic energy of the fluid at section (2) is partially lost because of viscous dissipation, so that the pressure does not return to the ideal value. An entrance head loss (pressure drop) is produced as is indicated in Fig. 8.23. The majority of this loss is due to inertia effects that are eventually dissipated by the shear stresses within the fluid. Only a small portion of the loss is due to the wall shear stress within the entrance region. The net effect is that the loss coefficient for a square-edged entrance is approximately $K_L = 0.50$. One-half of a velocity head is lost as the fluid enters the pipe. If the pipe protrudes into the tank (a reentrant entrance) as is shown in Fig. 8.22*a*, the losses are even greater.

V8.4

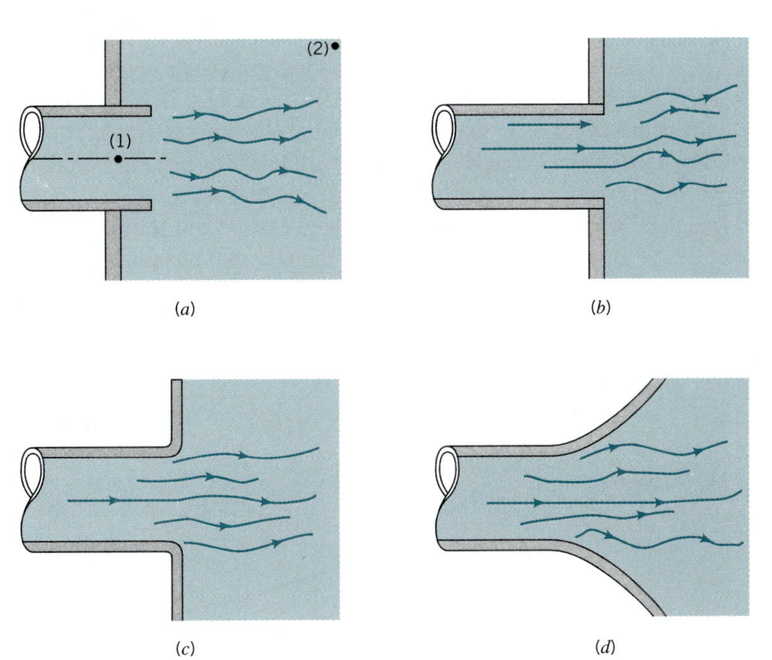

■ **FIGURE 8.25** **Exit flow conditions and loss coefficient. (*a*) Reentrant, $K_L = 1.0$, (*b*) sharp-edged, $K_L = 1.0$, (*c*) slightly rounded, $K_L = 1.0$, (*d*) well-rounded, $K_L = 1.0$.**

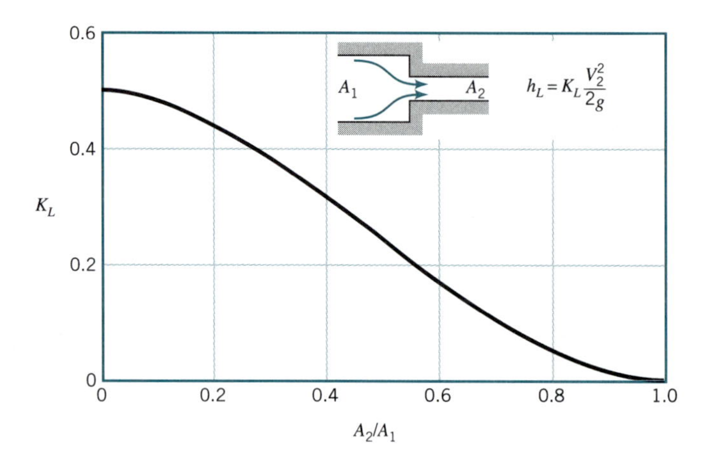

■ FIGURE 8.26 **Loss coefficient for a sudden contraction (Ref. 10).**

Pipe entrance losses can be relatively easily reduced by rounding the inlet.

An obvious way to reduce the entrance loss is to round the entrance region as is shown in Fig. 8.22*c*, thereby reducing or eliminating the vena contracta effect. Typical values for the loss coefficient for entrances with various amounts of rounding of the lip are shown in Fig. 8.24. A significant reduction in K_L can be obtained with only slight rounding.

A head loss (the exit loss) is also produced when a fluid flows from a pipe into a tank as is shown in Fig. 8.25. In these cases the entire kinetic energy of the exiting fluid (velocity V_1) is dissipated through viscous effects as the stream of fluid mixes with the fluid in the tank and eventually comes to rest ($V_2 = 0$). The exit loss from points (1) and (2) is therefore equivalent to one velocity head, or $K_L = 1$.

Losses also occur because of a change in pipe diameter as is shown in Figs. 8.26 and 8.27. The sharp-edged entrance and exit flows discussed in the previous paragraphs are limiting cases of this type of flow with either $A_1/A_2 = \infty$, or $A_1/A_2 = 0$, respectively. The loss coefficient for a sudden contraction, $K_L = h_L/(V_2^2/2g)$, is a function of the area ratio, A_2/A_1, as is shown in Fig. 8.26. The value of K_L changes gradually from one extreme of a sharp-edged entrance ($A_2/A_1 = 0$ with $K_L = 0.50$) to the other extreme of no area change ($A_2/A_1 = 1$ with $K_L = 0$).

In many ways, the flow in a sudden expansion is similar to exit flow. As is indicated in Fig. 8.28, the fluid leaves the smaller pipe and initially forms a jet-type structure as it enters the larger pipe. Within a few diameters downstream of the expansion, the jet becomes dispersed across the pipe, and fully developed flow becomes established again. In this process [between sections (2) and (3)] a portion of the kinetic energy of the fluid is dissipated as a result of viscous effects. A square-edged exit is the limiting case with $A_1/A_2 = 0$.

A sudden expansion is one of the few components (perhaps the only one) for which

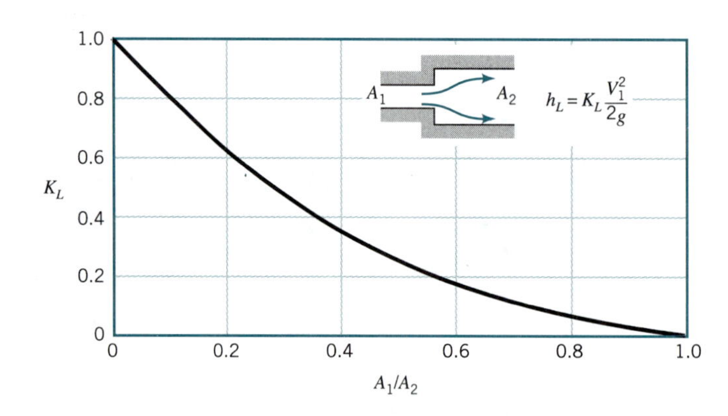

■ FIGURE 8.27 **Loss coefficient for a sudden expansion (Ref. 10).**

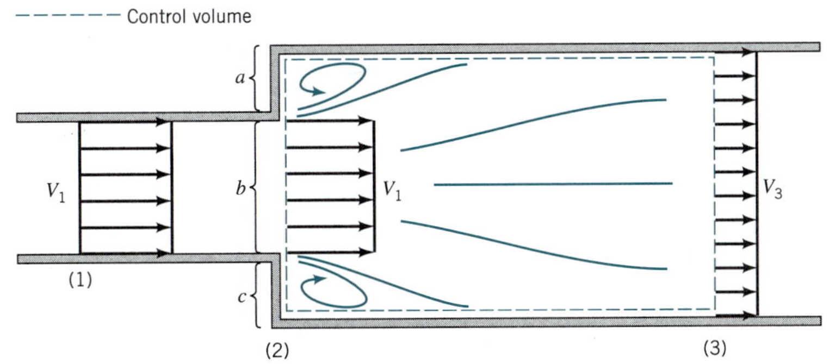

■ FIGURE 8.28 **Control volume used to calculate the loss coefficient for a sudden expansion.**

the loss coefficient can be obtained by means of a simple analysis. To do this we consider the continuity and momentum equations for the control volume shown in Fig. 8.28 and the energy equation applied between (2) and (3). We assume that the flow is uniform at sections (1), (2), and (3) and the pressure is constant across the left-hand side of the control volume ($p_a = p_b = p_c = p_1$). The resulting three governing equations (mass, momentum, and energy) are

$$A_1V_1 = A_3V_3$$

$$p_1A_3 - p_3A_3 = \rho A_3V_3(V_3 - V_1)$$

and

The loss coefficient for a sudden expansion can be theoretically calculated.

$$\frac{p_1}{\gamma} + \frac{V_1^2}{2g} = \frac{p_3}{\gamma} + \frac{V_3^2}{2g} + h_L$$

These can be rearranged to give the loss coefficient, $K_L = h_L/(V_1^2/2g)$, as

$$K_L = \left(1 - \frac{A_1}{A_2}\right)^2$$

where we have used the fact that $A_2 = A_3$. This result, plotted in Fig. 8.27, is in good agreement with experimental data. As with so many minor loss situations, it is not the viscous effects directly (i.e., the wall shear stress) that cause the loss. Rather, it is the dissipation of kinetic energy (another type of viscous effect) as the fluid decelerates inefficiently.

The losses may be quite different if the contraction or expansion is gradual. Typical results for a conical *diffuser* with a given area ratio, A_2/A_1, are shown in Fig. 8.29. (A diffuser is a device shaped to decelerate a fluid.) Clearly the included angle of the diffuser, θ, is a very important parameter. For very small angles, the diffuser is excessively long and most of the head loss is due to the wall shear stress as in fully developed flow. For moderate or large angles, the flow separates from the walls and the losses are due mainly to a dissipation of the kinetic energy of the jet leaving the smaller diameter pipe. In fact, for moderate or large values of θ (i.e., $\theta > 35°$ for the case shown in Fig. 8.29), the conical diffuser is, perhaps unexpectedly, less efficient than a sharp-edged expansion which has $K_L = 1$. There is an optimum angle($\theta \approx 8°$ for the case illustrated) for which the loss coefficient is a minimum. The relatively small value of θ for the minimum K_L results in a long diffuser and is an indication of the fact that it is difficult to efficiently decelerate a fluid.

It must be noted that the conditions indicated in Fig. 8.29 represent typical results only. Flow through a diffuser is very complicated and may be strongly dependent on the area ratio

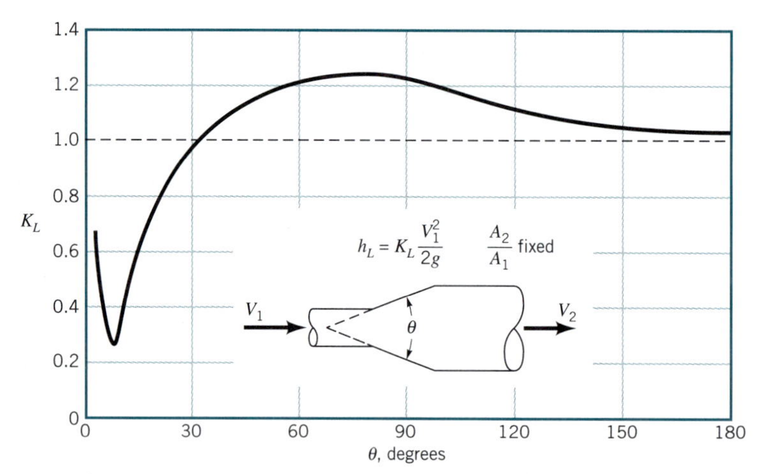

■ **FIGURE 8.29** **Loss coefficient for a typical conical diffuser (Ref. 5).**

A_2/A_1, specific details of the geometry, and the Reynolds number. The data are often presented in terms of a *pressure recovery coefficient*, $C_p = (p_2 - p_1)/(\rho V_1^2/2)$, which is the ratio of the static pressure rise across the diffuser to the inlet dynamic pressure. Considerable effort has gone into understanding this important topic (Refs. 11, 12).

Flow in a conical contraction (a nozzle; reverse the flow direction shown in Fig. 8.29) is less complex than that in a conical expansion. Typical loss coefficients based on the downstream (high-speed) velocity can be quite small, ranging from $K_L = 0.02$ for $\theta = 30°$, to $K_L = 0.07$ for $\theta = 60°$, for example. It is relatively easy to accelerate a fluid efficiently.

Bends in pipes produce a greater head loss than if the pipe were straight. The losses are due to the separated region of flow near the inside of the bend (especially if the bend is sharp) and the swirling secondary flow that occurs because of the imbalance of centripetal forces as a result of the curvature of the pipe centerline. These effects and the associated values of K_L for large Reynolds number flows through a 90° bend are shown in Fig. 8.30. The friction loss due to the axial length of the pipe bend must be calculated and added to that given by the loss coefficient of Fig. 8.30.

For situations in which space is limited, a flow direction change is often accomplished by use of miter bends, as is shown in Fig. 8.31, rather than smooth bends. The considerable losses in such bends can be reduced by the use of carefully designed guide vanes that help direct the flow with less unwanted swirl and disturbances.

V8.5

Extensive tables are available for loss coefficients of standard pipe components.

Another important category of pipe system components is that of commercially available pipe fittings such as elbows, tees, reducers, valves, and filters. The values of K_L for such components depend strongly on the shape of the component and only very weakly on the Reynolds number for typical large Re flows. Thus, the loss coefficient for a 90° elbow depends on whether the pipe joints are threaded or flanged but is, within the accuracy of the data, fairly independent of the pipe diameter, flow rate, or fluid properties (the Reynolds number effect). Typical values of K_L for such components are given in Table 8.2. These typical components are designed more for ease of manufacturing and costs than for reduction of the head losses that they produce. The flowrate from a faucet in a typical house is sufficient whether the value of K_L for an elbow is the typical $K_L = 1.5$, or it is reduced to $K_L = 0.2$ by use of a more expensive long-radius, gradual bend (Fig. 8.30).

Valves control the flowrate by providing a means to adjust the overall system loss coefficient to the desired value. When the valve is closed, the value of K_L is infinite and no

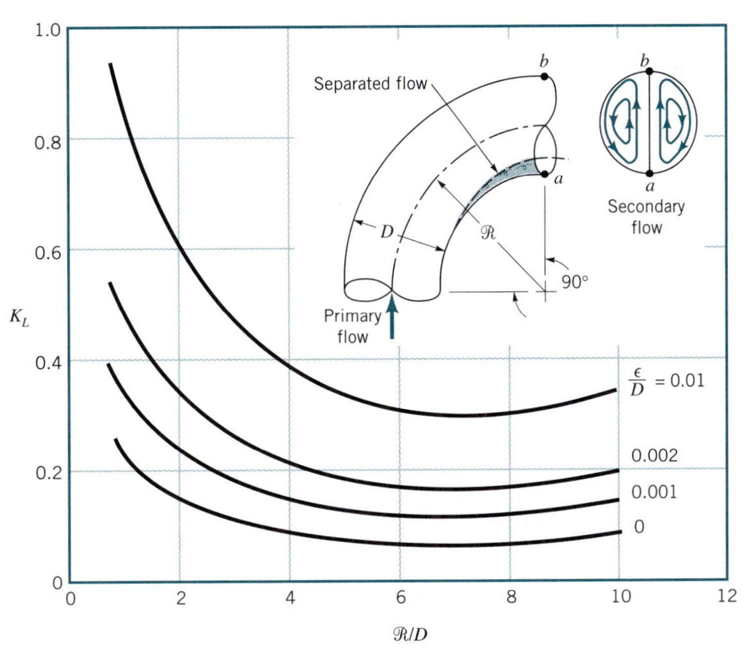

■ **FIGURE 8.30** **Character of the flow in a 90° bend and the associated loss coefficient (Ref. 5).**

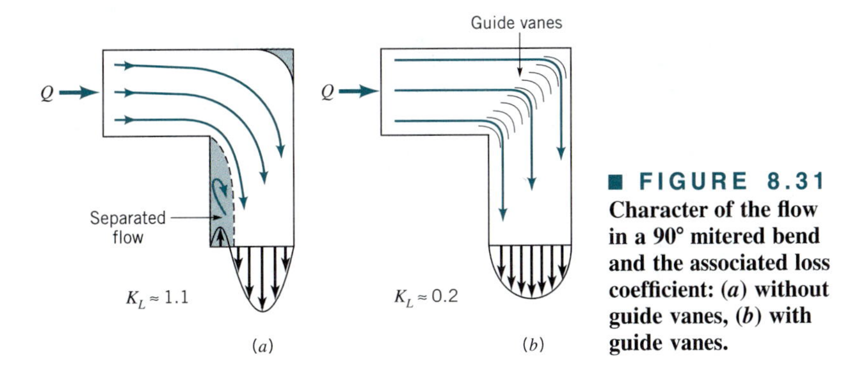

■ **FIGURE 8.31** **Character of the flow in a 90° mitered bend and the associated loss coefficient: (*a*) without guide vanes, (*b*) with guide vanes.**

A valve is a variable resistance element in a pipe circuit.

fluid flows. Opening of the valve reduces K_L, producing the desired flowrate. Typical cross sections of various types of valves are shown in Fig. 8.32. Some valves (such as the conventional globe valve) are designed for general use, providing convenient control between the extremes of fully closed and fully open. Others (such as a needle valve) are designed to provide very fine control of the flowrate. The check valve provides a diode type operation that allows fluid to flow in one direction only.

Loss coefficients for typical valves are given in Table 8.2. As with many system components, the head loss in valves is mainly a result of the dissipation of kinetic energy of a high-speed portion of the flow. This is illustrated in Fig. 8.33.

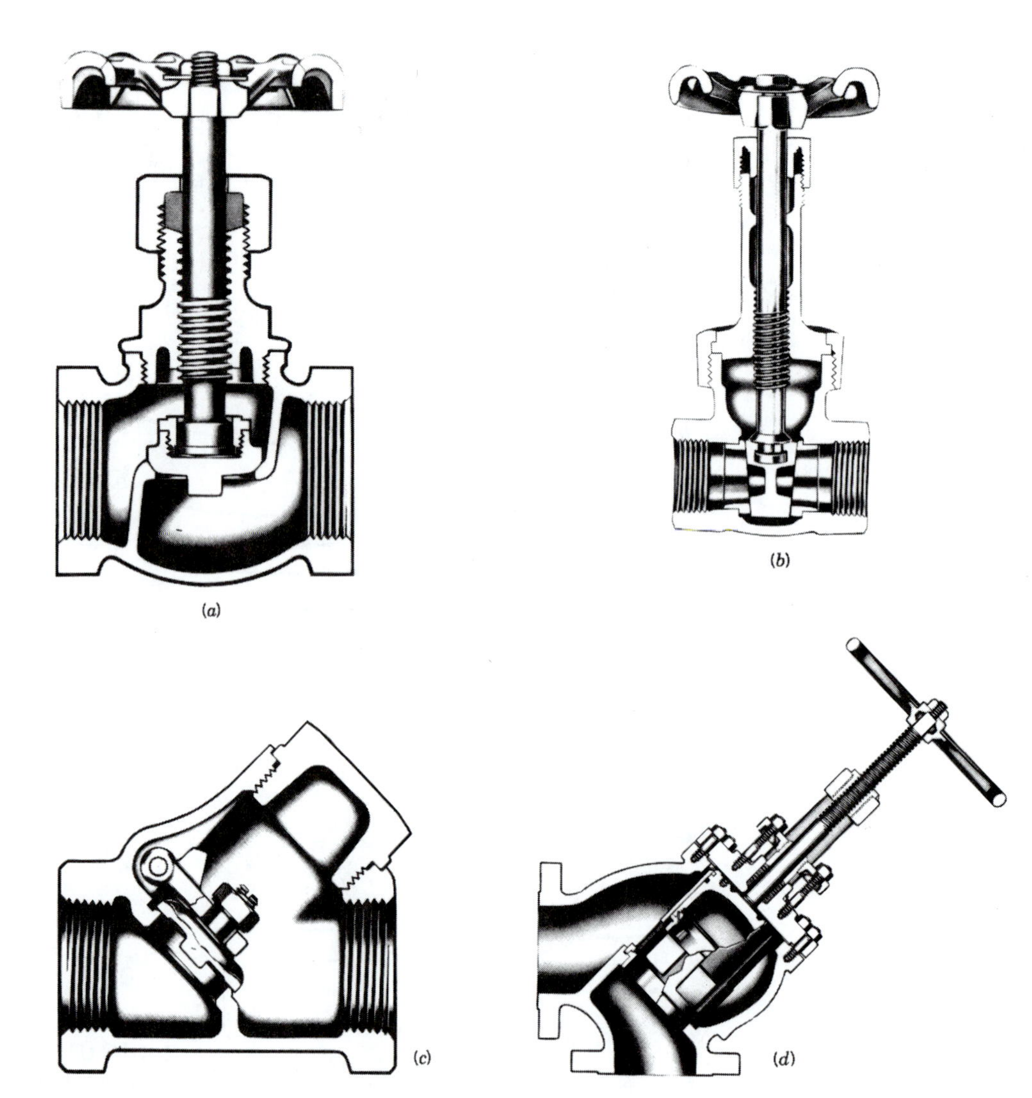

■ **FIGURE 8.32** Internal structure of various valves: (a) globe valve, (b) gate valve, (c) swing check valve, (d) stop check valve. (Courtesy of Crane Co., Valve Division.)

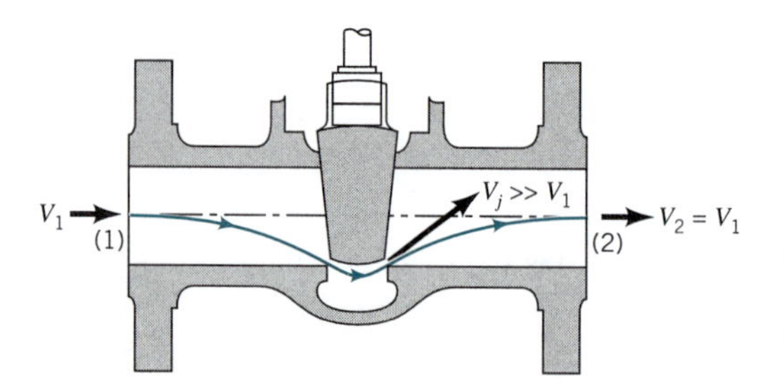

■ **FIGURE 8.33** Head loss in a valve is due to dissipation of the kinetic energy of the large-velocity fluid near the valve seat.

■ **TABLE 8.2**

Loss Coefficients for Pipe Components $\left(h_L = K_L \frac{V^2}{2g}\right)$ **(Data from Refs. 5, 10, 27)**

Component	K_L
a. Elbows	
Regular 90°, flanged	0.3
Regular 90°, threaded	1.5
Long radius 90°, flanged	0.2
Long radius 90°, threaded	0.7
Long radius 45°, flanged	0.2
Regular 45°, threaded	0.4
b. 180° return bends	
180° return bend, flanged	0.2
180° return bend, threaded	1.5
c. Tees	
Line flow, flanged	0.2
Line flow, threaded	0.9
Branch flow, flanged	1.0
Branch flow, threaded	2.0
d. Union, threaded	0.08
***e. Valves**	
Globe, fully open	10
Angle, fully open	2
Gate, fully open	0.15
Gate, $\frac{1}{4}$ closed	0.26
Gate, $\frac{1}{2}$ closed	2.1
Gate, $\frac{3}{4}$ closed	17
Swing check, forward flow	2
Swing check, backward flow	∞
Ball valve, fully open	0.05
Ball valve, $\frac{1}{3}$ closed	5.5
Ball valve, $\frac{2}{3}$ closed	210

*See Fig. 8.36 for typical valve geometry

Air at standard conditions is to flow through the test section [between sections (5) and (6)] of the closed-circuit wind tunnel shown in Fig. E8.6 with a velocity of 200 ft/s. The flow is driven by a fan that essentially increases the static pressure by the amount $p_1 - p_9$ that is needed to overcome the head losses experienced by the fluid as it flows around the circuit. Estimate the value of $p_1 - p_9$ and the horsepower supplied to the fluid by the fan.

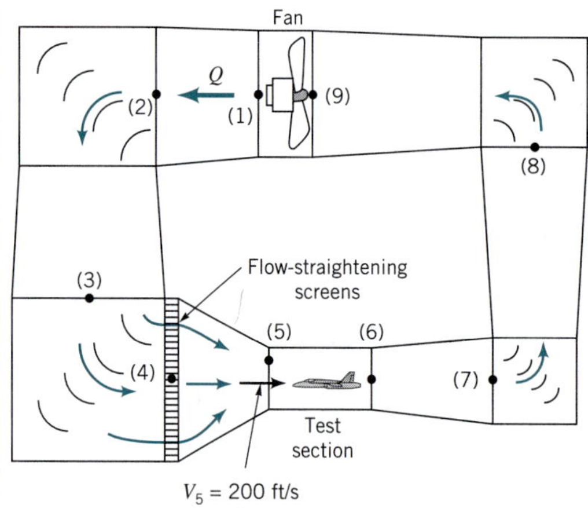

■ FIGURE E8.6

Location	Area (ft^2)	Velocity (ft/s)
1	22.0	36.4
2	28.0	28.6
3	35.0	22.9
4	35.0	22.9
5	4.0	200.0
6	4.0	200.0
7	10.0	80.0
8	18.0	44.4
9	22.0	36.4

SOLUTION

The maximum velocity within the wind tunnel occurs in the test section (smallest area). Thus, the maximum Mach number of the flow is $\text{Ma}_5 = V_5/c_5$, where $V_5 = 200$ ft/s and from Eq. 1.20 the speed of sound is $c_5 = (kRT_5)^{1/2} = \{1.4(1716 \text{ ft}\cdot\text{lb/slug}\cdot{}^\circ\text{R})[(460 + 59)^\circ\text{R}]\}^{1/2} = 1117$ ft/s. Thus, $\text{Ma}_5 = 200/1117 = 0.179$. As was indicated in Chapter 3 and discussed fully in Chapter 11, most flows can be considered as incompressible if the Mach number is less than about 0.3. Hence, we can use the incompressible formulas for this problem.

The purpose of the fan in the wind tunnel is to provide the necessary energy to overcome the net head loss experienced by the air as it flows around the circuit. This can be found from the energy equation between points (1) and (9) as

$$\frac{p_1}{\gamma} + \frac{V_1^2}{2g} + z_1 = \frac{p_9}{\gamma} + \frac{V_9^2}{2g} + z_9 + h_{L_{1-9}}$$

where $h_{L_{1-9}}$ is the total head loss from (1) to (9). With $z_1 = z_9$ and $V_1 = V_9$ this gives

$$\frac{p_1}{\gamma} - \frac{p_9}{\gamma} = h_{L_{1-9}} \tag{1}$$

Similarly, by writing the energy equation (Eq. 5.84) across the fan, from (9) to (1), we obtain

$$\frac{p_9}{\gamma} + \frac{V_9^2}{2g} + z_9 + h_p = \frac{p_1}{\gamma} + \frac{V_1^2}{2g} + z_9$$

where h_p is the actual head rise supplied by the pump (fan) to the air. Again since $z_9 = z_1$ and $V_9 = V_1$ this, when combined with Eq. 1, becomes

$$h_p = \frac{(p_1 - p_9)}{\gamma} = h_{L_{1-9}}$$

The actual power supplied to the air (horsepower, $\mathcal{P}_a$) is obtained from the fan head by

$$\mathcal{P}_a = \gamma Q h_p = \gamma A_5 V_5 h_p = \gamma A_5 V_5 h_{L_{1\text{-}9}} \tag{2}$$

Thus, the power that the fan must supply to the air depends on the head loss associated with the flow through the wind tunnel. To obtain a reasonable, approximate answer we make the following assumptions. We treat each of the four turning corners as a mitered bend with guide vanes so that from Fig. 8.31 $K_{L_{\text{corner}}} = 0.2$. Thus, for each corner

$$h_{L_{\text{corner}}} = K_L \frac{V^2}{2g} = 0.2 \frac{V^2}{2g}$$

where, because the flow is assumed incompressible, $V = V_5 A_5/A$. The values of A and the corresponding velocities throughout the tunnel are given in Table E8.6.

We also treat the enlarging sections from the end of the test section (6) to the beginning of the nozzle (4) as a conical diffuser with a loss coefficient of $K_{L_{dif}} = 0.6$. This value is larger than that of a well-designed diffuser (see Fig. 8.29, for example). Since the wind tunnel diffuser is interrupted by the four turning corners and the fan, it may not be possible to obtain a smaller value of $K_{L_{dif}}$ for this situation. Thus,

$$h_{L_{dif}} = K_{L_{dif}} \frac{V_6^2}{2g} = 0.6 \frac{V_6^2}{2g}$$

The loss coefficients for the conical nozzle between section (4) and (5) and the flow-straightening screens are assumed to be $K_{L_{noz}} = 0.2$ and $K_{L_{scr}} = 4.0$ (Ref. 13), respectively. We neglect the head loss in the relatively short test section.

Thus, the total head loss is

$$h_{L_{1\text{-}9}} = h_{L_{corner7}} + h_{L_{corner8}} + h_{L_{corner2}} + h_{L_{corner3}} + h_{L_{dif}} + h_{L_{noz}} + h_{L_{scr}}$$

or

$$h_{L_{1\text{-}9}} = [0.2(V_7^2 + V_8^2 + V_2^2 + V_3^2) + 0.6V_6^2 + 0.2V_5^2 + 4.0V_4^2]/2g$$

$$= [0.2(80.0^2 + 44.4^2 + 28.6^2 + 22.9^2) + 0.6(200)^2 + 0.2(200)^2 + 4.0(22.9)^2] \text{ ft}^2/\text{s}^2/[2(32.2 \text{ ft/s}^2)]$$

or

$$h_{L_{1\text{-}9}} = 560 \text{ ft}$$

Hence, from Eq. 1 we obtain the pressure rise across the fan as

$$p_1 - p_9 = \gamma h_{L_{1\text{-}9}} = (0.0765 \text{ lb/ft}^3)(560 \text{ ft})$$
$$= 42.8 \text{ lb/ft}^2 = 0.298 \text{ psi} \qquad \textbf{(Ans)}$$

From Eq. 2 we obtain the power added to the fluid as

$$\mathcal{P}_a = (0.0765 \text{ lb/ft}^3)(4.0 \text{ ft}^2)(200 \text{ ft/s})(560 \text{ ft}) = 34{,}300 \text{ ft}\cdot\text{lb/s}$$

or

$$\mathcal{P}_a = \frac{34{,}300 \text{ ft}\cdot\text{lb/s}}{550 \text{ (ft}\cdot\text{lb/s)/hp}} = 62.3 \text{ hp} \qquad \textbf{(Ans)}$$

With a closed-return wind tunnel of this type, all of the power required to maintain the flow is dissipated through viscous effects, with the energy remaining within the closed tunnel. If heat transfer across the tunnel walls is negligible, the air temperature within the tunnel will increase in time. For steady state operations of such tunnels, it is often necessary to provide some means of cooling to maintain the temperature at acceptable levels.

It should be noted that the actual size of the motor that powers the fan must be greater than the calculated 62.3 hp because the fan is not 100% efficient. The power calculated above is that needed by the fluid to overcome losses in the tunnel, excluding those in the fan. If the fan were 60% efficient, it would require a shaft power of $\mathcal{P} = 62.3 \text{ hp}/(0.60) = 104$ hp to run the fan. Determination of fan (or pump) efficiencies can be a complex problem that depends on the specific geometry of the fan. Introductory material about fan performance is presented in Chapter 12; additional material can be found in various references (Refs. 14, 15, 16, for example).

It should also be noted that the above results are only approximate. Clever, careful design of the various components (corners, diffuser, etc.) may lead to improved (i.e., lower) values of the various loss coefficients, and hence lower power requirements. Since h_L is proportional to V^2, the components with the larger V tend to have the larger head loss. Thus, even though $K_L = 0.2$ for each of the four corners, the head loss for corner (7) is $(V_7/V_3)^2 = (80/22.9)^2 = 12.2$ times greater than it is for corner (3).

8.4.3 Noncircular Conduits

Many of the conduits that are used for conveying fluids are not circular in cross section. Although the details of the flows in such conduits depend on the exact cross-sectional shape, many round pipe results can be carried over, with slight modification, to flow in conduits of other shapes.

Theoretical results can be obtained for fully developed laminar flow in noncircular ducts, although the detailed mathematics often becomes rather cumbersome. For an arbitrary cross section, as is shown in Fig. 8.34, the velocity profile is a function of both y and z [$\mathbf{V} = u(y, z)\hat{\mathbf{i}}$]. This means that the governing equation from which the velocity profile is obtained (either the Navier-Stokes equations of motion or a force balance equation similar to that used for circular pipes, Eq. 8.6) is a partial differential equation rather than an ordinary differential equation. Although the equation is linear (for fully developed flow the convective acceleration is zero), its solution is not as straightforward as for round pipes. Typically the velocity profile is given in terms of an infinite series representation (Ref. 17).

The hydraulic diameter is used for noncircular duct calculations.

Practical, easy-to-use results can be obtained as follows. Regardless of the cross-sectional shape, there are no inertia effects in fully developed laminar pipe flow. Thus, the friction factor can be written as $f = C/\text{Re}_h$, where the constant C depends on the particular shape of the duct, and Re_h is the Reynolds number, $\text{Re}_h = \rho V D_h/\mu$, based on the hydraulic diameter. The *hydraulic diameter* defined as $D_h = 4A/P$ is four times the ratio of the cross-sectional flow area divided by the wetted perimeter, P, of the pipe as is illustrated in Fig. 8.34. It represents a characteristic length that defines the size of a cross section of a specified shape. The factor of 4 is included in the definition of D_h so that for round pipes the diameter and hydraulic diameter are equal [$D_h = 4A/P = 4(\pi D^2/4)/(\pi D) = D$]. The hydraulic diameter is also used in the definition of the friction factor, $h_L = f(\ell/D_h)V^2/2g$, and the relative roughness, ε/D_h.

The values of $C = f\,\text{Re}_h$ for laminar flow have been obtained from theory and/or experiment for various shapes. Typical values are given in Table 8.3 along with the hydraulic diameter. Note that the value of C is relatively insensitive to the shape of the conduit. Unless the cross section is very "thin" in some sense, the value of C is not too different from its circular pipe value, $C = 64$. Once the friction factor is obtained, the calculations for noncircular conduits are identical to those for round pipes.

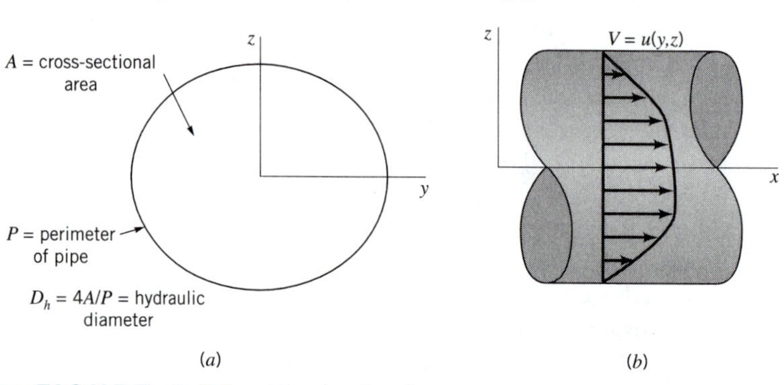

■ FIGURE 8.34 Noncircular duct.

TABLE 8.3
Friction Factors for Laminar Flow in Noncircular Ducts (Data from Ref. 18)

Shape	Parameter	$C = f\mathrm{Re}_h$
I. Concentric Annulus $D_h = D_2 - D_1$	D_1/D_2	
	0.0001	71.8
	0.01	80.1
	0.1	89.4
	0.6	95.6
	1.00	96.0
II. Rectangle $D_h = \dfrac{2ab}{a+b}$	a/b	
	0	96.0
	0.05	89.9
	0.10	84.7
	0.25	72.9
	0.50	62.2
	0.75	57.9
	1.00	56.9

The Moody chart, developed for round pipes, can also be used for noncircular ducts.

Calculations for fully developed turbulent flow in ducts of noncircular cross section are usually carried out by using the Moody chart data for round pipes with the diameter replaced by the hydraulic diameter and the Reynolds number based on the hydraulic diameter. Such calculations are usually accurate to within about 15%. If greater accuracy is needed, a more detailed analysis based on the specific geometry of interest is needed.

Air at a temperature of 120 °F and standard pressure flows from a furnace through an 8-in.-diameter pipe with an average velocity of 10 ft/s. It then passes through a transition section and into a square duct whose side is of length a. The pipe and duct surfaces are smooth ($\varepsilon = 0$). Determine the duct size, a, if the head loss per foot is to be the same for the pipe and the duct.

SOLUTION

We first determine the head loss per foot for the pipe, $h_L/\ell = (f/D)\, V^2/2g$, and then size the square duct to give the same value. For the given pressure and temperature we obtain (from Table B.3) $\nu = 1.89 \times 10^{-4}\ \mathrm{ft^2/s}$ so that

$$\mathrm{Re} = \frac{VD}{\nu} = \frac{(10\ \mathrm{ft/s})(\frac{8}{12}\ \mathrm{ft})}{1.89 \times 10^{-4}\ \mathrm{ft^2/s}} = 35{,}300$$

With this Reynolds number and with $\varepsilon/D = 0$ we obtain the friction factor from Fig. 8.20 as $f = 0.022$ so that

$$\frac{h_L}{\ell} = \frac{0.022}{(\frac{8}{12}\ \mathrm{ft})} \frac{(10\ \mathrm{ft/s})^2}{2(32.2\ \mathrm{ft/s^2})} = 0.0512$$

Thus, for the square duct we must have

$$\frac{h_L}{\ell} = \frac{f}{D_h}\frac{V_s^2}{2g} = 0.0512 \tag{1}$$

where

$$D_h = 4A/P = 4a^2/4a = a \quad \text{and}$$

$$V_s = \frac{Q}{A} = \frac{\frac{\pi}{4}\left(\frac{8}{12}\text{ ft}\right)^2 (10 \text{ ft/s})}{a^2} = \frac{3.49}{a^2} \tag{2}$$

is the velocity in the duct.

By combining Eqs. 1 and 2 we obtain

$$0.0512 = \frac{f}{a}\frac{(3.49/a^2)^2}{2(32.2)}$$

or

$$a = 1.30\, f^{1/5} \tag{3}$$

where a is in feet. Similarly, the Reynolds number based on the hydraulic diameter is

$$\text{Re}_h = \frac{V_s D_h}{\nu} = \frac{(3.49/a^2)a}{1.89 \times 10^{-4}} = \frac{1.85 \times 10^4}{a} \tag{4}$$

We have three unknowns (a, f, and Re_h) and three equations (Eqs. 3, 4, and the third equation in graphical form, Fig. 8.20, the Moody chart). Thus, a trial and error solution is required.

As an initial attempt, assume the friction factor for the duct is the same as for the pipe. That is, assume $f = 0.022$. From Eq. 3 we obtain $a = 0.606$ ft, while from Eq. 4 we have $\text{Re}_h = 3.05 \times 10^4$. From Fig. 8.20, with this Reynolds number and the given smooth duct we obtain $f = 0.023$, which does not quite agree with the assumed value of f. Hence, we do not have the solution. We try again, using the latest calculated value of $f = 0.023$ as our guess. The calculations are repeated until the guessed value of f agrees with the value obtained from Fig. 8.20. The final result (after only two iterations) is $f = 0.023$, $\text{Re}_h = 3.03 \times 10^4$, and

$$a = 0.611 \text{ ft} = 7.34 \text{ in.} \tag{Ans}$$

Note that the length of the side of the equivalent square duct is $a/D = 7.34/8 = 0.918$, or approximately 92% of the diameter of the equivalent duct. It can be shown that this value, 92%, is a very good approximation for any pipe flow—laminar or turbulent. The cross-sectional area of the duct ($A = a^2 = 53.9 \text{ in.}^2$) is greater than that of the round pipe ($A = \pi D^2/4 = 50.3 \text{ in.}^2$). Also, it takes less material to form the round pipe (perimeter $= \pi D = 25.1$ in.) than the square duct (perimeter $= 4a = 29.4$ in.). Circles are very efficient shapes.

8.5 Pipe Flow Examples

Pipe systems may contain a single pipe with components or multiple interconnected pipes.

In the previous sections of this chapter, we discussed concepts concerning flow in pipes and ducts. The purpose of this section is to apply these ideas to the solutions of various practical problems. The application of the pertinent equations is straightforward, with rather simple calculations that give answers to problems of engineering importance. The main idea involved is to apply the energy equation between appropriate locations within the flow system, with the head loss written in terms of the friction factor and the minor loss coefficients. We will consider two classes of pipe systems: those containing a single pipe (whose length may be interrupted by various components), and those containing multiple pipes in parallel, series, or network configurations.

8.5.1 Single Pipes

The nature of the solution process for pipe flow problems can depend strongly on which of the various parameters are independent parameters (the "given") and which is the dependent parameter (the "determine"). The three most common types of problems are shown in Table 8.4 in terms of the parameters involved. We assume the pipe system is defined in terms of the length of pipe sections used and the number of elbows, bends, and valves needed to convey the fluid between the desired locations. In all instances we assume the fluid properties are given.

Pipe flow problems can be categorized by what parameters are given and what is to be calculated.

In a Type I problem we specify the desired flowrate or average velocity and determine the necessary pressure difference or head loss. For example, if a flowrate of 2.0 gal/min is required for a dishwasher that is connected to the water heater by a given pipe system, what pressure is needed in the water heater?

In a Type II problem we specify the applied driving pressure (or, alternatively, the head loss) and determine the flowrate. For example, how many gal/min of hot water are supplied to the dishwasher if the pressure within the water heater is 60 psi and the pipe system details (length, diameter, roughness of the pipe; number of elbows; etc.) are specified?

In a Type III problem we specify the pressure drop and the flowrate and determine the diameter of the pipe needed. For example, what diameter of pipe is needed between the water heater and dishwasher if the pressure in the water heater is 60 psi (determined by the city water system) and the flowrate is to be not less than 2.0 gal/min (determined by the manufacturer)?

Several examples of these types of problems follow.

TABLE 8.4
Pipe Flow Types

Variable	Type I	Type II	Type III
a. Fluid			
Density	Given	Given	Given
Viscosity	Given	Given	Given
b. Pipe			
Diameter	Given	Given	Determine
Length	Given	Given	Given
Roughness	Given	Given	Given
c. Flow			
Flowrate or Average Velocity	Given	Determine	Given
d. Pressure			
Pressure Drop or Head Loss	Determine	Given	Given

EXAMPLE 8.8 (TYPE I, DETERMINE PRESSURE DROP)

Water at 60 °F flows from the basement to the second floor through the 0.75-in. (0.0625-ft)-diameter copper pipe (a drawn tubing) at a rate of $Q = 12.0 \text{ gal/min} = 0.0267 \text{ ft}^3\text{/s}$ and exits through a faucet of diameter 0.50 in. as shown in Fig. E8.8*a*. Determine the pressure at point (1) if: (a) all losses are neglected, (b) the only losses included are major losses, or (c) all losses are included.

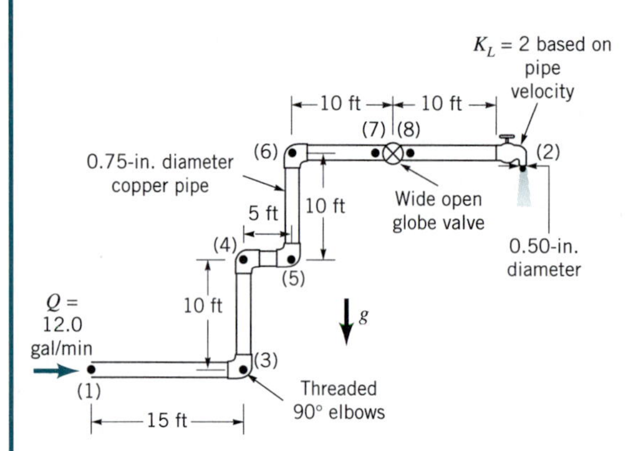

■ FIGURE E8.8a

SOLUTION

Since the fluid velocity in the pipe is given by $V_1 = Q/A_1 = Q/(\pi D^2/4) = (0.0267 \text{ ft}^3\text{/s})/[\pi(0.0625 \text{ ft})^2/4] = 8.70 \text{ ft/s}$, and the fluid properties are $\rho = 1.94 \text{ slugs/ft}^3$ and $\mu = 2.34 \times 10^{-5} \text{ lb·s/ft}^2$ (see Table B.1), it follows that $\text{Re} = \rho VD/\mu = (1.94 \text{ slugs/ft}^3)(8.70 \text{ ft/s})(0.0625 \text{ ft})/(2.34 \times 10^{-5} \text{ lb·s/ft}^2) = 45{,}000$. Thus, the flow is turbulent. The governing equation for either case (a), (b), or (c) is Eq. 8.21,

$$\frac{p_1}{\gamma} + \alpha_1 \frac{V_1^2}{2g} + z_1 = \frac{p_2}{\gamma} + \alpha_2 \frac{V_2^2}{2g} + z_2 + h_L$$

where $z_1 = 0$, $z_2 = 20$ ft, $p_2 = 0$ (free jet), $\gamma = \rho g = 62.4 \text{ lb/ft}^3$, and the outlet velocity is $V_2 = Q/A_2 = (0.0267 \text{ ft}^3\text{/s})/[\pi(0.50/12)^2\text{ft}^2/4] = 19.6 \text{ ft/s}$. We assume that the kinetic energy coefficients α_1 and α_2 are unity. This is reasonable because turbulent velocity profiles are nearly uniform across the pipe. Thus,

$$p_1 = \gamma z_2 + \tfrac{1}{2}\rho(V_2^2 - V_1^2) + \gamma h_L \quad \textbf{(1)}$$

where the head loss is different for each of the three cases.

(a) If all losses are neglected ($h_L = 0$), Eq. 1 gives

$$p_1 = (62.4 \text{ lb/ft}^3)(20 \text{ ft}) + \frac{1.94 \text{ slugs/ft}^3}{2}\left[\left(19.6 \frac{\text{ft}}{\text{s}}\right)^2 - \left(8.70 \frac{\text{ft}}{\text{s}}\right)^2\right]$$
$$= (1248 + 299) \text{ lb/ft}^2 = 1547 \text{ lb/ft}^2$$

or

$$p_1 = 10.7 \text{ psi} \quad \textbf{(Ans)}$$

Note that for this pressure drop, the amount due to elevation change (the hydrostatic effect) is $\gamma(z_2 - z_1) = 8.67$ psi and the amount due to the increase in kinetic energy is $\rho(V_2^2 - V_1^2)/2 = 2.07$ psi.

(b) If the only losses included are the major losses, the head loss is

$$h_L = f \frac{\ell}{D} \frac{V_1^2}{2g}$$

From Table 8.1 the roughness for a 0.75-in.-diameter copper pipe (drawn tubing) is $\varepsilon = 0.00005$ ft so that $\varepsilon/D = 8 \times 10^{-5}$. With this ε/D and the calculated Reynolds number (Re $= 45{,}000$), the value of f is obtained from the Moody chart as $f = 0.0215$. Note that the Colebrook equation (Eq. 8.35) would give the same value of f. Hence, with the total length of the pipe as $\ell = (15 + 5 + 10 + 10 + 20)$ ft $= 60$ ft and the elevation and kinetic energy portions the same as for part (a), Eq. 1 gives

$$\begin{aligned} p_1 &= \gamma z_2 + \frac{1}{2} \rho (V_2^2 - V_1^2) + \rho f \frac{\ell}{D} \frac{V_1^2}{2} \\ &= (1248 + 299) \text{ lb/ft}^2 \\ &\quad + (1.94 \text{ slugs/ft}^3)(0.0215) \left(\frac{60 \text{ ft}}{0.0625 \text{ ft}} \right) \frac{(8.70 \text{ ft/s})^2}{2} \\ &= (1248 + 299 + 1515) \text{ lb/ft}^2 = 3062 \text{ lb/ft}^2 \end{aligned}$$

or

$$p_1 = 21.3 \text{ psi} \qquad \textbf{(Ans)}$$

Of this pressure drop, the amount due to pipe friction is approximately $(21.3 - 10.7)$ psi $= 10.6$ psi.

(c) If major and minor losses are included, Eq. 1 becomes

$$p_1 = \gamma z_2 + \frac{1}{2} \rho (V_2^2 - V_1^2) + f \gamma \frac{\ell}{D} \frac{V_1^2}{2g} + \sum \rho K_L \frac{V^2}{2}$$

or

$$p_1 = 21.3 \text{ psi} + \sum \rho K_L \frac{V^2}{2} \qquad \textbf{(2)}$$

where the 21.3 psi contribution is due to elevation change, kinetic energy change, and major losses [part (b)], and the last term represents the sum of all of the minor losses. The loss coefficients of the components ($K_L = 1.5$ for each elbow and $K_L = 10$ for the wide-open globe valve) are given in Table 8.2 (except for the loss coefficient of the faucet, which is given in Fig. E8.8*a* as $K_L = 2$). Thus,

$$\begin{aligned} \sum \rho K_L \frac{V^2}{2} &= (1.94 \text{ slugs/ft}^3) \frac{(8.70 \text{ ft})^2}{2} [10 + 4(1.5) + 2] \\ &= 1321 \text{ lb/ft}^2 \end{aligned}$$

or

$$\sum \rho K_L \frac{V^2}{2} = 9.17 \text{ psi} \qquad \textbf{(3)}$$

Note that we did not include an entrance or exit loss because points (1) and (2) are located within the fluid streams, not within an attaching reservoir where the kinetic energy is zero. Thus, by combining Eqs. 2 and 3 we obtain the entire pressure drop as

$$p_1 = (21.3 + 9.17) \text{ psi} = 30.5 \text{ psi} \qquad \textbf{(Ans)}$$

This pressure drop calculated by including all losses should be the most realistic answer of the three cases considered.

More detailed calculations will show that the pressure distribution along the pipe is as illustrated in Fig. E8.8*b* for cases (a) and (c)—neglecting all losses or including all losses. Note that not all of the pressure drop, $p_1 - p_2$, is a "pressure loss." The pressure change due to the elevation and velocity changes are completely reversible. The portion due to the major and minor losses are irreversible.

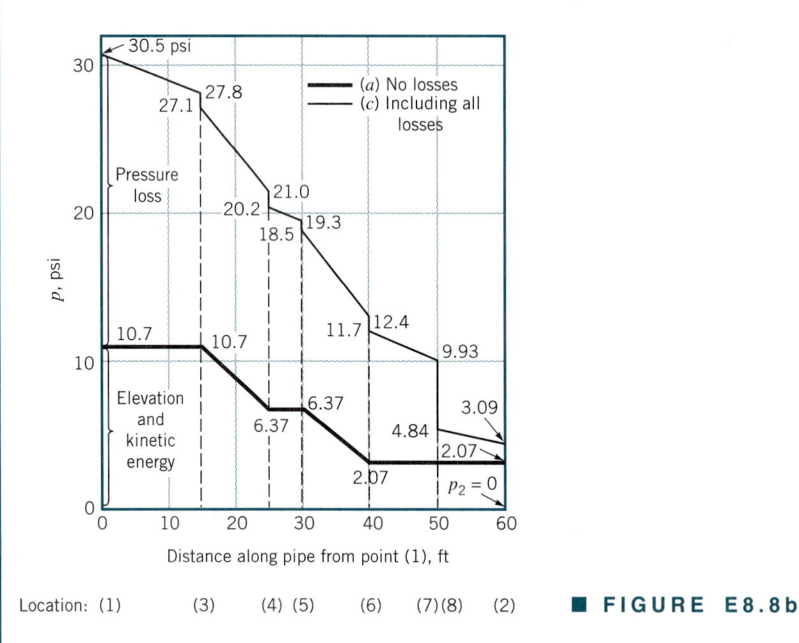

■ **FIGURE E8.8b**

This flow can be illustrated in terms of the energy line and hydraulic grade line concepts introduced in Section 3.7. As is shown in Fig. E8.8*c*, for case (a) there are no losses and the energy line (EL) is horizontal, one velocity head ($V^2/2g$) above the hydraulic grade line (HGL), which is one pressure head (γz) above the pipe itself. For case (c) the energy line is not horizontal. Each bit of friction in the pipe or loss in a component reduces the available energy, thereby lowering the energy line. Thus, for

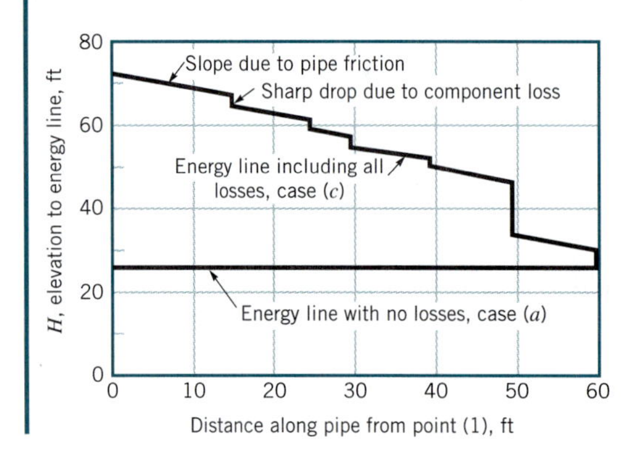

■ **FIGURE E8.8c**

case (a) the total head remains constant throughout the flow with a value of

$$H = \frac{p_1}{\gamma} + \frac{V_1^2}{2g} + z_1 = \frac{(1547 \text{ lb/ft}^2)}{(62.4 \text{ lb/ft}^3)} + \frac{(8.70 \text{ ft/s})^2}{2(32.2 \text{ ft/s}^2)} + 0$$

$$= 26.0 \text{ ft}$$

$$= \frac{p_2}{\gamma} + \frac{V_2^2}{2g} + z_2 = \frac{p_3}{\gamma} + \frac{V_3^3}{2g} + z_3 = \dots$$

For case (c) the energy line starts at

$$H_1 = \frac{p_1}{\gamma} + \frac{V_1^2}{2g} + z_1 = \frac{(30.5 \times 144)\text{lb/ft}^2}{(62.4 \text{ lb/ft}^3)} + \frac{(8.70 \text{ ft/s})^2}{2(32.2 \text{ ft/s}^2)} + 0 = 71.6 \text{ ft}$$

and falls to a final value of

$$H_2 = \frac{p_2}{\gamma} + \frac{V_2^2}{2g} + z_2 = 0 + \frac{(19.6 \text{ ft/s})^2}{2(32.2 \text{ ft/s}^2)} + 20 \text{ ft} = 26.0 \text{ ft}$$

The elevation of the energy line can be calculated at any point along the pipe. For example, at point (7), 50 ft from point (1),

$$H_7 = \frac{p_7}{\gamma} + \frac{V_7^2}{2g} + z_7 = \frac{(9.93 \times 144) \text{ lb/ft}^2}{(62.4 \text{ lb/ft}^3)} + \frac{(8.70 \text{ ft/s})^2}{2(32.2 \text{ ft/s}^2)} + 20 \text{ ft} = 44.1 \text{ ft}$$

The head loss per foot of pipe is the same all along the pipe. That is,

$$\frac{h_L}{\ell} = f\frac{V^2}{2gD} = \frac{0.0215(8.70 \text{ ft/s})^2}{2(32.2 \text{ ft/s}^2)(0.0625 \text{ ft})} = 0.404 \text{ ft/ft}$$

Thus, the energy line is a set of straight line segments of the same slope separated by steps whose height equals the head loss of the minor component at that location. As is seen from Fig. E8.8*c*, the globe valve produces the largest of all the minor losses.

Although the governing pipe flow equations are quite simple, they can provide very reasonable results for a variety of applications, as is shown in the next example.

EXAMPLE 8.9 (TYPE I, DETERMINE HEAD LOSS)

Crude oil at 140 °F with $\gamma = 53.7 \text{ lb/ft}^3$ and $\mu = 8 \times 10^{-5} \text{ lb·s/ft}^2$ (about four times the viscosity of water) is pumped across Alaska through the Alaskan pipeline, a 799-mile-long, 4-ft-diameter steel pipe, at a maximum rate of $Q = 2.4$ million barrels/day $= 117 \text{ ft}^3/\text{s}$, or $V = Q/A = 9.31 \text{ ft/s}$. Determine the horsepower needed for the pumps that drive this large system.

SOLUTION

From the energy equation (Eq. 8.21) we obtain

$$\frac{p_1}{\gamma} + \frac{V_1^2}{2g} + z_1 + h_p = \frac{p_2}{\gamma} + \frac{V_2^2}{2g} + z_2 + h_L$$

where points (1) and (2) represent locations within the large holding tanks at either end of the line and h_p is the head provided to the oil by the pumps. We assume that $z_1 = z_2$ (pumped from sea level to sea level), $p_1 = p_2 = V_1 = V_2 = 0$ (large, open tanks) and $h_L = (f\ell/D)V^2/2g$). Minor losses are negligible because of the large length-to-diameter ratio of the relatively straight, uninterrupted pipe; $\ell/D = (799 \text{ mi})(5280 \text{ ft/mi})/(4 \text{ ft}) = 1.05 \times 10^6$. Thus,

$$h_p = h_L = f\frac{\ell}{D}\frac{V^2}{2g}$$

where from Fig. 8.20, $f = 0.0125$ since $\varepsilon/D = (0.00015 \text{ ft})/(4 \text{ ft}) = 0.0000375$ (see Table 8.1) and $\text{Re} = \rho VD/\mu = [(53.7/32.2) \text{ slugs/ft}^3](9.31 \text{ ft/s})(4.0 \text{ ft})/(8 \times 10^{-5} \text{ lb·s/ft}^2) = 7.76 \times 10^5$. Thus,

$$h_p = 0.0125(1.05 \times 10^6)\frac{(9.31 \text{ ft/s})^2}{2(32.2 \text{ ft/s}^2)} = 17{,}700 \text{ ft}$$

and the actual power supplied to the fluid, $\mathcal{P}_a$, is

$$\mathcal{P}_a = \gamma Q h_p = (53.7 \text{ lb/ft}^3)(117 \text{ ft}^3\text{/s})(17{,}700 \text{ ft})$$

$$= 1.11 \times 10^8 \text{ ft·lb/s}\left(\frac{1 \text{ hp}}{550 \text{ ft·lb/s}}\right)$$

$$= 202{,}000 \text{ hp}$$ **(Ans)**

There are many reasons why it is not practical to drive this flow with a single pump of this size. First, there are no pumps this large! Second, the pressure at the pump outlet would need to be $p = \gamma h_L = (53.7 \text{ lb/ft}^3)(17{,}700 \text{ ft})(1 \text{ ft}^2/144 \text{ in.}^2) = 6600$ psi. No practical 4-ft-diameter pipe would withstand this pressure. An equally unfeasible alternative would be to place the holding tank at the beginning of the pipe on top of a hill of height $h_L = 17{,}700$ ft and let gravity force the oil through the 799-mi pipe!

To produce the desired flow, the actual system contains 12 pumping stations positioned at strategic locations along the pipeline. Each station contains four pumps, three of which operate at any one time (the fourth is in reserve in case of emergency). Each pump is driven by a 13,500-hp motor, thereby producing a total horsepower of $\mathcal{P}$ = 12 stations (3 pumps/station)(13,500 hp/pump) = 486,000 hp. If we assume that the pump/motor combination is approximately 60% efficient, there is a total of 0.60 (486,000) hp = 292,000 hp available to drive the fluid. This number compares favorably with the 202,000-hp answer calculated above.

The assumption of a 140 °F oil temperature may not seem reasonable for flow across Alaska. Note, however, that the oil is warm when it is pumped from the ground and that the 202,000 hp needed to pump the oil is dissipated as a head loss (and therefore a temperature rise) along the pipe. However, if the oil temperature were 70 °F rather than 140 °F, the viscosity would be approximately 16×10^{-5} lb·s/ft^2 (twice as large), but the friction factor would only increase from $f = 0.0125$ at 140 °F ($\text{Re} = 7.76 \times 10^5$) to $f = 0.0140$ at 70 °F ($\text{Re} = 3.88 \times 10^5$). This doubling of viscosity would result in only an 11% increase in power (from 202,000 to 226,000 hp). Because of the large Reynolds numbers involved, the shear stress is due mostly to the turbulent nature of the flow. That is, the value of Re for this flow is large enough (on the relatively flat part of the Moody chart) so that f is nearly independent of Re (or viscosity).

Pipe flow problems in which it is desired to determine the flowrate for a given set of conditions (Type II problems) often require trial-and-error solution techniques. This is because

Some pipe flow problems require a trial-and-error solution technique.

it is necessary to know the value of the friction factor to carry out the calculations, but the friction factor is a function of the unknown velocity (flowrate) in terms of the Reynolds number. The solution procedure is indicated in Example 8.10.

EXAMPLE 8.10 (TYPE II, DETERMINE FLOWRATE)

According to an appliance manufacturer, the 4-in.-diameter galvanized iron vent on a clothes dryer is not to contain more than 20 ft of pipe and four 90° elbows. Under these conditions determine the air flowrate if the pressure within the dryer is 0.20 inches of water. Assume a temperature of 100 °F and standard pressure.

SOLUTION

Application of the energy equation (Eq. 8.21) between the inside of the dryer, point (1), and the exit of the vent pipe, point (2), gives

$$\frac{p_1}{\gamma} + \frac{V_1^2}{2g} + z_1 = \frac{p_2}{\gamma} + \frac{V_2^2}{2g} + z_2 + f\frac{\ell}{D}\frac{V^2}{2g} + \sum K_L \frac{V^2}{2g} \qquad \textbf{(1)}$$

where K_L for the entrance is assumed to be 0.5 and that for each elbow is assumed to be 1.5. In addition we assume that $V_1 = 0$ and $z_1 = z_2$. (The change in elevation is often negligible for gas flows.) Also, $p_2 = 0$, and $p_1/\gamma_{H_2O} = 0.2$ in., or

$$p_1 = (0.2 \text{ in.})\left(\frac{1 \text{ ft}}{12 \text{ in.}}\right)(62.4 \text{ lb/ft}^3) = 1.04 \text{ lb/ft}^2$$

Thus, with $\gamma = 0.0709 \text{ lb/ft}^3$ (see Table B.3) and $V_2 = V$ (the air velocity in the pipe), Eq. 1 becomes

$$\frac{(1.04 \text{ lb/ft}^2)}{(0.0709 \text{ lb/ft}^3)} = \left[1 + f\frac{(20 \text{ ft})}{(\frac{4}{12} \text{ ft})} + 0.5 + 4(1.5)\right]\frac{V^2}{2(32.2 \text{ ft/s}^2)}$$

or

$$945 = (7.5 + 60f)V^2 \qquad \textbf{(2)}$$

where V is in ft/s.

The value of f is dependent on Re, which is dependent on V, an unknown. However, from Table B.3, $\nu = 1.79 \times 10^{-4} \text{ ft}^2/\text{s}$ and we obtain

$$\text{Re} = \frac{VD}{\nu} = \frac{(\frac{4}{12} \text{ ft})\, V}{1.79 \times 10^{-4} \text{ ft}^2/\text{s}}$$

or

$$\text{Re} = 1860\, V \qquad \textbf{(3)}$$

where again V is in ft/s.

Also, since $\varepsilon/D = (0.0005 \text{ ft})/(4/12 \text{ ft}) = 0.0015$ (see Table 8.1 for the value of ε), we know which particular curve of the Moody chart is pertinent to this flow. Thus, we have three relationships (Eqs. 2, 3, and the $\varepsilon/D = 0.0015$ curve of Fig. 8.20) from which we can solve for the three unknowns f, Re, and V. This is done easily by an iterative scheme as follows.

It is usually simplest to assume a value of f, calculate V from Eq. 2, calculate Re from Eq. 3, and look up the appropriate value of f in the Moody chart for this value of Re. If the assumed f and the new f do not agree, the assumed answer is not correct—we do not have the solution to the three equations. Although values of either f, V, or Re could be assumed as starting values, it is usually simplest to assume a value of f because the correct value often lies on the relatively flat portion of the Moody chart for which f is quite insensitive to Re.

Thus, we assume $f = 0.022$, approximately the large Re limit for the given relative roughness. From Eq. 2 we obtain

$$V = \left[\frac{945}{7.5 + 60(0.022)}\right]^{1/2} = 10.4 \text{ ft/s}$$

and from Eq. 3

$$\text{Re} = 1860(10.4) = 19{,}300$$

With this Re and ε/D, Fig. 8.20 gives $f = 0.029$, which is not equal to the assumed solution $f = 0.022$ (although it is close!). We try again, this time with the newly obtained value of $f = 0.029$, which gives $V = 10.1$ ft/s and Re $= 18{,}800$. With these values, Fig. 8.20 gives $f = 0.029$, which agrees with the assumed value. Thus, the solution is $V = 10.1$ ft/s, or

$$Q = AV = \frac{\pi}{4}\left(\tfrac{4}{12} \text{ ft}\right)^2(10.1 \text{ ft/s}) = 0.881 \text{ ft}^3\text{/s} \qquad \textbf{(Ans)}$$

Note that the need for the iteration scheme is because one of the equations, $f = \phi(\text{Re}, \varepsilon/D)$, is in graphical form (the Moody chart). If the dependence of f on Re and ε/D is known in equation form, this graphical dependency is eliminated, and the solution technique may be easier. Such is the case if the flow is laminar so that the friction factor is simply $f = 64/\text{Re}$. For turbulent flow, we can use the Colebrook equation rather than the Moody chart, although this will normally require an iterative scheme also because of the complexity of the equation. As is shown below, such a formulation is ideally suited for an iterative computer solution.

We keep Eqs. 2 and 3 and use the Colebrook equation (Eq. 8.35, rather than the Moody chart) with $\varepsilon/D = 0.0015$ to give

$$\frac{1}{\sqrt{f}} = -2.0 \log\left(\frac{\varepsilon/D}{3.7} + \frac{2.51}{\text{Re}\sqrt{f}}\right) = -2.0 \log\left(4.05 \times 10^{-4} + \frac{2.51}{\text{Re}\sqrt{f}}\right) \qquad \textbf{(4)}$$

From Eq. 2 we have $V = [945/(7.5 + 60\,f)]^{1/2}$, which can be combined with Eq. 3 to give

$$\text{Re} = \frac{57{,}200}{\sqrt{7.5 + 60\,f}} \qquad \textbf{(5)}$$

The combination of Eqs. 4 and 5 provides a single equation for the determination of f

$$\frac{1}{\sqrt{f}} = -2.0 \log\left(4.05 \times 10^{-4} + 4.39 \times 10^{-5}\sqrt{60 + \frac{7.5}{f}}\right) \qquad \textbf{(6)}$$

A simple iterative solution of this equation gives $f = 0.029$, in agreement with the above solution which used the Moody chart. [This iterative solution using the Colebrook equation can be done as follows: (a) assume a value of f, (b) calculate a new value by using the assumed value in the right-hand side of Eq. 6, (c) use this new f to recalculate another value of f, and (d) repeat until the successive values agree.]

Note that unlike the Alaskan pipeline example (Example 8.9) in which we assumed minor losses are negligible, minor losses are of importance in this example because of the relatively small length-to-diameter ratio: $\ell/D = 20/(4/12) = 60$. The ratio of minor to major losses in this case is $K_L/(f\ell/D) = 6.5/[0.029\,(60)] = 3.74$. The elbows and entrance produce considerably more loss than the pipe itself.

EXAMPLE 8.11 (TYPE II, DETERMINE FLOWRATE)

The turbine shown in Fig. E8.11 extracts 50 hp from the water flowing through it. The 1-ft-diameter, 300-ft-long pipe is assumed to have a friction factor of 0.02. Minor losses are negligible. Determine the flowrate through the pipe and turbine.

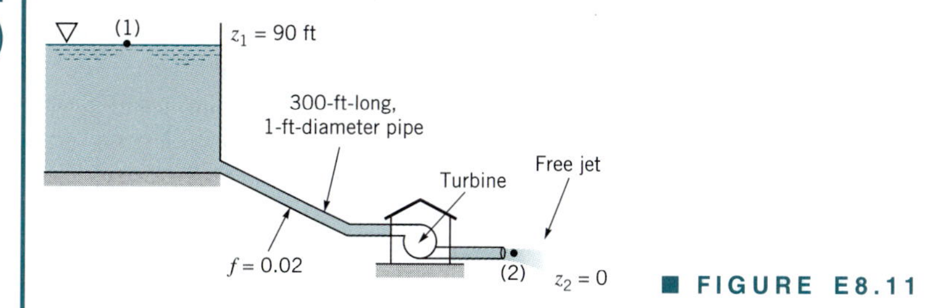

■ FIGURE E8.11

SOLUTION

The energy equation (Eq. 8.21) can be applied between the surface of the lake (point (1)) and the outlet of the pipe as

$$\frac{p_1}{\gamma} + \frac{V_1^2}{2g} + z_1 = \frac{p_2}{\gamma} + \frac{V_2^2}{2g} + z_2 + h_L + h_T \tag{1}$$

where $p_1 = V_1 = p_2 = z_2 = 0$, $z_1 = 90$ ft, and $V_2 = V$, the fluid velocity in the pipe. The head loss is given by

$$h_L = f\frac{\ell}{D}\frac{V^2}{2g} = 0.02\frac{(300 \text{ ft})}{(1 \text{ ft})}\frac{V^2}{2(32.2 \text{ ft/s}^2)} = 0.0932V^2 \text{ ft}$$

where V is in ft/s. Also, the turbine head is

$$h_T = \frac{\mathcal{P}_a}{\gamma Q} = \frac{\mathcal{P}_a}{\gamma(\pi/4)D^2V} = \frac{(50 \text{ hp})[(550 \text{ ft}\cdot\text{lb/s})/\text{hp}]}{(62.4 \text{ lb/ft}^3)[(\pi/4)(1 \text{ ft})^2V]} = \frac{561}{V} \text{ ft}$$

Thus, Eq. 1 can be written as

$$90 = \frac{V^2}{2(32.2)} + 0.0932V^2 + \frac{561}{V}$$

or

$$0.109V^3 - 90V + 561 = 0 \tag{2}$$

where V is in ft/s. The velocity of the water in the pipe is found as the solution of Eq. 2. Surprisingly, there are two real, positive roots: $V = 6.58$ ft/s or $V = 24.9$ ft/s. The third root is negative ($V = -31.4$ ft/s) and has no physical meaning for this flow. Thus, the two acceptable flowrates are

$$Q = \frac{\pi}{4}D^2V = \frac{\pi}{4}(1 \text{ ft})^2(6.58 \text{ ft/s}) = 5.17 \text{ ft}^3/\text{s} \qquad \textbf{(Ans)}$$

or

$$Q = \frac{\pi}{4}(1 \text{ ft})^2(24.9 \text{ ft/s}) = 19.6 \text{ ft}^3/\text{s} \qquad \textbf{(Ans)}$$

Either of these two flowrates gives the same power, $\mathcal{P}_a = \gamma Q h_T$. The reason for two possible solutions can be seen from the following. With the low flowrate ($Q = 5.17 \text{ ft}^3/\text{s}$), we obtain the head loss and turbine head as $h_L = 4.04$ ft and $h_T = 85.3$ ft. Because of the relatively low velocity there is a relatively small head loss and, therefore, a large head available for the turbine. With the large flowrate ($Q = 19.6 \text{ ft}^3/\text{s}$), we find $h_L = 57.8$ ft and $h_T = 22.5$ ft. The high-speed flow in the pipe produces a relatively large loss due to friction, leaving a relatively small head for the turbine. However, in either case the product of the turbine head times the flowrate is the same. That is, the power extracted ($\mathcal{P}_a = \gamma Q h_T$) is identical for each case. Although either flowrate will allow the extraction of 50 hp from the water, the details of the design of the turbine itself will depend strongly on which flowrate is to be used. Such information can be found in Chapter 12 and various references about turbomachines (Refs. 14, 19, 20).

If the friction factor were not given, the solution to the problem would be much more lengthy. A trial-and-error solution similar to that in Example 8.10 would be required along with the solution of a cubic equation.

In pipe flow problems for which the diameter is the unknown (Type III), an iterative technique is required. This is, again, because the friction factor is a function of the diameter—through both the Reynolds number and the relative roughness. Thus, neither $\text{Re} = \rho VD/\mu = 4\rho Q/\pi\mu D$ nor ε/D are known unless D is known. Examples 8.12 and 8.13 illustrate this.

EXAMPLE 8.12 (TYPE III WITHOUT MINOR LOSSES, DETERMINE DIAMETER)

Air at standard temperature and pressure flows through a horizontal, galvanized iron pipe ($\varepsilon = 0.0005$ ft) at a rate of $2.0 \text{ ft}^3/\text{s}$. Determine the minimum pipe diameter if the pressure drop is to be no more than 0.50 psi per 100 ft of pipe.

SOLUTION

We assume the flow to be incompressible with $\rho = 0.00238 \text{ slugs/ft}^3$ and $\mu = 3.74 \times 10^{-7} \text{ lb}\cdot\text{s/ft}^2$. Note that if the pipe were too long, the pressure drop from one end to the other, $p_1 - p_2$, would not be small relative to the pressure at the beginning, and compressible flow considerations would be required. For example, a pipe length of 200 ft gives $(p_1 - p_2)/p_1 = [(0.50 \text{ psi})/(100 \text{ ft})](200 \text{ ft})/14.7 \text{ psi} = 0.068 = 6.8\%$, which is probably small enough to justify the incompressible assumption.

With $z_1 = z_2$ and $V_1 = V_2$ the energy equation (Eq. 8.21) becomes

$$p_1 = p_2 + f\frac{\ell}{D}\frac{\rho V^2}{2} \tag{1}$$

where $V = Q/A = 4Q/(\pi D^2) = 4(2.0 \text{ ft}^3/\text{s})/\pi D^2$, or

$$V = \frac{2.55}{D^2}$$

where D is in feet. Thus, with $p_1 - p_2 = (0.5 \text{ lb/in.}^2)(144 \text{ in.}^2/\text{ft}^2)$ and $\ell = 100$ ft, Eq. 1

becomes

$$p_1 - p_2 = (0.5)(144)\ \text{lb/ft}^2$$
$$= f\,\frac{(100\ \text{ft})}{D}\,(0.00238\ \text{slugs/ft}^3)\,\frac{1}{2}\left(\frac{2.55}{D^2}\,\frac{\text{ft}}{\text{s}}\right)^2$$

or

$$D = 0.404 f^{1/5} \tag{2}$$

where D is in feet. Also $\text{Re} = \rho VD/\mu = (0.00238\ \text{slugs/ft}^3)\,[(2.55/D^2)\ \text{ft/s}]D/(3.74 \times 10^{-7}\ \text{lb·s/ft}^2)$, or

$$\text{Re} = \frac{1.62 \times 10^4}{D} \tag{3}$$

and

$$\frac{\varepsilon}{D} = \frac{0.0005}{D} \tag{4}$$

Thus, we have four equations (Eqs. 2, 3, 4, and either the Moody chart or the Colebrook equation) and four unknowns (f, D, ε/D, and Re) from which the solution can be obtained by trial-and-error methods.

If we use the Moody chart, it is probably easiest to assume a value of f, use Eqs. 2, 3, and 4 to calculate D, Re, and ε/D, and then compare the assumed f with that from the Moody chart. If they do not agree, try again. Thus, we assume $f = 0.02$, a typical value, and obtain $D = 0.404(0.02)^{1/5} = 0.185$ ft, which gives $\varepsilon/D = 0.0005/0.185 = 0.0027$ and $\text{Re} = 1.62 \times 10^4/0.185 = 8.76 \times 10^4$. From the Moody chart we obtain $f = 0.027$ for these values of ε/D and Re. Since this is not the same as our assumed value of f, we try again. With $f = 0.027$, we obtain $D = 0.196$ ft, $\varepsilon/D = 0.0026$, and $\text{Re} = 8.27 \times 10^4$, which in turn give $f = 0.027$, in agreement with the assumed value. Thus, the diameter of the pipe should be

$$D = 0.196\ \text{ft} \qquad \textbf{(Ans)}$$

If we use the Colebrook equation (Eq. 8.35) with $\varepsilon/D = 0.0005/0.404 f^{1/5} = 0.00124/f^{1/5}$ and $\text{Re} = 1.62 \times 10^4/0.404 f^{1/5} = 4.01 \times 10^4/f^{1/5}$, we obtain

$$\frac{1}{\sqrt{f}} = -2.0 \log\left(\frac{\varepsilon/D}{3.7} + \frac{2.51}{\text{Re}\sqrt{f}}\right)$$

or

$$\frac{1}{\sqrt{f}} = -2.0 \log\left(\frac{3.35 \times 10^{-4}}{f^{1/5}} + \frac{6.26 \times 10^{-5}}{f^{3/10}}\right)$$

An iterative scheme (see solution of Eq. 6 in Example 8.10) to solve this equation for f gives $f = 0.027$, and hence $D = 0.196$ ft, in agreement with the Moody chart method.

In the previous example we only had to consider major losses. In some instances the inclusion of major and minor losses can cause a slightly more lengthy solution procedure, even though the governing equations are essentially the same. This is illustrated in Example 8.13.

EXAMPLE 8.13 (TYPE III WITH MINOR LOSSES, DETERMINE DIAMETER)

Water at 10 °C ($\nu = 1.307 \times 10^{-6}$ m^2/s, see Table B.2) is to flow from reservoir A to reservoir B through a cast-iron pipe ($\varepsilon = 0.26$ mm) of length 20 m at a rate of $Q = 0.0020$ m^3/s as shown in Fig. E8.13. The system contains a sharp-edged entrance and six regular threaded 90° elbows. Determine the pipe diameter needed.

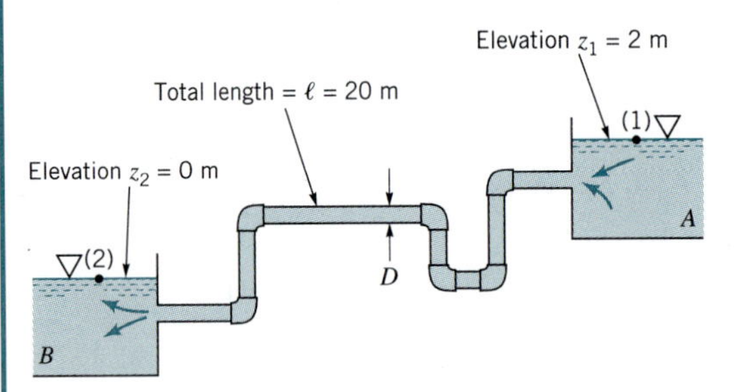

■ FIGURE E8.13

SOLUTION

The energy equation (Eq. 8.21) can be applied between two points on the surfaces of the reservoirs ($p_1 = p_2 = V_1 = V_2 = z_2 = 0$) as follows:

$$\frac{p_1}{\gamma} + \frac{V_1^2}{2g} + z_1 = \frac{p_2}{\gamma} + \frac{V_2^2}{2g} + z_2 + h_L$$

or

$$z_1 = \frac{V^2}{2g}\left(f\frac{\ell}{D} + \sum K_L\right) \tag{1}$$

where $V = Q/A = 4Q/\pi D^2 = 4(2 \times 10^{-3} \text{ m}^3/\text{s})/\pi D^2$, or

$$V = \frac{2.55 \times 10^{-3}}{D^2} \tag{2}$$

is the velocity within the pipe. (Note that the units on V and D are m/s and m, respectively.) The loss coefficients are obtained from Table 8.2 and Figs. 8.22 and 8.25 as $K_{L_{\text{ent}}} = 0.5$, $K_{L_{\text{elbow}}} = 1.5$, and $K_{L_{\text{exit}}} = 1$. Thus, Eq. 1 can be written as

$$2 \text{ m} = \frac{V^2}{2(9.81 \text{ m/s}^2)}\left\{\frac{20}{D}f + [6(1.5) + 0.5 + 1]\right\}$$

or, when combined with Eq. 2 to eliminate V,

$$6.03 \times 10^6 D^5 - 10.5D - 20f = 0 \tag{3}$$

To determine D we must know f, which is a function of Re and ε/D, where

$$\text{Re} = \frac{VD}{\nu} = \frac{[(2.55 \times 10^{-3})/D^2]D}{1.307 \times 10^{-6}} = \frac{1.95 \times 10^3}{D} \tag{4}$$

and

$$\frac{\varepsilon}{D} = \frac{2.6 \times 10^{-4}}{D} \tag{5}$$

where D is in meters. Again, we have four equations (Eqs. 3, 4, 5, and the Moody chart or the Colebrook equation) for the four unknowns D, f, Re, and ε/D.

Consider the solution by using the Moody chart. Although it is often easiest to assume a value of f and make calculations to determine if the assumed value is the correct one, with the inclusion of minor losses this may not be the simplest method. For example, if we assume $f = 0.02$ and calculate D from Eq. 3, we would have to solve a fifth-order equation. With only major losses (see Example 8.12), the term proportional to D in Eq. 3 is absent, and it is easy to solve for D if f is given. With both major and minor losses included (represented by the second and third terms in Eq. 3), this solution for D (given f) would require a trial-and-error or iterative technique.

Thus, for this type of problem it is perhaps easier to assume a value of D, calculate the corresponding f from Eq. 3, and with the values of Re and ε/D determined from Eqs. 4 and 5, look up the value of f in the Moody chart (or the Colebrook equation). The solution is obtained when the two values of f are in agreement. For example, assume $D = 0.05$ m, so that Eq. 3 gives $f = 0.0680$ and Eqs. 4 and 5 give $\text{Re} = 3.90 \times 10^4$ and $\varepsilon/D = 5.2 \times 10^{-3}$. With these values of Reynolds number and relative roughness, the Moody chart gives $f = 0.033$, which does not coincide with that obtained from Eq. 3 ($f = 0.0680$). Thus, $D \neq 0.05$ m.

A few more rounds of calculation will reveal that the solution is given by $D \approx 0.045$ m with $f = 0.032$.

$$D \approx 45 \text{ mm} \qquad \textbf{(Ans)}$$

It is interesting to attempt to solve this example if all losses are neglected so that Eq. 1 becomes $z_1 = 0$. Clearly from Fig. E8.13, $z_1 = 2$ m. Obviously something is wrong. A fluid cannot flow from one elevation, beginning with zero pressure and velocity, and end up at a lower elevation with zero pressure and velocity unless energy is removed (i.e., a head loss or a turbine) somewhere between the two locations. If the pipe is short (negligible friction) and the minor losses are negligible, there is still the kinetic energy of the fluid as it leaves the pipe and enters the reservoir. After the fluid meanders around in the reservoir for some time, this kinetic energy is lost and the fluid is stationary. No matter how small the viscosity is, the exit loss cannot be neglected. The same result can be seen if the energy equation is written from the free surface of the upstream tank to the exit plane of the pipe, at which point the kinetic energy is still available to the fluid. In either case the energy equation becomes $z_1 = V^2/2g$ in agreement with the inviscid results of Chapter 3 (the Bernoulli equation).

8.5.2 Multiple Pipe Systems

In many pipe systems there is more than one pipe involved. The complex system of tubes in our lungs (beginning with the relatively large-diameter trachea and ending in minute bronchi after numerous branchings) and the maze of pipes in a city's water distribution system are typical of such systems. The governing mechanisms for the flow in multiple pipe systems are the same as for the single pipe systems discussed in this chapter. However, because of the numerous unknowns involved, additional complexities may arise in solving for the flow in multiple pipe systems. Some of these complexities are discussed in this section.

An analogy between pipe systems and electrical circuits can be made.

The simplest multiple pipe systems can be classified into series or parallel flows, as are shown in Fig. 8.35. The nomenclature is similar to that used in electrical circuits. Indeed, an analogy between fluid and electrical circuits is often made as follows. In a simple electrical circuit, there is a balance between the voltage (e), current (i), and resistance (R) as given by Ohm's law: $e = iR$. In a fluid circuit there is a balance between the pressure drop (Δp), the flowrate or velocity (Q or V), and the flow resistance as given in terms of the friction factor and minor loss coefficients (f and K_L). For a simple flow [$\Delta p = f(\ell/D)(\rho V^2/2)$], it follows that $\Delta p = Q^2\tilde{R}$, where $\tilde{R}$, a measure of the resistance to the flow, is proportional to f.

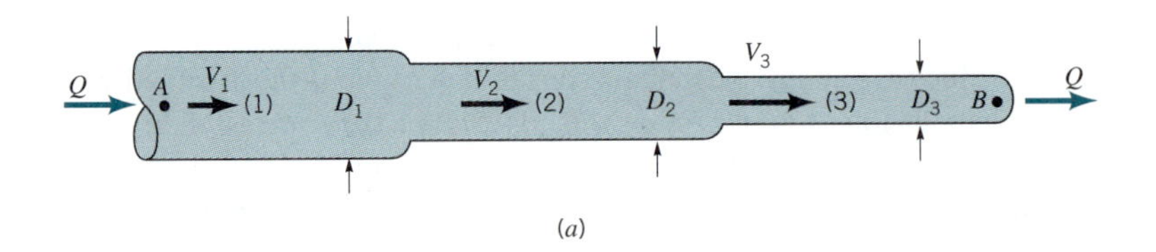

(a)

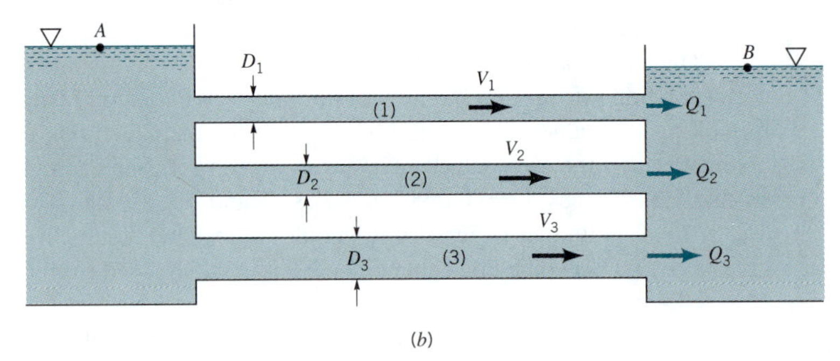

(b)

■ **FIGURE 8.35 Series (*a*) and parallel (*b*) pipe systems.**

The main differences between the solution methods used to solve electrical circuit problems and those for fluid circuit problems lie in the fact that Ohm's law is a linear equation (doubling the voltage doubles the current), while the fluid equations are generally nonlinear (doubling the pressure drop does not double the flowrate unless the flow is laminar). Thus, although some of the standard electrical engineering methods can be carried over to help solve fluid mechanics problems, others cannot.

One of the simplest multiple pipe systems is that containing pipes in *series*, as is shown in Fig. 8.35*a*. Every fluid particle that passes through the system passes through each of the pipes. Thus, the flowrate (but not the velocity) is the same in each pipe, and the head loss from point A to point B is the sum of the head losses in each of the pipes. The governing equations can be written as follows

$$Q_1 = Q_2 = Q_3$$

and

$$h_{L_{A-B}} = h_{L_1} + h_{L_2} + h_{L_3}$$

Series and parallel pipe systems are often encountered.

where the subscripts refer to each of the pipes. In general, the friction factors will be different for each pipe because the Reynolds numbers ($\text{Re}_i = \rho V_i D_i/\mu$) and the relative roughnesses (ε_i/D_i) will be different. If the flowrate is given, it is a straightforward calculation to determine the head loss or pressure drop (Type I problem). If the pressure drop is given and the flowrate is to be calculated (Type II problem), an iteration scheme is needed. In this situation none of the friction factors, f_i, are known, so the calculations may involve more trial-and-error attempts than for corresponding single pipe systems. The same is true for problems in which the pipe diameter (or diameters) is to be determined (Type III problems).

Another common multiple pipe system contains pipes in *parallel*, as is shown in Fig. 8.35*b*. In this system a fluid particle traveling from A to B may take any of the paths available, with the total flowrate equal to the sum of the flowrates in each pipe. However, by writing the energy equation between points A and B it is found that the head loss experienced by any fluid particle traveling between these locations is the same, independent of the path taken. Thus, the governing equations for parallel pipes are

$$Q = Q_1 + Q_2 + Q_3$$

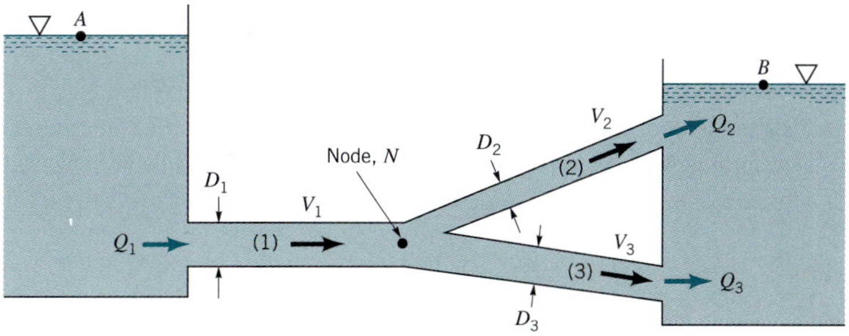

■ FIGURE 8.36 **Multiple pipe loop system.**

and

$$h_{L_1} = h_{L_2} = h_{L_3}$$

Again, the method of solution of these equations depends on what information is given and what is to be calculated.

Another type of multiple pipe system called a *loop* is shown in Fig. 8.36. In this case the flowrate through pipe (1) equals the sum of the flowrates through pipes (2) and (3), or $Q_1 = Q_2 + Q_3$. As can be seen by writing the energy equation between the surfaces of each reservoir, the head loss for pipe (2) must equal that for pipe (3), even though the pipe sizes and flowrates may be different for each. That is,

$$\frac{p_A}{\gamma} + \frac{V_A^2}{2g} + z_A = \frac{p_B}{\gamma} + \frac{V_B^2}{2g} + z_B + h_{L_1} + h_{L_2}$$

for a fluid particle traveling through pipes (1) and (2), while

$$\frac{p_A}{\gamma} + \frac{V_A^2}{2g} + z_A = \frac{p_B}{\gamma} + \frac{V_B^2}{2g} + z_B + h_{L_1} + h_{L_3}$$

for fluid that travels through pipes (1) and (3). These can be combined to give $h_{L_2} = h_{L_3}$. This is a statement of the fact that fluid particles that travel through pipe (2) and particles that travel through pipe (3) all originate from common conditions at the junction (or node, *N*) of the pipes and all end up at the same final conditions.

The three-reservoir problem can be quite complex.

The flow in a relatively simple looking multiple pipe system may be more complex than it appears initially. The branching system termed the *three-reservoir problem* shown in Fig. 8.37 is such a system. Three reservoirs at known elevations are connected together with three pipes of known properties (lengths, diameters, and roughnesses). The problem is to determine the flowrates into or out of the reservoirs. If valve (1) were closed, the fluid would flow from reservoir *B* to *C*, and the flowrate could be easily calculated. Similar calculations could be carried out if valves (2) or (3) were closed with the others open.

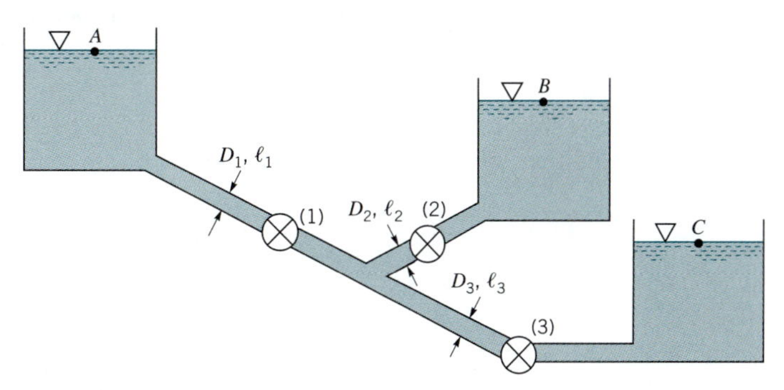

■ FIGURE 8.37 **A three-reservoir system.**

For some pipe systems, the direction of flow is not known a priori.

With all valves open, however, it is not necessarily obvious which direction the fluid flows. For the conditions indicated in Fig. 8.37, it is clear that fluid flows from reservoir A because the other two reservoir levels are lower. Whether the fluid flows into or out of reservoir B depends on the elevation of reservoirs B and C and the properties (length, diameter, roughness) of the three pipes. In general, the flow direction is not obvious, and the solution process must include the determination of this direction. This is illustrated in Example 8.14.

EXAMPLE 8.14

Three reservoirs are connected by three pipes as are shown in Fig. E8.14. For simplicity we assume that the diameter of each pipe is 1 ft, the friction factor for each is 0.02, and because of the large length-to-diameter ratio, minor losses are negligible. Determine the flowrate into or out of each reservoir.

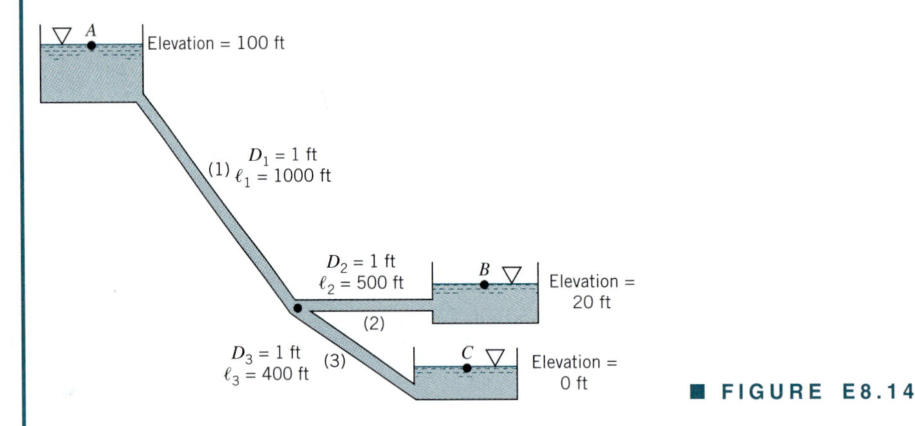

■ FIGURE E8.14

SOLUTION

It is not obvious which direction the fluid flows in pipe (2). However, we assume that it flows out of reservoir B, write the governing equations for this case, and check our assumption. The continuity equation requires that $Q_1 + Q_2 = Q_3$, which, since the diameters are the same for each pipe, becomes simply

$$V_1 + V_2 = V_3 \tag{1}$$

The energy equation for the fluid that flows from A to C in pipes (1) and (3) can be written as

$$\frac{p_A}{\gamma} + \frac{V_A^2}{2g} + z_A = \frac{p_C}{\gamma} + \frac{V_C^2}{2g} + z_C + f_1 \frac{\ell_1}{D_1}\frac{V_1^2}{2g} + f_3 \frac{\ell_3}{D_3}\frac{V_3^2}{2g}$$

By using the fact that $p_A = p_C = V_A = V_C = z_C = 0$, this becomes

$$z_A = f_1 \frac{\ell_1}{D_1}\frac{V_1^2}{2g} + f_3 \frac{\ell_3}{D_3}\frac{V_3^2}{2g}$$

For the given conditions of this problem we obtain

$$100 \text{ ft} = \frac{0.02}{2(32.2 \text{ ft/s}^2)} \frac{1}{(1 \text{ ft})} [(1000 \text{ ft})V_1^2 + (400 \text{ ft})V_3^2]$$

or

$$322 = V_1^2 + 0.4V_3^2 \tag{2}$$

where V_1 and V_3 are in ft/s. Similarly the energy equation for fluid flowing from B to C is

$$\frac{p_B}{\gamma} + \frac{V_B^2}{2g} + z_B = \frac{p_C}{\gamma} + \frac{V_C^2}{2g} + z_C + f_2 \frac{\ell_2}{D_2} \frac{V_2^2}{2g} + f_3 \frac{\ell_3}{D_3} \frac{V_3^2}{2g}$$

or

$$z_B = f_2 \frac{\ell_2}{D_2} \frac{V_2^2}{2g} + f_3 \frac{\ell_3}{D_3} \frac{V_3^2}{2g}$$

For the given conditions this can be written as

$$64.4 = 0.5V_2^2 + 0.4V_3^2 \qquad \textbf{(3)}$$

Equations 1, 2, and 3 (in terms of the three unknowns V_1, V_2, and V_3) are the governing equations for this flow, provided the fluid flows from reservoir B. It turns out, however, that there is no solution for these equations with positive, real values of the velocities. Although these equations do not appear to be complicated, there is no simple way to solve them directly. Thus, a trial-and-error solution is suggested. This can be accomplished as follows. Assume a value of $V_1 > 0$, calculate V_3 from Eq. 2, and then V_2 from Eq. 3. It is found that the resulting V_1, V_2, V_3 trio does not satisfy Eq. 1 for any value of V_1 assumed. There is no solution to Eqs. 1, 2, and 3 with real, positive values of V_1, V_2, and V_3. Thus, our original assumption of flow out of reservoir B must be incorrect.

To obtain the solution, assume the fluid flows into reservoirs B and C and out of A. For this case the continuity equation becomes

$$Q_1 = Q_2 + Q_3$$

or

$$V_1 = V_2 + V_3 \qquad \textbf{(4)}$$

Application of the energy equation between points A and B and A and C gives

$$z_A = z_B + f_1 \frac{\ell_1}{D_1} \frac{V_1^2}{2g} + f_2 \frac{\ell_2}{D_2} \frac{V_2^2}{2g}$$

and

$$z_A = z_C + f_1 \frac{\ell_1}{D_1} \frac{V_1^2}{2g} + f_3 \frac{\ell_3}{D_3} \frac{V_3^2}{2g}$$

which, with the given data, become

$$258 = V_1^2 + 0.5\, V_2^2 \qquad \textbf{(5)}$$

and

$$322 = V_1^2 + 0.4\, V_3^2 \qquad \textbf{(6)}$$

Equations 4, 5, and 6 can be solved as follows. By subtracting Eq. 5 from 6 we obtain

$$V_3 = \sqrt{160 + 1.25V_2^2}$$

Thus, Eq. 5 can be written as

$$258 = (V_2 + V_3)^2 + 0.5V_2^2 = (V_2 + \sqrt{160 + 1.25V_2^2})^2 + 0.5V_2^2$$

or

$$2V_2\sqrt{160 + 1.25V_2^2} = 98 - 2.75V_2^2 \qquad \textbf{(7)}$$

which, upon squaring both sides, can be written as

$$V_2^4 - 460\, V_2^2 + 3748 = 0$$

By using the quadratic formula we can solve for V_2^2 to obtain either $V_2^2 = 452$ or $V_2^2 = 8.30$. Thus, either $V_2 = 21.3$ ft/s or $V_2 = 2.88$ ft/s. The value $V_2 = 21.3$ ft/s is not a root of the original equations. It is an extra root introduced by squaring Eq. 7, which with $V_2 = 21.3$ becomes "1140 = −1140." Thus, $V_2 = 2.88$ ft/s and from Eq. 5, $V_1 = 15.9$ ft/s. The corresponding flowrates are

$$Q_1 = A_1V_1 = \frac{\pi}{4}D_1^2V_1 = \frac{\pi}{4}(1 \text{ ft})^2 (15.9 \text{ ft/s})$$

$$= 12.5 \text{ ft}^3\text{/s from } A \qquad \textbf{(Ans)}$$

$$Q_2 = A_2V_2 = \frac{\pi}{4}D_2^2V_2 = \frac{\pi}{4}(1 \text{ ft})^2 (2.88 \text{ ft/s}) \qquad \textbf{(Ans)}$$

$$= 2.26 \text{ ft}^3\text{/s into } B$$

and

$$Q_3 = Q_1 - Q_2 = (12.5 - 2.26) \text{ ft}^3\text{/s} = 10.2 \text{ ft}^3\text{/s into } C \qquad \textbf{(Ans)}$$

Note the slight differences in the governing equations depending on the direction of the flow in pipe (2)—compare Eqs. 1, 2, and 3 with Eqs. 4, 5, and 6.

If the friction factors were not given, a trial-and-error procedure similar to that needed for Type II problems (see Section 8.5.1) would be required.

The ultimate in multiple pipe systems is a *network* of pipes such as that shown in Fig. 8.38. Networks like these often occur in city water distribution systems and other systems that may have multiple "inlets" and "outlets." The direction of flow in the various pipes is by no means obvious—in fact, it may vary in time, depending on how the system is used from time to time.

Pipe network problems can be solved using node and loop concepts.

The solution for pipe network problems is often carried out by use of node and loop equations similar in many ways to that done in electrical circuits. For example, the continuity equation requires that for each *node* (the junction of two or more pipes) the net flowrate is zero. What flows into a node must flow out at the same rate. In addition, the net pressure difference completely around a *loop* (starting at one location in a pipe and returning to that location) must be zero. By combining these ideas with the usual head loss and pipe flow equations, the flow throughout the entire network can be obtained. Of course, trial-and-error solutions are usually required because the direction of flow and the friction factors may not be known. Such a solution procedure using matrix techniques is ideally suited for computer use (Refs. 21, 22).

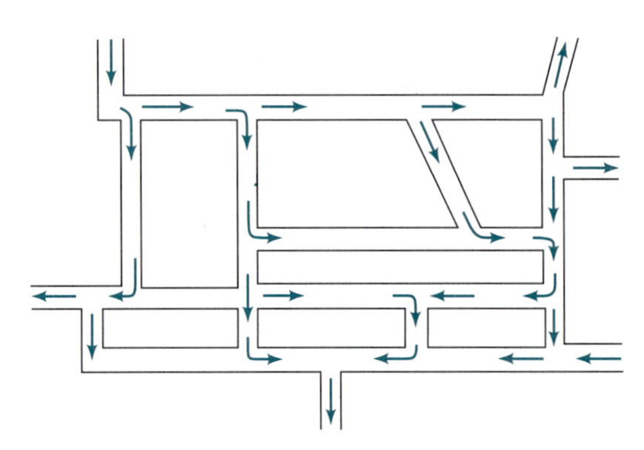

■ **FIGURE 8.38 A general pipe network.**

8.6 Pipe Flowrate Measurement

It is often necessary to determine experimentally the flowrate in a pipe. In Chapter 3 we introduced various types of flow-measuring devices (Venturi meter, nozzle meter, orifice meter, etc.) and discussed their operation under the assumption that viscous effects were not important. In this section we will indicate how to account for the ever-present viscous effects in these flow meters. We will also indicate other types of commonly used flow meters.

8.6.1 Pipe Flowrate Meters

Three of the most common devices used to measure the instantaneous flowrate in pipes are the orifice meter, the nozzle meter, and the Venturi meter. As was discussed in Section 3.6.3, each of these meters operates on the principle that a decrease in flow area in a pipe causes an increase in velocity that is accompanied by a decrease in pressure. Correlation of the pressure difference with the velocity provides a means of measuring the flowrate. In the absence of viscous effects and under the assumption of a horizontal pipe, application of the Bernoulli equation (Eq. 3.7) between points (1) and (2) shown in Fig. 8.39 gave

$$Q_{\text{ideal}} = A_2V_2 = A_2\sqrt{\frac{2(p_1 - p_2)}{\rho(1 - \beta^4)}} \qquad \textbf{(8.37)}$$

where $\beta = D_2/D_1$. Based on the results of the previous sections of this chapter, we anticipate that there is a head loss between (1) and (2) so that the governing equations become

$$Q = A_1V_1 = A_2V_2$$

and

$$\frac{p_1}{\gamma} + \frac{V_1^2}{2g} = \frac{p_2}{\gamma} + \frac{V_2^2}{2g} + h_L$$

The ideal situation has $h_L = 0$ and results in Eq. 8.37. The difficulty in including the head loss is that there is no accurate expression for it. The net result is that empirical coefficients are used in the flowrate equations to account for the complex real world effects brought on by the nonzero viscosity. The coefficients are discussed below.

An orifice discharge coefficient is used to account for nonideal effects.

A typical *orifice meter* is constructed by inserting between two flanges of a pipe a flat plate with a hole, as shown in Fig. 8.40. The pressure at point (2) within the vena contracta is less than that at point (1). Nonideal effects occur for two reasons. First, the vena contracta area, A_2, is less than the area of the hole, A_o, by an unknown amount. Thus, $A_2 = C_cA_o$, where C_c is the contraction coefficient ($C_c < 1$). Second, the swirling flow and turbulent motion near the orifice plate introduce a head loss that cannot be calculated theoretically. Thus, an *orifice discharge coefficient*, C_o, is used to take these effects into account. That is,

$$Q = C_oQ_{\text{ideal}} = C_oA_o\sqrt{\frac{2(p_1 - p_2)}{\rho(1 - \beta^4)}} \qquad \textbf{(8.38)}$$

where $A_o = \pi d^2/4$ is the area of the hole in the orifice plate. The value of C_o is a function of $\beta = d/D$ and the Reynolds number $\text{Re} = \rho VD/\mu$, where $V = Q/A_1$. Typical values of

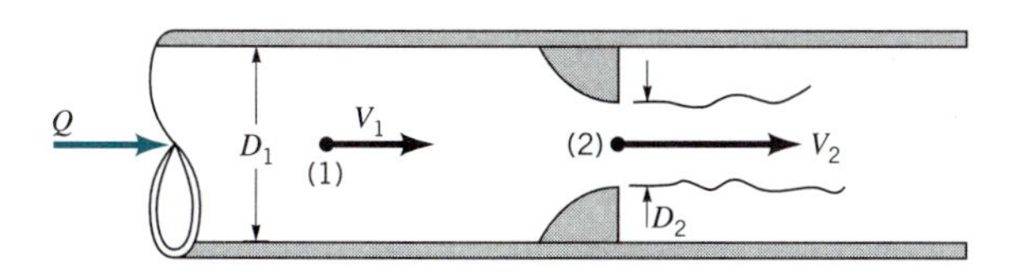

■ FIGURE 8.39 Typical pipe flow meter geometry.

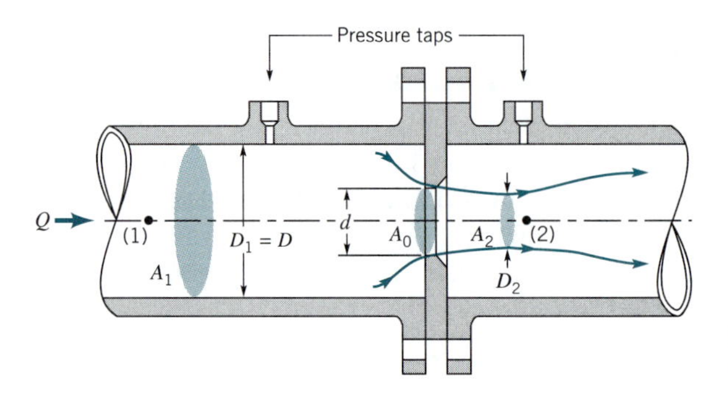

■ **FIGURE 8.40 Typical orifice meter construction.**

C_o are given in Fig. 8.41. Note that the value of C_o depends on the specific construction of the orifice meter (i.e., the placement of the pressure taps, whether the orifice plate edge is square or beveled, etc.). Very precise conditions governing the construction of standard orifice meters have been established to provide the greatest accuracy possible (Refs. 23, 24).

The nozzle meter is more efficient than the orifice meter.

Another type of pipe flow meter that is based on the same principles used in the orifice meter is the *nozzle meter*, three variations of which are shown in Fig. 8.42. This device uses a contoured nozzle (typically placed between flanges of pipe sections) rather than a simple (and less expensive) plate with a hole as in an orifice meter. The resulting flow pattern for the nozzle meter is closer to ideal than the orifice meter flow. There is only a slight vena contracta and the secondary flow separation is less severe, but there still are viscous effects. These are accounted for by use of the *nozzle discharge coefficient*, C_n, where

$$Q = C_n Q_{\text{ideal}} = C_n A_n \sqrt{\frac{2(p_1 - p_2)}{\rho(1 - \beta^4)}} \qquad \textbf{(8.39)}$$

with $A_n = \pi d^2/4$. As with the orifice meter, the value of C_n is a function of the diameter ratio, $\beta = d/D$, and the Reynolds number, $\text{Re} = \rho VD/\mu$. Typical values obtained from experiments are shown in Fig. 8.43. Again, precise values of C_n depend on the specific details of the nozzle design. Accepted standards have been adopted (Ref. 24). Note that $C_n > C_o$; the nozzle meter is more efficient (less energy dissipated) than the orifice meter.

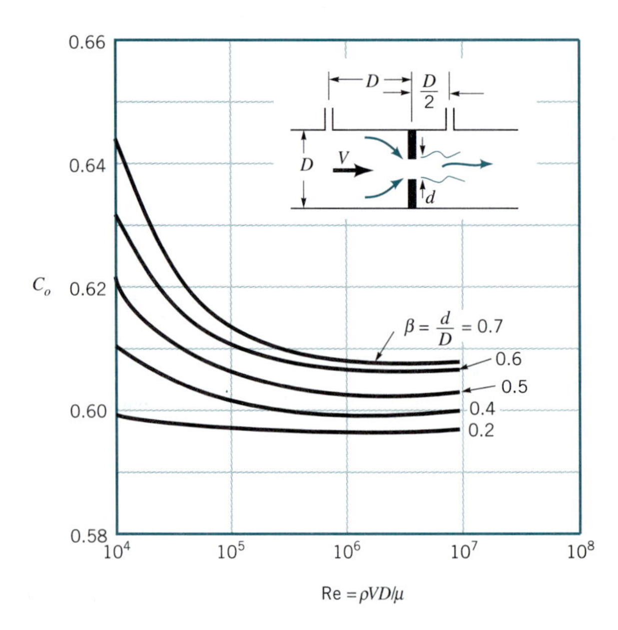

■ **FIGURE 8.41 Orifice meter discharge coefficient (Ref. 24).**

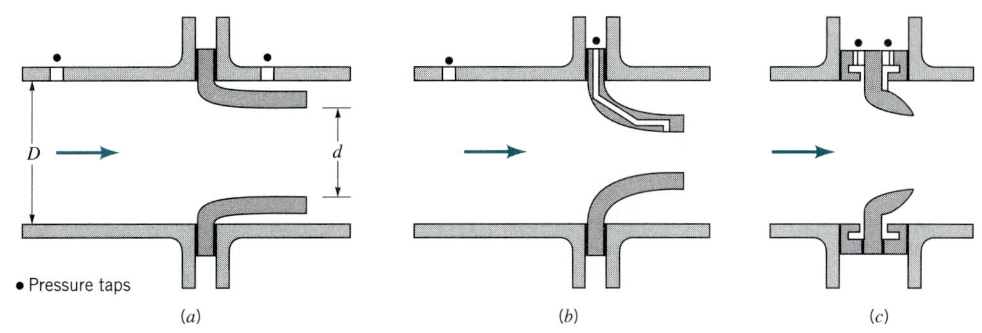

■ FIGURE 8.42 **Typical nozzle meter construction.**

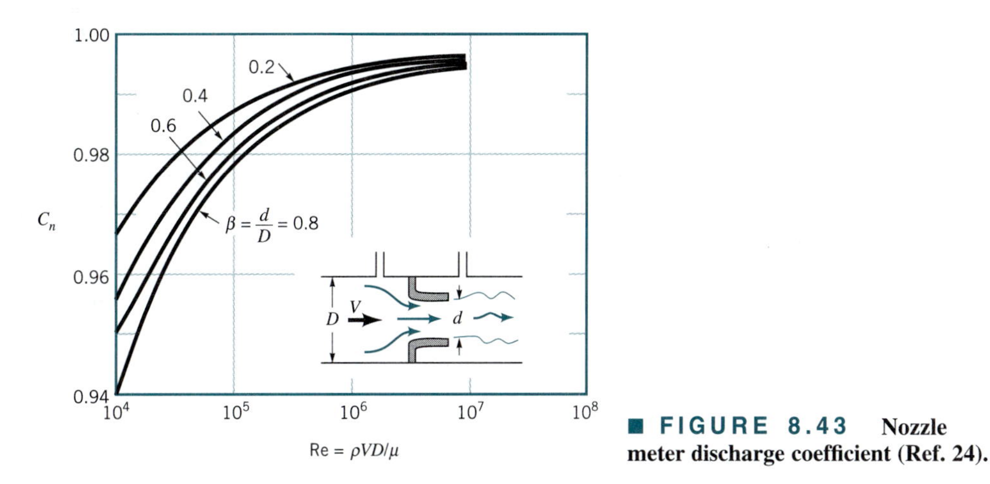

■ FIGURE 8.43 **Nozzle meter discharge coefficient (Ref. 24).**

The most precise and most expensive of the three obstruction-type flow meters is the *Venturi meter* shown in Fig. 8.44. Although the operating principle for this device is the same as for the orifice or nozzle meters, the geometry of the Venturi meter is designed to reduce head losses to a minimum. This is accomplished by providing a relatively streamlined contraction (which eliminates separation ahead of the throat) and a very gradual expansion downstream of the throat (which eliminates separation in this decelerating portion of the device). Most of the head loss that occurs in a well-designed Venturi meter is due to friction losses along the walls rather than losses associated with separated flows and the inefficient mixing motion that accompanies such flow.

Thus, the flowrate through a Venturi meter is given by

$$Q = C_v Q_{\text{ideal}} = C_v A_T \sqrt{\frac{2(p_1 - p_2)}{\rho(1 - \beta^4)}}$$

The Venturi discharge coefficient is a function of the specific geometry of the meter.

where $A_T = \pi d^2/4$ is the throat area. The range of values of C_v, the *Venturi discharge coefficient*, is given in Fig. 8.45. The throat-to-pipe diameter ratio ($\beta = d/D$), the Reynolds

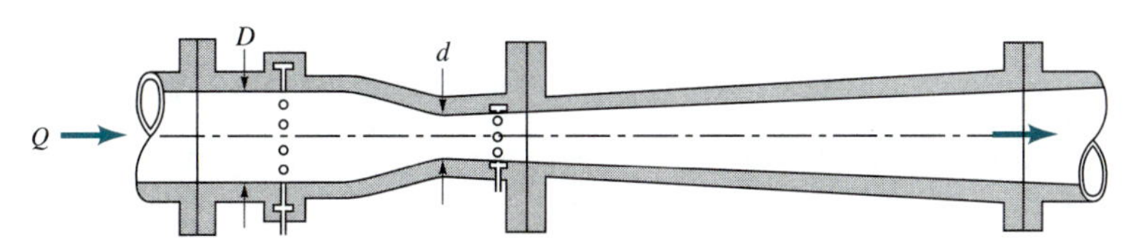

■ FIGURE 8.44 **Typical Venturi meter construction.**

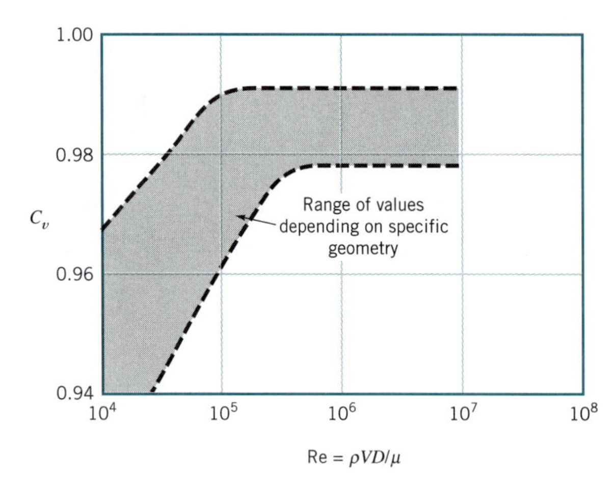

■ **FIGURE 8.45 Venturi meter discharge coefficient (Ref. 23).**

number, and the shape of the converging and diverging sections of the meter are among the parameters that affect the value of C_v.

Again, the precise values of C_n, C_o, and C_v depend on the specific geometry of the devices used. Considerable information concerning the design, use, and installation of standard flow meters can be found in various books (Refs. 23, 24, 25, 26, 31).

EXAMPLE 8.15

Ethyl alcohol flows through a pipe of diameter $D = 60$ mm in a refinery. The pressure drop across the nozzle meter used to measure the flowrate is to be $\Delta p = 4.0$ kPa when the flowrate is $Q = 0.003 \text{ m}^3/\text{s}$. Determine the diameter, d, of the nozzle.

SOLUTION

From Table 1.6 the properties of ethyl alcohol are $\rho = 789 \text{ kg/m}^3$ and $\mu = 1.19 \times 10^{-3}$ N·s/m^2. Thus,

$$\text{Re} = \frac{\rho VD}{\mu} = \frac{4\rho Q}{\pi D\mu} = \frac{4(789 \text{ kg/m}^3)(0.003 \text{ m}^3/\text{s})}{\pi(0.06 \text{ m})(1.19 \times 10^{-3} \text{ N·s/m}^2)} = 42{,}200$$

From Eq. 8.39 the flowrate through the nozzle is

$$Q = 0.003 \text{ m}^3/\text{s} = C_n \frac{\pi}{4} d^2 \sqrt{\frac{2(4 \times 10^3 \text{ N/m}^2)}{789 \text{ kg/m}^3(1 - \beta^4)}}$$

or

$$1.20 \times 10^{-3} = \frac{C_n d^2}{\sqrt{1 - \beta^4}} \qquad \textbf{(1)}$$

where d is in meters. Note that $\beta = d/D = d/0.06$. Equation 1 and Fig. 8.43 represent two equations for the two unknowns d and C_n that must be solved by trial and error.

As a first approximation we assume that the flow is ideal, or $C_n = 1.0$, so that Eq. 1 becomes

$$d = (1.20 \times 10^{-3} \sqrt{1 - \beta^4})^{1/2} \qquad \textbf{(2)}$$

In addition, for many cases $1 - \beta^4 \approx 1$, so that an approximate value of d can be obtained from Eq. 2 as

$$d = (1.20 \times 10^{-3})^{1/2} = 0.0346 \text{ m}$$

Hence, with an initial guess of $d = 0.0346$ m or $\beta = d/D = 0.0346/0.06 = 0.577$, we obtain from Fig. 8.43 (using Re = 42,200) a value of $C_n = 0.972$. Clearly this does not agree with our initial assumption of $C_n = 1.0$. Thus, we do not have the solution to Eq. 1 and Fig. 8.43. Next we assume $\beta = 0.577$ and $C_n = 0.972$ and solve for d from Eq. 1 to obtain

$$d = \left(\frac{1.20 \times 10^{-3}}{0.972} \sqrt{1 - 0.577^4} \right)^{1/2}$$

or $d = 0.0341$ m. With the new value of $\beta = 0.0341/0.060 = 0.568$ and Re = 42,200, we obtain (from Fig. 8.43) $C_n \approx 0.972$ in agreement with the assumed value. Thus,

$$d = 34.1 \text{ mm} \quad \textbf{(Ans)}$$

If numerous cases are to be investigated, it may be much easier to replace the discharge coefficient data of Fig. 8.43 by the equivalent equation, $C_n = \phi(\beta, \text{Re})$, and use a computer to iterate for the answer. Such equations are available in the literature (Ref. 24). This would be similar to using the Colebrook equation rather than the Moody chart for pipe friction problems.

There are many types of flow meters.

Numerous other devices are used to measure the flowrate in pipes. Many of these devices use principles other than the high-speed/low-pressure concept of the orifice, nozzle, and Venturi meters.

A quite common, accurate, and relatively inexpensive flow meter is the *rotameter*, or variable area meter as is shown in Fig. 8.46. In this device a float is contained within a tapered, transparent metering tube that is attached vertically to the pipeline. As fluid flows

V8.6

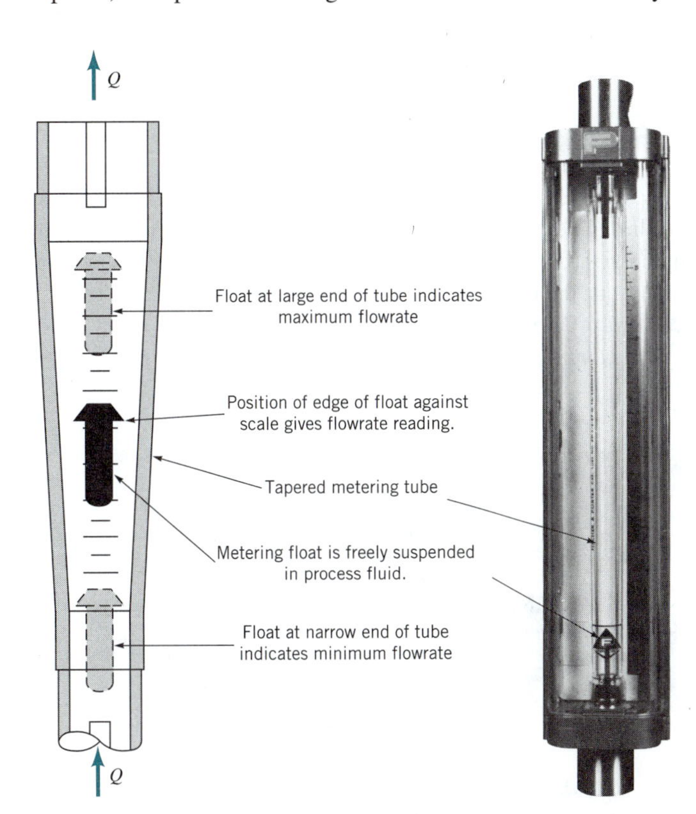

■ **FIGURE 8.46**
Rotameter-type flow meter. (Courtesy of Fischer & Porter Co.).

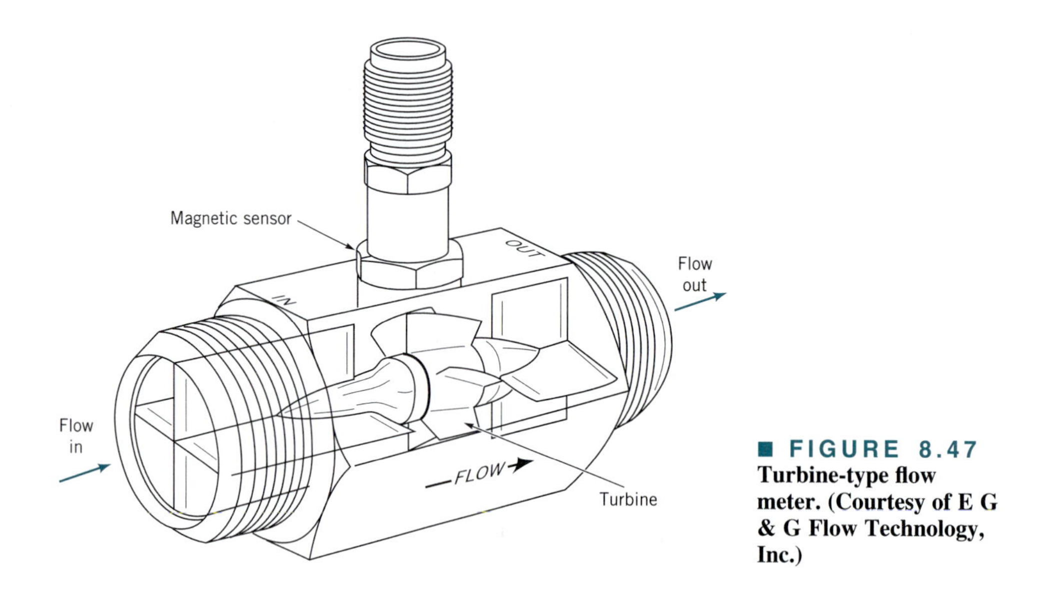

■ FIGURE 8.47 **Turbine-type flow meter. (Courtesy of E G & G Flow Technology, Inc.)**

through the meter (entering at the bottom), the float will rise within the tapered tube and reach an equilibrium height that is a function of the flowrate. This height corresponds to an equilibrium condition for which the net force on the float (buoyancy, float weight, fluid drag) is zero. A calibration scale in the tube provides the relationship between the float position and the flowrate.

Another useful pipe flowrate meter is a *turbine meter* as is shown in Fig. 8.47. A small, freely rotating propeller or turbine within the turbine meter rotates with an angular velocity that is a function of (nearly proportional to) the average fluid velocity in the pipe. This angular velocity is picked up magnetically and calibrated to provide a very accurate measure of the flowrate through the meter.

8.6.2 Volume Flow Meters

Volume flow meters measure volume rather than volume flowrate.

In many instances it is necessary to know the amount (volume or mass) of fluid that has passed through a pipe during a given time period, rather than the instantaneous flowrate. For example, we are interested in how many gallons of gasoline are pumped into the tank in our car rather than the rate at which it flows into the tank. There are numerous quantity-measuring devices that provide such information.

The *nutating disk meter* shown in Fig. 8.48 is widely used to measure the net amount

V8.7

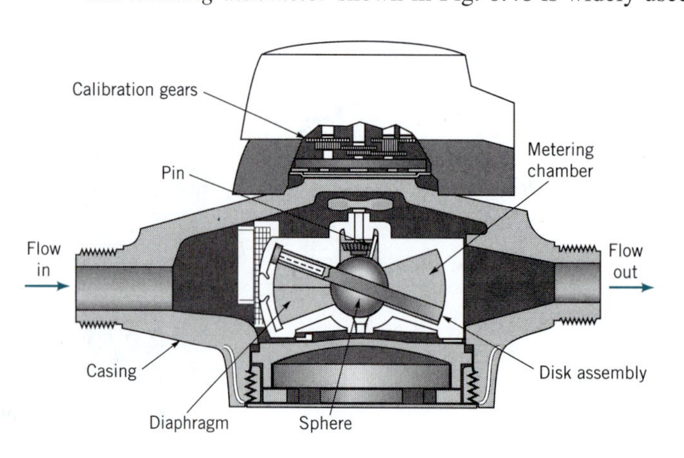

■ FIGURE 8.48 **Nutating disk flow meter. (Courtesy of Badger Meter, Inc.)**

of water used in domestic and commercial water systems as well as the amount of gasoline delivered to your gas tank. This meter contains only one essential moving part and is relatively inexpensive and accurate. Its operating principle is very simple, but it may be difficult to understand its operation without actually inspecting the device firsthand. The device consists of a metering chamber with spherical sides and conical top and bottom. A disk passes through a central sphere and divides the chamber into two portions. The disk is constrained to be at an angle not normal to the axis of symmetry of the chamber. A radial plate (diaphragm) divides the chamber so that the entering fluid causes the disk to wobble (nutate), with fluid flowing alternately above or below the disk. The fluid exits the chamber after the disk has completed one wobble, which corresponds to a specific volume of fluid passing through the chamber. During each wobble of the disk, the pin attached to the tip of the center sphere, normal to the disk, completes one circle. The volume of fluid that has passed through the meter can be obtained by counting the number of revolutions completed.

The nutating disk meter is very simple in design, using one moving part.

Another quantity-measuring device that is used for gas flow measurements is the *bellows meter* as shown in Fig. 8.49. It contains a set of bellows that alternately fill and empty as a result of the pressure of the gas and the motion of a set of inlet and outlet valves. The

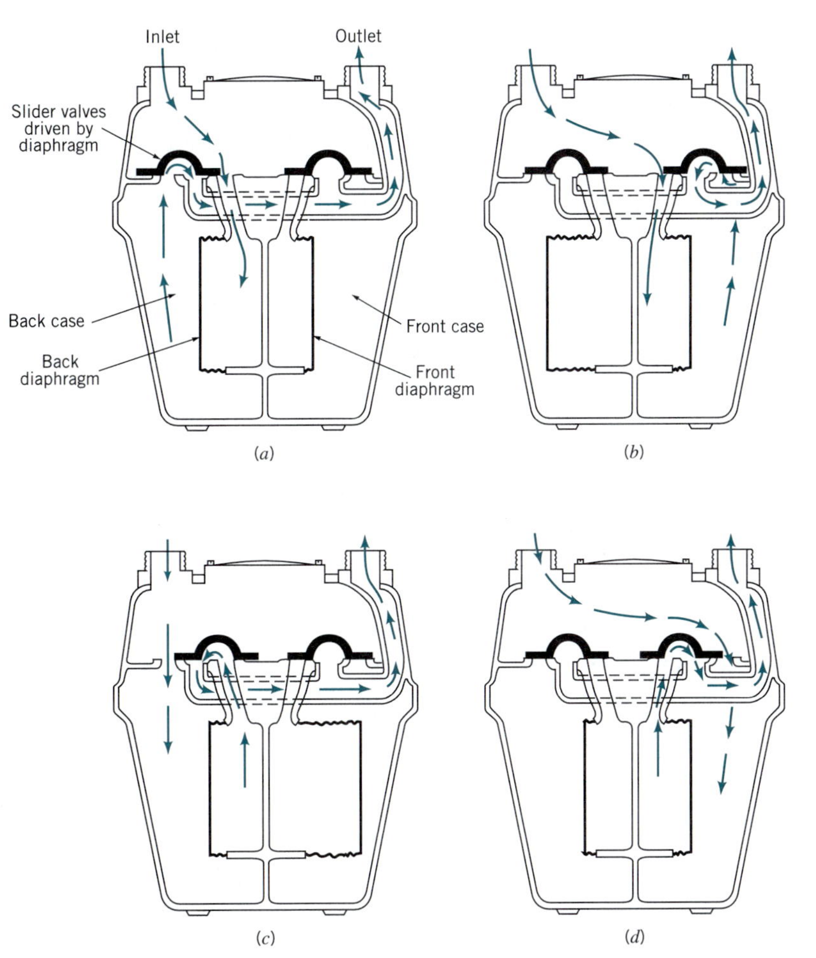

■ **FIGURE 8.49** **Bellows-type flow meter. (Courtesy of BTR—Rockwell Gas Products). (*a*) Back case emptying, back diaphragm filling. (*b*) Front diaphragm filling, front case emptying. (*c*) Back case filling, back diaphragm emptying. (*d*) Front diaphragm emptying, front case filling.**

The bellows meter contains a complex set of moving parts.

common household natural gas meter is of this type. For each cycle [(*a*) through (*d*)] a known volume of gas passes through the meter.

The nutating disk meter (water meter) is an example of extreme simplicity—one cleverly designed moving part. The bellows meter (gas meter), on the other hand, is relatively complex—it contains many moving, interconnected parts. This difference is dictated by the application involved. One measures a common, safe-to-handle, relatively high-pressure liquid, whereas the other measures a relatively dangerous, low-pressure gas. Each device does its intended job very well.

There are numerous devices used to measure fluid flow, only a few of which have been discussed here. The reader is encouraged to review the literature to gain familiarity with other useful, clever devices (Refs. 25, 26).

References

1. Hinze, J. O., *Turbulence*, 2nd Ed., McGraw-Hill, New York, 1975.
2. Panton, R. L., *Incompressible Flow*, Wiley, New York, 1984.
3. Schlichting, H., *Boundary Layer Theory*, 7th Ed., McGraw-Hill, New York, 1979.
4. Gleick, J., *Chaos: Making a New Science*, Viking Penguin, New York, 1987.
5. White, F. M., *Fluid Mechanics*, McGraw-Hill, New York, 1979.
6. Nikuradse, J., ''Stomungsgesetz in Rauhen Rohren,'' *VDI-Forschungsch*, No. 361, 1933; or see NACA Tech Memo 1922.
7. Moody, L. F., ''Friction Factors for Pipe Flow,'' *Transactions of the ASME*, Vol. 66, 1944.
8. Colebrook, C. F., ''Turbulent Flow in Pipes with Particular Reference to the Transition Between the Smooth and Rough Pipe Laws,'' *Journal of the Institute of Civil Engineers London*, Vol. 11, 1939.
9. *ASHRAE Handbook of Fundamentals*, ASHRAE, Atlanta, 1981.
10. Streeter, V. L., ed., *Handbook of Fluid Dynamics*, McGraw-Hill, New York, 1961.
11. Sovran, G., and Klomp, E. D., ''Experimentally Determined Optimum Geometries for Rectilinear Diffusers with Rectangular, Conical, or Annular Cross Sections,'' in *Fluid Mechanics of Internal Flow*, Sovran, G., ed., Elsevier, Amsterdam, 1967.
12. Runstadler, P. W., ''Diffuser Data Book,'' Technical Note 186, Creare, Inc., Hanover, NH, 1975.
13. Laws, E. M., and Livesey, J. L., ''Flow Through Screens,'' *Annual Review of Fluid Mechanics*, Vol. 10, Annual Reviews, Inc., Palo Alto, CA, 1978.
14. Balje, O. E., *Turbomachines: A Guide to Design, Selection and Theory*, Wiley, New York, 1981.
15. Wallis, R. A., *Axial Flow Fans and Ducts*, Wiley, New York, 1983.
16. Karassick, I. J. et al., *Pump Handbook*, 2nd Ed., McGraw-Hill, New York, 1985.
17. White, F. M., *Viscous Fluid Flow*, McGraw-Hill, New York, 1974.
18. Olson, R. M., *Essentials of Engineering Fluid Mechanics*, 4th Ed., Harper & Row, New York, 1980.
19. Dixon, S. L., *Fluid Mechanics of Turbomachinery*, 3rd Ed., Pergamon, Oxford, 1978.
20. Daugherty, R. L., and Franzini, J. R., *Fluid Mechanics*, 7th Ed., McGraw-Hill, New York, 1977.

21. Streeter, V. L., and Wylie, E. B., *Fluid Mechanics*, 8th Ed., McGraw-Hill, New York, 1985.
22. Jeppson, R. W., *Analysis of Flow in Pipe Networks*, Ann Arbor Science Publishers, Ann Arbor, Mich., 1976.
23. Bean, H. S., ed., *Fluid Meters: Their Theory and Application*, 6th Ed., American Society of Mechanical Engineers, New York, 1971.
24. ''Measurement of Fluid Flow by Means of Orifice Plates, Nozzles, and Venturi Tubes Inserted in Circular Cross Section Conduits Running Full,'' Int. Organ. Stand. Rep. DIS-5167, Geneva, 1976.
25. Goldstein, R. J., ed., *Fluid Mechanics Measurements*, Hemisphere Publishing, New York, 1983.
26. Benedict, R. P., *Measurement of Temperature, Pressure, and Flow*, 2nd Ed., Wiley, New York, 1977.
27. Hydraulic Institute, *Engineering Data Book*, 1st Ed., Cleveland Hydraulic Institute, 1979.
28. Harris, C. W., *University of Washington Engineering Experimental Station Bulletin*, 48, 1928.
29. Hamilton, J. B., *University of Washington Engineering Experimental Station Bulletin*, 51, 1929.
30. Miller, D. S., *Internal Flow Systems*, 2nd Ed., BHRA, Cranfield, UK, 1990.
31. Spitzer, D. W., ed., *Flow Measurement: Practical Guides for Measurement and Control*, Instrument Society of America, Research Triangle Park, North Carolina, 1991.
32. Wilcox, D. C., *Turbulence Modeling for CFD*, DCW Industries, Inc., La Canada, California, 1994.
33. Gleick, J., *Chaos, Making a New Science*, Penguin Books, New York, 1988.
34. Mullin, T., ed., *The Nature of Chaos*, Oxford University Press, Oxford, 1993.

Review Problems

Note: Problems designated with (R) are review problems. The phrases within parentheses refer to the main topics to be used in solving the problems. Complete, detailed solutions to these review problems can be found in the supplement titled *Student Solution Manual for Fundamentals of Fluid Mechanics*, by Munson, Young, and Okiishi (John Wiley and Sons, New York, 1997).

8.1R (Laminar flow) Asphalt at 120 °F, considered to be a Newtonian fluid with a viscosity 80,000 times that of water and a specific gravity of 1.09, flows through a pipe of diameter 2.0 in. If the pressure gradient is 1.6 psi/ft determine the flowrate assuming the pipe is **(a)** horizontal; **(b)** vertical with flow up.

(ANS: 4.69×10^{-3} ft^3/s; 3.30×10^{-3} ft^3/s)

8.2R (Laminar flow) A fluid flows through two horizontal pipes of equal length which are connected together to form a pipe of length 2ℓ. The flow is laminar and fully developed. The pressure drop for the first pipe is 1.44 times greater than it is for the second pipe. If the diameter of the first pipe is D, determine the diameter of the second pipe.

(ANS: 1.095 D)

8.3R (Velocity profile) A fluid flows through a pipe of radius R with a Reynolds number of 100,000. At what location, r/R, does the fluid velocity equal the average velocity? Repeat if the Reynolds number is 1000.

(ANS: 0.758; 0.707)

8.4R (Turbulent velocity profile) Water at 80 °C flows through a 120-mm-diameter pipe with an average velocity of 2 m/s. If the pipe wall roughness is small enough so that it does not protrude through the laminar sublayer, the pipe can be considered as smooth. Approximately what is the largest roughness allowed to classify this pipe as smooth?

(ANS: 2.31×10^{-5} m)

8.5R (Moody chart) Water flows in a smooth plastic pipe of 200-mm diameter at a rate of 0.10 m^3/s. Determine the friction factor for this flow.

(ANS: 0.0128)

8.6R (Moody chart) After a number of years of use, it is noted that to obtain a given flowrate, the head loss is increased to 1.6 times its value for the originally smooth pipe. If the Reynolds number is 10^6, determine the relative roughness of the old pipe.

(ANS: 0.00070)

8.7R (Minor losses) Air flows through the fine mesh gauze shown in Fig. P8.7R with an average velocity of 1.50 m/s in the pipe. Determine the loss coefficient for the gauze.

(ANS: 56.7)

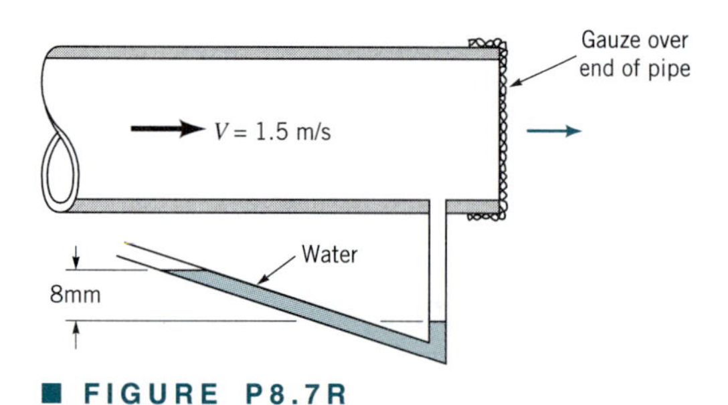

■ FIGURE P8.7R

8.8R (Noncircular conduits) A manufacturer makes two types of drinking straws: one with a square cross-sectional shape, and the other type the typical round shape. The amount of material in each straw is to be the same. That is, the length of the perimeter of the cross section of each shape is the same. For a given pressure drop, what is the ratio of the flowrates through the straws? Assume the drink is viscous enough to ensure laminar flow and neglect gravity.

(ANS: $Q_{round} = 1.83\ Q_{square}$)

8.9R (Single pipe—determine pressure drop) Determine the pressure drop per 300-m length of new 0.20-m-diameter horizontal cast iron water pipe when the average velocity is 1.7 m/s.

(ANS: 47.6 kN/m^2)

8.10R (Single pipe—determine pressure drop) A fire protection association code requires a minimum pressure of 65 psi at the outlet end of a 250-ft-long, 4-in.-diameter hose when the flowrate is 500 gal/min. What is the minimum pressure allowed at the pumper truck that supplies water to the hose? Assume a roughness of $\varepsilon = 0.03$ in.

(ANS: 94.0 psi)

8.11R (Single pipe—determine flowrate) An above ground swimming pool of 30 ft diameter and 5 ft depth is to be filled from a garden hose (smooth interior) of length 100 ft and diameter 5/8 in. If the pressure at the faucet to which the hose is attached remains at 55 psi, how long will it take to fill the pool? The water exits the hose as a free jet 6 ft above the faucet.

(ANS: 32.0 hr)

8.12R (Single pipe—determine pipe diameter) Water is to flow at a rate of 1.0 m^3/s through a rough concrete pipe (ε = 3 mm) that connects two ponds. Determine the pipe diameter if the elevation difference between the two ponds is 10 m and the pipe length is 1000 m. Neglect minor losses.

(ANS: 0.748 m)

8.13R (Single pipe with pump) Without the pump shown in Fig. P8.13R it is determined that the flowrate is too small. Determine the horsepower added to the fluid if the pump causes the flowrate to be doubled. Assume the friction factor remains at 0.020 in either case.

(ANS: 1.51 hp)

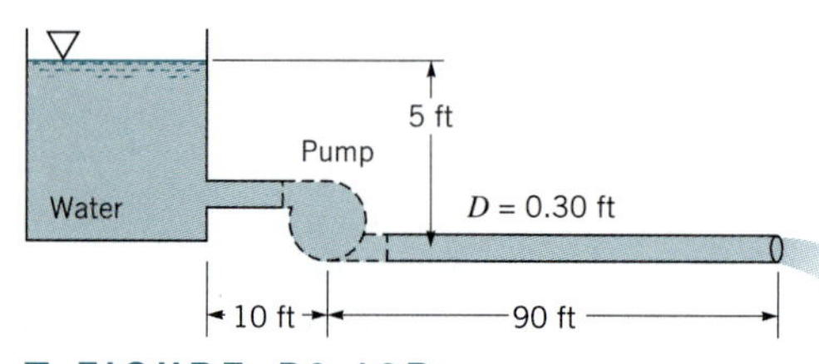

■ FIGURE P8.13R

8.14R (Single pipe with pump) The pump shown in Fig. P8.14R adds a 15-ft head to the water being pumped when the flowrate is 1.5 ft^3/s. Determine the friction factor for the pipe.

(ANS: 0.0306)

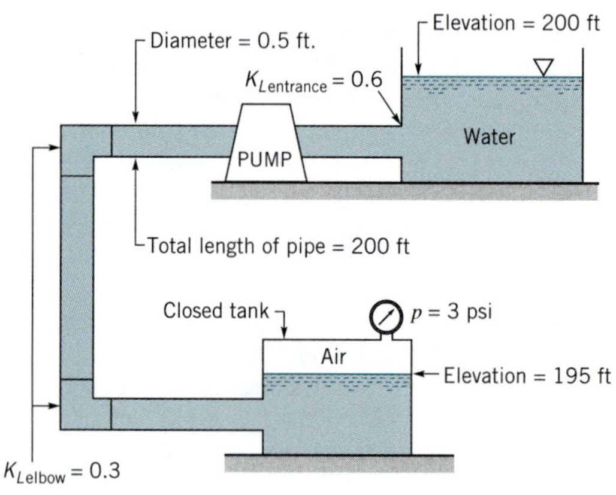

■ FIGURE P8.14R

8.15R (Single pipe with turbine) Water drains from a pressurized tank through a pipe system as shown in Fig. P8.15R. The head of the turbine is equal to 116 m. If entrance effects are negligible, determine the flow rate.

(ANS: 3.71×10^{-2} m^3/s)

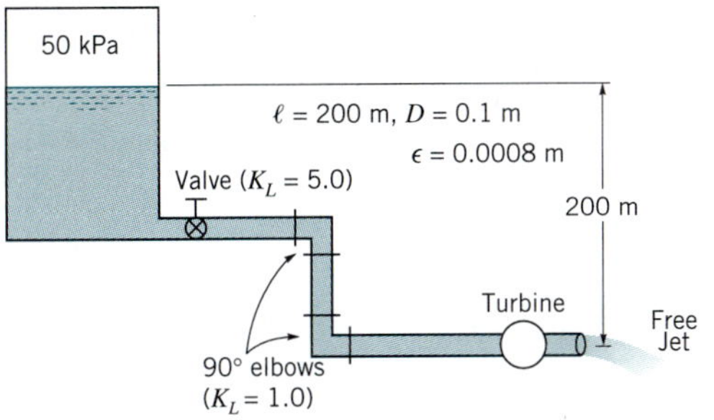

■ FIGURE P8.15R

8.16R (Multiple pipes) The three tanks shown in Fig. P8.16R are connected by pipes with friction factors of 0.03 for each pipe. Determine the water velocity in each pipe. Neglect minor losses.

(ANS: (A) 4.73 ft/s, (B) 8.35 ft/s, (C) 10.3 ft/s)

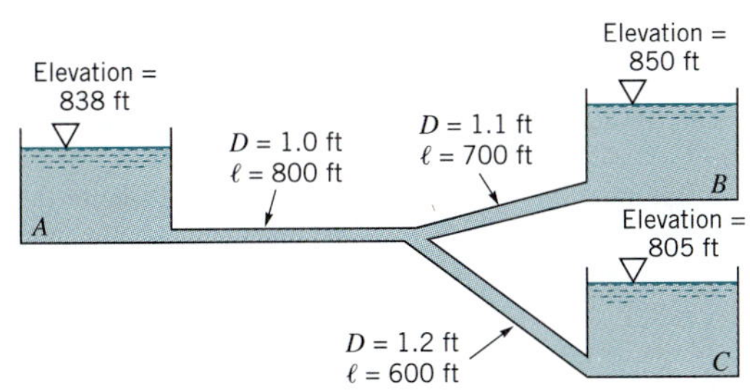

■ FIGURE P8.16R

8.17R (Flow meters) Water flows in a 0.10-m-diameter pipe at a rate of 0.02 m^3/s. If the pressure difference across the orifice meter in the pipe is to be 28 kPa, what diameter orifice is needed?

(ANS: 0.070 m)

8.18R (Flow meters) A 2.5-in.-diameter flow nozzle is installed in a 3.8-in.-diameter pipe that carries water at 160 °F. If the flowrate is 0.78 cfs, determine the reading on the inverted air-water U-tube manometer used to measure the pressure difference across the meter.

(ANS: 6.75 ft)

Problems

Note: Unless otherwise indicated use the values of fluid properties found in the tables on the inside of the front cover. Problems designated with an (*) are intended to be solved with the aid of a programmable calculator or a computer. Problems designated with a (†) are "open-ended" problems and require critical thinking in that to work them one must make various assumptions and provide the necessary data. There is not a unique answer to these problems.

8.1 Rainwater runoff from a parking lot flows through a 3-ft-diameter pipe, completely filling it. Would you expect the flow to be laminar or turbulent? Support your answer with appropriate calculations.

† **8.2** Under normal circumstances is the air flow through your trachea (your windpipe) laminar or turbulent? List all assumptions and show all calculations.

8.3 The flow of water in a 3-mm-diameter pipe is to remain laminar. Plot a graph of the maximum flowrate allowed as a function of temperature for $0 < T < 100$ °C.

8.4 Air at 100 °F flows at standard atmospheric pressure in a pipe at a rate of 0.08 lb/s. Determine the minimum diameter allowed if the flow is to be laminar.

8.5 Carbon dioxide at 20 °C and a pressure of 550 kPa (abs) flows in a pipe at a rate of 0.04 N/s. Determine the maximum diameter allowed if the flow is to be turbulent.

† **8.6** List some typical everyday pipe flow situations and discuss whether the flow is laminar or turbulent; fully developed or entrance flow.

8.7 To cool a given room it is necessary to supply 5 ft^3/s of air through an 8-in.-diameter pipe. Approximately how long is the entrance length in this pipe?

8.8 The wall shear stress in a fully developed flow portion of a 12-in.-diameter pipe carrying water is 1.85 lb/ft^2. Determine the pressure gradient, $\partial p/\partial x$, where x is in the flow direction, if the pipe is **(a)** horizontal, **(b)** vertical with flow up, or **(c)** vertical with flow down.

8.9 The pressure drop needed to force water through a horizontal 1-in.-diameter pipe is 0.60 psi for every 12-ft length of pipe. Determine the shear stress on the pipe wall. Determine the shear stress at distances 0.3 and 0.5 in. away from the pipe wall.

8.10 Repeat Problem 8.9 if the pipe is on a 20° hill. Is the flow up or down the hill? Explain.

8.11 Water flows in a constant diameter pipe with the following conditions measured: At section (a) $p_a = 32.4$ psi and $z_a = 56.8$ ft; at section (b) $p_b = 29.7$ psi and $z_b = 68.2$ ft. Is the flow from (a) to (b) or from (b) to (a)? Explain.

8.12 Repeat Problem 8.11 if the specific gravity of the fluid is 0.50.

8.13 Some fluids behave as a non-Newtonian power-law fluid characterized by $\tau = -C(du/dr)^n$, where $n = 1, 3, 5$, and so on, and C is a constant. (If $n = 1$, the fluid is the customary Newtonian fluid.) For flow in a round pipe of a diameter D, integrate the force balance equation (Eq. 8.3) to obtain the velocity profile

$$u(r) = \frac{-n}{(n+1)}\left(\frac{\Delta p}{2\ell C}\right)^{1/n}\left[r^{(n+1)/n} - \left(\frac{D}{2}\right)^{(n+1)/n}\right]$$

***8.14** For the flow discussed in Problem 8.13, plot the dimensionless velocity profile u/V_c, where V_c is the centerline velocity (at $r = 0$), as a function of the dimensionless radial coordinate $r/(D/2)$, where D is the pipe diameter. Consider values of $n = 1, 3, 5$, and 7.

8.15 A fluid of specific gravity 0.96 flows steadily in a long, vertical 1-in.-diameter pipe with an average velocity of 0.50 ft/s. If the pressure is constant throughout the fluid, what is the viscosity of the fluid? Determine the shear stress on the pipe wall.

8.16 Water is pumped between two tanks as shown in Fig. P8.16. The energy line is as indicated. Is the fluid being pumped from A to B or B to A? Explain. Which pipe has the larger diameter: A to the pump or B to the pump? Explain.

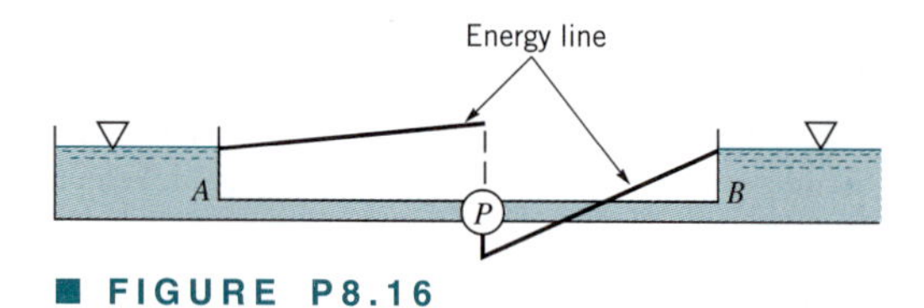

■ FIGURE P8.16

8.17 Glycerin at 20 °C flows upward in a vertical 75-mm-diameter pipe with a centerline velocity of 1.0 m/s. Determine the head loss and pressure drop in a 10-m length of the pipe.

8.18 A fluid flows through a horizontal 0.1-in.-diameter pipe. When the Reynolds number is 1500, the head loss over a 20-ft length of the pipe is 6.4 ft. Determine the fluid velocity.

8.19 A viscous fluid flows in a 0.10-m-diameter pipe such that its velocity measured 0.012 m away from the pipe wall is 0.8 m/s. If the flow is laminar, determine the centerline velocity and the flowrate.

8.20 Oil (specific weight = 8900 N/m^3, viscosity = 0.10 N·s/m^2) flows through a horizontal 23-mm-diameter tube as shown in Fig. P8.20. A differential U-tube manometer is used to measure the pressure drop along the tube. Detrmine the range of values for h for laminar flow.

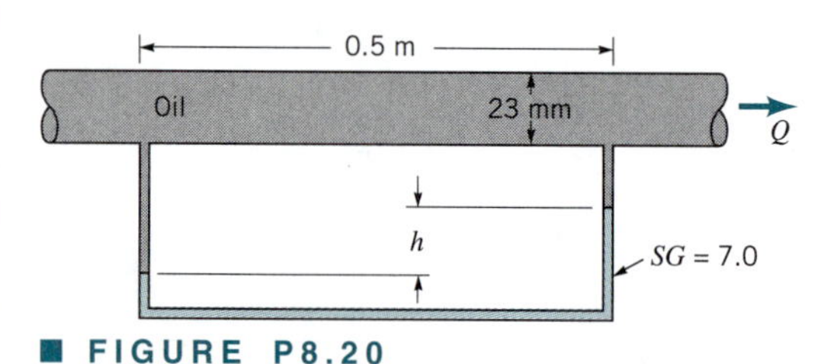

■ FIGURE P8.20

8.21 A fluid flows in a smooth pipe with a Reynolds number of 6000. By what percent would the head loss be reduced if the flow could be maintained as laminar flow rather than the expected turbulent flow?

8.22 Oil of $SG = 0.87$ and a kinematic viscosity $\nu = 2.2 \times 10^{-4}$ m^2/s flows through the vertical pipe shown in Fig. P8.22 at a rate of 4×10^{-4} m^3/s. Determine the manometer reading, h.

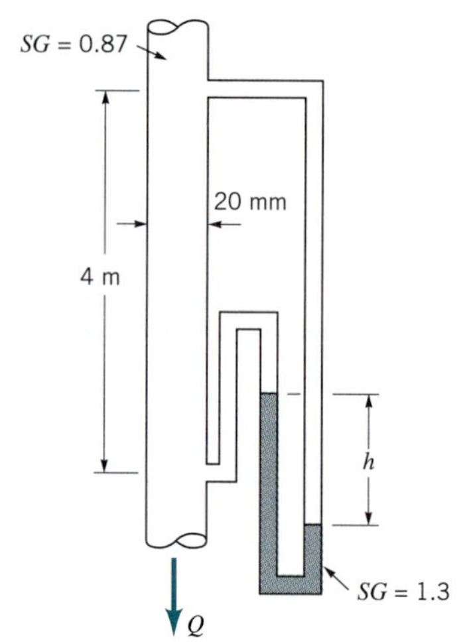

■ FIGURE P8.22

8.23 Determine the manometer reading, h, for Problem 8.22 if the flow is up rather than down the pipe. Note: The manometer reading will be reversed.

8.24 For Problem 8.22, what flowrate (magnitude and direction) will cause $h = 0$?

8.25 The kinetic energy coefficient, α, is defined in Eq. 5.86. Show that its value for a power-law turbulent velocity profile (Eq. 8.31) is given by $\alpha = (n+1)^3(2n+1)^3/[4n^4(n+3)(2n+3)]$.

8.26 If the velocity profile for turbulent flow in a pipe is approximated by the power-law profile (Eq. 8.31), at what radial location should a Pitot tube be placed if it is to measure the average velocity in the pipe? Assume $n = 7, 8,$ or 9.

8.27 Water at 80 °F flows in a 6-in.-diameter pipe with a flowrate of 2.0 cfs. What is the approximate velocity at a distance 2.0 in. away from the wall? Determine the centerline velocity.

8.28 During a heavy rainstorm, water from a parking lot completely fills an 18-in.-diameter, smooth, concrete storm sewer. If the flowrate is 10 ft^3/s, determine the pressure drop in a 100-ft horizontal section of the pipe. Repeat the problem if there is a 2-ft change in elevation of the pipe per 100 ft of its length.

8.29 Carbon dioxide at a temperature of 0 °C and a pressure of 600 kPa (abs) flows through a horizontal 40-mm-diameter pipe with an average velocity of 2 m/s. Determine the friction factor if the pressure drop is 235 N/m^2 per 10-m length of pipe.

8.30 Water flows through a 6-in.-diameter horizontal pipe at a rate of 2.0 cfs and a pressure drop of 4.2 psi per 100 ft of pipe. Determine the friction factor.

8.31 Air flows through the 0.108-in.-diameter, 24-in.-long tube shown in Fig. P8.31. Determine the friction factor if the flowrate is $Q = 0.00191$ cfs when $h = 1.70$ in. Compare your results with the expression $f = 64/\text{Rd}$. Is the flow laminar or turbulent?

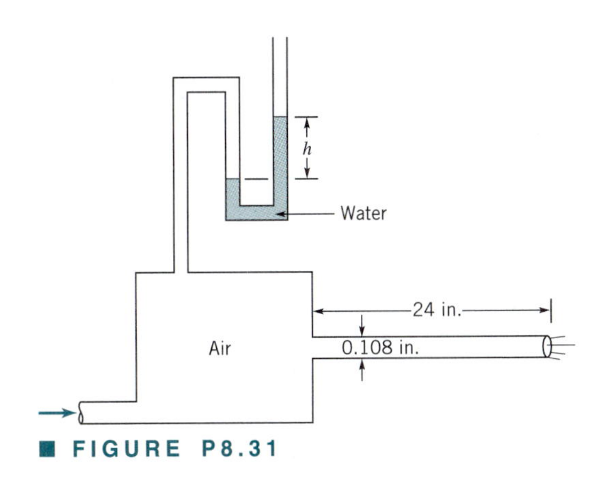

■ **FIGURE P8.31**

8.32 Water at 10 °C flows through a smooth 60-mm-diameter pipe with an average velocity of 8 m/s. Would a scratch of height 0.005 mm on the pipe wall protrude through the viscous sublayer? Explain.

8.33 Determine the thickness of the viscous sublayer in a smooth 8-in.-diameter pipe if the Reynolds number is 25,000.

8.34 Water at 60 °F flows through a 6-in.-diameter pipe with an average velocity of 15 ft/s. Approximately what is the height of the largest roughness element allowed if this pipe is to be classified as smooth?

8.35 A 70-ft-long, 0.5-in.-diameter hose with a roughness of $\varepsilon = 0.0009$ ft is fastened to a water faucet where the pressure is p_1. Determine p_1 if there is no nozzle attached and the average velocity in the hose is 6 ft/s. Neglect minor losses and elevation changes.

8.36 Repeat Problem 8.35 if there is a nozzle of diameter 0.25 in. attached to the end of the hose.

***8.37** The following equation is sometimes used in place of the Colebrook equation (Eq. 8.35):

$$f = \frac{1.325}{\{\ln[(\varepsilon/3.7D) + (5.74/\text{Re}^{0.9})]\}^2}$$

for $10^{-6} < \varepsilon/D < 10^{-2}$ and $5000 < \text{Re} < 10^{+8}$ (Ref. 22, pg. 220). An advantage of this equation is that given Re and ε/D, it does not require an iteration procedure to obtain f. Plot a graph of the percent difference in f as given by this equation and the original Colebrook equation for Reynolds numbers in the range of validity of the above equation, with $\varepsilon/D = 10^{-4}$.

8.38 Water flows at a rate of 10 gallons per minute in a new horizontal 0.75-in.-diameter galvanized iron pipe. Determine the pressure gradient, $\Delta p/\ell$, along the pipe.

8.39 For a given head loss per unit length, what effect on the flowrate does doubling the pipe diameter have if the flow is **(a)** laminar, or **(b)** completely turbulent?

8.40 A garden hose is attached to a faucet that is fully opened. Without a nozzle on the end of the hose, the water does not shoot very far. However, if you place your thumb over a portion of the end of the hose, it is possible to shoot the water a considerable distance. Explain this phenomenon. (*Note:* The flowrate decreases as the area covered by your thumb increases.)

8.41 Air at standard temperature and pressure flows through a 1-in.-diameter galvanized iron pipe with an average velocity of 8 ft/s. What length of pipe produces a head loss equivalent to **(a)** a flanged 90° elbow, **(b)** a wide-open angle valve, or **(c)** a sharp-edged entrance?

***8.42** Water at 40 °C flows through drawn tubings with diameters of 0.025, 0.050, or 0.075 m. Plot the head loss in each meter length of pipe for flowrates between 5×10^{-4} m^3/s and 50×10^{-4} m^3/s. In your solution obtain the friction factor from the Colebrook formula.

8.43 Air at standard temperature and pressure flows at a rate of 7.0 cfs through a horizontal, galvanized iron duct that has a rectangular cross-sectional shape of 12 in. by 6 in. Estimate the pressure drop per 200 ft of duct.

8.44 A viscous oil with a specific gravity $SG = 0.85$ and a viscosity of 0.10 Pa·s flows from tank *A* to tank *B* through the six rectangular slots indicated in Fig. P8.44. If the total flowrate is 30 mm^3/s and minor losses are negligible, determine the pressure in tank *A*.

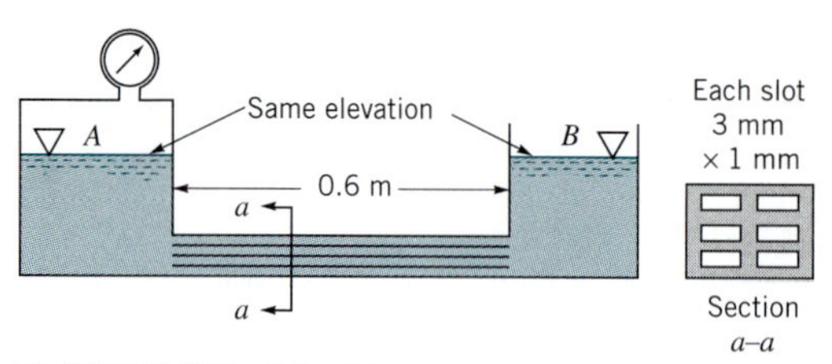

■ **FIGURE P8.44**

† **8.45** Consider the process of donating blood. Blood flows from a vein in which the pressure is greater than atmospheric, through a long small-diameter tube, and into a plastic bag that is essentially at atmospheric pressure. Based on fluid mechanics principles, estimate the amount of time it takes to donate a pint of blood. List all assumptions and show calculations.

8.46 To conserve water and energy, a "flow reducer" is installed in the shower head as shown in Fig. P8.46. If the pressure at point (1) remains constant and all losses except for that in the "flow reducer" are neglected, determine the value of the loss coefficient (based on the velocity in the pipe) of the "flow reducer" if its presence is to reduce the flowrate by a factor of 2. Neglect gravity.

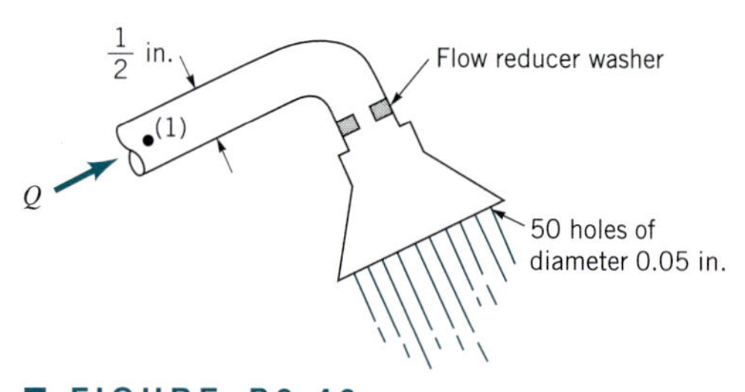

■ **FIGURE P8.46**

8.47 Water flows at a rate of 0.040 m^3/s in a 0.12-m-diameter pipe that contains a sudden contraction to a 0.06-m-diameter pipe. Determine the pressure drop across the contraction section. How much of this pressure difference is due to losses and how much is due to kinetic energy changes?

† **8.48** Except during periods of high energy demand, the flow of blood in a human's veins and arteries is laminar rather than turbulent. Discuss why this is beneficial.

8.49 At time $t = 0$ the level of water in tank *A* shown in Fig. P8.49 is 2 ft above that in tank *B*. Plot the elevation of the water in tank *A* as a function of time until the free surfaces in both tanks are at the same elevation. Assume quasisteady conditions—that is, the steady pipe flow equations are assumed valid at any time, even though the flowrate does change (slowly) in time. Neglect minor losses. *Note:* Verify and use the fact that the flow is laminar.

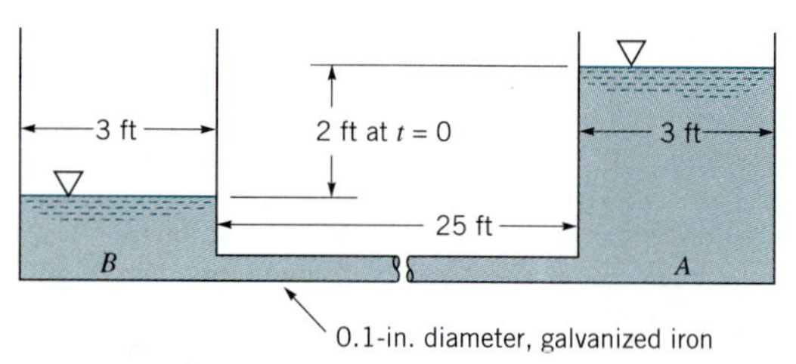

■ **FIGURE P8.49**

***8.50** Repeat Problem 8.49 if the pipe diameter is changed to 0.1 ft rather than 0.1 in. *Note:* The flow may not be laminar for this case.

† **8.51** Your garden hose has developed a leak in the form of a small hole through which water sprays vertically upward. Explain why the water from the nozzle at the end of the hose can go higher than that leaking through the small hole.

8.52 Gasoline flows in a smooth pipe of 40-mm diameter at a rate of 0.001 m^3/s. If it were possible to prevent turbulence from occurring, what would be the ratio of the head loss for the actual turbulent flow compared to that if it were laminar flow?

8.53 A 3-ft-diameter duct is used to carry ventilating air into a vehicular tunnel at a rate of 9000 ft^3/min. Tests show that the pressure drop is 1.5 in. of water per 1500 ft of duct. What is the value of the friction factor for this duct and the approximate size of the equivalent roughness of the surface of the duct?

8.54 Natural gas ($\rho = 0.0044$ slugs/ft^3 and $\nu = 5.2 \times 10^{-5}$ ft^2/s) is pumped through a horizontal 6-in.-diameter cast-iron pipe at a rate of 800 lb/hr. If the pressure at section (1) is 50 psi (abs), determine the pressure at section (2) 8 mi downstream if the flow is assumed incompressible. Is the incompressible assumption reasonable? Explain.

***8.55** Water flows in a 20-mm-diameter galvanized iron pipe with average velocities between 0.01 and 10.0 m/s. Plot the head loss per meter of pipe length over this velocity range. Discuss.

8.56 A fluid flows through a smooth horizontal 2-m-long tube of diameter 2 mm with an average velocity of 2.1 m/s.

Determine the head loss and the pressure drop if the fluid is **(a)** air, **(b)** water, or **(c)** mercury.

8.57 Air at standard temperature and pressure flows through a horizontal 2 ft by 1.3 ft rectangular galvanized iron duct with a flowrate of 8.2 cfs. Determine the pressure drop in inches of water per 200-ft length of duct.

8.58 Air flows through a rectangular galvanized iron duct of size 0.30 m by 0.15 m at a rate of 0.068 m^3/s. Determine the head loss in 12 m of this duct.

8.59 Air at standard conditions flows through a horizontal 1 ft by 1.5 ft rectangular wooden duct at a rate of 5000 ft^3/min. Determine the head loss, pressure drop, and power supplied by the fan to overcome the flow resistance in 500 ft of the duct.

8.60 When the valve is closed, the pressure throughout the horizontal pipe shown in Fig. P8.60 is 400 kPa, and the water level in the closed, air-filled surge chamber is $h = 0.4$ m. If the valve is fully opened and the pressure at point (1) remains 400 kPa, determine the new level of the water in the surge chamber. Assume the friction factor is $f = 0.02$ and the fittings are threaded fittings.

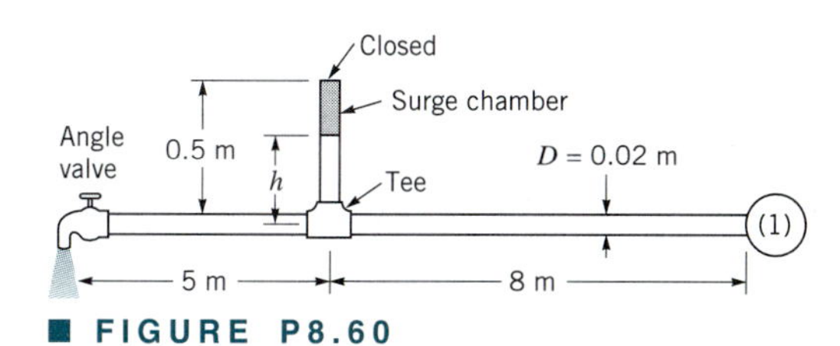

■ **FIGURE P8.60**

8.61 What horsepower is added to water to pump it vertically through a 200-ft-long, 1.0-in.-diameter drawn tubing at a rate of 0.060 ft^3/s if the pressures at the inlet and outlet are the same?

8.62 Water flows from a lake as is shown in Fig. P8.62 at a rate of 4.0 cfs. Is the device inside the building a pump or a turbine? Explain and determine the horsepower of the device. Neglect all minor losses and assume the friction factor is 0.025.

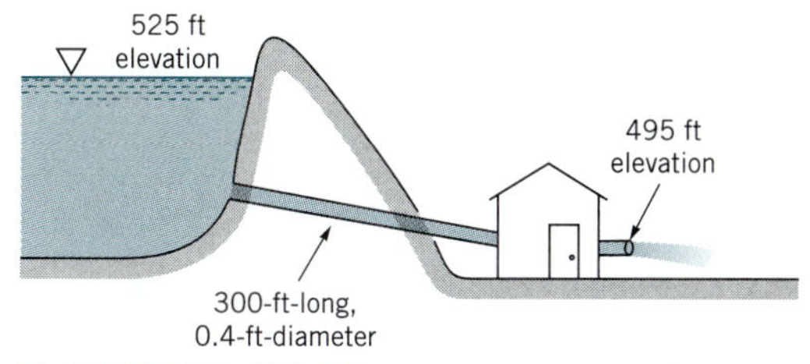

■ **FIGURE P8.62**

8.63 Repeat Problem 8.62 if the flowrate is 1.0 cfs.

8.64 At a ski resort, water at 40 °F is pumped through a 3-in.-diameter, 2000-ft-long steel pipe from a pond at an elevation of 4286 ft to a snow-making machine at an elevation of 4623 ft at a rate of 0.26 ft^3/s. If it is necessary to maintain a pressure of 180 psi at the snow-making machine, determine the horsepower added to the water by the pump. Neglect minor losses.

8.65 Water flows through the screen in the pipe shown in Fig. P8.65 as indicated. Determine the loss coefficient for the screen.

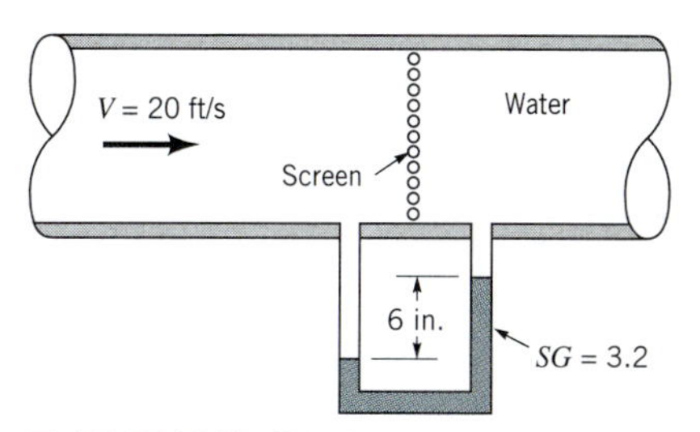

■ **FIGURE P8.65**

† **8.66** The needle valve shown in Fig. P8.66 consists of a cone-shaped plunger that can be positioned accurately by turning the handle. Obtain an equation for the loss coefficient for this valve as a function of the number of turns the valve stem is rotated from its closed position. List all assumptions and show calculations.

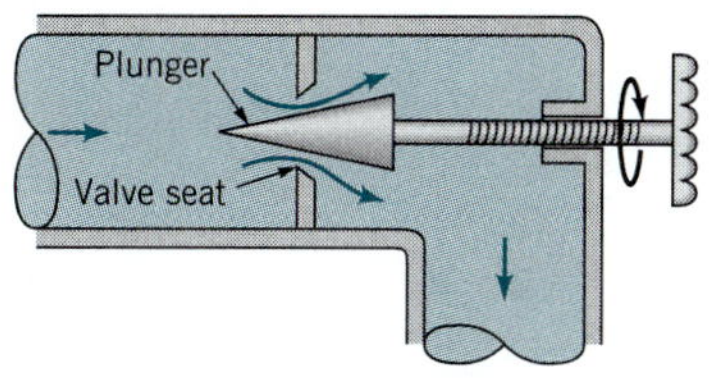

■ **FIGURE P8.66**

8.67 Air at 80 °F and standard atmospheric pressure flows through a furnace filter with an average velocity of 2.4 ft/s. If the pressure drop across the filter is 0.06 in. of water, what is the loss coefficient for the filter?

8.68 Assume a car's exhaust system can be approximated as 14 ft of 0.125-ft-diameter cast-iron pipe with the equivalent of six 90° flanged elbows and a muffler. The muffler acts as a resistor with a loss coefficient of $K_L = 8.5$. Determine the pressure at the beginning of the exhaust system if the flowrate is 0.10 cfs, the temperature is 250 °F, and the flow is incompressible.

8.69 Air is to flow through a smooth, horizontal, rectangular duct at a rate of 100 m^3/s with a pressure drop of not more than 40 mm of water per 50 m of duct. If the aspect ratio (width to height) is 3 to 1, determine the size of the duct.

8.70 Repeat Problem 3.14 if all head losses are included. The pipes are 1-in. copper pipes with regular flanged fittings. The faucets are globe valves.

8.71 Water at 40 °F flows through the coils of the heat exchanger as shown in Fig. P8.71 at a rate of 0.9 gal/min. Determine the pressure drop between the inlet and outlet of the horizontal device.

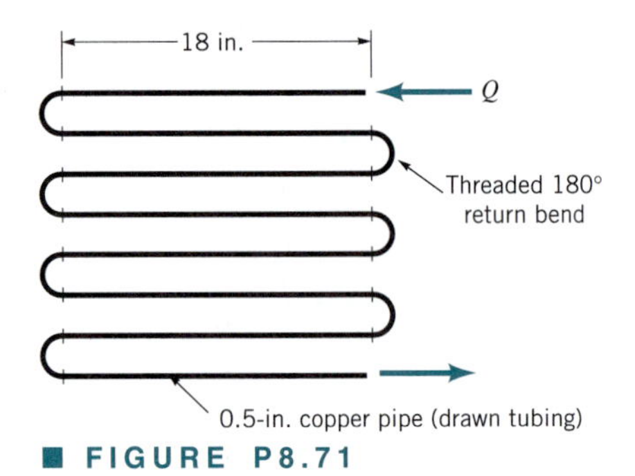

■ FIGURE P8.71

8.72 Water at 40 °F is pumped from a lake as shown in Fig. P8.72. What is the maximum flowrate possible without cavitation occurring?

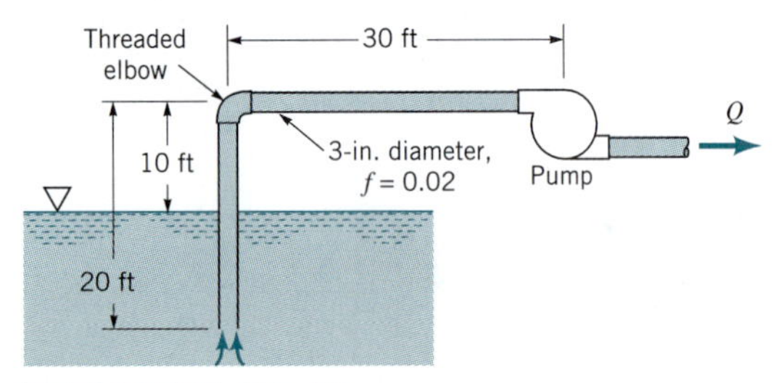

■ FIGURE P8.72

8.73 The $\frac{1}{2}$-in.-diameter hose shown in Fig. P8.73 can withstand a maximum pressure of 200 psi without rupturing. Determine the maximum length, ℓ, allowed if the friction factor is 0.022 and the flowrate is 0.010 cfs. Neglect minor losses.

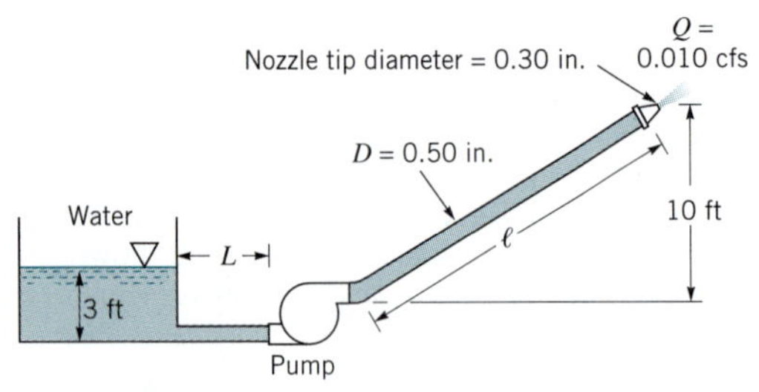

■ FIGURE P8.73

8.74 The hose shown in Fig. P8.73 will collapse if the pressure within it is lower than 10 psi below atmospheric pressure. Determine the maximum length, L, allowed if the friction factor is 0.015 and the flowrate is 0.010 cfs. Neglect minor losses.

8.75 The contents of the fire extinguisher shown in Fig. P8.75 are pressurized to 60 psi. Determine the flowrate if the friction factor of the tube is 0.018 and the loss coefficients for the entrance and the valve/nozzle assembly (based on the velocity in the tube) are 1.0 and 7.8, respectively. Repeat the problem if pipe friction were neglected but minor losses included. Repeat the problem if minor losses were neglected but pipe friction included. Comment on the importance of the various losses for this flow.

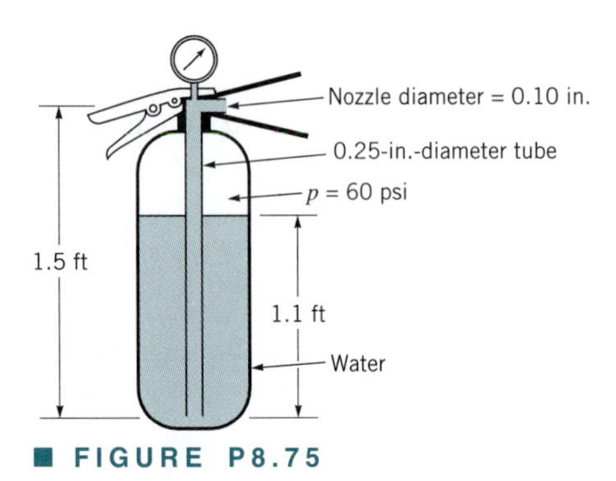

■ FIGURE P8.75

8.76 Water flows from the container shown in Fig. P8.76. Determine the loss coefficient needed in the valve if the water is to "bubble up" 3 in. above the outlet pipe. The entrance is slightly rounded.

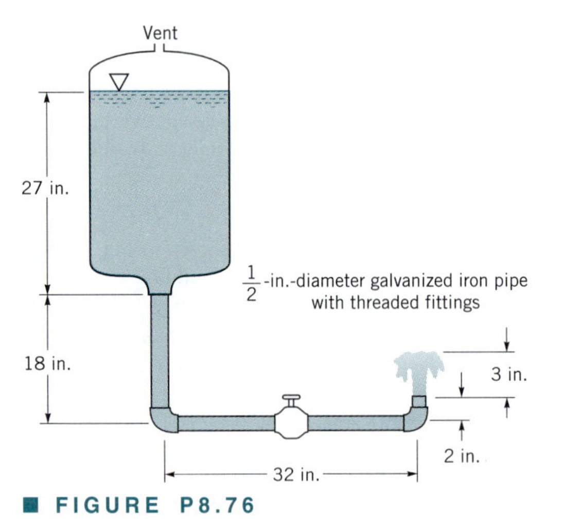

■ FIGURE P8.76

8.77 The pressure at section (2) shown in Fig. P8.77 is not to fall below 60 psi when the flowrate from the tank varies from 0 to 1.0 cfs and the branch line is shut off. Determine the minimum height, h, of the water tank under the assumption that **(a)** minor losses are negligible, **(b)** minor losses are not negligible.

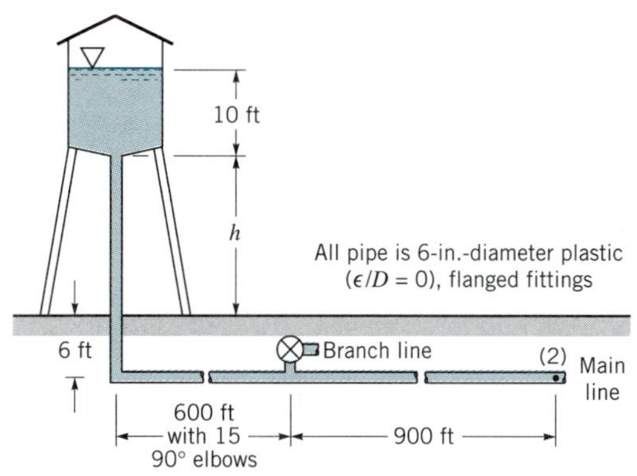

■ FIGURE P8.77

8.78 Repeat Problem 8.77 with the assumption that the branch line is open so that half of the flow from the tank goes into the branch, and half continues in the main line.

8.79 Repeat Problem 3.43 if head losses are included.

***8.80** Water flows from a large, open tank through a sharp-edged entrance and into a galvanized iron pipe of length 100 m and diameter 10 mm. The water exits the pipe as a free jet at a distance h below the free surface of the tank. Plot a log–log graph of the flowrate, Q, as a function of h for $0.1 \le h \le 10$ m.

8.81 Water flows at a rate of 0.50 cfs from tank A to tank B through a horizontal 3-in.-diameter cast-iron pipe of length 200 ft. If minor losses are neglected, determine the difference in elevation of the free surfaces of the tanks.

8.82 A flowrate of 3.5 ft³/s is to be maintained in a horizontal aluminum pipe ($\varepsilon = 5 \times 10^{-6}$ ft). The inlet and outlet pressures are 65 psi and 30 psi, respectively, and the pipe length is 500 ft. Determine the diameter of this water pipe.

8.83 Water flows downward through a vertical smooth pipe. When the flowrate is 0.5 ft³/s there is no change in pressure along the pipe. Determine the diameter of the pipe.

† **8.84** Estimate the increase in flowrate from your lawn sprinkler if you were to replace your present $\frac{1}{2}$-in.-diameter hose with a $\frac{5}{8}$-in. diameter hose. List all assumptions and show all calculations.

***8.85** Water drains from the bathtub shown in Fig. P8.85 through the drain indicated. Determine the length of time taken for the tub to drain if the water is initially 8 in. deep and the friction factor is $f = 0.02$. Assume quasisteady flow conditions and that the drain pipe remains full. See table below.

A (ft²)	h (in.)
0	0
4.1	1
6.7	2
8.5	3
9.1	4
9.4	5
9.6	6
9.8	7
10.0	8

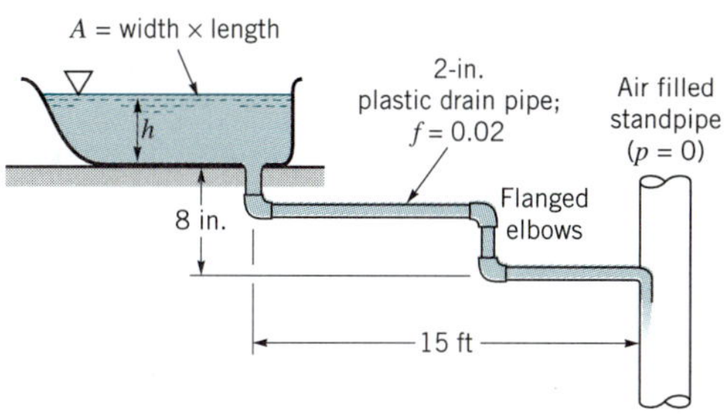

■ FIGURE P8.85

8.86 Water flows from a tank and through the new pipe shown in Fig. P8.86. The flowrate is 63 ft³/s when the valve is adjusted to produce the water level in the piezometer tube as indicated. Is the pipe made from concrete, galvanized iron, or plastic? Show all calculations.

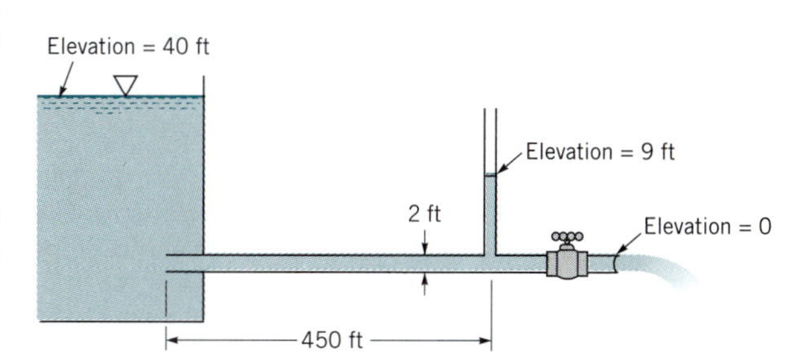

■ FIGURE P8.86

8.87 Water flows through the pipe shown in Fig. P8.87. Determine the net tension in the bolts if minor losses are neglected and the wheels on which the pipe rests are frictionless.

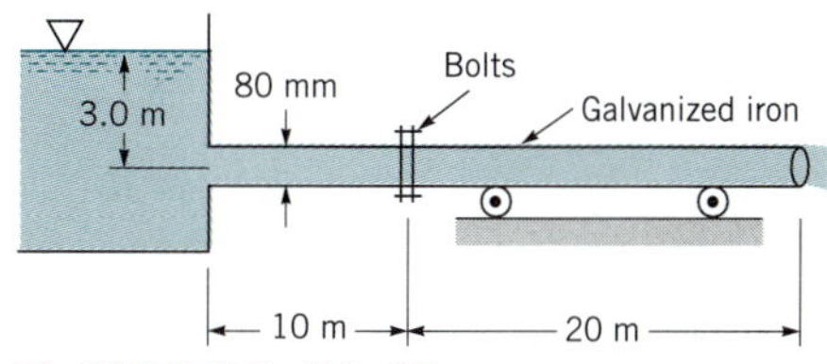

■ FIGURE P8.87

8.88 Water flows through two sections of the vertical pipe shown in Fig. P8.88. The bellows connection cannot support any force in the vertical direction. The 0.4-ft-diameter pipe weighs 0.2 lb/ft and the friction factor is assumed to be 0.02. At what velocity will the force, F, required to hold the pipe be zero?

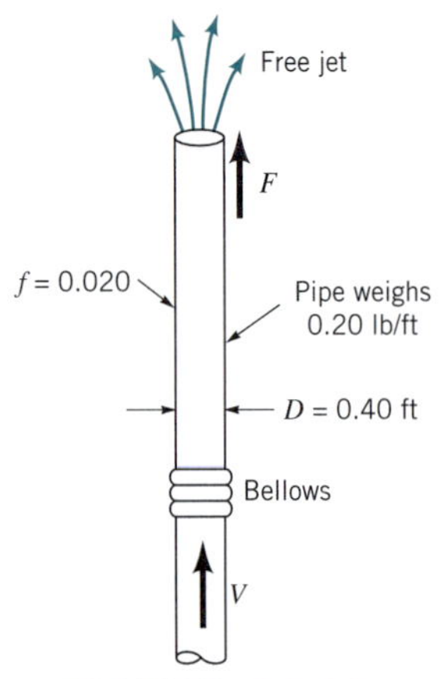

■ **FIGURE P8.88**

8.89 The pump shown in Fig. P8.89 adds 25 kW to the water and causes a flowrate of 0.04 m^3/s. Determine the flowrate expected if the pump is removed from the system. Assume $f = 0.016$ for either case and neglect minor losses.

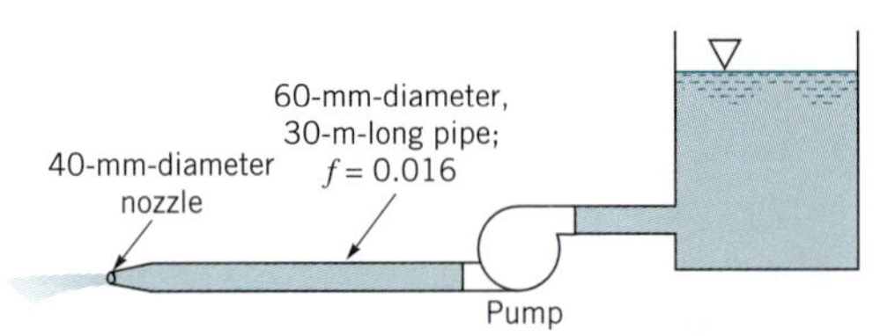

■ **FIGURE P8.89**

8.90 A certain process requires 2.3 cfs of water to be delivered at a pressure of 30 psi. This water comes from a large diameter supply main in which the pressure remains at 60 psi. If the galvanized iron pipe connecting the two locations is 200 ft long and contains six threaded 90° elbows, determine the pipe diameter. Elevation differences are negligible.

8.91 The turbine shown in Fig. P8.91 develops 400 kW. Determine the flowrate if **(a)** head losses are negligible or **(b)** head loss due to friction in the pipe is considered. Assume $f = 0.02$. *Note:* There may be more than one solution or there may be no solution to this problem.

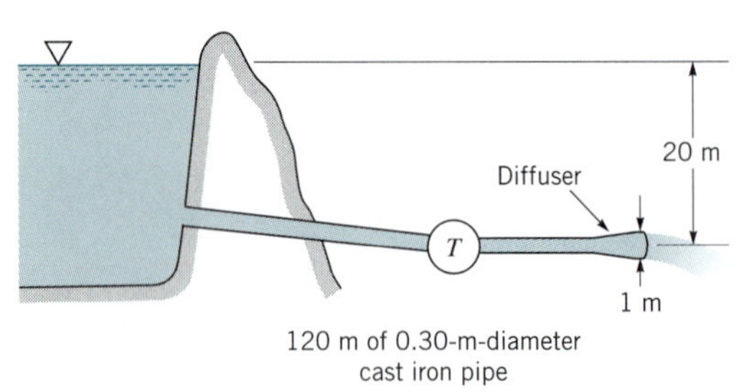

■ **FIGURE P8.91**

8.92 Water is pumped from a large, closed, pressurized tank as shown in Fig. P8.92. The friction factor for the constant diameter pipe is 0.025, and minor losses are negligible. **(a)** Determine the flowrate if the pump adds 1 horsepower to the water. **(b)** Repeat the problem if the outlet of the pipe is 10 ft above the air–water interface rather than 8 ft as shown in the figure. Comment on the differences between that of Part (a) and Part (b).

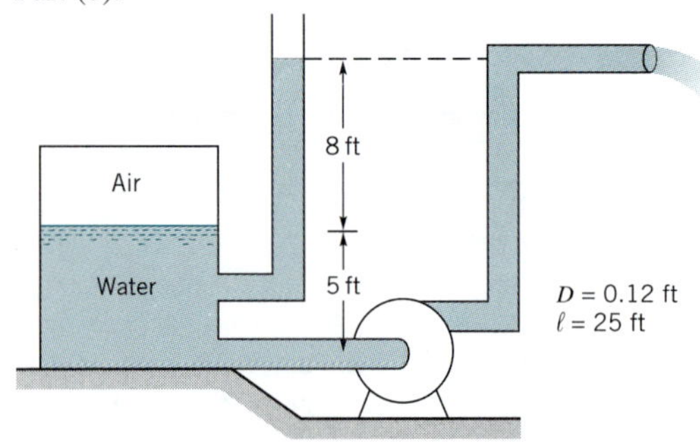

■ **FIGURE P8.92**

8.93 Water is circulated from a large tank, through a filter, and back to the tank as shown in Fig. P8.93. The power added to the water by the pump is 200 ft·lb/s. Determine the flowrate through the filter.

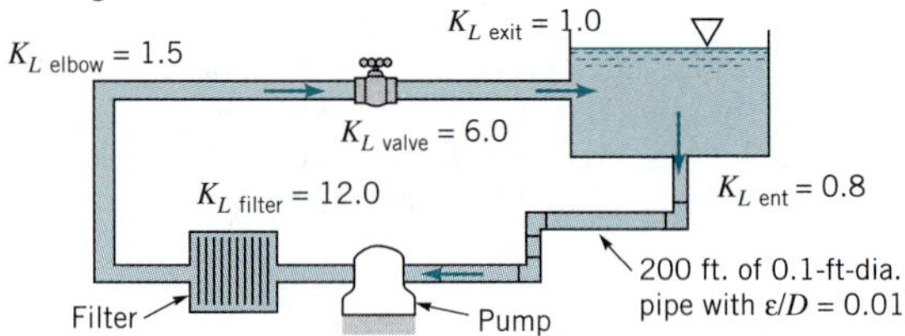

■ **FIGURE P8.93**

8.94 Water is to be moved from a large, closed tank in which the air pressure is 20 psi into a large, open tank through 2000 ft of smooth pipe at the rate of 3 ft^3/s. The fluid level in the open tank is 150 ft below that in the closed tank. Determine the required diameter of the pipe. Neglect minor losses.

8.95 Rainwater flows through the galvanized iron downspout shown in Fig. P8.95 at a rate of 0.006 m^3/s. Determine the size of the downspout cross section if it is a rectangle with an aspect ratio of 1.7 to 1 and it is completely filled with water. Neglect the velocity of the water in the gutter at the free surface and the head loss associated with the elbow.

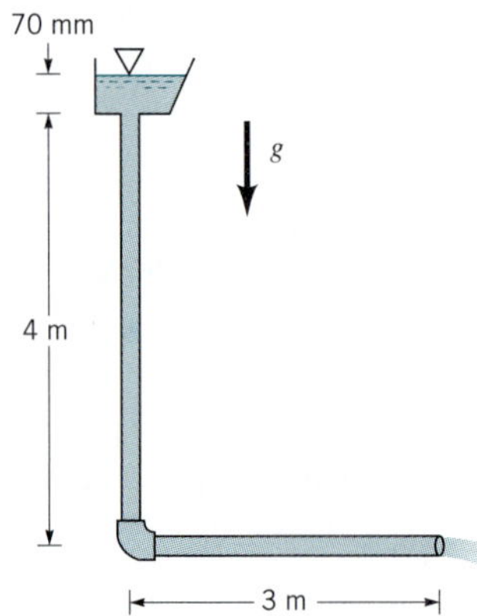

■ **FIGURE P8.95**

*8.96 Repeat Problem 8.95 if the downspout is circular.

8.97 Air, assumed incompressible, flows through the two pipes shown in Fig. P8.97. Determine the flowrate if minor losses are neglected and the friction factor in each pipe is 0.015. Determine the flowrate if the 0.5-in.-diameter pipe were replaced by a 1-in.-diameter pipe. Comment on the assumption of incompressibility.

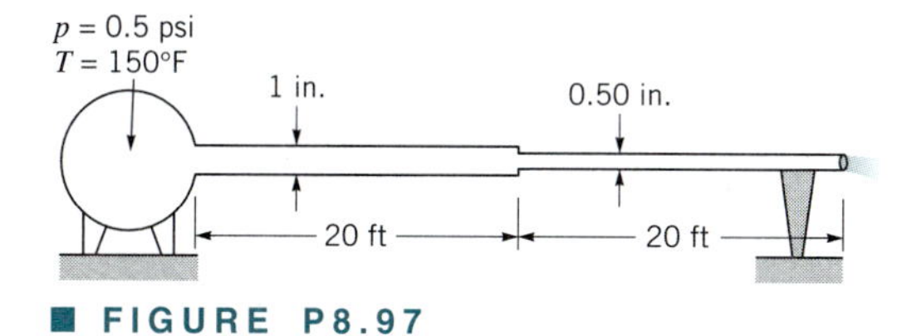

■ FIGURE P8.97

*8.98 Repeat Problem 8.97 if the pipes are galvanized iron and the friction factors are not known a priori.

† 8.99 A clothes washer in the basement, a dishwasher in the first floor kitchen, and the shower in the second floor bathroom are all attached to the same hot water feeder line coming from the water heater. Discuss and show with appropriate equations how the operation of one of these three devices could affect the operation of another.

8.100 With the valve closed, water flows from tank *A* to tank *B* as shown in Fig. P8.100. What is the flowrate into tank *B* when the valve is opened to allow water to flow into tank *C* also? Neglect all minor losses and assume that the friction factor is 0.02 for all pipes.

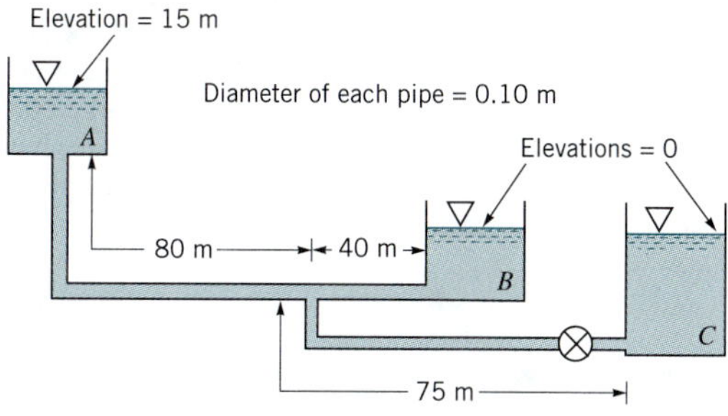

■ FIGURE P8.100

*8.101 Repeat Problem 8.100 if the friction factors are not known, but the pipes are steel pipes.

8.102 The three water-filled tanks shown in Fig. P8.102 are connected by pipes as indicated. If minor losses are neglected, determine the flowrate in each pipe.

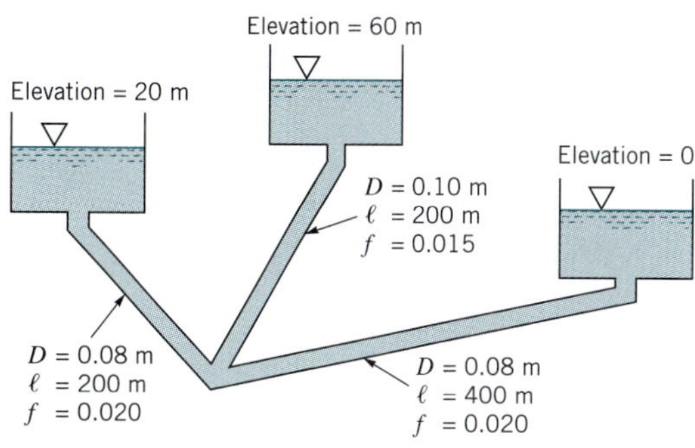

■ FIGURE P8.102

*8.103 Repeat Problem 8.102 if the friction factors are not known, but the pipes are steel pipes.

8.104 A 2-in.-diameter orifice plate is inserted in a 3-in.-diameter pipe. If the water flowrate through the pipe is 0.90 cfs, determine the pressure difference indicated by a manometer attached to the flow meter.

8.105 Air to ventilate an underground mine flows through a large 2-m-diameter pipe. A crude flowrate meter is constructed by placing a sheet metal "washer" between two sections of the pipe. Estimate the flowrate if the hole in the sheet metal has a diameter of 1.6 m and the pressure difference across the sheet metal is 8.0 mm of water.

8.106 Gasoline flows through a 35-mm-diameter pipe at a rate of 0.0032 m^3/s. Determine the pressure drop across a flow nozzle placed in the line if the nozzle diameter is 20 mm.

8.107 Air at 200 °F and 60 psia flows in a 4-in.-diameter pipe at a rate of 0.52 lb/s. Determine the pressure at the 2-in-diameter throat of a Venturi meter placed in the pipe.

8.108 A 50-mm-diameter nozzle meter is installed at the end of a 80-mm-diameter pipe through which air flows. A manometer attached to the static pressure tap just upstream from the nozzle indicates a pressure of 7.3 mm of water. Determine the flowrate.

8.109 A 2.5-in.-diameter nozzle meter is installed in a 3.8-in.-diameter pipe that carries water at 160 °F. If the inverted air-water U-tube manometer used to measure the pressure difference across the meter indicates a reading of 3.1 ft, determine the flowrate.

8.110 Water flows through the Venturi meter shown in Fig. P8.110. The specific gravity of the manometer fluid is 1.52. Determine the flowrate.

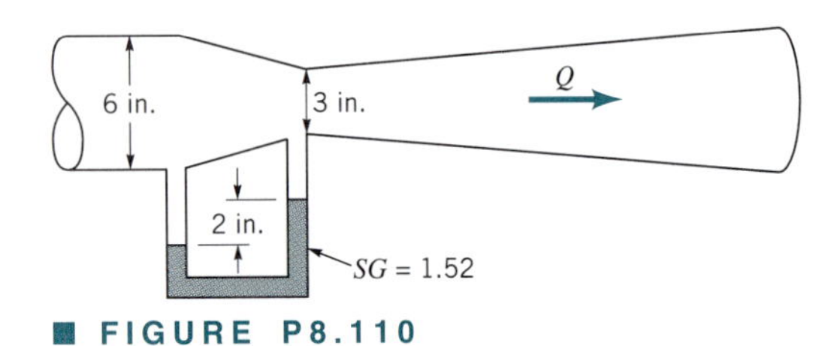

■ FIGURE P8.110

8.111 If the fluid flowing in Problem 8.110 were air, what would the flowrate be? Would compressibility effects be important? Explain.

8.112 Water flows through the orifice meter shown in Fig. P8.112 at a rate of 0.10 cfs. If $d = 0.1$ ft, determine the value of h.

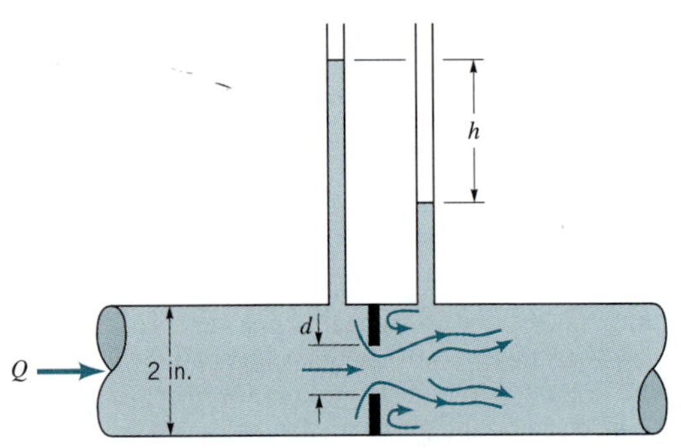

■ FIGURE P8.112

8.113 Water flows through the orifice meter shown in Fig. P8.112 at a rate of 0.10 cfs. If $h = 3.8$ ft, determine the value of d.

8.114 Water flows through the orifice meter shown in Fig. P8.112 such that $h = 1.6$ ft with $d = 1.5$ in. Determine the flowrate.

8.115 The device shown in Fig. P8.115 is used to investigate laminar flow through a pipe and to determine the Reynolds number for transition from laminar to turbulent flow. Air at a temperature of 73 °F and an absolute pressure of 29.93 in. of mercury flows with an average velocity of V through a small diameter, $D = 0.108$ in., tube of length $\ell = 24.0$ in. as indicated. The flowrate, Q, is determined by a rotameter and the pressure within the tank to which the tube is attached is given by the water manometer reading, h.

Experimentally determined values of Q and h are shown in the following table. Use these results to plot a graph on log–log paper of the friction factor, f, for this tube as a function of the Reynolds number based on the tube diameter, $\text{Re} = \rho VD/\mu$. On the same graph, plot the theoretical curve for laminar pipe flow. From the graphed results, determine the value of the Reynolds number for which the flow becomes turbulent in this pipe.

Compare the experimental and theoretical results and discuss some possible reasons for any differences between them.

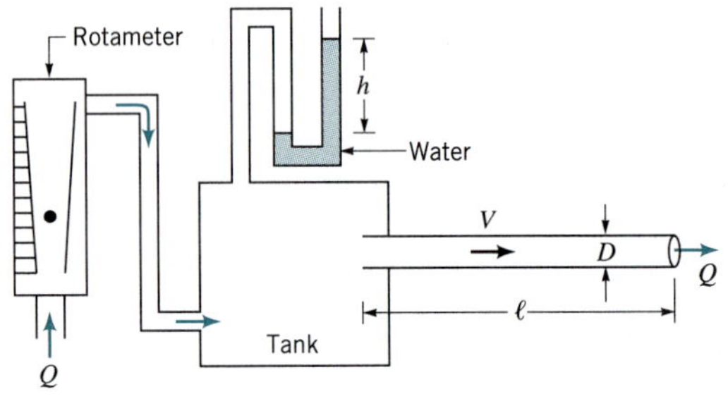

■ FIGURE P8.115

Q (cm³/min)	h (in.)
1,100	0.60
1,800	1.08
2,400	1.49
2,900	1.89
3,700	2.70
4,500	3.75
4,600	4.06
4,860	4.57
5,000	5.01
5,150	5.43
5,650	6.47
6,000	7.31
6,200	7.89

8.116 The device shown in Fig. P8.116 is used to calibrate a Venturi flow meter and an orifice flow meter. Water pumped from a tank flows through each of the meters. The flowrate, Q, is obtained by determining the time, t, it takes for a given volume, $\forall$, of water to be pumped from the tank. That is, $Q = \forall/t$. For this experiment, $\forall = 2$ gal. The pressure difference across each flow meter is determined by use of an inverted air-filled manometer as indicated. The manometer readings are h_v and h_o.

Experimentally determined values of t, h_v, and h_o are listed in the following table. Use these results to plot a graph on log–log paper of the flowrate as a function of the manometer reading. (There will be two separate curves—one for the Venturi meter and one for the orifice meter.) On the same graph, plot the theoretical flowrate assuming that the discharge coefficient is equal to one. (Note that both of these flow meters have the same pipe and throat diameters.) From the experimental results determine the discharge coefficient for each of these flow meters.

Discuss some possible sources of error in the results.

t (s)	h_v (in.)	h_o (in.)
34.0	1.9	5.5
27.0	3.8	9.3
25.0	4.2	10.1
20.4	6.2	14.7
17.3	8.7	21.4
15.7	11.0	26.7
13.2	14.5	37.0
12.0	18.1	43.2

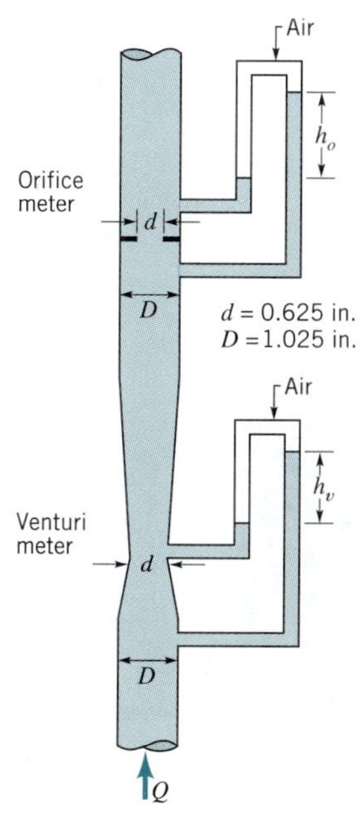

■ FIGURE P8.116

8.117 The apparatus shown in Fig. P8.117 is used to investigate the flowrate of water from a tank. Galvanized iron pipes of diameter $D = 0.595$ in. are used to drain water from the tank as indicated. The tank has a rectangular cross-section 12.0 by 9.0 in. in size. As the water flows from the tank with the valve wide open, the water depth, h, in the tank is measured as a function of time, t. The flowrate can therefore be determined as $Q = A\, dh/dt$, where A is the cross-sectional area of the tank.

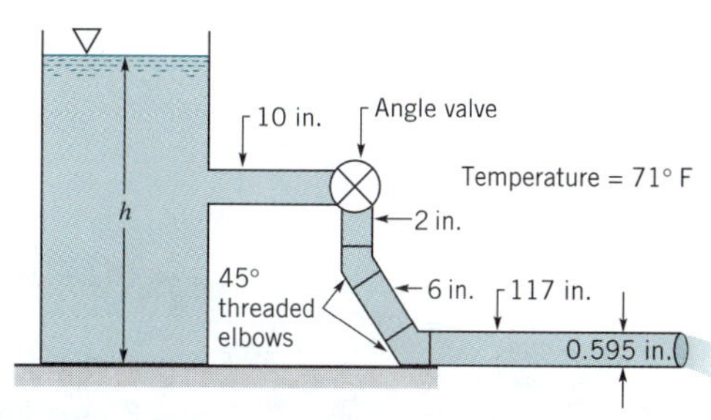

■ FIGURE P8.117

Experimentally determined values of h and t are shown in the following table. Use these results to determine the flowrate from the tank. For the given pipe system shown in Fig. P8.117, determine the theoretical flowrate from the tank corresponding to the following situations: (**a**) all losses are negligible, (**b**) major losses are important, and (**c**) all losses (major and minor) are important.

Compare the experimental and theoretical results and discuss some possible reasons for any differences between them.

h **(ft)**	***t*** **(s)**
2.0	0
1.9	13
1.8	26
1.7	40
1.6	54

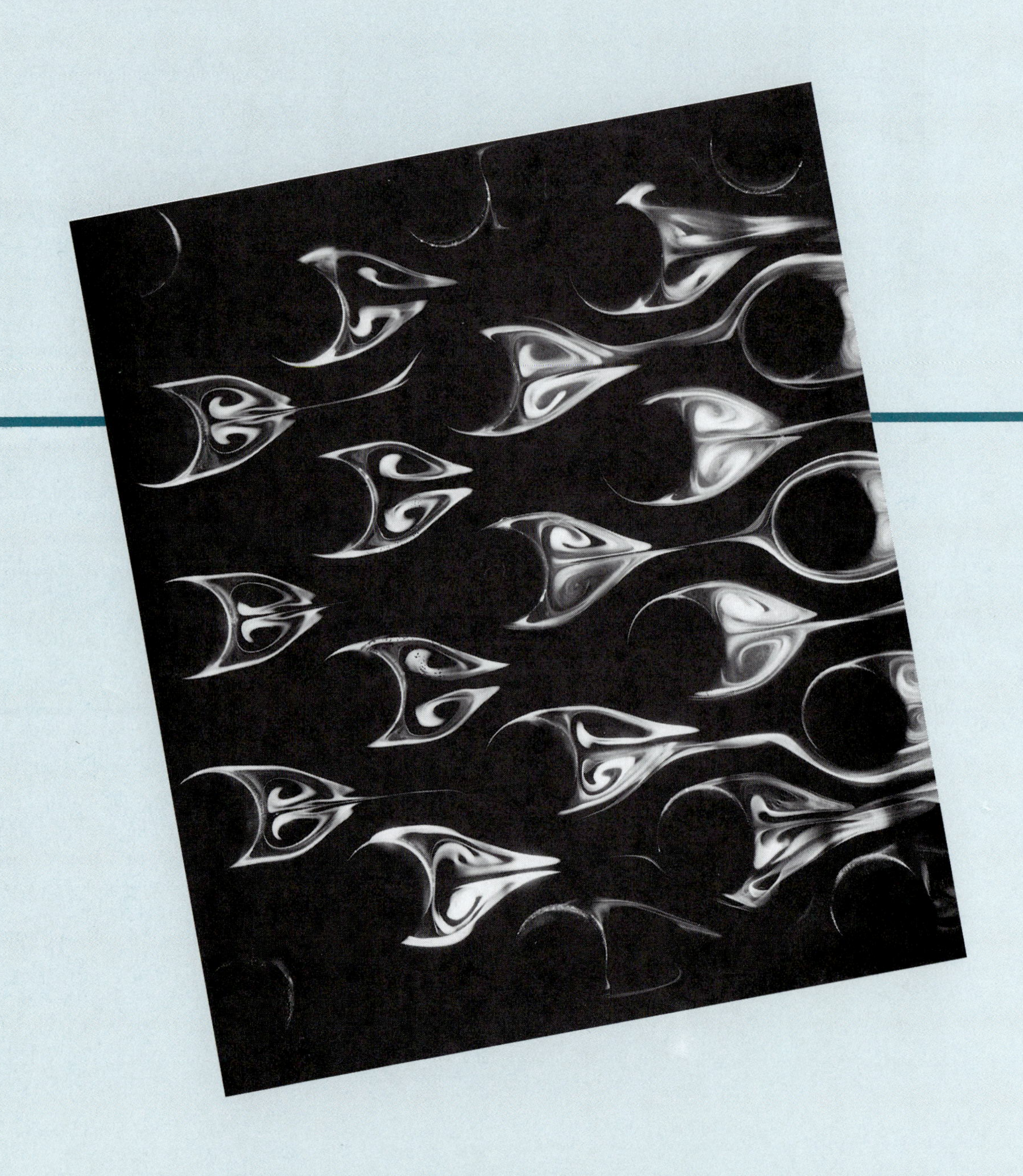

Impulsive start of flow past an array of cylinders: The complex structure of laminar flow past a relatively simple geometric structure illustrates why it is often difficult to obtain exact analytical results for external flows. (Dye in water.) (Photograph courtesy of ONERA, France.)

9 Flow Over Immersed Bodies

In this chapter we consider various aspects of the flow over bodies that are immersed in a fluid. Examples include the flow of air around airplanes, automobiles, and falling snow flakes, or the flow of water around submarines and fish. In these situations the object is completely surrounded by the fluid and the flows are termed *external flows*.

External flows involving air are often termed aerodynamics in response to the important external flows produced when an object such as an airplane flies through the atmosphere. Although this field of external flows is extremely important, there are many other examples that are of equal importance. The fluid force (lift and drag) on surface vehicles (cars, trucks, bicycles) has become a very important topic. By correctly designing cars and trucks, it has become possible to greatly decrease the fuel consumption and improve the handling characteristics of the vehicle. Similar efforts have resulted in improved ships, whether they are surface vessels (surrounded by two fluids, air and water) or submersible vessels (surrounded completely by water).

Many practical situations involve flow past objects.

Other applications of external flows involve objects that are not completely surrounded by fluid, although they are placed in some external-type flow. For example, the proper design of a building (whether it is your house or a tall skyscraper) must include consideration of the various wind effects involved.

As with other areas of fluid mechanics, two approaches (theoretical and experimental) are used to obtain information on the fluid forces developed by external flows. Theoretical (i.e., analytical and numerical) techniques can provide much of the needed information about such flows. However, because of the complexities of the governing equations and the complexities of the geometry of the objects involved, the amount of information obtained from purely theoretical methods is limited. With current and anticipated advancements in the area of computational fluid mechanics, it is likely that computer prediction of forces and complicated flow patterns will become more readily available.

Much of the information about external flows comes from experiments carried out, for the most part, on scale models of the actual objects. Such testing includes the obvious wind tunnel testing of model airplanes, buildings, and even entire cities. In some instances the actual device, not a model, is tested in wind tunnels. Figure 9.1 shows tests of vehicles in wind tunnels. Better performance of cars, bikes, skiers, and numerous other objects has resulted from testing in wind tunnels. The use of water tunnels and towing tanks also provides useful information about the flow around ships and other objects.

In this chapter we consider characteristics of external flow past a variety of objects. We investigate the qualitative aspects of such flows and learn how to determine the various forces on objects surrounded by a moving liquid.

9.1 General External Flow Characteristics

For external flows it is usually easiest to use a coordinate system fixed to the object.

A body immersed in a moving fluid experiences a resultant force due to the interaction between the body and the fluid surrounding it. In some instances (such as an airplane flying through still air) the fluid far from the body is stationary and the body moves through the fluid with velocity U. In other instances (such as the wind blowing past a building) the body is stationary and the fluid flows past the body with velocity U. In any case, we can fix the coordinate system in the body and treat the situation as fluid flowing past a stationary body

(*a*)

(*b*)

■ **FIGURE 9.1** **(*a*) Flow past a full-sized streamlined vehicle in the GM aerodynamics laboratory wind tunnel, an 18-ft by 34-ft test section facility driven by a 4000-hp, 43-ft-diameter fan. (Photograph courtesy of General Motors Corporation.) (*b*) Surface flow on a model vehicle as indicated by tufts attached to the surface. (Reprinted with permission from Society of Automotive Engineers, Ref. 28.)**

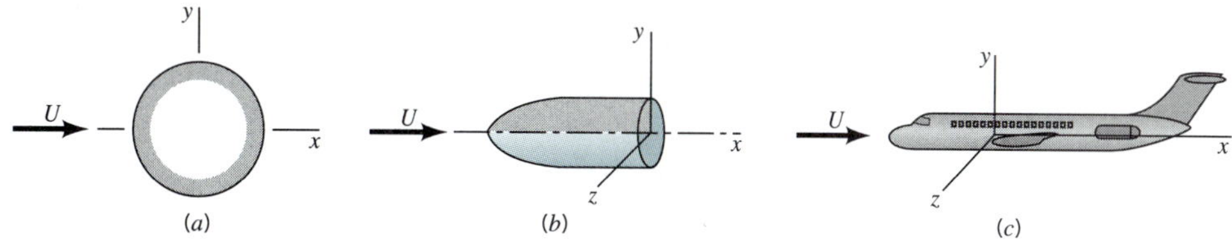

■ FIGURE 9.2 **Flow classification: (*a*) two-dimensional, (*b*) axisymmetric, (*c*) three-dimensional.**

with velocity U, the *upstream velocity*. For the purposes of this book, we will assume that the upstream velocity is constant in both time and location. That is, there is a uniform, constant velocity fluid flowing past the object. In actual situations this is often not true. For example, the wind blowing past a smokestack is nearly always turbulent and gusty (unsteady) and probably not of uniform velocity from the top to the bottom of the stack. Usually the unsteadiness and nonuniformity are of minor importance.

V9.1

Even with a steady, uniform upstream flow, the flow in the vicinity of an object may be unsteady. Examples of this type of behavior include the flutter that is sometimes found in the flow past airfoils (wings), the regular oscillation of telephone wires that ''sing'' in a wind, and the irregular turbulent fluctuations in the wake regions behind bodies.

The structure of an external flow and the ease with which the flow can be described and analyzed often depend on the nature of the body in the flow. Three general categories of bodies are shown in Fig. 9.2. They include (a) two-dimensional objects (infinitely long and of constant cross-sectional size and shape), (b) axisymmetric bodies (formed by rotating their cross-sectional shape about the axis of symmetry), and (c) three-dimensional bodies that may or may not possess a line or plane of symmetry. In practice there can be no truly two-dimensional bodies—nothing extends to infinity. However, many objects are sufficiently long so that the end effects are negligibly small.

The shape of a body affects the flow characteristics.

Another classification of body shape can be made depending on whether the body is streamlined or blunt. The flow characteristics depend strongly on the amount of streamlining present. In general, *streamlined bodies* (i.e., airfoils, racing cars, etc.) have little effect on the surrounding fluid, compared with the effect that *blunt bodies* (i.e., parachutes, buildings, etc.) have on the fluid. Usually, but not always, it is easier to force a streamlined body through a fluid than it is to force a similar-sized blunt body at the same velocity. There are important exceptions to this basic rule.

9.1.1 Lift and Drag Concepts

When any body moves through a fluid, an interaction between the body and the fluid occurs; this effect can be described in terms of the forces at the fluid–body interface. This can be described in terms of the stresses—wall shear stresses, τ_w, due to viscous effects and normal stresses due to the pressure, p. Typical shear stress and pressure distributions are shown in Figs. 9.3*a* and 9.3*b*. Both τ_w and p vary in magnitude and direction along the surface.

It is often useful to know the detailed distribution of shear stress and pressure over the surface of the body, although such information is difficult to obtain. Many times, however, only the integrated or resultant effects of these distributions are needed. The resultant force in the direction of the upstream velocity is termed the *drag*, $\mathscr{D}$, and the resultant force normal to the upstream velocity is termed the *lift*, $\mathscr{L}$, as is indicated in Fig. 9.3*c*. For some

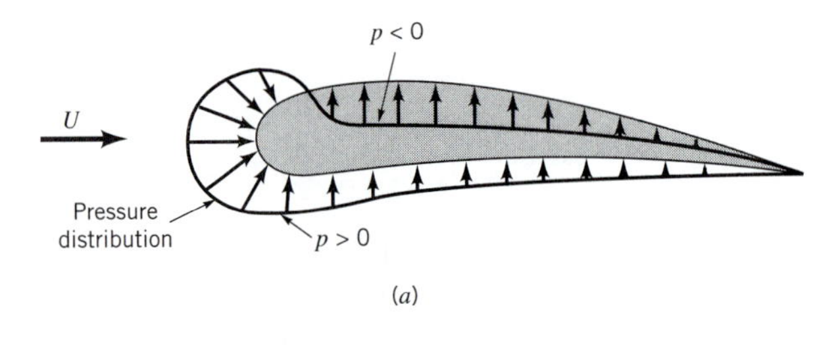

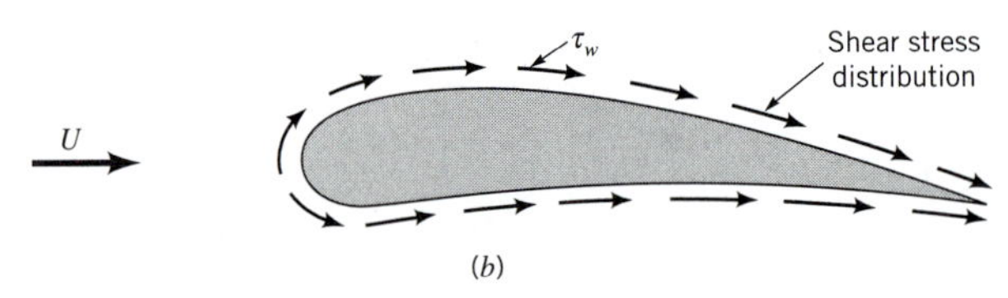

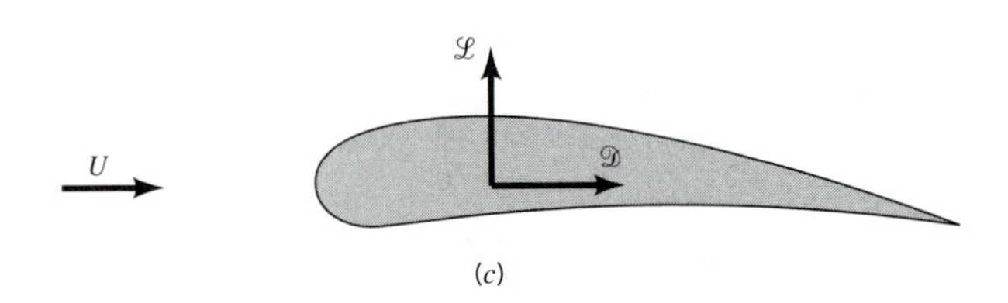

■ **FIGURE 9.3 Forces from the surrounding fluid on a two-dimensional object: (*a*) pressure force, (*b*) viscous force, (*c*) resultant force (lift and drag).**

three-dimensional bodies there may also be a side force that is perpendicular to the plane containing $\mathscr{D}$ and $\mathscr{L}$.

A body interacts with the surrounding fluid through pressure and shear stresses.

The resultant of the shear stress and pressure distributions can be obtained by integrating the effect of these two quantities on the body surface as is indicated in Fig. 9.4. The x and y components of the fluid force on the small area element dA are

$$dF_x = (p\ dA)\cos\theta + (\tau_w\ dA)\sin\theta$$

and

$$dF_y = -(p\ dA)\sin\theta + (\tau_w\ dA)\cos\theta$$

Thus, the net x and y components of the force on the object are

$$\mathscr{D} = \int dF_x = \int p\cos\theta\ dA + \int \tau_w \sin\theta\ dA \tag{9.1}$$

and

$$\mathscr{L} = \int dF_y = -\int p\sin\theta\ dA + \int \tau_w \cos\theta\ dA \tag{9.2}$$

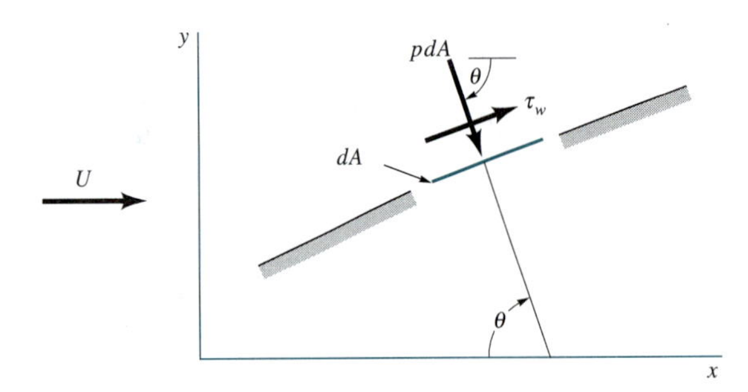

■ **FIGURE 9.4 Pressure and shear forces on a small element of the surface of a body.**

Lift and drag on a section of a body depend on the orientation of the surface.

Of course, to carry out the integrations and determine the lift and drag, we must know the body shape (i.e., θ as a function of location along the body) and the distribution of τ_w and p along the surface. These distributions are often extremely difficult to obtain, either experimentally or theoretically. The pressure distribution can be obtained experimentally without too much difficulty by use of a series of static pressure taps along the body surface. On the other hand, it is usually quite difficult to measure the wall shear stress distribution.

It is seen that both the shear stress and pressure force contribute to the lift and drag, since for an arbitrary body θ is neither zero nor 90° along the entire body. The exception is a flat plate aligned either parallel to the upstream flow ($\theta = 90°$) or normal to the upstream flow ($\theta = 0$) as is discussed in Example 9.1.

EXAMPLE 9.1

Air at standard conditions flows past a flat plate as is indicated in Fig. E9.1. In case (a) the plate is parallel to the upstream flow, and in case (b) it is perpendicular to the upstream flow. If the pressure and shear stress distributions on the surface are as indicated (obtained either by experiment or theory), determine the lift and drag on the plate.

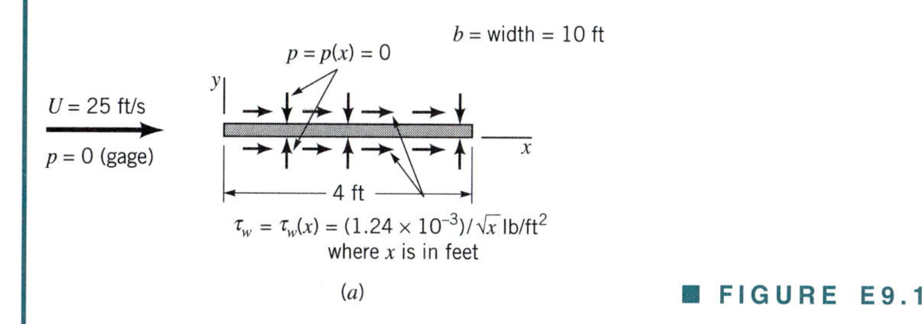

■ FIGURE E9.1

SOLUTION

For either orientation of the plate, the lift and drag are obtained from Eqs. 9.1 and 9.2. With the plate parallel to the upstream flow we have $\theta = 90°$ on the top surface and $\theta = 270°$ on the bottom surface so that the lift and drag are given by

$$\mathscr{L} = -\int_{\text{top}} p\, dA + \int_{\text{bottom}} p\, dA = 0$$

and

$$\mathscr{D} = \int_{\text{top}} \tau_w\, dA + \int_{\text{bottom}} \tau_w\, dA = 2\int_{\text{top}} \tau_w\, dA \qquad (1)$$

where we have used the fact that because of symmetry the shear stress distribution is the same on the top and the bottom surfaces, as is the pressure also [whether we use gage ($p = 0$) or absolute ($p = p_{\text{atm}}$) pressure]. There is no lift generated—the plate does not know up

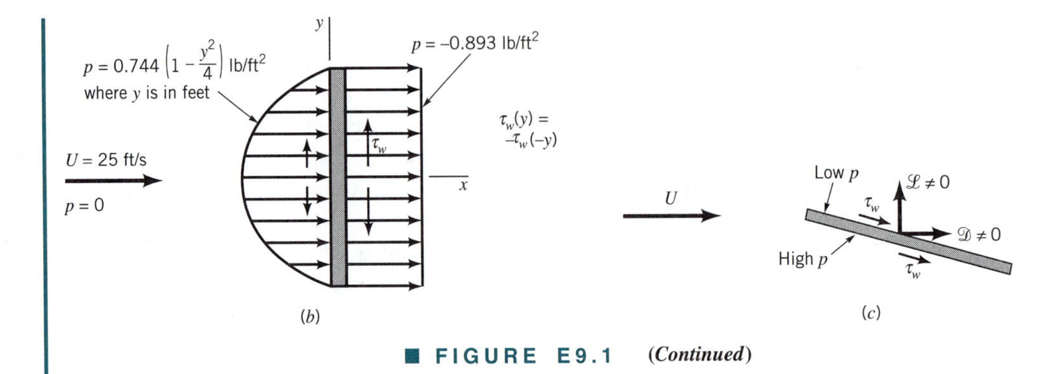

■ **FIGURE E9.1** ***(Continued)***

from down. With the given shear stress distribution, Eq. 1 gives

$$\mathscr{D} = 2 \int_{x=0}^{4\text{ ft}} \left(\frac{1.24 \times 10^{-3}}{x^{1/2}} \text{ lb/ft}^2 \right) (10 \text{ ft})\, dx$$

or

$$\mathscr{D} = 0.0992 \text{ lb} \qquad \textbf{(Ans)}$$

With the plate perpendicular to the upstream flow, we have $\theta = 0°$ on the front and $\theta = 180°$ on the back. Thus, from Eqs. 9.1 and 9.2

$$\mathscr{L} = \int_{\text{front}} \tau_w \, dA - \int_{\text{back}} \tau_w \, dA = 0$$

and

$$\mathscr{D} = \int_{\text{front}} p \, dA - \int_{\text{back}} p \, dA$$

Again there is no lift because the pressure forces act parallel to the upstream flow (in the direction of $\mathscr{D}$ not $\mathscr{L}$) and the shear stress is symmetrical about the center of the plate. With the given relatively large pressure on the front of the plate (the center of the plate is a stagnation point) and the negative pressure (less than the upstream pressure) on the back of the plate, we obtain the following drag

$$\mathscr{D} = \int_{y=-2}^{2\text{ ft}} \left[0.744 \left(1 - \frac{y^2}{4} \right) \text{lb/ft}^2 - (-0.893) \text{ lb/ft}^2 \right] (10 \text{ ft})\, dy$$

or

$$\mathscr{D} = 55.6 \text{ lb} \qquad \textbf{(Ans)}$$

Clearly there are two mechanisms responsible for the drag. On the ultimately streamlined body (a zero thickness flat plate parallel to the flow) the drag is entirely due to the shear stress at the surface and, in this example, is relatively small. For the ultimately blunted body (a flat plate normal to the upstream flow) the drag is entirely due to the pressure difference between the front and back portions of the object and, in this example, is relatively large.

If the flat plate were oriented at an arbitrary angle relative to the upstream flow as indicated in Fig. E9.1*c*, there would be both a lift and a drag, each of which would be dependent on both the shear stress and the pressure. Both the pressure and shear stress distributions would be different for the top and bottom surfaces.

Although Eqs. 9.1 and 9.2 are valid for any body, the difficulty in their use lies in obtaining the appropriate shear stress and pressure distributions on the body surface. Considerable effort has gone into determining these quantities, but because of the various complexities involved, such information is available only for certain simple situations.

Without detailed information concerning the shear stress and pressure distributions on a body, Eqs. 9.1 and 9.2 cannot be used. The widely used alternative is to define dimensionless lift and drag coefficients and determine their approximate values by means of either a simplified analysis, some numerical technique, or an appropriate experiment. The *lift coefficient*, C_L, and *drag coefficient*, C_D, are defined as

Lift coefficients and drag coefficients are dimensionless forms of lift and drag.

$$C_L = \frac{\mathscr{L}}{\frac{1}{2}\rho U^2 A}$$

and

$$C_D = \frac{\mathscr{D}}{\frac{1}{2}\rho U^2 A}$$

where A is a characteristic area of the object (see Chapter 7). Typically, A is taken to be *frontal area*—the projected area seen by a person looking toward the object from a direction parallel to the upstream velocity, U. It would be the area of the shadow of the object projected onto a screen normal to the upstream velocity as formed by a light shining along the upstream flow. In other situations A is taken to be the *planform area*—the projected area seen by an observer looking toward the object from a direction normal to the upstream velocity (i.e., from "above" it). Obviously, which characteristic area is used in the definition of the lift and drag coefficients must be clearly stated.

9.1.2 Characteristics of Flow Past an Object

External flows past objects encompass an extremely wide variety of fluid mechanics phenomena. Clearly the character of the flow field is a function of the shape of the body. Flows past relatively simple geometric shapes (i.e., a sphere or circular cylinder) are expected to have less complex flow fields than flows past a complex shape such as an airplane or a tree. However, even the simplest-shaped objects produce rather complex flows.

For a given-shaped object, the characteristics of the flow depend very strongly on various parameters such as size, orientation, speed, and fluid properties. As is discussed in Chapter 7, according to dimensional analysis arguments, the character of the flow should depend on the various dimensionless parameters involved. For typical external flows the most important of these parameters are the Reynolds number, $\text{Re} = \rho U \ell / \mu = U\ell/\nu$, the Mach number, $\text{Ma} = U/c$, and for flows with a free surface (i.e., flows with an interface between two fluids, such as the flow past a surface ship), the Froude number, $\text{Fr} = U/\sqrt{g\ell}$. (Recall that ℓ is some characteristic length of the object and c is the speed of sound.)

For the present, we consider how the external flow and its associated lift and drag vary as a function of Reynolds number. Recall that the Reynolds number represents the ratio of inertial effects to viscous effects. In the absence of all viscous effects ($\mu = 0$), the Reynolds number is infinite. On the other hand, in the absence of all inertial effects (negligible mass or $\rho = 0$), the Reynolds number is zero. Clearly, any actual flow will have a Reynolds number between (but not including) these two extremes. The nature of the flow past a body depends strongly on whether $\text{Re} \gg 1$ or $\text{Re} \ll 1$.

The character of flow past an object is dependent on the value of the Reynolds number.

Most external flows with which we are familiar are associated with moderately sized objects with a characteristic length on the order of $0.01 \text{ m} < \ell < 10 \text{ m}$. In addition, typical upstream velocities are on the order of $0.01 \text{ m/s} < U < 100 \text{ m/s}$ and the fluids involved are typically water or air. The resulting Reynolds number range for such flows is approximately $10 < \text{Re} < 10^9$. As a rule of thumb, flows with $\text{Re} > 100$ are dominated by inertial effects, whereas flows with $\text{Re} < 1$ are dominated by viscous effects. Hence, most familiar external flows are dominated by inertia.

On the other hand, there are many external flows in which the Reynolds number is considerably less than 1, indicating in some sense that viscous forces are more important than inertial forces. The gradual settling of small particles of dirt in a lake or stream is governed by low Reynolds number flow principles because of the small diameter of the particles and their small settling speed. Similarly, the Reynolds number for objects moving through large viscosity oils is small because μ is large. The general differences between small and large Reynolds number flow past streamlined and blunt objects can be illustrated by considering flows past two objects—one a flat plate parallel to the upstream velocity and the other a circular cylinder.

Flows past three flat plates of length ℓ with $\text{Re} = \rho U \ell / \mu = 0.1, 10,$ and 10^7, are shown in Fig. 9.5. If the Reynolds number is small, the viscous effects are relatively strong and the plate affects the uniform upstream flow far ahead, above, below, and behind the plate. To reach that portion of the flow field where the velocity has been altered by less than 1% of its undisturbed value (i.e., $U - u < 0.01U$) we must travel relatively far from the plate. In low Reynolds number flows the viscous effects are felt far from the object in all directions.

As the Reynolds number is increased (by increasing U, for example), the region in which viscous effects are important becomes smaller in all directions except downstream, as is shown in Fig. 9.5*b*. One does not need to travel very far ahead, above, or below the plate to reach areas in which the viscous effects of the plate are not felt. The streamlines are displaced from their original uniform upstream conditions, but the displacement is not as great as for the $\text{Re} = 0.1$ situation shown in Fig. 9.5*a*.

If the Reynolds number is large (but not infinite), the flow is dominated by inertial effects and the viscous effects are negligible everywhere except in a region very close to the plate and in the relatively thin *wake region* behind the plate, as shown in Fig. 9.5*c*. Since the fluid viscosity is not zero ($\text{Re} < \infty$), it follows that the fluid must stick to the solid surface (the no-slip boundary condition). There is a thin *boundary layer* region of thickness $\delta = \delta(x) \ll \ell$ (i.e., thin relative to the length of the plate) next to the plate in which the fluid velocity changes from the upstream value of $u = U$ to zero velocity on the plate. The thickness of this layer increases in the direction of flow, starting from zero at the forward or leading edge of the plate. The flow within the boundary layer may be laminar or turbulent, depending on various parameters involved.

The streamlines of the flow outside of the boundary layer are nearly parallel to the plate. As we will see in the next section, the slight displacement of the external streamlines that are outside of the boundary layer is due to the thickening of the boundary layer in the direction of flow. The existence of the plate has very little effect on the streamlines outside of the boundary layer—either ahead, above, or below the plate. On the other hand, the wake region is due entirely to the viscous interaction between the fluid and the plate.

One of the great advancements in fluid mechanics occurred in 1904 as a result of the insight of Ludwig Prandtl (1875–1953), a German physicist and aerodynamicist. He con-

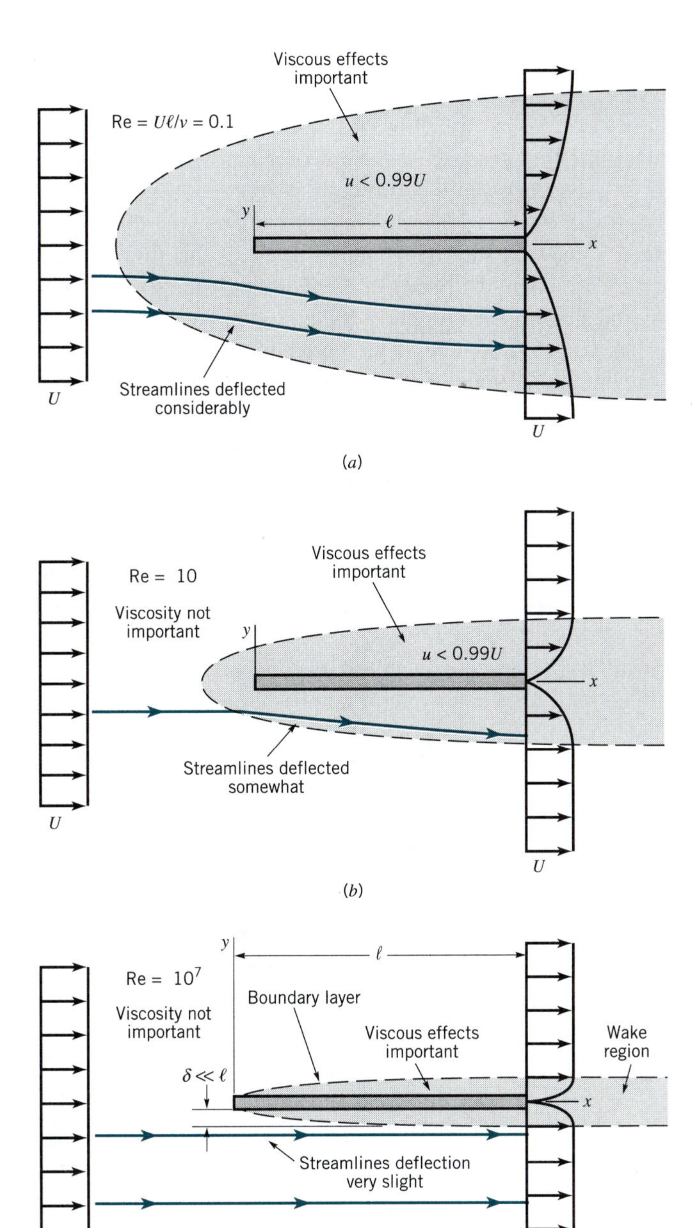

■ FIGURE 9.5 Character of the steady, viscous flow past a flat plate parallel to the upstream velocity: (*a*) low Reynolds number flow, (*b*) moderate Reynolds number flow, (*c*) large Reynolds number flow.

Thin boundary layers may develop in large Reynolds number flows.

ceived of the idea of the boundary layer—a thin region on the surface of a body in which viscous effects are very important and outside of which the fluid behaves essentially as if it were inviscid. Clearly the actual fluid viscosity is the same throughout; only the relative importance of the viscous effects (due to the velocity gradients) is different within or outside of the boundary layer. As is discussed in the next section, by using such a hypothesis it is possible to simplify the analysis of large Reynolds number flows, thereby allowing solution to external flow problems that are otherwise still unsolvable.

As with the flow past the flat plate described above, the flow past a blunt object (such as a circular cylinder) also varies with Reynolds number. In general, the larger the Reynolds number, the smaller the region of the flow field in which viscous effects are important. For objects that are not sufficiently streamlined, however, an additional characteristic of the flow is observed. This is termed *flow separation* and is illustrated in Fig. 9.6.

Low Reynolds number flow ($\text{Re} = UD/\nu < 1$) past a circular cylinder is characterized by the fact that the presence of the cylinder and the accompanying viscous effects are felt throughout a relatively large portion of the flow field. As is indicated in Fig. 9.6*a*, for $\text{Re} = UD/\nu = 0.1$, the viscous effects are important several diameters in any direction from the cylinder. A somewhat surprising characteristic of this flow is that the streamlines are essentially symmetric about the center of the cylinder—the streamline pattern is the same in front of the cylinder as it is behind the cylinder.

As the Reynolds number is increased, the region ahead of the cylinder in which viscous effects are important becomes smaller, with the viscous region extending only a short distance ahead of the cylinder. The viscous effects are convected downstream and the flow loses its symmetry. Another characteristic of external flows becomes important—the flow separates from the body at the *separation location* as indicated in Fig. 9.6*b*. With the increase in Reynolds number, the fluid inertia becomes more important and at some location on the body, denoted the separation location, the fluid's inertia is such that it cannot follow the curved path around to the rear of the body. The result is a separation bubble behind the cylinder in which some of the fluid is actually flowing upstream, against the direction of the upstream flow.

Flow separation may occur behind blunt objects.

At still larger Reynolds numbers, the area affected by the viscous forces is forced farther downstream until it involves only a thin ($\delta \ll D$) boundary layer on the front portion of the

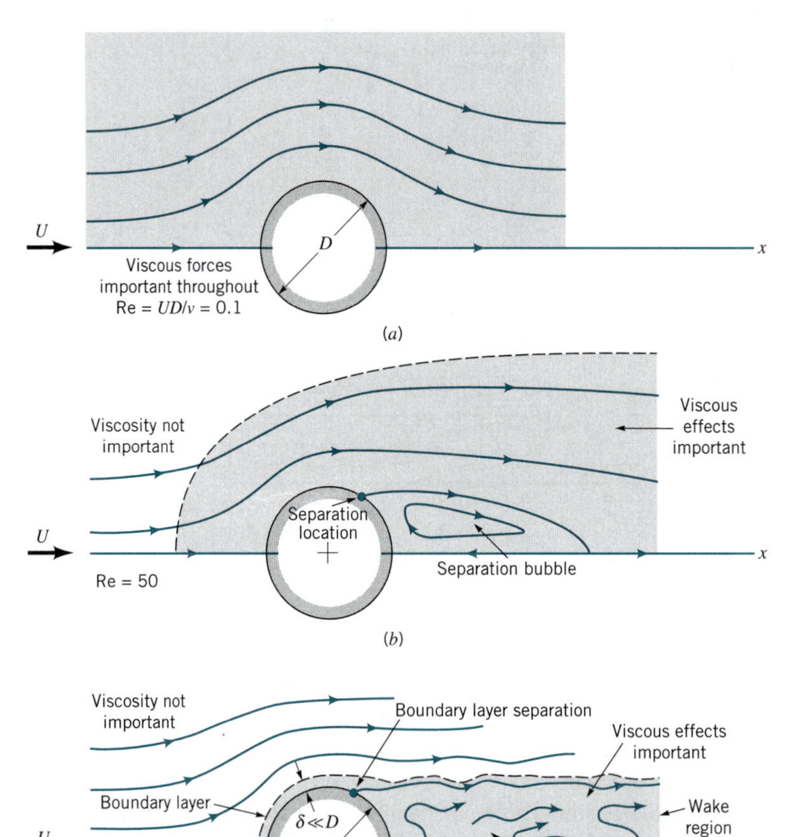

■ **FIGURE 9.6 Character of the steady, viscous flow past a circular cylinder: (*a*) low Reynolds number flow, (*b*) moderate Reynolds number flow, (*c*) large Reynolds number flow.**

cylinder and an irregular, unsteady (perhaps turbulent) wake region that extends far downstream of the cylinder. The fluid in the region outside of the boundary layer and wake region flows as if it were inviscid. Of course, the fluid viscosity is the same throughout the entire flow field. Whether viscous effects are important or not depends on which region of the flow field we consider. The velocity gradients within the boundary layer and wake regions are much larger than those in the remainder of the flow field. Since the shear stress (i.e., viscous effect) is the product of the fluid viscosity and the velocity gradient, it follows that viscous effects are confined to the boundary layer and wake regions.

V9.2

Most familiar flows involve large Reynolds numbers.

The characteristics described in Figs. 9.5 and 9.6 for flow past a flat plate and a circular cylinder are typical of flows past streamlined and blunt bodies, respectively. The nature of the flow depends strongly on the Reynolds number. Most familiar flows are similar to the large Reynolds number flows depicted in Figs. 9.5*c* and 9.6*c*, rather than the low Reynolds number flow situations. In the remainder of this chapter we will investigate more thoroughly these ideas and determine how to calculate the forces on immersed bodies.

EXAMPLE 9.2

It is desired to determine the various characteristics of flow past a car. The following tests could be carried out: (a) $U = 20$ mm/s flow of glycerin past a scale model that is 34-mm tall, 100-mm long and 40-mm wide, (b) $U = 20$ mm/s air flow past the scale model, or (c) $U = 25$ m/s air flow past the actual car, which is 1.7-m tall, 5-m long, and 2-m wide. Would the flow characteristics for these three situations be similar? Explain.

SOLUTION

The characteristics of flow past an object depend on the Reynolds number. For this instance we could pick the characteristic length to be the height, h, width, b, or length, ℓ, of the car to obtain three possible Reynolds numbers, $\text{Re}_h = Uh/\nu$, $\text{Re}_b = Ub/\nu$, and $\text{Re}_\ell = U\ell/\nu$. These numbers will be different because of the different values of h, b, and ℓ. Once we arbitrarily decide on the length we wish to use as the characteristic length, we must stick with it for all calculations when using comparisons between model and prototype.

With the values of kinematic viscosity for air and glycerin obtained from Tables 1.8 and 1.6 as $\nu_{\text{air}} = 1.46 \times 10^{-5}\ \text{m}^2/\text{s}$ and $\nu_{\text{glycerin}} = 1.19 \times 10^{-3}\ \text{m}^2/\text{s}$, we obtain the following Reynolds numbers for the flows described.

Reynolds Number	(a) Model in Glycerin	(b) Model in Air	(c) Car in Air
Re_h	0.571	46.6	2.91×10^6
Re_b	0.672	54.8	3.42×10^6
Re_ℓ	1.68	137.0	8.56×10^6

Clearly, the Reynolds numbers for the three flows are quite different (regardless of which characteristic length we choose). Based on the previous discussion concerning flow past a flat plate or flow past a circular cylinder, we would expect that the flow past the actual car would behave in some way similar to the flows shown in Figs. 9.5*c* or 9.6*c*. That is, we would expect some type of boundary layer characteristic in which viscous effects would be confined to relatively thin layers near the surface of the car and the wake region behind it. Whether the car would act more like a flat plate or a cylinder would depend on the amount of streamlining incorporated into the car's design.

Because of the small Reynolds number involved, the flow past the model car in glycerin would be dominated by viscous effects, in some way reminiscent of the flows depicted in Figs. 9.5*a* or 9.6*a*. Similarly, with the moderate Reynolds number involved for the air flow past the model, a flow with characteristics similar to those indicated in Figs. 9.5*b* and 9.6*b* would be expected. Viscous effects would be important—not as important as with the glycerin flow, but more important than with the full-sized car.

It would not be a wise decision to expect the flow past the full-sized car to be similar to the flow past either of the models. The same conclusions result regardless of whether we use Re_h, Re_b, or Re_ℓ. As is indicated in Chapter 7, the flows past the model car and the full-sized prototype will not be similar unless the Reynolds numbers for the model and prototype are the same. It is not always an easy task to ensure this condition. One (expensive) solution is to test full-sized prototypes in very large wind tunnels (see Fig. 9.1).

9.2 Boundary Layer Characteristics

Large Reynolds number flow fields may be divided into viscous and inviscid regions.

As was discussed in the previous section, it is often possible to treat flow past an object as a combination of viscous flow in the boundary layer and inviscid flow elsewhere. If the Reynolds number is large enough, viscous effects are important only in the boundary layer regions near the object (and in the wake region behind the object). The boundary layer is needed to allow for the no-slip boundary condition that requires the fluid to cling to any solid surface that it flows past. Outside of the boundary layer the velocity gradients normal to the flow are relatively small, and the fluid acts as if it were inviscid, even though the viscosity is not zero. A necessary condition for this structure of the flow is that the Reynolds number be large.

9.2.1 Boundary Layer Structure and Thickness on a Flat Plate

There can be a wide variety in the size of a boundary layer and the structure of the flow within it. Part of this variation is due to the shape of the object on which the boundary layer forms. In this section we consider the simplest situation, one in which the boundary layer is formed on an infinitely long flat plate along which flows a viscous, incompressible fluid as is shown in Fig. 9.7. If the surface were curved (i.e., a circular cylinder or an airfoil), the boundary layer structure would be more complex. Such flows are discussed in Section 9.2.6.

If the Reynolds number is sufficiently large, only the fluid in a relatively thin boundary layer on the plate will feel the effect of the plate. That is, except in the region next to the plate the flow velocity will be essentially $\mathbf{V} = U\hat{\mathbf{i}}$, the upstream velocity. For the infinitely long flat plate extending from $x = 0$ to $x = \infty$, it is not obvious how to define the Reynolds

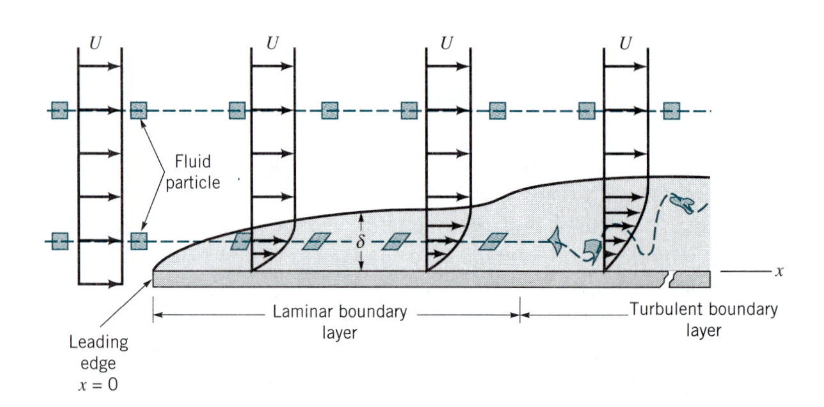

■ **FIGURE 9.7 Distortion of a fluid particle as it flows within the boundary layer.**

number because there is no characteristic length. The plate has no thickness and is not of finite length!

For a finite length plate, it is clear that the plate length, ℓ, can be used as the characteristic length. For an infinitely long plate we use x, the coordinate distance along the plate from the leading edge, as the characteristic length and define the Reynolds number as $\text{Re}_x = Ux/\nu$. Thus, for any fluid or upstream velocity the Reynolds number will be sufficiently large for boundary layer type flow (i.e., Fig. 9.5*c*) if the plate is long enough. Physically, this means that the flow situations illustrated in Fig. 9.5 could be thought of as occurring on the same plate, but should be viewed by looking at longer portions of the plate as we step away from the plate to see the flows in Fig. 9.5*a*, 9.5*b*, and 9.5*c*, respectively.

If the plate is sufficiently long, the Reynolds number $\text{Re} = U\ell/\nu$ is sufficiently large so that the flow takes on its boundary layer character (except very near the leading edge). The details of the flow field near the leading edge are lost to our eyes because we are standing so far from the plate that we cannot make out these details. On this scale (Fig. 9.5*c*) the plate has negligible effect on the fluid ahead of the plate. The presence of the plate is felt only in the relatively thin boundary layer and wake regions. As previously noted, Prandtl in 1904 was the first to hypothesize such a concept. It has become one of the major turning points in fluid mechanics analysis.

A better appreciation of the structure of the boundary layer flow can be obtained by considering what happens to a fluid particle that flows into the boundary layer. As is indicated in Fig. 9.7, a small rectangular particle retains its original shape as it flows in the uniform flow outside of the boundary layer. Once it enters the boundary layer, the particle begins to distort because of the velocity gradient within the boundary layer—the top of the particle has a larger speed than its bottom. The fluid particles do not rotate as they flow along outside the boundary layer, but they begin to rotate once they pass through the fictitious boundary layer surface and enter the world of viscous flow. The flow is said to be irrotational outside the boundary layer and rotational within the boundary layer. (In terms of the kinematics of fluid particles as is discussed in Section 6.1, the flow outside the boundary layer has zero vorticity, and the flow within the boundary layer has nonzero vorticity.)

Fluid particles within the boundary layer experience viscous effects.

At some distance downstream from the leading edge, the boundary layer flow becomes turbulent and the fluid particles become greatly distorted because of the random, irregular nature of the turbulence. One of the distinguishing features of turbulent flow is the occurrence of irregular mixing of fluid parcels that range in size from the smallest fluid particles up to those comparable in size with the object of interest. For laminar flow, mixing occurs only on the molecular scale. This molecular scale is orders of magnitude smaller in size than typical size scales for turbulent flow mixing. The transition from laminar to turbulent flow occurs at a critical value of the Reynolds number, $\text{Re}_{x\text{cr}}$, on the order of 2×10^5 to 3×10^6, depending on the roughness of the surface and the amount of turbulence in the upstream flow, as is discussed in Section 9.2.4.

V9.3

The purpose of the boundary layer on the plate is to allow the fluid to change its velocity from the upstream value of U to zero on the plate. Thus, $\mathbf{V} = 0$ at $y = 0$ and $\mathbf{V} \approx U\hat{\mathbf{i}}$ at $y = \delta$, with the velocity profile, $u = u(x, y)$ bridging the boundary layer thickness. In actuality (both mathematically and physically), there is no sharp ''edge'' to the boundary layer. That is, $u \rightarrow U$ as we get farther from the plate; it is not precisely $u = U$ at $y = \delta$. We define the *boundary layer thickness*, δ, as that distance from the plate at which the fluid velocity is within some arbitrary value of the upstream velocity. Typically, as indicated in Fig. 9.8*a*,

$$\delta = y \quad \text{where} \quad u = 0.99U$$

To remove this arbitrariness (i.e., what is so special about 99%; why not 98%?), the following definitions are introduced. Shown in Fig. 9.8*b* are two velocity profiles for flow past a flat plate—one if there were no viscosity (a uniform profile) and the other if there is

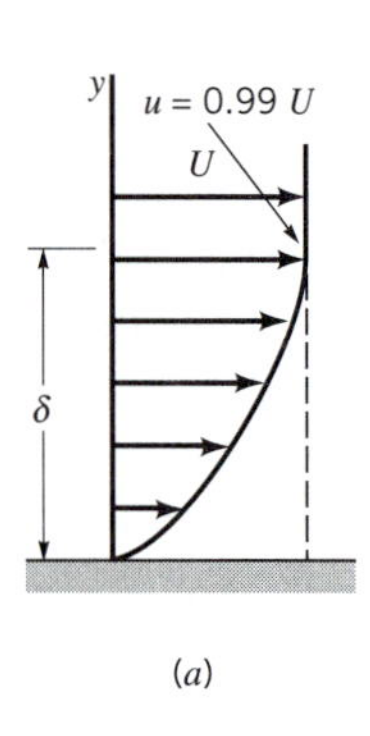

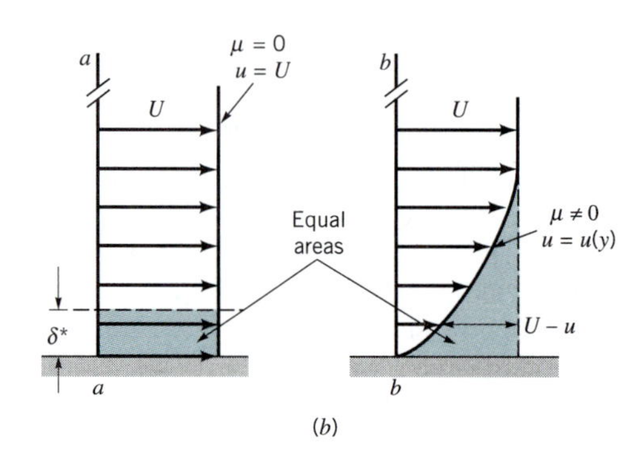

■ **FIGURE 9.8** **Boundary layer thickness: (*a*) standard boundary layer thickness, (*b*) boundary layer displacement thickness.**

viscosity and zero slip at the wall (the boundary layer profile). Because of the velocity deficit, $U - u$, within the boundary layer, the flowrate across section b–b is less than that across section a–a. However, if we displace the plate at section a–a by an appropriate amount δ^*, the *boundary layer displacement thickness*, the flowrates across each section will be identical. This is true if

The boundary layer displacement thickness is defined in terms of volumetric flowrate.

$$\delta^* bU = \int_0^\infty (U - u)b \, dy$$

where b is the plate width. Thus,

$$\delta^* = \int_0^\infty \left(1 - \frac{u}{U}\right) dy \qquad \textbf{(9.3)}$$

The displacement thickness represents the amount that the thickness of the body must be increased so that the fictitious uniform inviscid flow has the same mass flowrate properties as the actual viscous flow. It represents the outward displacement of the streamlines caused by the viscous effects on the plate. This idea allows us to simulate the presence that the boundary layer has on the flow outside of the boundary layer by adding the displacement thickness to the actual wall and treating the flow over the thickened body as an inviscid flow. The displacement thickness concept is illustrated in Example 9.3.

EXAMPLE 9.3

Air flowing into a 2-ft-square duct with a uniform velocity of 10 ft/s forms a boundary layer on the walls as shown in Fig. E9.3. The fluid within the core region (outside the boundary layers) flows as if it were inviscid. From advanced calculations it is determined that for this flow the boundary layer displacement thickness is given by

$$\delta^* = 0.0070(x)^{1/2} \qquad \textbf{(1)}$$

where δ^* and x are in feet. Determine the velocity $U = U(x)$ of the air within the duct but outside of the boundary layer.

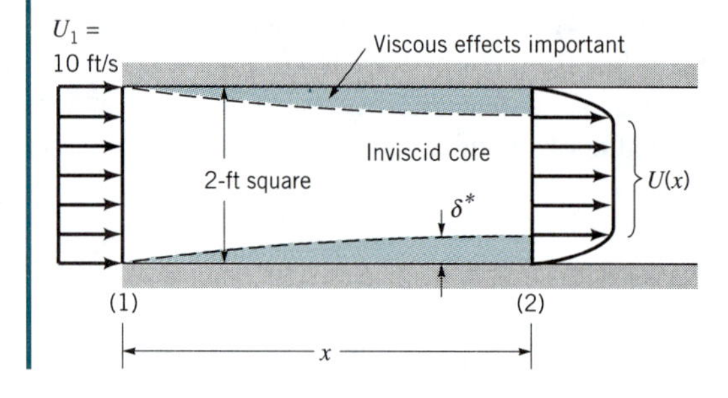

■ **FIGURE E9.3**

SOLUTION

If we assume incompressible flow (a reasonable assumption because of the low velocities involved), it follows that the volume flowrate across any section of the duct is equal to that at the entrance (i.e., $Q_1 = Q_2$). That is,

$$U_1 A_1 = 10 \text{ ft/s } (2 \text{ ft})^2 = 40 \text{ ft}^3/\text{s} = \int_{(2)} u \, dA$$

According to the definition of the displacement thickness, δ^*, the flowrate across section (2) is the same as that for a uniform flow with velocity U through a duct whose walls have been moved inward by δ^*. That is,

$$40 \text{ ft}^3/\text{s} = \int_{(2)} u \, dA = U(2 \text{ ft} - 2\delta^*)^2 \tag{2}$$

By combining Eqs. 1 and 2 we obtain

$$40 \text{ ft}^3/\text{s} = 4U(1 - 0.0070x^{1/2})^2$$

or

$$U = \frac{10}{(1 - 0.0070x^{1/2})^2} \text{ ft/s} \qquad \textbf{(Ans)}$$

Note that U increases in the downstream direction. For example, $U = 11.6$ ft/s at $x = 100$ ft. The viscous effects that cause the fluid to stick to the walls of the duct reduce the effective size of the duct, thereby (from conservation of mass principles) causing the fluid to accelerate. The pressure drop necessary to do this can be obtained by using the Bernoulli equation (Eq. 3.7) along the inviscid streamlines from section (1) to (2). (Recall that this equation is not valid for viscous flows within the boundary layer. It is, however, valid for the inviscid flow outside the boundary layer.) Thus,

$$p_1 + \tfrac{1}{2}\rho U_1^2 = p + \tfrac{1}{2}\rho U^2$$

Hence, with $\rho = 2.38 \times 10^{-3}$ slugs/ft^3 and $p_1 = 0$ we obtain

$$\begin{aligned} p &= \frac{1}{2}\rho(U_1^2 - U^2) \\ &= \frac{1}{2}(2.38 \times 10^{-3} \text{ slugs/ft}^3)\left[(10 \text{ ft/s})^2 - \frac{10^2}{(1 - 0.0079x^{1/2})^4} \text{ ft}^2/\text{s}^2\right] \end{aligned}$$

or

$$p = 0.119\left[1 - \frac{1}{(1 - 0.0070x^{1/2})^4}\right] \text{ lb/ft}^2$$

For example, $p = -0.0401$ lb/ft^2 at $x = 100$ ft.

If it were desired to maintain a constant velocity along the centerline of this entrance region of the duct, the walls could be displaced outward by an amount equal to the boundary layer displacement thickness, δ^*.

Another boundary layer thickness definition, the *boundary layer momentum thickness*, Θ, is often used when determining the drag on an object. Again because of the velocity deficit, $U - u$, in the boundary layer, the momentum flux across section b–b in Fig. 9.8 is less than that across section a–a. This deficit in momentum flux for the actual boundary layer flow is given by

$$\int \rho u(U - u)\, dA = \rho b \int_0^\infty u(U - u)\, dy$$

The boundary layer momentum thickness is defined in terms of momentum flux.

which by definition is the momentum flux in a layer of uniform speed U and thickness Θ. That is,

$$\rho b U^2 \Theta = \rho b \int_0^\infty u(U - u)\, dy$$

or

$$\Theta = \int_0^\infty \frac{u}{U}\left(1 - \frac{u}{U}\right) dy \qquad \textbf{(9.4)}$$

All three boundary layer thickness definitions, δ, δ^*, and Θ, are of use in boundary layer analyses.

The boundary layer concept is based on the fact that the boundary layer is thin. For the flat plate flow this means that at any location x along the plate, $\delta \ll x$. Similarly, $\delta^* \ll x$ and $\Theta \ll x$. Again, this is true if we do not get too close to the leading edge of the plate (i.e., not closer than $\mathrm{Re}_x = Ux/\nu = 1000$ or so).

The structure and properties of the boundary layer flow depend on whether the flow is laminar or turbulent. As is illustrated in Fig. 9.9 and discussed in Sections 9.2.2 through 9.2.5, both the boundary layer thickness and the wall shear stress are different in these two regimes.

9.2.2 Prandtl/Blasius Boundary Layer Solution

In theory, the details of viscous, incompressible flow past any object can be obtained by solving the governing Navier-Stokes equations discussed in Section 6.8.2. For steady, two-dimensional laminar flows with negligible gravitational effects, these equations (Eqs. 6.127*a*,

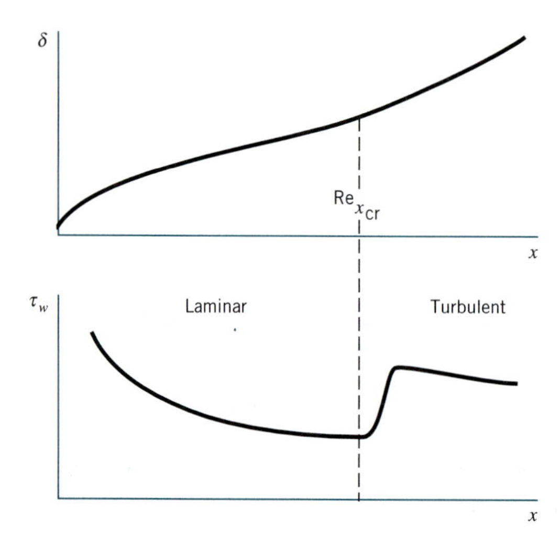

■ **FIGURE 9.9 Typical characteristics of boundary layer thickness and wall shear stress for laminar and turbulent boundary layers.**

b, and c) reduce to the following

$$u \frac{\partial u}{\partial x} + v \frac{\partial u}{\partial y} = -\frac{1}{\rho} \frac{\partial p}{\partial x} + \nu \left(\frac{\partial^2 u}{\partial x^2} + \frac{\partial^2 u}{\partial y^2} \right) \tag{9.5}$$

$$u \frac{\partial v}{\partial x} + v \frac{\partial v}{\partial y} = -\frac{1}{\rho} \frac{\partial p}{\partial y} + \nu \left(\frac{\partial^2 v}{\partial x^2} + \frac{\partial^2 v}{\partial y^2} \right) \tag{9.6}$$

which express Newton's second law. In addition, the conservation of mass equation, Eq. 6.31, for incompressible flow is

$$\frac{\partial u}{\partial x} + \frac{\partial v}{\partial y} = 0 \tag{9.7}$$

The appropriate boundary conditions are that the fluid velocity far from the body is the upstream velocity and that the fluid sticks to the solid body surfaces. Although the mathematical problem is well-posed, no one has obtained an analytical solution to these equations for flow past any shaped body! Currently much work is being done to obtain numerical solutions to these governing equations for many flow geometries.

By using boundary layer concepts introduced in the previous sections, Prandtl was able to impose certain approximations (valid for large Reynolds number flows), and thereby to simplify the governing equations. In 1908, H. Blasius (1883–1970), one of Prandtl's students, was able to solve these simplified equations for the boundary layer flow past a flat plate parallel to the flow. A brief outline of this technique and the results are presented below. Additional details may be found in the literature (Refs. 1, 2, 3).

Since the boundary layer is thin, it is expected that the component of velocity normal to the plate is much smaller than that parallel to the plate and that the rate of change of any parameter across the boundary layer should be much greater than that along the flow direction. That is,

$$v \ll u \quad \text{and} \quad \frac{\partial}{\partial x} \ll \frac{\partial}{\partial y}$$

Physically, the flow is primarily parallel to the plate and any fluid property is convected downstream much more quickly than it is diffused across the streamlines.

With these assumptions it can be shown that the governing equations (Eqs. 9.5, 9.6, and 9.7) reduce to the following boundary layer equations

The Navier–Stokes equations can be simplified for boundary layer flow analysis.

$$\frac{\partial u}{\partial x} + \frac{\partial v}{\partial y} \tag{9.8}$$

$$u \frac{\partial u}{\partial x} + v \frac{\partial u}{\partial y} = \nu \frac{\partial^2 u}{\partial y^2} \tag{9.9}$$

Although both these boundary layer equations and the original Navier–Stokes equations are nonlinear partial differential equations, there are considerable differences between them. For one, the y momentum equation has been eliminated, leaving only the original, unaltered continuity equation and a modified x momentum equation. One of the variables, the pressure, has been eliminated, leaving only the x and y components of velocity as unknowns. For boundary layer flow over a flat plate the pressure is constant throughout the fluid. The flow represents a balance between viscous and inertial effects, with pressure playing no role.

The boundary conditions for the governing boundary layer equations are that the fluid sticks to the plate

$$u = v = 0 \quad \text{on} \quad y = 0 \tag{9.10}$$

and that outside of the boundary layer the flow is the uniform upstream flow $u = U$. That is,

$$u \to U \quad \text{as} \quad y \to \infty \tag{9.11}$$

Mathematically, the upstream velocity is approached asymptotically as one moves away from the plate. Physically, the flow velocity is within 1% of the upstream velocity at a distance of δ from the plate.

In mathematical terms, the Navier–Stokes equations (Eqs. 9.5, 9.6) and the continuity equation (Eq. 9.7) are elliptic equations, whereas the equations for boundary layer flow (Eqs. 9.8 and 9.9) are parabolic equations. The nature of the solutions to these two sets of equations, therefore, is different. Physically, this fact translates to the idea that what happens downstream of a given location in a boundary layer cannot affect what happens upstream of that point. That is, whether the plate shown in Fig. 9.5*c* ends with length ℓ or is extended to length 2ℓ, the flow within the first segment of length ℓ will be the same. In addition, the presence of the plate has no effect on the flow ahead of the plate.

In general, the solutions of nonlinear partial differential equations (such as the boundary layer equations, Eqs. 9.8 and 9.9) are extremely difficult to obtain. However, by applying a clever coordinate transformation and change of variables, Blasius reduced the partial differential equations to an ordinary differential equation that he was able to solve. A brief description of this process is given below. Additional details can be found in standard books dealing with boundary layer flow (Refs. 1, 2).

It can be argued that in dimensionless form the boundary layer velocity profiles on a flat plate should be similar regardless of the location along the plate. That is,

$$\frac{u}{U} = g\left(\frac{y}{\delta}\right)$$

where $g(y/\delta)$ is an unknown function to be determined. In addition, by applying an order of magnitude analysis of the forces acting on fluid within the boundary layer, it can be shown that the boundary layer thickness grows as the square root of x and inversely proportional to the square root of U. That is,

$$\delta \sim \left(\frac{\nu x}{U}\right)^{1/2}$$

Such a conclusion results from a balance between viscous and inertial forces within the boundary layer and from the fact that the velocity varies much more rapidly in the direction across the boundary layer than along it.

The boundary layer equations can be written in terms of a similarity variable.

Thus, we introduce the dimensionless *similarity variable* $\eta = (U/\nu x)^{1/2} y$ and the stream function $\psi = (\nu x U)^{1/2} f(\eta)$, where $f = f(\eta)$ is an unknown function. Recall from Section 6.2.3 that the velocity components for two-dimensional flow are given in terms of the stream function as $u = \partial\psi/\partial y$ and $v = -\partial\psi/\partial x$, which for this flow become

$$u = Uf'(\eta) \tag{9.12}$$

and

$$v = \left(\frac{\nu U}{4x}\right)^{1/2} (\eta f' - f) \tag{9.13}$$

with the notation $(\;)' = d/d\eta$. We substitute Eqs. 9.12 and 9.13 into the governing equations, Eqs. 9.8 and 9.9, to obtain (after considerable manipulation) the following nonlinear, third-order ordinary differential equation:

$$2f''' + ff'' = 0 \tag{9.14a}$$

The boundary conditions given in Eqs. 9.10 and 9.11 can be written as

$$f = f' = 0 \text{ at } \eta = 0 \quad \text{and} \quad f' \to 1 \text{ as } \eta \to \infty \tag{9.14b}$$

The original partial differential equation and boundary conditions have been reduced to an ordinary differential equation by use of the similarity variable η. The two independent variables, x and y, were combined into the similarity variable in a fashion that reduced the partial differential equation (and boundary conditions) to an ordinary differential equation. This type of reduction is not generally possible. For example, this method does not work on the full Navier-Stokes equations, although it does on the boundary layer equations (Eqs. 9.8 and 9.9).

Although there is no known analytical solution to Eq. 9.14, it is relatively easy to integrate this equation on a computer. The dimensionless boundary layer profile, $u/U = f'(\eta)$, obtained by numerical solution of Eq. 9.14 (termed the Blasius solution), is sketched in Fig. 9.10*a* and is tabulated in Table 9.1. The velocity profiles at different x locations are similar in that there is only one curve necessary to describe the velocity at any point in the boundary layer. Because the similarity variable η contains both x and y, it is seen from Fig. 9.10*b* that the actual velocity profiles are a function of both x and y. The profile at location x_1 is the same as that at x_2 except that the y coordinate is stretched by a factor of $(x_2/x_1)^{1/2}$.

From the solution it is found that $u/U \approx 0.99$ when $\eta = 5.0$. Thus,

Laminar, flat plate boundary layer thickness grows as the square root of the distance from the leading edge.

$$\delta = 5\sqrt{\frac{\nu x}{U}} \tag{9.15}$$

or

$$\frac{\delta}{x} = \frac{5}{\sqrt{\mathrm{Re}_x}}$$

where $\mathrm{Re}_x = Ux/\nu$. It can also be shown that the displacement and momentum thicknesses are given by

$$\frac{\delta^*}{x} = \frac{1.721}{\sqrt{\mathrm{Re}_x}} \tag{9.16}$$

and

$$\frac{\Theta}{x} = \frac{0.664}{\sqrt{\mathrm{Re}_x}} \tag{9.17}$$

As postulated, the boundary layer is thin provided that Re_x is large (i.e., $\delta/x \to 0$ as $\mathrm{Re}_x \to \infty$).

With the velocity profile known, it is an easy matter to determine the wall shear stress, $\tau_w = \mu(\partial u/\partial y)_{y=0}$, where the velocity gradient is evaluated at the plate. The value of $\partial u/\partial y$ at $y = 0$ can be obtained from the Blasius solution to give

$$\tau_w = 0.332U^{3/2}\sqrt{\frac{\rho\mu}{x}} \tag{9.18}$$

Note that the shear stress decreases with increasing x because of the increasing thickness of the boundary layer—the velocity gradient at the wall decreases with increasing x. Also, τ_w varies as $U^{3/2}$, not as U as it does for fully developed laminar pipe flow. These variations are discussed in Section 9.2.3.

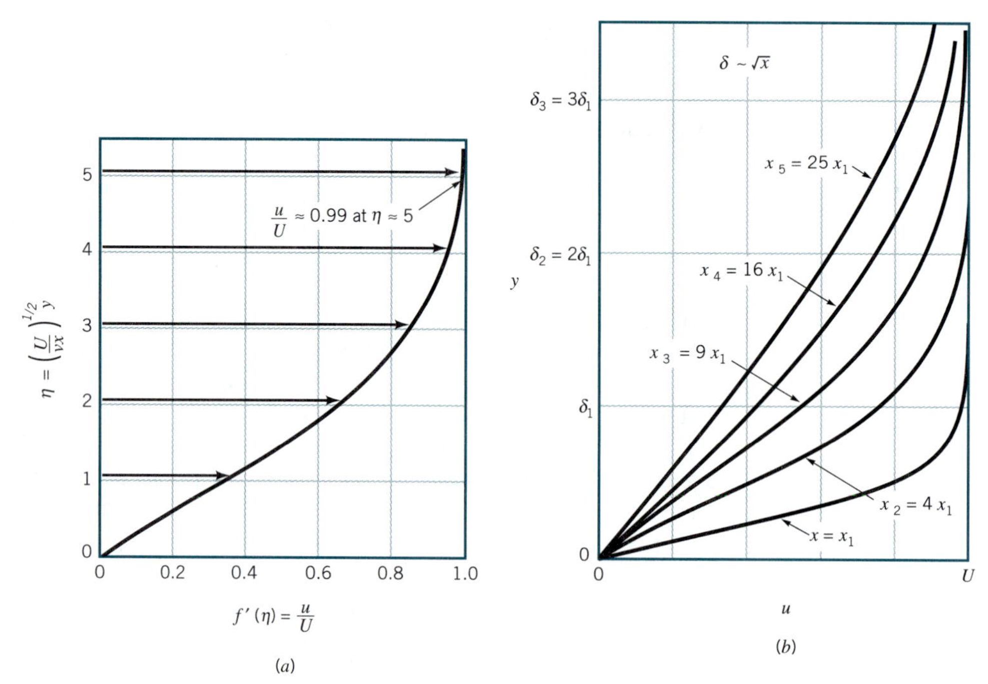

■ **FIGURE 9.10 Blasius boundary layer profile: (*a*) boundary layer profile in dimensionless form using the similarity variable η, (*b*) similar boundary layer profiles at different locations along the flat plate.**

9.2.3 Momentum-Integral Boundary Layer Equation for a Flat Plate

The momentum integral method provides an approximate technique to analyze boundary layer flow.

One of the important aspects of boundary layer theory is the determination of the drag caused by shear forces on a body. As was discussed in the previous section, such results can be obtained from the governing differential equations for laminar boundary layer flow. Since these solutions are extremely difficult to obtain, it is of interest to have an alternative approximate method. The momentum integral method described in this section provides such an alternative.

We consider the uniform flow past a flat plate and the fixed control volume as shown in Fig. 9.11. In agreement with advanced theory and experiment, we assume that the pressure is constant throughout the flow field. The flow entering the control volume at the leading edge of the plate [section (1)] is uniform, while the velocity of the flow exiting the control volume [section (2)] varies from the upstream velocity at the edge of the boundary layer to zero velocity on the plate.

The fluid adjacent to the plate makes up the lower portion of the control surface. The upper surface coincides with the streamline just outside the edge of the boundary layer at section (2). It need not (in fact, does not) coincide with the edge of the boundary layer except at section (2). If we apply the x component of the momentum equation (Eq. 5.22) to the steady flow of fluid within this control volume we obtain

$$\sum F_x = \rho \int_{(1)} u\mathbf{V} \cdot \hat{\mathbf{n}}\, dA + \rho \int_{(2)} u\mathbf{V} \cdot \hat{\mathbf{n}}\, dA$$

■ **TABLE 9.1**
Laminar Flow Along a Flat Plate (the Blasius Solution)

$\eta = y(U/\nu x)^{1/2}$	$f'(\eta) = u/U$	η	$f'(\eta)$
0	0	3.6	0.9233
0.4	0.1328	4.0	0.9555
0.8	0.2647	4.4	0.9759
1.2	0.3938	4.8	0.9878
1.6	0.5168	5.0	0.9916
2.0	0.6298	5.2	0.9943
2.4	0.7290	5.6	0.9975
2.8	0.8115	6.0	0.9990
3.2	0.8761	∞	1.0000

Flat plate drag is directly related to wall shear stress.

where for a plate of width b

$$\sum F_x = -\mathfrak{D} = -\int_{\text{plate}} \tau_w \, dA = -b \int_{\text{plate}} \tau_w \, dx \tag{9.19}$$

and $\mathfrak{D}$ is the drag that the plate exerts on the fluid. Note that the net force caused by the uniform pressure distribution does not contribute to this flow. Since the plate is solid and the upper surface of the control volume is a streamline, there is no flow through these areas. Thus,

$$-\mathfrak{D} = \rho \int_{(1)} U(-U) \, dA + \rho \int_{(2)} u^2 \, dA$$

or

$$\mathfrak{D} = \rho U^2 bh - \rho b \int_0^\delta u^2 \, dy \tag{9.20}$$

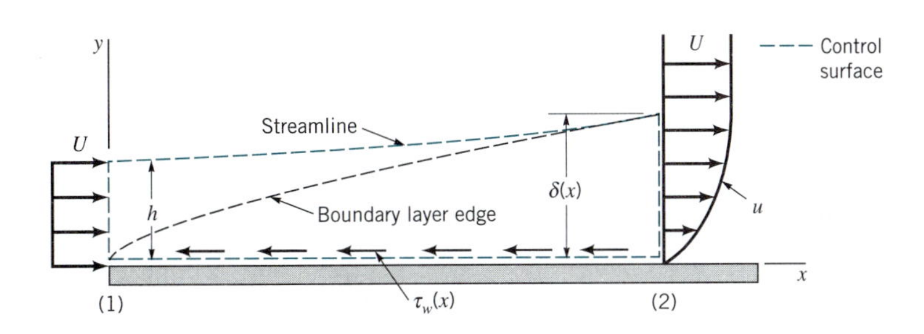

■ **FIGURE 9.11 Control volume used in the derivation of the momentum integral equation for boundary layer flow.**

Although the height h is not known, it is known that for conservation of mass the flowrate through section (1) must equal that through section (2), or

$$Uh = \int_0^\delta u\, dy$$

which can be written as

$$\rho U^2 bh = \rho b \int_0^\delta Uu\, dy \tag{9.21}$$

Thus, by combining Eqs. 9.20 and 9.21 we obtain the drag in terms of the deficit of momentum flux across the outlet of the control volume as

Drag on a flat plate is related to momentum deficit within the boundary layer.

$$\mathcal{D} = \rho b \int_0^\delta u(U - u)\, dy \tag{9.22}$$

If the flow were inviscid, the drag would be zero, since we would have $u \equiv U$ and the right-hand side of Eq. 9.22 would be zero. (This is consistent with the fact that $\tau_w = 0$ if $\mu = 0$.) Equation 9.22 points out the important fact that boundary layer flow on a flat plate is governed by a balance between shear drag (the left-hand side of Eq. 9.22) and a decrease in the momentum of the fluid (the right-hand side of Eq. 9.22). As x increases, δ increases and the drag increases. The thickening of the boundary layer is necessary to overcome the drag of the viscous shear stress on the plate. This is contrary to horizontal fully developed pipe flow in which the momentum of the fluid remains constant and the shear force is overcome by the pressure gradient along the pipe.

The development of Eq. 9.22 and its use was first put forth in 1921 by T. von Karman (1881–1963), a Hungarian/German aerodynamicist. By comparing Eqs. 9.22 and 9.4 we see that the drag can be written in terms of the momentum thickness, Θ, as

$$\mathcal{D} = \rho b U^2\, \Theta \tag{9.23}$$

Note that this equation is valid for laminar or turbulent flows.

The shear stress distribution can be obtained from Eq. 9.23 by differentiating both sides with respect to x to obtain

$$\frac{d\mathcal{D}}{dx} = \rho b U^2 \frac{d\Theta}{dx} \tag{9.24}$$

The increase in drag per length of the plate, $d\mathcal{D}/dx$, occurs at the expense of an increase of the momentum boundary layer thickness, which represents a decrease in the momentum of the fluid.

Since $d\mathcal{D} = \tau_w\, b\, dx$ (see Eq. 9.19) it follows that

$$\frac{d\mathcal{D}}{dx} = b\tau_w \tag{9.25}$$

Hence, by combining Eqs. 9.24 and 9.25 we obtain the *momentum integral equation* for the boundary layer flow on a flat plate

Shear stress on a flat plate is proportional to the rate of boundary layer growth.

$$\tau_w = \rho U^2 \frac{d\Theta}{dx} \tag{9.26}$$

The usefulness of this relationship lies in the ability to obtain approximate boundary layer results easily by using rather crude assumptions. For example, if we knew the detailed velocity profile in the boundary layer (i.e., the Blasius solution discussed in the previous section), we could evaluate either the right-hand side of Eq. 9.23 to obtain the drag, or the right-hand side of Eq. 9.26 to obtain the shear stress. Fortunately, even a rather crude guess at the velocity profile will allow us to obtain reasonable drag and shear stress results from Eq. 9.26. This method is illustrated in Example 9.4.

EXAMPLE 9.4

Consider the laminar flow of an incompressible fluid past a flat plate at $y = 0$. The boundary layer velocity profile is approximated as $u = Uy/\delta$ for $0 \leq y \leq \delta$ and $u = U$ for $y > \delta$, as is shown in Fig. E9.4. Determine the shear stress by using the momentum integral equation. Compare these results with the Blasius results given by Eq. 9.18.

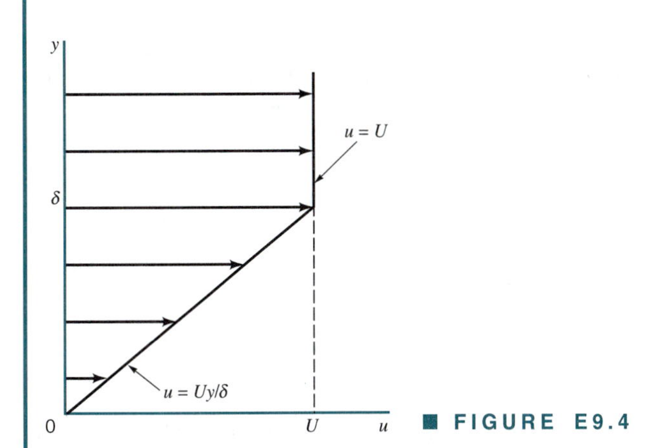

■ FIGURE E9.4

SOLUTION

From Eq. 9.26 the shear stress is given by

$$\tau_w = \rho U^2 \frac{d\Theta}{dx} \tag{1}$$

while for laminar flow we know that $\tau_w = \mu(\partial u/\partial y)_{y=0}$. For the assumed profile we have

$$\tau_w = \mu \frac{U}{\delta} \tag{2}$$

and from Eq. 9.4

$$\Theta = \int_0^\infty \frac{u}{U}\left(1 - \frac{u}{U}\right) dy = \int_0^\delta \frac{u}{U}\left(1 - \frac{u}{U}\right) dy = \int_0^\delta \left(\frac{y}{\delta}\right)\left(1 - \frac{y}{\delta}\right) dy$$

or

$$\Theta = \frac{\delta}{6} \tag{3}$$

Note that as yet we do not know the value of δ (but suspect that it should be a function of x).

By combining Eqs. 1, 2, and 3 we obtain the following differential equation for δ:

$$\frac{\mu U}{\delta} = \frac{\rho U^2}{6} \frac{d\delta}{dx}$$

or

$$\delta \, d\delta = \frac{6\mu}{\rho U} dx$$

This can be integrated from the leading edge of the plate, $x = 0$ (where $\delta = 0$) to an arbitrary location x where the boundary layer thickness is δ. The result is

$$\frac{\delta^2}{2} = \frac{6\mu}{\rho U} x$$

or

$$\delta = 3.46 \sqrt{\frac{\nu x}{U}} \tag{4}$$

Note that this approximate result (i.e., the velocity profile is not actually the simple straight line we assumed) compares favorably with the (much more laborious to obtain) Blasius result given by Eq. 9.15.

The wall shear stress can also be obtained by combining Eqs. 1, 3, and 4 to give

$$\tau_w = 0.289 U^{3/2} \sqrt{\frac{\rho \mu}{x}} \qquad \textbf{(Ans)}$$

Again this approximate result is close (within 13%) to the Blasius value of τ_w given by Eq. 9.18.

As is illustrated in Example 9.4, the momentum integral equation, Eq. 9.26, can be used along with an assumed velocity profile to obtain reasonable, approximate boundary layer results. The accuracy of these results depends on how closely the shape of the assumed velocity profile approximates the actual profile.

Thus, we consider a general velocity profile

$$\frac{u}{U} = g(Y) \quad \text{for} \quad 0 \leq Y \leq 1$$

Approximate velocity profiles are used in the momentum integral equation.

and

$$\frac{u}{U} = 1 \quad \text{for} \quad Y > 1$$

where the dimensionless coordinate $Y = y/\delta$ varies from 0 to 1 across the boundary layer. The dimensionless function $g(Y)$ can be any shape we choose, although it should be a reasonable approximation to the boundary layer profile. In particular, it should certainly satisfy the boundary conditions $u = 0$ at $y = 0$ and $u = U$ at $y = \delta$. That is,

$$g(0) = 0 \quad \text{and} \quad g(1) = 1$$

The linear function $g(Y) = Y$ used in Example 9.4 is one such possible profile. Other conditions, such as $dg/dY = 0$ at $Y = 1$ (i.e., $\partial u/\partial y = 0$ at $y = \delta$), could also be incorporated into the function $g(Y)$ to more closely approximate the actual profile.

For a given $g(Y)$, the drag can be determined from Eq. 9.22 as

$$\mathcal{D} = \rho b \int_0^\delta u(U - u)\, dy = \rho b U^2 \delta \int_0^1 g(Y)[1 - g(Y)]\, dY$$

or

$$\mathcal{D} = \rho b U^2 \delta C_1 \tag{9.27}$$

where the dimensionless constant C_1 has the value

$$C_1 = \int_0^1 g(Y)[1 - g(Y)]\, dY$$

Also, the wall shear stress can be written as

$$\tau_w = \mu \left.\frac{\partial u}{\partial y}\right|_{y=0} = \left.\frac{\mu U}{\delta}\frac{dg}{dY}\right|_{Y=0} = \frac{\mu U}{\delta} C_2 \tag{9.28}$$

where the dimensionless constant C_2 has the value

$$C_2 = \left.\frac{dg}{dY}\right|_{Y=0}$$

By combining Eqs. 9.25, 9.27, and 9.28 we obtain

$$\delta\, d\delta = \frac{\mu C_2}{\rho U C_1}\, dx$$

which can be integrated from $\delta = 0$ at $x = 0$ to give

$$\delta = \sqrt{\frac{2\nu C_2 x}{U C_1}}$$

or

$$\frac{\delta}{x} = \frac{\sqrt{2C_2/C_1}}{\sqrt{\mathrm{Re}_x}} \tag{9.29}$$

By substituting this expression back into Eqs. 9.28 we obtain

$$\tau_w = \sqrt{\frac{C_1 C_2}{2}}\, U^{3/2} \sqrt{\frac{\rho\mu}{x}} \tag{9.30}$$

Approximate boundary layer results are obtained from the momentum integral equation.

To use Eqs. 9.29 and 9.30 we must determine the values of C_1 and C_2. Several assumed velocity profiles and the resulting values of δ are given in Fig. 9.12 and Table 9.2. The more closely the assumed shape approximates the actual (i.e., Blasius) profile, the more accurate the final results. For any assumed profile shape, the functional dependence of δ and τ_w on the physical parameters ρ, μ, U, and x is the same. Only the constants are different. That is, $\delta \sim (\mu x/\rho U)^{1/2}$ or $\delta \mathrm{Re}_x^{1/2}/x =$ constant, and $\tau_w \sim (\rho\mu U^3/x)^{1/2}$, where $\mathrm{Re}_x = \rho U x/\mu$.

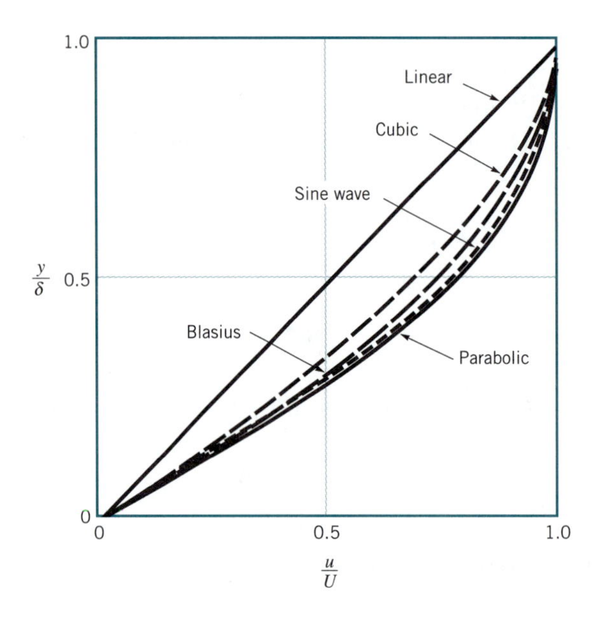

■ **FIGURE 9.12 Typical approximate boundary layer profiles used in the momentum integral equation.**

It is often convenient to use the dimensionless *local friction coefficient*, c_f, defined as

The local friction coefficient is the dimensionless wall shear stress.

$$c_f = \frac{\tau_w}{\frac{1}{2}\rho U^2} \tag{9.31}$$

to express the wall shear stress. From Eq. 9.30 we obtain the approximate value

$$c_f = \sqrt{2C_1C_2}\sqrt{\frac{\mu}{\rho U x}} = \frac{\sqrt{2C_1C_2}}{\sqrt{\text{Re}_x}}$$

while the Blasius solution result is given by

$$c_f = \frac{0.644}{\sqrt{\text{Re}_x}} \tag{9.32}$$

These results are also indicated in Table 9.2.

■ **TABLE 9.2**
Flat Plate Momentum-Integral Results for Various Assumed Laminar Flow Velocity Profiles

Profile Character	$\delta\text{Re}_x^{1/2}/x$	$c_f\text{Re}_x^{1/2}$	$C_{Df}\text{Re}_\ell^{1/2}$
a. Blasius solution	5.00	0.664	1.328
b. Linear $u/U = y/\delta$	3.46	0.578	1.156
c. Parabolic $u/U = 2y/\delta - (y/\delta)^2$	5.48	0.730	1.460
d. Cubic $u/U = 3(y/\delta)/2 - (y/\delta)^3/2$	4.64	0.646	1.292
e. Sine wave $u/U = \sin[\pi(y/\delta)/2]$	4.79	0.655	1.310

For a flat plate of length ℓ and width b, the net friction drag, $\mathcal{D}_f$, can be expressed in terms of the *friction drag coefficient*, C_{Df}, as

$$C_{Df} = \frac{\mathcal{D}_f}{\frac{1}{2}\rho U^2 b\ell} = \frac{b\int_0^\ell \tau_w\, dx}{\frac{1}{2}\rho U^2 b\ell}$$

The friction drag coefficient is an integral of the local friction coefficient.

or

$$C_{Df} = \frac{1}{\ell}\int_0^\ell c_f\, dx \tag{9.33}$$

We use the above approximate value of $c_f = (2C_1C_2\mu/\rho U x)^{1/2}$ to obtain

$$C_{Df} = \frac{\sqrt{8C_1C_2}}{\sqrt{\text{Re}_\ell}}$$

where $\text{Re}_\ell = U\ell/\nu$ is the Reynolds number based on the plate length. The corresponding value obtained from the Blasius solution (Eq. 9.32) gives

$$C_{Df} = \frac{1.328}{\sqrt{\text{Re}_\ell}}$$

These results are also indicated in Table 9.2.

The momentum-integral boundary layer method provides a relatively simple technique to obtain useful boundary layer results. As is discussed in Sections 9.2.5 and 9.2.6, this technique can be extended to boundary layer flows on curved surfaces (where the pressure and fluid velocity at the edge of the boundary layer are not constant) and to turbulent flows.

9.2.4 Transition from Laminar to Turbulent Flow

The analytical results given in Table 9.2 are restricted to laminar boundary layer flows along a flat plate with zero pressure gradient. They agree quite well with experimental results up to the point where the boundary layer flow becomes turbulent, which will occur for any free stream velocity and any fluid provided the plate is long enough. This is true because the parameter that governs the transition to turbulent flow is the Reynolds number—in this case the Reynolds number based on the distance from the leading edge of the plate, $\text{Re}_x = Ux/\nu$.

The value of the Reynolds number at the transition location is a rather complex function of various parameters involved, including the roughness of the surface, the curvature of the surface (e.g., a flat plate or a sphere), and some measure of the disturbances in the flow outside the boundary layer. On a flat plate with a sharp leading edge in a typical air stream, the transition takes place at a distance x from the leading edge given by $\text{Re}_{x\text{cr}} = 2 \times 10^5$ to 3×10^6. Unless otherwise stated, we will use $\text{Re}_{x\text{cr}} = 5 \times 10^5$ in our calculations.

The actual transition from laminar to turbulent boundary layer flow may occur over a region of the plate, not at a specific single location. This occurs, in part, because of the spottiness of the transition. Typically, the transition begins at random locations on the plate in the vicinity of $\text{Re}_x = \text{Re}_{x\text{cr}}$. These spots grow rapidly as they are convected downstream until the entire width of the plate is covered with turbulent flow. The photo shown in Fig. 9.13 illustrates this transition process.

The complex process of transition from laminar to turbulent flow involves the instability of the flow field. Small disturbances imposed on the boundary layer flow (i.e., from a vibration of the plate, a roughness of the surface, or a ''wiggle'' in the flow past the plate) will either grow (instability) or decay (stability), depending on where the disturbance is introduced into the flow. If these disturbances occur at a location with $\text{Re}_x < \text{Re}_{x\text{cr}}$ they will die out, and the

■ **FIGURE 9.13 Turbulent spots and the transition from laminar to turbulent boundary layer flow on a flat plate. Flow from left to right. (Photograph courtesy of B. Cantwell, Stanford University.)**

The boundary layer on a flat plate will become turbulent if the plate is long enough.

boundary layer will return to laminar flow at that location. Disturbances imposed at a location with $Re_x > Re_{xcr}$ will grow and transform the boundary layer flow downstream of this location into turbulence. The study of the initiation, growth, and structure of these turbulent bursts or spots is an active area of fluid mechanics research.

Transition from laminar to turbulent flow also involves a noticeable change in the shape of the boundary layer velocity profile. Typical profiles obtained in the neighborhood of the transition location are indicated in Fig. 9.14. The turbulent profiles are flatter, have a larger velocity gradient at the wall, and produce a larger boundary layer thickness than do the laminar profiles.

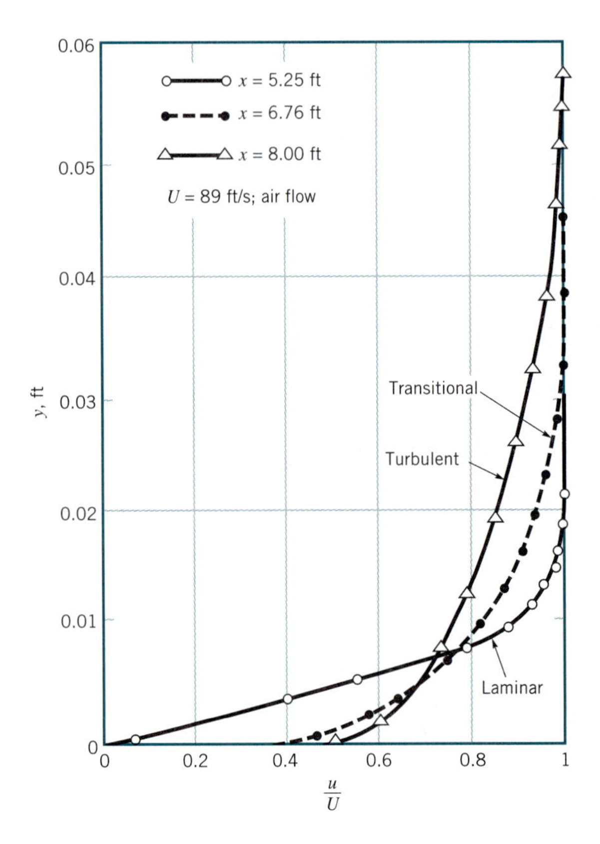

■ **FIGURE 9.14 Typical boundary layer profiles on a flat plate for laminar, transitional, and turbulent flow (Ref. 1).**

EXAMPLE 9.5

A fluid flows steadily past a flat plate with a velocity of $U = 10$ ft/s. At approximately what location will the boundary layer become turbulent, and how thick is the boundary layer at that point if the fluid is (a) water at 60 °F, (b) standard air, or (c) glycerin at 68 °F?

SOLUTION

For any fluid, the laminar boundary layer thickness is found from Eq. 9.15 as

$$\delta = 5\sqrt{\frac{\nu x}{U}}$$

The boundary layer remains laminar up to

$$x_{cr} = \frac{\nu \text{Re}_{xcr}}{U}$$

Thus, if we assume $\text{Re}_{xcr} = 5 \times 10^5$ we obtain

$$x_{cr} = \frac{5 \times 10^5}{10 \text{ ft/s}} \nu = 5 \times 10^4 \nu$$

and

$$\delta_{cr} \equiv \delta|_{x=x_{cr}} = 5 \left[\frac{\nu}{10}(5 \times 10^4 \ \nu)\right]^{1/2} = 354 \ \nu$$

where ν is in ft^2/s and x_{cr} and δ_{cr} are in feet. The values of the kinematic viscosity obtained from Tables 1.5 and 1.7 are listed in Table E9.5 along with the corresponding x_{cr} and δ_{cr}.

TABLE E9.5

Fluid	ν (ft^2/s)	x_{cr} (ft)	δ_{cr} (ft)
a. Water	1.21×10^{-5}	0.605	0.00428
b. Air	1.57×10^{-4}	7.85	0.0556
c. Glycerin	1.28×10^{-2}	640.0	4.53

(Ans)

Laminar flow can be maintained on a longer portion of the plate if the viscosity is increased. However, the boundary layer flow eventually becomes turbulent, provided the plate is long enough. Similarly, the boundary layer thickness is greater if the viscosity is increased.

9.2.5 Turbulent Boundary Layer Flow

Random transport of finite-sized fluid particles occurs within turbulent boundary layers.

The structure of turbulent boundary layer flow is very complex, random, and irregular. It shares many of the characteristics described for turbulent pipe flow in Section 8.3. In particular, the velocity at any given location in the flow is unsteady in a random fashion. The flow can be thought of as a jumbled mix of intertwined eddies (or swirls) of different sizes (diameters and angular velocities). The various fluid quantities involved (i.e., mass, momentum, energy) are convected downstream in the free-stream direction as in a laminar boundary layer. For turbulent flow they are also convected across the boundary layer (in the direction perpendicular to the plate) by the random transport of finite-sized fluid particles associated with the turbulent eddies. There is considerable mixing involved with these finite-sized eddies—

considerably more than is associated with the mixing found in laminar flow where it is confined to the molecular scale. Although there is considerable random motion of fluid particles perpendicular to the plate, there is very little net transfer of mass across the boundary layer—the largest flowrate by far is parallel to the plate.

There is, however, a considerable net transfer of x component of momentum perpendicular to the plate because of the random motion of the particles. Fluid particles moving toward the plate (in the negative y direction) have some of their excess momentum (they come from areas of higher velocity) removed by the plate. Conversely, particles moving away from the plate (in the positive y direction) gain momentum from the fluid (they come from areas of lower velocity). The net result is that the plate acts as a momentum sink, continually extracting momentum from the fluid. For laminar flows, such cross-stream transfer of these properties takes place solely on the molecular scale. For turbulent flow the randomness is associated with fluid particle mixing. Consequently, the shear force for turbulent boundary layer flow is considerably greater than it is for laminar boundary layer flow (see Section 8.3.2).

There are no exact solutions available for turbulent boundary layer flows.

There are no "exact" solutions for turbulent boundary layer flow. As is discussed in Section 9.2.2, it is possible to solve the Prandtl boundary layer equations for laminar flow past a flat plate to obtain the Blasius solution (which is "exact" within the framework of the assumptions involved in the boundary layer equations). Since there is no precise expression for the shear stress in turbulent flow (see Section 8.3), solutions are not available for turbulent flow. However, considerable headway has been made in obtaining numerical (computer) solutions for turbulent flow by using approximate shear stress relationships. Also, progress is being made in the area of direct, full numerical integration of the basic governing equations, the Navier-Stokes equations.

Approximate turbulent boundary layer results can also be obtained by use of the momentum integral equation, Eq. 9.26, which is valid for either laminar or turbulent flow. What is needed for the use of this equation are reasonable approximations to the velocity profile $u = U\,g(Y)$, where $Y = y/\delta$ and u is the time-averaged velocity (the overbar notation, $\bar{u}$, of Section 8.3.2 has been dropped for convenience) and a functional relationship describing the wall shear stress. For laminar flow the wall shear stress was used as $\tau_w = \mu(\partial u/\partial y)_{y=0}$. In theory, such a technique should work for turbulent boundary layers also. However, as is discussed in Section 8.3, the details of the velocity gradient at the wall are not well understood for turbulent flow. Thus, it is necessary to use some empirical relationship for the wall shear stress. This is illustrated in Example 9.6.

Consider turbulent flow of an incompressible fluid past a flat plate. The boundary layer velocity profile is assumed to be $u/U = (y/\delta)^{1/7} = Y^{1/7}$ for $Y = y/\delta \le 1$ and $u = U$ for $Y > 1$ as shown in Fig. E9.6. This is a reasonable approximation of experimentally observed profiles, except very near the plate where this formula gives $\partial u/\partial y = \infty$ at $y = 0$. Note the differences between the assumed turbulent profile and the laminar profile. Also assume that the shear stress agrees with the experimentally determined formula:

$$\tau_w = 0.0225\rho U^2 \left(\frac{\nu}{U\delta}\right)^{1/4} \qquad \textbf{(1)}$$

Determine the boundary layer thicknesses δ, δ^*, and Θ and the wall shear stress, τ_w, as a function of x. Determine the friction drag coefficient, C_{Df}.

SOLUTION

Whether the flow is laminar or turbulent, it is true that the drag force is accounted for by a reduction in the momentum of the fluid flowing past the plate. The shear is obtained from

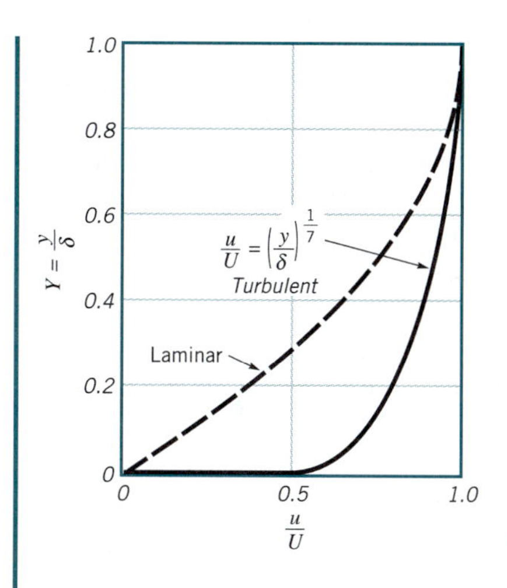

■ FIGURE E9.6

Eq. 9.26 in terms of the rate at which the momentum boundary layer thickness, Θ, increases with distance along the plate as

$$\tau_w = \rho U^2 \frac{d\Theta}{dx}$$

For the assumed velocity profile, the boundary layer momentum thickness is obtained from Eq. 9.4 as

$$\Theta = \int_0^\infty \frac{u}{U}\left(1 - \frac{u}{U}\right) dy = \delta \int_0^1 \frac{u}{U}\left(1 - \frac{u}{U}\right) dY$$

or by integration

$$\Theta = \delta \int_0^1 Y^{1/7}(1 - Y^{1/7})\, dY = \frac{7}{72}\,\delta \qquad \textbf{(2)}$$

where δ is an unknown function of x. By combining the assumed shear force dependence (Eq. 1) with Eq. 2, we obtain the following differential equation for δ:

$$0.0225\rho U^2 \left(\frac{\nu}{U\delta}\right)^{1/4} = \frac{7}{72}\rho U^2 \frac{d\delta}{dx}$$

or

$$\delta^{1/4}\, d\delta = 0.231 \left(\frac{\nu}{U}\right)^{1/4} dx$$

This can be integrated from $\delta = 0$ at $x = 0$ to obtain

$$\delta = 0.370 \left(\frac{\nu}{U}\right)^{1/5} x^{4/5} \qquad \textbf{(3)} \quad \textbf{(Ans)}$$

or in dimensionless form

$$\frac{\delta}{x} = \frac{0.370}{\mathrm{Re}_x^{1/5}}$$

Strictly speaking, the boundary layer near the leading edge of the plate is laminar, not turbulent, and the precise boundary condition should be the matching of the initial turbulent boundary layer thickness (at the transition location) with the thickness of the laminar boundary layer at that point. In practice, however, the laminar boundary layer often exists over a relatively short portion of the plate, and the error associated with starting the turbulent boundary layer with $\delta = 0$ at $x = 0$ can be negligible.

The displacement thickness, δ^*, and the momentum thickness, Θ, can be obtained from Eqs. 9.3 and 9.4 by integrating as follows:

$$\delta^* = \int_0^\infty \left(1 - \frac{u}{U}\right) dy = \delta \int_0^1 \left(1 - \frac{u}{U}\right) dY$$

$$= \delta \int_0^1 (1 - Y^{1/7})\, dY = \frac{\delta}{8}$$

Thus, by combining this with Eq. 3 we obtain

$$\delta^* = 0.0463 \left(\frac{\nu}{U}\right)^{1/5} x^{4/5} \qquad \textbf{(Ans)}$$

Similarly, from Eq. 2,

$$\Theta = \tfrac{7}{72}\delta = 0.0360 \left(\frac{\nu}{U}\right)^{1/5} x^{4/5} \qquad \textbf{(4)} \quad \textbf{(Ans)}$$

The functional dependence for δ, δ^*, and Θ is the same; only the constants of proportionality are different. Typically, $\Theta < \delta^* < \delta$.

By combining Eqs. 1 and 3, we obtain the following result for the wall shear stress

$$\tau_w = 0.0225\rho U^2 \left[\frac{\nu}{U(0.370)(\nu/U)^{1/5} x^{4/5}}\right]^{1/4} = \frac{0.0288\rho U^2}{\mathrm{Re}_x^{1/5}} \qquad \textbf{(Ans)}$$

This can be integrated over the length of the plate to obtain the friction drag on one side of the plate, $\mathcal{D}_f$, as

$$\mathcal{D}_f = \int_0^\ell b\tau_w\, dx = b(0.0288\rho U^2) \int_0^\ell \left(\frac{\nu}{Ux}\right)^{1/5} dx$$

or

$$\mathcal{D}_f = 0.0360\rho U^2 \frac{A}{\mathrm{Re}_\ell^{1/5}}$$

where $A = b\ell$ is the area of the plate. (This result can also be obtained by combining Eq. 9.23 and the expression for the momentum thickness given in Eq. 4.) The corresponding friction drag coefficient, C_{Df}, is

$$C_{Df} = \frac{\mathcal{D}_f}{\frac{1}{2}\rho U^2 A} = \frac{0.0720}{\mathrm{Re}_\ell^{1/5}} \qquad \textbf{(Ans)}$$

Note that for the turbulent boundary layer flow the boundary layer thickness increases with x as $\delta \sim x^{4/5}$ and the shear stress decreases as $\tau_w \sim x^{-1/5}$. For laminar flow these dependencies are $x^{1/2}$ and $x^{-1/2}$, respectively. The random character of the turbulent flow causes a different structure of the flow.

Obviously the results presented in this example are valid only in the range of validity of the original data—the assumed velocity profile and shear stress. This range covers smooth flat plates with $5 \times 10^5 < \mathrm{Re}_\ell < 10^7$.

The flat plate drag coefficient is a function of relative roughness and Reynolds number.

In general, the drag coefficient for a flat plate of length ℓ is a function of the Reynolds number, Re_ℓ, and the relative roughness, ε/ℓ. The results of numerous experiments covering a wide range of the parameters of interest are shown in Fig. 9.15. For laminar boundary layer flow the drag coefficient is a function of only the Reynolds number—surface roughness is not important. This is similar to laminar flow in a pipe. However, for turbulent flow, the surface roughness does affect the shear stress and, hence, the drag coefficient. This is similar to turbulent pipe flow in which the surface roughness may protrude into or through the viscous sublayer next to the wall and alter the flow in this thin, but very important, layer (see Section 8.4.1). Values of the roughness, ε, for different materials can be obtained from Table 8.1.

The drag coefficient diagram of Fig. 9.15 (boundary layer flow) shares many characteristics in common with the familiar Moody diagram (pipe flow) of Fig. 8.23, even though the mechanisms governing the flow are quite different. Fully developed horizontal pipe flow is governed by a balance between pressure forces and viscous forces. The fluid inertia remains constant throughout the flow. Boundary layer flow on a horizontal flat plate is governed by a balance between inertia effects and viscous forces. The pressure remains constant throughout the flow. (As is discussed in Section 9.2.6, for boundary layer flow on curved surfaces, the pressure is not constant.)

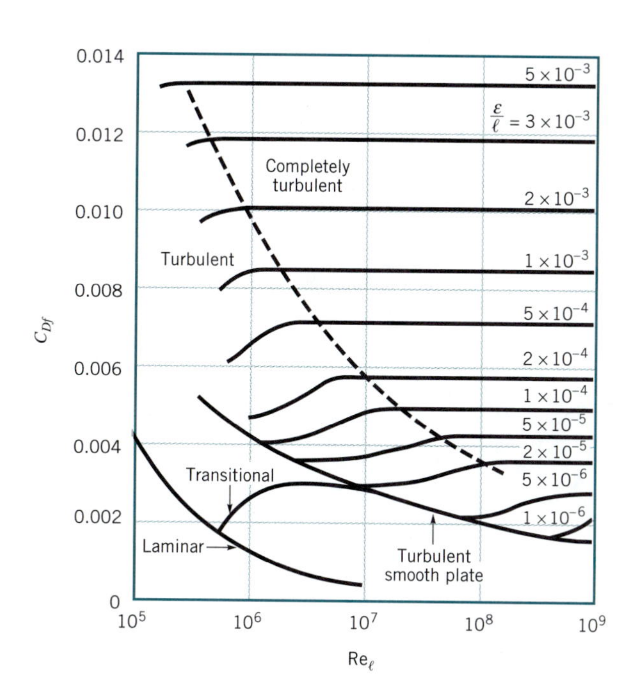

■ **FIGURE 9.15** **Friction drag coefficient for a flat plate parallel to the upstream flow (Ref. 18, with permission).**

TABLE 9.3

Empirical Equations for the Flat Plate Drag Coefficient (Ref. 1)

Equation	Flow Conditions
$C_{Df} = 1.328/(\mathrm{Re}_\ell)^{0.5}$	Laminar flow
$C_{Df} = 0.455/(\log \mathrm{Re}_\ell)^{2.58} - 1700/\mathrm{Re}_\ell$	Transitional with $\mathrm{Re}_{xcr} = 5 \times 10^5$
$C_{Df} = 0.455/(\log \mathrm{Re}_\ell)^{2.58}$	Turbulent, smooth plate
$C_{Df} = [1.89 - 1.62 \log(\varepsilon/\ell)]^{-2.5}$	Completely turbulent

Various equations are available for flat plate drag coefficients.

It is often convenient to have an equation for the drag coefficient as a function of the Reynolds number and relative roughness rather than the graphical representation given in Fig. 9.15. Although there is not one equation valid for the entire $\mathrm{Re}_\ell - \varepsilon/\ell$ range, the equations presented in Table 9.3 do work well for the conditions indicated.

EXAMPLE 9.7

The water ski shown in Fig. E9.7*a* moves through 70 °F water with a velocity U. Estimate the drag caused by the shear stress on the bottom of the ski for $0 < U < 30$ ft/s.

SOLUTION

Clearly the ski is not a flat plate, and it is not aligned exactly parallel to the upstream flow. However, we can obtain a reasonable approximation to the shear force by using the flat plate results. That is, the friction drag, $\mathfrak{D}_f$, caused by the shear stress on the bottom of the ski (the wall shear stress) can be determined as

$$\mathfrak{D}_f = \tfrac{1}{2}\rho U^2 \ell b C_{Df}$$

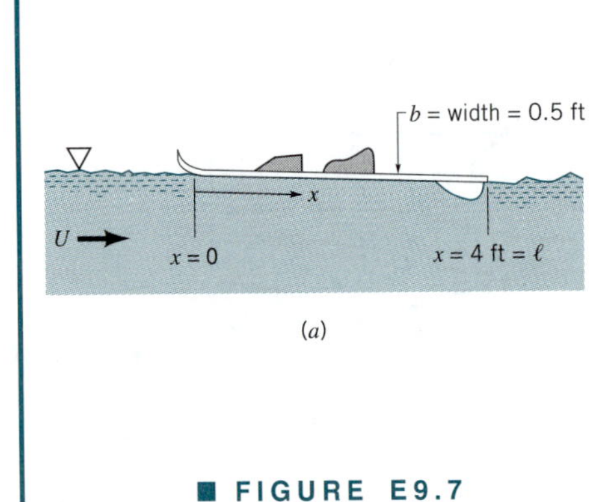

(*a*)

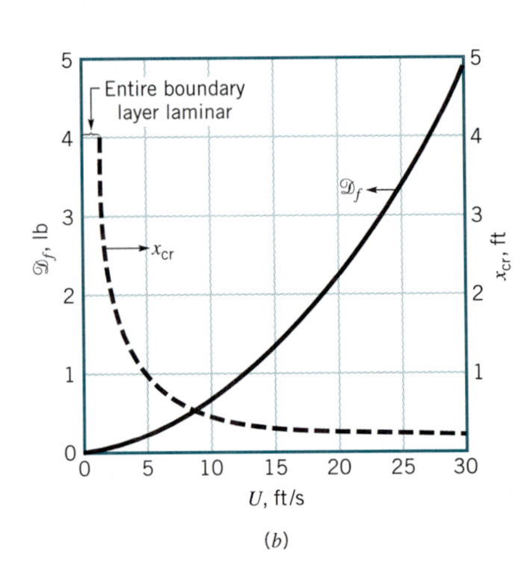

(*b*)

■ **FIGURE E9.7**

With $A = \ell b = 4 \text{ ft} \times 0.5 \text{ ft} = 2 \text{ ft}^2$, $\rho = 1.94 \text{ slugs/ft}^3$, and $\mu = 2.04 \times 10^{-5} \text{ lb·s/ft}^2$ (see Table B.1) we obtain

$$\mathcal{D}_f = \tfrac{1}{2}(1.94 \text{ slugs/ft}^3)(2.0 \text{ ft}^2)U^2 C_{Df}$$
$$= 1.94\, U^2 C_{Df} \quad \textbf{(1)}$$

where $\mathcal{D}_f$ and U are in pounds and ft/s, respectively.

The friction coefficient, C_{Df}, can be obtained from Fig. 9.15 or from the appropriate equations given in Table 9.3. As we will see, for this problem, much of the flow lies within the transition regime where both the laminar and turbulent portions of the boundary layer flow occupy comparable lengths of the plate. We choose to use the values of C_{Df} from the table.

For the given conditions we obtain

$$\text{Re}_\ell = \frac{\rho U \ell}{\mu} = \frac{(1.94 \text{ slugs/ft}^3)(4 \text{ ft})U}{2.04 \times 10^{-5} \text{ lb·s/ft}^2} = 3.80 \times 10^5\, U$$

where U is in ft/s. With $U = 10$ ft/s, or $\text{Re}_\ell = 3.80 \times 10^6$, we obtain from Table 9.3 $C_{Df} = 0.455/(\log \text{Re}_\ell)^{2.58} - 1700/\text{Re}_\ell = 0.00308$. From Eq. 1 the corresponding drag is

$$\mathcal{D}_f = 1.94(10)^2(0.00308) = 0.598 \text{ lb}$$

By covering the range of upstream velocities of interest we obtain the results shown in Fig. E9.7*b*.

If Re $\lesssim$ 1000, the results of boundary layer theory are not valid—inertia effects are not dominant enough and the boundary layer is not thin compared with the length of the plate. For our problem this corresponds to $U = 2.63 \times 10^{-3}$ ft/s. For all practical purposes U is greater than this value, and the flow past the ski is of the boundary layer type.

The approximate location of the transition from laminar to turbulent boundary layer flow as defined by $\text{Re}_{\text{cr}} = \rho U x_{\text{cr}}/\mu = 5 \times 10^5$ is indicated in Fig. E9.7*b*. Up to $U = 1.31$ ft/s the entire boundary layer is laminar. The fraction of the boundary layer that is laminar decreases as U increases until only the front 0.18 ft is laminar when $U = 30$ ft/s.

For anyone who has water skied, it is clear that it can require considerably more force to be pulled along at 30 ft/s than the 2 × 4.88 lb = 9.76 lb (two skis) indicated in Fig. E9.7*b*. As is discussed in Section 9.3, the total drag on an object such as a water ski consists of more than just the friction drag. Other components, including pressure drag and wave-making drag, add considerably to the total resistance.

9.2.6 Effects of Pressure Gradient

The free-stream velocity on a curved surface is not constant.

The boundary layer discussions in the previous parts of Section 9.2 have dealt with flow along a flat plate in which the pressure is constant throughout the fluid. In general, when a fluid flows past an object other than a flat plate, the pressure field is not uniform. As shown in Fig. 9.6, if the Reynolds number is large, relatively thin boundary layers will develop along the surfaces. Within these layers the component of the pressure gradient in the streamwise direction (i.e., along the body surface) is not zero, although the pressure gradient normal to the surface is negligibly small. That is, if we were to measure the pressure while moving across the boundary layer from the body to the boundary layer edge, we would find that the pressure is essentially constant. However, the pressure does vary in the direction along the body surface if the body is curved. The variation in the *free-stream velocity*, U_{fs}, the fluid velocity at the edge of the boundary layer, is the cause of the pressure gradient in the boundary layer. The characteristics of the entire flow (both within and outside of the boundary layer) are often highly dependent on the pressure gradient effects on the fluid within the boundary layer.

For a flat plate parallel to the upstream flow, the upstream velocity (that far ahead of the plate) and the free-stream velocity (that at the edge of the boundary layer) are equal—$U = U_{fs}$. This is a consequence of the negligible thickness of the plate. For bodies of nonzero thickness, these two velocities are different. This can be seen in the flow past a circular cylinder of diameter D. The upstream velocity and pressure are U and p_0, respectively. If the fluid were completely inviscid ($\mu = 0$), the Reynolds number would be infinite ($\text{Re} = \rho UD/\mu = \infty$) and the streamlines would be symmetrical, as are shown in Fig. 9.16*a*. The fluid velocity along the surface would vary from $U_{fs} = 0$ at the very front and rear of the cylinder (points A and F are stagnation points) to a maximum of $U_{fs} = 2U$ at the top and bottom of the cylinder (point C). The pressure on the surface of the cylinder would be symmetrical about the vertical midplane of the cylinder, reaching a maximum value of $p_0 + \rho U^2/2$ (the stagnation pressure) at both the front and back of the cylinder, and a minimum of $p_0 - 3\rho U^2/2$ at the top and bottom of the cylinder. The pressure and free-stream velocity distributions are shown in Figs. 9.16*b* and 9.16*c*. These characteristics can be obtained from potential flow analysis of Section 6.6.3.

If there were no viscosity, there would be no pressure or friction drag on a cylinder.

Because of the absence of viscosity (therefore, $\tau_w = 0$) and the symmetry of the pressure distribution for inviscid flow past a circular cylinder, it is clear that the drag on the cylinder is zero. Although it is not obvious, it can be shown that the drag is zero for any object that does not produce a lift (symmetrical or not) in an inviscid fluid (Ref. 4). Based on experimental evidence, however, we know that there must be a net drag. Clearly, since there is no purely inviscid fluid, the reason for the observed drag must lie on the shoulders of the viscous effects.

To test this hypothesis, we could conduct an experiment by measuring the drag on an object (such as a circular cylinder) in a series of fluids with decreasing values of viscosity. To our initial surprise we would find that no matter how small we make the viscosity (provided

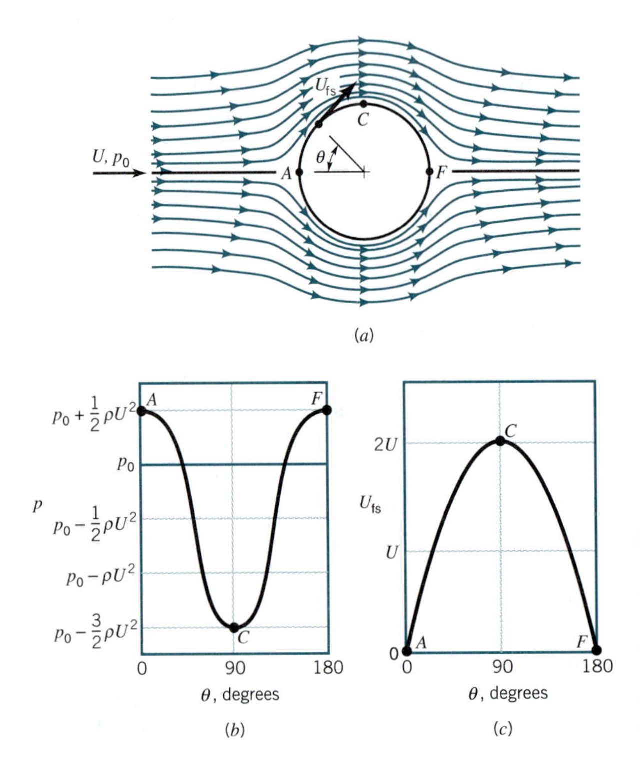

■ **FIGURE 9.16**
Inviscid flow past a circular cylinder: (*a*) streamlines for the flow if there were no viscous effects, (*b*) pressure distribution on the cylinder's surface, (*c*) free stream velocity on the cylinder's surface.

it is not precisely zero) we would measure a finite drag, essentially independent of the value of μ. As was noted in Section 6.6.3, this leads to what has been termed *d'Alembert's paradox*—the drag on an object in an inviscid fluid is zero, but the drag on an object in a fluid with vanishingly small (but nonzero) viscosity is not zero.

The reason for the above paradox can be described in terms of the effect of the pressure gradient on boundary layer flow. Consider large Reynolds number flow of a real (viscous) fluid past a circular cylinder. As was discussed in Section 9.1.2, we expect the viscous effects to be confined to thin boundary layers near the surface. This allows the fluid to stick ($\mathbf{V} = 0$) to the surface—a necessary condition for any fluid, provided $\mu \neq 0$. The basic idea of boundary layer theory is that the boundary layer is thin enough so that it does not greatly disturb the flow outside the boundary layer. Based on this reasoning, for large Reynolds numbers the flow throughout most of the flow field would be expected to be as is indicated in Fig. 9.16*a*, the inviscid flow field.

The pressure gradient in the external flow is imposed throughout the boundary layer fluid.

The pressure distribution indicated in Fig. 9.16*b* is imposed on the boundary layer flow along the surface of the cylinder. In fact, there is negligible pressure variation across the thin boundary layer so that the pressure within the boundary layer is that given by the inviscid flow field. This pressure distribution along the cylinder is such that the stationary fluid at the nose of the cylinder ($U_{fs} = 0$ at $\theta = 0$) is accelerated to its maximum velocity ($U_{fs} = 2U$ at $\theta = 90°$) and then is decelerated back to zero velocity at the rear of the cylinder ($U_{fs} = 0$ at $\theta = 180°$). This is accomplished by a balance between pressure and inertia effects; viscous effects are absent for the inviscid flow outside the boundary layer.

Physically, in the absence of viscous effects, a fluid particle traveling from the front to the back of the cylinder coasts down the ''pressure hill'' from $\theta = 0$ to $\theta = 90°$ (from point *A* to *C* in Fig. 9.16*b*) and then back up the hill to $\theta = 180°$ (from point *C* to *F*) without any loss of energy. There is an exchange between kinetic and pressure energy, but there are no energy losses. The same pressure distribution is imposed on the viscous fluid within the boundary layer. The decrease in pressure in the direction of flow along the front half of the cylinder is termed a *favorable pressure gradient.* The increase in pressure in the direction of flow along the rear half of the cylinder is termed an *adverse pressure gradient.*

Consider a fluid particle within the boundary layer indicated in Fig. 9.17. In its attempt to flow from *A* to *F* it experiences the same pressure distribution as the particles in the free stream immediately outside the boundary layer—the inviscid flow field pressure. However, because of the viscous effects involved, the particle in the boundary layer experiences a loss of energy as it flows along. This loss means that the particle does not have enough energy to coast all of the way up the pressure hill (from *C* to *F*) and to reach point *F* at the rear of the cylinder. This kinetic energy deficit is seen in the velocity profile detail at point *C*, shown in Fig. 9.17*a*. Because of friction, the boundary layer fluid cannot travel from the front to the rear of the cylinder. (This conclusion can also be obtained from the concept that due to viscous effects the particle at *C* does not have enough momentum to allow it to coast up the pressure hill to *F*.)

The situation is similar to a bicyclist coasting down a hill and up the other side of the valley. If there were no friction the rider starting with zero speed could reach the same height from which he or she started. Clearly friction (rolling resistance, aerodynamic drag, etc.) causes a loss of energy (and momentum), making it impossible for the rider to reach the height from which he or she started without supplying additional energy (i.e., peddling). The fluid within the boundary layer does not have such an energy supply. Thus, the fluid flows against the increasing pressure as far as it can, at which point the boundary layer separates from (lifts off) the surface. This *boundary layer separation* is indicated in Fig. 9.17*a*. Typical velocity profiles at representative locations along the surface are shown in Fig. 9.17*b*. At the separation location (profile *D*), the velocity gradient at the wall and the wall shear stress are zero. Beyond that location (from *D* to *E*) there is reverse flow in the boundary layer.

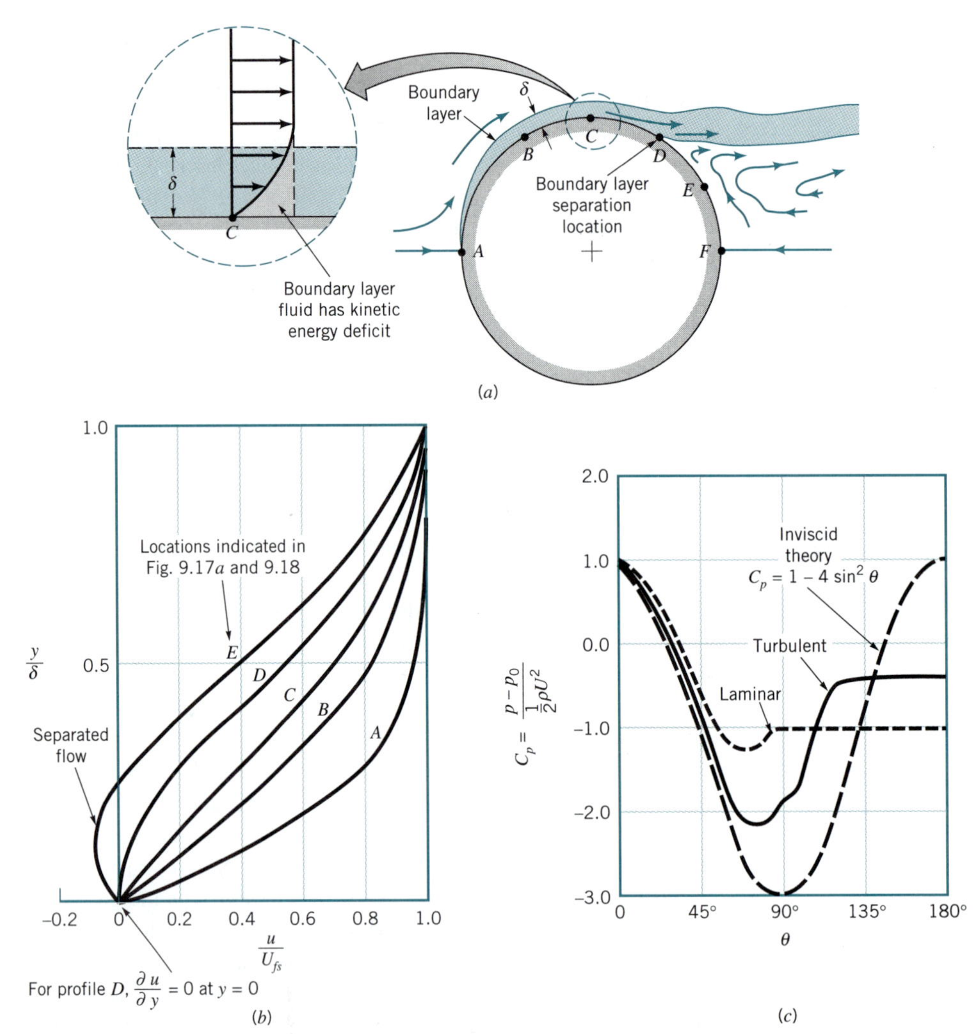

■ **FIGURE 9.17 Boundary layer characteristics on a circular cylinder: (*a*) boundary layer separation location, (*b*) typical boundary layer velocity profiles at various locations on the cylinder, (*c*) surface pressure distributions for inviscid flow and boundary layer flow.**

V9.4

Viscous effects within the boundary layer cause boundary layer separation.

As is indicated in Fig. 9.17*c*, because of the boundary layer separation, the average pressure on the rear half of the cylinder is considerably less than that on the front half. Thus, a large pressure drag is developed, even though (because of small viscosity) the viscous shear drag may be quite small. D'Alembert's paradox is explained. No matter how small the viscosity, provided it is not zero, there will be a boundary layer that separates from the surface, giving a drag that is, for the most part, independent of the value of μ.

The location of separation, the width of the wake region behind the object, and the pressure distribution on the surface depend on the nature of the boundary layer flow. Compared with a laminar boundary layer, a turbulent boundary layer flow has more kinetic energy and momentum associated with it because (1) as is indicated in Fig. E9.6, the velocity profile is fuller, more nearly like the ideal uniform profile, and (2) there can be considerable energy associated with the swirling, random components of the velocity that do not appear in the time-averaged x component of velocity. Thus, as is indicated in Fig. 9.17c, the turbulent boundary layer can flow farther around the cylinder (farther up the pressure hill) before it separates than can the laminar boundary layer. This idea is discussed in detail in Section 9.3.2.

The structure of the flow field past a circular cylinder is completely different for a zero viscosity fluid than it is for a viscous fluid, no matter how small the viscosity is, provided it is not zero. This is due to boundary layer separation. Similar concepts hold for other shaped bodies as well. The flow past an airfoil at zero *angle of attack* (the angle between the upstream flow and the axis of the object) is shown in Fig. 9.18a; flow past the same airfoil at a 5° angle of attack is shown in Fig. 9.18b. Over the front portion of the airfoil the pressure decreases in the direction of flow—a favorable pressure gradient. Over the rear portion the pressure increases in the direction of flow—an adverse pressure gradient. The boundary layer velocity profiles at representative locations are similar to those indicated in Fig. 9.17b for flow past a circular cylinder. If the adverse pressure gradient is not too great (because the body is not too ''thick'' in some sense), the boundary layer fluid can flow into the slightly increasing pressure region (i.e., from C to the trailing edge in Fig. 9.18a) without separating from the surface. However, if the pressure gradient is too adverse (because the angle of attack is too large), the boundary layer will separate from the surface as indicated in Fig. 9.18b. Such situations can lead to the catastrophic loss of lift called *stall*, which is discussed in Section 9.4.

Boundary layer separation causes airplane wings to stall.

Streamlined bodies are generally those designed to eliminate (or at least to reduce) the effects of separation, whereas nonstreamlined bodies generally have relatively large drag due to the low pressure in the separated regions (the wake). Although the boundary layer may be quite thin, it can appreciably alter the entire flow field because of boundary layer separation. These ideas are discussed in Section 9.3.

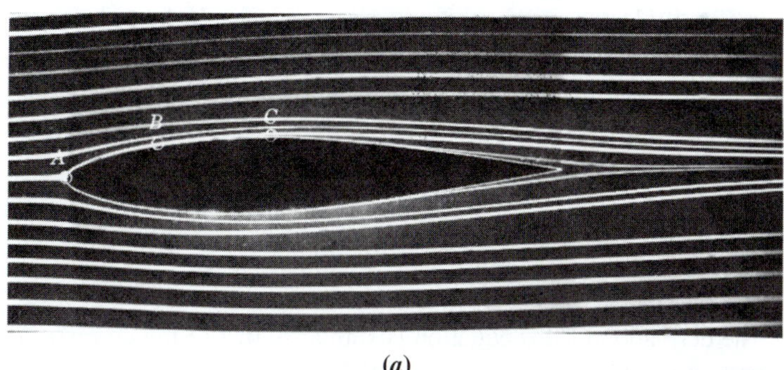

(*a*)

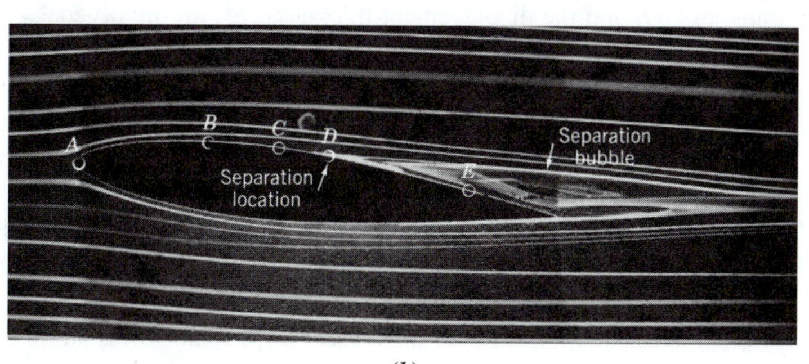

(*b*)

■ FIGURE 9.18 Flow visualization photographs of flow past an airfoil (the boundary layer velocity profiles for the points indicated are similar to those indicated in Fig. 9.17*b*): (*a*) zero angle of attack, no separation, (*b*) 5° angle of attack, flow separation. Dye in water. (Photograph courtesy of ONERA, France.)

9.2.7 Momentum-Integral Boundary Layer Equation with Nonzero Pressure Gradient

The boundary layer results discussed in Sections 9.2.2 and 9.2.3 are valid only for boundary layers with zero pressure gradients. They correspond to the velocity profile labeled C in Fig. 9.17*b*. Boundary layer characteristics for flows with nonzero pressure gradients can be obtained from nonlinear, partial differential boundary layer equations similar to Eqs. 9.8 and 9.9, provided the pressure gradient is appropriately accounted for. Such an approach is beyond the scope of this book (Refs. 1, 2).

An alternative approach is to extend the momentum integral boundary layer equation technique (Section 9.2.3) so that it is applicable for flows with nonzero pressure gradients. The momentum integral equation for boundary layer flows with zero pressure gradient, Eq. 9.26, is a statement of the balance between the shear force on the plate (represented by τ_w) and rate of change of momentum of the fluid within the boundary layer [represented by $\rho U^2 (d\Theta/dx)$]. For such flows the free-stream velocity is constant ($U_{fs} = U$). If the free-stream velocity is not constant [$U_{fs} = U_{fs}(x)$, where x is the distance measured along the curved body], the pressure will not be constant. This follows from the Bernoulli equation with negligible gravitational effects, since $p + \rho U_{fs}^2/2$ is constant along the streamlines outside the boundary layer. Thus,

$$\frac{dp}{dx} = -\rho U_{fs} \frac{dU_{fs}}{dx} \tag{9.34}$$

For a given body the free-stream velocity and the corresponding pressure gradient on the surface can be obtained from inviscid flow techniques (potential flow) discussed in Section 6.7. (This is how the circular cylinder results of Fig. 9.16 were obtained.)

Pressure gradient effects can be included in the momentum integral equation.

Flow in a boundary layer with nonzero pressure gradient is very similar to that shown in Fig. 9.11, except that the upstream velocity, U, is replaced by the free-stream velocity, $U_{fs}(x)$, and the pressures at sections (1) and (2) are not necessarily equal. By using the x component of the momentum equation (Eq. 5.22) with the appropriate shear forces and pressure forces acting on the control surface indicated in Fig. 9.11, the following integral momentum equation for boundary layer flows is obtained:

$$\tau_w = \rho \frac{d}{dx} (U_{fs}^2 \, \Theta) + \rho \delta^* \, U_{fs} \frac{dU_{fs}}{dx} \tag{9.35}$$

The derivation of this equation is similar to that of the corresponding equation for constant-pressure boundary layer flow, Eq. 9.26, although the inclusion of the pressure gradient effect brings in additional terms (Refs. 1, 2, 3). For example, both the boundary layer momentum thickness, Θ, and the displacement thickness, δ^*, are involved.

Equation 9.35, the general momentum integral equation for two-dimensional boundary layer flow, represents a balance between viscous forces (represented by τ_w), pressure forces (represented by $\rho U_{sf} \, dU_{fs}/dx = -dp/dx$), and the fluid momentum (represented by Θ, the boundary layer momentum thickness). In the special case of a flat plate, $U_{fs} = U =$ constant, and Eq. 9.35 reduces to Eq. 9.26.

Equation 9.35 can be used to obtain boundary layer information in a manner similar to that done for the flat plate boundary layer (Section 9.2.3). That is, for a given body shape the free-stream velocity, U_{fs}, is determined, and a family of approximate boundary layer profiles is assumed. Equation 9.35 is then used to provide information about the boundary layer thickness, wall shear stress, and other properties of interest. The details of this technique are not within the scope of this book (Refs. 1, 3).

9.3 Drag

As was discussed in Section 9.1, any object moving through a fluid will experience a drag, $\mathcal{D}$—a net force in the direction of flow due to the pressure and shear forces on the surface of the object. This net force, a combination of flow direction components of the normal and tangential forces on the body, can be determined by use of Eqs. 9.1 and 9.2, provided the distributions of pressure, p, and wall shear stress, τ_w, are known. Only in very rare instances can these distributions be determined analytically. The boundary layer flow past a flat plate parallel to the upstream flow as is discussed in Section 9.2 is one such case. Current advances in computational fluid dynamics (i.e., the use of computers to solve the governing equations of the flow field) have provided encouraging results for more complex shapes. However, much work in this area remains.

Most of the information pertaining to drag on objects is a result of numerous experiments with wind tunnels, water tunnels, towing tanks, and other ingenious devices that are used to measure the drag on scale models. As was discussed in Chapter 7, these data can be put into dimensionless form and the results can be appropriately ratioed for prototype calculations. Typically, the result for a given-shaped object is a drag coefficient, C_D, where

$$C_D = \frac{\mathcal{D}}{\frac{1}{2}\rho U^2 A} \tag{9.36}$$

The drag coefficient is a function of other dimensionless parameters.

and C_D is a function of other dimensionless parameters such as Reynolds number, Re, Mach number, Ma, Froude number, Fr, and relative roughness of the surface, ε/ℓ. That is,

$$C_D = \phi(\text{shape, Re, Ma, Fr}, \varepsilon/\ell)$$

The character of C_D as a function of these parameters is discussed in this section.

9.3.1 Friction Drag

Friction drag, $\mathcal{D}_f$, is that part of the drag that is due directly to the shear stress, τ_w, on the object. It is a function of not only the magnitude of the wall shear stress, but also of the orientation of the surface on which it acts. This is indicated by the factor $\tau_w \sin\theta$ in Eq. 9.1. If the surface is parallel to the upstream velocity, the entire shear force contributes directly to the drag. This is true for the flat plate parallel to the flow as was discussed in Section 9.2. If the surface is perpendicular to the upstream velocity, the shear stress contributes nothing to the drag. Such is the case for a flat plate normal to the upstream velocity as was discussed in Section 9.1.

In general, the surface of a body will contain portions parallel to and normal to the upstream flow, as well as any direction in between. A circular cylinder is such a body. Because the viscosity of most common fluids is small, the contribution of the shear force to the overall drag on a body is often quite small. Such a statement should be worded in dimensionless terms. That is, because the Reynolds number of most familiar flows is quite large, the percent of the drag caused directly by the shear stress is often quite small. For highly streamlined bodies or for low Reynolds number flow, however, most of the drag may be due to friction drag.

The friction drag on a flat plate of width b and length ℓ oriented parallel to the upstream flow can be calculated from

$$\mathcal{D}_f = \tfrac{1}{2}\rho U^2 b\ell C_{Df}$$

where C_{Df} is the friction drag coefficient. The value of C_{Df}, given as a function of Reynolds number, $\text{Re}_\ell = \rho U\ell/\mu$, and relative surface roughness, ε/ℓ, in Fig. 9.15 and Table 9.3, is a

Friction (viscous) drag is that drag produced by viscous shear stresses.

result of boundary layer analysis and experiments (see Section 9.2). Typical values of roughness, ε, for various surfaces are given in Table 8.1. As with the pipe flow discussed in Chapter 8, the flow is divided into two distinct categories—laminar or turbulent, with a transitional regime connecting them. The drag coefficient (and, hence, the drag) is not a function of the plate roughness if the flow is laminar. However, for turbulent flow the roughness does considerably affect the value of C_{Df}. As with pipe flow, this dependence is a result of the surface roughness elements protruding into or through the laminar sublayer (see Section 8.3).

Most objects are not flat plates parallel to the flow; instead, they are curved surfaces along which the pressure varies. As was discussed in Section 9.2.6, this means that the boundary layer character, including the velocity gradient at the wall, is different for most objects from that for a flat plate. This can be seen in the change of shape of the boundary layer profile along the cylinder in Fig. 9.17*b*.

The precise determination of the shear stress along the surface of a curved body is quite difficult to obtain. Although approximate results can be obtained by a variety of techniques (Refs. 1, 2), these are outside the scope of this text. As is shown by the following example, if the shear stress is known, its contribution to the drag can be determined.

EXAMPLE 9.8

A viscous, incompressible fluid flows past the circular cylinder shown in Fig. E9.8*a*. According to a more advanced theory of boundary layer flow, the boundary layer remains attached to the cylinder up to the separation location at $\theta \approx 108.8°$, with the dimensionless wall shear stress as is indicated in Fig. E9.8*b* (Ref. 1). The shear stress on the cylinder in the wake region, $108.8 < \theta < 180°$, is negligible. Determine C_{Df}, the drag coefficient for the cylinder based on the friction drag only.

SOLUTION

The friction drag, $\mathscr{D}_f$, can be determined from Eq. 9.1 as

$$\mathscr{D}_f = \int \tau_w \sin\theta \, dA = 2\left(\frac{D}{2}\right) b \int_0^{\pi} \tau_w \sin\theta \, d\theta$$

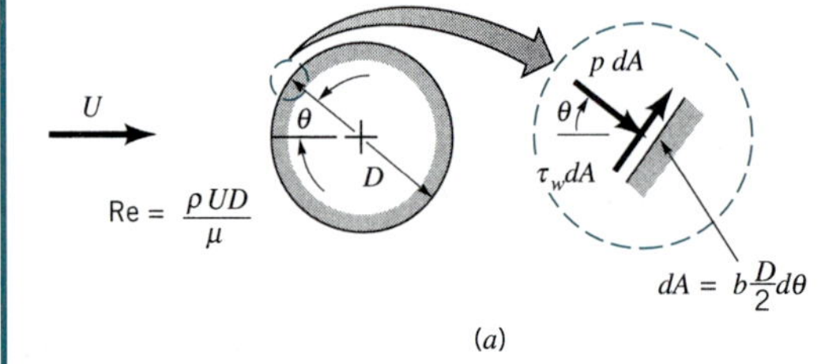

(*a*)

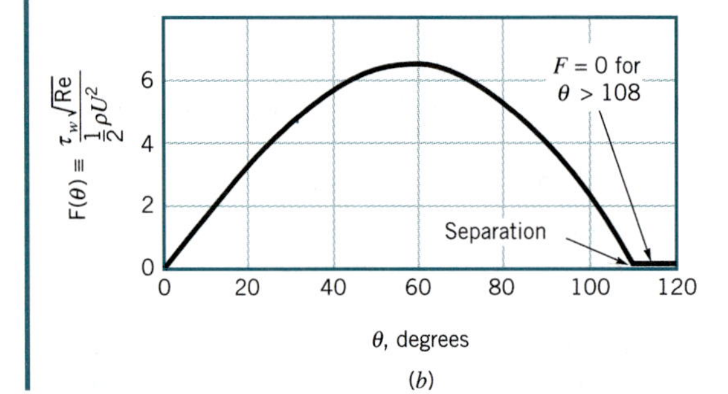

(*b*)

■ FIGURE E9.8

(c)

■ FIGURE E9.8 *(Continued)*

where b is the length of the cylinder. Note that θ is in radians (not degrees) to ensure the proper dimensions of $dA = 2\,(D/2)\,b\,d\theta$. Thus,

$$C_{Df} = \frac{\mathscr{D}_f}{\frac{1}{2}\rho U^2 bD} = \frac{2}{\rho U^2}\int_0^{\pi} \tau_w \sin\theta\, d\theta$$

This can be put into dimensionless form by using the dimensionless shear stress parameter, $F(\theta) = \tau_w\sqrt{\text{Re}}/(\rho U^2/2)$, given in Fig. E9.8*b* as follows:

$$C_{Df} = \int_0^{\pi} \frac{\tau_w}{\frac{1}{2}\rho U^2}\sin\theta\, d\theta = \frac{1}{\sqrt{\text{Re}}}\int_0^{\pi}\frac{\tau_w\sqrt{\text{Re}}}{\frac{1}{2}\rho U^2}\sin\theta\, d\theta$$

where $\text{Re} = \rho UD/\mu$. Thus,

$$C_{Df} = \frac{1}{\sqrt{\text{Re}}}\int_0^{\pi} F(\theta)\sin\theta\, d\theta \qquad (1)$$

The function $F(\theta)\sin\theta$, obtained from Fig. E9.8*b*, is plotted in Fig. E9.8*c*. The necessary integration to obtain C_{Df} from Eq. 1 can be done by an appropriate numerical technique or by an approximate graphical method to determine the area under the given curve.

The result is $\int_0^{\pi} F(\theta)\sin\theta\, d\theta = 5.93$, or

$$C_{Df} = \frac{5.93}{\sqrt{\text{Re}}} \qquad \textbf{(Ans)}$$

Note that the total drag must include both the shear stress (friction) drag and the pressure drag. As we will see in Example 9.9, for the circular cylinder most of the drag is due to the pressure force.

The above friction drag result is valid only if the boundary layer flow on the cylinder is laminar. As is discussed in Section 9.3.3, for a smooth cylinder this means that $\text{Re} = \rho UD/\mu < 3 \times 10^5$. It is also valid only for flows that have a Reynolds number sufficiently large to ensure the boundary layer structure to the flow. For the cylinder this means $\text{Re} > 100$.

9.3.2 Pressure Drag

Pressure (form) drag is that drag produced by normal stresses.

Pressure drag, $\mathscr{D}_p$, is that part of the drag that is due directly to the pressure, p, on an object. It is often referred to as *form drag* because of its strong dependency on the shape or form of the object. Pressure drag is a function of the magnitude of the pressure and the orientation of the surface element on which the pressure force acts. For example, the pressure force on either side of a flat plate parallel to the flow may be very large, but it does not contribute to the drag because it acts in the direction normal to the upstream velocity. On the other hand, the pressure force on a flat plate normal to the flow provides the entire drag.

As previously noted, for most bodies, there are portions of the surface that are parallel to the upstream velocity, others normal to the upstream velocity, and the majority of which are at some angle in between. The pressure drag can be obtained from Eq. 9.1 provided a detailed description of the pressure distribution and the body shape is given. That is,

$$\mathfrak{D}_p = \int p \cos \theta \, dA$$

which can be rewritten in terms of the *pressure drag coefficient*, C_{Dp}, as

$$C_{Dp} = \frac{\mathfrak{D}_p}{\frac{1}{2}\rho U^2 A} = \frac{\int p \cos \theta \, dA}{\frac{1}{2}\rho U^2 A} = \frac{\int C_p \cos \theta \, dA}{A} \qquad \textbf{(9.37)}$$

The pressure coefficient is a dimensionless form of the pressure.

Here $C_p = (p - p_0)/(\rho U^2/2)$ is the *pressure coefficient*, where p_0 is a reference pressure. The level of the reference pressure does not influence the drag directly because the net pressure force on a body is zero if the pressure is constant (i.e., p_0) on the entire surface.

For flows in which inertial effects are large relative to viscous effects (i.e., large Reynolds number flows), the pressure difference, $p - p_0$, scales directly with the dynamic pressure, $\rho U^2/2$, and the pressure coefficient is independent of Reynolds number. In such situations we expect the drag coefficient to be relatively independent of Reynolds number.

For flows in which viscous effects are large relative to inertial effects (i.e., very small Reynolds number flows), it is found that both the pressure difference and wall shear stress scale with the characteristic viscous stress, $\mu U/\ell$, where ℓ is a characteristic length. In such situations we expect the drag coefficient to be proportional to 1/Re. That is, $C_D \sim \mathfrak{D}/(\rho U^2/2) \sim (\mu U/\ell)/(\rho U^2/2) \sim \mu/\rho U\ell = 1/\text{Re}$. These characteristics are similar to the friction factor dependence of $f \sim 1/\text{Re}$ for laminar pipe flow and $f \sim$ constant for large Reynolds number flow (see Section 8.4).

If the viscosity were zero, the pressure drag on any shaped object (symmetrical or not) in a steady flow would be zero. There perhaps would be large pressure forces on the front portion of the object, but there would be equally large (and oppositely directed) pressure forces on the rear portion. If the viscosity is not zero, the net pressure drag may be nonzero because of boundary layer separation as is discussed in Section 9.2.6. Example 9.9 illustrates this.

A viscous, incompressible fluid flows past the circular cylinder shown in Fig. E9.8*a*. The pressure coefficient on the surface of the cylinder (as determined from experimental measurements) is as indicated in Fig. E9.9*a*. Determine the pressure drag coefficient for this flow. Combine the results of Examples 9.8 and 9.9 to determine the drag coefficient for a circular cylinder. Compare your results with those given in Fig. 9.21.

SOLUTION

The pressure (form) drag coefficient, C_{Dp}, can be determined from Eq. 9.37 as

$$C_{Dp} = \frac{1}{A}\int C_p \cos \theta \, dA = \frac{1}{bD}\int_0^{2\pi} C_p \cos \theta \, b \left(\frac{D}{2}\right) d\theta$$

or because of symmetry

$$C_{Dp} = \int_0^{\pi} C_p \cos \theta \, d\theta$$

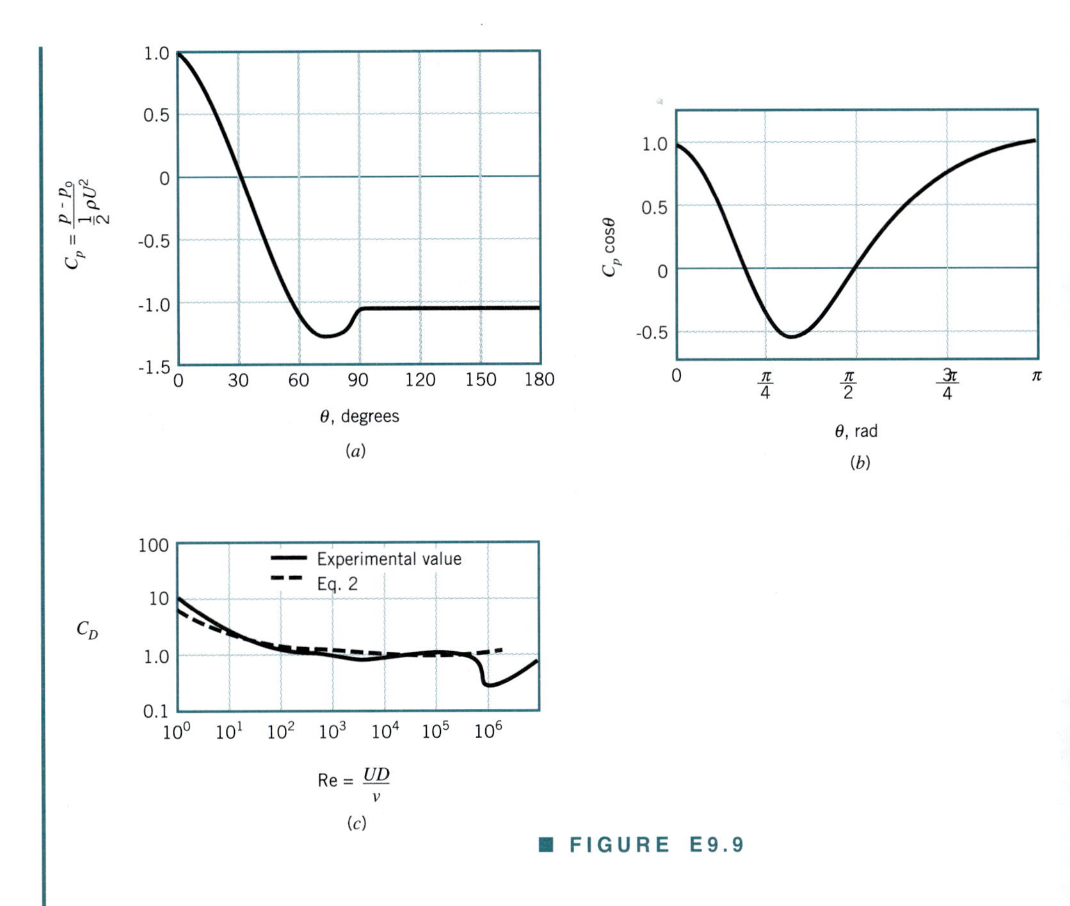

■ **FIGURE E9.9**

where b and D are the length and diameter of the cylinder. To obtain C_{Dp}, we must integrate the C_p cos θ function from $\theta = 0$ to $\theta = \pi$ radians. Again, this can be done by some numerical integration scheme or by determining the area under the curve shown in Fig. E9.9*b*. The result is

$$C_{Dp} = 1.17 \qquad \textbf{(1)} \quad \textbf{(Ans)}$$

Note that the positive pressure on the front portion of the cylinder ($0 \leq \theta \leq 30°$) and the negative pressure (less than the upstream value) on the rear portion ($90 \leq \theta \leq 180°$) produce positive contributions to the drag. The negative pressure on the front portion of the cylinder ($30 < \theta < 90°$) reduces the drag by pulling on the cylinder in the upstream direction. The positive area under the C_p cos θ curve is greater than the negative area—there is a net pressure drag. In the absence of viscosity, these two contributions would be equal—there would be no pressure (or friction) drag.

The net drag on the cylinder is the sum of friction and pressure drag. Thus, from Eq. 1 of Example 9.8 and Eq. 1 of this example, we obtain the drag coefficient

$$C_D = C_{Df} + C_{Dp} = \frac{5.93}{\sqrt{\text{Re}}} + 1.17 \qquad \textbf{(2)} \quad \textbf{(Ans)}$$

This result is compared with the standard experimental value (obtained from Fig. 9.21) in Fig. E9.9*c*. The agreement is very good over a wide range of Reynolds numbers. For $\text{Re} < 10$ the curves diverge because the flow is not a boundary layer type flow—the shear stress and pressure distributions used to obtain Eq. 2 are not valid in this range. The drastic divergence in the curves for $\text{Re} > 3 \times 10^5$ is due to the change from a laminar to turbulent boundary layer, with the corresponding change in the pressure distribution. This is discussed in Section 9.3.3.

It is of interest to compare the friction drag to the total drag on the cylinder. That is,

$$\frac{\mathfrak{D}_f}{\mathfrak{D}} = \frac{C_{Df}}{C_D} = \frac{5.93/\sqrt{\text{Re}}}{(5.93/\sqrt{\text{Re}}) + 1.17} = \frac{1}{1 + 0.197\sqrt{\text{Re}}}$$

For $\text{Re} = 10^3$, 10^4, and 10^5 this ratio is 0.138, 0.0483, and 0.0158, respectively. Most of the drag on the blunt cylinder is pressure drag—a result of the boundary layer separation.

9.3.3 Drag Coefficient Data and Examples

As was discussed in previous sections, the net drag is produced by both pressure and shear stress effects. In most instances these two effects are considered together and an overall drag coefficient, C_D, as defined in Eq. 9.36 is used. There is an abundance of such drag coefficient data available in the literature. This information covers incompressible and compressible viscous flows past objects of almost any shape of interest—both manmade and natural objects. In this section we consider a small portion of this information for representative situations. Additional data can be obtained from various sources (Refs. 5, 6).

Shape Dependence. Clearly the drag coefficient for an object depends on the shape of the object, with shapes ranging from those that are streamlined to those that are blunt. The drag on an ellipse with aspect ratio ℓ/D, where D and ℓ are the thickness and length parallel to the flow, illustrates this dependence. The drag coefficient $C_D = \mathfrak{D}/(\rho U^2 bD/2)$, based on the frontal area, $A = bD$, where b is the length normal to the flow, is as shown in Fig. 9.19. The more blunt the body, the larger the drag coefficient. With $\ell/D = 0$ (i.e., a flat plate normal to the flow) we obtain the flat plate value of $C_D = 1.9$. With $\ell/D = 1$ the corresponding value for a circular cylinder is obtained. As ℓ/D becomes larger the value of C_D decreases.

V9.5

For very large aspect ratios ($\ell/D \to \infty$) the ellipse behaves as a flat plate parallel to the flow. For such cases, the friction drag is greater than the pressure drag, and the value of C_D based on the frontal area, $A = bD$, would increase with increasing ℓ/D. (This occurs for larger ℓ/D values than those shown in the figure.) For such extremely thin bodies (i.e., an ellipse with $\ell/D \to \infty$, a flat plate, or very thin airfoils) it is customary to use the planform area, $A = b\ell$, in defining the drag coefficient. After all, it is the planform area on which the shear stress acts, rather than the much smaller (for thin bodies) frontal area. The ellipse drag coefficient based on the planform area, $C_D = \mathfrak{D}/(\rho U^2 b\ell/2)$, is also shown in Fig. 9.19. Clearly the drag obtained by using either of these drag coefficients would be the same. They merely represent two different ways to package the same information.

The drag coefficient may be based on the frontal area or the planform area.

The amount of streamlining can have a considerable effect on the drag. Incredibly, the drag on the two two-dimensional objects drawn to scale in Fig. 9.20 is the same. The width of the wake for the streamlined strut is very thin, on the order of that for the much smaller diameter circular cylinder.

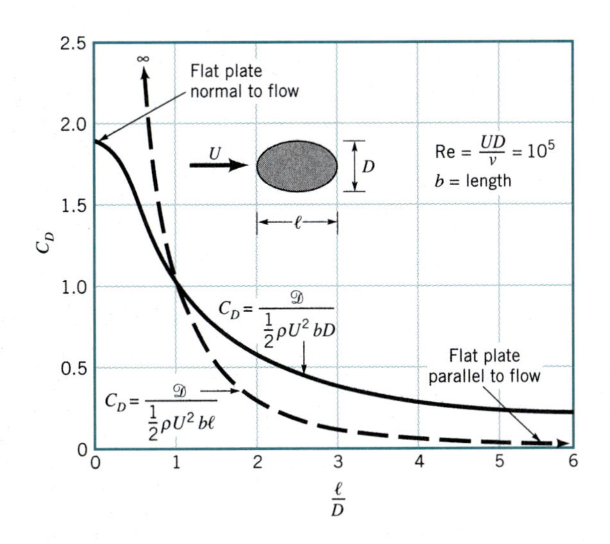

■ **FIGURE 9.19 Drag coefficient for an ellipse with the characteristic area either the frontal area, $A = bD$, or the planform area, $A = b\ell$ (Ref. 5).**

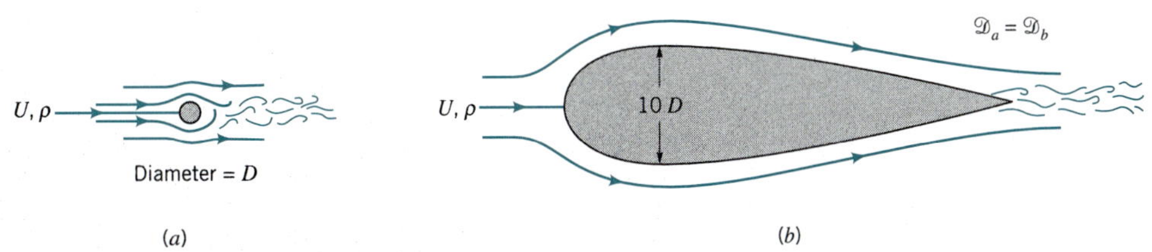

■ **FIGURE 9.20 Two objects of considerably different size that have the same drag force: (*a*) circular cylinder $C_D = 1.2$, (*b*) streamlined strut $C_D = 0.12$.**

Reynolds Number Dependence. Another parameter on which the drag coefficient can be very dependent is the Reynolds number. The main categories of Reynolds number dependence are (1) very low Reynolds number flow, (2) moderate Reynolds number flow (laminar boundary layer), and (3) very large Reynolds number flow (turbulent boundary layer). Examples of these three situations are discussed below.

For very low Reynolds number flows inertia is negligible.

Low Reynolds number flows ($\mathrm{Re} < 1$) are governed by a balance between viscous and pressure forces. Inertia effects are negligibly small. In such instances the drag is expected to be a function of the upstream velocity, U, the body size, ℓ, and the viscosity, μ. That is,

$$\mathcal{D} = f(U, \ell, \mu)$$

From dimensional considerations (see Section 7.7.1)

$$\mathcal{D} = C\mu\ell U \tag{9.38}$$

where the value of the constant C depends on the shape of the body. If we put Eq. 9.38 into dimensionless form using the standard definition of the drag coefficient, we obtain

$$C_D = \frac{\mathcal{D}}{\frac{1}{2}\rho U^2 \ell^2} = \frac{2C\mu\ell U}{\rho U^2 \ell^2} = \frac{2C}{\mathrm{Re}}$$

where $\mathrm{Re} = \rho U\ell/\mu$. The use of the dynamic pressure, $\rho U^2/2$, in the definition of the drag coefficient is somewhat misleading in the case of creeping flows ($\mathrm{Re} < 1$) because it

■ **TABLE 9.4**
Low Reynolds Number Drag Coefficients (Ref. 7) ($\text{Re} = \rho UD/\mu$, $A = \pi D^2/4$)

Object	$C_D = \mathcal{D}/(\rho U^2 A/2)$ (for Re $\lesssim$ 1)	Object	C_D
a. Circular disk normal to flow	20.4/Re	c. Sphere	24.0/Re
b. Circular disk parallel to flow	13.6/Re	d. Hemisphere	22.2/Re

introduces the fluid density, which is not an important parameter for such flows (inertia is not important). Use of this standard drag coefficient definition gives the 1/Re dependence for small Re drag coefficients.

For very small Reynolds number flows, the drag coefficient varies inversely with the Reynolds number.

Typical values of C_D for low Reynolds number flows past a variety of objects are given in Table 9.4. It is of interest that the drag on a disk normal to the flow is only 1.5 times greater than that on a disk parallel to the flow. For large Reynolds number flows this ratio is considerably larger (see Example 9.1). Streamlining (i.e., making the body slender) can produce a considerable drag reduction for large Reynolds number flows; for very small Reynolds number flows it can actually increase the drag because of an increase in the area on which shear forces act. For most objects, the low Reynolds number flow results are valid up to a Reynolds number of about 1.

A small grain of sand, diameter $D = 0.10$ mm and specific gravity $SG = 2.3$, settles to the bottom of a lake after having been stirred up by a passing boat. Determine how fast it falls through the still water.

SOLUTION

A free-body diagram of the particle (relative to the moving particle) is shown in Fig. E9.10. The particle moves downward with a constant velocity U that is governed by a balance between the weight of the particle, $\mathcal{W}$, the buoyancy force of the surrounding water, F_B, and the drag of the water on the particle, $\mathcal{D}$.

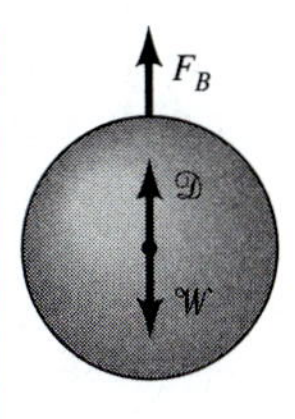

■ FIGURE E9.10

From the free-body diagram we obtain

$$\mathscr{W} = \mathscr{D} + F_B$$

where

$$\mathscr{W} = \gamma_{\text{sand}} \forall = SG\ \gamma_{H_2O} \frac{\pi}{6} D^3 \tag{1}$$

and

$$F_B = \gamma_{H_2O} \forall = \gamma_{H_2O} \frac{\pi}{6} D^3 \tag{2}$$

We assume (because of the smallness of the object) that the flow will be creeping flow ($\text{Re} < 1$) with $C_D = 24/\text{Re}$ (see Table 9.4) so that

$$\mathscr{D} = \frac{1}{2} \rho_{H_2O} U^2 \frac{\pi}{4} D^2 C_D = \frac{1}{2} \rho_{H_2O} U^2 \frac{\pi}{4} D^2 \left(\frac{24}{\rho_{H_2O} UD/\mu_{H_2O}} \right)$$

or

$$\mathscr{D} = 3\pi\mu_{H_2O} UD \tag{3}$$

We must eventually check to determine if this assumption is valid or not. Equation 3 is called Stokes law in honor of G. G. Stokes (1819–1903), a British mathematician and physicist. By combining Eqs. 1, 2, and 3, we obtain

$$SG\ \gamma_{H_2O} \frac{\pi}{6} D^3 = 3\pi\mu_{H_2O} UD + \gamma_{H_2O} \frac{\pi}{6} D^3$$

or, since $\gamma = \rho g$,

$$U = \frac{(SG\rho_{H_2O} - \rho_{H_2O}) g D^2}{18\mu} \tag{4}$$

From Table 1.6 for water at 15.6 °C we obtain $\rho_{H_2O} = 999 \text{ kg/m}^3$ and $\mu_{H_2O} = 1.12 \times 10^{-3}$ N·s/m^2. Thus, from Eq. 4 we obtain

$$U = \frac{(2.3 - 1)(999 \text{ kg/m}^3)(9.81 \text{ m/s}^2)(0.10 \times 10^{-3} \text{ m})^2}{18(1.12 \times 10^{-3} \text{ N·s/m}^2)}$$

or

$$U = 6.32 \times 10^{-3} \text{ m/s} \qquad \textbf{(Ans)}$$

Since

$$\text{Re} = \frac{\rho D U}{\mu} = \frac{(999 \text{ kg/m}^3)(0.10 \times 10^{-3} \text{ m})(0.00632 \text{ m/s})}{1.12 \times 10^{-3} \text{ N·s/m}^2} = 0.564$$

we see that $\text{Re} < 1$, and the form of the drag coefficient used is valid.

Note that if the density of the particle were the same as the surrounding fluid, from Eq. 4 we would obtain $U = 0$. This is reasonable since the particle would be neutrally buoyant and there would be no force to overcome the motion-induced drag. Note also that we have assumed that the particle falls at its steady terminal velocity. That is, we have neglected the acceleration of the particle from rest to its terminal velocity. Since the terminal velocity is small, this acceleration time is quite small. For faster objects (such as a free-falling sky diver) it may be important to consider the acceleration portion of the fall.

Moderate Reynolds number flows tend to take on a boundary layer flow structure. For such flows past streamlined bodies, the drag coefficient tends to decrease slightly with Reynolds number. The $C_D \sim \text{Re}^{-1/2}$ dependence for a laminar boundary layer on a flat plate (see Table 9.3) is such an example. Moderate Reynolds number flows past blunt bodies generally produce drag coefficients that are relatively constant. The C_D values for the spheres and circular cylinders shown in Fig. 9.21*a* indicate this character in the range $10^3 < \text{Re} < 10^5$.

Flow past a cylinder can take on a variety of different structures.

The structure of the flow field at selected Reynolds numbers indicated in Fig. 9.21*a* is shown in Fig. 9.21*b*. For a given object there is a wide variety of flow situations, depending on the Reynolds number involved. The curious reader is strongly encouraged to study the many beautiful photographs of these (and other) flow situations found in Ref. 8.

V9.6

For many shapes there is a sudden change in the character of the drag coefficient when the boundary layer becomes turbulent. This is illustrated in Fig. 9.15 for the flat plate and in Fig. 9.21 for the sphere and the circular cylinder. The Reynolds number at which this transition takes place is a function of the shape of the body.

For streamlined bodies, the drag coefficient increases when the boundary layer becomes turbulent because most of the drag is due to the shear force, which is greater for turbulent flow than for laminar flow. On the other hand, the drag coefficient for a relatively blunt object, such as a cylinder or sphere, actually decreases when the boundary layer becomes turbulent. As is discussed in Section 9.2.6, a turbulent boundary layer can travel further along the surface into the adverse pressure gradient on the rear portion of the cylinder before separation occurs.

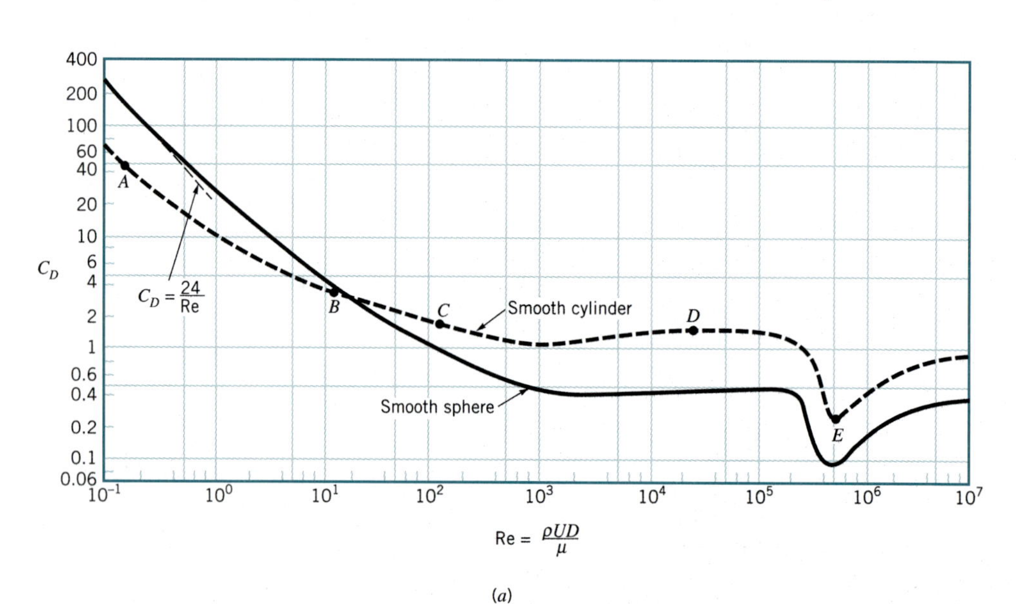

FIGURE 9.21 (*a*) **Drag coefficient as a function of Reynolds number for a smooth circular cylinder and a smooth sphere.**

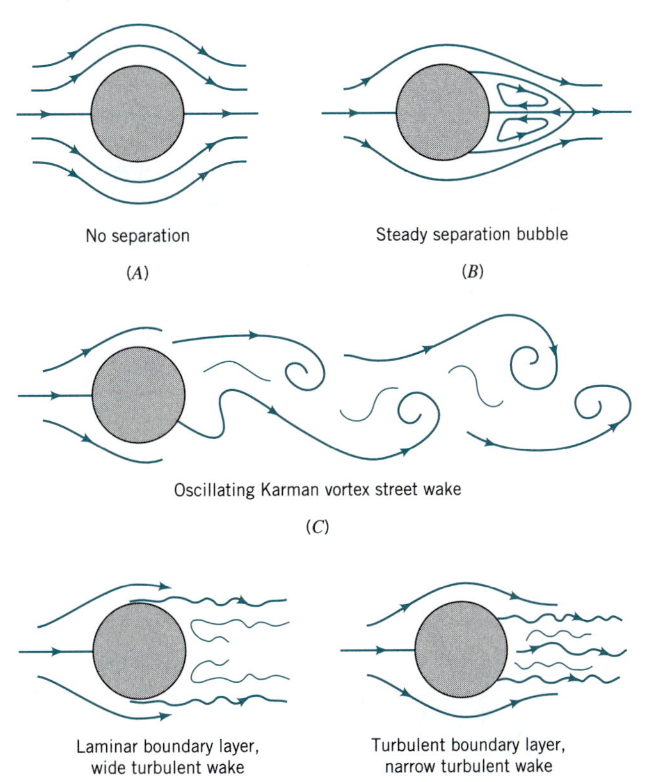

■ **FIGURE 9.21** **(*Continued*) (*b*) Typical flow patterns for flow past a circular cylinder at various Reynolds numbers as indicated in (*a*).**

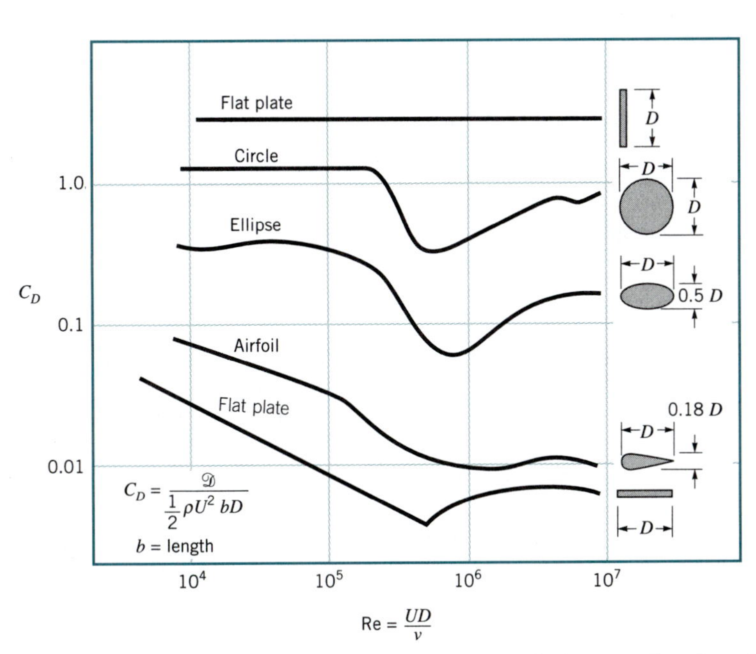

■ **FIGURE 9.22** **Character of the drag coefficient as a function of Reynolds number for objects with various degrees of streamlining, from a flat plate normal to the upstream flow to a flat plate parallel to the flow (two-dimensional flow) (Ref. 5).**

The result is a thinner wake and smaller pressure drag for turbulent boundary layer flow. This is indicated in Fig. 9.21 by the sudden decrease in C_D for $10^5 < \text{Re} < 10^6$. In a portion of this range the actual drag (not just the drag coefficient) decreases with increasing speed. It would be very difficult to control the steady flight of such an object in this range—an increase in velocity requires a decrease in thrust (drag). In all other Reynolds number ranges the drag increases with an increase in the upstream velocity (even though C_D may decrease with Re).

The drag coefficient may change considerably when the boundary layer becomes turbulent.

For extremely blunt bodies, like a flat plate perpendicular to the flow, the flow separates at the edge of the plate regardless of the nature of the boundary layer flow. Thus, the drag coefficient shows very little dependence on the Reynolds number.

The drag coefficients for a series of two-dimensional bodies of varying bluntness are given as a function of Reynolds number in Fig. 9.22. The characteristics described above are evident.

EXAMPLE 9.11

Hail is produced by the repeated rising and falling of ice particles in the updraft of a thunderstorm, as is indicated in Fig. E9.11. When the hail becomes large enough, the aerodynamic drag from the updraft can no longer support the weight of the hail, and it falls from the storm cloud. Estimate the velocity, U, of the updraft needed to make $D = 1.5$-in.-diameter (i.e., "golf ball-sized") hail.

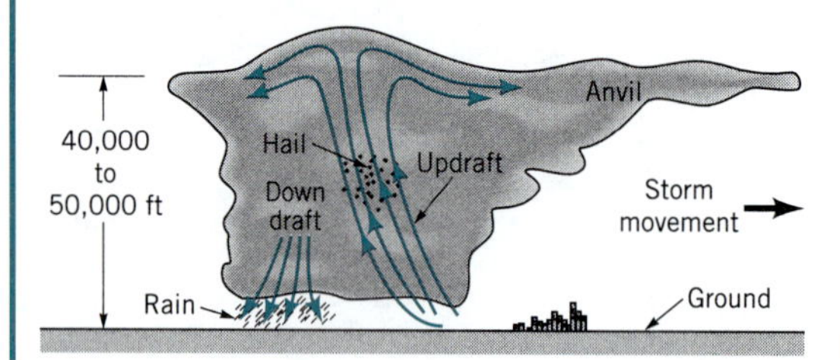

■ FIGURE E9.11

SOLUTION

As is discussed in Example 9.10, for steady state conditions a force balance on an object falling through a fluid gives

$$\mathcal{W} = \mathcal{D} + F_B$$

where $F_B = \gamma_{\text{air}} \forall$ is the buoyant force of the air on the particle, $\mathcal{W} = \gamma_{\text{ice}} \forall$ is the particle weight, and $\mathcal{D}$ is the aerordynamic drag. This equation can be rewritten as

$$\tfrac{1}{2}\rho_{\text{air}} U^2 \frac{\pi}{4} D^2 C_D = \mathcal{W} - F_B \tag{1}$$

With $\forall = \pi D^3/6$ and since $\gamma_{\text{ice}} \gg \gamma_{\text{air}}$ (i.e., $\mathcal{W} \gg F_B$), Eq. 1 can be simplified to

$$U = \left(\frac{4}{3}\frac{\rho_{\text{ice}}}{\rho_{\text{air}}}\frac{gD}{C_D}\right)^{1/2} \tag{2}$$

By using $\rho_{\text{ice}} = 1.84 \text{ slugs/ft}^3$, $\rho_{\text{air}} = 2.38 \times 10^{-3} \text{ slugs/ft}^3$, and $D = 1.5 \text{ in.} = 0.125 \text{ ft}$, Eq. 2 becomes

$$U = \left[\frac{4(1.84 \text{ slugs/ft}^3)(32.2 \text{ ft/s}^2)(0.125 \text{ ft})}{3(2.38 \times 10^{-3} \text{ slugs/ft}^3)C_D}\right]^{1/2}$$

or

$$U = \frac{64.5}{\sqrt{C_D}} \tag{3}$$

where U is in ft/s. To determine U, we must know C_D. Unfortunately, C_D is a function of the Reynolds number (see Fig. 9.21), which is not known unless U is known. Thus, we must use an iterative technique similar to that done with the Moody chart for certain types of pipe flow problems (see Section 8.5).

From Fig. 9.21 we expect that C_D is on the order of 0.5. Thus, we assume $C_D = 0.5$ and from Eq. 3 obtain

$$U = \frac{64.5}{\sqrt{0.5}} = 91.2 \text{ ft/s}$$

The corresponding Reynolds number (assuming $\nu = 1.57 \times 10^{-4} \text{ ft}^2/\text{s}$) is

$$\text{Re} = \frac{UD}{\nu} = \frac{91.2 \text{ ft/s } (0.125 \text{ ft})}{1.57 \times 10^{-4} \text{ ft}^2/\text{s}} = 7.26 \times 10^4$$

For this value of Re we obtain from Fig. 9.21, $C_D = 0.5$. Thus, our assumed value of $C_D = 0.5$ was correct. The corresponding value of U is

$$U = 91.2 \text{ ft/s} = 62.2 \text{ mph} \qquad \textbf{(Ans)}$$

This result was obtained by using standard sea level properties for the air. If conditions at 20,000 ft altitude are used (i.e., from Table C.1, $\rho_{\text{air}} = 1.267 \times 10^{-3} \text{ slugs/ft}^3$ and $\mu = 3.324 \times 10^{-7} \text{ lb}\cdot\text{s/ft}^2$), the corresponding result is $U = 125 \text{ ft/s} = 85.2 \text{ mph}$.

Clearly, an airplane flying through such an updraft would feel its effects (even if it were able to dodge the hail). As seen from Eq. 2, the larger the hail, the stronger the necessary updraft. Hailstones greater than 6 in. in diameter have been reported. In reality, a hailstone is seldom spherical and often not smooth. However, the calculated updraft velocities are in agreement with measured values.

Compressibility Effects. The above discussion is restricted to incompressible flows. If the velocity of the object is sufficiently large, compressibility effects become important and the drag coefficient becomes a function of the Mach number, $\text{Ma} = U/c$, where c is the speed of sound in the fluid. The introduction of Mach number effects complicates matters because the drag coefficient for a given object is then a function of both Reynolds number and Mach number—$C_D = \phi(\text{Re}, \text{Ma})$. The Mach number and Reynolds number effects are often closely connected because both are directly proportional to the upstream velocity. For example, both Re and Ma increase with increasing flight speed of an airplane. The changes in C_D due to a change in U are due to changes in both Re and Ma.

The precise dependence of the drag coefficient on Re and Ma is generally quite complex (Ref. 13). However, the following simplifications are often justified. For low Mach numbers, the drag coefficient is essentially independent of Ma as is indicated in Fig. 9.23. For this situation, if $\text{Ma} < 0.5$ or so, compressibility effects are unimportant. On the other hand, for larger Mach number flows, the drag coefficient can be strongly dependent on Ma, with only secondary Reynolds number effects.

The drag coefficient is usually independent of Mach number for Mach numbers up to approximately 0.5.

For most objects, values of C_D increase dramatically in the vicinity of $\text{Ma} = 1$ (i.e., sonic flow). This change in character, indicated by Fig. 9.24, is due to the existence of shock waves (extremely narrow regions in the flow field across which the flow parameters change in a nearly discontinuous manner), which are discussed in Chapter 11. Shock waves, which cannot exist in subsonic flows, provide a mechanism for the generation of drag that is not present in the relatively low-speed subsonic flows.

The character of the drag coefficient as a function of Mach number is different for blunt bodies than for sharp bodies. As is shown in Fig. 9.24, sharp-pointed bodies develop their

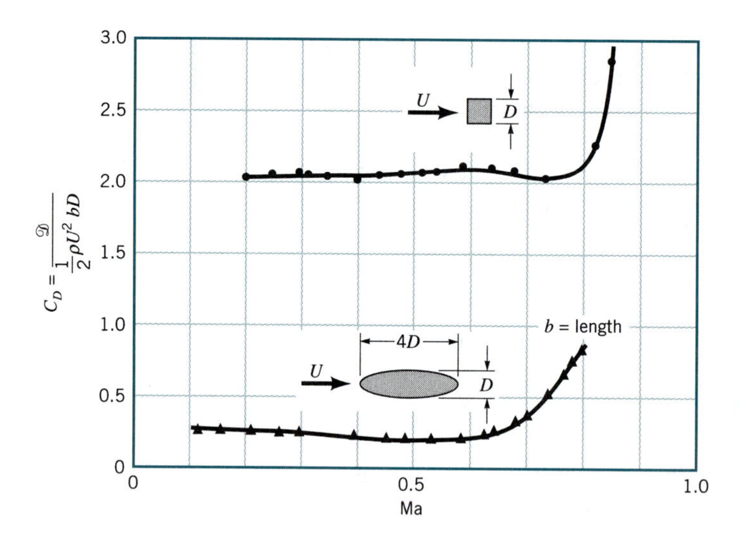

■ **FIGURE 9.23** **Drag coefficient as a function of Mach number for two-dimensional objects in subsonic flow (Ref. 5).**

Compressibility effects can significantly increase the drag coefficient.

maximum drag coefficient in the vicinity of Ma = 1 (sonic flow), whereas the drag coefficient for blunt bodies increases with Ma far above Ma = 1. This behavior is due to the nature of the shock wave structure and the accompanying flow separation. The leading edges of wings for subsonic aircraft are usually quite rounded and blunt, while those of supersonic aircraft tend to be quite pointed and sharp. More information on these important topics can be found in standard texts about compressible flow and aerodynamics (Refs. 9, 10, 29).

Surface Roughness. As is indicated in Fig. 9.15, the drag on a flat plate parallel to the flow is quite dependent on the surface roughness, provided the boundary layer flow is

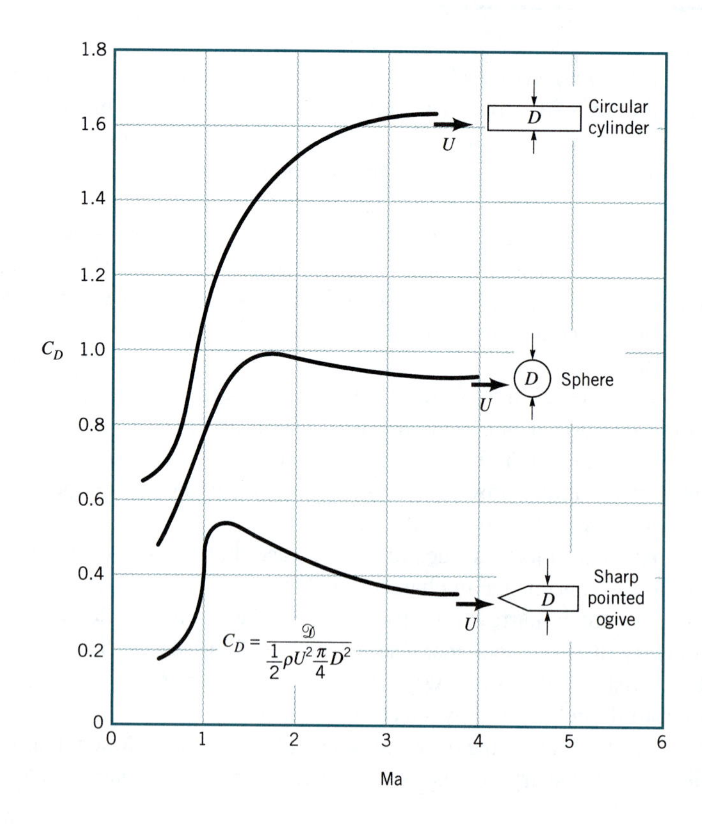

■ **FIGURE 9.24** **Drag coefficient as a function of Mach number for supersonic flow (adapted from Ref. 19).**

turbulent. In such cases the surface roughness protrudes through the laminar sublayer adjacent to the surface (see Section 8.4) and alters the wall shear stress. In addition to the increased turbulent shear stress, surface roughness can alter the Reynolds number at which the boundary layer flow becomes turbulent. Thus, a rough flat plate may have a larger portion of its length covered by a turbulent boundary layer than does the corresponding smooth plate. This also acts to increase the net drag on the plate.

Depending on the body shape, an increase in surface roughness may increase or decrease drag.

In general, for streamlined bodies, the drag increases with increasing surface roughness. Great care is taken to design the surfaces of airplane wings to be as smooth as possible, since protruding rivets or screw heads can cause a considerable increase in the drag. On the other hand, for an extremely blunt body, such as a flat plate normal to the flow, the drag is independent of the surface roughness, since the shear stress is not in the upstream flow direction and contributes nothing to the drag.

For blunt bodies like a circular cylinder or sphere, an increase in surface roughness can actually cause a decrease in the drag. This is illustrated for a sphere in Fig. 9.25. As is discussed in Section 9.2.6, when the Reynolds number reaches the critical value ($\text{Re} = 3 \times 10^5$ for a smooth sphere), the boundary layer becomes turbulent and the wake region behind the sphere becomes considerably narrower than if it were laminar (see Fig. 9.17). The result is a considerable drop in pressure drag with a slight increase in friction drag, combining to give a smaller overall drag (and C_D).

The boundary layer can be tripped into turbulence at a smaller Reynolds number by using a rough-surfaced sphere. For example, the critical Reynolds number for a golf ball is approximately $\text{Re} = 4 \times 10^4$. In the range $4 \times 10^4 < \text{Re} < 4 \times 10^5$, the drag on the standard rough (i.e., dimpled) golf ball is considerably less ($D_{D\text{rough}}/C_{D\text{smooth}} \approx 0.25/0.5 = 0.5$) than for the smooth ball. As is shown in Example 9.12, this is precisely the Reynolds number range for well-hit golf balls—hence, the reason for dimples on golf balls. The Reynolds number range for well-hit table tennis balls is less than $\text{Re} = 4 \times 10^4$. Thus, table tennis balls are smooth.

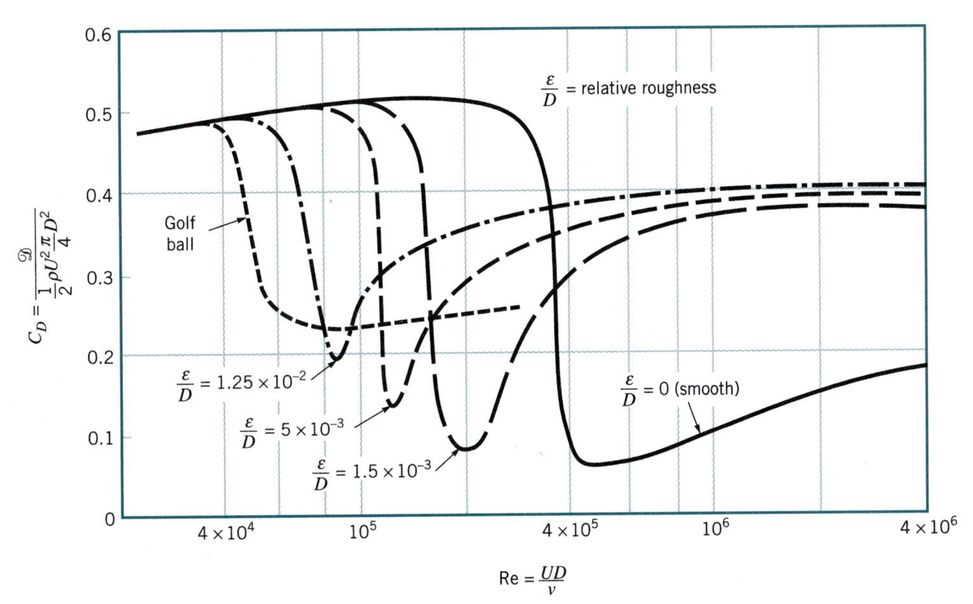

■ **FIGURE 9.25** **The effect of surface roughness on the drag coefficient of a sphere in the Reynolds number range for which the laminar boundary layer becomes turbulent (Ref. 5).**

A well-hit golf ball (diameter $D = 1.69$ in., weight $\mathcal{W} = 0.0992$ lb) can travel at $U = 200$ ft/s as it leaves the tee. A well-hit table tennis ball (diameter $D = 1.50$ in., weight $\mathcal{W} = 0.00551$ lb) can travel at $U = 60$ ft/s as it leaves the paddle. Determine the drag on a standard golf ball, a smooth golf ball, and a table tennis ball for the conditions given. Also determine the deceleration of each ball for these conditions.

SOLUTION

For either ball, the drag can be obtained from

$$\mathfrak{D} = \frac{1}{2}\rho U^2 \frac{\pi}{4} D^2 C_D \tag{1}$$

where the drag coefficient, C_D, is given in Fig. 9.25 as a function of the Reynolds number and surface roughness. For the golf ball in standard air

$$\text{Re} = \frac{UD}{\nu} = \frac{(200 \text{ ft/s})(1.69/12 \text{ ft})}{1.57 \times 10^{-4} \text{ ft}^2\text{/s}} = 1.79 \times 10^5$$

while for the table tennis ball

$$\text{Re} = \frac{UD}{\nu} = \frac{(60 \text{ ft/s})(1.50/12 \text{ ft})}{1.57 \times 10^{-4} \text{ ft}^2\text{/s}} = 4.78 \times 10^4$$

The corresponding drag coefficients are $C_D = 0.25$ for the standard golf ball, $C_D = 0.51$ for the smooth golf ball, and $C_D = 0.50$ for the table tennis ball. Hence, from Eq. 1 for the standard golf ball

$$\mathfrak{D} = \frac{1}{2}(2.38 \times 10^{-3} \text{ slugs/ft}^3)(200 \text{ ft/s})^2 \frac{\pi}{4}\left(\frac{1.69}{12} \text{ ft}\right)^2 (0.25) = 0.185 \text{ lb} \qquad \textbf{(Ans)}$$

for the smooth golf ball

$$\mathfrak{D} = \frac{1}{2}(2.38 \times 10^{-3} \text{ slugs/ft}^3)(200 \text{ ft/s})^2 \frac{\pi}{4}\left(\frac{1.69}{12} \text{ ft}\right)^2 (0.51) = 0.378 \text{ lb} \qquad \textbf{(Ans)}$$

and for the table tennis ball

$$\mathfrak{D} = \frac{1}{2}(2.38 \times 10^{-3} \text{ slugs/ft}^3)(60 \text{ ft/s})^2 \frac{\pi}{4}\left(\frac{1.50}{12} \text{ ft}\right)^2 (0.50) = 0.0263 \text{ lb} \qquad \textbf{(Ans)}$$

The corresponding decelerations are $a = \mathfrak{D}/m = g\mathfrak{D}/\mathcal{W}$, where m is the mass of the ball. Thus, the deceleration relative to the acceleration of gravity, a/g (i.e., the number of g's deceleration) is $a/g = \mathfrak{D}/\mathcal{W}$ or

$$\frac{a}{g} = \frac{0.185 \text{ lb}}{0.0992 \text{ lb}} = 1.86 \text{ for the standard golf ball} \qquad \textbf{(Ans)}$$

$$\frac{a}{g} = \frac{0.378 \text{ lb}}{0.0992 \text{ lb}} = 3.81 \text{ for the smooth golf ball} \qquad \textbf{(Ans)}$$

and

$$\frac{a}{g} = \frac{0.0263 \text{ lb}}{0.00551 \text{ lb}} = 4.77 \text{ for the table tennis ball} \qquad \textbf{(Ans)}$$

Note that there is a considerably smaller deceleration for the rough golf ball than for the smooth one. Because of its much larger drag-to-mass ratio, the table tennis ball slows down relatively quickly and does not travel as far as the golf ball. (Note that with $U = 60$ ft/s the standard golf ball has a drag of $\mathfrak{D} = 0.0200$ lb and a deceleration of $a/g = 0.202$, considerably less than the $a/g = 4.77$ of the table tennis ball. Conversely, a table tennis ball hit from a tee at 200 ft/s would decelerate at a rate of $a = 1740$ ft/s^2, or $a/g = 54.1$. It would not travel nearly as far as the golf ball.)

The Reynolds number range for which a rough golf ball has smaller drag than a smooth one (i.e., 4×10^4 to 4×10^5) corresponds to a flight velocity range of $45 < U < 450$ ft/s. This is comfortably within the range of most golfers. As is discussed in Section 9.4.2, the dimples (roughness) on a golf ball also help produce a lift (due to the spin of the ball) that allows the ball to travel farther than a smooth ball.

V9.7

The drag coefficient for surface ships is a function of the Froude number.

Froude Number Effects. Another parameter on which the drag coefficient may be strongly dependent is the Froude number, $\text{Fr} = U/\sqrt{g\ell}$. As is discussed in Chapter 10, the Froude number is a ratio of the free-stream speed to a typical wave speed on the interface of two fluids, such as the surface of the ocean. An object moving on the surface, such as a ship, often produces waves that require a source of energy to generate. This energy comes from the ship and is manifest as a drag. [Recall that the rate of energy production (power) equals speed times force.] The nature of the waves produced often depends on the Froude number of the flow and the shape of the object—the waves generated by a water skier ''plowing'' through the water at a low speed (low Fr) are different than those generated by the skier ''planing'' along the surface at high speed (large Fr).

Thus, the drag coefficient for surface ships is a function of Reynolds number (viscous effects) and Froude number (wave-making effects); $C_D = \phi(\text{Re, Fr})$. As was discussed in Chapter 7, it is often quite difficult to run model tests under conditions similar to those of

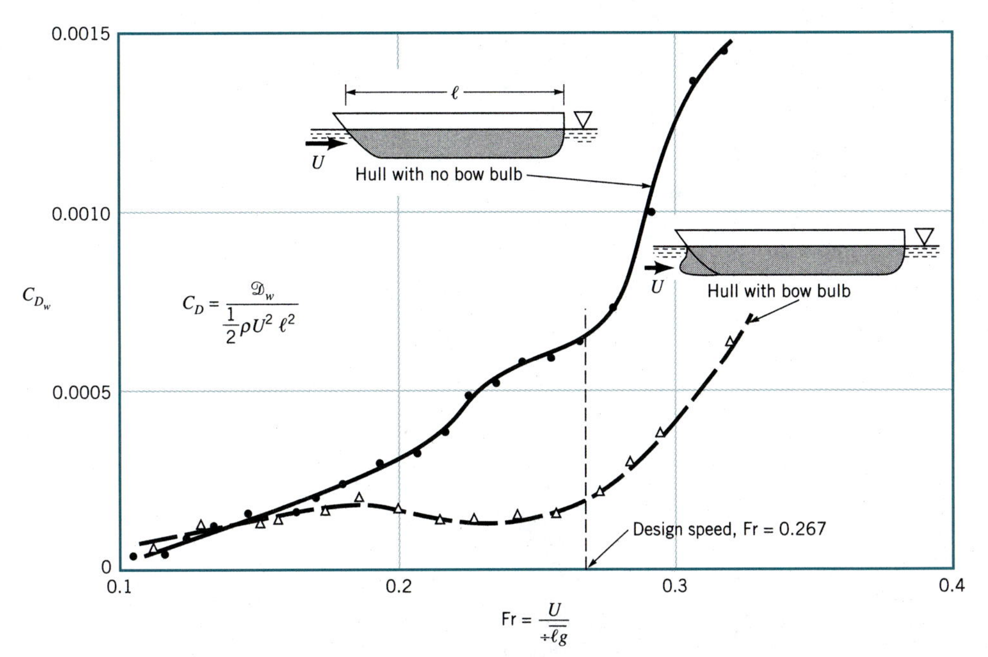

■ **FIGURE 9.26** **Typical drag coefficient data as a function of Froude number and hull characteristics for that portion of the drag due to the generation of waves (adapted from Ref. 25).**

the prototype (i.e., same Re and Fr for surface ships). Fortunately, the viscous and wave effects can often be separated, with the total drag being the sum of the drag of these individual effects. A detailed account of this important topic can be found in standard texts (Ref. 11).

As is indicated in Fig. 9.26, the wave-making drag, $\mathfrak{D}_w$, can be a complex function of the Froude number and the body shape. The rather "wiggly" dependence of wave drag coefficient, $C_{Dw} = \mathfrak{D}_w/(\rho U^2\ell^2/2)$, on the Froude number shown is typical. It results from the fact that the structure of the waves produced by the hull is a strong function of the ship speed or, in dimensionless form, the Froude number. This wave structure is also a function of the body shape. For example, the bow wave, which is often the major contributor to the wave drag, can be reduced by use of an appropriately designed bulb on the bow, as is indicated in Fig. 9.26. In this instance the streamlined body (hull without a bulb) has more drag than the less streamlined one.

Composite Body Drag. Approximate drag calculations for a complex body can often be obtained by treating the body as a composite collection of its various parts. For example, the drag on an airplane can be approximated by adding the drag produced by its various components—the wings, fuselage, tail section, and so on. Considerable care must be used in such an approach because of the interactions between the various parts. For example, the flow past the wing root (near the wing-fuselage intersection) is considerably altered by the fuselage. Hence, it may not be correct to merely add the drag of the components to obtain the drag of the entire object, although such approximations are often reasonable.

The drag on a complex body can be approximated as the sum of the drag on its parts.

EXAMPLE 9.13

A 60-mph (i.e., 88-fps) wind blows past the water tower shown in Fig. E9.13*a*. Estimate the moment (torque), M, needed at the base to keep the tower from tipping over.

SOLUTION

We treat the water tower as a sphere resting on a circular cylinder and assume that the total drag is the sum of the drag from these parts. The free-body diagram of the tower is shown in Fig. E.9.13*b*. By summing moments about the base of the tower, we obtain

$$M = \mathfrak{D}_s\left(b + \frac{D_s}{2}\right) + \mathfrak{D}_c\left(\frac{b}{2}\right) \tag{1}$$

where

$$\mathfrak{D}_s = \frac{1}{2}\rho U^2 \frac{\pi}{4} D_s^2 C_{Ds} \tag{2}$$

and

$$\mathfrak{D}_c = \frac{1}{2}\rho U^2 b D_c C_{Dc} \tag{3}$$

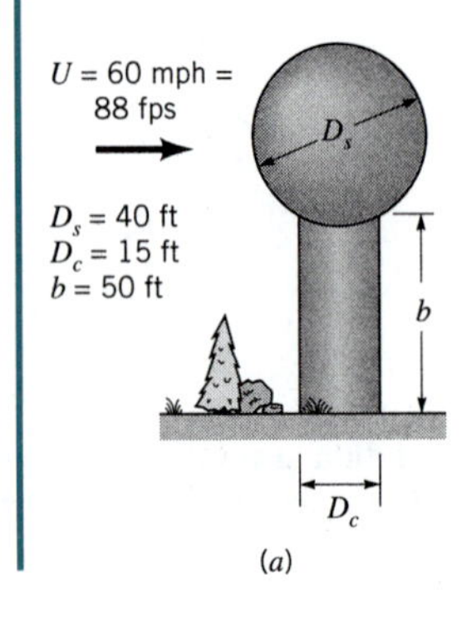

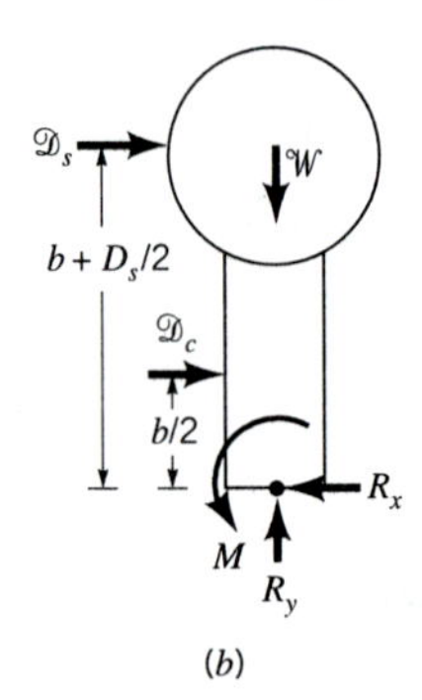

■ FIGURE E9.13

are the drag on the sphere and cylinder, respectively. For standard atmospheric conditions, the Reynolds numbers are

$$\mathrm{Re}_s = \frac{UD_s}{\nu} = \frac{(88 \text{ ft/s})(40 \text{ ft})}{1.57 \times 10^{-4} \text{ ft}^2\text{/s}} = 2.24 \times 10^7$$

and

$$\mathrm{Re}_c = \frac{UD_c}{\nu} = \frac{(88 \text{ ft/s})(15 \text{ ft})}{1.57 \times 10^{-4} \text{ ft}^2\text{/s}} = 8.41 \times 10^6$$

The corresponding drag coefficients, C_{Ds} and C_{Dc}, can be approximated from Fig. 9.21 as

$$C_{Ds} \approx 0.3 \quad \text{and} \quad C_{Dc} \approx 0.7$$

Note that the value of C_{Ds} was obtained by an extrapolation of the given data to Reynolds numbers beyond those given (a potentially dangerous practice!). From Eqs. 2 and 3 we obtain

$$\mathcal{D}_s = 0.5(2.38 \times 10^{-3} \text{ slugs/ft}^3)(88 \text{ ft/s})^2 \frac{\pi}{4}(40 \text{ ft})^2(0.3) = 3470 \text{ lb}$$

and

$$\mathcal{D}_c = 0.5(2.38 \times 10^{-3} \text{ slugs/ft}^3)(88 \text{ ft/s})^2(50 \text{ ft} \times 15 \text{ ft})(0.7) = 4840 \text{ lb}$$

From Eq. 1 the corresponding moment needed to prevent the tower from tipping is

$$M = 3470 \text{ lb}\left(50 \text{ ft} + \frac{40}{2} \text{ ft}\right) + 4840 \text{ lb}\left(\frac{50}{2} \text{ ft}\right) = 3.64 \times 10^5 \text{ ft·lb} \qquad \textbf{(Ans)}$$

The above result is only an estimate because (a) the wind is probably not uniform from the top of the tower to the ground, (b) the tower is not exactly a combination of a smooth sphere and a circular cylinder, (c) the cylinder is not of infinite length, (d) there will be some interaction between the flow past the cylinder and that past the sphere so that the net drag is not exactly the sum of the two, and (e) a drag coefficient value was obtained by extrapolation of the given data. However, such approximate results are often quite accurate.

V9.8

Considerable effort has gone into reducing the aerodynamic drag of automobiles.

The aerodynamic drag on automobiles provides an example of the use of composite bodies. The power required to move a car along a level street is used to overcome the rolling resistance and the aerodynamic drag. For speeds above approximately 30 mph, the aerodynamic drag becomes a significant contribution to the net propulsive force needed. The contribution of the drag due to various portions of car (i.e., front end, windshield, roof, rear end, windshield peak, rear roof/trunk, and cowl) have been determined by numerous model and full-sized tests as well as by numerical calculations. As a result it is possible to predict the aerodynamic drag on cars of a wide variety of body styles.

As is indicated in Fig. 9.27, the drag coefficient for cars has decreased rather continuously over the years. This reduction is a result of careful design of the shape and the details (such as window molding, rear view mirrors, etc.). An additional reduction in drag has been accomplished by a reduction of the projected area. The net result is a considerable increase in the gas mileage, especially at highway speeds. Considerable additional information about the aerodynamics of road vehicles can be found in the literature (Ref. 30).

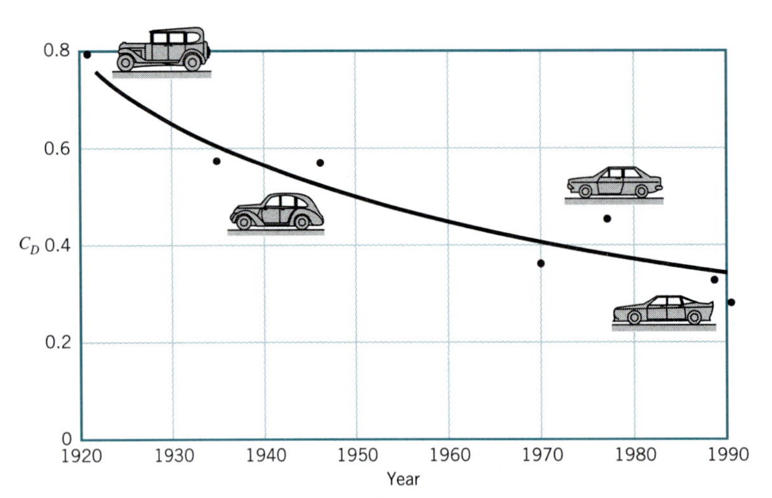

■ **FIGURE 9.27** **The historical trend of streamlining automobiles to reduce their aerodynamic drag and increase their miles per gallon (adapted from Ref. 5).**

The effect of several important parameters (shape, Re, Ma, Fr, and roughness) on the drag coefficient for various objects has been discussed in this section. As stated previously, drag coefficient information for a very wide range of objects is available in the literature. Some of this information is given in Figs. 9.28, 9.29, and 9.30 for a variety of two- and three-dimensional, natural and manmade objects. Recall that a drag coefficient of unity is equivalent to the drag produced by the dynamic pressure acting on an area of size A. That is, $\mathscr{D} = \frac{1}{2}\rho U^2 A C_D = \frac{1}{2}\rho U^2 A$ if $C_D = 1$. Typical nonstreamlined objects have drag coefficients on this order.

9.4 Lift

As is indicated in Section 9.1, any object moving through a fluid will experience a net force of the fluid on the object. For symmetrical objects, this force will be in the direction of the free stream—a drag, $\mathscr{D}$. If the object is not symmetrical (or if it does not produce a symmetrical flow field, such as the flow around a rotating sphere), there may also be a force normal to the free stream—a lift, $\mathscr{L}$. Considerable effort has been put forth to understand the various properties of the generation of lift. Some objects, such as an airfoil, are designed to generate lift. Other objects are designed to reduce the lift generated. For example, the lift on a car tends to reduce the contact force between the wheels and the ground, causing reduction in traction and cornering ability. It is desirable to reduce this lift.

9.4.1 Surface Pressure Distribution

The lift can be determined from Eq. 9.2 if the distributions of pressure and wall shear stress around the entire body are known. As is indicated in Section 9.1, such data are usually not known. Typically, the lift is given in terms of the lift coefficient.

The lift coefficient is a dimensionless form of the lift.

$$C_L = \frac{\mathscr{L}}{\frac{1}{2}\rho U^2 A} \tag{9.39}$$

which is obtained from experiments, advanced analysis, or numerical considerations. The lift coefficient is a function of the appropriate dimensionless parameters and, as the drag

Shape	Reference area A (b = length)	Drag coefficient $C_D = \dfrac{\mathfrak{D}}{\frac{1}{2}\rho U^2 A}$	Reynolds number $\mathrm{Re} = \rho UD/\mu$
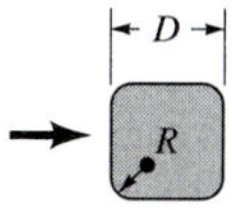 Square rod with rounded corners (D, R)	$A = bD$	R/D | C_D 0 | 2.2 0.02 | 2.0 0.17 | 1.2 0.33 | 1.0	$\mathrm{Re} = 10^5$
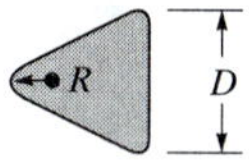 Rounded equilateral triangle (R, D)	$A = bD$	R/D | C_D (→) | C_D (←) 0 | 1.4 | 2.1 0.02 | 1.2 | 2.0 0.08 | 1.3 | 1.9 0.25 | 1.1 | 1.3	$\mathrm{Re} = 10^5$
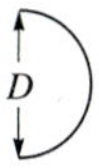 Semicircular shell (D)	$A = bD$	→ 2.3 ← 1.1	$\mathrm{Re} = 2 \times 10^4$
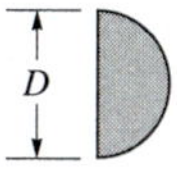 Semicircular cylinder (D)	$A = bD$	→ 2.15 ← 1.15	$\mathrm{Re} > 10^4$
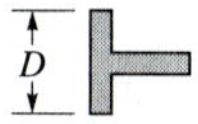 T-beam (D)	$A = bD$	→ 1.80 ← 1.65	$\mathrm{Re} > 10^4$
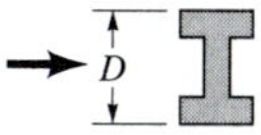 I-beam (D)	$A = bD$	2.05	$\mathrm{Re} > 10^4$
 Angle (D)	$A = bD$	→ 1.98 ← 1.82	$\mathrm{Re} > 10^4$
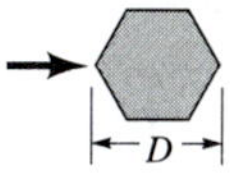 Hexagon (D)	$A = bD$	1.0	$\mathrm{Re} > 10^4$
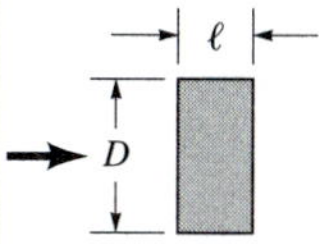 Rectangle (ℓ, D)	$A = bD$	ℓ/D | C_D ≤0.1 | 1.9 0.5 | 2.5 0.65 | 2.9 1.0 | 2.2 2.0 | 1.6 3.0 | 1.3	$\mathrm{Re} = 10^5$

■ **FIGURE 9.28** **Typical drag coefficients for regular two-dimensional objects (Refs. 5 and 6).**

coefficient, can be written as

$$C_L = \phi(\text{shape, Re, Ma, Fr}, \varepsilon/\ell)$$

The lift coefficient is a function of other dimensionless parameters.

The Froude number, Fr, is important only if there is a free surface present, as with an underwater "wing" used to support a high-speed hydrofoil surface ship. Often the surface roughness, ε, is relatively unimportant in terms of lift—it has more of an effect on the drag. The Mach number, Ma, is of importance for relatively high-speed subsonic and supersonic flows (i.e., $\mathrm{Ma} > 0.8$), and the Reynolds number effect is often not great. The most important parameter that affects the lift coefficient is the shape of the object. Considerable effort has gone into designing optimally shaped lift-producing devices. We will emphasize the effect

Shape	Reference area A	Drag coefficient C_D	Reynolds number $\text{Re} = \rho UD/\mu$
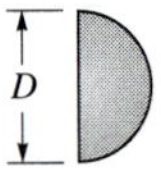 Solid hemisphere	$A = \frac{\pi}{4}D^2$	→ 1.17 ← 0.42	$\text{Re} > 10^4$
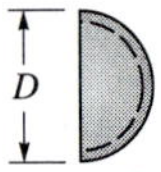 Hollow hemisphere	$A = \frac{\pi}{4}D^2$	→ 1.42 ← 0.38	$\text{Re} > 10^4$
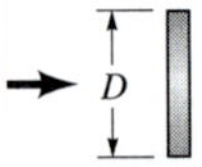 Thin disk	$A = \frac{\pi}{4}D^2$	1.1	$\text{Re} > 10^3$
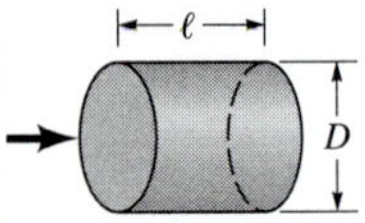 Circular rod parallel to flow	$A = \frac{\pi}{4}D^2$	ℓ/D | C_D: 0.5 | 1.1; 1.0 | 0.93; 2.0 | 0.83; 4.0 | 0.85	$\text{Re} > 10^5$
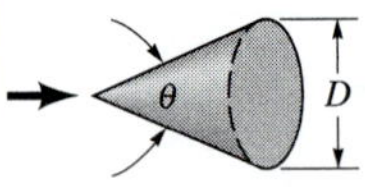 Cone	$A = \frac{\pi}{4}D^2$	θ, degrees | C_D: 10 | 0.30; 30 | 0.55; 60 | 0.80; 90 | 1.15	$\text{Re} > 10^4$
Cube	$A = D^2$	1.05	$\text{Re} > 10^4$
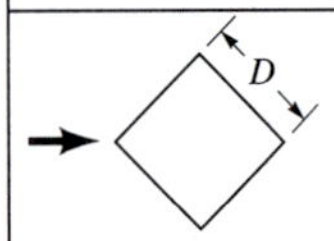 Cube	$A = D^2$	0.80	$\text{Re} > 10^4$
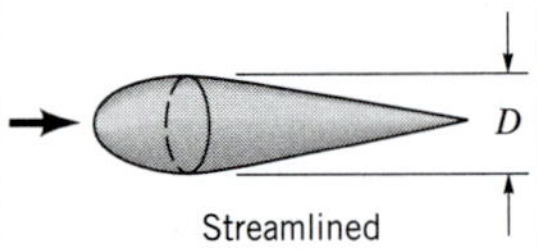 Streamlined body	$A = \frac{\pi}{4}D^2$	0.04	$\text{Re} > 10^5$

■ **FIGURE 9.29** **Typical drag coefficients for regular three-dimensional objects (Ref. 5).**

of the shape on lift—the effects of the other dimensionless parameters can be found in the literature (Refs. 13, 14, 29).

Usually most lift comes from pressure forces, not viscous forces.

Most common lift-generating devices (i.e., airfoils, fans, spoilers on cars, etc.) operate in the large Reynolds number range in which the flow has a boundary layer character, with viscous effects confined to the boundary layers and wake regions. For such cases the wall shear stress, τ_w, contributes little to the lift. Most of the lift comes from the surface pressure distribution. A typical pressure distribution on a moving car is shown in Fig. 9.31. The distribution, for the most part, is consistent with simple Bernoulli equation analysis. Locations with high-speed flow (i.e., over the roof and hood) have low pressure, while locations with low-speed flow (i.e., on the grill and windshield) have high pressure. It is easy to believe that the integrated effect of this pressure distribution would provide a net upward force.

Shape	Reference area	Drag coefficient C_D
 Parachute	Frontal area $A = \frac{\pi}{4}D^2$	1.4
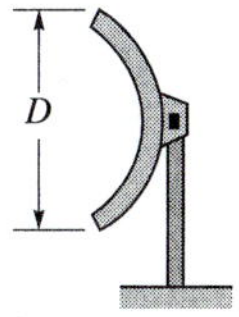 Porous parabolic dish	Frontal area $A = \frac{\pi}{4}D^2$	Porosity: 0, 0.2, 0.5; → : 1.42, 1.20, 0.82; ← : 0.95, 0.90, 0.80; Porosity = open area/total area
 Average person	Standing Sitting Crouching	$C_DA = 9\ \text{ft}^2$ $C_DA = 6\ \text{ft}^2$ $C_DA = 2.5\ \text{ft}^2$
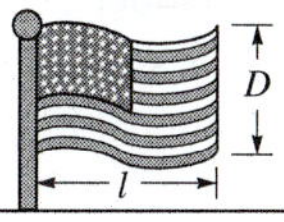 Fluttering flag	$A = \ell D$	ℓ/D = 1: C_D = 0.07 ℓ/D = 2: C_D = 0.12 ℓ/D = 3: C_D = 0.15
 Empire State Building	Frontal area	1.4
 Six-car passenger train	Frontal area	1.8
Bikes		
 Upright commuter	$A = 5.5\ \text{ft}^2$	1.1
 Racing	$A = 3.9\ \text{ft}^2$	0.88
 Drafting	$A = 3.9\ \text{ft}^2$	0.50
 Streamlined	$A = 5.0\ \text{ft}^2$	0.12
Tractor-trailor tucks		
 Standard	Frontal area	0.96
 With fairing	Frontal area	0.76
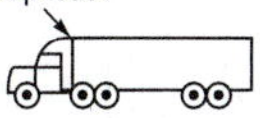 With fairing and gap seal	Frontal area	0.70
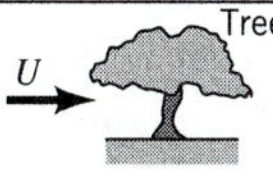 Tree, U = 10 m/s U = 20 m/s U = 30 m/s	Frontal area	0.43 0.26 0.20
 Dolphin	Wetted area	0.0036 at Re = 6×10^6 (flat plate has $C_{Df} = 0.0031$)
 Large birds	Frontal area	0.40

■ **FIGURE 9.30** **Typical drag coefficients for objects of interest (Refs. 5, 6, 15, and 20).**

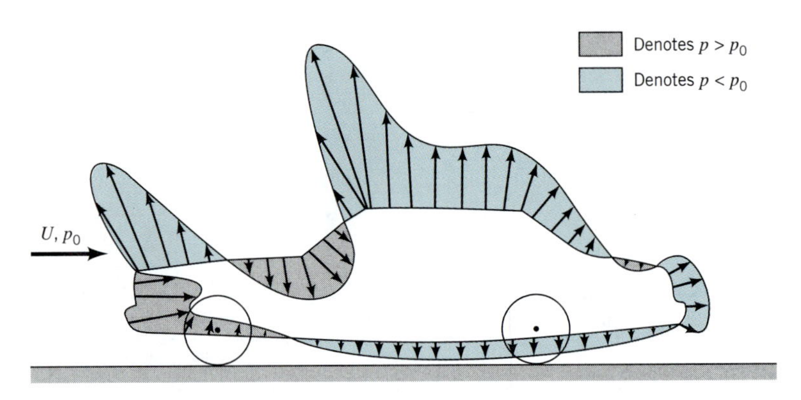

■ FIGURE 9.31 **Pressure distribution on the surface of an automobile.**

The relative importance of shear stress and pressure effects depends strongly on the Reynolds number.

For objects operating in very low Reynolds number regimes (i.e., Re < 1), viscous effects are important, and the contribution of the shear stress to the lift may be as important as that of the pressure. Such situations include the flight of minute insects and the swimming of microscopic organisms. The relative importance of τ_w and p in the generation of lift in a typical large Reynolds number flow is shown in Example 9.14.

EXAMPLE 9.14

When a uniform wind of velocity U blows past the semicircular building shown in Fig. E9.14*a*, the wall shear stress and pressure distributions on the outside of the building are as given previously in Figs. E9.8*b* and E9.9*a*, respectively. If the pressure in the building is atmospheric (i.e., the value, p_0, far from the building), determine the lift coefficient and the lift on the roof.

SOLUTION

From Eq. 9.2 we obtain the lift as

$$\mathscr{L} = -\int p \sin\theta \, dA + \int \tau_w \cos\theta \, dA \qquad (1)$$

As is indicated in Fig. E9.14*a*, we assume that on the inside of the building the pressure is uniform, $p = p_0$, and that there is no shear stress. Thus, Eq. 1 can be written as

$$\mathscr{L} = -\int_0^\pi (p - p_0) \sin\theta \, b\left(\frac{D}{2}\right) d\theta + \int_0^\pi \tau_w \cos\theta \, b\left(\frac{D}{2}\right) d\theta$$

or

$$\mathscr{L} = \frac{bD}{2}\left[-\int_0^\pi (p - p_0)\sin\theta \, d\theta + \int_0^\pi \tau_w \cos\theta \, d\theta\right] \qquad (2)$$

where b and D are the length and diameter of the building, respectively, and $dA = b(D/2)d\theta$. Equation 2 can be put into dimensionless form by using the dynamic pressure, $\rho U^2/2$, planform area, $A = bD$, and dimensionless shear stress

$$F(\theta) = \tau_w(\text{Re})^{1/2}/(\rho U^2/2)$$

to give

$$\mathscr{L} = \frac{1}{2}\rho U^2 A\left[-\frac{1}{2}\int_0^\pi \frac{(p - p_0)}{\frac{1}{2}\rho U^2}\sin\theta \, d\theta + \frac{1}{2\sqrt{\text{Re}}}\int_0^\pi F(\theta)\cos\theta \, d\theta\right] \qquad (3)$$

The values of the two integrals in Eq. 3 can be obtained by determining the area under the curves of $[(p - p_0)/(\rho U^2/2)] \sin\theta$ versus θ and $F(\theta) \cos\theta$ versus θ plotted in Figs. E9.14*b*

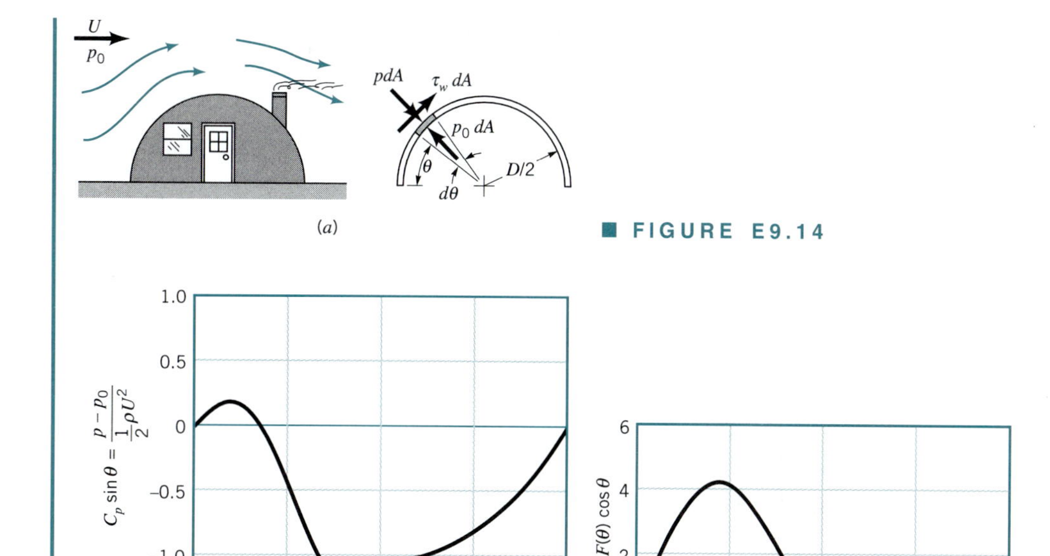

■ FIGURE E9.14

and E9.14*c*. The results are

$$\int_0^{\pi} \frac{(p - p_0)}{\frac{1}{2}\rho U^2} \sin\theta \, d\theta = -1.76$$

and

$$\int_0^{\pi} F(\theta) \cos\theta \, d\theta = 3.92$$

Thus, the lift is

$$\mathscr{L} = \frac{1}{2}\rho U^2 A \left[\left(-\frac{1}{2}\right)(-1.76) + \frac{1}{2\sqrt{\text{Re}}}(3.92) \right]$$

or

$$\mathscr{L} = \left(0.88 + \frac{1.96}{\sqrt{\text{Re}}}\right)\left(\frac{1}{2}\rho U^2 A\right) \qquad \text{(Ans)}$$

and

$$C_L = \frac{\mathscr{L}}{\frac{1}{2}\rho U^2 A} = 0.88 + \frac{1.96}{\sqrt{\text{Re}}} \qquad (4) \quad \text{(Ans)}$$

Consider a typical situation with $D = 20$ ft, $U = 30$ ft/s, $b = 50$ ft, and standard atmospheric conditions ($\rho = 2.38 \times 10^{-3}$ slugs/ft^3 and $\nu = 1.57 \times 10^{-4}$ ft^2/s), which gives a Reynolds number of

$$\text{Re} = \frac{UD}{\nu} = \frac{(30 \text{ ft/s})(20 \text{ ft})}{1.57 \times 10^{-4} \text{ ft}^2/\text{s}} = 3.82 \times 10^6$$

Hence, the lift coefficient is

$$C_L = 0.88 + \frac{1.96}{(3.82 \times 10^6)^{1/2}} = 0.88 + 0.001 = 0.881$$

Note that the pressure contribution to the lift coefficient is 0.88 whereas that due to the wall shear stress is only $1.96/(\text{Re}^{1/2}) = 0.001$. The Reynolds number dependency of C_L is quite minor. The lift is pressure dominated. Recall from Example 9.9 that this is also true for the drag on a similar shape.

From Eq. 4, we obtain the lift for the assumed conditions as

$$\mathscr{L} = \tfrac{1}{2}\rho U^2 A C_L = \tfrac{1}{2}(2.38 \times 10^{-3} \text{ slugs/ft}^3)(30 \text{ ft/s})^2(20 \text{ ft} \times 50 \text{ ft})(0.881)$$

or

$$\mathscr{L} = 944 \text{ lb}$$

There is a considerable tendency for the building to lift off the ground. Clearly this is due to the object being nonsymmetrical. The lift force on a complete circular cylinder is zero, although the fluid forces do tend to pull the upper and lower halves apart.

Most lift-producing objects are not symmetrical.

A typical device designed to produce lift does so by generating a pressure distribution that is different on the top and bottom surfaces. For large Reynolds number flows these pressure distributions are usually directly proportional to the dynamic pressure, $\rho U^2/2$, with viscous effects being of secondary importance. Two airfoils used to produce lift are indicated in Fig. 9.32. Clearly the symmetrical one cannot produce lift unless the angle of attack, α, is nonzero. Because of the asymmetry of the nonsymmetric airfoil, the pressure distributions on the upper and lower surfaces are different, and a lift is produced even with $\alpha = 0$. Of course, there will be a certain value of α (less than zero for this case) for which the lift is zero. For this situation, the pressure distributions on the upper and lower surfaces are different, but their resultant (integrated) pressure forces will be equal and opposite.

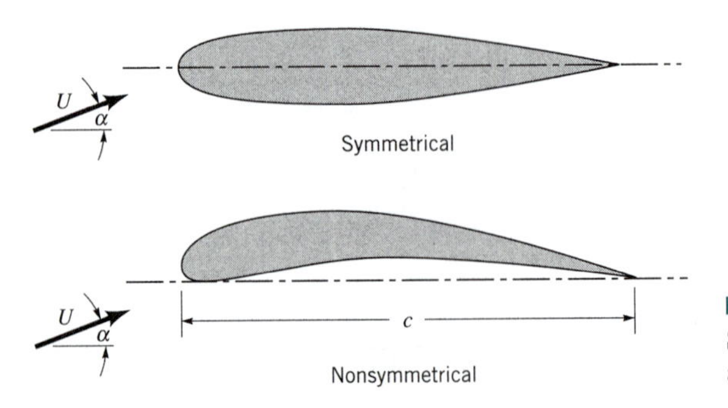

■ **FIGURE 9.32** **Symmetrical and nonsymmetrical airfoils.**

Since most airfoils are thin, it is customary to use the planform area, $A = bc$, in the definition of the lift coefficient. Here b is the length of the airfoil and c is the *chord length*—the length from the leading edge to the trailing edge as indicated in Fig. 9.32. Typical lift coefficients so defined are on the order of unity. That is, the lift force is on the order of the dynamic pressure times the planform area of the wing, $\mathscr{L} \approx (\rho U^2/2)A$. The *wing loading*, defined as the average lift per unit area of the wing, $\mathscr{L}/A$, therefore, increases with speed. For example, the wing loading of the 1903 Wright Flyer aircraft was 1.5 lb/ft^2, while for the present-day Boeing 747 aircraft it is 150 lb/ft^2. The wing loading for a bumble bee is approximately 1 lb/ft^2 (Ref. 15).

Lift and drag coefficients for wings are functions of the angle of attack.

Typical lift and drag coefficient data as a function of angle of attack, α, and *aspect ratio*, $\mathscr{A}$, are indicated in Figs. 9.33*a* and 9.33*b*. The aspect ratio is defined as the ratio of the square of the wing length to the planform area, $\mathscr{A} = b^2/A$. If the chord length, c, is constant along the length of the wing (a rectangular planform wing), this reduces to $\mathscr{A} = b/c$.

In general, the lift coefficient increases and the drag coefficient decreases with an increase in aspect ratio. Long wings are more efficient because their wing tip losses are relatively

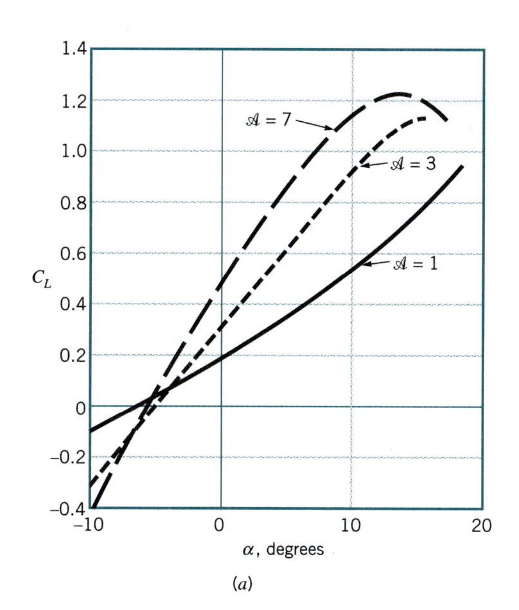

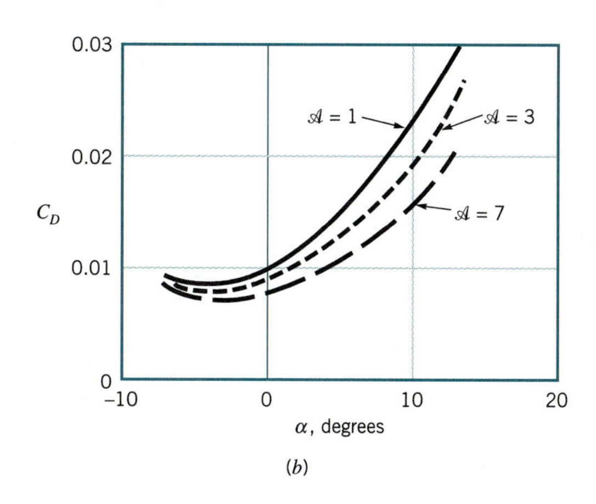

■ **FIGURE 9.33 Typical lift and drag coefficient data as a function of angle of attack and the aspect ratio of the airfoil: (*a*) lift coefficient, (*b*) drag coefficient.**

more minor than for short wings. The increase in drag due to the finite length ($\mathcal{A} < \infty$) of the wing is often termed induced drag. It is due to the interaction of the complex swirling flow structure near the wing tips (see Fig. 9.37) and the free stream (Ref. 13). High-performance soaring airplanes and highly efficient soaring birds (i.e., the albatross and sea gull) have long, narrow wings. Such wings, however, have considerable inertia that inhibits rapid maneuvers. Thus, highly maneuverable fighter or acrobatic airplanes and birds (i.e., the falcon) have small-aspect-ratio wings.

Although viscous effects and the wall shear stress contribute little to the direct generation of lift, they play an extremely important role in the design and use of lifting devices. This is because of the viscosity-induced boundary layer separation that can occur on nonstreamlined bodies such as airfoils that have too large an angle of attack (see Fig. 9.18). As is indicated in Fig. 9.33, up to a certain point, the lift coefficient increases rather steadily with the angle of attack. If α is too large, the boundary layer on the upper surface separates, the flow over the wing develops a wide, turbulent wake region, the lift decreases, and the drag increases. The airfoil *stalls*. Such conditions are extremely dangerous if they occur while the airplane is flying at a low altitude where there is not sufficient time and altitude to recover from the stall.

At large angles of attack the boundary layer separates and the wing stalls.

In many lift-generating devices the important quantity is the ratio of the lift to drag developed, $\mathcal{L}/\mathcal{D} = C_L/C_D$. Such information is often presented in terms of C_L/C_D versus α, as is shown in Fig. 9.34*a*, or in a *lift-drag polar* of C_L versus C_D with α as a parameter, as is shown in Fig. 9.34*b*. The most efficient angle of attack (i.e., largest C_L/C_D) can be found by drawing a line tangent to the $C_L - C_D$ curve from the origin, as is shown in Fig. 9.34*b*.

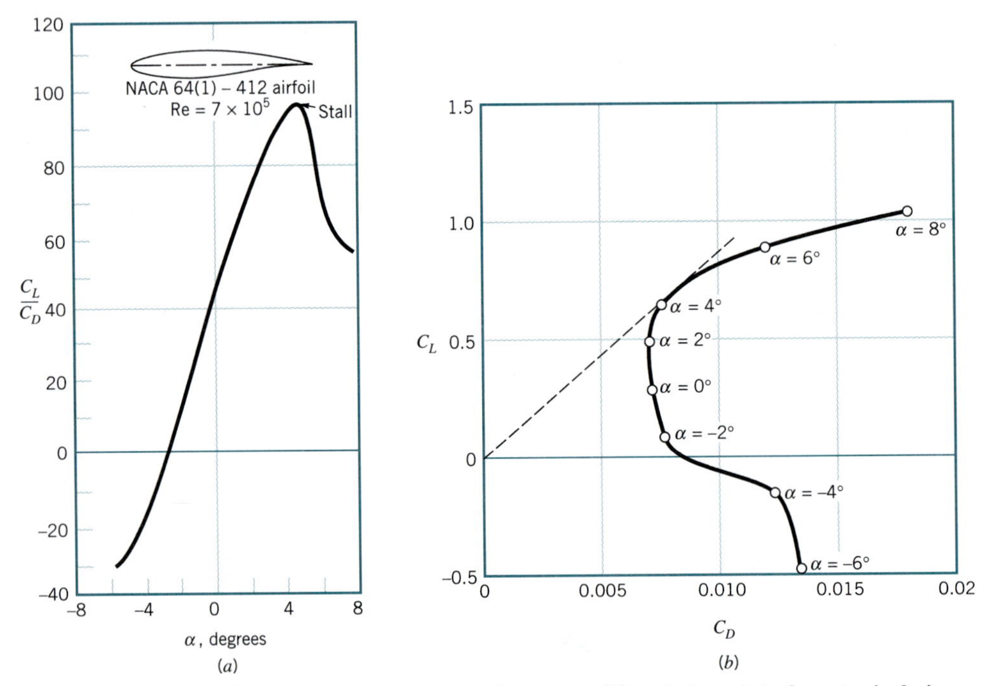

■ **FIGURE 9.34** **Two representations of the same lift and drag data for a typical airfoil: (*a*) lift-to-drag ratio as a function of angle of attack, with the onset of boundary layer separation on the upper surface indicated by the occurrence of stall, (*b*) the lift and drag polar diagram with the angle of attack indicated (Ref. 27).**

High-performance airfoils generate lift that is perhaps 100 or more times greater than their drag. This translates into the fact that in still air they can glide a horizontal distance of 100 m for each 1 m drop in altitude.

As is indicated above, the lift and drag on an airfoil can be altered by changing the angle of attack. This actually represents a change in the shape of the object. Other shape changes can be used to alter the lift and drag when desirable. In modern airplanes it is common to utilize leading edge and trailing edge flaps as is shown in Fig. 9.35. To generate the necessary lift during the relatively low-speed landing and takeoff procedures, the airfoil shape is altered by extending special flaps on the front and/or rear portions of the wing. Use of the flaps considerably enhances the lift, although it is at the expense of an increase in the drag (the airfoil is in a "dirty" configuration). This increase in drag is not of much concern during landing and takeoff operations—the decrease in landing or takeoff speed is more important than is a temporary increase in drag. During normal flight with the flaps retracted (the "clean" configuration), the drag is relatively small, and the needed lift force is achieved with the smaller lift coefficient and the larger dynamic pressure (higher speed).

Flaps alter the lift and drag characteristics of a wing.

The use of the complex flap systems for modern aircraft has proved to be an important breakthrough in aeronautics. Actually, certain birds use the leading edge flap concept. Some species have special feathers on the leading edge of their wings that extend as a leading edge flap when low-speed flight is required (such as when their wings are fully extended during landing) (Ref. 15).

A wide variety of lift and drag information for airfoils can be found in standard aerodynamics books (Ref. 13, 14, 29).

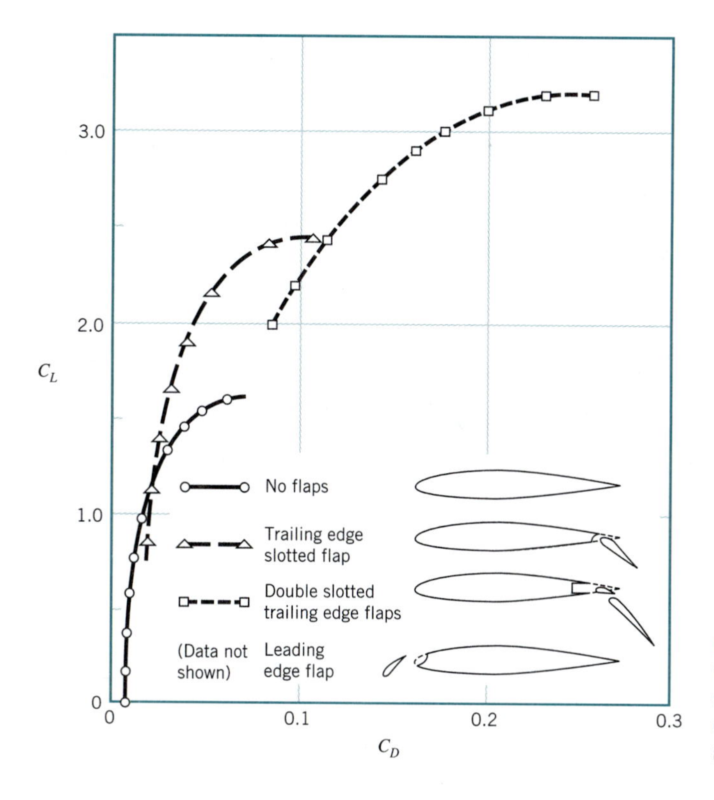

■ **FIGURE 9.35**
Typical lift and drag alterations possible with the use of various types of flap designs (Ref. 21).

EXAMPLE 9.15

In 1977 the *Gossamer Condor* won the Kremer prize by being the first human-powered aircraft to complete a prescribed figure-of-eight course around two turning points 0.5 mi apart (Ref. 22). The following data pertain to this aircraft:

$$\text{flight speed} = U = 15 \text{ ft/s}$$

$$\text{wing size} = b = 96 \text{ ft}, c = 7.5 \text{ ft (average)}$$

$$\text{weight (including pilot)} = \mathcal{W} = 210 \text{ lb}$$

$$\text{drag coefficient} = C_D = 0.046 \text{ (based on planform area)}$$

$$\text{power train efficiency} = \eta = \text{power to overcome drag/pilot power} = 0.8$$

Determine the lift coefficient, C_L, and the power, $\mathcal{P}$, required by the pilot.

SOLUTION

For steady flight conditions the lift must be exactly balanced by the weight, or

$$\mathcal{W} = \mathcal{L} = \tfrac{1}{2}\rho U^2 A C_L$$

Thus,

$$C_L = \frac{2\mathcal{W}}{\rho U^2 A}$$

where $A = bc = 96 \text{ ft} \times 7.5 \text{ ft} = 720 \text{ ft}^2$, $\mathcal{W} = 210$ lb, and $\rho = 2.38 \times 10^{-3}$ slugs/ft^3 for standard air. This gives

$$C_L = \frac{2(210 \text{ lb})}{(2.38 \times 10^{-3} \text{ slugs/ft}^3)(15 \text{ ft/s})^2(720 \text{ ft}^2)} = 1.09 \qquad \textbf{(Ans)}$$

a reasonable number. The overall-lift-to drag ratio for the aircraft is $C_L/C_D = 1.09/0.046 = 23.7$.

The product of the power that the pilot supplies and the power train efficiency equals the useful power needed to overcome the drag, $\mathcal{D}$. That is,

$$\eta\mathcal{P} = \mathcal{D}U$$

where

$$\mathcal{D} = \tfrac{1}{2}\rho U^2 A C_D$$

Thus,

$$\mathcal{P} = \frac{\mathcal{D}U}{\eta} = \frac{\tfrac{1}{2}\rho U^2 A C_D U}{\eta} = \frac{\rho A C_D U^3}{2\eta} \qquad \textbf{(1)}$$

or

$$\mathcal{P} = \frac{(2.38 \times 10^{-3} \text{ slugs/ft}^3)(720 \text{ ft}^2)(0.046)(15 \text{ ft/s})^3}{2(0.8)}$$

$$\mathcal{P} = 166 \text{ ft}\cdot\text{lb/s}\left(\frac{1 \text{ hp}}{550 \text{ ft}\cdot\text{lb/s}}\right) = 0.302 \text{ hp} \qquad \textbf{(Ans)}$$

This power level is obtainable by a well-conditioned athlete (as is indicated by the fact that the flight was successfully completed). Note that only 80% of the pilot's power (i.e., $0.8 \times 0.302 = 0.242$ hp, which corresponds to a drag of $\mathfrak{D} = 8.86$ lb) is needed to force the aircraft through the air. The other 20% is lost because of the power train inefficiency. Note from Eq. 1 that for a constant drag coefficient the power required increases as U^3—a doubling of the speed to 30 ft/s would require an eightfold increase in power (i.e., 2.42 hp, well beyond the range of any human).

9.4.2 Circulation

Inviscid flow analysis can be used to obtain ideal flow past airfoils.

Since viscous effects are of minor importance in the generation of lift, it should be possible to calculate the lift force on an airfoil by integrating the pressure distribution obtained from the equations governing inviscid flow past the airfoil. That is, the potential flow theory discussed in Chapter 6 should provide a method to determine the lift. Although the details are beyond the scope of this book, the following is found from such calculations (Ref. 4).

The calculation of the inviscid flow past a two-dimensional airfoil gives a flow field as indicated in Fig. 9.36. The predicted flow field past an airfoil with no lift (i.e., a symmetrical airfoil at zero angle of attack, Fig. 9.36*a*) appears to be quite accurate (except for the absence of thin boundary layer regions). However, as is indicated in Fig. 9.36*b*, the calculated flow

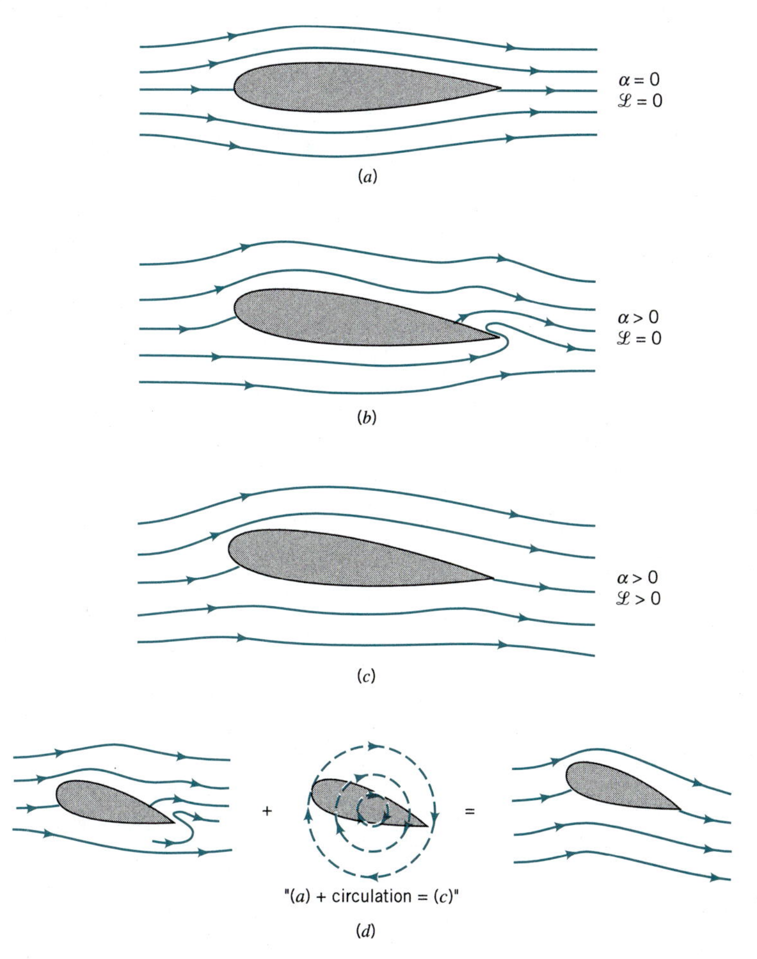

■ FIGURE 9.36 Inviscid flow past an airfoil: (*a*) symmetrical flow past the symmetrical airfoil at a zero angle of attack, (*b*) same airfoil at a nonzero angle of attack—no lift, flow near trailing edge not realistic, (*c*) same conditions as for (*b*) except circulation has been added to the flow–nonzero lift, realistic flow, (*d*) superposition of flows to produce the final flow past the airfoil.

past the same airfoil at a nonzero angle of attack (but one small enough so that boundary layer separation would not occur) is not proper near the trailing edge. In addition, the calculated lift for a nonzero angle of attack is zero—in conflict with the known fact that such airfoils produce lift.

In reality, the flow should pass smoothly over the top surface as is indicated in Fig. 9.36*c*, without the strange behavior indicated near the trailing edge in Fig. 9.36*b*. As is shown in Fig. 9.36*d*, the unrealistic flow situation can be corrected by adding an appropriate clockwise swirling flow around the airfoil. The results are twofold: (1) The unrealistic behavior near the trailing edge is eliminated (i.e., the flow pattern of Fig. 9.36*b* is changed to that of Fig. 9.36*c*), and (2) the average velocity on the upper surface of the airfoil is increased while that on the lower surface is decreased. From the Bernoulli equation concepts (i.e., $p/\gamma + V^2/2g + z =$ constant), the average pressure on the upper surface is decreased and that on the lower surface is increased. The net effect is to change the original zero lift condition to that of a lift-producing airfoil.

Lift generated by wings can be explained in terms of circulation.

The addition of the clockwise swirl is termed the addition of *circulation*. The amount of swirl (circulation) needed to have the flow leave the trailing edge smoothly is a function of the airfoil size and shape and can be calculated from potential flow (inviscid) theory (see Section 6.6.3 and Ref. 29). Although the addition of circulation to make the flow field physically realistic may seem artificial, it has well-founded mathematical and physical grounds. For example, consider the flow past a finite length airfoil, as is indicated in Fig. 9.37. For lift-generating conditions the average pressure on the lower surface is greater than that on the upper surface. Near the tips of the wing this pressure difference will cause some of the fluid to attempt to migrate from the lower to the upper surface, as is indicated in Fig. 9.37*b*. At the same time, this fluid is swept downstream, forming a *trailing vortex* (swirl) from each wing tip (see Fig. 4.3). It is speculated that the reason some birds migrate in vee-formation is to take advantage of the updraft produced by the trailing vortex of the preceding bird. [It

V9.9

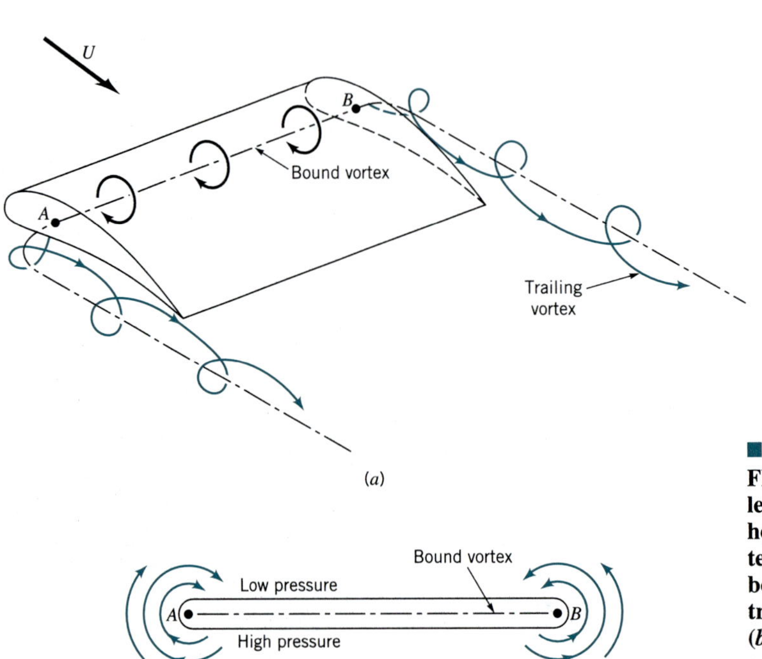

■ **FIGURE 9.37** **Flow past a finite length wing: (*a*) the horseshoe vortex system produced by the bound vortex and the trailing vortices; (*b*) the leakage of air around the wing tips produces the trailing vortices.**

is calculated that for a given expenditure of energy, a flock of 25 birds flying in vee-formation could travel 70% farther than if each bird were to fly separately (Ref. 15.)]

The trailing vortices from the right and left wing tips are connected by the *bound vortex* along the length of the wing. It is this vortex that generates the circulation that produces the lift. The combined vortex system (the bound vortex and the trailing vortices) is termed a horseshoe vortex. The strength of the trailing vortices (which is equal to the strength of the bound vortex) is proportional to the lift generated. Large aircraft (for example, a Boeing 747) can generate very strong trailing vortices that persist for a long time before viscous effects finally cause them to die out. Such vortices are strong enough to flip smaller aircraft out of control if they follow too closely behind the large aircraft.

As is indicated above, the generation of lift is directly related to the production of a swirl or vortex flow around the object. A nonsymmetric airfoil, by design, generates its own prescribed amount of swirl and lift. A symmetric object like a circular cylinder or sphere, which normally provides no lift, can generate swirl and lift if it rotates.

As is discussed in Section 6.6.3, the inviscid flow past a circular cylinder has the symmetrical flow pattern indicated in Fig. 9.38*a*. By symmetry the lift and drag are zero. However, if the cylinder is rotated about its axis in a stationary real ($\mu \neq 0$) fluid, the rotation will drag some of the fluid around, producing circulation about the cylinder as in Fig. 9.38*b*. When this circulation is combined with an ideal, uniform upstream flow, the flow pattern indicated in Fig. 9.38*c* is obtained. The flow is no longer symmetrical about the horizontal plane through the center of the cylinder; the average pressure is greater on the lower half of the cylinder than on the upper half, and a lift is generated. This effect is called the *Magnus effect*, after Heinrich Magnus (1802–1870), a German chemist and physicist who first investigated this phenomenon. A similar lift is generated on a rotating sphere. It accounts for the various types of pitches in baseball (i.e., curve ball, floater, sinker, etc.), the ability of a soccer player to hook the ball, and the hook or slice of a golf ball.

A spinning sphere or cylinder can generate lift.

Typical lift and drag coefficients for a smooth, spinning sphere are shown in Fig. 9.39. Although the drag coefficient is fairly independent of the rate of rotation, the lift coefficient is strongly dependent on it. In addition (although not indicated in the figure), both C_L and C_D are dependent on the roughness of the surface. As was discussed in Section 9.3, in a certain Reynolds number range an increase in surface roughness actually decreases the drag coefficient. Similarly, an increase in surface roughness can increase the lift coefficient because the roughness helps drag more fluid around the sphere increasing the circulation for a given angular velocity. Thus, a rotating, rough golf ball travels farther than a smooth one because the drag is less and the lift is greater. However, do not expect a severely roughed up (cut)

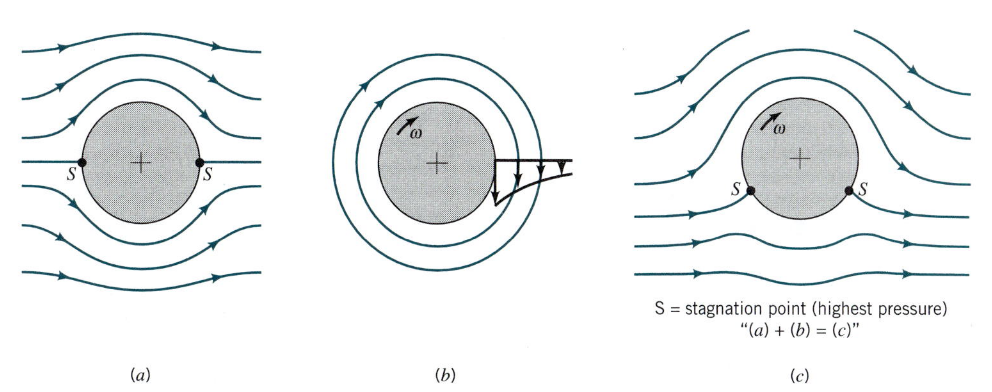

■ **FIGURE 9.38 Inviscid flow past a circular cylinder: (*a*) uniform upstream flow without circulation, (*b*) free vortex at the center of the cylinder, (*c*) combination of free vortex and uniform flow past a circular cylinder giving nonsymmetric flow and a lift.**

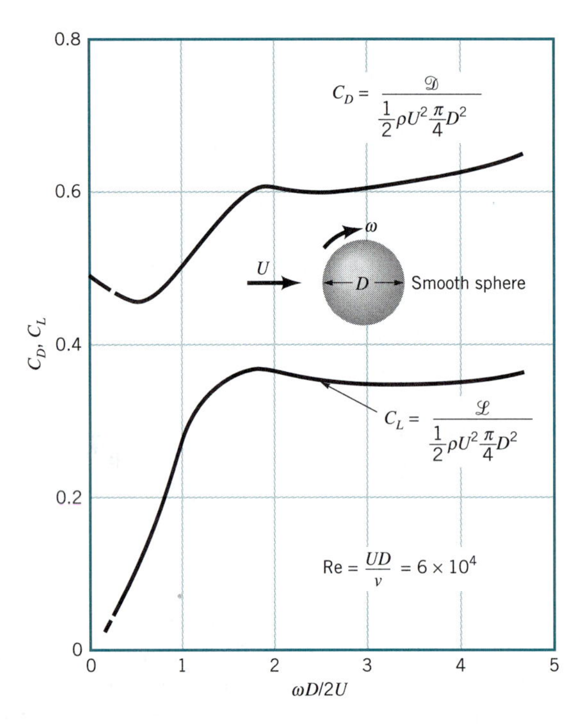

■ **FIGURE 9.39 Lift and drag coefficients for a spinning smooth sphere (Ref. 23).**

ball to work better—extensive testing has gone into obtaining the optimum surface roughness for golf balls.

EXAMPLE 9.16

A table tennis ball weighing 2.45×10^{-2} N with diameter $D = 3.8 \times 10^{-2}$ m is hit at a velocity of $U = 12$ m/s with a back spin of angular velocity ω as is shown in Fig. E9.16. What is the value of ω if the ball is to travel on a horizontal path, not dropping due to the acceleration of gravity?

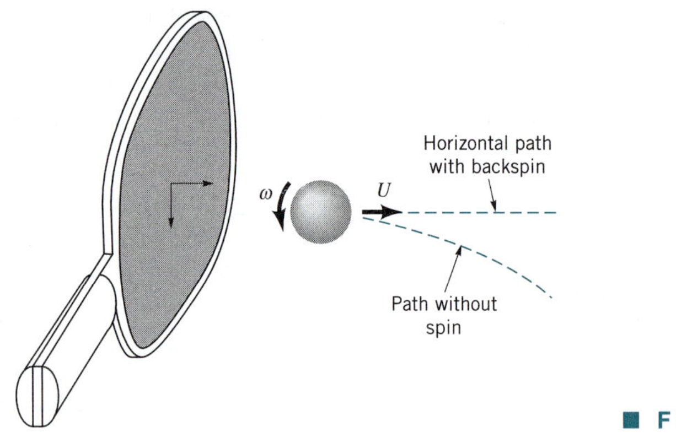

■ **FIGURE E9.16**

SOLUTION

For horizontal flight, the lift generated by the spinning of the ball must exactly balance the weight, $\mathscr{W}$, of the ball so that

$$\mathscr{W} = \mathscr{L} = \tfrac{1}{2}\rho U^2 A C_L$$

or

$$C_L = \frac{2\mathcal{W}}{\rho U^2(\pi/4)D^2}$$

where the lift coefficient, C_L, can be obtained from Fig. 9.39. For standard atmospheric conditions with $\rho = 1.23 \text{ kg/m}^3$ we obtain

$$C_L = \frac{2(2.45 \times 10^{-2} \text{ N})}{(1.23 \text{ kg/m}^3)(12 \text{ m/s})^2(\pi/4)(3.8 \times 10^{-2} \text{ m})^2} = 0.244$$

which, according to Fig. 9.39, can be achieved if

$$\frac{\omega D}{2U} = 0.9$$

or

$$\omega = \frac{2U(0.9)}{D} = \frac{2(12 \text{ m/s})(0.9)}{3.8 \times 10^{-2} \text{ m}} = 568 \text{ rad/s}$$

Thus,

$$\omega = (568 \text{ rad/s})(60 \text{ s/min})(1 \text{ rev}/2\pi \text{ rad}) = 5420 \text{ rpm}$$ **(Ans)**

Is it possible to impart this angular velocity to the ball? With larger angular velocities the ball will rise and follow an upward curved path. Similar trajectories can be produced by a well-hit golf ball—rather than falling like a rock, the golf ball trajectory is actually curved up and the spinning ball travels a greater distance than one without spin. However, if top spin is imparted to the ball (as in an improper tee shot) the ball will curve downward more quickly than under the action of gravity alone—the ball is ''topped'' and a negative lift is generated. Similarly, rotation about a vertical axis will cause the ball to hook or slice to one side or the other.

References

1. Schlichting, H., *Boundary Layer Theory*, 7th Ed., McGraw-Hill, New York, 1979.
2. Rosenhead, L., *Laminar Boundary Layers*, Oxford University Press, London, 1963.
3. White, F. M., *Viscous Fluid Flow*, McGraw-Hill, New York, 1974.
4. Currie, I. G., *Fundamental Mechanics of Fluids*, McGraw-Hill, New York, 1974.
5. Blevins, R. D., *Applied Fluid Dynamics Handbook*, Van Nostrand Reinhold, New York, 1984.
6. Hoerner, S. F., *Fluid-Dynamic Drag*, published by the author, Library of Congress No. 64,19666, 1965.
7. Happel, J., *Low Reynolds Number Hydrodynamics*, Prentice Hall, Englewood Cliffs, NJ, 1965.
8. Van Dyke, M., *An Album of Fluid Motion*, Parabolic Press, Stanford, Calif., 1982.
9. Thompson, P. A., *Compressible-Fluid Dynamics*, McGraw-Hill, New York, 1972.
10. Zucrow, M. J., and Hoffman, J. D., *Gas Dynamics, Vol. I*, Wiley, New York, 1976.
11. Clayton, B. R., and Bishop, R. E. D., *Mechanics of Marine Vehicles*, Gulf Publishing Co., Houston, 1982.
12. *CRC Handbook of Tables for Applied Engineering Science*, 2nd Ed., CRC Press, 1973.

13. Shevell, R. S., *Fundamentals of Flight*, 2nd Ed., Prentice Hall, Englewood Cliffs, NJ, 1989.
14. Kuethe, A. M. and Chow, C. Y., *Foundations of Aerodynamics, Bases of Aerodynamics Design*, 4th Ed., Wiley, 1986.
15. Vogel, J., *Life in Moving Fluids*, 2nd Ed., Willard Grant Press, Boston, 1994.
16. Kreider, J. F., *Principles of Fluid Mechanics*, Allyn and Bacon, Newton, Mass., 1985.
17. Dobrodzicki, G. A., Flow Visualization in the National Aeronautical Establishment's Water Tunnel, National Research Council of Canada, Aeronautical Report LR-557, 1972.
18. White, F. M., *Fluid Mechanics*, McGraw-Hill, New York, 1986.
19. Vennard, J. K., and Street, R. L., *Elementary Fluid Mechanics*, 6th Ed., Wiley, New York, 1982.
20. Gross, A. C., Kyle, C. R., and Malewicki, D. J., The Aerodynamics of Human Powered Land Vehicles, *Scientific American*, Vol. 249, No. 6, 1983.
21. Abbott, I. H., and Von Doenhoff, A. E., *Theory of Wing Sections*, Dover Publications, New York, 1959.
22. MacReady, P. B., "Flight on 0.33 Horsepower: The Gossamer Condor," *Proc. AIAA 14th Annual Meeting* (Paper No. 78-308), Washington, DC, 1978.
23. Goldstein, S., *Modern Developments in Fluid Dynamics*, Oxford Press, London, 1938.
24. Achenbach, E., Distribution of Local Pressure and Skin Friction around a Circular Cylinder in Cross-Flow up to Re $= 5 \times 10^6$, *Journal of Fluid Mechanics*, Vol. 34, Pt. 4, 1968.
25. Inui, T., Wave-Making Resistance of Ships, *Transactions of the Society of Naval Architects and Marine Engineers*, Vol. 70, 1962.
26. Sovran, G., et al. (ed.), *Aerodynamic Drag Mechanisms of Bluff Bodies and Road Vehicles*, Plenum Press, New York, 1978.
27. Abbott, I. H., von Doenhoff, A. E. and Stivers, L. S., Summary of Airfoil Data, NACA Report No. 824, Langley Field, Va., 1945.
28. Society of Automotive Engineers Report HSJ1566, "Aerodynamic Flow Visualization Techniques and Procedures," 1986.
29. Anderson, J. D., *Fundamentals of Aerodynamics*, 2nd Ed., McGraw-Hill, New York, 1991.
30. Hucho, W. H., *Aerodynamics of Road Vehicles*, Butterworth–Heinemann, 1987.

Review Problems

Note: Problems designated with (R) are review problems. The phrases within parentheses refer to the main topics to be used in solving the problems. Complete, detailed solutions to these review problems can be found in the supplement titled *Student Solution Manual for Fundamentals of Fluid Mechanics*, by Munson, Young, and Okiishi (John Wiley and Sons, New York, 1997).

9.1R (Lift/drag calculation) Determine the lift and drag coefficients (based on frontal area) for the triangular two-dimensional object shown in Fig. P9.1R. Neglect shear forces.

(ANS: 0; 1.70)

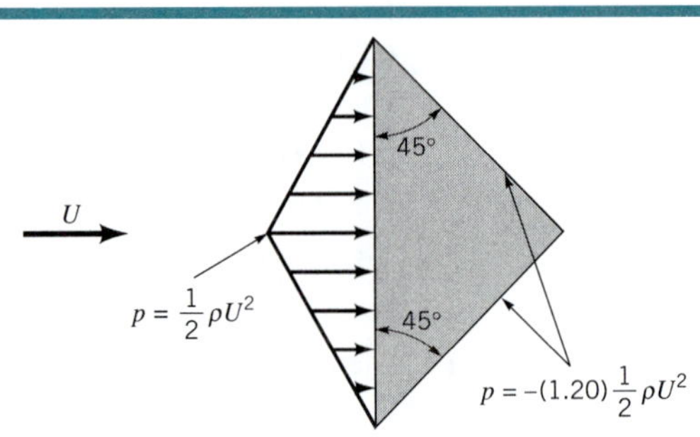

■ FIGURE P9.1R

9.2R (External flow character) A 0.23-m-diameter soccer ball moves through the air with a speed of 10 m/s. Would the flow around the ball be classified as low, moderate, or large Reynolds number flow? Explain.

(ANS: Large Reynolds number flow)

9.3R (External flow character) A small 15-mm-long fish swims with a speed of 20 mm/s. Would a boundary layer type flow be developed along the sides of the fish? Explain.

(ANS: No)

9.4R (Boundary layer flow) Air flows over a flat plate of length $\ell = 2$ ft such that the Reynolds number based on the plate length is $\text{Re} = 2 \times 10^5$. Plot the boundary layer thickness, δ, for $0 \leq x \leq \ell$.

9.5R (Boundary layer flow) At a given location along a flat plate the boundary layer thickness is $\delta = 45$ mm. At this location, what would be the boundary layer thickness if it were defined as the distance from the plate where the velocity is 97% of the upstream velocity rather than the standard 99%? Assume laminar flow.

(ANS: 38.5 mm)

9.6R (Friction drag) A laminar boundary layer formed on one side of a plate of length ℓ produces a drag $\mathfrak{D}$. How much must the plate be shortened if the drag on the new plate is to be $\mathfrak{D}/4$? Assume the upstream velocity remains the same. Explain your answer physically.

(ANS: $\ell_{\text{new}} = \ell/16$)

9.7R (Momentum integral equation) As is indicated in Table 9.2, the laminar boundary layer results obtained from the momentum integral equation are relatively insensitive to the shape of the assumed velocity profile. Consider the profile given by $u = U$ for $y > \delta$, and $u = U\{1 - [(y - \delta)/\delta]^2\}^{1/2}$ for $y \leq \delta$ as shown in Fig. P9.7R. Note that this satisfies the conditions $u = 0$ at $y = 0$ and $u = U$ at $y = \delta$. However, show that such a profile produces meaningless results when used with the momentum integral equation. Explain.

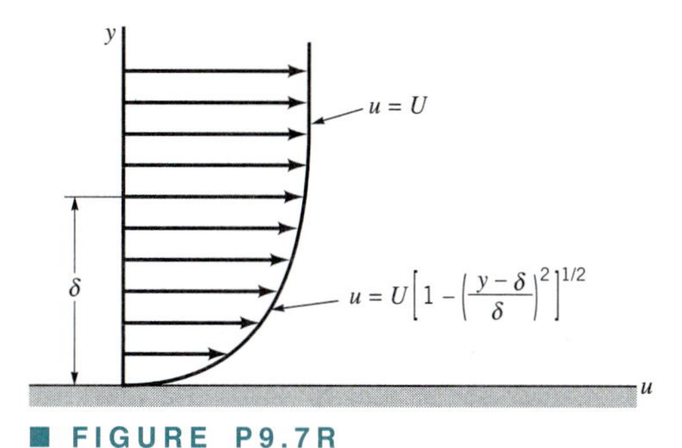

■ FIGURE P9.7R

9.8R (Drag—low Reynolds number) How fast do small water droplets of 0.06-μm (6×10^{-8} m) diameter fall through the air under standard sea-level conditions? Assume the drops do not evaporate. Repeat the problem for standard conditions at 5000-m altitude.

(ANS: 1.10×10^{-7} m/s; 1.20×10^{-7} m/s)

9.9R (Drag) A 12-mm-diameter cable is strung between a series of poles that are 40 m apart. Determine the horizontal force this cable puts on each pole if the wind velocity is 30 m/s.

(ANS: 372 N)

9.10R (Drag) How much less power is required to pedal a racing-style bicycle at 20 mph with a 10-mph tail wind than at the same speed with a 10-mph head wind? (See Fig. 9.30.)

(ANS: 0.375 hp)

9.11R (Drag) A rectangular car-top carrier of 1.6-ft height, 5.0-ft length (front to back), and a 4.2-ft width is attached to the top of a car. Estimate the additional power required to drive the car with the carrier at 60 mph through still air compared with the power required to drive only the car at 60 mph.

(ANS: 12.9 hp)

9.12R (Drag) Estimate the wind velocity necessary to blow over the 250-kN boxcar shown in Fig. P9.12R.

(ANS: approximately 32.6 m/s to 35.1 m/s)

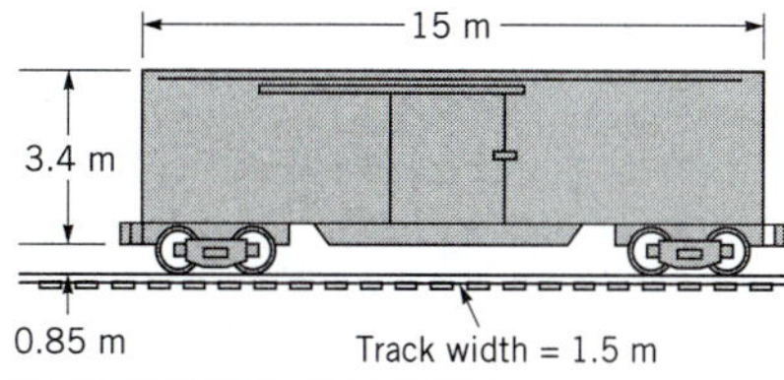

■ FIGURE P9.12R

9.13R (Drag) A 200-N rock (roughly spherical in shape) of specific gravity $SG = 1.93$ falls at a constant speed U. Determine U if the rock falls through **(a)** air; **(b)** water.

(ANS: 176 m/s; 5.28 m/s)

9.14R (Drag—composite body) A shortwave radio antenna is constructed from circular tubing, as is illustrated in Fig. P9.14R. Estimate the wind force on the antenna in a 100 km/hr wind.

(ANS: 180 N)

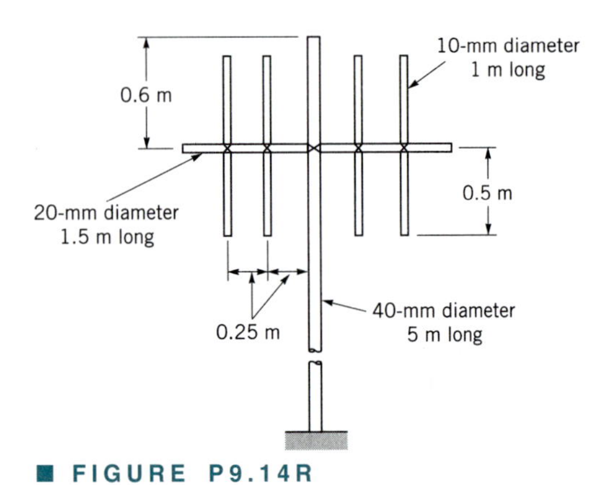

■ FIGURE P9.14R

9.15R (Lift) Show that for level flight the drag on a given airplane is independent of altitude if the lift and drag coefficients remain constant. Note that with C_L constant the airplane must fly faster at a higher altitude.

9.16R (Lift) The wing area of a small airplane weighing 6.22 kN is 10.2 m^2. **(a)** If the cruising speed of the plane is 210 km/hr, determine the lift coefficient of the wing. **(b)** If the engine delivers 150 kW at this speed, and if 60% of this power represents propeller loss and body resistance, what is the drag coefficient of the wing.

(ANS: 0.292; 0.0483)

Problems

Note: Unless otherwise indicated use the values of fluid properties found in the tables on the inside of the front cover. Problems designated with an (*) are intended to be solved with the aid of a programmable calculator or a computer. Problems designated with a (†) are "open ended" problems and require critical thinking in that to work them one must make various assumptions and provide the necessary data. There is not a unique answer to these problems.

9.1 Assume that water flowing past the equilateral triangular bar shown in Fig. P9.1 produces the pressure distributions indicated. Determine the lift and drag on the bar and the corresponding lift and drag coefficients (based on frontal area). Neglect shear forces.

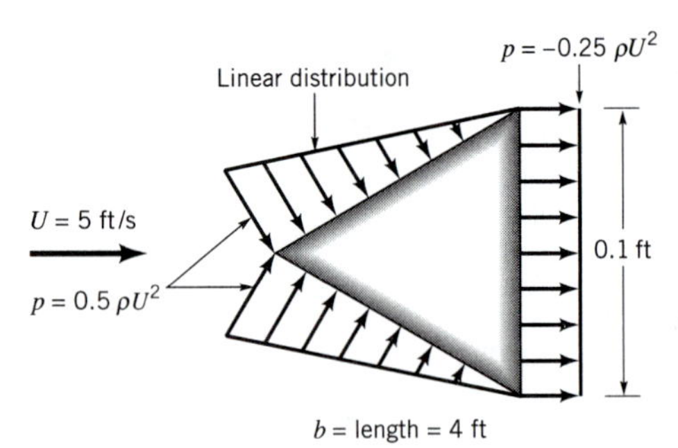

■ FIGURE P9.1

9.2 Fluid flows past the two-dimensional bar shown in Fig. P9.2. The pressures on the ends of the bar are as shown, and the average shear stress on the top and bottom of the bar is τ_{avg}. Assume that the drag due to pressure is equal to the drag due to viscous effects. **(a)** Determine τ_{avg} in terms of the dynamic pressure, $\rho U^2/2$. **(b)** Determine the drag coefficient for this object.

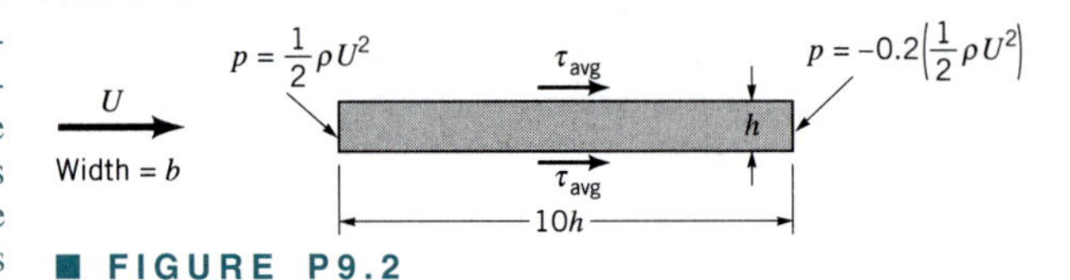

■ FIGURE P9.2

***9.3** The pressure distribution on the 1-m-diameter circular disk in Fig. P9.3 is given in the table below. Determine the drag on the disk.

r (m)	p (kN/m^2)
0	4.34
0.05	4.28
0.10	4.06
0.15	3.72
0.20	3.10
0.25	2.78
0.30	2.37
0.35	1.89
0.40	1.41
0.45	0.74
0.50	0.0

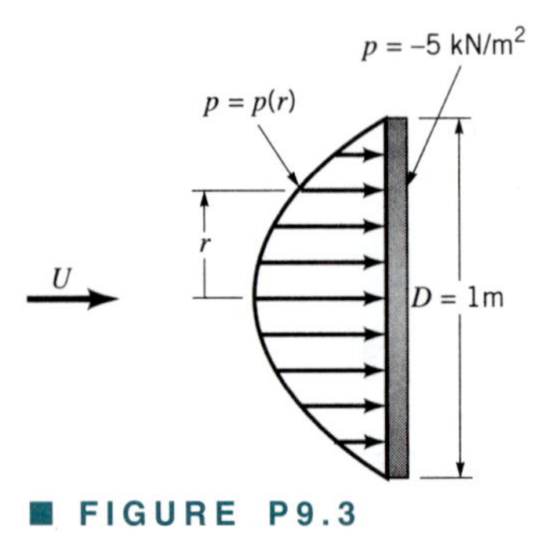

■ FIGURE P9.3

9.4 The pressure distribution on a cylinder is approximated by the two straight line segments shown in Fig. P9.4. Determine the drag coefficient for the cylinder. Neglect shear forces.

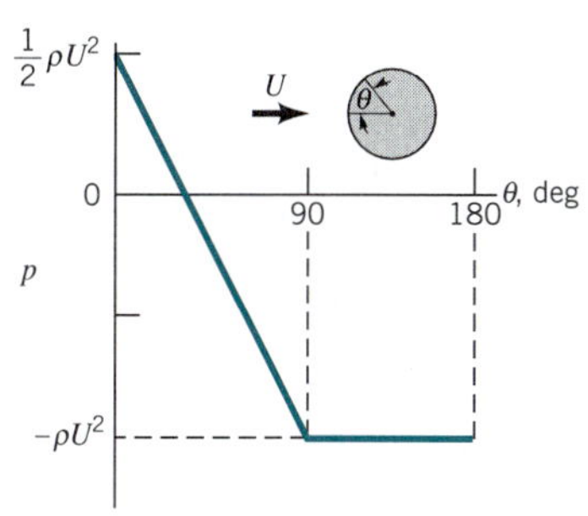

■ **FIGURE P9.4**

9.5 Repeat Problem 9.1 if the object is a cone (made by rotating the equilateral triangle about the horizontal axis through its tip) rather than a triangular bar.

9.6 A 12-ft-long kayak moves with a speed of 5 ft/s. Would a boundary layer type flow be developed along the sides of the boat? Explain.

9.7 Typical values of the Reynolds number for various animals moving through air or water are listed below. For which cases is inertia of the fluid important? For which cases do viscous effects dominate? For which cases would the flow be laminar; turbulent? Explain.

Animal	Speed	Re
(a) large whale	10 m/s	300,000,000
(b) flying duck	20 m/s	300,000
(c) large dragonfly	7 m/s	30,000
(d) invertebrate larva	1 mm/s	0.3
(e) bacterium	0.01 mm/s	0.00003

9.8 When you walk through still air, would you expect the character of the air flow around you to be most like that depicted in Fig. 9.6a, b, or c? Explain.

9.9 Approximately how fast can the wind blow past a 0.25-in.-diameter twig if viscous effects are to be of importance throughout the entire flow field (i.e., Re < 1)? Explain. Repeat for a 0.004-in.-diameter hair and a 6-ft-diameter smokestack.

9.10 A viscous fluid flows past a flat plate such that the boundary layer thickness at a distance 1.3 m from the leading edge is 12 mm. Determine the boundary layer thickness at distances of 0.20, 2.0, and 20 m from the leading edge. Assume laminar flow.

9.11 If the upstream velocity of the flow in Problem 9.10 is $U = 1.5$ m/s, determine the kinematic viscosity of the fluid.

9.12 Water flows past a flat plate with an upstream velocity of $U = 0.02$ m/s. Determine the water velocity a distance of 10 mm from the plate at distances of $x = 1.5$ m and $x = 15$ m from the leading edge.

***9.13** A Pitot tube connected to a water-filled U-tube manometer is used to measure the total pressure within a boundary layer. Based on the data given in the table below, determine the boundary layer thickness, δ, the displacement thickness, δ^*, and the momentum thickness, Θ.

y (mm), distance above plate	h (mm), manometer reading
0	0
2.1	10.6
4.3	21.1
6.4	25.6
10.7	32.5
15.0	36.9
19.3	39.4
23.6	40.5
26.8	41.0
29.3	41.0
32.7	41.0

9.14 Because of the velocity deficit, $U - u$, in the boundary layer, the streamlines for flow past a flat plate are not exactly parallel to the plate. This deviation can be determined by use of the displacement thickness, δ^*. For air blowing past the flat plate shown in Fig. P9.14, plot the streamline A–B that passes through the edge of the boundary layer ($y = \delta_B$ at $x = \ell$) at point B. That is, plot $y = y(x)$ for streamline A–B. Assume laminar boundary layer flow.

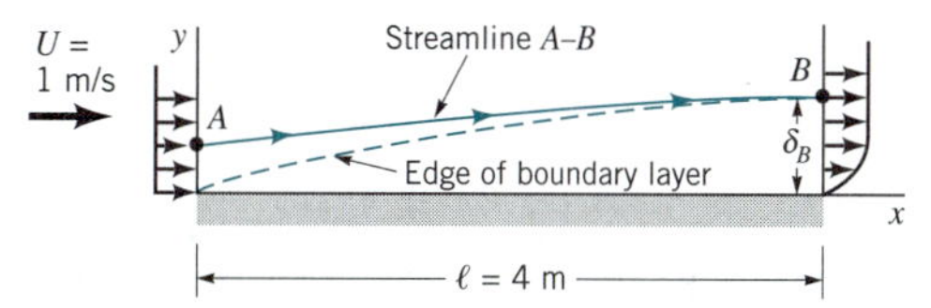

■ **FIGURE P9.14**

9.15 Air enters a square duct through a 1-ft opening as is shown in Fig. P9.15. Because the boundary layer displacement thickness increases in the direction of flow, it is necessary to increase the cross-sectional size of the duct if a constant $U = 2$ ft/s velocity is to be maintained outside the boundary layer. Plot a graph of the duct size, d, as a function of x for $0 \leq x \leq 10$ ft if U is to remain constant. Assume laminar flow.

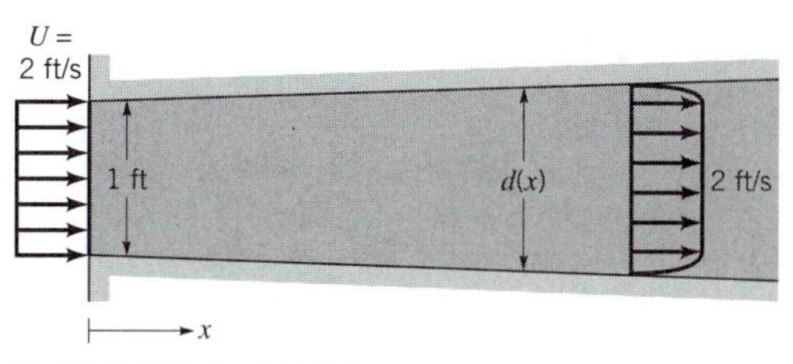

■ **FIGURE P9.15**

9.16 A smooth flat plate of length $\ell = 6$ m and width $b = 4$ m is placed in water with an upstream velocity of $U = 0.5$ m/s. Determine the boundary layer thickness and the wall shear stress at the center and the trailing edge of the plate. Assume a laminar boundary layer.

9.17 An atmospheric boundary layer is formed when the wind blows over the earth's surface. Typically, such velocity profiles can be written as a power law: $u = ay^n$, where the constants a and n depend on the roughness of the terrain. As is indicated in Fig. P9.17, typical values are $n = 0.40$ for urban areas, $n = 0.28$ for woodland or suburban areas, and $n = 0.16$ for flat open country (Ref. 23). **(a)** If the velocity is 20 ft/s at the bottom of the sail on your boat ($y = 4$ ft), what is the velocity at the top of the mast ($y = 30$ ft)? **(b)** If the average velocity is 10 mph on the tenth floor of an urban building, what is the average velocity on the sixtieth floor?

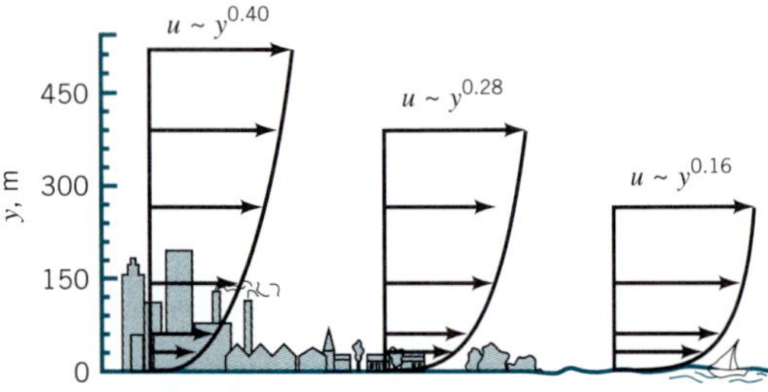

■ **FIGURE P9.17**

9.18 A 30-story office building (each story is 12 ft tall) is built in a suburban industrial park. Plot the dynamic pressure, $\rho u^2/2$, as a function of elevation if the wind blows at hurricane strength (75 mph) at the top of the building. Use the atmospheric boundary layer information of Problem 9.17.

9.19 The typical shape of small cumulus clouds is as indicated in Fig. P9.19. Based on boundary layer ideas, explain why it is clear that the wind is blowing from right to left as indicated.

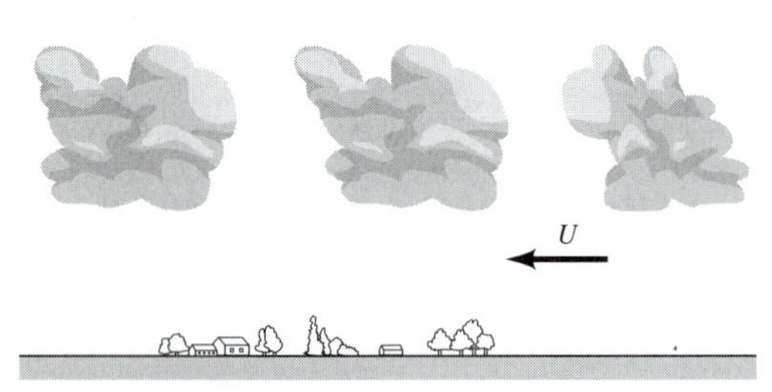

■ **FIGURE P9.19**

9.20 Show that by writing the velocity in terms of the similarity variable η and the function $f(\eta)$, the momentum equation for boundary layer flow on a flat plate (Eq. 9.9) can be written as the ordinary differential equation given by Eq. 9.14.

***9.21** Integrate the Blasius equation (Eq. 9.14) numerically to determine the boundary layer profile for laminar flow past a flat plate. Compare your results with those of Table 9.1.

9.22 An airplane flies at a speed of 400 mph at an altitude of 10,000 ft. If the boundary layers on the wing surfaces behave as those on a flat plate, estimate the extent of laminar boundary layer flow along the wing. Assume a transitional Reynolds number of $\text{Re}_{x\text{cr}} = 5 \times 10^5$. If the airplane maintains its 400-mph speed but descends to sea level elevation, will the portion of the wing covered by a laminar boundary layer increase or decrease compared with its value at 10,000 ft? Explain.

† **9.23** If the boundary layer on the hood of your car behaves as one on a flat plate, estimate how far from the front edge of the hood the boundary layer becomes turbulent. How thick is the boundary layer at this location?

9.24 A laminar boundary layer velocity profile is approximated by $u/U = 2(y/\delta) - 2(y/\delta)^3 + (y/\delta)^4$ for $y \leq \delta$, and $u = U$ for $y > \delta$. **(a)** Show that this profile satisfies the appropriate boundary conditions. **(b)** Use the momentum integral equation to determine the boundary layer thickness, $\delta = \delta(x)$.

9.25 A laminar boundary layer velocity profile is approximated by the two straight-line segments indicated in Fig. P9.25. Use the momentum integral equation to determine the boundary layer thickness, $\delta = \delta(x)$, and wall shear stress, $\tau_w = \tau_w(x)$. Compare these results with those in Table 9.2.

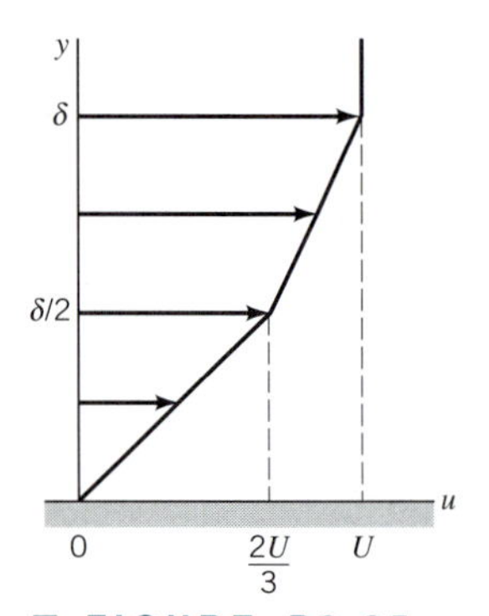

■ **FIGURE P9.25**

***9.26** An assumed, dimensionless laminar boundary layer profile for flow past a flat plate is given in the table below. Use the momentum integral equation to determine $\delta = \delta(x)$. Compare your result with the exact Blasius solution result (see Table 9.2).

y/δ	u/U
0	0
0.080	0.133
0.16	0.265
0.24	0.394
0.32	0.517
0.40	0.630
0.48	0.729
0.56	0.811
0.64	0.876
0.72	0.923
0.80	0.956
0.88	0.976
0.96	0.988
1.00	1.000

***9.27** For a fluid of specific gravity $SG = 0.86$ flowing past a flat plate with an upstream velocity of $U = 5$ m/s, the wall shear stress on a flat plate was determined to be as indicated in the table below. Use the momentum integral equation to determine the boundary layer momentum thickness, $\Theta = \Theta(x)$. Assume $\Theta = 0$ at the leading edge, $x = 0$.

x (m)	τ_w (N/m^2)
0	—
0.2	13.4
0.4	9.25
0.6	7.68
0.8	6.51
1.0	5.89
1.2	6.57
1.4	6.75
1.6	6.23
1.8	5.92
2.0	5.26

9.28 The net drag on one side of the two plates (each of size ℓ by $\ell/2$) parallel to the free stream shown in Fig. P9.28*a* is $\mathfrak{D}$. Determine the drag (in terms of $\mathfrak{D}$) on the same two plates when they are connected together as indicated in Fig. P9.28*b*. Assume laminar boundary flow. Explain your answer physically.

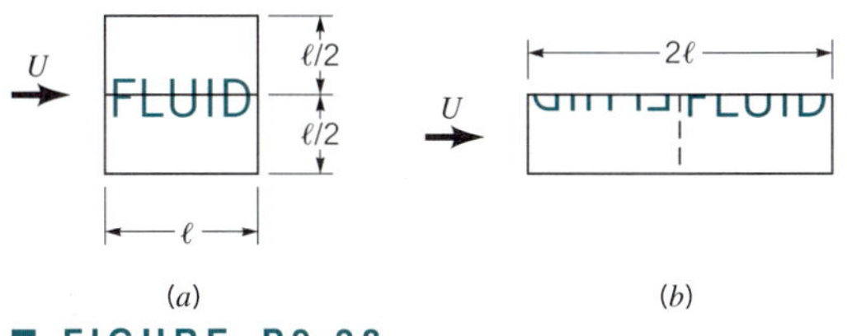

■ FIGURE P9.28

9.29 A plate is oriented parallel to the free stream as is indicated in Fig. 9.29. If the boundary layer flow is laminar, determine the ratio of the drag for case *a* to that for case *b*. Explain your answer physically.

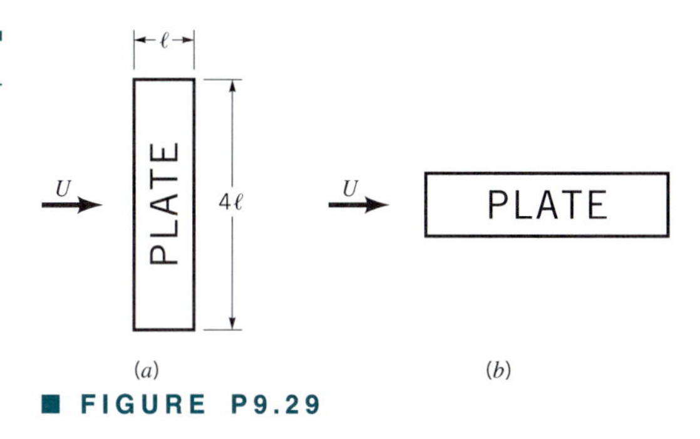

■ FIGURE P9.29

9.30 If the drag on one side of a flat plate parallel to the upstream flow is $\mathfrak{D}$ when the upstream velocity is U, what will the drag be when the upstream velocity is $2U$; or $U/2$? Assume laminar flow.

9.31 Air flows past a parabolic-shaped flat plate oriented parallel to the free stream shown in Fig. P9.31. Integrate the wall shear stress over the plate to determine the friction drag on one side of the plate. Assume laminar flow.

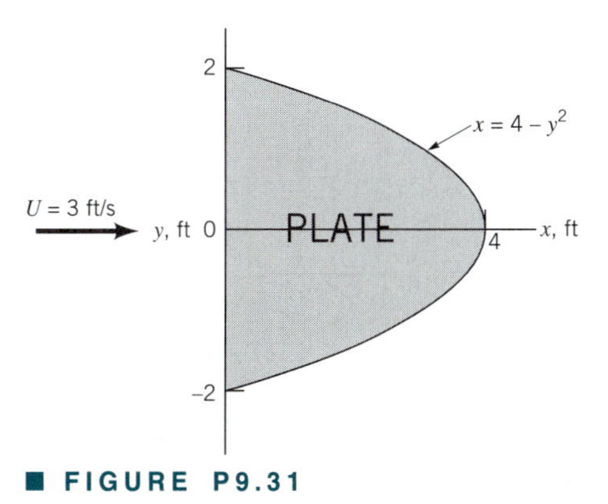

■ FIGURE P9.31

9.32 It is often assumed that "sharp objects can cut through the air better than blunt ones." Based on this assumption, the drag on the object shown in Fig. P9.32 should be less when the wind blows from right to left than when it blows from left to right. Experiments show that the opposite is true. Explain.

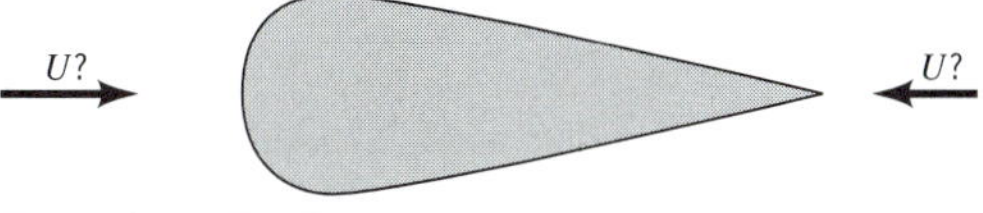

■ FIGURE P9.32

***9.33** The device shown in Fig. P9.33 is to be designed to measure the wall shear stress as air flows over the smooth surface with an upstream velocity U. It is proposed that τ_w can be obtained by measuring the bending moment, M, at the base [point (1)] of the support that holds the small surface element which is free from contact with the surrounding surface. Plot a graph of M as a function of U for $5 \leq U \leq 50$ m/s, with $\ell = 2, 3, 4$, and 5 m.

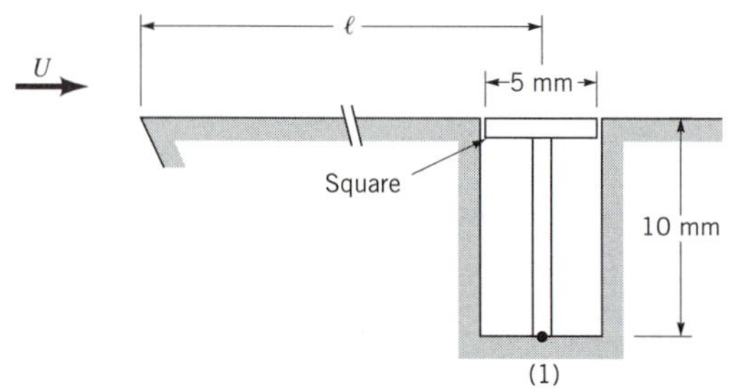

■ FIGURE P9.33

9.34 Repeat Problem 9.33 if the fluid is water, the upstream velocity is $U = 5$ m/s, and $\ell = 3$ m.

9.35 A three-bladed helicopter blade rotates at 200 rpm. If each blade is 12 ft long and 1.5 ft wide, estimate the torque needed to overcome the friction on the blades if they act as flat plates.

9.36 A ceiling fan consists of five blades of 0.80-m length and 0.10-m width which rotate at 100 rpm. Estimate the torque needed to overcome the friction on the blades if they act as flat plates.

9.37 A thin smooth sign is attached to the side of a truck as is indicated in Fig. P9.37. Estimate the friction drag on the sign when the truck is driven at 55 mph.

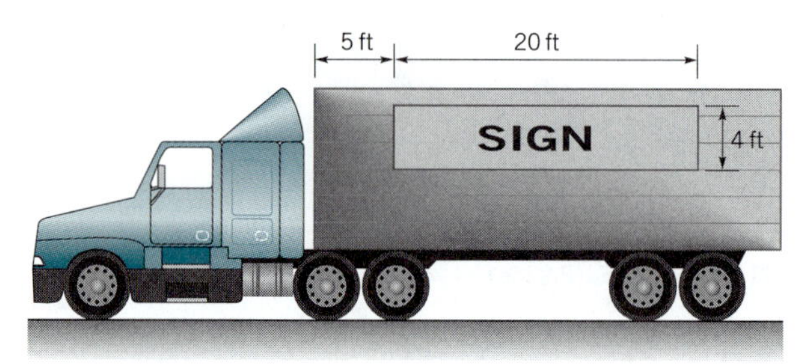

■ FIGURE P9.37

9.38 A sphere of diameter D and density ρ_s falls at a steady rate through a liquid of density ρ and viscosity μ. If the Reynolds number, $\text{Re} = \rho DU/\mu$, is less than 1, show that the viscosity can be determined from $\mu = gD^2(\rho_s - \rho)/18\ U$.

9.39 Determine the drag on a small circular disk of 0.01-ft diameter moving 0.01 ft/s through oil with a specific gravity of 0.87 and a viscosity 10,000 times that of water. The disk is oriented normal to the upstream velocity. By what percent is the drag reduced if the disk is oriented parallel to the flow?

9.40 For small Reynolds number flows the drag coefficient of an object is given by a constant divided by the Reynolds number (see Table 9.4). Thus, as the Reynolds number tends to zero, the drag coefficient becomes infinitely large. Does this mean that for small velocities (hence, small Reynolds numbers) the drag is very large? Explain.

9.41 Compare the rise velocity of an $\frac{1}{8}$-in.-diameter air bubble in water to the fall velocity of an $\frac{1}{8}$-in.-diameter water drop in air. Assume each to behave as a solid sphere.

9.42 A 38.1-mm-diameter, 0.0245-N table tennis ball is released from the bottom of a swimming pool. With what velocity does it rise to the surface? Assume it has reached its terminal velocity.

† **9.43** How fast will a toy balloon filled with helium rise through still air? List all of your assumptions.

9.44 A hot air balloon roughly spherical in shape has a volume of 70,000 ft^3 and a weight of 500 lb (including passengers, basket, balloon fabric, etc.). If the outside air temperature is 80 °F and the temperature within the balloon is 165 °F, estimate the rate at which it will rise under steady-state conditions if the atmospheric pressure is 14.7 psi.

9.45 A 500-N cube of specific gravity $SG = 1.8$ falls through water at a constant speed U. Determine U if the cube falls **(a)** as oriented in Fig. P9.45*a*, **(b)** as oriented in Fig. P9.45*b*.

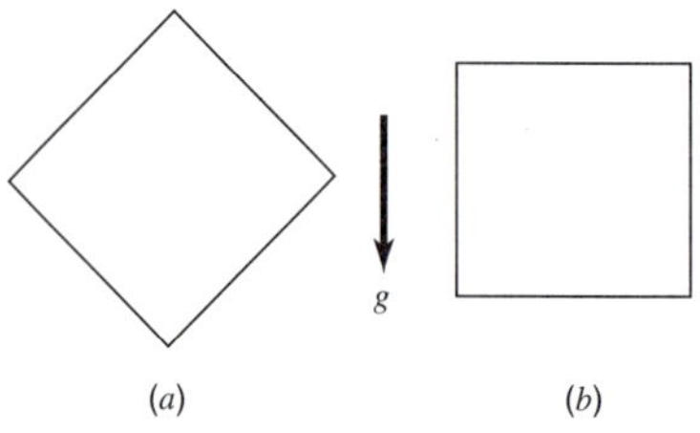

■ FIGURE P9.45

9.46 A 50-lb box shaped like a 1-ft cube falls from the cargo hold of an airplane at an altitude of 30,000 ft. If the drag coefficient of the falling box is 1.2, determine the time it takes for the box to hit the ocean. Assume that it falls at the terminal velocity corresponding to its current altitude and use a standard atmosphere (see Table C.1).

9.47 A 60-mph wind blows against an outdoor movie screen that is 70 ft wide and 20 ft tall. Estimate the wind force on the screen.

9.48 Determine the moment needed at the base of 20-m-tall, 0.12-m-diameter flag pole to keep it in place in a 20 m/s wind.

9.49 Repeat Problem 9.48 if a 2-m by 2.5-m flag is attached to the top of the pole. See Fig. 9.30 for drag coefficient data for flags.

9.50 An object falls at a rate of 100 ft/s immediately prior to the time that the parachute attached to it opens. The final descent rate with the chute open is 10 ft/s. Calculate and plot the speed of falling as a function of time from when the chute opens. Assume that the chute opens instantly, that the drag coefficient and air density remain constant, and that the flow is quasisteady.

9.51 If for a given vehicle it takes 20 hp to overcome aerodynamic drag while being driven at 65 mph, estimate the horsepower required at 75 mph.

9.52 Two bicycle racers ride 30 km/hr through still air. By what percentage is the power required to overcome aerodynamic drag for the second cyclist reduced if she drafts closely behind the first cyclist rather than riding alongside her? Neglect any forces other than aerodynamic drag. (See Fig. 9.30.)

† **9.53** Estimate the wind speed needed to tip over a garbage can. List all assumptions and show all calculations.

9.54 It is suggested that the power, $\mathcal{P}$, needed to overcome the aerodynamic drag on a vehicle traveling at a speed U varies as $\mathcal{P} \sim U^n$. What is an appropriate value for the constant n? Explain.

9.55 A 25-ton (50,000-lb) truck coasts down a steep 7% mountain grade without brakes, as shown in Fig. P9.55. The truck's ultimate steady-state speed, V, is determined by a balance between weight, rolling resistance, and aerodynamic drag. Determine V if the rolling resistance for a truck on concrete is 1.2% of the weight and the drag coefficient based on frontal area is 0.76.

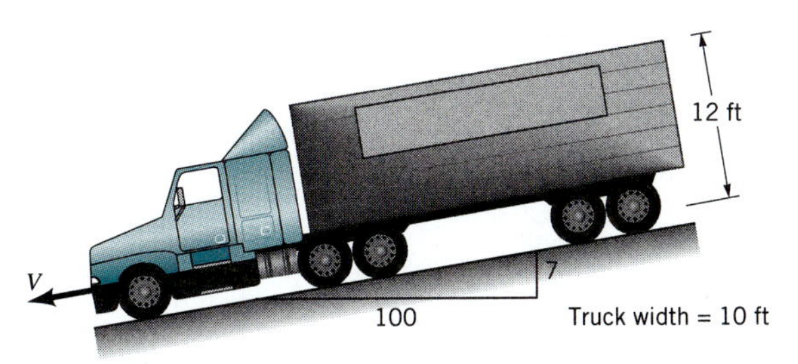

■ **FIGURE P9.55**

9.56 As shown in Fig. P9.56, the drag coefficient for a car depends on whether the windows and/or sun roof are open or closed. Assume it takes 20 horsepower to overcome aerodynamic drag at a given speed with the windows and roof closed. How much additional horsepower is needed to overcome aerodynamic drag at the same speed if **(a)** the windows are opened; **(b)** the windows and roof are opened?

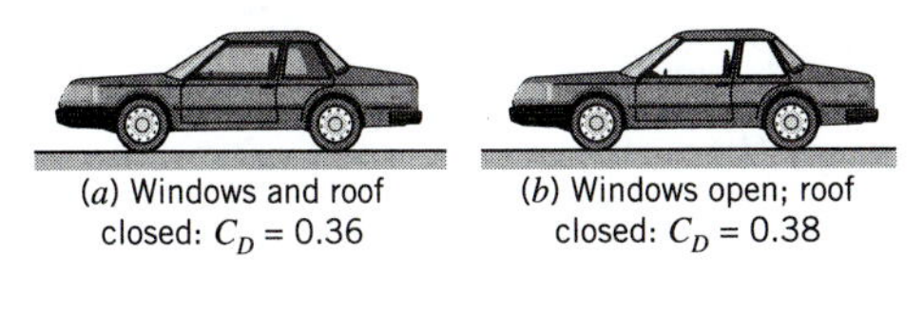

■ **FIGURE P9.56**

9.57 The large, newly planted tree shown in Fig. P9.57 is kept from tipping over in a wind by use of a rope as shown. It is assumed that the sandy soil cannot support any moment about the center of the soil ball, point A. Estimate the tension in the rope if the wind is 80 km/hr. (See Fig. 9.30.)

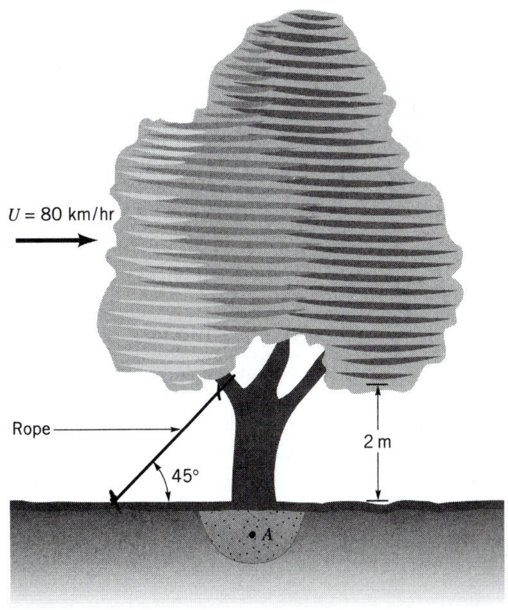

■ **FIGURE P9.57**

† **9.58** Estimate the maximum wind velocity in which you can stand without holding on to something. List your assumptions.

9.59 Estimate the velocity with which you would contact the ground if you jumped from an airplane at an altitude of 5,000 ft and **(a)** air resistance is negligible, **(b)** air resistance is important, but you forgot your parachute, or **(c)** you use a 25-ft-diameter parachute.

***9.60** The helium-filled balloon shown in Fig. P9.60 is to be used as a wind speed indicator. The specific weight of the helium is $\gamma = 0.011$ lb/ft^3, the weight of the balloon material is 0.20 lb, and the weight of the anchoring cable is negligible. Plot a graph of θ as a function of U for $1 \leq U \leq 50$ mph. Would this be an effective device over the range of U indicated? Explain.

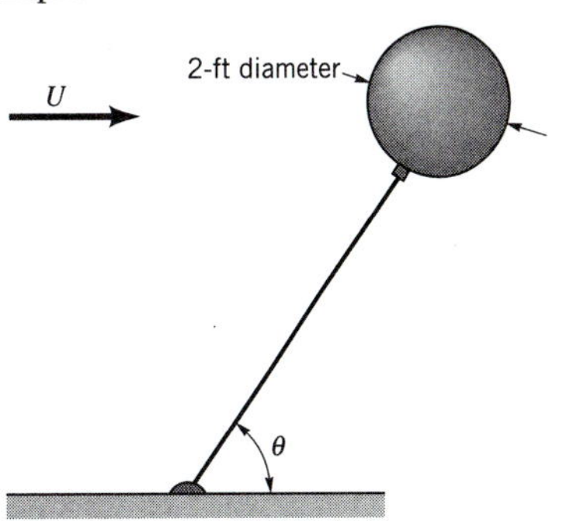

■ **FIGURE P9.60**

9.61 A 2-in.-diameter cork sphere (specific weight = 13 lb/ft^3) is attached to the bottom of a river with a thin cable, as is illustrated in Fig. P9.61. If the sphere has a drag coefficient of 0.5, determine the river velocity. Both the drag on the cable and its weight are negligible.

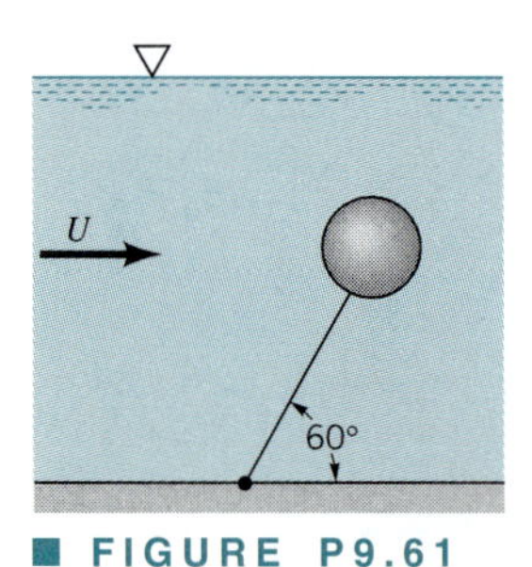

■ FIGURE P9.61

9.62 Two smooth spheres are attached to a thin rod that is free to rotate in the horizontal plane about point O as shown in Fig. P9.62. The rod is held stationary until the air speed reaches 50 ft/s. Which direction will the rod rotate (clockwise or counterclockwise) when the holding force is released? Explain your answer.

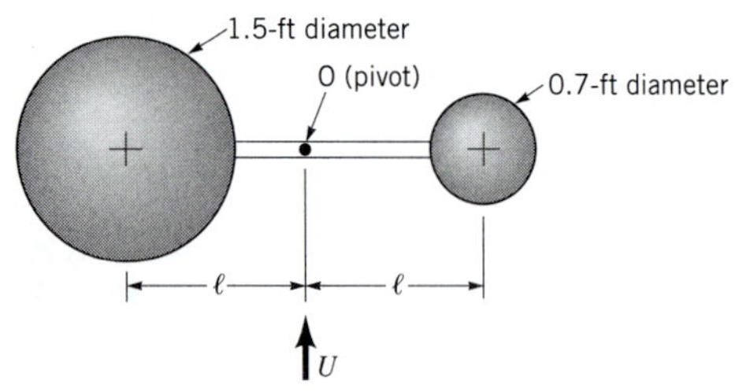

■ FIGURE P9.62

9.63 A radio antenna on a car consists of a circular cylinder $\frac{1}{4}$ in. in diameter and 4 ft long. Determine the bending moment at the base of the antenna if the car is driven 55 mph through still air.

† **9.64** Estimate the energy that a runner expends to overcome aerodynamic drag while running a complete marathon race. This expenditure of energy is equivalent to climbing a hill of what height? List all assumptions and show all calculations.

9.65 Estimate the wind force on your hand when you hold it out of your car window while driving 55 mph. Repeat your calculations if you were to hold your hand out of the window of an airplane flying 550 mph.

***9.66** Let $\mathcal{P}_0$ be the power required to fly a particular airplane at 500 mph at sea level conditions. Plot a graph of the ratio $\mathcal{P}/\mathcal{P}_0$, where $\mathcal{P}$ is the power required at a speed of U, for 500 mph $\leq U \leq$ 3000 mph at altitudes of sea level, 10,000 ft, 20,000 ft, and 30,000 ft. Assume that the drag coefficient for the aircraft behaves similarly to that of the sharp-pointed ogive indicated in Fig. 9.24.

9.67 Estimate the deceleration (in terms of the number of g's) of a 6-ft-diameter, 12,500-lb blunt reentry vehicle traveling 15,000 mph through the atmosphere at an altitude of 30 mi.

9.68 A 30-ft-tall tower is constructed of equal 1-ft segments as is indicated in Fig. P9.68. Each of the four sides is similar. Estimate the drag on the tower when a 75-mph wind blows against it.

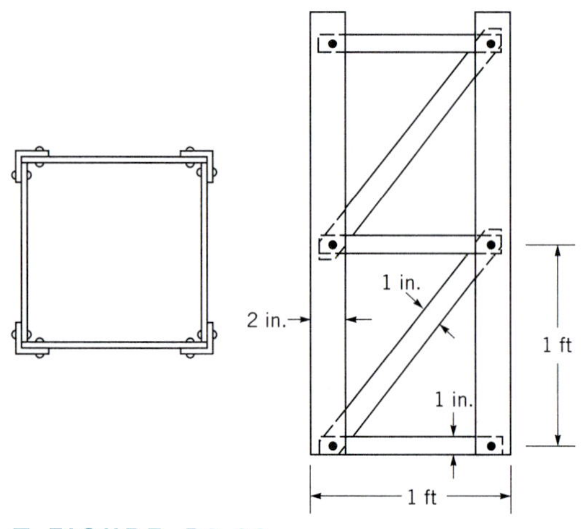

■ FIGURE P9.68

9.69 A 2-in.-diameter sphere weighing 0.14 lb is suspended by the jet of air shown in Fig. P9.69. The drag coefficient for the sphere is 0.5. Determine the reading on the pressure gage if friction and gravity effects can be neglected for the flow between the pressure gage and the nozzle exit.

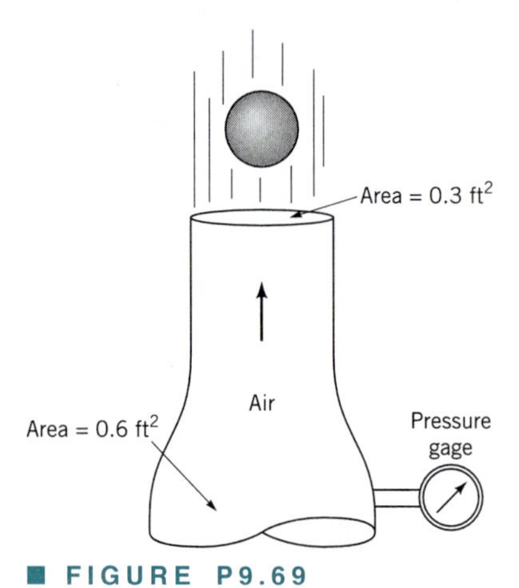

■ FIGURE P9.69

9.70 The United Nations Building in New York is approximately 87.5-m wide and 154-m tall. **(a)** Determine the drag on this building if the drag coefficient is 1.3 and the wind speed is a uniform 20 m/s. **(b)** Repeat your calculations if the velocity profile against the building is a typical profile for an urban area (see Problem 9.17) and the wind speed halfway up the building is 20 m/s.

† **9.71** An "air-popper" popcorn machine blows hot air past the kernels with speed U so that the unpopped ones remain in the holder ($U < U_{max}$), but the popped kernels are blown out of the holder ($U > U_{min}$). Estimate the range of air velocity allowed for proper operation of the machine ($U_{min} < U < U_{max}$). List all assumptions and show all calculations.

9.72 A 1.2-lb kite with an area of 6 ft^2 flies in a 20 ft/s wind such that the weightless string makes an angle of 55° relative to the horizontal. If the pull on the string is 1.5 lb, determine the lift and drag coefficients based on the kite area.

9.73 A regulation football is 6.78 in. in diameter and weighs 0.91 lb. If its drag coefficient is $C_D = 0.2$, determine its deceleration if it has a speed of 20 ft/s at the top of its trajectory.

9.74 Explain how the drag on a given smokestack could be the same in a 2 mph wind as in a 4 mph wind. Assume the values of ρ and μ are the same for each case.

9.75 As is discussed in Section 9.3, the drag on a rough golf ball is less than that on an equal-sized smooth ball. Does it follow that a 10-m-diameter spherical water tank resting on a 20-m-tall support should have a rough surface so as to reduce the moment needed at the base of the support when a wind blows? Explain.

† **9.76** If the wind becomes strong enough, it is "impossible" to paddle a canoe into the wind. Estimate the wind speed at which this will happen. List all assumptions and show all calculations.

9.77 A strong wind can blow a golf ball off the tee by pivoting it about point 1 as shown in Fig. P9.77. Determine the wind speed necessary to do this.

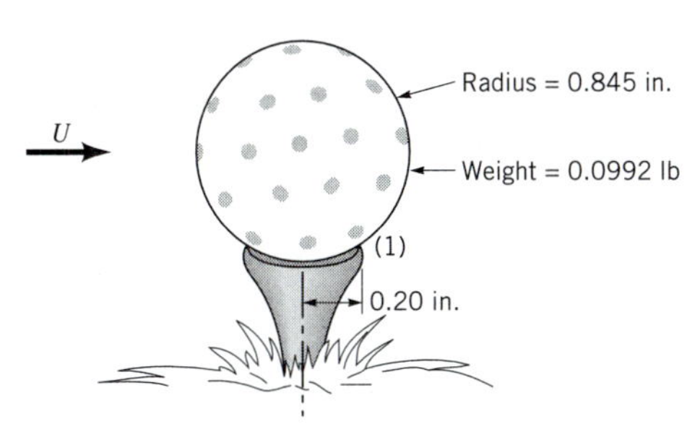

■ FIGURE P9.77

9.78 An airplane tows a banner that is $b = 0.8$ m tall and $\ell = 25$ m long at a speed of 150 km/hr. If the drag coefficient based on the area $b\ell$ is $C_D = 0.06$, estimate the power required to tow the banner. Compare the drag force on the banner with that on a rigid flat plate of the same size. Which has the larger drag force and why?

9.79 By appropriate streamlining, the drag coefficient for an airplane is reduced by 12% while the frontal area remains the same. For the same power output, by what percentage is the flight speed increased?

9.80 The dirigible Akron had a length of 239 m and a maximum diameter of 40.2 m. Estimate the power required at its maximum speed of 135 km/hr if the drag coefficient based on frontal area is 0.060.

9.81 Estimate the power needed to overcome the aerodynamic drag of a person who runs at a rate of 100 yds in 10 s in still air. Repeat the calculations if the race is run into a 20-mph headwind; a 20-mph tailwind. Explain.

† **9.82** Skydivers often join together to form patterns during the free-fall portion of their jump. The current *Guiness Book of World Records* record is 297 skydivers joined hand-to-hand. Given that they can't all jump from the same airplane at the same time, describe how they manage to get together. Use appropriate fluid mechanics equations and principles in your answer.

9.83 A fishnet consists of 0.10-in.-diameter strings tied into squares 4 in. per side. Estimate the force needed to tow a 15-ft by 30-ft section of this net through seawater at 5 ft/s.

9.84 An iceberg floats with approximately $\frac{1}{7}$ of its volume in the air as is shown in Fig. P9.84. If the wind velocity is U and the water is stationary, estimate the speed at which the wind forces the iceberg through the water.

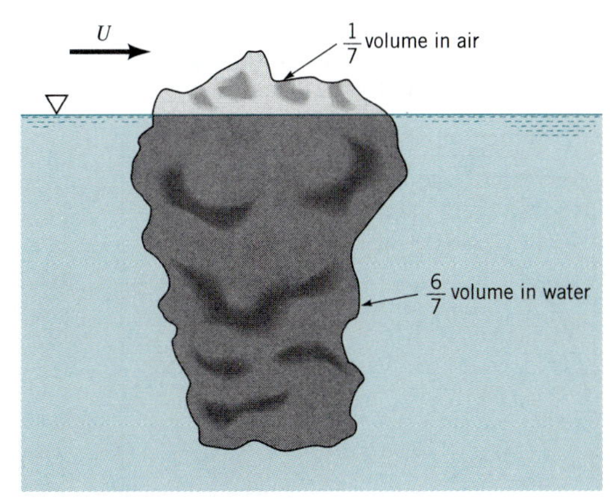

■ FIGURE P9.84

9.85 A Piper Cub airplane has a gross weight of 1750 lb, a cruising speed of 115 mph, and a wing area of 179 ft^2. Determine the lift coefficient of this airplane for these conditions.

9.86 A light aircraft with a wing area of 200 ft^2 and a weight of 2000 lb has a lift coefficient of 0.40 and a drag coefficient of 0.05. Determine the power required to maintain level flight.

† **9.87** Fold a piece of paper into a paper airplane and observe its flight characteristics. Based on your observations, estimate the lift-to-drag ratio for this airplane. List all assumptions and show all calculations.

9.88 The wings of old airplanes are often strengthened by the use of wires that provided cross-bracing as shown in Fig. P9.88. If the drag coefficient for the wings was 0.020 (based on the planform area), determine the ratio of the drag from the wire bracing to that from the wings.

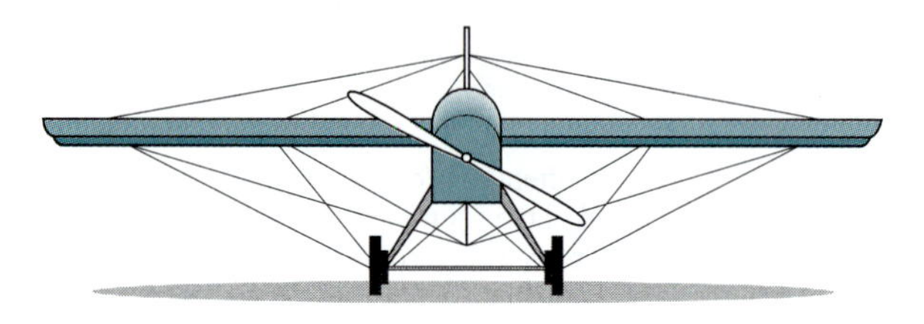

Speed: 70 mph
Wing area: 148 ft^2
Wire: length = 160 ft
diameter = 0.05 in.

■ **FIGURE P9.88**

9.89 For a given airplane, compare the power to maintain level flight at a 5000-ft altitude with that at 30,000 ft at the same velocity. Assume C_D remains constant.

9.90 A wing generates a lift $\mathscr{L}$ when moving through sea level air with a velocity U. How fast must the wing move through the air at an altitude of 35,000 ft with the same lift coefficient if it is to generate the same lift?

*__9.91__ When air flows past the airfoil shown in Fig. P9.91, the velocity just outside the boundary layer, u, is as indicated. Estimate the lift coefficient for these conditions.

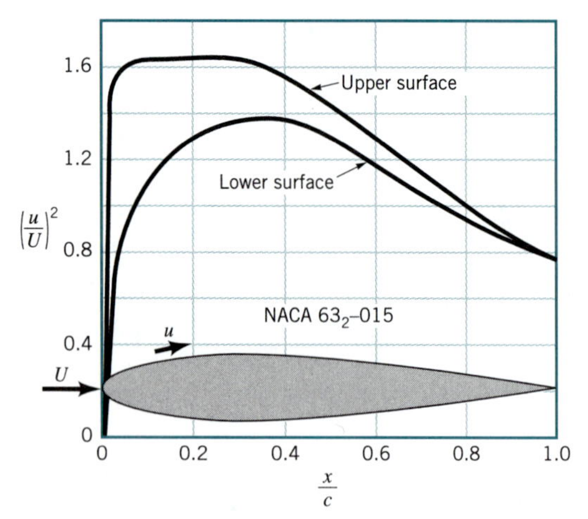

■ **FIGURE P9.91**

9.92 A certain airfoil has the measured lift and drag coefficient characteristics listed in the table below.

α, deg	C_L	C_D
−3.3	−0.182	0.0098
−1.8	−0.025	0.0094
−0.5	0.135	0.0087
1.2	0.286	0.0087
2.6	0.441	0.0093
5.8	0.746	0.0104
8.6	1.05	0.0141
11.8	1.33	0.0203
13.6	1.46	0.0264
15.3	1.58	0.0361
15.9	1.60	0.0436
19.7	1.33	0.231
25.9	1.03	0.426

Plot a graph of C_L and C_D as a function of angle of attack. Plot the polar diagram for this airfoil. Determine the stall angle.

9.93 A Boeing 747 aircraft weighing 580,000 lb when loaded with fuel and 100 passengers takes off with an airspeed of 140 mph. With the same configuration (i.e., angle of attack, flap settings, etc.), what is its takeoff speed if it is loaded with 372 passengers? Assume each passenger with luggage weighs 200 lb.

9.94 Show that for unpowered flight (for which the lift, drag, and weight forces are in equilibrium) the glide slope angle, θ, is given by $\tan\theta = C_D/C_L$.

9.95 If the lift coefficient for a Boeing 777 aircraft is 15 times greater than its drag coefficient, can it glide from an altitude of 30,000 ft to an airport 80 mi away if it loses power from its engines? Explain. (See Problem 9.94.)

9.96 On its final approach to the airport, an airplane flies on a flight path that is 3.0° relative to the horizontal. What lift-to-drag ratio is needed if the airplane is to land with its engines idled back to zero power? (See Problem 9.94.)

9.97 A sail plane with a lift-to-drag ratio of 25 flies with a speed of 50 mph. It maintains or increases its altitude by flying in thermals, columns of vertically rising air produced by buoyancy effects of nonuniformly heated air. What vertical airspeed is needed if the sail plane is to maintain a constant altitude?

9.98 The stall speed, U_{stall}, is the minimum landing speed for an aircraft. Show that $U_{stall} = [2\mathscr{W}/(\rho C_{Lmax} A)]^{1/2}$, where $\mathscr{W}$ is the weight of the airplane. (Normally, aircraft land at a higher speed to allow a margin of safety.) By what percentage can the landing speed be reduced by using double-slotted trailing edge flaps (see Fig. 9.35) rather than no flaps?

9.99 If the takeoff speed of a particular airplane is 120 mi/hr at sea level, what will it be at Denver (elevation 5000 ft)? Use properties of the U.S. Standard Atmosphere.

9.100 Commercial airliners normally cruise at relatively high altitudes (30,000 to 35,000 ft). Discuss how flying at this high altitude (rather than 10,000 ft, for example) can save fuel costs.

9.101 A pitcher can pitch a ''curve ball'' by putting sufficient spin on the ball when it is thrown. A ball that has absolutely no spin will follow a ''straight'' path. A ball that is pitched with a very small amount of spin (on the order of one revolution during its flight between the pitcher's mound and home plate) is termed a knuckle ball. A ball pitched this way tends to ''jump around'' and ''zig-zag'' back and forth. Explain this phenomenon. Note: A baseball has seams.

9.102 Repeated controversy regarding the ability of a baseball to curve appeared in the literature for years. According to a test (*Life*, July 27, 1953), a baseball (assume the diameter is 2.9 in. and weight is 5.25 oz) spinning 1400 rpm while traveling 43 mph was observed to follow a path with an 800-ft horizontal radius of curvature. Based on the data of Fig. 9.39, do you agree with this test result? Explain.

9.103 The velocity profile within the boundary layer on a flat plate can be determined by means of the device shown in Fig. P9.103. Air at a temperature of 80 °F and an absolute pressure of 29.09 in. of mercury blows steadily along a flat plate. A small-diameter, open-ended tube located at various distances, y, above the plate is used to measure the stagnation pressure of the flow at these locations. The static pressure is measured by means of a static pressure tap on the plate as indicated. The difference between the stagnation and static pressures is determined from the inclined manometer readings, ℓ. The manometer is filled with a liquid of specific gravity 0.812.

Values of ℓ and y obtained experimentally at a distance of 15.0 in. from the leading edge of the plate are shown in the following table. Use these results to plot a graph of the air speed, u, as a function of the distance above the plate, y. Determine the boundary layer thickness at this location. Also calculate the theoretical boundary layer thickness for the conditions corresponding to the above experimental data.

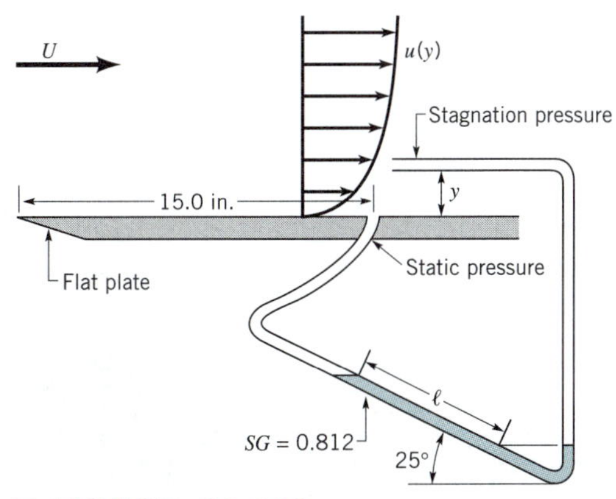

■ FIGURE P9.103

Compare the experimental and theoretical results and discuss some possible reasons for any differences between them.

y (in.)	ℓ (in.)
0.020	0.15
0.035	0.35
0.044	0.40
0.060	0.70
0.096	0.90
0.110	1.30
0.138	1.45
0.178	1.65
0.230	1.95
0.270	2.00
0.322	2.00

Motion of water induced by surface waves: As a wave passes along the surface of the water, the water particles follow elliptical paths. There is no net motion of the water, just a periodic, cyclic trajectory (neutrally buoyant particles in water). (Photograph by A. Wallet and F. Ruellan, Ref. 13, courtesy of M. C. Vasseur, Sogreah.)

10 Open-Channel Flow

Open-channel flow involves the flow of a liquid in a channel or conduit that is not completely filled. There exists a free surface between the flowing fluid (usually water) and fluid above it (usually the atmosphere). The main driving force for such flows is the fluid weight—gravity forces the fluid to flow downhill. Under steady, fully developed flow conditions, the component of the weight force in the direction of flow is balanced by the equal and opposite shear force between the fluid and the channel surfaces. For unsteady or nonfully developed situations, the inertia of the following fluid is also important.

Such flows are different from the pipe flows discussed in Chapter 8 in that there can be no pressure force driving the fluid through the channel or conduit. Any attempt to impose a pressure gradient in the flow direction is met with failure because of the negligible inertial and viscous effects of the gas (atmosphere) above the flowing fluid. For steady, fully developed channel flow, the pressure distribution within the fluid is merely hydrostatic.

Open-channel flows are essential to the world as we know it.

Open-channel flows are essential to the world as we know it. The natural drainage of water through the numerous creek and river systems is a complex example of open-channel flow. Although the flow geometry for these systems is extremely complex, the resulting flow properties are of considerable economic, ecological, and recreational importance. Other examples of open-channel flows include the flow of rainwater in the gutters of our houses; the flow in canals, drainage ditches, sewers, and gutters along roads; the flow of small rivulets and sheets of water across fields or parking lots; and the flow in the chutes of water rides in amusement parks.

Clearly the character, description, and complexity of open-channel flow geometry is quite variable. The bounding geometry for flow in a sewer pipe laid on a constant slope and running half-full is much simpler than the geometry of the Mississippi River with its variable cross-sectional shape, bends, bottom slope variation, and character of its bounding surfaces. Because of complexities like these, most open-channel flow results are based on correlations obtained from model and full-scale experiments. Additional information can be gained from various analytical and numerical efforts.

The purpose of this chapter is to investigate the concepts of open-channel flow. Because of the amount and variety of material available, only a brief introduction to the topic can be presented. Further information can be obtained from the references indicated.

10.1 General Characteristics of Open-Channel Flow

In our study of pipe flow (Chapter 8) we found that there are many ways to classify a flow—developing, fully developed, laminar, turbulent, and so on. For open-channel flow, the existence of a free surface allows additional types of flow. The extra freedom that allows the fluid to select its free surface location and configuration (because it does not completely fill a pipe or conduit) allows important phenomena in open-channel flow that cannot occur in pipe flow. Some of the classifications of the flows are described below.

The manner in which the fluid depth, y, varies with time, t, and distance along the channel, x, is used to partially classify a flow. For example, the flow is *unsteady* or *steady* depending on whether the depth at a given location does or does not change with time. Some unsteady flows can be viewed as steady flows if the reference frame of the observer is changed. For example, a tidal bore moving up a river is unsteady to an observer standing on the bank, but steady to an observer moving along the bank with the speed of the wave front of the bore. Other flows are unsteady regardless of the reference frame used. The complex, time-dependent, wind-generated waves on a lake are in this category. In this book we will consider only steady open-channel flows.

An open-channel flow is classified as *uniform flow* (UF) if the depth of flow does not vary along the channel ($dy/dx = 0$). Conversely, it is *nonuniform flow* or *varied flow* if the depth varies with distance ($dy/dx \neq 0$). Nonuniform flows are further classified as *rapidly varying flow* (RVF) if the flow depth changes considerably over a relatively short distance; $dy/dx \sim 1$. *Gradually varying flows* (GVF) are those in which the flow depth changes slowly with distance along the channel; $dy/dx \ll 1$. Examples of these types of flow are illustrated in Fig. 10.1. The relative importance of the various types of forces involved (pressure, weight, shear, inertia) is different for the different types of flows.

Open-channel flow can have a variety of characteristics.

As for any flow geometry, open-channel flow may be *laminar*, *transitional*, or *turbulent*, depending on various conditions involved. Which type of flow occurs depends on the Reynolds number, $\text{Re} = \rho V R_h/\mu$, where V is the average velocity of the fluid and R_h is the hydraulic radius of the channel (see Section 10.4). A general rule is that open-channel flow is laminar if $\text{Re} < 500$, turbulent if $\text{Re} > 12{,}500$, and transitional otherwise. The values of these dividing Reynolds numbers are only approximate—a precise knowledge of the channel geometry is necessary to obtain specific values. Since most open-channel flows involve water (which has a fairly small viscosity) and have relatively large characteristic lengths, it is uncommon to have laminar open-channel flows. For example, flow of 50 °F water ($\nu = 1.41 \times 10^{-5}$ ft^2/s) with an average velocity of $V = 1$ ft/s in a river with a hydraulic radius

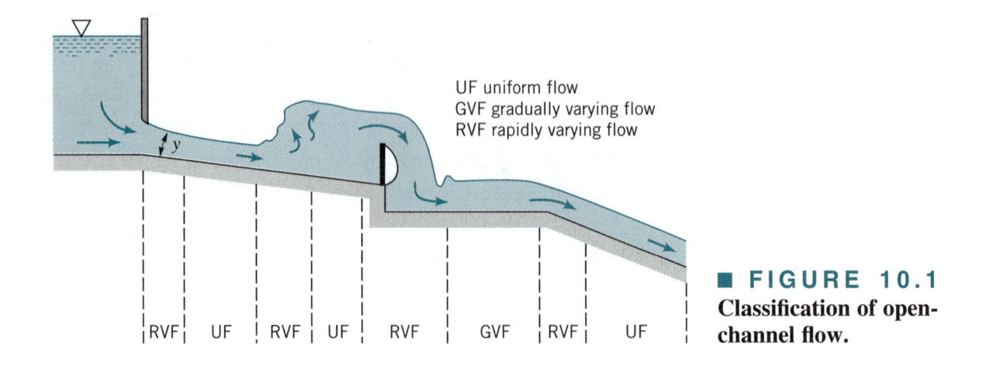

■ **FIGURE 10.1** **Classification of open-channel flow.**

of $R_h = 10$ ft has $\text{Re} = VR_h/\nu = 7.1 \times 10^5$. The flow is turbulent. However, flow of a thin sheet of water down a driveway with an average velocity of $V = 0.25$ ft/s such that $R_h = 0.02$ ft (in such cases the hydraulic radius is approximately equal to the fluid depth; see Section 10.4) has $\text{Re} = 355$. The flow is laminar.

In some cases *stratified flows* are important. In such situations layers of two or more fluids of different densities flow in a channel. A layer of oil on water is one example of this type of flow. All of the open-channel flows considered in this book are *homogeneous flows.* That is, the fluid has uniform properties throughout.

The Froude number is important in open-channel flows.

Open-channel flows involve a free surface that can deform from its undisturbed relatively flat configuration to form waves. Such waves move across the surface at speeds that depend on their size (height, length) and properties of the channel (depth, fluid velocity, etc.). The character of an open-channel flow may depend strongly on how fast the fluid is flowing relative to how fast a typical wave moves relative to the fluid. The dimensionless parameter that describes this behavior is termed the Froude number, $\text{Fr} = V/(g\ell)^{1/2}$, where ℓ is an appropriate characteristic length of the flow. This dimensionless parameter was introduced in Chapter 7 and is discussed more fully in Section 10.2. The special case of a flow with a Froude number of unity, $\text{Fr} = 1$, is termed a *critical flow.* If the Froude number is less than 1, the flow is *subcritical* (or *tranquil*). A flow with the Froude number greater than 1 is termed *supercritical* (or *rapid*).

10.2 Surface Waves

The distinguishing feature of flows involving a free surface (as in open-channel flows) is the opportunity for the free surface to distort into various shapes. The surface of a lake or the ocean is seldom "smooth as a mirror." It is usually distorted into ever-changing patterns associated with surface waves. Some of these waves are very high, some barely ripple the surface; some waves are very long (the distance between wave crests), some are short; some are breaking waves that form whitecaps, others are quite smooth. Although a general study of this wave motion is beyond the scope of this book, an understanding of certain fundamental properties of simple waves is necessary for open-channel flow considerations. The interested reader is encouraged to use some of the excellent references available for further study about wave motion (Refs. 1, 2, 3).

10.2.1 Wave Speed

Consider the situation illustrated in Fig. 10.2*a* in which a single elementary wave of small height, δy, is produced on the surface of a channel by suddenly moving the initially stationary end wall with speed δV. The water in the channel was stationary at the initial time, $t = 0$. A stationary observer will observe a single wave move down the channel with a *wave speed c*, with no fluid motion ahead of the wave and a fluid velocity of δV behind the wave. The motion is unsteady for such an observer. For an observer moving along the channel with speed c, the flow will appear steady as shown in Fig. 10.2*b*. To this observer, the fluid velocity will be $\mathbf{V} = -c\hat{\mathbf{i}}$ on the observer's right and $\mathbf{V} = (-c + \delta V)\hat{\mathbf{i}}$ to the left of the observer.

The relationship between the various parameters involved for this flow can be obtained by application of the continuity and momentum equations to the control volume shown in Fig. 10.2*b* as follows. With the assumption of uniform one-dimensional flow, the continuity equation (Eq. 5.12) becomes

$$-cyb = (-c + \delta V)(y + \delta y)b$$

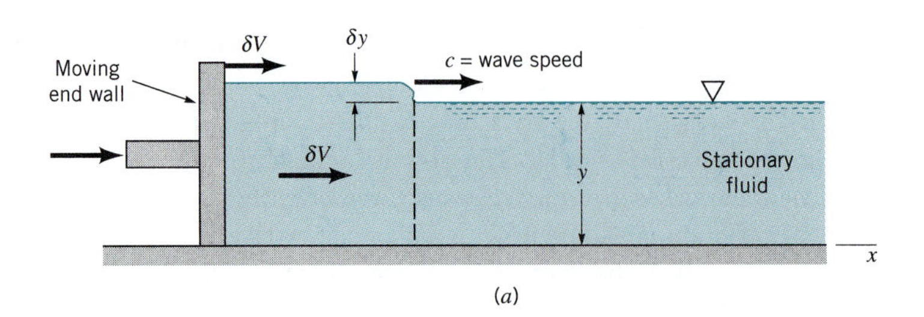

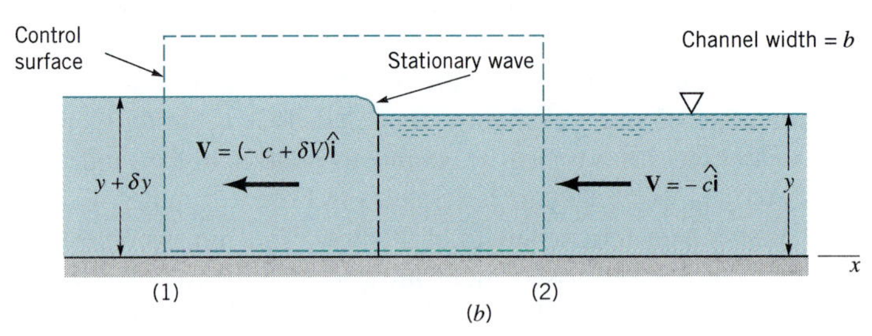

■ **FIGURE 10.2** **(*a*) Production of a single elementary wave in a channel as seen by a stationary observer. (*b*) Wave as seen by an observer moving with a speed equal to the wave speed.**

where b is the channel width. This simplifies to

$$c = \frac{(y + \delta y)\,\delta V}{\delta y}$$

or in the limit of small amplitude waves with $\delta y \ll y$

$$c = y\,\frac{\delta V}{\delta y} \tag{10.1}$$

Similarly, the momentum equation (Eq. 5.22) is

$$\tfrac{1}{2}\gamma y^2 b - \tfrac{1}{2}\gamma (y + \delta y)^2 b = \rho b c y[(c - \delta V) - c]$$

where we have written the mass flowrate as $\dot{m} = \rho b c y$ and have assumed that the pressure variation is hydrostatic within the fluid. That is, the pressure forces on the channel cross sections (1) and (2) are $F_1 = \gamma y_{c1} A_1 = \gamma (y + \delta y)^2 b/2$ and $F_2 = \gamma y_{c2} A_2 = \gamma y^2 b/2$, respectively. If we again impose the assumption of small amplitude waves [that is, $(\delta y)^2 \ll y\,\delta y$], the momentum equation reduces to

The wave speed can be obtained from the continuity and momentum equations.

$$\frac{\delta V}{\delta y} = \frac{g}{c} \tag{10.2}$$

Combination of Eqs. 10.1 and 10.2 gives the wave speed as

$$c = \sqrt{gy} \tag{10.3}$$

The speed of a small amplitude solitary wave as is indicated in Fig. 10.2 is proportional to the square root of the fluid depth, y, and independent of the wave amplitude, δy. The fluid density is not an important parameter, although the acceleration of gravity is. This is a result of the fact that such wave motion is a balance between inertial effects (proportional to ρ) and weight or hydrostatic pressure effects (proportional to $\gamma = \rho g$). A ratio of these forces eliminates the common factor ρ but retains g.

The wave speed can also be calculated by using the energy and continuity equations rather than the momentum and continuity equations as is done above. A simple wave on the surface is shown in Fig. 10.3. As seen by an observer moving with the wave speed, c, the

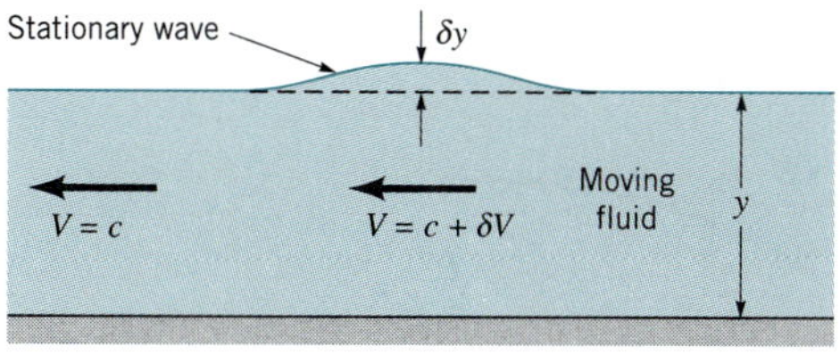

■ FIGURE 10.3 **Stationary simple wave in a flowing fluid.**

flow is steady. Since the pressure is constant at any point on the free surface, the Bernoulli equation for this frictionless flow is simply

$$\frac{V^2}{2g} + y = \text{constant}$$

or by differentiating

$$\frac{V\,\delta V}{g} + \delta y = 0$$

Also, by differentiating the continuity equation, $Vy = \text{constant}$, we obtain

$$y\,\delta V + V\,\delta y = 0$$

The wave speed can be obtained from the continuity and energy equations.

We combine these two equations to eliminate δV and δy and use the fact that $V = c$ for this situation (the observer moves with speed c) to obtain the wave speed given by Eq. 10.3. The above results are restricted to waves of small amplitude because we have assumed one-dimensional flow. That is, $\delta y/y \ll 1$. More advanced analysis and experiments show that the wave speed for finite-sized solitary waves exceeds that given by Eq. 10.3. To a first approximation, one obtains (Ref. 4)

$$c \approx \sqrt{gy}\left(1 + \frac{\delta y}{y}\right)^{1/2}$$

The larger the amplitude, the faster the wave travels.

A more general description of wave motion can be obtained by considering continuous (not solitary) waves of sinusoidal shape as is shown in Fig. 10.4. By combining waves of various wavelengths, λ, and amplitudes, δy, it is possible to describe very complex surface patterns found in nature, such as the wind-driven waves on a lake. Mathematically, such a process consists of using a Fourier series (each term of the series represented by a wave of different wavelength and amplitude) to represent an arbitrary function (the free-surface shape).

A more advanced analysis of such sinusoidal surface waves of small amplitude shows that the wave speed varies with both the wavelength and fluid depth as (Ref. 1)

$$c = \left[\frac{g\lambda}{2\pi}\tanh\left(\frac{2\pi y}{\lambda}\right)\right]^{1/2} \qquad \textbf{(10.4)}$$

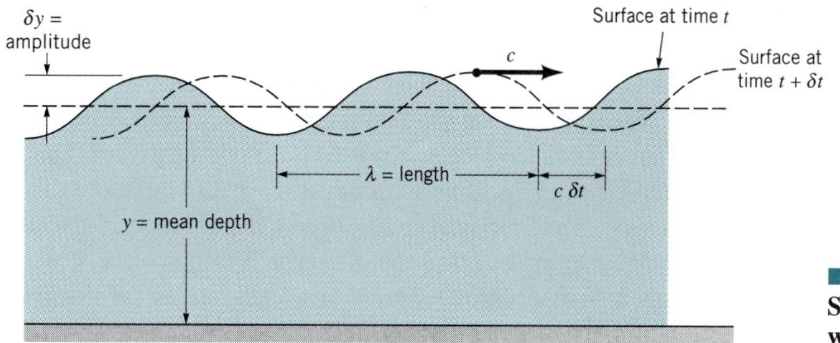

■ FIGURE 10.4 **Sinusoidal surface wave.**

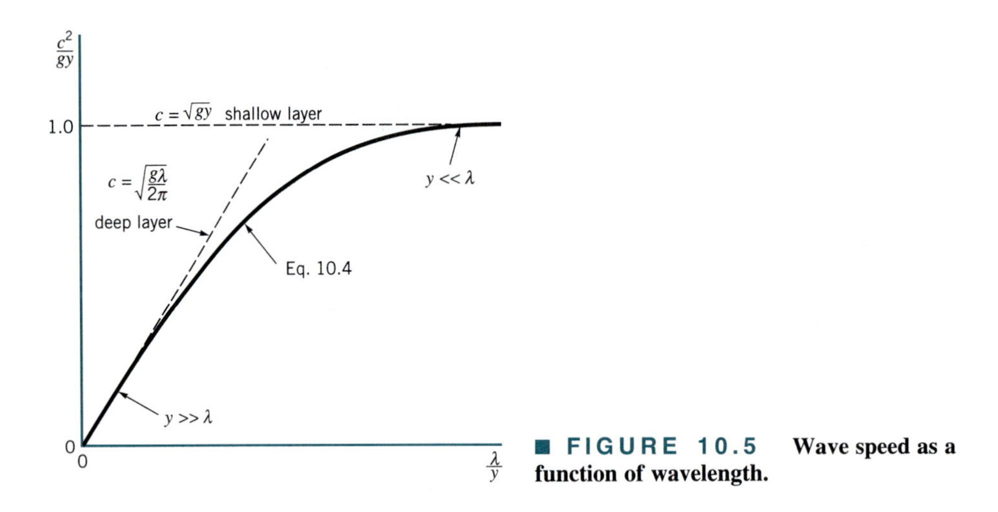

■ **FIGURE 10.5** **Wave speed as a function of wavelength.**

where $\tanh(2\pi y/\lambda)$ is the hyperbolic tangent of the argument $2\pi y/\lambda$. This result is plotted in Fig. 10.5. For conditions for which the water depth is much greater than the wavelength ($y \gg \lambda$, as in the ocean), the wave speed is independent of y and given by

$$c = \sqrt{\frac{g\lambda}{2\pi}}$$

This result follows from Eq. 10.4, since $\tanh(2\pi y/\lambda) \to 1$ as $y/\lambda \to \infty$. On the other hand, if the fluid layer is shallow ($y \ll \lambda$, as often happens in open channels), the wave speed is given by $c = (gy)^{1/2}$, as derived for the solitary wave in Fig. 10.2. This result also follows from Eq. 10.4, since $\tanh(2\pi y/\lambda) \to 2\pi y/\lambda$ as $y/\lambda \to 0$. These two limiting cases are shown in Fig. 10.5. For moderate depth layers ($y \sim \lambda$), the results are given by the complete Eq. 10.4. Note that for a given fluid depth, the long wave travels the fastest. Hence, for our purposes we will consider the wave speed to be this limiting situation, $c = (gy)^{1/2}$.

10.2.2 Froude Number Effects

The wave speed is measured relative to the flowing fluid, not the fixed ground.

Consider an elementary wave traveling on the surface of a fluid, as is shown in Fig. 10.2*a*. If the fluid layer is stationary, the wave moves to the right with speed c relative to the fluid and the stationary observer. If the fluid is flowing to the left with velocity $V < c$, the wave (which travels with speed c relative to the fluid) will travel to the right with a speed of $c - V$ relative to a fixed observer. If the fluid flows to the left with $V = c$, the wave will remain stationary, but if $V > c$ the wave will be washed to the left with speed $V - c$.

The above ideas can be expressed in dimensionless form by use of the Froude number, $\text{Fr} = V/(gy)^{1/2}$, where we take the characteristic length to be the fluid depth, y. Thus, the Froude number, $\text{Fr} = V/(gy)^{1/2} = V/c$, is the ratio of the fluid velocity to the wave speed.

The following characteristics are observed when a wave is produced on the surface of a moving stream, as happens when a rock is thrown into a river. If the stream is not flowing, the wave spreads equally in all directions. If the stream is nearly stationary or moving in a tranquil manner (i.e., $V < c$), the wave can move upstream. Upstream locations are said to be in hydraulic communication with the downstream locations. That is, an observer upstream of a disturbance can tell that there has been a disturbance on the surface because that disturbance can propagate upstream to the observer. Viscous effects, which have been neglected in this discussion, will eventually damp out such waves far upstream. Such flow conditions, $V < c$, or $\text{Fr} < 1$, are termed subcritical.

On the other hand, if the stream is moving rapidly so that the flow velocity is greater than the wave speed (i.e., $V > c$), no upstream communication with downstream locations is possible. Any disturbance on the surface downstream from the observer will be washed farther downstream. Such conditions, $V > c$ or $\text{Fr} > 1$, are termed supercritical. For the special case of $V = c$ or $\text{Fr} = 1$, the upstream propagating wave remains stationary and the flow is termed critical.

V10.1

Subcritical flows may behave differently than supercritical flows.

The character of an open-channel flow may depend strongly on whether the flow is subcritical or supercritical. The characteristics of the flow may be completely opposite for subcritical flow than for supercritical flow. For example, as is discussed in Section 10.3, a "bump" on the bottom of a river (such as a submerged log) may cause the surface of the river to dip below the level it would have had if the log were not there, or it may cause the surface level to rise above its undisturbed level. Which situation will happen depends on the value of Fr. Similarly, for supercritical flows it is possible to produce steplike discontinuities in the fluid depth (called a hydraulic jump; see Section 10.6.1). For subcritical flows, however, changes in depth must be smooth and continuous. Certain open-channel flows, such as the broad-crested weir (Section 10.6.3), depend on the existence of critical flow conditions for their operation.

As strange as it may seem, there exist many similarities between the open-channel flow of a liquid and the compressible flow of a gas. The governing dimensionless parameter in each case is the fluid velocity, V, divided by a wave speed, the surface wave speed for open-channel flow or sound wave speed for compressible flow. Many of the differences between subcritical ($\text{Fr} < 1$) and supercritical ($\text{Fr} > 1$) open-channel flows have analogs in subsonic ($\text{Ma} < 1$) and supersonic ($\text{Ma} > 1$) compressible gas flow, where Ma is the Mach number. Some of these similarities are discussed in this chapter and in Chapter 11.

10.3 Energy Considerations

A typical segment of an open-channel flow is shown in Fig. 10.6. The slope of the channel bottom (or *bottom slope*), $S_0 = (z_1 - z_2)/\ell$, is assumed constant over the segment shown. The fluid depths and velocities are y_1, y_2, V_1, and V_2 as indicated. Note that the fluid depth is measured in the vertical direction and the distance x is horizontal. For most open-channel flows the value of S_0 is very small (the bottom is nearly horizontal). For example, the Mississippi River drops a distance of 1470 ft in its 2350-mi length to give an average value of $S_0 = 0.000118$. In such circumstances the values of x and y are often taken as the distance along the channel bottom and the depth normal to the bottom, with negligibly small differences introduced by the two coordinate schemes.

With the assumption of a uniform velocity profile across any section of the channel, the one-dimensional energy equation for this flow (Eq. 5.84) becomes

$$\frac{p_1}{\gamma} + \frac{V_1^2}{2g} + z_1 = \frac{p_2}{\gamma} + \frac{V_2^2}{2g} + z_2 + h_L \tag{10.5}$$

where h_L is the head loss due to viscous effects between sections (1) and (2) and $z_1 - z_2 = S_0\ell$. Since the pressure is essentially hydrostatic at any cross section, we find that $p_1/\gamma = y_1$ and $p_2/\gamma = y_2$ so that Eq. 10.5 becomes

$$y_1 + \frac{V_1^2}{2g} + S_0\ell = y_2 + \frac{V_2^2}{2g} + h_L \tag{10.6}$$

One of the difficulties of analyzing open-channel flow, similar to that discussed in Chapter 8 for pipe flow, is associated with the determination of the head loss in terms of other physical parameters. Without getting into such details at present, we write the head loss in terms of

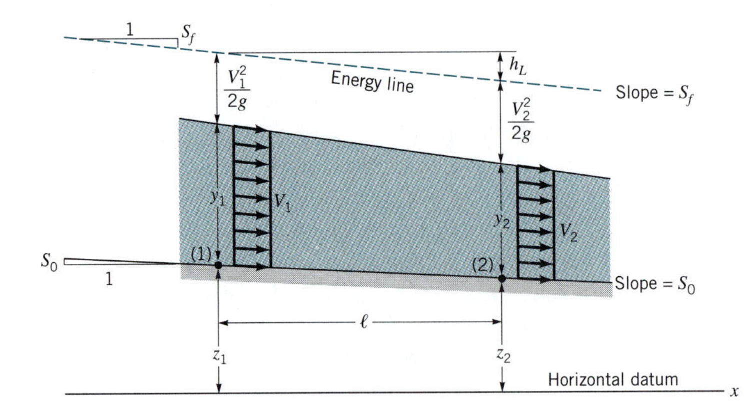

■ **FIGURE 10.6** **Typical open-channel geometry.**

the slope of the energy line, $S_f = h_L/\ell$ (often termed the *friction slope*), as is indicated in Fig. 10.6. Recall from Chapter 3 that the energy line is located a distance z (the elevation from some datum to the channel bottom) plus the pressure head (p/γ) plus the velocity head $(V^2/2g)$ above the datum. Therefore, Eq. 10.6 can be written as

$$y_1 - y_2 = \frac{(V_2^2 - V_1^2)}{2g} + (S_f - S_0)\ell \tag{10.7}$$

If there is no head loss, the energy line is horizontal ($S_f = 0$), and the total energy of the flow is free to shift between kinetic energy and potential energy in a conservative fashion. In the specific instance of a horizontal channel bottom ($S_0 = 0$) and negligible head loss ($S_f = 0$), Eq. 10.7 simply becomes

$$y_1 - y_2 = \frac{(V_2^2 - V_1^2)}{2g}$$

10.3.1 Specific Energy

The specific energy is the sum of potential energy and kinetic energy (per unit weight).

The concept of the *specific energy*, E, defined as

$$E = y + \frac{V^2}{2g} \tag{10.8}$$

is often useful in open-channel flow considerations. The energy equation, Eq. 10.7, can be written in terms of E as

$$E_1 = E_2 + (S_f - S_0)\ell \tag{10.9}$$

If head losses are negligible, then $S_f = 0$ so that $(S_f - S_0)\ell = -S_0\ell = z_2 - z_1$ and the sum of the specific energy and the elevation of the channel bottom remains constant (i.e., $E_1 + z_1 = E_2 + z_2$, a statement of the Bernoulli equation).

If we consider a simple channel whose cross-sectional shape is a rectangle of width b, the specific energy can be written in terms of the flowrate per unit width, $q = Q/b = Vyb/b = Vy$, as

$$E = y + \frac{q^2}{2gy^2} \tag{10.10}$$

For a given channel of constant width, the value of q remains constant along the channel, although the depth, y, may vary. To gain insight into the flow processes involved, we consider the *specific energy diagram*, a graph of $E = E(y)$ with q fixed, as shown in Fig. 10.7.

For given q and E, Eq. 10.10 is a cubic equation with three solutions, y_{sup}, y_{sub}, and

y_{neg}. If the specific energy is large enough (i.e., $E > E_{min}$, where E_{min} is a function of q), two of the solutions are positive and the other, y_{neg}, is negative. The negative root, represented by the curved dashed line in Fig. 10.7, has no physical meaning and can be ignored. Thus, for a given flowrate and specific energy there are two possible depths, unless the vertical line from the E axis does not intersect the specific energy curve corresponding to the value of q given (i.e., $E < E_{min}$). These two depths are termed *alternate depths.*

For a given value of specific energy, a flow may have alternate depths.

For large values of E the upper and lower branches of the specific energy diagram (y_{sub} and y_{sup}) approach $y = E$ and $y = 0$, respectively. These limits correspond to a very deep channel flowing very slowly ($E = y + V^2/2g \to y$ as $y \to \infty$ with $q = Vy$ fixed), or a very high-speed flow in a shallow channel ($E = y + V^2/2g \to V^2/2g$ as $y \to 0$).

As is indicated in Fig. 10.7, $y_{sup} < y_{sub}$. Thus, since $q = Vy$ is constant along the curve, it follows that $V_{sup} > V_{sub}$, where the subscripts "sub" and "sup" on the velocities correspond to the depths so labeled. The specific energy diagram consists of two portions divided by the E_{min} "nose" of the curve. We will show that the flow conditions at this location correspond to critical conditions (Fr = 1), those on the upper portion of the curve correspond to subcritical conditions (hence, the "sub" subscript), and those on the lower portion of the curve correspond to supercritical conditions (hence, the "sup" subscript).

To determine the value of E_{min}, we use Eq. 10.10 and set $dE/dy = 0$ to obtain

$$\frac{dE}{dy} = 1 - \frac{q^2}{gy^3} = 0$$

or

$$y_c = \left(\frac{q^2}{g}\right)^{1/3} \tag{10.11}$$

where the subscript "c" denotes conditions at E_{min}. By substituting this back into Eq. 10.10 we obtain

$$E_{min} = \frac{3y_c}{2}$$

By combining Eq. 10.11 and $V_c = q/y_c$ we obtain

$$V_c = \frac{q}{y_c} = \frac{(y_c^{3/2}\, g^{1/2})}{y_c} = \sqrt{gy_c}$$

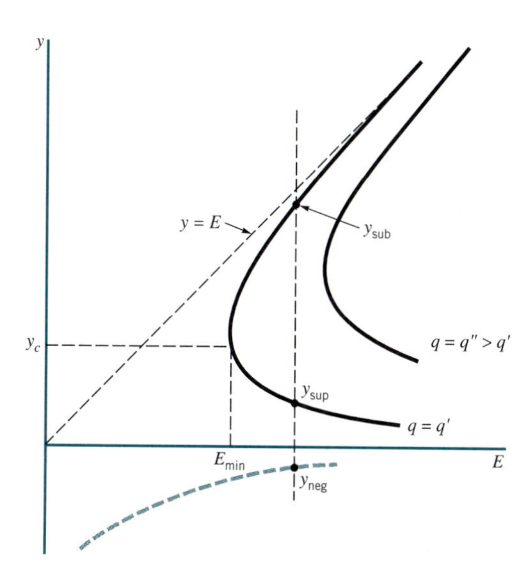

■ **FIGURE 10.7 Specific energy diagram.**

or $\text{Fr}_c \equiv V_c/(gy_c)^{1/2} = 1$. Thus, critical conditions (Fr $= 1$) occur at the location of E_{min}. Since the layer is deeper and the velocity smaller for the upper part of the specific energy diagram (compared with the conditions at E_{min}), such flows are subcritical (Fr < 1). Conversely, flows for the lower part of the diagram are supercritical. Thus, for a given flowrate, q, if $E > E_{\text{min}}$ there are two possible depths of flow, one subcritical and the other supercritical.

EXAMPLE 10.1

Water flows under the sluice gate in a constant width rectangular channel as shown in Fig. E10.1*a*. Describe this flow in terms of the specific energy diagram. Assume inviscid flow.

SOLUTION

For this inviscid ($S_f = 0$) flow the channel bottom is horizontal, $z_1 = z_2$ (or $S_0 = 0$), so that the energy equation (Eq. 10.9) reduces to $E_1 = E_2$. Although the flowrate is not given, we do know that $q_1 = q_2$ and that the specific energy diagram for this flow is as shown in Fig. E10.1*b*. The flow upstream of the gate is subcritical while the flow downstream is supercritical. The particular value of q obtained, $q = q_0$, and hence the specific curve of Fig. E10.1*b* which is valid for this flow, are illustrated by those that give $E_1 = E_2$ for the given y_1 and y_2.

The flowrate can remain the same for this channel even if the upstream depth is increased. This is indicated by depths $y_{1'}$ and $y_{2'}$ in Fig. E10.1*c*. Of course, to do this the distance between the bottom of the gate and the channel bottom must be decreased to give a smaller flow area ($y_{2'} < y_2$), and the upstream depth must be increased to give a bigger head ($y_{1'} > y_1$). On the other hand, if the gate remains fixed so that the downstream depth remains fixed ($y_{2''} = y_2$), the flowrate will increase as the upstream depth increases to $y_{1''} > y_1$. This is indicated in Fig. E10.1*c* by the curve with flowrate $q'' > q_0$.

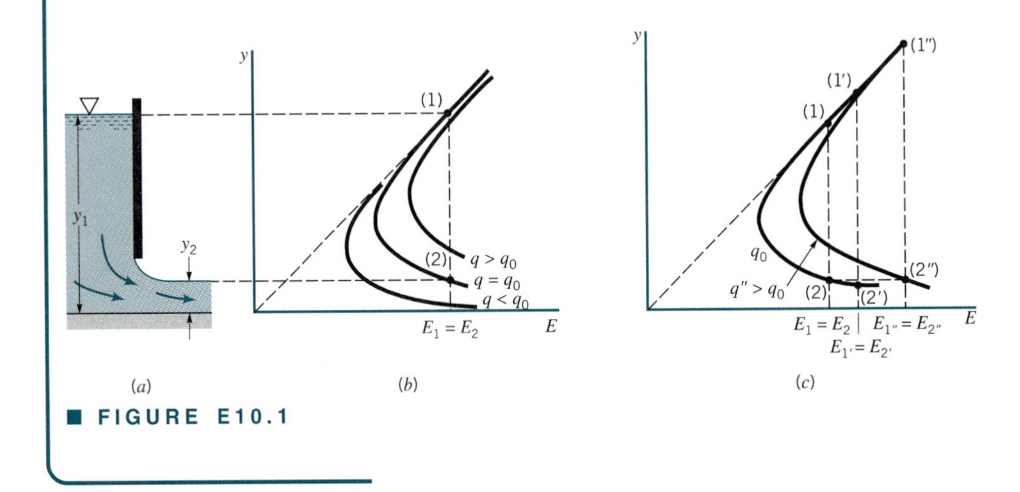

■ FIGURE E10.1

It is often possible to determine various characteristics of a flow by considering the specific energy diagram. Example 10.2 illustrates this for a situation in which the channel bottom elevation is not constant.

EXAMPLE 10.2

Water flows up a 0.5-ft-tall ramp in a constant width rectangular channel at a rate $q = 5.75$ ft^2/s as is shown in Fig. E10.2*a*. (For now disregard the "bump.") If the upstream depth is 2.3 ft, determine the elevation of the water surface downstream of the ramp, $y_2 + z_2$. Neglect viscous effects.

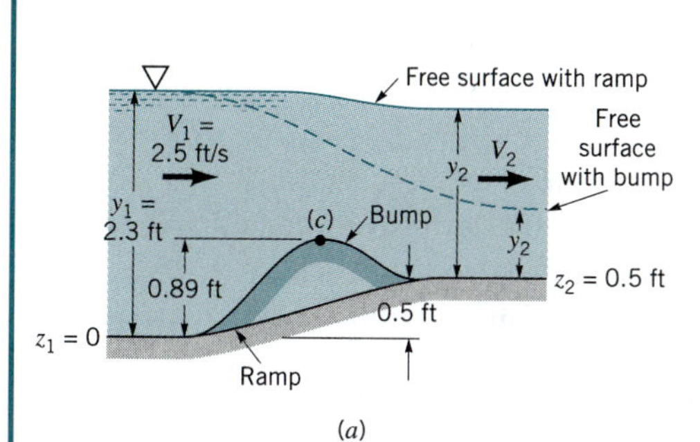

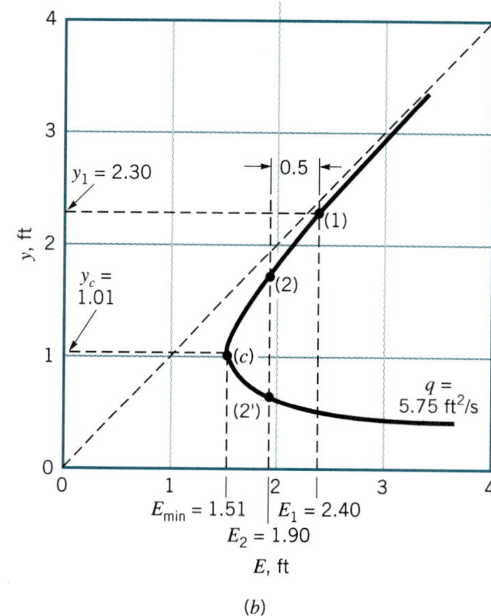

■ FIGURE E10.2

SOLUTION

With $S_0\ell = z_1 - z_2$ and $h_L = 0$, conservation of energy (Eq. 10.6 which, under these conditions, is actually the Bernoulli equation) requires that

$$y_1 + \frac{V_1^2}{2g} + z_1 = y_2 + \frac{V_2^2}{2g} + z_2$$

For the conditions given ($z_1 = 0$, $z_2 = 0.5$ ft, $y_1 = 2.3$ ft, and $V_1 = q/y_1 = 2.5$ ft/s), this becomes

$$1.90 = y_2 + \frac{V_2^2}{64.4} \quad \text{(1)}$$

where V_2 and y_2 are in ft/s and feet, respectively. The continuity equation provides the second equation

$$y_2V_2 = y_1V_1$$

or

$$y_2V_2 = 5.75 \text{ ft}^2/\text{s} \quad \text{(2)}$$

Equations 1 and 2 can be combined to give

$$y_2^3 - 1.90y_2^2 + 0.513 = 0$$

which has solutions

$$y_2 = 1.72 \text{ ft}, \qquad y_2 = 0.638 \text{ ft}, \quad \text{or} \quad y_2 = -0.466 \text{ ft}$$

Note that two of these solutions are physically realistic, but the negative solution is meaningless. This is consistent with the previous discussions concerning the specific energy (recall the three roots indicated in Fig. 10.7). The corresponding elevations of the free surface are either

$$y_2 + z_2 = 1.72 \text{ ft} + 0.50 \text{ ft} = 2.22 \text{ ft}$$

or

$$y_2 + z_2 = 0.638 \text{ ft} + 0.50 \text{ ft} = 1.14 \text{ ft}$$

The question is which of these two flows is to be expected? This can be answered by use of the specific energy diagram obtained from Eq. 10.10, which for this problem is

$$E = y + \frac{0.513}{y^2}$$

where E and y are in feet. The diagram is shown in Fig. E10.2*b*. The upstream condition corresponds to subcritical flow; the downstream condition is either subcritical or supercritical, corresponding to points 2 or 2′. Note that since $E_1 = E_2 + (z_2 - z_1) = E_2 + 0.5$ ft, it follows that the downstream conditions are located 0.5 ft to the left of the upstream conditions on the diagram.

With a constant-width channel, the value of q remains the same for any location along the channel. That is, all points for the flow from (1) to (2) or (2′) must lie along the $q = 5.75 \text{ ft}^2/\text{s}$ curve shown. Any deviation from this curve would imply either a change in q or a relaxation of the one-dimensional flow assumption. To stay on the curve and go from (1) around the critical point (point c) to point (2′) would require a reduction in specific energy to E_{min}. As is seen from Fig. E10.2*a*, this would require a specified elevation (bump) in the channel bottom so that critical conditions would occur above this bump. In particular, since $E_1 = y_1 + 0.513/y_1^2 = 2.40$ ft and $E_{\text{min}} = 3y_c/2 = 3(q^2/g)^{1/3}/2 = 1.51$ ft, the top of this bump would need to be $z_c - z_1 = E_1 - E_{\text{min}} = 2.40 \text{ ft} - 1.51 \text{ ft} = 0.89$ ft above the channel bottom at section (1). The flow could then accelerate to supercritical conditions ($\text{Fr}_{2'} > 1$) as is shown by the free surface represented by the dashed line in Fig. E10.2*a*.

Since the actual elevation change (a ramp) shown in Fig. E10.2*a* does not contain a bump, the downstream conditions will correspond to the subcritical flow denoted by (2), not the supercritical condition (2′). Without a bump on the channel bottom, the state (2′) is

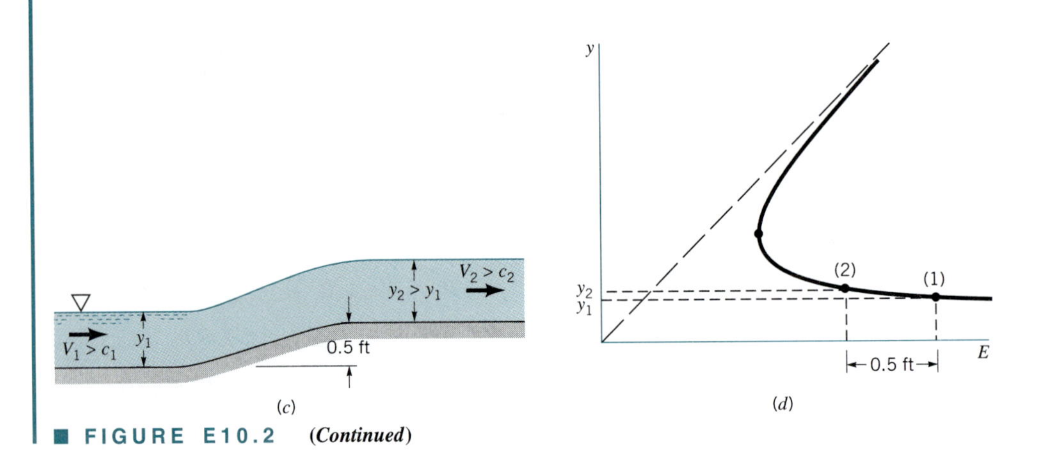

■ FIGURE E10.2 (*Continued*)

inaccessible from the upstream condition state (1). Such considerations are often termed the *accessibility of flow regimes*. Thus, the surface elevation is

$$y_2 + z_2 = 2.22 \text{ ft} \qquad \textbf{(Ans)}$$

Note that since $y_1 + z_1 = 2.30$ ft and $y_2 + z_2 = 2.22$ ft, the elevation of the free surface decreases as it goes across the ramp.

If the flow conditions upstream of the ramp were supercritical, the free-surface elevation and fluid depth would increase as the fluid flows up the ramp. This is indicated in Fig. E10.2*c* along with the corresponding specific energy diagram, as is shown in Fig. E10.2*d*. For this case the flow starts at (1) on the lower (supercritical) branch of the specific energy curve and ends at (2) on the same branch with $y_2 > y_1$. Since both y and z increase from (1) to (2), the surface elevation, $y + z$, also increases. Thus, flow up a ramp is different for subcritical than it is for supercritical conditions.

10.3.2 Channel Depth Variations

By using the concepts of the specific energy and critical flow conditions (Fr = 1), it is possible to determine how the depth of a flow in an open channel changes with distances along the channel. In some situations the depth change is very rapid so that the value of dy/dx is of the order one. Complex effects involving two- or three-dimensional flow phenomena are often involved in such flows.

In this section we consider only gradually varying flows. For such flows, $dy/dx \ll 1$ and it is reasonable to impose the one-dimensional velocity assumption. At any section the total head is $H = V^2/2g + y + z$ and the energy equation (Eq. 10.5) becomes

$$H_1 = H_2 + h_L$$

where h_L is the head loss between sections (1) and (2).

As is discussed in the previous section, the slope of the energy line is $dH/dx = dh_L/dx = S_f$ and the slope of the channel bottom is $dz/dx = S_0$. Thus, since

$$\frac{dH}{dx} = \frac{d}{dx}\left(\frac{V^2}{2g} + y + z\right) = \frac{V}{g}\frac{dV}{dx} + \frac{dy}{dx} + \frac{dz}{dx}$$

we obtain

The total head is not constant because the head loss is not zero.

$$\frac{dh_L}{dx} = \frac{V}{g}\frac{dV}{dx} + \frac{dy}{dx} + S_0$$

or

$$\frac{V}{g}\frac{dV}{dx} + \frac{dy}{dx} = S_f - S_0 \qquad \textbf{(10.12)}$$

For a given flowrate per unit width, q, in a rectangular channel of constant width b, we have $V = q/y$ or by differentiation

$$\frac{dV}{dx} = -\frac{q}{y^2}\frac{dy}{dx} = -\frac{V}{y}\frac{dy}{dx}$$

so that the kinetic energy term in Eq. 10.12 becomes

$$\frac{V}{g}\frac{dV}{dx} = -\frac{V^2}{gy}\frac{dy}{dx} = -\text{Fr}^2\frac{dy}{dx} \quad \textbf{(10.13)}$$

where $\text{Fr} = V/(gy)^{1/2}$ is the local Froude number of the flow. Substituting Eq. 10.13 into Eq. 10.12 and simplifying gives

$$\frac{dy}{dx} = \frac{(S_f - S_0)}{(1 - \text{Fr}^2)} \quad \textbf{(10.14)}$$

It is seen that the rate of change of fluid depth, dy/dx, depends on the local slope of the channel bottom, the slope of the energy line, and the Froude number. The value of dy/dx can be either negative, zero, or positive, depending on the values of these three parameters. That is, the channel flow depth may be constant or it may increase or decrease in the flow direction, depending on the values of S_0, S_f, and Fr. The behavior of subcritical flow may be the opposite of that for supercritical flow, as seen by the denominator, $1 - \text{Fr}^2$, of Eq. 10.14.

Although in the derivation of Eq. 10.14 we assumed q is constant (i.e., a rectangular channel), Eq. 10.14 is valid for channels of any constant cross-sectional shape, provided the Froude number is interpreted properly (Ref. 3). In this book we will consider only rectangular cross-section channels when using this equation.

10.4 Uniform Depth Channel Flow

V10.2

For uniform depth flow, the loss in potential energy equals the energy dissipated by friction.

Many channels are designed to carry fluid at a uniform depth all along their length. Irrigation canals are frequently of uniform depth and cross section for considerable lengths. Natural channels such as rivers and creeks are seldom of uniform shape, although a reasonable approximation to the flowrate in such channels can often be obtained by assuming uniform flow. In this section we will discuss various aspects of such flows.

Uniform depth flow ($dy/dx = 0$) can be accomplished by adjusting the bottom slope, S_0, so that it precisely equals the slope of the energy line, S_f. That is, $S_0 = S_f$. This can be seen from Eq. 10.14. From an energy point of view, uniform depth flow is achieved by a balance between the potential energy lost by the fluid as it coasts downhill and the energy that is dissipated by viscous effects (head loss) associated with shear stresses throughout the fluid. Similar conclusions can be reached from a force balance analysis as discussed in the following section.

10.4.1 Uniform Flow Approximations

We consider fluid flowing in an open channel of constant cross-sectional size and shape such that the depth of flow remains constant as is indicated in Fig. 10.8. The area of the section is A and the *wetted perimeter* (i.e., the length of the perimeter of the cross section in contact with the fluid) is P. The interaction between the fluid and the atmosphere at the free surface is assumed negligible so that this portion of the perimeter is not included in the definition of the wetted perimeter.

Since the fluid must adhere to the solid surfaces, the actual velocity distribution in an open channel is not uniform. Some typical velocity profiles measured in channels of various

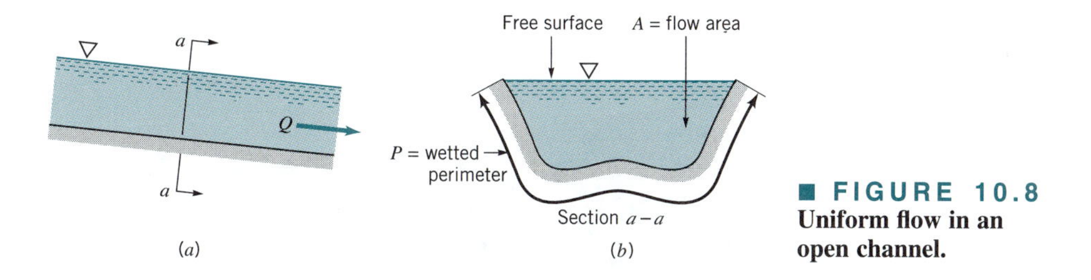

■ **FIGURE 10.8 Uniform flow in an open channel.**

shapes are indicated in Fig. 10.9*a*. The maximum velocity is often found somewhat below the free surface, and the fluid velocity is zero on the wetted perimeter, where a wall shear stress, τ_w, is developed. This shear stress is seldom uniform along the wetted perimeter, with typical variations as are indicated in Fig. 10.9*b*.

The wall shear stress acts on the wetted perimeter of the channel.

Fortunately, reasonable analytical results can be obtained by assuming a uniform velocity profile, V, and a constant wall shear stress, τ_w. Similar assumptions were made for pipe flow situations (Chapter 8), with the friction factor being used to obtain the head loss.

10.4.2 The Chezy and Manning Equations

The basic equations used to determine the uniform flowrate in open channels were derived many years ago. Continual refinements have taken place to obtain better values of the empirical coefficients involved. The result is a semiempirical equation that provides reasonable engineering results. A more refined analysis is perhaps not warranted because of the complexity and uncertainty of the flow geometry (i.e., channel shape and the irregular makeup of the wetted perimeter, particularly for natural channels).

Under the assumptions of steady uniform flow, the x component of the momentum

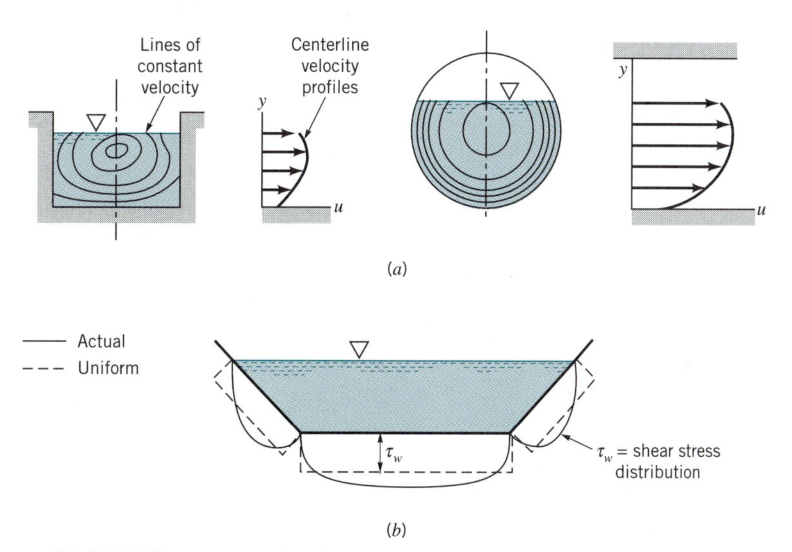

■ **FIGURE 10.9 Typical velocity and shear stress distributions in an open channel: (*a*) velocity distribution throughout the cross section, (*b*) shear stress distribution on the wetted perimeter.**

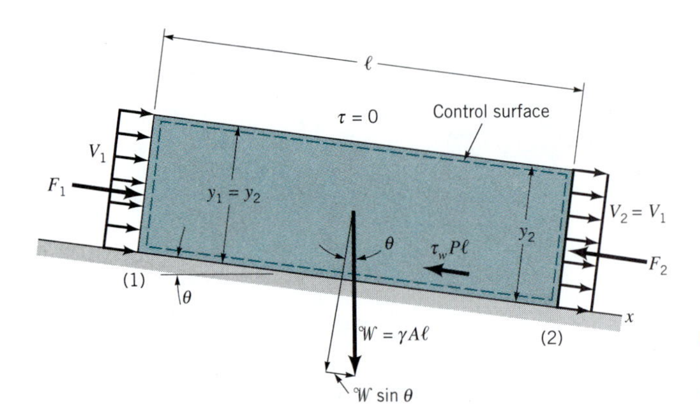

■ **FIGURE 10.10** **Control volume for uniform flow in an open channel.**

equation (Eq. 5.22) applied to the control volume indicated in Fig. 10.10 simply reduces to

$$\Sigma F_x = \rho Q(V_2 - V_1) = 0$$

since $V_1 = V_2$. There is no acceleration of the fluid, and the momentum flux across section (1) is the same as that across section (2). The flow is governed by a simple balance between the forces in the direction of the flow. Thus, $\Sigma F_x = 0$, or

$$F_1 - F_2 - \tau_w P\ell + \mathcal{W} \sin\theta = 0 \qquad \textbf{(10.15)}$$

For uniform depth, channel flow is governed by a balance between friction and weight.

where F_1 and F_2 are the hydrostatic pressure forces across either end of the control volume. Because the flow is at a uniform depth $(y_1 = y_2)$, it follows that $F_1 = F_2$ so that these two forces do not contribute to the force balance. The term $\mathcal{W} \sin\theta$ is the component of the fluid weight that acts down the slope, and $\tau_w P\ell$ is the shear force on the fluid, acting up the slope as a result of the interaction of the water and the channel's wetted perimeter. Thus, Eq. 10.15 becomes

$$\tau_w = \frac{\mathcal{W}\sin\theta}{P\ell} = \frac{\mathcal{W}S_0}{P\ell}$$

where we have used the approximation that $\sin\theta \approx \tan\theta = S_0$, since the bottom slope is typically very small (i.e., $S_0 \ll 1$). Since $\mathcal{W} = \gamma A\ell$ and the *hydraulic radius* is defined as $R_h = A/P$, the force balance equation becomes

$$\tau_w = \frac{\gamma A\ell S_0}{P\ell} = \gamma R_h S_0 \qquad \textbf{(10.16)}$$

Most open-channel flows are turbulent rather than laminar. In fact, typical Reynolds numbers are quite large, well above the transitional value and into the wholly turbulent regime. As was discussed in Chapter 8, for very large Reynolds number pipe flows (wholly turbulent flows), the friction factor, f, is found to be independent of Reynolds number, dependent only on the relative roughness of the pipe surface. For such cases, the wall shear stress is proportional to the dynamic pressure, $\rho V^2/2$, and independent of the viscosity. That is,

$$\tau_w = K\rho\frac{V^2}{2}$$

where K is a constant dependent upon the roughness of the pipe.

It is not unreasonable that similar shear stress dependencies occur for the large Reynolds number open-channel flows. In such situations, Eq. 10.16 becomes

$$K\rho\frac{V^2}{2} = \gamma R_h S_0$$

or

$$V = C\sqrt{R_h S_0} \qquad \textbf{(10.17)}$$

where the constant C is termed the Chezy coefficient and Eq. 10.17 is termed the *Chezy equation*. This equation, one of the oldest in the area of fluid mechanics, was developed in 1768 by A. Chezy (1718–1798), a French engineer who designed a canal for the Paris water supply. The value of the Chezy coefficient, which must be determined by experiments, is not dimensionless but has the dimensions of $(\text{length})^{1/2}$ per time (i.e., the square root of the units of acceleration).

From a series of experiments it was found that the slope dependence of Eq. 10.17 ($V \sim S_0^{1/2}$) is reasonable, but that the dependence on the hydraulic radius is actually not as given. In 1889, R. Manning (1816–1897), an Irish engineer, developed the following somewhat modified equation for open-channel flow to more accurately describe the R_h dependence:

$$V = \frac{R_h^{2/3} S_0^{1/2}}{n} \qquad \textbf{(10.18)}$$

The Manning equation is used to obtain the velocity or flowrate in an open channel.

Equation 10.18 is termed the *Manning equation*, and the parameter n is the *Manning resistance coefficient*. Its value is dependent on the surface material of the channel's wetted perimeter and is obtained from experiments. It is not dimensionless, having the units of $\text{s/m}^{1/3}$ or $\text{s/ft}^{1/3}$.

As is discussed in Chapter 7, any correlation should be expressed in dimensionless form, with the coefficients that appear being dimensionless coefficients, such as the friction factor for pipe flow or the drag coefficient for flow past objects. Thus, Eq. 10.18 should be expressed in dimensionless form. Unfortunately, the Manning equation is so widely used and has been used for so long that it will continue to be used in its dimensional form with a coefficient, n, that is not dimensionless. The values of n found in the literature (such as Table 10.1) were developed for SI units. Standard practice is to use the same value of n when using the BG system of units, and to insert a conversion factor into the equation.

Thus, uniform flow in an open channel is obtained from the Manning equation written as

$$\boxed{V = \frac{\kappa}{n} R_h^{2/3} S_0^{1/2}} \qquad \textbf{(10.19)}$$

and

$$\boxed{Q = \frac{\kappa}{n} A R_h^{2/3} S_0^{1/2}} \qquad \textbf{(10.20)}$$

where $\kappa = 1$ if SI units are used, and $\kappa = 1.49$ if BG units are used. The value 1.49 is the cube root of the number of feet per meter: $(3.281 \text{ ft/m})^{1/3} = 1.49$. Thus, by using R_h in meters, A in m^2, and $\kappa = 1$, the average velocity is m/s and the flowrate m^3/s. By using R_h in feet, A in ft^2, and $\kappa = 1.49$, the average velocity is ft/s and the flowrate ft^3/s.

Typical values of the Manning coefficient are indicated in Table 10.1. As expected, the rougher the wetted perimeter, the larger the value of n. For example, the roughness of floodplain surfaces increases from pasture to brush to tree conditions. So does the corresponding value of the Manning coefficient. Precise values of n are often difficult to obtain. Except for artificially lined channel surfaces like those found in new canals or flumes, the channel surface structure may be quite complex and variable. These are various methods used to obtain a reasonable estimate of the value of n for a given situation (Ref. 5). For the purpose of this

■ **TABLE 10.1**
Values of the Manning Coefficient, n (Ref. 6)

Wetted Perimeter	n	Wetted Perimeter	n
A. *Natural channels*		D. *Artificially lined channels*	
Clean and straight	0.030	Glass	0.010
Sluggish with deep pools	0.040	Brass	0.011
Major rivers	0.035	Steel, smooth	0.012
		Steel, painted	0.014
B. *Floodplains*		Steel, riveted	0.015
Pasture, farmland	0.035	Cast iron	0.013
Light brush	0.050	Concrete, finished	0.012
Heavy brush	0.075	Concrete, unfinished	0.014
Trees	0.15	Planed wood	0.012
		Clay tile	0.014
C. *Excavated earth channels*		Brickwork	0.015
Clean	0.022	Asphalt	0.016
Gravelly	0.025	Corrugated metal	0.022
Weedy	0.030	Rubble masonry	0.025
Stony, cobbles	0.035		

V10.3

book, the values from Table 10.1 are sufficient. Note that the error in Q is directly proportional to the error in n. A 10% error in the value of n produces a 10% error in the flowrate. Considerable effort has been put forth to obtain the best estimate of n, with extensive tables of values covering a wide variety of surfaces (Ref. 7). It should be noted that the values of n given in Table 10.1 are valid only for water as the flowing fluid.

The value of the Manning coefficient depends on the nature of the channel surface.

Both the friction factor for pipe flow and the Manning coefficient for channel flow are parameters that relate the wall shear stress to the makeup of the bounding surface. Thus, various results are available that describe n in terms of the equivalent pipe friction factor, f, and the surface roughness, ε (Ref. 8). For our purposes we will use the values of n from Table 10.1.

10.4.3 Uniform Depth Examples

A variety of interesting and useful results can be obtained from the Manning equation. The following examples illustrate some of the typical considerations.

The main parameters involved in uniform depth open-channel flow are the size and shape of the channel cross section (A, R_h), the slope of the channel bottom (S_0), the character of the material lining the channel bottom and walls (n), and the average velocity or flowrate (V or Q). Although the Manning equation is a rather simple equation, the ease of using it depends in part on which variables are given and which are to be determined.

Determination of the flowrate for a given channel with flow at a given depth (often termed the *normal flowrate* or *normal depth*, sometimes denoted y_n) is obtained from a straightforward calculation as is shown in Example 10.3.

Water flows in the canal of trapezoidal cross section shown in Fig. E10.3. The bottom drops 1.4 ft per 1000 ft of length. Determine the flowrate if the canal is lined with new smooth concrete, or if weeds cover the wetted perimeter. Determine the Froude number for each of these flows.

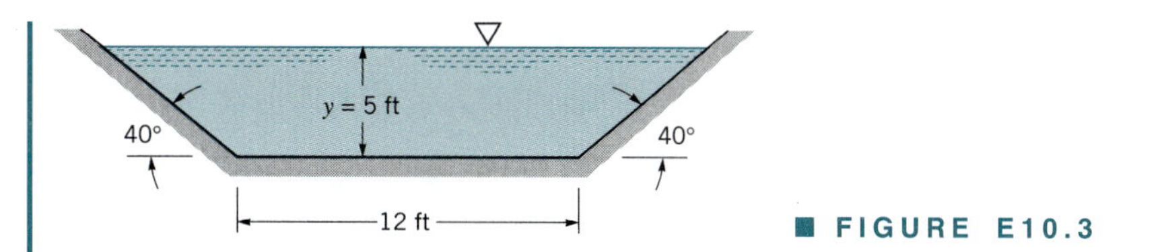

■ FIGURE E10.3

SOLUTION

From Eq. 10.20,

$$Q = \frac{1.49}{n} A R_h^{2/3} S_0^{1/2} \tag{1}$$

where we have used $\kappa = 1.49$, since the dimensions are given in BG units. For a depth of $y = 5$ ft, the flow area is

$$A = 12 \text{ ft } (5 \text{ ft}) + 5 \text{ ft} \left(\frac{5}{\tan 40°} \text{ ft} \right) = 89.8 \text{ ft}^2$$

so that with a wetted perimeter of $P = 12 \text{ ft} + 2(5/\sin 40° \text{ ft}) = 27.6$ ft, the hydraulic radius is determined to be $R_h = A/P = 3.25$ ft. Note that even though the channel is quite wide (the free-surface width is 23.9 ft), the hydraulic radius is only 3.25 ft, which is less than the depth.

Thus, with $S_0 = 1.4 \text{ ft}/1000 \text{ ft} = 0.0014$, Eq. 1 becomes

$$Q = \frac{1.49}{n} (89.8 \text{ ft}^2)(3.25 \text{ ft})^{2/3}(0.0014)^{1/2} = \frac{10.98}{n}$$

where Q is in ft^3/s.

From Table 10.1, the values of n are estimated to be $n = 0.012$ for the smooth concrete and $n = 0.030$ for the weedy conditions. Thus,

$$Q = \frac{10.98}{0.012} = 915 \text{ cfs} \qquad \textbf{(Ans)}$$

for the new concrete lining and

$$Q = \frac{10.98}{0.030} = 366 \text{ cfs} \qquad \textbf{(Ans)}$$

for the weedy lining. The corresponding average velocities, $V = Q/A$, are 10.2 ft/s and 4.08 ft/s, respectively. It does not take a very steep slope ($S_0 = 0.0014$ or $\theta = \tan^{-1}(0.0014) = 0.080°$) for this velocity.

Note that the increased roughness causes a decrease in the flowrate. This is an indication that for the turbulent flows involved, the wall shear stress increases with surface roughness. [For water at 50 °F, the Reynolds number based on the 3.25-ft hydraulic radius of the channel is $\text{Re} = R_h V/\nu = 3.25 \text{ ft } (4.08 \text{ ft/s})/(1.41 \times 10^{-5} \text{ ft}^2/\text{s}) = 9.40 \times 10^5$, well into the turbulent regime.]

The Froude numbers based on the maximum depths for the two flows can be determined from $\text{Fr} = V/(gy)^{1/2}$. For the new concrete case,

$$\mathrm{Fr} = \frac{10.2 \text{ ft/s}}{(32.2 \text{ ft/s}^2 \times 5 \text{ ft})^{1/2}} = 0.804 \qquad \textbf{(Ans)}$$

while for the weedy case

$$\mathrm{Fr} = \frac{4.08 \text{ ft/s}}{32.2 \text{ ft/s}^2 \times 5 \text{ ft})^{1/2}} = 0.322 \qquad \textbf{(Ans)}$$

In either case the flow is subcritical.

The same results would be obtained for the channel if its size were given in meters. We would use the same value of n but set $\kappa = 1$ for this SI units situation.

In some instances a trial-and-error or iteration method must be used to solve for the dependent variable. This is often encountered when the flowrate, channel slope, and channel material are given, and the flow depth is to be determined as illustrated in the following examples.

Water flows in the channel shown in Fig. E10.3 at a rate of $Q = 10.0 \text{ m}^3\text{/s}$. If the canal lining is weedy, determine the depth of the flow.

SOLUTION

In this instance neither the flow area nor the hydraulic radius are known, although they can be written in terms of the depth, y, as

$$A = 1.19y^2 + 3.66y$$

where the bottom width is $(12 \text{ ft})(1 \text{ m}/3.281 \text{ ft}) = 3.66 \text{ m}$ and A and y are in square meters and meters, respectively. Also, the wetted perimeter is

$$P = 3.66 + 2\left(\frac{y}{\sin 40°}\right) = 3.11y + 3.66$$

so that

$$R_h = \frac{A}{P} = \frac{1.19y^2 + 3.66y}{3.11y + 3.66}$$

where R_h and y are in meters. Thus, with $n = 0.030$ (from Table 10.1), Eq. 10.20 can be written as

$$\begin{aligned} Q = 10 &= \frac{\kappa}{n} A R_h^{2/3} S_0^{1/2} \\ &= \frac{1.0}{0.030}(1.19y^2 + 3.66y)\left(\frac{1.19y^2 + 3.66y}{3.11y + 3.66}\right)^{2/3}(0.0014)^{1/2} \end{aligned}$$

which can be rearranged into the form

$$(1.19y^2 + 3.66y)^5 - 515(3.11y + 3.66)^2 = 0 \quad \textbf{(1)}$$

where y is in meters. The solution of Eq. 1 can be easily obtained by use of a simple root-finding numerical technique or by trial-and-error methods. The only physically meaningful root of Eq. 1 (i.e., a positive, real number) gives the solution for the normal flow depth at this flowrate as

$$y = 1.50 \text{ m} \quad \textbf{(Ans)}$$

In Example 10.4 we found the flow depth for a given flowrate. Since the equation for this depth is a nonlinear equation, it may be that there is more than one solution to the problem. For a given channel there may be two or more depths that carry the same flowrate. Although this is not normally so, it can and does happen, as is illustrated by Example 10.5.

EXAMPLE 10.5

Water flows in a round pipe of diameter D at a depth of $0 \leq y \leq D$, as is shown in Fig. E10.5*a*. The pipe is laid on a constant slope of S_0, and the Manning coefficient is n. At what depth does the maximum flowrate occur? Show that for certain flowrates there are two depths possible with the same flowrate. Explain this behavior.

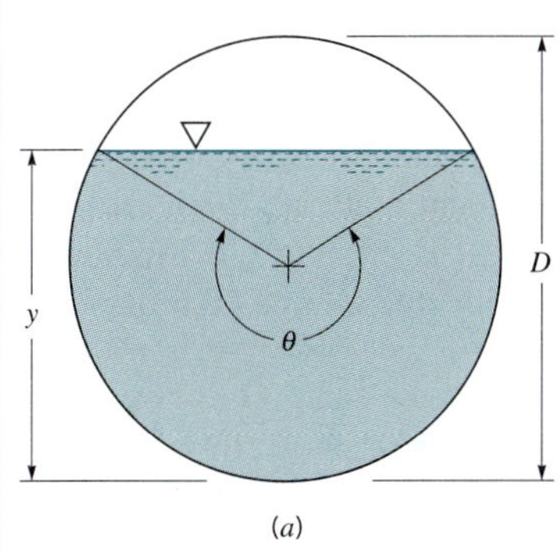

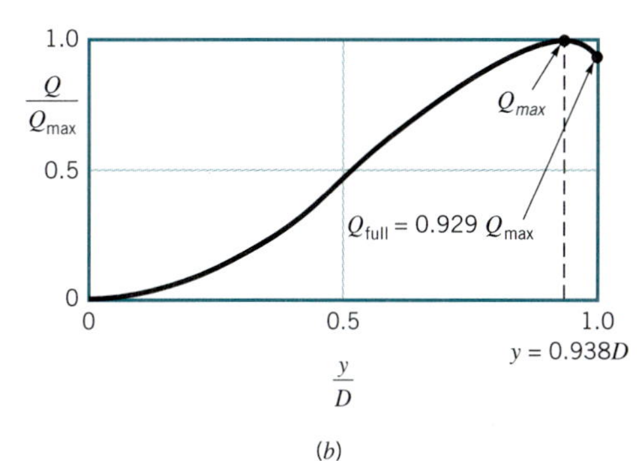

■ FIGURE E10.5

SOLUTION

According to the Manning equation (Eq. 10.20) the flowrate is

$$Q = \frac{\kappa}{n} A R_h^{2/3} S_0^{1/2} \quad \textbf{(1)}$$

where S_0, n, and κ are constants for this problem. From geometry it can be shown that

$$A = \frac{D^2}{8}(\theta - \sin\theta)$$

where θ, the angle indicated in Fig. E10.5*a*, is in radians. Similarly, the wetted perimeter is

$$P = \frac{D\theta}{2}$$

so that the hydraulic radius is

$$R_h = \frac{A}{P} = \frac{D(\theta - \sin\theta)}{4\theta}$$

Therefore, Eq. 1 becomes

$$Q = \frac{\kappa}{n} S_0^{1/2} \frac{D^{8/3}}{8(4)^{2/3}} \left[\frac{(\theta - \sin\theta)^{5/3}}{\theta^{2/3}}\right]$$

This can be written in terms of the flow depth by using $y = (D/2)[1 - \cos(\theta/2)]$.

A graph of flowrate versus flow depth, $Q = Q(y)$, has the characteristic indicated in Fig. E10.5*b*. In particular, the maximum flowrate, Q_{max}, does not occur when the pipe is full; $Q_{full} = 0.929 Q_{max}$. It occurs when $y = 0.938D$, or $\theta = 5.28$ rad $= 303°$. Thus,

$$Q = Q_{max} \text{ when } y = 0.938D \qquad \textbf{(Ans)}$$

For any $0.929 < Q/Q_{max} < 1$ there are two possible depths that give the same Q. The reason for this behavior can be seen by considering the gain in flow area, A, compared to the increase in wetted perimeter, P, for $y \approx D$. The flow area increase for an increase in y is very slight in this region, whereas the increase in wetted perimeter, and hence the increase in shear force holding back the fluid, is relatively large. The net result is a decrease in flowrate as the depth increases. For most practical problems, the slight difference between the maximum flowrate and full pipe flowrates is negligible, particularly in light of the usual inaccuracy of the value of n.

Sometimes it is necessary to determine the slope needed to produce a desired flowrate in a channel of specified shape, size, and material. Such a calculation is a straightforward application of Eq. 10.20. In addition, it is often of use to know whether the flow obtained will be subcritical, critical, or supercritical. Example 10.6 illustrates these ideas.

EXAMPLE 10.6

Water flows in a rectangular channel of width $b = 10$ m that has a Manning coefficient of $n = 0.025$. Plot a graph of flowrate, Q, as a function of slope, S_0, indicating lines of constant depth and lines of constant Froude number.

SOLUTION

For this channel the flow area is $A = by = 10y$, and the hydraulic radius is $R_h = A/P = by/(b + 2y)$, where y is the flow depth. Thus, the Manning equation (Eq. 10.19) becomes

$$V = \frac{\kappa}{n} R_h^{2/3} S_0^{1/2} = \frac{1.0}{0.025}\left(\frac{10y}{10 + 2y}\right)^{2/3} S_0^{1/2} \qquad \textbf{(1)}$$

where we have set $\kappa = 1$ because we are using SI units. Since we wish to plot Q versus S_0 at constant values of Fr, we use the fact that for a rectangular channel $\text{Fr} = V/(gy)^{1/2}$ and write Eq. 1 as

$$(gy)^{1/2}\text{Fr} = \frac{1}{0.025}\left(\frac{10y}{10 + 2y}\right)^{2/3} S_0^{1/2}$$

which simplifies to

$$S_0 = 0.00613(\text{Fr})^2\, y \left(\frac{5 + y}{5y}\right)^{4/3} \qquad \textbf{(2)}$$

For a given value of Fr, we pick various values of y, determine the corresponding values of S_0 from Eq. 2, and then calculate $Q = VA$, with V from either Eq. 1 or $V = (gy)^{1/2}$ Fr. The results are indicated in Fig. E10.6.

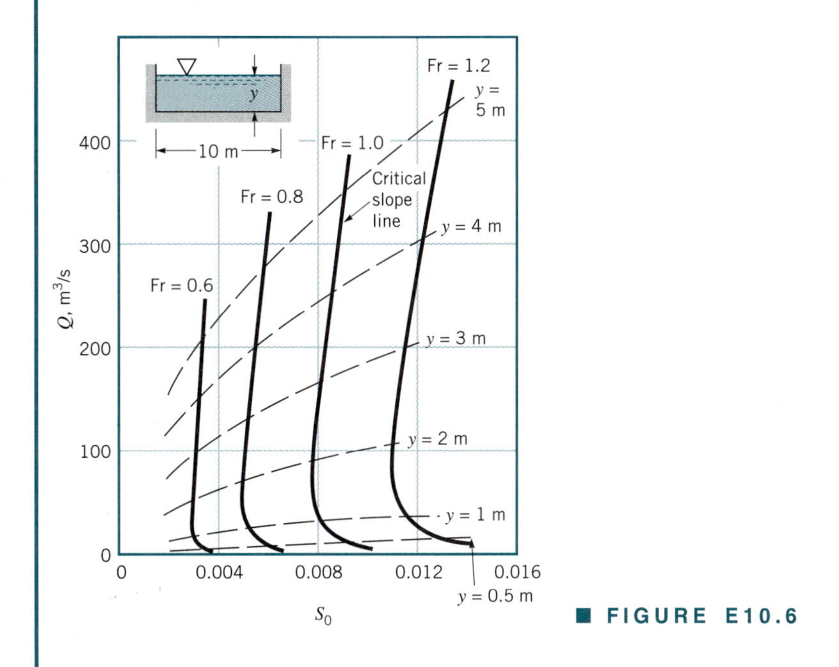

■ **FIGURE E10.6**

Note that for a given flowrate, there is a specific value of the slope that gives the critical condition, Fr = 1. This slope, denoted S_{0c}, is called the *critical slope*. The critical slope line divides the graph into two regions—one subcritical and the other supercritical. The dependence of S_{0c} on Q is rather weak over a large range of Q. If the slope is such that the flow is critical or nearly critical, it will remain so over a wide range of flowrates (and depths).

Lines of constant depth, y, are also indicated in Fig. E10.6. A figure like this allows one to easily see what effects are to be expected by varying the parameters involved.

For many open channels, the surface roughness varies across the channel.

In many man-made channels and in most natural channels, the surface roughness (and hence the Manning coefficient) varies along the wetted perimeter of the channel. A drainage ditch, for example, may have a rocky bottom surface with concrete side walls to prevent erosion. Thus, the effective n will be different for shallow depths than for deep depths of flow. Similarly, a river channel may have one value of n appropriate for its normal channel and another very different value of n during its flood stage when a portion of the flow occurs across fields or through floodplain woods. An ice-covered channel usually has a different value of n for the ice than for the remainder of the wetted perimeter (Ref. 7). (Strictly speaking, such ice-covered channels are not "open" channels, although analysis of their flow is often based on open-channel flow equations. This is acceptable, since the ice covered is often thin enough so that it represents a fixed boundary in terms of the shear stress resistance, but it cannot support a significant pressure differential as in pipe flow situations.)

A variety of methods has been used to determine an appropriate value of the effective roughness of channels that contain subsections with different values of n. Which method gives the most accurate, easy-to-use results is not firmly established, since the results are nearly the same for each method (Ref. 5). A reasonable approximation is to divide the channel cross section into N subsections, each with its own wetted perimeter, P_i, area, A_i, and Manning coefficient, n_i. The P_i values do not include the imaginary boundaries between the different subsections. The total flowrate is assumed to be the sum of the flowrates through each section. This technique is illustrated by Example 10.7.

EXAMPLE 10.7

Water flows along the drainage canal having the properties shown in Fig. E10.7. If the bottom slope is $S_0 = 1 \text{ ft}/500 \text{ ft} = 0.002$, estimate the flowrate.

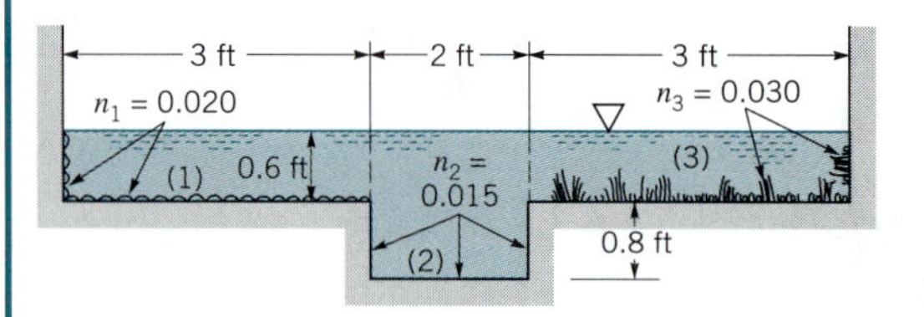

■ FIGURE E10.7

SOLUTION

We divide the cross section into three subsections as is indicated in Fig. E10.7 and write the flowrate as $Q = Q_1 + Q_2 + Q_3$, where for each section

$$Q_i = \frac{1.49}{n_i} A_i R_{h_i}^{2/3} S_0^{1/2}$$

The appropriate values of A_i, P_i, R_{hi}, and n_i are listed in Table E10.7. Note that the imaginary portions of the perimeters between sections (denoted by the dashed lines in Fig. E10.7) are not included in the P_i. That is, for section (2)

$$A_2 = 2 \text{ ft } (0.8 + 0.6) \text{ ft} = 2.8 \text{ ft}^2$$

and

$$P_2 = 2 \text{ ft} + 2(0.8 \text{ ft}) = 3.6 \text{ ft}$$

so that

$$R_{h_2} = \frac{A_2}{P_2} = \frac{2.8 \text{ ft}^2}{3.6 \text{ ft}} = 0.778 \text{ ft}$$

■ TABLE E10.7

i	A_i (ft²)	P_i (ft)	R_{hi} (ft)	n_i
1	1.8	3.6	0.500	0.020
2	2.8	3.6	0.778	0.015
3	1.8	3.6	0.500	0.030

Thus, the total flowrate is

$$Q = Q_1 + Q_2 + Q_3 = 1.49(0.002)^{1/2}$$
$$\times \left[\frac{(1.8 \text{ ft}^2)(0.500 \text{ ft})^{2/3}}{0.020} + \frac{(2.8 \text{ ft}^2)(0.778 \text{ ft})^{2/3}}{0.015} + \frac{(1.8 \text{ ft}^2)(0.500 \text{ ft})^{2/3}}{0.030} \right]$$

or

$$Q = 16.8 \text{ ft}^3/\text{s} \qquad \textbf{(Ans)}$$

If the entire channel cross section were considered as one flow area, then $A = A_1 + A_2 + A_3 = 6.4 \text{ ft}^2$ and $P = P_1 + P_2 + P_3 = 10.8$ ft, or $R_h = A/P = 6.4 \text{ ft}^2/10.8 \text{ ft} = 0.593$ ft. The flowrate is given by Eq. 10.20, which can be written as

$$Q = \frac{1.49}{n_{\text{eff}}} A R_h^{2/3} S_0^{1/2}$$

where n_{eff} is the effective value of n for this channel. With $Q = 16.8 \text{ ft}^3/\text{s}$ as determined above, the value of n_{eff} is found to be

$$n_{\text{eff}} = \frac{1.49 A R_h^{2/3} S_0^{1/2}}{Q}$$

$$= \frac{1.49(6.4)(0.593)^{2/3}(0.002)^{1/2}}{16.8} = 0.0179$$

As expected, the effective roughness (Manning n) is between the minimum ($n_2 = 0.015$) and maximum ($n_3 = 0.030$) values for the individual subsections.

For a given flowrate, the channel of minimum area is denoted as the best hydraulic cross section.

One type of problem often encountered in open-channel flows is that of determining the *best hydraulic cross section* defined as the section of the minimum area for a given flowrate, Q, slope, S_0, and roughness coefficient, n. By using $R_h = A/P$ we can write Eq. 10.20 as

$$Q = \frac{\kappa}{n} A \left(\frac{A}{P}\right)^{2/3} S_0^{1/2} = \frac{\kappa}{n} \frac{A^{5/3} S_0^{1/2}}{P^{2/3}}$$

which can be rearranged as

$$A = \left(\frac{nQ}{\kappa S_0^{1/2}}\right)^{3/5} P^{2/5}$$

where the quantity in the parentheses is a constant. Thus, a channel with minimum A is one with a minimum P, so that both the amount of excavation needed and the amount of material to line the surface are minimized by the best hydraulic cross section.

The best hydraulic cross section possible is that of a semicircular channel. No other shape has as small a wetted perimeter for a given area. It is often desired to determine the best shape for a class of cross sections such as rectangles, trapezoids, or triangles. Example 10.8 illustrates the concept of the best hydraulic cross section for rectangular channels.

EXAMPLE 10.8

Water flows uniformly in a rectangular channel of width b and depth y. Determine the aspect ratio, b/y, for the best hydraulic cross section.

SOLUTION

The uniform flow is given by Eq. 10.20 as

$$Q = \frac{\kappa}{n} A R_h^{2/3} S_0^{1/2} \quad (1)$$

where $A = by$ and $P = b + 2y$, so that $R_h = A/P = by/(b + 2y)$. We rewrite the hydraulic radius in terms of A as

$$R_h = \frac{A}{(2y + b)} = \frac{A}{(2y + A/y)} = \frac{Ay}{(2y^2 + A)}$$

so that Eq. 1 becomes

$$Q = \frac{\kappa}{n} A \left(\frac{Ay}{2y^2 + A}\right)^{2/3} S_0^{1/2}$$

This can be rearranged to give

$$A^{5/2}y = K(2y^2 + A) \tag{2}$$

where $K = (nQ/\kappa S_0^{1/2})^{3/2}$ is a constant. The best hydraulic section is the one that gives the minimum A for all y. That is, $dA/dy = 0$. By differentiating Eq. 2 with respect to y, we obtain

$$\frac{5}{2} A^{3/2} \frac{dA}{dy} y + A^{5/2} = K\left(4y + \frac{dA}{dy}\right)$$

which, with $dA/dy = 0$, reduces to

$$A^{5/2} = 4Ky \tag{3}$$

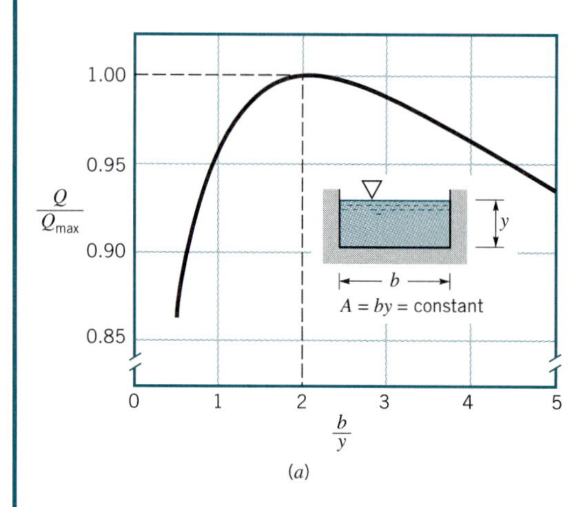

(a)

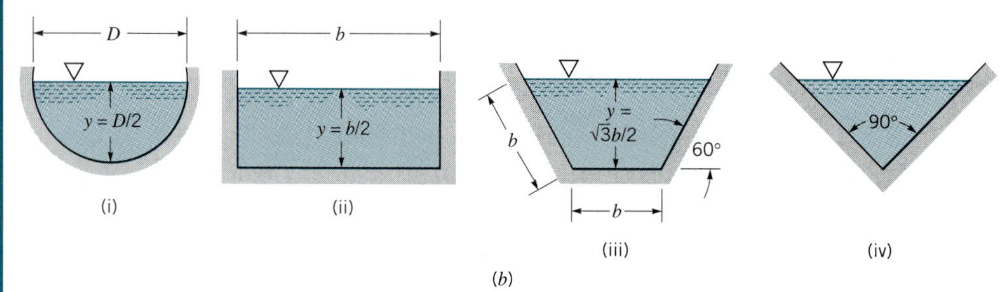

(b)

■ FIGURE E10.8

With $K = A^{5/2} y/(2y^2 + A)$ from Eq. 2, Eq. 3 can be written in the form

$$A^{5/2} = \frac{4A^{5/2}y^2}{(2y^2 + A)}$$

which simplifies to $y = (A/2)^{1/2}$. Thus, since $A = by$, the best hydraulic cross section for a rectangular shape has a width b and a depth

$$y = \left(\frac{A}{2}\right)^{1/2} = \left(\frac{by}{2}\right)^{1/2}$$

or

$$2y^2 = by$$

That is, the rectangle with the best hydraulic cross section is twice as wide as it is deep, or

$$b/y = 2 \qquad \textbf{(Ans)}$$

A rectangular channel with $b/y = 2$ will give the smallest area (and smallest wetted perimeter) for a given flowrate. Conversely, for a given area, the largest flowrate in a rectangular channel will occur when $b/y = 2$. For $A = by =$ constant, if $y \rightarrow 0$ then $b \rightarrow \infty$, and the flowrate is small because of the large wetted perimeter $P = b + 2y \rightarrow \infty$. Similarly, if $y \rightarrow \infty$ then $b \rightarrow 0$, and the flowrate is small because $P = b + 2y \rightarrow \infty$. The maximum Q occurs when $y = b/2$. However, as is seen in Fig. E10.8*a*, the maximum represented by this optimal configuration is a rather weak one. For example, for aspect ratios between 1 and 4, the flowrate is within 96% of the maximum flowrate obtained with the same area and $b/y = 2$.

An alternate but equivalent method to obtain the above answer is to use the fact that $dR_h/dy = 0$, which follows from Eq. 1 using $dQ/dy = 0$ (constant flowrate) and $dA/dy = 0$ (best hydraulic cross section has minimum area). Differentiation of $R_h = Ay/(2y^2 + A)$ with constant A gives $b/y = 2$ when $dR_h/dy = 0$.

The best hydraulic cross section can be calculated for other shapes in a similar fashion. The results (given here without proof) for circular, rectangular, trapezoidal (with 60° sides), and triangular shapes are shown in Fig. E10.8*b*.

10.5 Gradually Varied Flow

Open-channel flows are classified as uniform depth, gradually varying, or rapidly varying.

In many situations the flow in an open channel is not of uniform depth ($y =$ constant) along the channel. This can occur because of several reasons: The bottom slope is not constant, the cross-sectional shape and area vary in the flow direction, or there is some obstruction across a portion of the channel. Such flows are classified as gradually varying flows (or gradually varied flows) if $dy/dx \ll 1$ or rapidly varying flows if $dy/dx \sim 1$. In this section we consider some of the properties of gradually varied flows.

In a complex open-channel flow situation, it is common practice to divide the channel into specific segments or reaches that possess certain qualities. One such method is to use a classification scheme denoting how the flow depth varies with distance along the channel.

As discussed in the previous section, if the channel bottom slope is equal to the slope of the energy line, $S_0 = S_f$, the flow depth is constant, $dy/dx = 0$. Physically, the loss in potential energy of the fluid as it flows downhill is exactly balanced by the dissipation of energy through viscous effects.

If the bottom slope and the energy line slope are not equal, the flow depth will vary along the channel, either increasing or decreasing in the flow direction. In such cases $dy/dx \neq 0$, and the left-hand side of Eq. 10.14 is not zero. Physically, the difference between the component of weight and the shear forces in the direction of flow produces a change in the fluid momentum which requires a change in velocity and, from continuity considerations, a change in depth. Whether the depth increases or decreases depends on various parameters of the flow, with 12 types of surface profile configurations [flow depth as a function of distance, $y = y(x)$] possible.

In some situations the water depth increases in the direction of flow; in other cases it decreases.

The sign of dy/dx, that is, whether the flow depth increases or decreases with distance along the channel, depends on the sign of both the numerator and denominator of Eq. 10.14. The sign of the denominator clearly depends on whether the flow is subcritical or supercritical. For a given channel (i.e., size, shape, Manning coefficient, and flowrate) there is a *critical slope*, $S_0 = S_{0c}$, and a corresponding critical depth, $y = y_c$, such that $\text{Fr} = 1$ under conditions of uniform flow. (See Example 10.6.)

For uniform depth flow, it follows that $S_0 = S_f$ and the numerator of Eq. 10.14 is zero. However, if S_0 is greater than S_f, the bottom slope is steeper than is required for uniform flow, and the numerator is negative. If S_0 is less than S_f, the numerator is positive. Thus, there are a number of combinations of numerator and denominator of Eq. 10.14 that give positive, zero, or negative values of dy/dx. These conditions result in the following flow configurations.

10.5.1 Classification of Surface Shapes

The character of a gradually varying flow is often classified in terms of the actual channel slope, S_0, compared with the slope required to produce uniform critical flow, S_{0c}. In addition, the character of the flow depends on whether the fluid depth is less than or greater than the uniform normal depth, y_n, that would occur in the channel with the given slope and flowrate. The 12 configurations indicated in Table 10.2 are possible. We will merely indicate the possible types of surface configurations without becoming involved in the details of how to calculate their shape. Additional discussion of this complex topic can be found in standard references about open-channel flows (Refs. 3, 9).

The five types of slopes indicated include: (1) mild slope with $S_0 < S_{0c}$ (the flow would be subcritical if it were uniform depth), (2) critical slope with $S_0 = S_{0c}$ (the flow would have

TABLE 10.2
Possible Free-Surface Configurations

Slope Type	Slope Notation	Froude No.	Surface Shape Designation
$S_0 < S_{0c}$	Mile (M)	$\text{Fr} < 1$	M-1
		$\text{Fr} < 1$	M-2
		$\text{Fr} > 1$	M-3
$S_0 = S_{0c}$	Critical (C)	$\text{Fr} < 1$	C-1
		$\text{Fr} > 1$	C-3
$S_0 > S_{0c}$	Steep (S)	$\text{Fr} < 1$	S-1
		$\text{Fr} > 1$	S-2
		$\text{Fr} > 1$	S-3
$S_0 = 0$	Horizontal (H)	$\text{Fr} < 1$	H-2
		$\text{Fr} > 1$	H-3
$S_0 < 0$	Adverse (A)	$\text{Fr} < 1$	A-2
		$\text{Fr} > 1$	A-3

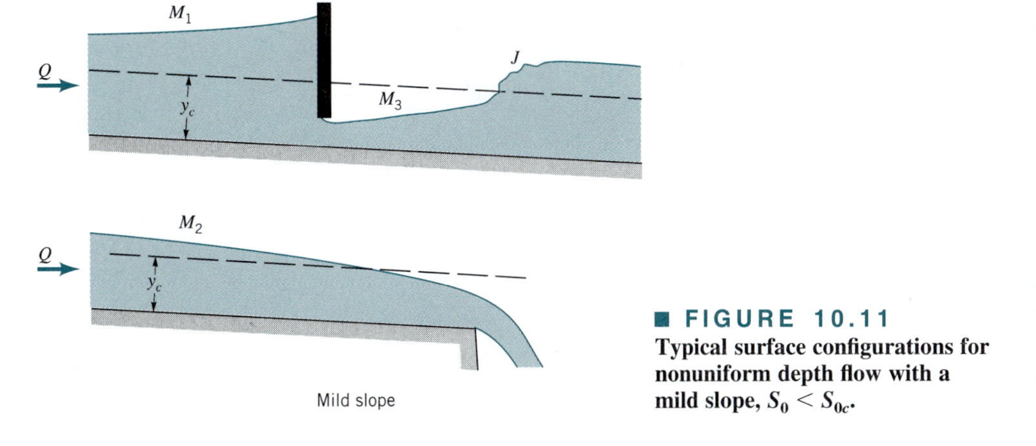

■ **FIGURE 10.11**
Typical surface configurations for nonuniform depth flow with a mild slope, $S_0 < S_{0c}$.

Fr $= 1$ if it were uniform depth), (3) steep slope with $S_0 > S_{0c}$ (the flow would be supercritical if it were uniform depth), (4) horizontal slope with $S_0 = 0$, and (5) adverse slope with $S_0 < 0$ (flow uphill). Since we are considering nonuniform depth flows, the flows within these categories may have Fr < 1 or Fr > 1, depending on whether $y > y_c$ or $y < y_c$, respectively. Here, y_c is the critical depth, the depth of flow for the given conditions that would give critical, uniform depth flow. The standard designations attached to these flows (i.e., *M*-1, *A*-2, etc.) are indicated in the table.

If the flow were uniform depth flow along the channel, the determination of whether the flow is sub- or supercritical would be dependent solely on whether $S_0 < S_{0c}$ or $S_0 > S_{0c}$, respectively. For gradually varying flow, inertia of the fluid becomes a governing mechanism (in addition to weight and shear force), and it is possible to have a variety of additional flow situations. Whether the actual flow is subcritical or supercritical depends on additional conditions, not just the value of S_0. For example, with $S_0 < S_{0c}$ (a mild slope indicating Fr < 1 if the flow were uniform depth), it is possible to have either Fr < 1 or Fr > 1 depending on the depth. That is, a balance of weight, shear, and inertia effects provides an additional freedom to the flow that is not present in uniform flow where inertial effects are absent.

Fluid inertia is important for nonuniform open-channel flows.

10.5.2 Examples of Gradually Varied Flows

Although calculation of the actual surface shapes is beyond the scope of this book, flow situations involving the 12 types of surface profiles possible are indicated in Figs. 10.11 through 10.15. In each instance the depth variation is caused by some obstruction or change in the channel geometry, such as a dam, a sluice gate, or a sudden change in the channel elevation. Gradually varied flows can also occur in the vicinity of the location where the channel bottom changes from one slope to another slope, either steeper or shallower.

Note that the horizontal length scale of Figs. 10.11 through 10.15 has been reduced. They should have $dy/dx \ll 1$ and $S_0 \ll 1$. Generally, the bottom slope is so small that it would

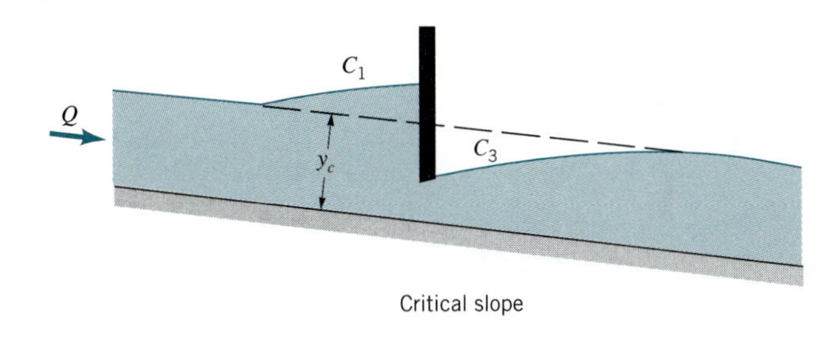

■ **FIGURE 10.12**
Typical surface configurations for nonuniform depth flow with a critical slope, $S_0 = S_{0c}$.

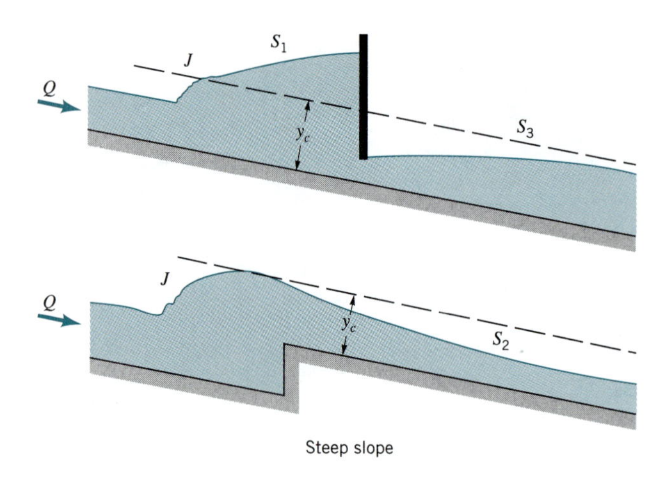

■ **FIGURE 10.13 Typical surface configurations for nonuniform depth flow with a steep slope, $S_0 > S_{0c}$.**

not be visible on a true scale drawing and the distances over which the depth changes occur are very large compared with the depth change. For rapidly varying flows, such as the hydraulic jumps indicated by the regions J in the figures and discussed in Section 10.6.1, the depth variation occurs in a relatively short distance.

The most common of the gradually varying flows is the M-1 type indicated in Fig. 10.11. Such flows occur upstream of a dam or sluice gate where the subcritical flow is slowed down by the obstruction across the channel. It is often of considerable importance to determine the extent to which the water level is raised above what it would assume if the dam were not present. The lake surface immediately upstream of a dam is horizontal—the water is stationary. However, the surface profile in the transition region between the lake and the unaffected flowing river is neither horizontal nor at an elevation it would have if the dam were not present. Such water surface profiles are called the *backwater curve*.

The free surface shape depends on the channel bottom slope and the Froude number.

The M-2 type curve is obtained by the reduction of the downstream resistance in a channel such as that produced by the outflow condition shown in Fig. 10.11. The subcritical flow accelerates as it approaches the end of the channel and produces the *drop-down* profile as indicated.

The M-3 type curve indicated in Fig. 10.11 results from the fact that the mild slope is not sufficient to maintain the supercritical conditions after the sluice gate. At a certain point downstream the supercritical flow passes through a hydraulic jump, J, and changes to subcritical flow.

The C-1 and C-3 critical slope profiles indicated in Fig. 10.12 can be produced by a sluice gate across the channel. However, they are infrequently encountered because the precise $\text{Fr} = 1$ or $S_0 = S_{0c}$ conditions are not often found.

The steep slope S-1 and S-3 conditions indicated in Fig. 10.13 can be produced by an obstruction such as a dam or sluice gate in a manner very similar to the M-1 and M-3 mild slope flows. Note that the sign of the free-surface curvature is opposite for the steep slope case compared with the mild slope case. The S-2 profile is obtained when a channel receives flow from a critical depth situation, as often occurs for the flow from a reservoir into a channel.

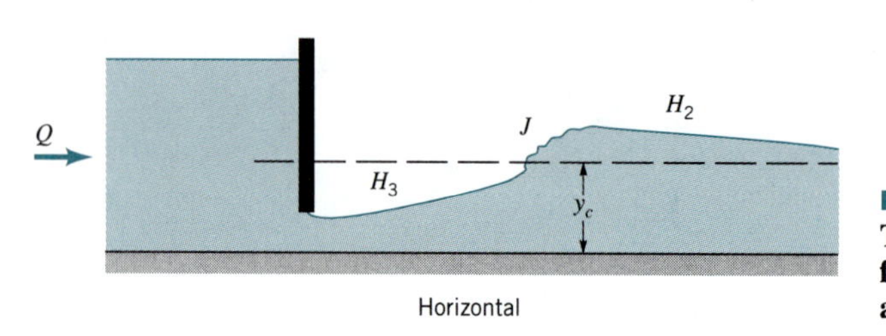

■ **FIGURE 10.14 Typical surface configurations for nonuniform depth flow with a horizontal slope, $S_0 = 0$.**

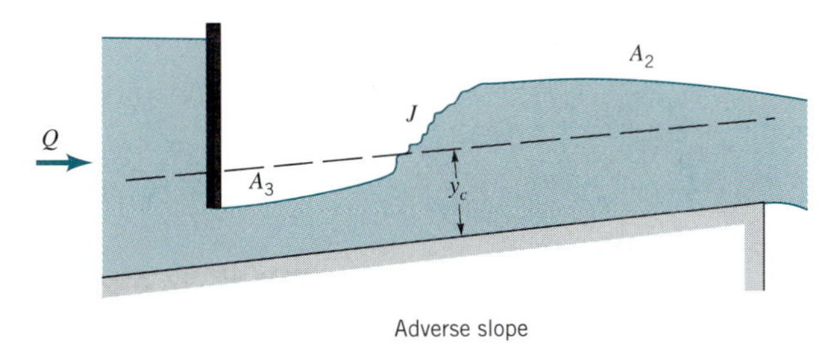

■ **FIGURE 10.15** **Typical surface configurations for nonuniform depth flow with an adverse slope, $S_0 < 0$.**

The horizontal or adverse slope situations are unique in that it is impossible to maintain uniform flow under such conditions. The fluid inertia is clearly an important mechanism in these flows, since the component of weight along the channel is either zero (horizontal slope) or in the direction opposite to the flow (adverse slope). Typical flows are indicated in Figs. 10.14 and 10.15.

Numerical techniques can be used to predict open-channel surface shapes.

As is discussed above, a variety of surface profiles is possible in gradually varied open-channel flow. The free surface is relatively free to conform to the shape that satisfies the governing mass, momentum, and energy equations. The actual shape of the surface is often very important in the design of open-channel devices or in the prediction of flood levels in natural channels. The surface shape, $y = y(x)$, can be calculated by solving the governing differential equation obtained from a combination of the Manning equation (Eq. 10.20) and the energy equation (Eq. 10.14). The result is a complex, nonlinear differential equation that has been the subject of many investigations over the past century. With the use of modern computers, a variety of numerical techniques has been developed and is used with considerable success in determining the various gradually varied flow surface profiles. Such results are beyond the scope of this book, but are readily available in the literature (Refs. 5, 9).

10.6 Rapidly Varied Flow

In many open channels, flow depth changes occur over a relatively short distance so that $dy/dx \sim 1$. Such rapidly varied flow conditions are often quite complex and difficult to analyze in a precise fashion. Fortunately, many useful approximate results can be obtained by using a simple one-dimensional model along with appropriate experimentally determined coefficients when necessary. In this section we discuss several of these flows.

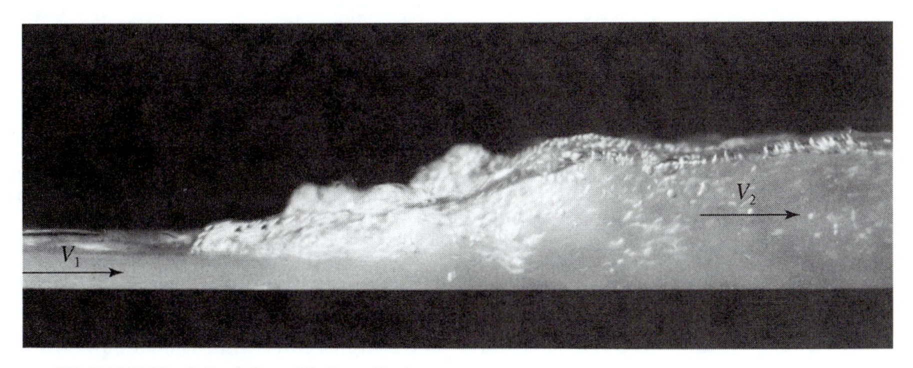

■ **FIGURE 10.16** **Hydraulic jump.**

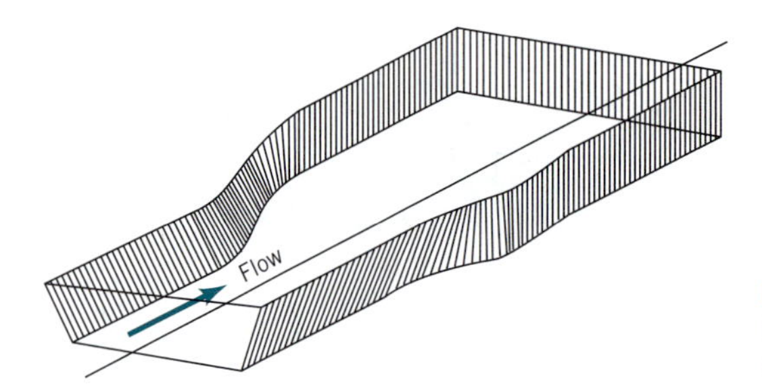

■ FIGURE 10.17 **Rapidly varied flow may occur in a channel transition section.**

In many cases the flow depth may change significantly in a short distance.

V10.4

Some rapidly varied flows occur in constant area channels for reasons that are not immediately obvious. The hydraulic jump is one such case. As is indicated in Fig. 10.16, the flow may change from a relatively shallow, high-speed condition into a relatively deep, low-speed condition within a horizontal distance of just a few channel depths. Other rapidly varied flows may be due to a sudden change in the channel geometry such as the flow in an expansion or contraction section of a channel as is indicated in Fig. 10.17.

In such situations the flow field is often two- or three-dimensional in character. There may be regions of flow separation, flow reversal, or unsteady oscillations of the free surface. For the purpose of some analyses, these complexities can be neglected and a simplified analysis can be undertaken. In other cases, however, it is the complex details of the flow that are the most important property of the flow; any analysis must include their effects. The scouring of a river bottom in the neighborhood of a bridge pier, as is indicated in Fig. 10.18, is such an example. A one- or two-dimensional model of this flow would not be sufficient to describe the complex structure of the flow that is responsible for the erosion near the foot of the bridge pier.

Many open-channel flow-measuring devices are based on principles associated with rapidly varied flows. Among these devices are broad-crested weirs, sharp-crested weirs, critical flow flumes, and sluice gates. The operation of such devices is discussed in the following sections.

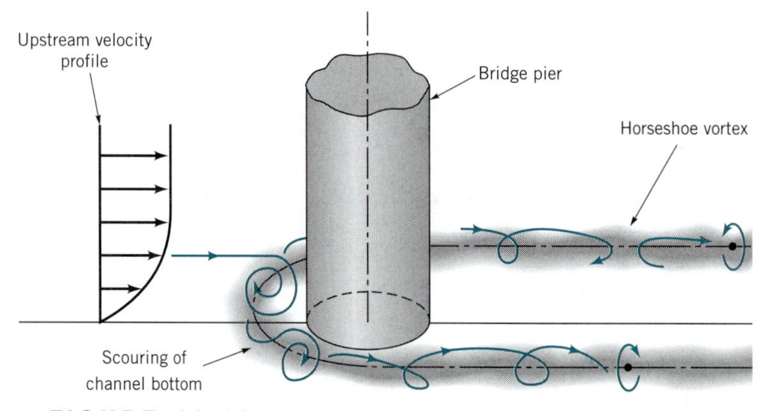

■ FIGURE 10.18 **The complex three-dimensional flow structure around a bridge pier.**

10.6.1 The Hydraulic Jump

Observations of flows in open channels show that under certain conditions it is possible that the fluid depth will change very rapidly over a short length of the channel without any change in the channel configuration. Such changes in depth can be approximated as a discontinuity in the free-surface elevation ($dy/dx = \infty$). For reasons discussed below, this step change in depth is always from a shallow to a deeper depth—always a step up, never a step down.

A hydraulic jump is a step-like increase in fluid depth in an open channel.

Physically, this near discontinuity, called a *hydraulic jump*, may result when there is a conflict between the upstream and downstream influences that control a particular section (or reach) of a channel. For example, a sluice gate may require that the conditions at the upstream portion of the channel (downstream of the gate) be supercritical flow, while obstructions in the channel on the downstream end of the reach may require that the flow be subcritical. The hydraulic jump provides the mechanism (a nearly discontinuous one at that) to make the transition between the two types of flow.

V10.5

The simplest type of hydraulic jump occurs in a horizontal, rectangular channel as is indicated in Fig. 10.19. Although the flow within the jump itself is extremely complex and agitated, it is reasonable to assume that the flow at sections (1) and (2) is nearly uniform, steady, and one-dimensional. In addition, we neglect any wall shear stresses, τ_w, within the relatively short segment between these two sections. Under these conditions the x component of the momentum equation (Eq. 5.22) for the control volume indicated can be written as

$$F_1 - F_2 = \rho Q(V_2 - V_1) = \rho V_1 y_1 b(V_2 - V_1)$$

where the pressure force at either section is hydrostatic. That is, $F_1 = p_{c1}A_1 = \gamma y_1^2 b/2$ and $F_2 = p_{c2}A_2 = \gamma y_2^2 b/2$, where $p_{c1} = \gamma y_1/2$ and $p_{c2} = \gamma y_2/2$ are the pressures at the centroids of the channel cross sections and b is the channel width. Thus, the momentum equation becomes

$$\frac{y_1^2}{2} - \frac{y_2^2}{2} = \frac{V_1 y_1}{g}(V_2 - V_1) \quad \textbf{(10.21)}$$

In addition to the momentum equation, we have the conservation of mass equation (Eq. 5.12)

$$y_1 b V_1 = y_2 b V_2 = Q \quad \textbf{(10.22)}$$

and the energy equation (Eq. 5.84)

$$y_1 + \frac{V_1^2}{2g} = y_2 + \frac{V_2^2}{2g} + h_L \quad \textbf{(10.23)}$$

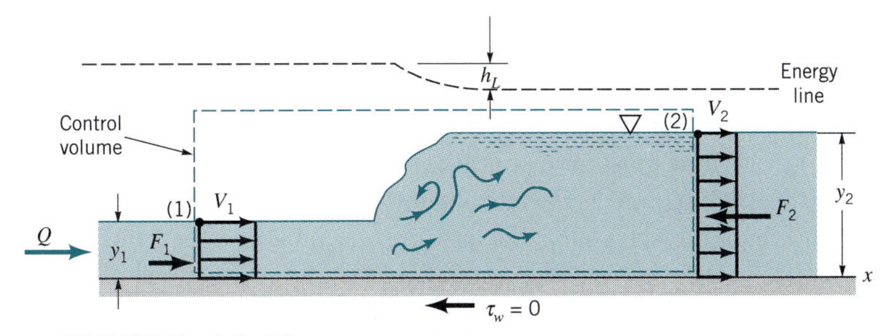

■ FIGURE 10.19 **Hydraulic jump geometry.**

The head loss, h_L, in Eq. 10.23 is due to the violent turbulent mixing and dissipation that occur within the jump itself. We have neglected any head loss due to wall shear stresses.

Clearly Eqs. 10.21, 10.22, and 10.23 have a solution $y_1 = y_2$, $V_1 = V_2$, and $h_L = 0$. This represents the trivial case of no jump. Since these are nonlinear equations, it may be possible that more than one solution exists. The other solutions can be obtained as follows. By combining Eqs. 10.21 and 10.22 to eliminate V_2 we obtain

$$\frac{y_1^2}{2} - \frac{y_2^2}{2} = \frac{V_1 y_1}{g}\left(\frac{V_1 y_1}{y_2} - V_1\right) = \frac{V_1^2 y_1}{g y_2}(y_1 - y_2)$$

which can be simplified by factoring out a common nonzero factor $y_1 - y_2$ from each side to give

$$\left(\frac{y_2}{y_1}\right)^2 + \left(\frac{y_2}{y_1}\right) - 2\,\mathrm{Fr}_1^2 = 0$$

where $\mathrm{Fr}_1 = V_1/\sqrt{g y_1}$ is the upstream Froude number. By using the quadratic formula we obtain

$$\frac{y_2}{y_1} = \frac{1}{2}\left(-1 \pm \sqrt{1 + 8\mathrm{Fr}_1^2}\right)$$

Clearly the solution with the minus sign is not possible (it would give a negative y_2/y_1). Thus,

The depth ratio across a hydraulic jump depends on the Froude number only.

$$\frac{y_2}{y_1} = \frac{1}{2}\left(-1 + \sqrt{1 + 8\mathrm{Fr}_1^2}\right) \tag{10.24}$$

This depth ratio, y_2/y_1, across the hydraulic jump is shown as a function of the upstream Froude number in Fig. 10.20. The portion of the curve for $\mathrm{Fr}_1 < 1$ is dashed in recognition of the fact that to have a hydraulic jump the flow must be supercritical. That is, the solution as given by Eq. 10.24 must be restricted to $\mathrm{Fr}_1 \geq 1$, for which $y_2/y_1 \geq 1$. This can be shown by consideration of the energy equation, Eq. 10.23, as follows. The dimensionless head loss, h_L/y_1, can be obtained from Eq. 10.23 as

$$\frac{h_L}{y_1} = 1 - \frac{y_2}{y_1} + \frac{\mathrm{Fr}_1^2}{2}\left[1 - \left(\frac{y_1}{y_2}\right)^2\right] \tag{10.25}$$

where, for given values of Fr_1, the values of y_2/y_1 are obtained from Eq. 10.24. As is indicated in Fig. 10.20, the head loss is negative if $\mathrm{Fr}_1 < 1$. Since negative head losses violate the second law of thermodynamics (viscous effects dissipate energy, they cannot create energy; see Section 5.3), it is not possible to produce a hydraulic jump with $\mathrm{Fr}_1 < 1$. The head loss across the jump is indicated by the lowering of the energy line shown in Fig. 10.19.

A flow must be supercritical (Froude number > 1) to produce the discontinuity called a hydraulic jump. This is analogous to the compressible flow ideas discussed in Chapter 11 in which it is shown that the flow of a gas must be supersonic (Mach number > 1) to produce the discontinuity called a normal shock wave. However, the fact that a flow is supercritical (or supersonic) does not guarantee the production of a hydraulic jump (or shock wave). The trivial solution $y_1 = y_2$ and $V_1 = V_2$ is also possible.

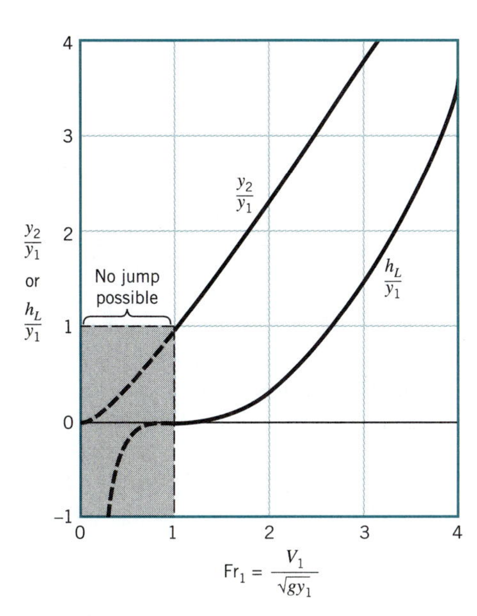

■ FIGURE 10.20 Depth ratio and dimensionless head loss across a hydraulic jump as a function of upstream Froude number.

The fact that there is an energy loss across a hydraulic jump is useful in many situations. For example, the relatively large amount of energy contained in the fluid flowing down the spillway of a dam could cause damage to the channel below the dam. By placing suitable flow control objects in the channel downstream of the spillway, it is possible (if the flow is supercritical) to produce a hydraulic jump on the apron of the spillway and thereby dissipate a considerable portion of the energy of the flow. That is, the dam spillway produces supercritical flow, and the channel downstream of the dam requires subcritical flow. The resulting hydraulic jump provides the means to change the character of the flow.

EXAMPLE 10.9

Water on the horizontal apron of the 100-ft-wide spillway shown in Fig. E10.9*a* has a depth of 0.60 ft and a velocity of 18 ft/s. Determine the depth, y_2, after the jump, the Froude numbers before and after the jump, Fr_1 and Fr_2, and the power dissipated, $\mathcal{P}_d$, within the jump.

SOLUTION

Conditions across the jump are determined by the upstream Froude number

$$Fr_1 = \frac{V_1}{\sqrt{gy_1}} = \frac{18 \text{ ft/s}}{[(32.2 \text{ ft/s}^2)(0.60 \text{ ft})]^{1/2}} = 4.10 \qquad \textbf{(Ans)}$$

Thus, the upstream flow is supercritical, and it is possible to generate a hydraulic jump as sketched.

From Eq. 10.24 we obtain the depth ratio across the jump as

$$\frac{y_2}{y_1} = \frac{1}{2}(-1 + \sqrt{1 + 8\,Fr_1^2}) = \frac{1}{2}[-1 + \sqrt{1 + 8(4.10)^2}] = 5.32$$

or

$$y_2 = 5.32\ (0.60\ \text{ft}) = 3.19\ \text{ft} \qquad \textbf{(Ans)}$$

Since $Q_1 = Q_2$, or $V_2 = (y_1 V_1)/y_2 = 0.60\ \text{ft}\ (18\ \text{ft/s})/3.19\ \text{ft} = 3.39\ \text{ft/s}$, it follows that

$$\text{Fr}_2 = \frac{V_2}{\sqrt{gy_2}} = \frac{3.39\ \text{ft/s}}{[(32.2\ \text{ft/s}^2)(3.19\ \text{ft})]^{1/2}} = 0.334 \qquad \textbf{(Ans)}$$

As is true for any hydraulic jump, the flow changes from supercritical to subcritical flow across the jump.

The power (energy per unit time) dissipated, $\mathcal{P}_d$, by viscous effects within the jump can be determined from the head loss as (see Eq. 5.85)

$$\mathcal{P}_d = \gamma Q h_L = \gamma b y_1 V_1 h_L \qquad \textbf{(1)}$$

where h_L is obtained from Eqs. 10.23 or 10.25 as

$$h_L = \left(y_1 + \frac{V_1^2}{2g}\right) - \left(y_2 + \frac{V_2^2}{2g}\right) = \left[0.60\ \text{ft} + \frac{(18.0\ \text{ft/s})^2}{2(32.2\ \text{ft/s}^2)}\right] - \left[3.19\ \text{ft} + \frac{(3.39\ \text{ft/s})^2}{2(32.2\ \text{ft/s}^2)}\right]$$

or

$$h_L = 2.26\ \text{ft}$$

Thus, from Eq. 1,

$$\mathcal{P}_d = (62.4\ \text{lb/ft}^3)(100\ \text{ft})(0.60\ \text{ft})(18.0\ \text{ft/s})(2.26\ \text{ft}) = 1.52 \times 10^5\ \text{ft}\cdot\text{lb/s}$$

or

$$\mathcal{P}_d = \frac{1.52 \times 10^5\ \text{ft}\cdot\text{lb/s}}{550[(\text{ft}\cdot\text{lb/s})/\text{hp}]} = 277\ \text{hp} \qquad \textbf{(Ans)}$$

This power, which is dissipated within the highly turbulent motion of the jump, is converted into an increase in water temperature, T. That is, $T_2 > T_1$. Although the power dissipated is considerable, the difference in temperature is not great because the flowrate is quite large.

The hydraulic jump flow process can be illustrated by use of the specific energy concept introduced in Section 10.3 as follows. Equation 10.23 can be written in terms of the specific energy, $E = y + V^2/2g$, as $E_1 = E_2 + h_L$, where $E_1 = y_1 + V_1^2/2g = 5.63$ ft and $E_2 = y_2 + V_2^2/2g = 3.37$ ft. As is discussed in Section 10.3, the specific energy diagram for this flow can be obtained by using $V = q/y$, where

$$q = q_1 = q_2 = \frac{Q}{b} = y_1 V_1 = 0.60\ \text{ft}\ (18.0\ \text{ft/s}) = 10.8\ \text{ft}^2/\text{s}$$

Thus,

$$E = y + \frac{q^2}{2gy^2} = y + \frac{(10.8\ \text{ft}^2/\text{s})^2}{2(32.2\ \text{ft/s}^2)y^2} = y + \frac{1.81}{y^2}$$

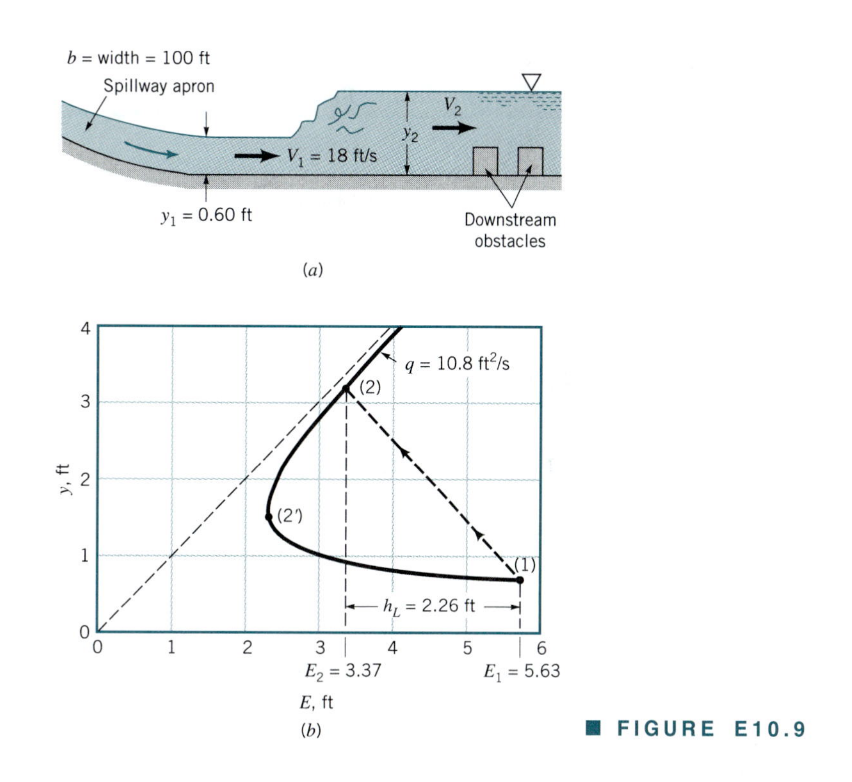

■ FIGURE E10.9

where y and E are in feet. The resulting specific energy diagram is shown in Fig. E10.9*b*. Because of the head loss across the jump, the upstream and downstream values of E are different. In going from state (1) to state (2) the fluid does not proceed along the specific energy curve and pass through the critical condition at state 2′. Rather, it jumps from (1) to (2) as is represented by the dashed line in the figure. From a one-dimensional consideration, the jump is a discontinuity. In actuality, the jump is a complex three-dimensional flow incapable of being represented on the one-dimensional specific energy diagram.

The actual structure of a hydraulic jump depends on the Froude number.

The actual structure of a hydraulic jump is a complex function of Fr_1, even though the depth ratio and head loss are given quite accurately by a simple one-dimensional flow analysis (Eqs. 10.24 and 10.25). A detailed investigation of the flow indicates that there are essentially five types of surface and jump conditions. The classification of these jumps is indicated in Table 10.3, along with sketches of the structure of the jump. For flows that are barely supercritical, the jump is more like a standing wave, without a nearly step change in depth. In some Froude number ranges the jump is unsteady, with regular periodic oscillations traveling downstream. (Recall that the wave cannot travel upstream against the supercritical flow.)

The length of a hydraulic jump (the distance between the nearly uniform upstream and downstream flows) may be of importance in the design of channels. Although its value cannot be determined theoretically, experimental results indicate that over a wide range of Froude numbers, the jump is approximately seven downstream depths long (Ref. 5).

V10.6

Hydraulic jumps can occur in a variety of channel flow configurations, not just in horizontal, rectangular channels as discussed above. Jumps in nonrectangular channels (i.e., circular pipes, trapezoidal canals, etc.) behave in a manner quite like those in rectangular

■ **TABLE 10.3**
Classification of Hydraulic Jumps (Ref. 12)

Fr_1	y_2/y_1	Classification	Sketch
<1	1	Jump impossible	
1 to 1.7	1 to 2.0	Standing wave or undulant jump	
1.7 to 2.5	2.0 to 3.1	Weak jump	
2.5 to 4.5	3.1 to 5.9	Oscillating jump	
4.5 to 9.0	5.9 to 12	Stable, well-balanced steady jump; insensitive to downstream conditions	
>9.0	>12	Rough, somewhat intermittent strong jump	

channels, although the details of the depth ratio and head loss are somewhat different from jumps in rectangular channels.

Other common types of hydraulic jumps include those that occur in sloping channels as is indicated in Fig. 10.21*a* and the submerged hydraulic jumps that can occur just downstream of a sluice gate as is indicated in Fig. 10.21*b*. Details of these and other jumps can be found in standard open-channel flow references (Refs. 3 and 5).

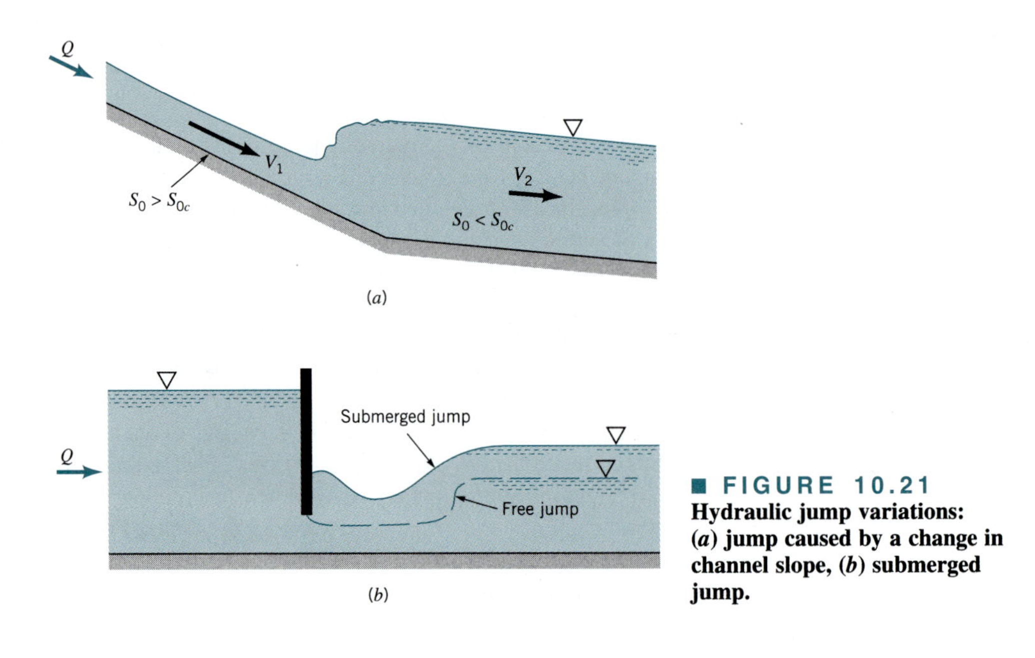

■ **FIGURE 10.21** **Hydraulic jump variations: (*a*) jump caused by a change in channel slope, (*b*) submerged jump.**

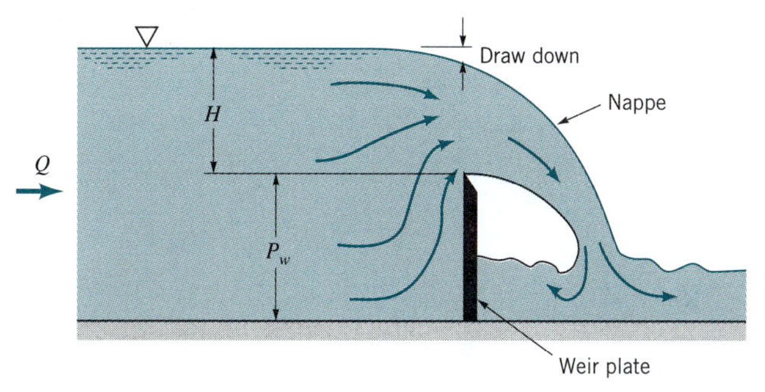

■ **FIGURE 10.22** **Sharp-crested weir geometry.**

10.6.2 Sharp-Crested Weirs

A sharp-crested weir can be used to determine the flowrate.

A weir is an obstruction on a channel bottom over which the fluid must flow. It provides a convenient method of determining the flowrate in an open channel in terms of a single depth measurement. A *sharp-crested weir* is essentially a vertical sharp-edged flat plate placed across the channel in a way such that the fluid must flow across the sharp edge and drop into the pool downstream of the weir plate, as is shown in Fig. 10.22. The specific shape of the flow area in the plane of the weir plate is used to designate the type of weir. Typical shapes include the rectangular weir, the triangular weir, and the trapezoidal weir, as indicated in Fig. 10.23.

The complex nature of the flow over a weir makes it impossible to obtain precise analytical expressions for the flow as a function of other parameters, such as the weir height, P_w, the head of the weir, H, the fluid depth upstream, and the geometry of the weir plate (angle θ for triangular weirs or aspect ratio, b/H, for rectangular weirs). The flow structure is far from one-dimensional, with a variety of interesting flow phenomena obtained.

The main mechanisms governing flow over a weir are gravity and inertia. From a highly simplified point of view, gravity accelerates the fluid from its free-surface elevation upstream of the weir to larger velocity as it flows down the hill formed by the nappe. Although viscous and surface tension effects are usually of secondary importance, such effects cannot be entirely neglected. Generally, appropriate experimentally determined coefficients are used to account for these effects.

As a first approximation, we assume that the velocity profile upstream of the weir plate is uniform and that the pressure within the nappe is atmospheric. In addition, we assume that the fluid flows horizontally over the weir plate with a nonuniform velocity profile, as indicated

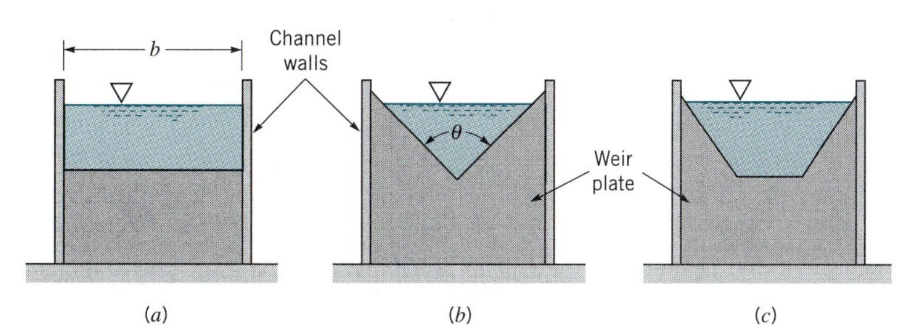

■ **FIGURE 10.23** **Sharp-crested weir plate geometry: (*a*) rectangular, (*b*) triangular, (*c*) trapezoidal.**

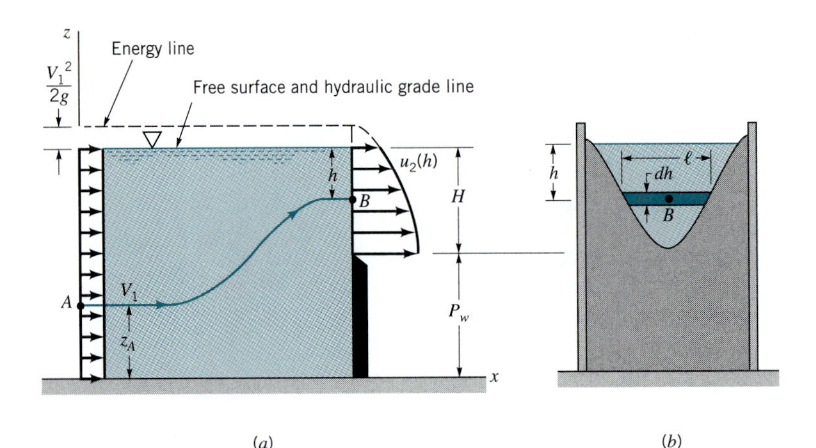

■ FIGURE 10.24 **Assumed flow structure over a weir.**

in Fig. 10.24. With $p_B = 0$ the Bernoulli equation for flow along the arbitrary streamline A–B indicated can be written as

$$\frac{p_A}{\gamma} + \frac{V_1^2}{2g} + z_A = (H + P_w - h) + \frac{u_2^2}{2g} \tag{10.26}$$

The Bernoulli equation can be used to determine the fluid velocity over a weir.

where h is the distance that point B is below the free surface. We do not know the location of point A from which came the fluid that passes over the weir at point B. However, since the total head for any particle along the vertical section (1) is the same, $z_A + p_A/\gamma + V_1^2/2g = H + P_w + V_1^2/2g$, the specific location of A is not needed, and the velocity of the fluid over the weir plate is obtained from Eq. 10.26 as

$$u_2 = \sqrt{2g\left(h + \frac{V_1^2}{2g}\right)}$$

The flowrate can be calculated from

$$Q = \int_{(2)} u_2 \, dA = \int_{h=0}^{h=H} u_2 \ell \, dh \tag{10.27}$$

where $\ell = \ell(h)$ is the cross-channel width of a strip of the weir area, as is indicated in Fig. 10.24*b*. For a rectangular weir ℓ is constant. For other weirs, such as triangular or circular weirs, the value of ℓ is known as a function of h.

For a rectangular weir, $\ell = b$, and the flowrate becomes

$$Q = \sqrt{2g}\, b \int_0^H \left(h + \frac{V_1^2}{2g}\right)^{1/2} dh$$

or

$$Q = \frac{2}{3}\sqrt{2g}\, b\left[\left(H + \frac{V_1^2}{2g}\right)^{3/2} - \left(\frac{V_1^2}{2g}\right)^{3/2}\right] \tag{10.28}$$

Equation 10.28 is a rather cumbersome expression that can be simplified by using the fact that with $P_w \gg H$ (as often happens in practical situations) the upstream velocity is negligibly small. That is, $V_1^2/2g \ll H$ and Eq. 10.28 simplifies to the basic rectangular weir equation

$$Q = \tfrac{2}{3}\sqrt{2g}\, b\, H^{3/2} \tag{10.29}$$

Note that the weir head, H, is the height of the upstream free surface above the crest of the weir. As is indicated in Fig. 10.22, because of the drawdown effect, H is not the distance of the free surface above the weir crest as measured directly above the weir plate.

Because of the numerous approximations made to obtain Eq. 10.29, it is not unexpected that an experimentally determined correction factor must be used to obtain the actual flowrate as a function of weir head. Thus, the final form is

A weir coefficient is used to account for nonideal conditions excluded in the simplified analysis.

$$Q = C_{wr} \tfrac{2}{3} \sqrt{2g}\; b\; H^{3/2} \tag{10.30}$$

where C_{wr} is the rectangular weir coefficient. From dimensional analysis arguments, it is expected that C_{wr} is a function of Reynolds number (viscous effects), Weber number (surface tension effects), and H/P_w (geometry). In most practical situations, the Reynolds and Weber number effects are negligible, and the following correlation can be used (Refs. 4 and 7):

$$C_{wr} = 0.611 + 0.075 \left(\frac{H}{P_w}\right) \tag{10.31}$$

More precise values of C_{wr} can be found in the literature, if needed (Refs. 3, 14).

V10.7

The triangular sharp-crested weir is often used for flow measurements, particularly for measuring flowrates over a wide range of values. For small flowrates, the head, H, for a rectangular weir would be very small and the flowrate could not be measured accurately. However, with the triangular weir, the flow area decreases as H decreases so that even for small flowrates, reasonable heads are developed. Accurate results can be obtained over a wide range of Q.

The triangular weir equation can be obtained from Eq. 10.27 by using

$$\ell = 2(H - h) \tan\left(\frac{\theta}{2}\right)$$

where θ is the angle of the V-notch (see Figs. 10.23 and 10.24). After carrying out the integration and again neglecting the upstream velocity ($V_1^2/2g \ll H$), we obtain

$$Q = \frac{8}{15} \tan\left(\frac{\theta}{2}\right) \sqrt{2g}\; H^{5/2}$$

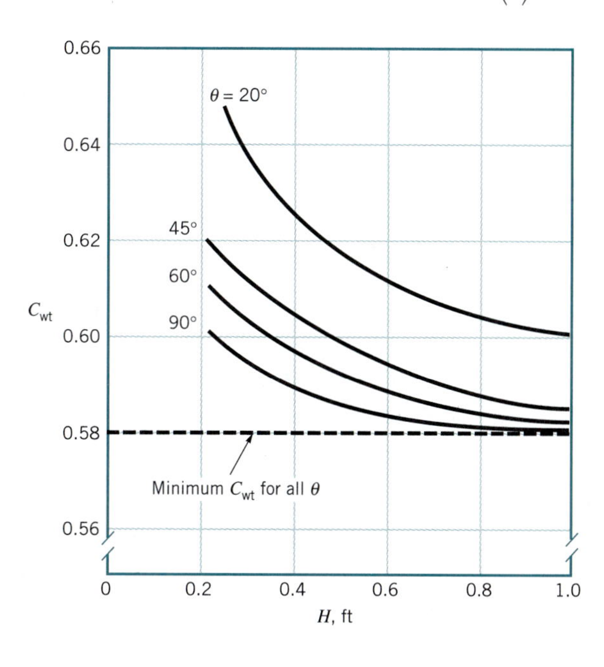

■ **FIGURE 10.25 Weir coefficient for triangular sharp-crested weirs (Ref. 10).**

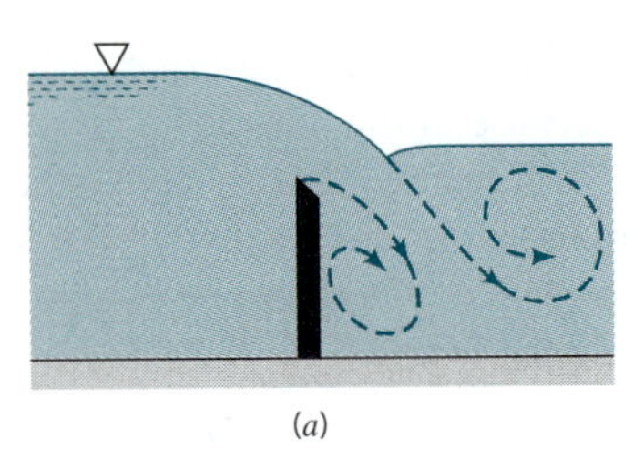

(a)

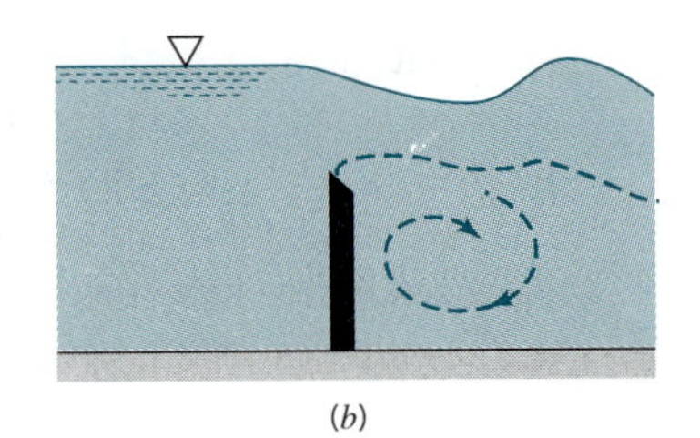

(b)

■ **FIGURE 10.26 Flow conditions over a weir without a free nappe: (*a*) plunging nappe, (*b*) submerged nappe.**

An experimentally determined triangular weir coefficient, C_{wt}, is used to account for the real world effects neglected in the analysis so that

$$Q = C_{wt} \frac{8}{15} \tan\left(\frac{\theta}{2}\right) \sqrt{2g}\, H^{5/2} \tag{10.32}$$

Typical values of C_{wt} for triangular weirs are in the range of 0.58 to 0.62, as is shown in Fig. 10.25. Note that although C_{wt} and θ are dimensionless, the value of C_{wt} is given as a function of the weir head, H, which is a dimensional quantity. Although using dimensional parameters is not recommended (see the dimensional analysis discussion in Chapter 7), such parameters are often used for open-channel flow.

V10.8

Flowrate over a weir depends on whether the nappe is free or submerged.

The above results for sharp-crested weirs are valid provided the area under the nappe is ventilated to atmospheric pressure. Although this is not a problem for triangular weirs, for rectangular weirs it is sometimes necessary to provide ventilation tubes to ensure atmospheric pressure in this region. In addition, depending on downstream conditions, it is possible to obtain submerged weir operation, as is indicated in Fig. 10.26. Clearly the flowrate will be different for these situations than that given by Eqs. 10.30 and 10.32.

10.6.3 Broad-Crested Weirs

A *broad-crested weir* is a structure in an open channel that has a horizontal crest above which the fluid pressure may be considered hydrostatic. A typical configuration is shown in Fig. 10.27. Generally, to ensure proper operation, these weirs are restricted to the range $0.08 < H/L_w < 0.50$. For long weir blocks (H/L_w less than 0.08), head losses across the weir cannot be neglected. On the other hand, for short weir blocks (H/L_w greater than 0.50) the streamlines of the flow over the weir block are not horizontal. Although broad-crested weirs can be used in channels of any cross-sectional shape, we restrict our attention to rectangular channels.

The operation of a broad-crested weir is based on the fact that nearly uniform critical flow is achieved in the short reach above the weir block. (If $H/L_w < 0.08$, viscous effects are important, and the flow is subcritical over the weir.) If the kinetic energy of the upstream flow is negligible, then $V_1^2/2g \ll y_1$ and the upstream specific energy is $E_1 = V_1^2/2g + y_1 \approx y_1$. Observations show that as the flow passes over the weir block, it accelerates and reaches critical conditions, $y_2 = y_c$ and $\text{Fr}_2 = 1$ (i.e., $V_2 = c_2$), corresponding to the nose of the specific energy curve (see Fig. 10.7). The flow does not accelerate to supercritical conditions ($\text{Fr}_2 > 1$). To do so would require the ability of the downstream fluid to communicate with the upstream fluid to let it know that there is an end of the weir block. Since waves cannot propagate upstream against a critical flow, this information cannot be transmitted. The flow remains critical, not supercritical, across the weir block.

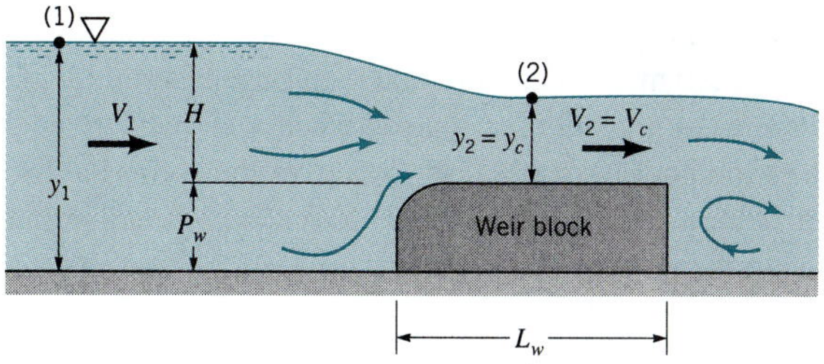

■ FIGURE 10.27 Broad-crested weir geometry.

The Bernoulli equation can be applied between point (1) upstream of the weir and point (2) over the weir where the flow is critical to obtain

$$H + P_w + \frac{V_1^2}{2g} = y_c + P_w + \frac{V_c^2}{2g}$$

or, if the upstream velocity head is negligible

$$H - y_c = \frac{(V_c^2 - V_1^2)}{2g} = \frac{V_c^2}{2g}$$

The broad-crested weir is governed by critical flow across the weir block.

However, since $V_2 = V_c = (gy_c)^{1/2}$, we find that $V_c^2 = gy_c$ so that we obtain

$$H - y_c = \frac{y_c}{2}$$

or

$$y_c = \frac{2H}{3}$$

Thus, the flowrate is

$$Q = by_2V_2 = by_cV_c = by_c(gy_c)^{1/2} = b\sqrt{g}\, y_c^{3/2}$$

or

$$Q = b\sqrt{g}\left(\frac{2}{3}\right)^{3/2} H^{3/2}$$

Again an empirical weir coefficient is used to account for the various real world effects not included in the above simplified analysis. That is

$$Q = C_{wb}\, b\sqrt{g}\left(\frac{2}{3}\right)^{3/2} H^{3/2} \quad \textbf{(10.33)}$$

where typical values of C_{wb}, the broad-crested weir coefficient, can be obtained from the equation (Ref. 6)

$$C_{wb} = \frac{0.65}{(1 + H/P_w)^{1/2}} \quad \textbf{(10.34)}$$

EXAMPLE 10.10

Water flows in a rectangular channel of width $b = 2$ m with flowrates between $Q_{\min} = 0.02$ m^3/s and $Q_{\max} = 0.60$ m^3/s. This flowrate is to be measured by using either (a) a rectangular sharp-crested weir, (b) a triangular sharp-crested weir with $\theta = 90°$, or (c) a broad-crested weir. In all cases the bottom of the flow area over the weir is a distance $P_w = 1$ m above the channel bottom. Plot a graph of $Q = Q(H)$ for each weir and comment on which weir would be best for this application.

SOLUTION

For the rectangular weir with $P_w = 1$ m, Eqs. 10.30 and 10.31 give

$$Q = C_{\text{wt}} \frac{2}{3} \sqrt{2g}\, bH^{3/2} = \left(0.611 + 0.075 \frac{H}{P_w}\right) \frac{2}{3} \sqrt{2g}\, bH^{3/2}$$

Thus,

$$Q = (0.611 + 0.075H) \frac{2}{3} \sqrt{2(9.81 \text{ m/s}^2)}\, (2 \text{ m})\, H^{3/2}$$

or

$$Q = 5.91(0.611 + 0.075H)H^{3/2} \tag{1}$$

where H and Q are in meters and m^3/s, respectively. The results from Eq. 1 are plotted in Fig. E10.10.

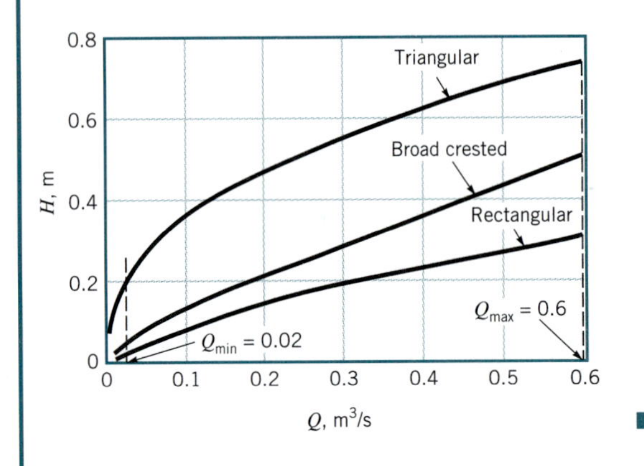

■ FIGURE E10.10

Similarly, for the triangular weir, Eq. 10.32 gives

$$Q = C_{\text{wt}} \frac{8}{15} \tan\left(\frac{\theta}{2}\right) \sqrt{2g}\, H^{5/2}$$

$$= C_{\text{wt}} \frac{8}{15} \tan(45°) \sqrt{2(9.81 \text{ m/s}^2)}\, H^{5/2}$$

or

$$Q = 2.36C_{\text{wt}}\, H^{5/2} \tag{2}$$

where H and Q are in meters and m^3/s and C_{wt} is obtained from Fig. 10.25. For example, with $H = 0.20$ m, we find $C_{\text{wt}} = 0.60$, or $Q = 2.36\,(0.60)(0.20)^{5/2} = 0.0253$ m^3/s. The triangular weir results are also plotted in Fig. E10.10.

For the broad-crested weir, Eqs. 10.33 and 10.34 give

$$Q = C_{wb}\, b\sqrt{g}\left(\frac{2}{3}\right)^{3/2} H^{3/2} = \frac{0.65}{(1 + H/P_w)^{1/2}}\, b\sqrt{g}\left(\frac{2}{3}\right)^{3/2} H^{3/2}$$

Thus, with $P_w = 1$ m

$$Q = \frac{0.65}{(1 + H)^{1/2}}\,(2\text{ m})\,\sqrt{9.81\text{ m/s}^2}\left(\frac{2}{3}\right)^{3/2} H^{3/2}$$

or

$$Q = \frac{2.22}{(1 + H)^{1/2}}\, H^{3/2} \tag{3}$$

where, again, H and Q are in meters and m^3/s. This result is also plotted in Fig. E10.10.

Although it appears as though any of the three weirs would work well for the upper portion of the flowrate range, neither the rectangular nor the broad-crested weir would be very accurate for small flowrates near $Q = Q_{min}$ because of the small head, H, at these conditions. The triangular weir, however, would allow reasonably large values of H at the lowest flowrates. The corresponding heads with $Q = Q_{min} = 0.02\text{ m}^3/\text{s}$ for rectangular, triangular, and broad-crested weirs are 0.0312 m, 0.182 m, and 0.0440 m, respectively.

In addition, as is discussed in this section, for proper operation the broad-crested weir geometry is restricted to $0.08 < H/L_w < 0.50$, where L_w is the weir block length. From Eq. 3 with $Q_{max} = 0.60\text{ m}^3/\text{s}$, we obtain $H_{max} = 0.476$. Thus, we must have $L_w > H_{max}/0.5 = 0.952$ m to maintain proper critical flow conditions at the largest flowrate in the channel. On the other hand, with $Q = Q_{min} = 0.02\text{ m}^3/\text{s}$, we obtain $H_{min} = 0.0440$ m. Thus, we must have $L_w < H_{min}/0.08 = 0.549$ m to ensure that frictional effects are not important. Clearly, these two constraints on the geometry of the weir block, L_w, are incompatible.

A broad-crested weir will not function properly under the wide range of flowrates considered in this example. The sharp-crested triangular weir would be the best of the three types considered, provided the channel can handle the $H_{max} = 0.719$-m head.

10.6.4 Underflow Gates

Various underflow gate geometries are available for flowrate control.

A variety of gate-type structures is available for flowrate control at the crest of an overflow spillway, or at the entrance of an irrigation canal or river from a lake. Three typical types are illustrated in Fig. 10.28. Each has certain advantages and disadvantages in terms of costs of construction, ease of use, and the like, although the basic fluid mechanics involved are the same in all instances.

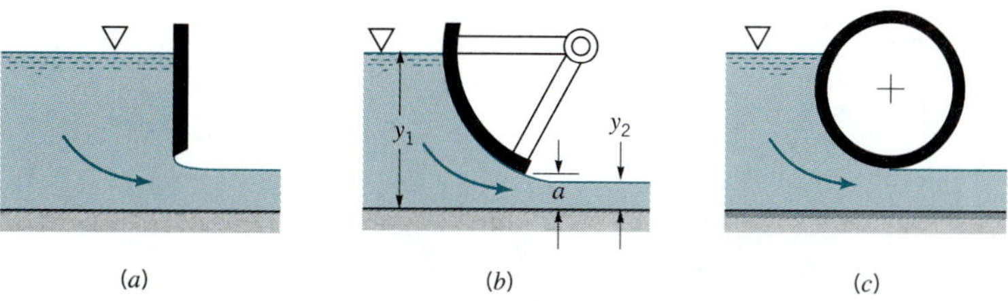

■ **FIGURE 10.28 Three variations of underflow gates: (*a*) vertical gate, (*b*) radial gate, (*c*) drum gate.**

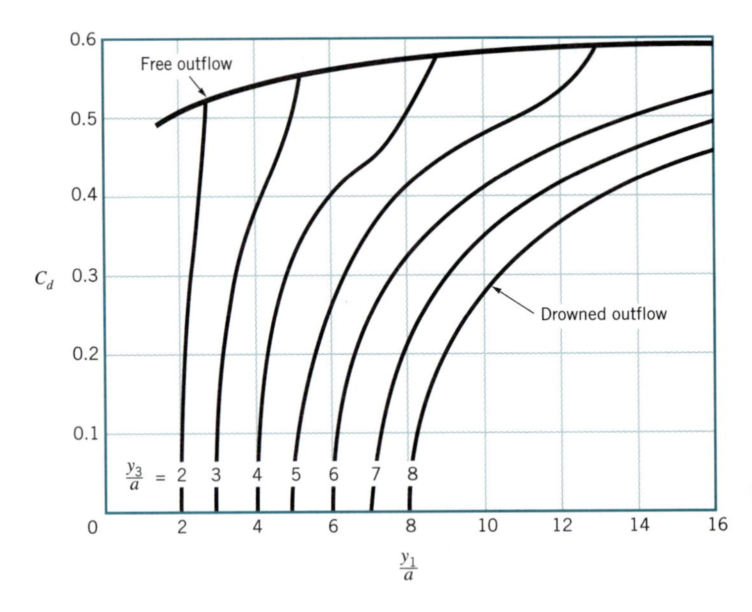

■ **FIGURE 10.29 Typical discharge coefficients for underflow gates (Ref. 3).**

The flow under a gate is said to be free outflow when the fluid issues as a jet of supercritical flow with a free surface open to the atmosphere as shown in Fig. 10.28. In such cases it is customary to write this flowrate as the product of the distance, a, between the channel bottom and the bottom of the gate times the convenient reference velocity $(2gy_1)^{1/2}$. That is

$$q = C_d a \sqrt{2gy_1} \tag{10.35}$$

where q is the flowrate per unit width. The discharge coefficient, C_d, is a function of the contraction coefficient, $C_c = y_2/a$, and the depth ratio y_1/a. Typical values of the discharge coefficient for free outflow (or free discharge) from a vertical sluice gate are on the order of 0.55 to 0.60 as indicated by the top line in Fig. 10.29 (Ref. 3).

As is indicated in Fig. 10.30, in certain situations the depth downstream of the gate is controlled by some downstream obstacle and the jet of water issuing from under the gate is overlaid by a mass of water that is quite turbulent.

The flowrate from an underflow gate depends on whether the outlet is free or drowned.

The flowrate for a submerged (or drowned) gate can be obtained from the same equation that is used for free outflow (Eq. 10.35), provided the discharge coefficient is modified appropriately. Typical values of C_d for drowned outflow cases are indicated as the series of lower curves in Fig. 10.29. Consider flow for a given gate and upstream conditions (i.e., given y_1/a) corresponding to a vertical line in the figure. With $y_3/a = y_1/a$ (i.e., $y_3 = y_1$) there is no head to drive the flow so that $C_d = 0$ and the fluid is stationary. For a given upstream depth (y_1/a fixed), the value of C_d increases with decreasing y_3/a until the maximum value

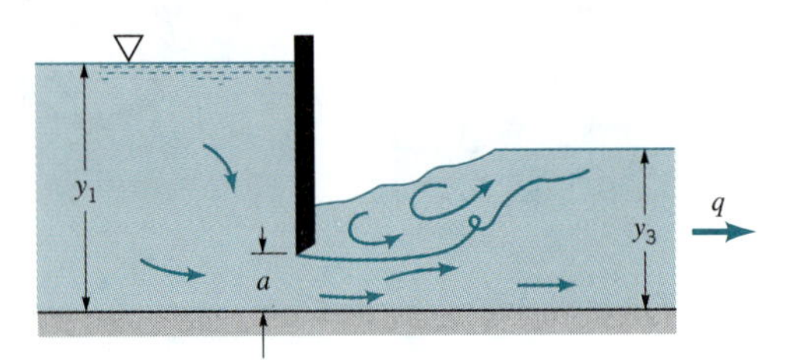

■ **FIGURE 10.30 Drowned outflow from a sluice gate.**

of C_d is reached. This maximum corresponds to the free discharge conditions and is represented by the free outflow line so labeled in Fig. 10.29. For values of y_3/a that give C_d values between zero and its maximum, the jet from the gate is overlaid (drowned) by the downstream water and the flowrate is therefore reduced when compared with a free discharge situation. Similar results are obtained for the radial gate and drum gate.

EXAMPLE 10.11

Water flows under the sluice gate shown in Fig. E10.11. The channel width is $b = 20$ ft, the upstream depth is $y_1 = 6$ ft, and the gate is $a = 1.0$ ft off the channel bottom. Plot a graph of flowrate, Q, as a function of y_3.

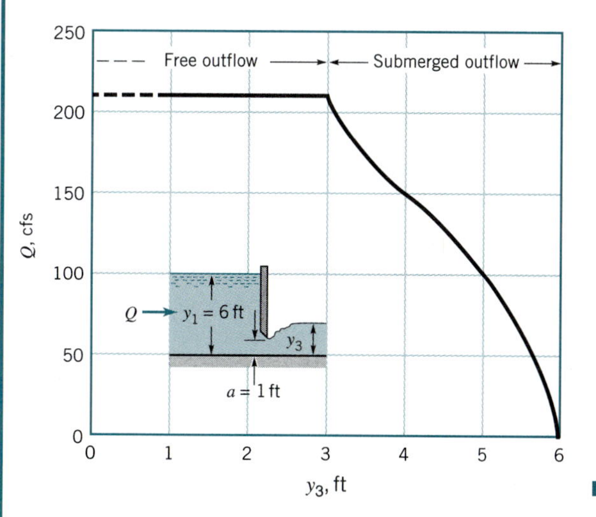

■ FIGURE E10.11

SOLUTION

From Eq. 10.35 we have

$$Q = bq = baC_d\sqrt{2gy_1} = 20 \text{ ft } (1.0 \text{ ft})\, C_d \sqrt{2(32.2 \text{ ft/s}^2)(6.0 \text{ ft})}$$

or

$$Q = 393C_d \text{ cfs} \tag{1}$$

The value of C_d is obtained from Fig. 10.29 along the vertical line $y_1/a = 6$ ft/1 ft $= 6$. For $y_3 = 6$ ft (i.e., $y_3/a = 6 = y_1/a$) we obtain $C_d = 0$ indicating that there is no flow when there is no head difference across the gate. The value of C_d increases as y_3/a decreases, reaching a maximum of $C_d = 0.56$ when $y_3/a = 3.2$ Thus, with $y_3 = 3.2a = 3.2$ ft

$$Q = 393\ (0.56) \text{ cfs} = 220 \text{ cfs}$$

For $y_3 < 3.2$ ft the flowrate is independent of y_3, and the outflow is a free (not submerged) outflow. For such cases the inertia of the water flowing under the gate is sufficient to produce free outflow even with $y_3 > a$.

The flowrate for 3.2 ft $\leq y_3 \leq 6$ ft is obtained from Eq. 1 and the C_d values of Fig. 10.28 with the results as indicated in Fig. E10.11.

References

1. Currie, I. G., *Fundamental Mechanics of Fluids*, McGraw-Hill, New York, 1974.
2. Stoker, J. J., *Water Waves*, Interscience, New York, 1957.
3. Henderson, F. M., *Open Channel Flow*, Macmillan, New York, 1966.
4. Rouse, H, *Elementary Fluid Mechanics*, Wiley, New York, 1946.
5. French, R. H., *Open Channel Hydraulics*, McGraw-Hill, New York, 1985.
6. Chow, V. T., *Open Channel Hydraulics*, McGraw-Hill, New York, 1959.
7. Blevins, R. D., *Applied Fluid Dynamics Handbook*, Van Nostrand Reinhold, New York, 1984.
8. Daugherty, R. L., and Franzini, J. B., *Fluid Mechanics with Engineering Applications*, McGraw-Hill, New York, 1977.
9. Vennard, J. K., and Street, R. L., *Elementary Fluid Mechanics*, Wiley, New York, 1976.
10. Lenz, A. T., ''Viscosity and Surface Tension Effects on V-Notch Weir Coefficients,'' *Transactions of the American Society of Chemical Engineers*, Vol. 108, 759–820, 1943.
11. White, F. M., *Fluid Mechanics*, McGraw-Hill, New York, 1986.
12. U.S. Bureau of Reclamation, Research Studies on Stilling Basins, Energy Dissipators, and Associated Appurtenances, Hydraulic Lab Report Hyd.-399, June 1, 1955.
13. Wallet, A., and Ruellan, F., *Houille Blanche*, Vol. 5, 1950.
14. Spitzer, D. W., ed., *Flow Measurement: Practical Guides for Measurement and Control*, Instrument Society of America, Research Triangle Park, 1991.

Review Problems

Note: Problems designated with (R) are review problems. The phrases within parentheses refer to the main topics to be used in solving the problems. Complete, detailed solutions to these review problems can be found in the supplement titled *Student Solution Manual for Fundamentals of Fluid Mechanics*, by Munson, Young, and Okiishi (John Wiley and Sons, New York, 1997).

10.1R (Surface waves) If the water depth in a pond is 15 ft, determine the speed of small amplitude, long wavelength ($\lambda \gg y$) waves on the surface.

(ANS: 22.0 ft/s)

10.2R (Surface waves) A small amplitude wave travels across a pond with a speed of 9.6 ft/s. Determine the water depth.

(ANS: $y \geq 2.86$ ft)

10.3R (Froude number) The average velocity and average depth of a river from its beginning in the mountains to its discharge into the ocean are given in the table below. Plot a graph of the Froude number as a function of distance along the river.

Distance (mi)	Average Velocity (ft/s)	Average Depth (ft)
0	13	1.5
5	10	2.0
10	9	2.3
30	5	3.7
50	4	5.4
80	4	6.0
90	3	6.2

(ANS: Fr = 1.87 at the beginning, 0.212 at the discharge)

10.4R (Froude number) Water flows in a rectangular channel at a depth of 4 ft and a flowrate of $Q = 200$ cfs. Determine the minimum channel width if the flow is to be subcritical.

(ANS: 4.41 ft)

10.5R (Specific energy) Plot the specific energy diagram for a wide channel carrying $q = 50$ ft^2/s. Determine **(a)** the critical depth, **(b)** the minimum specific energy, **(c)** the alternate depth corresponding to a depth of 2.5 ft, and **(d)** the possible flow velocities if $E = 10$ ft.

(ANS: 4.27 ft; 6.41 ft; 8.12 ft; 5.22 ft/s or 22.3 ft/s)

10.6R (Specific energy) Water flows at a rate of 1000 ft^3/s in a horizontal rectangular channel 30 ft wide with a 2-ft depth. Determine the depth if the channel contracts to a width of 25 ft. Explain.

(ANS: 2.57 ft)

10.7R (Wall shear stress) Water flows in a 10-ft-wide rectangular channel with a flowrate of 150 cfs and a depth of 3 ft. If the slope is 0.005, determine the Manning coefficient, n, and the average shear stress at the sides and bottom of the channel.

(ANS: 0.0320; 0.585 lb/ft^2)

10.8R (Manning equation) The triangular flume shown in Fig. P10.8R is built to carry its design flowrate, Q_0, at a depth of 0.90 m as is indicated. If the flume is to be able to carry up to twice its design flowrate, $Q = 2Q_0$, determine the freeboard, ℓ, needed.

(ANS: 0.378 m)

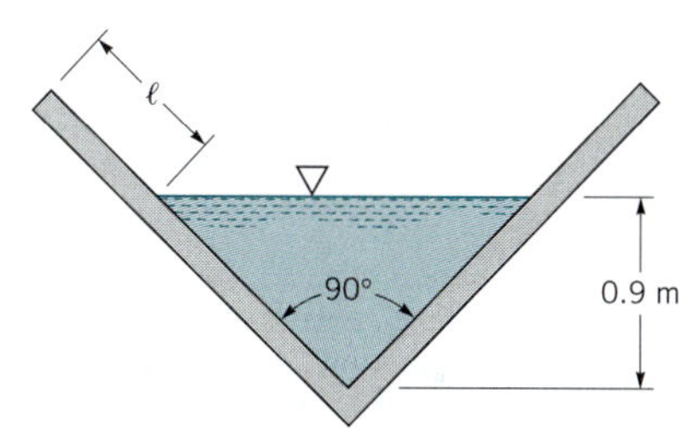

■ **FIGURE P10.8R**

10.9R (Manning equation) Water flows in a rectangular channel of width b at a depth of $b/3$. Determine the diameter of a circular channel (in terms of b) that carries the same flowrate when it is half-full. Both channels have the same Manning coefficient, n, and slope.

(ANS: 0.889 b)

10.10R (Manning equation) A weedy irrigation canal of trapezoidal cross section is to carry 20 m^3/s when built on a slope of 0.56 m/km. If the sides are at a 45° angle and the bottom is 8 m wide, determine the width of the waterline at the free surface.

(ANS: 12.0 m)

10.11R (Manning equation) Determine the maximum flowrate possible for the creek shown in Fig. P10.11R if it is not to overflow onto the floodplain. The creek bed drops an average of 5 ft/half mile of length. Determine the flowrate during a flood if the depth is 8 ft.

(ANS: 182 ft^3/s; 1517 ft^3/s)

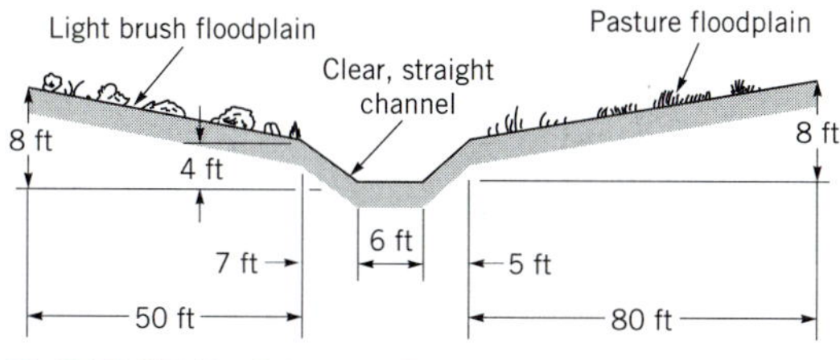

■ **FIGURE P10.11R**

10.12R (Best hydraulic cross section) Show that the triangular channel with the best hydraulic cross section (i.e., minimum area, A, for a given flowrate) is a right triangle as is shown in Fig. E10.8b.

10.13R (Hydraulic jump) At the bottom of a water ride in an amusement park, the water in the rectangular channel has a depth of 1.2 ft and a velocity of 15.6 ft/s. Determine the height of the "standing wave" (a hydraulic jump) that the boat passes through for its final "splash."

(ANS: 2.50 ft)

10.14R (Hydraulic jump) Water flows in a rectangular channel with velocity V = 6 m/s. A gate at the end of the channel is suddenly closed so that a wave (a moving hydraulic jump) travels upstream with velocity V_w = 2 m/s as is indicated in Fig. P10.14R. Determine the depths ahead of and behind the wave. Note that this is an unsteady problem for a stationary observer. However, for an observer moving to the left with velocity V_w, the flow appears as a steady hydraulic jump.

(ANS: 0.652 m; 2.61 m)

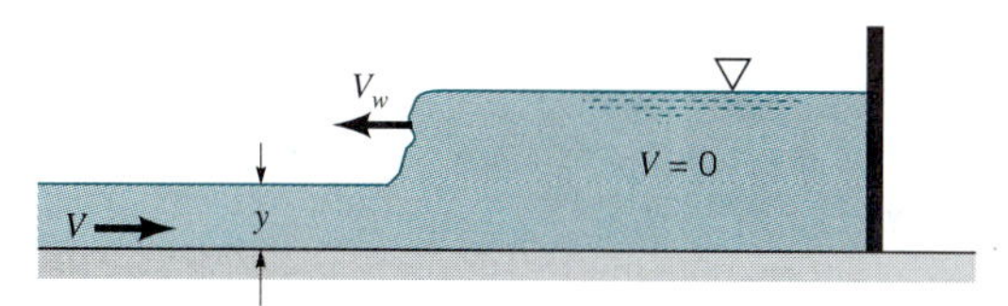

■ **FIGURE P10.14R**

10.15R (Sharp-crested weir) Determine the head, H, required to allow a flowrate of 600 m^3/hr over a sharp-crested triangular weir with θ = 60°.

(ANS: 0.536 m)

10.16R (Broad-crested weir) The top of a broad-crested weir block is at an elevation of 724.5 ft, which is 4 ft above the channel bottom. If the weir is 20-ft wide and the flowrate is 400 cfs, determine the elevation of the reservoir upstream of the weir.

(ANS: 730.86 ft)

10.17R (Underflow gate) Water flows under a sluice gate in a 60-ft-wide finished concrete channel as is shown in Fig. P10.17R. Determine the flowrate. If the slope of the channel is 2.5 ft/100 ft, will the water depth increase or decrease downstream of the gate? Assume $C_c = y_2/a = 0.65$. Explain.

(ANS: 1670 ft^3/s; decrease)

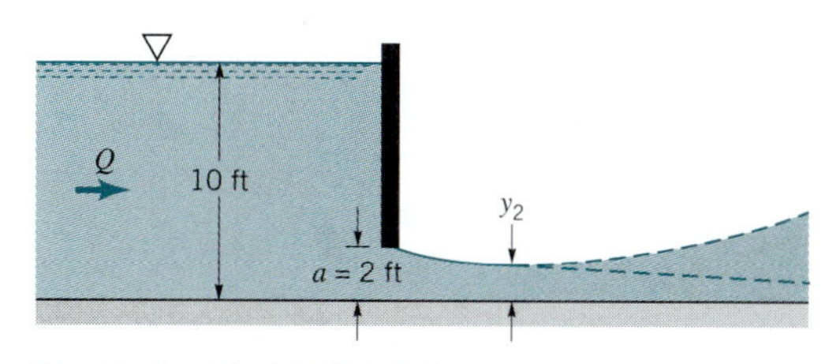

■ **FIGURE P10.17R**

Problems

Note: Unless otherwise indicated, use the values of fluid properties found in the tables on the inside of the front cover. Problems designated with an (*) are intended to be solved with the aid of a programmable calculator or a computer. Problems designated with a (†) are "open ended" problems and require critical thinking in that to work them one must make various assumptions and provide the necessary data. There is not a unique answer to these problems.

10.1 The flowrate in a 10-ft-wide, 2-ft-deep river is $Q = 190$ cfs. Is the flow subcritical or supercritical?

10.2 The flowrate per unit width in a wide channel is $q = 2.3\ m^2/s$. Is the flow subcritical or supercritical if the depth is **(a)** 0.2 m, **(b)** 0.8 m, or **(c)** 2.5 m?

10.3 Water flows in a canal at a depth of 2.8 ft and a velocity of 5.3 ft/s. Will waves produced by throwing a stick into the canal travel both upstream and downstream, or will they all be washed downstream? Explain.

10.4 A trout jumps, producing waves on the surface of a 0.8-m-deep mountain stream. If it is observed that the waves do not travel upstream, what is the minimum velocity of the current?

10.5 Waves on the surface of a tank are observed to travel at a speed of 2 m/s. How fast would these waves travel if **(a)** the tank were in an elevator accelerating upward at a rate of 4 m/s^2, **(b)** the tank accelerates horizontally at a rate of 9.81 m/s^2, **(c)** the tank were aboard the orbiting Space Shuttle. Explain.

10.6 In flowing from section (1) to section (2) along an open channel, the water depth decreases by a factor of two and the Froude number changes from a subcritical value of 0.5 to a supercritical value of 3.0. Determine the channel width at (2) if it is 12 ft wide at (1).

10.7 Observations at a shallow sandy beach show that even though the waves several hundred yards out from the shore are not parallel to the beach, the waves often "break" on the beach nearly parallel to the shore as is indicated in Fig. P10.7. Explain this behavior based on the wave speed $c = (gy)^{1/2}$.

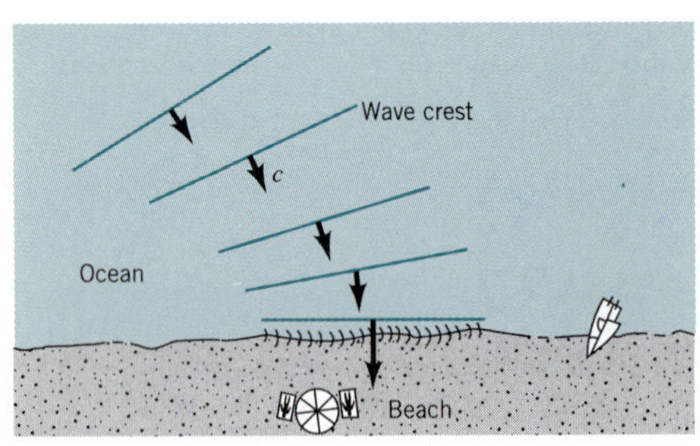

■ FIGURE P10.7

10.8 Waves on the surface of a tank containing water are observed to move with a velocity of 1.8 m/s. If the water is replaced by mercury, with all other conditions the same, determine the wave speed expected. Determine the wave speed if the tank were in a laboratory on the surface of a planet where the acceleration of gravity is 4 times that on earth.

10.9 Often when an earthquake shifts a segment of the ocean floor, a relatively small amplitude wave of very long wavelength is produced. Such waves go unnoticed as they move across the open ocean; only when they approach the shore do they become dangerous (a tsunami or "tidal wave"). Determine the wave speed if the wavelength, λ, is 6000 ft and the ocean depth is 15,000 ft.

† **10.10** It is interesting to observe waves on the surface of the ocean. Usually there are waves of various wave lengths and amplitudes moving across the surface. Do the short wavelength waves overtake the longer waves, or is the opposite true? Explain.

10.11 Water flows in a rectangular channel with a flowrate per unit width of $q = 2.5\ m^2/s$. Plot the specific energy diagram for this flow. Determine the two possible depths of flow if $E = 2.5$ m.

10.12 Water flows radially outward on a horizontal round disk as is shown in Fig. P10.12. **(a)** Show that the specific energy can be written in terms of the flowrate, Q, the radial distance from the axis of symmetry, r, and the fluid depth, y, as

$$E = y + \left(\frac{Q}{2\pi r}\right)^2 \frac{1}{2gy^2}$$

(b) For a constant flowrate, sketch the specific energy diagram. Recall Fig. 10.7, but note that for the present case r is a variable. Explain the important characteristics of your sketch. **(c)** Based on the results of Part (b), show that the water depth increases in the flow direction if the flow is subcritical, but that it decreases in the flow direction if the flow is supercritical.

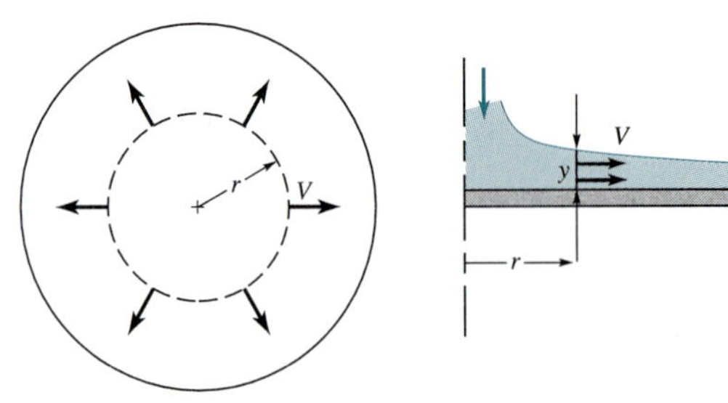

■ FIGURE P10.12

***10.13** Water flows in a rectangular channel with a specific energy of $E = 5$ ft. If the flowrate per unit width is $q = 30\ ft^2/s$, determine the two possible flow depths and the corresponding Froude numbers. Plot the specific energy diagram for this flow. Repeat the problem for $E = 1, 2, 3$, and 4 ft.

10.14 Water flows in a rectangular channel at a rate of q = 20 cfs/ft. When a Pitot tube is placed in the stream, water in the tube rises to a level of 4.5 ft above the channel bottom. Determine the two possible flow depths in the channel. Illustrate this flow on a specific energy diagram.

10.15 Water flows in a 5-ft-wide rectangular channel with a flowrate of $Q = 30\ \text{ft}^3/\text{s}$ and an upstream depth of $y_1 = 2.5$ ft as is shown in Fig. P10.15. Determine the flow depth and the surface elevation at section (2).

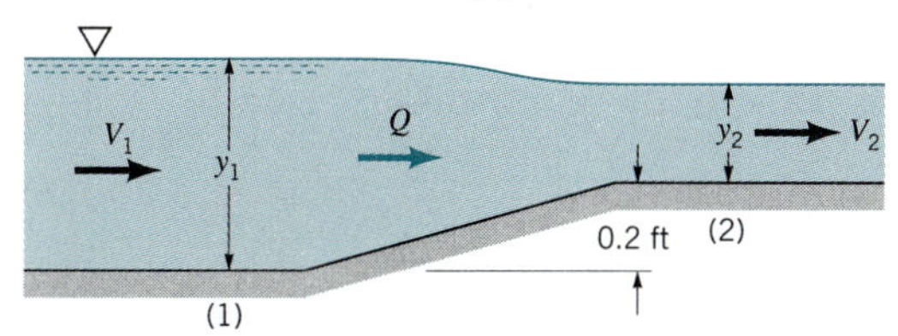

■ **FIGURE P10.15**

10.16 Repeat Problem 10.15 if the upstream depth is $y_1 = 0.5$ ft.

***10.17** Water flows over the bump in the bottom of the rectangular channel shown in Fig. P10.17 with a flowrate per unit width of $q = 4\ \text{m}^2/\text{s}$. The channel bottom contour is given by $z_B = 0.2e^{-x^2}$, where z_B and x are in meters. The water depth far upstream of the bump is $y_1 = 2$ m. Plot a graph of the water depth, $y = y(x)$, and the surface elevation, $z = z(x)$, for -4 m $\leq x \leq 4$ m. Assume one-dimensional flow.

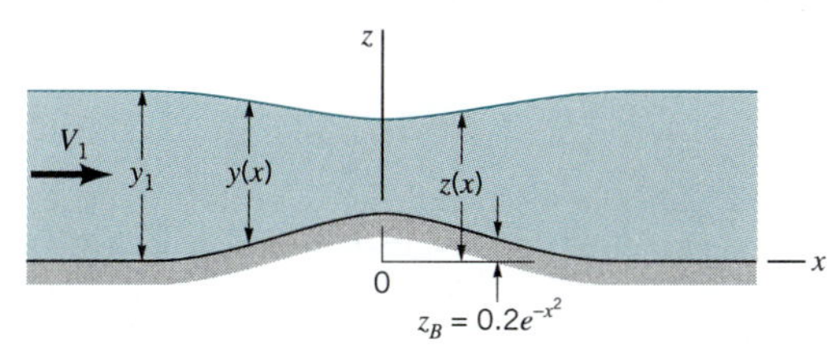

■ **FIGURE P10.17**

***10.18** Repeat Problem 10.17 if the upstream depth is 0.4 m.

10.19 Water in a rectangular channel flows into a gradual contraction section as is indicated in Fig. P10.19. If the flowrate is $Q = 25\ \text{ft}^3/\text{s}$ and the upstream depth is $y_1 = 2$ ft, determine the downstream depth, y_2.

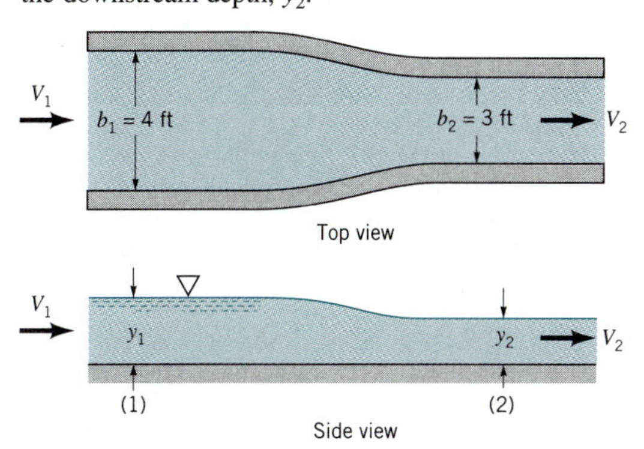

■ **FIGURE P10.19**

10.20 Sketch the specific energy diagram for the flow of Problem 10.19 and indicate its important characteristics. Note that $q_1 \neq q_2$.

10.21 Repeat Problem 10.19 if the upstream depth is $y_1 = 0.5$ ft. Assume that there are no losses between sections (1) and (2).

10.22 Water flows in a rectangular channel with a flowrate per unit width of $q = 1.5\ \text{m}^2/\text{s}$ and a depth of 0.5 m at section (1). The head loss between sections (1) and (2) is 0.03 m. Plot the specific energy diagram for this flow and locate states (1) and (2) on this diagram. Is it possible to have a head loss of 0.06 m? Explain.

10.23 Water flows in a horizontal rectangular channel with a flowrate per unit width of $q = 10\ \text{ft}^2/\text{s}$ and a depth of 1.0 ft at the downstream section (2). The head loss between section (1) upstream and section (2) is 0.2 ft. Plot the specific energy diagram for this flow and locate states (1) and (2) on this diagram.

10.24 Water flows in a horizontal, rectangular channel with an initial depth of 2 ft and initial velocity of 12 ft/s. Determine the depth downstream if losses are negligible. Note that there may be more than one solution. Repeat the problem if the initial depth remains the same, but the initial velocity is 6 ft/s.

10.25 A smooth transition section connects two rectangular channels as shown in Fig. P10.25. The channel width increases from 6.0 to 7.0 ft and the water surface elevation is the same in each channel. If the upstream depth of flow is 3.0 ft, determine h, the amount the channel bed needs to be raised across the transition section to maintain the same surface elevation.

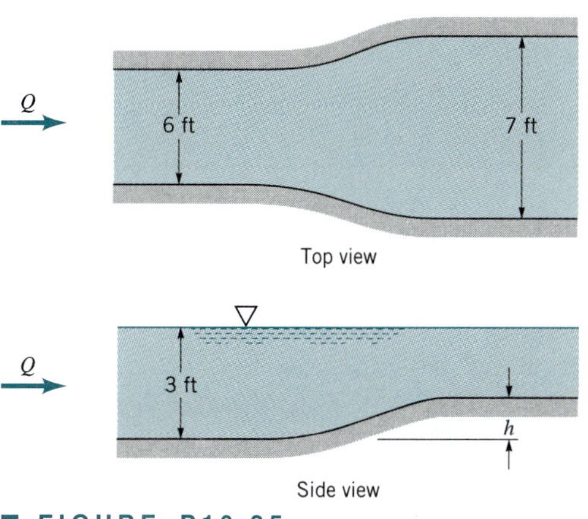

■ **FIGURE P10.25**

10.26 Water flows over a bump of height $h = h(x)$ on the bottom of a wide rectangular channel as is indicated in Fig. P10.26. If energy losses are negligible, show that the slope of the water surface is given by $dy/dx = -(dh/dx)/[1 - (V^2/gy)]$, where $V = V(x)$ and $y = y(x)$ are the local velocity and depth of flow. Comment on the sign (i.e., <0, $=0$, or >0) of dy/dx relative to the sign of dh/dx.

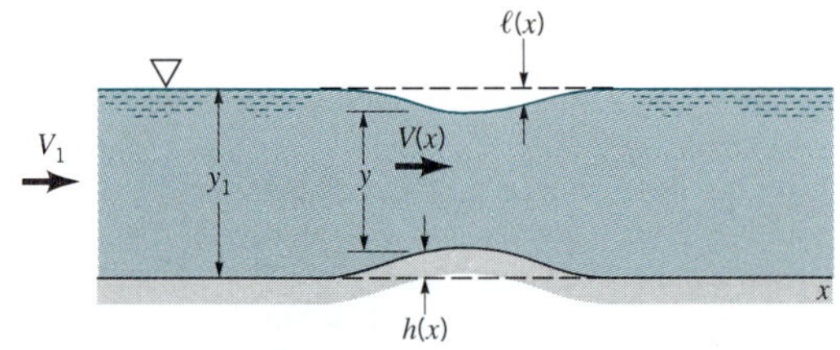

■ FIGURE P10.26

10.27 Integrate the differential equation obtained in Problem 10.26 to determine the draw-down distance, $\ell = \ell(x)$, indicated in Fig. P10.26. Comment on your results.

10.28 Determine the maximum depth in a 3-m-wide rectangular channel if the flow is to be supercritical with a flowrate of $Q = 60\ m^3/s$.

† **10.29** The Manning equation is used to determine the flowrate in an open channel. Is this formula appropriate for the flow of lava in a ditch on the flanks of a volcano? Is it appropriate for open-channel flow on another planet? Explain.

10.30 The following data are taken from measurements on Indian Fork Creek: $A = 26\ m^2$, $P = 16$ m, and $S_0 = 0.02$ m/62 m. Determine the average shear stress on the wetted perimeter of this channel.

10.31 A viscous oil flows down a wide plate with a uniform depth of 8 mm and an average velocity of 50 mm/s. The plate is on a 3° hill and the specific gravity of the oil is 0.85. Determine the average shear stress between the oil and the plate.

10.32 The following data are obtained for a particular reach of the Provo River in Utah: $A = 183\ ft^2$, free-surface width = 55 ft, average depth = 3.3 ft, $R_h = 3.22$ ft, $V = 6.56$ ft/s, length of reach = 116 ft, and elevation drop of reach = 1.04 ft. Determine **(a)** the average shear stress on the wetted perimeter, **(b)** the Manning coefficient, n, and **(c)** the Froude number of the flow.

10.33 By what percent is the flowrate reduced in the rectangular channel shown in Fig. P10.33 because of the addition of the thin center board? All surfaces are of the same material.

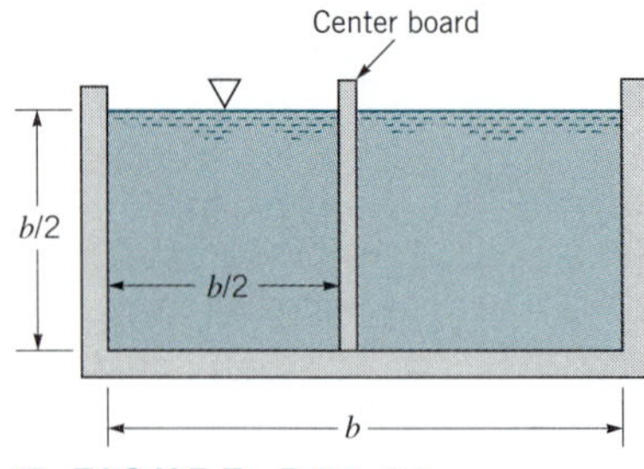

■ FIGURE P10.33

10.34 Water flows in an unfinished concrete channel at a rate of $30\ m^3/s$. What flowrate can be expected if the concrete were finished and the depth remains constant?

10.35 The Blue Ridge Flume, constructed in Shasta County, California, in 1872 to supply water for gold mining operations, was built of planed wood and had a cross section as shown in Fig. P10.35. Determine the water velocity in a section of this flume if it was full and the elevation change was 27 ft per mile.

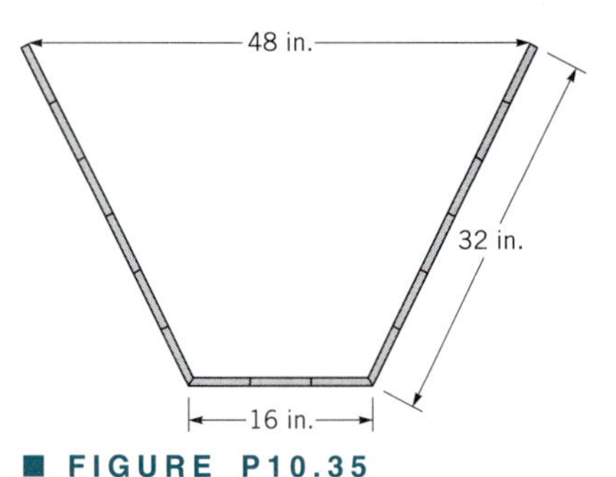

■ FIGURE P10.35

10.36 It is claimed that on steep portions of the Blue Ridge Flume (see Problem 10.35 for the flume cross section), the water would rush along at 50 mph. Is this possible? If so, determine the slope; if not, explain why not. Note: Use a flowrate of $52\ ft^3/s$ which is that obtained for the conditions (slope) of Problem 10.35.

10.37 At a particular location the cross section of the Columbia River is as indicated in Fig. P10.37. If on a day without wind it takes 5 min to float 0.5 mi along the river, which drops 0.46 ft in that distance, determine the value of the Manning coefficient, n.

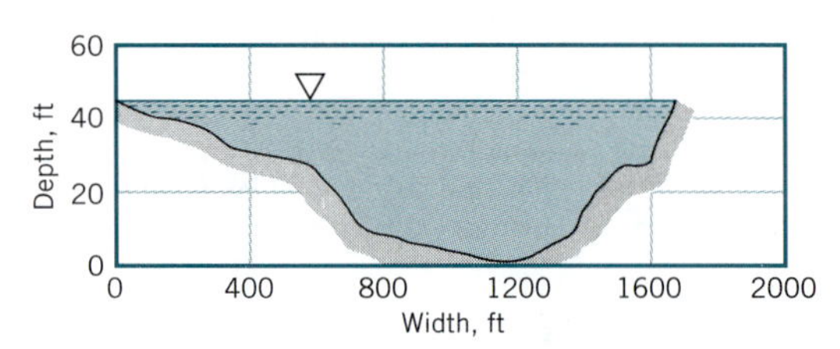

■ FIGURE P10.37

10.38 If the free surface of the Columbia River shown in Fig. P10.37 were 20 ft above the bottom rather than 44 ft, as is indicated in the figure, how long would it take to float the 0.5-mi stretch considered in Problem 10.37? Assume the elevation change remains 0.46 ft.

10.39 Rainwater runoff from a 200-ft by 500-ft parking lot is to drain through a circular concrete pipe that is laid on a slope of 3 ft/mi. Determine the pipe diameter if it is to be full with a steady rainfall of 1.5 in./hr.

10.40 To prevent weeds from growing in a clean earthen-lined canal, it is recommended that the velocity be no less than 2.5 ft/s. For the symmetrical canal shown in Fig. P10.40, determine the minimum slope needed.

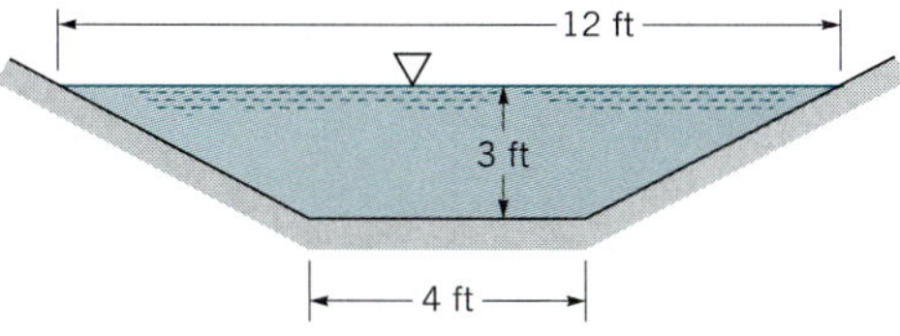

■ FIGURE P10.40

10.41 The smooth concrete-lined symmetrical channel shown in Fig. P10.40 carries water from the silt-laden Colorado River. If the velocity must be 4.0 ft/s to prevent the silt from settling out (and eventually clogging the channel), determine the minimum slope needed.

10.42 The symmetrical channel shown in Fig. P10.40 is dug in sandy loam soil with $n = 0.020$. For such surface material it is recommended that to prevent scouring of the surface the average velocity be no more than 1.75 ft/s. Determine the maximum slope allowed.

10.43 The flowrate in the clay-lined channel ($n = 0.025$) shown in Fig. P10.43 is to be 300 ft^3/s. To prevent erosion of the sides, the velocity must not exceed 5 ft/s. For this maximum velocity, determine the width of the bottom, b, and the slope, S_0.

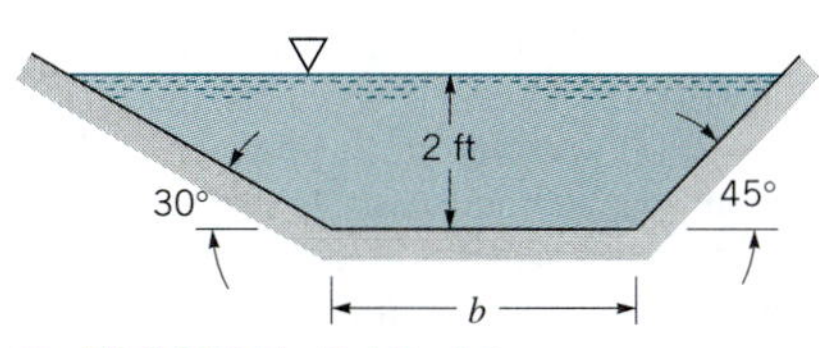

■ FIGURE P10.43

10.44 A trapezoidal channel with a bottom width of 3.0 m and sides with a slope of 2:1 (horizontal:vertical) is lined with fine gravel ($n = 0.020$) and is to carry 10 m^3/s. Can this channel be built with a slope of $S_0 = 0.00010$ if it is necessary to keep the velocity below 0.75 m/s to prevent scouring of the bottom? Explain.

10.45 Water flows in a 2-m-diameter finished concrete pipe so that it is completely full and the pressure is constant all along the pipe. If the slope is $S_0 = 0.005$, determine the flowrate by using open-channel flow methods. Compare this result with that obtained by using pipe flow methods of Chapter 8.

† **10.46** The Manning equation used to estimate the flow rate in an open channel is not capable of producing highly accurate answers. Moreover, it is not even in dimensionless form. However, not much effort has gone into developing a "better" equation. Why do you think this apparent lack of effort to obtain better results is acceptable?

10.47 A natural riverbed has an approximate parabolic cross section such that the free-surface width is 200 ft when the center depth is 5 ft. Determine the flowrate if $n = 0.035$ and the slope is 1.5 ft/mi.

10.48 The flowrate through the trapezoidal canal shown in Fig. P10.48 is Q. If it is desired to double the flowrate to $2Q$ without changing the depth, determine the additional width, L, needed. The bottom slope, surface material, and the slope of the walls are to remain the same.

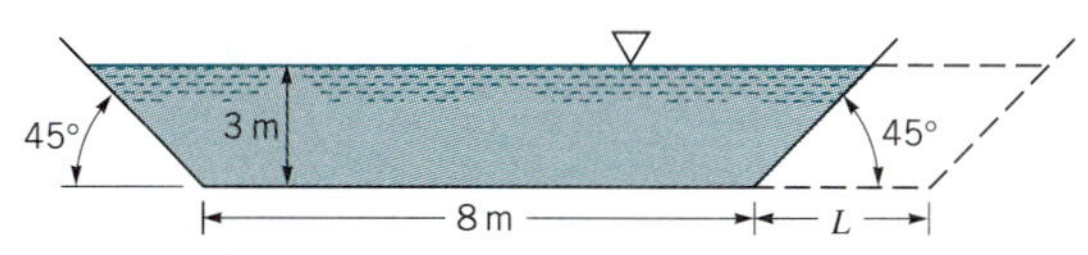

■ FIGURE P10.48

10.49 A circular, finished concrete culvert is to carry a discharge of 50 ft^3/s on a slope of 0.0010. It is to flow not more than half-full. The culvert pipes are available from the manufacturer with diameters that are multiples of 1 ft. Determine the smallest suitable culvert diameter.

10.50 A rectangular, unfinished concrete channel of 28-ft-width is laid on a slope of 8 ft/mi. Determine the flow depth and Froude number of the flow if the flowrate is 400 ft^3/s.

10.51 Rainwater flows down a street whose cross section is shown in Fig. P10.51. The street is on a hill at an angle of 2°. Determine the maximum flowrate possible if the water is not to overflow onto the sidewalk.

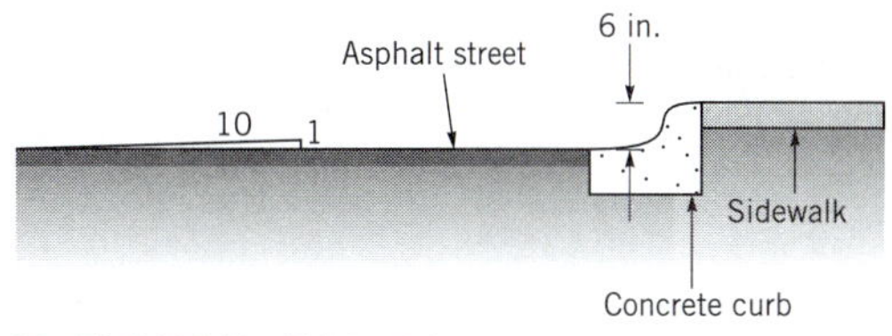

■ FIGURE P10.51

10.52 An engineer is to design a channel lined with planed wood to carry water at a flowrate of 2 m^3/s on a slope of 10 m/800m. The channel cross section can be either a 90° triangle or a rectangle with a cross section twice as wide as its depth. Which would require less wood and by what percent?

10.53 An 8-ft-diameter concrete drainage pipe which flows half-full is to be replaced by a concrete-lined V-shaped open channel having an interior angle of 90°. Determine the depth of fluid which will exist in the V-shaped channel if it is laid on the same slope and carries the same discharge as the drainage pipe.

10.54 Water flows in a channel with an equilateral triangle cross section as shown in Fig. P10.54. For a given Manning coefficient, n, and channel slope, determine the depth that gives the maximum flowrate.

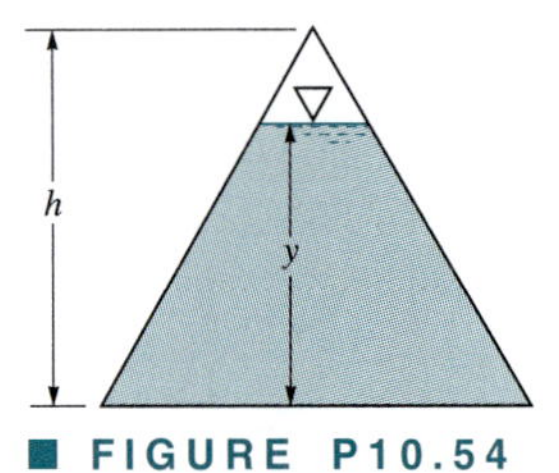

■ FIGURE P10.54

10.55 At what depth will 50 ft^3/s of water flow in a 6-ft-wide rectangular channel lined with rubble masonry set on a slope of 1 ft in 500 ft? Is a hydraulic jump possible under these conditions? Explain.

10.56 Water flows in the symmetrical, unfinished concrete trapezoidal channel shown in Fig. P10.56 at a rate of 120 ft^3/s. The slope is 4.2 ft/2000 ft. Determine the number of cubic yards of concrete needed to line each 1000 ft of the channel.

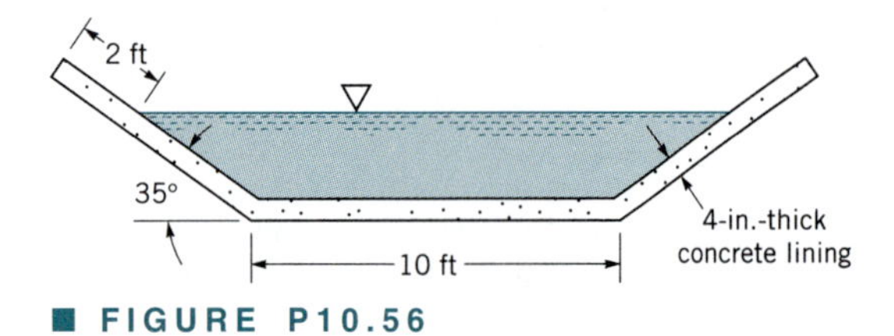

■ FIGURE P10.56

10.57 Determine the critical depth for a flow of 200 m^3/s through a rectangular channel of 10-m width. If the water flows 3.8-m deep, is the flow supercritical? Explain.

10.58 Water flows in a rectangular, brick-lined aqueduct of width 1.2 m at a rate of 73,000 m^3/day. Determine the water depth if the change in elevation over the 16-km length of this channel is 9.6 m.

10.59 A smooth steel water slide at an amusement park is of semicircular cross section with a diameter of 2.5 ft. The slide descends a vertical distance of 35 ft in its 420 ft length. If pumps supply water to the slide at a rate of 6 cfs, determine the depth of flow. Neglect the effects of the curves and bends of the slide.

10.60 Three sewer pipes of diameter D_1 join to form one pipe of diameter D. If the Manning coefficient, n, and the slope are the same for all of the pipes, and if each pipe flows half-full, determine D.

***10.61** Water flows in the painted steel rectangular channel with rounded corners shown in Fig. P10.61. The bottom slope is 1 ft/200 ft. Plot a graph of flowrate as a function of water depth for $0 \leq y \leq 1$ ft with corner radii of $r = 0, 0.2, 0.4, 0.6, 0.8$, and 1.0 ft.

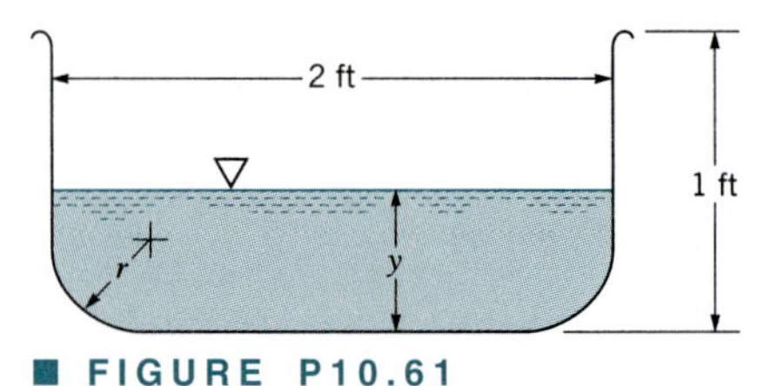

■ FIGURE P10.61

***10.62** Water flows in the fiberglass ($n = 0.014$) triangular channel with a round bottom shown in Fig. P10.62. The channel slope is 0.1 m/90 m. Plot a graph of flowrate as a function of water depth for $0 \leq y \leq 0.50$ m with bottom radii of $r = 0$, 0.25, 0.50, 0.75, and 1.0 m.

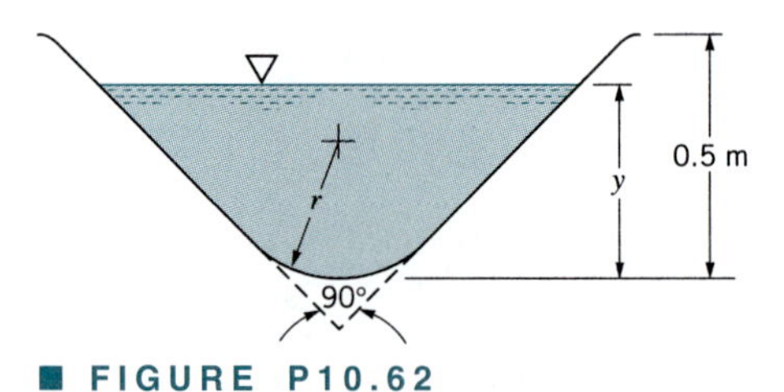

■ FIGURE P10.62

10.63 The cross section of an ancient Roman aqueduct is drawn to scale in Fig. P10.63. When it was new the channel was essentially rectangular and for a flowrate of 100,000 m^3/day, the water depth was as indicated. Archeological evidence indicates that after many years of use, calcium carbonate deposits on the sides and bottom modified the shape to that shown in the figure. Estimate the flowrate for the modified shape if the slope and surface roughness did not change.

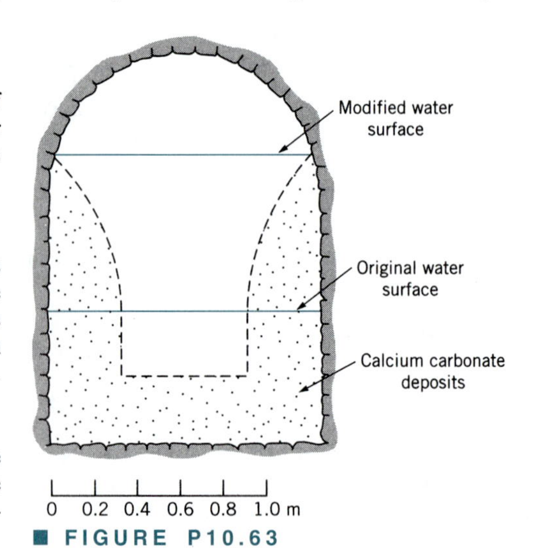

■ FIGURE P10.63

10.64 The smooth concrete-lined channel shown in Fig. P10.64 is built on a slope of 2 m/km. Determine the flowrate if the depth is $y = 1.5$ m.

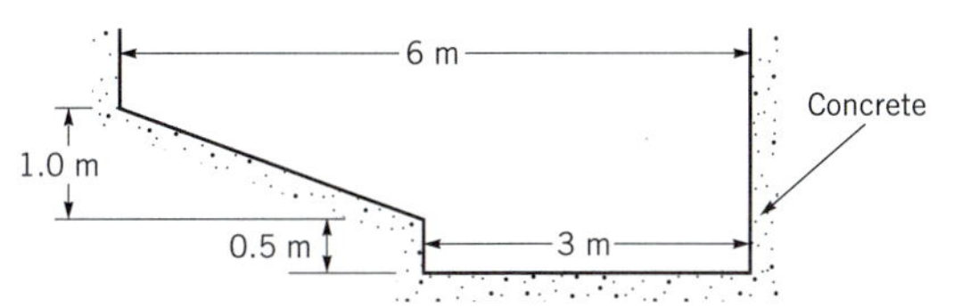

■ **FIGURE P10.64**

10.65 Determine the flow depth for the channel shown in Fig. P10.64 if the flowrate is 15 m^3/s.

***10.66** The cross section of a creek valley is shown in Fig. P10.66. Plot a graph of flowrate as a function of depth, y, for $0 \leq y \leq 10$ ft. The slope is 5 ft/mi.

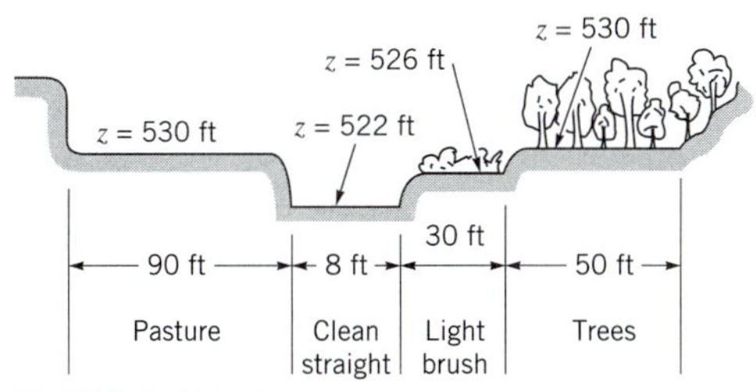

■ **FIGURE P10.66**

10.67 Repeat Problem 10.64 if the surfaces are smooth concrete as is indicated except for the diagonal surface, which is gravelly with $n = 0.025$.

***10.68** Water flows through the storm sewer shown in Fig. P10.68. The slope of the bottom is 2 m/400 m. Plot a graph of the flowrate as a function of depth for $0 \leq y \leq 1.7$ m. On the same graph, plot the flowrate expected if the entire surface were lined with material similar to that of a clay tile.

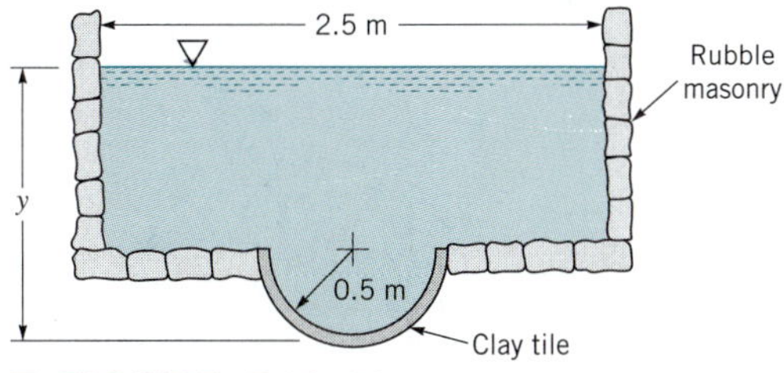

■ **FIGURE P10.68**

10.69 Determine the flowrate for the symmetrical channel shown in Fig. P10.40 if the bottom is smooth concrete and the sides are weedy. The bottom slope is $S_0 = 0.001$.

10.70 Water in a painted steel rectangular channel of width $b = 1$ ft and depth y is to flow at critical conditions, $\text{Fr} = 1$. Plot a graph of the critical slope, S_{0c}, as a function of y for 0.05 ft $\leq y \leq 5$ ft. What is the maximum slope allowed if critical flow is not to occur regardless of the depth?

10.71 Water flows in a rectangular channel of width b and depth y with a Froude number of unity. The slope, S_{0c}, of the channel needed to produce this critical flow is a function of y. Show that as $y \to \infty$ the slope becomes proportional to y (i.e., $S_{0c} = C_1 y$, where C_1 is a constant) and that as $y \to 0$ the slope becomes proportional to $y^{-1/3}$ (i.e., $S_{0c} = C_2/y^{1/3}$, where C_2 is a constant). Show that the channel with an aspect ratio of $b/y = 6$ gives the minimum value of S_{0c}.

10.72 Water flows in a rectangular channel with a bottom slope of 4.2 ft/mi and a head loss of 2.3 ft/mi. At a section where the depth is 5.8 ft and the average velocity 5.9 ft/s, does the flow depth increase or decrease in the direction of flow? Explain.

10.73 Water flows in the river shown in Fig. P10.73 with a uniform bottom slope. The total head at each section is measured by using Pitot tubes as indicated. Determine the value of dy/dx at a location where the Froude number is 0.357.

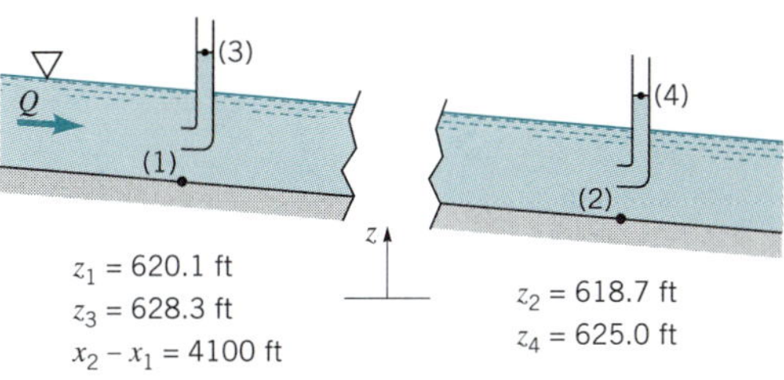

■ **FIGURE P10.73**

10.74 Repeat Problem 10.73 if the Froude number is 2.75.

10.75 Assume that the conditions given in Fig. P10.73 are as indicated except that the value of z_4 is not known. Determine the value of z_4 if the flow is uniform depth.

10.76 A 2.0-ft standing wave is produced at the bottom of the rectangular channel in an amusement park water ride. If the water depth upstream of the wave is estimated to be 1.5 ft, determine how fast the boat is traveling when it passes through this standing wave (hydraulic jump) for its final ''splash.''

10.77 The water depths upstream and downstream of a hydraulic jump are 0.3 and 1.2 m, respectively. Determine the upstream velocity and the power dissipated if the channel is 50 m wide.

10.78 Water flows under a sluice gate and forms a hydraulic jump as shown in Fig. P10.78. The head loss, h_L, across the jump is measured by the difference in elevation of the water levels in the tubes connected to the two Pitot tubes shown. Measured values of head loss as a function of upstream water depth, y_0, are given in the table below. Based on this data, plot a graph of dimensionless head loss, h_L/y_1, as a function of

Froude number, $Fr_1 = V_1/(gy_1)^{1/2}$. Also plot the theoretical result on the same graph.

y_0, ft	h_L, ft
0.855	0.383
0.759	0.325
0.691	0.268
0.578	0.201
0.492	0.144
0.414	0.117
0.289	0.058

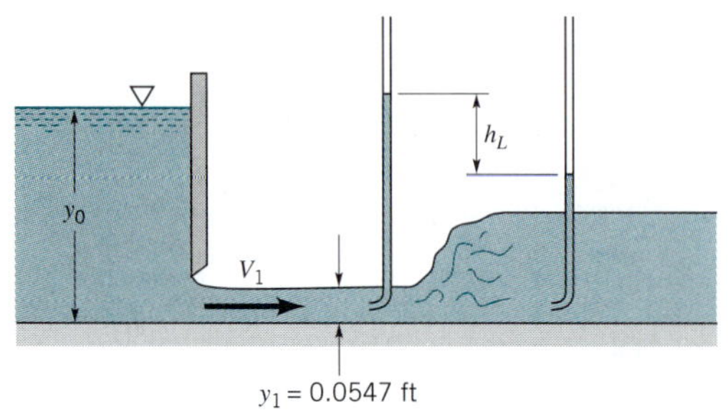

■ FIGURE P10.78

10.79 Show that for a hydraulic jump in a rectangular channel, the Froude number upstream, Fr_1, and the Froude number downstream, Fr_2, are related by

$$Fr_2^2 = \frac{8Fr_1^2}{[(1 + 8\,Fr_1^2)^{1/2} - 1]^3}$$

Plot Fr_2 as a function of Fr_1 and show that the flow downstream of a jump is subcritical.

10.80 Water flows in a 2-ft-wide rectangular channel at a rate of 10 ft^3/s. If the water depth downstream of a hydraulic jump is 2.5 ft, determine **(a)** the water depth upstream of the jump, **(b)** the upstream and downstream Froude numbers, and **(c)** the head loss across the jump.

10.81 A hydraulic jump at the base of a spillway of a dam is such that the depths upstream and downstream of the jump are 0.90 and 3.6 m, respectively. If the spillway is 100 m wide, what is the flowrate over the spillway?

10.82 Determine the head loss and power dissipated by the hydraulic jump of Problem 10.81.

10.83 Water flowing radially outward along a circular plate forms a circular hydraulic jump as is shown in Fig. P10.83*a*. This is shown easily by holding a dinner plate under the faucet of the kitchen sink. **(a)** Sketch a typical specific energy diagram for this flow (see Problem 10.12) and locate points 1, 2, 3, and 4 on the diagram. **(b)** Which of the water depth profiles shown in Fig. P10.83*b* represents the actual situation? Explain.

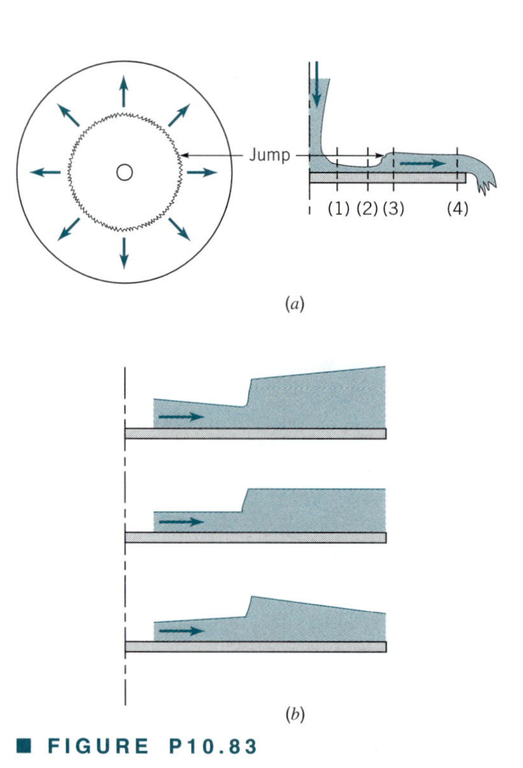

■ FIGURE P10.83

10.84 Water flows in a wide, finished concrete channel as is shown in Fig. P10.84 such that a hydraulic jump occurs at the transition of the change in slope of the channel bottom. If the upstream Froude number and depth are 4.0 and 0.2 ft, respectively, determine the slopes upstream, S_{01}, and downstream, S_{02}, of the jump to maintain uniform flows in those regions. The jump can be treated as a jump on a horizontal surface.

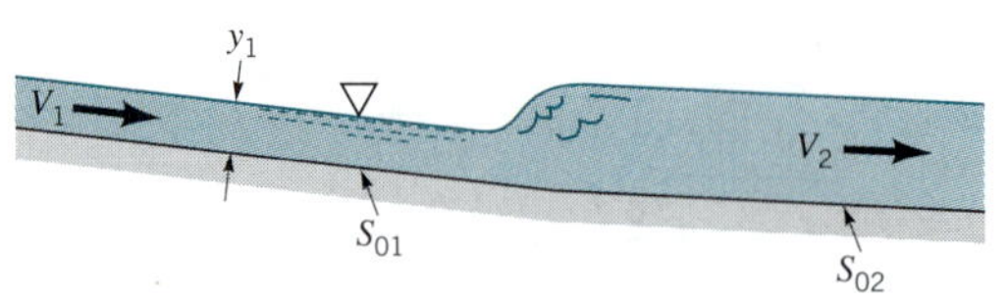

■ FIGURE P10.84

***10.85** A rectangular channel of width b is to carry water at flowrates from $30 \leq Q \leq 600$ cfs. The water depth upstream of the hydraulic jump that occurs (if one does occur) is to remain 1.5 ft for all cases. Plot the power dissipated in the jump as a function of flowrate for channels of width b = 10, 20, 30, and 40 ft.

10.86 Water flows in a rectangular channel at a depth of $y = 1$ ft and a velocity of $V = 20$ ft/s. When a gate is suddenly placed across the end of the channel, a wave (a moving hydraulic jump) travels upstream with velocity V_w as is indicated in Fig. P10.86. Determine V_w. Note that this is an unsteady problem for a stationary observer. However, for an observer moving to the left with velocity V_w, the flow appears as a steady hydraulic jump.

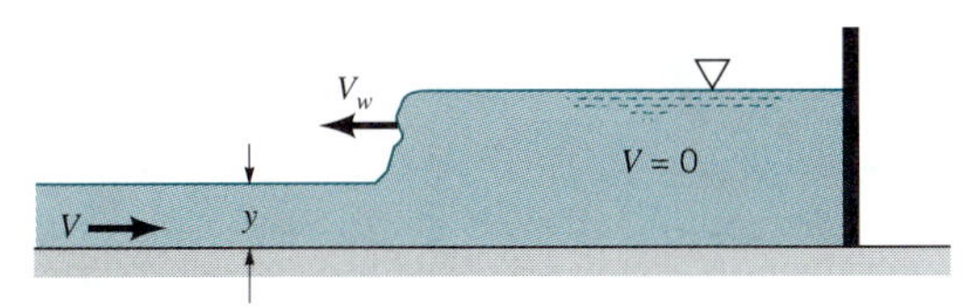

■ **FIGURE P10.86**

† **10.87** The depth ratio across a hydraulic jump can be obtained by using the momentum and continuity equations. In doing so the drag force between the channel bottom and the flowing water (i.e., the wall shear stress) is usually neglected. Redo the hydraulic jump depth ratio analysis with this friction force somehow included. Discuss the differences in the analysis and results for the cases with and without friction.

10.88 Water flows over a 5-ft-wide, rectangular sharp-crested weir that is $P_w = 4.5$ ft tall. If the depth upstream is 5 ft, determine the flowrate.

10.89 A rectangular sharp-crested weir is used to measure the flowrate in a channel of width 10 ft. It is desired to have the channel flow depth be 6 ft when the flowrate is 50 cfs. Determine the height, P_w, of the weir plate.

10.90 Water flows over a sharp-crested triangular weir with $\theta = 90°$. The head range covered is $0.2 \leq H \leq 1.0$ ft, and the accuracy in the measurement of the head, H, is $\delta H = \pm 0.01$ ft. Plot a graph of the percent error expected in Q as a function of Q.

10.91 Water flows over a broad-crested weir that has a width of 4 m and a height of $P_w = 1.5$ m. The free-surface well upstream of the weir is at a height of 0.5 m above the surface of the weir. Determine the flowrate in the channel and the minimum depth of the water above the weir block.

10.92 Determine the flowrate per unit width, q, over a broad-crested weir that is 2.0 m tall if the head, H, is 0.50 m.

10.93 Water flows under a sluice gate in a channel of 10-ft width. If the upstream depth remains constant at 5 ft, plot a graph of flowrate as a function of the distance between the gate and the channel bottom as the gate is slowly opened. Assume free outflow.

10.94 Water flows over the rectangular sharp-crested weir in a wide channel as shown in Fig. P10.94. If the channel is lined with unfinished concrete with a bottom slope of 2 m/300 m, will it be possible to produce a hydraulic jump in the channel downstream of the weir? Explain.

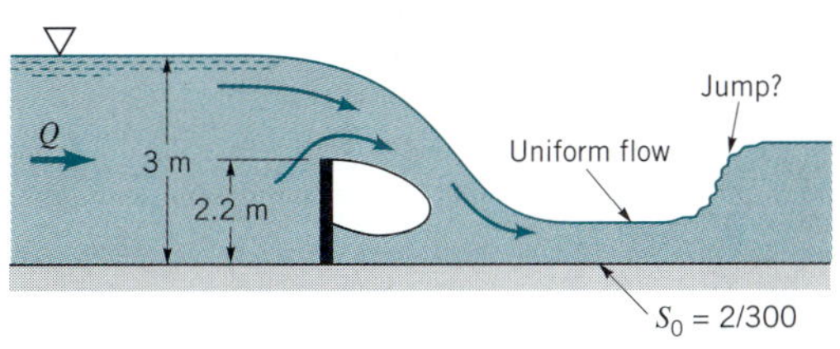

■ **FIGURE P10.94**

10.95 Water flows in a rectangular channel of width $b = 20$ ft at a rate of 100 ft^3/s. The flowrate is to be measured by using either a rectangular weir of height $P_w = 4$ ft or a triangular ($\theta = 90°$) sharp-crested weir. Determine the head, H, necessary. If measurement of the head is accurate to only ± 0.04 ft, determine the accuracy of the measured flowrate expected for each of the weirs. Which weir would be the most accurate? Explain.

10.96 Water flows over the sharp-crested weir shown in Fig. P10.96. The weir plate cross section consists of a semicircle and a rectangle. Plot a graph of the estimated flowrate, Q, as a function of head, H. List all assumptions and show all calculations.

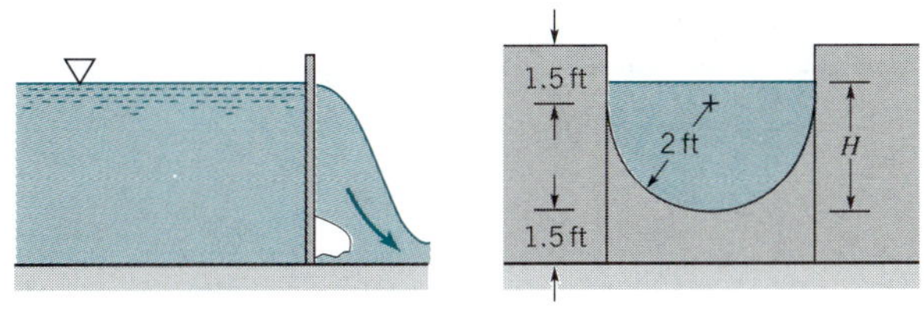

■ **FIGURE P10.96**

10.97 A water-level regulator (not shown) maintains a depth of 2.0 m downstream from a 50-ft-wide drum gate as shown in Fig. P10.97. Plot a graph of flowrate, Q, as a function of water depth upstream of the gate, y_1, for $2.0 \leq y_1 \leq 5.0$ m.

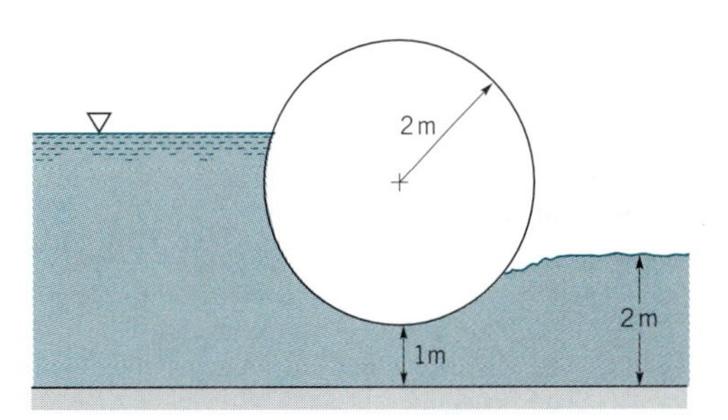

■ **FIGURE P10.97**

10.98 The device shown in Fig. P10.98 is used to calibrate the flow of water over a 90-degree *V*-notch triangular weir. For a given flowrate, the water in the channel upstream of the weir is at a uniform depth of y_1, and the weir head is H. The average velocity of the water in the channel, V_1, is determined by measuring the time, t, it takes a float to travel a known distance, $\ell = 2$ ft. That is, $V_1 = \ell/t$. The flowrate is $Q = V_1 A_1$, where A_1 is the cross-sectional area of the channel.

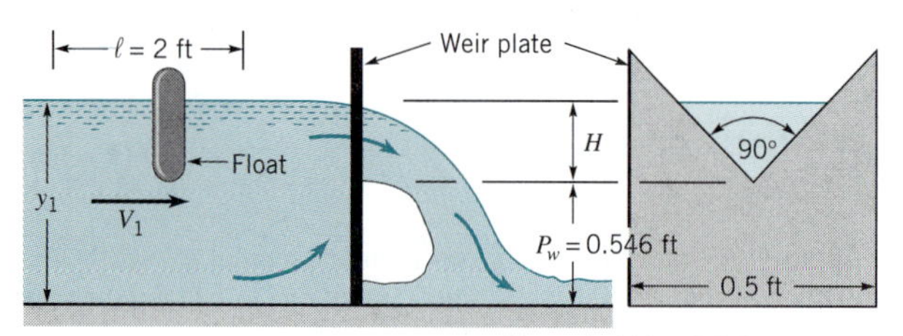

■ FIGURE P10.98

Experimentally determined values of y_1 and t for various flowrates are shown in the table below. Use these results to plot a graph on log–log paper of the flowrate as a function of the weir head. On the same graph, plot the theoretical curve assuming that the weir coefficient, C_{wt}, is equal to one. From the experimental results, determine the value of the weir coefficient for this weir.

Discuss some possible sources of error in the results.

y_1, (ft)	t (s)
0.777	10.2
0.770	10.5
0.757	13.3
0.738	15.5
0.722	21.2
0.702	24.2
0.682	33.5
0.652	59.5
0.636	78.1
0.634	84.3

10.99 The device shown in Fig. P10.99 is used to investigate the flowrate of water under a sluice gate. For a given flowrate, the water upstream of the sluice gate is at a uniform depth of y_1, while that downstream of the gate is y_2. The bottom of the gate is distance $a = 1.0$ in. from the channel bottom. The average velocity in the channel, V_1, is determined by measuring the time, t, it takes a float to travel a known distance, $\ell = 2$ ft. That is, $V_1 = \ell/t$. The flowrate is $Q = V_1 A_1$, where A_1 is the cross-sectional area of the channel.

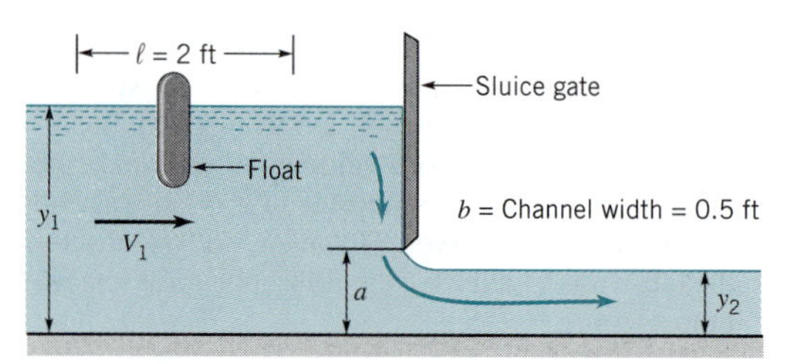

■ FIGURE P10.99

Experimentally determined values of y_1, y_2, and t for various flowrates are shown in the table below. Use these results to plot a graph on log–log paper of the flowrate as a function of the upstream depth. On the same graph, plot the theoretical curve assuming that the discharge coefficient, C_d, is equal to one. From the experimental results, determine the value of the discharge coefficient and the contraction coefficient, $C_c = y_2/a$, for this sluice gate.

Discuss some possible sources of error in the results.

y_1 (ft)	y_2 (ft)	t (s)
0.877	0.057	5.1
0.725	0.057	4.5
0.569	0.059	4.3
0.453	0.059	3.5
0.343	0.060	3.3
0.267	0.060	3.0
0.183	0.060	2.9

10.100 The device shown in Fig. P10.100 is used to investigate hydraulic jumps. Water supplied to an open channel is regulated by a supply pump and the sluice gate upstream of the hydraulic jump. For a given water depth upstream of the sluice gate, y_0, the tail gate is adjusted so that a hydraulic jump is produced as indicated. The water depths upstream and downstream of the jump are y_1 and y_2, respectively. With the bottom of the sluice gate fixed at a distance of $a = 0.0912$ ft above the bottom of the channel, the value of y_1 was determined to be essentially constant at $y_1 = 0.0547$ ft, independent of y_0. The water speed upstream of the hydraulic jump, V_1, can be determined from the measured values of y_0 and y_1 and use of the Bernoulli and continuity equations applied between points (0) and (1).

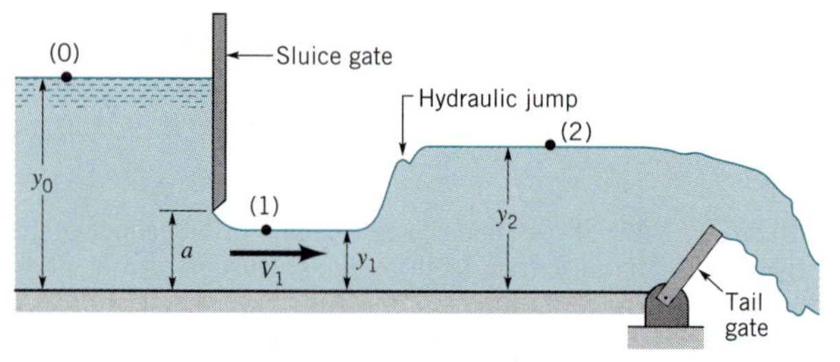

■ FIGURE P10.100

Experimentally determined values of y_0 and y_2 corresponding to different flowrates are shown in the following table. Use these results to plot a graph of the depth ratio, y_2/y_1, as a function of the Froude number upstream of the hydraulic jump, $Fr_1 = V_1\sqrt{gy_1}$. On the same graph, plot the theoretical curve.

Compare the experimental and theoretical results and discuss some possible reasons for any differences between them.

y_0 (ft)	y_2 (ft)
0.855	0.404
0.759	0.375
0.691	0.367
0.578	0.337
0.492	0.308
0.414	0.288
0.289	0.233
0.248	0.211

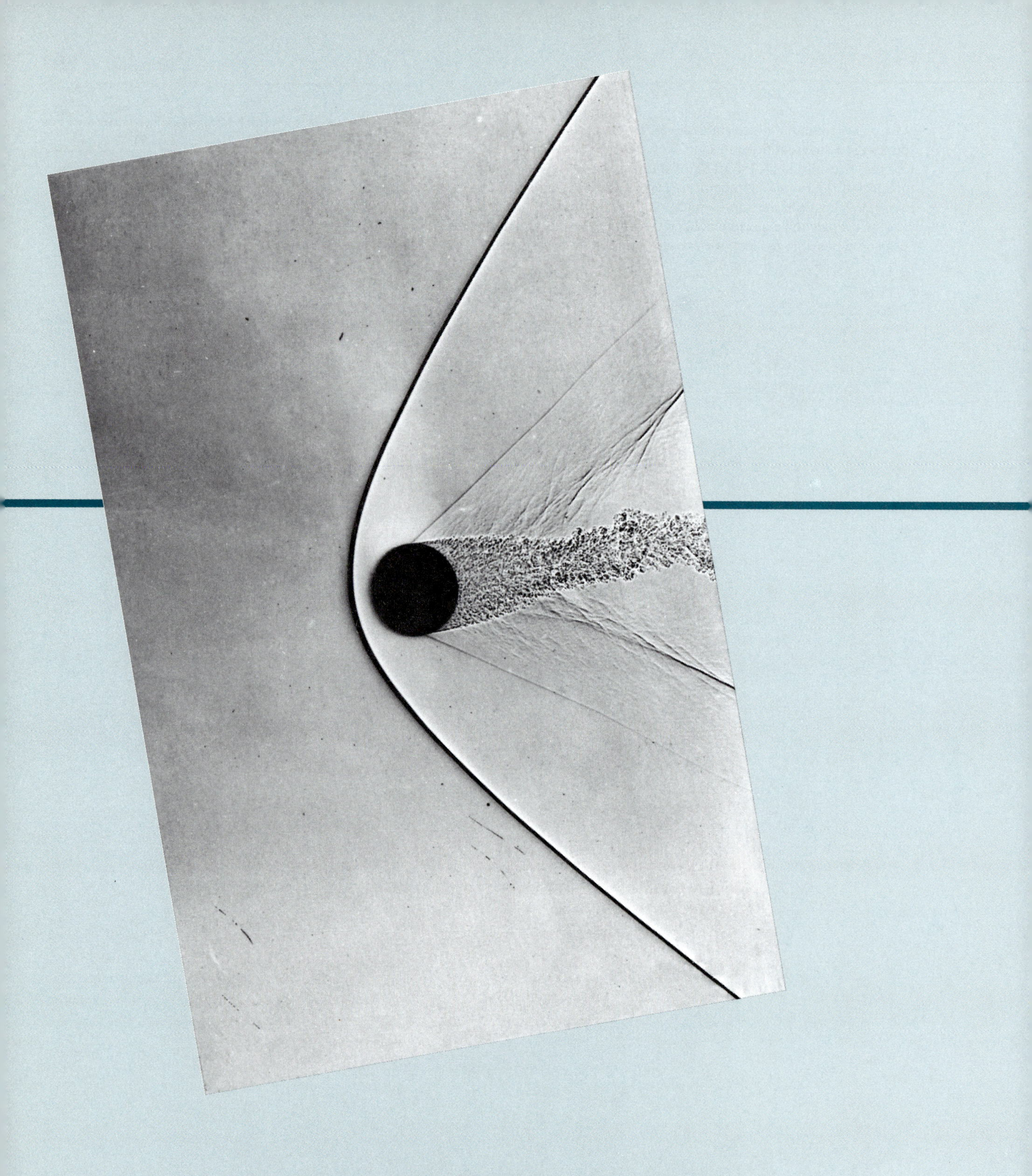

Flow past a sphere at Mach 1.53: An object moving through a fluid at supersonic speed (Mach number greater than one) creates a shock wave (a discontinuity in flow conditions shown by the dark curved line), which is heard as a sonic boom as the object passes overhead. The turbulent wake is also shown (shadowgraph technique used in air). (Photography courtesy of A. C. Charters.)

11 Compressible Flow

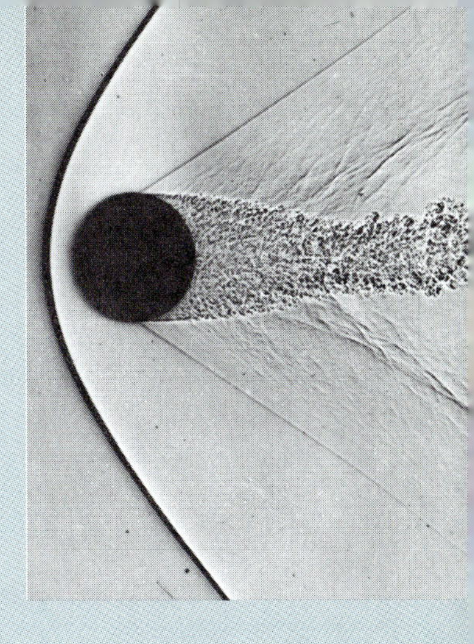

Most first courses in fluid mechanics concentrate on constant density (incompressible) flows. In earlier chapters of this book, we mainly considered incompressible flow behavior. In a few instances, variable density (compressible) flow effects were covered briefly. The notion of an incompressible fluid is convenient because when constant density and constant (including zero) viscosity are assumed, problem solutions are greatly simplified. Also, fluid incompressibility allows us to build on the Bernoulli equation as was done, for example, in Chapter 5. Preceding examples should have convinced us that nearly incompressible flows are common in everyday experiences.

Any study of fluid mechanics would, however, be incomplete without a brief introduction to compressible flow behavior. Fluid compressibility is a very important consideration in numerous engineering applications of fluid mechanics. For example, the measurement of high-speed flow velocities requires compressible flow theory. The flows in gas turbine engine components are generally compressible. Many aircraft fly fast enough to involve a compressible flow field.

The variation of fluid density for compressible flows requires attention to density and other fluid property relationships. The fluid equation of state, often unimportant for incompressible flows, is vital in the analysis of compressible flows. Also, temperature variations for compressible flows are usually significant and thus the energy equation is important. Curious phenomena can occur with compressible flows. For example, with compressible flows we can have fluid acceleration because of friction, fluid deceleration in a converging duct, fluid temperature decrease with heating, and the formation of abrupt discontinuities in flows across which fluid properties change appreciably.

Compressible flow phenomena are sometimes surprising; for example, friction can accelerate fluid.

For simplicity, in this introductory study of compressibility effects we mainly consider the steady, one-dimensional, constant (including zero) viscosity, compressible flow of an ideal gas. In this chapter, one-dimensional flow refers to flow involving uniform distributions of fluid properties over any flow cross-section area. Both frictionless ($\mu = 0$) and frictional ($\mu \neq 0$) compressible flows are considered. If the change in volume associated with a change of pressure is considered a measure of compressibility, our experience suggests that gases and vapors are much more compressible than liquids. We focus our attention on the compressible flow of a gas because such flows occur often. We limit our discussion to ideal gases, since the equation of state for an ideal gas is uncomplicated, yet representative of actual gases at pressures and temperatures of engineering interest, and because the flow trends associated with an ideal gas are generally applicable to other compressible fluids.

An excellent film about compressible flow is available (see Ref. 1). This resource is a useful supplement to the material covered in this chapter.

11.1 Ideal Gas Relationships

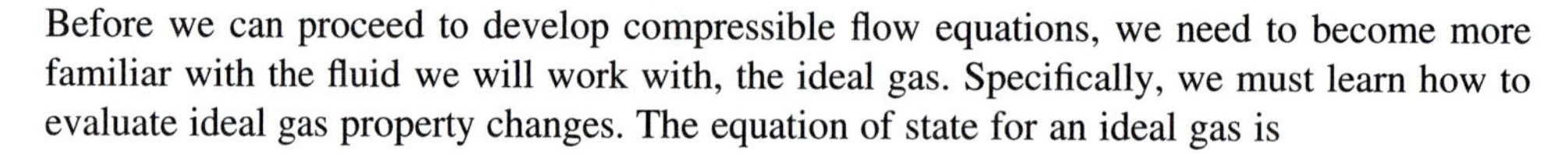

Before we can proceed to develop compressible flow equations, we need to become more familiar with the fluid we will work with, the ideal gas. Specifically, we must learn how to evaluate ideal gas property changes. The equation of state for an ideal gas is

V11.1

$$\boxed{p = \rho R T} \tag{11.1}$$

We have already discussed fluid pressure, p, density, ρ, and temperature, T, in earlier chapters. The gas constant, R, represents a constant for each distinct ideal gas or mixture of ideal gases, where

$$R = \frac{\lambda}{M_{\text{gas}}} \tag{11.2}$$

We consider ideal gas flows only.

With this notation, λ is the universal gas constant and M_{gas} is the molecular weight of the ideal gas or gas mixture. Listed in Tables 1.7 and 1.8 are values of the gas constants of some commonly used gases. Nonideal gas state equations are beyond the scope of this text, and those interested in this topic are directed to texts on engineering thermodynamics, for example, Refs. 2 and 3. Note that the trends of ideal gas flows are generally good indicators of what nonideal gas flow behavior is like. Nonideal gas flows are also discussed briefly in Refs. 2 and 3.

For an ideal gas, internal energy, $\check{u}$, is considered to be a function of temperature only (Refs. 2 and 3). Thus, the ideal gas specific heat at constant volume, c_v, can be expressed as

$$c_v = \left(\frac{\partial \check{u}}{\partial T}\right)_v = \frac{d\check{u}}{dT} \tag{11.3}$$

where the subscript v on the partial derivative refers to differentiation at constant specific volume, $v = 1/\rho$. From Eq. 11.3 we conclude that for a particular ideal gas, c_v is a function of temperature only. Equation 11.3 can be rearranged to yield

$$d\check{u} = c_v \, dT$$

Thus,

$$\check{u}_2 - \check{u}_1 = \int_{T_1}^{T_2} c_v \, dT \tag{11.4}$$

Equation 11.4 is useful because it allows us to evaluate the change in internal energy, $\check{u}_2 - \check{u}_1$, associated with ideal gas flow from section (1) to section (2) in a flow. For simplicity, we can assume that c_v is constant for a particular ideal gas and obtain from Eq. 11.4

$$\boxed{\check{u}_2 - \check{u}_1 = c_v(T_2 - T_1)} \tag{11.5}$$

Actually, c_v for a particular gas varies with temperature (see Refs. 2 and 3). However, for moderate changes in temperature, the constant c_v assumption is reasonable.

The fluid property enthalpy, $\check{h}$, is defined as

$$\check{h} = \check{u} + \frac{p}{\rho} \tag{11.6}$$

For an ideal gas, we have already stated that

$$\check{u} = \check{u}(T)$$

From the equation of state (Eq. 11.1)

$$\frac{p}{\rho} = RT$$

Thus, it follows that

$$\check{h} = \check{h}(T)$$

Since for an ideal gas, enthalpy is a function of temperature only, the ideal gas specific heat at constant pressure, c_p, can be expressed as

$$c_p = \left(\frac{\partial \check{h}}{\partial T}\right)_p = \frac{d\check{h}}{dT} \tag{11.7}$$

where the subscript p on the partial derivative refers to differentiation at constant pressure, and c_p is a function of temperature only. The rearrangement of Eq. 11.7 leads to

$$d\check{h} = c_p \, dT$$

and

$$\check{h}_2 - \check{h}_1 = \int_{T_1}^{T_2} c_p \, dT \tag{11.8}$$

Equation 11.8 is useful because it allows us to evaluate the change in enthalpy, $\check{h}_2 - \check{h}_1$, associated with ideal gas flow from section (1) to section (2) in a flow. For simplicity, we can assume that c_p is constant for a specific ideal gas and obtain from Eq. 11.8

$$\boxed{\check{h}_2 - \check{h}_1 = c_p(T_2 - T_1)} \tag{11.9}$$

For moderate temperature changes, specific heat values can be considered constant.

As is true for c_v, the value of c_p for a given gas varies with temperature. Nevertheless, for moderate changes in temperature, the constant c_p assumption is reasonable.

From Eqs. 11.5 and 11.9 we see that changes in internal energy and enthalpy are related to changes in temperature by values of c_v and c_p. We turn our attention now to developing useful relationships for determining c_v and c_p. Combining Eqs. 11.6 and 11.1 we get

$$\check{h} = \check{u} + RT \tag{11.10}$$

Differentiating Eq. 11.10 leads to

$$d\check{h} = d\check{u} + R \, dT$$

or

$$\frac{d\check{h}}{dT} = \frac{d\check{u}}{dT} + R \tag{11.11}$$

From Eqs. 11.3, 11.7, and 11.11 we conclude that

$$c_p - c_v = R \tag{11.12}$$

Equation 11.12 indicates that the difference between c_p and c_v is constant for each ideal gas regardless of temperature. Also $c_p > c_v$. If the specific heat ratio, k, is defined as

$$k = \frac{c_p}{c_v} \tag{11.13}$$

then combining Eqs. 11.12 and 11.13 leads to

$$c_p = \frac{Rk}{k - 1} \tag{11.14}$$

and

$$c_v = \frac{R}{k - 1} \tag{11.15}$$

The gas constant is related to the specific heat values.

Actually, c_p, c_v, and k are all somewhat temperature dependent for any ideal gas. We will assume constant values for these variables in this book. Values of k and R for some commonly used gases at nominal temperatures are listed in Tables 1.7 and 1.8. These tabulated values can be used with Eqs. 11.13 and 11.14 to determine the values of c_p and c_v. Example 11.1 demonstrates how internal energy and enthalpy changes can be calculated for a flowing ideal gas having constant c_p and c_v.

EXAMPLE 11.1

Air flows steadily between two sections in a long straight portion of 4-in.-diameter pipe as is indicated in Fig. E11.1. The uniformly distributed temperature and pressure at each section are $T_1 = 540$ °R, $p_1 = 100$ psia, and $T_2 = 453$ °R, $p_2 = 18.4$ psia. Calculate the (a) change in internal energy between sections (1) and (2), (b) change in enthalpy between sections (1) and (2), and (c) change in density between sections (1) and (2).

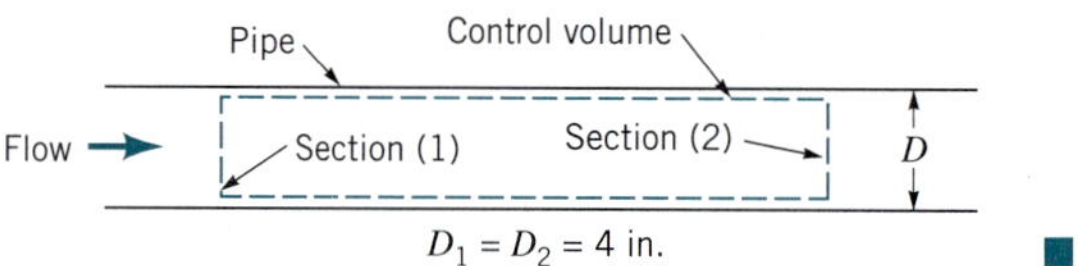

■ FIGURE E11.1

SOLUTION

(a) Assuming air behaves as an ideal gas, we can use Eq. 11.5 to evaluate the change in internal energy between sections (1) and (2). Thus

$$\check{u}_2 - \check{u}_1 = c_v(T_2 - T_1) \tag{1}$$

From Eq. 11.15 we have

$$c_v = \frac{R}{k - 1} \tag{2}$$

and from Table 1.7, $R = 1716$ (ft·lb)/(slug· °R) and $k = 1.4$. Throughout this book, we use the nominal values of k for common gases listed in Tables 1.7 and 1.8 and consider these values as being representative. From Eq. 2 we obtain

$$c_v = \frac{1716}{(1.4 - 1)} \text{ (ft·lb)/(slug· °R)} = 4290 \text{ (ft·lb)/(slug· °R)} \qquad \textbf{(3)}$$

Combining Eqs. 1 and 3 yields

$$\check{u}_2 - \check{u}_1 = c_v(T_2 - T_1) = 4290 \text{ (ft·lb)/(slug· °R)}$$
$$\times (453 \text{ °R} - 540 \text{ °R}) = -3.73 \times 10^5 \text{ ft·lb/slug} \qquad \textbf{(Ans)}$$

(b) For enthalpy change we use Eq. 11.9. Thus

$$\check{h}_2 - \check{h}_1 = c_p(T_2 - T_1) \qquad \textbf{(4)}$$

where since $k = c_p/c_v$ we obtain

$$c_p = kc_v = (1.4)[4290 \text{ (ft·lb)/(slug· °R)}] = 6006 \text{ (ft·lb)/(slug· °R)} \qquad \textbf{(5)}$$

From Eqs. 4 and 5 we obtain

$$\check{h}_2 - \check{h}_1 = c_p(T_2 - T_1) = 6006 \text{ (ft·lb)/(slug· °R)}$$
$$\times (453 \text{ °R} - 540 \text{ °R}) = -5.22 \times 10^5 \text{ ft·lb/slug} \qquad \textbf{(Ans)}$$

(c) For density change we use the ideal gas equation of state (Eq. 11.1) to get

$$\rho_2 - \rho_1 = \frac{p_2}{RT_2} - \frac{p_1}{RT_1} = \frac{1}{R}\left(\frac{p_2}{T_2} - \frac{p_1}{T_1}\right) \qquad \textbf{(6)}$$

Using the pressures and temperatures given in the problem statement we calculate from Eq. 6

$$\rho_2 - \rho_1 = \frac{1}{1716 \text{ (ft·lb)/(slug· °R)}}$$
$$\times \left[\frac{(18.4 \text{ psia})(144 \text{ in.}^2/\text{ft}^2)}{453 \text{ °R}} - \frac{(100 \text{ psia})(144 \text{ in.}^2/\text{ft}^2)}{540 \text{ °R}}\right]$$

or

$$\rho_2 - \rho_1 = -0.0121 \text{ slug/ft}^3 \qquad \textbf{(Ans)}$$

This is a significant change in density when compared with the upstream density

$$\rho_1 = \frac{p_1}{RT_1} = \frac{(100 \text{ psia})(144 \text{ in.}^2/\text{ft}^2)}{[1716 \text{ (ft·lb)/(slug· °R)}](540 \text{ °R})} = 0.0155 \text{ slug/ft}^3$$

Compressibility effects are important for this flow.

For compressible flows, changes in the thermodynamic property *entropy*, s, are important. For any pure substance including ideal gases, the "first $T\,ds$ equation" is (see Ref. 2 or Ref. 3)

$$T\,ds = d\check{u} + pd\left(\frac{1}{\rho}\right) \qquad \textbf{(11.16)}$$

where T is absolute temperature, s is entropy, $\check{u}$ is internal energy, p is absolute pressure, and ρ is density. Differentiating Eq. 11.6 leads to

$$d\check{h} = d\check{u} + pd\left(\frac{1}{\rho}\right) + \left(\frac{1}{\rho}\right) dp \quad \textbf{(11.17)}$$

By combining Eqs. 11.16 and 11.17, we obtain

$$T\, ds = d\check{h} - \left(\frac{1}{\rho}\right) dp \quad \textbf{(11.18)}$$

Equation 11.18 is often referred to as the ''second $T\, ds$ equation.'' For an ideal gas, Eqs. 11.1, 11.3, and 11.16 can be combined to yield

$$ds = c_v \frac{dT}{T} + \frac{R}{1/\rho} d\left(\frac{1}{\rho}\right) \quad \textbf{(11.19)}$$

and Eqs. 11.1, 11.7, and 11.18 can be combined to yield

$$ds = c_p \frac{dT}{T} - R \frac{dp}{p} \quad \textbf{(11.20)}$$

If c_p and c_v are assumed to be constant for a given gas, Eqs. 11.19 and 11.20 can be integrated to get

$$\boxed{s_2 - s_1 = c_v \ln \frac{T_2}{T_1} + R \ln \frac{\rho_1}{\rho_2}} \quad \textbf{(11.21)}$$

and

$$\boxed{s_2 - s_1 = c_p \ln \frac{T_2}{T_1} - R \ln \frac{p_2}{p_1}} \quad \textbf{(11.22)}$$

Changes in entropy can be important in compressible flows.

Equations 11.21 and 11.22 allow us to calculate the change of entropy of an ideal gas flowing from one section to another with constant specific heat values (c_p and c_v).

For the air flow of Example 11.1, calculate the change in entropy, $s_2 - s_1$, between sections (1) and (2).

SOLUTION

Assuming that the flowing air in Fig. E11.1 behaves as an ideal gas, we can calculate the entropy change between sections by using either Eq. 11.21 or Eq. 11.22. We use both to demonstrate that the same result is obtained either way.

From Eq. 11.21,

$$s_2 - s_1 = c_v \ln \frac{T_2}{T_1} + R \ln \frac{\rho_1}{\rho_2} \quad \textbf{(1)}$$

To evaluate $s_2 - s_1$ from Eq. 1 we need the density ratio, ρ_1/ρ_2, which can be obtained from the ideal gas equation of state (Eq. 11.1) as

$$\frac{\rho_1}{\rho_2} = \left(\frac{p_1}{T_1}\right)\left(\frac{T_2}{p_2}\right) \quad \textbf{(2)}$$

and thus from Eqs. 1 and 2,

$$s_2 - s_1 = c_v \ln \frac{T_2}{T_1} + R \ln \left[\left(\frac{p_1}{T_1} \right) \left(\frac{T_2}{p_2} \right) \right] \tag{3}$$

By substituting values already identified in the Example 11.1 problem statement and solution into Eq. 3 we get

$$s_2 - s_1 = [4290 \text{ (ft·lb)/(slug· °R)}] \ln \left(\frac{453 \text{ °R}}{540 \text{ °R}} \right) + [1716 \text{ (ft·lb)/(slug· °R)}] \ln \left[\left(\frac{100 \text{ psia}}{540 \text{ °R}} \right) \left(\frac{453 \text{ °R}}{18.4 \text{ psia}} \right) \right]$$

or

$$s_2 - s_1 = 1850 \text{ (ft·lb)/(slug· °R)} \qquad \textbf{(Ans)}$$

From Eq. 11.22,

$$s_2 - s_1 = c_p \ln \frac{T_2}{T_1} - R \ln \frac{p_2}{p_1} \tag{4}$$

By substituting known values into Eq. 4 we obtain

$$s_2 - s_1 = [6006 \text{ (ft·lb)/(slug· °R)}] \ln \left(\frac{453 \text{ °R}}{540 \text{ °R}} \right) - [1716 \text{ (ft·lb)/(slug· °R)}] \ln \left(\frac{18.4 \text{ psia}}{100 \text{ psia}} \right)$$

or

$$s_2 - s_1 = 1850 \text{ (ft·lb)/(slug· °R)} \qquad \textbf{(Ans)}$$

Absolute values of pressure and temperature must be used in entropy difference calculations.

As anticipated, both Eqs. 11.21 and 11.22 yield the same result for the entropy change, $s_2 - s_1$.

Note that since the ideal gas equation of state was used in the derivation of the entropy difference equations, both the pressures and temperatures used must be absolute.

If internal energy, enthalpy, and entropy changes for ideal gas flow with variable specific heats are desired, Eqs. 11.4, 11.8, and 11.19 or 11.20 must be used as explained in Refs. 2 and 3. Detailed tables (see, for example, Ref. 4) are available for variable specific heat calculations.

The second law of thermodynamics requires that the adiabatic and frictionless flow of any fluid results in $ds = 0$ or $s_2 - s_1 = 0$. Constant entropy flow is called *isentropic* flow. For the isentropic flow of an ideal gas with constant c_p and c_v, we get from Eqs. 11.21 and 11.22

$$c_v \ln \frac{T_2}{T_1} + R \ln \frac{\rho_1}{\rho_2} = c_p \ln \frac{T_2}{T_1} - R \ln \frac{p_2}{p_1} = 0 \tag{11.23}$$

By combining Eq. 11.23 with Eqs. 11.14 and 11.15 we obtain

$$\left(\frac{T_2}{T_1}\right)^{k/(k-1)} = \left(\frac{\rho_2}{\rho_1}\right)^k = \left(\frac{p_2}{p_1}\right) \quad \textbf{(11.24)}$$

which is a useful relationship between temperature, density, and pressure for the isentropic flow of an ideal gas. From Eq. 11.24 we can conclude that

$$\boxed{\frac{p}{\rho^k} = \text{constant}} \quad \textbf{(11.25)}$$

for an ideal gas with constant c_p and c_v flowing isentropically, a result already used without proof earlier in Chapters 1, 3, and 5.

11.2 Mach Number and Speed of Sound

Mach number is the ratio of local flow and sound speeds.

The Mach number, Ma, was introduced in Chapters 1 and 7 as a dimensionless measure of compressibility in a fluid flow. In this and subsequent sections, we develop some useful relationships involving the Mach number. The Mach number is defined as the ratio of the value of the local flow velocity, V, to the local speed of sound, c. In other words,

$$\text{Ma} = \frac{V}{c}$$

What we perceive as sound generally consists of weak pressure pulses that move through air. When our ear drums respond to a succession of moving pressure pulses, we hear sounds.

To better understand the notion of speed of sound, we analyze the one-dimensional fluid mechanics of an infinitesimally thin, weak pressure pulse moving at the speed of sound through a fluid at rest (see Fig. 11.1*a*). Ahead of the pressure pulse, the fluid velocity is zero and the fluid pressure and density are p and ρ. Behind the pressure pulse, the fluid velocity has changed by an amount δV, and the pressure and density of the fluid have also changed by amounts δp and $\delta \rho$. We select an infinitesimally thin control volume that moves with the pressure pulse as is sketched in Fig. 11.1*a*. The speed of the weak pressure pulse is considered constant and in one direction only; thus, our control volume is inertial.

For an observer moving with this control volume (Fig. 11.1*b*), it appears as if fluid is entering the control volume through surface area A with speed c at pressure p and density ρ and leaving the control volume through surface area A with speed $c - \delta V$, pressure $p + \delta p$, and density $\rho + \delta \rho$. When the continuity equation (Eq. 5.16) is applied to the flow through this control volume, the result is

$$\rho A c = (\rho + \delta\rho) A (c - \delta V) \quad \textbf{(11.26)}$$

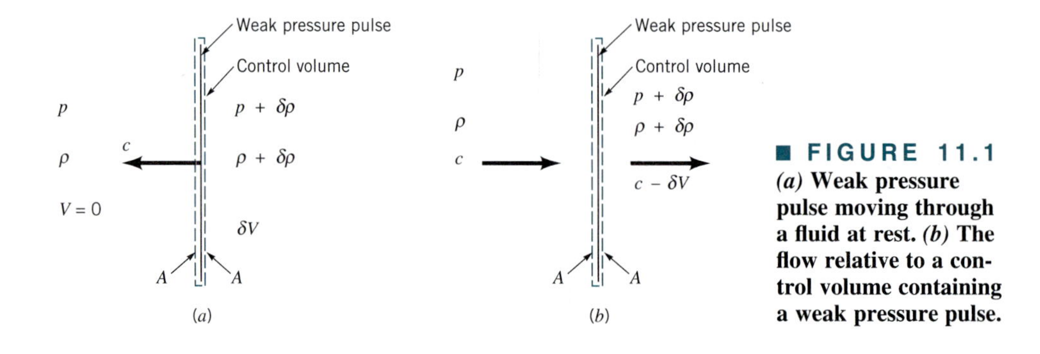

■ **FIGURE 11.1** ***(a)* Weak pressure pulse moving through a fluid at rest. *(b)* The flow relative to a control volume containing a weak pressure pulse.**

or

$$\rho c = \rho c - \rho\ \delta V + c\ \delta\rho - (\delta\rho)(\delta V) \tag{11.27}$$

Since $(\delta\rho)(\delta V)$ is much smaller than the other terms in Eq. 11.27, we drop it from further consideration and keep

$$\rho\ \delta V = c\ \delta\rho \tag{11.28}$$

The linear momentum equation (Eq. 5.29) can also be applied to the flow through the control volume of Fig. 11.1*b*. The result is

$$-c\rho cA + (c - \delta V)(\rho + \delta\rho)(c - \delta V)A = pA - (p + \delta p)A \tag{11.29}$$

Note that any frictional forces are considered as being negligibly small. We again neglect higher order terms [such as $(\delta V)^2$ compared to $c\ \delta V$, for example] and combine Eqs. 11.26 and 11.29 to get

$$-c\rho cA + (c - \delta V)\rho Ac = -\delta pA$$

or

$$\rho\delta V = \frac{\delta p}{c} \tag{11.30}$$

From Eqs. 11.28 (continuity) and 11.30 (linear momentum) we obtain

$$c^2 = \frac{\delta p}{\delta\rho}$$

or

$$c = \sqrt{\frac{\delta p}{\delta\rho}} \tag{11.31}$$

This expression for the speed of sound results from application of the conservation of mass and conservation of linear momentum principles to the flow through the control volume of Fig. 11.1*b*. These principles were similarly used in Section 10.2.1 to obtain an expression for the speed of surface waves traveling on the surface of fluid in a channel.

Mass conservation and the energy or momentum law lead to the sound speed formula.

The conservation of energy principle can also be applied to the flow through the control volume of Fig. 11.1*b*. If the energy equation (Eq. 5.103) is used for the flow through this control volume, the result is

$$\frac{\delta p}{\rho} + \delta\left(\frac{V^2}{2}\right) + g\ \delta z = \delta(\text{loss}) \tag{11.32}$$

For gas flow we can consider $g\ \delta z$ as being negligibly small in comparison to the other terms in the equation. Also, if we assume that the flow is frictionless, then $\delta(\text{loss}) = 0$ and Eq. 11.32 becomes

$$\frac{\delta p}{\rho} + \frac{(c - \delta V)^2}{2} - \frac{c^2}{2} = 0$$

or, neglecting $(\delta V)^2$ compared to $c\ \delta V$, we obtain

$$\rho\ \delta V = \frac{\delta p}{c} \tag{11.33}$$

By combining Eqs. 11.28 (continuity) and 11.33 (energy) we again find that

$$c = \sqrt{\frac{\delta p}{\delta\rho}}$$

which is identical to Eq. 11.31. Thus, the conservation of linear momentum and the conservation of energy principles lead to the same result. If we further assume that the frictionless flow through the control volume of Fig. 11.1b is adiabatic (no heat transfer), then the flow is isentropic. In the limit, as δp becomes vanishingly small ($\delta p \rightarrow \partial p \rightarrow 0$)

$$c = \sqrt{\left(\frac{\partial p}{\partial \rho}\right)_s} \tag{11.34}$$

where the subscript s is used to designate that the partial differentiation occurs at constant entropy.

Equation 11.34 suggests to us that we can calculate the speed of sound by determining the partial derivative of pressure with respect to density at constant entropy. For the isentropic flow of an ideal gas (with constant c_p and c_v), we learned earlier (Eq. 11.25) that

$$p = (\text{constant})(\rho^k)$$

and thus

$$\left(\frac{\partial p}{\partial \rho}\right)_s = (\text{constant})\, k\rho^{k-1} = \frac{p}{\rho^k}\, k\rho^{k-1} = \frac{p}{\rho}\, k = RTk \tag{11.35}$$

Thus, for an ideal gas

$$\boxed{c = \sqrt{RTk}} \tag{11.36}$$

More generally, the bulk modulus of elasticity, E_v, of any fluid including liquids is defined as (see Section 1.7.1)

$$E_v = \frac{dp}{d\rho/\rho} = \rho \left(\frac{\partial p}{\partial \rho}\right)_s \tag{11.37}$$

Thus, in general, from Eqs. 11.34 and 11.37,

$$c = \sqrt{\frac{E_v}{\rho}} \tag{11.38}$$

Speed of sound is larger in fluids that are more difficult to compress.

Values of the speed of sound are tabulated in Tables B.1 and B.2 for water and in Tables B.3 and B.4 for air. From experience we know that air is more easily compressed than water. Note from the values of c in Tables B.1 through B.4 that the speed of sound in air is much less than it is in water. From Eq. 11.37, we can conclude that if a fluid is truly incompressible, its bulk modulus would be infinitely large, as would be the speed of sound in that fluid. Thus, an incompressible flow must be considered an idealized approximation of reality.

EXAMPLE 11.3

Verify the speed of sound for air at 0 °C listed in Table B.4.

SOLUTION

In Table B.4, we find the speed of sound of air at 0 °C given as 331.4 m/s. Assuming that air behaves as an ideal gas, we can calculate the speed of sound from Eq. 11.36 as

$$c = \sqrt{RTk} \tag{1}$$

The value of the gas constant is obtained from Table 1.8 as

$$R = 286.9 \text{ J/(kg·K)}$$

and the specific heat ratio is listed in Table B.4 as

$$k = 1.401$$

By substituting values of R, k, and T into Eq. 1 we obtain

$$c = \sqrt{[(286.9)\ \text{J}/(\text{kg}\cdot\text{K})](273.15\text{K})(1.401)[1(\text{kg}\cdot\text{m})/(\text{N}\cdot\text{s}^2)][1(\text{N}\cdot\text{m})/\text{J}]}$$
$$= 331.4\ \text{m/s} \qquad \textbf{(Ans)}$$

The value of the speed of sound calculated with Eq. 11.36 agrees very well with the value of c listed in Table B.4.

11.3 Categories of Compressible Flow

In Section 3.8.1, we learned that the effects of compressibility become more significant as the Mach number increases. For example, the error associated with using $\rho V^2/2$ in calculating the stagnation pressure of an ideal gas increases at larger Mach numbers. From Fig. 3.24 we can conclude that incompressible flows can only occur at low Mach numbers.

Experience has also demonstrated that compressibility can have a large influence on other important flow variables. For example, in Fig. 11.2 the variation of drag coefficient with Reynolds number and Mach number is shown for air flow over a sphere. Compressibility effects can be of considerable importance.

Compressibility effects are more important at higher Mach numbers.

To further illustrate some curious features of compressible flow, a simplified example is considered. Imagine the emission of weak pressure pulses from a point source. These pressure waves are spherical and expand radially outward from the point source at the speed of sound, c. If a pressure wave is emitted at different times, t_{wave}, we can determine where

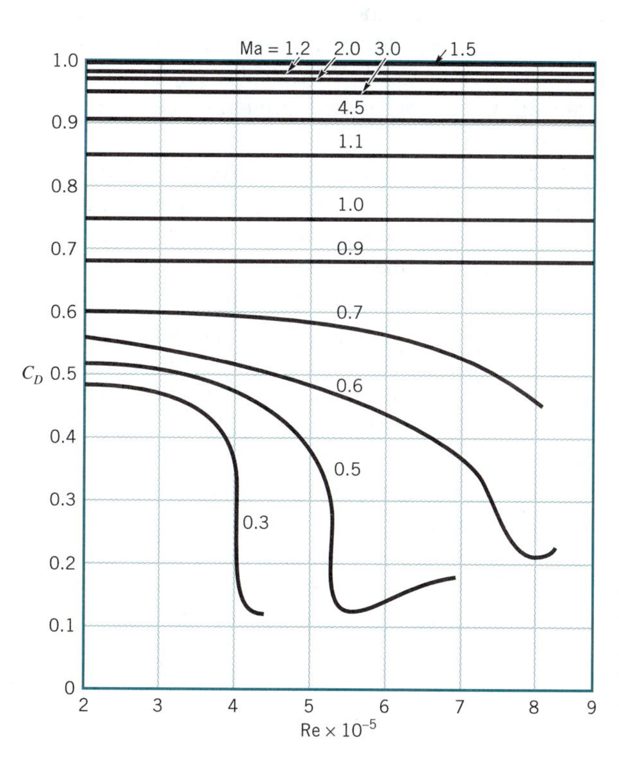

■ **FIGURE 11.2** **The variation of the drag coefficient of a sphere with Reynolds number and Mach number. (Adapted from Fig. 1.8 in Ref. 1 of Chapter 9.)**

several waves will be at a common instant of time, t, by using the relationship

$$r = (t - t_{\text{wave}})c$$

where r is the radius of the sphere-shaped wave emitted at time $= t_{\text{wave}}$. For a stationary point source, the symmetrical wave pattern shown in Fig. 11.3*a* is involved.

When the point source moves to the left with a constant velocity, V, the wave pattern is no longer symmetrical. In Figs. 11.3*b*, 11.3*c*, and 11.3*d* are illustrated the wave patterns at $t = 3$ s for different values of V. Also shown with a "+" are the positions of the moving point source at values of time, t, equal to 0 s, 1 s, 2 s, and 3 s. Knowing where the point source has been at different instances is important because it indicates to us where the different waves originated.

From the pressure wave patterns of Fig. 11.3, we can draw some useful conclusions. Before doing this we should recognize that if instead of moving the point source to the left, we held the point source stationary and moved the fluid to the right with velocity V, the resulting pressure wave patterns would be identical to those indicated in Fig. 11.3.

The speed of sound is infinite in a truly incompressible fluid.

When the point source and the fluid are stationary, the pressure wave pattern is symmetrical (Fig. 11.3*a*) and an observer anywhere in the pressure field would hear the same sound frequency from the point source. When the velocity of the point source (or the fluid) is very small in comparison with the speed of sound, the pressure wave pattern will still be nearly symmetrical. The speed of sound in an incompressible fluid is infinitely large. Thus, the stationary point source and stationary fluid situation are representative of incompressible flows. For truly incompressible flows, the communication of pressure information throughout the flow field is unrestricted and instantaneous ($c = \infty$).

When the point source moves in fluid at rest (or when fluid moves past a stationary point source), the pressure wave patterns vary in asymmetry, with the extent of asymmetry depending on the ratio of the point source (or fluid) velocity and the speed of sound. When $V/c < 1$, the wave pattern is similar to the one shown in Fig. 11.3*b*. This flow is considered *subsonic* and compressible. A stationary observer will hear a different sound frequency coming from the point source depending on where the observer is relative to the source because the wave pattern is asymmetrical. We call this phenomenon the Doppler effect. Pressure information can still travel unrestricted throughout the flow field, but not symmetrically or instantaneously.

When $V/c = 1$, pressure waves are not present ahead of the moving point source. The flow is *sonic*. If you were positioned to the left of the moving point source, you would not hear the point source until it was coincident with your location. For flow moving past a stationary point source at the speed of sound ($V/c = 1$), the pressure waves are all tangent to a plane that is perpendicular to the flow and that passes through the point source. The concentration of pressure waves in this tangent plane suggests the formation of a significant pressure variation across the plane. This plane is often called a Mach wave. Note that communication of pressure information is restricted to the region of flow downstream of the Mach wave. The region of flow upstream of the Mach wave is called the *zone of silence* and the region of flow downstream of the tangent plane is called the *zone of action*.

When $V > c$, the flow is *supersonic* and the pressure wave pattern resembles the one depicted in Fig. 11.3*d*. A cone (*Mach cone*) that is tangent to the pressure waves can be constructed to represent the Mach wave that separates the zone of silence from the zone of action in this case. The communication of pressure information is restricted to the zone of action. From the sketch of Fig. 11.3*d*, we can see that the angle of this cone, α, is given by

$$\sin \alpha = \frac{c}{V} = \frac{1}{\text{Ma}} \qquad \textbf{(11.39)}$$

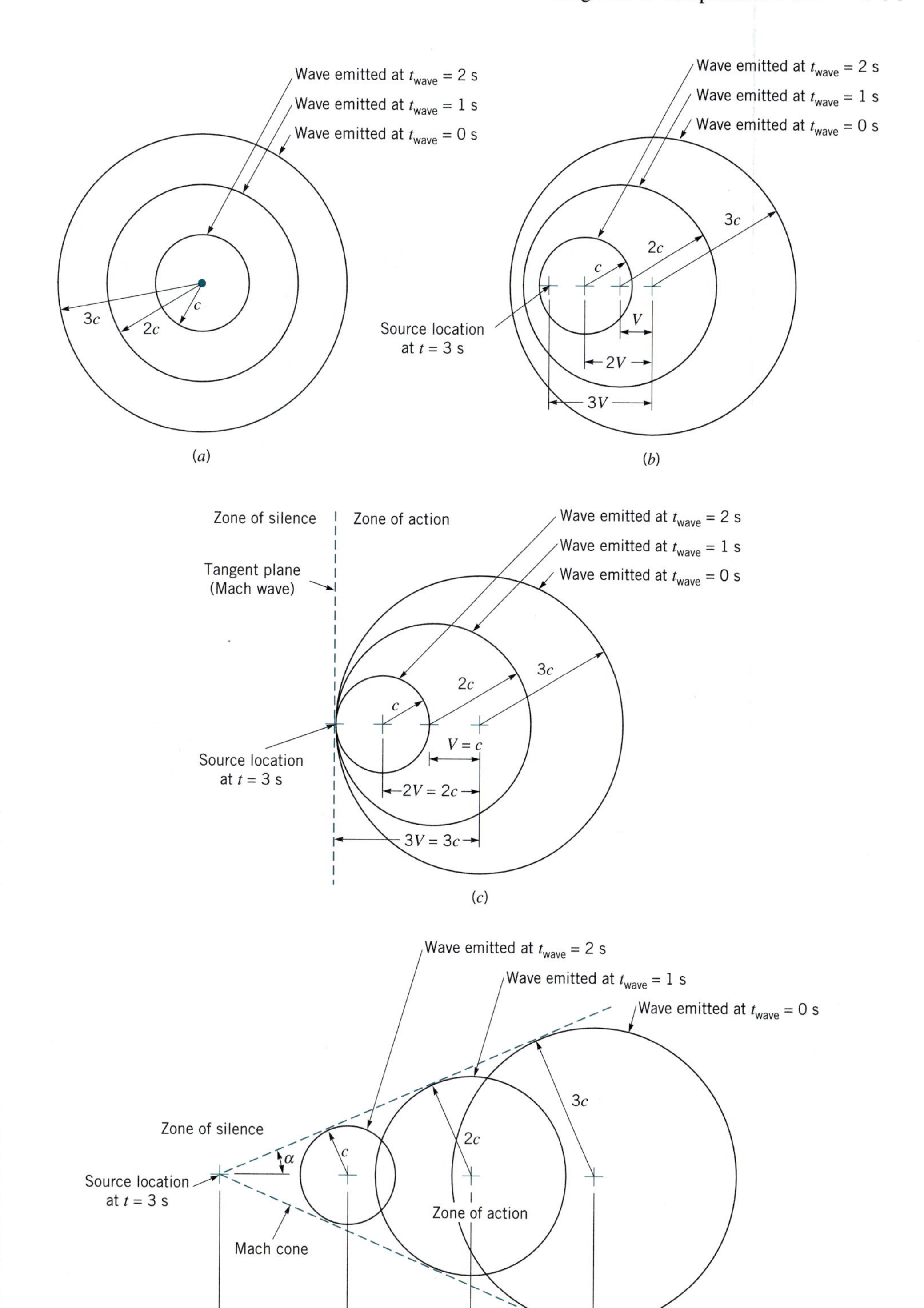

■ **FIGURE 11.3** ***(a)* Pressure waves at $t = 3$ s, $V = 0$; *(b)* Pressure waves at $t = 3$ s, $V < c$; *(c)* Pressure waves at $t = 3$ s, $V = c$; *(d)* Pressure waves at $t = 3$ s, $V > c$.**

■ FIGURE 11.4 The schlieren visualization of flow (supersonic to subsonic) through a row of compressor airfoils. (Photograph provided by Dr. Hans Starken of the DLR Köln-Porz, Germany.)

V11.2

Abrupt changes in fluid properties can occur in supersonic flows.

Equation 11.39 is often used to relate the Mach cone angle, α, and the flow Mach number, Ma, when studying flows involving $V/c > 1$. The concentration of pressure waves at the surface of the Mach cone suggests a significant pressure, and thus density, variation across the cone surface. An abrupt density change can be visualized in a flow field by using special optics. Examples of flow visualization methods include the schlieren, shadowgraph, and interferometer techniques (see Ref. 5). A schlieren photo of a flow for which $V > c$ is shown in Fig. 11.4. The air flow through the row of compressor blade airfoils is as shown with the arrow. The flow enters supersonically ($\text{Ma}_1 = 1.14$) and leaves subsonically ($\text{Ma}_2 = 0.86$). The center two airfoils have pressure tap hoses connected to them. Regions of significant changes in fluid density appear in the supersonic portion of the flow. Also, the region of separated flow on each airfoil is visible.

This discussion about pressure wave patterns suggests the following categories of fluid flow:

1. Incompressible flow: $\text{Ma} \leq 0.3$. Unrestricted, nearly symmetrical and instantaneous pressure communication.
2. Compressible subsonic flow: $0.3 < \text{Ma} < 1.0$. Unrestricted but noticeably asymmetrical pressure communication.
3. Compressible supersonic flow: $\text{Ma} \geq 1.0$. Formation of Mach wave; pressure communication restricted to zone of action.

In addition to the above-mentioned categories of flows, two other regimes are commonly referred to: namely, transonic flows ($0.9 \leq \text{Ma} \leq 1.2$) and hypersonic flows ($\text{Ma} > 5$). Modern aircraft are mainly powered by gas turbine engines that involve transonic flows. When a space shuttle reenters the earth's atmosphere, the flow is hypersonic. Future aircraft may be expected to operate from subsonic to hypersonic flow conditions.

EXAMPLE 11.4

An aircraft cruising at 1000-m elevation, z, above you moves past in a flyby. How many seconds after the plane passes overhead do you expect to wait before you hear the aircraft if it is moving with a Mach number equal to 1.5 and the ambient temperature is 20 °C?

SOLUTION

Since the aircraft is moving supersonically ($\text{Ma} > 1$), we can imagine a Mach cone originating from the forward tip of the craft as is illustrated in Fig. E11.4. When the surface of the cone

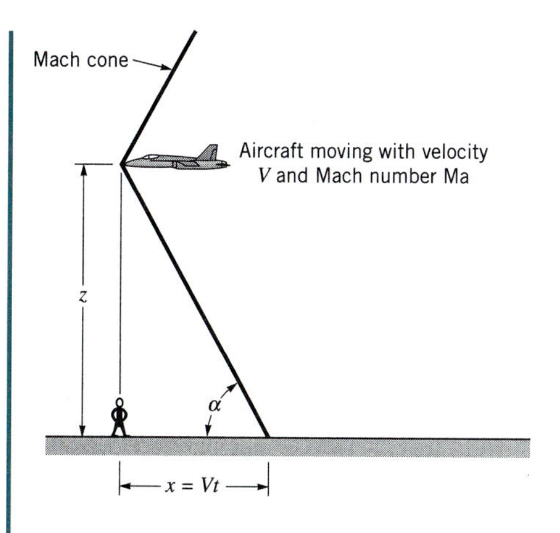

■ FIGURE E11.4

reaches the observer, the "sound" of the aircraft is perceived. The angle α in Fig. E11.4 is related to the elevation of the plane, z, and the ground distance, x, by

$$\alpha = \tan^{-1}\frac{z}{x} = \tan^{-1}\frac{1000}{Vt} \tag{1}$$

Also, assuming negligible change of Mach number with elevation, we can use Eq. 11.39 to relate Mach number to the angle α. Thus,

$$\text{Ma} = \frac{1}{\sin\alpha} \tag{2}$$

Combining Eqs. 1 and 2 we obtain

$$\text{Ma} = \frac{1}{\sin\left[\tan^{-1}(1000/Vt)\right]} \tag{3}$$

The speed of the aircraft can be related to the Mach number with

$$V = (\text{Ma})c \tag{4}$$

where c is the speed of sound. From Table B.4, $c = 343.3$ m/s. Using Ma = 1.5, we get from Eqs. 3 and 4

$$1.5 = \frac{1}{\sin\left\{\tan^{-1}\left[\dfrac{1000 \text{ m}}{(1.5)(343.3 \text{ m/s})t}\right]\right\}}$$

or

$$t = 2.17 \text{ s} \qquad \textbf{(Ans)}$$

11.4 Isentropic Flow of an Ideal Gas

In this section, we consider in further detail the steady, one-dimensional, isentropic flow of an ideal gas with constant specific heat values (c_p and c_v). Because the flow is steady throughout, shaft work cannot be involved. Also, as explained earlier, the one-dimensionality of flows we discuss in this chapter implies velocity and fluid property changes in the streamwise direction only. We consider flows through finite control volumes with uniformly distributed velocities and fluid properties at each section of flow. Much of what we develop can also apply to the flow of a fluid particle along its pathline.

An important class of isentropic flow involves no heat transfer and zero friction.

Isentropic flow involves constant entropy and was discussed earlier in Section 11.1, where we learned that adiabatic and frictionless (reversible) flow is one form of isentropic flow. Some ideal gas relationships for isentropic flows were developed in Section 11.1. An isentropic flow is not achievable with actual fluids because of friction. Nonetheless, the study of isentropic flow trends is useful because it helps us to gain an understanding of actual compressible flow phenomena.

11.4.1 Effect of Variations in Flow Cross-Section Area

When fluid flows steadily through a conduit that has a flow cross-section area that varies with axial distance, the conservation of mass (continuity) equation

$$\dot{m} = \rho A V = \text{constant} \tag{11.40}$$

can be used to relate the flow rates at different sections. For incompressible flow, the fluid density remains constant and the flow velocity from section to section varies inversely with cross-section area. However, when the flow is compressible, density, cross-section area, and flow velocity can all vary from section to section. We proceed to determine how fluid density and flow velocity change with axial location in a variable area duct when the fluid is an ideal gas and the flow through the duct is steady and isentropic.

In Chapter 3, Newton's second law was applied to the inviscid (frictionless) and steady flow of a fluid particle. For the streamwise direction, the result (Eq. 3.5) for either compressible or incompressible flows is

$$dp + \tfrac{1}{2}\rho\, d(V^2) + \gamma\, dz = 0 \tag{11.41}$$

The frictionless flow from section to section through a finite control volume is also governed by Eq. 11.41, if the flow is one-dimensional, because every particle of fluid involved will have the same experience. For ideal gas flow, the potential energy difference term, $\gamma\, dz$, can be dropped because of its small size in comparison to the other terms, namely, dp and $d(V^2)$. Thus, an appropriate equation of motion in the streamwise direction for the steady, one-dimensional, and isentropic (adiabatic and frictionless) flow of an ideal gas is obtained from Eq. 11.41 as

$$\frac{dp}{\rho V^2} = -\frac{dV}{V} \tag{11.42}$$

If we form the logarithm of both sides of the continuity equation (Eq. 11.40), the result is

$$\ln \rho + \ln A + \ln V = \text{constant} \tag{11.43}$$

Differentiating Eq. 11.43 we get

$$\frac{d\rho}{\rho} + \frac{dA}{A} + \frac{dV}{V} = 0$$

or

$$-\frac{dV}{V} = \frac{d\rho}{\rho} + \frac{dA}{A} \tag{11.44}$$

Now we combine Eqs. 11.42 and 11.44 to obtain

$$\frac{dp}{\rho V^2}\left(1 - \frac{V^2}{dp/d\rho}\right) = \frac{dA}{A} \tag{11.45}$$

Since the flow being considered is isentropic, the speed of sound is related to variations of pressure with density by Eq. 11.34, repeated here for convenience as

$$c = \sqrt{\left(\frac{\partial p}{\partial \rho}\right)_s}$$

Equation 11.34, combined with the definition of Mach number

$$\text{Ma} = \frac{V}{c} \tag{11.46}$$

and Eq. 11.45 yields

$$\frac{dp}{\rho V^2}(1 - \text{Ma}^2) = \frac{dA}{A} \tag{11.47}$$

Equations 11.42 and 11.47 merge to form

$$\frac{dV}{V} = -\frac{dA}{A}\frac{1}{(1 - \text{Ma}^2)} \tag{11.48}$$

We can use Eq. 11.48 to conclude that when the flow is subsonic ($\text{Ma} < 1$), velocity and section area changes are in opposite directions. In other words, the area increase associated with subsonic flow through a diverging duct like the one shown in Fig. 11.5*a* is accompanied by a velocity decrease. Subsonic flow through a converging duct (see Fig. 11.5*b*) involves an increase of velocity. These trends are consistent with incompressible flow behavior, which we described earlier in this book, for instance, in Chapters 3 and 8.

Equation 11.48 also serves to show us that when the flow is supersonic ($\text{Ma} > 1$), velocity and area changes are in the same direction. A diverging duct (Fig. 11.5*a*) will accelerate a supersonic flow. A converging duct (Fig. 11.5*b*) will decelerate a supersonic flow. These trends are the opposite of what happens for incompressible and subsonic compressible flows.

A converging duct will decelerate a supersonic flow and accelerate a subsonic flow.

To better understand why subsonic and supersonic duct flows are so different, we combine Eqs. 11.44 and 11.48 to form

$$\frac{d\rho}{\rho} = \frac{dA}{A}\frac{\text{Ma}^2}{(1 - \text{Ma}^2)} \tag{11.49}$$

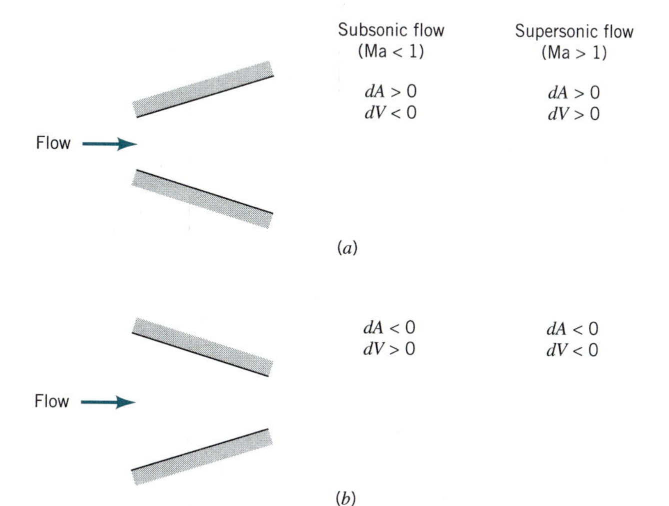

■ **FIGURE 11.5**
***(a)* A diverging duct. *(b)* A converging duct.**

Using Eq. 11.49, we can conclude that for subsonic flows (Ma < 1), density and area changes are in the same direction, whereas for supersonic flows (Ma > 1), density and area changes are in opposite directions. Since ρAV must remain constant (Eq. 11.40), when the duct diverges and the flow is subsonic, density and area both increase and thus flow velocity must decrease. However, for supersonic flow through a diverging duct, when the area increases, the density decreases enough so that the flow velocity has to increase to keep ρAV constant.

By rearranging Eq. 11.48, we can obtain

$$\frac{dA}{dV} = -\frac{A}{V}(1 - \text{Ma}^2) \tag{11.50}$$

Equation 11.50 gives us some insight into what happens when Ma $= 1$. For Ma $= 1$, Eq. 11.50 requires that $dA/dV = 0$. This result suggests that the area associated with Ma $= 1$ is either a minimum or a maximum amount.

A converging-diverging duct (Fig. 11.6*a*) involves a minimum area. If the flow entering such a duct were subsonic, Eq. 11.48 discloses that the fluid velocity would increase in the converging portion of the duct, and achievement of a sonic condition (Ma $= 1$) at the minimum area location appears possible. If the flow entering the converging-diverging duct is supersonic, Eq. 11.48 states that the fluid velocity would decrease in the converging portion of the duct and the sonic condition at the minimum area is possible.

A diverging-converging duct (Fig. 11.6*b*), on the other hand, would involve a maximum area. If the flow entering this duct were subsonic, the fluid velocity would decrease in the diverging portion of the duct and the sonic condition could not be attained at the maximum area location. For supersonic flow in the diverging portion of the duct, the fluid velocity would increase and thus Ma $= 1$ at the maximum area is again impossible.

For the steady isentropic flow of an ideal gas, we conclude that the sonic condition (Ma $= 1$) can be attained in a converging-diverging duct at the minimum area location. This minimum area location is often called the *throat* of the converging-diverging duct. Furthermore, to achieve supersonic flow from a subsonic state in a duct, a converging-diverging area variation is necessary. For this reason, we often refer to such a duct as a *converging-diverging nozzle*. Note that a converging-diverging duct can also decelerate a supersonic flow to subsonic conditions. Thus, a converging-diverging duct can be a nozzle or a diffuser depending on whether the flow in the converging portion of the duct is subsonic or supersonic. A supersonic wind tunnel test section is generally preceded by a converging-diverging nozzle and followed by a converging-diverging diffuser (see Ref. 1). Further details about steady, isentropic, ideal gas flow through a converging-diverging duct are discussed in the next section.

A converging-diverging duct is required to accelerate a flow from subsonic to supersonic flow conditions.

11.4.2 Converging-Diverging Duct Flow

In the preceding section, we discussed the variation of density and velocity of the steady isentropic flow of an ideal gas through a variable area duct. We proceed now to develop equations that help us determine how other important flow properties vary in these flows.

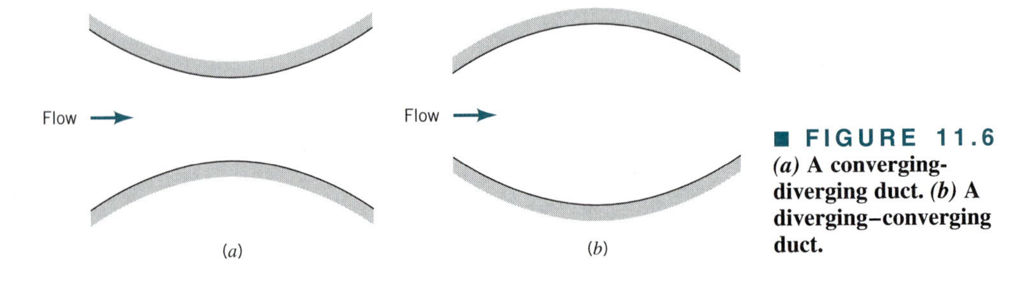

■ **FIGURE 11.6** ***(a)* A converging-diverging duct. *(b)* A diverging–converging duct.**

The stagnation state is associated with zero flow velocity that is attained isentropically.

It is convenient to use the stagnation state of the fluid as a reference state for compressible flow calculations. The stagnation state is associated with zero flow velocity and an entropy value that corresponds to the entropy of the flowing fluid. The subscript 0 is used to designate the stagnation state. Thus, stagnation temperature and pressure are T_0 and p_0. For example, if the fluid flowing through the converging-diverging duct of Fig. 11.6*a* were drawn isentropically from the atmosphere, the atmospheric pressure and temperature would represent the stagnation state of the flowing fluid. The stagnation state can also be achieved by isentropically decelerating a flow to zero velocity. This can be accomplished with a diverging duct for subsonic flows or a converging-diverging duct for supersonic flows. Also, as discussed earlier in Chapter 3, an approximately isentropic deceleration can be accomplished with a Pitot-static tube (see Fig. 3.6). It is thus possible to measure, with only a small amount of uncertainty, values of stagnation pressure, p_0, and stagnation temperature, T_0, of a flowing fluid.

In Section 11.1, we demonstrated that for the isentropic flow of an ideal gas (see Eq. 11.25)

$$\frac{p}{\rho^k} = \text{constant} = \frac{p_0}{\rho_0^k}$$

The streamwise equation of motion for steady and frictionless flow (Eq. 11.41) can be expressed for an ideal gas as

$$\frac{dp}{\rho} + d\left(\frac{V^2}{2}\right) = 0 \quad \textbf{(11.51)}$$

since the potential energy term, $\gamma\, dz$ can be considered as being negligibly small in comparison with the other terms involved.

By incorporating Eq. 11.25 into Eq. 11.51 we obtain

$$\frac{p_0^{1/k}}{\rho_0} \frac{dp}{(p)^{1/k}} + d\left(\frac{V^2}{2}\right) = 0 \quad \textbf{(11.52)}$$

Consider the steady, one-dimensional, isentropic flow of an ideal gas with constant c_p and c_v through the converging-diverging nozzle of Fig. 11.6*a*. Equation 11.52 is valid for this flow and can be integrated between the common stagnation state of the flowing fluid to the state of the gas at any location in the converging-diverging duct to give

$$\frac{k}{k-1}\left(\frac{p_0}{\rho_0} - \frac{p}{\rho}\right) - \frac{V^2}{2} = 0 \quad \textbf{(11.53)}$$

By using the ideal gas equation of state (Eq. 11.1) with Eq. 11.53 we obtain

$$\frac{kR}{k-1}(T_0 - T) - \frac{V^2}{2} = 0 \quad \textbf{(11.54)}$$

It is of interest to note that combining Eqs. 11.14 and 11.54 leads to

$$c_p(T_0 - T) - \frac{V^2}{2} = 0$$

which, when merged with Eq. 11.9, results in

$$\check{h}_0 - \left(\check{h} + \frac{V^2}{2}\right) = 0 \quad \textbf{(11.55)}$$

where $\check{h}_0$ is the stagnation enthalpy. If the steady flow energy equation (Eq. 5.69) is applied to the flow situation we are presently considering, the resulting equation will be identical to

Eq. 11.55. Further, we conclude that the stagnation enthalpy is constant. The conservation of momentum and energy principles lead to the same equation (Eq. 11.55) for steady isentropic flows.

The definition of Mach number (Eq. 11.46) and the speed of sound relationship for ideal gases (Eq. 1.36) can be combined with Eq. 11.54 to yield

$$\frac{T}{T_0} = \frac{1}{1 + [(k-1)/2]\mathrm{Ma}^2} \tag{11.56}$$

With Eq. 11.56 we can calculate the temperature of an ideal gas anywhere in the converging-diverging duct of Fig. 11.6*a* if the flow is steady, one-dimensional, and isentropic, provided we know the value of the local Mach number and the stagnation temperature.

We can also develop an equation for pressure variation. Since $p/\rho = RT$, then

$$\left(\frac{p}{p_0}\right)\left(\frac{\rho_0}{\rho}\right) = \frac{T}{T_0} \tag{11.57}$$

From Eqs. 11.57 and 11.25 we obtain

$$\left(\frac{p}{p_0}\right) = \left(\frac{T}{T_0}\right)^{k/(k-1)} \tag{11.58}$$

Combining Eqs. 11.58 and 11.56 leads to

$$\frac{p}{p_0} = \left\{\frac{1}{1 + [(k-1)/2]\mathrm{Ma}^2}\right\}^{k/(k-1)} \tag{11.59}$$

For density variation we consolidate Eqs. 11.56, 11.57, and 11.59 to get

$$\frac{\rho}{\rho_0} = \left\{\frac{1}{1 + [(k-1)/2]\mathrm{Ma}^2}\right\}^{1/(k-1)} \tag{11.60}$$

The temperature-entropy diagram is a useful concept for mapping compressible flows.

A very useful means of keeping track of the states of an isentropic flow of an ideal gas involves a temperature-entropy ($T - s$) diagram, as is shown in Fig. 11.7. Experience has shown (see, for example, Refs. 2 and 3) that lines of constant pressure are generally as are sketched in Fig. 11.7. An isentropic flow is confined to a vertical line on a $T - s$ diagram. The vertical line in Fig. 11.7 is representative of flow between the stagnation state and any state within the converging-diverging nozzle. Equation 11.56 shows that fluid temperature

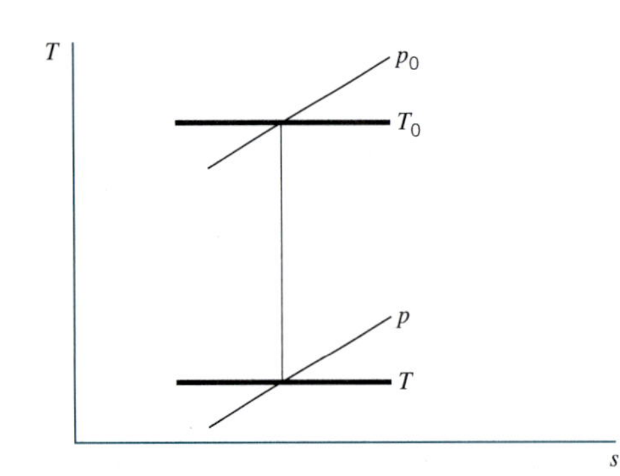

■ FIGURE 11.7 The T–s diagram relating stagnation and static states.

decreases with an increase in Mach number. Thus, the lower temperature levels on a $T - s$ diagram correspond to higher Mach numbers. Equation 11.59 suggests that fluid pressure also decreases with an increase in Mach number. Thus, lower fluid temperatures and pressures are associated with higher Mach numbers in our isentropic converging-diverging duct example.

One way to produce flow through a converging-diverging duct like the one in Fig. 11.6*a* is to connect the downstream end of the duct to a vacuum pump. When the pressure at the downstream end of the duct (the back pressure) is decreased slightly, air will flow from the atmosphere through the duct and vacuum pump. Neglecting friction and heat transfer and considering the air to act as an ideal gas, Eqs. 11.56, 11.59, and 11.60 and a $T - s$ diagram can be used to describe steady flow through the converging-diverging duct.

If the pressure in the duct is only slightly less than atmospheric pressure, we predict with Eq. 11.59 that the Mach number levels in the duct will be low. Thus, with Eq. 11.60 we conclude that the variation of fluid density in the duct is also small. The continuity equation (Eq. 11.40) leads us to state that there is a small amount of fluid flow acceleration in the converging portion of the duct followed by flow deceleration in the diverging portion of the duct. We considered this type of flow when we discussed the Venturi meter in Section 3.6.3. The $T - s$ diagram for this flow is sketched in Fig. 11.8.

We next consider what happens when the back pressure is lowered further. Since the flow starts from rest upstream of the converging portion of the duct of Fig. 11.6*a*, Eqs. 11.48 and 11.50 reveal to us that flow up to the nozzle throat can be accelerated to a maximum allowable Mach number of 1 at the throat. Thus, when the duct back pressure is lowered sufficiently, the Mach number at the throat of the duct will be 1. Any further decrease of the back pressure will not affect the flow in the converging portion of the duct because, as is discussed in Section 11.3, information about pressure cannot move upstream when Ma = 1. When Ma = 1 at the throat of the converging-diverging duct, we have a condition called *choked flow*. Some useful equations for choked flow are developed below.

Choked flow occurs when the Mach number is 1.0 at the minimum cross-section area.

We have already used the stagnation state for which Ma = 0 as a reference condition. It will prove helpful to us to use the state associated with Ma = 1 and the same entropy level as the flowing fluid as another reference condition we shall call the *critical state*, denoted $(\ )^*$.

The ratio of pressure at the converging-diverging duct throat for choked flow, p^*, to stagnation pressure, p_0, is referred to as the *critical pressure ratio*. By substituting Ma = 1 into Eq. 11.59 we obtain

$$\frac{p^*}{p_0} = \left(\frac{2}{k+1}\right)^{k/(k-1)} \quad \textbf{(11.61)}$$

For $k = 1.4$, the nominal value of k for air, Eq. 11.61 yields

$$\left(\frac{p^*}{p_0}\right)_{k=1.4} = 0.528 \quad \textbf{(11.62)}$$

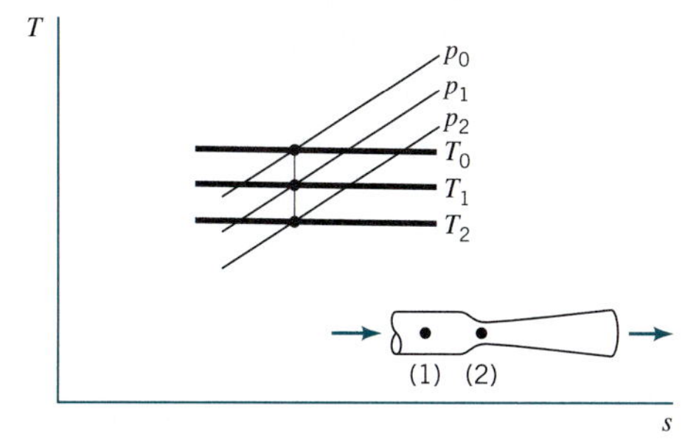

■ **FIGURE 11.8** **The *T–s* diagram for Venturi meter flow.**

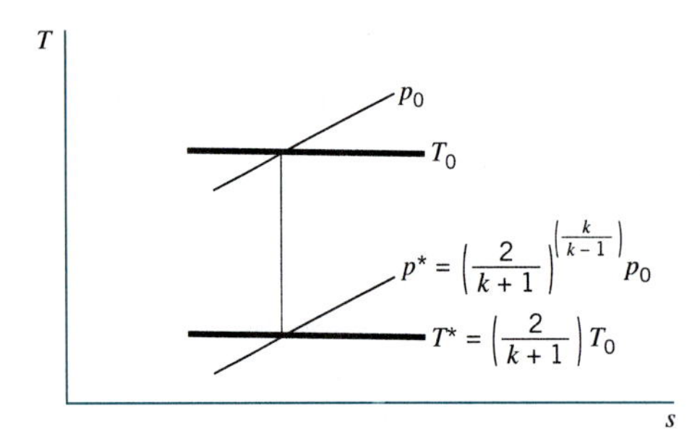

■ **FIGURE 11.9 The relationship between the stagnation and critical states.**

Because the stagnation pressure for our converging-diverging duct example is the atmospheric pressure, p_{atm}, the throat pressure for choked air flow is, from Eq. 11.62

$$p^*_{k=1.4} = 0.528 p_{atm}$$

We can get a relationship for the critical temperature ratio, T^*/T_0, by substituting $Ma = 1$ into Eq. 11.56. Thus,

$$\frac{T^*}{T_0} = \frac{2}{k+1} \tag{11.63}$$

or for $k = 1.4$

$$\left(\frac{T^*}{T_0}\right)_{k=1.4} = 0.833 \tag{11.64}$$

For the duct of Fig. 11.6*a*, Eq. 11.64 yields

$$T^*_{k=1.4} = 0.833 T_{atm}$$

The stagnation and critical pressures and temperatures are shown on the $T - s$ diagram of Fig. 11.9.

The stagnation and critical states are at the same entropy level.

When we combine the ideal gas equation of state (Eq. 11.1) with Eqs. 11.61 and 11.63, for $Ma = 1$ we get

$$\frac{\rho^*}{\rho_0} = \left(\frac{p^*}{T^*}\right)\left(\frac{T_0}{p_0}\right) = \left(\frac{2}{k+1}\right)^{k/(k-1)}\left(\frac{k+1}{2}\right) = \left(\frac{2}{k+1}\right)^{1/(k-1)} \tag{11.65}$$

For air ($k = 1.4$), Eq. 11.65 leads to

$$\left(\frac{\rho^*}{\rho_0}\right)_{k=1.4} = 0.634 \tag{11.66}$$

and we see that when the converging-diverging duct flow is choked, the density of the air at the duct throat is 63.4% of the density of atmospheric air.

EXAMPLE 11.5

A converging duct passes air steadily from standard atmospheric conditions to a receiver pipe as illustrated in Fig. E11.5*a*. The throat (minimum) flow cross-section area of the converging

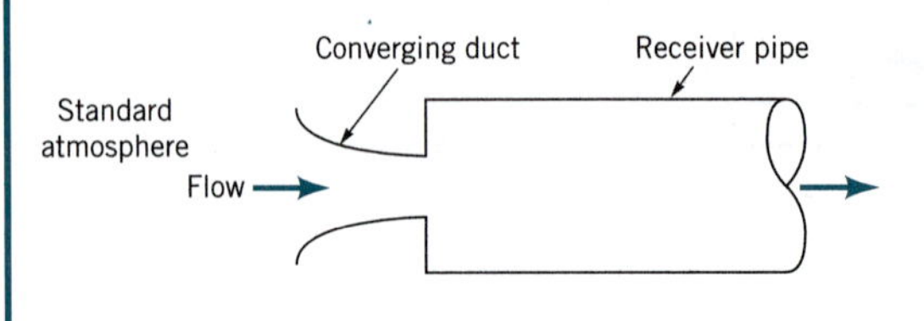

(*a*)

■ **FIGURE E11.5**

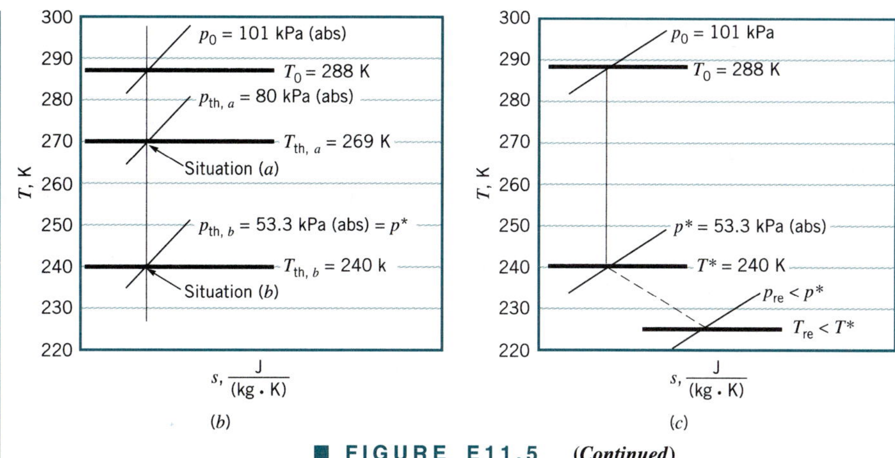

■ **FIGURE E11.5** ***(Continued)***

duct is 1×10^{-4} m^2. Determine the mass flowrate through the duct if the receiver pressure is (a) 80 kPa (abs), (b) 40 kPa (abs). Sketch temperature-entropy diagrams for situations (a) and (b).

SOLUTION

To determine the mass flowrate through the converging duct we use Eq. 11.40. Thus,

$$\dot{m} = \rho AV = \text{constant}$$

or in terms of the given throat area, A_{th},

$$\dot{m} = \rho_{th} A_{th} V_{th} \tag{1}$$

We assume that the flow through the converging duct is isentropic and that the air behaves as an ideal gas with constant c_p and c_v. Then, from Eq. 11.60

$$\frac{\rho_{th}}{\rho_0} = \left\{\frac{1}{1 + [(k-1)/2]\mathrm{Ma}_{th}^2}\right\}^{1/(k-1)} \tag{2}$$

The stagnation density, ρ_0, for standard atmosphere is 1.23 kg/m^3 and the specific heat ratio is 1.4. To determine the throat Mach number, Ma_{th}, we can use Eq. 11.59,

$$\frac{p_{th}}{p_0} = \left\{\frac{1}{1 + [(k-1)/2]\mathrm{Ma}_{th}^2}\right\}^{k/(k-1)} \tag{3}$$

The critical pressure, p^*, is obtained from Eq. 11.62 as

$$p^* = 0.528 p_0 = 0.528 p_{atm} = (0.528)[101 \text{ kPa(abs)}] = 53.3 \text{ kPa(abs)}$$

If the receiver pressure, p_{re}, is greater than or equal to p^*, then $p_{th} = p_{re}$. If $p_{re} < p^*$, then $p_{th} = p^*$ and the flow is choked. With p_{th}, p_0, and k known, Ma_{th} can be obtained from Eq. 3, and ρ_{th} can be determined from Eq. 2.

The flow velocity at the throat can be obtained from Eqs. 11.36 and 11.46 as

$$V_{th} = \mathrm{Ma}_{th} c_{th} = \mathrm{Ma}_{th} \sqrt{RT_{th} k} \tag{4}$$

The value of temperature at the throat, T_{th}, can be calculated from Eq. 11.56,

$$\frac{T_{th}}{T_0} = \frac{1}{1 + [(k - 1)/2]\text{Ma}_{th}^2} \quad \textbf{(5)}$$

Since the flow through the converging duct is assumed to be isentropic, the stagnation temperature is considered constant at the standard atmosphere value of $T_0 = 15\text{ K} + 273\text{ K} = 288\text{ K}$. Note that absolute pressures and temperatures are used.

(a) For $p_{re} = 80\text{ kPa(abs)} > 53.3\text{ kPa(abs)} = p^*$, we have $p_{th} = 80\text{ kPa(abs)}$. Then from Eq. 3

$$\frac{80\text{ kPa(abs)}}{101\text{ kPa(abs)}} = \left\{\frac{1}{1 + [(1.4 - 1)/2]\text{Ma}_{th}^2}\right\}^{1.4/(1.4-1)}$$

or

$$\text{Ma}_{th} = 0.587$$

From Eq. 2

$$\frac{\rho_{th}}{1.23\text{ kg/m}^3} = \left\{\frac{1}{1 + [(1.4 - 1)/2](0.587)^2}\right\}^{1/(1.4-1)}$$

or

$$\rho_{th} = 1.04\text{ kg/m}^3$$

From Eq. 5

$$\frac{T_{th}}{288\text{ K}} = \frac{1}{1 + [(1.4 - 1)/2](0.587)^2}$$

or

$$T_{th} = 269\text{ K}$$

Substituting $\text{Ma}_{th} = 0.587$ and $T_{th} = 269\text{ K}$ into Eq. 4 we obtain

$$V_{th} = 0.587\sqrt{[286.9\text{ J/(kg}\cdot\text{K)}](269\text{ K})(1.4)[1(\text{kg}\cdot\text{m})/(\text{N}\cdot\text{s}^2)][1(\text{N}\cdot\text{m})/\text{J}]}$$

or

$$V_{th} = 193\text{ m/s}$$

Finally from Eq. 1 we have

$$\dot{m} = (1.04\text{ kg/m}^3)(1 \times 10^{-4}\text{ m}^2)(193\text{ m/s}) = 0.0201\text{ kg/s} \quad \textbf{(Ans)}$$

(b) For $p_{re} = 40\text{ kPa(abs)} < 53.3\text{ kPa(abs)} = p^*$, we have $p_{th} = p^* = 53.3\text{ kPa(abs)}$ and $\text{Ma}_{th} = 1$. The converging duct is choked. From Eq. 2 (see also Eq. 11.66)

$$\frac{\rho_{th}}{1.23\text{ kg/m}^3} = \left\{\frac{1}{1 + [(1.4 - 1)/2](1)^2}\right\}^{1/(1.4-1)}$$

or

$$\rho_{th} = 0.780\text{ kg/m}^3$$

From Eq. 5 (see also Eq. 11.64),

$$\frac{T_{th}}{288\ K} = \frac{1}{1 + [(1.4 - 1)/2](1)^2}$$

or

$$T_{th} = 240\ K$$

From Eq. 4,

$$V_{th} = (1)\sqrt{[286.9\ J/(kg \cdot K)](240\ K)(1.4)[1(kg \cdot m)/(N \cdot s^2)][1(N \cdot m)/J]} = 310\ m/s$$

Finally from Eq. 1

$$\dot{m} = (0.780\ kg/m^3)(1 \times 10^{-4}\ m^2)(310\ m/s) = 0.0242\ kg/s \qquad \textbf{(Ans)}$$

From the values of throat temperature and throat pressure calculated above for flow situations (a) and (b), we can construct the temperature–entropy diagram shown in Fig. E11.5*b*.

Note that the flow from standard atmosphere to the receiver for receiver pressure, p_{re}, greater than or equal to the critical pressure, p^*, is isentropic. When the receiver pressure is less than the critical pressure as in situation (b) above, what is the flow like downstream from the exit of the converging duct? Experience suggests that this flow, when $p_{re} < p^*$, is three-dimensional and nonisentropic and involves a drop in pressure from p_{th} to p_{re}, a drop in temperature, and an increase of entropy as are indicated in Fig. E11.5*c*.

Isentropic flow Eqs. 11.56, 11.59, and 11.60 have been used to construct Fig. D.1 in Appendix D for air ($k = 1.4$). Examples 11.6 and 11.7 illustrate how these graphs of T/T_0, p/p_0, and ρ/ρ_0 as a function of Mach number, Ma, can be used to solve compressible flow problems.

Solve Example 11.5 using Fig. D.1 of Appendix D.

SOLUTION

We still need the density and velocity of the air at the converging duct throat to solve for mass flowrate from

$$\dot{m} = \rho_{th} A_{th} V_{th} \qquad \textbf{(1)}$$

(a) Since the receiver pressure, $p_{re} = 80$ kPa(abs), is greater than the critical pressure, $p^* = 53.3$ kPa(abs), the throat pressure, p_{th}, is equal to the receiver pressure. Thus

$$\frac{p_{th}}{p_0} = \frac{80\ kPa(abs)}{101\ kPa(abs)} = 0.792$$

From Fig. D.1, for $p/p_0 = 0.79$, we get from the graph

$$\text{Ma}_{\text{th}} = 0.59$$

$$\frac{T_{\text{th}}}{T_0} = 0.94 \quad \textbf{(2)}$$

$$\frac{\rho_{\text{th}}}{\rho_0} = 0.85 \quad \textbf{(3)}$$

Thus, from Eqs. 2 and 3

$$T_{\text{th}} = (0.94)(288 \text{ K}) = 271 \text{ K}$$

and

$$\rho_{\text{th}} = (0.85)(1.23 \text{ kg/m}^3) = 1.04 \text{ kg/m}^3$$

Furthermore, using Eqs. 11.36 and 11.46 we get

$$\begin{aligned} V_{\text{th}} &= \text{Ma}_{\text{th}} \sqrt{RT_{\text{th}}k} \\ &= (0.59) \sqrt{[286.9 \text{ J/(kg}\cdot\text{K)}](269 \text{ K})(1.4)[1(\text{kg}\cdot\text{m})/(\text{N}\cdot\text{s}^2)][1(\text{N}\cdot\text{m})/\text{J}]} = 194 \text{ m/s} \end{aligned}$$

Finally, from Eq. 1

$$\dot{m} = (1.04 \text{ kg/m}^3)(1 \times 10^{-4} \text{ m}^2)(194 \text{ m/s}) = 0.0202 \text{ kg/s} \quad \textbf{(Ans)}$$

(b) For $p_{\text{re}} = 40$ kPa(abs) < 53.3 kPa(abs) $= p^*$, the throat pressure is equal to 53.3 kPa(abs) and the duct is choked with $\text{Ma}_{\text{th}} = 1$. From Fig. D.1, for Ma $= 1$ we get

$$\frac{T_{\text{th}}}{T_0} = 0.83 \quad \textbf{(4)}$$

and

$$\frac{\rho_{\text{th}}}{\rho_0} = 0.64 \quad \textbf{(5)}$$

From Eqs. 4 and 5 we obtain

$$T_{\text{th}} = (0.83)(288 \text{ K}) = 240 \text{ K}$$

and

$$\rho_{\text{th}} = (0.64)(1.23 \text{ kg/m}^3) = 0.79 \text{ kg/m}^3$$

Also, from Eqs. 11.36 and 11.46 we conclude that

$$\begin{aligned} V_{\text{th}} &= \text{Ma}_{\text{th}} \sqrt{RT_{\text{th}}k} \\ &= (1) \sqrt{[286.9 \text{ J/(kg}\cdot\text{K)}](240 \text{ K})(1.4)[1(\text{kg}\cdot\text{m})/(\text{N}\cdot\text{s}^2)][1(\text{N}\cdot\text{m})/\text{J}]} = 310 \text{ m/s} \end{aligned}$$

Then, from Eq. 1

$$\dot{m} = (0.79 \text{ kg/m}^3)(1 \times 10^{-4} \text{ m}^2)(310 \text{ m/s}) = 0.024 \text{ kg/s} \quad \textbf{(Ans)}$$

The values from Fig. D.1 resulted in answers for mass flowrate that are close to those using the ideal gas equations (see Example 11.5).

The temperature–entropy diagrams remain the same as those provided in the solution of Example 11.5.

The static pressure to stagnation pressure ratio at a point in a flow stream is measured with a Pitot-static tube (see Fig. 3.6) as being equal to 0.82. The stagnation temperature of the fluid is 68 °F. Determine the flow velocity if the fluid is (a) air, (b) helium.

SOLUTION

We consider both air and helium, flowing as described above, to act as ideal gases with constant specific heats. Then, we can use any of the ideal gas relationships developed in this chapter. To determine the flow velocity, we can combine Eqs. 11.36 and 11.46 to obtain

$$V = \text{Ma}\sqrt{RTk} \tag{1}$$

By knowing the value of static to stagnation pressure ratio, p/p_0, and the specific heat ratio we can obtain the corresponding Mach number from Eq. 11.59, or for air, from Fig. D.1. Fig. D.1 cannot be used for helium, since k for helium is 1.66 and Fig. D.1 is for $k = 1.4$ only. With Mach number, specific heat ratio, and stagnation temperature known, the value of static temperature can be subsequently ascertained from Eq. 11.56 (or Fig. D.1 for air).

(a) For air, $p/p_0 = 0.82$; thus from Fig. D.1,

$$\text{Ma} = 0.54 \tag{2}$$

and

$$\frac{T}{T_0} = 0.94 \tag{3}$$

Then, from Eq. 3

$$T = (0.94)[(68 + 460)\ °\text{R}] = 496\ °\text{R} \tag{4}$$

and using Eqs. 1, 2, and 4 we get

$$V = (0.54)\sqrt{[1.716 \times 10^3\ (\text{ft}\cdot\text{lb})/(\text{slug}\cdot°\text{R})](496\ °\text{R})(1.4)[1\ (\text{slug}\cdot\text{ft})/(\text{lb}\cdot\text{s}^2)]}$$

or

$$V = 590\ \text{ft/s} \qquad \textbf{(Ans)}$$

(b) For helium, $p/p_0 = 0.82$ and $k = 1.66$. By substituting these values into Eq. 11.59 we get

$$0.82 = \left\{\frac{1}{1 + [(1.66 - 1)/2]\ \text{Ma}^2}\right\}^{1.66/(1.66-1)}$$

or

$$\text{Ma} = 0.499$$

From Eq. 11.56 we obtain

$$\frac{T}{T_0} = \frac{1}{1 + [(k - 1)/2]\text{Ma}^2}$$

Thus,

$$T = \left\{\frac{1}{1 + [(1.66 - 1)/2](0.499)^2}\right\}[(68 + 460)\ °\text{R}] = 488\ °\text{R}$$

From Eq. 1 we obtain

$$V = (0.499)\sqrt{[1.242 \times 10^4(\text{ft}\cdot\text{lb})/(\text{slug}\cdot°\text{R})](488°\text{R})(1.66)[1(\text{slug}\cdot\text{ft})/(\text{lb}\cdot\text{s}^2)]}$$

or

$$V = 1580 \text{ ft/s} \qquad \textbf{(Ans)}$$

Note that the isentropic flow equations and Fig. D.1 for $k = 1.4$ were used presently to describe fluid particle isentropic flow along a pathline in a stagnation process. Even though these equations and graph were developed for one-dimensional duct flows, they can be used for frictionless, adiabatic pathline flows also.

Furthermore, while the Mach numbers calculated above are of similar size for the air and helium flows, the flow speed is much larger for helium than for air because the speed of sound in helium is much larger than it is in air.

Also included in Fig. D.1 is a graph of the ratio of local area, A, to critical area, A^*, for different values of local Mach number. The importance of this area ratio is clarified below.

For choked flow through the converging-diverging duct of Fig. 11.6*a*, the conservation of mass equation (Eq. 11.40) yields

$$\rho AV = \rho^* A^* V^*$$

or

$$\frac{A}{A^*} = \left(\frac{\rho^*}{\rho}\right)\left(\frac{V^*}{V}\right) \tag{11.67}$$

From Eqs. 11.36 and 11.46, we obtain

$$V^* = \sqrt{RT^*k} \tag{11.68}$$

and

$$V = \text{Ma}\,\sqrt{RTk} \tag{11.69}$$

By combining Eqs. 11.67, 11.68, and 11.69 we get

$$\frac{A}{A^*} = \frac{1}{\text{Ma}}\left(\frac{\rho^*}{\rho_0}\right)\left(\frac{\rho_0}{\rho}\right)\sqrt{\frac{(T^*/T_0)}{(T/T_0)}} \tag{11.70}$$

The incorporation of Eqs. 11.56, 11.60, 11.63, 11.65, and 11.70 results in

$$\frac{A}{A^*} = \frac{1}{\text{Ma}}\left\{\frac{1 + [(k-1)/2]\text{Ma}^2}{1 + [(k-1)/2]}\right\}^{(k+1)/[2(k-1)]} \tag{11.71}$$

The ratio of flow area to the critical area is a useful concept for isentropic duct flow.

Equation 11.71 was used to generate the values of A/A^* for air ($k = 1.4$) in Fig. D.1. These values of A/A^* are graphed as a function of Mach number in Fig. 11.10. As is demonstrated in the following examples, whether or not the critical area, A^*, is physically present in the flow, the area ratio, A/A^*, is still a useful concept for the isentropic flow of an ideal gas through a converging-diverging duct.

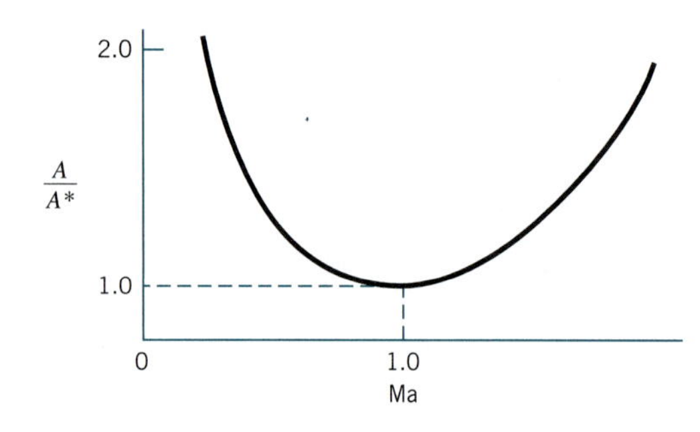

■ **FIGURE 11.10 The variation of area ratio with Mach number for isentropic flow of an ideal gas ($k = 1.4$, linear coordinate scales).**

EXAMPLE 11.8

Air enters subsonically from standard atmosphere and flows isentropically through a choked converging-diverging duct having a circular cross-section area, A, that varies with axial distance from the throat, x, according to the formula

$$A = 0.1 + x^2$$

where A is in square meters and x is in meters. The duct extends from $x = -0.5$ m to $x = +0.5$ m. For this flow situation, sketch the side view of the duct and graph the variation of Mach number, static temperature to stagnation temperature ratio, T/T_0, and static pressure to stagnation pressure ratio, p/p_0, through the duct from $x = -0.5$ m to $x = +0.5$ m. Also show the possible fluid states at $x = -0.5$ m, 0 m, and $+0.5$ m using temperature-entropy coordinates.

SOLUTION

The side view of the converging-diverging duct is a graph of radius r from the duct axis as a function of axial distance. For a circular flow cross section we have

$$A = \pi r^2 \tag{1}$$

where

$$A = 0.1 + x^2 \tag{2}$$

Thus, combining Eqs. 1 and 2, we have

$$r = \left(\frac{0.1 + x^2}{\pi}\right)^{1/2} \tag{3}$$

and a graph of radius as a function of axial distance can be easily constructed (see Fig. E11.8*a*).

Since the converging-diverging duct in this example is choked, the throat area is also the critical area, A^*. From Eq. 2 we see that

$$A^* = 0.1 \text{ m}^2 \tag{4}$$

For any axial location, from Eqs. 2 and 4 we get

$$\frac{A}{A^*} = \frac{0.1 + x^2}{0.1} \tag{5}$$

Values of A/A^* from Eq. 5 can be used in Eq. 11.71 to calculate corresponding values of Mach number, Ma. For air with $k = 1.4$, we could enter Fig. D.1 with values of A/A^* and read off values of the Mach number. With values of Mach number ascertained, we could use Eqs. 11.56 and 11.59 to calculate related values of T/T_0 and p/p_0. For air with $k = 1.4$, Fig. D.1 could be entered with A/A^* or Ma to get values of T/T_0 and p/p_0. To solve this example, we elect to use values from Fig. D.1.

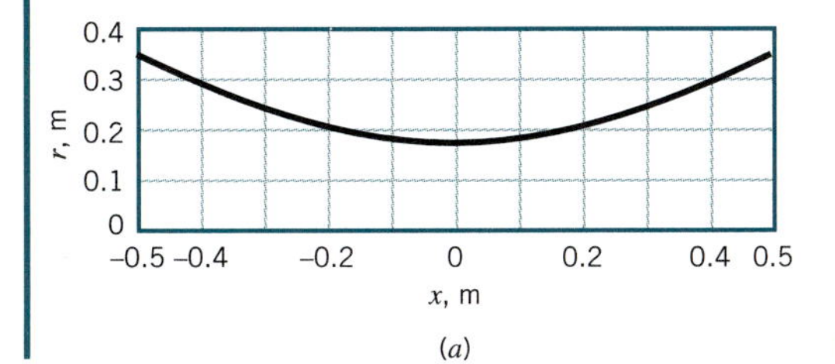

■ FIGURE E11.8

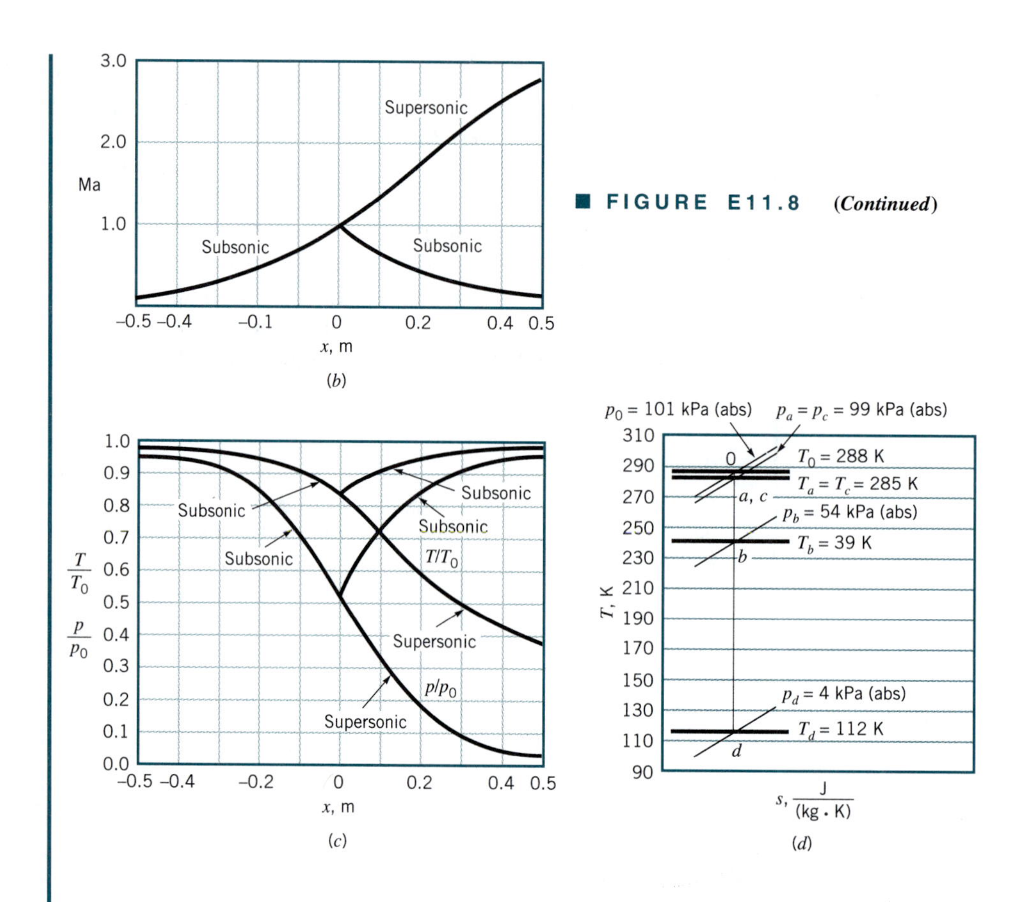

■ **FIGURE E11.8** *(Continued)*

The following table was constructed by using Eqs. 3 and 5 and Fig. D.1.

With the air entering the choked converging-diverging duct subsonically, only one isentropic solution exists for the converging portion of the duct. This solution involves an accelerating flow that becomes sonic ($\text{Ma} = 1$) at the throat of the passage. Two isentropic flow solutions are possible for the diverging portion of the duct—one subsonic, the other supersonic. If the pressure ratio, p/p_0, is set at 0.98 at $x = +0.5$ m (the outlet), the subsonic flow will occur. Alternatively, if p/p_0 is set at 0.04 at $x = +0.5$ m, the supersonic flow field will exist. These conditions are illustrated in Fig. E11.8. An unchoked subsonic flow through the converging-diverging duct of this example is discussed in Example 11.10. Choked flows involving flows other than the two isentropic flows in the diverging portion of the duct of this example are discussed after Example 11.10.

x (m)	From Eq. 3, r (m)	From Eq. 5, A/A^*	From Fig. D.1 Ma	T/T_0	p/p_0	State
Subsonic Solution						
−0.5	0.334	3.5	0.17	0.99	0.98	*a*
−0.4	0.288	2.6	0.23	0.99	0.96	
−0.3	0.246	1.9	0.32	0.98	0.93	
−0.2	0.211	1.4	0.47	0.96	0.86	
−0.1	0.187	1.1	0.69	0.91	0.73	
0	0.178	1	1.00	0.83	0.53	*b*
+0.1	0.187	1.1	0.69	0.91	0.73	
+0.2	0.211	1.4	0.47	0.96	0.86	
+0.3	0.246	1.9	0.32	0.98	0.93	
+0.4	0.288	2.6	0.23	0.99	0.97	
+0.5	0.344	3.5	0.17	0.99	0.98	*c*
Supersonic Solution						
+0.1	0.187	1.1	1.37	0.73	0.33	
+0.2	0.211	1.4	1.76	0.62	0.18	
+0.3	0.246	1.9	2.14	0.52	0.10	
+0.4	0.288	2.6	2.48	0.45	0.06	
+0.5	0.334	3.5	2.80	0.39	0.04	*d*

EXAMPLE 11.9

Air enters supersonically with T_0 and p_0 equal to standard atmosphere values and flows isentropically through the choked converging-diverging duct described in Example 11.8. Graph the variation of Mach number, Ma, static temperature to stagnation temperature ratio, T/T_0, and static pressure to stagnation pressure ratio, p/p_0, through the duct from $x = -0.5$ m to $x = +0.5$ m. Also show the possible fluid states at $x = -0.5$ m, 0 m, and $+0.5$ m by using temperature-entropy coordinates.

SOLUTION

With the air entering the converging-diverging duct of Example 11.8 supersonically instead of subsonically, a unique isentropic flow solution is obtained for the converging portion of the duct. Now, however, the flow decelerates to the sonic condition at the throat. The two solutions obtained previously in Example 11.8 for the diverging portion are still valid. Since the area variation in the duct is symmetrical with respect to the duct throat, we can use the supersonic flow values obtained from Example 11.8 for the supersonic flow in the converging portion of the duct. The supersonic flow solution for the converging passage is summarized in the following table. The solution values for the entire duct are graphed in Fig. E11.9.

x (m)	A/A^*	From Fig. D.1 Ma	T/T_0	p/p_0	State
−0.5	3.5	2.8	0.39	0.04	e
−0.4	2.6	2.5	0.45	0.06	
−0.3	1.9	2.1	0.52	0.10	
−0.2	1.4	1.8	0.62	0.18	
−0.1	1.1	1.4	0.73	0.33	
0	1	1.0	0.83	0.53	b

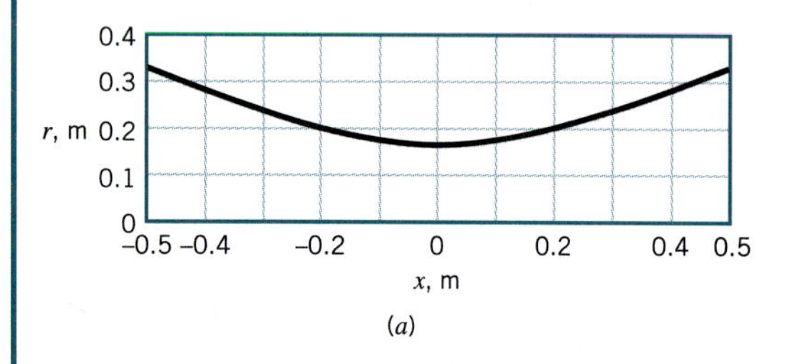

(a)

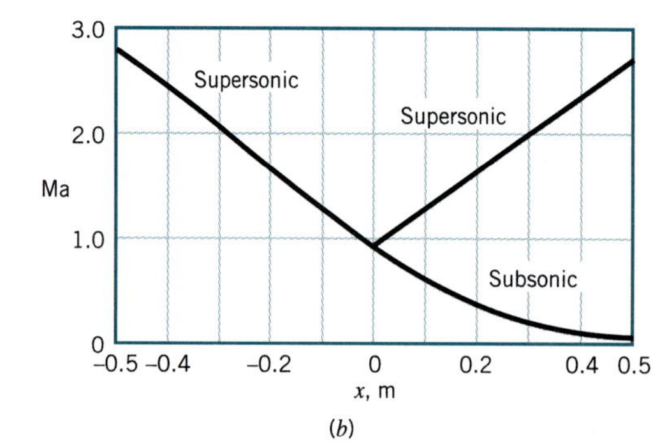

(b)

■ FIGURE E11.9

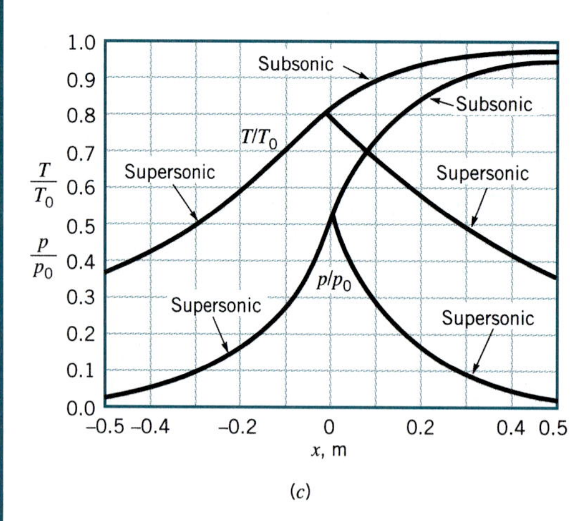

(c)

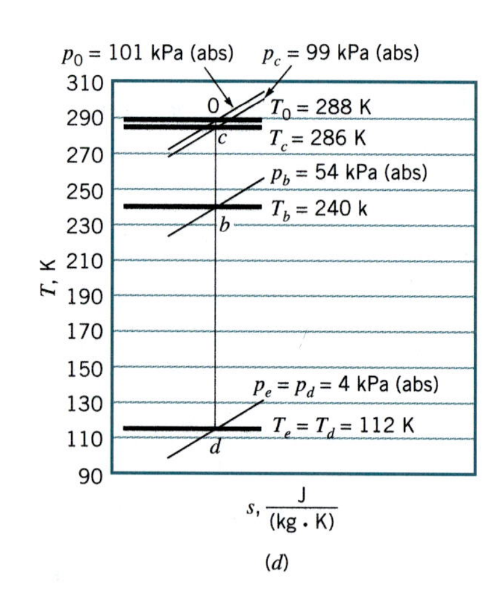

(d)

EXAMPLE 11.10

Air flows subsonically and isentropically through the converging-diverging duct of Example 11.8. Graph the variation of Mach number, Ma, static temperature to stagnation temperature ratio, T/T_0, the static pressure to stagnation pressure ratio, p/p_0, through the duct from $x = -0.5$ m to $x = +0.5$ m for Ma $= 0.48$ at $x = 0$ m. Show the corresponding temperature-entropy diagram.

SOLUTION

Since for this example, Ma $= 0.48$ at $x = 0$ m, the isentropic flow through the converging-diverging duct will be entirely subsonic and not choked. For air ($k = 1.4$) flowing isentropically through the duct, we can use Fig. D.1 for flow field quantities. Entering Fig. D.1 with Ma $= 0.48$ we read off $p/p_0 = 0.85$, $T/T_0 = 0.96$, and $A/A^* = 1.4$. Even though the duct flow is not choked in this example and A^* does not therefore exist physically, it still represents a valid reference. For a given isentropic flow, p_0, T_0, and A^* are constants. Since A at $x = 0$ m is equal to 0.10 m^2 (from Eq. 2 of Example 11.8), A^* for this example is

$$A^* = \frac{A}{(A/A^*)} = \frac{0.10 \text{ m}^2}{1.4} = 0.07 \text{ m}^2 \tag{1}$$

With known values of duct area at different axial locations, we can calculate corresponding area ratios, A/A^*, knowing $A^* = 0.07$ m^2. Having values of the area ratio, we can use Fig. E.1 and obtain related values of Ma, T/T_0, and p/p_0. The following table summarizes flow quantities obtained in this manner. The results are graphed in Fig. E11.10.

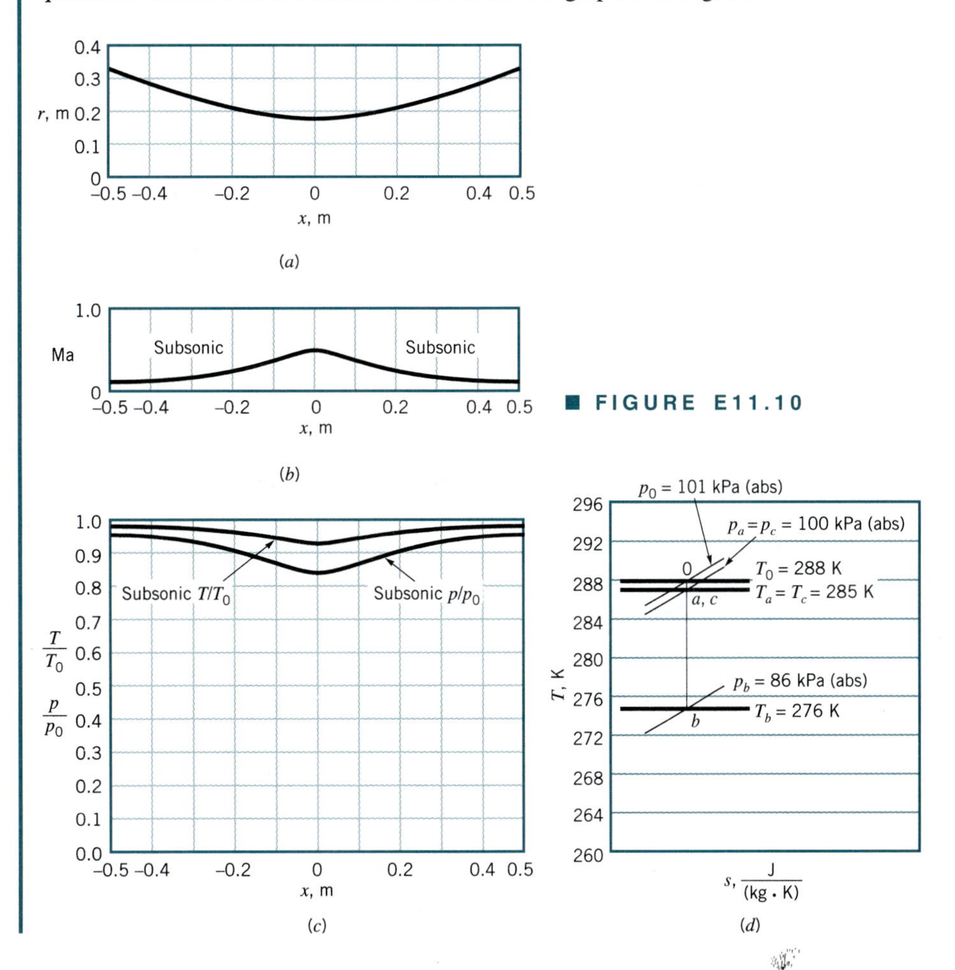

■ FIGURE E11.10

A more precise solution for the flow of this example could have been obtained with isentropic flow equations by following the steps outlined below.

1. Use Eq. 11.59 to get p/p_0 at $x = 0$ knowing k and Ma $= 0.48$.
2. From Eq. 11.71, obtain value of A/A^* at $x = 0$ knowing k and Ma.
3. Determine A^* knowing A and A/A^* at $x = 0$.
4. Determine A/A^* at different axial locations, x.
5. Use Eq. 11.71 and A/A^* from step 4 above to get values of Mach numbers at different axial locations.
6. Use Eqs. 11.56 and 11.59 and Ma from step 5 above to obtain T/T_0 and p/p_0 at different axial locations, x.

There are an infinite number of subsonic, isentropic flow solutions for the converging-diverging duct considered in this example.

x (m)	Calculated, A/A^*	From Fig. D.1 Ma	T/T_0	p/p_0	State
−0.5	5.0	0.12	0.99	0.99	a
−0.4	3.7	0.16	0.99	0.98	
−0.3	2.7	0.23	0.99	0.96	
−0.2	2.0	0.31	0.98	0.94	
−0.1	1.6	0.40	0.97	0.89	
0	1.4	0.48	0.96	0.85	b
+0.1	1.6	0.40	0.97	0.89	
+0.2	2.0	0.31	0.98	0.94	
+0.3	2.7	0.23	0.99	0.96	
+0.4	3.7	0.16	0.99	0.98	
+0.5	5.0	0.12	0.99	0.99	c

A variety of flow situations can occur for flow in a converging-diverging duct.

The isentropic flow behavior for the converging-diverging duct discussed in Examples 11.8, 11.9, and 11.10 is summarized in the area ratio–Mach number graphs sketched in Fig. 11.11. The points a, b, and c represent states at axial distance $x = -0.5$ m, 0 m, and $+0.5$ m. In Fig. 11.11a, the isentropic flow through the converging-diverging duct is subsonic without choking at the throat. This situation was discussed in Example 11.10. Figure 11.11b represents subsonic to subsonic choked flow (Example 11.8) and Fig. 11.11c is for subsonic to supersonic choked flow (also Example 11.8). The states in Fig. 11.11d are related to the supersonic to supersonic choked flow of Example 11.9; the states in Fig. 11.11e are for the supersonic to subsonic choked flow of Example 11.9. Not covered by an example but also possible are the isentropic flow states a, b, and c shown in Fig. 11.11f for supersonic to supersonic flow without choking. These six categories generally represent the possible kinds of isentropic, ideal gas flow through a converging-diverging duct.

For a given stagnation state (i.e., T_0 and p_0 fixed), ideal gas (k = constant), and converging-diverging duct geometry, an infinite number of isentropic subsonic to subsonic (not choked) and supersonic to supersonic (not choked) flow solutions exist. In contrast, the isentropic subsonic to supersonic (choked), subsonic to subsonic (choked), supersonic to subsonic (choked), and supersonic to supersonic (choked) flow solutions are each unique. The

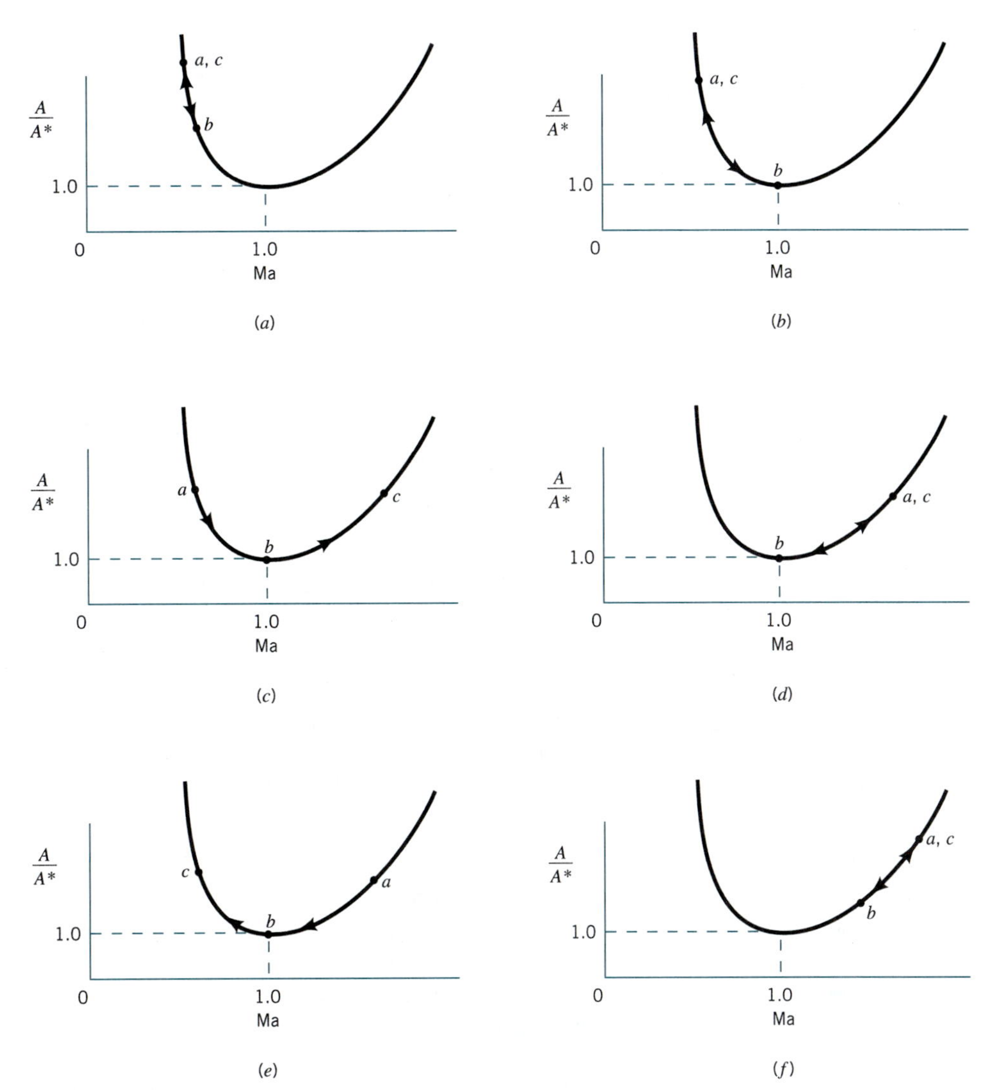

■ **FIGURE 11.11** ***(a)*** **Subsonic to subsonic isentropic flow (not choked).** ***(b)*** **Subsonic to subsonic isentropic flow (choked).** ***(c)*** **Subsonic to supersonic isentropic flow (choked)** ***(d)*** **Supersonic to supersonic isentropic flow (choked).** ***(e)*** **Supersonic to subsonic isentropic flow (choked).** ***(f)*** **Supersonic to supersonic isentropic flow (not choked).**

above-mentioned isentropic flow solutions are represented in Fig. 11.12. When the pressure at $x = +0.5$ (exit) is greater than or equal to p_{I} indicated in Fig. 11.12*d*, an isentropic flow is possible. When the pressure at $x = +0.5$ is equal to or less than p_{II}, isentropic flows in the duct are possible. However, when the exit pressure is less than p_{I} and greater than p_{III} as indicated in Fig. 11.13, isentropic flows are no longer possible in the duct. Determination of the value of P_{III} is discussed in Example 11.20.

A shock wave can occur in supersonic flows.

Some possible nonisentropic choked flows through our converging-diverging duct are represented in Fig. 11.13. Each abrupt pressure rise shown within and at the exit of the flow passage occurs across a very thin discontinuity in the flow called a *normal shock wave*. Except for flow across the normal shock wave, the flow is isentropic. The nonisentropic flow equations that describe the changes in fluid properties that take place across a normal shock wave

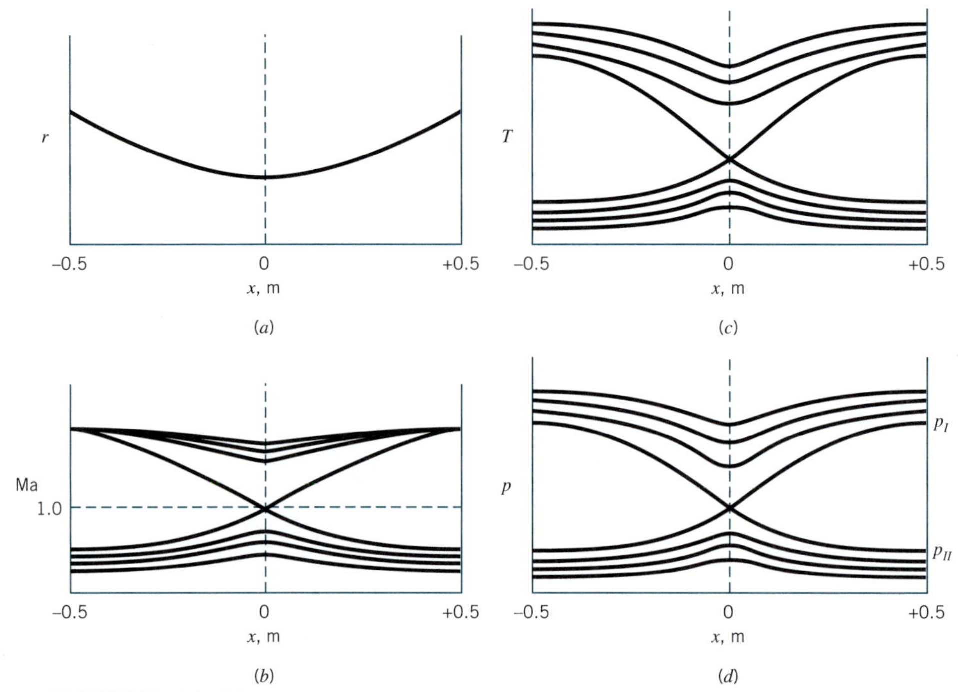

■ **FIGURE 11.12** ***(a)* The variation of duct radius with axial distance. *(b)* The variation of Mach number with axial distance. *(c)* The variation of temperature with axial distance. *(d)* The variation of pressure with axial distance.**

V11.4

Unlike incompressible flow, the pressure at the end of a supersonic nozzle need not equal the surrounding pressure.

are developed in Section 11.5.3. The less abrupt pressure rise or drop that occurs after the flow leaves the duct is nonisentropic and attributable to three-dimensional *oblique shock waves*. If the pressure rises downstream of the duct exit, the flow is considered *overexpanded*. If the pressure drops downstream of the duct exit, the flow is called *underexpanded*. Further details about over- and underexpanded flows and oblique shock waves are beyond the scope of this text. Interested readers are referred to texts on compressible flows and gas dynamics (for example, Refs. 5, 6, 7, and 8) for additional material on this subject.

11.4.3 Constant-Area Duct Flow

For steady, one-dimensional, isentropic flow of an ideal gas through a constant-area duct (see Fig. 11.14), Eq. 11.50 suggests that $dV = 0$ or that flow velocity remains constant. With the energy equation (Eq. 5.69) we can conclude that since flow velocity is constant, the fluid

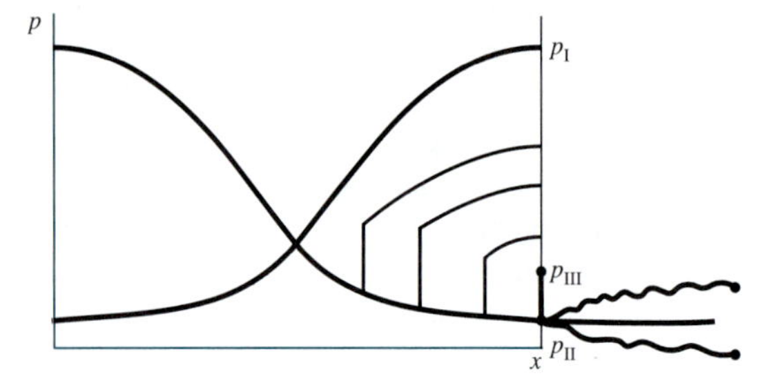

■ **FIGURE 11.13 Shock formation in converging-diverging duct flows.**

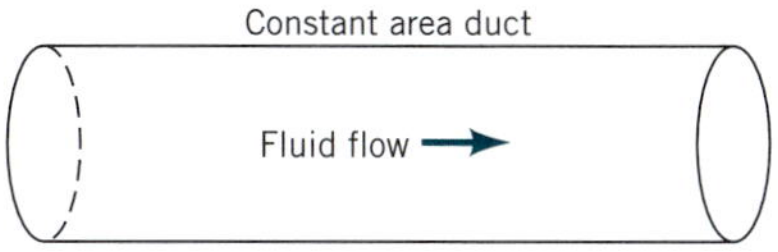

Constant area duct

Fluid flow

■ **FIGURE 11.14 Constant-area duct flow.**

enthalpy and thus temperature are also constant for this flow. This information and Eqs. 11.36 and 11.46 indicate that the Mach number is constant for this flow also. This being the case, Eqs. 11.59 and 11.60 tell us that fluid pressure and density also remain unchanged. Thus, we see that a steady, one-dimensional, isentropic flow of an ideal gas does not involve varying velocity or fluid properties unless the flow cross-section area changes.

In Section 11.5 we discuss nonisentropic, steady, one-dimensional flows of an ideal gas through a constant-area duct and also a normal shock wave. We learn that friction and/or heat transfer can also accelerate or decelerate a fluid.

11.5 Nonisentropic Flow of an Ideal Gas

Actual fluid flows are generally nonisentropic. An important example of nonisentropic flow involves adiabatic (no heat transfer) flow with friction. Flows with heat transfer (diabatic flows) are generally nonisentropic also. In this section we consider the adiabatic flow of an ideal gas through a constant-area duct with friction. This kind of flow is often referred to as *Fanno flow*. We also analyze the diabatic flow of an ideal gas through a constant-area duct without friction (*Rayleigh flow*). The concepts associated with Fanno and Rayleigh flows lead to further discussion of normal shock waves.

Fanno flow involves wall friction with no heat transfer and constant cross-section area.

11.5.1 Adiabatic Constant-Area Duct Flow with Friction (Fanno Flow)

Consider the steady, one-dimensional, and adiabatic flow of an ideal gas through the constant area duct shown in Fig 11.15. This is Fanno flow. For the control volume indicated, the energy equation (Eq. 5.69) leads to

0(negligibly small for gas flow)   0(flow is adiabatic)   0(flow is steady throughout)

$$\dot{m}\left[\check{h}_2 - \check{h}_1 + \frac{V_2^2 - V_1^2}{2} + g(z_2 - z_1)\right] = \dot{Q}_{\substack{\text{net}\\ \text{in.}}} + \dot{W}_{\substack{\text{shaft}\\ \text{net in}}}$$

or

$$\check{h} + \frac{V^2}{2} = \check{h}_0 = \text{constant} \tag{11.72}$$

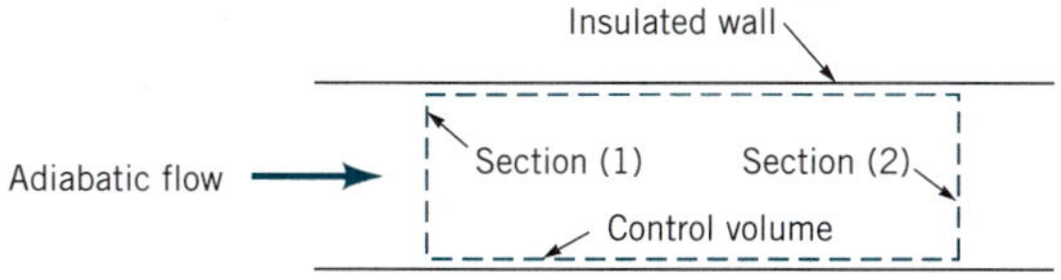

Insulated wall

Adiabatic flow

Section (1)   Section (2)

Control volume

■ **FIGURE 11.15 Adiabatic constant-area flow.**

where h_0 is the stagnation enthalpy. For an ideal gas we gather from Eq. 11.9 that

$$\check{h} - \check{h}_0 = c_p(T - T_0) \tag{11.73}$$

so that by combining Eqs. 11.72 and 11.73 we get

$$T + \frac{V^2}{2c_p} = T_0 = \text{constant}$$

or

$$T + \frac{(\rho V)^2}{2c_p\rho^2} = T_0 = \text{constant} \tag{11.74}$$

By substituting the ideal gas equation of state (Eq. 11.1) into Eq. 11.74 we obtain

$$T + \frac{(\rho V)^2 T^2}{2c_p(p^2/R^2)} = T_0 = \text{constant} \tag{11.75}$$

From the continuity equation (Eq. 11.40) we can conclude that the density-velocity product, ρV, is constant for a given Fanno flow since the area, A, is constant. Also, for a particular Fanno flow, the stagnation temperature, T_0, is fixed. Thus, Eq. 11.75 allows us to calculate values of fluid temperature corresponding to values of fluid pressure in the Fanno flow. We postpone our discussion of how pressure is determined until later.

As with earlier discussions in this chapter, it is helpful to describe Fanno flow with a temperature-entropy diagram. From the second $T\,ds$ relationship, an expression for entropy variation was already derived (Eq. 11.22). If the temperature, T_1, pressure, p_1, and entropy, s_1, at the entrance of the Fanno flow duct are considered as reference values, then Eq. 11.22 yields

$$s - s_1 = c_p \ln \frac{T}{T_1} - R \ln \frac{p}{p_1} \tag{11.76}$$

Entropy increases in Fanno flows because of wall friction.

Equations 11.75 and 11.76 taken together result in a curve with $T - s$ coordinates as is illustrated in Fig. 11.16. This curve involves a given gas (c_p and R) with fixed values of stagnation temperature, density-velocity product, and inlet temperature, pressure, and entropy. Curves like the one sketched in Fig. 11.16 are called Fanno lines.

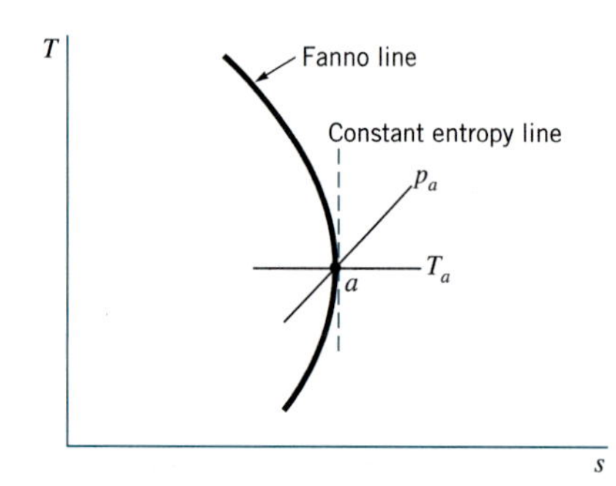

■ FIGURE 11.16 **The T–s diagram for Fanno flow.**

EXAMPLE 11.11

Air ($k = 1.4$) enters [section (1)] an insulated, constant cross-section area duct with the following properties:

$$T_0 = 518.67\ °\text{R}$$

$$T_1 = 514.55\ °\text{R}$$

$$p_1 = 14.3\ \text{psia}$$

For Fanno flow, determine corresponding values of fluid temperature and entropy change for various values of downstream pressures and plot the related Fanno line.

SOLUTION

To plot the Fanno line we can use Eq. 11.75

$$T + \frac{(\rho V)^2 T^2}{2c_p p^2/R^2} = T_0 = \text{constant} \quad \textbf{(1)}$$

and Eq. 11.76

$$s - s_1 = c_p \ln \frac{T}{T_1} - R \ln \frac{p}{p_1} \quad \textbf{(2)}$$

to construct a table of values of temperature and entropy change corresponding to different levels of pressure in the Fanno flow.

We need values of the ideal gas constant and the specific heat at constant pressure to use in Eqs. 1 and 2. From Table 1.7 we read for air

$$R = 1716\ (\text{ft}\cdot\text{lb})/(\text{slug}\cdot°\text{R})$$

From Eq. 11.14 we obtain

$$c_p = \frac{Rk}{k - 1} \quad \textbf{(3)}$$

or

$$c_p = \frac{[1716\ (\text{ft}\cdot\text{lb})/(\text{slug}\cdot°\text{R})](1.4)}{1.4 - 1} = 6006\ (\text{ft}\cdot\text{lb})/(\text{slug}\cdot°\text{R})$$

From Eqs. 11.1 and 11.69 we obtain

$$\rho V = \frac{p}{RT}\text{Ma}\sqrt{RTk}$$

and ρV is constant for this flow

$$\rho V = \rho_1 V_1 = \frac{p_1}{RT_1}\text{Ma}_1\sqrt{RT_1 k} \quad \textbf{(4)}$$

But

$$\frac{T_1}{T_0} = \frac{514.55\ °\text{F}}{518.67\ °\text{R}} = 0.99206$$

and from Eq. 11.56

$$\text{Ma}_1 = \sqrt{\left(\frac{1}{0.99202} - 1\right)/.02} = 0.2$$

Thus with Eq. 4

$$\rho V = \frac{(14.3\ \text{psia})(144\ \text{in.}^2/\text{ft}^2)0.2\sqrt{(1.4)[1716(\text{ft}\cdot\text{lb})/(\text{slug}\cdot°\text{R})](514.55\ °\text{R})[1(\text{slug}\cdot\text{ft})/(\text{lb}\cdot\text{s}^2)]}}{[1716(\text{ft}\cdot\text{lb})/(\text{slug}\cdot°\text{R})](514.55\ °\text{R})}$$

or

$$\rho V = 0.519 \text{ slug/(ft}^2\text{·s)}$$

For $p = 7$ psia we have from Eq. 1

$$T + \frac{[0.519 \text{ slug/(ft}^2\text{·s)}]^2 T^2}{2[6006 \text{ (ft·lb)/(slug·°R)}] \dfrac{(7 \text{ psia})^2(144 \text{ in.}^2/\text{ft}^2)^2}{[1716 \text{ (ft·lb)/(slug·°R)}]^2}} = 518.67 \text{ °R}$$

or

$$6.5 \times 10^{-5} T^2 + T - 518.67 = 0$$

Thus,

$$T = 502.3 \text{ °R} \qquad \textbf{(Ans)}$$

From Eq. 2, we obtain

$$s - s_1 = [6006 \text{ (ft·lb)/(slug·°R)}] \ln\left(\frac{502.3 \text{ °R}}{514.55 \text{ °R}}\right) - [1716 \text{ (ft·lb)/(slug·°R)}] \ln\left(\frac{7 \text{ psia}}{14.3 \text{ psia}}\right)$$

or

$$s - s_1 = 1081 \text{ (ft·lb)/(slug·°R)} \qquad \textbf{(Ans)}$$

Proceeding as outlined above, we construct the table of values shown below and graphed as the Fanno line in Fig. E11.11. The maximum entropy difference occurs at a pressure of 2.62 psia and a temperature of 432.1 °R.

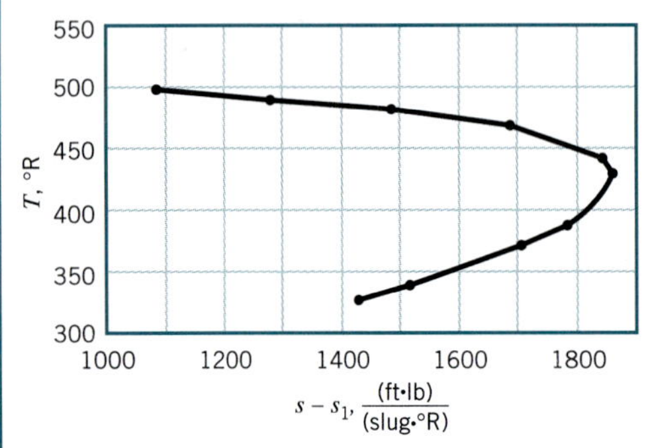

■ **FIGURE E11.11**

p (psia)	T (°R)	$s - s_1$ [(ft·lb)/(slug·°R)]
7	502.3	1081
6	496.8	1280
5	488.3	1489
4	474.0	1693
3	447.7	1844
2.62	432.1	1863
2	394.7	1783
1.8	378.1	1706
1.5	347.6	1513
1.4	335.6	1421

We can learn more about Fanno lines by further analyzing the equations that describe the physics involved. For example, the second $T\,ds$ equation (Eq. 11.18) is

$$T\,ds = d\check{h} - \frac{dp}{\rho} \qquad \textbf{(11.18)}$$

For an ideal gas

$$d\check{h} = c_p \, dT \tag{11.7}$$

and

$$p = \rho R T \tag{11.1}$$

or

$$\frac{dp}{p} = \frac{d\rho}{\rho} + \frac{dT}{T} \tag{11.77}$$

Thus, consolidating Eqs. 11.1, 11.7, 11.18, and 11.77 we obtain

$$T \, ds = c_p \, dT - RT \left(\frac{d\rho}{\rho} + \frac{dT}{T} \right) \tag{11.78}$$

Also, from the continuity equation (Eq. 11.40), we get for Fanno flow

$$\rho V = \text{constant}$$

or

$$\frac{d\rho}{\rho} = -\frac{dV}{V} \tag{11.79}$$

Substituting Eq. 11.79 into Eq. 11.78 yields

$$T \, ds = c_p \, dT - RT \left(-\frac{dV}{V} + \frac{dT}{T} \right)$$

or

$$\frac{ds}{dT} = \frac{c_p}{T} - R \left(-\frac{1}{V} \frac{dV}{dT} + \frac{1}{T} \right) \tag{11.80}$$

By differentiating the energy equation (11.74) obtained earlier, we obtain

$$\frac{dV}{dT} = -\frac{c_p}{V} \tag{11.81}$$

which, when substituted into Eq. 11.80, results in

$$\frac{ds}{dT} = \frac{c_p}{T} - R \left(\frac{c_p}{V^2} + \frac{1}{T} \right) \tag{11.82}$$

The maximum entropy state on the Fanno diagram corresponds to sonic conditions.

The Fanno line in Fig. 11.16 goes through a state (labeled state a) for which $ds/dT = 0$. At this state, we can conclude from Eqs. 11.14 and 11.82 that

$$V_a = \sqrt{R T_a k} \tag{11.83}$$

However, by comparing Eqs. 11.83 and 11.36 we see that the Mach number at state a is 1. Since the stagnation temperature is the same for all points on the Fanno line [see energy equation (Eq. 11.74)], the temperature at point a is the critical temperature, T^*, for the entire Fanno line. Thus, Fanno flow corresponding to the portion of the Fanno line above the critical temperature must be subsonic, and Fanno flow on the line below T^* must be supersonic.

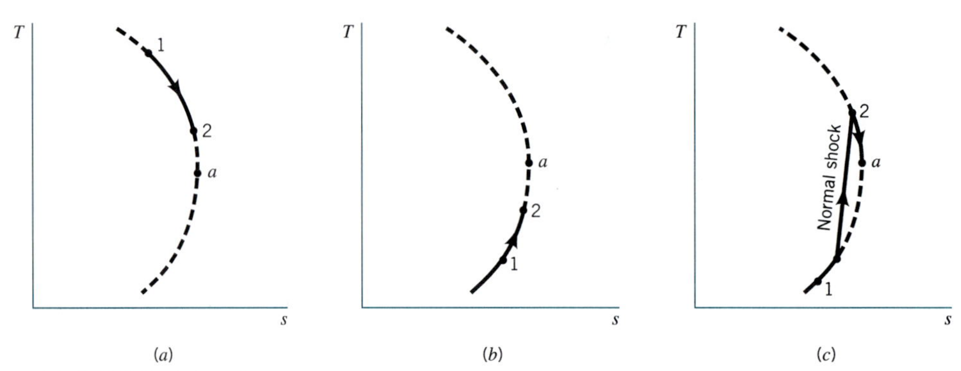

■ **FIGURE 11.17** ***(a)* Subsonic Fanno flow. *(b)* Supersonic Fanno flow. *(c)* Normal shock occurrence in Fanno flow.**

Friction accelerates a subsonic Fanno flow.

The second law of thermodynamics states that, based on all past experience, entropy can only remain constant or increase for adiabatic flows. For Fanno flow to be consistent with the second law of thermodynamics, flow can only proceed along the Fanno line toward state a, the critical state. The critical state may or may not be reached by the flow. If it is, the Fanno flow is *choked*. Some examples of Fanno flow behavior are summarized in Fig. 11.17. A case involving subsonic Fanno flow that is accelerated by friction to a higher Mach number without choking is illustrated in Fig. 11.17*a*. A supersonic flow that is decelerated by friction to a lower Mach number without choking is illustrated in Fig. 11.17*b*. In Fig. 11.17*c*, an abrupt change from supersonic to subsonic flow in the Fanno duct is represented. This sudden deceleration occurs across a standing *normal shock wave* that is described in more detail in Section 11.5.3.

The qualitative aspects of Fanno flow that we have already discussed are summarized in Table 11.1 and Fig. 11.18. To quantify Fanno flow behavior we need to combine a relationship that represents the linear momentum law with the set of equations already derived in this chapter.

If the linear momentum equation (Eq. 5.22) is applied to the Fanno flow through the control volume sketched in Fig. 11.19*a*, the result is

$$p_1A_1 - p_2A_2 - R_x = \dot{m}(V_2 - V_1)$$

where R_x is the frictional force exerted by the inner pipe wall on the fluid. Since $A_1 = A_2 = A$ and $\dot{m} = \rho AV = $ constant, we obtain

$$p_1 - p_2 - \frac{R_x}{A} = \rho V(V_2 - V_1) \qquad \textbf{(11.84)}$$

■ **TABLE 11.1**
Summary of Fanno Flow Behavior

	Flow	
Parameter	**Subsonic Flow**	**Supersonic Flow**
Stagnation temperature	Constant	Constant
Ma	Increases (maximum is 1)	Decreases (minimum is 1)
Friction	Accelerates flow	Decelerates flow
Pressure	Decreases	Increases
Temperature	Decreases	Increases

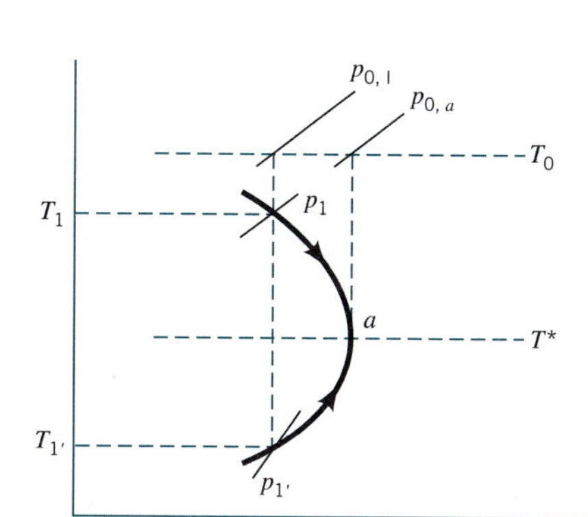

■ **FIGURE 11.18**
Fanno flow.

The differential form of Eq. 11.84, which is valid for Fanno flow through the semi-infinitesimal control volume shown in Fig. 11.19*b*, is

$$-dp - \frac{\tau_w \pi D\, dx}{A} = \rho V\, dV \tag{11.85}$$

Friction forces in Fanno flow are given in terms of the friction factor.

The wall shear stress, τ_w, is related to the wall friction factor, f, by Eq. 8.20 as

$$f = \frac{8\tau_w}{\rho V^2} \tag{11.86}$$

By substituting Eq. 11.86 and $A = \pi D^2/4$ into Eq. 11.85, we obtain

$$-dp - f\rho \frac{V^2}{2} \frac{dx}{D} = \rho V\, dV \tag{11.87}$$

or

$$\frac{dp}{p} + \frac{f}{p} \frac{\rho V^2}{2} \frac{dx}{D} + \frac{\rho}{p} \frac{d(V^2)}{2} = 0 \tag{11.88}$$

Combining the ideal gas equation of state (Eq. 11.1), the ideal gas speed-of-sound equation (Eq. 11.36), and the Mach number definition (Eq. 11.46) with Eq. 11.88 leads to

$$\frac{dp}{p} + \frac{fk}{2} \text{Ma}^2 \frac{dx}{D} + k \frac{\text{Ma}^2}{2} \frac{d(V^2)}{V^2} = 0 \tag{11.89}$$

Since $V = \text{Ma}\, c = \text{Ma}\sqrt{RTk}$, then

$$V^2 = \text{Ma}^2 RTk$$

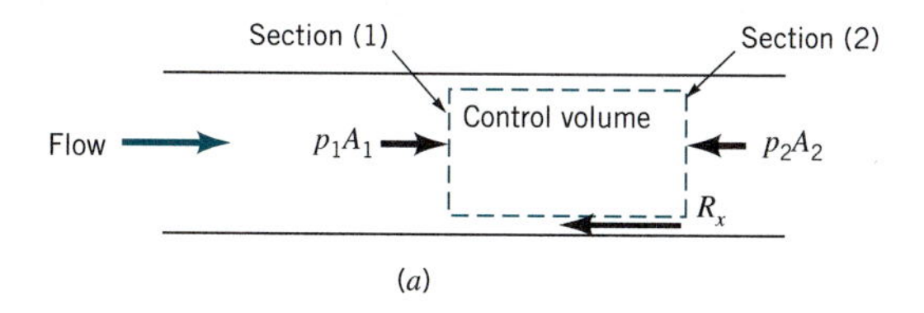

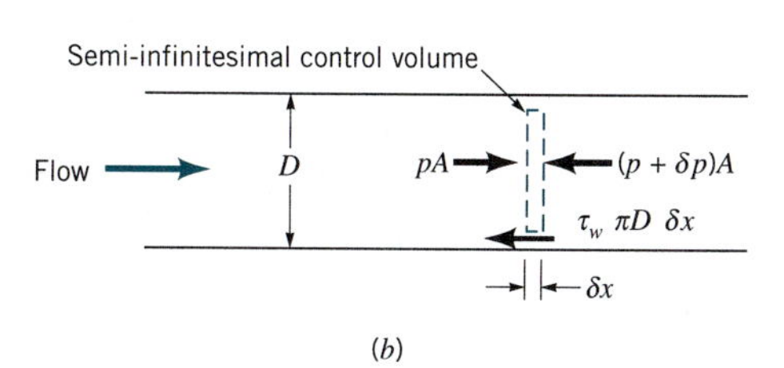

■ **FIGURE 11.19** ***(a)* Finite control volume. *(b)* Semi-infinitesimal control volume.**

or

$$\frac{d(V^2)}{V^2} = \frac{d(\mathrm{Ma}^2)}{\mathrm{Ma}^2} + \frac{dT}{T} \tag{11.90}$$

The application of the energy equation (Eq. 5.69) to Fanno flow gave Eq. 11.74. If Eq. 11.74 is differentiated and divided by temperature, the result is

$$\frac{dT}{T} + \frac{d(V^2)}{2c_pT} = 0 \tag{11.91}$$

Substituting Eqs. 11.14, 11.36, and 11.46 into Eq. 11.91 yields

$$\frac{dT}{T} + \frac{k-1}{2}\mathrm{Ma}^2\frac{d(V^2)}{V^2} = 0 \tag{11.92}$$

which can be combined with Eq. 11.90 to form

$$\frac{d(V^2)}{V^2} = \frac{d(\mathrm{Ma}^2)/\mathrm{Ma}^2}{1 + [(k-1)/2]\mathrm{Ma}^2} \tag{11.93}$$

We can merge Eqs. 11.77, 11.79, and 11.90 to get

$$\frac{dp}{p} = \frac{1}{2}\frac{d(V^2)}{V^2} - \frac{d(\mathrm{Ma}^2)}{\mathrm{Ma}^2} \tag{11.94}$$

Consolidating Eqs. 11.94 and 11.89 leads to

$$\frac{1}{2}(1 + k\mathrm{Ma}^2)\frac{d(V^2)}{V^2} - \frac{d(\mathrm{Ma}^2)}{\mathrm{Ma}^2} + \frac{fk}{2}\mathrm{Ma}^2\frac{dx}{D} = 0 \tag{11.95}$$

Finally, incorporating Eq. 11.93 into Eq. 11.95 yields

$$\frac{(1-\mathrm{Ma}^2)\,d(\mathrm{Ma}^2)}{\{1 + [(k-1)/2]\mathrm{Ma}^2\}k\mathrm{Ma}^4} = f\frac{dx}{D} \tag{11.96}$$

Equation 11.96 can be integrated from one section to another in a Fanno flow duct. We elect to use the critical (*) state as a reference and to integrate Eq. 11.96 from an upstream state to the critical state. Thus

$$\int_{\mathrm{Ma}}^{\mathrm{Ma}^*=1}\frac{(1-\mathrm{Ma}^2)\,d(\mathrm{Ma}^2)}{\{1 + [(k-1)/2]\,\mathrm{Ma}^2\}k\mathrm{Ma}^4} = \int_{\ell}^{\ell^*} f\frac{dx}{D} \tag{11.97}$$

where ℓ is length measured from an arbitrary but fixed upstream reference location to a section in the Fanno flow. For an approximate solution, we can assume that the friction factor is constant at an average value over the integration length, $\ell^* - \ell$. We also consider a constant value of k. Thus, we obtain from Eq. 11.97

For Fanno flow, the Mach number is a function of the distance to the critical state.

$$\frac{1}{k}\frac{(1-\mathrm{Ma}^2)}{\mathrm{Ma}^2} + \frac{k+1}{2k}\ln\left\{\frac{[(k+1)/2]\mathrm{Ma}^2}{1 + [(k-1)/2]\mathrm{Ma}^2}\right\} = \frac{f(\ell^* - \ell)}{D} \tag{11.98}$$

For a given gas, values of $f(\ell^* - \ell)/D$ can be tabulated as a function of Mach number for Fanno flow. For example, values of $f(\ell^* - \ell)/D$ for air ($k = 1.4$) Fanno flow are graphed as a function of Mach number in Fig. D.2 in Appendix D. Note that the critical state does not have to exist in the actual Fanno flow being considered, since for any two sections in a given Fanno flow

$$\frac{f(\ell^* - \ell_2)}{D} - \frac{f(\ell^* - \ell_1)}{D} = \frac{f}{D}(\ell_1 - \ell_2) \tag{11.99}$$

The sketch in Fig. 11.20 illustrates the physical meaning of Eq. 11.99.

For a given Fanno flow (constant specific heat ratio, duct diameter, and friction factor) the length of duct required to change the Mach number from Ma_1 to Ma_2 can be determined from Eqs. 11.98 and 11.99 or a graph such as Fig. D.2. To get the values of other fluid properties in the Fanno flow field we need to develop more equations.

For Fanno flow, the length of duct needed to produce a given change in Mach number can be determined.

By consolidating Eqs. 11.90 and 11.92 we obtain

$$\frac{dT}{T} = -\frac{(k-1)}{2\{1 + [(k-1)/2]\text{Ma}^2\}}\, d(\text{Ma}^2) \tag{11.100}$$

Integrating Eq. 11.100 from any state upstream in a Fanno flow to the critical (*) state leads to

$$\frac{T}{T^*} = \frac{(k+1)/2}{1 + [(k-1)/2]\text{Ma}^2} \tag{11.101}$$

Equations 11.68 and 11.69 allow us to write

$$\frac{V}{V^*} = \frac{\text{Ma}\sqrt{RTk}}{\sqrt{RT^*k}} = \text{Ma}\sqrt{\frac{T}{T^*}} \tag{11.102}$$

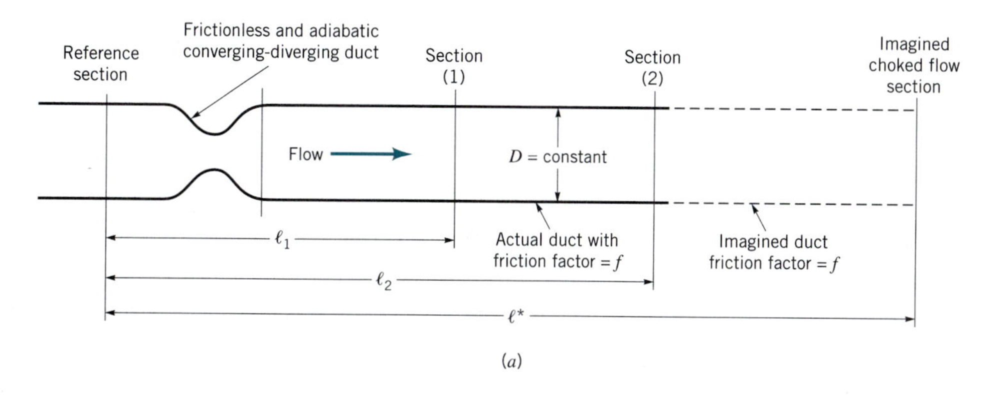

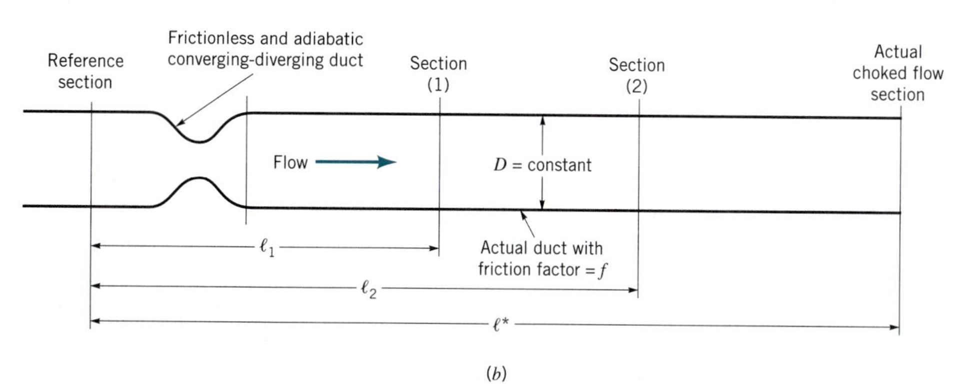

FIGURE 11.20 ***(a) Unchoked Fanno flow. (b) Choked Fanno flow.***

Substituting Eq. 11.101 into Eq. 11.102 yields

$$\frac{V}{V^*} = \left\{ \frac{[(k+1)/2]\text{Ma}^2}{1 + [(k-1)/2]\text{Ma}^2} \right\}^{1/2} \tag{11.103}$$

From the continuity equation (Eq. 11.40) we get for Fanno flow

$$\frac{\rho}{\rho^*} = \frac{V^*}{V} \tag{11.104}$$

Combining 11.104 and 11.103 results in

$$\frac{\rho}{\rho^*} = \left\{ \frac{1 + [(k-1)/2]\text{Ma}^2}{[(k+1)/2]\text{Ma}^2} \right\}^{1/2} \tag{11.105}$$

The ideal gas equation of state (Eq. 11.1) leads to

$$\frac{p}{p^*} = \frac{\rho}{\rho^*}\frac{T}{T^*} \tag{11.106}$$

For Fanno flow, thermodynamic and flow properties can be calculated as a function of Mach number.

and merging Eqs. 11.106, 11.105, and 11.101 gives

$$\frac{p}{p^*} = \frac{1}{\text{Ma}} \left\{ \frac{(k+1)/2}{1 + [(k-1)/2]\text{Ma}^2} \right\}^{1/2} \tag{11.107}$$

Finally, the stagnation pressure ratio can be written as

$$\frac{p_0}{p_0^*} = \left(\frac{p_0}{p}\right)\left(\frac{p}{p^*}\right)\left(\frac{p^*}{p_0^*}\right) \tag{11.108}$$

which by use of Eqs. 11.59 and 11.107 yields

$$\frac{p_0}{p_0^*} = \frac{1}{\text{Ma}} \left[\left(\frac{2}{k+1}\right)\left(1 + \frac{k-1}{2}\text{Ma}^2\right) \right]^{[(k+1)/2(k-1)]} \tag{11.109}$$

Values of $f(\ell^* - \ell)/D$, T/T^*, V/V^*, p/p^*, and p_0/p_0^* for Fanno flow of air ($k = 1.4$) are graphed as a function of Mach number (using Eqs. 11.99, 11.101, 11.103, 11.107, and 11.109) in Fig. D.2 of Appendix D. The usefulness of Fig. D.2 is illustrated in Examples 11.12, 11.13, and 11.14.

Standard atmospheric air [$T_0 = 288$ K, $p_0 = 101$ kPa(abs)] is drawn steadily through a frictionless, adiabatic converging nozzle into an adiabatic, constant-area duct as shown in Fig. E11.12*a*. The duct is 2-m long and has an inside diameter of 0.1 m. The average friction factor for the duct is estimated as being equal to 0.02. What is the maximum mass flowrate through the duct? For this maximum flowrate, determine the values of static temperature, static pressure, stagnation temperature, stagnation pressure, and velocity at the inlet [section (1)] and exit [section (2)] of the constant-area duct. Sketch a temperature-entropy diagram for this flow.

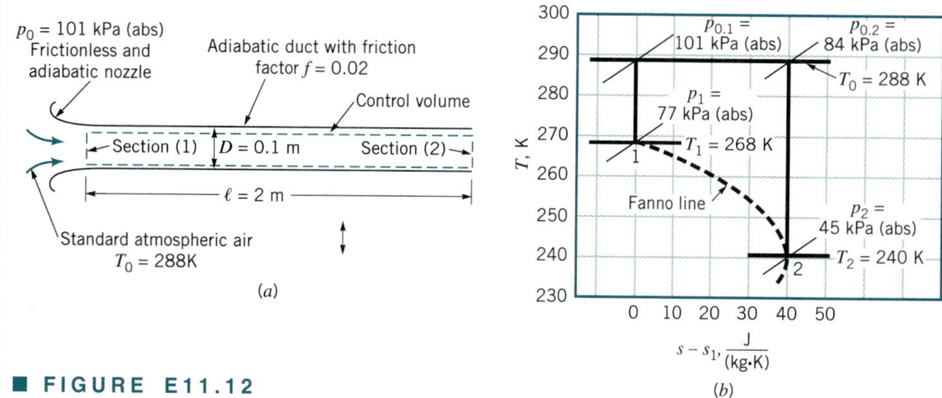

■ **FIGURE E11.12**

SOLUTION

We consider the flow through the converging nozzle to be isentropic and the flow through the constant-area duct to be Fanno flow. A decrease in the pressure at the exit of the constant-area duct (back pressure) causes the mass flowrate through the nozzle and the duct to increase. The flow throughout is subsonic. The maximum flowrate will occur when the back pressure is lowered to the extent that the constant-area duct chokes and the Mach number at the duct exit is equal to 1. Any further decrease of back pressure will not affect the flowrate through the nozzle-duct combination.

For the maximum flowrate condition, the constant-area duct must be choked, and

$$\frac{f(\ell^* - \ell_1)}{D} = \frac{f(\ell_2 - \ell_1)}{D} = \frac{(0.02)(2 \text{ m})}{(0.1 \text{ m})} = 0.4 \tag{1}$$

With $k = 1.4$ for air and the above calculated value of $f(\ell^* - \ell_1)/D = 0.4$, we could use Eq. 11.98 to determine a value of Mach number at the entrance of the duct [section (1)]. With $k = 1.4$ and Ma_1 known, we could then rely on Eqs. 11.101, 11.103, 11.107, and 11.109 to obtain values of T_1/T^*, V_1/V^*, p_1/p^*, and $p_{0,1}/p_0^*$. Alternatively, for air ($k = 1.4$), we can use Fig. D.2 with $f(\ell^* - \ell_1)/D = 0.4$ and read off values of Ma_1, T_1/T^*, V_1/V^*, p_1/p^*, and $p_{0,1}/p_0^*$.

The pipe entrance Mach number, Ma_1, also represents the Mach number at the throat (and exit) of the isentropic, converging nozzle. Thus, the isentropic flow equations of Section 11.4 or Fig. D.1 can be used with Ma_1. We use Fig. D.1 in this example.

With Ma_1 known, we can enter Fig. D.1 and get values of T_1/T_0, p_1/p_0, and ρ_1/ρ_0. Through the isentropic nozzle, the values of T_0, p_0, and ρ_0 are each constant, and thus T_1, p_1, and ρ_1 can be readily obtained.

Since T_0 also remains constant through the constant-area duct (see Eq. 11.75), we can use Eq. 11.63 to get T^*. Thus,

$$\frac{T^*}{T_0} = \frac{2}{k + 1} = \frac{2}{1.4 + 1} = 0.8333 \tag{2}$$

Since $T_0 = 288$ K, we get from Eq. 2,

$$T^* = (0.8333)(288 \text{ K}) = 240 \text{ K} = T_2 \tag{3}$$ **(Ans)**

With T^* known, we can calculate V^* from Eq. 11.36 as

$$V^* = \sqrt{RT^*k}$$
$$= \sqrt{[(286.9 \text{ J})/(\text{kg}\cdot\text{K})](240 \text{ K})(1.4)[1(\text{kg}\cdot\text{m})/(\text{N}\cdot\text{s}^2)][1(\text{N}\cdot\text{m})/\text{J}]} \quad \textbf{(4)} \quad \text{(Ans)}$$
$$= 310 \text{ m/s} = V_2$$

Now V_1 can be obtained from V^* and V_1/V^*. Having A_1, ρ_1, and V_1 we can get the mass flowrate from

$$\dot{m} = \rho_1 A_1 V_1 \quad \textbf{(5)}$$

Values of the other variables asked for can be obtained from the ratios mentioned above.
Entering Fig. D.2 with $f(\ell^* - \ell)/D = 0.4$ we read

$$\text{Ma}_1 = 0.63 \quad \textbf{(7)}$$

$$\frac{T_1}{T^*} = 1.1 \quad \textbf{(8)}$$

$$\frac{V_1}{V^*} = 0.66 \quad \textbf{(9)}$$

$$\frac{p_1}{p^*} = 1.7 \quad \textbf{(10)}$$

$$\frac{p_{0,1}}{p_0^*} = 1.16 \quad \textbf{(11)}$$

Entering Fig. D.1 with $\text{Ma}_1 = 0.63$ we read

$$\frac{T_1}{T_0} = 0.93 \quad \textbf{(12)}$$

$$\frac{p_1}{p_{0,1}} = 0.76 \quad \textbf{(13)}$$

$$\frac{\rho_1}{\rho_{0,1}} = 0.83 \quad \textbf{(14)}$$

Thus, from Eqs. 4 and 9 we obtain

$$V_1 = (0.66)(310 \text{ m/s}) = 205 \text{ m/s} \quad \text{(Ans)}$$

From Eq. 14 we get

$$\rho_1 = 0.83\rho_{0,1} = (0.83)(1.23 \text{ kg/m}^3) = 1.02 \text{ kg/m}^3$$

and from Eq. 5 we conclude that

$$\dot{m} = (1.02 \text{ kg/m}^3)\left[\frac{\pi(0.1 \text{ m})^2}{4}\right](206 \text{ m/s}) = 1.65 \text{ kg/s} \quad \text{(Ans)}$$

From Eq. 12, it follows that

$$T_1 = (0.93)(288 \text{ K}) = 268 \text{ K} \quad \text{(Ans)}$$

Equation 13 yields

$$p_1 = (0.76)[101 \text{ kPa (abs)}] = 77 \text{ kPa (abs)} \quad \textbf{(Ans)}$$

The stagnation temperature, T_0, remains constant through this adiabatic flow at a value of

$$T_{0,1} = T_{0,2} = 288 \text{ K} \quad \textbf{(Ans)}$$

The stagnation pressure, p_0, at the entrance of the constant-area duct is the same as the constant value of stagnation pressure through the isentropic nozzle. Thus

$$p_{0,1} = 101 \text{ kPa(abs)} \quad \textbf{(Ans)}$$

To obtain the duct exit pressure ($p_2 = p^*$) we can use Eqs. 10 and 13. Thus,

$$p_2 = \left(\frac{p^*}{p_1}\right)\left(\frac{p_1}{p_{0,1}}\right)(p_{0,1}) = \left(\frac{1}{1.7}\right)(0.76)[101 \text{ kPa(abs)}] = 45 \text{ kPa(abs)} \quad \textbf{(Ans)}$$

For the duct exit stagnation pressure ($p_{0,2} = p_0^*$) we can use Eq. 11 as

$$p_{0,2} = \left(\frac{p_0^*}{p_{0,1}}\right)(p_{0,1}) = \left(\frac{1}{1.2}\right)[101 \text{ kPa(abs)}] = 84 \text{ kPa(abs)} \quad \textbf{(Ans)}$$

The stagnation pressure, p_0, decreases in a Fanno flow because of friction.

Use of graphs such as Figs. D.1 and D.2 illustrates the solution of a problem involving Fanno flow. The $T - s$ diagram for this flow is shown in Fig. E.11.12*b*, where the entropy difference, $s_2 - s_1$, is obtained from Eq. 11.22.

EXAMPLE 11.13

The duct in Example 11.12 is shortened by 50%, but the duct discharge pressure is maintained at the choked flow value for Example 11.12, namely,

$$p_d = 45 \text{ kPa(abs)}$$

Will shortening the duct cause the mass flowrate through the duct to increase or decrease? Assume that the average friction factor for the duct remains constant at a value of $f = 0.02$.

SOLUTION

We guess that the shortened duct will still choke and check our assumption by comparing p_d with p^*. If $p_d < p^*$, the flow is choked; if not, another assumption has to be made. For choked flow we can calculate the mass flowrate just as we did for Example 11.12. For unchoked flow, we will have to devise another strategy.

For choked flow

$$\frac{f(\ell^* - \ell_1)}{D} = \frac{(0.02)(1 \text{ m})}{0.1 \text{ m}} = 0.2$$

and from Fig. D.2, we read the values $Ma_1 = 0.70$ and $p_1/p^* = 1.5$. With $Ma_1 = 0.70$, we use Fig. D.1 and get

$$\frac{p_1}{p_0} = 0.72$$

Now the duct exit pressure ($p_2 = p^*$) can be obtained from

$$p_2 = p^* = \left(\frac{p^*}{p_1}\right)\left(\frac{p_1}{p_{0,1}}\right)(p_{0,1})$$

$$= \left(\frac{1}{1.5}\right)(0.72)[101 \text{ kPa(abs)}] = 48.5 \text{ kPa(abs)}$$

and we see that $p_d < p^*$. Our assumption of choked flow is justified. The pressure at the exit plane is greater than the surrounding pressure outside the duct exit. The final drop of pressure from 48.5 kPa(abs) to 45 kPa(abs) involves complicated three-dimensional flow downstream of the exit.

To determine the mass flowrate we use

$$\dot{m} = \rho_1 A_1 V_1 \quad \textbf{(1)}$$

The density at section (1) is obtained from

$$\frac{\rho_1}{\rho_{0,1}} = 0.79 \quad \textbf{(2)}$$

which is read in Fig. D.1 for $Ma_1 = 0.7$. Thus,

$$\rho_1 = (0.79)(1.23 \text{ kg/m}^3) = 0.97 \text{ kg/m}^3 \quad \textbf{(3)}$$

We get V_1 from

$$\frac{V_1}{V^*} = 0.73 \quad \textbf{(4)}$$

from Fig. D.2 for $Ma_1 = 0.7$. The value of V^* is the same as it was in Example 11.12, namely,

$$V^* = 310 \text{ m/s} \quad \textbf{(5)}$$

Thus, from Eqs. 4 and 5 we obtain

$$V_1 = (0.73)(310) = 226 \text{ m/s} \quad \textbf{(6)}$$

and from Eqs. 1, 3, and 6 we get

$$\dot{m} = (0.97 \text{ kg/m}^3)\left[\frac{\pi(0.1)^2}{4}\right](226 \text{ m/s}) = 1.73 \text{ kg/s} \quad \textbf{(Ans)}$$

The mass flowrate associated with a shortened tube is larger than the mass flowrate for the longer tube, $\dot{m} = 1.65$ kg/s. This trend is general for subsonic Fanno flow. For the same upstream stagnation state and downstream pressure, the mass flowrate for the Fanno flow will decrease with increase in length of duct for subsonic flow. Equivalently, if the length of the duct remains the same but the wall friction is increased, the mass flowrate will decrease.

EXAMPLE 11.14

If the same flowrate obtained in Example 11.12 ($\dot{m} = 1.65$ kg/s) is desired through the shortened duct of Example 11.13 ($\ell_2 - \ell_1 = 1$ m), determine the Mach number at the exit of the duct, M_2, and the back pressure, p_2, required. Assume f remains constant at a value of 0.02.

SOLUTION

Since the mass flowrate of Example 11.12 is desired, the Mach number and other properties at the entrance of the constant-area duct remain at the values determined in Example 11.12. Thus, from Example 11.12, $\mathrm{Ma}_1 = 0.63$ and from Fig. D.2

$$\frac{f(\ell^* - \ell_1)}{D} = 0.4$$

For this example,

$$\frac{f(\ell_2 - \ell_1)}{D} = \frac{f(\ell^* - \ell_1)}{D} - \frac{f(\ell^* - \ell_2)}{D}$$

or

$$\frac{(0.02)(1 \text{ m})}{0.1 \text{ m}} = 0.4 - \frac{f(\ell^* - \ell_2)}{D}$$

so that

$$\frac{f(\ell^* - \ell_2)}{D} = 0.2 \qquad \textbf{(1)}$$

By using the value from Eq. 1 and Fig. D.2, we get

$$\mathrm{Ma}_2 = 0.70 \qquad \textbf{(Ans)}$$

and

$$\frac{p_2}{p^*} = 1.5 \qquad \textbf{(2)}$$

We obtain p_2 from

$$p_2 = \left(\frac{p_2}{p^*}\right)\left(\frac{p^*}{p_1}\right)\left(\frac{p_1}{p_{0,1}}\right)(p_{0,1})$$

where p_2/p^* is given in Eq. 2 and p^*/p_1, $p_1/p_{0,1}$, and $p_{0,1}$ are the same as they were in Example 11.12. Thus,

$$p_2 = (1.5)\left(\frac{1}{1.7}\right)(0.76)[101 \text{ kPa(abs)}] = 68.0 \text{ kPa(abs)} \qquad \textbf{(Ans)}$$

A larger back pressure [68.0 kPa(abs)] than the one associated with choked flow through a Fanno duct [45 kPa(abs)] will maintain the same flowrate through a shorter Fanno duct with the same friction coefficient. The flow through the shorter duct is not choked. It would not be possible to maintain the same flowrate through a Fanno duct longer than the choked one with the same friction coefficient, regardless of what back pressure is used.

11.5.2 Frictionless Constant-Area Duct Flow with Heat Transfer (Rayleigh Flow)

Rayleigh flow involves heat transfer with no wall friction and constant cross-section area.

Consider the steady, one-dimensional, and frictionless flow of an ideal gas through the constant-area duct with heat transfer illustrated in Fig. 11.21. This is *Rayleigh flow*. Application of the linear momentum equation (Eq. 5.22) to the Rayleigh flow through the finite control volume sketched in Fig. 11.21 results in

$$p_1 A_1 + \dot{m} V_1 = p_2 A_2 + \dot{m} V_2 + \overset{0\text{(frictionless flow)}}{\cancel{R_x}}$$

or

$$p + \frac{(\rho V)^2}{\rho} = \text{constant} \qquad \textbf{(11.110)}$$

Use of the ideal gas equation of state (Eq. 11.1) in Eq. 11.110 leads to

$$p + \frac{(\rho V)^2 \, RT}{p} = \text{constant} \qquad \textbf{(11.111)}$$

Since the flow cross-section area remains constant for Rayleigh flow, from the continuity equation (Eq.11.40) we conclude that

$$\rho V = \text{constant}$$

For a given Rayleigh flow, the constant in Eq. 11.111, the density–velocity product, ρV, and the ideal gas constant are all fixed. Thus, Eq. 11.111 can be used to determine values of fluid temperature corresponding to the local pressure in a Rayleigh flow.

To construct a temperature-entropy diagram for a given Rayleigh flow, we can use Eq. 11.76, which was developed earlier from the second $T\,ds$ relationship. Equations 11.111 and 11.76 can be solved simultaneously to obtain the curve sketched in Fig. 11.22. Curves like the one in Fig. 11.22 are called *Rayleigh lines*.

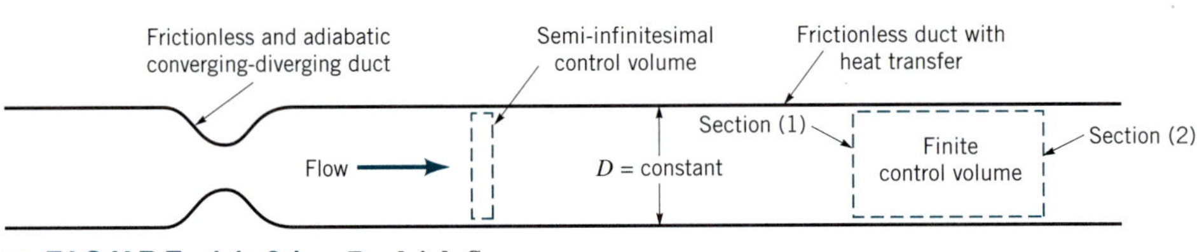

■ **FIGURE 11.21** **Rayleigh flow.**

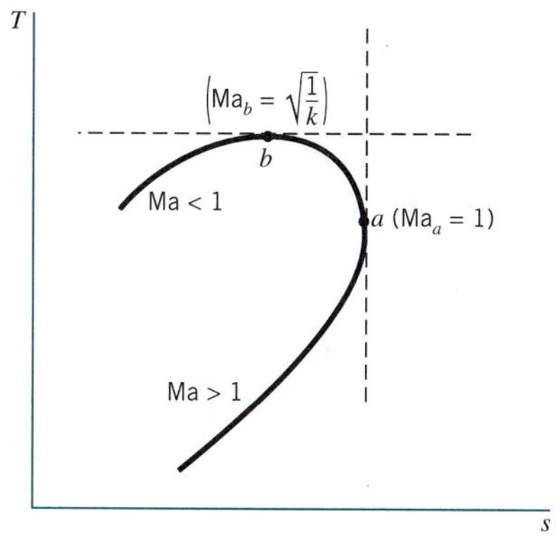

T
$\left(\text{Ma}_b = \sqrt{\frac{1}{k}}\right)$
b
Ma < 1
a ($\text{Ma}_a = 1$)
Ma > 1
s

■ **FIGURE 11.22** **Rayleigh line.**

EXAMPLE 11.15

Air ($k = 1.4$) enters [section (1)] a frictionless, constant flow cross-section area duct with the following properties (the same as in Example 11.11):

$$T_0 = 518.67\ °\text{R}$$

$$T_1 = 514.55\ °\text{R}$$

$$p_1 = 14.3\ \text{psia}$$

For Rayleigh flow, determine corresponding values of fluid temperature and entropy change for various levels of downstream pressure and plot the related Rayleigh line.

SOLUTION

To plot the Rayleigh line asked for, use Eq. 11.111

$$p + \frac{(\rho V)^2\, RT}{p} = \text{constant} \tag{1}$$

and Eq. 11.76

$$s - s_1 = c_p \ln \frac{T}{T_1} - R \ln \frac{p}{p_1} \tag{2}$$

to construct a table of values of temperature and entropy change corresponding to different levels of pressure downstream in a Rayleigh flow.

Use the value of ideal gas constant for air from Table 1.7

$$R = 1716\ (\text{ft}\cdot\text{lb})/(\text{slug}\cdot°\text{R})$$

and the value of specific heat at constant pressure for air from Example 11.11, namely,

$$c_p = 6006\ (\text{ft}\cdot\text{lb})/\text{slug}\cdot°\text{R})$$

Also, from Example 11.11, $\rho V = 0.519\ \text{slug}/(\text{ft}^2\cdot\text{s})$. For the given inlet [section (1)] conditions, we get from Eq. 1

$$p + \frac{(\rho V)^2\, RT}{p} = 14.3\ \text{psia}$$

$$+ \frac{[0.519\ \text{slug}/(\text{ft}^2\cdot\text{s})]^2[1716\ (\text{ft}\cdot\text{lb})/(\text{slug}\cdot°\text{R})](514.55\ °\text{R})}{(144\ \text{in.}^2/\text{ft}^2)^2\ 14.3\ \text{psia}} = \text{constant}$$

or

$$p + \frac{(\phi V)^2 RT}{p} = 15.10 \text{ psia} = \text{constant} \tag{3}$$

From Eq. 3, with the downstream pressure $p = 13.5$ psia, we obtain

$$13.5 \text{ psia} + \frac{[0.519 \text{ slug}/(\text{ft}^2/\text{s})]^2[1716 \text{ (ft·lb)}/(\text{slug·°R})]T}{(144 \text{ in.}^2/\text{ft}^2)^2 \; 13.5 \text{ psia}} = 15.10 \text{ psia}$$

or

$$T = 969 \text{ °R}$$

From Eq. 2 with the downstream pressure $p = 13.5$ psia and temperature $T = 969$ °R we get

$$s - s_1 = [6006 \text{ (ft·lb)}/(\text{slug·°R})] \ln\left(\frac{969 \text{ °R}}{514.55 \text{ °R}}\right) - [1716 \text{ (ft·lb)}/(\text{slug·°R})] \ln\left(\frac{13.5 \text{ psia}}{14.3 \text{ psia}}\right)$$

$$s - s_1 = 3900 \text{ (ft·lb)}/(\text{slug·°R})$$

By proceeding as outlined above, we can construct the table of values shown below and graph the Rayleigh line of Fig. E11.15.

p (psia)	T (°R)	$s - s_1$ [(ft·lb)/(slug·°R)]
13.5	969	3,900
12.5	1459	6,490
11.5	1859	8,089
10.5	2168	9,168
9.0	2464	10,202
8.0	2549	10,607
7.6	2558	10,716
7.5	2558	10,739
7.0	2544	10,825
6.3	2488	10,872
6.0	2450	10,863
5.5	2369	10,810
5.0	2266	10,707
4.5	2140	10,544
4.0	1992	10,315
2.0	1175	8,335
1.0	633	5,809

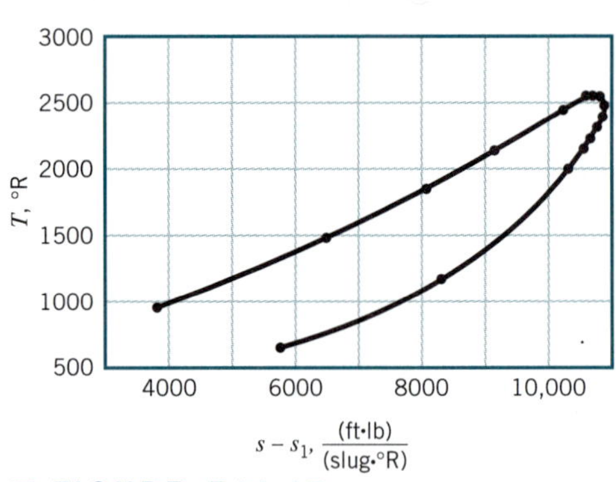

■ FIGURE E11.15

At point a on the Rayleigh line of Fig. 11.22, $ds/dT = 0$. To determine the physical importance of point a, we analyze further some of the governing equations. By differentiating the linear momentum equation for Rayleigh flow (Eq. 11.110) we obtain

$$dp = -\rho V\, dV$$

or

$$\frac{dp}{\rho} = -V\, dV \qquad \textbf{(11.112)}$$

Combining Eq. 11.112 with the second $T\,ds$ equation (Eq. 11.18) leads to

$$T\,ds = d\check{h} + V\,dV \qquad \textbf{(11.113)}$$

For an ideal gas (Eq. 11.7) $d\check{h} = c_p\,dT$. Thus, substituting Eq. 11.7 into Eq. 11.113 gives

$$T\,ds = c_p\,dT + V\,dV$$

or

$$\frac{ds}{dT} = \frac{c_p}{T} + \frac{V}{T}\frac{dV}{dT} \qquad \textbf{(11.114)}$$

Consolidation of Eqs. 11.114, 11.112 (linear momentum), 11.1, 11.77 (differentiated equation of state), and 11.79 (continuity) leads to

$$\frac{ds}{dT} = \frac{c_p}{T} + \frac{V}{T}\frac{1}{[(T/V) - (V/R)]} \qquad \textbf{(11.115)}$$

Hence, at state a where $ds/dT = 0$, Eq. 11.115 reveals that

$$V_a = \sqrt{RT_a k} \qquad \textbf{(11.116)}$$

Comparison of Eqs. 11.116 and 11.36 tells us that the Mach number at state a is equal to 1,

$$\text{Ma}_a = 1 \qquad \textbf{(11.117)}$$

At point b on the Rayleigh line of Fig. 11.22, $dT/ds = 0$. From Eq. 11.115 we get

$$\frac{dT}{ds} = \frac{1}{ds/dT} = \frac{1}{(c_p/T) + (V/T)[(T/V) - (V/R)]^{-1}}$$

which for $dT/ds = 0$ (point b) gives

$$\text{Ma}_b = \sqrt{\frac{1}{k}} \qquad \textbf{(11.118)}$$

The flow at point b is subsonic ($\text{Ma}_b < 1.0$). Recall that $k > 1$ for any gas.

To learn more about Rayleigh flow, we need to consider the energy equation in addition to the equations already used. Application of the energy equation (Eq. 5.69) to the Rayleigh flow through the finite control volume of Fig. 11.21 yields

$$\dot{m}\left[\check{h}_2 - \check{h}_1 + \frac{V_2^2 - V_1^2}{2} + \underbrace{g(z_2 - z_1)}_{\substack{0\text{(negligibly small}\\\text{for gas flow)}}}\right] = \dot{Q}_{\substack{\text{net}\\\text{in}}} + \underbrace{\dot{W}_{\substack{\text{shaft}\\\text{net in}}}}_{\substack{0\text{(flow is steady}\\\text{throughout)}}}$$

■ **TABLE 11.2**
Summary of Rayleigh Flow Characteristics

	Heating	Cooling
Ma < 1	Acceleration	Deceleration
Ma > 1	Deceleration	Acceleration

or in differential form for Rayleigh flow through the semi-infinitesimal control volume of Fig. 11.21

$$d\check{h} + V\,dV = \delta q \tag{11.119}$$

where δq is the heat transfer per unit mass of fluid in the semi-infinitesimal control volume.

By using $d\check{h} = c_p\,dT = Rk\,dT/(k - 1)$ in Eq. 11.119, we obtain

$$\frac{dV}{V} = \frac{\delta q}{c_p T}\left[\frac{V}{T}\frac{dT}{dV} + \frac{V^2(k-1)}{kRT}\right]^{-1} \tag{11.120}$$

Thus, by combining Eqs. 11.36 (ideal gas speed of sound), 11.46 (Mach number), 11.1 and 11.77 (ideal gas equation of state), 11.79 (continuity), and 11.112 (linear momentum) with Eq. 11.120 (energy) we get

$$\frac{dV}{V} = \frac{\delta q}{c_p T}\frac{1}{(1 - \text{Ma}^2)} \tag{11.121}$$

With the help of Eq. 11.121, we see clearly that when the Rayleigh flow is subsonic (Ma < 1), fluid heating ($\delta q > 0$) increases fluid velocity while fluid cooling ($\delta q < 0$) decreases fluid velocity. When Rayleigh flow is supersonic (Ma > 1), fluid heating decreases fluid velocity and fluid cooling increases fluid velocity.

The second law of thermodynamics states that, based on experience, entropy increases with heating and decreases with cooling. With this additional insight provided by the conservation of energy principle and the second law of thermodynamics, we can say more about the Rayleigh line in Fig. 11.22. A summary of the qualitative aspects of Rayleigh flow is outlined in Table 11.2 and Fig. 11.23. Along the upper portion of the line, which includes point b, the flow is subsonic. Heating the fluid results in flow acceleration to a maximum Mach number of 1 at point a. Note that between points b and a along the Rayleigh line, heating the fluid results in a temperature decrease and cooling the fluid leads to a temperature increase. This trend is not surprising if we consider the stagnation temperature and fluid

Fluid temperature reduction can accompany heating a subsonic Rayleigh flow.

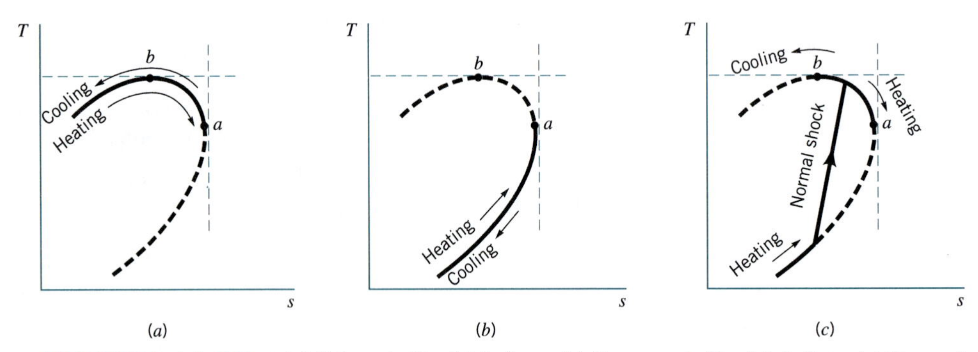

■ **FIGURE 11.23** ***(a)*** **Subsonic Rayleigh flow.** ***(b)*** **Supersonic Rayleigh flow.** ***(c)*** **Normal shock in a Rayleigh flow.**

velocity changes that occur between points a and b when the fluid is heated or cooled. Along the lower portion of the Rayleigh curve the flow is supersonic. Rayleigh flows may or may not be choked. The amount of heating or cooling involved determines what will happen in a specific instance. As with Fanno flows, an abrupt deceleration from supersonic flow to subsonic flow across a normal shock wave can also occur in Rayleigh flows.

To quantify Rayleigh flow behavior we need to develop appropriate forms of the governing equations. We elect to use the state of the Rayleigh flow fluid at point a of Fig. 11.22 as the reference state. As shown earlier, the Mach number at point a is 1. Even though the Rayleigh flow being considered may not choke and state a is not achieved by the flow, this reference state is useful.

The Mach number is unity at the reference state for Rayleigh flow.

If we apply the linear momentum equation (Eq. 11.110) to Rayleigh flow between any upstream section and the section, actual or imagined, where state a is attained, we get

$$p + \rho V^2 = p_a + \rho_a V_a^2$$

or

$$\frac{p}{p_a} + \frac{\rho V^2}{p_a} = 1 + \frac{\rho_a}{p_a} V_a^2 \quad (11.122)$$

By substituting the ideal gas equation of state (Eq. 11.1) into Eq. 11.122 and making use of the ideal gas speed-of-sound equation (Eq. 11.36) and the definition of Mach number (Eq. 11.46), we obtain

$$\boxed{\frac{p}{p_a} = \frac{1 + k}{1 + k\text{Ma}^2}} \quad (11.123)$$

From the ideal gas equation of state (Eq. 11.1) we conclude that

$$\frac{T}{T_a} = \frac{p}{p_a}\frac{\rho_a}{\rho} \quad (11.124)$$

Conservation of mass (Eq. 11.40) with constant A gives

$$\frac{\rho_a}{\rho} = \frac{V}{V_a} \quad (11.125)$$

which when combined with Eqs. 11.36 (ideal gas speed of sound) and 11.46 (Mach number definition) gives

$$\frac{\rho_a}{\rho} = \text{Ma}\sqrt{\frac{T}{T_a}} \quad (11.126)$$

Combining Eqs. 11.124 and 11.126 leads to

$$\frac{T}{T_a} = \left(\frac{p}{p_a}\text{Ma}\right)^2 \quad (11.127)$$

which when combined with Eq. 11.123 gives

$$\boxed{\frac{T}{T_a} = \left[\frac{(1 + k)\text{Ma}}{1 + k\text{Ma}^2}\right]^2} \quad (11.128)$$

From Eqs. 11.125, 11.126, and 11.128 we see that

$$\frac{\rho_a}{\rho} = \frac{V}{V_a} = \text{Ma}\left[\frac{(1 + k)\text{Ma}}{1 + k\text{Ma}^2}\right] \tag{11.129}$$

The energy equation (Eq. 5.69) tells us that because of the heat transfer involved in Rayleigh flows, the stagnation temperature varies. We note that

$$\frac{T_0}{T_{0,a}} = \left(\frac{T_0}{T}\right)\left(\frac{T}{T_a}\right)\left(\frac{T_a}{T_{0,a}}\right) \tag{11.130}$$

Unlike Fanno flow, the stagnation temperature in Rayleigh flow varies.

We can use Eq. 11.56 (developed earlier for steady, isentropic, ideal gas flow) to evaluate T_0/T and T_a/T_{0a} because these two temperature ratios, by definition of the stagnation state, involve isentropic processes. Equation 11.128 can be used for T/T_a. Thus, consolidating Eqs. 11.130, 11.56, and 11.128 we obtain

$$\frac{T_0}{T_{0,a}} = \frac{2(k + 1)\text{Ma}^2\left(1 + \frac{k - 1}{2}\text{Ma}^2\right)}{(1 + k\text{Ma}^2)^2} \tag{11.131}$$

Finally, we observe that

$$\frac{p_0}{p_{0,a}} = \left(\frac{p_0}{p}\right)\left(\frac{p}{p_a}\right)\left(\frac{p_a}{p_{0,a}}\right) \tag{11.132}$$

We can use Eq. 11.59 developed earlier for steady, isentropic, ideal gas flow to evaluate p_0/p and $p_a/p_{0,a}$ because these two pressure ratios, by definition, involve isentropic processes. Equation 11.123 can be used for p/p_a. Together, Eqs. 11.59, 11.123, and 11.132 give

$$\frac{p_0}{p_{0a}} = \frac{(1 + k)}{(1 + k\text{Ma}^2)}\left[\left(\frac{2}{k + 1}\right)\left(1 + \frac{k - 1}{2}\text{Ma}^2\right)\right]^{k/(k-1)} \tag{11.133}$$

Values of p/p_a, T/T_a, ρ_a/ρ, or V/V_a, $T_0/T_{0,a}$, and $p_0/p_{0,a}$ are graphed in Fig. D.3 of Appendix D as a function of Mach number for Rayleigh flow of air ($k = 1.4$). The values in Fig. D.3 were calculated from Eqs. 11.123, 11.128, 11.129, 11.131, and 11.133. The usefulness of Fig. D.3 is illustrated in Example 11.16.

EXAMPLE 11.16

The information in Fig. 11.2 shows us that subsonic Rayleigh flow accelerates when heated and decelerates when cooled. Supersonic Rayleigh flow behaves just opposite to subsonic Rayleigh flow; it decelerates when heated and accelerates when cooled. Using Fig. D.3 for air ($k = 1.4$), state whether velocity, Mach number, static temperature, stagnation temperature, static pressure, and stagnation pressure increase or decrease as subsonic and supersonic Rayleigh flow is (a) heated, (b) cooled.

SOLUTION

Acceleration occurs when V/V_a in Fig. D.3 increases. For deceleration, V/V_a decreases. From Fig. D.3 and Table 11.2 the following chart can be constructed.

	Heating		Cooling	
	Subsonic	Supersonic	Subsonic	Supersonic
V	Increase	Decrease	Decrease	Increase
Ma	Increase	Decrease	Decrease	Increase
T	Increase for $0 \leq \text{Ma} \leq \sqrt{1/k} = 0.845$ Decrease for $\sqrt{1/k} \leq \text{Ma} \leq 1$	Increase	Decrease for $0 \leq \text{Ma} \leq \sqrt{1/k} = 0.845$ Increase for $\sqrt{1/k} \leq \text{Ma} \leq 1$	Decrease
T_0	Increase	Increase	Decrease	Decrease
p	Decrease	Increase	Increase	Decrease
p_0	Decrease	Decrease	Increase	Increase

From the Rayleigh flow trends summarized in the table above, we note that heating affects Rayleigh flows much like friction affects Fanno flows. Heating and friction both accelerate subsonic flows and decelerate supersonic flows. More importantly, both heating and friction cause the stagnation pressure to decrease. Since stagnation pressure loss is considered undesirable in terms of fluid mechanical efficiency, heating a fluid flow must be accomplished with this loss in mind.

11.5.3 Normal Shock Waves

V11.5

Normal shock waves are assumed to be infinitesimally thin discontinuities.

As mentioned earlier, normal shock waves can occur in supersonic flows through converging-diverging and constant-area ducts. Past experience suggests that normal shock waves involve deceleration from a supersonic flow to a subsonic flow, a pressure rise, and an increase of entropy. To develop the equations that verify this observed behavior of flows across a normal shock, we apply first principles to the flow through a control volume that completely surrounds a normal shock wave (see Fig. 11.24). We consider the normal shock and thus the control volume to be infinitesimally thin and stationary.

For steady flow through the control volume of Fig. 11.24, the conservation of mass principle yields

$$\rho V = \text{constant} \tag{11.134}$$

because the flow cross-section area remains essentially constant within the infinitesimal thick-

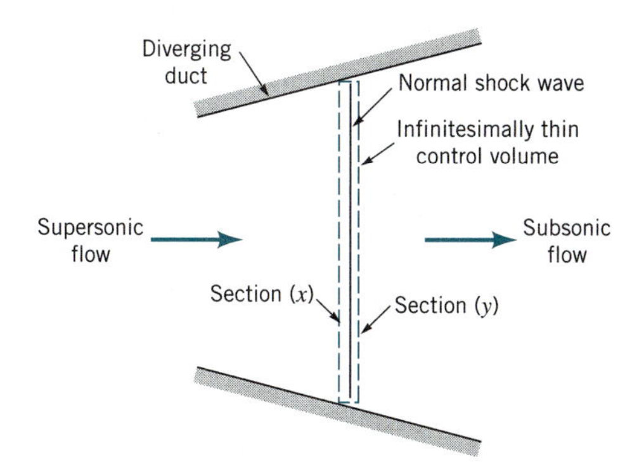

■ FIGURE 11.24 Normal shock control volume.

ness of the normal shock. Note that Eq. 11.134 is identical to the continuity equation used for Fanno and Rayleigh flows considered earlier.

The friction force acting on the contents of the infinitesimally thin control volume surrounding the normal shock is considered to be negligibly small. Also for ideal gas flow, the effect of gravity is neglected. Thus, the linear momentum equation (Eq. 5.22) describing steady gas flow through the control volume of Fig. 11.24 is

$$p + \rho V^2 = \text{constant}$$

or for an ideal gas for which $p = \rho RT$,

$$p + \frac{(\rho V)^2 RT}{p} = \text{constant} \tag{11.135}$$

Equation 11.135 is the same as the linear momentum equation for Rayleigh flow, which was derived earlier (Eq. 11.111).

For the control volume containing the normal shock, no shaft work is involved and the heat transfer is assumed negligible. Thus, the energy equation (Eq. 5.69) can be applied to steady gas flow through the control volume of Fig. 11.24 to obtain

$$\check{h} + \frac{V^2}{2} = \check{h}_0 = \text{constant}$$

or, for an ideal gas, since $\check{h} - \check{h}_0 = c_p(T - T_0)$ and $p = \rho RT$

$$T + \frac{(\rho V)^2 T^2}{2c_p(p^2/R^2)} = T_0 = \text{constant} \tag{11.136}$$

Equation 11.136 is identical to the energy equation for Fanno flow analyzed earlier (Eq. 11.75).

The $T\,ds$ relationship previously used for ideal gas flow (Eq. 11.22) is valid for the flow through the normal shock (Fig. 11.24) because it (Eq. 11.22) is an ideal gas property relationship.

The energy equation for Fanno flow and the momentum equation for Rayleigh flow are valid for flow across normal shocks.

From the analyses in the previous paragraphs, it is apparent that the steady flow of an ideal gas across a normal shock is governed by some of the same equations used for describing Fanno and Rayleigh flows (energy equation for Fanno flows and momentum equation for Rayleigh flow). Thus, for a given density-velocity product (ρV), gas (R,k), and conditions at the inlet of the normal shock (T_x, p_x, and s_x), the conditions downstream of the shock (state y) will be on both a Fanno line and a Rayleigh line that pass through the inlet state (state x), as is illustrated in Fig. 11.25. To conform with common practice we designate the states

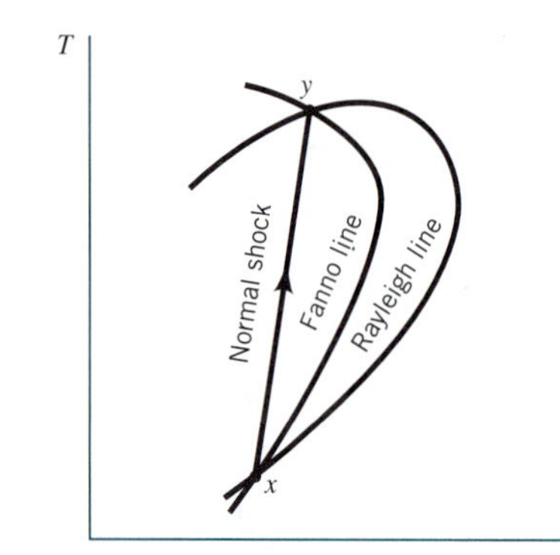

■ **FIGURE 11.25** **The relationship between a normal shock and Fanno and Rayleigh lines.**

upstream and downstream of the normal shock with x and y instead of numerals 1 and 2. The Fanno and Rayleigh lines describe more of the flow field than just in the vicinity of the normal shock when Fanno and Rayleigh flows are actually involved (solid lines in Figs. 11.26*a* and 11.26*b*). Otherwise, these lines (dashed lines in Figs. 11.26*a*, 11.26*b*, and 11.26*c*) are useful mainly as a way to better visualize how the governing equations combine to yield a solution to the normal shock flow problem.

The second law of thermodynamics requires that entropy must increase across a normal shock wave. This law and sketches of the Fanno line and Rayleigh line intersections, like those of Figs. 11.25 and 11.26, persuade us to conclude that flow across a normal shock can only proceed from supersonic to subsonic flow. Similarly, in open-channel flows (see Chapter 10) the flow across a hydraulic jump proceeds from supercritical to subcritical conditions.

The flow across a normal shock can only proceed from supersonic to subsonic flow.

Since the states upstream and downstream of a normal shock wave are represented by the supersonic and subsonic intersections of actual and/or imagined Fanno and Rayleigh lines, we should be able to use equations developed earlier for Fanno and Rayleigh flows to quantify normal shock flow. For example, for the Rayleigh line of Fig. 11.26*b*

$$\frac{p_y}{p_x} = \left(\frac{p_y}{p_a}\right)\left(\frac{p_a}{p_x}\right) \tag{11.137}$$

But from Eq. 11.123 for Rayleigh flow we get

$$\frac{p_y}{p_a} = \frac{1 + k}{1 + k\text{Ma}_y^2} \tag{11.138}$$

and

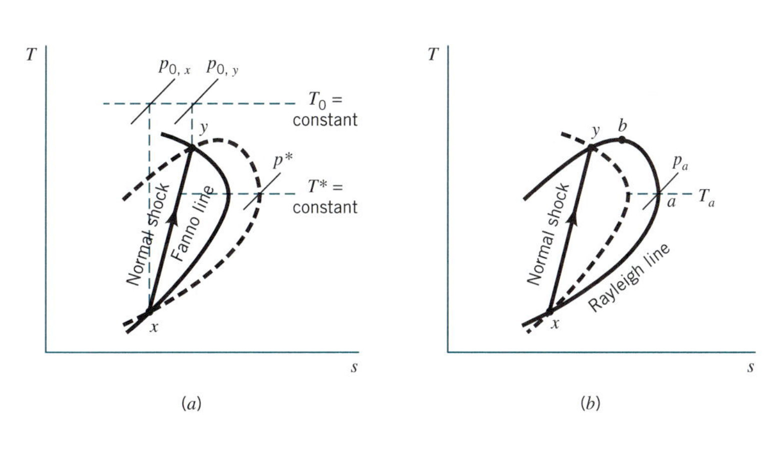

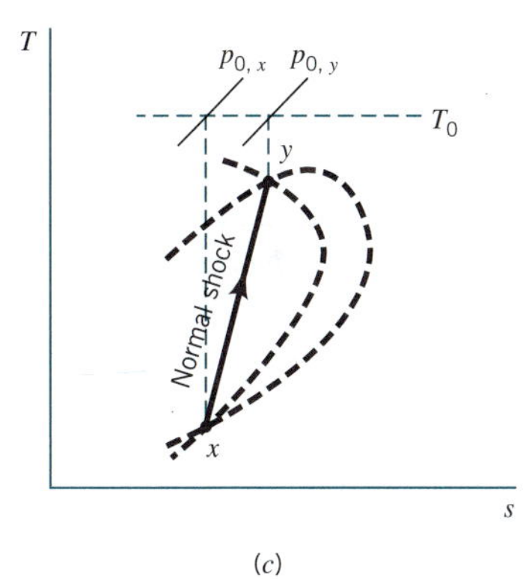

■ **FIGURE 11.26** ***(a)*** **The normal shock in a Fanno flow.** ***(b)*** **The normal shock in a Rayleigh flow.** ***(c)*** **The normal shock in a frictionless and adiabatic flow.**

$$\frac{p_x}{p_a} = \frac{1 + k}{1 + k\text{Ma}_x^2} \tag{11.139}$$

Thus, by combining Eqs. 11.137, 11.138, and 11.139 we get

$$\boxed{\frac{p_y}{p_x} = \frac{1 + k\text{Ma}_x^2}{1 + k\text{Ma}_y^2}} \tag{11.140}$$

Equation 11.140 can also be derived starting with

$$\frac{p_y}{p_x} = \left(\frac{p_y}{p^*}\right)\left(\frac{p^*}{p_x}\right)$$

and using the Fanno flow equation (Eq. 11.107)

$$\frac{p}{p^*} = \frac{1}{\text{Ma}}\left\{\frac{(k + 1)/2}{1 + [(k - 1)/2]\text{Ma}^2}\right\}^{1/2}$$

As might be expected, Eq. 11.140 can be obtained directly from the linear momentum equation

$$p_x + \rho_x V_x^2 = p_y + \rho_y V_y^2$$

since $\rho V^2/p = V^2/RT = kV^2/RTk = k\,\text{Ma}^2$.

For the Fanno flow of Fig. 11.26*a*

$$\frac{T_y}{T_x} = \left(\frac{T_y}{T^*}\right)\left(\frac{T^*}{T_x}\right) \tag{11.141}$$

From Eq. 11.101 for Fanno flow we get

$$\frac{T_y}{T^*} = \frac{(k + 1)/2}{1 + [(k - 1)/2]\text{Ma}_y^2} \tag{11.142}$$

Ratios of thermodynamic properties across a normal shock are functions of the Mach numbers.

and

$$\frac{T_x}{T^*} = \frac{(k + 1)/2}{1 + [(k - 1)/2]\text{Ma}_x^2} \tag{11.143}$$

A consolidation of Eqs. 11.141, 11.142, and 11.143 gives

$$\boxed{\frac{T_y}{T_x} = \frac{1 + [(k - 1)/2]\text{Ma}_x^2}{1 + [(k - 1)/2]\text{Ma}_y^2}} \tag{11.144}$$

We seek next to develop an equation that will allow us to determine the Mach number downstream of the normal shock, Ma_y, when the Mach number upstream of the normal shock, Ma_x, is known. From the ideal gas equation of state (Eq. 11.1), we can form

$$\frac{p_y}{p_x} = \left(\frac{T_y}{T_x}\right)\left(\frac{\rho_y}{\rho_x}\right) \tag{11.145}$$

Using the continuity equation

$$\rho_x V_x = \rho_y V_y$$

with Eq. 11.145 we obtain

$$\frac{p_y}{p_x} = \left(\frac{T_y}{T_x}\right)\left(\frac{V_x}{V_y}\right) \tag{11.146}$$

When combined with the Mach number definition (Eq. 11.46) and the ideal gas speed-of-sound equation (Eq. 11.36), Eq. 11.146 becomes

$$\frac{p_y}{p_x} = \left(\frac{T_y}{T_x}\right)^{1/2}\left(\frac{\mathrm{Ma}_x}{\mathrm{Ma}_y}\right) \tag{11.147}$$

Thus, Eqs. 11.147 and 11.144 lead to

$$\frac{p_y}{p_x} = \left\{\frac{1 + [(k-1)/2]\mathrm{Ma}_x^2}{1 + [(k-1)/2]\mathrm{Ma}_y^2}\right\}^{1/2} \frac{\mathrm{Ma}_x}{\mathrm{Ma}_y} \tag{11.148}$$

which can be merged with Eq. 11.140 to yield

$$\mathrm{Ma}_y^2 = \frac{\mathrm{Ma}_x^2 + [2/(k-1)]}{[2k/(k-1)]\mathrm{Ma}_x^2 - 1} \tag{11.149}$$

The flow changes from supersonic to subsonic across a normal shock.

Thus, we can use Eq. 11.149 to calculate values of Mach number downstream of a normal shock from a known Mach number upstream of the shock. As suggested by Fig. 11.26, to have a normal shock we must have $\mathrm{Ma}_x > 1$. From Eq. 11.149 we find that $\mathrm{Ma}_y < 1$.

If we combine Eqs. 11.149 and 11.140, we get

$$\frac{p_y}{p_x} = \frac{2k}{k+1}\mathrm{Ma}_x^2 - \frac{k-1}{k+1} \tag{11.150}$$

which allows us to calculate the pressure ratio across a normal shock from a known upstream Mach number. Similarly, taking Eqs. 11.149 and 11.144 together we obtain

$$\frac{T_y}{T_x} = \frac{\{1 + [(k-1)/2]\mathrm{Ma}_x^2\}\{[2k/(k-1)]\mathrm{Ma}_x^2 - 1\}}{\{(k+1)^2/[2(k-1)]\}\mathrm{Ma}_x^2} \tag{11.151}$$

From the continuity equation (Eq. 11.40), we have for flow across a normal shock

$$\frac{\rho_y}{\rho_x} = \frac{V_x}{V_y} \tag{11.152}$$

and from the ideal gas equation of state (Eq. 11.1)

$$\frac{\rho_y}{\rho_x} = \left(\frac{p_y}{p_x}\right)\left(\frac{T_x}{T_y}\right) \tag{11.153}$$

Thus, by combining Eqs. 11.152, 11.153, 11.150, and 11.151, we get

$$\frac{\rho_y}{\rho_x} = \frac{V_x}{V_y} = \frac{(k+1)\mathrm{Ma}_x^2}{(k-1)\mathrm{Ma}_x^2 + 2} \tag{11.154}$$

The stagnation pressure ratio across the shock can be determined by combining

■ TABLE 11.3
Summary of Normal Shock Wave Characteristics

Variable	Change Across Normal Shock Wave
Mach number	Decrease
Static pressure	Increase
Stagnation pressure	Decrease
Static temperature	Increase
Stagnation temperature	Constant
Density	Increase
Velocity	Decrease

Across a normal shock the values of some parameters increase, some remain constant, and some decrease.

$$\frac{p_{0,y}}{p_{0,x}} = \left(\frac{p_{0,y}}{p_y}\right)\left(\frac{p_y}{p_x}\right)\left(\frac{p_x}{p_{0,x}}\right) \tag{11.155}$$

with Eqs. 11.59, 11.149, and 11.150 to get

$$\frac{p_{0,y}}{p_{0,x}} = \frac{\left(\dfrac{k+1}{2}\,\mathrm{Ma}_x^2\right)^{k/(k-1)}\left(1+\dfrac{k-1}{2}\,\mathrm{Ma}_x^2\right)^{k/(1-k)}}{\left(\dfrac{2k}{k+1}\,\mathrm{Ma}_x^2-\dfrac{k-1}{k+1}\right)^{1/(k-1)}} \tag{11.156}$$

Fig. D.4 in Appendix D graphs values of downstream Mach numbers, Ma_y, pressure ratio, p_y/p_x, temperature ratio, T_y/T_x, density ratio, ρ_y/ρ_x or velocity ratio, V_x/V_y, and stagnation pressure ratio, $p_{0,y}/p_{0,x}$, as a function of upstream Mach number, Ma_x, for the steady flow across a normal shock wave of an ideal gas having a specific heat ratio $k = 1.4$. These values were calculated from Eqs. 11.149, 11.150, 11.151, 11.154, and 11.156.

Important trends associated with the steady flow of an ideal gas across a normal shock wave can be determined by studying Fig. D.4. These trends are summarized in Table 11.3.

Examples 11.17 and 11.18 illustrate how Fig. D.4 can be used to solve fluid flow problems involving normal shock waves.

Designers involved with fluid mechanics work hard at minimizing loss of available energy in their designs. Adiabatic, frictionless flows involve no loss in available energy. Entropy remains constant for these idealized flows. Adiabatic flows with friction involve available energy loss and entropy increase. Generally, larger entropy increases imply larger losses. For normal shocks, show that the stagnation pressure drop (and thus loss) is larger for higher Mach numbers.

SOLUTION

We assume that air ($k = 1.4$) behaves as a typical gas and use Fig. D.4 to respond to the above-stated requirements. Since

$$1 - \frac{p_{0,y}}{p_{0,x}} = \frac{p_{0,x} - p_{0,y}}{p_{0,x}}$$

we can construct the following table with values of $p_{0,y}/p_{0,x}$ from Fig. D.4.

When the Mach number of the flow entering the shock is low, say $\text{Ma}_x = 1.2$, the flow across the shock is nearly isentropic and the loss in stagnation pressure is small. However, at larger Mach numbers, the entropy change across the normal shock rises dramatically and the stagnation pressure drop across the shock is appreciable. If a shock occurs at $\text{Ma}_x = 2.5$, only about 50% of the upstream stagnation pressure is recovered.

In devices where supersonic flows occur, for example, high-performance aircraft engine inlet ducts and high-speed wind tunnels, designers attempt to prevent shock formation, or if shocks must occur, they design the flow path so that shocks are positioned where they are weak (small Mach number).

Ma_x	$p_{0,y}/p_{0,x}$	$\frac{p_{0,x} - p_{0,y}}{p_{0,x}}$
1.0	1.0	0
1.2	0.99	0.01
1.5	0.93	0.07
2.0	0.72	0.28
2.5	0.50	0.50
3.0	0.33	0.67
3.5	0.21	0.79
4.0	0.14	0.86
5.0	0.06	0.94

Ma_x	p_y/p_x
1.0	1.0
1.2	1.5
1.5	2.5
2.0	4.5
3.0	10
4.0	18
5.0	29

Of interest also is the static pressure rise that occurs across a normal shock. These static pressure ratios, p_y/p_x, obtained from Fig. D.4 are shown in the table above for a few Mach numbers. For a developing boundary layer, any pressure rise in the flow direction is considered as an adverse pressure gradient that can possibly cause flow separation (see Section 9.2.6). Thus, shock–boundary layer interactions are of great concern to designers of high-speed flow devices.

A total pressure probe is inserted into a supersonic air flow. A shock wave forms just upstream of the impact hole and head as illustrated in Fig. E11.18. The probe measures a total pressure of 60 psia. The stagnation temperature at the probe head is 1000 °R. The static pressure upstream of the shock is measured with a wall tap to be 12 psia. From these data determine the Mach number and velocity of the flow.

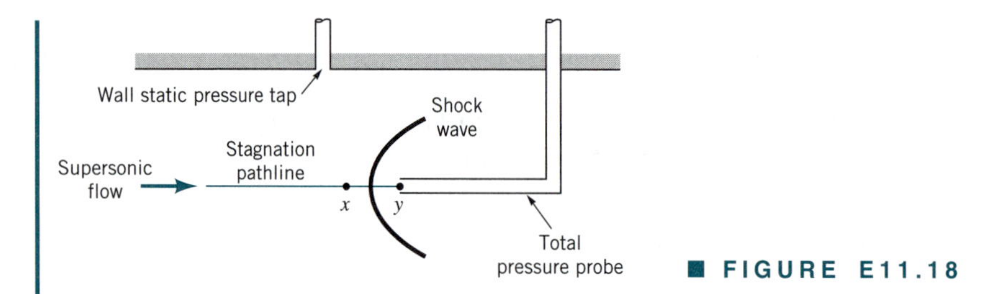

■ FIGURE E11.18

SOLUTION

We assume that the flow along the stagnation pathline is isentropic except across the shock. Also, the shock is treated as a normal shock. Thus, in terms of the data we have

$$\frac{p_{0,y}}{p_x} = \left(\frac{p_{0,y}}{p_{0,x}}\right)\left(\frac{p_{0,x}}{p_x}\right) \quad \textbf{(1)}$$

where $p_{0,y}$ is the stagnation pressure measured by the probe, and p_x is the static pressure measured by the wall tap. The stagnation pressure upstream of the shock, $p_{0,x}$, is not measured.

Combining Eqs. 1, 11.156, and 11.59 we obtain

$$\frac{p_{0,y}}{p_x} = \frac{\{[(k+1)/2]\text{Ma}_x^2\}^{k/(k-1)}}{\{[2k/(k+1)]\text{Ma}_x^2 - [(k-1)/(k+1)]\}^{1/(k-1)}} \quad \textbf{(2)}$$

which is called the *Rayleigh Pitot-tube formula*. Values of $p_{0,y}/p_x$ from Eq. 2 are considered important enough to be included in Fig. D.4 for $k = 1.4$. Thus, for $k = 1.4$ and

$$\frac{p_{0,y}}{p_x} = \frac{60 \text{ psia}}{12 \text{ psia}} = 5$$

we use Fig. D.4 to ascertain that

$$\text{Ma}_x = 1.9 \quad \textbf{(Ans)}$$

To determine the flow velocity we need to know the static temperature upstream of the shock, since Eqs. 11.36 and 11.46 can be used to yield

$$V_x = Ma_x c_x = \text{Ma}_x \sqrt{\text{R}T_x k} \quad \textbf{(3)}$$

The stagnation temperature downstream of the shock was measured and found to be

$$T_{0,y} = 1000 \text{ °R}$$

Since the stagnation temperature remains constant across a normal shock (see Eq. 11.136),

$$T_{0,x} = T_{0,y} = 1000 \text{ °R}$$

For the isentropic flow upstream of the shock, Eq. 11.56 or Fig. D.1 can be used. For $\text{Ma}_x = 1.9$,

$$\frac{T_x}{T_{0,x}} = 0.59$$

or

$$T_x = (0.59)(1000 \text{ °R}) = 590 \text{ °R}$$

With Eq. 3 we obtain

$$V_x = 1.87 \sqrt{[1716(\text{ft}\cdot\text{lb})/(\text{slug}\cdot{}^\circ\text{R})](590\ {}^\circ\text{R})(1.4)[1(\text{slug}\cdot\text{ft})/(\text{lb}\cdot\text{s}^2)]} = 2200\ \text{ft/s} \quad \textbf{(Ans)}$$

Application of the incompressible flow Pitot tube results (see Section 3.5) would give highly inaccurate results because of the large pressure and density changes involved.

EXAMPLE 11.19

Determine, for the converging-diverging duct of Example 11.8, the ratio of back pressure to inlet stagnation pressure, $p_{\text{III}}/p_{0,x}$ (see Fig. 11.13), that will result in a standing normal shock at the exit ($x = +0.5$ m) of the duct. What value of the ratio of back pressure to inlet stagnation pressure would be required to position the shock at $x = +0.3$ m? Show related temperature-entropy diagrams for these flows.

SOLUTION

For supersonic, isentropic flow through the nozzle to just upstream of the standing normal shock at the duct exit, we have from the table of Example 11.8 at $x = +0.5$ m

$$\text{Ma}_x = 2.8$$

and

$$\frac{p_x}{p_{0,x}} = 0.04$$

From Fig. D.4 for $\text{Ma}_x = 2.8$ we obtain

$$\frac{p_y}{p_x} = 9.0$$

Thus,

$$\frac{p_y}{p_{0,x}} = \left(\frac{p_y}{p_x}\right)\left(\frac{p_x}{p_{0,x}}\right) = (9.0)(0.04) = 0.36 = \frac{p_{\text{III}}}{p_{0,x}} \quad \textbf{(Ans)}$$

When the ratio of duct back pressure to inlet stagnation pressure, $p_{\text{III}}/p_{0,x}$, is set equal to 0.36, the air will accelerate through the converging-diverging duct to a Mach number of 2.8 at the duct exit. The air will subsequently decelerate to a subsonic flow across a normal shock at the duct exit. The stagnation pressure ratio across the normal shock, $p_{0,y}/p_{0,x}$, is 0.38 (Fig. D.4 for $M_x = 2.8$). A considerable amount of available energy is lost across the shock.

For a normal shock at $x = +0.3$ m, we note from the table of Example 11.8 that $\text{Ma}_x = 2.14$ and

$$\frac{p_x}{p_{0,x}} = 0.10 \quad \textbf{(1)}$$

From Fig. D.4 for $\text{Ma}_x = 2.14$ we obtain $p_y/p_x = 5.2$, $\text{Ma}_y = 0.56$ and

$$\frac{p_{0,y}}{p_{0,x}} = 0.66 \quad \textbf{(2)}$$

From Fig. D.1 for $\text{Ma}_y = 0.56$ we get

$$\frac{A_y}{A^*} = 1.24 \quad \textbf{(3)}$$

For $x = +0.3$ m, the ratio of duct exit area to local area (A_2/A_y) is, using the area equation from Example 11.8,

$$\frac{A_2}{A_y} = \frac{0.1 + (0.5)^2}{0.1 + (0.3)^2} = 1.842 \tag{4}$$

Using Eqs. 3 and 4 we get

$$\frac{A_2}{A^*} = \left(\frac{A_y}{A^*}\right)\left(\frac{A_2}{A_y}\right) = (1.24)(1.842) = 2.28$$

Note that for the isentropic flow upstream of the shock, $A^* = 0.10\ \text{m}^2$ (the actual throat area), while for the isentropic flow downstream of the shock, $A^* = A_2/2.28 = 0.35\ \text{m}^2/2.28 = 0.15\ \text{m}^2$. With $A_2/A^* = 2.28$ we use Fig. D.1 and find $\text{Ma}_2 = 0.26$ and

$$\frac{p_2}{p_{0,y}} = 0.95 \tag{5}$$

Combining Eqs. 2 and 5 we obtain

$$\frac{p_2}{p_{0,x}} = \left(\frac{p_2}{p_{0,y}}\right)\left(\frac{p_{0,y}}{p_{0,x}}\right) = (0.95)(0.66) = 0.63 \qquad \textbf{(Ans)}$$

When the back pressure, p_2, is set equal to 0.63 times the inlet stagnation pressure, $p_{0,x}$, the normal shock will be positioned at $x = +0.3$ m. Note that $p_2/p_{0,x} = 0.63$ is less than the value of this ratio for subsonic isentropic flow through the converging-diverging duct, $p_2/p_0 = 0.98$ (from Example 11.8) and is larger than $p_{\text{III}}/p_{0,x} = 0.36$, for duct flow with a normal shock at the exit (see Fig. 11.13). Also the stagnation pressure ratio with the shock at $x = +0.3$ m, $p_{0,y}/p_{0,x} = 0.66$, is much greater than the stagnation pressure ratio, 0.38, when the shock occurs at the exit ($x = +0.5$ m) of the duct. The corresponding $T - s$ diagrams are shown in Figs. E11.19*a* and E11.19*b*.

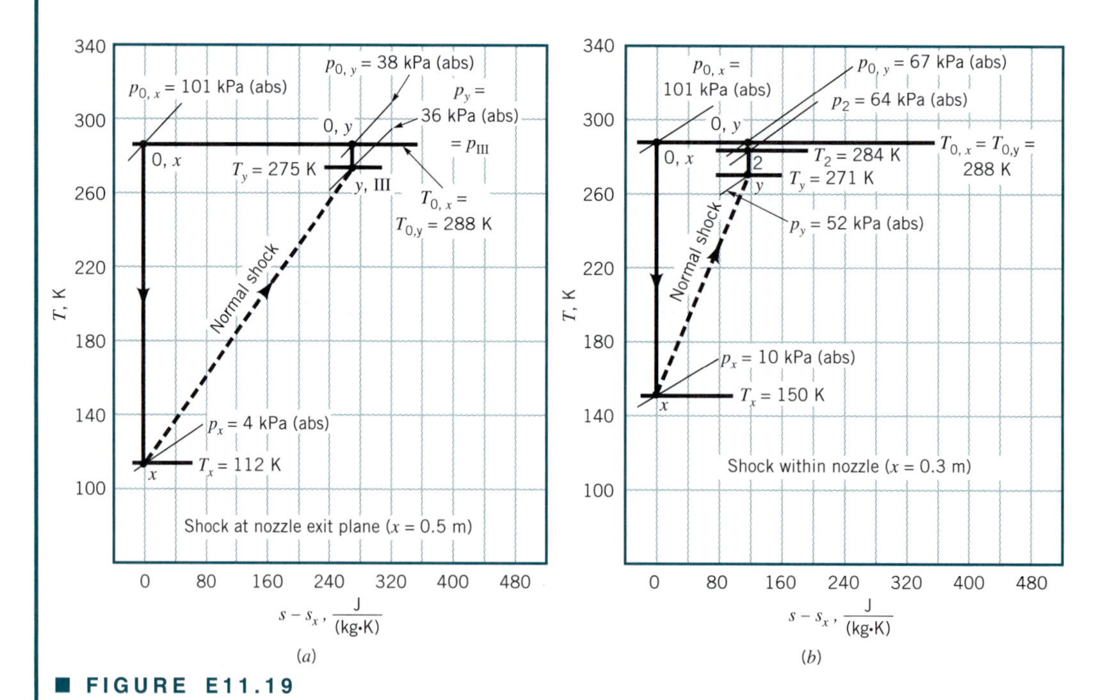

■ **FIGURE E11.19**

11.6 Analogy Between Compressible and Open-Channel Flows

During a first course in fluid mechanics, students rarely study both open-channel flows (Chapter 10) and compressible flows. This is unfortunate because these two kinds of flows are strikingly similar in several ways. Furthermore, the analogy between open-channel and compressible flows is useful because important two-dimensional compressible flow phenomena can be simply and inexpensively demonstrated with a shallow, open-channel flow field in a *ripple tank* or *water table.*

The propagation of weak pressure pulses (sound waves) in a compressible flow can be considered to be comparable to the movement of small amplitude waves on the surface of an open-channel flow. In each case—two-dimensional compressible flow and open-channel flow—the influence of flow velocity on wave pattern is similar. When the flow velocity is less than the wave speed, wave fronts can move upstream of the wave source and the flow is subsonic (compressible flow) or subcritical (open-channel flow). When the flow velocity is equal to the wave speed, wave fronts cannot move upstream of the wave source and the flow is sonic (compressible flow) or critical (open-channel flow). When the flow velocity is greater than the wave speed, the flow is supersonic (compressible flow) or supercritical (open-channel flow). Normal shocks can occur in supersonic compressible flows. Hydraulic jumps can occur in supercritical open-channel flows. Comparison of the characteristics of normal shocks (Section 11.5.3) and hydraulic jumps (Section 10.6.1) suggests a strong resemblance and thus analogy between the two phenomena.

Compressible gas flows and open-channel liquid flows are strikingly similar in several ways.

For compressible flows a meaningful dimensionless variable is the Mach number, where

$$\text{Ma} = \frac{V}{c} \tag{11.46}$$

In open-channel flows, an important dimensionless variable is the Froude number, where

$$\text{Fr} = \frac{V_{oc}}{\sqrt{gy}} \tag{11.157}$$

The velocity of the channel flow is V_{oc}, the acceleration of gravity is g, and the depth of the flow is y. Since the speed of a small amplitude wave on the surface of an open-channel flow, c_{oc}, is (see Section 10.2.1)

$$c_{oc} = \sqrt{gy} \tag{11.158}$$

we conclude that

$$\text{Fr} = \frac{V_{oc}}{c_{oc}} \tag{11.159}$$

From Eqs. 11.46 and 11.159 we see the similarity between Mach number (compressible flow) and Froude number (open-channel flow).

For compressible flow, the continuity equation is

$$\rho AV = \text{constant} \tag{11.160}$$

where V is the flow velocity, ρ is the fluid density, and A is the flow cross-section area. For an open-channel flow, conservation of mass leads to

$$ybV_{oc} = \text{constant} \tag{11.161}$$

where V_{oc} is the flow velocity, and y and b are the depth and width of the open-channel flow. Comparing Eqs. 11.160 and 11.161 we note that if flow velocities are considered similar and flow area, A, and channel width, b, are considered similar, then compressible flow density, ρ, is analogous to open-channel flow depth, y.

It should be pointed out that the similarity between Mach number and Froude number is generally not exact. If compressible flow and open-channel flow velocities are considered to be similar, then it follows that for Mach number and Froude number similarity the wave speeds c and c_{oc} must also be similar.

From the development of the equation for the speed of sound in an ideal gas (see Eqs. 11.34 and 11.35) we have for the compressible flow

$$c = \sqrt{(\text{constant})k\rho^{k-1}} \tag{11.162}$$

From Eqs. 11.162 and 11.158, we see that if y is to be similar to ρ as suggested by comparing Eq. 11.160 and 11.161, then k should be equal to 2. Typically $k = 1.4$ or 1.67, not 2. This limitation to exactness is, however, usually not serious enough to compromise the benefits of the analogy between compressible and open-channel flows.

11.7 Two-Dimensional Compressible Flow

A brief introduction to two-dimensional compressible flow is included here for those who are interested. We begin with a consideration of supersonic flow over a wall with a small change of direction as sketched in Fig. 11.27.

We apply the component of the linear momentum equation (Eq. 5.22) parallel to the Mach wave to the flow across the Mach wave. (See Eq. 11.39 for the definition of a Mach wave.) The result is that the component of velocity parallel to the Mach wave is constant across the Mach wave. That is, $V_{t1} = V_{t2}$. Thus, from the simple velocity triangle construction indicated in Fig. 11.27, we conclude that the flow accelerates because of the change in direction of the flow. If several changes in wall direction are involved as shown in Fig. 11.28, then the supersonic flow accelerates (expands) because of the changes in flow direction across the Mach waves (also called *expansion* waves). Each Mach wave makes an appropriately smaller angle α with the upstream wall because of the increase in Mach number that occurs with each direction change (see Section 11.3). A rounded expansion corner may be considered as a series of infinitesimal changes in direction. Conversely, even sharp corners are actually rounded when viewed on a small enough scale. Thus, expansion fans as illustrated in Fig. 11.29 are commonly used for supersonic flow around a ''sharp'' corner. If the flow across the Mach waves is considered to be isentropic, then Eq. 11.42 suggests that the increase in flow speed is accompanied by a decrease in static pressure.

Supersonic flows accelerate across expansion Mach waves.

When the change in supersonic flow direction involves the change in wall orientation sketched in Fig. 11.30, compression rather than expansion occurs. The flow decelerates and the static pressure increases across the Mach wave. For several changes in wall direction, as

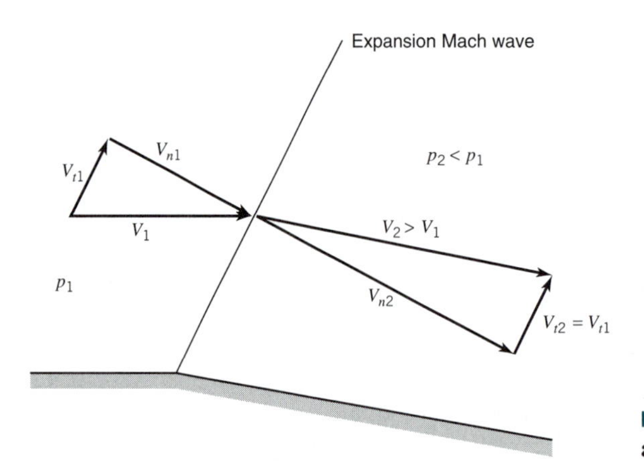

■ **FIGURE 11.27 Flow acceleration across a Mach wave.**

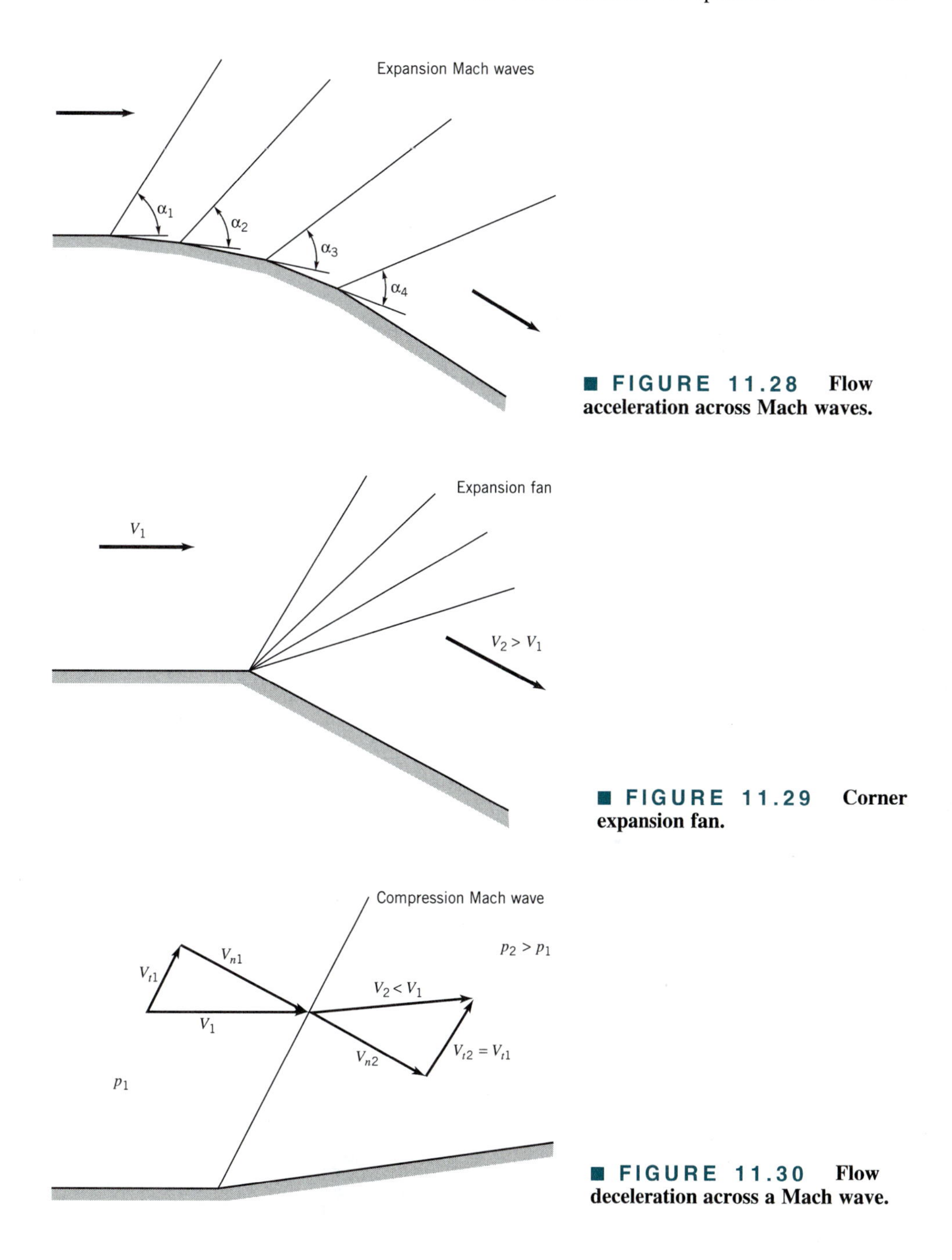

■ FIGURE 11.28 Flow acceleration across Mach waves.

■ FIGURE 11.29 Corner expansion fan.

■ FIGURE 11.30 Flow deceleration across a Mach wave.

Supersonic flows decelerate across compression Mach waves.

indicated in Fig. 11.31, several Mach waves occur, each at an appropriately larger angle α with the upstream wall. A rounded compression corner may be considered as a series of infinitesimal changes in direction and even sharp corners are actually rounded. Mach waves or compression waves can coalesce to form an oblique shock wave as shown in Fig. 11.32.

The above discussion of compression waves can be usefully extended to supersonic flow impinging on an object. For example, for supersonic flow incident on a wedge-shaped leading edge (see Fig. 11.33), an attached oblique shock can form as suggested in Fig. 11.33*a*. For the same incident Mach number but with a larger wedge angle, a detached curved shock as sketched in Fig. 11.33*b* can result. A detached, curved shock ahead of a blunt object (a

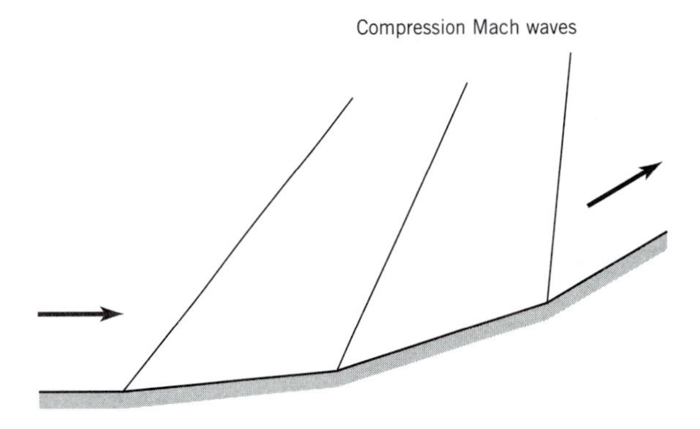

■ FIGURE 11.31 **Flow deceleration across Mach waves.**

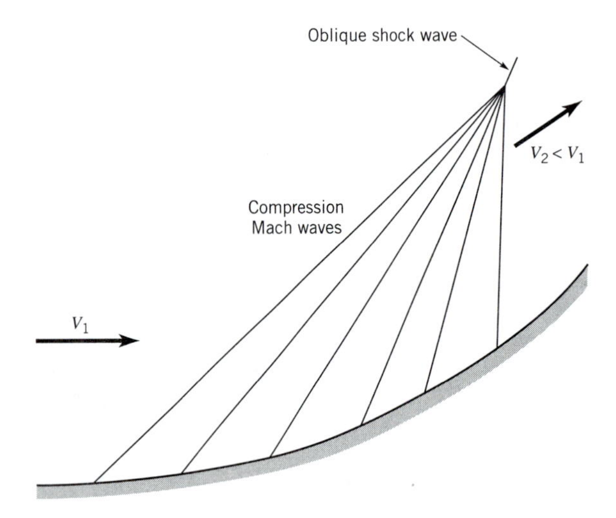

■ FIGURE 11.32 **Oblique shock wave.**

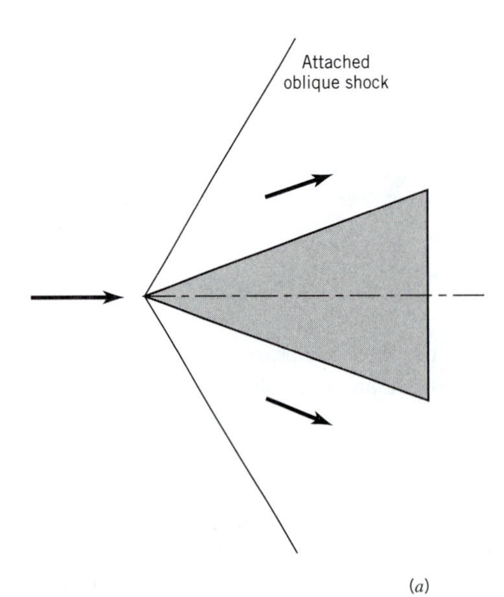

■ FIGURE 11.33 **Supersonic flow over a wedge: *(a)* Smaller wedge angle results in attached oblique shock.**

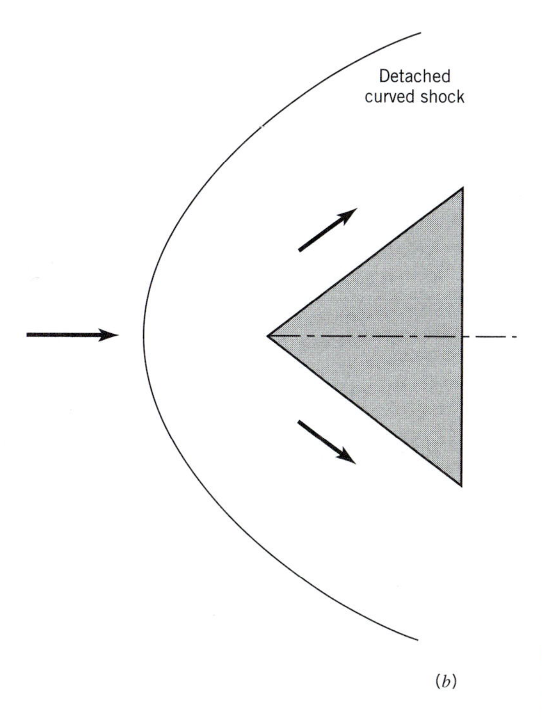

■ FIGURE 11.33 *(b)* **Large wedge angle results in detached curved shock.**

sphere) is shown in the photograph on page 698. In Example 11.19, we considered flow along a stagnation pathline across a detached curved shock to be identical to flow across a normal shock wave.

From this brief look at two-dimensional supersonic flow, one can easily conclude that the extension of these concepts to flows over immersed objects and within ducts can be exciting, especially if three-dimensional effects are considered. Some references that provide much more on this subject than could be included here are Refs. 6, 7, 8, and 9.

References

1. Coles, D., ''Channel Flow of a Compressible Fluid,'' Summary description of film in *Illustrated Experiments in Fluid Mechanics, The NCFMF Book of Film Notes*, MIT Press, Cambridge, Mass., 1972.
2. Jones, J. B., and Hawkins, G. A., *Engineering Thermodynamics*, 2nd Ed., Wiley, New York, 1986.
3. Moran, M. J., and Shapiro, H. N., *Fundamentals of Engineering Thermodynamics*, 2nd Ed., Wiley, New York, 1992.
4. Keenan, J. H., Chao, J., and Kaye, J., *Gas Tables*, 2nd Ed., Wiley, New York, 1980.
5. Shapiro, A. H., *The Dynamics and Thermodynamics of Compressible Fluid Flow*, Vol. 1, Ronald Press, New York, 1953.
6. Thompson, P. A., *Compressible-Fluid Dynamics*, McGraw-Hill, New York, 1972.
7. Zuchrow, M. J., and Hoffman, J. D., *Gas Dynamics*, Vol. 1, Wiley, New York, 1976.
8. Saad, M. A., *Compressible Fluid Flow*, 2nd Ed., Prentice-Hall, Englewood Cliffs, N.J., 1993.
9. Anderson, J. D., Jr., *Modern Compressible Flow with Historical Perspective*, 2nd Ed., McGraw-Hill, New York, 1990.

Review Problems

Note: Problems designated with (R) are review problems. The phrases within parentheses refer to the main topics to be used in solving the problems. Complete, detailed solutions to these review problems can be found in the supplement titled *Student Solution Manual for Fundamentals of Fluid Mechanics*, by Munson, Young, and Okiishi (John Wiley and Sons, New York, 1997).

11.1R (Speed of sound) Determine the speed of sound in air for a hot summer day when the temperature is 100 °F; for a cold winter day when the temperature is −20 °F.

(ANS: 1160 ft/s; 1028 ft/s)

11.2R (Speed of sound) Compare values of the speed of sound in ft/s in the following liquids at 68 °F: **(a)** ethyl alcohol, **(b)** glycerin, **(c)** mercury.

(ANS: 3810 ft/s; 6220 ft/s; 4760 ft/s)

11.3R (Sound waves) A stationary point source emits weak pressure pulses in a flow that moves uniformly from left to right with a Mach number of 0.5. Sketch the instantaneous outline at time = 10s of pressure waves emitted earlier at time = 5s and time = 8s. Assume the speed of sound is 1000 ft/s.

11.4R (Mach number) An airplane moves forward in air with a speed of 500 mph at an altitude of 40,000 ft. Determine the Mach number involved if the air is considered as U.S. standard atmosphere (see Table C.1).

(ANS: 0.757)

11.5R (Isentropic flow) At section (1) in the isentropic flow of carbon dioxide, p_1 = 40 kPa(abs), T_1 = 60 °C, and V_1 = 350 m/s. Determine the flow velocity, V_2, in m/s, at another section, section (2), where the Mach number is 2.0. Also calculate the section area ratio, A_2/A_1.

(ANS: 500 m/s; 1.71)

11.6R (Isentropic flow) An ideal gas in a large storage tank at 100 psia and 60 °F flows isentropically through a converging duct to the atmosphere. The throat area of the duct is 0.1 ft^2. Determine the pressure, temperature, velocity, and mass flowrate of the gas at the duct throat if the gas is **(a)** air, **(b)** carbon dioxide, **(c)** helium.

(ANS: **(a)** 52.8 psia; 433 °R; 1020 ft/s; 1.04 slugs/s; **(b)** 54.6 psia; 452 °R; 815 ft/s; 1.25 slugs/s; **(c)** 48.8 psia; 391 °R; 2840 ft/s; 0.411 slugs/s)

11.7R (Fanno flow) A long, smooth wall pipe (f = 0.01) is to deliver 8000 ft^3/min of air at 60 °F and 14.7 psia. The inside diameter of the pipe is 0.5 ft and the length of the pipe is 100 ft. Determine the static temperature and pressure required at the pipe entrance if the flow through the pipe is adiabatic.

(ANS: 539 °R; 23.4 psia)

11.8R (Rayleigh flow) Air enters a constant-area duct that may be considered frictionless with T_1 = 300 K and V_1 = 300 m/s. Determine the amount of heat transfer in kJ/kg required to choke the Rayleigh flow involved.

(ANS: 5020 J/kg)

11.9R (Normal shock waves) Standard atmospheric air enters subsonically and accelerates isentropically to supersonic flow in a duct. If the ratio of duct exit to throat cross-section areas is 3.0, determine the ratio of back pressure to inlet stagnation pressure that will result in a standing normal shock at the duct exit. Determine also the stagnation pressure loss across the normal shock in kPa.

(ANS: 0.375; 56.1 kPa)

Problems

Note: Unless otherwise indicated, use the values of fluid properties found in the tables on the inside of the front cover. Problems designated with an (*) are intended to be solved with the aid of a programmable calculator or a computer. If k = 1.4 and the figures of Appendix D can be used to simplify a problem solution. Problems designated with a (†) are ''open-ended'' problems and require critical thinking in that to work them one must make various assumptions and provide the necessary data. There is not a unique answer to these problems.

11.1 Determine, in SI units, nominal values of c_p and c_v for: **(a)** air, **(b)** carbon dioxide, **(c)** helium, **(d)** hydrogen, **(e)** methane, **(f)** nitrogen, **(g)** oxygen. Use information provided in Table 1.8.

11.2 Determine, in EE units, nominal values of c_p and c_v for: **(a)** air, **(b)** carbon dioxide, **(c)** helium, **(d)** hydrogen, **(e)** methane, **(f)** nitrogen, **(g)** oxygen. Use information provided in Table 1.7.

11.3 Air flows steadily between two sections in a duct. At section (1), the temperature and pressure are T_1 = 80 °C, p_1 = 301 kPa(abs), and at section (2), the temperature and pressure are T_2 = 180 °C, p_2 = 181 kPa(abs). Calculate the **(a)** change in internal energy between sections (1) and (2), **(b)** change in enthalpy between sections (1) and (2), **(c)** change in density between sections (1) and (2), **(d)** change in entropy between sections (1) and (2).

11.4 Three kilograms of hydrogen contained in a nondeforming sealed vessel are cooled from 400 °C, 400 kPa(abs) until the hydrogen pressure is 100 kPa(abs). Calculate the change in internal energy, enthalpy, and entropy associated with this cooling process.

11.5 Five pounds mass of air are heated in a closed, rigid container from 80 °F and 15 psia to 500 °F. Determine the final pressure of the air and the entropy rise involved.

11.6 Helium is compressed isothermally from 121 kPa(abs) to 301 kPa(abs). Determine the entropy change associated with this process.

11.7 Air at 14.7 psia and 70 °F is compressed adiabatically by a centrifugal compressor to a pressure of 60 psia. What is the minimum temperature rise possible? Explain.

11.8 Methane is compressed adiabatically from 100 kPa(abs) and 25 °C to 200 kPa(abs). What is the minimum compressor exit temperature possible? Explain.

11.9 Air expands adiabatically through a turbine from a pressure and temperature of 80 psia, 1600 °R to a pressure of 14.7 psia. If the actual temperature change is 85% of the ideal temperature change, determine the actual temperature of the expanded air and the actual enthalpy and entropy differences across the turbine.

11.10 An expression for the value of c_p for carbon dioxide as a function of temperature is

$$c_p = 9210 - \frac{3.71 \times 10^6}{T} + \frac{8.02 \times 10^8}{T^2}$$

where c_p is in (ft·lb)/(slug·°R) and T is in °R. Compare the change in enthalpy of carbon dioxide using the constant value of c_p from Problem 11.2 with the change in enthalpy of carbon dioxide using the expression above, for $T_2 - T_1$ equal to **(a)** 10 °R, **(b)** 1000 °R, **(c)** 3000 °R. Set $T_1 = 540$ °R.

11.11 Confirm the speed of sound for air at 50 °F listed in Table B.3.

11.12 From Table B.1 we can conclude that the speed of sound in water at 60 °F is 4814 ft/s. Is this value of c consistent with the value of bulk modulus, E_v, listed in Table 1.5?

11.13 Determine the Mach number of a car moving in standard air at a speed of **(a)** 25 mph, **(b)** 55 mph, and **(c)** 100 mph.

11.14 Using information provided in Table C.2, develop a table of speed of sound in m/s as a function of elevation for U.S. standard atmosphere.

† **11.15** Estimate the Mach number for swimming fish and discuss your results.

11.16 If a high-performance aircraft is able to cruise at a Mach number of 3.0 at an altitude of 80,000 ft, how fast is this in **(a)** mph, **(b)** ft/s, **(c)** m/s?

11.17 Compare values of the speed of sound in m/s at 20°C in the following gases: **(a)** air, **(b)** carbon dioxide, **(c)** helium, **(d)** hydrogen, **(e)** methane.

11.18 If a person inhales helium and then talks, his or her voice sounds like "Donald Duck." Explain why this happens.

11.19 Explain how you could vary the Mach number but not the Reynolds number in air flow past a sphere. For a constant Reynolds number of 300,000, estimate how much the drag coefficient will increase as the Mach number is increased from 0.3 to 1.0.

11.20 The flow of an ideal gas may be considered incompressible if the Mach number is less than 0.3. Determine the velocity level in ft/s and in m/s for Ma = 0.3 in the following gases: **(a)** standard air, **(b)** hydrogen at 68 °F.

11.21 At the seashore, you observe a high-speed aircraft moving overhead at an elevation of 10,000 ft. You hear the plane 8 s after it passes directly overhead. Using a nominal air temperature of 40 °F, estimate the Mach number and speed of the aircraft.

11.22 A schlieren photo of a bullet moving through air at 14.7 psia and 68 °F indicates a Mach cone angle of 28°. How fast was the bullet moving in: **(a)** m/s, **(b)** ft/s, **(c)** mph?

11.23 At a given instant of time, two of the pressure waves, each moving at the speed of sound, emitted by a point source moving with constant velocity in a fluid at rest are shown in Fig. P11.23. Determine the Mach number involved and indicate with a sketch the instantaneous location of the point source.

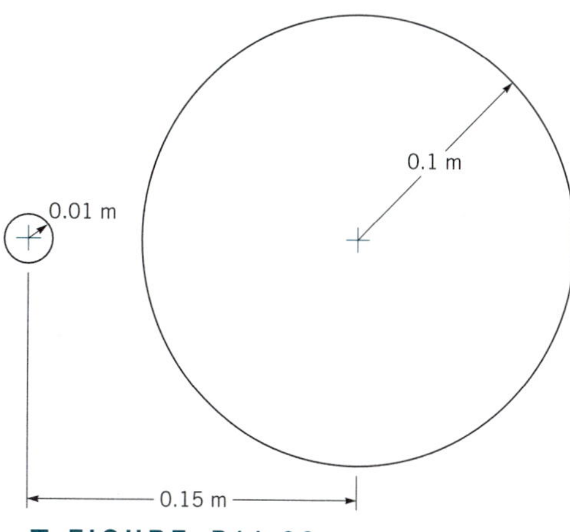

■ **FIGURE P11.23**

11.24 At a given instant of time, two of the pressure waves, each moving at the speed of sound, emitted by a point source moving with constant velocity in a fluid at rest, are shown in Fig. P11.24. Determine the Mach number involved and indicate with a sketch the instantaneous location of the point source.

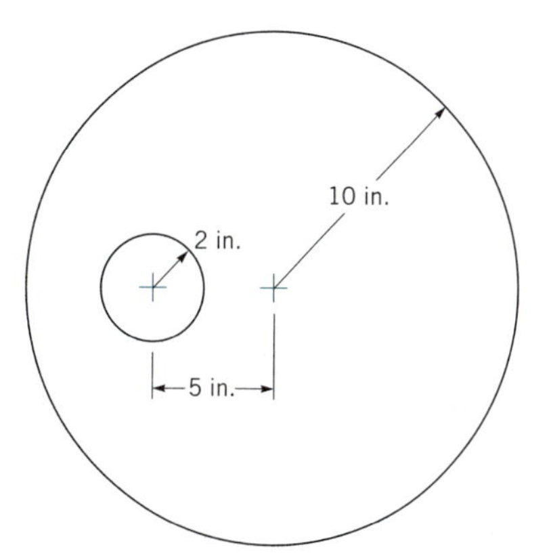

■ **FIGURE P11.24**

11.25 At a given instant of time, a pressure wave moving at the speed of sound, which is emitted by a point source moving with constant velocity along the line A–B in a fluid at rest, is shown in Fig. P11.25. If the point source Mach number is 0.8, sketch, for the given instant, the pressure wave emitted by the point source when it was at location B.

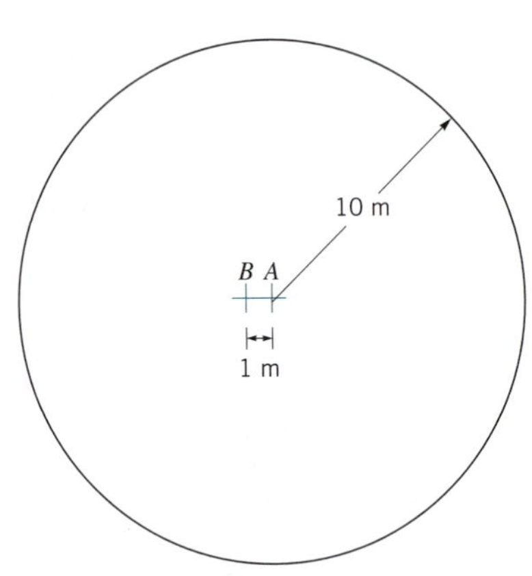

■ FIGURE P11.25

11.26 Sound waves are very small amplitude pressure pulses that travel at the ''speed of sound.'' Do very large amplitude waves such as a blast wave caused by an explosion travel less than, equal to, or greater than the speed of sound? Explain.

11.27 Starting with Eq. 11.52, prove that the stagnation enthalpy (and temperature) of an ideal gas remains constant during isentropic flow. Does this result fix the stagnation state for the flow? Explain. Using Eq. 5.69, comment on heat transfer and shaft work during any constant stagnation state process.

11.28 Explain how fluid pressure varies with cross-section area change for the isentropic flow of an ideal gas when the flow is **(a)** subsonic, **(b)** supersonic.

11.29 For any ideal gas, prove that the slope of constant pressure lines on a temperature–entropy diagram is positive and that higher pressure lines are above lower pressure lines.

11.30 Determine the critical pressure and temperature ratios for **(a)** air, **(b)** carbon dioxide, **(c)** helium, **(d)** hydrogen, **(e)** methane, **(f)** nitrogen, **(g)** oxygen.

11.31 Air flows steadily and isentropically from standard atmospheric conditions to a receiver pipe through a converging duct. The cross-section area of the throat of the converging duct is 0.05 ft^2. Determine the mass flowrate through the duct if the receiver pressure is **(a)** 10 psia, **(b)** 5 psia. Sketch temperature–entropy diagrams for situations (a) and (b). Verify results obtained with values from the appropriate graph in Appendix D with calculations involving ideal gas equations.

11.32 Helium at 68 °F and 14.7 psia in a large tank flows steadily and isentropically through a converging nozzle to a receiver pipe. The cross-section area of the throat of the converging passage is 0.05 ft^2. Determine the mass flowrate through the duct if the receiver pressure is **(a)** 10 psia, **(b)** 5 psia. Sketch temperature–entropy diagrams for situations (a) and (b).

11.33 What is the static pressure to stagnation pressure ratio associated with the following motion in standard air: **(a)** a runner moving at the rate of 20 mph, **(b)** a cyclist moving at the rate of 40 mph, **(c)** a car moving at the rate of 65 mph, **(d)** an airplane moving at the rate of 500 mph.

11.34 The static pressure to stagnation pressure ratio at a point in an ideal gas flow field is measured with a Pitot-static probe as being equal to 0.6. The stagnation temperature of the gas is 20 °C. Determine the flow speed in m/s and the Mach number if the gas is **(a)** air, **(b)** carbon dioxide, **(c)** hydrogen.

11.35 The stagnation pressure and temperature of air flowing past a probe are 120 kPa(abs) and 100 °C, respectively. The air pressure is 80 kPa(abs). Determine the air speed and the Mach number considering the flow to be **(a)** incompressible, **(b)** compressible.

11.36 The stagnation pressure indicated by a Pitot tube mounted on an airplane in flight is 45 kPa(abs). If the aircraft is cruising in standard atmosphere at an altitude of 10,000 m, determine the speed and Mach number involved.

11.37 According to the energy equation, the stagnation temperature increases with speed (see Eq. 11.56). Thus, as you ride your bike on a cold day the temperature of the end of your nose (a stagnation point) should increase as you ride faster. Obviously the opposite occurs. Explain why this happens.

***11.38** An ideal gas enters subsonically and flows isentropically through a choked converging-diverging duct having a circular cross-section area A that varies with axial distance from the throat, x, according to the formula

$$A = 0.1 + x^2$$

where A is in square feet and x is in feet. For this flow situation, sketch the side view of the duct and graph the variation of Mach number, static temperature to stagnation temperature ratio, T/T_0, and static pressure to stagnation pressure ratio, p/p_0, through the duct from $x = -0.6$ ft to $x = +0.6$ ft. Also show the possible fluid states at $x = -0.6$ ft, 0 ft, and $+0.6$ ft using temperature-entropy coordinates. Consider the gas as being helium (use $0.051 \leq \text{Ma} \leq 5.193$).

***11.39** An ideal gas enters supersonically and flows isentropically through the choked converging-diverging duct described in Problem 11.38. Graph the variation of Ma, T/T_0, and p/p_0 from the entrance to the exit sections of the duct for helium (use $0.051 \leq \text{Ma} \leq 5.193$). Show the possible fluid states at $x = -0.6$ ft, 0 ft, and $+0.6$ ft using temperature–entropy coordinates.

***11.40** Helium enters supersonically and flows isentropically through the choked converging-diverging duct described in Example 11.8. Compare the variation of Ma, T/T_0, and p/p_0 for helium with the variation of these parameters for air throughout the duct. Use $0.163 \leq \text{Ma} \leq 3.221$.

*11.41 Helium enters subsonically and flows isentropically through the converging-diverging duct of Example 11.8. Compare the values of Ma, T/T_0, and p/p_0 for helium with those for air at several locations in the duct. Use $0.163 \leq \text{Ma} \leq 3.221$.

*11.42 Helium flows subsonically and isentropically through the converging-diverging duct of Example 11.8. Graph the variation of Ma, T/T_0, and p/p_0 through the duct from $x = -0.5$ m to $x = +0.5$ m for $p/p_0 = 0.99$ at $x = -0.5$ m. Sketch the corresponding $T - s$ diagram. Use $0.110 \leq \text{Ma} \leq 0.430$.

11.43 An ideal gas flows subsonically and isentropically through the converging-diverging duct described in Problem 11.38. Graph the variation of Ma, T/T_0, and p/p_0 from the entrance to the exit sections of the duct for **(a)** air, **(b*)** helium (use $0.047 \leq \text{Ma} \leq 0.722$). The value of p/p_0 is 0.6708 at $x = 0$ ft. Sketch important states on a $T - s$ diagram.

11.44 An ideal gas contained in a large storage container at a constant temperature and pressure of 59 °F and 25 psia is to be expanded isentropically through a duct to standard atmospheric discharge conditions. Describe in general terms the kind of duct required and determine the duct exit cross-section area if the discharge mass flowrate required is 1.0 lbm/s and the gas is **(a)** air, **(b)** carbon dioxide, **(c)** helium.

11.45 An ideal gas is to flow isentropically from a large tank where the air is maintained at a temperature and pressure of 59 °F and 80 psia to standard atmospheric discharge conditions. Describe in general terms the kind of duct involved and determine the duct exit Mach number and velocity in ft/s if the gas is **(a)** air, **(b)** methane, **(c)** helium.

11.46 An ideal gas flows isentropically through a converging-diverging nozzle. At a section in the converging portion of the nozzle, $A_1 = 0.1$ m^2, $p_1 = 600$ kPa(abs), $T_1 = 20$ °C, and $\text{Ma}_1 = 0.6$. For section (2) in the diverging part of the nozzle, determine A_2, p_2, and T_2 if $\text{Ma}_2 = 3.0$ and the gas is **(a)** air, **(b)** helium.

11.47 Upstream of the throat of an isentropic converging–diverging nozzle at section (1), $V_1 = 150$ m/s, $p_1 = 100$ kPa(abs), and $T_1 = 20$ °C. If the discharge flow is supersonic and the throat area is 0.1 m^2, determine the mass flowrate in kg/s for the flow of **(a)** air, **(b)** methane, **(c)** helium.

11.48 The flow blockage associated with the use of an intrusive probe can be important. Determine the percentage increase in section velocity corresponding to a 0.5% reduction in flow area due to probe blockage for air flow if the section area is 1.0 m^2, $T_0 = 20$ °C, and the unblocked flow Mach numbers are **(a)** Ma = 0.2, **(b)** Ma = 0.8, **(c)** Ma = 1.5, **(d)** Ma = 30.

11.49 An ideal gas enters [section (1)] an insulated, constant cross-section area duct with the following properties:

$$T_0 = 293 \text{ K}$$

$$p_0 = 101 \text{ kPa(abs)}$$

$$\text{Ma}_1 = 0.2$$

For Fanno flow, determine corresponding values of fluid temperature and entropy change for various levels of pressure and plot the Fanno line if the gas is helium.

11.50 For Fanno flow, prove that

$$\frac{dV}{V} = \frac{fk(\text{Ma}^2/2)(dx/D)}{1 - \text{Ma}^2}$$

and in so doing show that when the flow is subsonic, friction accelerates the fluid, and when the flow is supersonic, friction decelerates the fluid.

11.51 Standard atmospheric air ($T_0 = 59$ °F, $p_0 = 14.7$ psia) is drawn steadily through a frictionless and adiabatic converging nozzle into an adiabatic, constant cross-section area duct. The duct is 10 ft long and has an inside diameter of 0.5 ft. The average friction factor for the duct may be estimated as being equal to 0.03. What is the maximum mass flowrate in slugs/s through the duct? For this maximum flowrate, determine the values of static temperature, static pressure, stagnation temperature, stagnation pressure, and velocity at the inlet [section (1)] and exit [section (2)] of the constant-area duct. Sketch a temperature-entropy diagram for this flow.

11.52 An ideal gas [$T_0 = 288$ K and $p_0 = 101$ kPa(abs)] is drawn steadily through the flow configuration of Example 11.12. Compare the maximum mass flowrate and constant-area duct inlet [section (1)] and exit [section (2)] values of static temperature, static pressure, stagnation temperature, stagnation pressure, and velocity for **(a)** helium and **(b)** carbon dioxide with the quantities for air from Example 11.12.

11.53 The duct in Problem 11.51 is shortened by 50%. The duct discharge pressure is maintained at the choked flow value determined in Problem 11.51. Determine the change in mass flowrate through the duct associated with the 50% reduction in length. The average friction factor remains constant at a value of 0.03.

11.54 If the same mass flowrate of air obtained in Problem 11.51 is desired through the shortened duct of Problem 11.53, determine the back pressure, p_2, required. Assume f remains constant at a value of 0.03.

11.55 If the average friction factor of the duct of Example 11.12 is changed to **(a)** 0.01 or **(b)** 0.03, determine the maximum mass flowrate of air through the duct associated with each new friction factor; compare with the maximum mass flowrate value of Example 11.12.

11.56 If the length of the constant-area duct of Example 11.12 is changed to **(a)** 1 m or **(b)** 3 m, and all other specifications in the problem statement remain the same, determine the maximum mass flowrate of air through the duct associated with each new length; compare with the maximum mass flowrate of Example 11.12.

11.57 The duct of Example 11.12 is lengthened by 50%. If the duct discharge pressure is set at a value of $p_d = 46.2$ kPa(abs), determine the mass flowrate of air through the lengthened duct. The average friction factor for the duct remains constant at a value of 0.02.

11.58 An ideal gas flows adiabatically with friction through a long, constant-area pipe. At upstream section (1), $p_1 = 60$ kPa(abs), $T_1 = 60$ °C, and $V_1 = 200$ m/s. At downstream section (2), $T_2 = 30$ °C. Determine p_2, V_2, and the stagnation pressure ratio $p_{0,2}/p_{0,1}$ if the gas is **(a)** air, **(b)** helium.

11.59 For the air flow of Problem 11.58, determine T, p, and V for the section halfway between sections (1) and (2).

11.60 An ideal gas flows adiabatically between two sections in a constant-area pipe. At upstream section (1), $p_{0,1} =$ 100 psia, $T_{0,1} = 600$ °R, and $\text{Ma}_1 = 0.5$. At downstream section (2), the flow is choked. Determine the magnitude of the force per unit cross-section area exerted by the inside wall of the pipe on the fluid between sections (1) and (2) if the gas is **(a)** air, **(b)** helium.

11.61 An ideal gas enters [section (1)] a frictionless, constant-area duct with the following properties:

$$T_0 = 293 \text{ K}$$

$$p_0 = 101 \text{ kPa(abs)}$$

$$\text{Ma}_1 = 0.2$$

For Rayleigh flow, determine corresponding values of fluid temperature and entropy change for various levels of pressure and plot the Rayleigh line if the gas is helium.

11.62 Standard atmospheric air [$T_0 = 288$ K, $p_0 = 101$ kPa(abs)] is drawn steadily through an isentropic converging nozzle into a frictionless diabatic ($q = 500$ kJ/kg) constant-area duct. For maximum flow, determine the values of static temperature, static pressure, stagnation temperature, stagnation pressure, and flow velocity at the inlet [section (1)] and exit [section (2)] of the constant-area duct. Sketch a temperature-entropy diagram for this flow.

11.63 An ideal gas enters a 0.5-ft inside diameter duct with $p_1 = 20$ psia, $T_1 = 80$ °F, and $V_1 = 200$ ft/s. What frictionless heat addition rate in Btu/s is necessary for an exit gas temperature $T_2 = 1500$ °F? Determine p_2, V_2, and Ma_2 also. The gas is **(a)** air, **(b)** helium.

11.64 Air enters a length of constant-area pipe with $p_1 =$ 200 kPa(abs), $T_1 = 500$ K, and $V_1 = 400$ m/s. If 500 kJ/kg of energy is removed from the air by frictionless heat transfer between sections (1) and (2), determine p_2, T_2, and V_2. Sketch a temperature-entropy diagram for the flow between sections (1) and (2).

11.65 Air flows through a constant-area pipe. At an upstream section (1), $p_1 = 15$ psia, $T_1 = 530$ °R, and $V_1 =$ 200 ft/s. Downstream at section (2), $p_2 = 10$ psia and $T_2 =$ 1760 °R. For this flow, determine the stagnation temperature and pressure ratios, $T_{0,2}/T_{0,1}$ and $p_{0,2}/p_{0,1}$, and the heat transfer per unit mass of air flowing between sections (1) and (2). Is the flow between sections (1) and (2) frictionless? Explain.

11.66 The Mach number and stagnation pressure of an ideal gas are 2.0 and 200 kPa(abs) just upstream of a normal shock. Determine the stagnation pressure loss across the shock for the following gases: **(a)** air, **(b)** helium. Comment on the effect of specific heat ratio, k, on shock loss.

11.67 The stagnation pressure ratio across a normal shock in an ideal gas flow is 0.8. Determine the Mach number of the flow entering the shock if the gas is air.

11.68 Just upstream of a normal shock in an ideal gas flow, $\text{Ma} = 3.0$, $T = 600$ °R, and $p = 30$ psia. Determine values of Ma, T_0, T, p_0, p, and V downstream of the shock if the gas is **(a)** air, **(b)** helium.

11.69 A total pressure probe is inserted into a supersonic air flow. A shock wave forms just upstream of the impact hole. The probe measures a total pressure of 500 kPa(abs). The stagnation temperature at the probe head is 500 K. The static pressure upstream of the shock is measured with a wall tap to be 100 kPa(abs). From these data, determine the Mach number and velocity of the flow.

11.70 The Pitot tube on a supersonic aircraft cruising at an altitude of 30,000 ft senses a stagnation pressure of 12 psia. If the atmosphere is considered standard, determine the air speed and Mach number of the aircraft. A shock wave is present just upstream of the probe impact hole.

11.71 An aircraft cruises at a Mach number of 2.0 at an altitude of 15 km. Inlet air is decelerated to a Mach number of 0.4 at the engine compressor inlet. A normal shock occurs in the inlet diffuser upstream of the compressor inlet at a section where the Mach number is 1.2. For isentropic diffusion, except across the shock, and for standard atmosphere, determine the stagnation temperature and pressure of the air entering the engine compressor.

11.72 Determine, for the air flow through the frictionless and adiabatic converging-diverging duct of Example 11.8, the ratio of duct exit pressure to duct inlet stagnation pressure that will result in a standing normal shock at: **(a)** $x = +0.1$ m, **(b)** $x = +0.2$ m, **(c)** $x = +0.4$ m. How large is the stagnation pressure loss in each case?

11.73 A normal shock is positioned in the diverging portion of a frictionless, adiabatic, converging-diverging air flow duct where the cross-section area is 0.1 ft^2 and the local Mach number is 2.0. Upstream of the shock, $p_0 = 200$ psia and $T_0 =$ 1200 °R. If the duct exit area is 0.15 ft^2, determine the exit area temperature and pressure and the duct mass flowrate.

11.74 Supersonic air flow enters an adiabatic, constant-area (inside diameter = 1 ft) 30-ft-long pipe with $\text{Ma}_1 = 3.0$. The pipe friction factor is estimated to be 0.02. What ratio of pipe exit pressure to pipe inlet stagnation pressure would result in a normal shock wave standing at **(a)** $x = 5$ ft, or **(b)** $x = 10$ ft, where x is the distance downstream from the pipe entrance? Determine also the duct exit Mach number and sketch the temperature-entropy diagram for each situation.

11.75 Supersonic ideal gas flow enters an adiabatic, constant-area pipe (inside diameter = 0.1 m) with $\text{Ma}_1 = 2.0$. The pipe friction factor is 0.02. If a standing normal shock is located right at the pipe exit, and the Mach number just upstream of the shock is 1.2, determine the length of the pipe if the gas is **(a)** air, **(b)** helium.

11.76 Air enters a frictionless, constant-area duct with $\text{Ma}_1 = 2.0$, $T_{0,1} = 59$ °F, and $p_{0,1} = 14.7$ psia. The air is decelerated by heating until a normal shock wave occurs where the local Mach number is 1.5. Downstream of the normal shock, the subsonic flow is accelerated with heating until it chokes at the duct exit. Determine the static temperature and pressure, the stagnation temperature and pressure, and the fluid velocity at the duct entrance, just upstream and downstream of the normal shock, and at the duct exit. Sketch the temperature-entropy diagram for this flow.

11.77 An ideal gas enters a frictionless, constant-area duct with $\text{Ma} = 2.5$, $T_0 = 20$ °C, and $p_0 = 101$ kPa(abs). The gas is decelerated by heating until a normal shock occurs where the local Mach number is 1.3. Downstream of the shock, the subsonic flow is accelerated with heating until it exits with a Mach number of 0.9. Determine the static temperature and pressure, the stagnation temperature and pressure, and the fluid velocity at the duct entrance, just upstream and downstream of the normal shock, and at the duct exit if the gas is **(a)** air or **(b)** helium. Sketch the temperature-entropy diagram for each flow.

11.78 Discuss the similarities between hydraulic jumps in open-channel flow and shock waves in compressible flow.

† **11.79** Estimate the surface temperature associated with the re-entry of the space shuttle into the earth's atmosphere. List all assumptions and show calculations.

† **11.80** Estimate the maximum temperature developed on the leading edge of the wings of a stealth fighter airplane. List all assumptions and show calculations.

Laser velocimeter measurements of the flow field in the rotor row of a low-speed research turbine. (Photograph courtesy of Dr. D. C. Wisler, Director, Aerodynamic Research Laboratory of GE Aircraft Engines.)

12 Turbomachines

In previous chapters we often used generic "black boxes" to represent fluid machines such as pumps or turbines. The purpose of this chapter is to understand from a fluid mechanics standpoint how these devices work.

Pumps and turbines (sometimes called *fluid machines*) occur in a wide variety of configurations. In general, pumps add energy to the fluid—they do work on the fluid; turbines extract energy from the fluid—the fluid does work on them. The term "pump" will be used to generically refer to all pumping machines, including *pumps*, *fans*, *blowers*, and *compressors*. Fluid machines can be divided into two main categories: *positive displacement machines* (denoted as the static type) and *turbomachines* (denoted as the dynamic type). The majority of this chapter deals with turbomachines.

Turbomachines are dynamic fluid machines that add (for pumps) or extract (for turbines) flow energy.

Positive displacement machines force a fluid into or out of a chamber by changing the volume of the chamber. The pressures developed and the work done are a result of essentially static forces rather than dynamic effects. Typical examples shown in Fig. 12.1 include the common tire pump used to fill bicycle tires, the human heart, and the gear pump. In these cases the device does work on the fluid (the container wall moves against the fluid pressure force on the moving wall). The internal combustion engine in your car is a positive-displacement machine in which the fluid does work on the machine, the opposite of what happens in a pump. In the car engine the piston moves in the direction of the fluid pressure force acting on the piston face during the power stroke.

Turbomachines, on the other hand, involve a collection of blades, buckets, flow channels, or passages arranged around an axis of rotation to form a rotor. Rotation of the rotor produces dynamic effects that either add energy to the fluid or remove energy from the fluid. Examples of turbomachine-type pumps include simple window fans, propellers on ships or airplanes, squirrel-cage fans on home furnaces, axial-flow water pumps used in deep wells, and compressors in automobile turbochargers. Examples of turbines include the turbine portion of gas turbine engines on aircraft, steam turbines used to drive generators at electrical generation stations, and the small, high-speed air turbines that power dentist drills.

Turbomachines serve in an enormous array of applications in our daily lives and thus play an important role in modern society. These machines can have a high power density

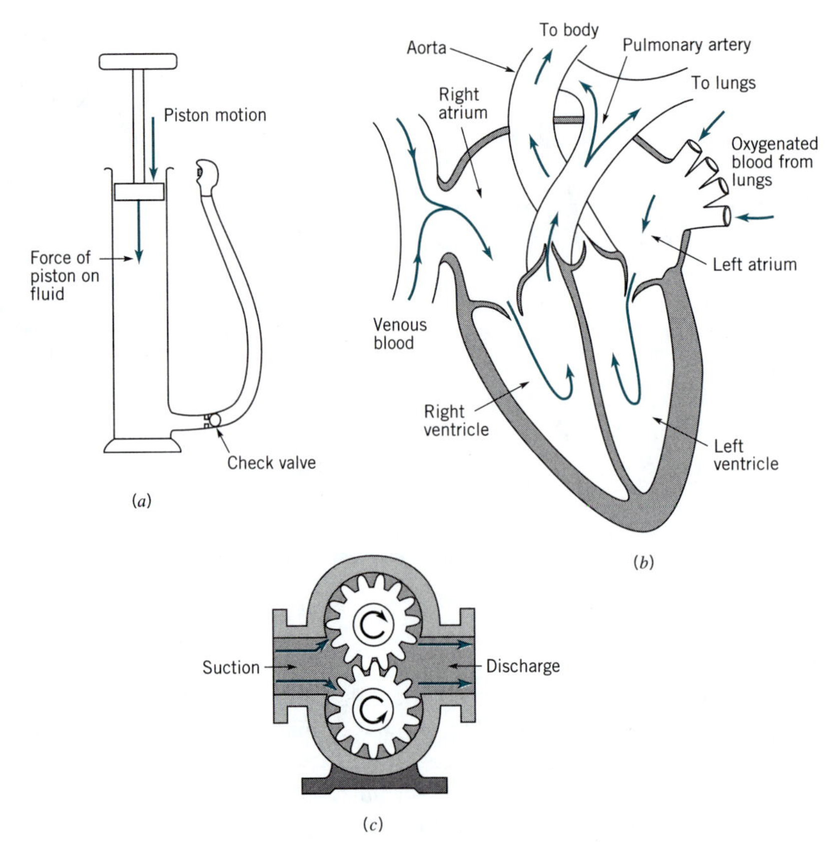

■ **FIGURE 12.1** **Typical positive displacement pumps: *(a)* tire pump, *(b)* human heart, *(c)* gear pump.**

(large power output per size), relatively few moving parts, and reasonable efficiency. The following sections provide a beginning to the understanding of these important machines. For additional information, the interested reader is encouraged to turn to the numerous references devoted entirely to turbomachines, including Refs. 1–6.

12.1 Introduction

Turbomachines are mechanical devices that either extract energy from a fluid (turbine) or add energy to a fluid (pump) as a result of dynamic interactions between the device and the fluid. While the actual design and construction of these devices often require considerable insight and effort, their basic operating principles are quite simple.

Turbomachines involve the related parameters of force, work, and power.

The dynamic interaction between a fluid and a solid is often based on flow and fluid/solid interaction forces. For example, it is clear that we do work with our muscles when we move a spoon through a cup of tea. The motion of the spoon through the tea causes a dynamic pressure difference between the front and back sides of the spoon, which produces a force on the spoon that we must overcome with our muscles. This force acting through a distance requires a specific amount of work from us; this work done in a given time period translates into a given power transfer. We do work on the fluid—we have a crude pump adding energy to the fluid, in this case to enhance tea steeping.

Conversely, the dynamic effect of the wind blowing past the sail on a boat creates pressure differences on the sail. The wind force on the moving sail in the direction of the boat's motion provides power to propel the boat. The sail and boat act as a machine extracting energy from the air.

Turbomachines operate on the principles described above. Rather than one spoon or one sail, a group of blades, airfoils, "buckets," flow channels, or passages is attached to a rotating shaft. Energy is either supplied to the rotating shaft (by a motor, for example) and transferred to the fluid by the blades (a pump), or the energy is transferred from the fluid to the blades and made available at the rotating shaft as shaft power (a turbine).

A group of blades moving with or against a lift force is the essence of a turbomachine.

The fluid used can be either a gas (as with a window fan or a gas turbine engine) or a liquid (as with the water pump on a car or a turbine at a hydroelectric power plant). While the basic operating principles are the same whether the fluid is a liquid or a gas, important differences in the fluid dynamics involved can occur. For example, cavitation may be an important design consideration when liquids are involved if the pressure at any point within the flow is reduced to the vapor pressure. Compressibility effects may be important when gases are involved if the Mach number becomes large enough.

Many turbomachines contain some type of housing or casing that surrounds the rotating blades or rotor, thus forming an internal flow passageway through which the fluid flows (see Fig. 12.2). Others, such as a windmill or a window fan, are unducted. Some turbomachines include stationary blades or vanes in addition to rotor blades. These stationary vanes can be arranged to accelerate the flow and thus serve as nozzles. Or, these vanes can be set to diffuse the flow and act as diffusers.

Turbomachines are classified as *axial-flow*, *mixed-flow*, or *radial-flow* machines depending on the predominant direction of the fluid motion relative to the rotor's axis as the fluid passes the blades (see Fig. 12.2). For an axial-flow machine the fluid maintains a significant axial-flow direction component from the inlet to outlet of the rotor. For a radial-flow machine the flow across the blades involves a substantial radial-flow component at the rotor inlet, exit, or both. In other machines, designated as mixed-flow machines, there may be significant radial- and axial-flow velocity components for the flow through the rotor row. Each type of machine has advantages and disadvantages for different applications and in terms of fluid-mechanical performance.

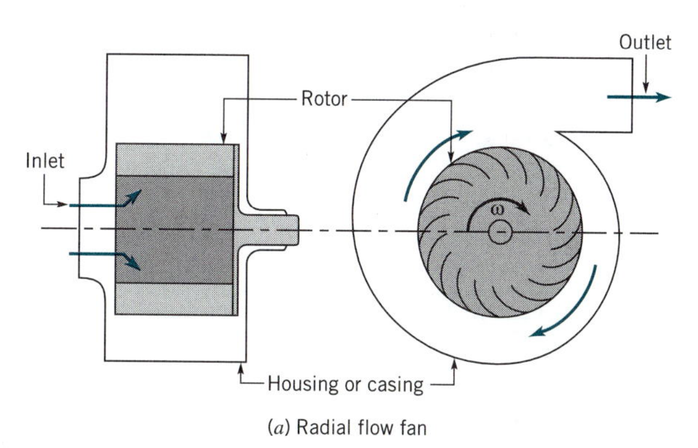

(*a*) Radial flow fan

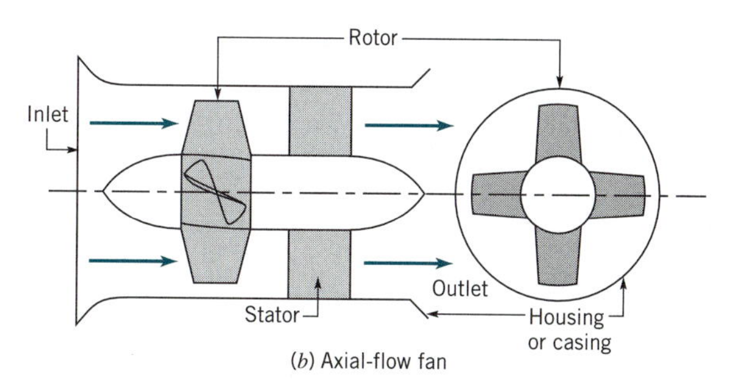

(*b*) Axial-flow fan

■ **FIGURE 12.2** ***(a)*** **A radial-flow turbomachine, *(b)* An axial-flow turbomachine.**

12.2 Basic Energy Considerations

An understanding of the work transfer in turbomachines can be obtained by considering the basic operation of a household fan (pump) and a windmill (turbine). Although the actual flows in such devices are very complex (i.e., three-dimensional and unsteady), the essential phenomena can be illustrated by use of simplified flow considerations and velocity triangles.

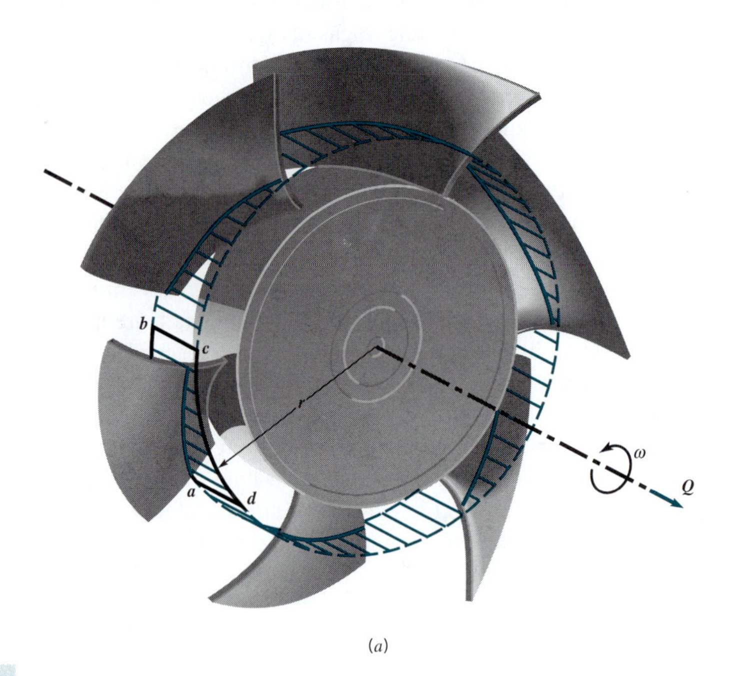

(a)

Absolute velocity is the vector sum of relative and blade velocities.

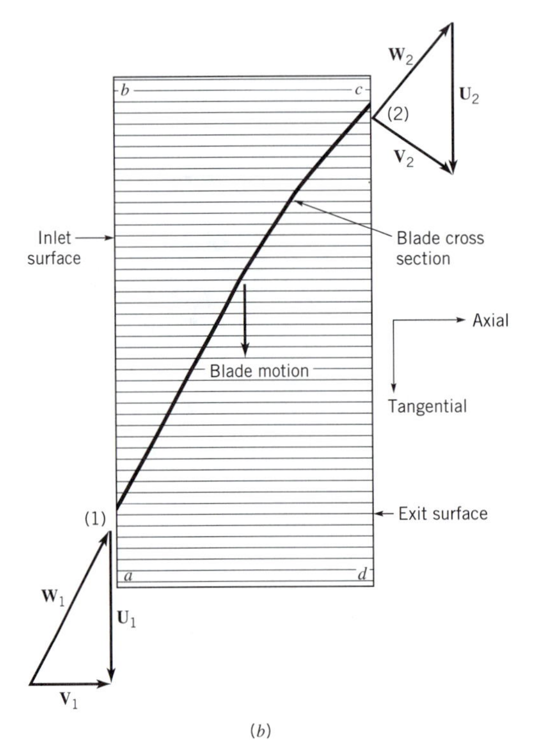

(b)

■ **FIGURE 12.3 Idealized flow through a fan: *(a)* fan blade geometry; *(b)* absolute velocity, V; relative velocity, W; and blade velocity, U at the inlet and exit of the fan blade section.**

Consider a fan blade driven at constant angular velocity, ω, by a motor as is shown in Fig. 12.3*a*. We denote the blade speed as $U = \omega r$, where r is the radial distance from the axis of the fan. The absolute fluid velocity (that seen by a person sitting stationary at the table on which the fan rests) is denoted **V**, and the relative velocity (that seen by a person riding on the fan blade) is denoted **W**. The actual (absolute) fluid velocity is the vector sum of the relative velocity and the blade velocity

$$\mathbf{V} = \mathbf{W} + \mathbf{U} \tag{12.1}$$

A simplified sketch of the fluid velocity as it ''enters'' and ''exits'' the fan at radius r is shown in Fig 12.3*b*. The shaded surface labeled *a-b-c-d* is a portion of the cylindrical surface (including a ''slice'' through the blade) shown in Fig. 12.3*a*. We assume for simplicity that the flow moves smoothly along the blade so that relative to the moving blade the velocity is parallel to the leading and trailing edges (points 1 and 2) of the blade. For now we assume that the fluid enters and leaves the fan at the same distance from the axis of rotation; thus, $U_1 = U_2 = \omega r$. In actual turbomachines, the entering and leaving flows are not necessarily tangent to the blades, and the fluid pathlines can involve changes in radius. These considerations are important at design and off-design operating conditions. Interested readers are referred to Refs. 7, 8, and 9 for more information about these aspects of turbomachine flows.

With this information we can construct the *velocity triangles* shown in Fig. 12.3*b*. Note that this view is from the top of the fan, looking radially down toward the axis of rotation. The motion of the blade is down; the motion of the incoming air is assumed to be directed along the axis of rotation. The important concept to grasp from this sketch is that the fan blade (because of its shape and motion) ''pushes'' the fluid, causing it to change direction. The absolute velocity vector, **V**, is turned during its flow across the blade from section (1) to section (2). Initially the fluid had no component of absolute velocity in the direction of the motion of the blade, the θ (or tangential) direction. When the fluid leaves the blade, this tangential component of absolute velocity is nonzero. For this to occur, the blade

Fan blades push fluid downstream.

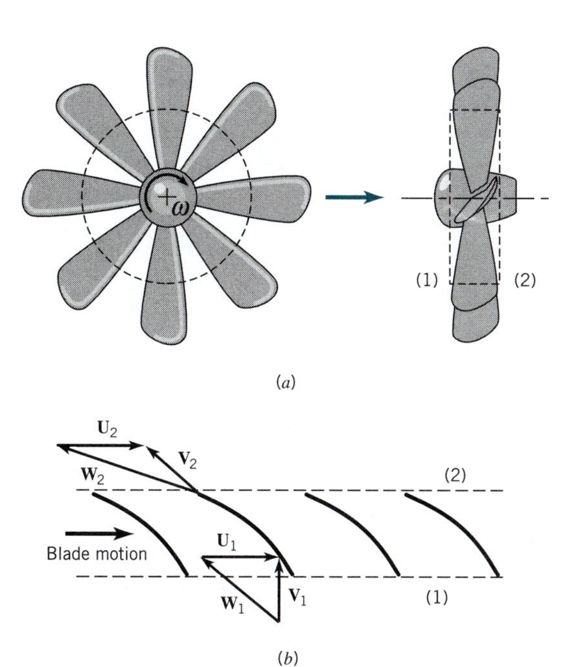

■ **FIGURE 12.4 Idealized flow through a windmill: *(a)* Windmill blade geometry; *(b)* Absolute velocity, V; relative velocity, W; and blade velocity, U; at the inlet and exit of the windmill blade section.**

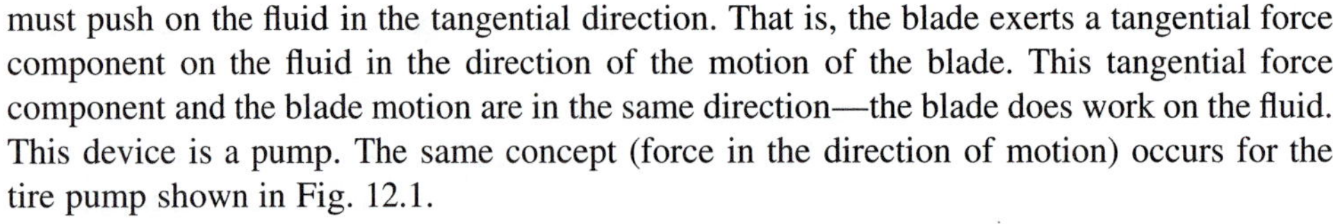

must push on the fluid in the tangential direction. That is, the blade exerts a tangential force component on the fluid in the direction of the motion of the blade. This tangential force component and the blade motion are in the same direction—the blade does work on the fluid. This device is a pump. The same concept (force in the direction of motion) occurs for the tire pump shown in Fig. 12.1.

V12.1

On the other hand, consider the windmill shown in Fig. 12.4*a*. Rather than the rotor being driven by a motor, it is rotated in the opposite direction (compared to the fan in Fig. 12.3) by the wind blowing through the rotor. We again note that because of the blade shape and motion, the absolute velocity vectors at sections (1) and (2), $\mathbf{V}_1$ and $\mathbf{V}_2$, have different directions. For this to happen, the blades must have pushed to the left on the fluid—opposite to the direction of their motion. Alternatively, because of equal and opposite forces (action/reaction) the fluid must have pushed on the blades in the direction of their motion—the fluid does work on the blades. This extraction of energy from the fluid is the purpose of a turbine.

When blades move because of the fluid force, we have a turbine; when blades are forced to move fluid, we have a pump.

These examples involve work transfer to or from a fluid using unducted flows in two axial-flow turbomachines. Similar concepts hold for other turbomachines including mixed-flow and radial-flow configurations.

EXAMPLE 12.1

The rotor shown in Fig. E12.1*a* rotates at a constant angular velocity of $\omega = 100 \text{ rad/s}$. Although the fluid initially approaches the rotor in an axial direction, the flow across the blades is primarily radial (see Fig. 12.2*a*). Measurements indicate that the absolute velocity at the inlet and outlet are $V_1 = 12 \text{ m/s}$ and $V_2 = 25 \text{ m/s}$, respectively. Is this device a pump or a turbine?

SOLUTION

To answer this question, we need to know if the tangential component of the force of the blade on the fluid is in the direction of the blade motion (a pump) or opposite to it (a turbine). We assume that the blades are tangent to the incoming relative velocity and that the relative flow leaving the rotor is tangent to the blades as shown in Fig. E12.1*b*. We can also calculate the inlet and outlet blade speeds as

$$U_1 = \omega r_1 = (100 \text{ rad/s})(0.1 \text{ m}) = 10 \text{ m/s}$$

and

$$U_2 = \omega r_2 = (100 \text{ rad/s})(0.2 \text{ m}) = 20 \text{ m/s}$$

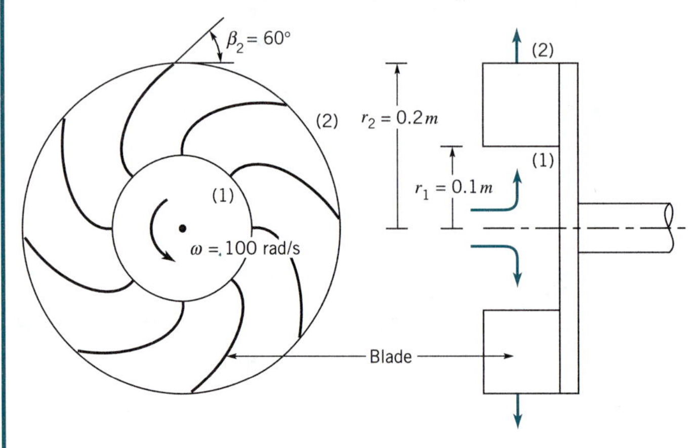

(*a*)

■ FIGURE E12.1

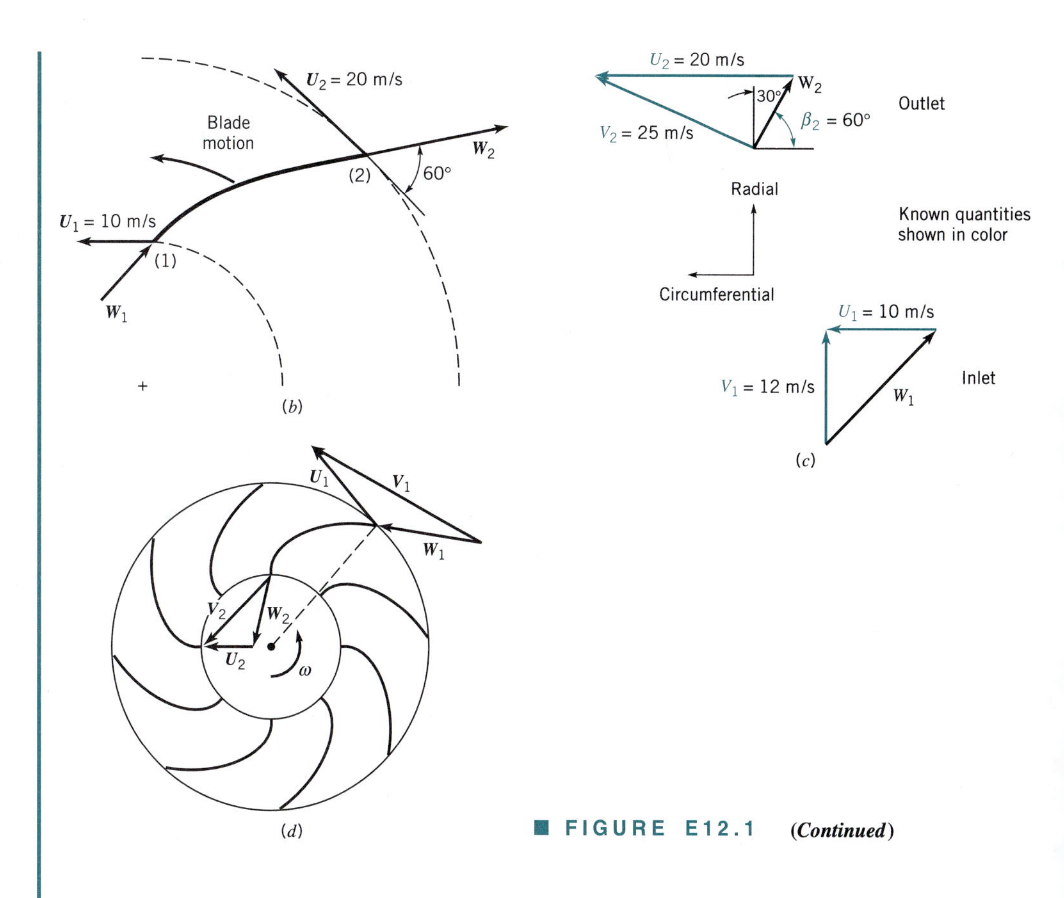

■ FIGURE E12.1 **(*Continued*)**

With the known, absolute fluid velocity and blade velocity at the inlet, we can draw the velocity triangle (the graphical representation of Eq. 12.1) at that location as shown in Fig. E12.1*c*. Note that we have assumed that the absolute flow at the blade row inlet is radial (i.e., the direction of $\mathbf{V}_1$ is radial). At the outlet we know the blade velocity, $\mathbf{U}_2$, the outlet speed, V_2, and the relative velocity direction, β_2, (because of the blade geometry). Therefore, we can graphically (or trigonometrically) construct the outlet velocity triangle as shown in the figure. By comparing the velocity triangles at the inlet and outlet, it can be seen that as the fluid flows across the blade row, the absolute velocity vector turns in the direction of the blade motion. At the inlet there is no component of absolute velocity in the direction of rotation; at the outlet this component is not zero. That is, the blade pushes the fluid in the direction of the blade motion, thereby doing work on the fluid, adding energy to it.

This device is a pump. **(Ans)**

On the other hand, by reversing the direction of flow from larger to smaller radii, this device can become a radial-flow turbine. In this case (Fig. 12.1*d*) the flow direction is reversed (compared to that in Fig. E12.1*a*, *b*, and *c*) and the velocity triangles are as indicated. Stationary vanes around the perimeter of the rotor would be needed to achieve $\mathbf{V}_1$ as shown. Note that the component of the absolute velocity, $\mathbf{V}$, in the direction of the blade motion is smaller at the outlet than at the inlet. The blade must push against the fluid in the direction opposite the motion of the blade to cause this. Hence (by equal and opposite forces), the fluid pushes against the blade in the direction of blade motion, thereby doing work on the blade. There is a transfer of work from the fluid to the blade—a turbine operation.

12.3 Basic Angular Momentum Considerations

When shaft torque and rotation are in the same direction, we have a pump; otherwise we have a turbine.

In the previous section we indicated how work transfer to or from a fluid flowing through a pump or a turbine occurs by interaction between moving rotor blades and the fluid. Since all of these turbomachines involve the rotation of an impeller or a rotor about a central axis, it is appropriate to discuss their performance in terms of torque and angular momentum.

Recall that work can be written as force times distance or as torque times angular displacement. Hence, if the shaft torque (the torque that the shaft applies to the rotor) and the rotation of the rotor are in the same direction, energy is transferred from the shaft to the rotor and from the rotor to the fluid—the machine is a pump. Conversely, if the torque exerted by the shaft on the rotor is opposite to the direction of rotation, the energy transfer is from the fluid to the rotor—a turbine. The amount of shaft torque (and hence shaft work) can be obtained from the moment-of-momentum equation derived formally in Section 5.2.3 and discussed as follows.

Consider a fluid particle traveling through the rotor in the radial-flow machine shown in Fig. E12.1*a*, *b*, and *c*. For now, assume that the particle enters the rotor with a radial velocity only (i.e., no "swirl"). After being acted upon by the rotor blades during its passage from the inlet [section (1)] to the outlet [section (2)], the particle exits with radial (r) and circumferential (θ) components of velocity. Thus, the particle enters with no angular momentum about the rotor axis of rotation but leaves with nonzero angular momentum about that axis. (Recall that the axial component of angular momentum for a particle is its mass times the distance from the axis times the θ component of absolute velocity.)

V12.2

A similar experience can occur at the neighborhood playground. Consider yourself as a particle and a merry-go-round as a rotor. Walk from the center to the edge of the spinning merry-go-round and note the forces involved. The merry-go-round does work on you—there is a "sideward force" on you. Another person must apply a torque (and power) to the merry-go-round to maintain a constant angular velocity, otherwise the angular momentum of the system (you and the merry-go-round) is conserved and the angular velocity decreases as you increase your distance from the axis of rotation. (Similarly, if the motor driving a pump is turned off, the pump will obviously slow down and stop.) Your friend is the motor supplying energy to the rotor that is transferred to you. Is the amount of energy your friend expends to keep the angular velocity constant dependent upon what path you follow along the merry-go-round (i.e., the blade shape); on how fast and in what direction you walk off the edge (i.e., the exit velocity); on how much you weigh (i.e., the density of the fluid)? What happens if you walk from the outside edge toward the center of the rotating merry-go-round? Recall that the opposite of a pump is a turbine.

In a turbomachine a series of particles (a continuum) passes through the rotor. Thus, the moment-of-momentum equation applied to a control volume as derived in Section 5.2.3 is valid. For steady flow (or for turbomachine rotors with steady-in-the-mean or steady-on-average cyclical flow), Eq. 5.42 gives

$$\sum (\mathbf{r} \times \mathbf{F}) = \int_{\text{cs}} (\mathbf{r} \times \mathbf{V})\, \rho \mathbf{V} \cdot \hat{\mathbf{n}}\, dA$$

Recall that the left-hand side of this equation represents the sum of the external torques (moments) acting on the contents of the control volume, and the right-hand side is the net rate of flow of moment-of-momentum (angular momentum) through the control surface.

The axial component of this equation applied to the one-dimensional simplification of flow through a turbomachine rotor with section (1) as the inlet and section (2) as the outlet results in

$$T_{\text{shaft}} = -\dot{m}_1(r_1 V_{\theta 1}) + \dot{m}_2(r_2 V_{\theta 2}) \quad \textbf{(12.2)}$$

where T_{shaft} is the shaft torque applied to the contents of the control volume. The "$-$" is associated with mass flowrate into the control volume and the "$+$" is used with the outflow. The sign of the V_θ component depends on the direction of V_θ and the blade motion, U. If V_θ and U are in the same direction, then V_θ is positive. The sign of the torque exerted by the shaft on the rotor, T_{shaft}, is positive if T_{shaft} is in the same direction as rotation, and negative otherwise.

As seen from Eq. 12.2, the shaft torque is directly proportional to the mass flowrate, $\dot{m} = \rho Q$. (It takes considerably more torque and power to pump water than to pump air with the same volume flowrate.) The torque also depends on the tangential component of the absolute velocity, V_θ. Equation 12.2 is often called the *Euler turbomachine equation.*

The Euler turbomachine equation is the axial component of the moment-of-momentum equation.

Also recall that the shaft power, $\dot{W}_{\text{shaft}}$, is related to the shaft torque and angular velocity by

$$\dot{W}_{\text{shaft}} = T_{\text{shaft}}\,\omega \quad \textbf{(12.3)}$$

By combining Eqs. 12.2 and 12.3 and using the fact that $U = \omega r$, we obtain

$$\dot{W}_{\text{shaft}} = -\dot{m}_1(U_1 V_{\theta 1}) + \dot{m}_2(U_2 V_{\theta 2}) \quad \textbf{(12.4)}$$

Again, the value of V_θ is positive when V_θ and U are in the same direction and negative otherwise. Also, $\dot{W}_{\text{shaft}}$ is positive when the shaft torque and ω are in the same direction and negative otherwise. Thus, $\dot{W}_{\text{shaft}}$ is positive when power is supplied to the contents of the control volume (pumps) and negative otherwise (turbines). This outcome is consistent with the sign convention involving the work term in the energy equation considered in Chapter 5 (see Eq. 5.67).

Finally, in terms of work per unit mass, $w_{\text{shaft}} = \dot{W}_{\text{shaft}}/\dot{m}$, we obtain

$$w_{\text{shaft}} = -U_1 V_{\theta 1} + U_2 V_{\theta 2} \quad \textbf{(12.5)}$$

where we have used the fact that by conservation of mass, $\dot{m}_1 = \dot{m}_2$. Equations 12.3, 12.4, and 12.5 are the basic governing equations for pumps or turbines whether the machines are radial-, mixed-, or axial-flow devices and for compressible and incompressible flows. Note that neither the axial nor the radial component of velocity enter into the specific work (work per unit mass) equation. [In the above merry-go-round example the amount of work your friend does is independent of how fast you jump "up" (axially) or "out" (radially) as you exit. The only thing that counts is your θ component of velocity.]

Another useful but more laborious form of Eq. 12.5 can be obtained by writing the right-hand side in a slightly different form based on the velocity triangles at the entrance or exit as shown generically in Fig. 12.5. The velocity component V_x is the generic through-

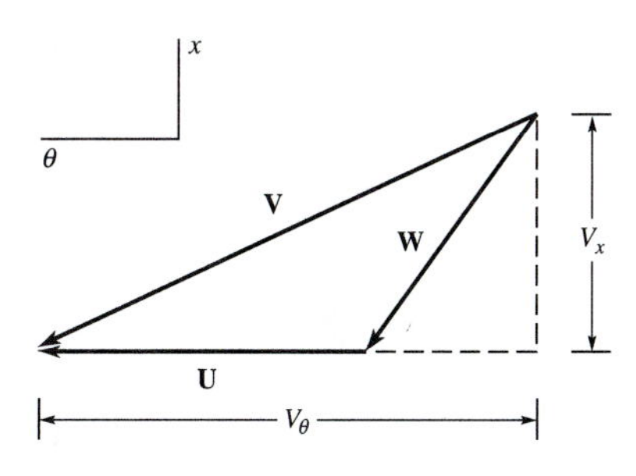

■ **FIGURE 12.5 Velocity triangle: V = absolute velocity, W = relative velocity, U = blade velocity.**

flow component of velocity and it can be axial, radial, or in-between depending on the rotor configuration. From the large right triangle we note that

$$V^2 = V_\theta^2 + V_x^2$$

or

$$V_x^2 = V^2 - V_\theta^2 \tag{12.6}$$

From the small right triangle we note that

$$V_x^2 + (V_\theta - U)^2 = W^2 \tag{12.7}$$

By combining Eqs. 12.6 and 12.7 we obtain

$$V_\theta U = \frac{V^2 + U^2 - W^2}{2}$$

which when written for the inlet and exit and combined with Eq. 12.5 gives

$$w_{\text{shaft}} = \frac{V_2^2 - V_1^2 + U_2^2 - U_1^2 - (W_2^2 - W_1^2)}{2} \tag{12.8}$$

Turbomachine work is related to changes in absolute, relative, and blade velocities.

Thus, the power and the shaft work per unit mass can be obtained from the speed of the blade, U, the absolute fluid speed, V, and the fluid speed relative to the blade, W. This is an alternative to using fewer components of the velocity as suggested by Eq. 12.5. Equation 12.8 contains more terms than Eq. 12.5; however, it is an important concept equation because it shows how the work transfer is related to absolute, relative, and blade velocity changes. Because of the general nature of the velocity triangle in Fig. 12.5, Eq. 12.8 is applicable for axial-, radial-, and mixed-flow rotors.

12.4 The Centrifugal Pump

One of the most common radial-flow turbomachines is the *centrifugal pump*. This type of pump has two main components: an *impeller* attached to a rotating shaft, and a stationary *casing*, *housing*, or *volute* enclosing the impeller. The impeller consists of a number of blades (usually curved), also sometimes called *vanes*, arranged in a regular pattern around the shaft. A sketch showing the essential features of a centrifugal pump is shown in Fig. 12.6. As the

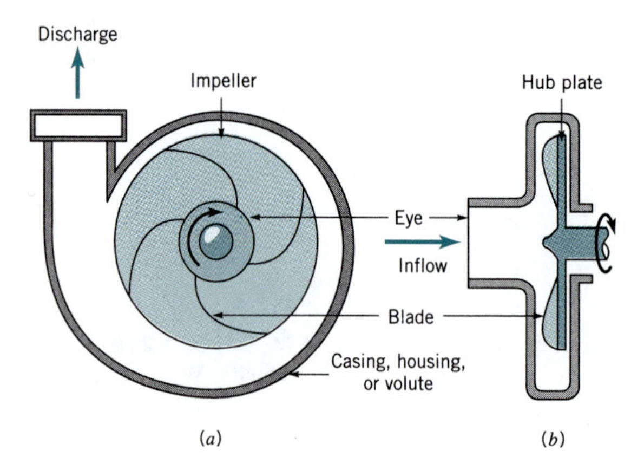

■ FIGURE 12.6 **Schematic diagram of basic elements of a centrifugal pump.**

impeller rotates, fluid is sucked in through the *eye* of the casing and flows radially outward. Energy is added to the fluid by the rotating blades, and both pressure and absolute velocity are increased as the fluid flows from the eye to the periphery of the blades. For the simplest type of centrifugal pump, the fluid discharges directly into a volute-shaped casing. The casing shape is designed to reduce the velocity as the fluid leaves the impeller, and this decrease in kinetic energy is converted into an increase in pressure. The volute-shaped casing, with its increasing area in the direction of flow, is used to produce an essentially uniform velocity distribution as the fluid moves around the casing into the discharge opening. For large centrifugal pumps, a different design is often used in which diffuser guide vanes surround the impeller. The diffuser vanes decelerate the flow as the fluid is directed into the pump casing. This type of centrifugal pump is referred to as a *diffuser* pump.

V12.3

Impellers are generally of two types. For one configuration the blades are arranged on a hub or backing plate and are open on the other (casing or shroud) side. A typical *open impeller* is shown in Fig. 12.7*a*. For the second type of impeller, called an *enclosed* or *shrouded* impeller, the blades are covered on both hub and shroud ends as shown in Fig. 12.7*b*.

Pump impellers can also be *single* or *double suction*. For the single-suction impeller the fluid enters through the eye on only one side of the impeller, whereas for the double-suction impeller the fluid enters the impeller along its axis from both sides. The double-suction arrangement reduces end thrust on the shaft, and also, since the net inlet flow area is larger, inlet velocities are reduced.

Pumps can be *single* or *multistage*. For a single-stage pump, only one impeller is mounted on the shaft, whereas for multistage pumps, several impellers are mounted on the same shaft. The stages operate in series, i.e., the discharge from the first stage flows into the eye of the second stage, the discharge from the second stage flows into the eye of the third stage, and so on. The flowrate is the same through all stages, but each stage develops an additional pressure rise. Thus, a very large discharge pressure, or head, can be developed by a multistage pump.

Centrifugal pumps involve radially outward flows.

Centrifugal pumps come in a variety of arrangements (open or shrouded impellers, volute or diffuser casings, single- or double-suction, single- or multistage), but the basic operating principle remains the same. Work is done on the fluid by the rotating blades (centrifugal action and tangential blade force acting on the fluid over a distance) creating a large increase in kinetic energy of the fluid flowing through the impeller. This kinetic energy is converted into an increase in pressure as the fluid flows from the impeller into the casing enclosing the impeller. A simplified theory describing the behavior of the centrifugal pump was introduced in the previous section and is expanded in the following section.

(a)

(b)

■ **FIGURE 12.7**
***(a)* Open impeller, *(b)* enclosed or shrouded impeller. (Courtesy of Ingersoll-Dresser Pump Company.)**

12.4.1 Theoretical Considerations

Although flow through a pump is very complex (unsteady and three-dimensional), the basic theory of operation of a centrifugal pump can be developed by considering the average one-dimensional flow of the fluid as it passes between the inlet and the outlet sections of the impeller as the blades rotate. As shown in Fig. 12.8, for a typical blade passage, the absolute velocity, $\mathbf{V}_1$, of the fluid entering the passage is the vector sum of the velocity of the blade, $\mathbf{U}_1$, rotating in a circular path with angular velocity ω, and the relative velocity, $\mathbf{W}_1$, within the blade passage so that $\mathbf{V}_1 = \mathbf{W}_1 + \mathbf{U}_1$. Similarly, at the exit $\mathbf{V}_2 = \mathbf{W}_2 + \mathbf{U}_2$. Note that $U_1 = r_1\omega$ and $U_2 = r_2\omega$. Fluid velocities are taken to be average velocities over the inlet and exit sections of the blade passage. The relationship between the various velocities is shown graphically in Fig. 12.8.

Centrifugal pump impellers involve an increase in blade velocity along the flow path.

As discussed in Section 12.3, the moment-of-momentum equation indicates that the shaft torque, T_{shaft}, required to rotate the pump impeller is given by equation Eq. 12.2 applied to a pump with $\dot{m}_1 = \dot{m}_2 = \dot{m}$. That is,

$$T_{\text{shaft}} = \dot{m}(r_2 V_{\theta 2} - r_1 V_{\theta 1}) \qquad \textbf{(12.9)}$$

or

$$T_{\text{shaft}} = \rho Q(r_2 V_{\theta 2} - r_1 V_{\theta 1}) \qquad \textbf{(12.10)}$$

where $V_{\theta 1}$ and $V_{\theta 2}$ are the tangential components of the absolute velocities, $\mathbf{V}_1$ and $\mathbf{V}_2$ (see Figs. 12.8*b*,*c*).

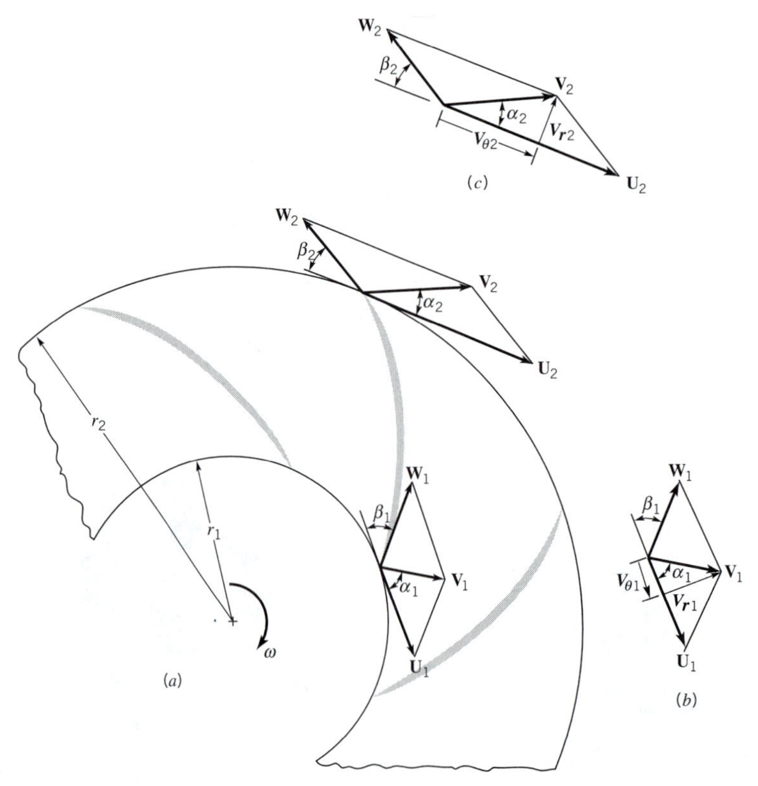

■ **FIGURE 12.8 Velocity diagrams at the inlet and exit of a centrifugal pump impeller.**

For a rotating shaft, the power transferred, $\dot{W}_{\text{shaft}}$, is given by

$$\dot{W}_{\text{shaft}} = T_{\text{shaft}}\omega$$

and therefore from Eq. 12.10

$$\dot{W}_{\text{shaft}} = \rho Q\omega(r_2 V_{\theta 2} - r_1 V_{\theta 1})$$

Since $U_1 = r_1\omega$ and $U_2 = r_2\omega$ we obtain

$$\dot{W}_{\text{shaft}} = \rho Q(U_2 V_{\theta 2} - U_1 V_{\theta 1}) \tag{12.11}$$

Equation 12.11 shows how the power supplied to the shaft of the pump is transferred to the flowing fluid. It also follows that the shaft power per unit mass of flowing fluid is

$$w_{\text{shaft}} = \frac{\dot{W}_{\text{shaft}}}{\rho Q} = U_2 V_{\theta 2} - U_1 V_{\theta 1} \tag{12.12}$$

Recall from Section 5.3.3 that the energy equation is often written in terms of heads—velocity head, pressure head, elevation head. The head that a pump adds to the fluid is an important parameter. The ideal or maximum head rise possible, h_i, is found from

$$\dot{W}_{\text{shaft}} = \rho g Q h_i$$

which is obtained from Eq. 5.84 by setting head loss (h_L) equal to zero and multiplying by the weight flowrate, $\rho g Q$. Combining this result with Eq. 12.12 we get

$$h_i = \frac{1}{g}(U_2 V_{\theta 2} - U_1 V_{\theta 1}) \tag{12.13}$$

The pump ideal head rise is the work per unit weight added to the fluid by the pump.

This ideal head rise, h_i, is the amount of energy per unit weight of fluid added to the fluid by the pump. The actual head rise realized by the fluid is less than the ideal amount by the head loss suffered. Some additional insight into the meaning of Eq. 12.13 can be obtained by using the following alternate version (see Eq. 12.8).

$$h_i = \frac{1}{2g}[(V_2^2 - V_1^2) + (U_2^2 - U_1^2) + (W_1^2 - W_2^2)] \tag{12.14}$$

A detailed examination of the physical interpretation of Eq. 12.14 would reveal the following. The first term in brackets on the right-hand side represents the increase in the kinetic energy of the fluid, and the other two terms represent the pressure head rise that develops across the impeller due to the centrifugal effect, $U_2^2 - U_1^2$, and the diffusion of relative flow in the blade passages, $W_1^2 - W_2^2$.

An appropriate relationship between the flowrate and the pump ideal head rise can be obtained as follows. Often the fluid has no tangential component of velocity $V_{\theta 1}$, or *swirl*, as it enters the impeller; i.e., the angle between the absolute velocity and the tangential direction is 90° ($\alpha_1 = 90°$ in Fig. 12.8). In this case, Eq. 12.13 reduces to

$$h_i = \frac{U_2 V_{\theta 2}}{g} \tag{12.15}$$

From Fig. 12.8*c*

$$\cot \beta_2 = \frac{U_2 - V_{\theta 2}}{V_{r2}}$$

so that Eq. 12.15 can be expressed as

$$h_i = \frac{U_2^2}{g} - \frac{U_2 V_{r2} \cot \beta_2}{g} \tag{12.16}$$

The flowrate, Q, is related to the radial component of the absolute velocity through the equation

$$Q = 2\pi r_2 b_2 V_{r2} \tag{12.17}$$

where b_2 is the impeller blade height at the radius r_2. Thus, combining Eqs. 12.16 and 12.17 yields

$$h_i = \frac{U_2^2}{g} - \frac{U_2 \cot \beta_2}{2\pi r_2 b_2 g} Q \tag{12.18}$$

The ideal head rise can be calculated with the Euler turbomachine equation.

This equation shows that the ideal or maximum head rise for a centrifugal pump varies linearly with Q for a given blade geometry and angular velocity. For actual pumps, the blade angle β_2 falls in the range of 15°–35°, with a normal range of $20° < \beta_2 < 25°$, and with $15° < \beta_1 < 50°$ (Ref. 10). Blades with $\beta_2 < 90°$ are called *backward curved*, whereas blades with $\beta_2 > 90°$ are called *forward curved*. Pumps are not usually designed with forward curved vanes since such pumps tend to suffer unstable flow conditions.

EXAMPLE 12.2

Water is pumped at the rate of 1400 gpm through a centrifugal pump operating at a speed of 1750 rpm. The impeller has a uniform blade height, b, of 2 in. with $r_1 = 1.9$ in. and $r_2 = 7.0$ in., and the exit blade angle β_2 is 23° (see Fig. 12.8). Assume ideal flow conditions and that the tangential velocity component, $V_{\theta 1}$, of the water entering the blade is zero ($\alpha_1 = 90°$). Determine **(a)** the tangential velocity component, $V_{\theta 2}$, at the exit, **(b)** the ideal head rise, h_i, and **(c)** the power, $\dot{W}_{\text{shaft}}$, transferred to the fluid.

SOLUTION

(a) At the exit the velocity diagram is as shown in Fig. 12.8*c*, where $\mathbf{V}_2$ is the absolute velocity of the fluid, $\mathbf{W}_2$ is the relative velocity, and $\mathbf{U}_2$ is the tip velocity of the impeller with

$$U_2 = r_2\omega = (7/12 \text{ ft})(2\pi \text{ rad/rev}) \frac{(1750 \text{ rpm})}{(60 \text{ s/min})}$$

$$= 107 \text{ ft/s}$$

Since the flowrate is given, it follows that

$$Q = 2\pi r_2 b_2 V_{r2}$$

or

$$V_{r2} = \frac{Q}{2\pi r_2 b_2} = \frac{1400 \text{ gpm}}{(7.48 \text{ gal/ft}^3)(60 \text{ s/min})(2\pi)(7/12 \text{ ft})(2/12 \text{ ft})}$$

$$= 5.11 \text{ ft/s}$$

From Fig. 12.8*c* we see that

$$\cot \beta_2 = \frac{U_2 - V_{\theta 2}}{V_{r2}}$$

so that

$$V_{\theta 2} = U_2 - V_{r2} \cot \beta_2$$
$$= (107 - 5.11 \cot 23°) \text{ ft/s}$$
$$= 95.0 \text{ ft/s} \qquad \textbf{(Ans)}$$

(b) From Eq. 12.15 the ideal head rise is given by

$$h_i = \frac{U_2 V_{\theta 2}}{g} = \frac{(107 \text{ ft/s})(95.0 \text{ ft/s})}{32.2 \text{ ft/s}^2}$$
$$= 316 \text{ ft} \qquad \textbf{(Ans)}$$

Alternatively, from Eq. 12.16, the ideal head rise is

$$h_i = \frac{U_2^2}{g} - \frac{U_2 V_{r2} \cot \beta_2}{g} = \frac{(107 \text{ ft/s})^2}{32.2 \text{ ft/s}^2} - \frac{(107 \text{ ft/s})(5.11 \text{ ft/s}) \cot 23°}{32.2 \text{ ft/s}^2}$$
$$= 316 \text{ ft} \qquad \textbf{(Ans)}$$

(c) From Eq. 12.11, with $V_{\theta 1} = 0$, the power transferred to the fluid is given by the equation

$$\dot{W}_{\text{shaft}} = \rho Q U_2 V_{\theta 2} = \frac{(1.94 \text{ slugs/ft}^3)(1400 \text{ gpm})(107 \text{ ft/s})(95.0 \text{ ft/s})}{[1(\text{slug}\cdot\text{ft/s}^2)/\text{lb}](7.48 \text{ gal/ft}^3)(60 \text{ s/min})}$$
$$= 61{,}500 \text{ ft}\cdot\text{lb/s} = 112 \text{ hp} \qquad \textbf{(Ans)}$$

Note that the ideal head rise and the power transferred are related through the relationship

$$\dot{W}_{\text{shaft}} = \rho g Q h_i$$

It should be emphasized that results given in the previous equation involve the ideal head rise. The actual head-rise performance characteristics of a pump are usually determined by experimental measurements obtained in a testing laboratory.

Ideal and actual head rise levels differ by the head loss.

Figure 12.9 shows the ideal head versus flowrate curve (Eq. 12.18) for a centrifugal pump with backward curved vanes ($\beta_2 < 90°$). Since there are simplifying assumptions (i.e., zero losses) associated with the equation for h_i, we would expect that the actual rise in head of fluid, h_a, would be less than the ideal head rise, and this is indeed the case. As shown in Fig. 12.9, the h_a versus Q curve lies below the ideal head-rise curve and shows a nonlinear variation with Q. The differences between the two curves (as represented by the shaded areas between the curves) arise from several sources. These differences include losses due to fluid skin friction in the blade passages, which vary as Q^2, and other losses due to such factors as flow separation, impeller blade-casing clearance flows, and other three-dimensional flow effects. Near the design flowrate, some of these other losses are minimized.

Centrifugal pump design is a highly developed field, with much known about pump theory and design procedures (see, e.g., Refs. 10, 11, 12, 13). However, due to the general complexity of flow through a centrifugal pump, the actual performance of the pump cannot be accurately predicted on a completely theoretical basis as indicated by the data of Fig. 12.9. Actual pump performance is determined experimentally through tests on the pump. From

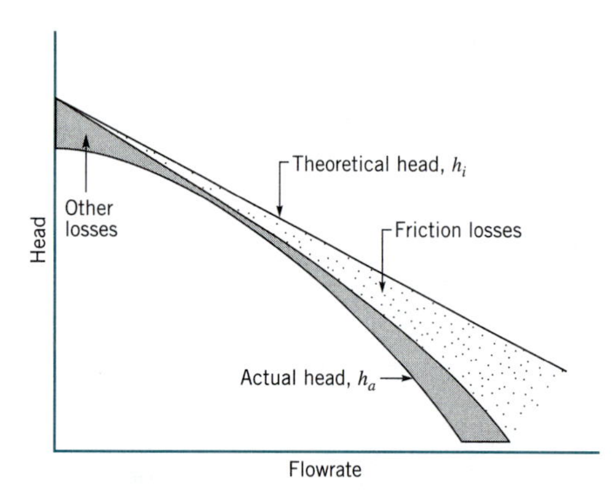

■ FIGURE 12.9 **Effect of losses on the pump head-flowrate curve.**

these tests, pump characteristics are determined and presented as *pump performance curves*. It is this information that is most helpful to the engineer responsible for incorporating pumps into a given flow system.

12.4.2 Pump Performance Characteristics

The energy equation can be used to calculate the pump actual head rise.

The actual head rise, h_a, gained by fluid flowing through a pump can be determined with an experimental arrangement of the type shown in Fig. 12.10, using the energy equation (Eq. 5.84 with $h_a = h_s - h_L$ where h_s is the shaft work head and is identical to h_i, and h_L is the pump head loss)

$$h_a = \frac{p_2 - p_1}{\gamma} + z_2 - z_1 + \frac{V_2^2 - V_1^2}{2g} \tag{12.19}$$

with sections (1) and (2) at the pump inlet and exit, respectively. The head, h_a, is the same as h_p used with the energy equation, Eq. 5.84, where h_p is interpreted to be the net head rise actually gained by the fluid flowing through the pump, i.e., $h_a = h_p = h_s - h_L$. Typically, the differences in elevations and velocities are small so that

$$h_a \approx \frac{p_2 - p_1}{\gamma} \tag{12.20}$$

The power, $\mathcal{P}_f$, gained by the fluid is given by the equation

$$\mathcal{P}_f = \gamma Q h_a \tag{12.21}$$

and this quantity, expressed in terms of horsepower is traditionally called the *water horsepower*. Thus,

$$\mathcal{P}_f = \text{water horsepower} = \frac{\gamma Q h_a}{550} \tag{12.22}$$

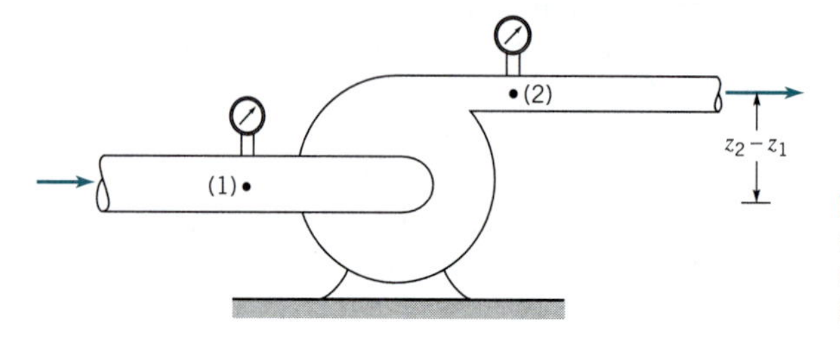

■ FIGURE 12.10 **Typical experimental arrangement for determining the head rise gained by a fluid flowing through a pump.**

with γ expressed in lb/ft^3, Q in ft^3/s, and h_a in ft. Note that if the pumped fluid is not water, the γ appearing in Eq. 12.22 must be the specific weight of the fluid moving through the pump.

In addition to the head or power added to the fluid, the *overall efficiency*, η, is of interest, where

$$\eta = \frac{\text{power gained by the fluid}}{\text{shaft power driving the pump}} = \frac{\mathcal{P}_f}{\dot{W}_{\text{shaft}}}$$

The denominator of this relationship represents the total power applied to the shaft of the pump and is often referred to as *brake horsepower* (bhp). Thus,

$$\eta = \frac{\gamma Q h_a/550}{\text{bhp}} \tag{12.23}$$

Pump overall efficiency is the ratio of power actually gained by the fluid to the shaft power supplied.

The overall pump efficiency is affected by the *hydraulic losses* in the pump, as previously discussed, and in addition, by the *mechanical losses* in the bearings and seals. There may also be some power loss due to leakage of the fluid between the back surface of the impeller hub plate and the casing, or through other pump components. This leakage contribution to the overall efficiency is called the *volumetric loss*. Thus, the overall efficiency arises from three sources, the *hydraulic efficiency*, η_h, the *mechanical efficiency*, η_m, and the *volumetric efficiency*, η_v, so that $\eta = \eta_h \eta_m \eta_v$.

Performance characteristics for a given pump geometry and operating speed are usually given in the form of plots of h_a, η, and bhp versus Q (commonly referred to as *capacity*) as illustrated in Fig. 12.11. Actually, only two curves are needed since h_a, η, and bhp are related through Eq. 12.23. For convenience, all three curves are usually provided. Note that for the pump characterized by the data of Fig. 12.11, the head curve continuously rises as the flowrate decreases, and in this case the pump is said to have a *rising head* curve. Pumps may also have $h_a - Q$ curves that initially rise as Q is decreased from the design value and then fall with a continued decrease in Q. These pumps have a *falling head* curve. The head developed by the pump at zero discharge is called the shutoff head, and it represents the rise in pressure head across the pump with the discharge valve closed. Since there is no flow with the valve closed, the related efficiency is zero, and the power supplied by the pump (bhp at $Q = 0$) is simply dissipated as heat. Although centrifugal pumps can be operated for short periods of time with the discharge valve closed, damage will occur due to overheating and large mechanical stress with any extended operation with the valve closed.

As can be seen from Fig. 12.11, as the discharge is increased from zero the brake horsepower increases, with a subsequent fall as the maximum discharge is approached. As

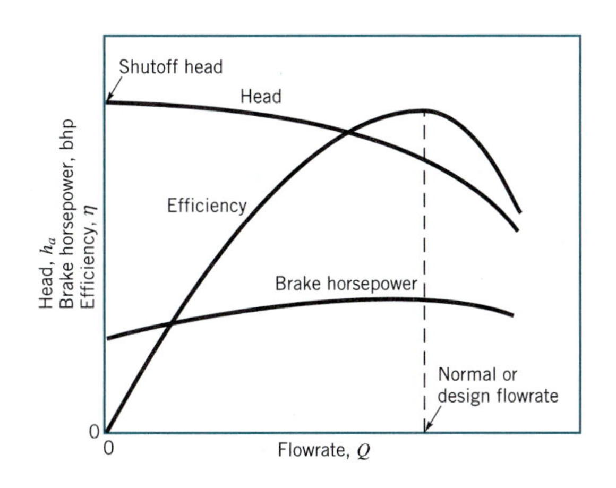

■ **FIGURE 12.11** **Typical performance characteristics for a centrifugal pump of a given size operating at a constant impeller speed.**

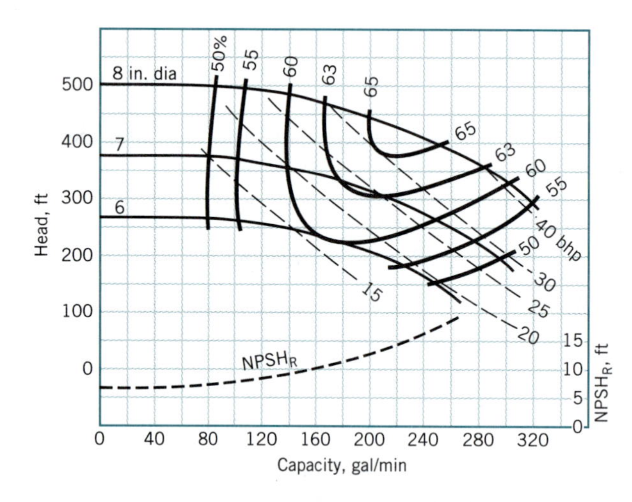

■ **FIGURE 12.12 Performance curves for a two-stage centrifugal pump operating at 3500 rpm. Data given for three different impeller diameters.**

previously noted, with h_a and bhp known, the efficiency can be calculated. As shown in Fig. 12.11, the efficiency is a function of the flowrate and reaches a maximum value at some particular value of the flowrate, commonly referred to as the *normal* or *design* flowrate or capacity for the pump. The points on the various curves corresponding to the maximum efficiency are denoted as the *best efficiency points* (BEP). It is apparent that when selecting a pump for a particular application, it is usually desirable to have the pump operate near its maximum efficiency. Thus, performance curves of the type shown in Fig. 12.11 are very important to the engineer responsible for the selection of pumps for a particular flow system. Matching the pump to a particular flow system is discussed in Section 12.4.4.

Pump efficiency varies with flowrate and has a single maximum value.

Pump performance characteristics are also presented in charts of the type shown in Fig. 12.12. Since impellers with different diameters may be used in a given casing, performance characteristics for several impeller diameters can be provided with corresponding lines of constant efficiency and brake horsepower as illustrated in Fig. 12.12. Thus, the same information can be obtained from this type of graph as from the curves shown in Fig. 12.11.

It is to be noted that an additional curve is given in Fig. 12.12, labeled NPSH_R, which stands for *required net positive suction head.* As discussed in the following section, the significance of this curve is related to conditions on the suction side of the pump, which must also be carefully considered when selecting and positioning a pump.

12.4.3 Net Positive Suction Head (NPSH)

On the suction side of a pump, low pressures are commonly encountered, with the concomitant possibility of cavitation occurring within the pump. As discussed in Section 1.8, cavitation occurs when the liquid pressure at a given location is reduced to the vapor pressure of the liquid. When this occurs, vapor bubbles form (the liquid starts to "boil"); this phenomenon can cause a loss in efficiency as well as structural damage to the pump. To characterize the potential for cavitation, the difference between the total head on the suction side, near the pump impeller inlet, $p_s/\gamma + V_s^2/2g$, and the liquid vapor pressure head, p_v/γ, is used. The position reference for the elevation head passes through the centerline of the pump impeller inlet. This difference is called the net positive suction head (NPSH) so that

$$\text{NPSH} = \frac{p_s}{\gamma} + \frac{V_s^2}{2g} - \frac{p_v}{\gamma} \tag{12.24}$$

There are actually two values of NPSH of interest. The first is the *required* NPSH, denoted NPSH_R, that must be maintained, or exceeded, so that cavitation will not occur. Since

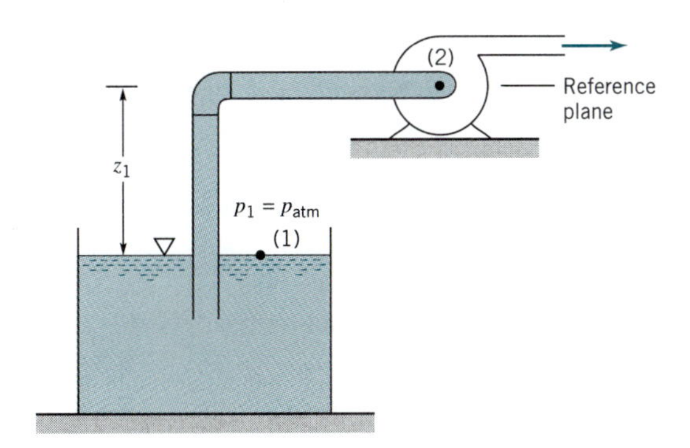

■ **FIGURE 12.13 Schematic of a pump installation in which the pump must lift fluid from one level to another.**

pressures lower than those in the suction pipe will develop in the impeller eye, it is usually necessary to determine experimentally, for a given pump, the required $NPSH_R$. This is the curve shown in Fig. 12.12. Pumps are tested to determine the value for $NPSH_R$, as defined by Eq. 12.24, by either directly detecting cavitation, or by observing a change in the head-flowrate curve (Ref. 14). The second value for NPSH of concern is the *available* NPSH, denoted $NPSH_A$, which represents the head that actually occurs for the particular flow system. This value can be determined experimentally, or calculated if the system parameters are known. For example, a typical flow system is shown in Fig. 12.13. The energy equation applied between the free liquid surface, where the pressure is atmospheric, p_{atm}, and a point on the suction side of the pump near the impeller inlet yields

$$\frac{p_{atm}}{\gamma} - z_1 = \frac{p_s}{\gamma} + \frac{V_s^2}{2g} + \sum h_L$$

where Σh_L represents head losses between the free surface and the pump impeller inlet. Thus, the head available at the pump impeller inlet is

$$\frac{p_s}{\gamma} + \frac{V_s^2}{2g} = \frac{p_{atm}}{\gamma} - z_1 - \sum h_L$$

so that

$$NPSH_A = \frac{p_{atm}}{\gamma} - z_1 - \sum h_L - \frac{p_v}{\gamma} \tag{12.25}$$

For this calculation, absolute pressures are normally used since the vapor pressure is usually specified as an absolute pressure. For proper pump operation it is necessary that

Cavitation, which may occur when pumping a liquid, is usually avoided.

$$NPSH_A \geq NPSH_R$$

It is noted from Eq. 12.25 that as the height of the pump impeller above the fluid surface, z_1, is increased, the $NPSH_A$ is decreased. Therefore, there is some critical value for z_1 above which the pump cannot operate without cavitation. The specific value depends on the head losses and the value of the vapor pressure. It is further noted that if the supply tank or reservoir is *above* the pump, z_1 will be negative in Eq. 12.25, and the $NPSH_A$ will increase as this height is increased.

A centrifugal pump is to be placed above a large, open water tank, as shown in Fig. 12.13, and is to pump water at a rate of 0.5 ft^3/s. At this flowrate the required net positive suction head, $NPSH_R$, is 15 ft, as specified by the pump manufacturer. If the water temperature is 80 °F and atmospheric pressure is 14.7 psi, determine the maximum height, z_1, that the pump can be located above the water surface without cavitation. Assume that the major head loss between the tank and the pump inlet is due to a filter at the pipe inlet having a minor loss coefficient $K_L = 20$. Other losses can be neglected. The pipe on the suction side of the pump has a diameter of 4 in.

SOLUTION

From Eq. 12.25 the available net positive suction head, NPSH_A, is given by the equation

$$\text{NPSH}_\text{A} = \frac{p_{\text{atm}}}{\gamma} - z_1 - \sum h_L - \frac{p_v}{\gamma}$$

and the maximum value for z_1 will occur when $\text{NPSH}_\text{A} = \text{NPSH}_\text{R}$. Thus,

$$(z_1)_{\max} = \frac{p_{\text{atm}}}{\gamma} - \sum h_L - \frac{p_v}{\gamma} - \text{NPSH}_\text{R} \tag{1}$$

Since the only head loss to be considered is the loss

$$\sum h_L = K_L \frac{V^2}{2g}$$

with

$$V = \frac{Q}{A} = \frac{0.5 \text{ ft}^3/\text{s}}{(\pi/4)(4/12 \text{ ft})^2} = 5.73 \text{ ft/s}$$

it follows that

$$\sum h_L = \frac{(20)(5.73 \text{ ft/s})^2}{2(32.2 \text{ ft/s}^2)} = 10.2 \text{ ft}$$

From Table B.1 the water vapor pressure at 80 °F is 0.5069 psia and $\gamma = 62.22 \text{ lb/ft}^3$. Equation (1) can now be written as

$$\begin{aligned}(z_1)_{\max} &= \frac{(14.7 \text{ lb/in.}^2)(144 \text{ in.}^2/\text{ft}^2)}{62.22 \text{ lb/ft}^3} - 10.2 \text{ ft} \\ &\quad - \frac{(0.5069 \text{ lb/in.}^2)(144 \text{ in.}^2/\text{ft}^2)}{62.22 \text{ lb/ft}^3} - 15 \text{ ft} \\ &= 7.65 \text{ ft}\end{aligned}$$

(Ans)

Thus, to prevent cavitation, with its accompanying poor pump performance, the pump should not be located higher than 7.65 ft above the water surface.

12.4.4 System Characteristics and Pump Selection

A typical flow system in which a pump is used is shown in Fig. 12.14. The energy equation applied between points (1) and (2) indicates that

$$h_p = z_2 - z_1 + \sum h_L \tag{12.26}$$

where h_p is the actual head gained by the fluid from the pump, and Σh_L represents all friction losses in the pipe and minor losses for pipe fittings and valves. From our study of pipe flow,

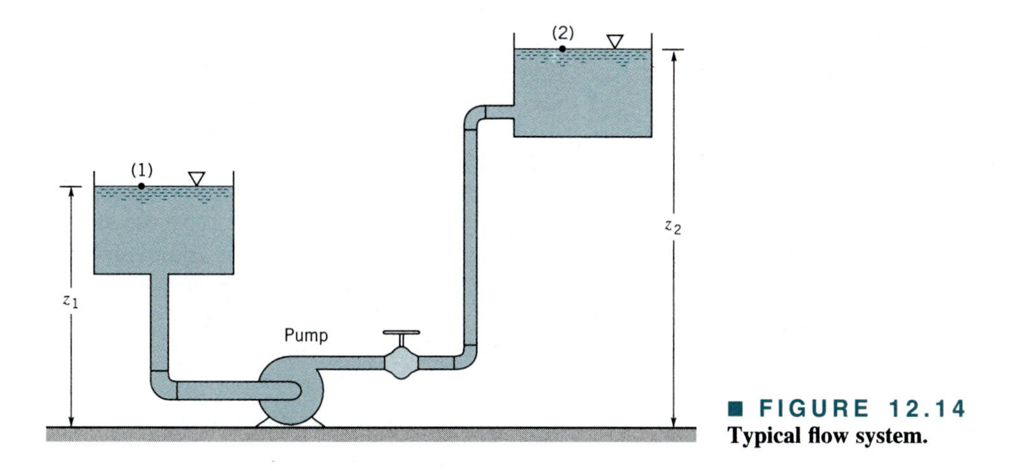

■ **FIGURE 12.14** **Typical flow system.**

we know that typically h_L varies approximately as the flowrate squared; that is, $h_L \propto Q^2$ (see Section 8.4). Thus, Eq. 12.26 can be written in the form

$$h_p = z_2 - z_1 + KQ^2 \tag{12.27}$$

where K depends on the pipe sizes and lengths, friction factors, and minor loss coefficients. Equation 12.27 is the *system equation* and shows how the actual head gained by the fluid from the pump is related to the system parameters. In this case the parameters include the change in elevation head, $z_2 - z_1$, and the losses due to friction as expressed by KQ^2. Each flow system has its own specific system equation. If the flow is laminar, the frictional losses will be proportional to Q rather than Q^2 (see Section 8.2).

The intersection of the pump performance curve and the system curve is the operating point.

There is also a unique relationship between the actual pump head gained by the fluid and the flowrate, which is governed by the pump design (as indicated by the pump performance curve). To select a pump for a particular application, it is necessary to utilize both the *system curve*, as determined by the system equation, and the pump performance curve. If both curves are plotted on the same graph, as illustrated in Fig. 12.15, their intersection (point A) represents the operating point for the system. That is, this point gives the head and flowrate that satisfies both the system equation and the pump equation. On the same graph the pump efficiency is shown. Ideally, we want the operating point to be near the best efficiency point (BEP) for the pump. For a given pump, it is clear that as the system equation changes, the operating point will shift. For example, if the pipe friction increases due to pipe wall fouling,

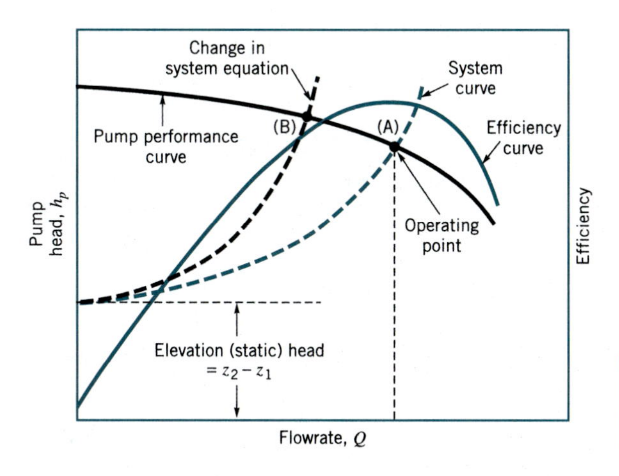

■ **FIGURE 12.15** **Utilization of the system curve and the pump performance curve to obtain the operating point for the system.**

the system curve changes, resulting in the operating point A shifting to point B in Fig. 12.15 with a reduction in flowrate and efficiency. The following example shows how the system and pump characteristics can be used to decide if a particular pump is suitable for a given application.

EXAMPLE 12.4

Water is to be pumped from one large, open tank to a second large, open tank as shown in Fig. E12.4*a*. The pipe diameter throughout is 6 in. and the total length of the pipe between the pipe entrance and exit is 200 ft. Minor loss coefficients for the entrance, exit, and the elbow are shown on the figure, and the friction factor for the pipe can be assumed constant and equal to 0.02. A certain centrifugal pump having the performance characteristics shown in Fig. E12.4*b* is suggested as a good pump for this flow system. With this pump, what would be the flowrate between the tanks? Do you think this pump would be a good choice?

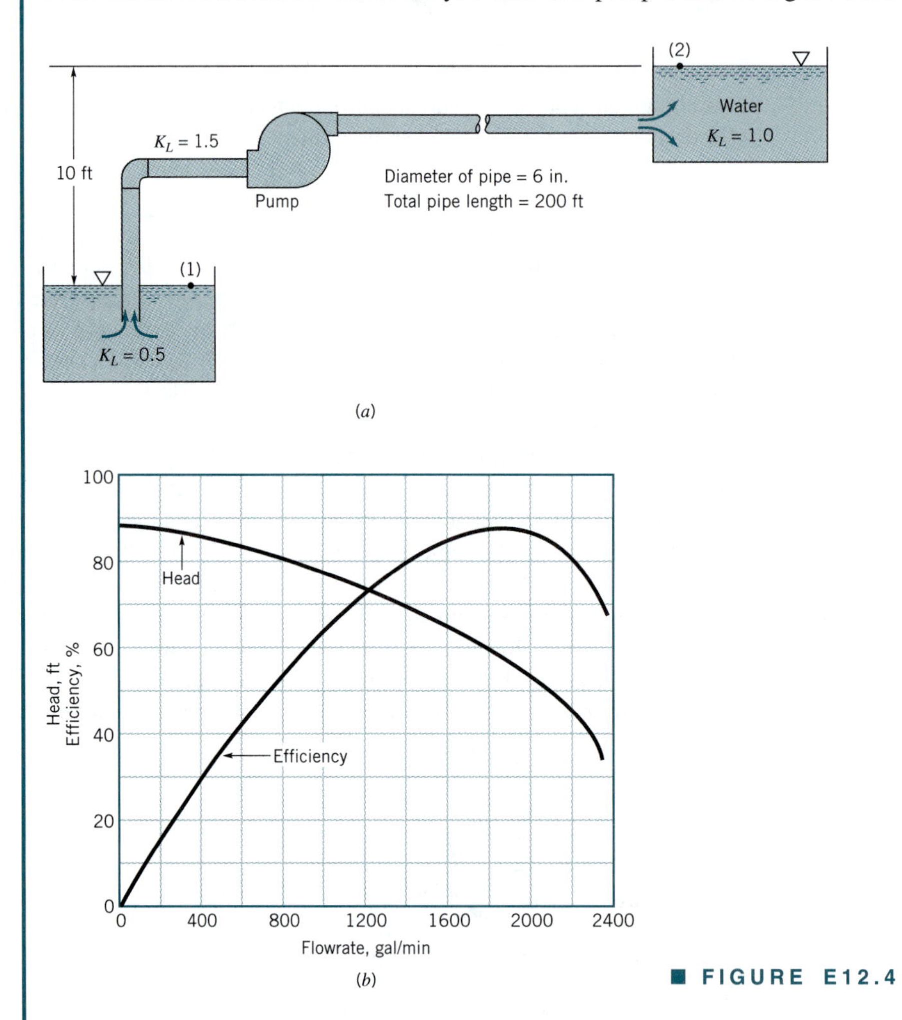

■ FIGURE E12.4

SOLUTION

Application of the energy equation between the two free surfaces, points (1) and (2) as indicated, gives

$$\frac{p_1}{\gamma} + \frac{V_1^2}{2g} + z_1 + h_p = \frac{p_2}{\gamma} + \frac{V_2^2}{2g} + z_2 + f\,\frac{\ell}{D}\,\frac{V^2}{2g} + \sum K_L\,\frac{V^2}{2g} \tag{1}$$

Thus, with $p_1 = p_2 = 0$, $V_1 = V_2 = 0$, $z_2 - z_1 = 10$ ft, $f = 0.02$, $D = 6/12$ ft, and $\ell =$ 200 ft, Eq. (1) becomes

$$h_p = 10 + \left[0.02 \frac{(200 \text{ ft})}{(6/12 \text{ ft})} + (0.5 + 1.5 + 1.0)\right] \frac{V^2}{2(32.2 \text{ ft/s}^2)} \qquad \textbf{(2)}$$

where the given minor loss coefficients have been used. Since

$$V = \frac{Q}{A} = \frac{Q(\text{ft}^3/\text{s})}{(\pi/4)(6/12 \text{ ft})^2}$$

Eq. 2 can be expressed as

$$h_p = 10 + 4.43\, Q^2 \qquad \textbf{(3)}$$

where Q is in ft^3/s, or with Q in gal/min,

$$h_p = 10 + 2.20 \times 10^{-5} Q^2 \qquad \textbf{(4)}$$

Equation 3 or 4 represents the system equation for this particular flow system and reveals how much actual head the fluid will need to gain from the pump to maintain a certain flowrate. The performance data shown in Fig. E12.4*b* indicate the actual head the fluid will gain from this particular pump when it operates at a certain flowrate. Thus, when Eq. 4 is plotted on the same graph with the performance data, the intersection of the two curves represents the operating point for the pump and the system. This combination is shown in Fig. E12.4*c* with the intersection (as obtained graphically) occurring at

$$Q = 1600 \text{ gal/min} \qquad \textbf{(Ans)}$$

with the corresponding actual head gained equal to 66.5 ft.

Another concern is whether or not the pump is operating efficiently at the operating point. As can be seen from Fig. E12.4*c*, although this is not peak efficiency, which is about 86%, it is close (about 84%). Thus, this pump would be a satisfactory choice, assuming the 1600 gal/min flowrate is at or near the desired flowrate.

The amount of pump head needed at the pump shaft is

$$\frac{66.5 \text{ ft}}{0.84} = 79.2 \text{ ft}$$

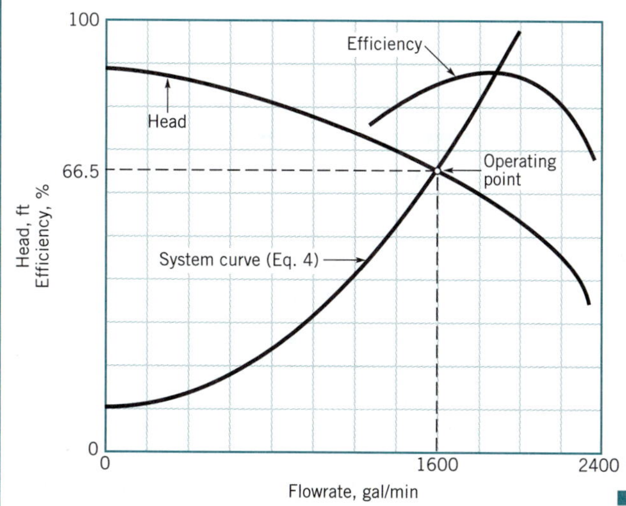

■ **FIGURE E12.4**
(Continued)

The power needed to drive the pump is

$$\dot{W}_{\text{shaft}} = \frac{\gamma Q h_a}{\eta}$$

$$= \frac{(62.4 \text{ lb/ft}^3)[(1600 \text{ gal/min})/(7.48 \text{ gal/ft}^3)(60 \text{ s/min})](66.5 \text{ ft})}{0.84}$$

$$= 17{,}600 \text{ ft}\cdot\text{lb/s} = 32.0 \text{ hp}$$

Pumps can be arranged in series or in parallel to provide for additional head or flow capacity. When two pumps are placed in *series*, the resulting pump performance curve is obtained by adding heads at the same flowrate. As illustrated in Fig. 12.16*a*, for two identical pumps in series, both the actual head gained by the fluid and the flowrate are increased, but neither will be doubled if the system curve remains the same. The operating point is at (A) for one pump and moves to (B) for two pumps in series. For two identical pumps in *parallel*, the combined performance curve is obtained by adding flowrates at the same head, as shown in Fig. 12.16*b*. As illustrated, the flowrate for the system will not be doubled with the addition of two pumps in parallel (if the same system curve applies). However, for a relatively flat system curve, as shown in Fig. 12.16*b*, a significant increase in flowrate can be obtained as the operating point moves from point (A) to point (B).

For two pumps in series, add heads; for two in parallel, add flowrates.

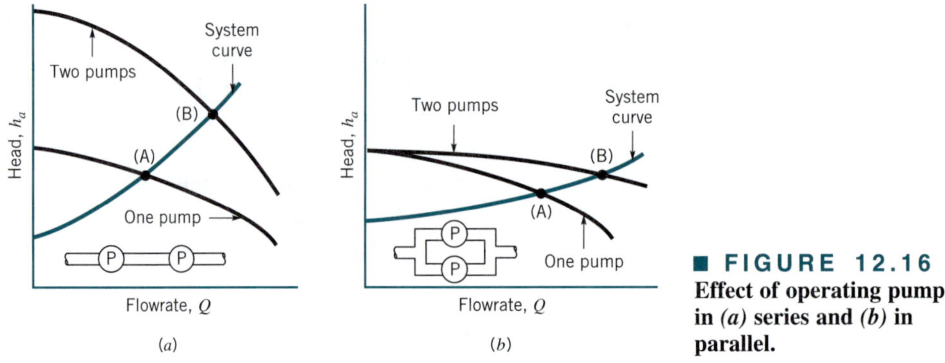

■ **FIGURE 12.16** **Effect of operating pumps in *(a)* series and *(b)* in parallel.**

12.5 Dimensionless Parameters and Similarity Laws

As discussed in Chapter 7, dimensional analysis is particularly useful in the planning and execution of experiments. Since the characteristics of pumps are usually determined experimentally, it is expected that dimensional analysis and similitude considerations will prove to be useful in the study and documentation of these characteristics.

From the previous section we know that the principal, dependent pump variables are the actual head rise, h_a, shaft power, $\dot{W}_{\text{shaft}}$, and efficiency, η. We expect that these variables will depend on the geometrical configuration, which can be represented by some characteristic diameter, D, other pertinent lengths, ℓ_i, and surface roughness, ε. In addition, the other important variables are flowrate, Q, the pump shaft rotational speed, ω, fluid viscosity, μ, and fluid density, ρ. We will only consider incompressible fluids presently, so compressibility effects need not concern us yet. Thus, any one of the dependent variables h_a, $\dot{W}_{\text{shaft}}$, and η

can be expressed as

$$\text{dependent variable} = f(D, \ell_i, \varepsilon, Q, \omega, \mu, \rho)$$

and a straightforward application of dimensional analysis leads to

$$\text{dependent pi term} = \phi\left(\frac{\ell_i}{D}, \frac{\varepsilon}{D}, \frac{Q}{\omega D^3}, \frac{\rho\omega D^2}{\mu}\right) \tag{12.28}$$

The dependent pi term involving the head is usually expressed as $C_H = gh_a/\omega^2 D^2$, where gh_a is the actual head rise in terms of energy per unit mass, rather than simply h_a, which is energy per unit weight. This dimensionless parameter is called the *head rise coefficient*. The dependent pi term involving the shaft power is expressed as $C_{\mathcal{P}} = \dot{W}_{\text{shaft}}/\rho\omega^3 D^5$, and this standard dimensionless parameter is termed the *power coefficient*. The power appearing in this dimensionless parameter is commonly based on the shaft (brake) horsepower, bhp, so that in BG units, $\dot{W}_{\text{shaft}} = 550 \times (\text{bhp})$. The rotational speed, ω, which appears in these dimensionless groups is expressed in rad/s. The final dependent pi term is the efficiency, η, which is already dimensionless. Thus, in terms of dimensionless parameters the performance characteristics are expressed as

$$C_H = \frac{gh_a}{\omega^2 D^2} = \phi_1\left(\frac{\ell_i}{D}, \frac{\varepsilon}{D}, \frac{Q}{\omega D^3}, \frac{\rho\omega D^2}{\mu}\right)$$

$$C_{\mathcal{P}} = \frac{\dot{W}_{\text{shaft}}}{\rho\omega^3 D^5} = \phi_2\left(\frac{\ell_i}{D}, \frac{\varepsilon}{D}, \frac{Q}{\omega D^3}, \frac{\rho\omega D^2}{\mu}\right)$$

$$\eta = \frac{\rho g Q h_a}{\dot{W}_{\text{shaft}}} = \phi_3\left(\frac{\ell_i}{D}, \frac{\varepsilon}{D}, \frac{Q}{\omega D^3}, \frac{\rho\omega D^2}{\mu}\right)$$

Dimensionless pi terms and similarity laws are important pump considerations.

The last pi term in each of the above equations is a form of Reynolds number that represents the relative influence of viscous effects. When the pump flow involves high Reynolds numbers, as is usually the case, experience has shown that the effect of the Reynolds number can be neglected. For simplicity, the relative roughness, ε/D, can also be neglected in pumps since the highly irregular shape of the pump chamber is usually the dominant geometric factor rather than the surface roughness. Thus, with these simplifications and for *geometrically similar* pumps (all pertinent dimensions, ℓ_i, scaled by a common length scale), the dependent pi terms are functions of only $Q/\omega D^3$, so that

$$\frac{gh_a}{\omega^2 D^2} = \phi_1\left(\frac{Q}{\omega D^3}\right) \tag{12.29}$$

$$\frac{\dot{W}_{\text{shaft}}}{\rho\omega^3 D^5} = \phi_2\left(\frac{Q}{\omega D^3}\right) \tag{12.30}$$

$$\eta = \phi_3\left(\frac{Q}{\omega D^3}\right) \tag{12.31}$$

The dimensionless parameter $C_Q = Q/\omega D^3$ is called the *flow coefficient*. These three equations provide the desired similarity relationships among a family of geometrically similar pumps. If two pumps from the family are operated at the same value of flow coefficient

$$\left(\frac{Q}{\omega D^3}\right)_1 = \left(\frac{Q}{\omega D^3}\right)_2 \tag{12.32}$$

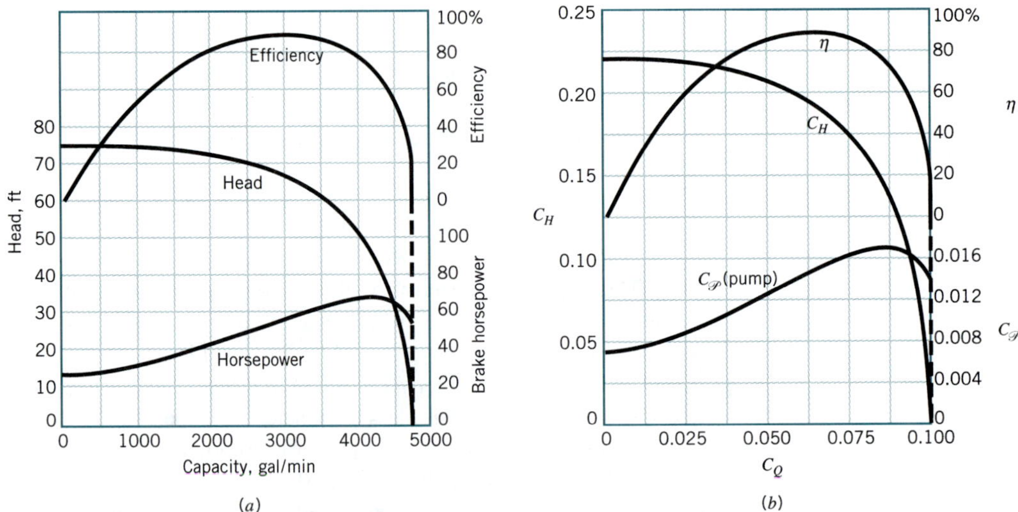

■ **FIGURE 12.17 Typical performance data for a centrifugal pump: *(a)* characteristic curves for a 12-in. centrifugal pump operating at 1000 rpm, *(b)* dimensionless characteristic curves. (Data from Ref. 16, used by permission.)**

it then follows that

$$\left(\frac{gh_a}{\omega^2 D^2}\right)_1 = \left(\frac{gh_a}{\omega^2 D^2}\right)_2 \tag{12.33}$$

$$\left(\frac{\dot{W}_{\text{shaft}}}{\rho\omega^3 D^5}\right)_1 = \left(\frac{\dot{W}_{\text{shaft}}}{\rho\omega^3 D^5}\right)_2 \tag{12.34}$$

$$\eta_1 = \eta_2 \tag{12.35}$$

where the subscripts 1 and 2 refer to any two pumps from the family of geometrically similar pumps.

Pump scaling laws relate geometrically similar pumps.

With these so-called *pump scaling laws* it is possible to experimentally determine the performance characteristics of one pump in the laboratory and then use these data to predict the corresponding characteristics for other pumps within the family under different operating conditions. Figure 12.17*a* shows some typical curves obtained for a centrifugal pump. Figure 12.17*b* shows the results plotted in terms of the dimensionless coefficients, C_Q, C_H, $C_{\mathcal{P}}$, and η. From these curves the performance of different-sized, geometrically similar pumps can be predicted, as can the effect of changing speeds on the performance of the pump from which the curves were obtained. It is to be noted that the efficiency, η, is related to the other coefficients through the relationship $\eta = C_Q C_H C_{\mathcal{P}}^{-1}$. This follows directly from the definition of η.

An 8-in.-diameter centrifugal pump operating at 1200 rpm is geometrically similar to the 12-in.-diameter pump having the performance characteristics of Figs. 12.17*a* and 12.17*b* while operating at 1000 rpm. For peak efficiency, predict the discharge, actual head rise, and shaft horsepower for this smaller pump. The working fluid is water at 60 °F.

SOLUTION

As is indicated by Eq. 12.31, for a given efficiency the flow coefficient has the same value for a given family of pumps. From Fig. 12.17*b* we see that at peak efficiency $C_Q = 0.0625$. Thus, for the 8-in. pump

$$Q = C_Q \omega D^3 = (0.0625)(1200/60 \text{ rev/s})(2\pi \text{ rad/rev})(8/12 \text{ ft})^3$$

$$Q = 2.33 \text{ ft}^3\text{/s} \qquad \textbf{(Ans)}$$

or in terms of gpm

$$Q = (2.33 \text{ ft}^3\text{/s})(7.48 \text{ gal/ft}^3)(60 \text{ s/min}) = 1046 \text{ gpm} \qquad \textbf{(Ans)}$$

The actual head rise and the shaft horsepower can be determined in a similar manner since at peak efficiency $C_H = 0.19$ and $C_{\mathcal{P}} = 0.014$, so that

$$h_a = \frac{C_H \omega^2 D^2}{g} = \frac{(0.19)(1200/60 \text{ rev/s})^2(2\pi \text{ rad/rev})^2(8/12 \text{ ft})^2}{32.2 \text{ ft/s}^2} = 41.4 \text{ ft} \qquad \textbf{(Ans)}$$

and

$$\begin{aligned} \dot{W}_{\text{shaft}} &= C_{\mathcal{P}} \rho \omega^3 D^5 \\ &= (0.014)(1.94 \text{ slugs/ft}^3)(1200/60 \text{ rev/s})^3(2\pi \text{ rad/rev})^3(8/12 \text{ ft})^5 \\ &= 7100 \text{ ft·lb/s} \end{aligned}$$

$$\dot{W}_{\text{shaft}} = \frac{7100 \text{ ft·lb/s}}{550 \text{ ft·lb/s/hp}} = 12.9 \text{ hp} \qquad \textbf{(Ans)}$$

This last result gives the shaft horsepower, which is the power supplied to the pump shaft. The power actually gained by the fluid is equal to $\gamma Q h_a$, which in this example is

$$\mathcal{P}_f = \gamma Q h_a = (62.4 \text{ lb/ft}^3)(2.33 \text{ ft}^3\text{/s})(41.4 \text{ ft}) = 6020 \text{ ft·lb/s}$$

Thus, the efficiency, η, is

$$\eta = \frac{\mathcal{P}_f}{\dot{W}_{\text{shaft}}} = \frac{6020}{7100} = 85\%$$

which checks with the efficiency curve of Fig. 12.17*b*.

12.5.1 Special Pump Scaling Laws

Two special cases related to pump similitude commonly arise. In the first case we are interested in how a change in the operating speed, ω, for a *given pump*, affects pump characteristics. It follows from Eq. 12.32 that for the same flow coefficient (and therefore the same efficiency) with $D_1 = D_2$ (the same pump)

$$\frac{Q_1}{Q_2} = \frac{\omega_1}{\omega_2} \qquad \textbf{(12.36)}$$

The subscripts 1 and 2 now refer to the same pump operating at two different speeds at the same flow coefficient. Also, from Eqs. 12.33 and 12.34 it follows that

$$\frac{h_{a1}}{h_{a2}} = \frac{\omega_1^2}{\omega_2^2} \tag{12.37}$$

and

$$\frac{\dot{W}_{\text{shaft1}}}{\dot{W}_{\text{shaft2}}} = \frac{\omega_1^3}{\omega_2^3} \tag{12.38}$$

Thus, for a given pump operating at a given flow coefficient, the flow varies directly with speed, the head varies as the speed squared, and the power varies as the speed cubed. These scaling laws are useful in estimating the effect of changing pump speed when some data are available from a pump test obtained by operating the pump at a particular speed.

In the second special case we are interested in how a change in the impeller diameter, D, of a geometrically similar family of pumps, operating at a *given speed,* affects pump characteristics. As before, it follows from Eq. 12.32 that for the same flow coefficient with $\omega_1 = \omega_2$

$$\frac{Q_1}{Q_2} = \frac{D_1^3}{D_2^3} \tag{12.39}$$

Similarly, from Eqs. 12.33 and 12.34

$$\frac{h_{a1}}{h_{a2}} = \frac{D_1^2}{D_2^2} \tag{12.40}$$

and

$$\frac{\dot{W}_{\text{shaft1}}}{\dot{W}_{\text{shaft2}}} = \frac{D_1^5}{D_2^5} \tag{12.41}$$

Thus, for a family of geometrically similar pumps operating at a given speed and the same flow coefficient, the flow varies as the diameter cubed, the head varies as the diameter squared, and the power varies as the diameter raised to the fifth power. These scaling relationships are based on the condition that, as the impeller diameter is changed, all other important geometric variables are properly scaled to maintain geometric similarity. This type of geometric scaling is not always possible due to practical difficulties associated with manufacturing the pumps. It is common practice for manufacturers to put impellers of different diameters in the same pump casing. In this case, complete geometric similarity is not maintained, and the scaling relationships expressed in Eqs. 12.39, 12.40, and 12.41 will not, in general, be valid. However, experience has shown that if the impeller diameter change is not too large, less than about 20%, these scaling relationships can still be used to estimate the effect of a change in the impeller diameter. The pump similarity laws expressed by Eqs. 12.36 through 12.41 are sometimes referred to as the *pump affinity laws.*

Pump affinity laws relate the same pump at different speeds or geometrically similar pumps at the same speed.

The effects of viscosity and surface roughness have been neglected in the foregoing similarity relationships. However, it has been found that as the pump size decreases these effects more significantly influence efficiency because of smaller clearances and blade size. An approximate, empirical relationship to estimate the influence of diminishing size on efficiency is (Ref. 15)

$$\frac{1 - \eta_2}{1 - \eta_1} \approx \left(\frac{D_1}{D_2}\right)^{1/5} \tag{12.42}$$

In general, it is to be expected that the similarity laws will not be very accurate if tests on a model pump with water are used to predict the performance of a prototype pump with a highly viscous fluid, such as oil, because at the much smaller Reynolds number associated with the oil flow, the fluid physics involved is different from the higher Reynolds number flow associated with water.

12.5.2 Specific Speed

Specific speed may be determined independent of pump size.

A useful pi term can be obtained by eliminating diameter D between the flow coefficient and the head rise coefficient. This is accomplished by raising the flow coefficient to an appropriate exponent (1/2) and dividing this result by the head coefficient raised to another appropriate exponent (3/4) so that

$$\frac{(Q/\omega D^3)^{1/2}}{(gh_a/\omega^2 D^2)^{3/4}} = \frac{\omega\sqrt{Q}}{(gh_a)^{3/4}} = N_s \qquad \textbf{(12.43)}$$

The dimensionless parameter N_s is called the *specific speed*. Specific speed varies with flow coefficient just as the other coefficients and efficiency discussed earlier do. However, for any pump it is customary to specify a value of specific speed at the flow coefficient corresponding to peak efficiency only. For pumps with low Q and high h_a, the specific speed is low compared to a pump with high Q and low h_a. Centrifugal pumps typically are low-capacity, high-head pumps, and therefore have low specific speeds.

Specific speed as defined by Eq. 12.43 is dimensionless, and therefore independent of the system of units used in its evaluation as long as a consistent unit system is used. However, in the United States a modified, dimensional form of specific speed, N_{sd}, is commonly used, where

$$N_{sd} = \frac{\omega(\text{rpm})\sqrt{Q(\text{gpm})}}{[h_a(\text{ft})]^{3/4}} \qquad \textbf{(12.44)}$$

In this case N_{sd} is said to be expressed in *U.S. customary units*. Typical values of N_{sd} are in the range $500 < N_{sd} < 4000$ for centrifugal pumps. Both N_s and N_{sd} have the same physical meaning, but their magnitudes will differ by a constant conversion factor ($N_{sd} = 2733\ N_s$) when ω in Eq. 12.43 is expressed in rad/s.

Each family or class of pumps has a particular range of values of specific speed associated with it. Thus, pumps that have low-capacity, high-head characteristics will have specific speeds that are smaller than pumps that have high-capacity, low-head characteristics. The concept of specific speed is very useful to engineers and designers, since if the required head, flowrate, and speed are specified, it is possible to select an appropriate (most efficient) type of pump for a particular application. As the specific speed, N_{sd}, increases beyond about 2000 the peak efficiency of the purely radial-flow centrifugal pump starts to fall off, and other types of more efficient pump design are preferred. In addition to the centrifugal pump, the *axial-flow* pump is widely used. As discussed in Section 12.6, in an axial-flow pump the direction of flow is primarily parallel to the rotating shaft rather than radial as in the centrifugal pump. Axial-flow pumps are essentially high-capacity, low-head pumps, and therefore have large specific speeds ($N_{sd} > 9000$) compared to centrifugal pumps. *Mixed-flow* pumps combine features of both radial-flow and axial-flow pumps and have intermediate values of specific speed. Figure 12.18 illustrates how the specific speed changes as the configuration of the pump changes from centrifugal or radial to axial.

12.5.3 Suction Specific Speed

With an analysis similar to that used to obtain the specific speed pi term, the *suction specific speed*, S_s, can be expressed as

$$S_s = \frac{\omega\sqrt{Q}}{[g(\text{NPSH}_\text{R})]^{3/4}} \qquad \textbf{(12.45)}$$

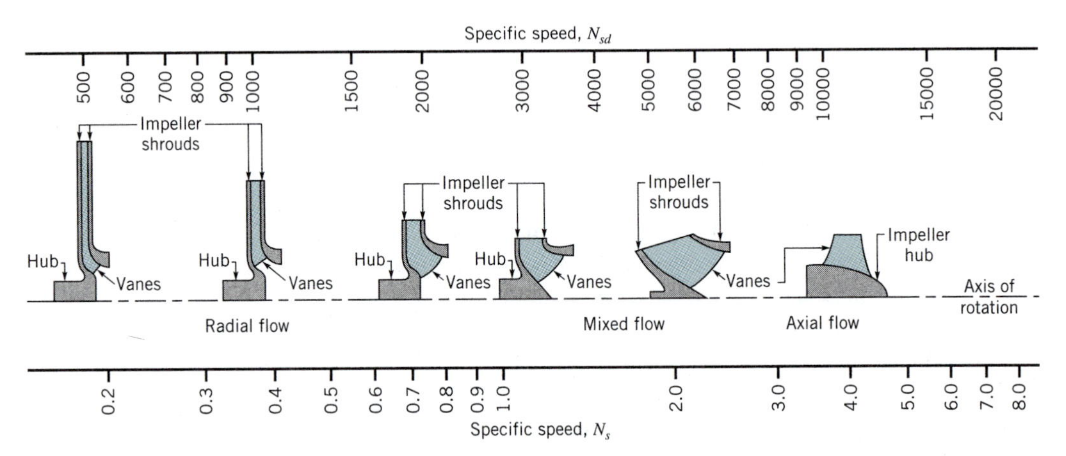

■ **FIGURE 12.18 Variation in specific speed with type of pump. (Adapted from Ref. 17, used with permission.)**

Specific speed may be used to approximate what general pump geometry (axial to radial) to use for maximum efficiency.

where h_a in Eq. 12.43 has been replaced by the required net positive suction head ($NPSH_R$). This dimensionless parameter is useful in determining the required operating conditions on the suction side of the pump. As was true for the specific speed, N_s, the value for S_s commonly used is for peak efficiency. For a family of geometrically similar pumps, S_s should have a fixed value. If this value is known, then the $NPSH_R$ can be estimated for other pumps within the same family operating at different values of ω and Q.

As noted for N_s, the suction specific speed as defined by Eq. 12.45 is also dimensionless, and the value for S_s is independent of the system of units used. However, as was the case for specific speed, in the United States a modified dimensional form for the suction specific speed, designated as S_{sd}, is commonly used, where

$$S_{sd} = \frac{\omega(\text{rpm})\sqrt{Q(\text{gpm})}}{[\text{NPSH}_\text{R}\ (\text{ft})]^{3/4}} \tag{12.46}$$

For double-suction pumps the discharge, Q, in Eq. 12.46 is one-half the total discharge.

Typical values for S_{sd} fall in the range of 7000 to 12,000 (Ref. 18). If S_{sd} is specified, Eq. 12.46 can be used to estimate the $NPSH_R$ for a given set of operating conditions. However, this calculation would generally only provide an approximate value for the $NPSH_R$, and the actual determination of the $NPSH_R$ for a particular pump should be made through a direct measurement whenever possible. Note that $S_{sd} = 2733\ S_s$, with ω expressed in rad/s in Eq. 12.45.

12.6 Axial-Flow and Mixed-Flow Pumps

As noted previously, centrifugal pumps are radial-flow machines that operate most efficiently for applications requiring high heads at relatively low flowrates. This head–flowrate combination typically yields specific speeds (N_{sd}) that are less than approximately 4000. For many applications, such as those associated with drainage and irrigation, high flowrates at low heads are required and centrifugal pumps are not suitable. In this case, axial-flow pumps are commonly used. This type of pump consists essentially of a propeller confined within a cylindrical casing. Axial-flow pumps are often called *propeller pumps*. For this type of pump the flow is primarily in the axial direction (parallel to the axis of rotation of the shaft), as opposed to the radial flow found in the centrifugal pump. Whereas the head developed by a centrifugal

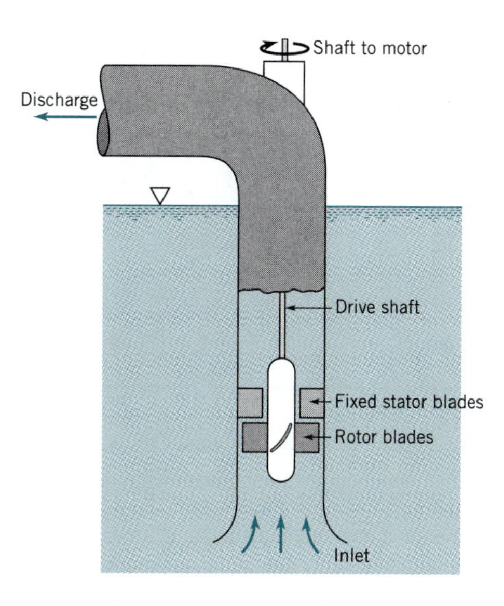

■ FIGURE 12.19 **Schematic diagram of an axial-flow pump arranged for vertical operation.**

pump includes a contribution due to centrifugal action, the head developed by an axial-flow pump is due primarily to the tangential force exerted by the rotor blades on the fluid. A schematic of an axial-flow pump arranged for vertical operation is shown in Fig. 12.19. The rotor is connected to a motor through a shaft, and as it rotates (usually at a relatively high speed) the fluid is sucked in through the inlet. Typically the fluid discharges through a row of fixed stator (guide) vanes used to straighten the flow leaving the rotor. Some axial-flow pumps also have inlet guide vanes upstream of the rotor row, and some are multistage in which pairs (*stages*) of rotating blades (*rotor blades*) and fixed vanes (*stator blades*) are arranged in series. Axial-flow pumps usually have specific speeds (N_{sd}) in excess of 9000.

Axial-flow pumps often have alternating rows of stator blades and rotor blades.

The definitions and broad concepts that were developed for centrifugal pumps are also applicable to axial-flow pumps. The actual flow characteristics, however, are quite different. In Fig. 12.20 typical head, power, and efficiency characteristics are compared for a centrifugal pump and an axial-flow pump. It is noted that at design capacity (maximum efficiency) the head and brake horsepower are the same for the two pumps. But as the flowrate decreases, the power input to the centrifugal pump falls to 180 hp at shutoff, whereas for the axial-flow

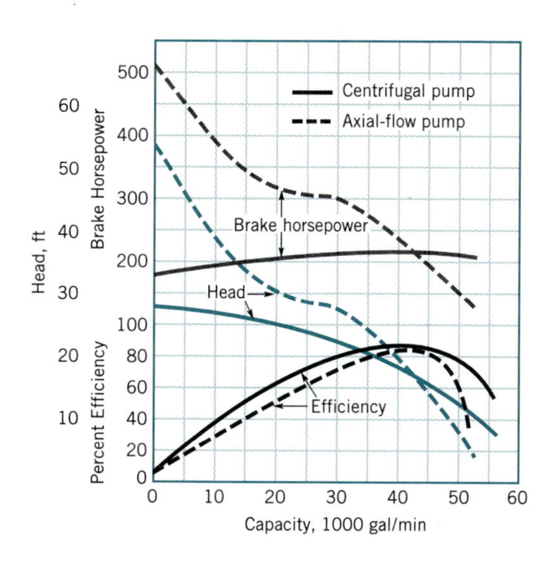

■ FIGURE 12.20 **Comparison of performance characteristics for a centrifugal pump and an axial-flow pump, each rated 42,000 gal/min at a 17-ft head. (Data from Ref. 19, used with permission.)**

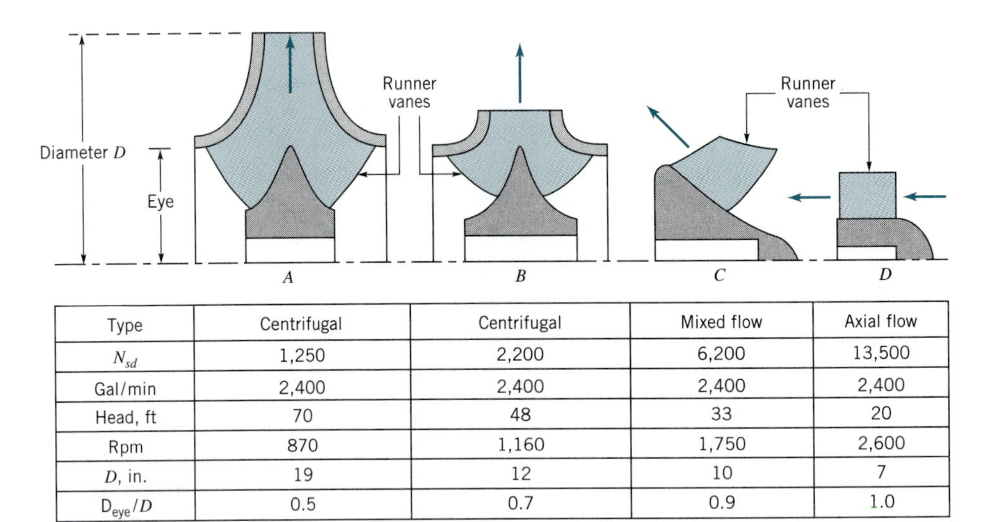

Type	Centrifugal	Centrifugal	Mixed flow	Axial flow
N_{sd}	1,250	2,200	6,200	13,500
Gal/min	2,400	2,400	2,400	2,400
Head, ft	70	48	33	20
Rpm	870	1,160	1,750	2,600
D, in.	19	12	10	7
D_{eye}/D	0.5	0.7	0.9	1.0

FIGURE 12.21 Comparison of different types of impellers. Specific speed for centrifugal pumps based on single suction. (Adapted from Ref. 19, used with permission.)

pump the power input increases to 520 hp at shutoff. This characteristic of the axial-flow pump can cause overloading of the drive motor if the flowrate is reduced significantly from the design capacity. It is also noted that the head curve for the axial-flow pump is much steeper than that for the centrifugal pump. Thus, with axial-flow pumps there will be a large change in head with a small change in the flowrate, whereas for the centrifugal pump, with its relatively flat head curve, there will be only a small change in head with large changes in the flowrate. It is further observed from Fig. 12.20 that, except at design capacity, the efficiency of the axial-flow pump is lower than that of the centrifugal pump. To improve operating characteristics, some axial-flow pumps are constructed with adjustable blades.

For applications requiring specific speeds intermediate to those for centrifugal and axial-flow pumps, mixed-flow pumps have been developed that operate efficiently in the specific speed range $4000 < N_{sd} < 9000$. As the name implies, the flow in a mixed-flow pump has both a radial and an axial component. Figure 12.21 shows some typical data for centrifugal, mixed-flow, and axial-flow pumps, each operating at the same design capacity. These data indicate that as we proceed from the centrifugal pump to the mixed-flow pump to the axial-flow pump, the specific speed increases, the head decreases, the speed increases, the impeller diameter decreases, and the eye diameter increases. These general trends are commonly found when these three types of pumps are compared.

The dimensionless parameters and scaling relationships developed in the previous sections apply to all three types of pumps—centrifugal, mixed-flow, and axial-flow—since the dimensional analysis used is not restricted to a particular type of pump. Additional information about pumps can be found in Refs. 3, 10, 14, 15, and 19.

12.7 Fans

Fans are used to pump air and other gases and vapors.

When the fluid to be moved is air, or some other gas or vapor, *fans* are commonly used. Types of fans vary from the small fan used for cooling desktop computers to large fans used in many industrial applications such as ventilating of large buildings. Fans typically operate at relatively low rotation speeds and are capable of moving large volumes of gas. Although the fluid of interest is a gas, the change in gas density through the fan does not usually exceed 7%, which for air represents a change in pressure of only about 1 psi (Ref. 20). Thus, in dealing with fans, the gas density is treated as a constant, and the flow analysis is based on incompressible flow concepts. Because of the low pressure rise involved, fans are often constructed of lightweight sheet metal. Fans are also called *blowers*, *boosters*, and *exhausters* depending on the location within the system; i.e., blowers are located at the system entrance, exhausters are at the system exit, and boosters are located at some intermediate position within the system. Turbomachines used to produce larger changes in gas density and pressure than possible with fans are called *compressors* (see Section 12.9.1).

As was the case for pumps, fan designs include centrifugal (radial-flow) fans, as well as mixed-flow and axial-flow (propeller) fans, and the analysis of fan performance closely follows that previously described for pumps. The shapes of typical performance curves for centrifugal and axial-flow fans are quite similar to those shown in Fig. 12.20 for centrifugal and axial-flow pumps. However, fan head-rise data are often given in terms of pressure rise, either static or total, rather than the more conventional head rise commonly used for pumps.

Scaling relationships for fans are the same as those developed for pumps, i.e., Eqs. 12.32 through 12.35 apply to fans as well as pumps. As noted above, for fans it is common to replace the head, h_a, in Eq. 12.33 with pressure head, $p_a/\rho g$, so that Eq. 12.33 becomes

$$\left(\frac{p_a}{\rho\omega^2 D^2}\right)_1 = \left(\frac{p_a}{\rho\omega^2 D^2}\right)_2 \tag{12.47}$$

where, as before, the subscripts 1 and 2 refer to any two fans from the family of geometrically similar fans. Equations 12.47, 12.32 and 12.34, are called the *fan laws* and can be used to scale performance characteristics between members of a family of geometrically similar fans. Additional information about fans can be found in Refs. 20, 21, 22, and 23.

12.8 Turbines

As discussed in Section 12.2, turbines are devices that extract energy from a flowing fluid. The geometry of turbines is such that the fluid exerts a torque on the rotor in the direction of its rotation. The shaft power generated is available to drive generators or other devices.

In the following sections we discuss mainly the operation of hydraulic turbines (those for which the working fluid is water) and to a lesser extent gas and steam turbines (those for which the density of the working fluid may be much different at the inlet than at the outlet).

The two basic types of hydraulic turbines are impulse and reaction.

Although there are numerous ingenious hydraulic turbine designs, most of these turbines can be classified into two basic types—*impulse turbines* and *reaction turbines*. (Reaction is related to the ratio of static pressure drop that occurs across the rotor to static pressure drop across the turbine stage, with larger rotor pressure drop corresponding to larger reaction.) For hydraulic impulse turbines, the pressure drop across the rotor is zero; all of the pressure drop across the turbine stage occurs in the nozzle row. The *Pelton wheel* shown in Fig. 12.22 is a classical example of an impulse turbine. In these machines the total head of the incoming fluid (the sum of the pressure head, velocity head, and elevation head) is converted into a large velocity head at the exit of the supply nozzle (or nozzles if a multiple nozzle configuration is used). Both the pressure drop across the bucket (blade) and the change in relative speed (i.e., fluid speed relative to the moving bucket) of the fluid across the bucket are negligible. The space surrounding the rotor is not completely filled with fluid. It is the impulse of the individual jets of fluid striking the buckets that generates the torque.

For reaction turbines, on the other hand, the rotor is surrounded by a casing (or volute), which is completely filled with the working fluid. There is both a pressure drop and a fluid

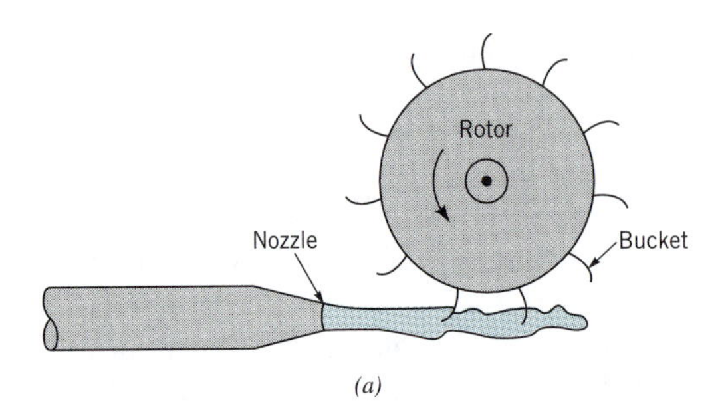

(a)

(b)

■ FIGURE 12.22 *(a)* **Schematic diagram of a Pelton wheel turbine,** *(b)* **photograph of a Pelton wheel turbine. (Courtesy of Voith Hydro, York, PA.)**

The Pelton wheel is an impulse turbine.

relative speed change across the rotor. As shown for the radial-inflow turbine in Fig 12.23, guide vanes act as nozzles to accelerate the flow and turn it in the appropriate direction as the fluid enters the rotor. Thus, part of the pressure drop occurs across the guide vanes and part occurs across the rotor. In many respects the operation of a reaction turbine is similar to that of a pump ''flowing backward,'' although such oversimplification can be quite misleading.

Both impulse and reaction turbines can be analyzed using the moment-of-momentum principles discussed in Section 12.3. In general, impulse turbines are high-head, low-flowrate devices, while reaction turbines are low-head, high-flowrate devices.

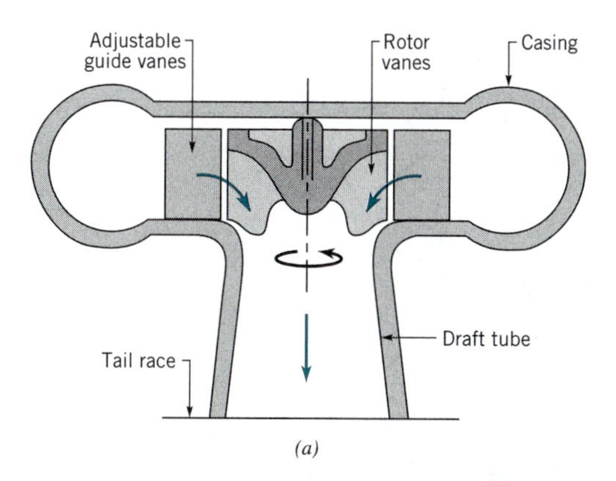

(a)

(b)

■ FIGURE 12.23 *(a)* **Schematic diagram of a reduction turbine,** *(b)* **photograph of a reaction turbine. (Courtesy of Voith Hydro, York, PA.)**

The Francis turbine involves reaction.

12.8.1 Impulse Turbines

Although there are various types of impulse turbine designs, perhaps the easiest to understand is the Pelton wheel (see Fig. 12.24). Lester Pelton (1829–1908), an American mining engineer during the California gold-mining days, is responsible for many of the still-used features of

■ FIGURE 12.24 Details of Pelton wheel turbine bucket.

this type of turbine. It is most efficient when operated with a large head (e.g., a water source from a lake located significantly above the turbine nozzle), which is converted into a relatively large velocity at the exit of the nozzle. Among the many design considerations for such a turbine are the head loss that occurs in the pipe (the penstock) transporting the water to the turbine, the design of the nozzle, and the design of the buckets on the rotor.

Pelton wheel turbines operate most efficiently with a larger head and lower flowrates.

As shown in Fig. 12.24, a high-speed jet of water strikes the Pelton wheel buckets and is deflected. The water enters and leaves the control volume surrounding the wheel as free jets (atmospheric pressure). In addition, a person riding on the bucket would note that the speed of the water does not change as it slides across the buckets (assuming viscous effects are negligible). That is, the magnitude of the relative velocity does not change, but its direction does. The change in direction of the velocity of the fluid jet causes a torque on the rotor, resulting in a power output from the turbine.

V12.4

Design of the optimum, complex shape of the buckets to obtain maximum power output is a very difficult matter. Ideally, the fluid enters and leaves the control volume shown in Fig. 12.25 with no radial component of velocity. (In practice there often is a small but negligible radial component.) In addition, the buckets would ideally turn the relative velocity vector through a 180° turn, but physical constraints dictate that β, the angle of the exit edge of the blade, is less than 180°. Thus, the fluid leaves with an axial component of velocity as shown in Fig. 12.26.

The inlet and exit velocity triangles at the arithmetic mean radius, r_m, are assumed to

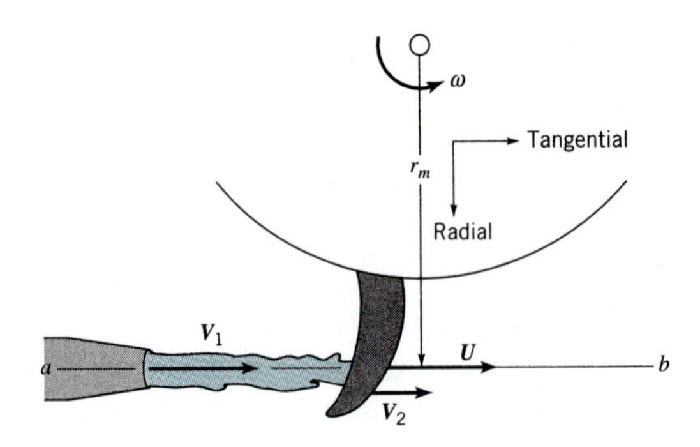

■ FIGURE 12.25 Ideal fluid velocities for a Pelton wheel turbine.

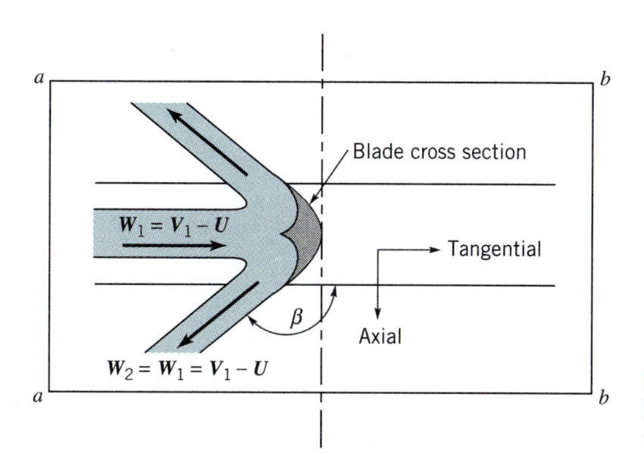

■ FIGURE 12.26 **Flow as viewed by an observer riding on the Pelton wheel—relative velocities.**

In Pelton wheel analyses, we assume the relative speed of the fluid is constant (no friction).

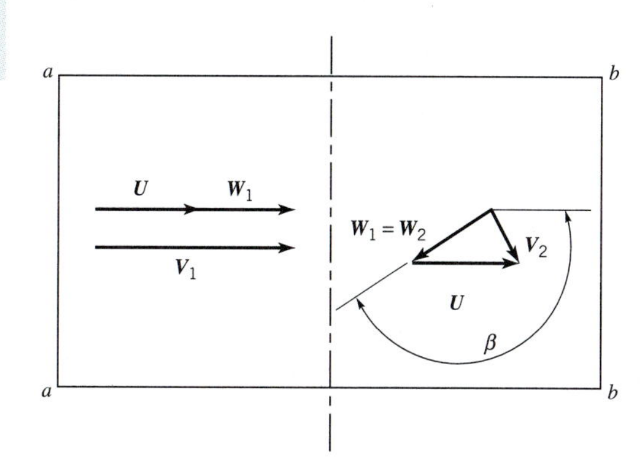

■ FIGURE 12.27 **Inlet and exit velocity triangles for a Pelton wheel turbine.**

be as shown in Fig. 12.27. To calculate the torque and power, we must know the tangential components of the absolute velocities at the inlet and exit. (Recall from the discussion in Section 12.3 that neither the radial nor the axial components of velocity enter into the torque or power equations.) From Fig. 12.27 we see that

$$V_{\theta 1} = V_1 = W_1 + U \tag{12.48}$$

and

$$V_{\theta 2} = W_2 \cos \beta + U \tag{12.49}$$

Thus, with the assumption that $W_1 = W_2$ (i.e., the relative speed of the fluid does not change as it is deflected by the buckets), we can combine Eqs. 12.48 and 12.49 to obtain

$$V_{\theta 2} - V_{\theta 1} = (U - V_1)(1 - \cos \beta) \tag{12.50}$$

This change in tangential component of velocity combined with the torque and power equations developed in Section 12.3 (i.e., Eqs. 12.2 and 12.4) gives

$$T_{\text{shaft}} = \dot{m} r_m (U - V_1)(1 - \cos \beta)$$

and since $U = \omega R_m$

$$\dot{W}_{\text{shaft}} = T_{\text{shaft}} \omega = \dot{m} U (U - V_1)(1 - \cos \beta) \tag{12.51}$$

These results are plotted in Fig. 12.28 along with typical experimental results. Note that $V_1 > U$ (i.e., the jet impacts the bucket), and $\dot{W}_{\text{shaft}} < 0$ (i.e., the turbine extracts power from the fluid).

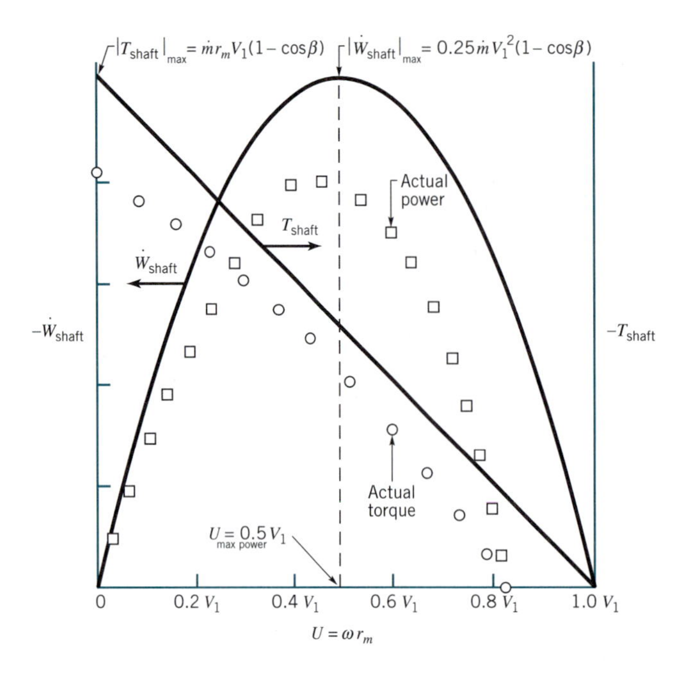

■ **FIGURE 12.28** **Typical theoretical and experimental power and torque for a Pelton wheel turbine as a function of bucket speed.**

Several interesting points can be noted from the above results. First, the power is a function of β. However, a typical value of $\beta = 165°$ (rather than the optimum 180°) results in a relatively small (less than 2%) reduction in power since $1 - \cos 165 = 1.966$, compared to $1 - \cos 180 = 2$. Second, although the torque is maximum when the wheel is stopped ($U = 0$), there is no power under this condition—to extract power one needs force and motion. On the other hand, the power output is a maximum when

$$U\big|_{\text{max power}} = \frac{V_1}{2} \tag{12.52}$$

This can be shown by using Eq. 12.51 and solving for U that gives $d\dot{W}_{shaft}/dU = 0$. A bucket speed of one-half the speed of the fluid coming from the nozzle gives the maximum power. Third, the maximum speed occurs when $T_{shaft} = 0$ (i.e., the load is completely removed from the turbine, as would happen if the shaft connecting the turbine to the generator were to break and frictional torques were negligible). For this case $U = \omega R = V_1$, the turbine is ''free wheeling,'' and the water simply passes across the rotor without putting any force on the buckets.

The maximum speed of a Pelton wheel occurs when the shaft torque is zero.

Although the actual flow through a Pelton wheel is considerably more complex than assumed in the above simplified analysis, reasonable results and trends are obtained by this simple application of the moment-of-momentum principle.

EXAMPLE 12.6

Water to drive a Pelton wheel is supplied through a pipe from a lake as indicated in Fig. E12.6*a*. Determine the nozzle diameter, D_1, that will give the maximum power output. Also determine this maximum power and the angular velocity of the rotor at this condition.

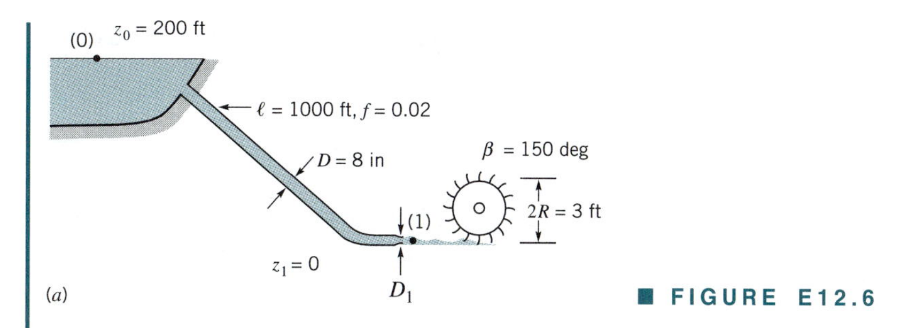

■ FIGURE E12.6

SOLUTION

As indicated by Eq. 12.51, the power output depends on the flowrate, $Q = \dot{m}/\rho$ and the jet speed at the nozzle exit, V_1, both of which depend on the diameter of the nozzle, D_1, and the head loss associated with the supply pipe. That is

$$\dot{W}_{\text{shaft}} = \rho Q U(U - V_1)(1 - \cos \beta) \tag{1}$$

The nozzle exit speed, V_1, can be obtained by applying the energy equation (Eq. 5.85) between a point on the lake surface (where $V_0 = p_0 = 0$) and the nozzle outlet (where $z_1 = p_1 = 0$) to give

$$z_0 = \frac{V_1^2}{2g} + h_L \tag{2}$$

where the head loss is given in terms of the friction factor, f, as (see Eq. 8.34)

$$h_L = f \frac{\ell}{D} \frac{V^2}{2g}$$

The speed, V, of the fluid in the pipe of diameter D is obtained from the continuity equation as

$$V = \frac{A_1 V_1}{A} = \left(\frac{D_1}{D}\right)^2 V_1$$

We have neglected minor losses associated with the pipe entrance and the nozzle. With the given data, Eq. 2 becomes

$$z_0 = \left[1 + f \frac{\ell}{D} \left(\frac{D_1}{D}\right)^4\right] \frac{V_1^2}{2g} \tag{3}$$

or

$$V_1 = \left[\frac{2gz_0}{1 + f \dfrac{\ell}{D}\left(\dfrac{D_1}{D}\right)^4}\right]^{1/2}$$

$$= \left[\frac{2(32.2 \text{ ft/s}^2)(200 \text{ ft})}{1 + 0.02\left(\dfrac{1000 \text{ ft}}{8/12 \text{ ft}}\right)\left(\dfrac{D_1}{8/12}\right)^4}\right]^{1/2} = \frac{113.5}{\sqrt{1 + 152\, D_1^4}} \tag{4}$$

where D_1 is in feet.

By combining Eqs. 1 and 4 and using $Q = \pi D_1^2 V_1/4$ we obtain the power as a function of D_1 and U as

$$\dot{W}_{\text{shaft}} = \frac{323\ UD_1^2}{\sqrt{1 + 152\ D_1^4}}\left[U - \frac{113.5}{\sqrt{1 + 152\ D_1^4}}\right] \quad \textbf{(5)}$$

where U is in ft/s and $\dot{W}_{\text{shaft}}$ is in ft·lb/s. These results are plotted as a function of U for various values of D_1 in Fig. E12.6*b*.

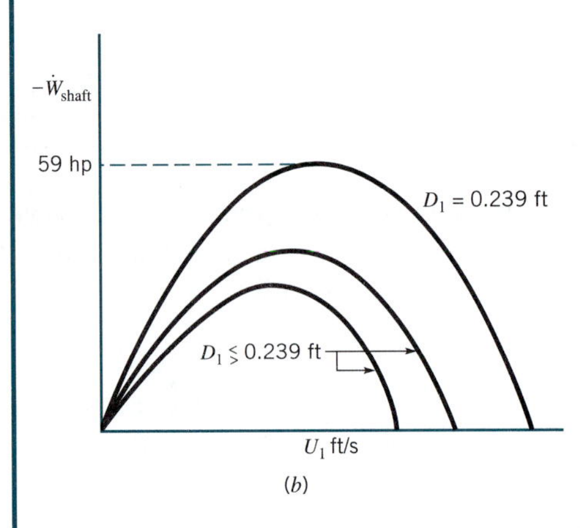

(*b*)

■ **FIGURE E12.6** ***(Continued)***

As shown by Eq. 12.52, the maximum power (in terms of its variation with U) occurs when $U = V_1/2$, which, when used with Eqs. 4 and 5, gives

$$\dot{W}_{\text{shaft}} = -\frac{1.04 \times 10^6\ D_1^2}{(1 + 152\ D_1^4)^{3/2}} \quad \textbf{(6)}$$

The maximum power possible occurs when $d\dot{W}_{\text{shaft}}/dD_1 = 0$, which according to Eq. 6 can be found as follows:

$$\frac{d\dot{W}_{\text{shaft}}}{dD_1} = -1.04 \times 10^6 \left[\frac{2\ D_1}{(1 + 152\ D_1^4)^{3/2}} - \left(\frac{3}{2}\right)\frac{4(152)\ D_1^5}{(1 + 152\ D_1^4)^{5/2}}\right] = 0$$

or

$$304\ D_1^4 = 1$$

Thus, the nozzle diameter for maximum power output is

$$D_1 = 0.239 \text{ ft} \quad \textbf{(Ans)}$$

The corresponding maximum power can be determined from Eq. 6 as

$$\dot{W}_{\text{shaft}} = -\frac{1.04 \times 10^6\ (0.239)^2}{[1 + 152(0.239)^4]^{3/2}} = -3.25 \times 10^4 \text{ ft·lb/s}$$

or

$$\dot{W}_{\text{shaft}} = -3.25 \times 10^4 \text{ ft·lb/s} \times \frac{1 \text{ hp}}{550 \text{ ft·lb/s}} = -59.0 \text{ hp} \quad \textbf{(Ans)}$$

The rotor speed at the maximum power condition can be obtained from

$$U = \omega R = \frac{V_1}{2}$$

where V_1 is given by Eq. 4. Thus,

$$\omega = \frac{V_1}{2R} = \frac{\dfrac{113.5}{\sqrt{1 + 152(0.239)^4}} \text{ ft/s}}{2\left(\dfrac{3}{2} \text{ ft}\right)}$$

$$= 30.9 \text{ rad/s} \times 1 \text{ rev}/2\pi \text{ rad} \times 60 \text{ s/min} = 295 \text{ rpm} \qquad \textbf{(Ans)}$$

The reason that an optimum diameter nozzle exists can be explained as follows. A larger diameter nozzle will allow a larger flowrate, but it will produce a smaller jet velocity because of the head loss within the supply side. A smaller diameter nozzle will reduce the flowrate but will produce a larger jet velocity. Since the power depends on a product combination of flowrate and jet velocity (see Eq. 1), there is an optimum-diameter nozzle that gives the maximum power.

The above results can be generalized (i.e., without regard to the specific parameter values of this problem) by considering Eqs. 1 and 3 and the condition that $U = V_1/2$ to obtain

$$\dot{W}_{\text{shaft}|U=V_1/2} = -\frac{\pi}{16}\rho(1 - \cos\beta)(2gz_0)^{3/2} D_1^2 \Big/ \left(1 + f\frac{\ell}{D^5} D_1^4\right)^{3/2}$$

By setting $d\dot{W}_{\text{shaft}}/dD_1 = 0$, it can be shown (see Problem 12.58) that the maximum power occurs when

$$D_1 = D \Big/ \left(2f\frac{\ell}{D}\right)^{1/4}$$

which gives the same results obtained above for the specific parameters of the example problem. Note that the optimum condition depends only on the friction factor and the length to diameter ratio of the supply pipe. What happens if the supply pipe is frictionless or of essentially zero length?

In previous chapters we mainly treated turbines (and pumps) as "black boxes" in the flow that removed (or added) energy to the fluid. We treated these devices as objects that removed a certain head (the turbine head, h_T) or added a certain head (the pump head, h_p) to the fluid. The relationship between the turbine head and the power output as described by the moment-of-momentum considerations is illustrated in Example 12.7.

Water flows through the Pelton wheel turbine shown in Fig. 12.24. For simplicity we assume that the water is turned 180° by the blade. Show, based on the energy equation (Eq. 5.84), that the maximum power output occurs when the absolute velocity of the fluid exiting the turbine is zero.

SOLUTION

As indicated by Eq. 12.51, the shaft power of the turbine is given by

$$\dot{W}_{\text{shaft}} = \rho Q U(U - V_1)(1 - \cos\beta) = 2\rho Q(U^2 - V_1 U) \qquad \textbf{(1)}$$

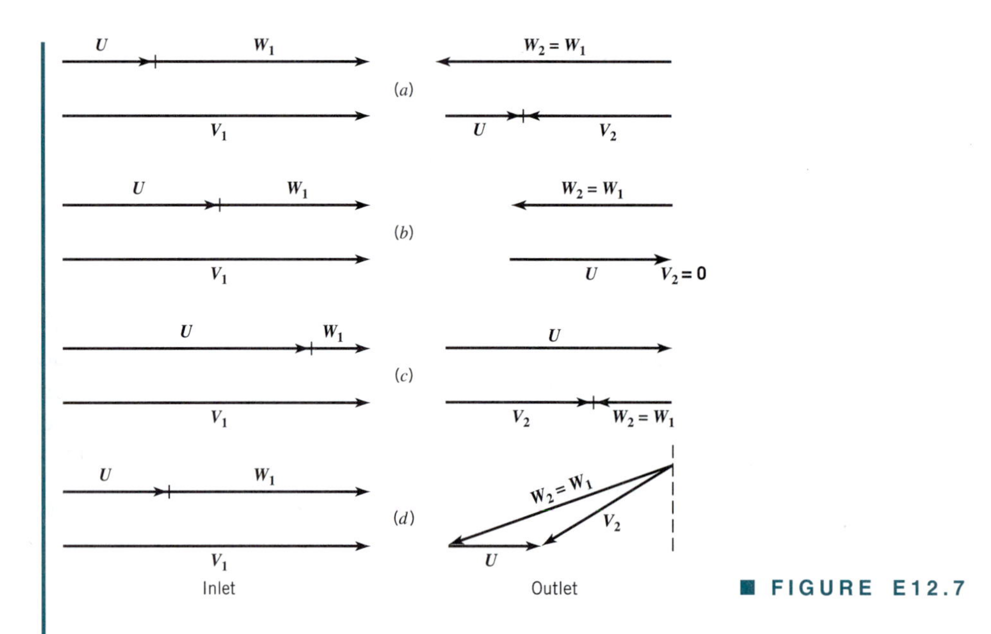

■ FIGURE E12.7

For this impulse turbine with $\beta = 180°$, the velocity triangles simplify into the diagram types shown in Fig. E12.7. Three possibilities are indicated: **(a)** the exit absolute velocity, $\mathbf{V}_2$, is directed back toward the nozzle, **(b)** the absolute velocity at the exit is zero, or **(c)** the exiting stream flows in the direction of the incoming stream.

According to Eq. 12.52, the maximum power occurs when $U = V_1/2$, which corresponds to the situation shown in Fig. E12.7*b*, That is, $U = V_1/2 = W_1$. If viscous effects are negligible, then $W_1 = W_2$ and we have $U = W_2$, which gives

$$V_2 = 0 \qquad \textbf{(Ans)}$$

If we consider the energy equation (Eq. 5.84) for flow across the rotor we have

$$\frac{p_1}{\gamma} + \frac{V_1^2}{2g} + z_1 = \frac{p_2}{\gamma} + \frac{V_2^2}{2g} + z_2 + h_T + h_L$$

where h_T is the turbine head. This simplifies to

$$h_T = \frac{V_1^2 - V_2^2}{2g} - h_L \qquad \textbf{(2)}$$

since $p_1 = p_2$ and $z_1 = z_2$. Note that the impulse turbine obtains its energy from a reduction in the velocity head. The largest turbine head possible (and therefore the largest power) occurs when all of the kinetic energy available is extracted by the turbine, giving

$$V_2 = 0 \qquad \textbf{(Ans)}$$

This is consistent with the maximum power condition represented by Fig. E12.7*b*.

As indicated by Eq. 1, if the exit absolute velocity is not in the plane of the rotor (i.e., $\beta < 180°$), there is a reduction in the power available (by a factor of $1 - \cos\beta$). This is also supported by the energy equation, Eq. 2, as follows. For $\beta < 180°$ the inlet and exit velocity triangles are as shown in Fig. E12.7*d*. Regardless of the bucket speed, U, it is not possible to reduce the value of V_2 to zero—there is always a component in the axial direction. Thus, according to Eq. 2, the turbine cannot extract the entire velocity head; the exiting fluid has some kinetic energy left in it.

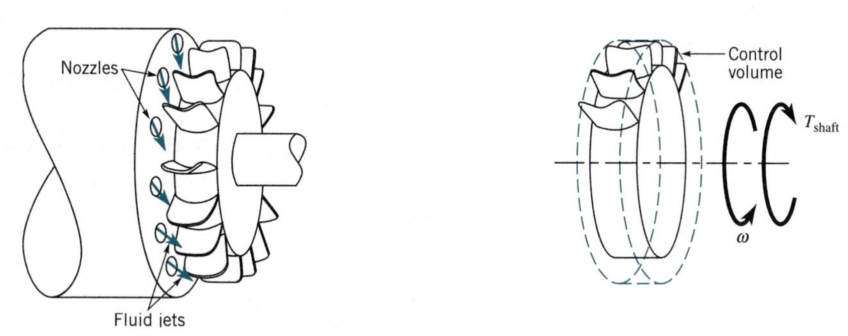

■ **FIGURE 12.29 A multinozzle, non-Pelton wheel impulse turbine commonly used with air as the working fluid.**

Dentist drill turbines are usually of the impulse class.

A second type of impulse turbine that is widely used (most often with gas as the working fluid) is indicated in Fig. 12.29. A circumferential series of fluid jets strikes the rotating blades which, as with the Pelton wheel, alter both the direction and magnitude of the absolute velocity. As with the Pelton wheel, the inlet and exit pressures (i.e., on either side of the rotor) are equal, and the magnitude of the relative velocity is unchanged as the fluid slides across the blades (if frictional effects are negligible).

Typical inlet and exit velocity triangles (absolute, relative, and blade velocities) are shown in Fig. 12.30. As discussed in Section 12.2, in order for the absolute velocity of the fluid to be changed as indicated during its passage across the blade, the blade must push on the fluid in the direction opposite of the blade motion. Hence, the fluid pushes on the blade in the direction of the blade's motion—the fluid does work on the blade (a turbine).

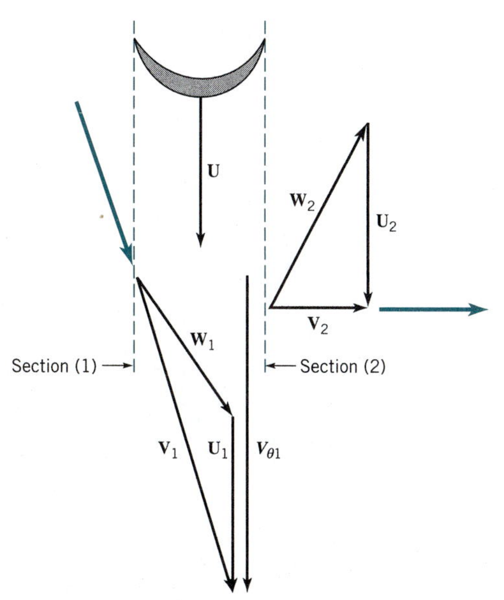

■ **FIGURE 12.30 Inlet and exit velocity triangles for the impulse turbine shown in Fig. 12.29.**

EXAMPLE 12.8

An air turbine used to drive the high-speed drill used by your dentist is shown in Fig. E12.8*a*. Air exiting from the upstream nozzle holes forces the turbine blades to move in the direction shown. Estimate the shaft energy per unit mass of air flowing through the turbine under the following conditions. The turbine rotor speed is 300,000 rpm, the tangential component of velocity out of the nozzle is twice the blade speed, and the tangential component of the absolute velocity out of the rotor is zero.

SOLUTION

We use the fixed, nondeforming control volume that includes the turbine rotor and the fluid in the rotor blade passages at an instant of time (see Fig. E12.8*b*). The only torque acting on this control volume is the shaft torque. For simplicity we analyze this problem using an arithmetic mean radius, r_m, where

$$r_m = \frac{1}{2}(r_0 + r_i)$$

A sketch of the velocity triangles at the rotor entrance and exit is shown in Fig. E12.8*c*.

Application of Eq. 12.5 (a form of the moment-of-momentum equation) gives

$$w_{\text{shaft}} = -U_1 V_{\theta 1} + U_2 V_{\theta 2} \tag{1}$$

where w_{shaft} is shaft energy per unit of mass flowing through the turbine. From the problem

V12.5

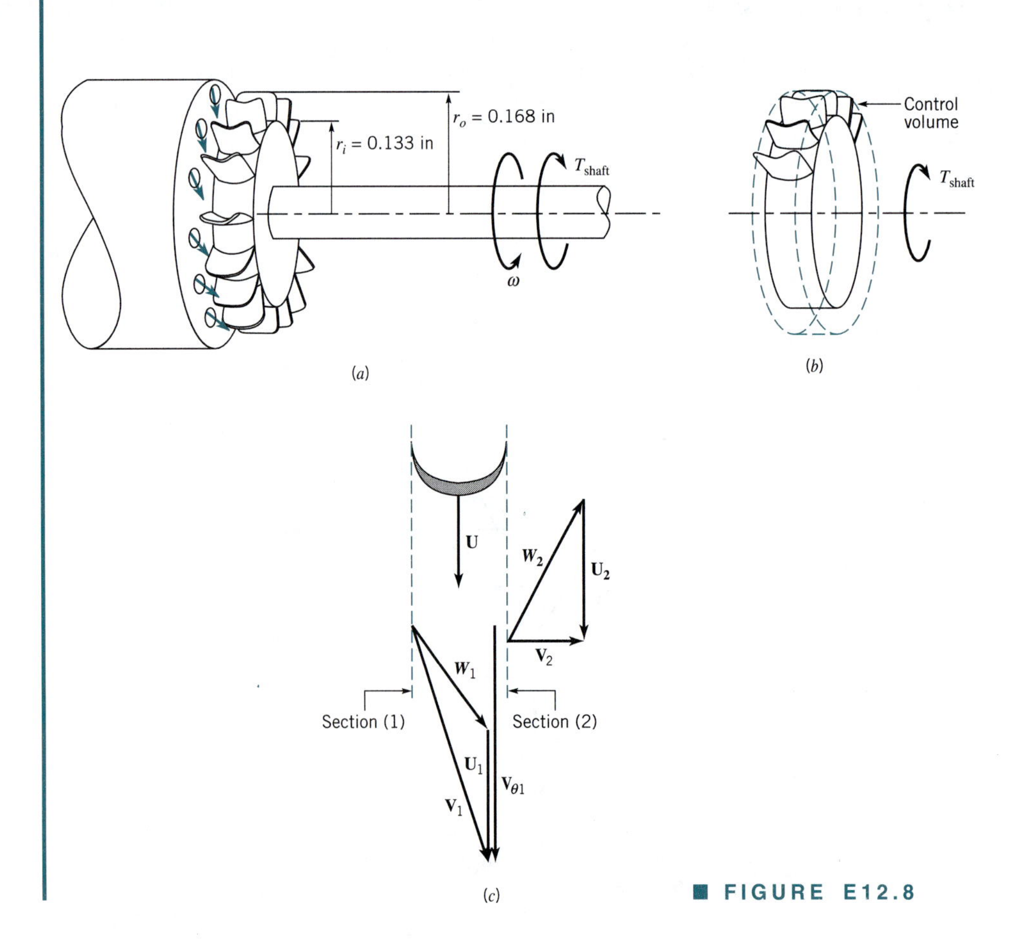

■ FIGURE E12.8

statement, $V_{\theta 1} = 2U$ and $V_{\theta 2} = 0$, where

$$U = \omega r_m = (300{,}000 \text{ rev/min})(1 \text{ min}/60 \text{ s})(2\pi \text{ rad/rev}) \times (0.168 \text{ in.} + 0.133 \text{ in.})/2(12 \text{ in./ft}) = 394 \text{ ft/s} \qquad \textbf{(2)}$$

is the mean-radius blade velocity. Thus, Eq. (1) becomes

$$w_{\text{shaft}} = -U_1 V_{\theta 1} = -2U^2 = -2(394 \text{ ft/s})^2 = -310{,}000 \text{ ft}^2/\text{s}^2$$
$$= (-310{,}000 \text{ ft}^2/\text{s}^2)(1 \text{ lb/slug·ft/s}^2) = -310{,}000 \text{ ft·lb/slug} \qquad \textbf{(Ans)}$$

For each slug of air passing through the turbine there is 310,000 ft·lb of energy available at the shaft to drive the drill. However, because of fluid friction, the actual amount of energy given up by each slug of air will be greater than the amount available at the shaft. How much greater depends on the efficiency of the fluid-mechanical energy transfer between the fluid and the turbine blades.

Recall that the shaft power, $\dot{W}_{\text{shaft}}$, is given by

$$\dot{W}_{\text{shaft}} = \dot{m} w_{\text{shaft}}$$

Hence, to determine the power we need to know the mass flowrate, $\dot{m}$, which depends on the size and number of the nozzles. Although the energy per unit mass is large (i.e., 310,000 ft·lb/slug), the flowrate is small, so the power is not "large."

12.8.2 Reaction Turbines

Reaction turbines are best suited for higher flowrate and lower head situations.

As indicated in the previous section, impulse turbines are best suited (i.e., most efficient) for lower flowrate and higher head operations. Reaction turbines, on the other hand, are best suited for higher flowrate and lower head situations such as are often encountered in hydroelectric power plants associated with a dammed river, for example.

In a reaction turbine the working fluid completely fills the passageways through which it flows (unlike an impulse turbine, which contains one or more individual unconfined jets of fluid). The angular momentum, pressure, and velocity of the fluid decrease as it flows through the turbine rotor—the turbine rotor extracts energy from the fluid.

As with pumps, turbines are manufactured in a variety of configurations—radial-flow, mixed-flow, and axial-flow. Typical radial- and mixed-flow hydraulic turbines are called *Francis turbines*, named after James B. Francis, an American engineer. At very low heads the most efficient type of turbine is the axial-flow or propeller turbine. The *Kaplan turbine*, named after Victor Kaplan, a German professor, is an efficient axial-flow hydraulic turbine with adjustable blades. Cross sections of these different turbine types are shown in Fig. 12.31.

As shown in Fig. 12.31*a*, flow across the rotor blades of a radial-inflow turbine has a major component in the radial direction. Inlet guide vanes (which may be adjusted to allow optimum performance) direct the water into the rotor with a tangential component of velocity. The absolute velocity of the water leaving the rotor is essentially without tangential velocity. Hence, the rotor decreases the angular momentum of the fluid, the fluid exerts a torque on the rotor in the direction of rotation, and the rotor extracts energy from the fluid. The Euler turbomachine equation (Eq. 12.2) and the corresponding power equation (Eq. 12.4) are equally valid for this turbine as they are for the centrifugal pump discussed in Section 12.4.

As shown in Fig. 12.31*b*, for an axial-flow Kaplan turbine, the fluid flows through the inlet guide vanes and achieves a tangential velocity in a vortex (swirl) motion before it reaches the rotor. Flow across the rotor contains a major axial component. Both the inlet guide vanes

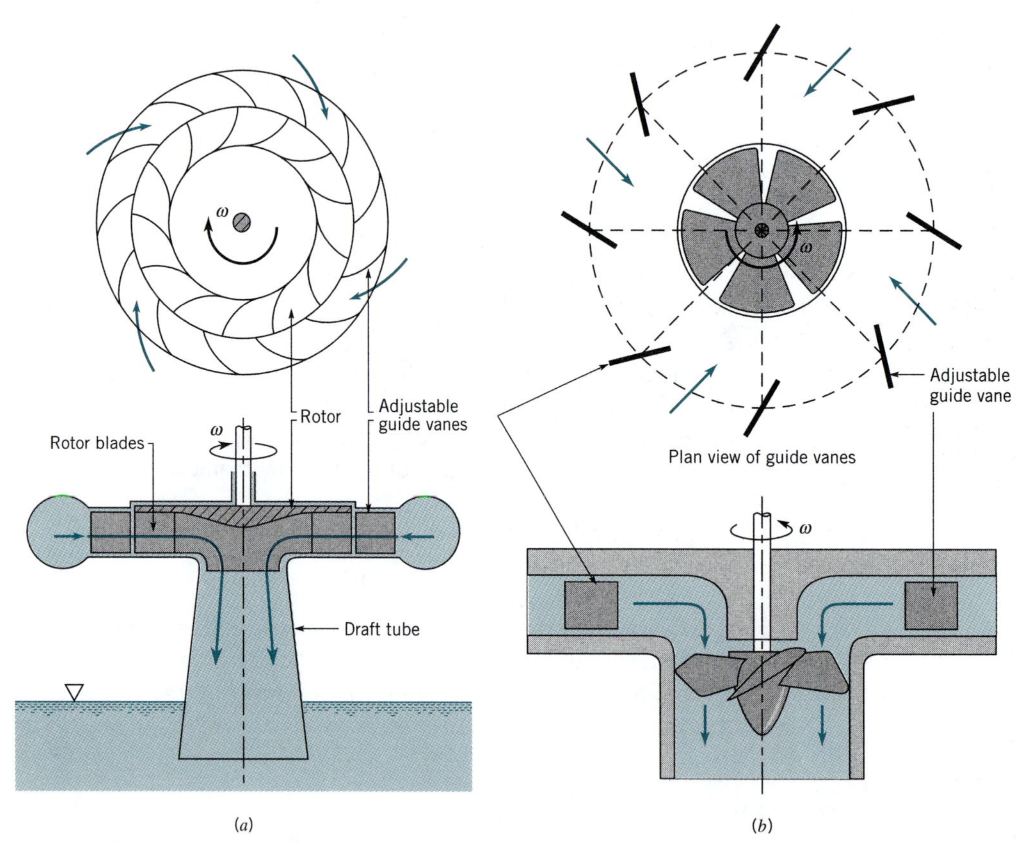

■ **FIGURE 12.31** ***(a)* Typical radial-flow Francis turbine, *(b)* typical axial-flow Kaplan turbine.**

and the turbine blades can be adjusted by changing their setting angles to produce the best match (optimum output) for the specific operating conditions. For example, the operating head available may change from season to season and/or the flow rate through the rotor may vary.

In a crude sense, pumps and turbines are the reverse flow versions of each other.

Pumps and turbines are often thought of as the ''inverse'' of each other. Pumps add energy to the fluid; turbines remove energy. The propeller on an outboard motor (a pump) and the propeller on a Kaplan turbine are in some ways geometrically similar, but they perform opposite tasks. Similar comparisons can be made for centrifugal pumps and Francis turbines. In fact, some large turbomachines at hydroelectric power plants are designed to be run as turbines during high-power demand periods (i.e., during the day) and as pumps to resupply the upstream reservoir from the downstream reservoir during low-demand times (i.e., at night). Thus, a pump type often has its corresponding turbine type. However, is it possible to have the ''inverse'' of a Pelton wheel turbine—an impulse pump?

As with pumps, incompressible flow turbine performance is often specified in terms of appropriate dimensionless parameters. The flow coefficient, $C_Q = Q/\omega D^3$, the head coefficient, $C_H = gh_T/\omega^2 D^2$, and the power coefficient, $C_{\mathcal{P}} = \dot{W}_{\text{shaft}}/\rho\omega^3 D^5$, are defined in the same way for pumps and turbines. On the other hand, turbine efficiency, η, is the inverse of pump efficiency. That is, the efficiency is the ratio of the shaft power output to the power available in the flowing fluid, or

$$\eta = \frac{\dot{W}_{\text{shaft}}}{\rho g Q h_T}$$

For geometrically similar turbines and for negligible Reynolds number and surface roughness difference effects, the relationships between the dimensionless parameters are given functionally by that shown in Eqs. 12.29, 30, and 31. That is,

$$C_H = \phi_1(C_Q), \quad C_{\mathcal{P}} = \phi_2(C_Q), \quad \text{and} \quad \eta = \phi_3(C_Q)$$

where the functions ϕ_1, ϕ_2, and ϕ_3 are dependent on the type of turbine involved. Also, for turbines the efficiency, η, is related to the other coefficients according to $\eta = C_{\mathcal{P}}/C_H C_Q$.

As indicated above, the design engineer has a variety of turbine types available for any given application. It is necessary to determine which type of turbine would best fit the job (i.e., be most efficient) before detailed design work is attempted. As with pumps, the use of a specific speed parameter can help provide this information. For hydraulic turbines, the rotor diameter D is eliminated between the flow coefficient and the power coefficient to obtain the *power specific speed*, N_s', where

$$N_s' = \frac{\omega\sqrt{\dot{W}_{\text{shaft}}/\rho}}{(gh_T)^{5/4}}$$

We use the more common, but not dimensionless, definition of specific speed

$$N_{sd}' = \frac{\omega(\text{rpm})\sqrt{\dot{W}_{\text{shaft}}\ (\text{bhp})}}{[h_T(\text{ft})]^{5/4}} \tag{12.53}$$

That is, N_{sd}' is calculated with angular velocity, ω, in rpm; shaft power, $\dot{W}_{\text{shaft}}$, in brake horsepower; and head, h_T, in feet. Optimum turbine efficiency (for large turbines) as a function of specific speed is indicated in Fig. 12.32. Also shown are representative rotor and casing cross sections. Note that impulse turbines are best at low specific speeds; that is, when operating with large heads and small flowrate. The other extreme is axial-flow turbines, which

Specific speed may be used to approximate what kind of turbine geometry (axial to radial) would operate most efficiently.

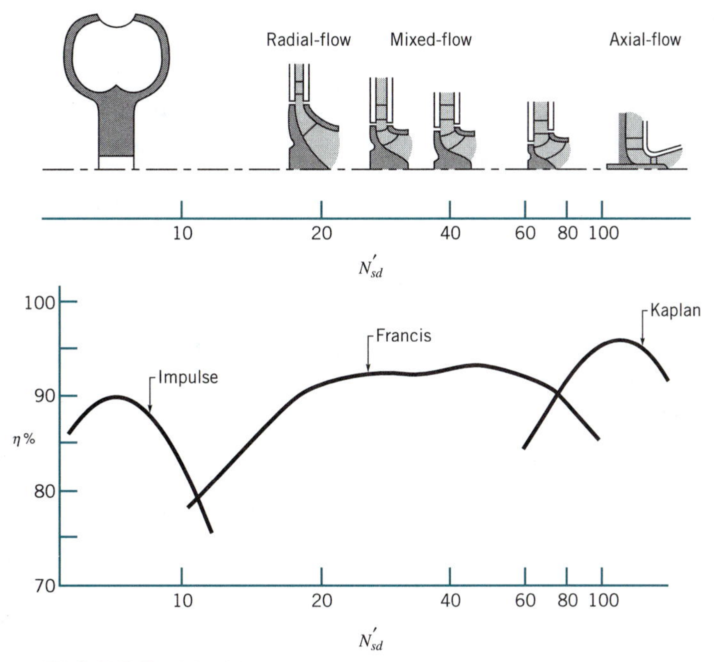

■ **FIGURE 12.32** **Typical turbine cross sections and maximum efficiencies as a function of specific speed.**

are the most efficient type if the head is low and if the flowrate is large. For intermediate values of specific speeds, radial- and mixed-flow turbines offer the best performance.

The data shown in Fig. 12.32 are meant only to provide a guide for turbine-type selection. The actual turbine efficiency for a given turbine depends very strongly on the detailed design of the turbine. Considerable analysis, testing, and experience are needed to produce an efficient turbine. However, the data of Fig. 12.32 are representative. Much additional information can be found in the literature (Ref. 1).

EXAMPLE 12.9

A hydraulic turbine is to operate at an angular velocity of 6 rev/s, a flowrate of 10 ft^3/s, and a head of 20 ft. What type of turbine should be selected? Explain.

SOLUTION

The most efficient type of turbine to use can be obtained by calculating the specific speed, N'_{sd} and using the information of Fig. 12.32. To use the dimensional form of the specific speed indicated in Fig. 12.32 we must convert the given data into the appropriate units. For the rotor speed we get

$$\omega = 6 \text{ rev/s} \times 60 \text{ s/min} = 360 \text{ rpm}$$

To estimate the shaft power, we assume all of the available head is converted into power and multiply this amount by an assumed efficiency (94%).

$$\dot{W}_{\text{shaft}} = \gamma Q z \eta = (62.4 \text{ lb/ft}^3)(10 \text{ ft}^3/\text{s}) \left[\frac{20 \text{ ft}(0.94)}{550 \text{ ft·lb/s·hp}} \right]$$

$$\dot{W}_{\text{shaft}} = 21.3 \text{ hp}$$

Thus for this turbine,

$$N'_{sd} = \frac{\omega\sqrt{\dot{W}_{\text{shaft}}}}{(h_T)^{5/4}} = \frac{(360 \text{ rpm})\sqrt{21.3 \text{ hp}}}{(20 \text{ ft})^{5/4}} = 39.3$$

According to the information of Fig. 12.32,

A mixed-flow Francis turbine would probably give the highest efficiency and an assumed efficiency of 0.94 is appropriate.

(Ans)

What would happen if we wished to use a Pelton wheel for this application? Note that with only a 20-ft head, the maximum jet velocity, V_1, obtainable (neglecting viscous effects) would be

$$V_1 = \sqrt{2\,gz} = \sqrt{2 \times 32.2 \text{ ft/s}^2 \times 20 \text{ ft}} = 35.9 \text{ ft/s}$$

As shown by Eq. 12.52, for maximum efficiency of a Pelton wheel the jet velocity is ideally two times the blade velocity. Thus, $V_1 = 2\omega R$, or the wheel diameter, $D = 2R$, is

$$D = \frac{V_1}{\omega} = \frac{35.9 \text{ ft/s}}{(6 \text{ rev/s} \times 2\pi \text{ rad/rev})} = 0.952 \text{ ft}$$

To obtain a flowrate of $Q = 10 \text{ ft}^3/\text{s}$ at a velocity of $V_1 = 35.9$ ft/s, the jet diameter, d_1, must be given by

$$Q = \frac{\pi}{4} d_1^2 V_1$$

or

$$d_1 = \left[\frac{4Q}{\pi V_1}\right]^{1/2} = \left[\frac{4(10 \text{ ft}^3/\text{s})}{\pi(35.9 \text{ ft/s})}\right]^{1/2} = 0.596 \text{ ft}$$

A Pelton wheel with a diameter of $D = 0.952$ ft supplied with water through a nozzle of diameter $d_1 = 0.596$ ft is not a practical design. Typically $d_1 << D$ (see Fig. 12.22). By using multiple jets it would be possible to reduce the jet diameter. However, even with 8 jets, the jet diameter would be 0.211 ft, which is still too large (relative to the wheel diameter) to be practical. Hence, the above calculations reinforce the results presented in Fig. 12.32—a Pelton wheel would not be practical for this application. If the flowrate were considerably smaller, the specific speed could be reduced to the range where a Pelton wheel would be the type to use (rather than a mixed-flow reaction turbine).

12.9 Compressible Flow Turbomachines

Compressors are pumps in which density changes are considerable.

Compressible flow turbomachines are in many ways similar to the incompressible flow pumps and turbines described in previous portions of this chapter. The main difference is that the density of the fluid (a gas or vapor) changes significantly from the inlet to the outlet of the compressible flow machines. This added feature has interesting consequences, benefits, and complications.

Compressors are pumps that add energy to the fluid, causing a significant pressure rise and a corresponding significant increase in density. Compressible flow turbines, on the other hand, remove energy from the fluid, causing a lower pressure and a smaller density at the outlet than at the inlet. The information provided earlier about basic energy considerations (Section 12.2) and basic angular momentum considerations (Section 12.3) is directly applicable to these turbomachines in the ways demonstrated earlier.

As discussed in Chapter 11, compressible flow study requires an understanding of the principles of thermodynamics. Similarly, an in-depth analysis of compressible flow turbomachines requires use of various thermodynamic concepts. In this section we provide only a brief discussion of some of the general properties of compressors and compressible flow turbines. The interested reader is encouraged to read some of the excellent references available for further information (Refs. 2, 4, 8, 26, 27).

12.9.1 Compressors

Turbocompressors operate with the continuous compression of gas flowing through the device. Since there is a significant pressure and density increase, there is also a considerable temperature increase.

Radial-flow (or centrifugal) compressors are essentially centrifugal pumps (see Section 12.4) that use a gas (rather than a liquid) as the working fluid. They are typically high pressure rise, low flowrate, and axially compact turbomachines. A photograph of the rotor of the centrifugal compressor in an automobile turbocharger is shown in Fig. 12.33.

The amount of compression is typically given in terms of the *total pressure ratio*, $PR = p_{02}/p_{01}$, where the pressures are absolute. Thus, a radial flow compressor with $PR = 3.0$ can compress standard atmospheric air from 14.7 psia to $3.0 \times 14.7 = 44.1$ psia.

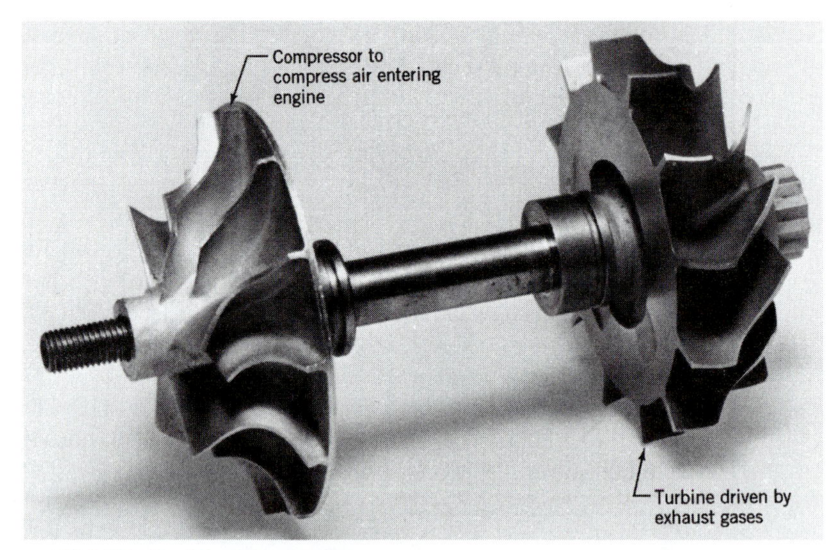

■ FIGURE 12.33 **Photograph of the rotor from an automobile turbocharger.**

Multistaging is common in high pressure ratio compressors.

Higher pressure ratios can be obtained by using *multiple stage* devices in which flow from the outlet of the preceding stage proceeds to the inlet of the following stage. If each stage has the same pressure ratio, *PR*, the overall pressure ratio after *n* stages is PR^n. Thus, a four-stage compressor with individual stage $PR = 2.0$ can compress standard air from 14.7 psia to $2^4 \times 14.7 = 235$ psia. Adiabatic (i.e., no heat transfer) compression of a gas causes an increase in temperature and requires more work than isothermal (constant temperature) compression of a gas. An interstage cooler (i.e., an intercooler heat exchanger) as shown in Fig. 12.34 can be used to reduce the compressed gas temperature and thus the work required.

Relative to centrifugal water pumps, radial compressors of comparable size rotate at much higher speeds. It is not uncommon for the rotor blade exit speed and the speed of the absolute flow leaving the impeller to be greater than the speed of sound. That such large

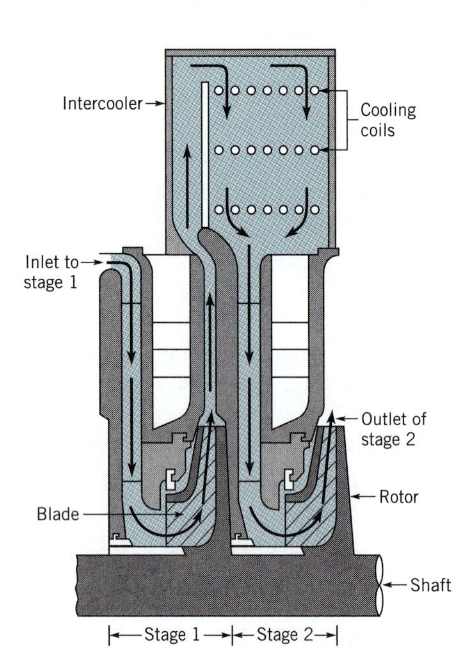

■ FIGURE 12.34 **Two-stage centrifugal compressor with an intercooler.**

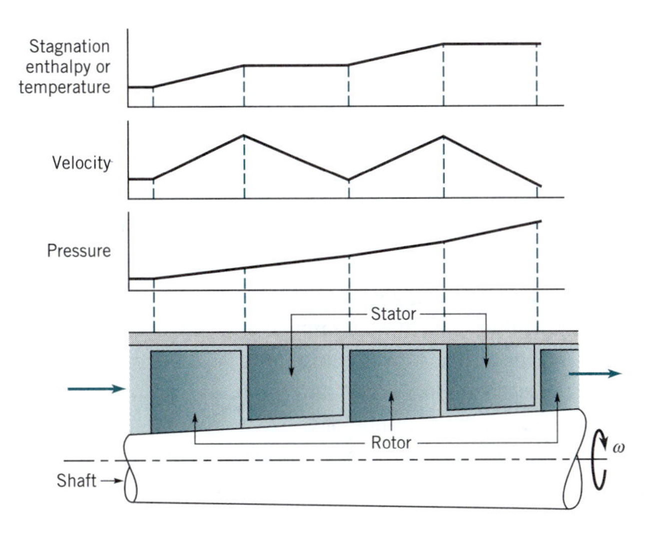

■ **FIGURE 12.35** **Enthalpy, velocity, and pressure distribution in an axial-flow compressor.**

speeds are necessary for compressors can be seen by noting that the large pressure rise designed for is related to the differences of several squared speeds (see Eq. 12.14).

Axial-flow compressor multistaging requires less space than centrifugal compressors.

The axial-flow compressor is the other widely used configuration. This type of turbomachine has a lower pressure rise per stage, a higher flowrate, and is more radially compact than a centrifugal compressor. As shown in Fig. 12.35, axial-flow compressors usually consist of several stages, with each stage containing a rotor/stator row pair. For an 11-stage compressor, a compression ratio of $PR = 1.2$ per stage gives an overall pressure ratio of $p_{02}/p_{01} = 1.2^{11} = 7.4$. As the gas is compressed and its density increases, a smaller annulus cross-section area is required and the flow channel size decreases from the inlet to the outlet of the compressor. The typical jet aircraft engine uses an axial-flow compressor as one of its main components (see Fig. 12.36).

An axial-flow compressor can include a set of *inlet guide vanes* upstream of the first rotor row. These guide vanes optimize the size of the relative velocity into the first rotor row by directing the flow away from the axial direction. *Rotor blades* push on the gas in the

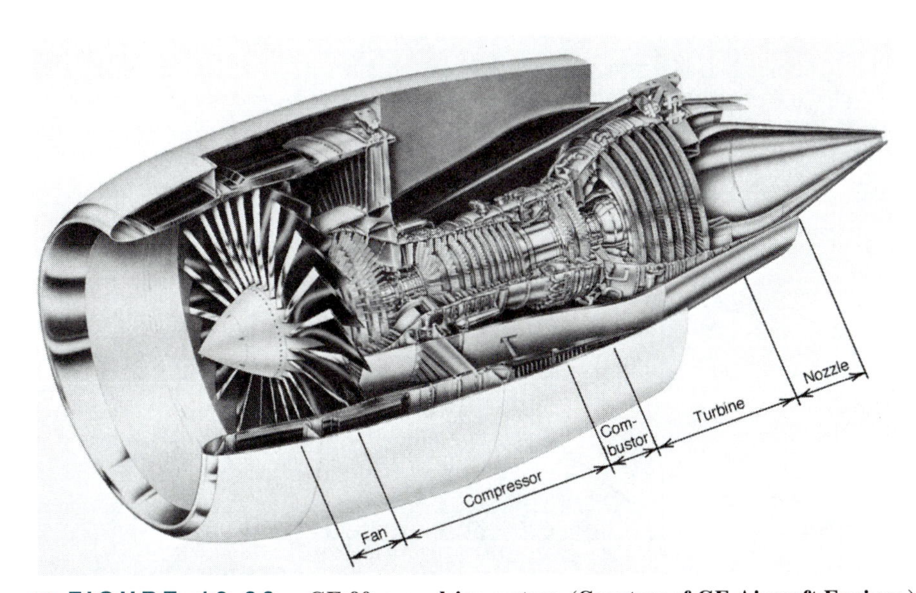

■ **FIGURE 12.36** **GE 90 propulsion system. (Courtesy of GE Aircraft Engines.)**

direction of blade motion and to the rear, adding energy (like in an axial-pump) and moving the gas through the compressor. The *stator blade* rows act as diffusers, turning the fluid back toward the axial direction and increasing the static pressure. The stator blades cannot add energy to the fluid because they are stationary. Typical pressure, velocity,and enthalpy distributions along the axial direction are shown in Fig. 12.35. [If you are not familiar with the thermodynamic concept of enthalpy (see Section 11.1), you may replace ''enthalpy'' by temperature as an approximation.] The reaction of the compressor stage is equal to the ratio of the rise in static enthalpy or temperature achieved across the rotor to the enthalpy or temperature rise across the stage. Most modern compressors involve 50% or higher reaction.

The blades in an axial-flow compressor are airfoils carefully designed to produce appropriate lift and drag forces on the flowing gas. As occurs with airplane wings, compressor blades can stall (see Section 9.4). When the flow rate is decreased from the design amount, the velocity triangle at the entrance of the rotor row indicates that the relative flow meets the blade leading edge at larger angles of incidence than the design value. When the angle of incidence becomes too large, blade stall can occur and the result is *compressor surge* or *stall*—unstable flow conditions that can cause excessive vibration, noise, poor performance, and possible damage to the machine. The lower flow rate bound of compressor operation is related to the beginning of these instabilities (see Fig. 12.37).

Compressor blades can stall, and unstable flow conditions can subsequently occur.

Other important compressible flow phenomena such as variations of the Mach cone (see Section 11.3), shock waves (see Section 11.5.3), and choked flow (see Section 11.4.2) occur commonly in compressible flow turbomachines. They must be carefully designed for. These phenomena are very sensitive to even very small changes or variations of geometry. Shock strength is kept low to minimize shock loss, and choked flows limit the upper flowrate boundary of machine operation (see Fig. 12.37).

The experimental performance data for compressors are systematically summarized with parameters prompted by dimensional analysis. As mentioned earlier, total pressure ratio, p_{02}/p_{01}, is used instead of the head-rise coefficient associated with pumps, blowers, and fans.

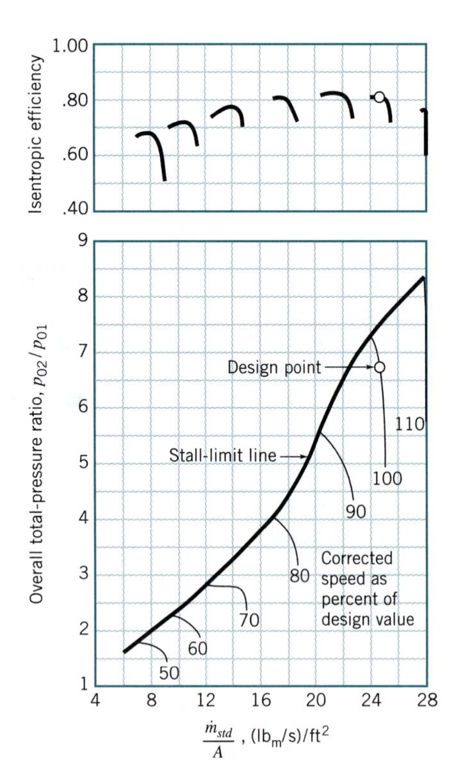

■ **FIGURE 12.37 Performance characteristics of an axial-flow compressor (Ref. 26).**

Either isentropic or polytropic efficiencies are used to characterize compressor performance. A detailed explanation of these efficiencies is beyond the scope of this text. Those interested in learning more about these parameters should study any of several available books on turbomachines (e.g., Refs. 7, 8, 9). Basically, each of these compressor efficiencies involves a ratio of ideal work to actual work required to accomplish the compression. The isentropic efficiency involves a ratio of the ideal work required with an adiabatic and frictionless (no loss) compression process to the actual work required to achieve the same total pressure rise. The polytropic efficiency involves a ratio of the ideal work required to achieve the actual end state of the compression with a polytropic and frictionless process between the actual beginning and end stagnation states across the compressor and the actual work involved between these same states.

The flow parameter commonly used for compressors is based on the following dimensionless grouping from dimensional analysis

As with incompressible flow pumps, dimensionless pi terms are useful for organizing compressor data.

$$\frac{R\dot{m}\sqrt{kRT_{01}}}{D^2\, p_{01}}$$

where R is the gas constant, $\dot{m}$ the mass flowrate, k the specific heat ratio, T_{01} the stagnation temperature at the compressor inlet, D a characteristic length, and p_{01} the stagnation pressure at the compressor inlet.

To account for variations in test conditions, the following strategy is employed. We set

$$\left(\frac{R\dot{m}\sqrt{kRT_{01}}}{D^2\, p_{01}}\right)_{\text{test}} = \left(\frac{R\dot{m}\sqrt{kRT_{01}}}{D^2\, p_{01}}\right)_{\text{std}}$$

where the subscript "test" refers to a specific test condition and "std" refers to the standard atmosphere ($p_0 = 14.7$ psia, $T_0 = 518.7\ ^\circ\text{R}$) condition. When we consider a given compressor operating on a given working fluid (so that R, k, and D are constant), the above equation reduces to

$$\dot{m}_{\text{std}} = \frac{\dot{m}_{\text{test}}\sqrt{T_{01\ \text{test}}/T_{0\ \text{std}}}}{p_{01\ \text{test}}/p_{0\ \text{std}}} \tag{12.54}$$

In essence, $\dot{m}_{\text{std}}$ is the compressor-test mass flowrate "corrected" to the standard atmosphere inlet condition. The *corrected compressor mass flowrate*, $\dot{m}_{\text{std}}$, is used instead of flow coefficient. Often, $\dot{m}_{\text{std}}$ is divided by A, the frontal area of the compressor flow path.

While for pumps, blowers, and fans, rotor speed was accounted for in the flow coefficient, it is not in the corrected mass flowrate derived above. Thus, for compressors, rotor speed needs to be accounted for with an additional group. This dimensionless group is

$$\frac{ND}{\sqrt{kRT_{01}}}$$

For the same compressor operating on the same gas, we eliminate D, k and R and, as with corrected mass flowrate, obtain a corrected speed, N_{std}, where

$$N_{\text{std}} = \frac{N}{\sqrt{T_{01}/T_{\text{std}}}} \tag{12.55}$$

Often, the percentage of the corrected speed design value is used.

An example of how compressor performance data are typically summarized is shown in Fig. 12.37.

12.9.2 Compressible Flow Turbines

Turbines that use a gas or vapor as the working fluid are in many respects similar to hydraulic turbines (see Section 12.8). Compressible flow turbines may be impulse or reaction turbines, and mixed-, radial-, or axial-flow turbines. The fact that the gas may expand (compressible flow) in coursing through the turbine can introduce some important phenomena that do not occur in hydraulic turbines. (Note: It is tempting to label turbines that use a gas as the working fluid as gas turbines. However, the terminology ''gas turbine'' is commonly used to denote a *gas turbine engine*, as employed, for example, for aircraft propulsion or stationary power generation. As shown in Fig. 12.36, these engines typically contain a compressor, combustion chamber, and turbine.)

A gas turbine engine generally consists of a compressor, a combustor, and a turbine.

Although for compressible flow turbines the axial-flow type is common, the radial-inflow type is also used for various purposes. As shown in Fig. 12.33, the turbine that drives the typical automobile turbocharger compressor is a radial-inflow type. The main advantages of the radial-inflow turbine are: (1) It is robust and durable, (2) it is axially compact, and (3) it can be relatively inexpensive. A radial-flow turbine usually has a lower efficiency than an axial-flow turbine, but lower initial costs may be the compelling incentive in choosing a radial-flow turbine over an axial-flow one.

Axial-flow turbines are widely used compressible flow turbines. Steam engines used in electrical generating plants and marine propulsion and the turbines used in gas turbine engines are usually of the axial-flow type. Often they are multistage turbomachines, although single-stage compressible turbines are also produced. They may be either an impulse type or a reaction type. With compressible flow turbines, the ratio of static enthalpy or temperature drop across the rotor to this drop across the stage, rather than the ratio of static pressure differences, is used to determine reaction. Strict impulse (zero pressure drop) turbines have slightly negative reaction; the static enthalpy or temperature actually increases across the rotor. Zero-reaction turbines involve no change of static enthalpy or temperature across the rotor but do involve a slight pressure drop.

A two-stage, axial-flow impulse turbine is shown in Fig. 12.38*a*. The gas accelerates

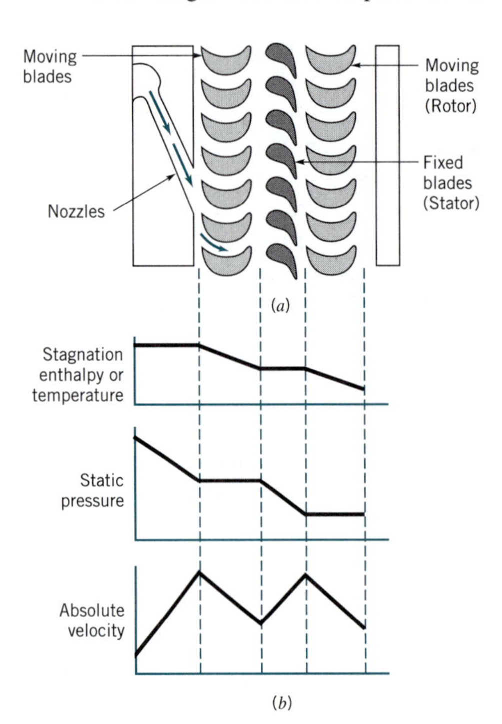

■ FIGURE 12.38 **Enthalpy, velocity, and pressure distribution in two-stage impulse turbine.**

through the supply nozzles, has some of its energy removed by the first-stage rotor blades, accelerates again through the second-stage nozzle row, and has additional energy removed by the second-stage rotor blades. As shown in Fig. 12.38*b*, the static pressure remains constant across the rotor rows. Across the second-stage nozzle row, the static pressure decreases, absolute velocity increases, and the stagnation enthalpy (temperature) is constant. Flow across the second rotor is similar to flow across the first rotor. Since the working fluid is a gas, the significant decrease in static pressure across the turbine results in a significant decrease in density—the flow is compressible. Hence, more detailed analysis of this flow must incorporate various compressible flow concepts developed in Chapter 11. Interesting phenomena such as shock waves and choking due to sonic conditions at the ''throat'' of the flow passage between blades can occur because of compressibility effects. The interested reader is encouraged to consult the various references available (Refs. 8, 9, and 27) for fascinating applications of compressible flow principles in turbines.

Static pressure decreases through all blade rows of a reaction turbine.

The rotor and nozzle blades in a three-stage, axial-flow reaction turbine are shown in Fig. 12.39*a*. The axial variations of pressure and velocity are shown in Fig. 12.39*c*. Both the stationary and rotor blade (passages) act as flow-accelerating nozzles. That is, the static pressure and enthalpy (temperature) decrease in the direction of flow for both the fixed and the rotating blade rows. This distinguishes the reaction turbine from the impulse turbine (see Fig. 12.38*b*). Energy is removed from the fluid by the rotors only (the stagnation enthalpy or temperature is constant across the adiabatic flow stators).

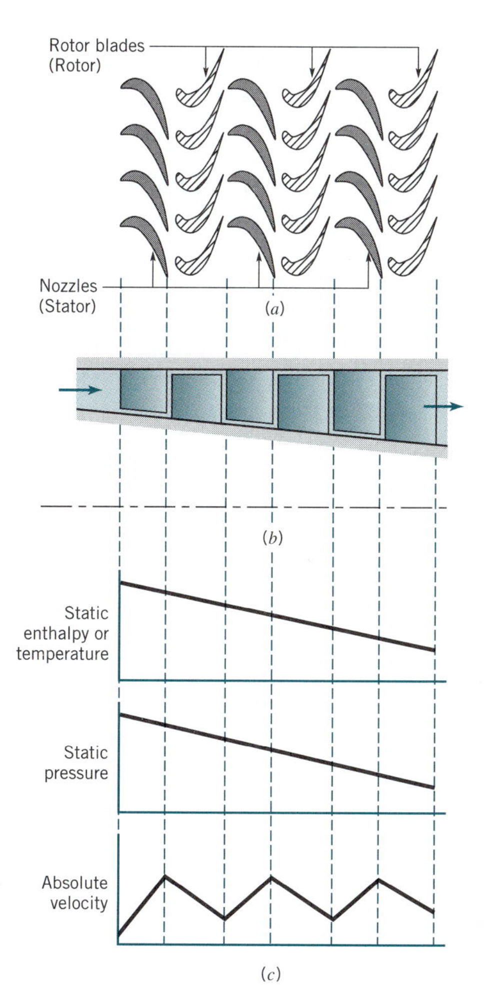

■ **FIGURE 12.39 Enthalpy, pressure, and velocity distribution in a three-stage reaction turbine.**

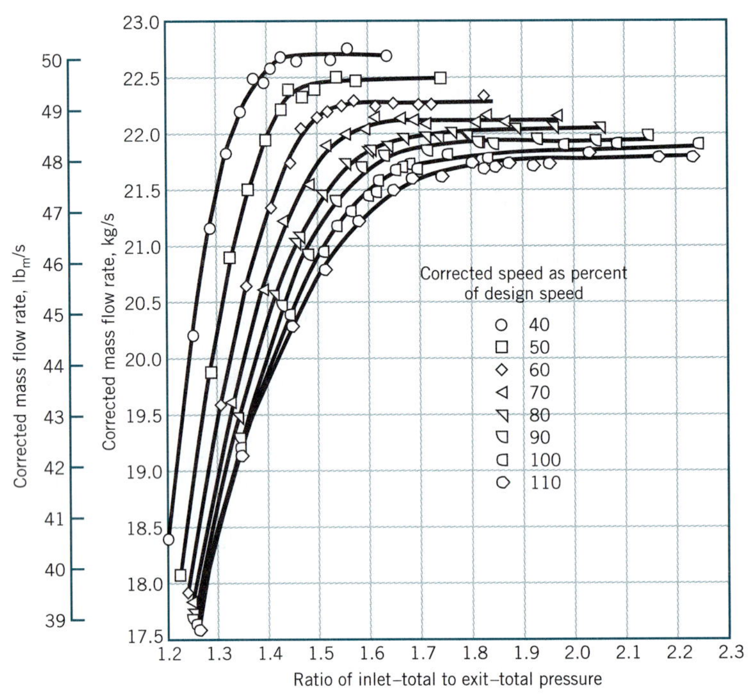

■ **FIGURE 12.40** **Typical compressible flow turbine performance "map." (Ref. 27)**

Because of the reduction of static pressure in the downstream direction, the gas expands, and the flow passage area must increase from the inlet to the outlet of this turbine. This is seen in Fig. 12.39*b*.

Turbine performance maps are used to display complex turbine characteristics.

Performance data for compressible flow turbines are summarized with the help of parameters derived from dimensional analysis. Isentropic and polytropic efficiencies (see Refs. 8, 9, and 27) are commonly used as are inlet-to-outlet total pressure ratios (p_{01}/p_{02}), corrected rotor speed (see Eq. 12.55), and corrected mass flowrate (see Eq. 12.54). In Fig. 12.40 is shown a compressible flow turbine performance "map."

References

1. Balje, O. E., *Turbomachines: A Guide to Design, Selection, and Theory*, Wiley, New York, 1981.
2. Bathie, W. W., *Fundamentals of Gas Turbines*, Wiley, New York, 1984.
3. Karassick, I. J., et al., *Pump Handbook*, McGraw-Hill, New York, 1985.
4. Boyce, M. P., *Gas Turbine Engineering Handbook*, Gulf Publishing, Houston, 1982.
5. Betz, A., *Introduction to the Theory of Flow Machines*, Pergamon, London, 1966.
6. Daugherty, R. L., Franzini, J. B., and Finnemore, E. J., *Fluid Mechanics with Engineering Applications*, 8th Ed., McGraw-Hill, New York, 1985.
7. Cumpsty, N. A., *Compressor Aerodynamics*, Longman Scientific & Technical, Essex, UK, and Wiley, New York, 1989.
8. Cohen, H., Rogers, G. F. C., and Saravanamuttoo, H. I. H., *Gas Turbine Theory*, 3rd Ed., Longman Scientific & Technical, Essex, UK, and Wiley, New York, 1987.

9. Wilson, D. G., *The Design of High-Efficiency Turbomachinery and Gas Turbines*, The MIT Press, Cambridge, 1984.
10. Stepanoff, H. J., *Centrifugal and Axial Flow Pumps*, 2nd Ed., Wiley, New York, 1957.
11. Shepherd, D. G., *Principles of Turbomachinery*, Macmillan, New York, 1956.
12. Wislicenus, G. F., *Preliminary Design of Turbopumps and Related Machinery*, NASA Reference Publication 1170, 1986.
13. Neumann, B., *The Interaction Between Geometry and Performance of a Centrifugal Pump*, Mechanical Engineering Publications Limited, London, 1991.
14. Garay, P. N., *Pump Application Desk Book*, Fairmont Press, Lilburn, Georgia, 1990.
15. Moody, L. F., and Zowski, T., ''Hydraulic Machinery,'' in *Handbook of Applied Hydraulics*, 3rd Ed., by C. V. Davis and K. E. Sorensen, McGraw-Hill, New York, 1969.
16. Rouse, H., *Elementary Mechanics of Fluids*, Wiley, New York, 1946.
17. Hydraulic Institute, *Hydraulic Institute Standards*, 14th Ed., Hydraulic Institute, Cleveland, Ohio, 1983.
18. Heald, C. C., Ed., *Cameron Hydraulic Data*, 17th Ed., Ingersoll-Rand, Woodcliff Lake, New Jersey, 1988.
19. Kristal, F. A., and Annett, F. A., *Pumps: Types, Selection, Installation, Operation, and Maintenance*, McGraw-Hill, New York, 1953.
20. Stepanoff, A. J., *Turboblowers*, Wiley, New York, 1955.
21. Berry, C. H., *Flow and Fan-Principles of Moving Air Through Ducts*, Industrial Press, New York, 1954.
22. Wallis, R. A., *Axial Flow Fans and Ducts*, Wiley, New York, 1983.
23. Reason, J., ''Fans,'' *Power*, Vol. 127, No. 9, 103–128, 1983.
24. Brown, R. N., *Compressors: Selection & Sizing*, Gulf Publishing, Houston, 1986.
25. Kováts, A., *Design and Performance of Centrifugal and Axial Flow Pumps and Compressors*, Pergamon Press, Oxford, 1964.
26. Johnson, I. A., and Bullock, R. D., Eds., *Aerodynamic Design of Axial-Flow Compressors*, NASA SP-36, National Aeronautics and Space Administration, Washington, 1965.
27. Glassman, A. J., Ed. *Turbine Design and Application*, Vol. 3, NASA SP-290, National Aeronautics and Space Administration, Washington, 1975.

Review Problems

Note: Problems designated with (R) are review problems. The phrases within parentheses refer to the main topics to be used in solving the problems. Complete, detailed solutions to these review problems can be found in the supplement titled *Student Solution Manual for Fundamentals of Fluid Mechanics*, by Munson, Young, and Okiishi (John Wiley and Sons, New York, 1997).

12.1R (Angular momentum) Water is supplied to a dishwasher through the manifold shown in Fig. P12.1R. Determine the rotational speed of the manifold if bearing friction and air resistance are neglected. The total flowrate of 2.3 gpm is divided evenly among the six outlets, each of which produces a 5/16-in.-diameter stream.

(ANS: 0.378 rev/s)

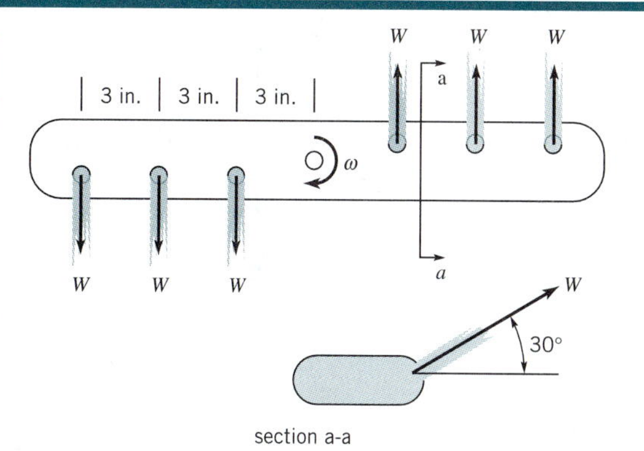

■ **FIGURE P12.1R**

12.2R (Velocity triangles) An axial-flow turbomachine rotor involves the upstream (1) and downstream (2) velocity triangles shown in Fig. P12.2R. Is this turbomachine a turbine or a fan? Sketch an appropriate blade section and determine the energy transferred per unit mass of fluid.

(ANS: turbine; $-36.9\ \text{ft}^2/\text{s}^2$)

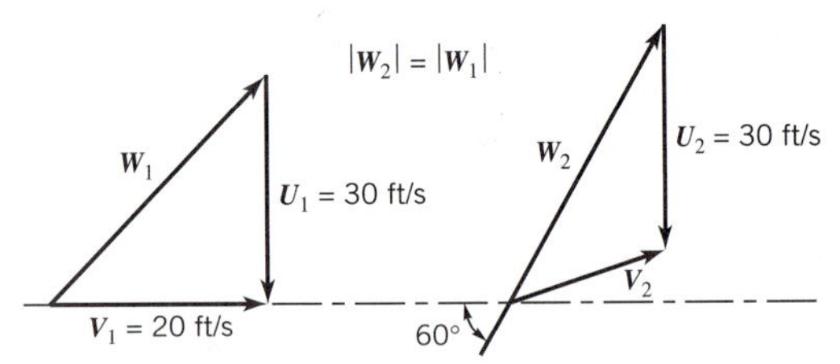

■ FIGURE P12.2R

12.3R (Centrifugal pump) Shown in Fig. P12.3R are front and side views of a centrifugal pump rotor or impeller. If the pump delivers 200 liters/s of water and the blade exit angle is 35° from the tangential direction, determine the power requirement associated with flow leaving at the blade angle. The flow entering the rotor blade row is essentially radial as viewed from a stationary frame.

(ANS: 348 kW)

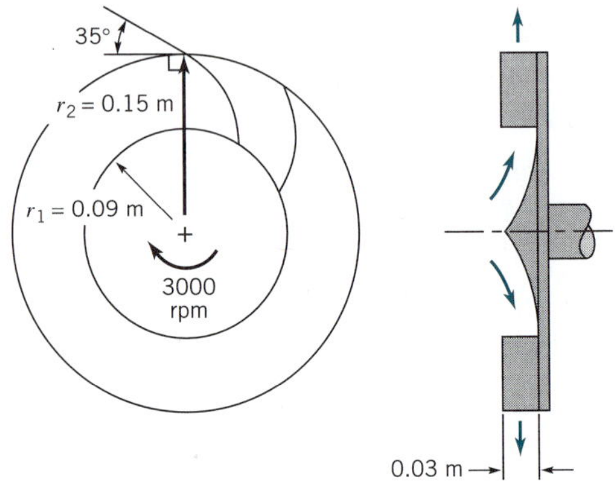

■ FIGURE P12.3R

12.4R (Centrifugal pump) The velocity triangles for water flow through a radial pump rotor are as indicated in Fig. P12.4R. **(a)** Determine the energy added to each unit mass (kg) of water as it flows through the rotor. **(b)** Sketch an appropriate blade section.

(ANS: 404 N·m/kg)

12.5R (Similarity laws) When the shaft horsepower supplied to a certain centrifugal pump is 25 hp, the pump discharges 700 gpm of water while operating at 1800 rpm with a head rise of 90 ft. **(a)** If the pump speed is reduced to 1200 rpm, determine

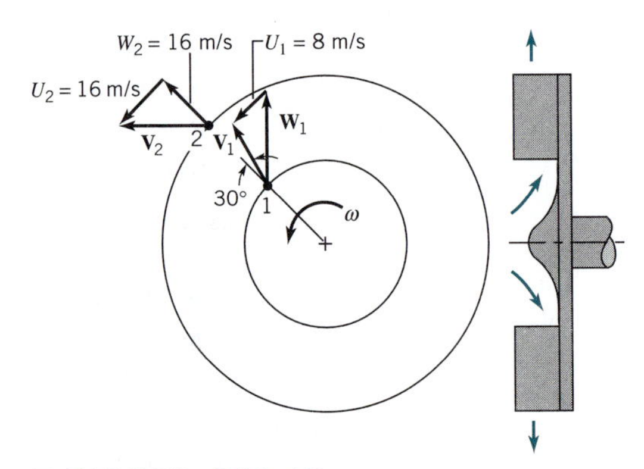

■ FIGURE P12.4R

the new head rise and shaft horsepower. Assume the efficiency remains the same. **(b)** What is the specific speed, N_{sd}, for this pump?

(ANS: 40 ft, 7.41 hp; 1630)

12.6R (Specific speed) An axial-flow turbine develops 10,000 hp when operating with a head of 40 ft. Determine the rotational speed if the efficiency is 88%.

(ANS: 65.4 rpm)

12.7R (Turbine) A water turbine with radial flow has the dimensions shown in Fig. P12.7R. The absolute entering velocity is 50 ft/s, and it makes an angle of 30° with the tangent to the rotor. The absolute exit velocity is directed radially inward. The angular speed of the rotor is 120 rpm. Find the power delivered to the shaft of the turbine.

(ANS: -1200 hp)

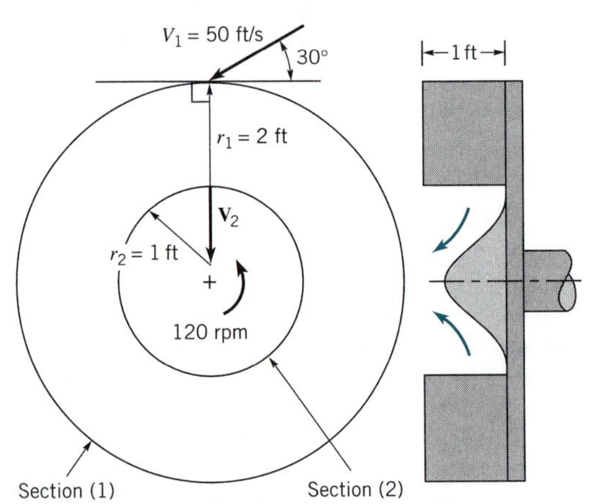

■ FIGURE P12.7R

12.8R (Turbine) Water enters an axial-flow turbine rotor with an absolute velocity tangential component, V_θ, of 30 ft/s. The corresponding blade velocity, U, is 100 ft/s. The water leaves the rotor blade row with no angular momentum. If the stagnation pressure drop across the turbine is 45 psi, determine the efficiency of the turbine.

(ANS: 0.898)

Problems

Notes: Unless otherwise indicated, use the values of fluid properties found in the tables on the inside of the front cover. Problems designated with a (†) are "open ended" problems and require critical thinking in that to work them one must make various assumptions and provide the necessary data. There is not a unique answer to these problems.

12.1 Water flows through a rotating sprinkler arm as shown in Fig. P12.1. Determine the flowrate if the angular velocity is 120 rpm. Friction is negligible. Is this a turbine or a pump?

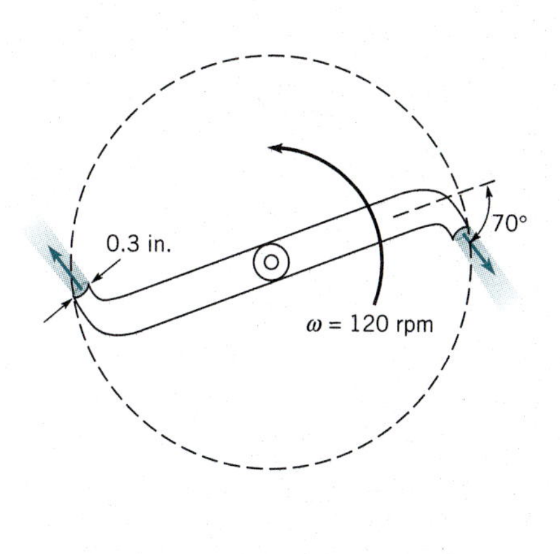

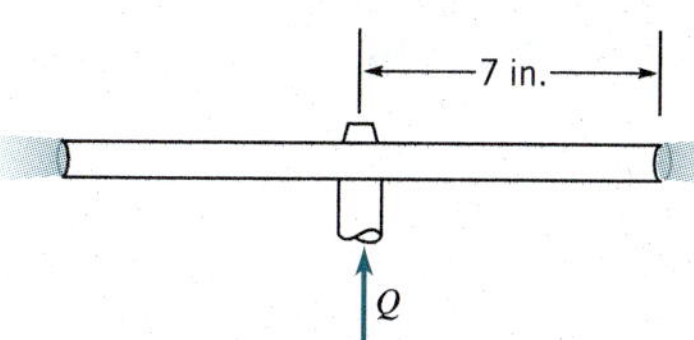

■ FIGURE P12.1

12.2 Uniform horizontal sheets of water of 3-mm thickness issue from the slits on the rotating manifold shown in Fig. P12.2. The velocity relative to the arm is a constant 3 m/s along each slit. Determine the torque needed to hold the manifold stationary. What would the angular velocity of the manifold be if the resisting torque is negligible?

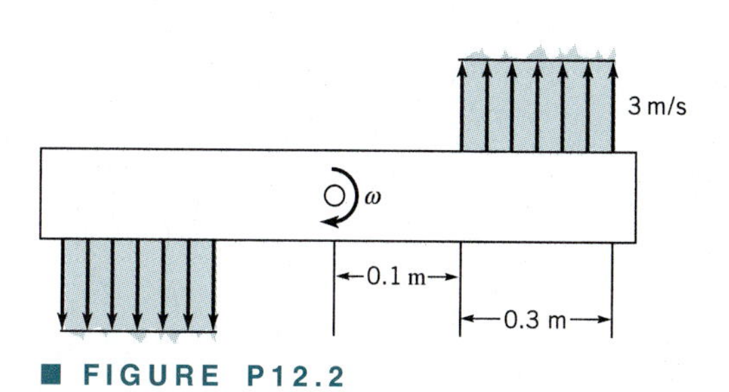

■ FIGURE P12.2

12.3 The rotor shown in Fig. P12.3 rotates with an angular velocity of 2000 rpm. Assume that the fluid enters in the radial direction and the relative velocity is tangent to the blades across the entire rotor. Is the device a pump or a turbine? Explain.

■ FIGURE P12.3

12.4 Air (assumed incompressible) flows across the rotor shown in Fig. P12.4 such that the magnitude of the absolute velocity increases from 15 m/s to 25 m/s. Measurements indicate that the absolute velocity at the inlet is in the direction shown. Determine the direction of the absolute velocity at the outlet if the fluid puts no torque on the rotor. Is the rotation CW or CCW? Is this device a pump or a turbine?

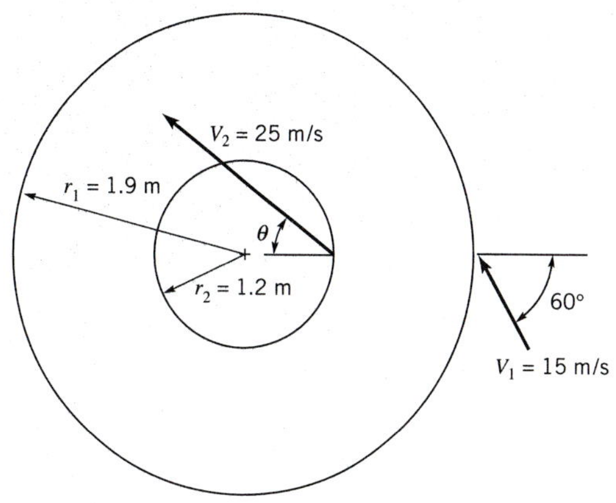

■ FIGURE P12.4

12.5 The measured shaft torque on the turbomachine shown in Fig. P12.5 is −60 N·m when the absolute velocities are as indicated. Determine the mass flowrate. What is the angular velocity if the magnitude of the shaft power is 1800 N·m/s? Is this machine a pump or a turbine? Explain.

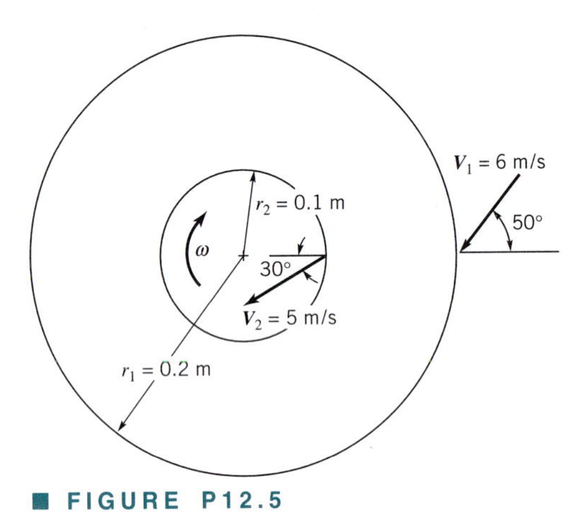

■ FIGURE P12.5

12.6 Sketched in Fig. P12.6 are the upstream [section (1)] and downstream [section (2)] velocity triangles at the arithmetic mean radius for flow through an axial-flow turbomachine rotor. The axial component of velocity is 100 ft/s at sections (1) and (2). **(a)** Label each velocity vector appropriately. Use **V** for absolute velocity, **W** for relative velocity, and **U** for blade velocity. **(b)** Are you dealing with a turbine or a fan? **(c)** Calculate the work per unit mass involved. **(d)** Sketch a reasonable blade section.

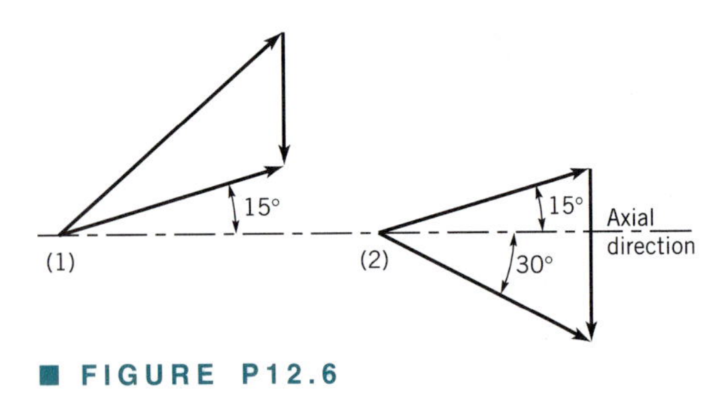

■ FIGURE P12.6

12.7 Shown in Fig. P12.7 is a toy "helicopter" powered by air escaping from a balloon. The air from the balloon flows radially through each of the three propeller blades and out small nozzles at the tips of the blades. The nozzles (along with the rotating propeller blades) are tilted at a small angle as indicated. Sketch the velocity triangle (i.e., blade, absolute, and relative velocities) for the flow from the nozzles. Explain why this toy tends to move upward. Is this a turbine? Pump?

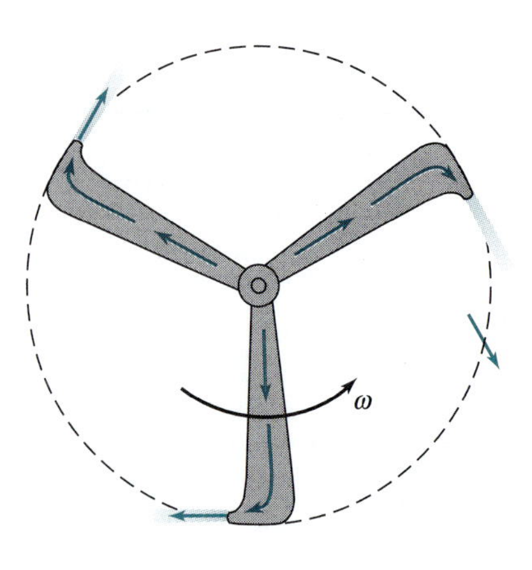

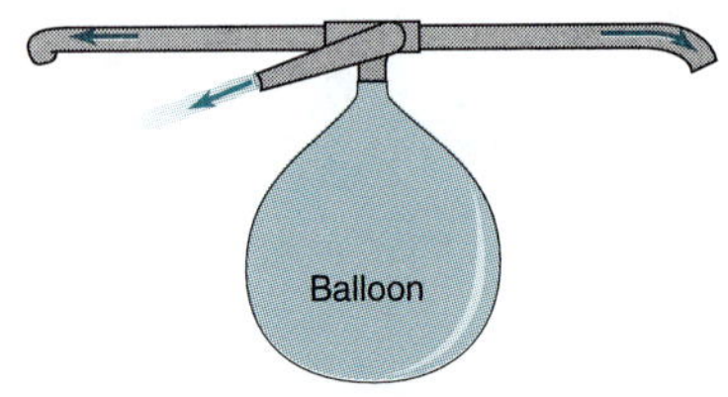

■ FIGURE P12.7

12.8 A centrifugal water pump having an impeller diameter of 0.5 m operates at 900 rpm. The water enters the pump parallel to the pump shaft. If the exit blade angle, β_2, (see Fig. 12.8) is 25°, determine the shaft power required to turn the impeller when the flow through the pump is 0.16 m^3/s. The uniform blade height is 50 mm.

12.9 The radial component of velocity of water leaving the centrifugal pump sketched in Fig. P12.9 is 45 ft/s. The magnitude of the absolute velocity at the pump exit is 90 ft/s. The fluid enters the pump rotor radially. Calculate the shaft work required per unit mass flowing through the pump. Show that Eqs. 12.5 and 12.8 both yield the same answer.

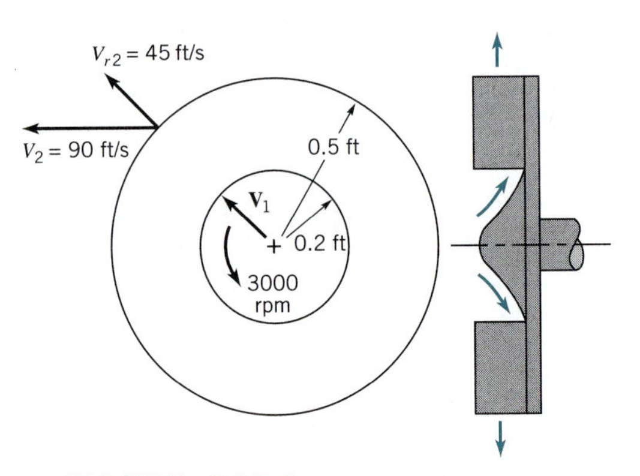

■ FIGURE P12.9

12.10 A centrifugal pump impeller is rotating at 1200 rpm in the direction shown in Fig. P12.10. The flow enters parallel to the axis of rotation and leaves at an angle of 30° to the radial direction. The absolute exit velocity, V_2, is 90 ft/s. **(a)** Draw the velocity triangle for the impeller exit flow. **(b)** Estimate the torque necessary to turn the impeller if the fluid density is 2.0 slugs/ft^3. What will the impeller rotation speed become if the shaft breaks?

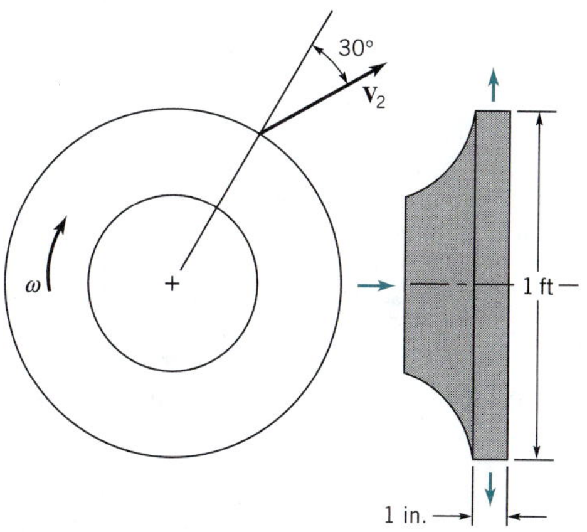

■ FIGURE P12.10

12.11 Discuss the main simplifying assumptions associated with Eq. 12.13 and explain why actual head rise is always less than ideal head rise. Discuss how ideal head rise is head "added" to the fluid and actual head rise is head "gained" by the fluid.

12.12 A centrifugal radial water pump has the dimensions shown in Fig. P12.12. The volume rate of flow is 0.25 ft^3/s, and the absolute inlet velocity is directed radially outward. The angular velocity of the impeller is 960 rpm. The exit velocity as seen from a coordinate system attached to the impeller can be assumed to be tangent to the vane at its trailing edge. Calculate the power required to drive the pump.

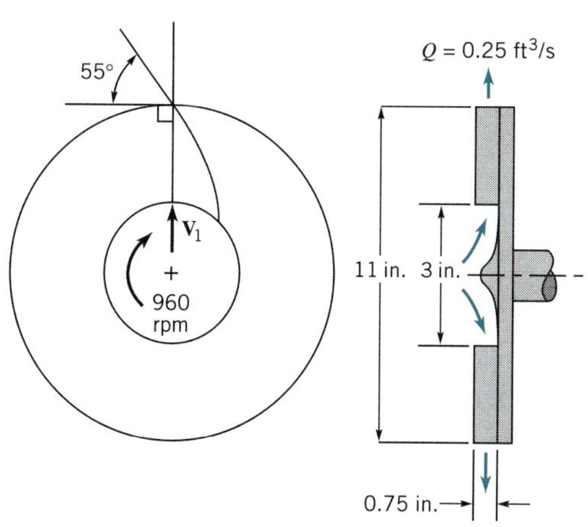

■ FIGURE P12.12

12.13 Water is pumped with a centrifugal pump, and measurements made on the pump indicate that for a flowrate of 240 gpm the required input power is 6 hp. For a pump efficiency of 62%, what is the actual head rise of the water being pumped?

12.14 The performance characteristics of a certain centrifugal pump are determined from an experimental setup similar to that shown in Fig. 12.10. When the flowrate of a liquid (SG = 0.9) through the pump is 120 gpm, the pressure gage at (1) indicates a vacuum of 95 mm of mercury and the pressure gage at (2) indicates a pressure of 80 kPa. The diameter of the pipe at the inlet is 110 mm and at the exit it is 55 mm. If $z_2 - z_1 = 0.5$ m, what is the actual head rise across the pump? Explain how you would estimate the pump motor power requirement.

12.15 The performance characteristics of a certain centrifugal pump having a 9-in.-diameter impeller and operating at 1750 rpm are determined using an experimental setup similar to that shown in Fig. 12.10. The following data were obtained during a series of tests in which $z_2 - z_1 = 0$, $V_2 = V_1$, and the fluid was water.

Q (gpm)	20	40	60	80	100	120	140
$p_2 - p_1$ (psi)	40.2	40.1	38.1	36.2	33.5	30.1	25.8
Power input (hp)	1.58	2.27	2.67	2.95	3.19	3.49	4.00

Based on these data, show or plot how the actual head rise, h_a, and the pump efficiency, η, vary with the flowrate. What is the design flowrate for this pump?

12.16 It is sometimes useful to have $h_a - Q$ pump performance curves expressed in the form of an equation. Fit the $h_a - Q$ data given in Problem 12.15 to an equation of the form $h_a = h_o - kQ^2$ and compare the values of h_a determined from the equation with the experimentally determined values. (Hint: Plot h_a versus Q^2 and use the method of least squares to fit the data to the equation.)

12.17 In Example 12.3, how will the maximum height, z_1, that the pump can be located above the water surface change if **(a)** the water temperature is increased to 120 °F, or **(b)** the fluid is changed from water to gasoline at 60 °F?

12.18 A centrifugal pump with a 7-in.-diameter impeller has the performance characteristics shown in Fig. 12.12. The pump is used to pump water at 100 °F, and the pump inlet is located 12 ft above the open water surface. When the flowrate is 200 gpm, the head loss between the water surface and the pump inlet is 6 ft of water. Would you expect cavitation in the pump to be a problem? Assume standard atmospheric pressure. Explain how you arrived at your answer.

12.19 Water at 40 °C is pumped from an open tank through 200 m of 50-mm-diameter smooth horizontal pipe as shown in Fig. P12.19 and discharges into the atmosphere with a velocity of 3 m/s. Minor losses are negligible. **(a)** If the efficiency of the pump is 70%, how much power is being supplied to the pump? **(b)** What is the $NPSH_A$ at the pump inlet? Neglect losses in the short section of pipe connecting the pump to the tank. Assume standard atmospheric pressure.

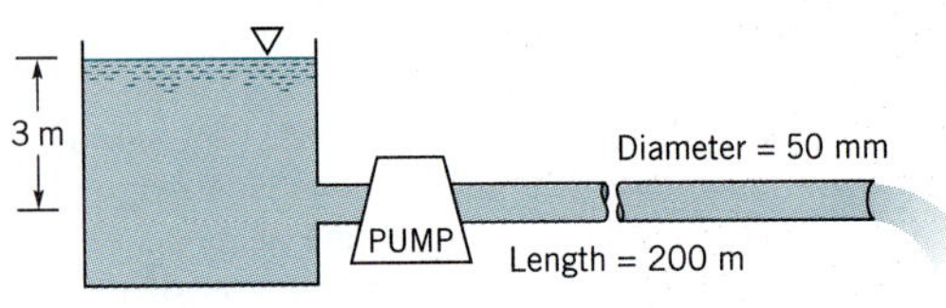

■ **FIGURE P12.19**

12.20 The centrifugal pump shown in Fig. P12.20 is not self-priming. That is, if the water is drained from the pump and pipe as shown in Fig. P12.20(*a*), the pump will not draw the water into the pump and start pumping when the pump is turned on. However, if the pump is primed [i.e., filled with water as in Fig. P12.20*b*)], the pump does start pumping water when turned on. Explain this behavior.

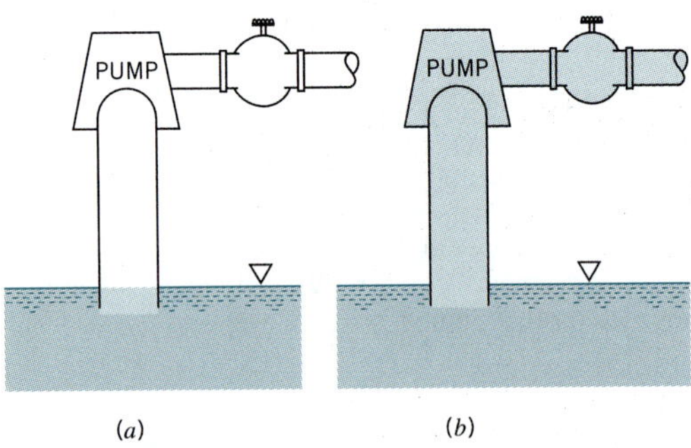

(*a*) (*b*)

■ **FIGURE P12.20**

12.21 Owing to fouling of the pipe wall, the friction factor for the pipe of Example 12.4 increases from 0.02 to 0.03. Determine the new flowrate, assuming all other conditions remain the same. What is the pump efficiency at this new flowrate? Explain how a line valve could be used to vary the flowrate through the pipe of Example 12.4. Would it be better to place the valve upstream or downstream of the pump? Why?

12.22 A centrifugal pump having a head-capacity relationship given by the equation $h_a = 180 - 6.10 \times 10^{-4}Q^2$, with h_a in feet when Q is in gpm, is to be used with a system similar to that shown in Fig. 12.14. For $z_2 - z_1 = 50$ ft, what is the expected flowrate if the total length of constant-diameter pipe is 600 ft and the fluid is water? Assume the pipe diameter to be 4 in. and the friction factor to be equal to 0.02. Neglect all minor losses.

12.23 A centrifugal pump having a 6-in.-diameter impeller and the characteristics shown in Fig. 12.12 is to be used to pump gasoline through 4000 ft of commercial steel 3-in.-diameter pipe. The pipe connects two reservoirs having open surfaces at the same elevation. Determine the flowrate. Do you think this pump is a good choice? Explain.

12.24 Determine the new flowrate for the system described in Problem 12.23 if the pipe diameter is increased from 3 in. to 4 in. Is this pump still a good choice? Explain.

12.25 A centrifugal pump having the characteristics shown in Example 12.4 is used to pump water between two large open tanks through 100 ft of 8-in.-diameter pipe. The pipeline contains 4 regular flanged 90° elbows, a check valve, and a fully open globe valve. Assume the friction factor $f = 0.02$ for the 100-ft section of pipe. Other minor losses are negligible. If the static head (difference in height of fluid surfaces in the two tanks) is 30 ft, what is the expected flowrate? Do you think this pump is a good choice? Explain.

12.26 In a chemical processing plant a liquid is pumped from an open tank, through a 0.1-m-diameter vertical pipe, and into another open tank as shown in Fig. P12.26(*a*). A valve is located in the pipe, and the minor loss coefficient for the valve as a function of the valve setting is shown in Fig. P12.26(*b*). The pump head-capacity relationship is given by the equation $h_a = 52.0 - 1.01 \times 10^3 Q^2$ with h_a in meters when Q is in m^3/s. Assume the friction factor $f = 0.02$ for the pipe, and all minor losses, except for the valve, are negligible. The fluid levels in the two tanks can be assumed to remain constant. **(a)** Determine the flowrate with the valve wide open. **(b)** Determine the required valve setting (percent open) to reduce the flowrate by 50%.

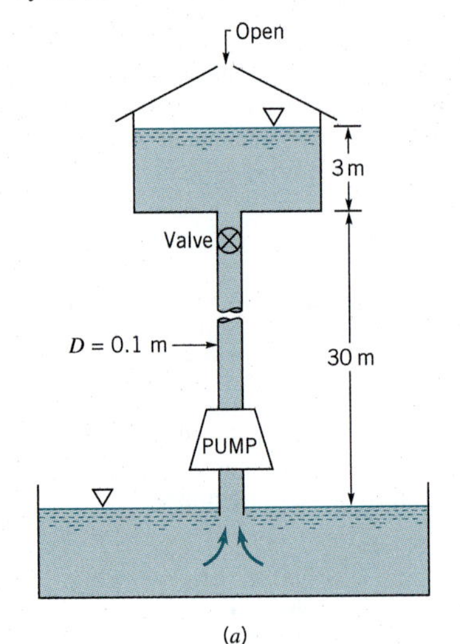

(*a*)

K_L: 0, 10, 20, 30, 40
Percent valve setting: 0 (Closed), 20, 40, 60, 80, 100 (Open)

(*b*)

■ **FIGURE P12.26**

12.27 A centrifugal pump having an impeller diameter of 1 m is to be constructed so that it will supply a head rise of 200 m at a flowrate of 4.1 m^3/s of water when operating at a speed of 1200 rpm. To study the characteristics of this pump, a 1/5 scale, geometrically similar model operated at the same speed is to be tested in the laboratory. Determine the required model discharge and head rise. Assume both model and prototype operate with the same efficiency (and therefore the same flow coefficient).

12.28 A small model of a pump is tested in the laboratory and found to have a specific speed, N_{sd}, equal to 1000 when operating at peak efficiency. Predict the discharge of a larger, geometrically similar pump operating at peak efficiency at a speed of 1800 rpm across an actual head rise of 200 ft.

12.29 Use the data given in Problem 12.15 and plot the dimensionless coefficients C_H, $C_{\mathcal{P}}$, η versus C_Q for this pump. Calculate a meaningful value of specific speed, discuss its usefulness, and compare the result with data of Fig. 12.18.

12.30 A centrifugal pump provides a flowrate of 500 gpm when operating at 1750 rpm against a 200-ft head. Determine the pump's flowrate and developed head if the pump speed is increased to 3500 rpm.

12.31 A centrifugal pump with a 12-in.-diameter impeller requires a power input of 60 hp when the flowrate is 3200 gpm against a 60-ft head. The impeller is changed to one with a 10-in. diameter. Determine the expected flowrate, head, and input power if the pump speed remains the same.

12.32 Do the head-flowrate data shown in Fig. 12.12 appear to follow the similarity laws as expressed by Eqs. 12.39 and 12.40? Explain.

12.33 A centrifugal pump has the performance characteristics of the pump with the 6-in.-diameter impeller described in Fig. 12.12. Note that the pump in this figure is operating at 3500 rpm. What is the expected head gained if the speed of this pump is reduced to 2800 rpm while operating at peak efficiency?

12.34 In a certain application a pump is required to deliver 5000 gpm against a 300-ft head when operating at 1200 rpm. What type of pump would you recommend?

12.35 A certain axial-flow pump has a specific speed of $N_S = 5.0$. If the pump is expected to deliver 3000 gpm when operating against a 15-ft head, at what speed (rpm) should the pump be run?

12.36 A certain pump is known to have a capacity of 3 m^3/s when operating at a speed of 60 rad/s against a head of 20 m. Based on the information in Fig. 12.18, would you recommend a radial-flow, mixed-flow, or axial-flow pump?

12.37 Fuel oil (sp. wt = 48.0 lb/ft^3, viscosity = 2.0×10^{-5} lb·s/ft^2) is pumped through the piping system of Fig. P12.37 with a velocity of 4.6 ft/s. The pressure 200 ft upstream from the pump is 5 psi. Pipe losses downstream from the pump are negligible, but minor losses are not (minor loss coefficients are given on the figure). **(a)** For a pipe diameter of 2 in. with a relative roughness $\varepsilon/D = 0.001$, determine the head that must be added by the pump. **(b)** For a pump operating speed of 1750 rpm, what type of pump (radial-flow, mixed-flow, or axial-flow) would you recommend for this application?

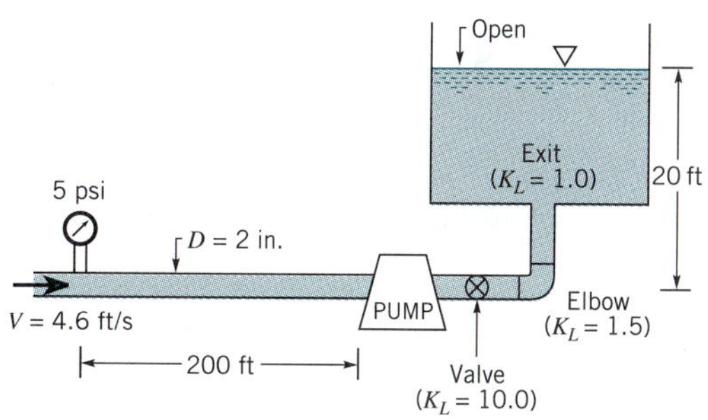

■ **FIGURE P12.37**

† **12.38** Water is pumped between the two tanks described in Example 12.4 once a day, 365 days a year, with each pumping period lasting two hours. The water levels in the two tanks remain essentially constant. Estimate the annual cost of the electrical power needed to operate the pump if it were located in your city. You will have to make a reasonable estimate for the efficiency of the motor used to drive the pump. Due to aging, it can be expected that the overall resistance of the system will increase with time. If the operating point shown in Fig. E12.4c changes to a point where the flowrate has been reduced to 1000 gpm, what will be the new annual cost of operating the pump? Assume the cost of electrical power remains the same.

† **12.39** Discuss the importance of boundary layer development in pumps.

† **12.40** Explain why the statement "pumps move fluids and moving fluids drive turbines" is true.

12.41 A Pelton wheel turbine is illustrated in Fig. P12.41. The radius to the line of action of the tangential reaction force on each vane is 1 ft. Each vane deflects fluid by an angle of 135° as indicated. Assume all of the flow occurs in a horizontal plane. Each of the four jets shown strikes a vane with a velocity of 100 ft/s and a stream diameter of 1 in. The magnitude of velocity of the jet remains constant along the vane surface.

(a) How much torque is required to hold the wheel stationary? **(b)** How fast will the wheel rotate if shaft torque is negligible and what practical situation is simulated by this condition?

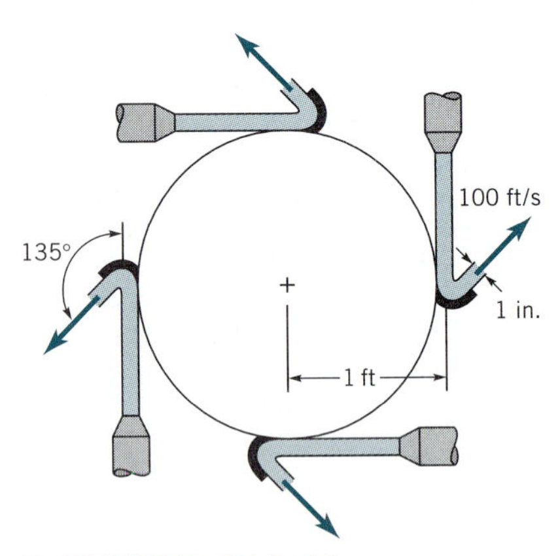

■ FIGURE P12.41

12.42 An inward-flow radial turbine (see Fig. P12.42) involves a nozzle angle, α_1, of 60° and an inlet rotor tip speed, U_1, of 3 m/s. The ratio of rotor inlet to outlet diameters is 2.0. The absolute velocity leaving the rotor at section (2) is radial with a magnitude of 6 m/s. Determine the energy transfer per unit mass of fluid flowing through this turbine if the fluid is **(a)** air, **(b)** water.

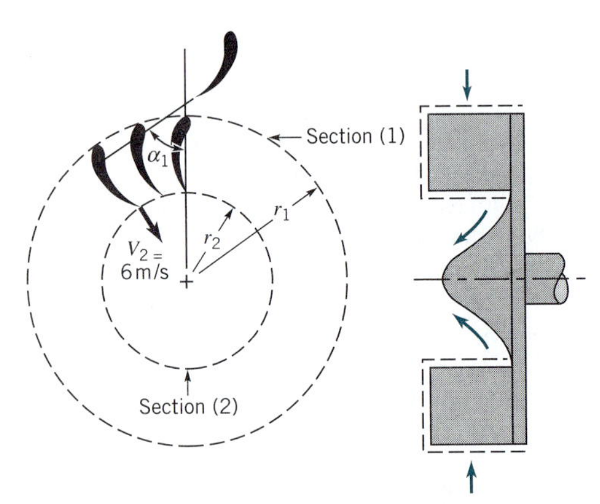

■ FIGURE P12.42

12.43 A simplified sketch of a hydraulic turbine runner is shown in Fig. P12.43. Relative to the rotating runner, water enters at section (1) (cylindrical cross section area A_1 at r_1 = 1.5 m) at an angle of 100° from the tangential direction and leaves at section (2) (cylindrical cross section area A_2 at r_2 = 0.85 m) at an angle of 50° from the tangential direction. The blade height at sections (1) and (2) is 0.45 m and the volume flowrate through the turbine is 30 m³/s. The runner speed is 130 rpm in the direction shown. Determine the shaft power developed. Is the shaft power greater or less than the power lost by the fluid? Explain.

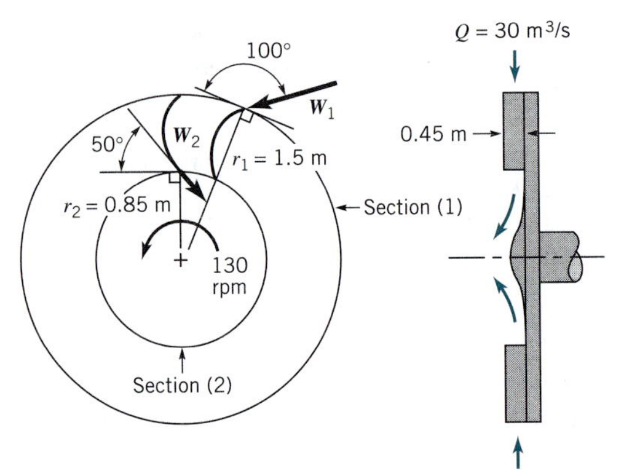

■ FIGURE P12.43

12.44 A water turbine wheel rotates at the rate of 100 rpm in the direction shown in Fig. P12.44. The inner radius, r_2, of the blade row is 1 ft, and the outer radius, r_1, is 2 ft. The absolute velocity vector at the turbine rotor entrance makes an angle of 20° with the tangential direction. The inlet blade angle is 60° relative to the tangential direction. The blade outlet angle is 120°. The flowrate is 10 ft³/s. For the flow tangent to the rotor blade surface at inlet and outlet, determine an appropriate constant blade height, b, and the corresponding power available at the rotor shaft. Is the shaft power greater or less than the power lost by the fluid? Explain.

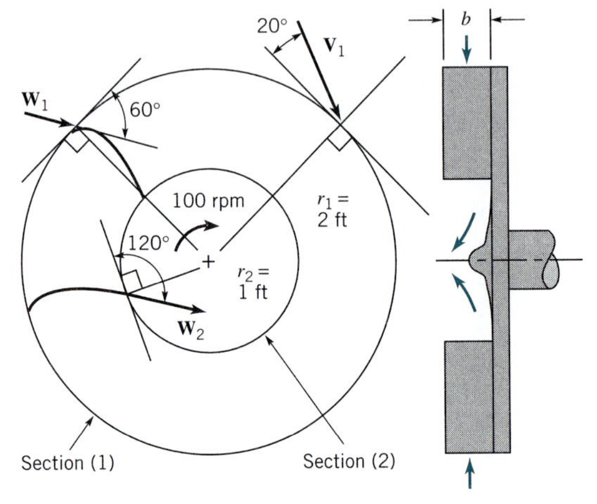

■ FIGURE P12.44

12.45 A sketch of the arithmetic mean radius blade sections of an axial-flow water turbine stage is shown in Fig. P12.45. The rotor speed is 1500 rpm. **(a)** Sketch and label velocity triangles for the flow entering and leaving the rotor row. Use **V** for absolute velocity, **W** for relative velocity, and **U** for blade velocity. Assume flow enters and leaves each blade row at the blade angles shown. **(b)** Calculate the work per unit mass delivered at the shaft.

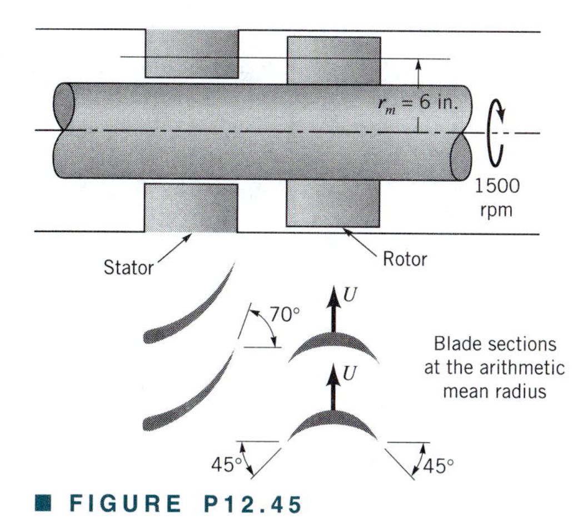

■ FIGURE P12.45

12.46 An inward flow radial turbine (see Fig. P12.46) involves a nozzle angle, α_1, of 60° and an inlet rotor tip speed, U_1, of 9 m/s. The ratio of rotor inlet to outlet diameters is 2.0. The radial component of velocity remains constant at 6 m/s through the rotor and the flow leaving the rotor at section (2) is without angular momentum. If the flowing fluid is water and the stagnation pressure drop across the rotor is 110 kPa, determine the loss of available energy across the rotor and the efficiency involved.

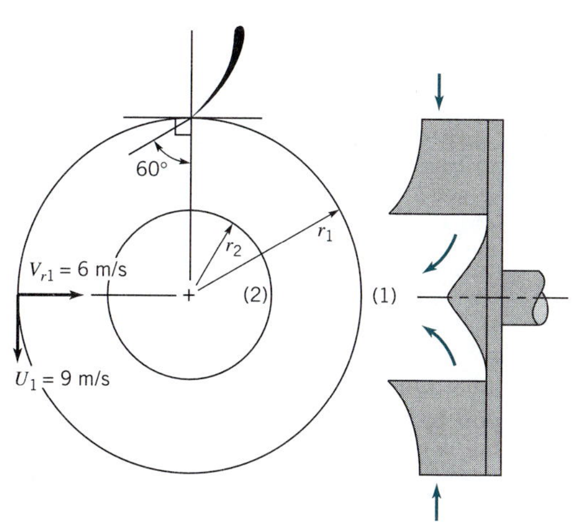

■ FIGURE P12.46

12.47 An inward flow radial turbine (see Fig. P12.46) involves a nozzle angle, α_1, of 60° and an inlet rotor tip speed of 9 m/s. The ratio of rotor inlet to outlet diameters is 2.0. The radial component of velocity remains constant at 6 m/s through the rotor and the flow leaving the rotor at section (2) is without angular momentum. If the flowing fluid is air and the static pressure drop across the rotor is 0.07 kPa, determine the loss of available energy across the rotor and the rotor efficiency.

12.48 A 10.9-ft-diameter Pelton wheel operates at 500 rpm with a total head just upstream of the nozzle of 5,330 ft. Estimate the diameter of the nozzle of the single-nozzle wheel if it develops 25,000 horsepower.

† **12.49** A Pelton wheel is to develop 10,000 hp under a head of 1,000 ft. Provide a preliminary design for such a system. List any assumptions made and show calculations.

12.50 A Pelton wheel has a diameter of 2 m and develops 500 kW when rotating 180 rpm. What is the average force of the water against the blades? If the turbine is operating at maximum efficiency, determine the speed of the water jet from the nozzle and the mass flowrate.

12.51 Water for a Pelton wheel turbine flows from the headwater and through the penstock as shown in Fig. P12.51. The effective friction factor for the penstock, control valves,and the like is 0.032 and the diameter of the jet is 0.20 m. Determine the maximum power output.

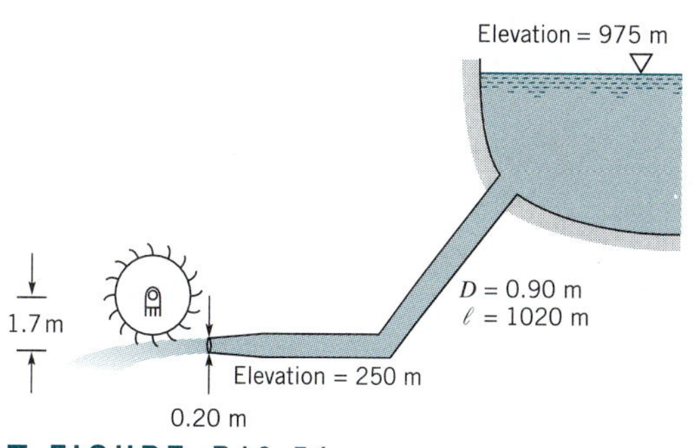

■ FIGURE P12.51

12.52 Water to run a Pelton wheel is supplied by a penstock of length ℓ and diameter D with a friction factor f. If the only losses associated with the flow in the penstock are due to pipe friction, shown that the maximum power output of the turbine occurs when the nozzle diameter, D_1, is given by $D_1 = D/(2f\,\ell/D)^{1/4}$.

12.53 A Pelton wheel is supplied with water from a lake at an elevation H above the turbine. The penstock that supplies the water to the wheel is of length ℓ, diameter D, and friction factor f. Minor losses are negligible. Show that the power developed by the turbine is maximum when the velocity head at the nozzle exit is $2H/3$. Note: The result of Problem 12.52 may be of use.

12.54 If there is negligible friction along the blades of a Pelton wheel, the relative speed remains constant as the fluid flows across the blades, and the maximum power output occurs when the blade speed is one-half the jet speed (see Eq. 12.52). Consider the case where friction is not negligible and the relative speed leaving the blade is some fraction, c, of the relative speed entering the blade. That is, $W_2 = cW_1$. Show that Eq. 12.52 is valid for this case also.

12.55 A hydraulic turbine operating at 180 rpm with a head of 170 feet develops 20,000 horsepower. Estimate the power and speed if the turbine were to operate under a head of 190 ft.

12.56 Draft tubes as shown in Fig. P12.56 are often installed at the exit of Kaplan and Francis turbines. Explain why such draft tubes are advantageous.

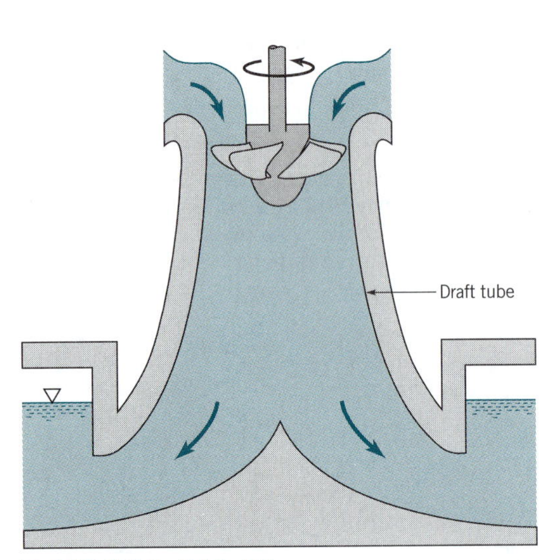

■ **FIGURE P12.56**

12.57 Turbines are to be designed to develop 30,000 horsepower while operating under a head of 70 ft and an angular velocity of 60 rpm. What type of turbines is best suited for this purpose? Estimate the flowrate needed.

12.58 A 1-m-diameter Pelton wheel rotates at 300 rpm. Which of the following heads (in meters) would be best suited for this turbine: **(a)** 2, **(b)** 5, **(c)** 40, **(d)** 70, or **(e)** 140? Explain.

12.59 Water at 400 psi is available to operate a turbine at 1750 rpm. What type of turbine would you suggest to use if the turbine should have an output of approximately 200 hp?

12.60 It is desired to produce 50,000 hp with a head of 50 ft and an angular velocity of 100 rpm. How many turbines would be needed if the specific speed is to be **(a)** 50, **(b)** 100?

† **12.61** What aircraft is powered by the gas turbine engine shown in Fig. 12.36 and what is the approximate market price of each engine and the entire airplane?

12.62 Test data for the small Francis turbine shown in Fig. P12.62 is given in the table below. The test was run at a constant 32.8 ft head just upstream of the turbine. The Prony brake on the turbine output shaft was adjusted to give various angular velocities, and the force on the brake arm, F, was recorded. Use the given data to plot curves of torque as a function of angular velocity and turbine efficiency as a function of angular velocity.

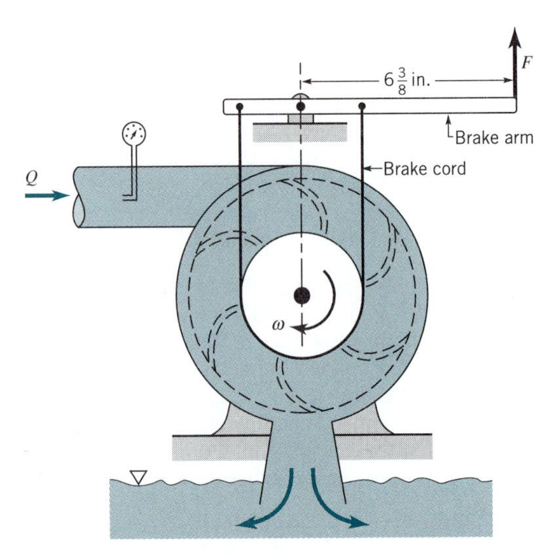

ω (rpm)	Q (ft^3/s)	F (lb)
0	0.129	2.63
1000	0.129	2.40
1500	0.129	2.22
1870	0.124	1.91
2170	0.118	1.49
2350	0.0942	0.876
2580	0.0766	0.337
2710	0.068	0.089

■ **FIGURE P12.62**

† **12.63** It is possible to generate power by using the water from your garden hose to drive a small Pelton wheel turbine. Provide a preliminary design of such a turbine and estimate the power output expected. List all assumptions and show calculations.

12.64 The device shown in Fig. P12.64 is used to investigate the power produced by a Pelton wheel turbine. Water supplied at a constant flowrate issues from a nozzle and strikes the turbine buckets as indicated. The angular velocity, ω, of the turbine wheel is varied by adjusting the tension on the Prony brake spring, thereby varying the torque, T_{shaft}, applied to the output shaft. This torque can be determined from the measured force, R, needed to keep the brake arm stationary as $T_{shaft} = F\ell$, where ℓ is the moment arm of the brake force.

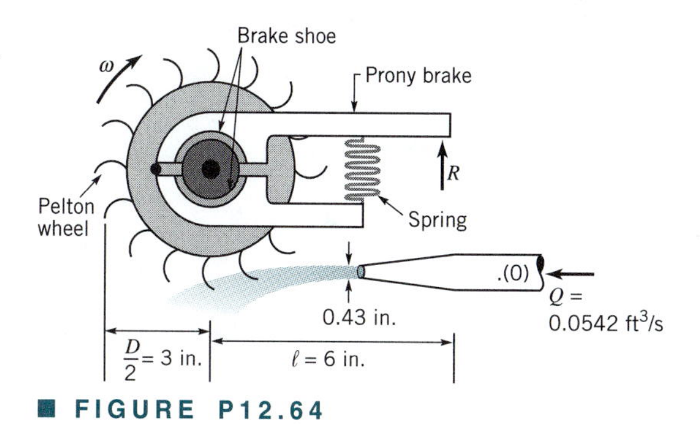

■ FIGURE P12.64

Experimentally determined values of ω and R are shown in the following table. Use these results to plot a graph of torque as a function of the angular velocity. On another graph plot the power output, $\dot{W}_{\text{shaft}} = T_{\text{shaft}}\,\omega$, as a function of the angular velocity. On each of these graphs plot the theoretical curves for this turbine, assuming 100 percent efficiency.

Compare the experimental and theoretical results and discuss some possible reasons for any differences between them.

ω (rpm)	R (lb)
0	2.47
360	1.91
450	1.84
600	1.69
700	1.55
940	1.17
1120	0.89
1480	0.16

Appendices

APPENDIX A

Unit Conversion Tables[1]

The following tables express the definitions of miscellaneous units of measure as exact numerical multiples of coherent SI units and provide multiplying factors for converting numbers and miscellaneous units to corresponding new numbers and SI units.

Conversion factors are expressed using computer exponential notation, and an asterisk follows each number which expresses an exact definition. For example, the entry "2.54 E − 2*" expresses the fact that 1 inch = 2.54×10^{-2} meter, exactly by definition. Numbers not followed by an asterisk are only approximate representations of definitions or are the results of physical measurements. In these tables pound-force is designated as lbf, whereas in the text pound-force is designated as lb.

■ **TABLE A.1**
Listing by Physical Quantity

To convert from	to	Multiply by
Acceleration		
foot/second^2	meter/second^2	3.048 E − 1*
free fall, standard	meter/second^2	9.806 65 E + 0*
gal (galileo)	meter/second^2	1.00 E − 2*
inch/second^2	meter/second^2	2.54 E − 2*
Area		
acre	meter^2	4.046 856 422 4 E + 3*
are	meter^2	1.00 E + 2*
barn	meter^2	1.00 E − 28*
foot^2	meter^2	9.290 304 E − 2*
hectare	meter^2	1.00 E + 4*
inch^2	meter^2	6.4516 E − 4*
mile^2 (U.S. statute)	meter^2	2.589 988 110 336 E + 6*
section	meter^2	2.589 988 110 336 E + 6*

[1]These tables abridged from Mechtly, E. A., *The International System of Units, 2nd Revision*, NASA SP-7012, 1973.

■ **TABLE A.1 (continued)**

To convert from	to	Multiply by
township	meter^2	9.323 957 2 E + 7
yard^2	meter^2	8.361 273 6 E − 1*
Density		
gram/centimeter^3	kilogram/meter^3	1.00 E + 3*
lbm/inch^3	kilogram/meter^3	2.767 990 5 E + 4
lbm/foot^3	kilogram/meter^3	1.601 846 3 E + 1
slug/foot^3	kilogram/meter^3	5.153 79 E + 2
Energy		
British thermal unit: (IST after 1956)	joule	1.055 056 E + 3
British thermal unit (thermochemical)	joule	1.054 350 E + 3
calorie (International Steam Table)	joule	4.1868 E + 0
calorie (thermochemical)	joule	4.184 E + 0*
calorie (kilogram, International Steam Table)	joule	4.1868 E + 3
calorie (kilogram, thermochemical)	joule	4.184 E + 3*
electron volt	joule	1.602 191 7 E − 19
erg	joule	1.00 E − 7*
foot lbf	joule	1.355 817 9 E + 0
foot poundal	joule	4.214 011 0 E − 2
joule (international of 1948)	joule	1.000 165 E + 0
kilocalorie (Internation Steam Table)	joule	4.1868 E + 3
kilocalorie (thermochemical)	joule	4.184 E + 3*
kilowatt hour	joule	3.60 E + 6*
watt hour	joule	3.60 E + 3*
Force		
dyne	newton	1.00 E − 5*
kilogram force (kgf)	newton	9.806 65 E + 0*
kilopound force	newton	9.806 65 E + 0*
kip	newton	4.448 221 615 260 5 E + 3*
lbf (pound force, avoirdupois)	newton	4.448 221 615 260 5 E + 0*
ounce force (avoirdupois)	newton	2.780 138 5 E − 1
pound force, lbf (avoirdupois)	newton	4.448 221 615 260 5 E + 0*
poundal	newton	1.382 549 543 76 E − 1*
Length		
angstrom	meter	1.00 E − 10*
astronomical unit (IAU)	meter	1.496 00 E + 11
cubit	meter	4.572 E − 1*
fathom	meter	1.8288 E + 0*
foot	meter	3.048 E − 1*
furlong	meter	2.011 68 E + 2*
hand	meter	1.016 E − 1*
inch	meter	2.54 E − 2*
league (international nautical)	meter	5.556 E + 3*
light year	meter	9.460 55 E + 15
meter	wavelengths Kr 86	1.650 763 73 E + 6*
micron	meter	1.00 E − 6*
mil	meter	2.54 E − 5*
mile (U.S. statute)	meter	1.609 344 E + 3*
nautical mile (U.S.)	meter	1.852 E + 3*
rod	meter	5.0292 E + 0*
yard	meter	9.144 E − 1*

■ **TABLE A.1 (continued)**

To convert from	to	Multiply by
Mass		
carat (metric)	kilogram	2.00 E − 4*
grain	kilogram	6.479 891 E − 5*
gram	kilogram	1.00 E − 3*
ounce mass (avoirdupois)	kilogram	2.834 952 312 5 E − 2*
pound mass, lbm (avoirdupois)	kilogram	4.535 923 7 E − 1*
slug	kilogram	1.459 390 29 E + 1
ton (long)	kilogram	1.016 046 908 8 E + 3*
ton (metric)	kilogram	1.00 E + 3*
ton (short, 2000 pound)	kilogram	9.071 847 4 E + 2*
tonne	kilogram	1.00 E + 3*
Power		
Btu (thermochemical)/second	watt	1.054 350 264 488 E + 3
calorie (thermochemical)/second	watt	4.184 E + 0*
foot lbf/second	watt	1.355 817 9 E + 0
horsepower (550 foot lbf/second)	watt	7.456 998 7 E + 2
kilocalorie (thermochemical)/second	watt	4.184 E + 3*
watt (international of 1948)	watt	1.000 165 E + 0
Pressure		
atmosphere	newton/meter^2	1.013 25 E + 5*
bar	newton/meter^2	1.00 E + 5*
barye	newton/meter^2	1.00 E − 1*
centimeter of mercury (0° C)	newton/meter^2	1.333 22 E + 3
centimeter of water (4° C)	newton/meter^2	9.806 38 E + 1
dyne/centimeter^2	newton/meter^2	1.00 E − 1*
foot of water (39.2° F)	newton/meter^2	2.988 98 E + 3
inch of mercury (32° F)	newton/meter^2	3.386 389 E + 3
inch of mercury (60° F)	newton/meter^2	3.376 85 E + 3
inch of water (39.2° F)	newton/meter^2	2.490 82 E + 2
inch of water (60° F)	newton/meter^2	2.4884 E + 2
kgf/centimeter^2	newton/meter^2	9.806 65 E + 4*
kgf/meter^2	newton/meter^2	9.806 65 E + 0*
lbf/foot^2	newton/meter^2	4.788 025 8 E + 1
lbf/inch^2 (psi)	newton/meter^2	6.894 757 2 E + 3
millibar	newton/meter^2	1.00 E + 2*
millimeter of mercury (0° C)	newton/meter^2	1.333 224 E + 2
pascal	newton/meter^2	1.00 E + 0*
psi (lbf/inch^2)	newton/meter^2	6.894 757 2 E + 3
torr (0° C)	newton/meter^2	1.333 22 E + 2
Speed		
foot/second	meter/second	3.048 E − 1*
inch/second	meter/second	2.54 E − 2*
kilometer/hour	meter/second	2.777 777 8 E − 1
knot (international)	meter/second	5.144 444 444 E − 1
mile/hour (U.S. statute)	meter/second	4.4704 E − 1*
Temperature		
Celsius	kelvin	$t_K = t_C + 273.15$
Fahrenheit	kelvin	$t_K = (5/9)(t_F + 459.67)$
Fahrenheit	Celsius	$t_C = (5/9)(t_F - 32)$
Rankine	kelvin	$t_K = (5/9)t_R$
Time		
day (mean solar)	second (mean solar)	8.64 E + 4*

TABLE A.1 (continued)

To convert from	to	Multiply by
hour (mean solar)	second (mean solar)	3.60 E + 3*
minute (mean solar)	second (mean solar)	6.00 E + 1*
year (calendar)	second (mean solar)	3.1536 E + 7*
Viscosity		
centistoke	meter^2/second	1.00 E − 6*
stoke	meter^2/second	1.00 E − 4*
foot^2/second	meter^2/second	9.290 304 E − 2*
centipoise	newton second/meter^2	1.00 E − 3*
lbm/foot second	newton second/meter^2	1.488 163 9 E + 0
lbf second/foot^2	newton second/meter^2	4.788 025 8 E + 1
poise	newton second/meter^2	1.00 E − 1*
poundal second/foot^2	newton second/meter^2	1.488 163 9 E + 0
slug/foot second	newton second/meter^2	4.788 025 8 E + 1
rhe	meter^2/newton second	1.00 E + 1*
Volume		
acre foot	meter^3	1.233 481 837 547 52 E + 3*
barrel (petroleum, 42 gallons)	meter^3	1.589 873 E − 1
board foot	meter^3	2.359 737 216 E − 3*
bushel (U.S.)	meter^3	3.523 907 016 688 E − 2*
cord	meter^3	3.624 556 3 E + 0
cup	meter^3	2.365 882 365 E − 4*
dram (U.S. fluid)	meter^3	3.696 691 195 312 5 E − 6*
fluid ounce (U.S.)	meter^3	2.957 352 956 25 E − 5*
foot^3	meter^3	2.831 684 659 2 E − 2*
gallon (U.K. liquid)	meter^3	4.546 087 E − 3
gallon (U.S. liquid)	meter^3	3.785 411 784 E − 3*
inch^3	meter^3	1.638 706 4 E − 5*
liter	meter^3	1.00 E − 3*
ounce (U.S. fluid)	meter^3	2.957 352 956 25 E − 5*
peck (U.S.)	meter^3	8.809 767 541 72 E − 3*
pint (U.S. liquid)	meter^3	4.731 764 73 E − 4*
quart (U.S. liquid)	meter^3	9.463 529 5 E − 4
stere	meter^3	1.00 E + 0*
tablespoon	meter^3	1.478 676 478 125 E − 5*
teaspoon	meter^3	4.928 921 593 75 E − 6*
yard^3	meter^3	7.645 548 579 84 E − 1*

APPENDIX B

Physical Properties of Fluids

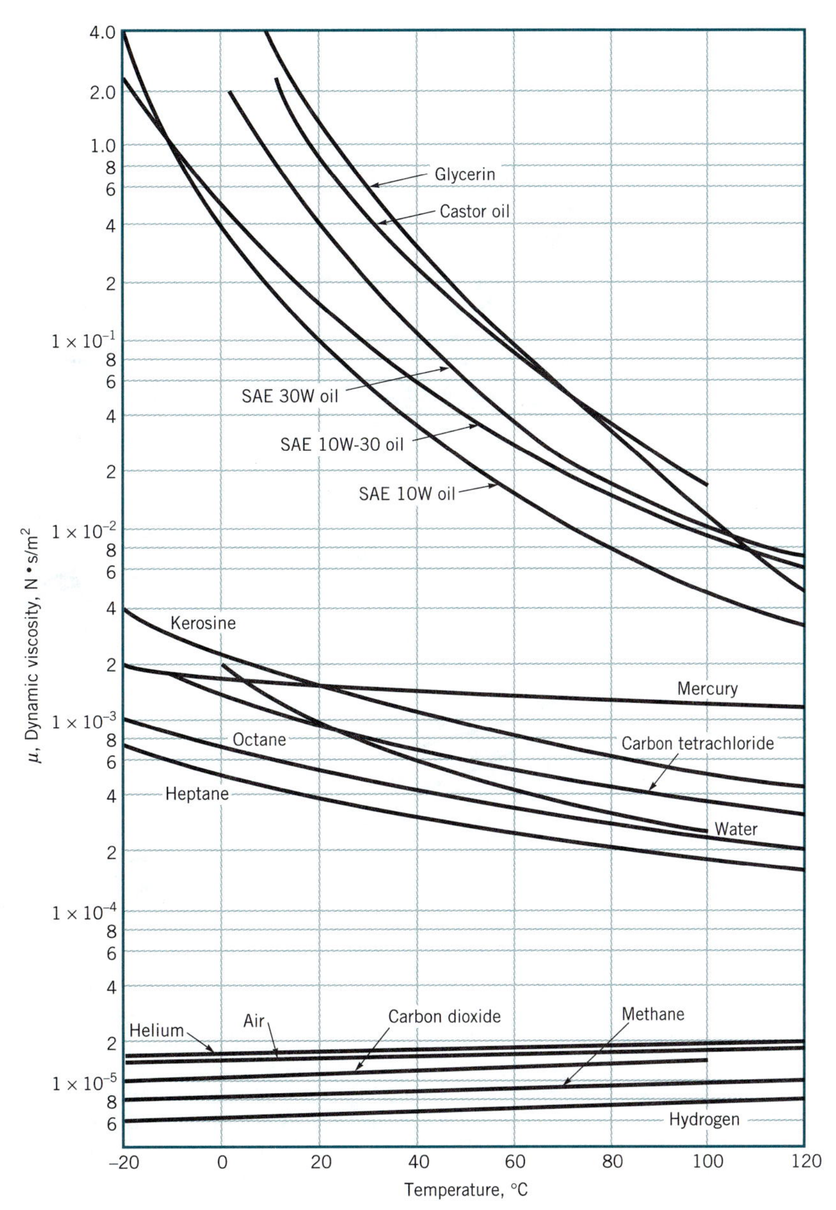

■ **FIGURE B.1** **Dynamic (absolute) viscosity of common fluids as a function of temperature. To convert to BG units of lb·s/ft² multiply N·s/m² by 2.089 × 10^{-2}. (Curves from R. W. Fox and A. T. McDonald, *Introduction to Fluid Mechanics*, 3rd Ed., Wiley, New York, 1985. Used by permission.)**

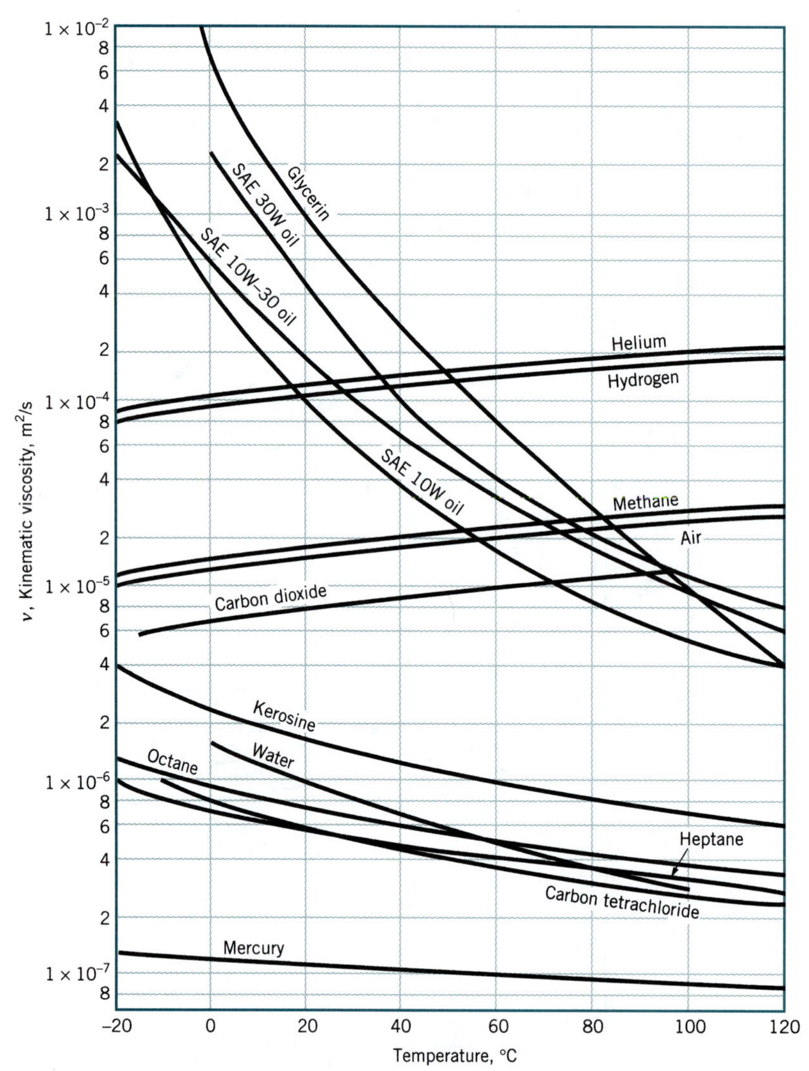

■ **FIGURE B.2** **Kinematic viscosity of common fluids (at atmospheric pressure) as a function of temperature. To convert to BG units of ft^2/s multiply m^2/s by 10.76. (Curves from R. W. Fox and A. T. McDonald, *Introduction to Fluid Mechanics*, 3rd Ed., Wiley, New York, 1985. Used by permission.)**

TABLE B.1
Physical Properties of Water (BG Units)[a]

Temperature (°F)	Density, ρ (slugs/ft^3)	Specific Weight[b], γ (lb/ft^3)	Dynamic Viscosity, μ (lb·s/ft^2)	Kinematic Viscosity, ν (ft^2/s)	Surface Tension[c], σ (lb/ft)	Vapor Pressure, p_v [lb/in.^{2}(abs)]	Speed of Sound[d], c (ft/s)
32	1.940	62.42	3.732 E − 5	1.924 E − 5	5.18 E − 3	8.854 E − 2	4603
40	1.940	62.43	3.228 E − 5	1.664 E − 5	5.13 E − 3	1.217 E − 1	4672
50	1.940	62.41	2.730 E − 5	1.407 E − 5	5.09 E − 3	1.781 E − 1	4748
60	1.938	62.37	2.344 E − 5	1.210 E − 5	5.03 E − 3	2.563 E − 1	4814
70	1.936	62.30	2.037 E − 5	1.052 E − 5	4.97 E − 3	3.631 E − 1	4871
80	1.934	62.22	1.791 E − 5	9.262 E − 6	4.91 E − 3	5.069 E − 1	4819
90	1.931	62.11	1.500 E − 5	8.233 E − 6	4.86 E − 3	6.979 E − 1	4960
100	1.927	62.00	1.423 E − 5	7.383 E − 6	4.79 E − 3	9.493 E − 1	4995
120	1.918	61.71	1.164 E − 5	6.067 E − 6	4.67 E − 3	1.692 E + 0	5049
140	1.908	61.38	9.743 E − 6	5.106 E − 6	4.53 E − 3	2.888 E + 0	5091
160	1.896	61.00	8.315 E − 6	4.385 E − 6	4.40 E − 3	4.736 E + 0	5101
180	1.883	60.58	7.207 E − 6	3.827 E − 6	4.26 E − 3	7.507 E + 0	5195
200	1.869	60.12	6.342 E − 6	3.393 E − 6	4.12 E − 3	1.152 E + 1	5089
212	1.860	59.83	5.886 E − 6	3.165 E − 6	4.04 E − 3	1.469 E + 1	5062

[a]Based on data from *Handbook of Chemistry and Physics*, 69th Ed., CRC Press, 1988. Where necessary, values obtained by interpolation.

[b]Density and specific weight are related through the equation $\gamma = \rho g$. For this table, $g = 32.174\ \text{ft/s}^2$.

[c]In contact with air.

[d]From R. D. Blevins, *Applied Fluid Dynamics Handbook*, Van Nostrand Reinhold Co., Inc., New York, 1984.

TABLE B.2
Physical Properties of Water (SI Units)[a]

Temperature (°C)	Density, ρ (kg/m^3)	Specific Weight[b], γ (kN/m^3)	Dynamic Viscosity, μ (N·s/m^2)	Kinematic Viscosity, ν (m^2/s)	Surface Tension[c], σ (N/m)	Vapor Pressure, p_v [N/m^2(abs)]	Speed of Sound[d], c (m/s)
0	999.9	9.806	1.787 E − 3	1.787 E − 6	7.56 E − 2	6.105 E + 2	1403
5	1000.0	9.807	1.519 E − 3	1.519 E − 6	7.49 E − 2	8.722 E + 2	1427
10	999.7	9.804	1.307 E − 3	1.307 E − 6	7.42 E − 2	1.228 E + 3	1447
20	998.2	9.789	1.002 E − 3	1.004 E − 6	7.28 E − 2	2.338 E + 3	1481
30	995.7	9.765	7.975 E − 4	8.009 E − 7	7.12 E − 2	4.243 E + 3	1507
40	992.2	9.731	6.529 E − 4	6.580 E − 7	6.96 E − 2	7.376 E + 3	1526
50	988.1	9.690	5.468 E − 4	5.534 E − 7	6.79 E − 2	1.233 E + 4	1541
60	983.2	9.642	4.665 E − 4	4.745 E − 7	6.62 E − 2	1.992 E + 4	1552
70	977.8	9.589	4.042 E − 4	4.134 E − 7	6.44 E − 2	3.116 E + 4	1555
80	971.8	9.530	3.547 E − 4	3.650 E − 7	6.26 E − 2	4.734 E + 4	1555
90	965.3	9.467	3.147 E − 4	3.260 E − 7	6.08 E − 2	7.010 E + 4	1550
100	958.4	9.399	2.818 E − 4	2.940 E − 7	5.89 E − 2	1.013 E + 5	1543

[a]Based on data from *Handbook of Chemistry and Physics*, 69th Ed., CRC Press, 1988.

[b]Density and specific weight are related through the equation $\gamma = \rho g$. For this table, $g = 9.807\ \text{m/s}^2$.

[c]In contact with air.

[d]From R. D. Blevins, *Applied Fluid Dynamics Handbook*, Van Nostrand Reinhold Co., Inc., New York, 1984.

■ **TABLE B.3**

Physical Properties of Air at Standard Atmospheric Pressure (BG Units)[a]

Temperature (°F)	Density, ρ (slugs/ft^3)	Specific Weight[b], γ (lb/ft^3)	Dynamic Viscosity, μ (lb·s/ft^2)	Kinematic Viscosity, ν (ft^2/s)	Specific Heat Ratio, k (—)	Speed of Sound, c (ft/s)
−40	2.939 E − 3	9.456 E − 2	3.29 E − 7	1.12 E − 4	1.401	1004
−20	2.805 E − 3	9.026 E − 2	3.34 E − 7	1.19 E − 4	1.401	1028
0	2.683 E − 3	8.633 E − 2	3.38 E − 7	1.26 E − 4	1.401	1051
10	2.626 E − 3	8.449 E − 2	3.44 E − 7	1.31 E − 4	1.401	1062
20	2.571 E − 3	8.273 E − 2	3.50 E − 7	1.36 E − 4	1.401	1074
30	2.519 E − 3	8.104 E − 2	3.58 E − 7	1.42 E − 4	1.401	1085
40	2.469 E − 3	7.942 E − 2	3.60 E − 7	1.46 E − 4	1.401	1096
50	2.420 E − 3	7.786 E − 2	3.68 E − 7	1.52 E − 4	1.401	1106
60	2.373 E − 3	7.636 E − 2	3.75 E − 7	1.58 E − 4	1.401	1117
70	2.329 E − 3	7.492 E − 2	3.82 E − 7	1.64 E − 4	1.401	1128
80	2.286 E − 3	7.353 E − 2	3.86 E − 7	1.69 E − 4	1.400	1138
90	2.244 E − 3	7.219 E − 2	3.90 E − 7	1.74 E − 4	1.400	1149
100	2.204 E − 3	7.090 E − 2	3.94 E − 7	1.79 E − 4	1.400	1159
120	2.128 E − 3	6.846 E − 2	4.02 E − 7	1.89 E − 4	1.400	1180
140	2.057 E − 3	6.617 E − 2	4.13 E − 7	2.01 E − 4	1.399	1200
160	1.990 E − 3	6.404 E − 2	4.22 E − 7	2.12 E − 4	1.399	1220
180	1.928 E − 3	6.204 E − 2	4.34 E − 7	2.25 E − 4	1.399	1239
200	1.870 E − 3	6.016 E − 2	4.49 E − 7	2.40 E − 4	1.398	1258
300	1.624 E − 3	5.224 E − 2	4.97 E − 7	3.06 E − 4	1.394	1348
400	1.435 E − 3	4.616 E − 2	5.24 E − 7	3.65 E − 4	1.389	1431
500	1.285 E − 3	4.135 E − 2	5.80 E − 7	4.51 E − 4	1.383	1509
750	1.020 E − 3	3.280 E − 2	6.81 E − 7	6.68 E − 4	1.367	1685
1000	8.445 E − 4	2.717 E − 2	7.85 E − 7	9.30 E − 4	1.351	1839
1500	6.291 E − 4	2.024 E − 2	9.50 E − 7	1.51 E − 3	1.329	2114

[a]Based on data from R. D. Blevins, *Applied Fluid Dynamics Handbook*, Van Nostrand Reinhold Co., Inc., New York, 1984.

[b]Density and specific weight are related through the equation $\gamma = \rho g$. For this table $g = 32.174$ ft/s^2.

TABLE B.4
Physical Properties of Air at Standard Atmospheric Pressure (SI Units)[a]

Temperature (°C)	Density, ρ (kg/m³)	Specific Weight[b], γ (N/m³)	Dynamic Viscosity, μ (N·s/m²)	Kinematic Viscosity, ν (m²/s)	Specific Heat Ratio, k (–)	Speed of Sound, c (m/s)
−40	1.514	14.85	1.57 E − 5	1.04 E − 5	1.401	306.2
−20	1.395	13.68	1.63 E − 5	1.17 E − 5	1.401	319.1
0	1.292	12.67	1.71 E − 5	1.32 E − 5	1.401	331.4
5	1.269	12.45	1.73 E − 5	1.36 E − 5	1.401	334.4
10	1.247	12.23	1.76 E − 5	1.41 E − 5	1.401	337.4
15	1.225	12.01	1.80 E − 5	1.47 E − 5	1.401	340.4
20	1.204	11.81	1.82 E − 5	1.51 E − 5	1.401	343.3
25	1.184	11.61	1.85 E − 5	1.56 E − 5	1.401	346.3
30	1.165	11.43	1.86 E − 5	1.60 E − 5	1.400	349.1
40	1.127	11.05	1.87 E − 5	1.66 E − 5	1.400	354.7
50	1.109	10.88	1.95 E − 5	1.76 E − 5	1.400	360.3
60	1.060	10.40	1.97 E − 5	1.86 E − 5	1.399	365.7
70	1.029	10.09	2.03 E − 5	1.97 E − 5	1.399	371.2
80	0.9996	9.803	2.07 E − 5	2.07 E − 5	1.399	376.6
90	0.9721	9.533	2.14 E − 5	2.20 E − 5	1.398	381.7
100	0.9461	9.278	2.17 E − 5	2.29 E − 5	1.397	386.9
200	0.7461	7.317	2.53 E − 5	3.39 E − 5	1.390	434.5
300	0.6159	6.040	2.98 E − 5	4.84 E − 5	1.379	476.3
400	0.5243	5.142	3.32 E − 5	6.34 E − 5	1.368	514.1
500	0.4565	4.477	3.64 E − 5	7.97 E − 5	1.357	548.8
1000	0.2772	2.719	5.04 E − 5	1.82 E − 4	1.321	694.8

[a]Based on data from R. D. Blevins, *Applied Fluid Dynamics Handbook*, Van Nostrand Reinhold Co., Inc., New York, 1984.

[b]Density and specific weight are related through the equation $\gamma = \rho g$. For this table $g = 9.807$ m/s².

APPENDIX C

Properties of the U.S. Standard Atmosphere

■ TABLE C.1

Properties of the U.S. Standard Atmosphere (BG Units)[a]

Altitude (ft)	Temperature (°F)	Acceleration of Gravity, g (ft/s^2)	Pressure, p [lb/in.^{2}(abs)]	Density, ρ (slugs/ft^3)	Dynamic Viscosity, μ (lb·s/ft^2)
−5,000	76.84	32.189	17.554	2.745 E − 3	3.836 E − 7
0	59.00	32.174	14.696	2.377 E − 3	3.737 E − 7
5,000	41.17	32.159	12.228	2.048 E − 3	3.637 E − 7
10,000	23.36	32.143	10.108	1.756 E − 3	3.534 E − 7
15,000	5.55	32.128	8.297	1.496 E − 3	3.430 E − 7
20,000	−12.26	32.112	6.759	1.267 E − 3	3.324 E − 7
25,000	−30.05	32.097	5.461	1.066 E − 3	3.217 E − 7
30,000	−47.83	32.082	4.373	8.907 E − 4	3.107 E − 7
35,000	−65.61	32.066	3.468	7.382 E − 4	2.995 E − 7
40,000	−69.70	32.051	2.730	5.873 E − 4	2.969 E − 7
45,000	−69.70	32.036	2.149	4.623 E − 4	2.969 E − 7
50,000	−69.70	32.020	1.692	3.639 E − 4	2.969 E − 7
60,000	−69.70	31.990	1.049	2.256 E − 4	2.969 E − 7
70,000	−67.42	31.959	0.651	1.392 E − 4	2.984 E − 7
80,000	−61.98	31.929	0.406	8.571 E − 5	3.018 E − 7
90,000	−56.54	31.897	0.255	5.610 E − 5	3.052 E − 7
100,000	−51.10	31.868	0.162	3.318 E − 5	3.087 E − 7
150,000	19.40	31.717	0.020	3.658 E − 6	3.511 E − 7
200,000	−19.78	31.566	0.003	5.328 E − 7	3.279 E − 7
250,000	−88.77	31.415	0.000	6.458 E − 8	2.846 E − 7

[a]Data abridged from *U.S. Standard Atmosphere*, 1976, U.S. Government Printing Office, Washington, D.C.

■ **TABLE C.2**
Properties of the U.S. Standard Atmosphere (SI Units)[a]

Altitude (m)	Temperature (°C)	Acceleration of Gravity, g (m/s²)	Pressure, p [N/m²(abs)]	Density, ρ (kg/m³)	Dynamic Viscosity, μ (N·s/m²)
−1,000	21.50	9.810	1.139 E + 5	1.347 E + 0	1.821 E − 5
0	15.00	9.807	1.013 E + 5	1.225 E + 0	1.789 E − 5
1,000	8.50	9.804	8.988 E + 4	1.112 E + 0	1.758 E − 5
2,000	2.00	9.801	7.950 E + 4	1.007 E + 0	1.726 E − 5
3,000	−4.49	9.797	7.012 E + 4	9.093 E − 1	1.694 E − 5
4,000	−10.98	9.794	6.166 E + 4	8.194 E − 1	1.661 E − 5
5,000	−17.47	9.791	5.405 E + 4	7.364 E − 1	1.628 E − 5
6,000	−23.96	9.788	4.722 E + 4	6.601 E − 1	1.595 E − 5
7,000	−30.45	9.785	4.111 E + 4	5.900 E − 1	1.561 E − 5
8,000	−36.94	9.782	3.565 E + 4	5.258 E − 1	1.527 E − 5
9,000	−43.42	9.779	3.080 E + 4	4.671 E − 1	1.493 E − 5
10,000	−49.90	9.776	2.650 E + 4	4.135 E − 1	1.458 E − 5
15,000	−56.50	9.761	1.211 E + 4	1.948 E − 1	1.422 E − 5
20,000	−56.50	9.745	5.529 E + 3	8.891 E − 2	1.422 E − 5
25,000	−51.60	9.730	2.549 E + 3	4.008 E − 2	1.448 E − 5
30,000	−46.64	9.715	1.197 E + 3	1.841 E − 2	1.475 E − 5
40,000	−22.80	9.684	2.871 E + 2	3.996 E − 3	1.601 E − 5
50,000	−2.50	9.654	7.978 E + 1	1.027 E − 3	1.704 E − 5
60,000	−26.13	9.624	2.196 E + 1	3.097 E − 4	1.584 E − 5
70,000	−53.57	9.594	5.221 E + 0	8.283 E − 5	1.438 E − 5
80,000	−74.51	9.564	1.052 E + 0	1.846 E − 5	1.321 E − 5

[a]Data abridged from *U.S. Standard Atmosphere*, 1976, U.S. Government Printing Office, Washington, D.C.

APPENDIX D

Compressible Flow Graphs for an Ideal Gas ($k = 1.4$)

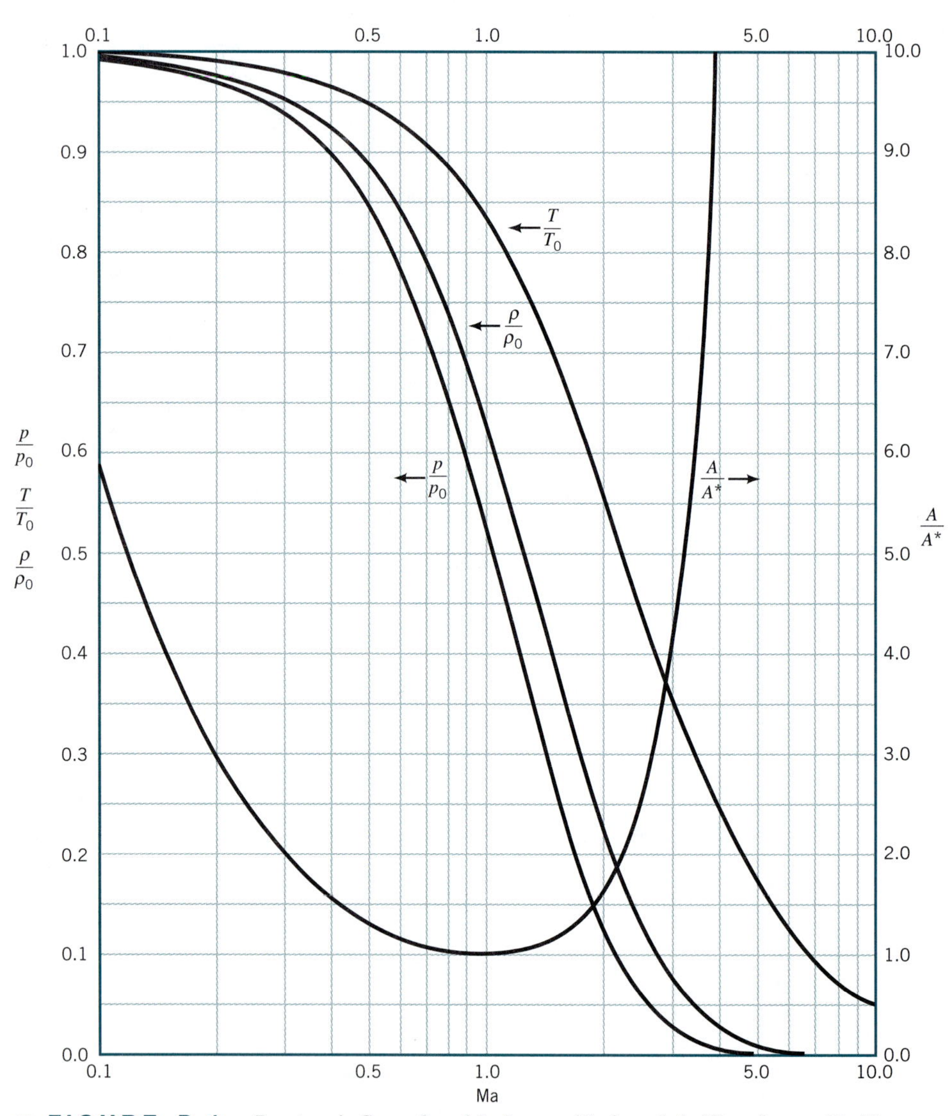

■ **FIGURE D.1** **Isentropic flow of an ideal gas with $k = 1.4$. (Graph provided by Professor Bruce A. Reichert of Kansas State University.)**

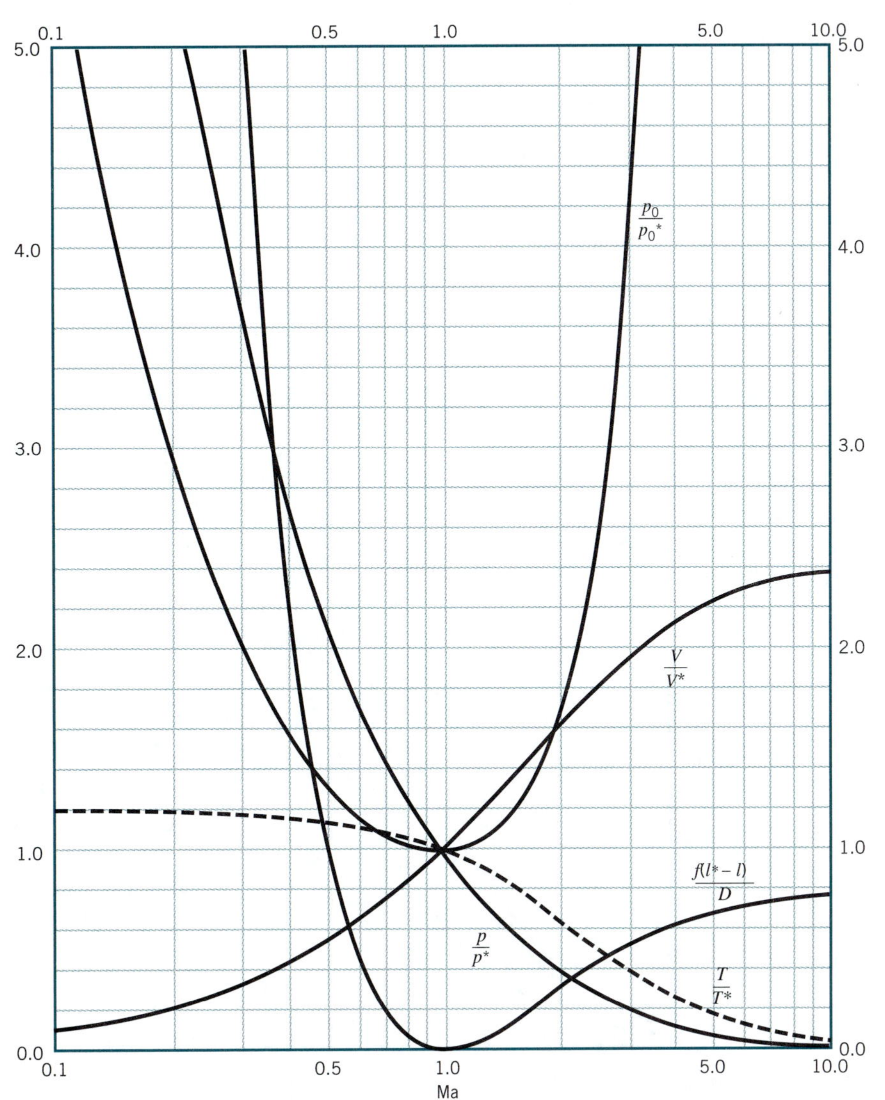

■ **FIGURE D.2** **Fanno flow of an ideal gas with $k = 1.4$. (Graph provided by Professor Bruce A. Reichert of Kansas State University.)**

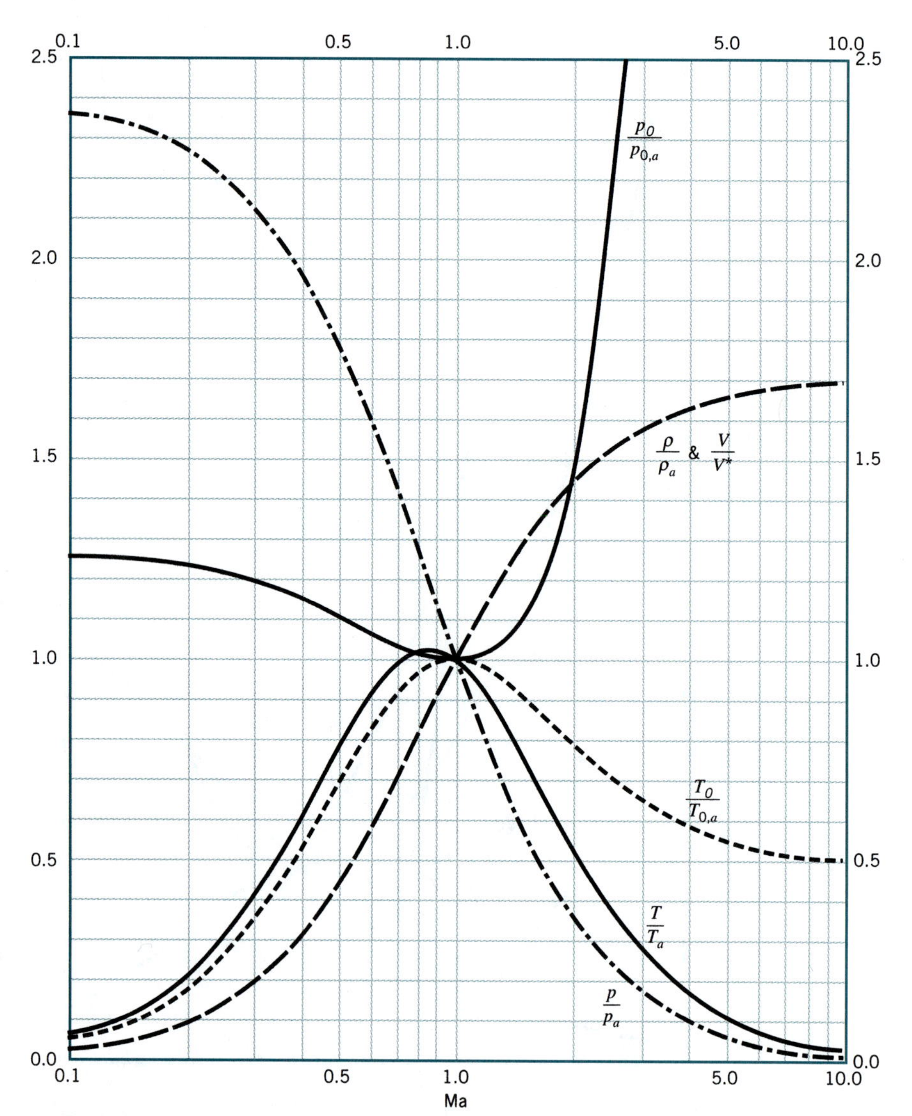

■ **FIGURE D.3** **Rayleigh flow of an ideal gas with $k = 1.4$. (Graph provided by Professor Bruce A. Reichert of Kansas State University.)**

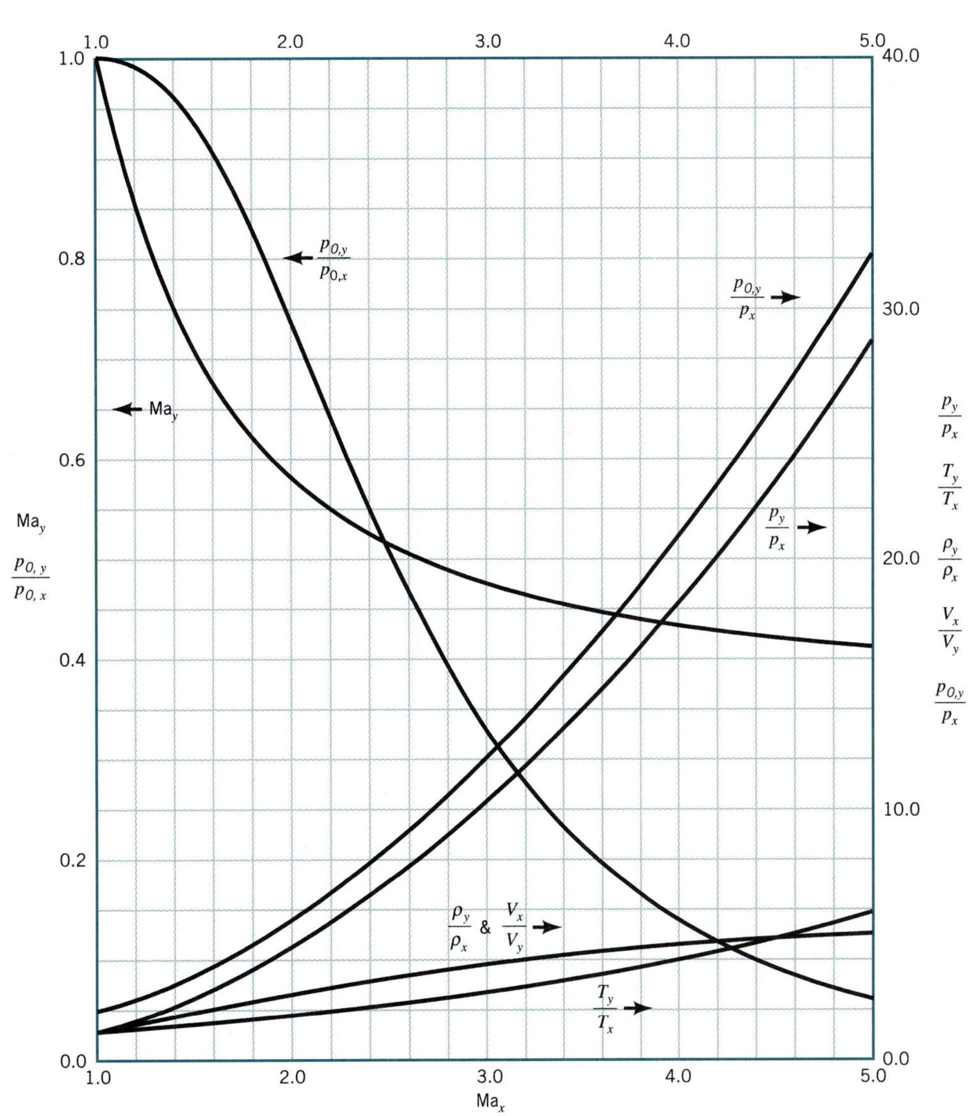

■ **FIGURE D.4** **Normal shock flow of an ideal gas with $k = 1.4$. (Graph provided by Professor Bruce A. Reichert of Kansas State University.)**

Answers to Selected Even Numbered Homework Problems

Chapter 1

- **1.4** (a) FL^{-2}; (b) FL^{-3}; (c) FL
- **1.6** (b)
- **1.8** dimensionless; yes
- **1.10** dimensionless; yes
- **1.12** no; no
- **1.14** (a) 4.32 mm/s; (b) 70.2 kg; (c) 13.4 N; (d) 22.3 m/s^2; (e) 1.12 $N \cdot s/m^2$
- **1.16** (a) 6.47×10^5 m^2; (b) 7.83×10^5 J; (c) 3.86×10^5 m; (d) 5.90×10^4 W; (e) 289 K
- **1.20** 9.46×10^{-2} m^3/s; 5.68×10^3 liters/min
- **1.22** 30.6 kg; 37.3 N
- **1.24** 2.65 $slugs/ft^3$; 1.37
- **1.26** 0.0186 ft^3; no
- **1.28** 16.0 kN/m^3; 1.63×10^3 kg/m^3; 1.63
- ***1.30** $\rho = 1001 - 0.05333\,T - 0.004095\,T^2$; 991.5 kg/m^3
- **1.32** 58.0 kPa
- **1.34** oxygen
- **1.36** 6.44×10^{-3} $slugs/ft^3$; 0.622 lb
- **1.40** 1.5×10^{-3} $N \cdot s/m^2$; 3.1×10^{-5} $lb \cdot s/ft^2$
- **1.42** 0.277 $N \cdot s/m^2$
- **1.44** 5×10^{-5} $N \cdot s/m^2$; 10.4×10^{-7} $lb \cdot s/ft^2$
- **1.46** $\mu = 1.1 \times 10^{-3}$ $N \cdot s/m^2$; $\mu = 2.3 \times 10^{-5}$ $lb \cdot s/ft^2$; $\nu = 1.4 \times 10^{-6}$ m^2/s; $\nu = 1.5 \times 10^{-5}$ ft^2/s
- **1.48** 11,200 (water); 564 (air)
- ***1.50** $C = 1.43 \times 10^{-6}$ $kg/(m \cdot s \cdot K^{1/2})$; $S =$ 107 K
- ***1.52** $D = 1.767 \times 10^{-6}$ $N \cdot s/m^2$; $B = 1.870 \times 10^3$ K; 5.76×10^{-4} $N \cdot s/m^2$
- **1.54** 0.268 ft/s
- **1.56** 0.0442 m/s
- **1.58** 6.72×10^{-2} N/m^2 acting in direction of flow
- ***1.60** (a) $C_1 = 153\ s^{-1}$; $C_2 = 4350\ ft^{-2}s^{-1}$ (b) 5.72×10^{-5} lb/ft^2 (y = 0); 6.94×10^{-5} lb/ft^2 (y = 0.05 ft)
- **1.62** 0.944 ft · lb; 17.8 ft · lb/s
- ***1.64** 2.45 $lb \cdot s/ft^2$
- **1.68** 3.00×10^5 psi
- **1.70** 104 psi
- **1.72** 1.52 kg/m^3
- **1.74** E_v (water) / E_v (air) $= 1.52 \times 10^4$
- **1.76** (a) 343 m/s; (b) 1010 m/s; (c) 446 m/s
- **1.78** 4.74 psia
- **1.80** 70.1 kPa (abs); 10.2 psia
- **1.82** 97.9 Pa
- **1.84** 0.111 lb/ft^2
- **1.86** 7.49 mm
- **1.88** water column
- **1.90** K (average) $= 0.527 \times 10^{-6}$ $m^2 \cdot ml / s^2$; K decreases with increasing temperature

Chapter 2

- **2.2** 10.4 m

2.6	60.6 M Pa; 8790 psi
2.8	(For 120 mm Hg) 16.0 k Pa; 2.32 psi (For 70 mm Hg) 9.31 k Pa; 1.35 psi
2.10	$p = K h^2/2 + \gamma_0 h$
2.12	36,000 lb
2.16	-7.7 psi
2.18	(a) 0.759 m; 0.759 m (without vapor pressure); (b) 10.1 m; 10.3 m (without vapor pressure); (c) 12.3 m; 13.0 m (without vapor pressure)
2.20	(a) 1240 lb/ft^2 (abs); (b) 1040 lb/ft^2 (abs); (c) 1270 lb/ft^2 (abs)
2.22	12.1 kPa; 0.195 kg/m^3
2.24	4.67 psi
2.26	22.4 psi
2.28	2.20 ft
2.32	0.424 psi
2.34	$h = (p_1 - p_2) / (\gamma_2 - \gamma_1)$
2.36	0.040 m
2.38	94.9 kPa
2.40	1.09
2.42	4.06 ft
2.44	27.8°
2.46	0.304 ft (down)
2.48	665 lb
2.50	-4.51 kPa
2.52	1350 lb
2.54	33,900 lb; 2.49 ft above base of triangle
2.56	482 kN; 0.987 m above bottom of conduit
2.58	436 kN
2.62	(a) 1060 kN (rectangle); 1010 kN (semicircle); (b) 1.37×10^6 N · m
2.64	26.5 kPa
***2.66**	426 kN; 2.46 m below fluid surface
2.68	0.146
2.70	$F_H = 392$ kN; $F_V = 437$ kN; yes
2.72	22,500 lb
2.74	3360 psi
2.76	60.8 kN; 0.100 m below center of tank end wall
2.78	203 kN
2.80	$F_H = 22{,}500$ lb $\leftarrow$; $F_V = 19{,}900$ lb $\uparrow$
2.82	1.22 ft; no change
2.84	368 lb directed vertically upward 8.40 ft right of point A
2.88	3130 kg
2.90	(a) 18.9 kPa; (b) 0.208 m^3
2.92	$dz/dy = 0.502$
2.94	4.91 m/s^2
2.96	158 lb/ft^2
2.98	10.5 rad/s
2.100	6.04 rad/s

Chapter 3

3.2	(a) $-194(1 + x) + 62.4$ lb/ft^3; (b) 41.2 psi
3.4	-30.0 kPa/m
3.6	-0.838 psi/ft; 0.0292 psi/ft
3.8	$p + \frac{1}{2}\rho V^2 + \rho g_0 z - \frac{1}{2}\rho c z^2 =$ constant
3.10	12.0 kPa; -20.1 kPa
3.14	34.2 ft/s; none
3.18	43.0 psi
3.20	$d = d_o/[1 - (2gz/V_o^2)]^{1/4}$; $d \equiv d_o$
3.22	10.8 lb/ft^2; 124 lb/ft^2
3.24	(a) -5.12 lb/ft^2; (b) 4.10 lb/ft^2
3.26	2.45×10^5 kN/m^2; 5.50×10^{-6} m^3/s
3.28	1.40 mph
3.30	$Q = 1.56\, D^2$, where $Q \sim$ m^3/s, $D \sim$ m
3.32	0
3.34	45.7 ft
3.36	$D = D_1/[1 - (\pi^2 g D_1^4\, z/8Q^2)]^{1/4}$
3.38	2.54×10^{-4} m^3/s
3.40	9.64×10^{-3} m^3/s; 29.4 kPa
3.42	0.188 in.
3.44	19.63 psi
3.46	2.50 in.
3.48	2.94 m
3.50	-8.11 psi
3.52	32.217 lb/ft^2
3.54	From 0.229 d_{10} to 1.00 d_{10}
3.58	15.4 m
3.60	1.98 ft
3.62	5.18 psi
3.64	0.303 ft^3/s; 499 lb/ft^2; 312 lb/ft^2; -187 lb/ft^2
3.66	0.37 m
3.68	7.53 ft
3.70	7.38 m^3/s
3.72	0.351 ft^3/s
3.74	8.68×10^{-3} m^3/s
3.76	$R = 0.998 z^{1/4}$
3.82	29.0 ft/s; 7.98 psi; 3.97 ft^3/s
3.84	53.4 psi; 48.1 psi
3.86	-7.14×10^{-5} psi
3.88	3000 ft^3/s; 2000 ft^3/s; 2120 ft^3/s
3.90	$Q = CH^2$
3.92	0.630 ft; 4.48 ft
3.94	0.174 m^3/s
3.96	6.51 m; 25.4 m; 6.51 m; -9.59 m

Chapter 4

4.2	$2\sqrt{2}\, x^2 t$ m/s; 45 deg from x axis
4.4	(2,2)
4.6	when $x = \pm\, y$; $x = y = 0$

4.10 $xy = C$
4.12 $x = -h(u_0/v_0)\, ln[1 - (y/h)]$
4.14 $2c^2x^3$; $2c^2y^3$; $x = y = 0$
4.18 -8 m/s^2; -2.0 m/s^2; -0.08 m/s^2
4.24 $-x/t^2$; x/t^2
4.26 $0.355\ V_0^2/\ell$
4.32 -18.9 °C/s; -16.8 °C/s
4.34 0; 100 N/m^2 · s
4.36 75 m/s^2
4.44 (a) 10 ft/s^2, 20 ft/s^2; (b) 20 ft/s^2; (c) 22.4 ft/s^2 at 63.4° from streamline
4.46 3.13×10^{-5} m/s^2; 2.00×10^{-3} m/s^2
4.48 $-25{,}600$ ft/s^2; -25.0 ft/s^2
4.50 3.75 m^3/s
4.52 $\rho_0 \forall_0 b e^{-bt}$
4.60 3000 kg/s; 3.00 m^3/s

Chapter 5

5.2 (a) 6280 kg/s
5.4 70.1 m/s
5.6 13.8 ft
5.8 3.18 ft
5.10 2.23 ft/s
5.12 0.064 m
5.14 1.00 m^3/s
5.16 (a) 0.711; (b) 0.791; (c) 0.837; (d) 0.866
5.18 0.0114 slug/s
5.20 (7/8) $U\ell\delta$
5.22 1260 min
5.24 (a) 15.6 gal/min; (b) 62.4 gal/min
5.30 352 lb to left
5.32 $F_{A,x}$ = 1890 lb to left; $F_{A,y} = 0$; R_x = 1890 lb to left; $R_y = 0$
5.34 7.01 ft^3/s; 674 lb down
5.36 30,800 N against flow
5.38 9.27 N to left
5.40 17.1 lb/ft
5.44 2.1%
5.48 213 lb to left
5.50 0.108 kg
5.52 1.82 kPa
5.54 7.07 ft/s at 45°
5.56 0
5.58 motion to right (a), (b), (c); motion to left (d)
5.60 3.97 m
5.62 2.96 lb
5.66 (a) 181 lb; (b) 146 lb
5.68 (a) 2.96 ft · lb; (b) 1.35 ft · lb; (c) 920 rpm
5.70 7.77 (N · m)/kg for (a) and (b)
5.74 (a) 43°; (b) 53.4 (ft · lb)/s
5.76 (b) 3130 (ft · lb)/slug
5.80 950 (ft · lb)/slug
5.82 84.2%
5.84 166 (ft · lb)/slug; 86%
5.86 778 ft
5.88 (a) 265 m/s; (b) 490 m/s
5.90 0.042 m^3/s
5.92 4.58×10^{-3} m^3/s
5.94 56 (ft · lb)/slug
5.96 (a) 392 kPa; (b) 422 kPa
5.102 (a) 1.65 psi; (b) 203 (ft · lb)/slug; (c) 77.2 lb
5.106 (a) 4.29 m/s, 17.2°; (b) 558 (N · m)/s
5.108 4.53 MW
5.110 301 hp
5.112 303°R actual, 267°R ideal, 88%
5.114 (a) 1.20 m^3/s; (b) 0.928 m^3/s; (c) 0.705 m^3/s
5.116 2.02 hp
5.118 79.9%
5.120 31.3 hp
5.122 14,500 (ft · lb)/s
5.124 (a) 1.11; (b) 1.08; (c) 1.06; (d) 1.05; (e) 1.04; (f) 1.03
5.126 0.60 lb experimental value, 0.69 lb theoretical value

Chapter 6

6.2 a_x (local) = 0; a_x (conv) = $18(x^3 + xy^2)$; a_y (local) = 0; a_y (conv) = $18(x^2 y + y^3)$; $\mathbf{V} = -24\hat{\mathbf{j}}$; $|\mathbf{V}| = 24$ ft/s; $\mathbf{a} = 288\hat{\mathbf{i}} + 288\hat{\mathbf{j}}$; $|\mathbf{a}| = 407$ ft/s^2
6.4 (a) 0; (b) $\omega = -(y/2 + z)\hat{\mathbf{i}} + (5z/2)\hat{\mathbf{j}} - (y/2)\hat{\mathbf{k}}$; No
6.6 No; None (except both equal to zero)
6.8 (a) 0; (b) $\omega = -(U/2b)\hat{\mathbf{k}}$; (c) $\zeta = -(U/b)\hat{\mathbf{k}}$; (d) $\dot{\gamma} = U/b$
6.10 $\omega = 4yz^2 - 6y^2z$
6.12 $\psi = x^2y - x^2/2 + C$
6.14 No
6.16 (a) $v_r = V \sin\theta$, $v_\theta\theta = V\cos\theta$; (b) $\psi = -Vx + C$, $\psi = -Vr\cos\theta + C$
6.18 (a) Yes; (b) Only for $A = B$; (c) $y^2 = (B/A)x^2 + C$
6.20 (a) $v = -2y$; (b) 1.41 ft/s
6.22 (b) -1 m^3/s (per unit width)
6.26 (a) $p_A = p_0$; (b) $p_B = p_0$
6.28 $\psi = 5x^2y - (5/3)y^3 + C$
6.30 $\psi = A\theta + Br\sin\theta + C$; $\theta = \pi$, $r = A/B$
6.32 60.5 psi
6.34 (a) Yes; (b) Yes, $\phi = 2(x + y) + C$; (c) 0
6.36 80.1 kPa
6.38 (a) $\psi = -Kr^2/2 + C$; (b) No

6.40 (a) $\partial p/\partial r = (1.60 \times 10^3)\, r^{-3}$ kPa/m; (b) 184 kPa
6.42 Yes
6.44 $C = 0.500$ m
6.46 $\Gamma = \omega\Delta\theta(b^2 - a^2)$
6.48 0.0185 m/s at 36.4° from x-axis
6.50 $\sin\theta = (\Gamma/2\pi rU)\, ln(r/4)$
6.54 8.49×10^{-4} m/s
6.60 (a) $p_{max} = p_0 + \rho U^2/2$ (at $\theta = 0$ and π); $p_{min} = p_0 - 3\rho U^2/2$ (at $\theta = \pi/2$); (b) $(2r/3a)(1 - a^2/r^2)\sin\theta = 1$
*6.62 $y/a \geq 10$
6.66 (b) $V_{wall} = (m/\pi)[y/(\ell^2 + y^2)]$; (c) $p_{wall} = p_0 - (\rho m^2/2\pi^2)[y/(\ell^2 + y^2)]^2$
6.68 $p = p_0 + (\rho U^2/2)[(\Gamma/\pi Ur)\sin\theta - (\Gamma/2\pi Ur)^2]$
6.70 -85.1×10^{-5} lb/ft^2
6.72 (a) $\partial v/\partial y = -2x$; (b) $\mathbf{a} = 2x^3\hat{\mathbf{i}}$; (c) $\partial p/\partial x = 2\mu - 2\rho x^3$
6.74 (a) 4.21×10^{-4} m^2/s; (b) 60 N/m^2 (acting in direction of flow); (c) 0.158 m/s
6.76 $q = (\rho g h^3 \sin\alpha)/3\mu$
6.78 $q = -2\rho g h^3/3\mu$
6.80 $u = [(U_1 + U_2)/b]y - U_2$
6.82 $U = (b^2/2\mu)\, \partial p/\partial x$
6.84 0.355 N · m
6.90 (a) Re = 320 < 2100; (b) 90.0 kPa; (c) 30.0 N/m^2
6.92 (b) 1.20 Pa
6.94 $v_\theta = R^2\,\omega/r$
6.98 0.165 ft

Chapter 7

7.4 (a) 103 m/s; (b) 444 m/s
7.6 $\mathcal{P}/\rho D^5\omega^3 = \phi(Q/D^3\omega)$
7.8 $q/b^{3/2}\sqrt{g} = \phi(H/b,\ \mu/\rho b^{3/2}\sqrt{g})$
7.10 $\mathcal{D}/d_1{}^2V^2\rho = \phi(d_2/d_1,\ \rho V d_1/\mu)$
7.12 $Q/A^{5/4}\sqrt{g} = \phi(\epsilon/\sqrt{A},\ S_0)$
7.14 $c\sqrt{\rho/E} = \phi(h/D)$
7.16 $\mathcal{D}/\rho V^2\ell_1{}^2 = \phi(V/c,\ \ell_i/\ell_1)$
7.18 $\Delta p \propto 1/D^2$ (for a given velocity)
7.20 $\omega = CD\sqrt{\gamma/m}$ (C = constant); decrease
7.22 $h/D = \phi(\sigma/\gamma D^2)$
*7.24 $\Delta p/\rho V^2 = -1.10(A_1/A_2)^2 + 1.07\,(A_1/A_2) - 0.0103$; 6.26 lb/ft^2
*7.26 $\theta = 298\,\omega\mu$
*7.28 1.16 mm
7.30 $Q/\sqrt{gH^5} = \phi(b/H)$
7.32 5.81×10^{-2} ft/s
7.34 129 m/s
7.36 $\sigma_m/\sigma = 4.44 \times 10^{-3}$
7.38 1170 km/hr
7.40 187 mph
7.42 4.62×10^{-9} m^2/s; no
7.44 (a) 400 km/hr; (b) 170 N
7.46 50.2 kPa (abs)
7.48 0.0762 (required length scale)
7.50 (a) $\mathcal{D}/\rho V^2 D^2 = \phi(d/D)$; (b) 31.1 lb
7.52 11.6 ft^3/s; 0.500 psi
7.54 (a) $D/d = \phi(\sigma/\rho g d^2,\ Q\sqrt{g}\,d^{5/2})$; (b) d_m = 0.791 in., $Q_m = 5.63 \times 10^{-3}$ ft^3/s, $D_m/D = \sqrt{10}$
7.56 100 to 450 mph; no
7.58 0.0647 to 0.0971
7.60 0.250 ft; 0.0586 ft^3/s
7.62 1.31 m/s; 1.25×10^5
7.64 -0.0809 lb/ft^2

Chapter 8

8.4 4.32 ft
8.8 (a) -7.40 lb/ft^3; (b) -69.8 lb/ft^3; (c) 55.0 lb/ft^3
8.10 Downhill; 0.596 lb/ft^2; 0.238 lb/ft^2; 0
8.12 (a) to (b)
8.16 B to A
8.18 2.01 ft/s
8.20 $h \leq 0.509$ m
8.22 18.5 m
8.24 0
8.26 $r/R = 0.758, 0.760, 0.762$
8.28 0.266 psi, 1.13 psi, -0.601 psi
8.30 0.0300
8.32 No
8.34 1.02×10^{-4} ft
8.36 25.1 psi
8.38 0.211 psi/ft
8.44 1.54 kPa
8.46 9.00
8.52 21.0
8.54 48.0 psi
8.56 (a) 50.2 m, 602 N/m^2; (b) 9.09 m, 8.91×10^4 N/m^2; (c) 4.95 m, 6.58×10^5 N/m^2
8.58 0.188 m
8.60 0.285 m
8.62 127 hp pump
8.64 24.4 hp
8.68 0.899 lb/ft^2
8.70 22.5 ft/s; 17.0 ft/s
8.72 0.710 ft^3/s
8.74 84.0 ft
8.76 5.68

8.78 (a) 135 ft; (b) 137 ft
8.82 0.476 ft
8.86 galvanized iron
8.88 5.73 ft/s
8.90 0.442 ft
8.92 (a) 0.227 ft^3/s; (b) 0.223 ft^3/s
8.94 0.491 ft
***8.96** 0.0445 m
8.98 0.0746 cfs; 0.339 cfs
8.100 0.0180 m^3/s
8.102 0.0284 m^3/s; 0.0143 m^3/s; 0.0141 m^3/s
8.104 24.9 psi
8.106 32.4 kPa
8.108 0.0221 m^3/s
8.110 0.115 ft^3/s
8.112 5.77 ft
8.114 0.0936 ft^3/s

Chapter 9
9.2 0.06 $(\rho U^2/2)$; 2.40
9.4 1.91
9.6 Yes
9.8 Fig. 9.6 (*c*)
9.10 4.70 mm; 14.8 mm; 47.0 mm
9.12 0.00718 m/s; 0.00229 m/s
9.16 0.0130 m; 0.0716 N/m^2; 0.0183 m; 0.0506 N/m^2
9.22 0.171 ft; 0.134 ft
9.24 $\delta = 5.83\ (\nu x/U)^{1/2}$
9.26 $\delta/x = 5.03/\mathrm{Re}_x^{1/2}$
9.28 0.707
9.30 $2.83\mathfrak{D}$; $0.354\mathfrak{D}$
9.34 6.78×10^{-6} N · m
9.36 0.0438 N · m
9.40 No
9.42 1.06 m/s
9.44 16.8 ft/s
9.46 143 s
9.48 7,080 N · m
9.52 43.2%
9.54 3
9.56 1.11 hp; 5.00 hp
9.62 counterclockwise
9.68 859 lb
9.70 4.31 MN; 4.17 MN
9.72 0.851; 0.301
9.78 53.5 kW; 4.46 kW
9.80 2.47×10^3 kW
9.84 0.0187 *U*
9.86 65.9 hp
9.88 22.5%
9.90 1.80 *U*
9.92 15 to 16 degrees
9.96 19.1
9.98 29.3%

Chapter 10
10.2 (a) supercritical; (b) supercritical; (c) subcritical
10.4 2.80 m/s
10.6 5.66 ft
10.8 3.60 m/s
10.14 4.14 ft or 1.42 ft
10.16 0.528 ft; 0.728 ft
10.22 No
10.24 2 ft, 3.51 ft; 2 ft, 1.38 ft
10.28 3.44 m
10.30 5.14 N/m^2
10.32 (a) 1.80 lb/ft^2; (b) 0.0469; (c) 0.636
10.34 35.0 m^3/s
10.36 Not possible
10.38 7.42 min
10.40 0.000664
10.42 0.000269
10.44 Yes
10.48 8.77 m
10.50 2.23 ft; 0.756
10.52 Same
10.54 0.856 *h*
10.56 244 yd^3
10.58 0.861 m
10.60 1.51 D_1
10.64 18.2 m^3/s
10.70 0.00757
10.72 Decreases
10.74 -7.07×10^{-5}
10.76 13.7 ft/s
10.80 (a) 0.228 ft; (b) 8.10, 0.223; (c) 5.15 ft
10.82 1.51 m; 12,500 kW
10.84 0.0577; 0.000240
10.86 4.36 ft/s
10.88 5.85 ft^3/s
10.92 0.350 m/s
10.94 Yes
10.98 0.699

Chapter 11
11.2 (a) $c_p = 187$ (ft · lb)/(lbm · °R); $c_v = 133$ (ft · lb)/(lbm · °R) (b) $c_p = 152$ (ft · lb)/(lbm · °R); $c_v = 117$ (ft · lb)/(lbm · °R) (c) $c_p = 971$ (ft · lb)/(lbm · °R); $c_v = 585$ (ft · lb)/(lbm · °R) (d) $c_p = 2636$ (ft ·

lb)/(lbm · °R); c_v = 1870 (ft · lb)/(lbm · °R) (e) c_p = 407 (ft · lb)/(lbm · °R); c_v = 311 ft · lb)/(lbm · °R) (f) c_p = 193 (ft · lb)/(lbm · °R); c_v = 138 (ft · lb)/(lbm · °R) (g) c_p = 169 (ft · lb)/(lbm · °R); c_v = 121 (ft · lb)/(lbm · °R)

11.4 $\check{u}_2 - \check{u}_1 = -5.08 \times 10^6$ (N · m)/kg; $\check{h}_2 - \check{h}_1 = -7.16 \times 10^6$ (N · m)/kg; $s_2 - s_1 = -1.40 \times 10^4$ (N · m)/(kg · K)

11.6 −1890 J/(kg · K)

11.8 351 K

11.10 (a) constant c_p, $\check{h}_2 - \check{h}_1$ = 49,000 (ft · lb)/slug; varying c_p, $\check{h}_2 - \check{h}_1$ = 51,000 (ft · lb)/slug; (b) constant c_p, $\check{h}_2 - \check{h}_1 = 4.9 \times 10^6$ (ft · lb)/slug; varying c_p, $\check{h}_2 - \check{h}_1 = 6.29 \times 10^6$ (ft · lb)/slug; (c) constant c_p, $\check{h}_2 - \check{h}_1 = 1.47 \times 10^7$ (ft · lb)/slug; varying c_p, $\check{h}_2 - \check{h}_1 = 2.19 \times 10^7$ (ft · lb)/slug

11.16 (a) 2000 mph; (b) 2930 ft/s; (c) 893 m/s

11.20 (a) 335 ft/s; 102 m/s; (b) 1280 ft/s; 390 m/s

11.22 (a) 732 m/s; (b) 2400 ft/s; (c) 1636 mph

11.24 0.625

11.30 (a) $p^*/p_0 = 0.5283$; $T^*/T_0 = 0.8333$; (b) $p^*/p_0 = 0.5457$; $T^*/T_0 = 0.8696$; (c) $p^*/p_0 = 0.4881$; $T^*/T_0 = 0.7519$; (d) $p^*/p_0 = 0.5266$; $T^*/T_0 = 0.8299$; (e) $p^*/p_0 = 0.5439$; $T^*/T_0 = 0.8658$; (f) $p^*/p_0 = 0.5283$; $T^*/T_0 = 0.8333$; (g) $p^*/p_0 = 0.5283$; $T^*/T_0 = 0.8333$

11.32 (a) 0.0277 slug/s; (b) 0.03 slug/s

11.34 (a) 283 m/s; 0.89; (b) 231 m/s; 0.913; (c) 1070 m/s; 0.884

11.36 269 m/s; 0.900

11.44 (a) 0.012 ft^2; (b) 0.00994 ft^2; (c) 0.0308 ft^2

11.46 (a) 0.36 m^2; 23 kPa (abs); 113 K; (b) 0.257 m^2; 24.8 kPa (abs); 82.6 K

11.48 (a) 0.0483%; (b) 1.43%; (c) −0.445%; (d) −0.0163%

11.54 11 psia

11.56 (a) 1.7 kg/s; (b) 1.52 kg/s

11.58 (a) 34.7 kPa (abs); 314 m/s; 0.8; (b) 18.4 kPa (abs); 593 m/s; 0.39

11.60 (a) 2830 lb/ft^2; (b) 2550 lb/ft^2

11.62 T_1 = 282K; p_1 = 95 kPa (abs); T_{01} = 288K; p_{01} = 101 kPa (abs); V_1 = 104 m/s; T_2 = 674K; p_2 = 45 kPa (abs); T_{02} = 786K; p_{02} = 84.9 kPa (abs); V_2 = 520 m/s

11.64 404 kPa (abs); 81.19 K; 31 m/s

11.66 (a) 56 kPa; (b) 47.6 kPa

11.68 (a) Ma_y = 0.475; $T_{0,y}$ = 1690 °R; T_y = 1620 °R; $p_{0,y}$ = 360 psia; p_y = 309 psia; V_y = 937 ft/s; (b) Ma_y = 0.521; $T_{0,y}$ = 2390 °R; T_y = 2190 °R; $p_{0,y}$ = 409 psia; p_y = 330 psia; V_y = 3500 ft/s

11.70 1240 ft/s; 1.25

11.72 (a) 0.94; 4 kPa; (b) 0.8; 17 kPa; (c) 0.47; 50 kPa

11.74 (a) $p_2/p_{0,1}$ = 0.213; Ma_2 = 0.62; (b) $p_2/p_{0,1}$ = 0.16; Ma_2 = 0.89

11.76 T_1 = 291 ° R; p_1 = 1.91 psia; $T_{0,1}$ = 519 °R; $p_{0,1}$ = 14.7 psia; V_1 = 1660 ft/s; T_x = 410 °R; p_x = 3 psia; $T_{0,x}$ = 598 °R; $p_{0,x}$ = 11 psia; V_x = 1490 ft/s; T_y = 533 °R; p_y = 7.5 psia; $T_{0,y}$ = 598 °R; $p_{0,y}$ = 10.2 psia; V_y = 784 ft/s; T_2 = 549 °R; p_2 = 5.31 psia; $T_{0,2}$ = 657 °R; $p_{0,2}$ = 9.8 psia; V_2 = 1140 ft/s

Chapter 12

12.2 4.05 N · m; 1.705 rev/s

12.4 55.4°; CCW; turbine

12.6 (b) fan; (c) 7160 ft^2/s^2

12.8 0.08 kW

12.10 (b) 918 ft · lb; 0 rpm

12.12 1.84 hp

12.14 11.5 m

12.18 No

12.22 365 gpm

12.24 255 gpm; no

12.26 0.0529 m^3/s; 13% open

12.28 873 gpm

12.30 1000 gpm; 800 ft

12.32 Yes

12.34 centrifugal pump

12.36 mixed-flow pump

12.44 0.0826 ft; 19.8 hp; less

12.46 16.6 m^2/s^2; 0.849

12.48 0.3 ft

12.50 26,600 N; 37.6 m/s; 707 kg/s

12.58 70 m

12.60 (a) 12; (b) 3

Index

■ **TABLE 1.3**

Conversion Factors from BG and EE Units to SI Units[a]

	To Convert from	to	Multiply by
Acceleration	ft/s^2	m/s^2	3.048 E − 1
Area	ft^2	m^2	9.290 E − 2
Density	lbm/ft^3	kg/m^3	1.602 E + 1
	$slugs/ft^3$	kg/m^3	5.154 E + 2
Energy	Btu	J	1.055 E + 3
	ft·lb	J	1.356
Force	lb	N	4.448
Length	ft	m	3.048 E − 1
	in.	m	2.540 E − 2
	mile	m	1.609 E + 3
Mass	lbm	kg	4.536 E − 1
	slug	kg	1.459 E + 1
Power	ft·lb/s	W	1.356
	hp	W	7.457 E + 2
Pressure	in. Hg (60 °F)	N/m^2	3.377 E + 3
	lb/ft^2 (psf)	N/m^2	4.788 E + 1
	$lb/in.^2$ (psi)	N/m^2	6.895 E + 3
Specific weight	lb/ft^3	N/m^3	1.571 E + 2
Temperature	°F	°C	$T_C = (5/9)(T_F - 32°)$
	°R	K	5.556 E − 1
Velocity	ft/s	m/s	3.048 E − 1
	mi/hr (mph)	m/s	4.470 E − 1
Viscosity (dynamic)	$lb{\cdot}s/ft^2$	$N{\cdot}s/m^2$	4.788 E + 1
Viscosity (kinematic)	ft^2/s	m^2/s	9.290 E − 2
Volume flowrate	ft^3/s	m^3/s	2.832 E − 2
	gal/min (gpm)	m^3/s	6.309 E − 5

[a]If more than four-place accuracy is desired, refer to Appendix A.